Introduction to VLSI Systems

A Logic, Circuit, and System Perspective

Introduction to VLSI Systems

A Logic, Circuit, and System Perspective

Ming-Bo Lin

CRC Press is an imprint of the
Taylor & Francis Group, an **informa** business

CRC Press
Taylor & Francis Group
6000 Broken Sound Parkway NW, Suite 300
Boca Raton, FL 33487-2742

Printed and bound in India by Replika Press Pvt. Ltd.
Version Date: 20110804

International Standard Book Number: 978-1-4398-6859-1 (Hardback)

Visit the Taylor & Francis Web site at
http://www.taylorandfrancis.com

and the CRC Press Web site at
http://www.crcpress.com

To Alfred, Fanny, Alice, and Frank
and in memory of my parents

Contents

Preface

With the advance of semiconductors and the prospering of the computer and communication industries, the use of system-on-a-chip (SoC) has become an essential technique to reduce product cost. With this progress and continuous reduction of feature sizes, it is important to understand the fundamentals of circuit and layout designs of very large-scale integration (VLSI) circuits for the following reasons. First, addressing the harder problems requires fundamental understanding of circuit and layout design issues. Second, distinguished engineers can often develop their physical intuition to estimate the behavior of circuits rapidly without relying predominantly on computer-aided design (CAD) tools. This book addresses the need for teaching such a topic in terms of a logic, circuit, and system design perspective.

To achieve these goals, this book will focus on building an understanding of integrated circuits from the bottom up and pay much attention to logic circuits, layout designs, and system designs. More precisely, this book has the following objectives. First, it will familiarize the reader with the process of implementing a digital system as a full-custom integrated circuit. Second, it covers the principle of switch logic design and provides useful paradigms, which may apply to various static and dynamic logic families. Third, it concretely copes with the fabrication and layout designs of complementary metal-oxide-semiconductor (CMOS) VLSI. Hence, the reader will be able to design a VLSI system with the full-custom skill after reading this book. Fourth, it intends to cover the important issues of modern CMOS processes, including deep submicron devices, circuit optimization, interconnect modeling and optimization, signal integrity, power integrity, clocking and timing, power dissipation, and electrostatic discharge (ESD). As a consequence, the reader not only can comprehensively understand the features and limitations of modern VLSI technologies but also have enough background to adapt to this ever-changing field.

The contents of this book stem largely from the courses VLSI System Designs and Digital Integrated Circuits Analysis and Design, offered yearly at our campus over the past 15 years. Both are elective courses for the undergraduate and first-year graduate. This book, *Introduction to VLSI Systems: A Logic, Circuit, and System Perspective*, is intended to be useful both as a text for students and as a reference book for practicing engineers, or as a self-study book for readers. For classroom use, an abundance of examples are provided throughout the book for helping readers understand the basic features of full-custom VLSI and grasp the essentials of digital logic designs, digital circuit analysis and design, and system design issues. In addition, many figures are used to illustrate the theme concepts of the topics. Abundant review questions are also included in each chapter to help readers test their understanding of the context.

Contents of This Book

The contents of this book are roughly partitioned into three parts. The first part covers Chapters 1 to 6 and focuses on the topics of hierarchical IC design, standard CMOS logic design, introductory physics of metal-oxide-semiconductor (MOS) transistors, device fabrication, physical layout, circuit simulation, and power dissipation and low-power design principles and techniques. The second part comprises Chapters 7 to 9 and deals with static logic and dynamic logic, as well as sequential logic. The third part concentrates on system design issues and covers Chapters 10 to 16 and an appendix. This part mainly takes into account the datapath subsystem designs, memory modules, design methodologies and implementation options, interconnect, power distribution and clock designs, input/output modules, and ESD protection networks, as well as testing and testable designs.

Chapter 1 introduces the features, capabilities, and the perspectives of VLSI systems. This chapter begins with the introduction of a brief history of integrated circuits and the challenges in VLSI design nowadays and in the near future. Then, it describes the VLSI design and fabrication, design principles and paradigms for CMOS logic circuits, as well as the design and implementation options of digital systems.

Chapter 2 deals with fundamental properties of semiconductors, characteristics of *pn* junctions, and features of MOS systems and MOS transistors. The topics of semiconductors include the differences between intrinsic and extrinsic semiconductors. The characteristics of *pn* junctions include basic structure, built-in potential, current equation, and junction capacitance. The basic operations of MOS transistors (MOSFETs) and their ideal current-voltage (I-V) characteristics are investigated in detail. In addition, the scaling theory of CMOS processes, nonideal features, threshold voltage effects, leakage current, short-channel I-V characteristics, temperature effects, and limitations of MOS transistors are also considered. Moreover, the simulation program with integrated circuit emphasis (SPICE) and related models for diodes and MOS transistors along with some examples are presented.

Chapter 3 addresses the basic manufacturing processes of semiconductor devices, including thermal oxidation, the doping process, photolithography, thin-film removal, and thin-film deposition, and their integration. In addition, the postfabrication processes, including the wafer test (or called the wafer probe), die separation, packaging, and the final test as well as the burn-in test, are discussed. Moreover, advanced packaging techniques, such as multichip module (MCM) and 3-D packaging, including system-in-package (SiP), system-on-package (SoP), and system-on-wafer (SoW), are described briefly.

Chapter 4 takes into account the interface between the fabrication process and circuit design, namely, layout design rules. The advanced layout considerations about signal integrity, manufacturability, and reliability in modern deep submicron (nanometer) processes are also considered. The latch-up problem inherent in CMOS processes is also considered in depth. After this, a widely used regular layout structure, the Euler path approach, is introduced with examples.

Chapter 5 concerns the resistance and capacitance parasitics of MOS transistors and their effects. To this end, we begin with the introduction of resistances and capacitances of MOS transistors and then consider the three common approaches to estimating the propagation delays, t_{pHL} and t_{pLH}. After this, we are concerned with cell delay and Elmore delay models. The path-delay optimization problems of logic chains composed of inverters or mixed-type logic gates follow. Finally, logical effect is defined, and its applications to the path-delay optimization problems are explored in great detail.

Chapter 6 presents the power dissipation and low-power designs. In this chapter, we first introduce the power dissipation of CMOS logic circuits and then focus on the low-power design principles and related issues. Finally, the techniques for designing power-manageable components and dynamic power management along with examples are dealt with in depth.

Chapter 7 investigates static logic circuits. For this purpose, we first consider CMOS inverters and its voltage transfer characteristics. NAND and NOR gates are then addressed. Finally, single-rail and dual-rail logic circuits along with typical examples are broadly examined. The logic design principles of these logic circuits are also investigated in depth.

Chapter 8 considers the dynamic logic circuits. In this chapter, we first examine the basic dynamic logic circuits and the partially discharged hazards and its avoidance. After this, nonideal effects of dynamic logic are described in more detail. Finally, we address three classes of dynamic logic: single-rail dynamic logic, dual-rail dynamic logic, and clocked CMOS logic.

Chapter 9 is concerned with the sequential logic circuits. This chapter begins to introduce the fundamentals of sequential logic circuits, covering the sequential logic models, basic bistable devices, and metastable states and hazards, as well as arbiters. A variety of CMOS static and dynamic memory elements, including latches, flip-flops, and pulsed latches, are then considered in depth. The timing issues of systems based on flip-flops, latches, and pulsed latches are also dealt with in detail. Finally, pipeline systems are discussed.

Chapter 10 explores the basic components widely used in datapaths. These include basic combinational and sequential components. The former comprises decoders, encoders, multiplexers, demultiplexers, magnitude comparators, and shifters, and the latter consists of registers and counters. In addition, arithmetic operations, including addition, subtraction, multiplication, and division, are also dealt with in depth. An arithmetic algorithm may usually be realized by using either a multiple-cycle or single-cycle structure. The shift, addition, multiplication, and division algorithms are used as examples to repeatedly manifest the essentials of these two structures.

Chapter 11 describes a broad variety of types of semiconductor memories. The semiconductor memory can be classified in terms of the type of data access and the information retention capability. According to the type of data access, semiconductor memory can be classified into serial access, content addressable, and random access. The random-access memory can be categorized into read/write memory and read-only memory. Read/write memory can be further subdivided into two types: static random access memory (SRAM) and dynamic random access memory (DRAM). According to the information retention capability, semiconductor memory may be classified into volatile and nonvolatile memory. The volatile memory includes static RAM and dynamic RAM while the nonvolatile memory contains ROM, ferroelectric RAM (FRAM), and magnetoresistance RAM (MRAM).

Chapter 12 is concerned with the design methodologies and implementation options of VLSI or digital systems. In this chapter, we begin to describe the related design approaches at both system and register-transfer (RT) levels. Design flows, including both RT and physical levels, and implementation options for digital systems are then addressed. Finally, a case study is given to illustrate how a real-world system can be designed and implemented with a variety of options, including the μP/DSP system, a field-programmable device, and an application-specific integrated circuit (ASIC) with a cell library.

Chapter 13 copes with interconnect and its related issues. Interconnect in a VLSI or digital system mainly provides power delivery paths, clock delivery paths, and signal delivery paths. It plays an important role in any VLSI or digital system because it controls timing, power, noise, design functionality, and reliability. All of these closely related issues are discussed in detail.

Chapter 14 deals with power distribution and clock designs. The design issues of power distribution, power distribution networks, and decoupling capacitors and their related issues are presented in detail. The main goal of a clock system is to generate and distribute one or more clocks to all sequential devices or dynamic logic circuitry in the system with as little skew as possible. For this purpose, the clock system architecture, methods used to generate clocks, and clock distribution networks are focused on. The phase-locked and delay-locked loops are also investigated.

Chapter 15 addresses input/output (I/O) modules and electrostatic discharge (ESD) protection networks. The I/O modules generally include input and output buffers. They play important roles for communicating with the outside of a chip or a VLSI system. Associated with I/O buffers are ESD protection networks that are used to create current paths for discharging the static charge caused by ESD events in order to protect the core circuits from being damaged.

The final chapter (Chapter 16) explores the topics of testability and testable design. The goal of testing is to find any existing faults in a system or a circuit. This chapter first takes a look at VLSI testing and then examines fault models, test vector generation, and testable circuit design or design for testability. After this, the boundary scan standard (IEEE 1149.1), the system-level testing, such as SRAM, a core-based system, system-on-a-chip (SoC), and IEEE 1500 standard, are briefly dealt with.

The appendix surveys some synthesizable features with examples of Verilog hardware description language (Verilog HDL) and SystemVerilog. Through the use of these examples, the reader can readily describe their own hardware modules in Verilog HDL/SystemVerilog. In addition, basic design approaches of test benches are also dealt with concisely. Finally, the complete description of the start/stop timer described in Chapter 12 is presented in the context of Verilog HDL.

Use of This Book for Courses

The author has been using the contents of this book for the following two courses for many years, VLSI Systems Design (or An Introduction to VLSI) and Digital Integrated Circuits Analysis and Design. The objectives of the "VLSI Systems Design" course are to familiarize the student with an understanding of how to implement a digital system as a full-custom integrated circuit, to cover the principles of switch logic design and provide useful paradigms for CMOS logic circuits, to cope concretely with the fabrication and layout designs of CMOS VLSI, and to provide the student with the fundamental understanding of modern VLSI techniques. Based on these, the reader not only can comprehensively understand the features and limitations of modern VLSI technologies but also have enough background to adapt to this ever-changing field. In this course, the following sections are covered. The details can be referred to in lecture notes.

- Sections 1.1 to 1.4
- Sections 3.1 to 3.3
- Sections 4.1 to 4.4
- Sections 5.1 to 5.3 and 6.1
- Sections 7.1 (7.1.2 to 7.1.4), 7.2, and 7.3
- Sections 8.1 (8.1.2 to 8.1.3), 8.3 (8.3.1 to 8.3.3), 8.4 (8.4.1 and 8.4.2), and 8.5

- Sections 9.1 (9.1.1 to 9.1.3), 9.2 (9.2.1 to 9.2.3), 9.3 , and 9.4
- Sections 10.1 to 10.6
- Optional Sections 12.1 to 12.3
- Optional Sections 16.1 to 16.5

The essential aims of the Digital Integrated Circuits Analysis and Design course are to introduce the student to the important issues of modern CMOS processes, including deep submicron devices, circuit optimization, memory designs, interconnect modeling and optimization, low-power designs, signal integrity, power integrity, clocking and timing, power distribution, and electrostatic discharge (ESD). To reach these goals, the following sections are covered. The details can also be referred to in lecture notes.

- Sections 2.1 to 2.5
- Sections 7.1 to 7.3
- Sections 8.1 and 8.2
- Sections 11.1 to 11.6
- Sections 13.1 to 13.4
- Sections 14.1 to 14.3
- Sections 15.1 to 15.4
- Sections 6.1 to 6.4

Of course, instructors who adopt this book are encouraged to customize their own course outlines based on the contents of this book.

Supplements

The instructor's supplements, containing a solution manual and lecture notes in PowerPoint slides, are available for all instructors who adopt this book.

Acknowledgments

Most materials of this book have been taken from the courses ET5302 and ET5006 offered yearly at our campus over the past 15 years. Many thanks are due to the students of these two courses, who suffered through many of the experimental class offerings based on the draft of this book. Valuable comments from the participants of both courses have helped in evolving the contents of this book and are greatly appreciated. Thanks are given to National Chip Implementation Center of National Applied Research Laboratories of Taiwan, R.O.C., for their support in VLSI education and related research in Taiwan over the past two decades. I also gratefully appreciate my mentor, Ben Chen, a cofounder of Chuan Hwa Book Ltd., for his invaluable support and continuous encouragement in my academic carrier over the past decades. I would like to extend special thanks to the people at CRC Press for their efforts in producing this book: in particular, Li-Ming Leong, Joselyn Banks-Kyle, and Jim McGovern. Finally, and most sincerely, I wish to thank my wife, Fanny, and my children, Alice and Frank, for their patience in enduring my absence from them during the writing of this book.

M. B. Lin
Taipei, Taiwan

1

Introduction

Although nowadays most application-specific integrated circuits (ASICs) are designed fully or partially based on a specific synthesis flow using hardware description languages (HDLs) and implemented by either field-programmable gate arrays (FPGAs) or cell libraries, it is increasingly important to understand the fundamentals of circuit and physical designs of very large-scale integration (VLSI) circuits at least due to the following reasons. First, addressing the harder problems requires a fundamental understanding of circuit and physical design issues. Second, distinguished engineers can often develop their physical intuition to estimate the behavior of circuits rapidly without predominantly relying on computer-aided design (CAD) tools.

To achieve the above-mentioned objectives, this book will focus on building an understanding of integrated circuits in a bottom-up fashion and pay much attention to physical layout, logic circuit designs, and system designs. After all, we always believe that the best way to learn VLSI design is by doing it.

To familiarize the reader with the process of implementing a digital system as a full-custom integrated circuit, in this tutorial chapter, we will address a brief history of integrated circuits, the challenges in VLSI design now and in the near future, the perspective on the VLSI design, the VLSI design and fabrication, principles and paradigms of complementary metal-oxide-semiconductor (CMOS) logic designs, as well as the design and implementation options of digital systems.

1.1 Introduction to VLSI

We first review in this section the history of VLSI technology in brief. Next, we introduce a silicon planar process on which modern CMOS processes are founded, and ultimate limitations of feature size. Then, we are concerned with the classification of VLSI circuits, the benefits of using VLSI circuits, the appropriate technologies for manufacturing VLSI circuits, and a brief introduction to scaling theory. Finally, design challenges in terms of deep-submicron (DSM) devices and wires are also dealt with in detail. Economics and future perspective of VLSI technology are explored briefly.

1.1.1 Introduction

In this subsection, we introduce the history of VLSI technology, a basic silicon planar process, and ultimate limitations of feature size.

1.1.1.1 A Brief History The history of VLSI can be dated back to the invention of transistors. In 1947, John Bardeen, Walter Brattain, and William Shockley (all at Bell laboratories, also known as Bell Labs) invented the first *point-contact transistor*. Because of this important contribution, they were awarded the Nobel Prize in Physics in 1956. After this invention, Bell Labs devoted itself to developing *bipolar junction transistors* (BJTs). Its efforts founded the basis of modern BJTs. BJTs soon replaced vacuum tubes and became the mainstream of electronic systems because that they are more reliable, less noisy, and consume less power.

Ten years later after the invention of the transistor, Jack Kilby at Texas Instruments (TI) built the first integrated circuit, which explored the potential of the miniaturization of building multiple transistors on a single piece of silicon. This work established the foundation of *transistor-transistor logic* (TTL) families, which were popular in the 1970s and 1980s. Some parts of them are still available and popular today although these circuits might not be still manufactured as their original design. Because of his invention of the integrated circuit, Jack Kilby was awarded the Nobel Prize in Physics in 2000.

Even though the *metal-oxide-semiconductor field-effect transistor* (MOSFET), or called the *metal-oxide-semiconductor* (MOS) transistor for short, was invented early before the BJT, it was not widely used until the 1970s. In 1925 and 1935, Julius Lilienfield (German) (US patent 1,745,175) and Oskar Heil (British patent 439,457), filed their patents separately. In 1963, Frank Wanlass at Fairchild built the first *complementary metal-oxide-semiconductor* (CMOS) logic gate with discrete components and demonstrated the ultra-low standby power feature of CMOS technology at the cost of two different types, n and p types, of MOS transistors. This circuit is the foundation for the current CMOS realm.

1.1.1.2 Silicon Planar Processes With the advent of the *silicon planar process*, MOS integrated circuits become popular due to their low cost since each transistor occupies less area than a BJT and the manufacturing process is much simpler. The p-type MOS (pMOS) transistors were used at the onset of the MOS process but soon were replaced by n-type MOS (nMOS) transistors due to their poor performance, reliability, and yield. One advantage of nMOS logic circuits is that they need less area than their pMOS counterparts. This advantage was demonstrated by Intel in its development in nMOS technology with 1101, a 256-bit static *random-access memory* (RAM), and 4004, the first 4-bit microprocessor. Nevertheless, in the 1980s, the demand on *integration density* was rapidly increased; the high standby power dissipation of nMOS logic circuits severely limited the degree of integration. Consequently, the CMOS process emerged and rapidly took the place of the nMOS process as the mainstream of VLSI technology although the logic circuits in the CMOS process needed much more area than in the nMOS process. Now the CMOS process has become the most mature and popular technology for VLSI designs.

The essential feature of the silicon planar process is the capability of depositing selective dopant atoms on the specific regions on the surface of silicon to change or modify the electrical properties of these regions. To reach this, as illustrated in Figure 1.1, the following basic steps are performed: (1) an *oxide* (*silicon dioxide*, SiO_2) layer is formed on the surface of silicon, (2) the selective region of the oxide layer is removed, (3) desired dopant atoms are deposited on the surface of silicon and oxide layer, and (4) the dopant atoms on the selective region are diffused onto the silicon. By repeatedly using these four steps, the desired integrated circuit (IC) can be made.

The basic silicon planar process has been refined over and over again since the 1960s. The *feature size*, that is, the allowed minimum line width or spacing between two lines

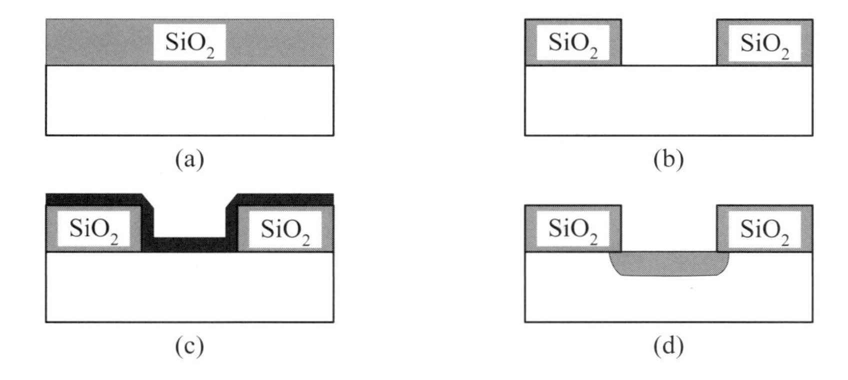

Figure 1.1: The essential steps of the silicon planar process: (a) oxide formation; (b) selective oxide removal; (c) deposition of dopant atoms; (d) diffusion of dopant atoms.

for a given process of an IC has continued to evolve at a rapid rate, from 8 μm in 1969 to 32 nm today (2010). With this evolution, the transistor count on a chip grows steeply, as depicted in Figure 1.2, which shows the evolution of Intel *microprocessors*. From the figure, we can see that the transistor count is almost doubled every two years. This exponential growth of transistor count per chip was first realized by Gorden Moore in 1965 and became the famous Moore's law, which states that *the transistor count per chip would double every 18 to 24 months.* Three years later, Robert Noyce and Gorden Moore founded Intel Corporation, which became the largest chip maker all over the world today.

Moore's law is actually driven by two major forces. First, the feature sizes are dramatically decreased with the refinement of equipment for integrated-circuit technology. Second, the die[1] area is increased to accommodate more transistors since the defect density of a wafer has been reduced profoundly. Nowadays, dies with an area of 2×2 cm^2 are not uncommon. However, we will see that the die yield and the cost of each die are strongly dependent on the die area. To achieve a good die yield, the die area has to be controlled under a bound. The details of die yield and the more general topic, VLSI economics, will be discussed later in this section.

1.1.1.3 Ultimate Limitations of Feature Size There is a limit to how much the feature size can be reduced. At least two factors will prevent feature sizes from being decreased without limit. First of all, as the device size decreases, the statistical fluctuations in the number ($\approx \sqrt{n}$) will become an important factor to limit the performance of the circuit, not only for analog circuits but finally also for digital circuits. This will make the circuit design more difficult than before. Eventually, each device may contain only a few electrons, and the entire concept of circuit design will be different from what we have been using today.

Another factor that limits the reduction of feature size indefinitely is the resolution and equipment of photolithography. Although resolution enhancement techniques such as *optical proximity correction* (OPC), *phase-shift masks*, *double patterning*, and *immersion photolithography* have enabled scaling to the present 45 nm and beyond, along the reduction of feature size, a more expensive photolithography equipment as well as the other manufacturing equipment are needed in the near future. The cost of

[1]A die usually means an integrated circuit on a wafer, while a chip denotes an integrated circuit that has been sliced from a wafer. Sometimes, they are used interchangeably.

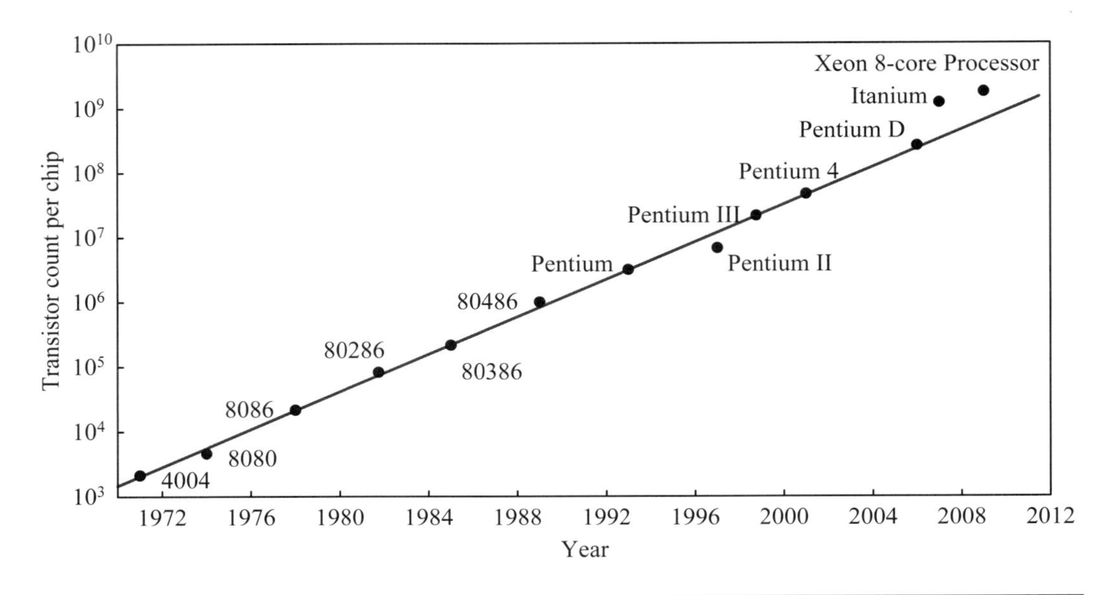

Figure 1.2: An illustration of Moore's law with the evolution of Intel microprocessors.

development and manufacturing equipment will likely limit the feature size available and used to manufacture ICs.

Review Questions

Q1-1. Describe the essential steps of the silicon planar process.

Q1-2. Describe the meaning and implication of Moore's law.

Q1-3. What are the ultimate limitations of feature size?

1.1.2 Basic Features of VLSI Circuits

In this section, we are concerned with the meaning of VLSI circuits, the benefits of using VLSI circuits, the appropriate technologies for manufacturing VLSI circuits, and scaling theory.

1.1.2.1 Classification of VLSI Circuits Roughly speaking, the word "VLSI" means any integrated circuit containing a lot of components, such as transistors, resistors, capacitors, and so on. More specifically, the *integrated circuits* (ICs) can be categorized into the following types in terms of the number of components that they contain.

- *Small-scale integration* (SSI) means an IC containing less than 10 gates.
- *Medium-scale integration* (MSI) means an IC containing up to 100 gates.
- *Large-scale integration* (LSI) means an IC containing up to 1000 gates.
- *Very large-scale integration* (VLSI) means an IC containing beyond 1000 gates.

Here the "gate" means a two-input basic gate, such as an AND gate or a NAND gate. Another common metric to measure the complexity of an IC is the *transistor count*. Both *gate count* and transistor count in an FPGA device, a cell-based design, and a bipolar full-custom design, can be converted from one to another by using the following rule-of-thumb factors.

- *Field programmable gate array* (FPGA): In FPGA devices, because of some unavoidable overhead, 1 gate is approximately equivalent to 7 to 10 transistors, depending on the type of FPGA device under consideration.
- *CMOS cell-based design*: Since a basic two-input NAND gate consists of two pMOS and two nMOS transistors, one basic gate in cell-based design is often counted as four transistors.
- *Bipolar full-custom design*: In BJT logic, such as TTL, each basic gate roughly comprises 10 basic components, including transistors and resistors.

Some popular VLSI circuit examples are as follows: microprocessors, *microcontrollers* (embedded systems), memory devices (static random access memory/SRAM, dynamic random access memory/DRAM, and *Flash memory*), various FPGA devices, and special-purpose processors, such as *digital signal processing* (DSP) and *graphics processing unit* (GPU) devices.

Recently, the term *ultra large-scale integration* (ULSI) is sometimes used to mean a single circuit capable of integrating billions of components. Nevertheless, we will not use this term in this book. Instead, we simply use the term VLSI to mean an IC with a high-degree integration of transistors and/or other components.

1.1.2.2 Benefits of Using VLSI Circuits Once we have classified the ICs, we are now in a position to consider the benefits of using VLSI circuits. In addition to the fact that integration may reduce manufacturing cost because virtually no manual assembly is required, integration significantly improves the design. This can be seen from the following consequences of integration. First, integration reduces parasitics, including capacitance, resistance, even inductance, and hence allows the resulting circuit to be operated at a higher speed. Second, integration reduces the power dissipation and hence generates less heat. A large amount of power dissipation of an IC is due to I/O circuits in which a lot of capacitors need to be charged and discharged when signals are switched. Most of these I/O circuits can be removed through proper integration. Third, the integrated system is physically smaller because of less chip area occupied than the original system. Thus, using VLSI technology to design an electronic system results in higher performance, consumes less power, and occupies less area. These factors reflect into the final product as smaller physical size, lower power dissipation, and reduced cost.

1.1.2.3 VLSI Technologies Many technologies may be used to design and implement an integrated circuit nowadays. The three most popular technologies are CMOS technology, BJT, and *gallium arsenide* (GaAs). Among these, the CMOS technology is the dominant one in the VLSI realm because of its attractive features of lowest power dissipation and highest integration density. The BJT has an inherent feature of higher-operating frequency than its CMOS counterpart. Hence, BJTs are often used in *radio-frequency* (RF) applications. Modern CMOS processes may also embed the fabrication of BJTs as an extension, thereby leading to a hybrid process known as the *Bipolar-CMOS* (BiCMOS) process. To improve the performance of BJT, most modern BiCMOS processes also provide a new type of transistor, called a *silicon-germanium* (SiGe) transistor, to be used in special high-frequency applications such as RF transceivers, which need to operate up to 5 to 10 GHz. The third technology is GaAs, which is special for microwave applications needing an operating frequency up to 100 GHz.

There are two criteria used to evaluate whether a technology is capable of providing high-degree integration. These criteria include power dissipation and the area of each

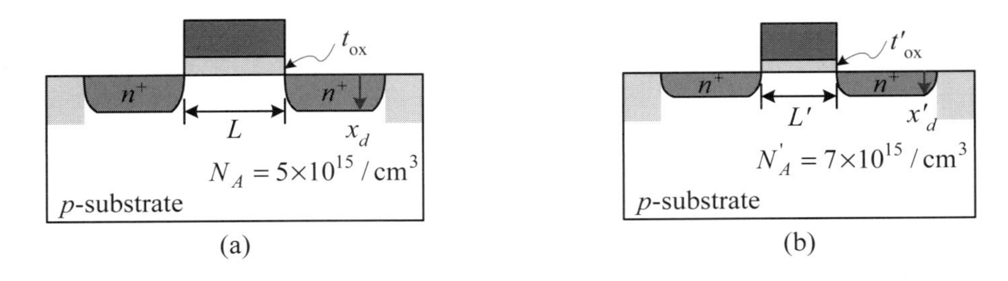

Figure 1.3: An illustration of the scaling principle (not drawn in scale): (a) original device; (b) scaled-down device.

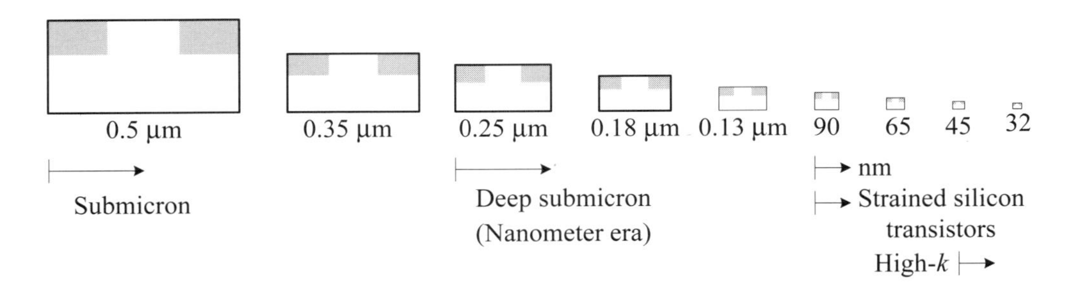

Figure 1.4: The feature-size trends of CMOS processes.

transistor (or logic gate). For the time being, only CMOS technology can meet these two criteria at the same time and hence is widely used in industry to produce a broad variety of chips in the fields of communications, computers, consumer products, multimedia, and so on. However, the power dissipation problem is still the major challenge in designing such CMOS chips since the power dissipation of a single chip may reach up to 100 or 120 watts. Beyond this, the heat-removing mechanism may complicate the entire system design. Consequently, the available commercial devices often limit their power dissipation below this value through low-power design and/or a complex on- and/or off-chip *power management module*.

1.1.2.4 Scaling Theory By the driving force of refinement of feature sizes, Dennard proposed the constant-field scaling theory in 1974. Based on this theory, each dimension, x, y, and z, of a device is scaled down by a factor k, where $0 < k < 1$. In order to maintain the constant field in the device, all operating voltages are required to be scaled down by the same factor of k and the charge density needs to be scaled up by a factor of $1/k$.

An illustration of the scaling principle is revealed in Figure 1.3. The channel length and width of the original device are scaled down by a factor k, namely, $L' = k \cdot L$ and $W' = k \cdot W$; the thickness of silicon dioxide (SiO_2) is also scaled down by the same factor, that is, $t'_{ox} = k \cdot t_{ox}$. The same rule is applied to the maximum depletion depth x_d, that is, $x'_d = k \cdot x_d$.

Figure 1.4 shows the historical scaling records from 0.5 μm down to today, 32 nm. From the figure, we can see that the scaling factor k of each generation is about 0.7. This means that the chip area is decreased by a factor of 2 for each generation, thereby doubling the transistor count provided that the chip area remains the same.

The advantages of scaling down a device are as follows. First, the device density is increased by a factor of $1/k^2$. Second, the circuit delay is shortened by a factor of k. Third, the power dissipation per device is reduced by a factor of k^2. Consequently, the scaled-down device occupies less area, has higher performance, and consumes less power than the original one.

■ Review Questions

Q1-4. What is VLSI?

Q1-5. Explain the reasons why the CMOS technology has become the mainstream of the current VLSI realm.

Q1-6. What are the benefits of using VLSI circuits?

1.1.3 Design Issues of VLSI Circuits

A VLSI manufacturing process is called a *submicron* (SM) process when the feature size is below 1 μm, and a *deep submicron* (DSM) process when the feature size is roughly below 0.25 μm.[2] The corresponding devices made by these two processes are denoted SM devices and DSM devices, respectively. At present, DSM devices are popular in the design of a large-scale system because they provide a more economical way to integrate a much more complicated system into a single chip. The resulting chip is often referred to as a *system-on-a-chip* (SoC) device.

Even though DSM processes allow us to design a very complicated large-scale system, many design challenges indeed exist, in particular, when the feature sizes are beyond 0.13 μm. The associated design issues can be subdivided into two main classes: DSM devices and DSM interconnect.[3] In the following sections, we address each of these briefly.

1.1.3.1 Design Issues of DSM Devices The design issues of DSM devices include *thin-oxide (gate-oxide) tunneling/breakdown*, *gate leakage current*, *subthreshold current*, *velocity saturation*, *short-channel effects* on V_T, *hot-carrier effects*, and *drain-induced barrier lowering* (DIBL) effect.

The device features of typical DSM processes are summarized in Table 1.1. From the table, we can see that the thin-oxide (gate-oxide, that is, silicon dioxide, SiO_2) thickness is reduced from 5.7 nm in a 0.25-μm process down to 1.65 nm in a 32-nm process. The side effects of this reduction are thin-oxide tunneling and breakdown. The thin-oxide tunneling may cause an extra gate leakage current. To avoid thin-oxide breakdown, the operating voltage applied to the gate has to be lowered. This means that the noise margins are reduced accordingly, and the subthreshold current may no longer be ignored. To reduce the gate leakage current, high-k MOS transistors are widely employed starting from a 45-nm process. In high-k MOS transistors, a high-k dielectric is used to replace the gate oxide. Hence, the gate-dielectric thickness may be increased significantly, thereby reducing the gate leakage current dramatically. The actual gate-dielectric thickness depends on the relative permittivity of gate-dielectric material, referring to Section 3.4.1.2 for more details.

In addition, as the channel length of a device is reduced, velocity saturation, short-channel effects on V_T, and hot-carrier effects may no longer be ignored as in the case of a long-channel device. The electron and hole velocities in the channel or silicon bulk

[2]Sometimes a process with a feature size below 100 nm is referred to as a *nanometer* (nm) process.

[3]The wires linking together transistors, circuits, cells, modules, and systems are called interconnected.

Table 1.1: The device features of typical DSM processes.

Process	0.25 μm	0.18 μm	0.13 μm	90 nm	65 nm	45 nm	32 nm
	1998	1999	2000	2002	2006	2008	2010
t_{ox} (nm)	5.7	4.1	3.1	2.5	1.85	1.75	1.65
V_{DD} (V)	2.5	1.8	1.2	1.0	0.80	0.80	0.80
V_T (V)	0.55	0.4	0.35	0.35	0.32	0.32	0.32

Table 1.2: The metal1-wire features of typical DSM processes.

Process	0.25 μm	0.18 μm	0.13 μm	90 nm	65 nm	45 nm	32 nm
Thickness	0.61 μm	0.48 μm	0.40 μm	150 nm	170 nm	144 nm	95 nm
Width/space	0.3 μm	0.23 μm	0.17 μm	110 nm	105 nm	80 nm	56 nm
R/sq (m$\Omega/\square$)	44	56	68	112	100	118	178

is proportional to the applied electric field when the electric field is below a critical value. However, these velocities will saturate at a value of about 8×10^6 cm/sec at 400 K, which is independent of the doping level and corresponds to an electric field with the strength of 6×10^4 V/cm for electrons and 2.4×10^5 V/cm for holes, respectively. When velocity saturation happens, the drain current of a MOS transistor will follow a linear rather than a quadratical relationship with applied gate-to-source voltage.

When the channel length is comparable to the drain *depletion-layer* thickness, the device is referred to as a *short-channel device*. The *short-channel effect* (SCE) refers to the reduction in V_T of a short-channel device and might result from the combination of the following effects: *charge sharing*, DIBL, and *subsurface punchthrough*. Charge sharing refers to the fact that the charge needed to induce a channel is not totally supported by the gate voltage but instead may obtain some help from others. DIBL refers to the influence of drain voltage on the threshold voltage. Subsurface punchthrough refers to the influence of drain voltage on the source *pn*-junction electron barrier.

A *hot electron* is an electron with kinetic energy greater than its thermal energy. Because of the high-enough electric field at the end of channel, electron-hole pairs may be generated in the space-charge region of the drain junction through *impact ionization* by hot electrons. These generated electron and hole carriers may have several effects. First, they may trap into the gate oxide and shift the threshold voltage V_T of the device gradually. Second, they may inject into the gate and become the gate current. Third, they may move into the substrate and become the substrate current. Fourth, they may cause the parasitic BJT to become forward-biased and cause the device to fail.

1.1.3.2 Design Issues of DSM Interconnect The design issues of DSM interconnect arise from *RLC* parasitics and include *IR drop*, *RC delays*, *capacitive coupling*, *inductive coupling*, *Ldi/dt noise*, electromigration, and *antenna effects*. In what follows, we briefly describe each of these.

The wires in a VLSI chip function as conductors for carrying signals, power, and clocks. Because of the inherent resistance and capacitance of wires, each wire has its definite *IR* drop. The metal-wire features of typical DSM processes are summarized in Table 1.2. We can see from the table that the thickness and width of wires are reduced with the evolution of feature sizes. Thereby, the *sheet resistance*, that is, resistance per square, of a wire is increased profoundly.

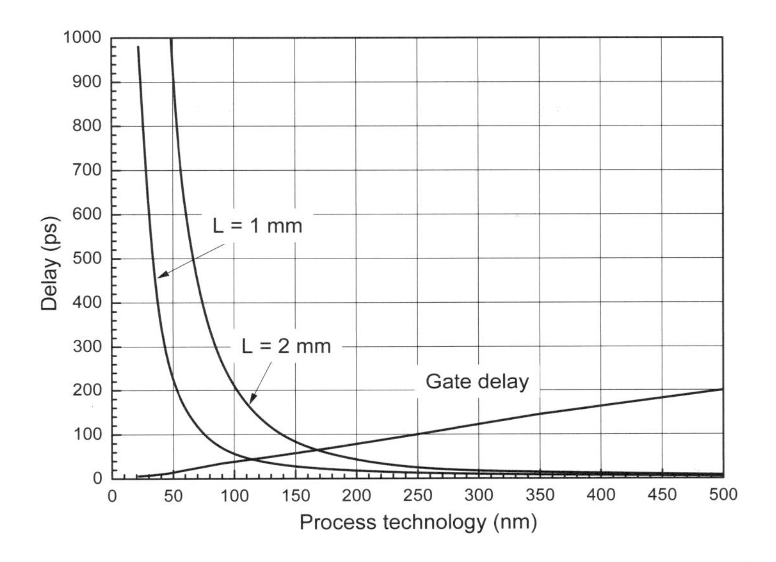

Figure 1.5: The gate versus wire delays in various processes.

The wire resistance leads to the IR drop, which may deteriorate the performance of a logic circuit, even making the logic circuit malfunction. The IR drop becomes more noticeable in DSM processes because of the following reasons. First, the wire resistance increases with the decreasing feature sizes. Second, the currents passing through a wire increase with the reduction of supply voltage. Third, the power dissipation in modern DSM chips increases because more complex functions are to be executed by these chips. This implies that the current is increased accordingly. As a consequence, an important system design issue in designing a modern DSM chip is to reduce the IR drop in the chip.

The combination of wire capacitance and resistance leads to the RC delay, which may deteriorate the performance of logic circuits. With the reduction of feature sizes, the RC delay of wire is increased quadratically, but the gate delay is reduced significantly, with each generation having a factor of $k \approx 0.7$. Consequently, the signal delay through a wire could be easily longer than that through the gate driving it. As illustrated in Figure 1.5, for the given 1-mm wire, the interconnect delays are less than gate delays and can be ignored when the feature sizes are above 130 nm. However, as feature sizes are below 90 nm, the interconnect delays are much longer than gate delays, even dominating the delays of the entire system.

In addition, in DSM processes, the spacing between two adjacent wires is narrowed, even much smaller than the thickness, as shown in Figure 1.6(a). An immediate consequence of this is that the capacitance between these two wires can no longer be overlooked and may result in a phenomenon called *capacitive coupling*, meaning that the signal in one wire is coupled into the other and may collapse the signal there. The capacitive coupling may cause a *signal-integrity problem.*

Besides capacitive coupling, the coupling effects may be caused by the wire inductance in DSM processes. *Faraday's law of induction* states that an induced voltage is produced in a conductor when it is immersed into a time-varying magnetic field produced by a nearby conductor carrying a current. This is called *inductive coupling*. Like capacitive coupling, inductive coupling may also make noises into adjacent wires. Nonetheless, unlike capacitive coupling which is confined between two conductors, in-

(a)

Process (nm)	Metal layers
500	3 (Al)
350	4 (Al)
250	5 (Al)
180	6 (Al, low-k)
130	7/8/9 (Cu, low-k)
90	7/8/9 (Cu, low-k)
65	8/9 (Cu, low-k)
45	8/9 (Cu, low-k)
32	8/9 (Cu, low-k)

(b)

Figure 1.6: The revolution of wires of modern CMOS processes: (a) Four-layer aluminum with tungsten vias in a 0.35-μm technology (Courtesy of International Business Corporation. Unauthorized use not permitted); (b) scaling of metal layers in typical processes.

ductive coupling may affect a large area of circuits in an unpredictable way. This is because the inductive effect is formed in a closed loop, and there may exist many possible *return paths* with different path lengths for a single signal path.

The induced voltage due to a time-varying current of a wire with self-inductance L can be expressed as Ldi/dt. The Ldi/dt effect often arises from the transitions of signals, especially clocks. Although the self-inductance of a wire is very small, the Ldi/dt effect could become pronounced if the current passing through it is changed rapidly. The Ldi/dt effect results in power supply spikes on the underlying circuit, leading to Ldi/dt *noises* and causing the *power-supply integrity problem* if they are large enough.

The *electromigration* is a phenomenon that electrons in a conductor migrates and dislodges the lattice of the conductor. This may result in breaking the conductor. To prevent this from occurring, the current density through a conductor must be controlled below a critical value determined by the conductor. The electromigration is more severely in aluminum than in copper. Hence, in practical processes using aluminum as conductors, the aluminum is usually alloyed with a small fraction of copper to reduce this effect.

A phenomenon that may cause a device to fail before the completion of a manufacture process is known as the *antenna effect*. The antenna effect is the phenomenon that charges are stored on metal wires during manufacturing. The charges stored on metal1 will cause the direct damage of gate oxide if the amount of charges is sufficient to produce a strong enough electric field. Consequently, it is necessary to limit the area of metal1 connected to polysilicon (or poly for short) directly.

1.1.3.3 Design Issues of VLSI Systems With the advance of CMOS processes, the feature sizes are reduced in a way much faster than what we can imagine. Along with this advance, many challenges in VLSI system designs follow immediately. These challenges mainly cover the following: *power distribution network*, *power management*, *clock distribution network*, and *design and test approach*.

A complete power distribution network needs to take into account the effects of IR drop, hot spots, Ldi/dt noises, ground bounce, and electromigration. In addition to IR drop and Ldi/dt noises, the *hot-spot problem* is essential in designing a VLSI system. The hot spots mean that temperatures of some small regions on a chip are curiously higher than the average value of the chip. These hot spots may deteriorate the performance of the chip, even causing the entire chip to fail eventually.

Careful analysis of power dissipation in all modules of a design is essential for keeping power dissipation within an acceptable limit. Since not all modules in a digital system need to be activated at all times, it is possible to only power the modules on demand at a time. Based on this, the power dissipation of a VLSI chip can be controlled under a desired bound. In fact, the power management has become an important issue for modern VLSI chip design. It may even determine whether the resulting product can be successful on the market.

The clock distribution network is another challenge in designing a VLSI system. Since in these systems a large area and a high operating frequency are required to carry out complex logic functions, the clock skew has to be controlled in a very narrow range. Otherwise, the systems may not function correctly. Hence, care must be taken in designing the clock distribution network. In addition, because a large capacitance needs to be driven by the clock distribution network, the power dissipation of the clock distribution network may no longer be ignored and has to be coped with very carefully.

With the advent of high-degree integration of a VLSI chip, the increasing complexity of systems has made the related design much more difficult. An efficient and effective design approach is to use the divide-and-conquer paradigm to limit the number of components that can be handled at a time. However, combining various different modules into a desired system becomes more difficult with the increasing complexity of the system to be designed. Moreover, testing for the combined system is more important and challenging.

■ Review Questions

Q1-7. Explain the meaning of Figure 1.5.

Q1-8. What is the hot-spot problem?

Q1-9. What are the design issues of DSM devices?

Q1-10. What are the design issues of DSM interconnect?

1.1.4 Economics of VLSI

The cost of an IC is roughly composed of two major factors: *fixed cost* and *variable cost*. The fixed cost, also referred to as the *nonrecurring engineering (NRE) cost*, is independent of the sales volume. It is mainly the start cost of the cost from a project until the first successful prototype is obtained. More precisely, the fixed cost covers direct and indirect costs. The direct cost includes the research and design (R&D) cost, manufacturing mask cost, as well as marketing and sales cost; the indirect cost comprises the investment of manufacturing equipments, the investment of CAD tools, building infrastructure cost, and so on. The variable cost is proportional to the product volume and is mainly the cost of manufacturing wafers, namely, *wafer price*, which is roughly in the range between 1,200 and 1,600 USD for a 300-mm wafer.

From the previous discussion, the cost per IC can be expressed as follows:

$$\text{Cost per IC} = \text{Variable cost of IC} + \frac{\text{Fixed cost}}{\text{Volume}} \tag{1.1}$$

The variable cost per IC can be formulated as the following equation:

$$\text{Variable cost of IC} = \frac{\text{Cost of die} + \text{Cost of testing die} + \text{Cost of packaging and final test}}{\text{Final test yield} \times \text{Dies per wafer}} \tag{1.2}$$

The cost of a die is the wafer price divided by the number of good dies and can be represented as the following formula:

$$\text{Cost of die} = \frac{\text{Wafer price}}{\text{Dies per wafer} \times \text{Die yield}} \tag{1.3}$$

The number of dies in a wafer, excluding fragmented dies on the boundary, can be approximated by the following equation:

$$\text{Dies per wafer} = \frac{3}{4}\frac{d^2}{A} - \frac{1}{2\sqrt{A}}d \tag{1.4}$$

where d is the diameter of the wafer and A is the area of square dies. The derivation of this equation is left to the reader as an exercise.

The die yield can be estimated by the following widely used function:

$$\text{Die yield} = \left(1 + \frac{D_0 A}{\alpha}\right)^{-\alpha} \tag{1.5}$$

where D_0 is the defect density, that is, the defects per unit area, in defects/cm^2, and α is a measure of manufacturing complexity. The typical values of D_0 and α are 0.3 to 1.3 and 4.0, respectively. From this equation, it is clear that the die yield is inversely proportional to the die area.

The following two examples exemplify the above concepts about the cost of an IC. In these two examples, we intend to ignore the fixed cost and only take into account the wafer price when calculating die cost.

■ Example 1-1: (Die area is 1 cm^2.) Assume that the diameter of the wafer is 30 cm and the die area is 1 cm^2. The defect density D_0 is 0.6 defects/cm^2 and the manufacturing complexity α is 4. Supposing that the wafer price is 1,500 USD, calculate the cost of each die without considering the fixed cost.

Solution: The number of dies per wafer can be estimated by using Equation (1.4).

$$\begin{aligned}\text{Dies per wafer} &= \frac{3}{4}\frac{d^2}{A} - \frac{1}{2\sqrt{A}}d \\ &= \frac{3}{4}\left(\frac{30^2}{1}\right) - \frac{1}{2\sqrt{1}}30 = 660\end{aligned}$$

The die yield is obtained from Equation (1.5) and is as follows:

$$\begin{aligned}\text{Die yield} &= \left(1 + \frac{D_0 A}{\alpha}\right)^{-\alpha} \\ &= \left(1 + \frac{0.6 \times 1}{4}\right)^{-4} = 0.57\end{aligned}$$

The cost of each die is calculated as follows by using Equation (1.3):

$$\begin{aligned}\text{Cost of die} &= \frac{\text{Wafer price}}{\text{Dies per wafer}\times\text{Die yield}}\\ &= \frac{1,500}{660\times 0.57} = 3.98 \text{ (USD)}\end{aligned}$$

■

■ Example 1-2: (Die area is 2.5 mm × 2.5 mm.) Assume that the diameter of the wafer is 30 cm and the die area is 2.5 mm × 2.5 mm. The defect density D_0 is 0.6 defects/cm^2 and the manufacturing complexity α is 4. The wafer price is 1,500 USD. Calculate the cost of each die without considering the fixed cost.

Solution: The number of dies per wafer can be estimated by using Equation (1.4).

$$\begin{aligned}\text{Dies per wafer} &= \frac{3}{4}\frac{d^2}{A} - \frac{1}{2\sqrt{A}}d\\ &= \frac{3}{4}\left(\frac{30^2}{0.0625}\right) - \frac{1}{2\sqrt{0.0625}}30 = 10,740\end{aligned}$$

The die yield is obtained from Equation (1.5) and is as follows:

$$\begin{aligned}\text{Die yield} &= \left(1+\frac{D_0A}{\alpha}\right)^{-\alpha}\\ &= \left(1+\frac{0.6\times 0.0625}{4}\right)^{-4} = 0.96\end{aligned}$$

The cost of each die is calculated as follows by using Equation (1.3).

$$\begin{aligned}\text{Cost of die} &= \frac{\text{Wafer price}}{\text{Dies per wafer}\times\text{Die yield}}\\ &= \frac{1,500}{10,740\times 0.96} = 0.15 \text{ (USD)}\end{aligned}$$

■

From the above two examples, it is apparent that the die yield is a strong function of die area, namely, inversely proportional to the die area. Thus, in practical applications, it is necessary to limit the die area to control the die yield and hence the die cost within an acceptable value. Note that, in practical chip design projects, the die area, power dissipation, and performance are usually the three major factors to be considered and traded off.

■ Review Questions

Q1-11. What are the two main factors determining the cost of an IC?

Q1-12. Describe the meaning of the nonrecurring engineering (NRE) cost.

Q1-13. What is the major factor of variable cost of an IC?

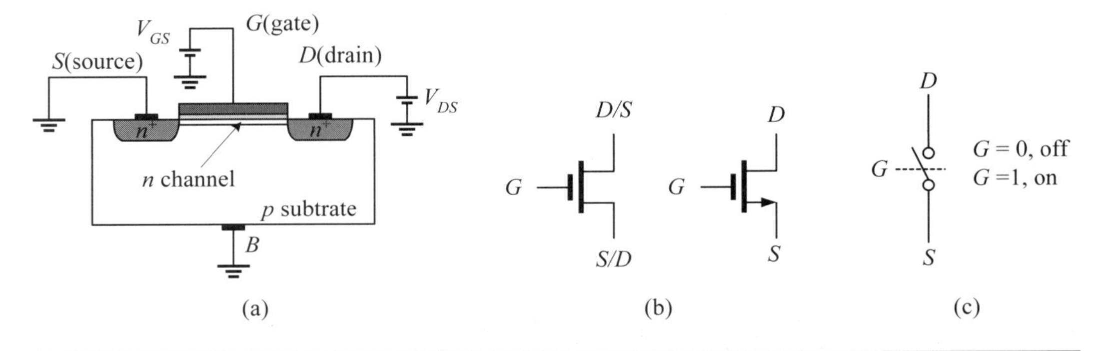

Figure 1.7: The (a) physical structure, (b) circuit symbols, and (c) switch circuit model of an nMOS transistor.

1.2 MOS Transistors as Switches

In this section, we begin to consider basic operations of both nMOS and pMOS transistors and regard these two types of transistors as switches, known as *nMOS switches* and *pMOS switches*, respectively. Then, we consider the combining effects of both nMOS and pMOS transistors as a combined switch, called a *transmission gate* (TG) or a *CMOS switch*. The features of these three types of switches used in logic circuits are also discussed. Finally, we deal with the basic design principles of *switch logic*. Such logic is also referred to as *steering logic* because the data inputs are routed directly to outputs under the control of some specific signals.

1.2.1 nMOS Transistors

The physical structure of an nMOS transistor is basically composed of a metal-oxide-silicon (MOS) system and two n^+ regions on the surface of a p-type silicon substrate, as depicted in Figure 1.7(a). The MOS system is a sandwich structure in which a dielectric (an insulator) is inserted between a metal or a polysilicon and a p-type substrate. The metal or polysilicon is called the *gate*. The two n^+ regions on the surface of the substrate are referred to as *drain* and *source*, respectively.

The operation of an nMOS transistor can be illustrated by Figure 1.7(a). When a large enough positive voltage V_{GS} is applied to the gate (electrode), electrons are attracted toward the silicon surface from the p-type substrate due to a positive electric field built on the silicon surface by the gate voltage. These electrons form a channel between the drain and source. The minimum voltage V_{GS} inducing the channel is defined as the *threshold voltage*, denoted V_{Tn}, of the nMOS transistor. The value of V_{Tn} ranges from 0.3 V to 0.7 V for the present submicron and deep-submicron processes, depending on a particular process of interest.

For digital applications, an nMOS transistor can be thought of as a simple switch element. The switch is turned on when the gate voltage is greater than or equal to its threshold voltage and turned off otherwise. Because of the symmetric structure of an nMOS transistor, either of the n^+ regions can be used as the source or drain, depending on how the operating voltage is applied. The one with more positive voltage is the drain, and the other is the source because the carriers on the nMOS transistor are electrons.

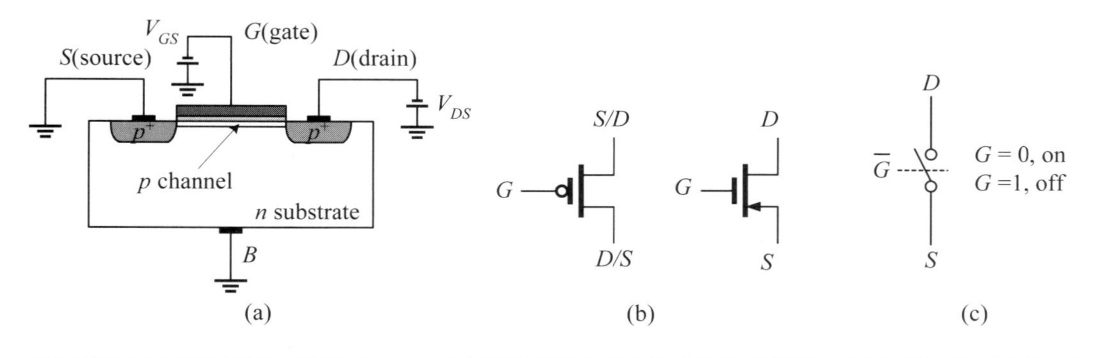

Figure 1.8: The (a) physical structure, (b) circuit symbols, and (c) switch circuit model of a pMOS transistor.

Figure 1.7(b) shows the circuit symbols often used in circuit designs. The one with an explicit arrow associated with the source electrode is often used in analog applications, where the roles of source and drain are fixed. The other without an explicit arrow is often used in digital applications because the roles of the drain and source will be dynamically determined by the actual operating conditions of the circuit. The switch circuit model is depicted in Figure 1.7(c).

1.2.2 pMOS Transistors

Likewise, the physical structure of a pMOS transistor comprises a MOS system and two p^+ regions on the surface of an n-type silicon substrate, as depicted in Figure 1.8(a). The MOS system is a sandwich structure in which a dielectric (an insulator) is inserted between a metal or polysilicon and an n-type substrate. The metal or polysilicon is called a gate. The two p^+ regions on the surface of substrate are referred to as the drain and source, respectively.

The operation of a pMOS transistor can be illustrated by Figure 1.8(a). When a large negative voltage V_{GS} is applied to the gate (electrode), holes are attracted toward the silicon surface from the n-type substrate due to a negative electric field being built on the silicon surface by the gate voltage. These holes form a channel between the drain and source. The minimum voltage $|V_{GS}|$ inducing the channel is defined as the threshold voltage, denoted V_{Tp}, of the pMOS transistor. The value of V_{Tp} ranges from -0.3 V to -0.7 V for the present submicron and deep-submicron processes, depending on a particular process of interest.

Like an nMOS transistor, a pMOS transistor can be regarded as a simple switch element for digital applications. The switch is turned on when the gate voltage is less than or equal to its threshold voltage and turned off otherwise. Because of the symmetric structure of a pMOS transistor, either of the p^+ regions can be used as the source or drain, depending on how the operating voltage is applied. The one with more positive voltage is the source, and the other is the drain because the carriers on the pMOS transistor are holes.

Figure 1.8(b) shows the circuit symbols often used in circuit designs. The symbol convention of pMOS transistors is exactly the same as that of nMOS transistors. The one with an explicit arrow associated with the source electrode is often used in analog applications in which the roles of source and drain are fixed. The other without an explicit arrow but with a circle at the gate is often used in digital applications because

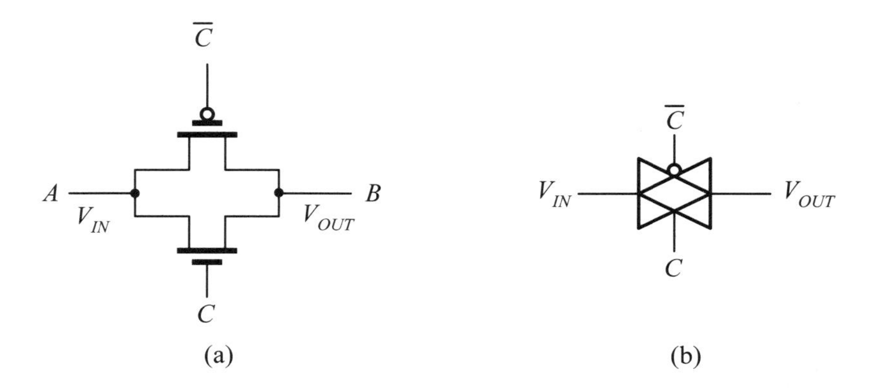

Figure 1.9: The (a) circuit structure and (b) logic symbol of a TG switch.

the roles of drain and source will be dynamically determined by the actual operating conditions of the circuit. The circle is used to distinguish it from the nMOS transistor and to indicate that the pMOS transistor is at active-low enable. The switch circuit model is depicted in Figure 1.8(c).

1.2.3 CMOS Transmission Gates

Since a voltage of magnitude V_{Tn} between the gate and source of an nMOS transistor is required to turn on the transistor, the maximum output voltage of an nMOS switch is equal to $V_{DD} - V_{Tn}$, provided that V_{DD} is applied to both gate and drain electrodes. Similarly, the minimum output voltage of a pMOS switch is equal to $|V_{Tp}|$, provided that 0 V is applied to both gate and drain electrodes. The above two statements can be restated in terms of information transfer by letting 0 V represent logic 0 and V_{DD} denote logic 1 as follows. The nMOS transistor can pass 0 perfectly but cannot pass 1 without degradation; the pMOS transistor can pass 1 perfectly but cannot pass 0 without degradation.

The aforementioned shortcomings of nMOS and pMOS transistors may be overcome by combining an nMOS transistor with a pMOS transistor as a parallel-connected switch, referred to as a transmission gate (TG) or a CMOS switch, as shown in Figure 1.9. Since both nMOS and pMOS transistors are connected in parallel, the imperfect feature of one transistor will be made up by the other. Figure 1.9(a) shows the circuit structure of a TG switch and Figure 1.9(b) shows the logic symbol often used in logic diagrams.

Even though using TG switches may overcome the degradation of information passing through them, each TG switch needs two transistors, one nMOS and one pMOS. This means that the use of TG switches needs more area than the use of nMOS switches or pMOS switches alone. In practice, for area-limited applications, the use of nMOS transistors is much more preferable to pMOS transistors since the electron mobility is much greater than hole mobility. Hence, nMOS transistors perform much better than pMOS transistors.

■ Review Questions

Q1-14. Describe the operation of nMOS transistors.

Q1-15. Describe the operation of pMOS transistors.

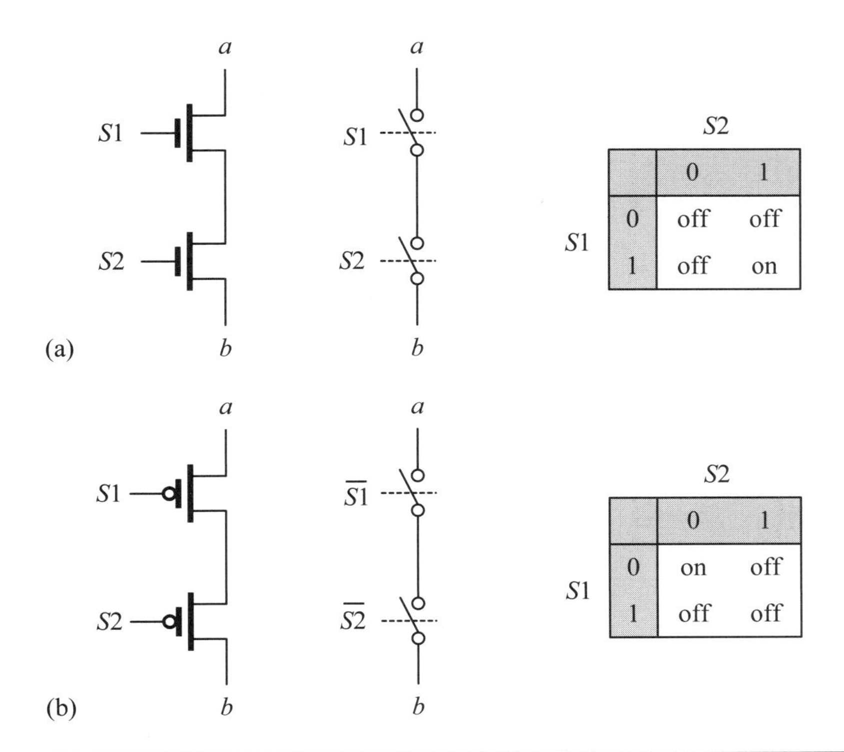

Figure 1.10: The operations of series switches: (a) using nMOS switches; (b) using pMOS switches.

Q1-16. Describe the operation of CMOS switches.
Q1-17. What is the drawback of an nMOS transistor when used as a switch?
Q1-18. What is the drawback of a pMOS transistor when used as a switch?

1.2.4 Simple Switch Logic Design

As introduced, any of three switches, nMOS, pMOS, and TG, may be used as a switch to control the close (on) or open (off) status of two points. Based on a proper combination of these switches, a switch logic circuit can be constructed. We begin with the discussion of compound switches and then introduce a systematic design methodology for constructing a switch logic circuit from a given switching function.

1.2.4.1 Compound Switches For many applications, we often combine two or more switches in a serial, parallel, or combined fashion to form a compound switch. For instance, the case of two switches being connected in series to form a compound switch is shown in Figure 1.10. The operation of the resulting switch is controlled by two control signals: $S1$ and $S2$. The compound switch is turned on only when both control signals $S1$ and $S2$ are asserted and remains in an off state otherwise.

Recall that to activate an nMOS switch we need to apply a high-level voltage to its gate, and to activate a pMOS switch we need to apply a low-level voltage to its gate. As a result, the compound nMOS switch shown in Figure 1.10(a) is turned on only when both control signals $S1$ and $S2$ are at high-level voltages (usually V_{DD}) and remains in an off state in all other combinations of control signals. The compound

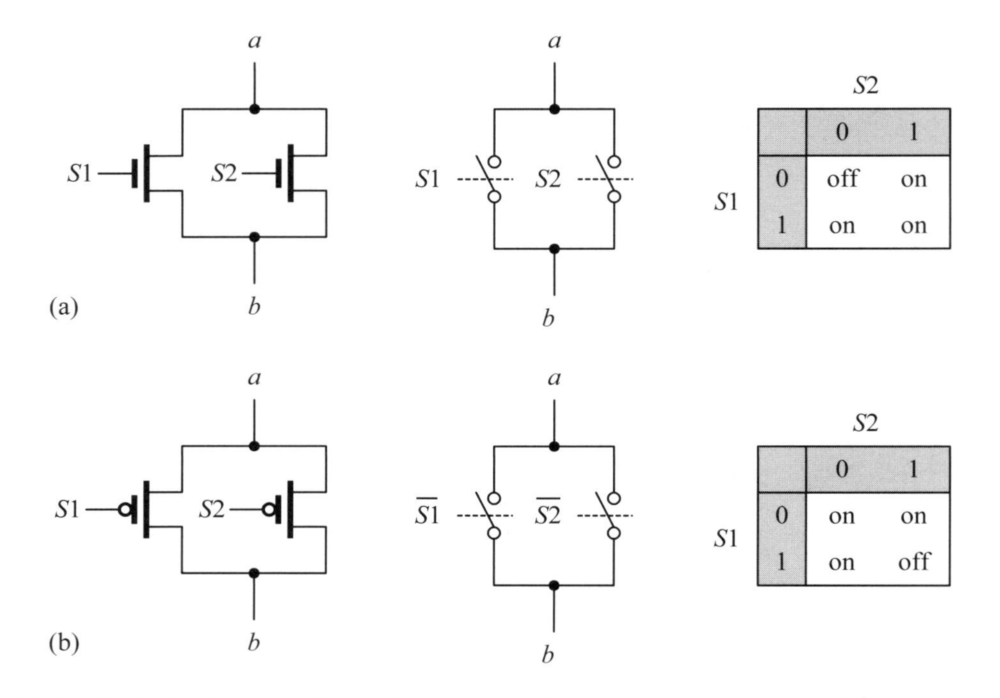

Figure 1.11: The operations of parallel switches: (a) using nMOS switches; (b) using pMOS switches.

pMOS switch depicted in Figure 1.10(b) is turned on only when both control signals $S1$ and $S2$ are at low-level voltages (usually at the ground level) and remains in an off state in all other combinations of control signals.

Figure 1.11 shows the case of two switches being connected in parallel to form a compound switch. The operation of the resulting switch is controlled by two control signals: $S1$ and $S2$. The compound switch is turned on whenever either switch is on. Therefore, the compound switch is turned off only if both control signals $S1$ and $S2$ are deasserted and remains in an on state otherwise.

In Figure 1.11(a), the compound nMOS switch is turned on whenever one control signal of $S1$ and $S2$ is at a high-level voltage (usually V_{DD}) and remains in an off state only when both control signals are at the ground level. In Figure 1.11(b), the compound pMOS switch is turned on whenever one control signal of $S1$ and $S2$ is at the ground level and remains in an off state only when both control signals $S1$ and $S2$ are at high-level voltages.

1.2.4.2 The f/$\bar{\text{f}}$ Paradigm A simple logic pattern utilizing the aforementioned features of both nMOS and pMOS switches to implement a switching function is referred to as the $f/\bar{f}$ *paradigm*, fully CMOS (FCMOS) logic, or CMOS logic for short. The rationale behind the $f/\bar{f}$ paradigm is on the observation that pMOS switches are ideal switches for transferring logic-1 signals, whereas nMOS switches are ideal switches for transferring logic-0 signals. Consequently, we apply pMOS switches to implement the true switching function f while applying nMOS switches to implement the complementary switching function $\bar{f}$.

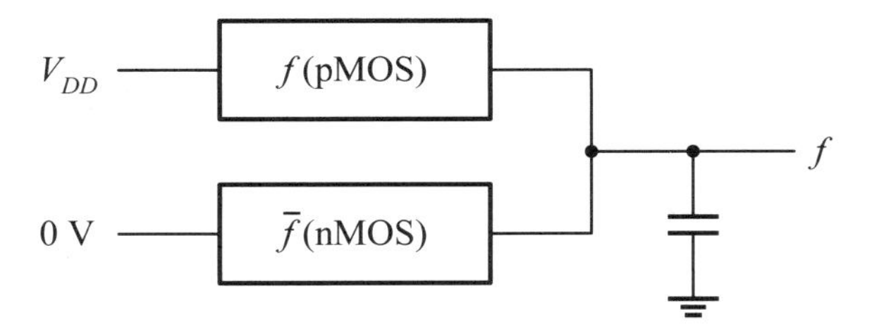

Figure 1.12: The block diagram of the $f/\bar{f}$ paradigm.

The block diagram of the $f/\bar{f}$ paradigm is shown in Figure 1.12, where two logic blocks are referred to as the f block and $\bar{f}$ block, respectively. The f block realizes the true switching function and connects the output to V_{DD} (namely, logic 1) when it is on, while the $\bar{f}$ block realizes the complementary switching function and connects the output to ground (namely, logic 0) when it is on. The capacitor at the output node denotes the parasitic capacitance inherent in the output node.

In what follows, we give two examples to illustrate how to design practical switch logic circuits using the $f/\bar{f}$ paradigm. The first example is to design a two-input NAND gate.

■ Example 1-3: (A two-input NAND gate.) The switching function of a two-input NAND gate is $f(x, y) = \overline{x \cdot y}$. Use the $f/\bar{f}$ paradigm to design and realize it.

Solution: By DeMorgan's law, $f(x, y) = \overline{x \cdot y} = \bar{x} + \bar{y}$. Hence, the f block is realized by two parallel pMOS transistors. The $\bar{f}$ block is realized by two series nMOS transistors because the complementary function of f is $\bar{f}(x, y) = x \cdot y$. The resulting switch and CMOS logic circuits are shown in Figures 1.13(a) and (b), respectively.

■

It is evident from Figure 1.13 that both functions of f and $\bar{f}$ blocks are dual with each other. The following example further demonstrates how a more complex logic circuit known as an *AND-OR-Inverter* (AOI) gate can be designed using the $f/\bar{f}$ paradigm and realized with both types of nMOS and pMOS switches.

■ Example 1-4: (An AOI gate.) Realize the following switching function $f(w, x, y, z) = \overline{w \cdot x + y \cdot z}$ using the $f/\bar{f}$ paradigm.

Solution: By DeMorgan's law, $f(w, x, y, z) = \overline{w \cdot x} \cdot \overline{y \cdot z} = (\bar{w} + \bar{x}) \cdot (\bar{y} + \bar{z})$. Hence, the f block is realized by two pairs of pMOS transistors connected in parallel and then these two pairs are connected in series. The $\bar{f}$ block is realized by two pairs of nMOS transistors connected in series and then these two pairs are connected in parallel because the complementary function of f is $\bar{f}(w, x, y, z) = w \cdot x + y \cdot z$. The resulting switch and CMOS logic circuits are shown in Figures 1.14(a) and (b), respectively.

■

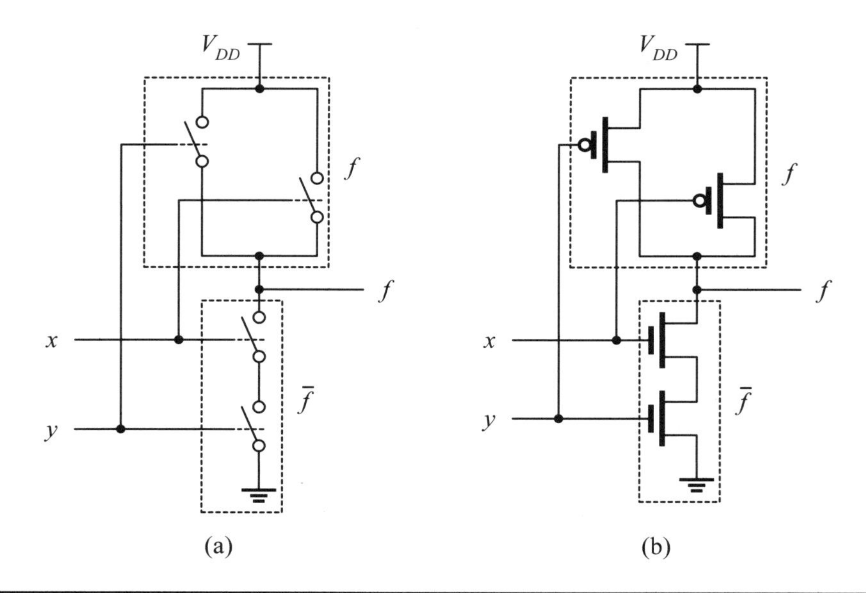

Figure 1.13: The (a) switch logic circuit and (b) CMOS logic circuit of a two-input NAND gate.

Review Questions

Q1-19. Using the $f/\bar{f}$ paradigm, design a CMOS two-input NOR gate.

Q1-20. Using the $f/\bar{f}$ paradigm, design a CMOS three-input NAND gate.

Q1-21. Using the $f/\bar{f}$ paradigm, design a CMOS three-input NOR gate.

1.2.5 Principles of CMOS Logic Design

The basic design principles of CMOS logic can be illustrated by exploring the difference between gate logic and switch logic circuits depicted in Figures 1.15(a) and (b), respectively. In Figure 1.15(a), the output f is 1 if both inputs x and y are 1 and is 0 otherwise; in Figure 1.15(b), the output f is 1 if both inputs x and y are 1 and is undefined otherwise. Consequently, the switch logic circuit shown in Figure 1.15(b) cannot realize the function of Figure 1.15(a) exactly because it only performs the function when both x and y are 1 but nothing else.

1.2.5.1 Two Fundamental Rules A general logic circuit has two definite values: logic 0 (ground) and logic 1 (V_{DD}). In order for a switch logic circuit to specify a definite value for every possible combination of input variables, the following two rules should be followed to correctly and completely realize a switching function using CMOS switches.

- **Rule 1 (node-value rule):** The signal summing point (such as f) must always be connected to 0 (ground) or 1 (V_{DD}) at any time.
- **Rule 2 (node-conflict-free rule):** The signal summing point (such as f) must never be connected to 0 (ground) and 1 (V_{DD}) at the same time.

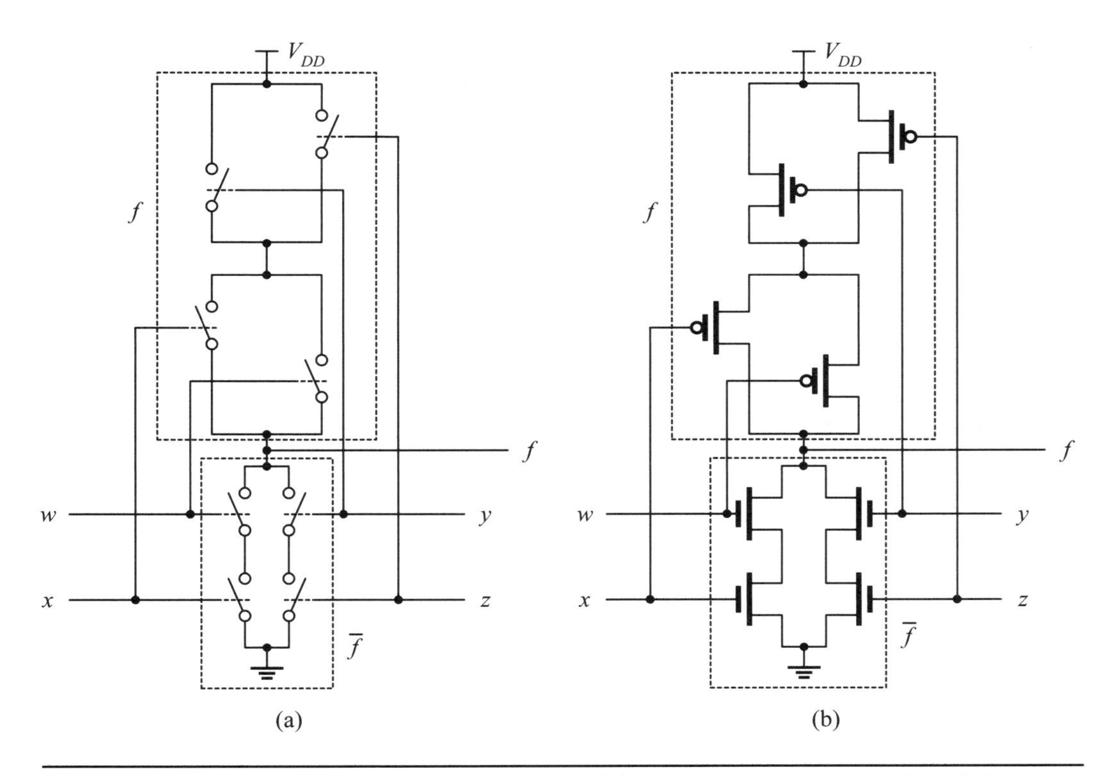

Figure 1.14: The (a) switch logic circuit and (b) CMOS logic circuit of an AOI gate.

Any logic circuit must always follow **Rule 1** to work correctly. **Rule 2** distinguishes a ratioless logic circuit from a ratioed one. A CMOS logic circuit is said to be *ratioless* if it follows both rules and is *ratioed* logic if **Rule 2** is violated, but the sizes of both pull-up and pull-down paths are set appropriately.

A ratioless logic circuit can always perform the designated switching function correctly regardless of the relative sizes of nMOS, pMOS, or TG switches. In contrast, for a ratioed logic circuit to function properly, the relative sizes of nMOS, pMOS, or TG switches used in the circuit must be set appropriately.

1.2.5.2 Residues of a Switching Function The principle of CMOS switch logic design is governed by Shannon's expansion (or decomposition) theorem, which is stated as follows.

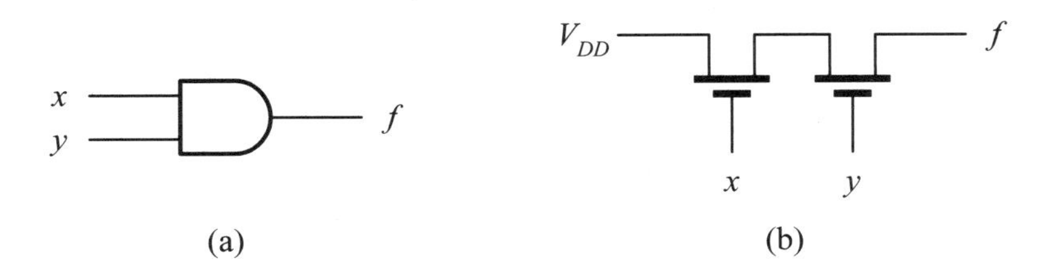

Figure 1.15: The difference between (a) gate logic circuit and (b) switch logic circuit.

Theorem 1.2.1 (Shannon's expansion theorem.)
Let $f(x_{n-1}, \cdots, x_{i+1}, x_i, x_{i-1}, \cdots, x_0)$ be an n-variable switching function. Then f can be decomposed with respect to variable x_i, where $0 \leq i < n$, as follows:

$$f(x_{n-1}, \cdots, x_{i+1}, x_i, x_{i-1}, \cdots, x_0) = x_i \cdot f(x_{n-1}, \cdots, x_{i+1}, 1, x_{i-1}, \cdots, x_0) + \bar{x}_i \cdot f(x_{n-1}, \cdots, x_{i+1}, 0, x_{i-1}, \cdots, x_0) \tag{1.6}$$

The proof of Shannon's expansion theorem is quite trivial; hence, we omit it here. For convenience, we will refer to the variable x_i as the *control variable* and the other variables as *function variables*.

Example 1-5: (Shannon's expansion theorem.) Consider the switching function $f(x, y, z) = xy + yz + xz$ and decompose it with respect to variable y.

$$\begin{aligned} f(x, y, z) &= xy + yz + xz \\ &= y \cdot f(x, 1, z) + \bar{y} \cdot f(x, 0, z) \\ &= y \cdot (x + z + xz) + \bar{y} \cdot (xz) \\ &= xy + yz + x\bar{y}z \end{aligned}$$

It is easy to show that the last line is indeed the same as the original switching function. Hence, we have shown the validity of Shannon's expansion theorem.

■

For convenience, we often distinguish a variable from a literal. A *variable* x is an object that can hold two values, 1 and 0, and can appear in one of two forms, x and $\bar{x}$. A *literal* is a variable in either true or complementary form. In other words, x and $\bar{x}$ denote two different literals but the same variable x. Once we have these, we define the residue of a switching function $f(X)$ with respect to a combination of a subset of $X = \{x_{n-1}, x_{n-2}, \cdots, x_1, x_0\}$ as follows.

Residues of a switching function. *Let $X = \{x_{n-1}, x_{n-2}, \cdots, x_1, x_0\}$ be the set of all n variables and $Y = \{y_{m-1}, y_{m-2}, \cdots, y_1, y_0\}$ be a subset of X, where $y_i \in X$ and $m \leq n$. The residue of $f(X)$ with respect to Y, denoted $f_Y(X)$, is defined to be the function value when all complementary literals in Y are set to 0 and all true literals in Y are set to 1.*

Based on this definition, Shannon's expansion theorem can be thought of as a combination of two residues of the switching function, namely, $f(X) = x_i f_{x_i}(X) + \bar{x}_i f_{\bar{x}_i}(X)$.

Example 1-6: (Residue of a switching function.) Compute the residue of the following switching function with respect to variables x:

$$f(x, y, z) = xy + yz + xz$$

Solution: By definition, the residue of f with respect to variable x is obtained by setting x to 1 and computing the value of the switching function f. That is,

$$f_x(1, y, z) = 1 \cdot y + yz + 1 \cdot z = x + y$$

Hence, the residue of f with respect to variable x is $x + y$.

■

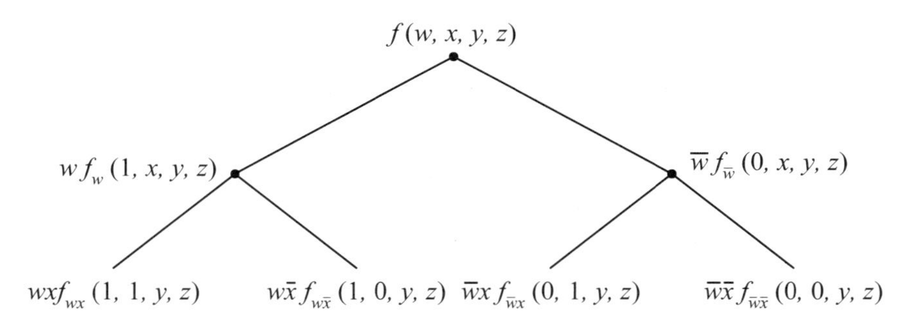

Figure 1.16: A decomposition tree for a four-variable switching function $f(w, x, y, z)$.

With the definition of residue, a switching function can be represented as a combination of two residues. For instance, $f(x, y, z) = x \cdot f_x(1, y, z) + \bar{x} \cdot f_{\bar{x}}(0, y, z)$. The interested reader is invited to verify this. By repeatedly applying Shannon's expansion theorem to further decompose residue functions, a binary tree is formed. Such a tree is often referred to as a *decomposition tree*. An example of a decomposition tree for a four-variable switching function $f(w, x, y, z)$ is exhibited in Figure 1.16, which is formed by decomposing the switching function with respect to variables w and x in order. The switching function f can then be expressed as a combination of the four residues as follows:

$$\begin{aligned} f(w,x,y,z) &= w \cdot f_w(1,x,y,z) + \bar{w} \cdot f_{\bar{w}}(0,x,y,z) \\ &= w \cdot [x \cdot f_{wx}(1,1,y,z) + \bar{x} \cdot f_{w\bar{x}}(1,0,y,z)] + \\ &\quad \bar{w} \cdot [x \cdot f_{\bar{w}x}(0,1,y,z) + \bar{x} \cdot f_{\bar{w}\bar{x}}(0,0,y,z)] \\ &= wxf_{wx}(1,1,y,z) + w\bar{x}f_{w\bar{x}}(1,0,y,z) + \\ &\quad \bar{w}xf_{\bar{w}x}(0,1,y,z) + \bar{w}\bar{x}f_{\bar{w}\bar{x}}(0,0,y,z) \end{aligned} \tag{1.7}$$

Each internal node of a decomposition tree can be realized by a 2-to-1 multiplexer with the control variable as its source selection variable and the residues as its two inputs. The resulting circuit is called a *tree network*.

The residues of a switching function f can also be found graphically. To this end, the residue map of a switching function f is defined as follows.

Residue map of a switching function. *A residue map is a two-dimensional graph with 2^m columns and 2^{n-m} rows. Each of 2^m columns corresponds to a combination of variables $Y = \{y_{m-1}, y_{m-2}, \cdots, y_1, y_0\}$, labeled in the increasing order, and each of 2^{n-m} rows corresponds to a combination of variables in the set of $X - Y$, labeled in Gray-code order. The cell at the intersection of each column and each row corresponds to a combination of variables $X = \{x_{n-1}, x_{n-2}, \cdots, x_1, x_0\}$ and is denoted the decimal value. To represent a switching function on the residue map, we circle the true minterm and leave the false minterm untouched. In addition, the don't care minterms are starred.*

The above definition is illustrated in Figure 1.17. Based on this definition, the variables in the set Y are referred to as *control variables* and the variables in the set $X - Y$ as called *function variables*. Of course, the set of function variables may be empty.

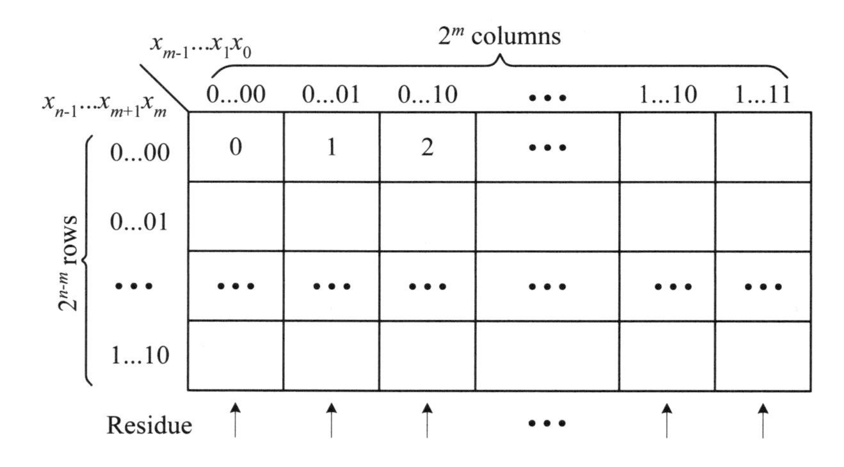

Figure 1.17: The general residue map of an n-variable switching function.

Once we have prepared a residue map for a switching function with respect to a subset Y of the set of variables X, we can further use it to find the residues of the switching function. The general procedure is as follows.

1. Examine each column of the residue map from left to right. The residue of the column is 0 if no cell in the column is circled. The residue of the column is 1 if at least one cell in the column is circled and the other cells are circled or starred.
2. The residue is a function of function variables (namely, variables not in the subset Y) if only a proper subset of cells in the column is circled. The starred cells in the same column may be used to help simplify the residue function in this case.

The following example illustrates how to apply the above procedure to find the residues of a switching function.

■ Example 1-7: (An example of using a residue map.) Assume that $f(w, x, y, z) = \sum(0, 1, 2, 3, 4, 11, 12, 15)$. Using the residue map, find the residues of $f(w, x, y, z)$ with respect to $Y = \{w, x, y\}$.

Solution: The residue map of f is depicted in Figure 1.18. According to the above procedure, the residues of columns 0 and 1 are 1 because all cells in these two columns are circled. The residues of columns 3 and 4 are 0 because no cells in these two columns are circled. The residues of columns 2 and 6 are $\bar{z}$ because only the cells corresponding to the row of $z = 0$ are circled. The residues of columns 5 and 7 are z because only the cells corresponding to the row of $z = 1$ are circled.

■

The following example shows how the don't care minterms may help find the residues of a switching function.

■ Example 1-8: (Another example of using a residue map.) Assume that $f(v, w, x, y, z) = \sum(1, 9, 11, 13, 25, 26, 27) + \sum_{\phi}(6, 14, 17, 19, 22, 30)$. Using the residue map, find the residues of $f(v, w, x, y, z)$ with respect to $Y = \{x, y, z\}$.

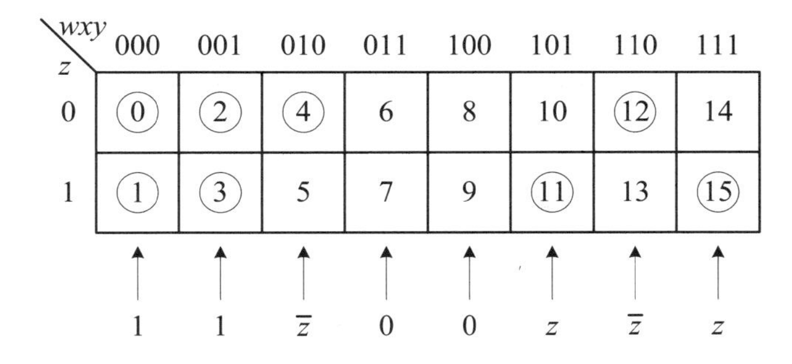

Figure 1.18: An example of using a residue map to find residues.

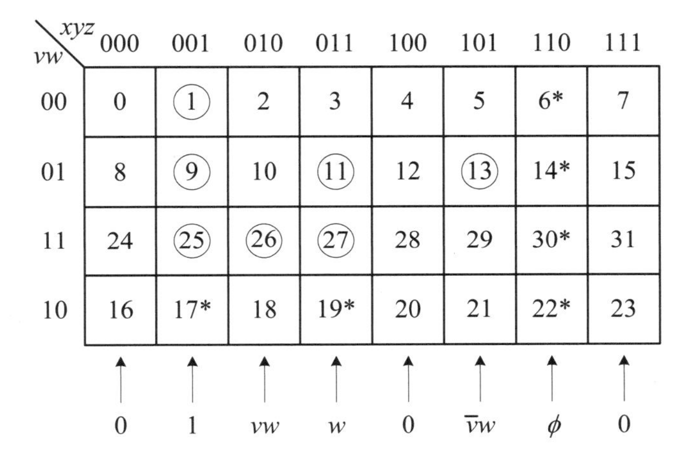

Figure 1.19: A more complicated example of using a residue map to find residues.

Solution: The residue map of f is depicted in Figure 1.19. According to the above procedure, the residues of columns 0, 4, and 7 are 0 because no cells in these three columns are circled. The residue of column 1 is 1 because three cells in this column are circled and the remaining cell is starred. The residue of column 6 is ϕ because all cells in the column are don't care (that is, no cell is circled). The residue of column 2 is vw since only the cell corresponding to row $vw = 11$ is circled. The residue of column 3 is w because the cells corresponding to rows $vw = 01$ and $vw = 11$ are circled. These two cells are combined into the result w. The residue of column 5 is $\bar{v}w$ because only the cell corresponding to the row of $vw = 01$ is circled.

■

1.2.5.3 Systematic Design Paradigms

By using Shannon's expansion theorem, we can derive two basic systematic paradigms that can be used to design a ratioless logic circuit. Depending on how the control variable is chosen at each decomposition, tree networks can be further categorized into a *freeform-tree network* and a *uniform-tree network*.

- **Freeform-tree networks.** In the freeform-tree network, each nontrivial node of the decomposition tree has its total freedom to select the control variable as it is

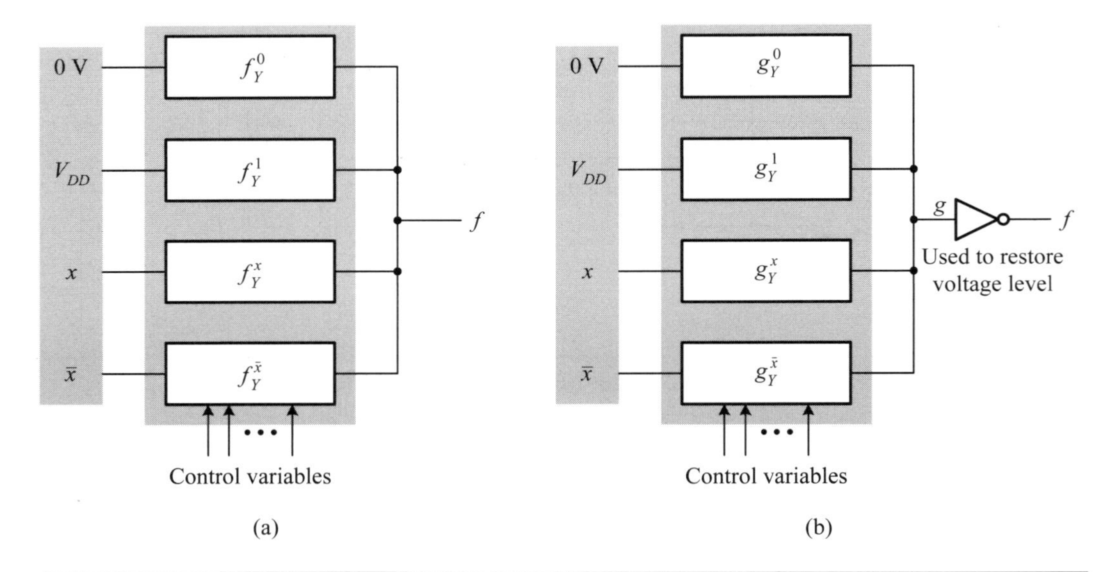

Figure 1.20: The general block diagrams of a $0/1$-$x/\bar{x}$-tree network (x is a function variable): (a) using TG switches; (b) using nMOS switches.

further decomposed. Hence, the survived literals in two leaf nodes need not be the same variable. Both sets of control and function variables need not be disjoint but are disjoint at the same branch.

- **Uniform-tree networks.** In the uniform-tree network, all nontrivial nodes at the same height of the decomposition tree use the same control variable when they are further decomposed. As in the freeform-tree network, the survived literals in two leaf nodes need not be the same variable. Two special cases are $0/1$-$x/\bar{x}$-tree and 0/1-tree networks.

These networks can generally be realized in any type of nMOS, pMOS, and TG switches, or their combinations. However, excepting in 0/1-tree networks, the pMOS switches are usually not used because of their poor performance compared with nMOS switches.

1.2.5.4 0/1-x/x̄-Tree Networks The $0/1$-$x/\bar{x}$-tree network is a special case of uniform-tree network in which $n-1$ variables are used as control variables and all leaf nodes have the same height in the decomposition tree. As a result, each residue can only have one of four values, 0, 1, x, and $\bar{x}$. The combinations of control variables (Y) generating the same residue are then combined into a switching function, denoted as f_Y^s, where $s \in \{0, 1, x, \bar{x}\}$. Each switching function may be further simplified prior to being realized by a TG or an nMOS switch network, as shown in Figure 1.20.

■ Example 1-9: (A simple example of a 0/1-x/x̄-tree network.) Realize the following switching function using a $0/1$-$x/\bar{x}$-tree network and nMOS switches:

$$f(x, y, z) = xy + yz + xz$$

Solution: Assume that the inverter at the output is not necessary; that is, we can tolerate the degradation of logic-1 signal. By computing the residues with respect to all combinations of $Y = \{y, z\}$, the functions of f_Y^0, f_Y^1, f_Y^x, and $f_Y^{\bar{x}}$ are $\{\bar{y}\bar{z}\}$,

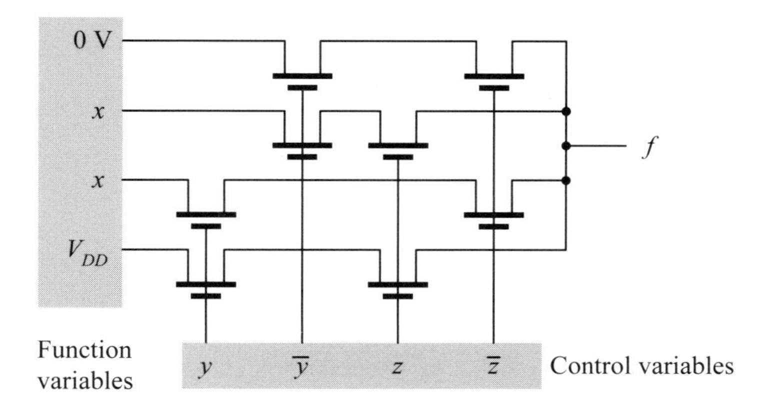

Figure 1.21: The realization of $f(x, y, z) = xy + yz + xz$ with a $0/1$-$x/\bar{x}$-tree network.

$\{yz\}$, $\{\bar{y}z, y\bar{z}\}$, and ϕ (empty set), respectively. The resulting logic circuit is shown in Figure 1.21.

■

The following example demonstrates the situation that the f_Y^s of $0/1$-$x/\bar{x}$-tree network can be further simplified.

■ Example 1-10: (A more complicated example of a 0/1-x/x̄-tree network.) Realize the following switching function using a $0/1$-$x/\bar{x}$-tree network and nMOS switches:

$$f(w, x, y, z) = \overline{\bar{w}\bar{x} + x\bar{y}\bar{z} + wyz}$$

Solution: According to the block diagram shown in Figure 1.20(b) and assuming that an inverter is used at the output to restore the output voltage level, the actual switching function to be realized is the complement of $f(w, x, y, z)$. Let this function be $g(w, x, y, z)$.

$$g(w, x, y, z) = \bar{f}(w, x, y, z) = \bar{w}\bar{x} + x\bar{y}\bar{z} + wyz$$

According to the definition of a $0/1$-$x/\bar{x}$-tree network and assuming that $Y = \{w, x, y\}$, the g_Y^0, g_Y^1, g_Y^z, and $g_Y^{\bar{z}}$ are as follows, respectively.

$$\begin{aligned}
g_Y^0 &= \bar{w}xy + w\bar{x}\bar{y} \\
g_Y^1 &= \bar{w}\bar{x}\bar{y} + \bar{w}\bar{x}y = \bar{w}\bar{x} \\
g_Y^z &= w\bar{x}y + wxy = wy \\
g_Y^{\bar{z}} &= \bar{w}x\bar{y} + wx\bar{y} = x\bar{y}
\end{aligned}$$

The resulting logic circuit is much simpler than an 8-to-1 multiplexer, as shown in Figure 1.22.

■

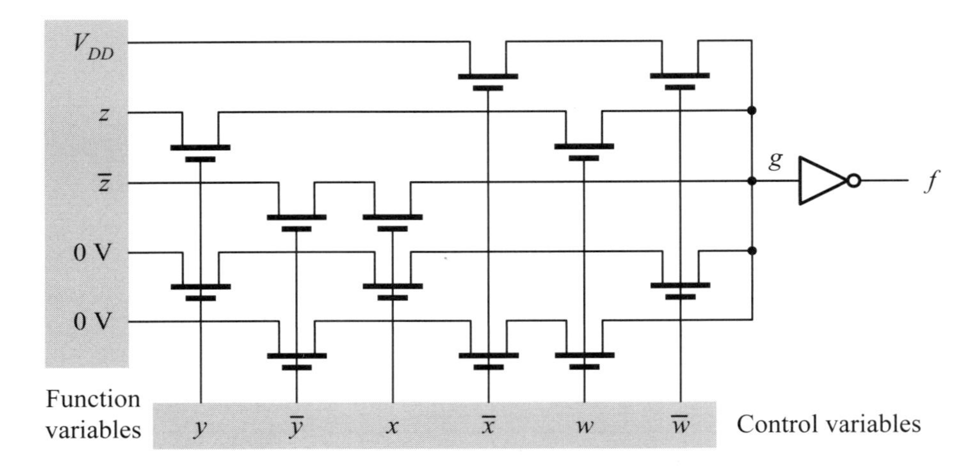

Figure 1.22: The realization of $f(w, x, y, z) = \overline{\bar{w}\bar{x} + x\bar{y}\bar{z} + wyz}$ with a $0/1$-$x/\bar{x}$-tree network.

1.2.5.5 0/1-Tree Networks The 0/1-tree network is also a special case of uniform-tree network in which all of n variables are set as control variables, and hence, each residue can only have one of two constant values, 0 and 1. The combinations of control variables (Y) generating the same residue are combined into a switching function, denoted f_Y^0 and f_Y^1. The switching function f_Y^0 corresponds to $\bar{f}$ and f_Y^1 corresponds to f. Each switching function can be further simplified prior to being realized by a TG or an nMOS switch network, or CMOS logic, as the general block diagrams shown in Figure 1.23.

■ **Example 1-11: (A simple example of 0/1-tree network.)** Realize the following switching function using a 0/1-tree network with nMOS switches:

$$f(w, x, y, z) = \overline{\bar{w} \cdot x + y \cdot z}$$

Solution: Assuming that an inverter is employed at the output, hence let

$$g(w, x, y, z) = \bar{f}(w, x, y, z) = \bar{w} \cdot x + y \cdot z = g_Y^1$$

Function g realizes the complementary function $\bar{f}$, and function $\bar{g}$ implements the truth function f.

$$\bar{g}(w, x, y, z) = f(w, x, y, z) = \overline{\bar{w} \cdot x + y \cdot z} = (w + \bar{x})(\bar{y} + \bar{z}) = g_Y^0$$

The resulting logic circuit is depicted in Figure 1.24.

■

Unlike the $0/1$-$x/\bar{x}$-tree networks, a mixed use of both nMOS and pMOS switches also finds its widespread use in the realizations of 0/1-tree networks. In this case, the 0/1-tree network is referred to as the $f/\bar{f}$ paradigm. The pMOS switches are used to implement the true function f, while nMOS switches are employed to realize the complementary function $\bar{f}$. In a CMOS logic circuit, the $f_Y^1(X)$ tree network is usually called a *pull-up network* (PUN) because it connects the output to V_{DD} when it is on, and the $f_Y^0(X)$ tree network is called a *pull-down network* (PDN) because it connects the output to ground when it is on.

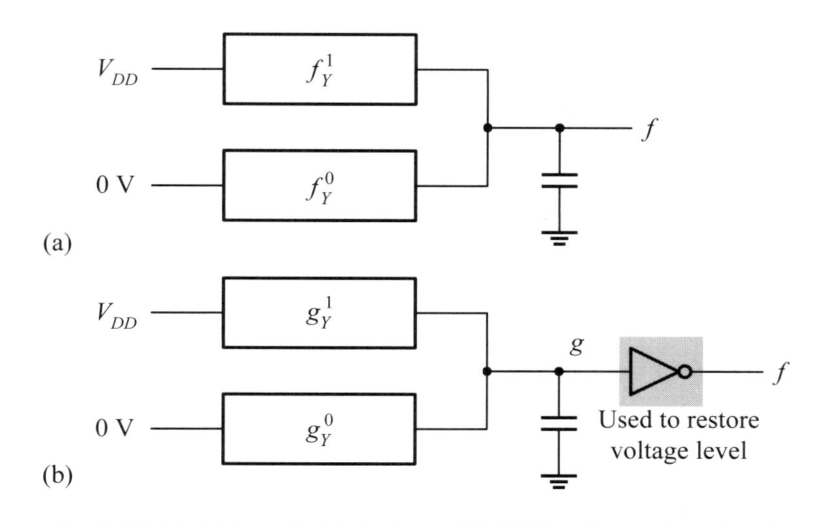

Figure 1.23: The general block diagrams of a 0/1-tree network: (a) using TG switches/CMOS logic; (b) using nMOS switches.

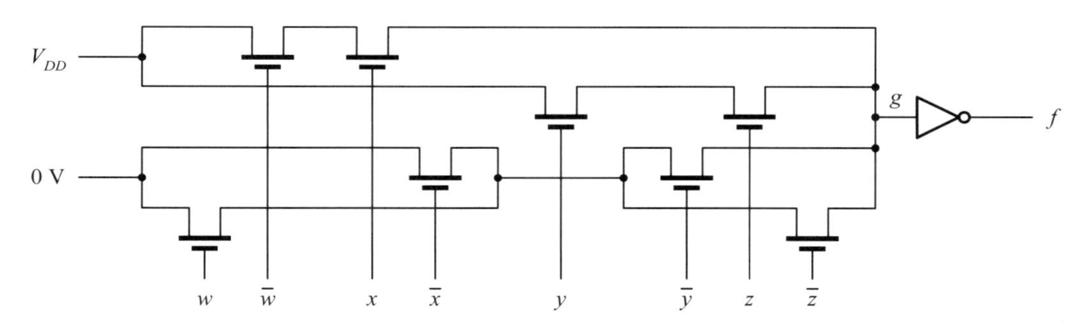

Figure 1.24: The realization of $f(w, x, y, z) = \overline{\overline{w} \cdot x + y \cdot z}$ with a 0/1-tree network.

■ Example 1-12: (An example of a 0/1-tree network with CMOS logic.)

Realize a two-input NAND gate using a 0/1-tree network with CMOS logic.

Solution: Since the output f of a two-input NAND gate is equal to 0 only when both inputs x and y are 1 and is 1, otherwise, the switching functions f_Y^0 and f_Y^1 are as follows.

$$\begin{aligned} f_Y^0 &= xy \\ f_Y^1 &= \bar{x}\bar{y} + \bar{x}y + x\bar{y} = \overline{xy} \end{aligned}$$

which are equal to $\bar{f}$ and f, respectively. Hence, it yields the same result as the $f/\bar{f}$ paradigm introduced in Section 1.2.4.

■

■ Review Questions

Q1-22. Describe Shannon's expansion theorem.

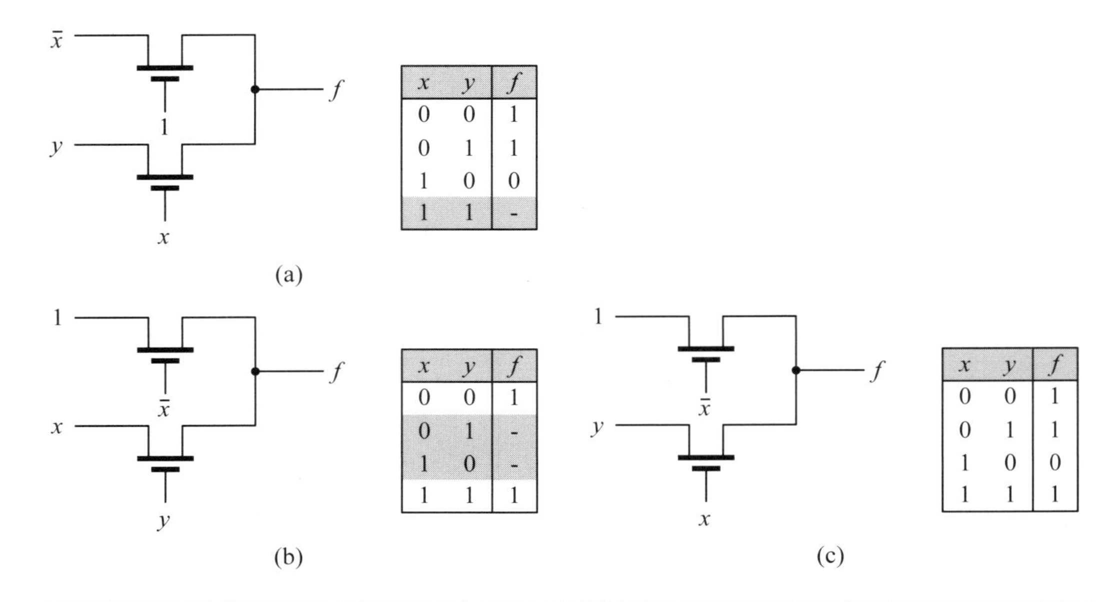

x	y	f
0	0	1
0	1	1
1	0	0
1	1	-

x	y	f
0	0	1
0	1	-
1	0	-
1	1	1

x	y	f
0	0	1
0	1	1
1	0	0
1	1	1

Figure 1.25: Pitfalls of realizing tree networks based on Shannon's expansion theorem: (a) wrong realization I; (b) wrong realization II; (c) right realization.

Q1-23. Define the residue of a switching function $f(x_{n-1}, \cdots, x_1, x_0)$.

Q1-24. Define the control variables and function variables.

Q1-25. What is the rationale behind the $f/\bar{f}$ paradigm?

Q1-26. What do pull-up and pull-down networks mean?

1.2.5.6 Pitfalls of Realizing Tree Networks When using CMOS switches to realize a tree network decomposed by Shannon's expansion theorem, we have to take care of the places where control and function variables are to be applied. As stated, control variables should be applied to the gates of MOS/TG switches while the function variables and constants should be applied at the inputs (namely, source/drain nodes) of MOS/TG switches. Otherwise, some undefined floating or conflict nodes might arise. Some instances are illustrated in the following example.

■ Example 1-13: (The effects of misplacing control and function variables.) Supposing that we want to implement the switching function, $f(x, y) = \bar{x} + y$, with nMOS switches, by using Shannon's expansion theorem, the switching function can be decomposed with respect to variable x into the following.

$$f(x, y) = \bar{x} \cdot 1 + x \cdot y$$

where variable x is the control variable and variable y is the function variable. As shown in Figure 1.25(a), since the variable $\bar{x}$ and constant 1 are interchanged, the resulting logic circuit can only correctly function in three out of four combinations of both inputs, x and y. In the fourth case, when both variables x and y are 1, the output node f will get both 1 and 0 at the same time, namely, a conflict situation.

A similar situation occurs when the switching function is realized with the circuit shown in Figure 1.25(b). The output node f will be in a conflict situation and will be floated when the variables x and y are 0 and 1, as well as 1 and 0, respectively.

Figure 1.25(c) shows the correct implementation, where the control variable x is applied to the gates of nMOS switches, and the constant 1 and function variable y are applied to the inputs of nMOS switches. It is easy to verify that it correctly functions in all four combinations of the input variables, x and y.

■

1.3 VLSI Design and Fabrication

The design of a VLSI chip is a complex process and is often proceeded with a hierarchical design process. The rationale behind this is the capability of repeatedly reusing some basic building blocks. To facilitate such reusable building blocks, logic circuits are designed and implemented as cells. Therefore, in the following, we first introduce the hierarchical design process and then address the concept of cell designs. After this, we briefly describe the CMOS fabrication process, the layouts of CMOS circuits, and the related layout design rules.

1.3.1 Design Techniques

The design of a VLSI chip is a complex process and is often proceeded with a hierarchical design process, including *top-down* and *bottom-up approaches*. In the design, the term *abstraction* is often used to mean the ability to generalize an object into a set of primitives that can describe the functionality of the object. We first cope with the concept of hierarchical design and then address the design abstraction of VLSI.

1.3.1.1 Hierarchical Design As in software programming, the *divide-and-conquer paradigm* is usually used to partition a large hardware system into several smaller subsystems. These subsystems are further partitioned into even smaller ones and this partition process is repeated until each subsystem can be easily handled. This system design methodology is known as a *hierarchical design approach*.

Generally speaking, the VLSI design methodology can be classified into two types: the top-down approach and bottom-up approach. In the top-down approach, functional details are progressively added into the system. That is, it creates lower-level abstractions from higher levels. These lower-level abstractions are in turn used to create even lower-level abstractions. The process is repeated until the created abstractions can be easily handled. An illustration is depicted in Figure 1.26. The 4-bit adder is decomposed into four full adders, with each consisting of two half adders and an OR gate. The half adder further comprises an XOR gate and an AND gate.

In contrast, in the bottom-up approach, the abstractions from lower-level behavior are created and added up to build its higher-level abstractions. These higher-level abstractions are then used to construct even higher-level abstractions. The process is repeated until the entire system is constructed.

However, a practical project usually needs the mixing efforts of both top-down and bottom-up approaches. During the design phase, the top-down approach is often used to partition a design into smaller subsystems; during the realization phase, the bottom-up approach is used to implement and verify the subsystems and their combinations.

The important features of hierarchical design are *modularity*, *locality*, and *regularity*. The modularity means that modules have well-defined functions and interfaces.

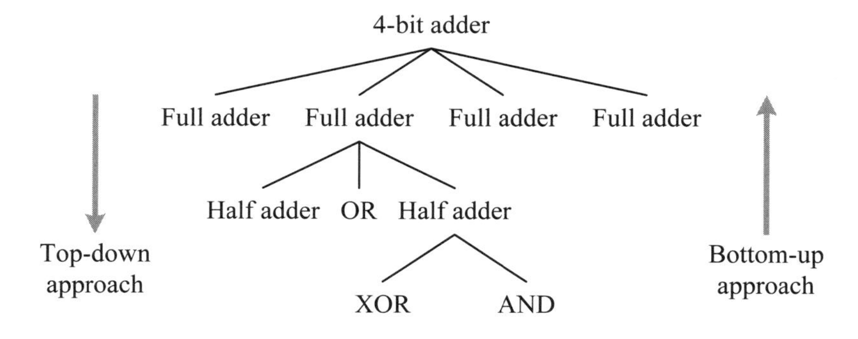

Figure 1.26: The concept of hierarchical design.

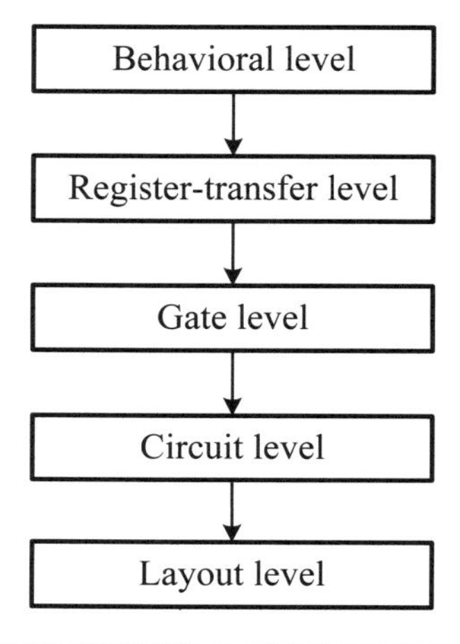

Figure 1.27: The design abstractions of VLSI systems.

Locality implies that the detailed implementation of each module is hidden from outside. The use of the module is completely through the well-defined interface. Regularity is sometimes synonymous with *reusability*; namely, modules can be reused many times in a design.

Although there are many factors that also need to be taken into account in designing a VLSI chip, area budget, power dissipation, and performance are the three most important factors that designers should always keep in mind. To make a practical project successful, the designer often needs to trade off among these three factors in some sense during the course of design and implementation.

1.3.1.2 Design Abstractions A VLSI chip is usually designed by following a design process or a synthesis flow. The general VLSI design abstraction levels are shown in Figure 1.27, which includes *behavioral level*, *register-transfer level* (*RTL* or *RT level*), *gate level*, *circuit level*, and *layout level*. We illustrate this design process with examples.

To begin with, a set of requirements, also called specifications, tell what the system should do, how fast it should run, what amount of power dissipation it could have, and so on. The specifications are not a design and are often incomplete and subject to be changed later. The following is an example of specifications.

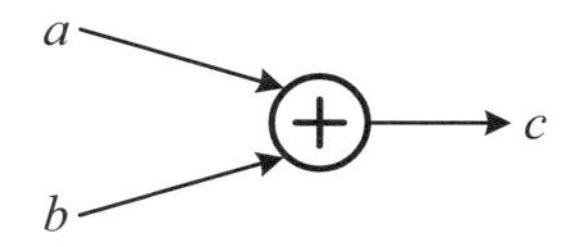

Figure 1.28: A behavioral level for an adder to calculate the sum of two numbers.

Specifications: Design an adder to calculate the sum of two numbers.

The abstraction levels of this design may be described in either a schematic or textual way.

Behavioral level. Once the specifications of a VLSI chip design have been set up, the next step is to model it at a behavioral level. At the behavioral level, the chip design is generally modeled as a set of executable programs, usually written in C, Verilog HDL, VHDL, or other high-level programming languages and is much more precise than the specifications. Through the careful verification of behavioral modeling, we could know whether the specifications meet the requirements. The simulation results of behavioral modeling are used to annotate the specifications of chip design.

Figure 1.28 shows the behavioral level for adding up two 4-bit numbers. Here, an adder is employed to calculate the sum of two numbers. At this behavioral level, only the behavior of design is verified. It leaves the area, delay, and power dissipation of the resulting adder untouched. A possible text description in Verilog HDL of the adder is as follows:

```
module adder_four_bit_behavior(
       input  [3:0] x, y,
       input  c_in,
       output [3:0] sum,
       output c_out);
// the body of an n-bit adder
always @(x or y)
   {c_out, sum} = x + y + c_in;
endmodule
```

Register-transfer level. When the specifications are assured, the next step is to describe the more detailed design at RTL. At the RTL, a design is described by a set of Boolean functions. The input and output values of the design on every cycle are exactly known. The components used in RTL are the basic combinational and sequential logic modules, such as full-adders, decoders, encoders, multiplexers, demultiplexers, registers, counters, and so on. At this level, the delay and area can be roughly estimated.

Figure 1.29 depicts the RTL structure of the 4-bit adder. From the figure, we can see that the 4-bit adder consists of four 1-bit adders in turn. A possible text description of the adder in Verilog HDL is as follows:

```
// gate-level description of 4-bit adder
module adder_four_bit_RTL (
       input  [3:0] x, y,
       input  c_in,
       output [3:0] sum,
       output c_out);
```

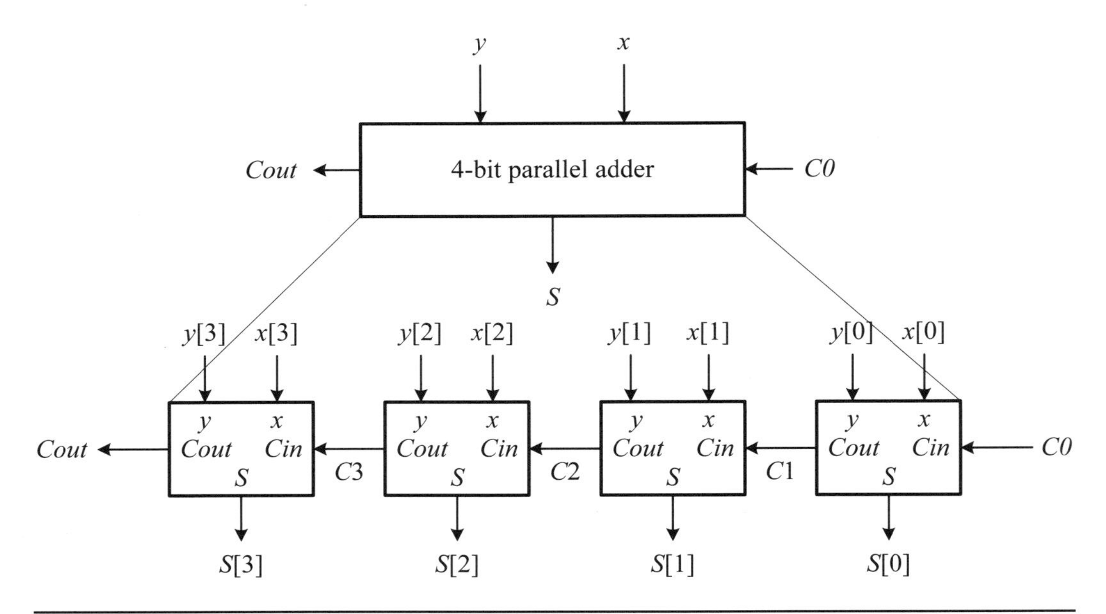

Figure 1.29: An RTL example of a 4-bit adder.

```
wire   c1, c2, c3; // intermediate carries
// -- four_bit adder body-- //
// instantiate the full adder
   full_adder fa_1 (x[0], y[0], c_in, c1, sum[0]);
   full_adder fa_2 (x[1], y[1], c1, c2, sum[1]);
   full_adder fa_3 (x[2], y[2], c2, c3, sum[2]);
   full_adder fa_4 (x[3], y[3], c3, c_out, sum[3]);
endmodule
```

Gate level. The gate level of a design is a structure of gates, flip-flops, and latches. At this level, more accurate delay and area can be estimated. To describe a design at the gate level, the switching functions of the design need to be derived. To illustrate this, let us consider the truth table of a full adder shown in Figure 1.30(a). From the Karnaugh maps given in Figure 1.30(b), we can derive the switching functions for c_{out} and sum as in the following:

$$c_{out}(x, y, c_{in}) = c_{in}(x \oplus y) + xy \quad (1.8)$$
$$sum(x, y, c_{in}) = x \oplus y \oplus c_{in} \quad (1.9)$$

If we define the function of a half adder as follows:

$$\begin{aligned} S_{ha}(x, y) &= x \oplus y \\ C_{ha}(x, y) &= x \cdot y \end{aligned} \quad (1.10)$$

then, a full adder can be represented as a function of two half adders and an OR gate. That is, the carry out and sum of a full adder can be expressed as the following two functions.

$$\begin{aligned} S_{fa} &= S_{ha2}(C_{in}, S_{ha1}(x, y)) \\ C_{fa} &= C_{ha2}(C_{in}, S_{ha1}(x, y)) + C_{ha1}(x, y) \end{aligned} \quad (1.11)$$

x	y	c_{in}	c_{out}	sum
0	0	0	0	0
0	0	1	0	1
0	1	0	0	1
0	1	1	1	0
1	0	0	0	1
1	0	1	1	0
1	1	0	1	0
1	1	1	1	1

(a)

c_{out}:

c_{in} \ xy	00	01	11	10
0	0 (0)	0 (2)	1 (6)	0 (4)
1	0 (1)	1 (3)	1 (7)	1 (5)

sum:

c_{in} \ xy	00	01	11	10
0	0 (0)	1 (2)	0 (6)	1 (4)
1	1 (1)	0 (3)	1 (7)	0 (5)

(b)

Figure 1.30: The (a) truth table and (b) Karnaugh maps for c_{out} and sum of a full adder.

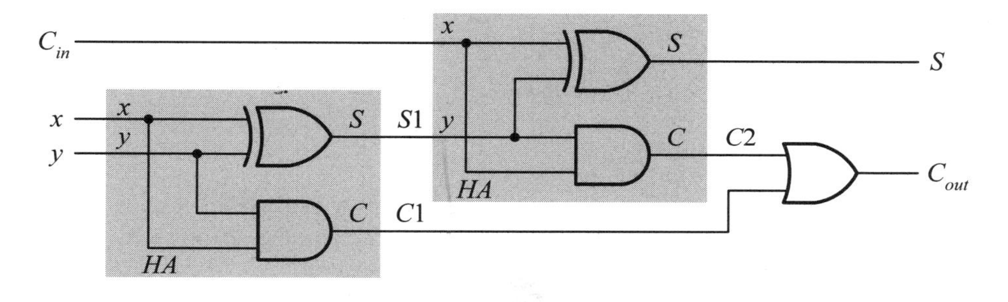

Figure 1.31: A gate-level example of a full adder.

The resulting gate-level full adder is shown in Figure 1.31.

A possible text description of the full adder in Verilog HDL at the gate level is as follows:

```
// gate-level description of full adder
module full_adder_gate (
       input  x, y, c_in,
       output sum, c_out);
wire s1, c1, c2;  // outputs of both half adders
// -- full adder body-- //
// instantiate the half adder
   half_adder ha_1 (x, y, c1, s1);
   half_adder ha_2 (c_in, s1, c2, sum);
   or (c_out, c1, c2);
endmodule

// gate-level description of half adder
module half_adder_gate (
       input  x, y,
       output c, s);
// -- half adder body-- //
// instantiate primitive gates
```

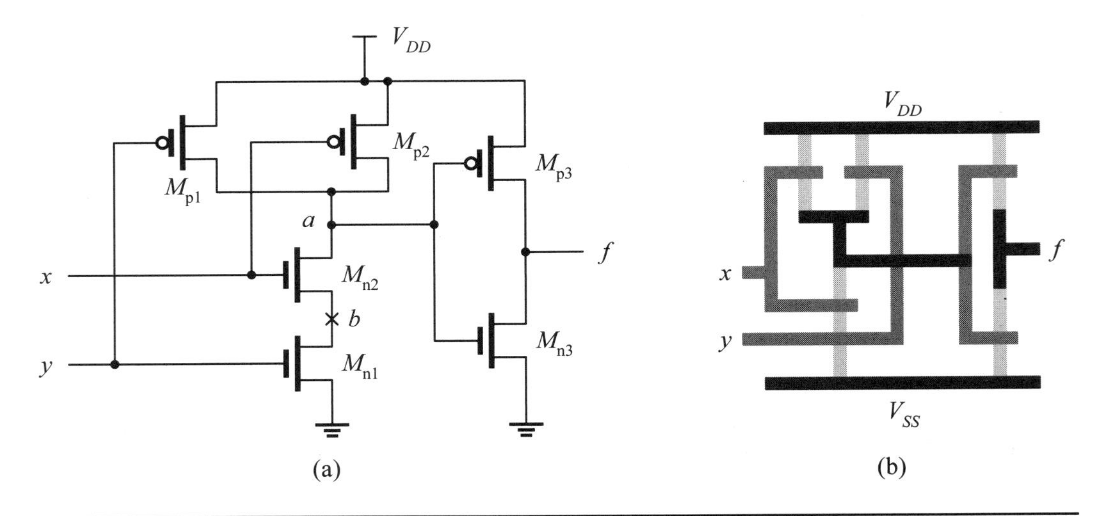

Figure 1.32: The (a) logic circuit and (b) stick diagram of a two-input AND gate.

```
  xor (s,x,y);
  and (c,x,y);
endmodule
```

Circuit level. At the circuit level, the design is implemented with transistors. The circuit level of a two-input AND gate is shown in Figure 1.32. In CMOS technology, the basic gate is an inverting circuit such as a NAND gate or a NOR gate. The simplest way to implement a noninverting gate, such as the AND gate or OR gate, is to add an *inverter* (NOT gate) at the output node of its corresponding basic gate, such as the combination of NAND+NOT or NOR+NOT. Hence, a two-input AND gate is implemented by cascading a two-input NAND gate with an inverter.

A possible text description of the two-input AND gate in SPICE at the circuit level is as follows:

```
AND Gate —— 0.35-um process
******** Parameters and model *********
.lib   '..\models\cmos35.txt' TT
.param Supply=3.3V   * Set value of Vdd
.opt    scale=0.175u
******** AND Subcircuit description **********
.global Vdd Gnd
.subckt AND x y f
Mn1   b   y   Gnd Gnd nmos L = 2 W = 6
Mn2   a   x     b Gnd nmos L = 2 W = 6
Mp1   a   y   Vdd Vdd pmos L = 2 W = 6
Mp2   a   x   Vdd Vdd pmos L = 2 W = 6
Mn3   f   a   Gnd Gnd nmos L = 2 W = 3
Mp3   f   a   Vdd Vdd pmos L = 2 W = 6
.ends
********* Circuit description ***********
Vdd Vdd Gnd 'Supply'
X1  t  w  z  AND  ** instantiate an AND gate
Vt  t Gnd pulse 0 'Supply' 0ps 100ps 100ps 2ns 4ns
```

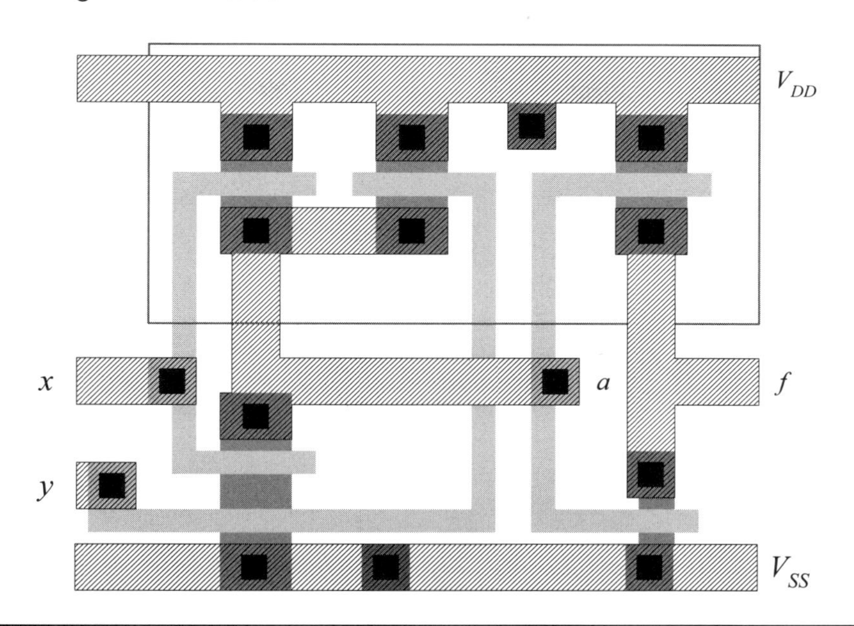

Figure 1.33: A layout example of the two-input AND gate shown in Figure 1.32.

```
Vw  w Gnd pulse 0 'Supply' 0ps 100ps 100ps 1ns 2ns
******** Analysis statement *******
.tran 1ps 4ns
******** Output statements *******
.probe V(t) V(w) V(z)
.option post
.end
```

Layout level. The last step of a VLSI chip design is to represent the circuit in mask layout layers. Before proceeding to the design of layout (that is, physical layout) of a logic circuit, it is useful to draw the logic circuit in a stick diagram. A *stick diagram* is a simplified layout, which only represents the relative relationship among wires and components without being confined to layout design rules. A stick diagram of the AND logic circuit shown in Figure 1.32(a) is depicted in Figure 1.32(b).

After a stick diagram of a circuit is completed, the layout of a logic circuit can be obtained by enforcing the layout design rules to all features of the stick diagram. A complete layout of the AND logic circuit exhibited in Figure 1.32(a) is given in Figure 1.33. This layout defines all mask layout layers required to fabricate the AND logic circuit.

Through these mask layout layers, the design can be manufactured in an IC foundry. After the layout is done, parasitics, including resistance, capacitance, and inductance, can be extracted, and the performance of layout can then be estimated through circuit-level simulators, such as the SPICE or its descendant programs, including HSPICE and Spectra.

A partial text description of the AND gate after extracting *RC* parasitics from the layout shown in Figure 1.33 is as follows:

```
.subckt PM_AND2   *** F 1 3 15 20 22
c5 20 0 0.333511f
c6 14 0 0.160047f
```

```
c7 12 0 0.172583f
r8 17 20 0.0938868
r9 22 16 0.6138
r10 15 16 1.83333
r11 13 17 0.0195
r12 13 14 0.0599444
r13 12 17 0.0195
r14 11 15 0.0342493
r15 11 12 0.0642778
r16 3 2 0.347907
r17 1 14 0.0314389
r18 1 2 0.833333
.ends
```

The parasitic capacitance and resistance of each node are extracted and represented as a SPICE data file. From this information, the characteristics of the layout can be quantified more accurately.

Review Questions

Q1-27. Describe the divide-and-conquer paradigm.

Q1-28. Distinguish the top-down approach from the bottom-up approach.

Q1-29. Describe the general VLSI design abstraction levels.

Q1-30. What are the three major factors that designers should always keep in mind when designing a VLSI system?

1.3.2 Cell Designs

The essence of hierarchical design is the capability of reusing some basic cells over and over again. During the course of bottom-up composition, a group of cells may be combined into a submodule, which in turn can be used as a submodule for another higher-level submodule. This process continues until the entire system is constructed. Therefore, the basic building blocks are cells. But what kind of circuit can be referred to as a cell? How complex a function should it contain? In fact, the function of a cell could range from a simple gate to a very complicated module, such as an MP3 decoder or even a 32-bit microprocessor.

Next, we will introduce the three most basic types of cells corresponding to those often introduced in basic texts: combinational cells, sequential cells, and subsystem cells. Their details will be treated in greater detail in dedicated chapters later in the book.

1.3.2.1 Combinational Cells Combinational cells are logic circuits whose outputs are only determined by the present inputs, that is, only functions of present inputs. The most basic CMOS circuit is the NOT gate (that is, inverter), which is composed of two MOS transistors, a pMOS transistor, and an nMOS transistor, as shown in Figure 1.34(a). The pMOS transistor connects the output to V_{DD} when it is on, whereas the nMOS transistor connects the output to ground when it is on. Since both MOS transistors cannot be turned on at the same time, there is no direct path from power (V_{DD}) to ground. Hence, the power dissipation is quite small, only due to a small leakage current and the transient current during switching. The average power

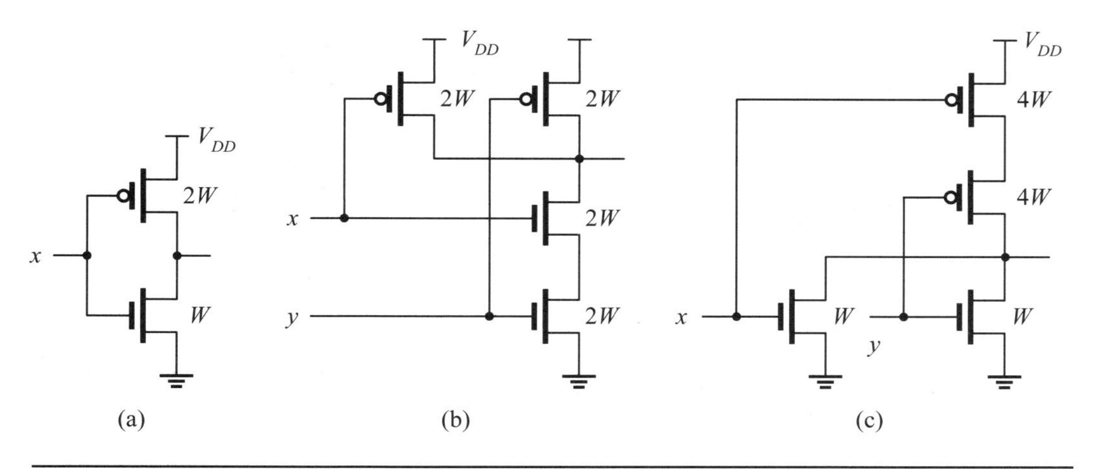

Figure 1.34: The three basic CMOS gates: (a) NOT gate; (b) NAND gate; (c) NOR gate.

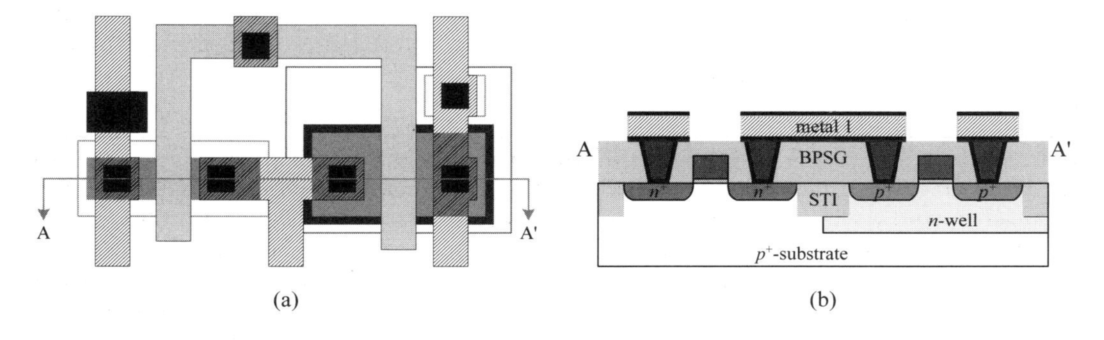

Figure 1.35: The (a) layout and (b) side view of a CMOS inverter.

dissipation is on the order of nanowatts. The layout and side views of a CMOS inverter are shown in Figure 1.35.

Combinational cells typically include basic gates: AND, OR, NOT, NAND, NOR, XOR, and XNOR. Also, these basic gates are often designed with various transistor sizes to provide different current-driving capabilities for different applications. The following examples give some representative combinational cells that are widely used in various digital system designs and will appear repeatedly throughout the book.

■ **Example 1-14: (Basic Gates.)** Three basic CMOS gates are depicted in Figure 1.34. These are a NOT gate (or inverter), a two-input NAND gate, and a two-output NOR gate. The reader may notice that a basic CMOS gate is an inverting gate; thus, it is necessary to append a NOT gate at the output when a noninverting function is needed. For example, a two-input AND gate is formed by appending an inverter at the output of a two-input NAND gate, as we have done in Figure 1.32. ■

In addition to basic gates, multiplexers and demultiplexers are often used in digital system designs. A multiplexer is a device capable of routing the input data from one

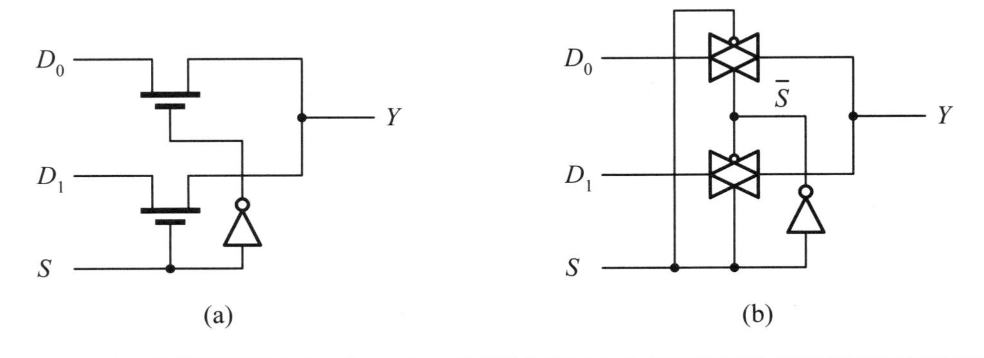

Figure 1.36: The examples of (a) nMOS-based and (b) TG-switch-based 2-to-1 multiplexers.

out of its many inputs specified by the value of a set of source selection inputs to its output. A demultiplexer operates in the reverse way; that is, it routes the input data to one of its many outputs selected by the value of a set of destination selection inputs.

The following two examples separately illustrate how a simple 2-to-1 multiplexer and demultiplexer can be constructed.

■ **Example 1-15: (A 2-to-1 multiplexer.)** Figure 1.36 shows two examples of a CMOS 2-to-1 multiplexer. Figure 1.36(a) is constructed by using two nMOS transistors and an inverter. The value of output Y is equal to D_0 or D_1 determined by the value of source selection input S. The upper nMOS transistor is turned on when the source selection input S is 0 and is turned off otherwise. The lower nMOS transistor operates in the opposite way; that is, it is on when the source selection input S is 1 and is turned off otherwise. As a result, it is operated as a 2-to-1 multiplexer. Figure 1.36(b) shows the case when transmission gates are used.

■

■ **Example 1-16: (A 1-to-2 demultiplexer.)** Figure 1.37 shows two examples of a CMOS 1-to-2 demultiplexer. Figure 1.37(a) is constructed by using two nMOS transistors and an inverter. The input D is routed to Y_0 or Y_1 determined by the value of destination selection input S. The input D is routed to Y_0 when the destination selection input S is 0 and routed to Y_1 otherwise. To facilitate this, the upper nMOS transistor is turned on only when the destination selection input S is 0 and the lower nMOS transistor is turned on only when the selection input S is 1. Hence, it results in a 1-to-2 demultiplexer. Figure 1.37(b) shows the case when transmission gates are used.

■

The 2-to-1 multiplexers and 1-to-2 demultiplexers virtually have the same circuit structure. In fact, they are interchangeable. That is, the 2-to-1 multiplexer may be used as a 1-to-2 demultiplexer and vice versa.

1.3.2.2 Sequential Cells Sequential cells are logic circuits whose outputs are not only determined by the current inputs but also dependent on their previous output

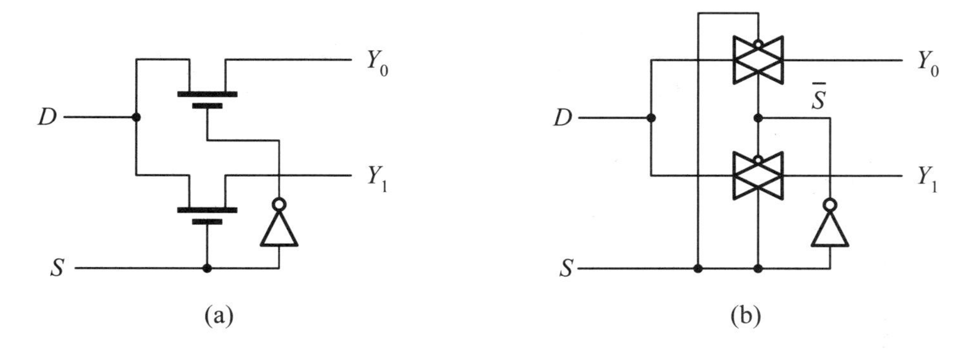

Figure 1.37: The examples of (a) nMOS-based and (b) TG-switch-based 1-to-2 demultiplexers.

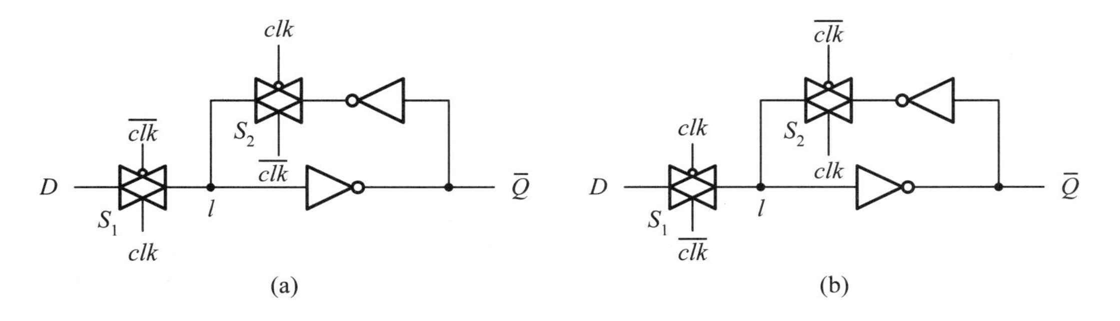

Figure 1.38: The examples of (a) positive and (b) negative D-type latches based on TG switches.

values; namely, their outputs are functions of both current inputs and previous outputs. Typical sequential cells include basic memory devices: *latches* and *flip-flops*.

The following examples first show how two transmission gates (TGs) and two inverters can be combined to build positive and negative latches, and then show how to use these two latches to construct a *master-slave flip-flop*.

■ Example 1-17: (A positive D-type latch.) Figure 1.38(a) shows a TG-based positive D-type latch. A latch is essentially a *bistable circuit* in which two stable states exist. These two states are called 0 and 1, respectively. To change the state of the bistable circuit, an external signal must be directed into the circuit to override the internal state. As shown in the figure, this operation is done by cutting off the feedback path and connecting the external signal to the bistable circuit; that is, it is achieved by turning off switch S_2 and turning on switch S_1 since the latch receives the input data at the high level of clock clk and retains the input data as the clock clk falls to 0. Such a type of latch is called a *positive latch*.

■

The following is an example of a negative D-type latch based on TGs.

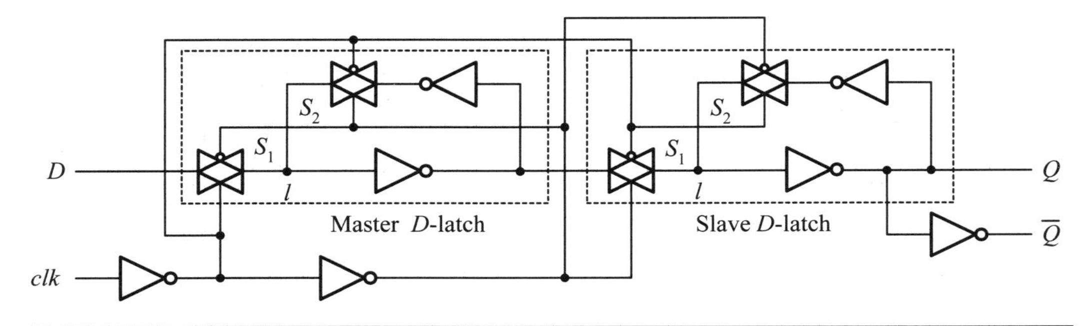

Figure 1.39: A master-slave positive edge-triggered D-type flip-flop.

Example 1-18: (A negative D-type latch.) Figure 1.38(b) shows a TG-based negative latch. It essentially operates in the same way as the positive latch except that now the latch receives the external data at the low level of clock *clk* and retains the input data as the clock *clk* rises high. Such a latch is called a *negative latch*.

The following example shows how to combine the above two latches into a master-slave flip-flop. The essential difference between latches and flip-flops is on the transparent property. For both positive and negative latches, their outputs exactly follow their inputs as the clocks are high and low, respectively. This property is known as *transparent property* since under this situation the latch sitting between the output and input seems not to exist at all. For flip-flops, their outputs are only a sample of their inputs at a particular instant of time, usually, the positive or negative edge of clock. Hence, flip-flops do not own the transparent property. The detailed discussion of latches and flip-flops with more examples is deferred to Chapter 9.

Example 1-19: (A master-slave positive edge-triggered D-type flip-flop.) In CMOS technology, a flip-flop is often created by cascading two latches with opposite polarities, that is, a negative latch followed by a positive latch, or the reverse order. Such a flip-flop is called a *master-slave flip-flop*. Figure 1.39 shows a master-slave D-type flip-flop created by cascading negative and positive D-type latches. The input data are routed to the master D-latch when the clock *clk* is low and then to the slave D-latch when the clock *clk* is high. Because of the transparent property of latches, the input data routed to the slave-D latch are the data just before the clock changes from low to high. As a consequence, it indeed functions as a positive edge-triggered D-type flip-flop.

Using the same principle, a negative edge-triggered D-type flip-flop can be built by cascading a positive and a negative D-type latches.

1.3.2.3 Subsystem Cells The third type of cells widely used in VLSI designs are *subsystem cells*. A subsystem cell is a logic circuit that is able to function as an independent module, such as an 8-bit *arithmetic logic unit* (ALU), an 8-bit or larger

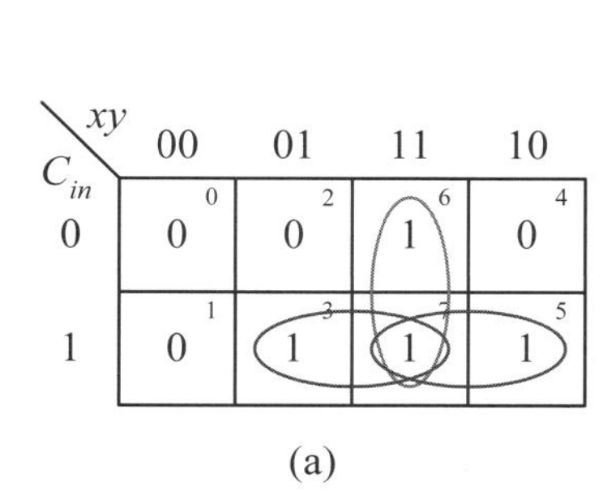

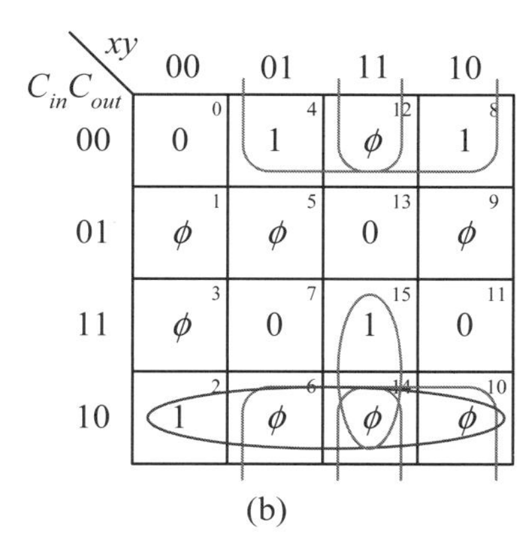

Figure 1.40: The Karnaugh maps for (a) c_{out} and (b) *sum* of a full adder.

multiplier, and so on. A subsystem cell is usually designed as a single primitive module to optimize an area, a performance, or both.

The following example gives the reader a sense for the subsystem cell showing how to directly design and implement a full adder from the circuit level without resorting to the combination of basic gates, such as the one shown in Figure 1.31.

■ Example 1-20: (A full adder.) At the circuit level, the designed logic circuit is implemented with transistors. The Karnaugh maps for c_{out} and *sum* of a full adder are depicted in Figures 1.40(a) and (b), respectively. From these two Karnaugh maps, the switching functions for c_{out} and *sum* can be obtained and expressed as follows.

$$\begin{aligned} c_{out} &= c_{in}(x + y) + xy \\ sum &= \overline{c_{out}}(x + y + c_{in}) + xyc_{in} \end{aligned}$$

The circuit-level realization of the above two functions is shown in Figure 1.41. The general approaches to realizing a switching function using CMOS transistors have been explored in Section 1.2.5 in depth.

■

In industry, cells with a broad variety of functions and driving capabilities are usually designed in advance. These cells along with their layouts and parameters, such as area and propagation delay, are then aggregated together in a *cell library* so that they can be reused whenever they are needed. The design based on a cell library is referred to as a *cell-based design*. It is often accomplished by using synthesis flow based on Verilog HDL or VHDL.

1.3.3 CMOS Processes

A *CMOS process* is a manufacturing technology capable of incorporating both pMOS and nMOS transistors in the same chip. Because of ultra-low power dissipation and high integration density, CMOS processes have become the major technology for modern VLSI circuits.

A specific CMOS circuit can be fabricated in a variety of ways. Figure 1.42 shows the three most commonly used CMOS fabrication structures. Figure 1.42(a) is an

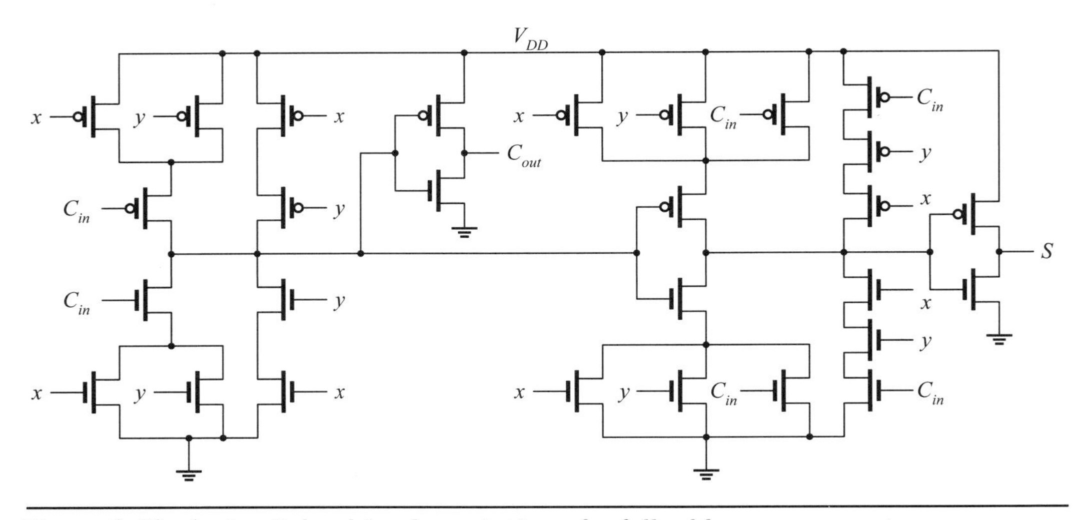

Figure 1.41: A circuit-level implementation of a full adder.

n-well (or n-tub) structure in which an n-well is first implanted and then a pMOS transistor is built on the surface of it. The nMOS transistor is built on top of the p-type substrate. Of course, a CMOS circuit can also be built with a p-well on an n-type substrate. However, this process yields an inferior performance to the n-well process on the p-type substrate. Hence, it is not widely used in industry.

In the n-well structure, the n-type dopant concentration must be high enough to overcompensate for the p-type substrate doping to form the required n-well. One major disadvantage of this structure is that the channel *mobility* is degraded because mobility is determined by the total concentration of dopant impurities, including both p and n types.

To improve channel mobility, most recent CMOS processes used in industry are the structure called a *twin-well structure* or a *twin-tub*, as shown in Figure 1.42(b). In this structure, a p-type epitaxial layer is first grown and then the desired p-type and n-type wells are separately grown on top of the epitaxial layer. Finally, the nMOS and pMOS transistors are then manufactured on the surface of p-type and n-type wells, respectively. The field-oxide is grown by using *local oxidation of silicon* (LOCOS). Because of no overcompensation problem, a high-channel mobility can be achieved.

The third CMOS structure shown in Figure 1.42(c) is still a twin-well structure but uses *shallow trench isolation* (STI) instead of LOCOS. Because of the lack of a *bird's beak effect* occurring in LOCOS, it can provide higher integration density. This structure is widely used in deep submicron (DSM) processes, in particular below 0.18 μm. The STI used here has a depth less than 1 μm. Some processes use a deep-trench isolation with a depth deeper than the depth of a well. In such a structure, an oxide layer is thermally grown on the bottom and side walls of the trench. The trench is then refilled by depositing polysilicon or silicon dioxide. The objective of this structure is to eliminate the inherent *latch-up problem* associated with CMOS processes. The latch-up problem is dealt with in more detail in Chapter 4.

1.3.4 CMOS Layout

As it can be seen from Figure 1.33, a layout is another view of the circuit. In other words, both layout and circuit denote the same thing just like a coin looked at from

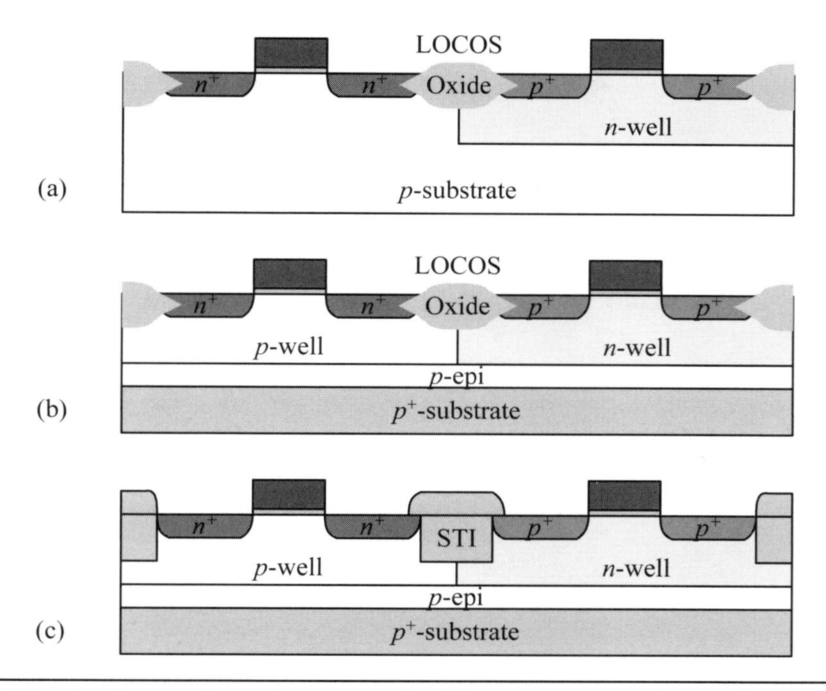

Figure 1.42: An illustration of various CMOS structures: (a) n-well structure; (b) twin-well structure; (c) twin-well structure with refilled trench.

both sides. Nevertheless, different viewpoints about a digital integrated circuit exist between circuit engineering and computer science. From the circuit engineering viewpoint, a digital integrated circuit is a system of circuits built on the surface of a silicon chip; from the computer science viewpoint, a digital integrated circuit is a set of geometrical patterns on the surface of a silicon chip.

As a consequence, from the circuit engineering/computer science viewpoint, a VLSI design is a system design discipline with the following two features. First, groups of circuits/patterns, called *modules*, represent different logic functions and can be repeated many times in a system. Second, complexity could be dealt with using the concept of repeated circuits/patterns put together hierarchically.

Fundamentally, a circuit is an abstract design that represents only the idea in one's mind. To implement the circuit, physical devices, either discrete devices or integrated circuits, must be used. As a result, a layout of a circuit is also only an abstract representation of the design. It denotes all information required for fabricating the circuit in an IC foundry. To illustrate the relationship between a layout and an actual IC fabrication, consider Figure 1.43, which shows a layout of a CMOS inverter along with the major steps for fabricating the inverter.

A layout of a circuit indeed defines the set of masks needed in manufacturing the circuit. As illustrated in Figure 1.43(a), seven mask layers are required for manufacturing such a simple CMOS inverter circuit. An IC is made in a layer-by-layer fashion from the bottom up. Since an n-well process is assumed to be used, the first mask is employed to define the n-well, where a pMOS transistor can be made, as depicted in Figure 1.43(b). After the n-well is defined, it is necessary to reserve areas needed by all MOS transistors and to fill the remaining part with *field oxide*, a thick silicon dioxide formed by the STI process, to isolate each MOS transistor electrically. This is

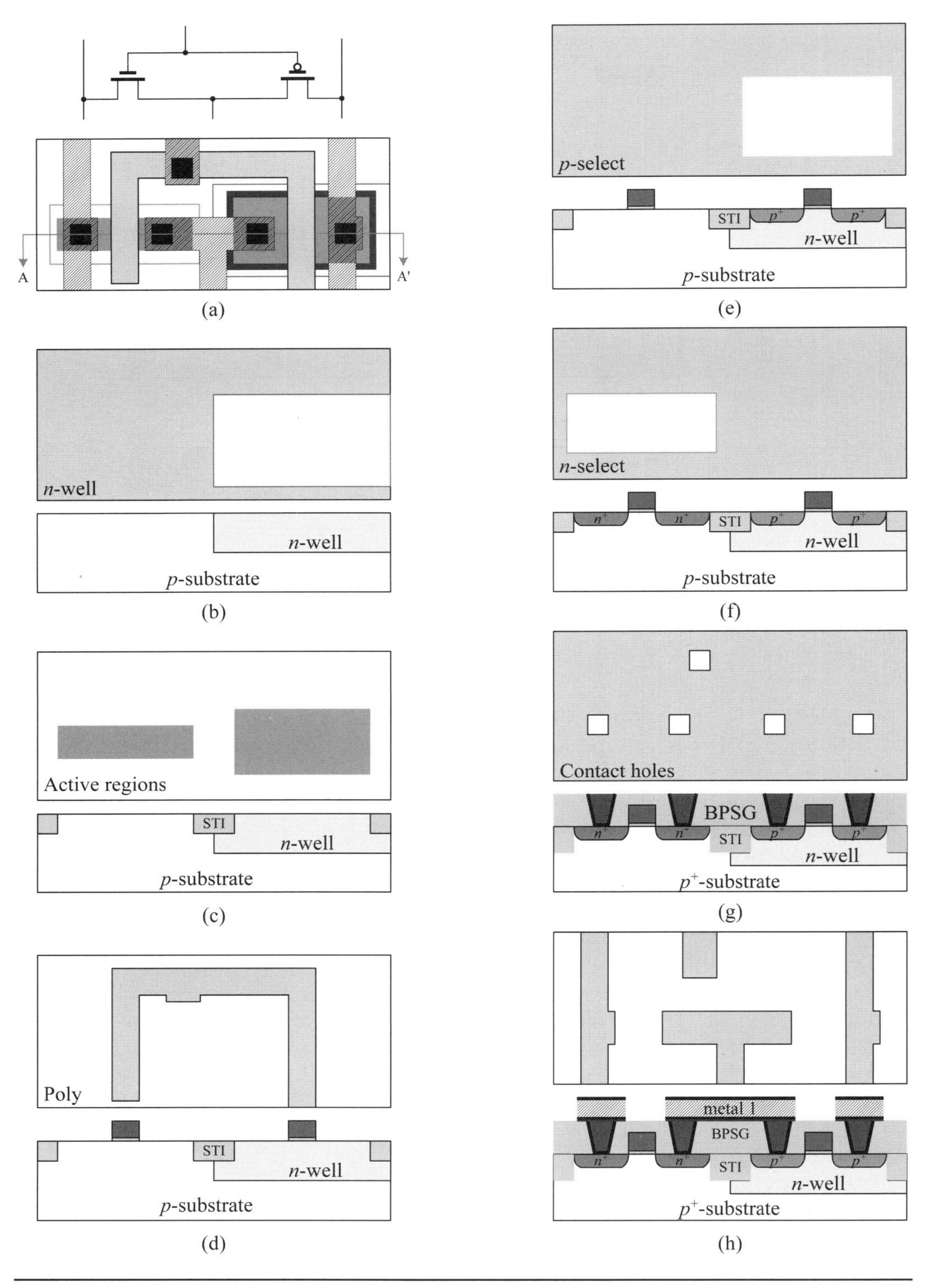

Figure 1.43: A CMOS inverter layout and its related mask set: (a) layout; (b) formation of n-well; (c) definition of nMOS transistors; (d) definition of polysilicon gates; (e) definition of p^+ diffusion; (f) definition of n^+ diffusion; (g) definition of contact holes; (h) definition of metal1 connections.

defined by the active mask, as shown in Figure 1.43(c).

Once the active regions have been defined, the next step is to use the polysilicon mask to make the gates of all MOS transistors, including both pMOS and nMOS transistors, and all wires using polysilicon as well. An illustration is exhibited in Figure 1.43(d). The next two masks are separately used to implant p^+ and n^+ diffusions needed in forming the drain and source regions of MOS transistors. These two masks are called p-select and n-select masks, respectively, and are derived masks obtained by bloating the size of active regions. These two masks and their effects are shown in Figures 1.43(e) and (f), respectively.

The last two masks are used to define contact holes and the metal layer, respectively. As depicted in Figure 1.43(g), all source and drain regions of MOS transistors need to be connected together in the same way as its original circuit to perform the same function. To achieve this, a mask is used to define the contact holes through which the required interconnect points can be prepared. The next step is to manufacture the required metal wires. This is done by using the metal mask, as shown in Figure 1.43(h).

Of course, in addition to the above steps, there are many steps that still need to be done. Generally, a CMOS circuit usually requires many masks, ranging from 7 to more than 20, depending how many metal layers are used. A later chapter is dedicated to the details of CMOS fabrication.

1.3.5 Layout Design Rules

A layout design must follow some predefined rules to specify geometrical objects, namely, polygons, that are either touching or overlapping on each masking layer. A set of such predefined rules is called *layout design rules*. The layout design rules may be specified as either of the following two forms:

- *Lambda* (λ) *rules*: In the λ rules, all rules are defined as a function of a single parameter λ. Typically, the feature size of a process is set to be 2λ. The λ rule is a scalable rule and widely used in the academic realm.
- *Micron* (μ) *rules*: In the μ rules, all rules are defined in absolute dimensions and can therefore exploit the features of a given process to its maximum degree. The μ rule is adopted in most IC foundries in industry.

The μ rules tend to differ from company to company and even from process to process. The λ rules are conservative in the sense that they use the integer multiples of λ. We will use λ rules throughout the book. Some sample design rules are shown in Figure 1.44, referring to Chapter 4 for more details.

■ Review Questions

Q1-31. Draw the logic circuit of a negative edge-triggered D-type flip-flop.

Q1-32. Design a gate-based 2-to-1 multiplexer.

Q1-33. What is the distinction between CMOS-based and gate-based 2-to-1 multiplexers?

Q1-34. Design a gate-based 1-to-2 demultiplexer.

Q1-35. What is the distinction between CMOS-based and gate-based 1-to-2 demultiplexers?

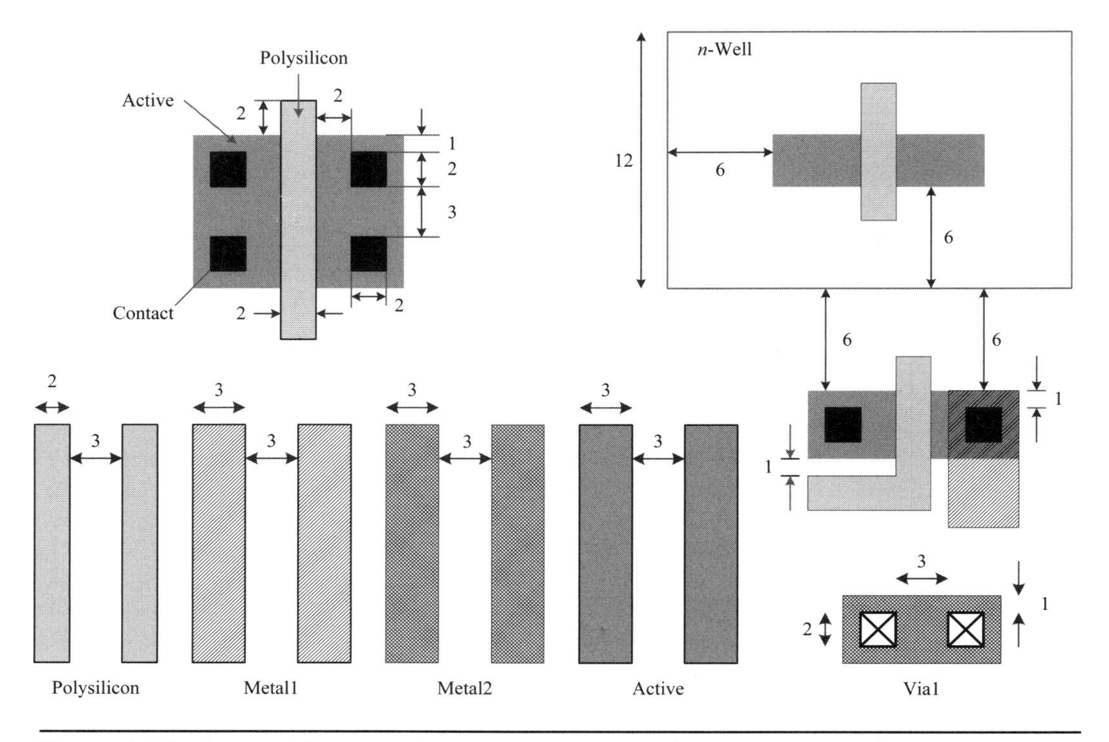

Figure 1.44: A simplified set of layout design rules (all units are in λ).

1.4 Implementation Options of Digital Systems

In this section, we first look at the future trends of VLSI (digital) systems design.[4] Then, we introduce a broad variety of implementation options currently available for digital systems.

1.4.1 Future Trends

The NRE cost (fixed cost) and *time to market* are two major factors that affect the future trends of VLSI designs. Recall that the cost of a VLSI chip is determined by the NRE and variable costs. The NRE cost is significantly increased with the decreasing feature sizes of manufacture processes due to the exponentially increased cost of related equipments, photolithography masks, CAD tools, and R&D. Recall three important issues in designing a VLSI system with DSM processes: IR drop, Ldi/dt effect, and hot-spot problems. To model and analyze these three issues accurately, it is inevitable to rely heavily on computer-aided design (CAD) tools. This means that the expensive CAD tools are indispensable for deep-submicron VLSI designs. To make the product more competitive or acceptable by end users, the NRE cost has to be profoundly reduced. Consequently, for a VLSI chip to be successful in the market, the product volume must be large enough to lower the amortized NRE cost to an acceptable level by the market.

[4]Even though the words *design* and *implementation* mean two different things, they are often strongly related. Hence, we will use these two words interchangeably when their meanings are unambiguous from the context.

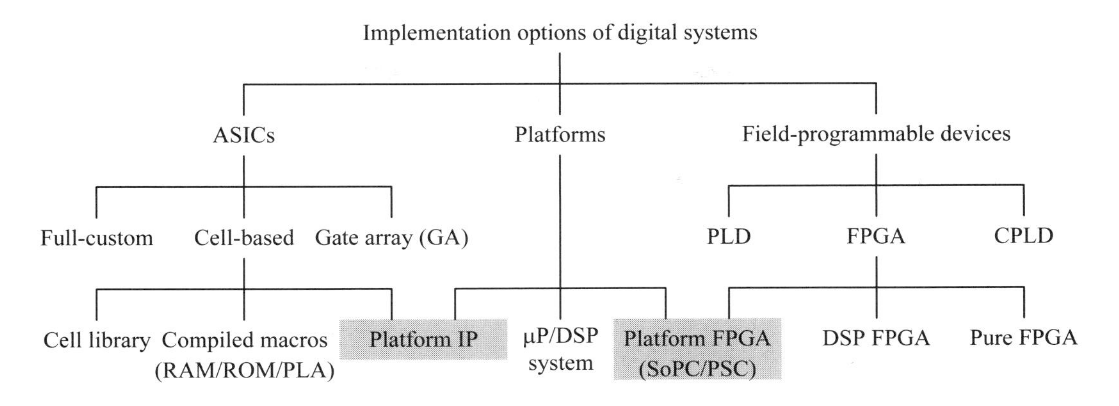

Figure 1.45: Implementation options of digital systems.

■ **Example 1-21: (Amortized NRE cost.)** Suppose that the NRE cost for getting the first prototype of a 50-M transistor chip is about 10 M dollars. Calculate the amortized NRE cost if the total product volume of the chip is estimated to be (a)100 k, and (b) 100 M.

Solution: (a) The amortized NRE cost is $10 \times 10^6/100 \times 10^3 = 100$ dollars. (b) The amortized NRE cost is $10 \times 10^6/100 \times 10^6 = 0.1$ dollars. Therefore, as the product volume increases to some point, the amortized NRE cost might be ignored.

■

The other factor that affects the future trends of VLSI design is the time to market. Late products are often irrelevant to modern consumer markets. In addition, all electronic products have increasing system complexity with the reduction of feature sizes and hence hardware cost. Although the divide-and-conquer paradigm can be used to partition the system into many smaller modules so that each module can be easily dealt with, the combination of these different functionality modules becomes more difficult and challenging, and the accompaning testing for the combined system is even more complicated. This means that to shorten the time to market of a product some effective design alternatives must be explored and used.

Based on the aforementioned factors, the future trends of VLSI (digital) system designs can be classified into three classes: ASICs, platforms, and field-programmable devices, as shown in Figure 1.45. One needs to choose an appropriate one from these options to meet the design specifications at the lowest cost and shortest time to market.

1.4.2 Implementation Options

The future trends of VLSI (digital) system implementations can be classified into three classes: *platforms*, *field-programmable devices*, and *ASICs*, as shown in Figure 1.45. We introduce these in the following briefly.

1.4.2.1 Platforms We are first concerned with the platform option. This option includes microprocessor and/or *digital-signal processor* (μP/DSP), platform IP, and platform FPGA. Many consumer products may be designed with a single μP and/or a simple DSP system. Depending on the desired performance, a variety of 8-bit to 32-bit μPs are available for use today.

A μP/DSP-based system is actually a microcontroller chip and is an embedded system built on silicon. Usually, the system contains an 8-bit or a 32-bit *center-processor unit* (CPU), nonvolatile memory (Flash memory or MRAM), static memory, and a variety of periphery modules, such as *universal serial bus* (USB), *general-purpose input-output* (GPIO), timer, and so on. Most digital systems are designed with this approach.

1.4.2.2 Field-Programmable Devices The field-programmable devices are the second design option. A field-programmable device is the one that contains many uncommitted logic modules that can be committed to the desired functions on-demand in laboratories. Currently, three major types of field-programmable devices are *field-programmable gate arrays* (FPGAs), *complex-programmable logic devices* (CPLDs), and *programmable logic devices* (PLDs). The FPGAs may be further subdivided into pure FPGAs, DSP FPGAs, and platform FPGAs. A platform FPGA combines features from both platforms and field-programmable devices into a single device; it is an FPGA device containing one or more CPUs in a hard, soft, or hardwired IP form, some periphery modules, and field-programmable logic modules.

Here, the IP is the acronym of *intellectual property* and is a predesigned component that can be reused in larger designs. It is often referred to as a *virtual component*. There are three common types of IP: *hard IP*, *soft IP*, and *hardwired IP*. A hard IP comes as a predesigned layout and routing but can be synthesized with other soft IP modules. The block size, performance, and power dissipation of a hard IP can be accurately measured. A soft IP is a synthesizable module in HDL, Verilog HDL or VHDL. Soft IPs are more flexible to new technologies but are harder to characterize, occupy more area, and perform more poorly than hard IPs. A hardwired IP is a circuit module already fabricated along with FPGA fabrics.

1.4.2.3 ASICs The third design option is *application-specific integrated circuits* (ASICs). Although the acronym ASIC represents any integrated circuit (IC) defined by a specific application, in industry the term ASIC is often reserved to identify an integrated circuit that needs to be processed in an IC foundry. That is, an ASIC denotes any IC designed and implemented with one of the following three methods: *full-custom*, *cell-based*, and *gate array*. One essential feature of an ASIC is that its final logic function needs to be committed through a partial or full set of mask layers in an IC foundry.

Full-custom design starts from scratch and needs to design the layouts of every transistor and wire. It is suitable for very high-regularity chips, such as DRAM/SRAM, and for very high-speed chips, such as CPUs, GPUs, or FPGAs, or other special-purpose chips, such as GPUs. The cell-based design combines various full-custom cells, blocks, and IPs into a chip. It is suitable for a product with a successful market only. The gate-array-based design combines a variety of IPs and builds the resulting design on a gate array, which is a wafer with prefabricated transistors. It is suitable for a product that has a successful market and needs a short time to market.

In summary, because of enough complexity, FPGA devices will replace ASICs and dominate the market of VLSI-size systems for the following reasons. First, the NRE cost of an ASIC profoundly increases with the reduction of feature sizes of CMOS processes. Second, the functionality of an ASIC chip increases with the progress of feature sizes. This makes the design and implementation more complicated, challenging, and difficult. Third, the increasing complexity of an ASIC design will become a difficult problem for most people.

■ Review Questions

Q1-36. What are the three major classes of options for designing a VLSI system?
Q1-37. What are the three types of platforms?
Q1-38. What are the three types of field-programmable devices?
Q1-39. What is the meaning of the term ASIC in industry?
Q1-40. What options can be used to design an ASIC?
Q1-41. Explain the reasons why FPGA devices will dominate the VLSI-size market.

1.5 Summary

In the introduction section, we first briefly reviewed the history of VLSI technology. Next, we introduced the silicon planar process on which, nowadays, CMOS processes are founded and ultimate limitations of feature size. Then, we were concerned with the classification of VLSI circuits, the benefits of using VLSI circuits, the appropriate technologies for manufacturing VLSI circuits, and the scaling theory. Finally, design challenges of DSM devices and wires were also dealt with in detail. Economics and future perspectives of VLSI technology were explored briefly.

Both nMOS and pMOS transistors are regarded as simple ideal switches, known as nMOS switches and pMOS switches, respectively. The combination of both nMOS and pMOS transistors as a combined switch is referred to as a *transmission gate* (TG), or a CMOS switch. The features of the three types of switches when used in logic circuits were also discussed, and the basic principles of switch logic design were coped with in detail along with examples.

The design of a VLSI chip is a complex process and is often proceeded with a hierarchical design process, including top-down and bottom-up approaches. To facilitate the reusable building blocks, logic circuits are designed and implemented as cells. After introducing the concepts of cell designs, we briefly described the CMOS fabrication process, the layout of CMOS circuits, and the related layout design rules.

The final section first looked at the future trends of VLSI (digital) systems design and then introduced the three major classes of implementation options of VLSI systems, including platforms, field-programmable devices, and ASICs.

References

1. B. Bailey, G. Martin, and A. Piziali, *ESL Design and Verification: A Prescription for Electronic System-Level Methodology.* San Francisco: Morgan Kaufmann, 2007.
2. J. A. Cunningham, "The use and evaluation of yield models in integrated circuit manufacturing," *IEEE Trans. on Semiconductor Manufacturing*, Vol. 3, No. 2, pp. 60–71, May 1990.
3. B. Davari,"CMOS technology: present and future," *Symposium VLSI Circuits Digest Technology Papers*, pp. 5-10, 1999.
4. R. Dennard et al.,"Design of ion-implanted MOSFET's with very small physical dimensions," *IEEE J. of Solid State Circuits*, Vol. 9, No. 5, pp. 256–268, October 1974.

5. P. Gelsinger, "Microprocessors for the new millennium: challenges, opportunities, and new frontiers," *Proc. of IEEE Int'l Solid-State Circuits Conf.*, pp. 22–25, 2001.

6. G. Gerosa et al., "A sub-2 W low power IA processor for mobile Internet devices in 45 nm high-k metal gate CMOS," *IEEE J. of Solid State Circuits*, Vol. 44, No. 1, pp. 73–82, January 2009.

7. D. A. Hodges, H. G. Jackson, and R. A. Saleh, *Analysis and Design of Digital Integrated Circuits: In Deep Submicron Technology*, 3rd ed. New York: McGraw-Hill Books, 2004.

8. J. S. Kilby, "Invention of the integrated circuit," *IEEE Trans. on Electronic Devices,* Vol. 23, No. 7, pp. 648–654, July 1976.

9. M. B. Lin, *Digital System Design: Principles, Practices, and Applications,* 4th ed. Taipei, Taiwan: Chuan Hwa Book Ltd., 2010.

10. M. B. Lin, *Basic Principles and Applications of Microprocessors: MCS-51 Embedded Microcomputer System, Software, and Hardware,* 2nd ed. Taipei, Taiwan: Chuan Hwa Book Ltd., 2003.

11. M. B. Lin, *Digital System Designs and Practices: Using Verilog HDL and FPGAs.* Singapore: John Wiley & Sons, 2008.

12. K. Mistry et al., "A 45nm logic technology with high-k+metal gate transistors, strained silicon, 9 Cu interconnect layers, 193 nm dry patterning, and 100% Pb-free packaging," *Proc. of IEEE Int'l Electron Devices Meeting (IEDM),* pp. 247–250, 2007.

13. G. Moore, "No exponential is forever: but 'forever' can be delayed!" *Proc. of IEEE Int'l Solid-State Circuits Conf.*, pp. 1–19, 2003.

14. G. Moore, "Cramming more components onto integrated circuits," *Electronics,* Vol. 38, No. 8, April 1965.

15. R. S. Muller and T. I. Kamins, *Device Electronics for Integrated Circuits,* 3rd ed. New York: John Wiley & Sons, Inc., 2003.

16. E. Nowak, "Maintaining the benefits of CMOS scaling when scaling bogs down," *IBM J. of Research and Development,* Vol. 46, No. 2/3, pp. 169–180, March/May 2002.

17. S. Parihar et al.,"A high density 0.10 μm CMOS technology using low-k dielectric and copper interconnect," *Proc. of IEEE Int'l Electron Devices Meeting (IEDM),* pp. 11.4.1–11.4.4, 2001.

18. S. Rusu et al., "A 45 nm 8-core enterprise Xeon processor," *IEEE J. of Solid State Circuits*, Vol. 45, No. 1, pp. 7–14, January 2010.

19. D. Sylvester and K. Keutzer,"Getting to the bottom of deep submicron," *Proc. of IEEE/ACM Int'l Conf. of Computer-Aided Design,* pp. 203–211, 1998.

20. S. Tyagi, "An advanced low power, high performance, strained channel 65 nm technology," *Proc. of IEEE Int'l Electron Devices Meeting (IEDM),* pp. 1070–1072, 2005.

21. J. P. Uyemura, *Introduction to VLSI Circuits and Systems.* New York: John Wiley & Sons, Inc., 2002.

22. F. Wanlass and C. Sah, "Nanowatt logic using field effect metal-oxide semiconductor triodes," *Proc. of IEEE Int'l Solid-State Circuits Conf.*, pp. 32–33, 1963.

Problems

1-1. Suppose that the diameter of a wafer is d and each die has an area of $A = a \times a$. Show that the number of dies in a wafer, excluding fragmented dies on the boundary, can be approximated as the following equation:

$$\text{Dies per wafer} = \frac{3}{4}\frac{d^2}{A} - \frac{1}{2\sqrt{A}}d$$

1-2. Assume that the diameter of a wafer is 30 cm and the die area is 0.65 cm^2. The defect density D_0 is 0.5 defects/cm^2 and the manufacturing complexity α is 4. The wafer price is 1,200 USD. Calculate the cost of each die without involving the fixed cost.

1-3. Assume that the diameter of a wafer is 30 cm and the die area is 0.86 cm^2. The defect density D_0 is 0.6 defects/cm^2 and the manufacturing complexity α is 4. The wafer price is 1,200 USD. Calculate the cost of each die without involving the fixed cost.

1-4. A half subtractor (also subtracter) is a device that accepts two inputs, x and y, and produces two outputs, b and d. The full subtractor is a device that accepts three inputs, x, y, and b_{in}, and produces two outputs, b_{out} and d, according to the truth table shown in Table 1.3.

Table 1.3: The truth table of a full subtractor.

x	y	b_{in}	b_{out}	d
0	0	0	0	0
0	0	1	1	1
0	1	0	1	1
0	1	1	1	0
1	0	0	0	1
1	0	1	0	0
1	1	0	0	0
1	1	1	1	1

(a) Derive the minimal expressions of both b_{out} and d of the full subtractor.

(b) Draw the logic diagram of switching expressions: b_{out} and d in terms of two half subtractors and one two-input OR gate.

1-5. Consider the logic circuit shown in Figure 1.46:

(a) Show that the logic circuit cannot correctly realize the switching function: $f(x, y, z) = xy + yz + xz$.

(b) Give a correct version using $f/\bar{f}$ design paradigm.

1-6. A universal logic module is a circuit that can realize any switching function by only using the circuit as many copies as needed. Show that both circuits shown in Figure 1.47 are universal logic modules if both true and complementary forms of control variable are available.

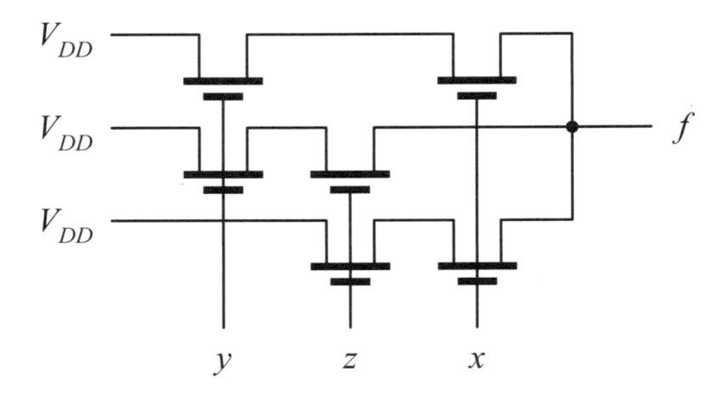

Figure 1.46: An example of an incorrect logic circuit.

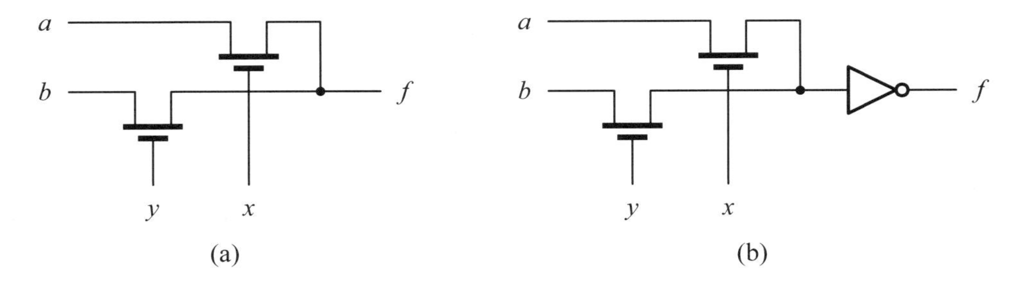

Figure 1.47: Two examples of universal logic modules: (a) unbuffered; (b) buffered.

1-7. Use a combination of CMOS logic gates to realize each of the following switching functions.

(a) $f(x, y, z) = \bar{x}\bar{y} + \bar{x}\bar{z} + \bar{y}\bar{z}$

(b) $f(w, x, y, z) = \overline{w \cdot (x + y) + y \cdot z}$

Suppose that only true literals are available. Your implementation needs to use the minimal number of transistors.

1-8. Using the $f/\bar{f}$ paradigm, implement the following logic gates and sketch their switch logic circuits.

(a) Three-input AND gate

(b) Three-input OR gate

1-9. Using the $f/\bar{f}$ paradigm, design a CMOS complex logic gate to realize each of the switching functions.

(a) $f(w, x, y, z) = \overline{wx + x \cdot y \cdot z}$

(b) $f(w, x, y, z) = \overline{w \cdot (x + y + z)}$

1-10. Suppose that both true and complementary forms of variables are available. Using the $f/\bar{f}$ paradigm, implement the following specified logic circuits.

(a) Two-input XOR gate

(b) Two-input XNOR gate

(c) 2-to-1 multiplexer

1-11. Using the following switching functions, verify the validity of Shannon's expansion theorem.

(a) $f(w,x,y,z) = \overline{w+x} \cdot (y \cdot z)$

(b) $f(w,x,y,z) = \bar{w} \cdot (\overline{x+y} + z)$

Assume that the control variable is x.

1-12. Compute the residues of the following switching functions with respect to all combinations of variables w, x, and y.

(a) $f(w,x,y,z) = \bar{w}\bar{x} + xy\bar{z} + wyz$

(b) $f(w,x,y,z) = \bar{w}y + xyz + \bar{w}\bar{y}z$

1-13. Compute the residues of the following switching functions with respect to all combinations of variables x, y, and z.

(a) $f(w,x,y,z) = \sum(0,1,4,6,7,9,11,13,15)$

(b) $f(w,x,y,z) = \sum(1,3,6,7,9,11,12) + \sum_{\phi}(2,5,13,15)$

1-14. Assuming that freeform-tree networks are used, plot the simplified decomposition trees for the following switching functions under the specified decomposition order.

(a) $f(w,x,y,z) = \bar{w}\bar{x}\bar{y}z + \bar{w}x\bar{y} + \bar{w}yz + w\bar{y}\bar{z} + wxy\bar{z} + wxz$. The function is first decomposed with respect to variable w. Then, the decomposition sequence of function f_w is variables z and y, and of function $f_{\bar{w}}$ is y and z.

(b) $f(w,x,y,z) = w\bar{x}\bar{z} + w\bar{x}y\bar{z} + w\bar{x}z + wx\bar{y}\bar{z} + \bar{w}x\bar{y}z + xyz$. The function is first decomposed with respect to variable x. Then, the decomposition sequence of function f_x is y and z, and of function $f_{\bar{x}}$ is variables z and w.

1-15. Assuming that uniform-tree networks are used, plot the simplified decomposition trees for the following switching functions under the specified decomposition order.

(a) $f(w,x,y,z) = \bar{x}\bar{y}z + w\bar{x}y + x\bar{y} + xy\bar{z}$. The decomposition order is x and y.

(b) $f(w,x,y,z) = \bar{x}\bar{y}z + \bar{w}\bar{x}y\bar{z} + w\bar{x}yz + wx\bar{y} + \bar{w}xyz + wxy\bar{z}$. The decomposition order is x, y, and w.

1-16. Using freeform-tree networks with nMOS switches, implement the following logic gates and sketch their switch logic circuits. Assume that the output inverter is not needed.

(a) Three-input NAND gate

(b) Three-input NOR gate

1-17. Realize each of the following switching functions using a freeform-tree network with nMOS switches.

(a) $f(x, y, z) = \bar{x}\bar{y} + \bar{x}\bar{z} + \bar{y}\bar{z}$

(b) $f(x, y, z) = xy + yz + xz$

1-18. Using a freeform-tree network with nMOS switches, design a complex logic gate to realize each of the switching functions.

(a) $f(w, x, y, z) = \overline{w + x \cdot y \cdot z}$

(b) $f(w, x, y, z) = \overline{(w + x) \cdot (y + z)}$

1-19. Using a uniform-tree network with nMOS switches, design a complex logic gate to realize each of the switching functions.

(a) $f(w, x, y, z) = \overline{w \cdot (x + y + z)}$

(b) $f(w, x, y, z) = \overline{(w \cdot x) + (y \cdot z)}$

1-20. Implement the following switching functions with nMOS switches using uniform-tree networks.

(a) $f(w, x, y, z) = \bar{w}x + x\bar{y}z + wy\bar{z}$

(b) $f(w, x, y, z) = w\bar{x} + \bar{x}\bar{y}z + wyz$

1-21. Assuming that the inverter at the output is not needed, use $0/1$-$x/\bar{x}$-tree networks with nMOS switches to implement the following logic gates.

(a) Three-input NAND gate

(b) Three-input NOR gate

1-22. Realize each of the following switching functions using a $0/1$-$x/\bar{x}$-tree network with TG switches.

(a) $f(x, y, z) = \bar{x}\bar{y} + \bar{x}\bar{z} + \bar{y}\bar{z}$

(b) $f(x, y, z) = xy + yz + xz$

1-23. Using a $0/1$-$x/\bar{x}$-tree network with nMOS switches, design a logic circuit to realize each of the switching functions.

(a) $f(w, x, y, z) = \overline{w + x \cdot y \cdot z}$

(b) $f(w, x, y, z) = \overline{w \cdot (x + y + z)}$

1-24. Implement the following switching functions with nMOS switches using $0/1$-$x/\bar{x}$-tree networks.

(a) $f(w, x, y, z) = \sum(3, 4, 5, 7, 9, 13, 14, 15)$

(b) $f(w, x, y, z) = \sum(3, 5, 6, 7, 9, 12, 13, 15)$

1-25. Consider a four-input NAND gate and answer the following questions.

(a) Sketch a CMOS logic circuit.

(b) Sketch a stick diagram.

(c) Estimate the area of your four-input NAND gate from the stick diagram.

(d) Using a CAD tool of your choice, construct a layout of your four-input NAND gate.

1-26. Consider a four-input NOR gate and answer the following questions.

(a) Sketch a CMOS logic circuit.
(b) Sketch a stick diagram.
(c) Estimate the area of your four-input NOR gate from the stick diagram.
(d) Using a CAD tool of your choice, construct a layout of your four-input NOR gate.

1-27. Consider a CMOS compound OR-AND-INVERT (OAI21) gate computing a switching function of $f(x, y, z) = \overline{(x + y) \cdot x}$.

(a) Sketch a CMOS logic circuit.
(b) Sketch a stick diagram.
(c) Estimate the area of your OAI21 gate from the stick diagram.
(d) Using a CAD tool of your choice, construct a layout of your OAI21 gate.

1-28. Consider a CMOS compound OR-AND-INVERT (OAI22) gate computing a switching function of $f(w, x, y, z) = \overline{(w + x) \cdot (y + z)}$.

(a) Sketch a CMOS logic circuit.
(b) Sketch a stick diagram.
(c) Estimate the area of your OAI22 gate from the stick diagram.
(d) Using a CAD tool of your choice, construct a layout of your OAI22 gate.

1-29. A majority circuit is a circuit that outputs 1 whenever more than half of its inputs are 1. Assume that only three inputs are considered.

(a) Sketch a CMOS logic circuit.
(b) Sketch a stick diagram.
(c) Estimate the area of your majority circuit from the stick diagram.
(d) Using a CAD tool of your choice, construct a layout of your majority circuit.

1-30. A minority circuit is a circuit that outputs 1 whenever less than half of its inputs are 1. Assume that only three inputs are considered.

(a) Sketch a CMOS logic circuit.
(b) Sketch a stick diagram.
(c) Estimate the area of your minority circuit from the stick diagram.
(d) Using a CAD tool of your choice, construct a layout of your minority circuit.

2

Fundamentals of MOS Transistors

Recall that a MOS transistor consists of a metal-oxide-semiconductor (MOS) system, two *pn* junctions, and two metal-semiconductor junctions. To understand the detailed operation of MOS transistors, in this chapter, we are first concerned with semiconductor fundamentals, including the intrinsic and extrinsic semiconductors coupled with Fermi-energy level, work function, and the relationships between impurity concentrations and Fermi-energy levels of both *p*-type and *n*-type materials. The carrier transport processes, drift and diffusion processes, are also discussed concisely.

Next, we address the characteristics of *pn* junctions, including basic structure, built-in potential, depletion-region width, current equation, and junction capacitance. In addition, we consider the effects of junction capacitance in digital circuits. The metal-semiconductor junctions are also discussed briefly.

The MOS transistor theory is then coped with in great detail. The MOS system is basically a capacitor, that is, consisting of a metal, a silicon dioxide, and a semiconductor. A MOS transistor can be constructed by adding an/a n^+/p^+ region on each side of a MOS system to serve as source and drain regions. The basic operations of MOS transistors and their ideal current-voltage (I-V) characteristics are examined in detail. After this, the scaling theory of complementary MOS (CMOS) processes is briefly explored. The nonideal features of MOS transistors, threshold voltage effects, and leakage currents are dealt with in great detail. The temperature effects on the drain currents of both *n*-tyep MOS (nMOS) and *p*-type MOS (pMOS) transistors and limitations of MOS transistors are addressed.

Finally, we conclude this chapter with the introduction of a simulation program with integrated circuit emphasis (SPICE) and look at how to model diodes and MOS transistors along with some examples.

2.1 Semiconductor Fundamentals

In this section, we begin with the introduction of semiconductors, including *intrinsic* and *extrinsic semiconductors* coupled with *Fermi-energy level*, *work function*, and the relationships between *impurity concentrations* and Fermi-energy levels of both *p*-type and *n*-type materials. Then, we address carrier transport processes: the *drift process* and *diffusion process*.

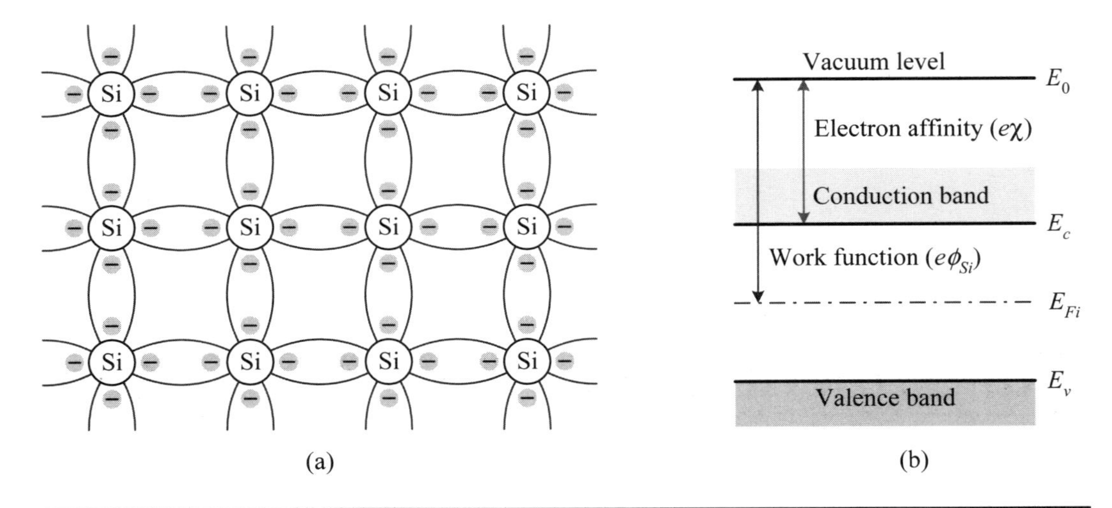

Figure 2.1: (a) Covalent bonding of the silicon atom. (b) Energy-band diagram of an intrinsic silicon crystal.

2.1.1 Intrinsic Semiconductors

The useful feature of a semiconductor is that its electrical properties can be altered by adding specific impurity atoms to it. A semiconductor that has been subjected to the addition of specific impurity atoms through some doping process is called an *extrinsic semiconductor*. In contrast, a pure semiconductor, without the addition of any specific impurity atoms, is called an *intrinsic semiconductor*.

It is well known that each silicon atom has four electrons in its outermost shell. These four electrons are referred to as *valence electrons*. In general, atoms having three, four, and five valence electrons are called *trivalent*, *tetravalent*, and *pentavalent*, respectively. For a silicon (or germanium) crystal, the four valence electrons of one atom form a bonding arrangement with its four adjacent atoms, as depicted in Figure 2.1(a). This kind of bonding is known as *covalent bonding*.

Figure 2.1(b) shows the energy-band diagram of the pure silicon crystal. The energy difference between the bottom of conduction band E_c and the top of valence band E_v is the *bandgap energy*, referred to as E_g. The bandgap energy E_g is a strong function of temperature and can be represented as follows:

$$E_g(T) = E_g(0) - \frac{\alpha T^2}{T + \beta} \qquad \text{(eV)} \tag{2.1}$$

For silicon, $E_g(0) = 1.17$ eV, $\alpha = 4.73 \times 10^{-4}$, and $\beta = 651$. At 300 K, $E_g(300) = 1.125$ eV.

The energy difference between the vacuum level, E_0, and the bottom of the conduction band is called the *electron affinity* ($e\chi$) of an electron. The work function ($e\phi_{Si}$) of an electron is the minimum energy of it required to break the covalent bonding to escape into the vacuum and become a free electron. The work function is defined as the energy difference between the vacuum level, E_0, and Fermi-energy level E_F. In intrinsic semiconductors, this is equal to the difference between E_0 and E_{Fi}, where E_{Fi} denotes the Fermi-energy level (E_F) of the intrinsic semiconductor.

2.1.1.1 The n_0 and p_0 Equations Let n_0 and p_0 denote the thermal-equilibrium electron and hole concentrations, respectively. The thermal-equilibrium electron con-

Table 2.1: Properties of three popular semiconductor materials.

Property	Si	Ge	GaAs	Unit
Energy gap (E_g) (300 K)	1.12	0.66	1.42	eV
Effective mass				
Electron	1.08	0.55	0.068	m_0
Hole	0.80	0.3	0.50	m_0
Intrinsic carrier concentration	1.45×10^{10}	2.4×10^{13}	9.0×10^{6}	cm^{-3}
Effective density of states				
Conduction band	2.82×10^{19}	1.04×10^{19}	4.7×10^{17}	cm^{-3}
Valence band	1.04×10^{19}	6.0×10^{18}	7.0×10^{18}	cm^{-3}

centration can be represented as follows:

$$n_0 = N_c \exp\left[-\frac{(E_c - E_F)}{kT}\right] \tag{2.2}$$

where the parameter N_c is called the *effective density of states* in the conduction band and is defined as

$$N_c = 2\left(\frac{2\pi m_n^* kT}{h^2}\right)^{3/2} \tag{2.3}$$

where m_n^* denotes the *electron effective mass*, T is the temperature in kelvin (K), h is the Planck's constant, equal to 4.135×10^{-15} eV-s, and k is the Boltzmann's constant, which is 8.62×10^{-5} eV/K.

Similarly, thermal-equilibrium hole concentration p_0 can be found as follows:

$$p_0 = N_v \exp\left[-\frac{(E_F - E_v)}{kT}\right] \tag{2.4}$$

where the parameter N_v is called the *effective density of states* in the valance band and is defined as

$$N_v = 2\left(\frac{2\pi m_p^* kT}{h^2}\right)^{3/2} \tag{2.5}$$

where m_p^* denotes the *hole effective mass*. Both n_0 and p_0 represent the electron and hole concentrations as a function of Fermi-energy level E_F. It is worth noting that the product of n_0 and p_0 is independent of Fermi-energy level and is a constant with a specific semiconductor material and a given temperature.

2.1.1.2 Positions of Fermi-Energy Levels In an intrinsic semiconductor, because of the feature of *charge neutrality*, the generation of electrons and holes due to thermal energy is always in pairs. Hence, the electron and hole concentrations are equal to each other. Because of this, the Fermi-energy level is approximately at the middle of bandgap energy, as shown in Figures 2.1(b). The Fermi-energy level in the intrinsic semiconductor is usually referred to as the *intrinsic Fermi-energy level* and is denoted as E_{Fi}. Similarly, the electron and hole concentrations, n_0 and p_0, in the intrinsic semiconductor are generally referred to as intrinsic electron and hole concentrations and denoted n_i and p_i, respectively. Some properties of three popular semiconductor materials are listed in Table 2.1.

2.1.1.3 Mass-Action Law The relationship between the electron and hole concentrations in a semiconductor can be described by the well-known *mass-action law*, which is stated as follows:

$$np = n_i^2 = BT^3 \exp\left(\frac{-E_g}{kT}\right) \tag{2.6}$$

where n and p denote the electron and hole concentrations in the material under consideration. n_i is the intrinsic carrier concentration and approximates to 1.45×10^{10} for silicon material at temperature 300 K. The constant B is a material-dependent parameter and equal to 6.2×10^{31}/K·cm^3 for silicon. In other words, the mass-action law states that, at a given temperature, the product of electron concentration and hole concentration of a given semiconductor is constant and equal to the square of intrinsic carrier concentration.

■ Example 2-1: (The intrinsic carrier concentration of silicon material.)
Using Equation (2.6), calculate the intrinsic carrier concentration of silicon at each of the following given temperatures.

1. $T = 300$ K
2. $T = 600$ K

Solution: From Equation (2.1), we obtain the values of bandgap energy E_g at temperatures 300 K and 600 K are 1.125 eV and 1.03 eV, respectively. Hence, the intrinsic carrier concentrations of silicon at these two temperatures can be calculated using Equation (2.6) and are as follows:

1. At $T = 300$ K

   $$n_i^2 = 6.2 \times 10^{31} \times 300^3 \exp\left(\frac{-1.125}{300 \times 8.62 \times 10^{-5}}\right) = 2.14 \times 10^{20}$$

   Therefore, the intrinsic carrier concentration n_i at temperature 300 K is 1.46×10^{10}/cm^3, which is very close to the most commonly accepted value: 1.45×10^{10}/cm^3.
2. At $T = 600$ K

   $$n_i^2 = 6.2 \times 10^{31} \times 600^3 \exp\left(\frac{-1.03}{600 \times 8.62 \times 10^{-5}}\right) = 3.00 \times 10^{31}$$

   That is, the intrinsic carrier concentration n_i at temperature 600 K is 5.48×10^{15}/cm^3. The intrinsic electron concentration n_i is therefore a very strong function of temperature.

■

■ Review Questions

Q2-1. Show that, at thermal equilibrium, the Fermi-energy level is not exactly at the middle of the energy gap.

Q2-2. Show the validity of the mass-action law and explain its meaning.

Q2-3. Distinguish between intrinsic and extrinsic semiconductors.

Q2-4. Give the expressions of n_i and p_i.

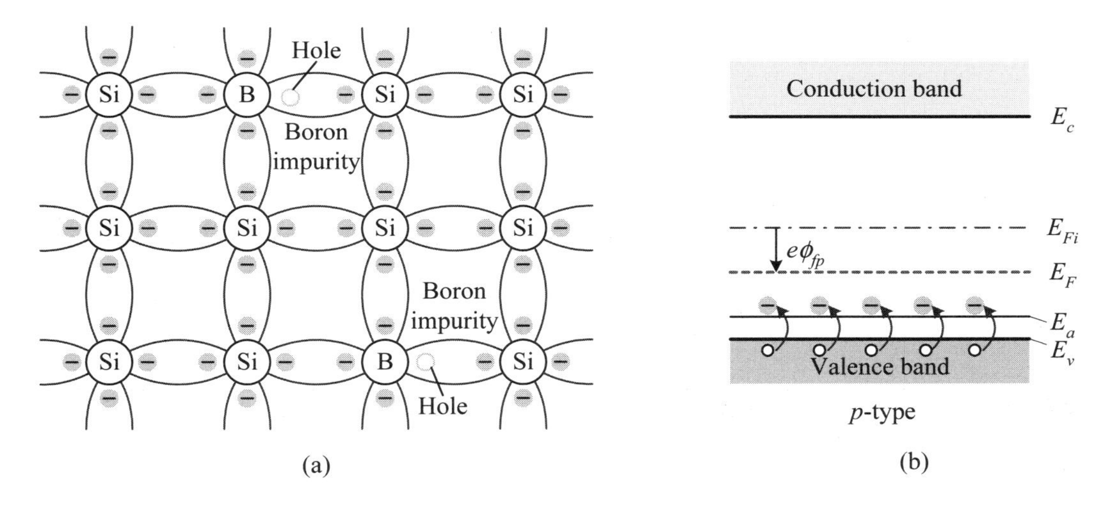

Figure 2.2: (a) Covalent bonding of boron impurity in p-type silicon. (b) Energy-band diagram of a p-type silicon semiconductor.

2.1.2 Extrinsic Semiconductors

In this subsection, we are concerned with extrinsic semiconductors and their related issues. We first introduce p- and n-type materials and then consider the *compensated semiconductors*.

2.1.2.1 The p-Type Semiconductors A p-type material is created by adding a predetermined number of trivalent impurity atoms, such as *boron* (B), *gallium* (Ga), and *indium* (In), to a pure silicon material. The effect of adding boron atoms into silicon material is shown in Figure 2.2(a), where holes are created due to the lack of one electron in the outermost shell of each boron atom. This results in a new lattice that has an insufficient number of electrons to complete the covalent bonds and leave vacancies. These vacancies are referred to as *holes*. The trivalent impurities are called *acceptor atoms* because they are ready to accept electrons. Each atom accepts one electron and leaves a hole in the covalent bond. Note that a trivalent atom becomes a negative ion after it accepts an electron because it is neutral prior to accepting an electron.

The energy-band diagram of the p-type silicon crystal is depicted in Figure 2.2(b). The acceptor impurity provides an acceptor energy level E_a with a bandgap energy E_g significantly less than the intrinsic material. The electrons in the valence band have less difficulty for absorbing sufficient thermal energy to move into the acceptor energy level at room temperature. This results in a large number of carriers (holes) in the valence band at room temperature and increases the conductivity of the material significantly.

The Fermi-energy level of a p-type material is determined by the number of impurity atoms added to the material. It is generally represented with respect to the intrinsic Fermi-energy level; namely, it can be expressed as

$$p_0 = n_i \exp\left[-\frac{(E_F - E_{Fi})}{kT}\right] \tag{2.7}$$

The difference between Fermi-energy level and intrinsic Fermi-energy level is often represented as a potential called *Fermi potential* of p-type material. This Fermi potential is denoted as ϕ_{fp} and is a function of acceptor and intrinsic carrier concentrations.

Assuming that all acceptor atoms are ionized, the Fermi potential of *p*-type material can be expressed as follows:

$$\phi_{fp} = \frac{E_F - E_{Fi}}{e} = -\frac{kT}{e}\ln\left(\frac{N_a}{n_i}\right) \tag{2.8}$$

where $(kT)/e$ is often defined as *thermal voltage* (V_t) or *thermal-equivalent voltage* and is equal to 25.86 mV at room temperature 300 K, e is the charge of an electron, that is, 1.6×10^{-19} C, and N_a is the concentration of acceptor impurity.

The following example illustrates the Fermi potential and thermal electron concentration of a *p*-type material with a given doping level.

Example 2-2: (A p-type material.) Consider a *p*-type silicon doped with boron impurity and assume that the concentration of boron impurity is 1.6×10^{16} cm^{-3}.

1. Calculate the Fermi potential of the *p*-type material.
2. Calculate the thermal-equilibrium electron concentration at $T = 300$ K.

Solution: The Fermi potential and thermal-equilibrium electron concentration are computed separately as follows:

1. The Fermi potential can be computed by using Equation (2.8).

$$\begin{aligned}\phi_{fp} &= -\frac{kT}{e}\ln\left(\frac{N_a}{n_i}\right)\\ &= -25.86 \times 10^{-3}\ln\left(\frac{1.6 \times 10^{16}}{1.45 \times 10^{10}}\right) = -0.36 \text{ V}\end{aligned}$$

2. Assume that at room temperature all acceptor atoms are ionized. Hence, $p_0 = N_a$. The thermal-equilibrium electron concentration can then be obtained from the mass-action law.

$$n_0 = \frac{n_i^2}{p_0} = \frac{(1.45 \times 10^{10})^2}{1.6 \times 10^{16}} = 1.31 \times 10^4 \text{ cm}^{-3}$$

The electron concentration is much less than the hole concentration. Because of the apparent difference in concentration, the electrons in the *p*-type material are called *minority carriers* while the holes are called *majority carriers*.

■

2.1.2.2 The n-Type Semiconductors Like the *p*-type material mentioned previously, an *n*-type material is created by adding a predetermined number of pentavalent impurity atoms, such as *phosphorous* (P), *antimony* (Sb), and *arsenic* (As), to a pure silicon material. The effect of adding antimony atoms into silicon material is shown in Figure 2.3(a), where free electrons are created due to an extra electron in the outermost shell of each antimony atom. The result of the newly formed lattice has some free electrons in addition to completing the covalent bonds. These pentavalent impurities are referred to as *donor atoms* because they donate electrons. Each atom donates one electron. Note that a pentavalent atom becomes a positive ion after it is donated an electron because it is neutral before donating an electron.

The energy-band diagram of the *n*-type silicon crystal is depicted in Figure 2.3(b). The donor impurity provides a donor energy level E_d with a bandgap energy E_g significantly less than the intrinsic material. The electrons in the donor energy level have

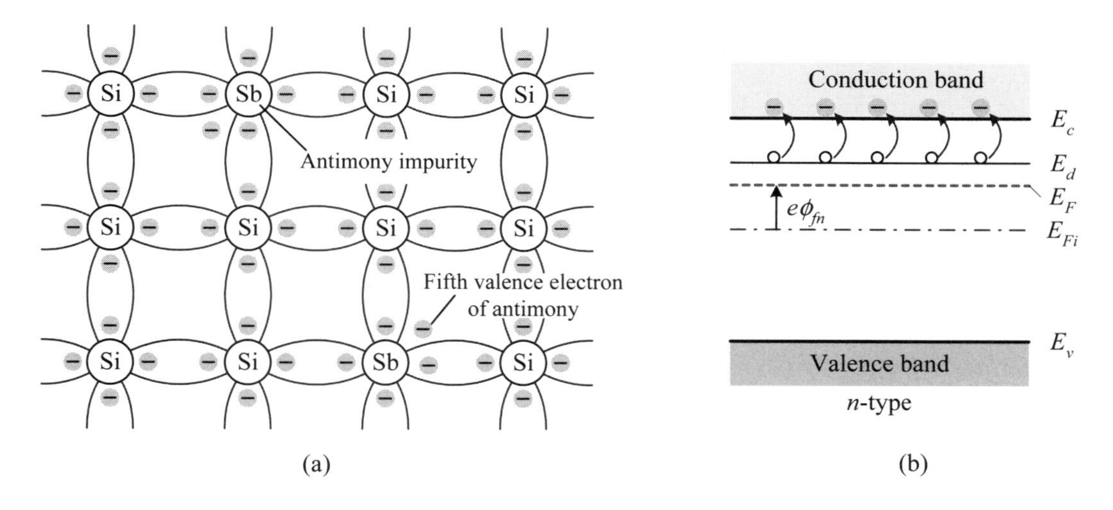

Figure 2.3: (a) Covalent bonding of antimony impurity in an n-type silicon. (b) Energy-band diagram of an n-type silicon semiconductor.

less difficulty for absorbing sufficient thermal energy to move into the conduction band at room temperature. This results in a large number of carriers (electrons) in the conduction band at room temperature and increases the conductivity of the material significantly.

The Fermi-energy level of an n-type material is determined by the number of impurity atoms added to the material. Like the p-type material, this Fermi-energy level is generally represented with respect to the intrinsic Fermi-energy level.

$$n_0 = n_i \exp\left[\frac{(E_F - E_{Fi})}{kT}\right] \tag{2.9}$$

The difference between Fermi-energy level and intrinsic Fermi-energy level is defined as a potential called *Fermi potential* of n-type material. This Fermi potential is denoted as ϕ_{fn} and is a function of donor and intrinsic carrier concentrations. Assuming that all donor atoms are ionized, the Fermi potential of n-type material can be expressed as follows:

$$\phi_{fn} = \frac{E_F - E_{Fi}}{e} = \frac{kT}{e}\ln\left(\frac{N_d}{n_i}\right) \tag{2.10}$$

where N_d is the concentration of donor impurity.

The following example illustrates the relationship between the Fermi potential and impurity concentration of an n-type material.

■ **Example 2-3: (An n-type material.)** Consider an n-type silicon doped with antimony impurity and assume that at $T = 300$ K, the Fermi-energy level is 0.25 eV below the conduction band.

1. Calculate the concentration of antimony impurity, assuming that all impurity atoms are ionized.
2. Calculate the thermal-equilibrium hole concentration.

Solution: The antimony impurity and thermal-equilibrium hole concentrations are computed separately as follows:

1. The bandgap energy of silicon at 300 K is 1.125 eV and $e\phi_{fn}$ is equal to $\frac{E_g}{2} - 0.25 = 0.3125$ eV. That is, $\phi_{fn} = 0.3125$ V. The concentration of antimony impurity can be computed by using Equation (2.10).

$$0.3125 = 25.86 \times 10^{-3} \ln \left(\frac{N_d}{1.45 \times 10^{10}} \right)$$

Solving it, the concentration of antimony impurity is found to be 2.57×10^{15} cm^{-3}.

2. The thermal-equilibrium hole concentration can be obtained by using mass-action law.

$$p_0 = \frac{n_i^2}{n_0} = \frac{(1.45 \times 10^{10})^2}{2.57 \times 10^{15}} = 8.18 \times 10^4 \text{ cm}^{-3}$$

The hole concentration is much less than the electron concentration. The electrons in the n-type material are called *majority carriers* while the holes are called *minority carriers*.

■

2.1.2.3 Compensated Semiconductors The above-mentioned p-type and n-type materials are simply formed by adding trivalent and pentavalent impurities into a pure semiconductor, such as silicon, respectively. In modern CMOS processes, it is not uncommon to start with a p-type or an n-type material and to form another type, n-type or p-type. For example, as depicted in Figure 2.5, an n-well is formed on top of a p-type substrate before doping trivalent impurities inside it to form a pMOS transistor. This kind of semiconductor containing different impurities in the same region is known as a *compensated semiconductor*. In a compensated semiconductor, the resulting type, p or n, is determined by the relative concentration of the added impurities.

■ Review Questions

Q2-5. What do acceptor atoms mean? What do donor atoms mean?
Q2-6. How would you form a p-type semiconductor?
Q2-7. What is the thermal (thermal-equivalent) voltage?
Q2-8. How would you form an n-type semiconductor?
Q2-9. What does the compensated semiconductor mean?

2.1.3 Carrier Transport Processes

Current is the net flow rate of charges. In a semiconductor, there are two carrier transport processes able to produce a current. These are *drift process* and *diffusion process*.

2.1.3.1 Drift Process and Current A drift process is the one that causes the net movement of charges due to an electric field. The net drift of charge particles gives rise to a *drift current*. The *electron drift velocity* can be expressed as follows:

$$v_d = -\mu_n E \tag{2.11}$$

where μ_n is called the *electron mobility*. Similarly, *hole drift velocity* can be expressed as

$$v_d = \mu_p E \tag{2.12}$$

Table 2.2: The electron and hole mobilities of three popular semiconductors.

Mobility	Si	Ge	GaAs	Unit
Electron (μ_n)	1350	3900	8800	cm^2/V-sec
Hole (μ_p)	480	1900	400	cm^2/V-sec

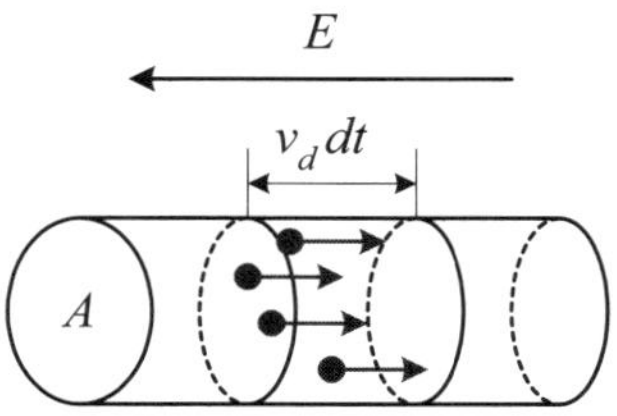

Figure 2.4: The drift current model of a conductor in an electric field.

where μ_p is called the *hole mobility*. Generally, mobility is a strong function of temperature and impurity concentrations.

The electron and hole mobilities at 300 K of three popular semiconductor materials are shown in Table 2.2. It is worth noting that these mobility values are in bulk material. The mobility values of MOS transistors are significantly less than these nominal values because the conducting channels of MOS transistors are on the surface of silicon material.

To describe the current quantitatively, that is, the net flow rate of charges, consider the drift current model shown in Figure 2.4. Let the cross-sectional area of the conductor be A and the electron concentration be n in cm^{-3}, then the drift current can be defined as the total amount of charge that flows through a cross-sectional area of A during a unit time dt and can be represented as follows:

$$I = \frac{dQ}{dt} = -\frac{d(enV)}{dt} = -\frac{en(Av_d dt)}{dt} = -enAv_d \tag{2.13}$$

where the volume V is equal to $v_d dtA$.

It is useful to denote the current as current density J so that the cross-sectional area A can be eliminated from the above equation. As a result, the *electron drift current density* is equal to

$$J_n = \frac{I}{A} = -env_d = en\mu_n E \tag{2.14}$$

Similarly, we may obtain the *hole drift current density* J_p as follows:

$$J_p = ep\mu_p E \tag{2.15}$$

Whenever both electrons and holes appear at the same time, the total drift current density J due to an external electric field is the sum of both electron and hole drift current densities and can be expressed as

$$J_{drift} = J_n + J_p = e(n\mu_n + p\mu_p)E = \sigma E \tag{2.16}$$

where σ is the *conductivity* of the material and is defined as

$$\sigma = e(n\mu_n + p\mu_p) \qquad \text{(Conductivity)} \tag{2.17}$$

The conductivity σ is given in units of $(\Omega\text{-cm})^{-1}$ and is a function of the electron and hole concentrations and mobilities. The reciprocal of conductivity is *resistivity*, which is denoted by ρ and is given in units of Ω-cm; namely,

$$\rho = \frac{1}{\sigma} = \frac{1}{e(n\mu_n + p\mu_p)} \qquad \text{(Resistivity)} \tag{2.18}$$

2.1.3.2 Diffusion Process and Current A diffusion process is the one that involves the particles flowing from a high-concentration region toward a low-concentration region. When such a process occurs, the net flow of charge particles results in a *diffusion current*. Assume that temperature is uniform over the region of interest so that the average terminal velocity of electrons is independent of x. In addition, we assume that there is no electric field existing in the system.

Let the electron concentration gradient be dn/dx. The *electron diffusion current density* J_n can be represented as follows:

$$J_{ndiff} = eD_n\frac{dn}{dx} \tag{2.19}$$

where $D_n = (kT/e)\mu_n$ is called the *electron diffusion coefficient*.

Similarly, let the hole concentration gradient be dp/dx. The *hole diffusion current density* J_p is found to be

$$J_{pdiff} = -eD_p\frac{dp}{dx} \tag{2.20}$$

where D_p is called the *hole diffusion coefficient* and is defined as $(kT/e)\mu_p$. Combining both definitions of electron and hole diffusion coefficients, Einstein's relation is the result.

$$\frac{D_n}{\mu_n} = \frac{D_p}{\mu_p} = \frac{kT}{e} \qquad \text{(Einstein's relation)} \tag{2.21}$$

Mobility indicates how well a charge carrier moves in a conductor or semiconductor as a result of an applied electric field; diffusion coefficient indicates how well a charge carrier moves in a conductor or semiconductor due to the concentration gradient of carriers.

2.1.3.3 Total Current Density The total current density is the sum of drift current and diffusion current components. When only electrons exist, the total current is

$$J_n = en\mu_n E + eD_n\frac{dn}{dx} \tag{2.22}$$

When only holes exist, the total current is

$$J_p = ep\mu_p E - eD_p\frac{dp}{dx} \tag{2.23}$$

In the situation in which both electrons and holes exist at the same time, the total current is the combination of the above four components. In other words, the total current can be generally expressed as the following equation:

$$J = J_n + J_p = en\mu_n E + eD_n\frac{dn}{dx} + ep\mu_p E - eD_p\frac{dp}{dx} \tag{2.24}$$

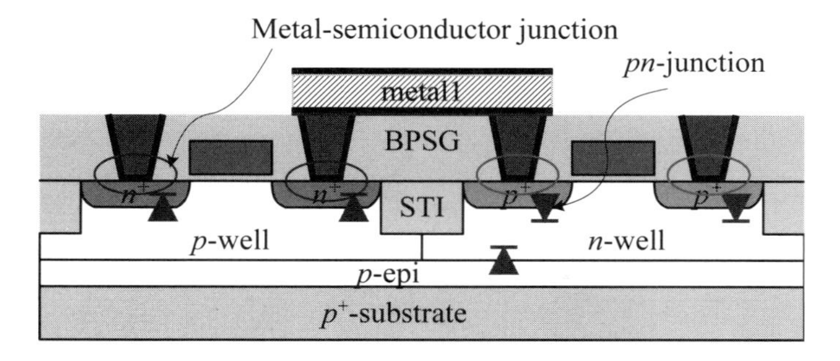

Figure 2.5: The *pn* and metal-semiconductor junctions in a CMOS inverter.

■ Review Questions

Q2-10. Define the drift process and diffusion process.
Q2-11. What is a current? Define it.
Q2-12. What is the electron mobility?
Q2-13. What is Einstein's relation?

2.2 The pn Junction

The *pn* junctions are ubiquitous in CMOS circuits. As shown in Figure 2.5, there are two pn^+ junctions for each nMOS transistor and two p^+n junctions of each pMOS transistor. In addition, an extra *pn* junction exists between an *n*-well and *p*-type substrate. Of course, all pn^+ and p^+n junctions must be in the reverse-bias condition such that they do not affect the normal operations of MOS transistors. To connect an nMOS or a pMOS transistor with others, metal wires are contacted with n^+ or p^+ diffusion regions. Hence, metal-semiconductor junctions are also formed in MOS transistors.

From the above description, we can see that a MOS transistor, regardless of *n* or *p* type, is composed of three major kinds of components: *pn*-junction, metal-semiconductor junction, and a MOS system. In this section, we will examine the structure and features of *pn*-junctions in detail and briefly describe the metal-semiconductor junctions. The MOS system and its applications to form MOS transistors will be dealt with later in this chapter.

2.2.1 The pn Junction

In this subsection, we are concerned with the characteristics of *pn* junctions. These include basic structure, built-in potential, depletion-region width, current equation, and junction capacitance. In addition, we consider the effects of junction capacitance in digital circuits.

2.2.1.1 Basic Structure A *pn* junction is formed by joining a *p*-type semiconductor with an *n*-type one, as shown in Figure 2.6(a). The interface separating the *p*-type and *n*-type regions is referred to as the *metallurgical junction*. The typical impurity concentrations in the *p*-type and *n*-type regions are also given in the figure. Here, we assume that the concentration of *p*-type impurity is much greater than that of *n*-type impurity. For simplicity, we are only concerned with the situation in which

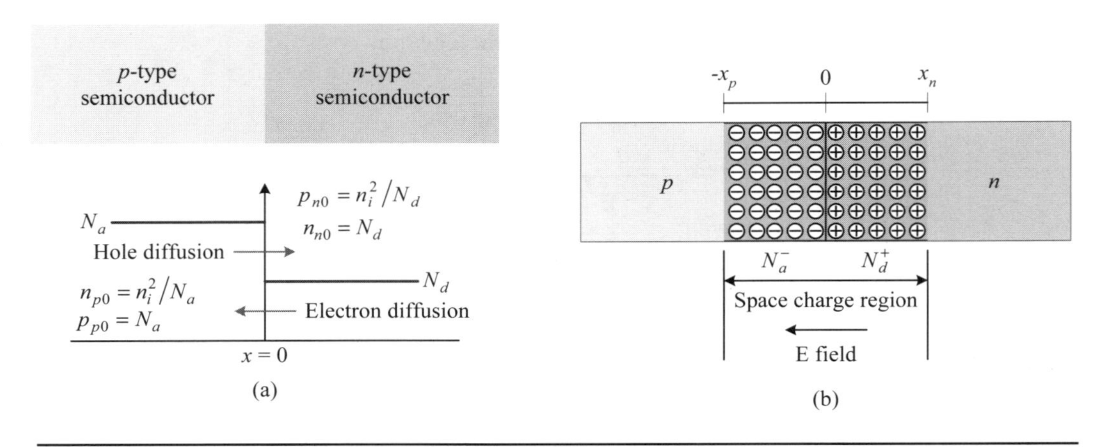

Figure 2.6: The (a) basic structure and (b) depletion region of a *pn*-junction.

the impurity concentration is uniform in each region and an abrupt change in the impurity concentration in the junction occurs. This type of junction is referred to as *step junction*, or *abrupt junction*.

Figure 2.6(b) shows the situation when both *p*-type and *n*-type materials are joined together. The electrons in the *n*-type region are diffused into the *p*-type region and leave behind positively charged donor ions. The holes in the *p*-type region are diffused into the *n*-type region and leave behind negatively charged acceptor ions. Because the charged ions are parts of the crystal lattice, they cannot move freely. As a consequence, the region near the junction is depleted of free carriers and called the *depletion region*, or *space-charge region*. In addition, both charged ions at the depletion region build up a potential barrier, called *built-in potential*, which tends to prevent electrons and holes from diffusing into the other side unlimitedly.

2.2.1.2 The Built-In Potential The energy-band diagrams of both *p*-type and *n*-type materials are shown in Figure 2.7(a). Recall that the Fermi-energy level E_F is determined by the impurity concentration and temperature for a given material. It can be shown that the position of Fermi-energy level E_F is constant throughout a system for a given impurity concentration and temperature.

The energy-band diagram after joining together both *p*-type and *n*-type materials is shown in Figure 2.7(b), which uses the fact that at thermal equilibrium the Fermi-energy level is constant throughout the system. Hence, an energy-level difference $e\phi_0$ exists between conduction bands of *p*-type and *n*-type materials. This difference is often represented as a potential known as the built-in potential, or *built-in potential difference*, or *built-in potential barrier*, and can be expressed as

$$\phi_0 = \phi_{fn} - \phi_{fp} = \frac{kT}{e}\left(\ln\frac{N_d}{n_i} + \ln\frac{N_a}{n_i}\right) = \frac{kT}{e}\ln\frac{N_dN_a}{n_i^2} \tag{2.25}$$

where the subscript 0 denotes the condition of thermal equilibrium and the subscripts p and n represent the *p*-type and *n*-type materials, respectively.

■ Example 2-4: (The built-in potential.) Assume that a *p*-type material with an impurity concentration of $N_a = 5 \times 10^{18}/\text{cm}^3$ and an *n*-type material with an impurity concentration of $N_d = 2 \times 10^{15}/\text{cm}^3$ are combined into a *pn* junction. Calculate the built-in potential of this *pn* junction at $T = 300$ K.

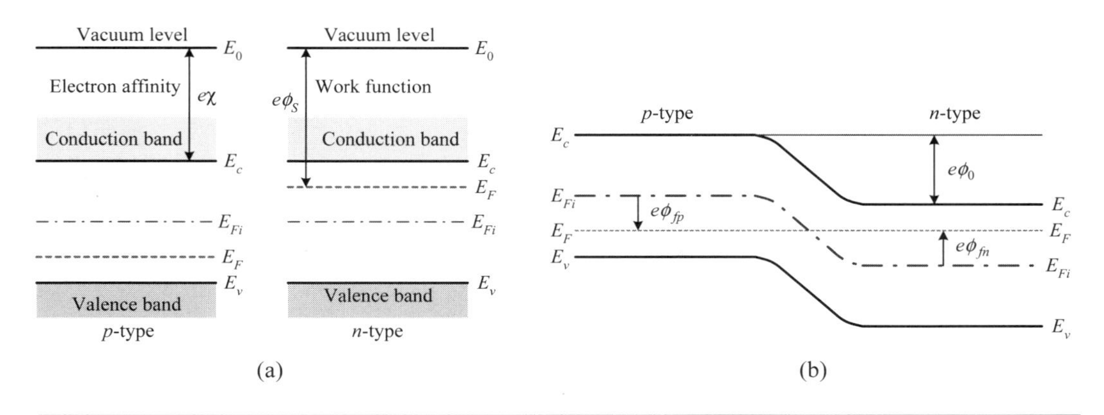

Figure 2.7: The energy-band diagrams of a typical *pn*-junction (a) before and (b) after combination.

Solution: From Equation (2.25), the built-in potential can be calculated as follows:

$$\begin{aligned}\phi_0 &= \frac{kT}{e}\ln\frac{N_dN_a}{n_i^2}\\ &= \frac{8.62\times 10^{-5}(\mathrm{eV/K})\times 300\ \mathrm{K}}{e}\ln\frac{2\times 10^{15}\times 5\times 10^{18}}{(1.45\times 10^{10})^2}\\ &= 0.81\ \mathrm{V}\end{aligned}$$

■

The built-in potential is a barrier that electrons in the conduction band of the n-type material must overcome to move into the conduction band of the p-type material. To reduce the built-in potential, an external voltage has to be applied across the pn junction in a way such that the positive terminal is connected to the p-type material and the negative terminal is connected to the n-type material. This bias condition is called *forward bias*. In a forward-biased pn junction, the resulting barrier height is reduced from ϕ_0 to $\phi_0 - V_a$, where V_a is the applied external voltage and is greater than or equal to 0 V.

On the contrary, if we reverse the polarity of applied external voltage, the resulting barrier height is increased from ϕ_0 to $\phi_0 + |V_a|$, where $V_a < 0$. Such a bias condition is called *reverse bias*.

2.2.1.3 The Depletion-Region Width One important feature of the depletion region is that both sides of it have the same amount of charge but of opposite polarities. That is,

$$x_nN_d = x_pN_a \tag{2.26}$$

where x_n and x_p represent the widths of n-type and p-type depletion regions accounting from the junction. For simplicity, the junction area is generally omitted in both sides. Referring to Figure 2.6(b), the depletion-region width of a pn junction is equal to

$$x_d = x_n - (-x_p) = x_n + x_p \tag{2.27}$$

and can be represented as a function of built-in potential and impurity concentrations.

$$x_d = \sqrt{\frac{2\varepsilon_{si}\phi_0}{e}\left(\frac{N_a + N_d}{N_a N_d}\right)} \tag{2.28}$$

where ϕ_0 is the built-in potential and the $\varepsilon_{si}(= \varepsilon_{r(si)}\varepsilon_0)$ is the permittivity of silicon. Generally, the value of $\varepsilon_{r(si)}$ is 11.7 and ε_0 is 8.854×10^{-14} F/cm.

The depletion-region width x_d is affected by the external voltage V_a being applied to the *pn* junction and can be expressed as follows:

$$x_d = \sqrt{\frac{2\varepsilon_{si}(\phi_0 - V_a)}{e}\left(\frac{N_a + N_d}{N_a N_d}\right)} \tag{2.29}$$

As V_a is greater than or equal to 0 V, that is, forward-biased condition, the depletion-region width x_d decreases; as V_a is less than 0 V, that is, reverse-biased condition, the depletion-region width x_d increases.

When the acceptor concentration N_a is much greater than the donor concentration N_d, the resulting depletion region is effectively extended into the n-type region. This results in a junction called a *p^+n one-sided junction.* Similarly, an n^+p one-sided junction is formed when the donor concentration N_d is much greater than the acceptor concentration N_a. In this case, the resulting depletion region is effectively extended into the p-type region. An illustration of p^+n one-sided junction is given in the following example.

■ Example 2-5: (A p$^+$n one-sided junction.) Consider a *pn* junction with acceptor and donor concentrations of $N_a = 5 \times 10^{18}/\text{cm}^3$ and $N_d = 2 \times 10^{15}/\text{cm}^3$, respectively. Calculate the depletion-region width at $T = 300$ K.

Solution: Since $N_a \gg N_d$, the resulting junction is a p^+n one-sided junction. The resulting depletion region is extended into the n-type material. The depletion-region width in the n-type side is obtained from Equation (2.28) and uses the fact that $N_a \gg N_d$.

$$\begin{aligned} x_n &= \sqrt{\frac{2\varepsilon_{si}\phi_0}{e}\left(\frac{1}{N_d}\right)} \\ &= \sqrt{\frac{2 \times 11.7 \times 8.85 \times 10^{-14} \times 0.81}{1.6 \times 10^{-19}}\left(\frac{1}{2 \times 10^{15}}\right)} = 0.724\ \mu\text{m} \end{aligned}$$

The built-in potential ϕ_0 is 0.81 V, obtained from the preceding example.

■

2.2.1.4 The Current Equation

The general current-voltage characteristic of a *pn*-junction diode is governed by the following *ideal-diode equation*:

$$I_D = I_s\left[\exp\left(\frac{V}{nV_t}\right) - 1\right] \tag{2.30}$$

where I_s is the reverse saturation current, V is the applied forward-bias voltage across the *pn* diode, n is an ideality factor and has a value between 1 and 2, and the voltage $V_t = kT/e$ is the thermal voltage.

2.2.1.5 Junction Capacitance As stated previously, both sides of the depletion region of a *pn* junction maintain the same amount of charge of opposite polarities, thereby forming a capacitor. Hence, the depletion region functions as a capacitor with capacitance C_j, which is referred to as *junction capacitance* or *depletion capacitance* and can be expressed as follows:

$$C_j = C_{j0}\left[1 - \frac{V}{\phi_0}\right]^{-m} \tag{2.31}$$

where C_{j0} is the zero-bias capacitance of the *pn* junction, V is the voltage across the junction, ϕ_0 is the built-in potential, and m is the *grading coefficient* with a value between 0.33 and 0.5. The C_{j0} can be expressed as

$$C_{j0} = A\left(\frac{e\varepsilon_{si}}{2\phi_0}\frac{N_a N_d}{N_a + N_d}\right)^{1/2} \tag{2.32}$$

where A is the junction area.

The following example illustrates the junction capacitance of the *pn* junction between an n-well and the p-substrate.

■ **Example 2-6: (Junction capacitance.)** Plot the junction capacitance versus the applied voltage V. The junction is formed between the n-well and p-type substrate. The area and depth of the n-well are $100 \times 50\ \mu\text{m}^2$ and $2\ \mu$m, respectively, as shown in Figure 2.8(a). Suppose that $N_a = 2 \times 10^{16}\ \text{cm}^{-3}$ and $N_d = 1 \times 10^{17}\ \text{cm}^{-3}$. The measured zero-bias junction capacitance is $100\ \text{aF}/\mu\text{m}^2$ and $m = 0.45$.

Solution: The built-in potential ϕ_0 is found to be

$$\begin{aligned}\phi_0 &= \frac{8.62 \times 10^{-5}(\text{eV/K}) \times 300\ \text{K}}{e}\ln\frac{2 \times 10^{16} \times 1 \times 10^{17}}{(1.45 \times 10^{10})^2} \\ &= 0.77\ \text{V}\end{aligned}$$

The zero-bias junction capacitance consists of two parts: bottom capacitance C_{j0b} and sidewall capacitance C_{j0sw}. The bottom capacitance under zero-bias condition can be calculated as follows:

$$\begin{aligned}C_{j0b} &= C_{j0} \times (100 \times 50\ \mu\text{m}^2) \\ &= 100\ \text{aF}/\mu\text{m}^2 \times (100 \times 50\ \mu\text{m}^2) \\ &= 0.5\ \text{pF}\end{aligned}$$

The sidewall capacitance under zero-bias condition can be calculated as in the following:

$$\begin{aligned}C_{j0sw} &= C_{j0} \times (300 \times 2\ \mu\text{m}^2) \\ &= 100\ \text{aF}/\mu\text{m}^2 \times (300 \times 2\ \mu\text{m}^2) \\ &= 0.06\ \text{pF}\end{aligned}$$

Therefore, the total zero-bias junction capacitance is 0.56 pF. Using Equation (2.31), the junction capacitance can be expressed as

$$C_j = 0.56\left[1 - \frac{V}{0.77}\right]^{-0.45} \quad (\text{pF})$$

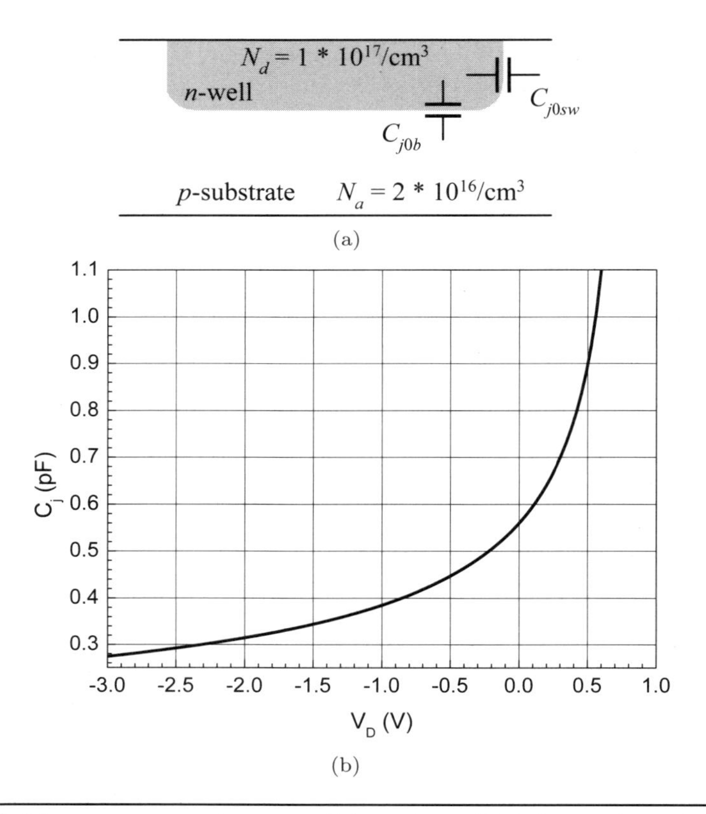

Figure 2.8: (a) The structure of an n-well junction. (b) The pn-junction capacitance between an n-well and the p-substrate.

The resulting junction capacitance C_j versus junction voltage V_D is plotted in Figure 2.8(b).

■

Although the junction capacitance of most MOS transistors is undesirable because it causes longer propagation delays of logic circuits, in modern CMOS radio-frequency (RF) oscillators the junction capacitance of parasitic diodes is often intentionally used along with the inductance formed by metal layers to form an LC resonance tank.

2.2.1.6 Large-Signal Equivalent Capacitance Since in digital circuits the signal excursions of both input and output are from 0 V to V_{DD} and vice versa, it is more convenient to estimate the junction capacitance of a reverse-biased diode at the two extreme points of operation, say, V_1 and V_2. By the definition of capacitance, the equivalent capacitance at these two voltages, V_1 and V_2, can be estimated as follows:

$$C_{eq} = \frac{\Delta Q}{\Delta V} = \frac{Q_j(V_2) - Q_j(V_1)}{V_2 - V_1} \tag{2.33}$$

where the amount of charge change is the integration of junction capacitance between two extreme voltages.

$$\Delta Q = \int_{V_1}^{V_2} C_j(V) dV = \int_{V_1}^{V_2} \frac{C_{j0}}{[1 - (V/\phi_0)]^m} dV \tag{2.34}$$

Completing the integration of the above equation and substituting into Equation (2.33), the equivalent junction capacitance is obtained.

$$\begin{aligned} C_{eq} &= -\frac{C_{j0}\phi_0}{(V_2 - V_1)(1-m)}\left[\left(1 - \frac{V_2}{\phi_0}\right)^{1-m} - \left(1 - \frac{V_1}{\phi_0}\right)^{1-m}\right] \\ &= K_{eq}C_{j0} \end{aligned} \tag{2.35}$$

where m is the grading coefficient. In the case of an abrupt junction, the grading coefficient $m = 1/2$, the *voltage equivalent factor* K_{eq} $(0 < K_{eq} < 1)$ is equal to

$$K_{eq} = \frac{C_{eq}}{C_{j0}} = -\frac{2\phi_0^{1/2}}{(V_2 - V_1)}\left[(\phi_0 - V_2)^{1/2} - (\phi_0 - V_1)^{1/2}\right] \tag{2.36}$$

In the case of a linearly graded junction, the grading coefficient $m = 1/3$, the voltage equivalent factor K_{eq} is

$$K_{eq} = \frac{C_{eq}}{C_{j0}} = -\frac{\frac{3}{2}\phi_0^{1/3}}{(V_2 - V_1)}\left[(\phi_0 - V_2)^{2/3} - (\phi_0 - V_1)^{2/3}\right] \tag{2.37}$$

A numeric example is given in the following.

■ Example 2-7: (Large-signal equivalent capacitance.) For an abrupt n^+p junction diode with $N_d = 5 \times 10^{18}$ cm^{-3} and $N_a = 10^{16}$ cm^{-3}, assume that the junction area is 20×20 μm^2.

1. Find ϕ_0 and C_{j0} at $T = 300$ K.
2. Find C_j for $V = -2.5$ V.
3. Find C_{eq} for $V_1 = -2.5$ V and $V_2 = 0$ V.

Solution:

1. The built-in potential at zero-bias condition, ϕ_0, can be calculated as follows:

$$\phi_0 = V_t \ln\left(\frac{N_a N_d}{n_i^2}\right) = 0.026 \ln\left(\frac{5 \times 10^{18} \times 10^{16}}{(1.45 \times 10^{10})^2}\right) = 0.86 \text{ V}$$

From Equation (2.32), the zero-bias junction capacitance, C_{j0}, is found to be

$$\begin{aligned} C_{j0} &= A\left(\frac{e\varepsilon_{si}}{2\phi_0}\frac{N_a N_d}{N_a + N_d}\right)^{1/2} \\ &= 4 \times 10^{-6}\left(\frac{1.6 \times 10^{-19} \times 11.7 \times 8.854 \times 10^{-14}}{2 \times 0.86}\frac{5 \times 10^{18} \times 10^{16}}{5 \times 10^{18} + 10^{16}}\right)^{1/2} \\ &= 0.12 \text{ pF} \end{aligned}$$

2. From Equation (2.31), the junction capacitance C_j at $V = -2.5$ V is

$$C_j = \frac{C_{j0}}{[1 - (V/\phi_0)]^m} = \frac{0.12 \text{ pF}}{[1 - (-2.5/0.86)]^{0.5}} = 60.7 \text{ fF}$$

3. From Equation (2.35), the equivalent capacitance C_{eq} for the two extreme voltages, $V_1 = -2.5$ V and $V_2 = 0$ V, is as follows:

$$C_{eq} = -\frac{C_{j0}\phi_0}{(V_2 - V_1)(1-m)}\left[\left(1 - \frac{V_2}{\phi_0}\right)^{1-m} - \left(1 - \frac{V_1}{\phi_0}\right)^{1-m}\right]$$

$$= -\frac{0.12 \times 0.86}{(0+2.5)(1-0.5)}\left[\left(1-\frac{0}{0.86}\right)^{0.5} - \left(1+\frac{2.5}{0.86}\right)^{0.5}\right]$$
$$= 80.6 \text{ fF}$$

■

■ Review Questions

Q2-14. What is the built-in potential of a *pn*-junction?
Q2-15. Explain the depletion region. Why is it so named?
Q2-16. What is an n^+p one-sided junction?
Q2-17. What is the junction capacitance of a *pn*-junction?
Q2-18. What is the large-signal equivalent capacitance of a *pn* junction?

2.2.2 Metal-Semiconductor Junctions

In the semiconductor realm, when an *n*-type material is physically joined with a *p*-type material, the resulting junction or contact is called a *rectifying junction* or *rectifying contact*. A rectifying junction is the one that provides a conduction with high resistance of current flow in one direction and with low resistance in the other direction. When the objective of a contact between a metal and a semiconductor is to connect the semiconductor to the outside world, as shown in Figure 2.5, the contact should provide a conduction with low resistance in both directions of current flow. Such a contact is referred to as a *nonrectifying*, or an *ohmic contact*.

Generally, when a metal is contacted with a semiconductor, the resulting junction or contact can be either a rectifying or an ohmic junction, depending on the work function difference between the metal and the semiconductor as well as the type of semiconductor. In theory, the resulting contact is an ohmic contact as a metal with work function ϕ_m contacts with an *n*-type semiconductor with work function ϕ_s such that $\phi_m < \phi_s$ or contacts with a *p*-type semiconductor with work function ϕ_s such that $\phi_m > \phi_s$. However, in practice, because of the imperfection of material, such as surface states, the above two types of junctions are not necessary to form good ohmic contacts.

The rectifying junction formed by contacting a metal with a semiconductor is referred to as a *Schottky junction*. The built-in potential and current-voltage characteristic of a Schottky junction are similar to those of a *pn* junction. The device built with a Schottky junction is called a *Schottky diode*. The current-voltage characteristic of a Schottky diode is basically the same as that of a *pn* diode except that it has a smaller cut-in or turn-on voltage than a *pn* junction diode. The *cut-in voltage* of a typical Schottky diode is around 0.3 to 0.5 V while that of a typical *pn* diode is about 0.6 to 0.8 V.

Because much more free electrons are in the metal, a Schottky junction is indeed a one-sided junction in which the depletion region is extended into the semiconductor side. Remember that the depletion-region width decreases with the increasing impurity concentration. As the depletion-region width reduces to a few nanometers, the effect of *barrier tunneling* becomes apparent. The barrier tunneling is a physical phenomenon that a nonzero probability of finding electrons from one side of a barrier exists if there are electrons appearing on the other side of the barrier. The probability is inversely proportional to an exponential function of the barrier width. The smaller the barrier width, the higher the tunneling probability. Hence, the tunneling probability increases profoundly with the increasing impurity concentration of the semiconductor.

Based on the aforementioned discussion, whenever an ohmic contact between a metal and a semiconductor is desired, a heavy impurity concentration, such as n^+ or p^+, must be used to avoid forming an undesirable Schottky junction. For instance, both source and drain areas shown in Figure 2.5 are heavily doped to avoid the formation of Schottky junctions between metals and these areas.

■ Review Questions

Q2-19. Distinguish between a rectifying junction and an ohmic contact.

Q2-20. What is the Schottky junction?

Q2-21. Explain the meaning of barrier tunneling.

Q2-22. Distinguish between a Schottky diode and a *pn*-junction diode.

Q2-23. Why are source/drain regions of MOS transistors heavily doped?

2.3 MOS Transistor Theory

In this section, we first introduce the metal-oxide-semiconductor (MOS) systems. Next, we address the basic operations of MOS transistors and their current-voltage (I-V) characteristics. Finally, we present the scaling theory.

2.3.1 MOS Systems

In the following discussion, a MOS system using aluminum as its top layer is used as an example to illustrate the features of MOS systems. The same principle can be applied to the case in which n^+ and p^+ polysilicon are used as the top layer conductor. This top layer conductor is often referred to as the gate of MOS system.

2.3.1.1 Basic Structure of MOS System The heart of MOS transistors is the MOS system, also known as a *MOS capacitor*. A typical MOS system consists of a metal (that is, a gate) and a semiconductor, and an insulator oxide (silicon dioxide, SiO_2) is sandwiched between them. The metal can be replaced by a heavily doped polysilicon (a low-resistance polysilicon of n-type or p-type) and the semiconductor can be a p-type or an n-type material. An example of a MOS system with p-type substrate is shown in Figure 2.9(a). For convenience, we will denote the metal as a gate and the semiconductor as the substrate. The energy-band diagram of these three components is depicted in Figure 2.9(b). From the figure, we can see that the work function of aluminum is 4.1 eV and the work function of the p-type substrate is 4.98 eV, supposing that the impurity concentration of the substrate is 2×10^{16} cm^{-3}. The work function difference between the aluminum and substrate is equal to

$$\phi_{GS} = \phi_G - \phi_{Si} = 4.1 - 4.98 = -0.88 \text{ V} \tag{2.38}$$

where $\phi_G = \phi_M$.

Remember that at thermal equilibrium the Fermi-energy level must be constant throughout the system consisting of many different materials. To reach this, electrons will transfer from the materials with a higher Fermi-energy level to the materials with a lower Fermi-energy level. Consequently, if we bring the three components, the

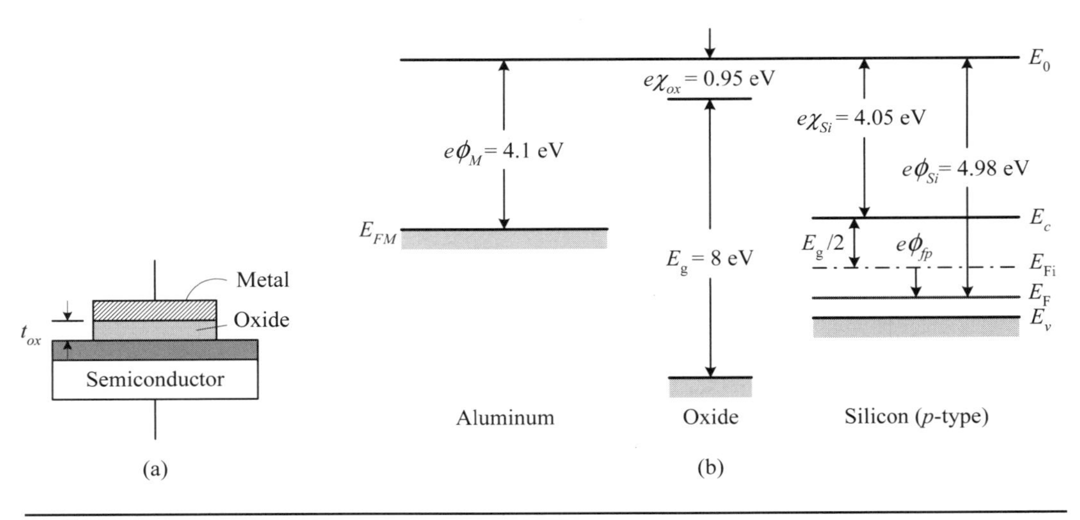

Figure 2.9: The (a) structure and (b) energy-band diagram of a basic MOS system.

metal, oxide, and silicon, into physical contact, some electrons from the metal will be transferred into the silicon and accumulated at the surface there. Because the oxide is free of mobile charges, an equal amount of positive charge is formed at the surface of the metal. Thus, the metal and the silicon form the two plates of a capacitor and a voltage drop is developed across them due to the stored charge on each side.

As depicted in Figure 2.10(a), the Fermi-energy level E_F at the surface of the silicon is shifted toward, even crosses, the intrinsic Fermi-energy level E_{Fi}. The negatively charged acceptors extend into the semiconductor from the surface. The *surface potential* ϕ_s is the potential difference across the depletion layer and is defined as the difference (in volts) between the Fermi potential E_{Fi}/e in the bulk semiconductor and the Fermi potential E_{Fi}/e at the surface. Formally, the surface potential ϕ_s can be expressed as follows:

$$\phi_s = \frac{E_{Fi(bulk)}}{e} - \frac{E_{Fi(surface)}}{e} \tag{2.39}$$

Let ϕ_{s0} and V_{ox0} be the surface potential and the voltage drop across the silicon dioxide when a zero voltage is applied to the metal (that is, a gate), respectively. From the figure, we are able to derive the following relation:

$$V_{ox0} + \phi_{s0} = -\phi_{GS} = -(\phi_G - \phi_{Si}) \tag{2.40}$$

where $e\phi_{Si}$ is the work function of the p-type silicon and is equal to

$$e\phi_{Si} = e\chi + \frac{E_g}{2} + e\phi_{fp} \tag{2.41}$$

where χ is the electron affinity in V, E_g is the energy gap of the silicon in eV, and ϕ_{fp} is the Fermi potential of the p-type silicon in V.

2.3.1.2 Flat-Band Voltage To compensate for the work function difference between the metal and silicon, an external voltage with the value of ϕ_{GS} should be applied across the metal and silicon. Once this is done, there is no band bending in the semiconductor and has a zero net depletion charge at the surface of the semiconductor. The voltage

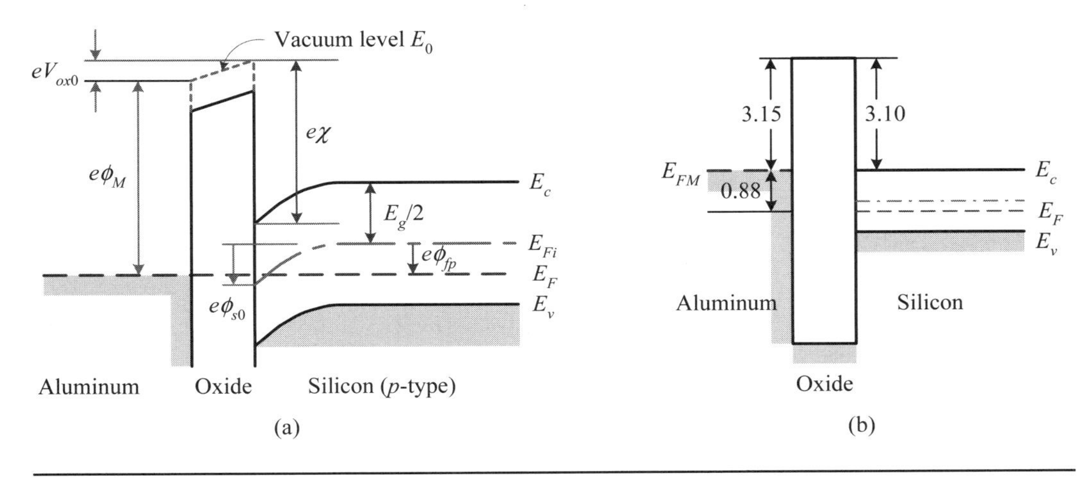

Figure 2.10: (a) An energy-band diagram showing the definition of surface potential. (b) The energy-band diagram of flat-band voltage.

being applied to a MOS system to yield the flat energy band in silicon is called the *flat-band voltage*, denoted as V_{FB}, and can be represented as

$$V_{FB} = \phi_G - \phi_{Si} = \phi_{GS} \tag{2.42}$$

The flat-band voltage is a function of the impurity concentration of the semiconductor and the specific gate material used for the MOS system. Figure 2.10(b) shows the energy-band diagram under the flat-band voltage.

Effects of fixed-oxide charge. For a real-world MOS system, a net fixed positive charge density may exist within the oxide and at the interface between the oxide and the semiconductor. This kind of surface charge with an amount of Q_{ss} results in a shift of the flat-band voltage V_{FB} to a more negative voltage. As a consequence, the flat-band voltage should take into account the effect of the fixed-oxide charge and is modified accordingly as follows:

$$V_{FB} = \phi_{GS} - \frac{Q_{ss}}{C_{ox}} \tag{2.43}$$

The C_{ox} in the above equation is called *oxide capacitance*, or *gate-oxide capacitance*, and is defined as

$$C_{ox} = \frac{\varepsilon_{ox}}{t_{ox}} = \frac{\varepsilon_{r(ox)}\varepsilon_0}{t_{ox}} \tag{2.44}$$

where $\varepsilon_{r(ox)}$ is the relative permittivity of silicon dioxide and ε_0 is the permittivity of free space. Generally, the value of $\varepsilon_{r(ox)}$ is 3.9 and ε_0 is 8.854×10^{-14} F/cm. The t_{ox} is the gate-oxide thickness.

2.3.1.3 Ideal MOS System

The real-world MOS system comes back to an ideal case whenever a flat-band voltage is applied to the gate, as shown in Figure 2.11(a). Under the ideal MOS system, when a negative voltage with respect to the substrate is applied to the gate, an amount of negative charge exists on the gate, that is, the top plate. Meanwhile, an accumulation layer of holes with the same amount of charge is induced at the surface of the *p*-type substrate near the oxide-semiconductor interface, as shown in Figure 2.11(b). This layer constitutes the bottom plate of the MOS

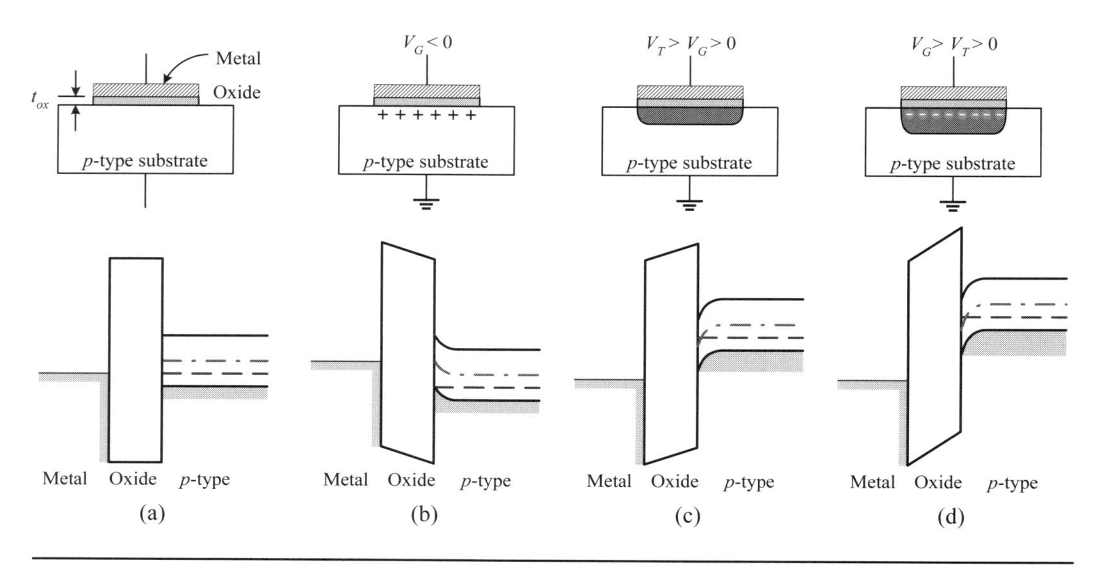

Figure 2.11: An ideal MOS system with a p-type substrate with various operation modes: (a) flat band; (b) accumulation; (c) depletion; (d) inversion.

system. In other words, the MOS system behaves as a parallel-plate capacitor. At the surface of substrate, because of the accumulation of a larger amount of positive charge, the Fermi-energy level goes more closely to the valance band and hence there exists energy-band bending near the surface.

The situation is different as we reverse the polarity of the applied voltage. As shown in Figure 2.11(c), when a moderate positive voltage is applied to the gate of the MOS system, a sheet of positive charge exists on the top metal plate and an equal amount of negative charge accumulates at the surface of the p-type substrate near the oxide-semiconductor junction. The holes at the surface of the substrate are repelled into the bulk of substrate by the induced electric field, leaving behind negative charged ions. As a consequence, a *depletion layer* is formed, a situation much like an n^+p one-sided junction. The Fermi-energy level goes toward, even crosses, the intrinsic Fermi-energy level.

A more interesting feature occurs when a larger positive voltage with respect to the substrate is applied to the gate of the MOS system. A larger positive voltage implies a larger induced electric field. Hence, more holes are repelled by the induced electric field into the bulk of silicon and the depletion-layer thickness is further increased, as shown in Figure 2.11(d). However, as the positive voltage is increased to some value, the *maximum depletion-layer thickness* is reached. At the same time, free electrons start to accumulate at the surface of the semiconductor and form an inversion layer. This process comes with the continually increasing positive voltage and hence the increasing concentration of the inversion layer. As the concentration of the inversion layer is equal to that of the substrate, the inversion layer is referred to as the *strong inversion layer* and the corresponding gate voltage is defined as the *threshold voltage* of the MOS system.

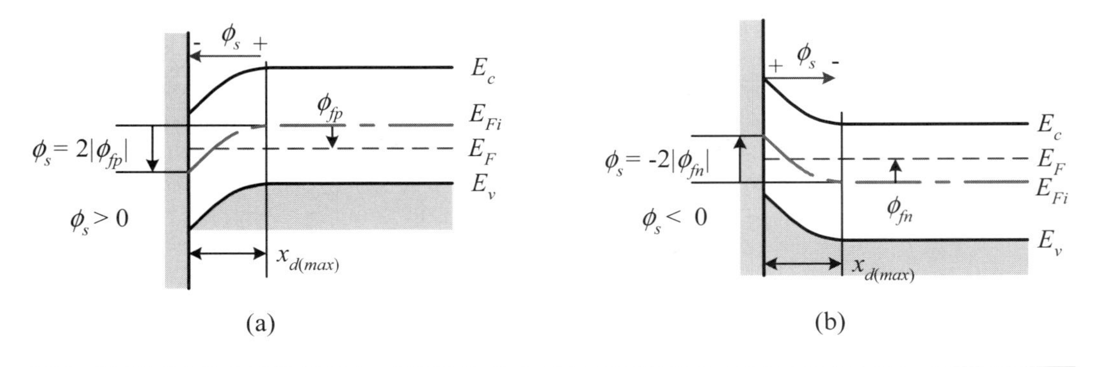

Figure 2.12: The definition of threshold voltages of ideal MOS systems with both (a) p-type and (b) n-type substrates, respectively.

As shown in Figure 2.12, the depletion-layer thickness can be obtained from the n^+p one-sided junction and is equal to

$$x_d = \sqrt{\frac{2\varepsilon_{si}\phi_s}{eN_a}}$$

The maximum depletion-layer thickness $x_{d(max)}$ occurs at the point where the surface potential ϕ_s is equal to $2|\phi_{fp}|$ and can be expressed as follows:

$$x_{d(max)} = \sqrt{\frac{4\varepsilon_{Si}|\phi_{fp}|}{eN_a}} \tag{2.45}$$

where the ϕ_{fp} is the Fermi potential and N_a is the impurity concentration of the p-type substrate.

■ Example 2-8: (Depletion-layer thickness.) Supposing that the p-type substrate has an impurity concentration of $N_a = 2 \times 10^{16}$ cm^{-3} and using $n_i = 1.45 \times 10^{10}$ cm^{-3}, find the maximum depletion-layer thickness $x_{d(max)}$ at $T = 300$ K.

Solution: From Equation (2.8), we obtain

$$|\phi_{fp}| = V_t \ln\left(\frac{N_a}{n_i}\right) = 0.026 \ln\left(\frac{2 \times 10^{16}}{1.45 \times 10^{10}}\right) = 0.37 \text{ V}$$

Substituting into Equation (2.45), the maximum depletion-layer thickness is

$$\begin{aligned} x_{d(max)} &= \sqrt{\frac{4\varepsilon_{si}|\phi_{fp}|}{eN_a}} = \sqrt{\frac{4 \times 11.7 \times 8.854 \times 10^{-14} \times 0.37}{1.6 \times 10^{-19} \times 2 \times 10^{16}}} \\ &= 0.22\ \mu\text{m} \end{aligned}$$

■

2.3.1.4 Threshold Voltage The threshold voltage V_T of a MOS system is defined as the applied gate voltage required to induce a strong inversion layer, with the same concentration as the impurity concentration in the bulk. When this condition occurs, the surface potential ϕ_s is equal to $2|\phi_{fp}|$ for the p-type substrate and $-2|\phi_{fn}|$ for the n-type substrate, as illustrated in Figure 2.12.

Suppose that the parasitic charge is located at the oxide-semiconductor interface and its value, denoted by Q_{ss}, is fixed. Ignoring the inversion-layer charge at the onset of the threshold inversion point, the voltage drop on the oxide V_{ox} is contributed by the charge Q_G, which is equal to $Q_{d(max)} - Q_{ss}$, where $Q_{d(max)}$ is the total charge as the depletion layer reaches its maximum thickness and is

$$Q_{d(max)} = eN_a x_{d(max)} = \sqrt{4eN_a\varepsilon_{si}|\phi_{fp}|} \tag{2.46}$$

Therefore, the threshold voltage for the p-type substrate can then be represented as in the following equation:

$$\begin{aligned} V_{Tn} &= 2|\phi_{fp}| + \frac{Q_G}{C_{ox}} + \phi_{GS} \\ &= 2|\phi_{fp}| + (Q_{d(max)} - Q_{ss})\frac{t_{ox}}{\varepsilon_{ox}} + \phi_{GS} \\ &= 2|\phi_{fp}| + Q_{d(max)}\frac{t_{ox}}{\varepsilon_{ox}} + \left(\phi_{GS} - Q_{ss}\frac{t_{ox}}{\varepsilon_{ox}}\right) \end{aligned} \tag{2.47}$$

The last term of Equation (2.47) is the flat-band voltage with the effects of a fixed-oxide charge.

Example 2-9: (Threshold-voltage of the p-type substrate.) For an aluminum-gate MOS system, the gate-oxide thickness t_{ox} is 5 nm and $\phi_{GS} = -0.88$ V. Assume that the acceptor concentration N_a is 2×10^{16} cm^{-3} and Q_{ss} is 5×10^{10} cm^{-2}. Find the threshold voltage V_{Tn} at $T = 300$ K.

Solution: At $T = 300$ K, the intrinsic carrier concentration n_i is 1.45×10^{10} cm^{-3}. From the result of the previous example and Equation (2.46), we have $Q_{d(max)}$ as follows:

$$\begin{aligned} Q_{d(max)} &= eN_a x_{d(max)} \\ &= 1.6 \times 10^{-19} \times 2 \times 10^{16} \times 0.22 \times 10^{-4} \\ &= 7.04 \times 10^{-8} \text{ C/cm}^2 \end{aligned}$$

The capacitance C_{ox} of the underlying MOS system is found to be

$$C_{ox} = \frac{\varepsilon_{ox}}{t_{ox}} = \frac{3.9 \times 8.854 \times 10^{-14}}{5 \times 10^{-7}} = 6.91 \times 10^{-7} \text{ F/cm}^2$$

The threshold voltage V_{Tn} can be calculated from Equation (2.47) as in the following:

$$\begin{aligned} V_{Tn} &= 2|\phi_{fp}| + Q_{d(max)}\frac{t_{ox}}{\varepsilon_{ox}} + \left(\phi_{GS} - Q_{ss}\frac{t_{ox}}{\varepsilon_{ox}}\right) \\ &= 2 \times 0.37 + \frac{7.04 \times 10^{-8}}{6.91 \times 10^{-7}} - 0.88 - \frac{5 \times 10^{10} \times (1.6 \times 10^{-19})}{6.91 \times 10^{-7}} \\ &= -0.05 \text{ V} \end{aligned}$$

Hence, the threshold voltage is a negative value. This means that an inversion layer already exists before a positive voltage is applied to the gate.

In summary, the threshold voltage of a p-type substrate may be negative, which means that an inversion layer (also called a *channel*) already exists before a positive voltage is applied to the gate. The MOS transistors stemming from such a MOS system are called *depletion mode* devices. At present, most modern CMOS applications require devices whose inversion layers are created only when a large enough positive voltage (for the p-type substrate) is applied to the gate of a MOS system. To achieve this, ion implantation may be used to adjust the threshold voltage to a desired value. The resulting devices, which have no channel existing with zero gate voltage, are known as *enhancement mode* devices. In the rest of this book, we only consider this type of MOS transistors, including both nMOS and pMOS transistors.

Threshold voltage adjustment. Assume that a delta function is used as the charge density model for simplicity. When an impurity density of D_I is implanted into the channel, the threshold voltage will be shifted by an amount of

$$\Delta V_T = \frac{eD_I}{C_{ox}} \tag{2.48}$$

The threshold voltage is shifted to a positive direction if an acceptor impurity is applied and to a negative direction if donor impurity is applied.

Combining Equation (2.48) with Equation (2.47), the following equation is obtained:

$$V_{Tn} = 2|\phi_{fp}| + \frac{Q_{d(max)}}{C_{ox}} + \left(\phi_{GS} - \frac{Q_{ss}}{C_{ox}}\right) + \frac{eD_I}{C_{ox}} \tag{2.49}$$

In summary, the threshold voltage of a MOS system is composed of four components:

1. The first component $2|\phi_{fp}|+Q_{d(max)}/C_{ox}$ is needed to induce the strong inversion layer and to offset the induced depletion charge $Q_{d(max)}$.
2. The term ϕ_{GS} is the work function difference between the gate material and the substrate on the side of the strong inversion layer.
3. The fixed charge term Q_{ss}/C_{ox} accounts for the charge effects appearing within the oxide and at the interface between the oxide and substrate.
4. The last term is the threshold voltage adjustment factor eD_i/C_{ox}, which adjusts the threshold voltage to a desired value in practical applications.

■ **Example 2-10: (Threshold voltage adjustment.)** Assuming that the delta function is used, find the impurity density required to adjust the threshold voltage of the preceding example to 0.45 V.

Solution: From Equation (2.48), we have

$$\Delta V_T = 0.45 - (-0.05) = 0.50 \text{ V} = \frac{eD_I}{C_{ox}}$$

The required impurity density of ion implantation is

$$D_I = \frac{C_{ox}}{e} \times \Delta V_T = \frac{6.91 \times 10^{-7}}{1.6 \times 10^{-19}} \times 0.50 = 2.16 \times 10^{12} \text{ cm}^{-2}$$

Hence, the impurity density of acceptors to be implanted is 2.16×10^{12} cm^{-2}. ■

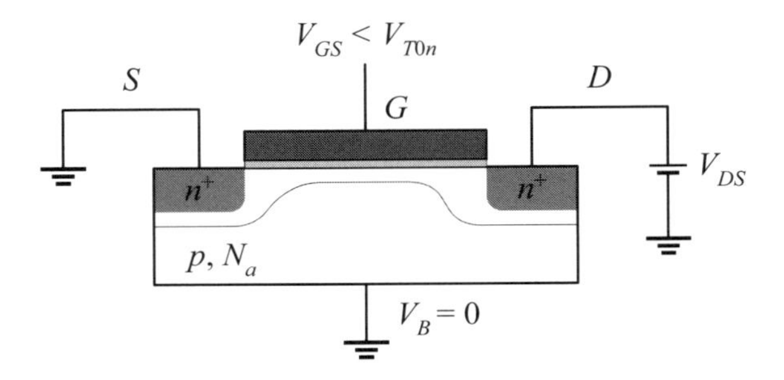

Figure 2.13: The cutoff mode of an nMOS transistor.

Review Questions

Q2-24. Define the strong inversion layer and threshold voltage of a MOS system.
Q2-25. Describe the components of the threshold voltage of a MOS system.
Q2-26. Explain the meaning of threshold voltage adjustment factor eD_i/C_{ox}.

2.3.2 The Operation of MOS Transistors

Now that we have understood the features of MOS systems, in this subsection we discuss the structure and operation of MOS transistors. A MOS transistor is a MOS system with the addition of two regions, referred to as the *drain* and *source*, on each side of the MOS system. The doping types of both drain and source regions are the same but opposite to the substrate. For instance, the basic structure of an nMOS transistor is shown in Figure 2.13.

At normal operation, the substrate is connected to a ground (that is, 0 V), a positive drain-to-source voltage is applied to the transistor, and a positive voltage is applied to the gate to induce an electron inversion layer at the surface of the substrate, thereby connecting both n-type source and drain regions. Such an inversion layer is called a *channel* since it provides a conductive path between source and drain regions. Because a MOS transistor has a symmetrical structure in both source and drain regions, the roles of source and drain in a circuit are determined by the voltage applied to them. One with more positive voltage is the drain and the other is the source. The source region is the source of carriers (here are electrons; for pMOS transistors, holes), which flow through the channel to the drain region.

There are three operation modes of a typical MOS transistor. These are *cut-off*, *linear*, and *saturation*. We will use an nMOS transistor as an example to illustrate these three operation modes. The nMOS transistor operated in the cut-off mode is shown in Figure 2.13. As mentioned above, at the normal operation, an external positive drain-to-source voltage is applied to the transistor. At the cut-off mode, a positive voltage smaller than the threshold voltage V_{T0n} is applied to the gate. Because of the induced electric field, a depletion layer appears at the surface of the p-type substrate under the silicon dioxide. Since there is no channel being induced between the source and drain regions, there is no current flowing from the drain to the source terminals.

The second operation mode of an nMOS transistor is the linear mode, as shown in Figure 2.14. In this mode, a positive voltage greater than the threshold voltage V_{T0n} is applied to the gate of an nMOS transistor. Thereby, an electron channel is induced at the surface of the p-type substrate under the silicon dioxide. This channel

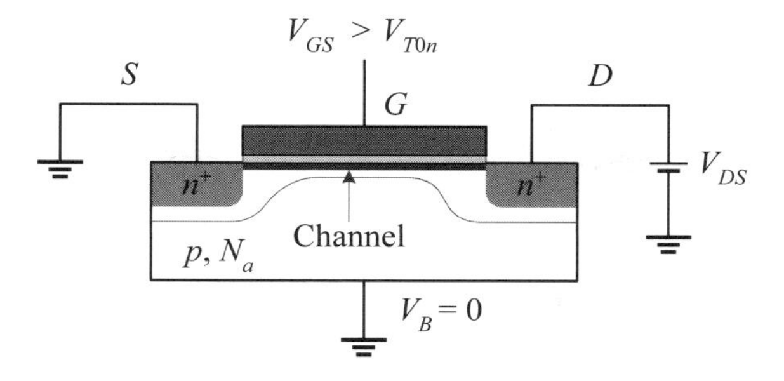

Figure 2.14: The linear mode of an nMOS transistor.

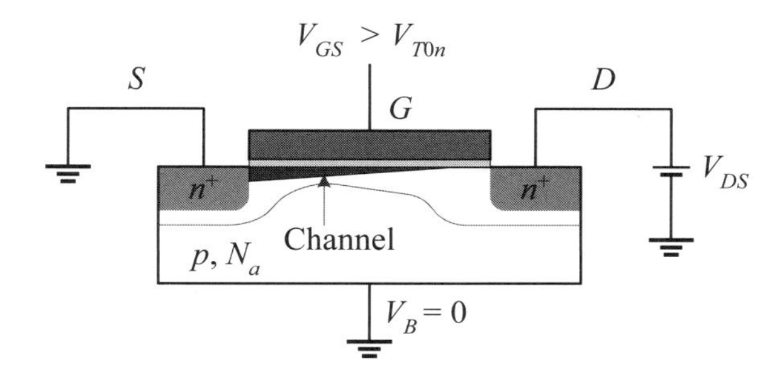

Figure 2.15: The saturation mode of an nMOS transistor.

directly connects the source to drain regions. Hence, electrons are able to flow from the source to drain regions; namely, a current flows from drain to source terminals. This current is called *drain current* I_{DS}, whose magnitude is significantly proportional to the drain-to-source voltage V_{DS}.

The third operation mode of an nMOS transistor is the saturation mode, as shown in Figure 2.15. As stated, when the gate-to-source voltage is above the threshold voltage V_{T0n}, an electron channel is induced and hence a drain current I_{DS} exists from drain to source terminals. This drain current initially increases with the increasing drain-to-source voltage V_{DS} but will virtually remain as a constant as the drain-to-source voltage V_{DS} reaches some value. Such a phenomenon that the drain current remains constant independent of drain-to-source voltage is referred to as the *saturation*.

The saturation phenomenon of an nMOS transistor can be explained as follows. From Figure 2.15, we can see that at the drain end, the n^+p junction is under reverse bias, with a magnitude much larger than that of the source end. Hence, the depletion region under the n^+ diffusion region at the drain end is also much deeper than that at the source end. This drain-end depletion-region depth is increased with the increasing drain-to-source voltage and will eventually reach the channel. After this, the channel is said to be *pinched-off* by the depletion region at the drain end. Therefore, the drain current no longer increases significantly with drain-to-source voltage and virtually remains a constant.

2.3.3 The I-V Characteristics of MOS Transistors

In the previous subsection, we have given a qualitative description of nMOS transistors. In this subsection, we explore the current versus voltage characteristics of MOS transistors.

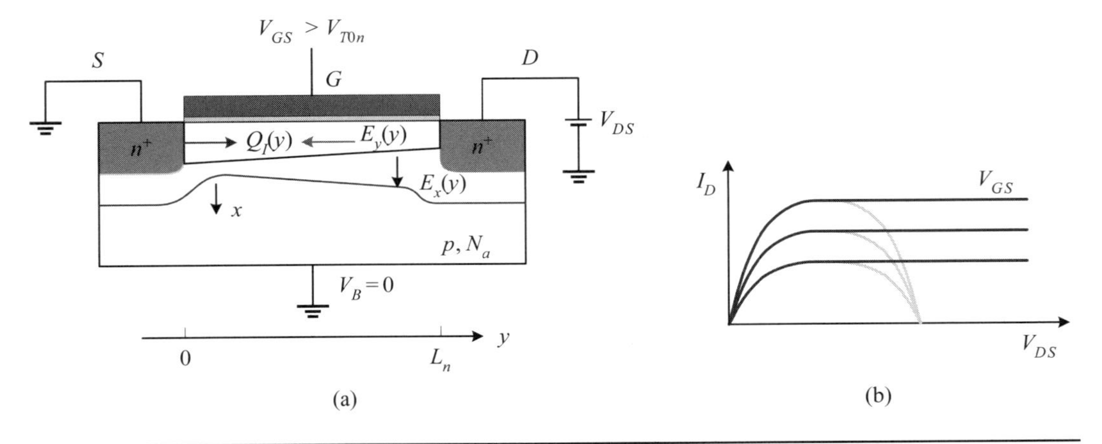

Figure 2.16: The (a) GCA model and (b) derived I-V characteristics of an nMOS transistor.

To derive the current versus voltage characteristics of an nMOS transistor shown in Figure 2.16(a), the *gradual-channel approximation* (GCA) model is widely used. In this model, the following two assumptions are made.

1. The vertical electric field $E_y(y)$ built by V_{GS} totally supports the depletion charge and the inversion layer. Here, we suppose that $V_{GS} \geq V_{Tn}$ and $V_{GD} = V_{GS} - V_{DS} \geq V_{Tn}$.
2. The channel electric field $E_x(y)$ is established by the drain-to-source voltage V_{DS}. As a result, the channel current is only caused by the drift current due to electrons in the channel. The electron mobility μ_n is assumed to be constant. The boundary conditions are $V(y = 0) = V_S = 0$ V and $V(y = L_n) = V_{DS}$.

Recall that an nMOS transistor is in the cut-off mode when its gate-to-source voltage V_{GS} is smaller than threshold voltage V_{Tn}. An nMOS transistor is in the linear mode when its gate-to-source voltage V_{GS} is larger than threshold voltage V_{Tn} but the drain-to-source voltage V_{DS} is a small positive voltage. In this mode, the drain current I_{DS} significantly increases with the increasing drain-to-source voltage V_{DS}.

To quantitatively describe the drain current as a function of gate-to-source voltage and drain-to-source voltage in the linear mode, we use the GCA model to quantify the drain current under the strong inversion condition. According to the definition of current,

$$I = \frac{dQ}{dt} \tag{2.50}$$

the drain current can be described as follows:

$$I_{DS} = W_n Q_I(y) \frac{dy}{dt} = W_n Q_I(y) v_d(y) \tag{2.51}$$

where $Q_I(y)$ is the *inversion layer charge* per unit area at a position y in the channel and $v_d(y)$ is the electron drift velocity at that position. From Equation (2.11), the electron drift velocity v_d can be related to the electric field E as $v_d = -\mu_n E$. The electric field E is established by a potential V across a distance d. Substituting v_d into Equation (2.51), we obtain

$$I_{DS} = -W_n Q_I(y) \mu_n \frac{dV(y)}{dy} \tag{2.52}$$

where μ_n denotes the electron mobility. Applying the basic relation $Q = CV$ to the inversion layer charge $Q_I(y)$, we obtain

$$Q_I(y) = -C_{ox}[V_{GS} - V_{Tn} - V(y)] \tag{2.53}$$

Substituting Equation (2.53) into Equation (2.52), rearranging the dy term, and integrating both sides as well as applying the boundary conditions, the following equation is obtained.

$$\int_0^{L_n} I_{DS} dy = \mu_n W_n C_{ox} \int_0^{V_{DS}} [V_{GS} - V_{Tn} - V(y)] dV \tag{2.54}$$

After integrating both sides, the drain current I_{DS} is found to be

$$I_{DS} = \mu_n C_{ox} \left(\frac{W_n}{L_n}\right) \left[(V_{GS} - V_{Tn}) V_{DS} - \frac{1}{2} V_{DS}^2\right] \tag{2.55}$$

In practical applications, it is convenient to define the *process transconductance* as $k'_n = \mu_n C_{ox}$ and the *device transconductance* as $k_n = \mu_n C_{ox}(W_n/L_n)$, respectively.

Equation (2.55) predicts the drain current I_{DS} as a parabolic function of V_{DS}. However, the experimental data show that the drain current virtually remains a constant value as it reaches its maximum value and is almost independent of the further increased drain-to-source voltage V_{DS}, as shown in Figure 2.16(b). The drain-to-source voltage causing the drain current to reach its maximum value is called the *saturation voltage* of nMOS transistor. The nMOS transistor is said to be saturated when this situation occurs.

To find the current characteristics of an nMOS transistor in the saturation mode, we have to find the saturation voltage. By using the fact that at the onset of the saturation mode, the drain current reaches its maximum value and then remains that value thereafter, the saturation voltage can be found from Equation (2.55) and can be expressed as

$$\frac{\partial I_{DS}}{\partial V_{DS}} = 0 = k_n[(V_{GS} - V_{Tn}) - V_{DS}] \tag{2.56}$$

Consequently, the saturation voltage is equal to

$$V_{DSsat} = V_{GS} - V_{Tn} \tag{2.57}$$

which is a function of gate-to-source voltage V_{GS}. Substituting this saturation voltage into Equation (2.55), the saturation drain current I_{DSsat} can be expressed as

$$I_{DSsat} = I_{DS}(V_{DS} = V_{DSsat}) = \frac{1}{2} \mu_n C_{ox} \left(\frac{W_n}{L_n}\right) (V_{GS} - V_{Tn})^2 \tag{2.58}$$

In summary, an nMOS transistor operates in the linear mode when its V_{GS} is larger than the threshold voltage V_{Tn} and its drain-to-source voltage V_{DS} is less than the $V_{GS} - V_{Tn}$. In this mode, the drain current I_{DS} is governed by Equation (2.55). An nMOS transistor operates in the saturation mode when its V_{GS} is larger than the threshold voltage V_{Tn} and its drain-to-source voltage V_{DS} is larger than or equal to $V_{GS} - V_{Tn}$. In this mode, the drain current I_{DS} is governed by Equation (2.58), namely, a quadratic function of V_{GS}.

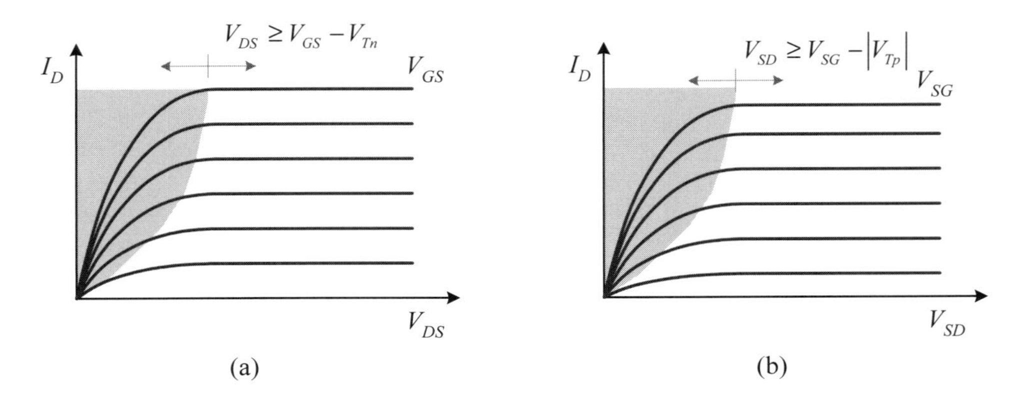

Figure 2.17: The typical characteristics of (a) nMOS and (b) pMOS transistors.

The GCA model can also be applied equally well to a pMOS transistor and obtain its current equations as follows:

$$I_{DS} = \mu_p C_{ox} \left(\frac{W_p}{L_p}\right) \left[(V_{SG} - |V_{Tp}|) V_{SD} - \frac{1}{2} V_{SD}^2\right] \quad (\text{for } V_{SD} < V_{SG} - |V_{Tp}|) \tag{2.59}$$

$$I_{DSsat} = \frac{1}{2} \mu_p C_{ox} \left(\frac{W_p}{L_p}\right) (V_{SG} - |V_{Tp}|)^2 \quad (\text{for } V_{SD} \geq V_{SG} - |V_{Tp}|) \tag{2.60}$$

As compared with the current equations of nMOS transistors, the current equations of pMOS transistors have the same form except the following: First, the order of source and drain in subscripts is exchanged. Second, the order of source and gate in subscripts is exchanged. Third, the absolute value of threshold voltage $|V_{Tp}|$ needs to be used.

The typical characteristics used to illustrate the operating modes of both nMOS and pMOS transistors are shown in Figures 2.17(a) and (b), respectively. The current-voltage characteristics of nMOS and pMOS transistors of a 0.18-μm process are shown in Figure 2.18(a) and (b), respectively. Here, both nMOS and pMOS transistors are assumed to have a size of $2\lambda \times 2\lambda$.

■ Review Questions

Q2-27. What are the three operation modes of a typical MOS transistor?

Q2-28. Describe the gradual-channel approximation (GCA) model.

Q2-29. Define process transconductance and device transconductance.

2.3.4 Scaling Theory

There are two scaling strategies that can be applied to reduce the sizes of MOS transistors: *constant-field scaling* (also known as *full scaling*) and *constant-voltage scaling*. Both scaling strategies have their unique effects on the operating characteristics of MOS transistors. We will deal with each scaling strategy and its effects.

2.3.4.1 Constant-Field Scaling In the constant-field scaling, all dimensions of a MOS transistor are reduced by a factor of k. The applied voltages are also scaled with the same scaling factor k as the device dimensions. Typically, the scaling factor is

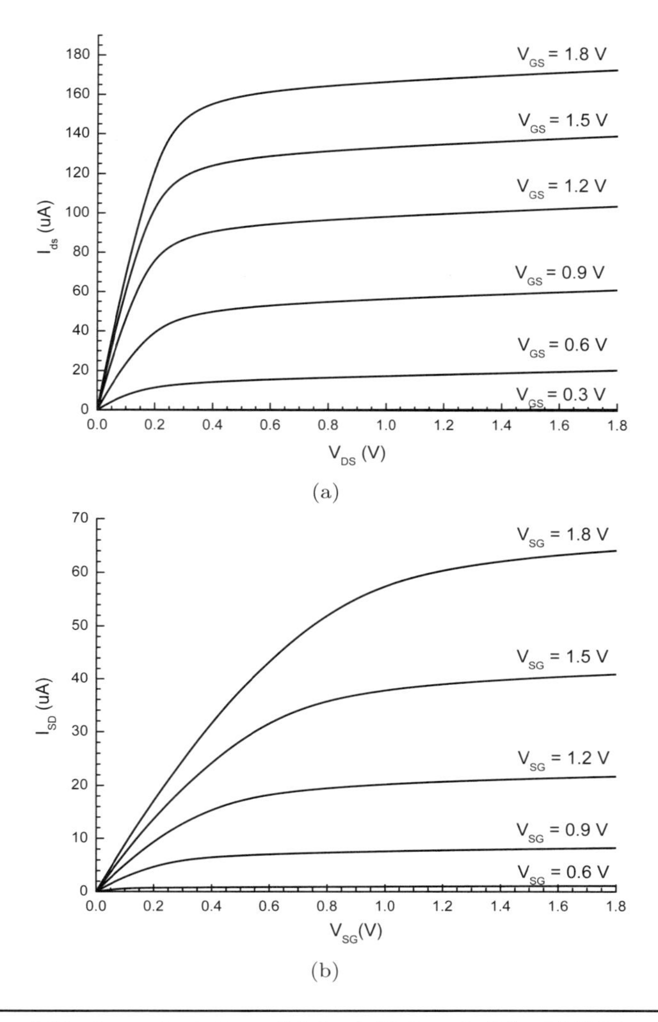

Figure 2.18: The I-V characteristics of (a) nMOS and (b) pMOS transistors.

about 0.7, namely, $k \approx 0.7$, per generation of a given CMOS process. A conceptual illustration of scaling theory is exhibited in Figure 2.19.

The essential feature of constant-field scaling is to keep the following relation unchanged.

$$\nabla^2 \phi(x, y, z) = -\frac{\rho(x, y, z)}{\varepsilon} \qquad \text{(Poisson's equation)} \tag{2.61}$$

Now if all dimensions are scaled by a factor k; namely, $x' = kx, y' = ky$, and $z' = kz$, then the above Poisson's equation becomes as

$$\frac{\partial^2 \phi'}{\partial (kx)^2} + \frac{\partial^2 \phi'}{\partial (ky)^2} + \frac{\partial^2 \phi'}{\partial (kz)^2} = -\frac{\rho'}{\varepsilon} \tag{2.62}$$

To preserve the same properties of the original electric field effect, the potential function needs to be scaled down by k and charge density needs to be increased by a factor of $1/k$; that is,

$$\phi' = k\phi \qquad \text{and} \qquad \rho' = \rho/k \tag{2.63}$$

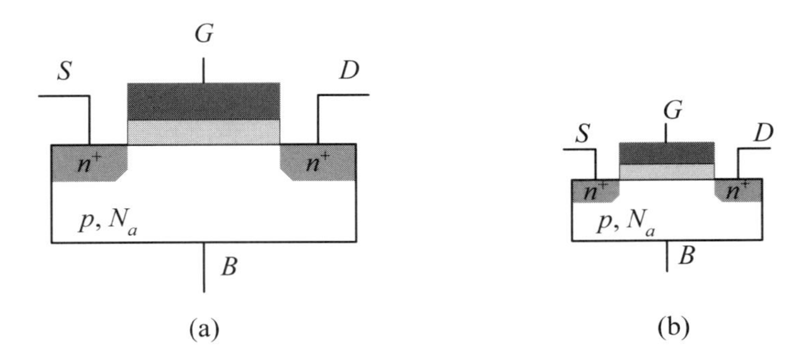

Figure 2.19: A conceptual illustration of scaling theory: (a) original; (b) scaled.

Representing the charge density as impurity concentrations, we have

$$N'_a = N_a/k \qquad \text{and} \qquad N'_d = N_d/k \tag{2.64}$$

Consequently, the impurity concentrations must be increased by a factor of $1/k$ to maintain the constant electric field.

In the constant-field scaling, both channel length L and width W are scaled down to kL and kW, respectively. The drain and gate voltages, V_{DS} and V_{GS}, are scaled down by a factor k. To maintain a constant vertical electric field, the gate-oxide thickness t_{ox} has to be scaled down by the same factor of k, namely, kt_{ox}. As a consequence, the maximum depletion-layer thickness

$$x_{d(max)} = \sqrt{\frac{2\varepsilon_{si}(\phi_0 + V_{DS})}{eN_a}} \tag{2.65}$$

is roughly reduced by k since N_a is increased by the factor $1/k$ and V_{DS} is reduced by k.

Now considering the gate-oxide capacitance $C_{ox} = \varepsilon_{ox}/t_{ox}$

$$C'_{ox} = \frac{\varepsilon_{ox}}{t'_{ox}} = \frac{1}{k}C_{ox} \tag{2.66}$$

As a result, the gate-oxide capacitance C'_{ox} is increased by a factor of $1/k$. However, the gate capacitance of a MOS transistor is reduced by a factor k, namely, $C'_g = kC_g$, because its channel width W and channel length L are reduced by the same factor.

Before considering the scaling of drain current in both linear and saturation regions, let us examine the scaling effect on the device transconductance, k_n, of nMOS transistors.

$$k'_n = \mu_n \frac{1}{k} C_{ox} \left(\frac{kW_n}{kL_n}\right) = \frac{1}{k} k_n \tag{2.67}$$

The same effect is applied to the pMOS transistors. As a result, device transconductance of MOS transistors is increased by a factor of $1/k$. The scaling effect on the drain current in a linear region is as follows:

$$\begin{aligned} I'_D &= \frac{1}{2} k'_n [2(V'_{GS} - V'_{T0})V'_{DS} - V'^2_{DS}] \\ &= \frac{1}{2}\left(\frac{1}{k}k_n\right) k^2 [2(V_{GS} - V_{T0})V_{DS} - V^2_{DS}] = kI_{DS} \end{aligned} \tag{2.68}$$

Table 2.3: Effects of constant-field and constant-voltage scaling ($0 < k < 1$).

Device and circuit parameters	Scaled factor	
	Constant field	Constant voltage
Device parameters (L, t_{ox}, W, x_j)	k	k
Impurity concentrations (N_a, N_d)	$1/k$	$1/k^2$
Voltages (V_{DD}, V_{DS}, and V_{GS})	k	1
Depletion-layer thickness ($x_{d(max)}$)	k	k
Capacitance ($C = \varepsilon A/d$)	k	k
Drain current (I_{DS})	k	$1/k$
Device density	$1/k^2$	$1/k^2$
Power density (P/A)	1	$1/k^3$
Power dissipation per device ($P_D = I_{DS}V_{DS}$)	k^2	$1/k$
Circuit propagation delay ($t_{pd} = R_{on}C_{load}$)	k	k^2
Power-delay product ($E = P_D t_{pd}$)	k^3	$1/k$.

The same result on the saturation region is also obtained.

$$\begin{aligned} I'_{D(sat)} &= \frac{1}{2}k'_n(V'_{GS} - V'_{T0})^2 \\ &= \frac{1}{2}\left(\frac{1}{k}k_n\right)k^2(V_{GS} - V_{T0})^2 = kI_{DSsat} \end{aligned} \tag{2.69}$$

Even though the drain current is decreased, the terminal voltage of the MOS transistor is also decreased by the same factor. Hence, the effective on-resistance of the MOS transistor is virtually unchanged. However, because of the decrease of gate capacitance, the circuit propagation delay $t_{pd} = R_{on}C_{load}$ is effectively reduced by a factor of k.

After constant-field scaling, the power dissipation of a single device is reduced by a quadratic factor of k.

$$P' = I'_D V'_{DS} = k^2 I_{DS} V_{DS} = k^2 P \tag{2.70}$$

thereby maintaining the power density of the chip due to the quadratic increase of the number of devices.

The scaling parameters and scaling effects on the device and circuit parameters under the constant-field scaling are listed in the first column of Table 2.3 for reference.

2.3.4.2 Constant-Voltage Scaling The salient feature of constant-field scaling is that the supply voltage and all terminal voltages are scaled down along the device dimensions. This results in new devices that may have different supply voltages from old ones. As a consequence, many different supply voltages are required in a system. This may cause trouble in some applications.

In constant-voltage scaling, all dimensions of a MOS transistor are reduced by a factor of k, as in the constant-field scaling. However, the supply voltage and all terminal voltages remain unchanged. To reach this, the charge density must be increased by a factor of $1/k^2$, according to Equation (2.62).

In constant-voltage scaling, the scaling effect on the device transconductance of nMOS transistors k_n is the same as in constant-field scaling.

$$k'_n = \mu_n \frac{1}{k} C_{ox}\left(\frac{kW_n}{kL_n}\right) = \frac{1}{k}k_n \tag{2.71}$$

The scaling effect on the drain current in a linear region is as follows:

$$\begin{aligned} I'_D &= \frac{1}{2}k'_n[2(V'_{GS} - V'_{T0})V'_{DS} - V'^2_{DS}] \\ &= \frac{1}{2}\left(\frac{1}{k}k_n\right)[2(V_{GS} - V_{T0})V_{DS} - V^2_{DS}] = \frac{1}{k}I_{DS} \end{aligned} \quad (2.72)$$

The same result is on the saturation region.

$$\begin{aligned} I'_{D(sat)} &= \frac{1}{2}k'_n(V'_{GS} - V'_{T0})^2 \\ &= \frac{1}{2}\left(\frac{1}{k}k_n\right)(V_{GS} - V_{T0})^2 = \frac{1}{k}I_{DSsat} \end{aligned} \quad (2.73)$$

The increase of drain current means that the effective on-resistance of MOS transistors is decreased. Combining this reduction of on-resistance with the reduction of gate capacitance, the result is that the circuit delay is shortened by a factor of k^2.

Because of the increase of drain current, the power dissipation of a device is increased by a factor of $1/k$. As a result, the power density increases by a factor of $1/k^3$, which is why the constant-voltage scaling theory cannot be applied to modern CMOS processes.

The scaling parameters and scaling effects on the device and circuit parameters under constant-voltage scaling are listed in the second column of Table 2.3 for reference.

■ Review Questions

Q2-30. Distinguish constant-field scaling from constant-voltage scaling.

Q2-31. Explain why the propagation delay is effectively reduced by a factor of k in constant-field scaling.

Q2-32. Explain why the propagation delay is effectively reduced by a factor of k^2 in constant-voltage scaling.

Q2-33. What is the power density in constant-field scaling?

2.4 Advanced Features of MOS Transistors

In this section, we first consider the nonideal features of MOS transistors, threshold voltage effects, and leakage currents. Next, we deal with the short-channel I-V characteristics. Then, we consider the temperature effects on the drain currents of both nMOS and pMOS transistors. Finally, we conclude this section with limitations of MOS transistors, which cover thin-oxide breakdown, avalanche breakdown, snapback breakdown, and punchthrough effects.

2.4.1 Nonideal Features of MOS Transistors

We are concerned with the nonideal features of MOS transistors: *channel-length modulation*, *velocity saturation*, and *hot carriers*.

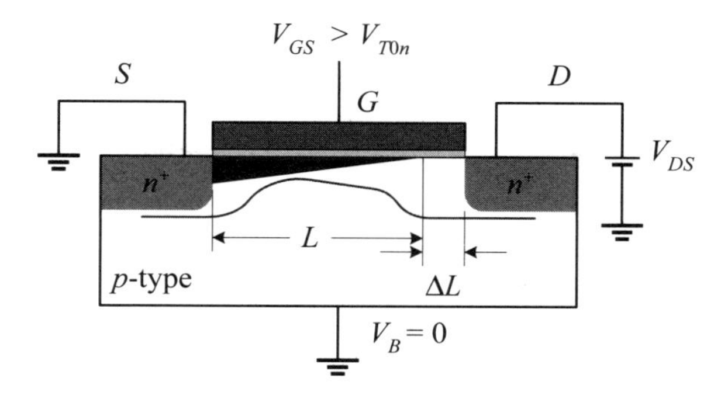

Figure 2.20: The effect of channel-length modulation of an nMOS transistor.

2.4.1.1 Channel-Length Modulation We have supposed that the drain current I_{DS} is constant when the drain-to-source voltage V_{DS} reaches its saturation value V_{DSsat} and thereafter. However, the drain current I_{DS} of actual devices is still slowly increased with the increasing drain-to-source voltage V_{DS} after $V_{DS} \geq V_{DSsat}$.

Referring to Figure 2.20, the depletion region at the drain end extends laterally into the channel when the MOS transistor is biased in the saturation mode, thereby reducing the effective channel length by an amount of ΔL. This results in the increase of drain current I_{DS} in accordance with the drain current equation. The phenomenon that the drain current is affected by the drain-to-source voltage V_{DS} in the saturation region is referred to as the *channel-length modulation* because the effective channel length is modulated (changed) by the drain-to-source voltage.

To further quantify the amount of the increase of drain current I_{DS} by the reduced channel length, we note that the reduced channel length ΔL can be related to ΔV_{DS}, where $\Delta V_{DS} = V_{DS} - V_{DSsat}$, by the following equation:

$$\Delta L = \sqrt{\frac{2\varepsilon_{si}}{eN_a}} \left(\sqrt{|\phi_{fp}| + V_{DSsat} + \Delta V_{DS}} - \sqrt{|\phi_{fp}| + V_{DSsat}} \right) \tag{2.74}$$

However, the above equation will make the current equation much more complicated. So in practice the following empirical relation is used instead.

$$I'_D = \left(\frac{L}{L - \Delta L} \right) I_{DS} = \frac{1}{1 - \lambda V_{DS}} I_{DS} \tag{2.75}$$

The resulting current equation taking into account the channel-length modulation is as follows:

$$I_{DS} = \frac{\mu_n C_{ox}}{2} \left(\frac{W_n}{L_n} \right) (V_{GS} - V_{T0n})^2 (1 + \lambda V_{DS}) \tag{2.76}$$

where λ is called the *channel-length modulation coefficient*, having a value in the range of 0.005 V^{-1} to 0.05 V^{-1}.

2.4.1.2 Velocity Saturation The total velocity of carriers (electrons and holes) in semiconductors is the sum of random thermal velocity and drift velocity due to an electric field. The carrier drift velocity is a function of the electric field; that is, it is proportional to the electric field and the proportional factor is defined as the carrier mobility, denoted as μ. The relationship between carrier drift velocity and electric field at 300 K is illustrated in Figure 2.21. At a low electric field, say, below

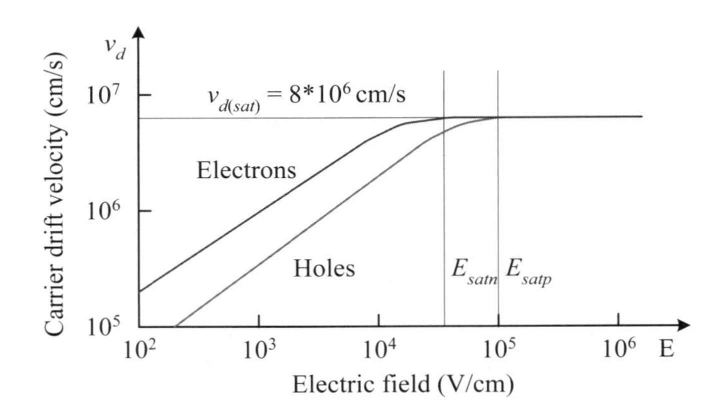

Figure 2.21: An illustration of velocity saturation of electrons and holes in a silicon semiconductor.

6×10^4 V/cm, the carrier drift velocity is linearly proportional to the electric field, that is, $v_d = \mu E$, and μ is a constant. However, when the electric field increases to a point called the *critical electric field*, or *saturation electric field*, E_{satn} for electrons and E_{satp} for holes, the carrier drift velocity no longer increases with the increasing electric field and remains at a constant velocity, called *carrier saturation velocity*, v_{sat}. This phenomenon is known as *carrier velocity saturation*. As shown in Figure 2.21, the carrier saturation velocity of both electrons and holes is about the same and has the value of 8×10^6 cm/s and the critical electric fields for electrons and holes in silicon are about 6×10^4 V/cm and 10^5 V/cm, respectively.

Under carrier velocity saturation, the drift current will saturate and become independent of the applied electric field. This effect will cause the drain current I_{DS} of a MOS transistor to follow a linear relationship rather than the ideal square-law dependence. As a result, the actual current flows in the MOS transistor is much smaller than the value estimated by the ideal square-law equation. The drain current I_{DS} under carrier velocity saturation can be expressed as follows:

$$I_{DSsat} = WC_{ox}(V_{GS} - V_T)v_{sat} \tag{2.77}$$

where v_{sat} is the saturation velocity of carriers, electrons, or holes.

2.4.1.3 Hot Carriers When carriers within a high electric field gain an energy much larger than the thermal energy that they have under thermal equilibrium, these carriers are called *hot carriers*. In MOS transistors, hot carriers are often generated by *impact ionization* occurring in the depletion region at the drain end (drain junction) by the channel electrons due to a high electric field. These carriers have energy far greater than thermal-equilibrium value and therefore are hot carriers. One useful feature is that the hot-carrier velocity may exceed the saturation value, an effect called *velocity overshoot*, which leads to currents greater than their predicted values, thereby enhancing the speed of MOS transistors. Another useful feature of hot carriers is that they may be used to program electrically erasable programmable read-only memory (EEPROM) and Flash memory devices. Such a programming mechanism is referred to as *hot-electron injection* (HEI).

However, hot carriers also cause many unwanted effects on MOS transistors. These effects include substrate current, device degradation, and gate leakage current. We will

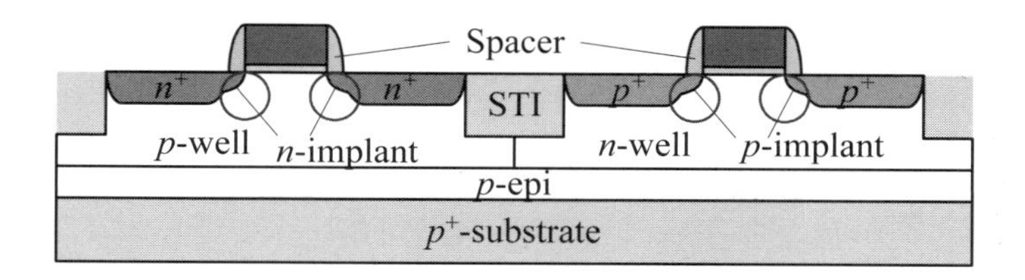

Figure 2.22: A cross-sectional view of a lightly doped drain (LDD) CMOS transistor.

deal with the first two effects in brief. The details of gate leakage current is deferred to Section 2.4.3.

Substrate current. The hot carriers repelled by the drain voltage can drift into the resistive substrate and become a substrate current I_{sub}. This current yields a voltage drop on the resistive substrate that causes a reverse-body effect, thereby reducing the threshold voltage and increasing the drain current of MOS transistors. This phenomenon is known as *substrate-current-induced body effect* (SCBE).

Device degradation. Some hot carriers generated by impact ionization can be trapped into the gate oxide and/or at the oxide-silicon interface and reside there forever. These trapped carriers amount to the quantity of surface state, Q_{ss}, thereby leading to the increase of the threshold voltage of the device. Because trapped carriers within the gate oxide cannot be removed, the hot-carrier charging effects are a continuous process, meaning that the performance of devices will be gradually degraded.

Lightly doped drain transistors. Recall that the high electric field occurs at the drain end of deep submicron devices. The high electric field will cause impact ionization to generate hot carriers. One way to alleviate such an effect is to lower the concentration near the drain end. This results in a device, referred to as a *lightly doped drain* (LDD) transistor, as shown in Figure 2.22.

The rationale behind the LDD transistor is to implant a shallow-n/p at the drain and source ends to provide a resistive buffer that would drop more voltage across a small region. Thus, the electric field gets reduced and keeps the carrier velocity from saturating.

■ Review Questions

Q2-34. What is the meaning of the channel-length modulation?

Q2-35. What is the meaning of the carrier velocity saturation?

Q2-36. Why does the carrier velocity saturation impact on drain current?

Q2-37. What are the hot carriers?

Q2-38. Explain the phenomenon of SCBE.

Q2-39. What is the objective of using LDD transistors?

2.4.2 Threshold Voltage Effects

Threshold voltages of MOS transistors are not constant. They are varied by many factors. In this section, we are concerned with these factors: *body effect*, *short-channel effect* (SCE), and *drain-induced barrier lowering* (DIBL).

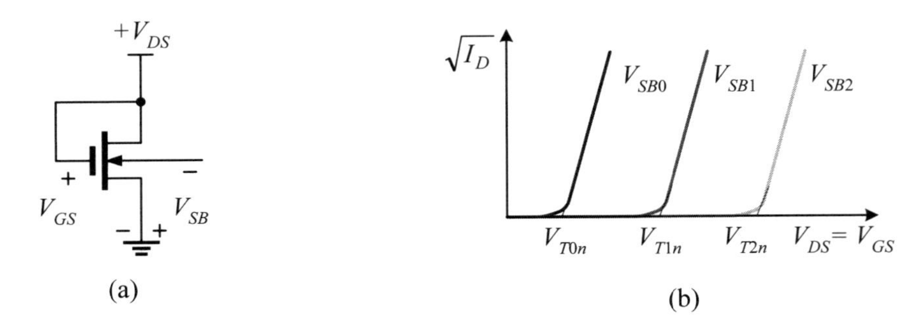

Figure 2.23: The body effects of a typical nMOS transistor: (a) circuit; (b) *I-V* characteristics.

2.4.2.1 Body Effect An important effect that must be considered in designing both digital and analog circuits is the *body effect*, or *body-bias effect*. Referring to Figure 2.23(b), the threshold voltage increases with the increasing voltage V_{SB} between the source and substrate, where subscripts S and B denote source and bulk, respectively. To see why this happens, let us go back to Figure 2.14. When the substrate is still grounded but a positive voltage V_{SB} is applied to the source terminal, part of the electrons in the inversion layer induced by the gate voltage V_{GS} will be attracted by the positive source-to-bulk voltage V_{SB} and flows into a grounded source. As a result, the gate needs to put forth more effort to induce the strong inversion layer. This means that the threshold voltage is increased with the rising source-to-bulk voltage V_{SB}. The phenomenon in which the threshold voltage is affected by the source-to-bulk voltage V_{SB} is called the *body effect*. In general, the increase of source-to-bulk voltage results in an increase of threshold voltage.

The relationship between the threshold voltage V_{Tn} of an nMOS transistor and its source-to-bulk voltage V_{SB} can be expressed as

$$V_{Tn} = V_{T0n} + \gamma\left(\sqrt{2|\phi_{fp}| + V_{SB}} - \sqrt{2|\phi_{fp}|}\right)$$

where V_{T0n} denotes the threshold voltage of an nMOS transistor without body effect, V_{Tn} denotes the threshold voltage of an nMOS transistor with body effect, and γ is called the *body-effect coefficient*. The body-effect coefficient (γ) can be expressed as follows:

$$\gamma = \frac{\sqrt{2e\varepsilon_{si}N_a}}{C_{ox}} \tag{2.78}$$

Similarly, the threshold voltage $|V_{Tp}|$ of a pMOS transistor can be related to its source-to-bulk voltage V_{SB} by the following equation:

$$|V_{Tp}| = |V_{T0p}| + \gamma\left(\sqrt{2|\phi_{fn}| + |V_{SB}|} - \sqrt{2|\phi_{fn}|}\right) \tag{2.79}$$

where V_{T0p} denotes the threshold voltage of a pMOS transistor without body effect and V_{Tp} denotes the threshold voltage of a pMOS transistor with body effect.

Figure 2.24 shows body effects of nMOS and pMOS transistors. The threshold voltage increases with the increasing source-to-bulk voltage (V_{SB}) in nMOS transistors. In pMOS transistors, the absolute value of threshold voltage increases with the increasing bulk-to-source voltage (V_{BS}).

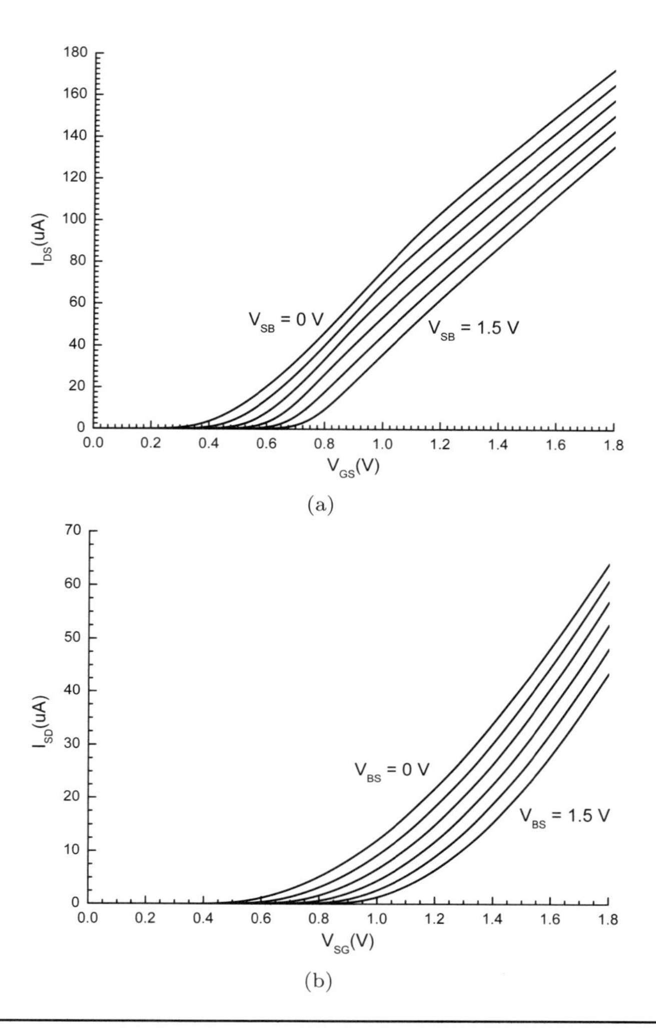

Figure 2.24: The body effects of (a) nMOS and (b) pMOS transistors.

2.4.2.2 Short-Channel Effect When the channel length (L) of a MOS transistor is comparable to the drain-depletion-layer thickness, the MOS transistor is referred to as a *short-channel device*. *The short-channel effect* (SCE) refers to all factors that cause threshold voltage reduction (hence, leakage current increase) when L decreases and the drain-to-source voltage V_{DS} increases. The reduction in threshold voltage (V_T) results from the combination of three effects: *charge sharing*, *subsurface punchthrough*, and *drain-induced barrier lowering* (DIBL).

Charge sharing. Charge sharing refers to the fact that the charge needed to induce a channel is not totally supported by the gate voltage but instead may obtain some help from others, such as *source/drain charge sharing*, *reverse short-channel effect* (RSCE), and *narrow-channel effect* (NCE) . The source/drain charge sharing accounts for the decrease of the depletion charge below the gate oxide due to the contribution of the drain-substrate junction under reverse bias. It results in the reduction of the threshold voltage. The RSCE is caused by the lateral nonuniformity of the doping across the channel. Its effect is the increase of the threshold voltage. The total short-channel effect is the sum of both SCE and RSCE. The NCE takes into account the extra depletion-layer region at both ends of the channel width. This effect results in the increase of the threshold voltage.

Subsurface punchthrough. Subsurface punchthrough refers to the phenomenon that the combination of channel length and reverse bias leads to the merging of both source and drain-depletion layers. In short-channel devices, the surface is more heavily doped than the bulk, causing a greater expansion of the depletion layer below the surface as compared with the surface. Thus, the punchthrough occurs below the surface and is often referred to as subsurface punchthrough. The subsurface punchthrough degrades the performance of MOS transistors, most notably by leading to an increase of the subthreshold current.

2.4.2.3 Drain-Induced Barrier Lowering Drain-induced barrier lowering (DIBL) refers to the influence of drain voltage on the threshold voltage. For long-channel devices, only the gate-to-source voltage lowers the source barrier; for short-channel devices, a sufficiently high drain voltage can pull down the source barrier and lower the threshold voltage by an amount of ηV_{DS}, where η is the DIBL coefficient, on the order of 0.1. Hence, the threshold voltage V_T decreases with the increasing V_{DS}. The DIBL can be measured at a constant V_{GS} as the change in I_{DS} for a change in V_{DS}, as shown in Figure 2.25.

Review Questions

Q2-40. What is the meaning of body effect?

Q2-41. What is the meaning of the short-channel effect?

Q2-42. Explain the phenomenon of DIBL.

Q2-43. What factors may affect the threshold voltages?

2.4.3 Leakage Currents

In processes above 0.18 μm, leakage current is so insignificant that it can be totally ignored. In processes between 90 nm and 65 nm, the threshold voltages are reduced to the point that the subthreshold current is in the range between several and tens nA and cannot be neglected as many million, even billion, devices are integrated on a chip. In processes below 65 nm, the gate-oxide thickness is reduced to the point that gate leakage current is comparable to the subthreshold current unless high-k dielectrics are used. Generally speaking, leakage current of MOS transistors comprises three components: junction-leakage current, subthreshold current, and gate leakage current.

2.4.3.1 Junction Leakage Current Recall that the current-voltage characteristic of a diode is governed by the ideal-diode equation, that is, Equation (2.30). As the junction is reverse biased, the leakage current is just the reverse saturation current I_s, which ranges from 0.1 to 0.01 fA/μm^2 and may be neglected compared with other leakage mechanisms, including *band-to-band tunneling* (BTBT) and *gate-induced drain leakage* (GIDL).

Band-to-band tunneling. Band-to-band tunneling is the phenomenon that electrons tunnel from the valence band of p-type material to the conduction band of n-type material when a high electric field ($> 10^6$ V/cm) is applied across the reverse-biased

pn junction. The tunneling current can be expressed as

$$I_{BTBT} = A\alpha \frac{EV}{E_g^{1/2}} \exp\left(-\beta \frac{E_g^{3/2}}{E}\right) \tag{2.80}$$

where A is the total area of the junction, α and β are constants, E_g is the bandgap voltage, and V is the applied forward-bias voltage. The electric field along the junction at a reverse bias of V is as follows:

$$E = \sqrt{\frac{2eN_aN_d(V+\phi_0)}{\varepsilon_{si}(N_a+N_d)}} \tag{2.81}$$

In deep submicron devices, high doping concentrations and abrupt doping profiles cause significant BTBT current through the drain-well junction.

Gate-induced drain leakage. Gate-induced drain leakage (GIDL) is due to high field effect in the pn junction of the gate-drain overlap area of an MOS transistor. The GIDL effect is most pronounced when the gate is at a negative voltage and the drain is at a high voltage (V_{DD}) in processes below 90 nm, as shown in Figure 2.25. The GIDL current is proportional to the gate-drain overlap area and hence to the channel width. It is a strong function of electric field; namely, thinner gate oxide and higher drain-to-gate voltage would increase the GIDL current.

2.4.3.2 Subthreshold Current Remember that the threshold voltage V_T is arbitrarily defined to be the gate-to-source voltage V_{GS} that induces under silicon dioxide a strong inversion layer with a concentration equal to that of the substrate. The drain current I_{DS} is not equal to 0 when the gate-to-source voltage V_{GS} is above threshold voltage V_T; otherwise, the drain current I_{DS} is zero.

However, the realistic devices indeed begin to conduct a current below the threshold voltage although the current is very small. This current is called the *subthreshold current*. Recall that the inversion layer starts to be formed after the depletion layer reaches its maximum thickness, and thereafter, the concentration of the inversion layer increases with the increasing gate-to-source voltage. Consequently, it is quite reasonable to have a current after the inversion layer is formed but before the strong inversion layer is reached. This operating region is called the *subthreshold region*.

Subthreshold conduction is so important in deep submicron devices because of the following reasons. First, in a modern VLSI design with millions, even billions, of transistors, the subthreshold current of each device aggregates to an appreciable power dissipation. Second, in dynamic circuits, the subthreshold current amounts to the *charge-loss effect* of soft nodes, thereby resulting in the deterioration of performance.

The current-voltage characteristics of the subthreshold region of MOS transistors are much like those of bipolar junction transistor (BJT); that is, the drain current I_{DS} is an exponential function of its gate-to-source voltage V_{GS} and can be approximated by the following equation:

$$I_{Dsub} = I_{D0sub}\left(\frac{W}{L}\right)\exp\left[\frac{V_{GS}-V_T}{nV_t}\right] \tag{2.82}$$

where I_{D0sub} is a constant and $V_t = kT/e$ is the thermal voltage. In bulk CMOS, the value of n is around 1.6.

The above equation can be expressed as a logarithmic function and is equal to

$$\log I_{Dsub} = \log I_{D0sub} + \log\left(\frac{W}{L}\right) - \frac{V_T}{nV_t}\log e + \frac{V_{GS}}{nV_t}\log e \tag{2.83}$$

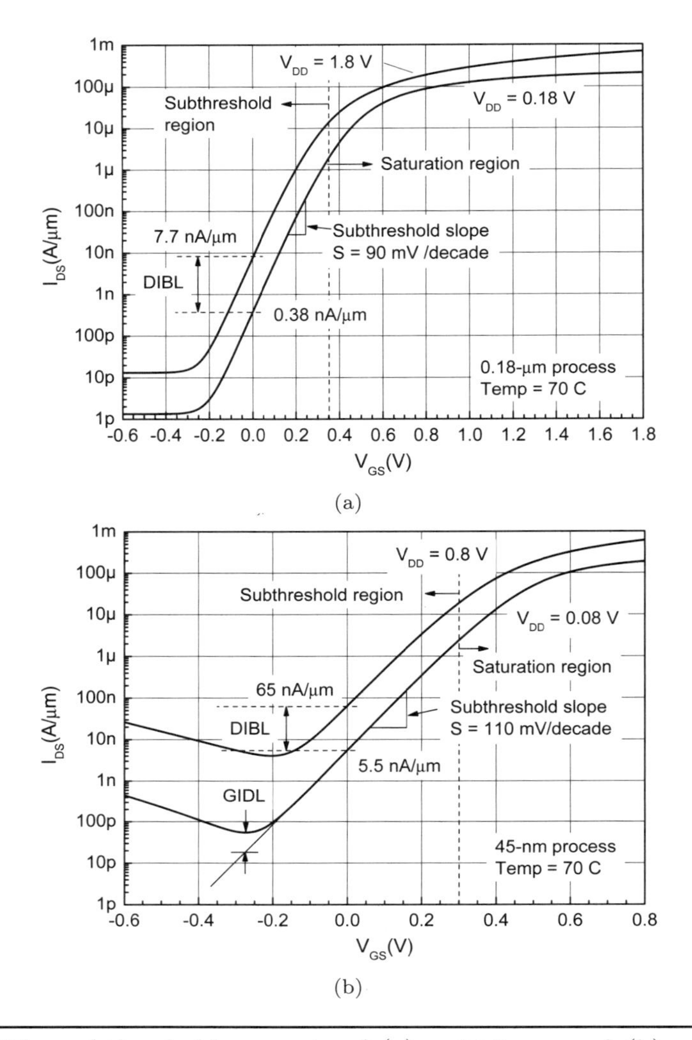

Figure 2.25: The subthreshold currents of (a) a 0.18-μm and (b) a 45-nm nMOS transistors at 70°C.

If $V_t = 26$ mV and $n = 1$, the subthreshold slope of Figure 2.25(a) is

$$\frac{nV_t}{\log e} = 90 \text{ mV/decade} \tag{2.84}$$

The subthreshold slope indicates how much gate voltage must drop to reduce the leakage current by an order of magnitude. The subthreshold currents of 0.18-μm and 45-nm nMOS transistors are given in Figure 2.25. The current-voltage (I-V) characteristics of typical 0.18-μm nMOS and pMOS transistors in subthreshold regions are shown in Figure 2.26.

The subthreshold operation has been proved useful for low-power applications such as solar-powered calculators, battery-operated watches, and CMOS image sensors. The major design challenges with subthreshold operation are on the issues of matching, noise, and bandwidth. Since the drain current is exponentially related to the gate-to-source voltage, any mismatch of the devices will result in significant differences in

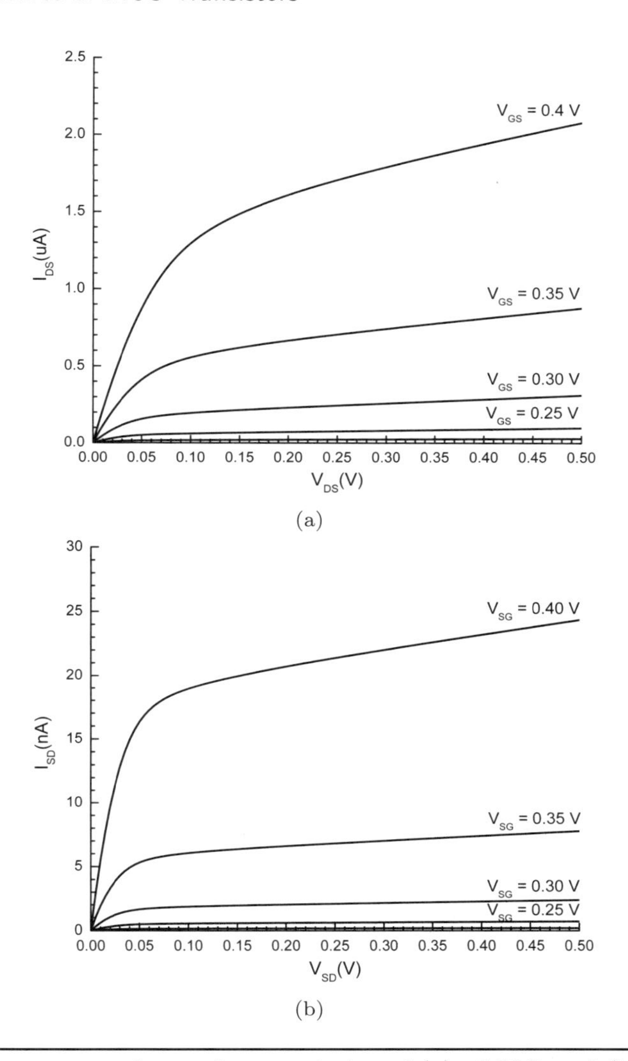

Figure 2.26: The current-voltage characteristics of (a) nMOS and (b) pMOS transistors in subthreshold regions.

drain currents. Discriminating the small signal from an almost equal amount of noise in the design is also a challenge. Because of limited bandwidth, subthreshold operation only finds its applications in low-frequency areas.

Before leaving this subsection, it is worthy to point out another important feature from Figure 2.25. As indicated in the figure, the drain currents of both nMOS transistors are not zero at zero V_{GS}. The drain current is 7.7 nA/μm for nMOS transistors in the given 0.18-μm process with minimum channel length under the $V_{DS} = 1.8$ V at 70°C and is 65 nA/μm for nMOS transistors in the given 45-nm process with minimum channel length under the $V_{DS} = 0.8$ V at 70°C. The drain current at zero V_{GS} is generally called the *off current*, or *off-state leakage current*, of a MOS transistor, even though the transistor is not indeed turned off.

2.4.3.3 Gate Leakage Current Gate leakage currents are caused by hot-carrier injection (HEI) and *gate-oxide tunneling*. These two currents are mostly found in deep

Table 2.4: Off-state and gate leakage currents of a typical triple gate-oxide process.

	HS-CMOS	LP-CMOS	I/O-CMOS
V_{DD} (V)	1.2	1.2	2.5
t_{ox} (nm)	1.9	2.5	5.0
I_{off} (A/μm)	8 n	$<$ 4 p	$<$ 1 p
I_{gate} (A/μm^2)	1 n	$<$ 40 p	—

submicron devices.

Hot-carrier injection. Some hot carriers generated by impact ionization can inject through the gate oxide and become gate current, constituting a part of the unwanted gate leakage current. For hot carriers in the channel to reach the gate, two conditions must be satisfied. First, the hot carriers must gain enough kinetic energy from the channel electric field to overcome the potential barrier between the bulk silicon and gate oxide. Second, the hot carriers must be redirected toward the silicon-oxide interface; that is, a vertical electric field must exist between the gate and substrate.

Gate tunneling. Gate tunneling current from a high electric field (E_{ox}) can be caused by either *Fowler-Nordheim* (FN) *tunneling* or *direct tunneling*. In the FN tunneling, electrons tunnel through a triangular potential barrier, whereas in the direct tunneling, electrons tunnel through a trapezoidal potential barrier. Because tunneling probability depends on the thickness of the barrier, the barrier height, and the structure of the barrier, the tunneling probabilities of a single electron in FN tunneling and direct tunneling are different, thereby resulting in different tunneling currents.

FN tunneling. FN tunneling occurs at high voltage with moderate oxide thickness and is most important in programming Flash memories. In FN tunneling, electrons tunnel into the conduction band of the oxide layer. For normal device operation, the measured value of FN tunneling current is very small and therefore can be negligible.

Direct tunneling. Direct tunneling occurs at lower voltage with thin (less than 3 to 4 nm) oxides. In direct tunneling, electrons from the inversion layer of silicon substrate directly tunnel to the gate through the forbidden energy gap of the gate oxide. For deep submicron devices, direct tunneling current is the dominant gate leakage current.

2.4.3.4 Reduction of Leakage Currents The most effective way to reduce gate and off-state leakage currents is to increase the gate-oxide thickness. To see this, consider a typical triple gate-oxide process shown in Table 2.4, where three MOS transistors with different gate-oxide thicknesses are provided by the process. The high-performance (HS) transistors have the thinnest gate oxide and hence the lowest threshold voltage, whereas the low-power (LP) transistors have a thicker gate oxide, thereby substantially reducing both gate leakage (I_{gate}) and off-state leakage (I_{off}) currents. The I/O transistors have the thickest gate oxide and hence the lowest performance but can withstand large breakdown voltages. Moreover, the I/O transistors have insignificant off-state and gate leakage currents.

Nowadays, many deep submicron processes often provide a combination of general-purpose (G) and low-power (LP) core transistors together with a 2.5-V I/O transistor as a triple gate-oxide process for optimizing operating speed, power dissipation, and leakage currents for a variety of applications. Some processes even offer 1.8-V, 2.5-V, and 3.3-V I/O options to meet different product requirements.

■ Review Questions

Q2-44. What are the ingredients of leakage current?
Q2-45. What is band-to-band tunneling (BTBT)?
Q2-46. What is gate-induced drain leakage (GIDL)?
Q2-47. Define the subthreshold current and subthreshold region?
Q2-48. What is the distinction between FN tunneling and direct tunneling?

2.4.4 Short-Channel I-V Characteristics

We begin to describe in the following the effective electron and hole mobilities of short-channel devices and then derive the current equations for such devices under both linear and saturation regions of operation.

2.4.4.1 Effective Mobility The saturation velocity of an electron in a surface channel is between 6×10^6 and 10×10^6 cm/s, as depicted in Figure 2.21. The saturation velocity of a hole in a surface channel is between 4×10^6 and 8×10^6 cm/s. The velocity of an electron or a hole in an electric field can be modeled as

$$\begin{aligned} v &= \frac{\mu_{eff} E}{1 + E/E_{sat}} & E < E_{sat} \\ &= v_{sat} & E \geq E_{sat} \end{aligned} \tag{2.85}$$

where μ_{eff} is the effective carrier mobility. From this equation, the saturation electric field E_{sat} is found to be

$$E_{sat} = \frac{2v_{sat}}{\mu_{eff}} \tag{2.86}$$

The effective carrier mobility can be estimated by an empirical formula, which can be expressed as follows:

$$\mu_{eff} = \frac{A}{1 + \left(\dfrac{E_{norm}}{B}\right)} \tag{2.87}$$

where A is 670 cm^2/V-s for electrons and 160 cm^2/V-s for holes, B is 6.6×10^5 V/cm for electrons and 7×10^5 V/cm for holes, and the factor E_{norm} can be calculated by the following equation:

$$E_{norm} = \frac{V_{GS} + V_T}{6t_{ox}} \tag{2.88}$$

where V_{GS} is the gate-to-source voltage, V_T is the threshold voltage, and t_{ox} is the thickness of gate oxide. The factor 6 in the expression E_{norm} accounts for the averaging of an integral and that $\varepsilon_{si}/\varepsilon_{ox} \approx 3$.

■ Example 2-11: (Effective electron and hole mobilities.) Assume that $t_{ox} = 4.1$ nm in a 0.18-μm process.

1. Calculate the effective electron mobility as $V_{GS} = 1.8$ V and $V_{Tn} = 0.4$ V.
2. Calculate the effective hole mobility as $V_{GS} = 1.8$ V and $V_{Tp} = -0.4$ V.

Solution: The effective electron and hole mobilities are calculated as follows:

1. Using Equation (2.88), we have $E_{norm} = 8.94 \times 10^5$ V/cm. By using Equation (2.87), the effective electron mobility is calculated as $\mu_{eff} = 285$ cm^2/V-s.
2. Using Equation (2.88), we have $E_{norm} = 5.69 \times 10^5$ V/cm. By using Equation (2.87), the effective hole mobility is calculated as $\mu_{eff} = 88$ cm^2/V-s.

■

2.4.4.2 The I-V Characteristics of the Linear Region Recall that the drain current in the linear region of operation can be expressed as follows:

$$I_{DS} = W \times Q_I \times v \tag{2.89}$$

Substituting the velocity relation given in Equation (2.85) into the above equation, the drain current in a short-channel MOS transistor can be represented as

$$I_{DS} = WC_{ox}[V_{GS} - V_T - V]\frac{\mu_{eff}E}{1 + E/E_{sat}} \tag{2.90}$$

where V is a function of y. Replacing E with dV/dy and rearranging terms, we obtain

$$I_{DS}dy = W\mu_{eff}\left[C_{ox}(V_{GS} - V_T - V(y)) - \frac{I_{DS}}{W\mu_{eff}E_{sat}}\right]dV \tag{2.91}$$

Integrating from $y = 0(V = 0)$ to $L(V = V_{DS})$, the drain current can be expressed as follows:

$$\begin{aligned} I_{DS} &= \mu_{eff}C_{ox}\left(\frac{W}{L}\right)\left(V_{GS} - V_T - \frac{1}{2}V_{DS}\right)V_{DS}\frac{1}{1 + V_{DS}/E_{sat}L} \\ &= \mu_n C_{ox}\left(\frac{W}{L}\right)\left(V_{GS} - V_T - \frac{1}{2}V_{DS}\right)V_{DS} \text{ (for } E_{sat}L \gg V_{DS}) \end{aligned} \tag{2.92}$$

The drain current converges back to the long-channel equation if $E_{sat}L \gg V_{DS}$.

2.4.4.3 The I-V Characteristics of the Saturation Region The drain current becomes saturated when the electrons arrive at the drain with their limiting velocity, saturation velocity v_{sat}. When this situation occurs, the drain current becomes

$$\begin{aligned} I_{DSsat} &= WQ_I v_{sat} \\ &= WC_{ox}[V_{GS} - V_T - V(y)]\, v_{sat} \\ &= WC_{ox}[V_{GS} - V_T - V_{DSsat}]\, v_{sat} \end{aligned} \tag{2.93}$$

The saturation voltage V_{DSsat} can be found by equating the drain currents at both linear and saturation regions, that is, Equations (2.92) and (2.93), and using the saturation electric field

$$E_{sat} = \frac{2v_{sat}}{\mu_{eff}}$$

The result V_{DSsat}, that is, the decision point between linear and saturation regions, can then be obtained as follows:

$$V_{DSsat} = \frac{(V_{GS} - V_T)E_{sat}L}{(V_{GS} - V_T) + E_{sat}L} \tag{2.94}$$

which converges to $V_{GS} - V_T$ when $E_{sat}L \gg V_{GS} - V_T$.

Once the saturation voltage V_{DSsat} is found, the drain current at the saturation region can then be represented as in the following:

$$\begin{aligned} I_{DSsat} &= WC_{ox}\left(V_{GS} - V_T - V_{DSsat}\right)v_{sat} \\ &= WC_{ox}v_{sat}\frac{(V_{GS} - V_T)^2}{(V_{GS} - V_T) - E_{sat}L} \quad (2.95) \\ &= \frac{1}{2}\mu_{eff}C_{ox}\left(\frac{W}{L}\right)(V_{GS} - V_T)^2 \quad \text{if } E_{sat}L \gg (V_{GS} - V_T) \quad (2.96) \\ &= Wv_{sat}C_{ox}(V_{GS} - V_T) \quad \text{if } E_{sat}L \ll (V_{GS} - V_T) \quad (2.97) \end{aligned}$$

Once again, it reduces to the long-channel case when $E_{sat}L \gg V_{GS} - V_T$.

Table 2.5 summarizes circuit design equations of MOS transistors, including both long-channel and short-channel devices.

2.4.5 Temperature Effects

The drain current I_{DSsat} is a strong function of temperature due to variations of threshold voltage and mobility with temperature. To see this, consider the current equation of an nMOS transistor.

$$I_{DSsat} = \frac{1}{2}\mu_n\left(\frac{\varepsilon_{ox}}{t_{ox}}\right)\left(\frac{W_n}{L_n}\right)(V_{GS} - V_{Tn})^2 \quad (2.98)$$

The dependence of V_{Tn} with temperature can be expressed as

$$\frac{dV_{Tn}}{dT} = -\frac{1}{T}\left(\frac{E_g}{2e} - |\phi_{fp}|\right)\left(2 + \frac{\gamma}{\sqrt{2|\phi_{fp}|}}\right) \quad (2.99)$$

The threshold voltage falls with increasing temperature if $|\phi_{fp}| < E_g/(2e)$ and increases with temperature, otherwise. The slope is usually in the range of -0.5 mV/°C to -4 mV/°C.

The mobility is also a function of temperature and can be modeled as follows:

$$\mu(T) = \mu(T_0)\left(\frac{T_0}{T}\right)^{1.5} \quad (2.100)$$

Therefore, the reduction of mobility with temperature is found to be

$$\frac{d\mu(T)}{dT} = -1.5\mu(T_0)\left(\frac{T_0}{T}\right)^{2.5} \quad (2.101)$$

From the previous discussion, we know that both threshold voltage and mobility decrease as temperature increases. From the current equation, the decrease in threshold voltage causes the drain current to go up while the decrease in mobility causes the drain current to go down. At low V_{GS}, the changes in V_{Tn} dominate and the drain current increases with increasing temperature. At higher V_{GS}, the mobility dominates and the drain current decreases with increasing temperature. At some V_{GS}, both effects cancel each other and the drain current does not change with temperature. The effects of temperature versus drain current are illustrated in Figure 2.27. Figure 2.27(a) gives the case when $V_{GS} = 0.01$ V, where the drain current increases with increasing temperature, while Figure 2.27(b) is the case when $V_{GS} = V_{DD}$, where the drain current decreases with increasing temperature.

Table 2.5: Circuit design equations for MOS transistors.

Long-channel devices

• nMOS transistor

$$V_{Tn} = V_{T0n} + \gamma\left(\sqrt{2|\phi_{fp}| + V_{SB}} - \sqrt{2|\phi_{fp}|}\right)$$

$$I_{DS} = \frac{\mu_n C_{ox}}{2}\left(\frac{W_n}{L_n}\right)\left[2\left(V_{GS} - V_{Tn}\right)V_{DS} - V_{DS}^2\right] \quad \text{for } V_{GS} \geq V_{Tn} \text{ and } V_{DS} < V_{GS} - V_{Tn}$$

$$I_{DSsat} = \frac{\mu_n C_{ox}}{2}\left(\frac{W_n}{L_n}\right)(V_{GS} - V_{Tn})^2(1 + \lambda V_{DS}) \quad \text{for } V_{GS} \geq V_{Tn} \text{ and } V_{DS} \geq V_{GS} - V_{Tn}$$

• pMOS transistor

$$|V_{Tp}| = |V_{T0p}| + \gamma\left(\sqrt{2|\phi_{fn}| + |V_{SB}|} - \sqrt{2|\phi_{fn}|}\right)$$

$$I_{DS} = \frac{\mu_p C_{ox}}{2}\left(\frac{W_p}{L_p}\right)\left[2\left(V_{SG} - |V_{Tp}|\right)V_{SD} - V_{SD}^2\right] \quad \text{for } V_{SG} \geq |V_{Tp}| \text{ and } V_{SD} < V_{SG} - |V_{Tp}|)$$

$$I_{DSsat} = \frac{\mu_p C_{ox}}{2}\left(\frac{W_p}{L_p}\right)\left(V_{SG} - |V_{Tp|}\right)^2(1 + \lambda V_{SD}) \quad \text{for } V_{SG} \geq |V_{Tp}| \text{ and } V_{SD} \geq V_{SG} - |V_{Tp}|)$$

Short-channel devices

• nMOS transistor

$$V_{DSsat} = \frac{(V_{GS} - V_{Tn})E_{satn}L_n}{(V_{GS} - V_{Tn}) + E_{satn}L_n}$$

$$I_{DS} = \frac{\mu_{effn} C_{ox}}{2}\left(\frac{W_n}{L_n}\right)\left[2(V_{GS} - V_{Tn})V_{DS} - V_{DS}^2\right]\frac{1}{1 + V_{DS}/E_{satn}L_n}$$

$$\text{for } V_{GS} \geq V_{Tn} \text{ and } V_{DS} < V_{DSsat}$$

$$I_{DSsat} = W_n v_{sat} C_{ox}\left(V_{GS} - V_{Tn} - V_{DSsat}\right)(1 + \lambda V_{DS}) \quad \text{for } V_{GS} \geq V_{Tn} \text{ and } V_{DS} \geq V_{DSsat}$$

• pMOS transistor

$$V_{SDsat} = \frac{(V_{SG} - |V_{Tp}|)E_{satp}L_p}{(V_{SG} - |V_{Tp}|) + E_{satp}L_p}$$

$$I_{DS} = \frac{\mu_{effn} C_{ox}}{2}\left(\frac{W_p}{L_p}\right)\left[2(V_{SG} - |V_{Tp}|)V_{SD} - V_{SD}^2\right]\frac{1}{1 + V_{SD}/E_{satp}L_p}$$

$$\text{for } V_{SG} \geq |V_{Tp}| \text{ and } V_{SD} < V_{SDsat}$$

$$I_{DSsat} = W_p v_{sat} C_{ox}\left(V_{SG} - |V_{Tp}| - V_{SDsat}\right)(1 + \lambda V_{SD}) \quad \text{for } V_{SG} \geq |V_{Tp}| \text{ and } V_{SD} \geq V_{SDsat}$$

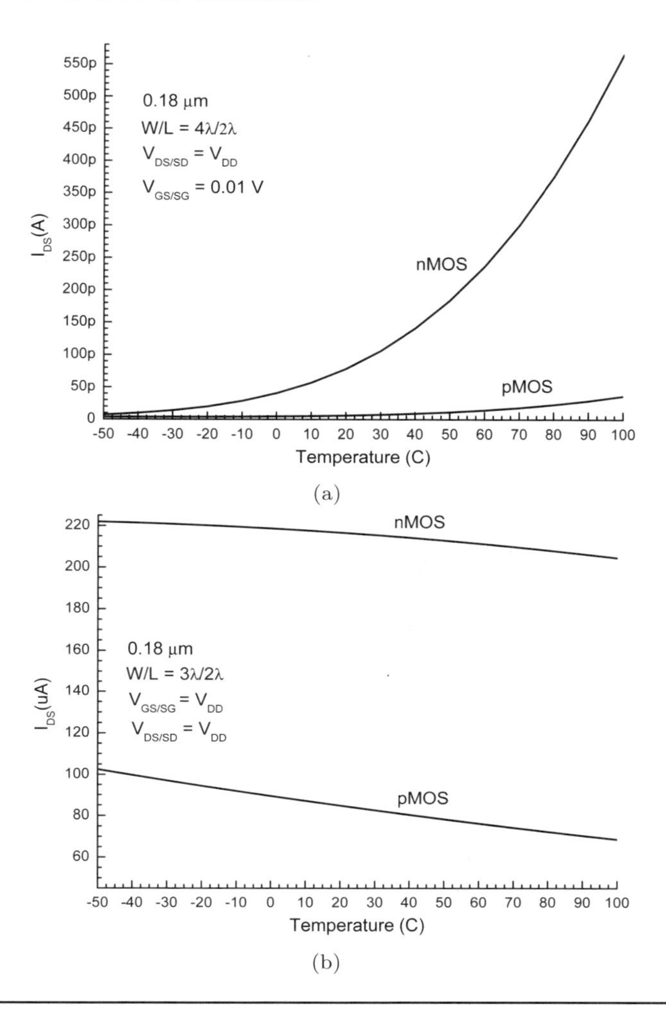

Figure 2.27: The temperature effects on the current-voltage characteristics of MOS transistors: (a) low V_{GS}; (b) $V_{GS} = V_{DD}$.

Figure 2.28 shows the detailed temperature effects on the current-voltage characteristics of MOS transistors. At the subthreshold voltage region, the drain current increases with the rising temperature due to the bipolar junction transistor in nature while at the above threshold region, the drain current decreases with the rising temperature due to the MOS transistor in nature.

2.4.6 Limitations of MOS Transistors

In this subsection, we address the limitations of MOS transistors, including thin-oxide breakdown, avalanche breakdown, snapback breakdown, and punchthrough effects.

2.4.6.1 Thin-Oxide Breakdown All MOS transistors must be protected against the vertical electric field built by the gate voltage. The electric field at breakdown of silicon dioxide is about 6×10^6 V/cm to 7×10^6 V/cm, corresponding to 0.6 V to 0.7 V applied from gate to channel with a 1-nm gate-oxide (namely, thin-oxide) thickness. As a result, the maximum gate voltage may be applied to a modern deep submicron

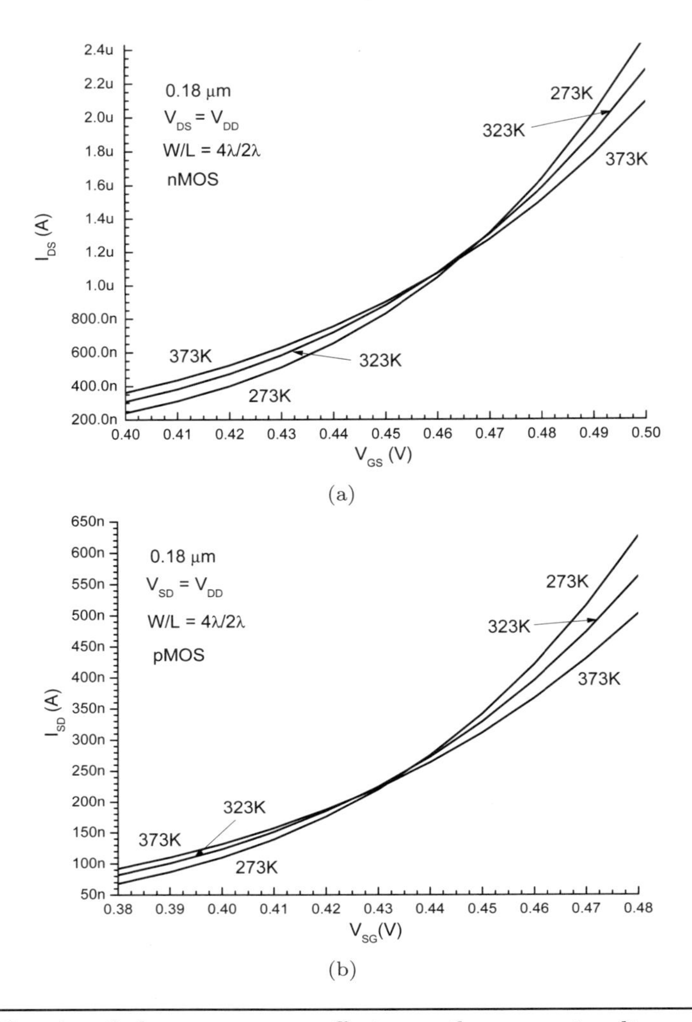

Figure 2.28: The detailed temperature effects on the current-voltage characteristics of MOS transistors: (a) nMOS transistor; (b) pMOS transistor.

device with a 1.5-nm gate-oxide thickness only about 0.9 V to 1.05 V. A very small gate voltage may cause the breakdown of gate oxide to occur.

The gate breakdown may also be caused by a static electric field. To avoid such a breakdown, especially in the input circuits of input/output pads, an electrostatic discharging circuit (ESD), such as the simplest clamped circuit built by diodes, is needed. Refer to Chapter 15 for more details of ESD.

2.4.6.2 Avalanche Breakdown As mentioned, a high electric field appears at the drain-substrate junction of an nMOS or a pMOS transistor. As the V_{DS} is small, the drain current is under the control of both V_{GS} and V_{DS}; however, as the V_{DS} exceeds the breakdown voltage of the drain-substrate junction, the drain current increases abruptly because of the avalanche breakdown which is caused by the impact ionization occurring in the depletion layer of the junction. The breakdown voltage is determined by the concentration of the low-doped region, that is, the substrate concentration.

2.4.6.3 Snapback Breakdown Once the avalanche breakdown occurs, the parasitic BJT comprising source, substrate, and drain, begins to operate in its active mode so that its collector current and current gain increase. This helps develop the avalanche breakdown condition. As a result, a lower V_{DS} is required for the increasing drain current and yields a negative resistance region. This phenomenon is known as the *snapback breakdown*.

2.4.6.4 Punchthrough Effects The punchthrough phenomenon in MOS transistors is the condition at which the depletion region around the drain end completely extends across the channel to reach the depletion region around the source end. This results in eliminating the potential barrier between the source and drain and a very large drain current would exist.

■ Review Questions

Q2-49. What is the maximum allowable gate voltage as the gate-oxide thickness is 2.0 nm?

Q2-50. Explain the avalanche breakdown.

Q2-51. What is the phenomenon of snapback breakdown?

Q2-52. Explain the punchthrough effects.

2.5 SPICE and Modeling

In this section, we first introduce a few features of the *Simulation Program with Integrated Circuit Emphasis* (SPICE), which will be used throughout the book. Then, we take a look at how to model diodes and MOS transistors along with some examples.

2.5.1 An Introduction to SPICE

SPICE is the acronym for Simulation Program with Integrated Circuit Emphasis. It was developed by the University of California, Berkeley (UCB), in the 1970s and was widely adopted as a circuit simulator in both industry and academia. Now, it has become the de facto standard of circuit simulation in industry. SPICE uses a numerical approach to carry out circuit simulations.

SPICE is generally a circuit analysis tool for the simulation of electrical circuits in steady-state, transient, and frequency domains. Many SPICE tools are available on the market, such as SBTSPICE, HSPICE, Spectre, TSPICE, Pspice, Smartspice, ISpice, and so on. Most of these stem from UCB SPICE program and therefore support the most common, original SPICE syntax. In addition, these SPICE tools have a similar basic algorithm scheme; only the control of time step, equation solver, and convergence control might be different.

SPICE is a circuit simulator and a circuit analysis tool rather than a circuit design tool. The role of SPICE is to verify the function and timing of a design. Therefore, before using SPICE to validate a circuit design, one should already understand the basic features of used devices, the functionality and specification of the circuit, the types of simulations of the circuit, the features of input signals, and the expected output responses as well.

■ Example 2-12: (A simple SPICE file example.) This example illustrates the major format and components of a typical SPICE program. The SPICE program describes an inverter circuit and analyzes its voltage-transfer characteristic (VTC).

```
Inverter VTC study —— 0.18 um process
******** Setting up various global parameters *********
.lib "..\cmos18.txt" cmos
.param Supply = 1.8V * Set value of Vdd
.opt scale = 0.09u   * Set value of lambda
******** Circuit description *************
MN  Vout Vin Gnd Gnd nmos L = 2 W = 4
MP  Vout Vin Vdd Vdd pmos L = 2 W = 8
Vdd Vdd  Gnd 'Supply'
Vin Vin  Gnd
******** Analysis statements ************
.dc Vin  0   'Supply'  'Supply/100'
******** Plotting and printing statements **
.probe V(Vout)
.option post
******** The .end statement *************
.end
```

■

A basic SPICE file will contain the following components:

1. A title: The first line of the file is always regarded as a title.

   ```
   Inverter VTC study —— 0.18 um process
   ```

2. Settings of various global parameters such as λ, V_{DD}, and the MOS device models to be used.

   ```
   .lib "..\cmos18.txt" cmos
   .param Supply = 1.8V * Set value of Vdd
   .opt scale = 0.09u   * Set value of lambda
   ```

3. A circuit description that lists sources, active elements, and passive elements.

   ```
   MP  Vout Vin Vdd Vdd pmos L = 2 W = 8
   Vdd Vdd  Gnd 'Supply'
   ```

4. Analysis statements such as direct current (DC) and transient.

   ```
   .dc Vin 0 'Supply' 'Supply/100'
   ```

5. Plotting and printing statements.

   ```
   .probe V(Vout)
   ```

6. An **.end** statement.

The three main types of global parameters that can be set are parameters, libraries, and lambda. Parameters are floating-point number variables. The syntax and an example are as follows:

```
.param parameter_name=real_number
.param Supply=2.5   * set value of Vdd
```

Table 2.6: The default instance naming scheme of SPICE.

Element type	Naming	Examples	Element type	Naming	Examples
Voltage sources			Passive elements		
Independent	Vxxx	V1, Vdd	Capacitor	Cxxx	C1, Cload
Dependent:VCVS	Exxx	E1	Inductor	Lxxx	L1, Lload
Dependent:CCVS	Hxxx	H12	Mutual inductor	Kxxx	K12
Current sources			Resistor	Rxxx	R1, Rload
Independent	Ixxx	V1, Vdd	Active elements		
Dependent:VCCS	Gxxx	G1	Diode	Dxxx	Dout
Dependent:CCCS	Fxxx	F2	JFET/MESFET	Jxxx	J1
Transmission line	Txxx	T12	MOSFET	Mxxx	Mn1, Mp1
Lossy transmission line	Uxxx	U2	BJT	Qxxx	Qa, Q2
	Wxxx	W1	Subcircuit call	Xxxx	X24

Libraries contain various parameters used in the MOS transistor model. The syntax and an example are as follows:

```
.lib "<file_path>" cmos      * CMOS is a corner name
.lib "..\cmos18.txt" cmos    * set 0.18 um library
```

Lambda defines the value (length) of the λ unit.

```
.opt scale=length
.opt scale=0.09u            * set value of lambda
MP Vout Vin Vdd Vdd pmos L=2 W=8 *L = 2 and W = 8
```

2.5.1.1 SPICE Data File In a SPICE program, a circuit is composed of a list of sources, elements, and associated connection nodes. The sources can be voltage and/or current sources; the elements can be active elements, including diodes and transistors, and passive elements, including resistors, capacitors, and inductors. Each type of source and element is specified on an individual line. Sources and elements are added to the list by specifying the following: a unique instance name, the node to which they are connected, specific values associated with the element, a model if an active element is being specified, and element specific parameters, such as the width and length of a MOS transistor. Unlike C programming language, SPICE is a case-insensitive language. Hence, both `gnd` and `GND` denote the same node.

Instance name. The instance name of an element is a unique identification that may contain up to eight alphanumeric characters. The first character of the name must correspond to the type of source or element that is being invoked. The default instance names are listed in Table 2.6.

Node naming convention. Most elements require two nodes while other elements such as BJT or MOS transistors need three or more nodes. Alphanumeric node names should be used whenever possible for a more descriptive and meaningful input. Each node name can be a name or number, for example, data1, n1, 21, and so on. The digit 0 (zero) is always reserved for the ground node; another commonly used name for the ground node is gnd (GND). All nodes default to be local except that a `.global` statement is used to declare specific nodes as global. Global nodes can be referred across all subcircuits.

Element values. Element values are specified using floating-point numbers. Standard units, such as ohms, farads, volts, and henrys, are not required since SPICE will

Table 2.7: The default multiplying factors used in SPICE.

Multiplying factor	Metric prefix	Notation	Exponential	SPICE Examples	
10^9	Giga	G, g	E9	2.4E9	2.4 GHz
10^6	Mega	Meg, meg	E6	100E6	100 Meg
10^3	Kilo	K, k	E3	22E3	22 kohms
10^{-3}	Milli	M, m	E-3	14E-3	14 mA
10^{-6}	Micro	U, u	E-6	2.4E-6	2.4 μV
10^{-9}	Nano	N, n	E-9	12E-9	12 nA
10^{-12}	Pico	P, p	E-12	21E-12	21 pF
10^{-15}	Femto	F, f	E-15	2E-15	2 fF

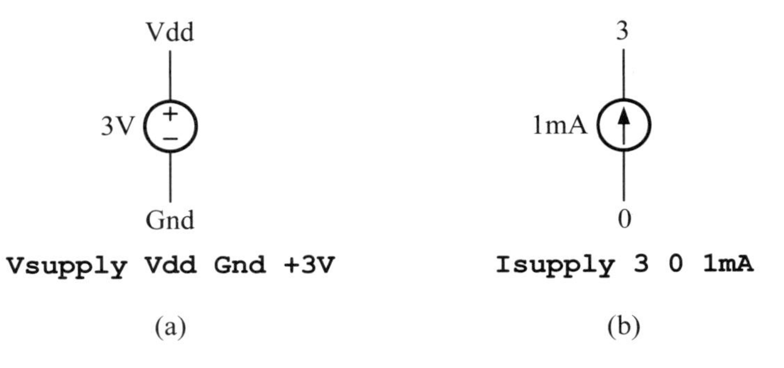

Figure 2.29: Examples of (a) voltage and (b) current sources.

infer the units based on the types of sources and elements, including resistors, inductors, capacitors, and current and voltage sources. However, they are allowed to be specified whenever possible to improve readability. The default multiplying factors used in SPICE are summarized in Table 2.7.

Voltage and current sources. The independent voltage and current sources have the following syntax:

```
Vxxx n+ n− voltage_value
Ixxx n+ n− current_value
```

where `Vxxx` and `Ixxx` are the names of voltage and current sources; `n+` is the positive node and `n-` is the negative node; `voltage_value` and `current_value` are values of the voltage and current sources, respectively. Examples of voltage and current sources are given in Figures 2.29(a) and (b), respectively.

Dependent voltage and current sources. SPICE supports the following four types of dependent sources:

- voltage-controlled voltage source (VCVS)
- voltage-controlled current source (VCCS)
- current-controlled voltage source (CCVS)
- current-controlled current source (CCCS)

Pulse voltage and current sources. The pulse voltage and current sources have the following syntax.

```
Vxxx n+ n− pulse <v1 v2 td tr tf pw per>
Ixxx n+ n− pulse <v1 v2 td tr tf pw per>
```

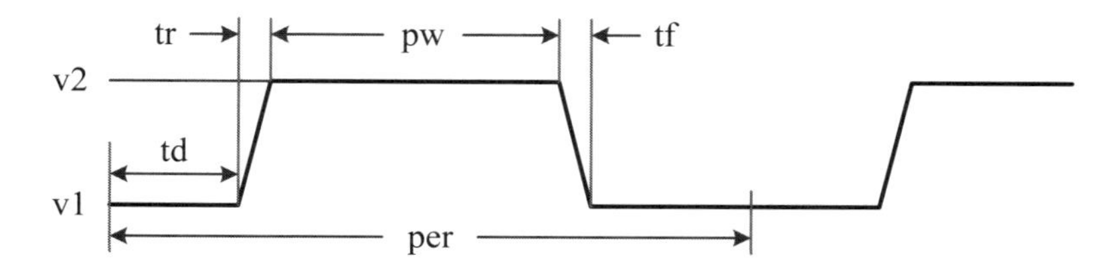

Figure 2.30: The specifications of a pulse source.

The keyword "`pulse`" designates a pulse to be specified. The other parameters characterize the pulse source. Parameters `tr` and `tf` specify rise and fall times, respectively; parameter `td` specifies delay time. The pulse width and period are separately specified by the other two parameters `pw` and `per`. An illustration of the relationship of various parameters associated with voltage and current pulse sources is shown in Figure 2.30.

Resistors, capacitors, and inductors. The three basic passive elements, resistor, capacitor, and inductor, can be specified as follows:

```
Rxxx n+ n- resistance_value
Cxxx n+ n- capacitance_value
Lxxx n+ n- inductance_value
```

where `Rxxx`, `Cxxx`, and `Lxxx` are the names of resistor, capacitor, and inductor, respectively. `n+` specifies the positive node while `n-` specifies the negative node. The values of the resistor, capacitor, and inductor are specified by `resistance_value`, `capacitance_value`, and `inductance_value`, respectively.

MOS transistor. A MOS transistor is a four-terminal element. It uses the following syntax:

```
Mxxx D G S B mname L = value W = value
+ AD = value AS = value PD = value PS = value
```

where `Mxxx` represents a MOS transistor element, including pMOS and nMOS transistors. The common practice uses the second character to differentiate a pMOS from an nMOS transistor. For instance, `Mnxx` denotes an nMOS transistor while `Mpxx` denotes a pMOS transistor. The `+` in the first column of the second line denotes that this line is a continuation of its previous line. The other fields of the syntax are listed in Table 2.8.

The area and perimeter of the source/drain of a MOS transistor are determined by the actual layout of the MOS transistor. Nevertheless, with the minimum allowable layout design rules, the area and perimeter of the source/drain of a MOS transistor can be estimated without carrying out the actual layout. According to the circuitry connections in practical circuits, three different situations are obtained: isolated contacted source/drain, contacted source/drain sharing, and uncontacted source/drain sharing. Their areas and perimeters of source/drain are shown in Figure 2.31.

■ **Example 2-13: (Specification of a MOS transistor.)** Suppose that the channel length and width of an nMOS transistor are 2λ and 8λ, respectively. Calculate the areas and perimeters of source and drain, and then specify the nMOS transistor in the SPICE data format.

Table 2.8: The specifications of a MOS transistor in SPICE.

Symbol	Description
Mxxx	The instance name of the MOSFET.
D	The drain node name.
G	The gate node name.
S	The source node name.
B	The bulk (substrate) node name.
mname	The model name of the pMOS or nMOS to be used.
L	The length of the transistor.
W	The width of the transistor.
AD	The bottom area of the drain diffusion region.
PD	The perimeter of the drain diffusion region.
AS	The bottom area of the source diffusion region.
PS	The perimeter of the source diffusion region.

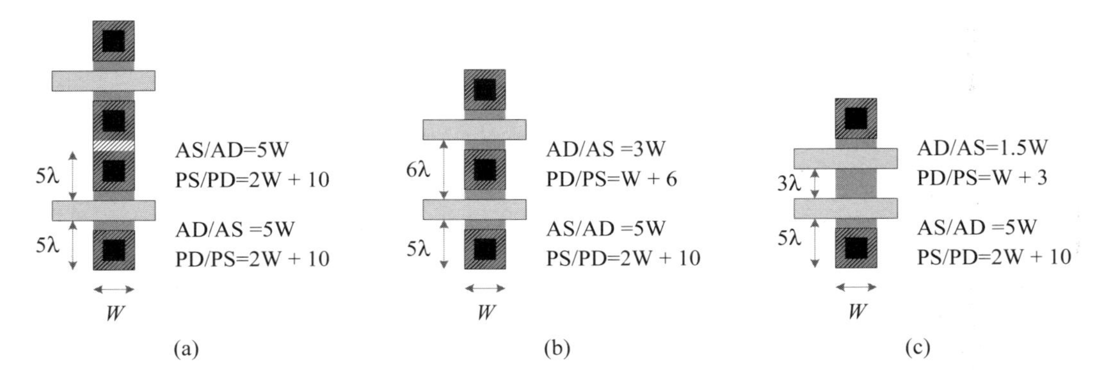

Figure 2.31: The specifications of the areas and perimeters of connected MOS transistors: (a) isolated contacted source/drain; (b) contacted source/drain sharing; (c) uncontacted source/drain sharing.

Solution: Referring to Figure 2.31(a), the areas of source and drain are the same and equal to $AS = AD = 40\lambda^2$; the perimeters of source and drain are $PS = PD = 26\lambda$. The resulting SPICE specification is as follows:

```
.opt scale = 0.09u * set value of lambda
Mp Vout Vin Vdd Vdd pmos L = 2 W = 8
+ AS = 40 AD = 40 PS = 26 PD = 26
```

∎

2.5.1.2 SPICE Analysis

SPICE supports many analysis modes. In digital circuits, only direct current (DC or dc) and transient (tran) analyses are fundamentally important. Hence, we only look at these two analyses.

DC analysis. DC analysis is used to measure the value of one variable while another is changing or sweeping from one value to another. The DC analysis has the following syntax.

```
.dc var1 start stop step <var2 start2 stop2 step2>
.dc var1 start stop step <sweep var2 type no_pts start2 stop2>
```

All DC statements must begin with the directive "`.dc`." At least one variable must be specified when a DC analysis is performed. `var1` specifies the variable to be swept. The initial and final values are specified by parameters `start` and `stop`, respectively. The amount of increment used to sweep the variable is specified by the `step` parameter.

Everything in <> is optional. In the second form of DC analysis, `no_pts` represents the number of points per decade or per octave, or just the number of points, specified by parameter `type`. Parameter `type` may be one of DEC (decade), OCT (octave), LIN (linear), and POI (list of points).

Transient analysis. Transient analysis is used to measure the value of a variable as it changes with time. In this analysis, we have to specify the interval and the time step. The interval specifies the duration over which the simulation will be carried out; the time step is used for the plotting purpose.

```
.tran tstep tstop <tstep2 tstop2 ... > <start=val><uic>
+ <sweep var tstart tstop tstep>
.tran tstep tstop <tstep2 tstop2 ... > <start=val><uic>
+ <sweep var type no_pts tstart tstop>
```

All transient statements must begin with the directive "`.tran`." The transient analysis starts at simulation time 0 and stops at `tstop` with an increment specified by parameter `tstep`.

Parameter `var` is the name of an independent voltage or current source, any element or model parameter, or the key name TEMP (a temperature sweep). Parameters `type` and `no_pts` have the same meaning and usage as DC analysis.

2.5.1.3 Output Statements

Printing and plotting statements are used to observe the results of transient or DC analysis. The `.plot` statement displays the results in graphical form, whereas `.print` statement displays them in tabular form.

```
.plot  analysis_type ov1 <ov2 ... ov32>
.print analysis_type ov1 <ov2 ... ov32>
.probe analysis_type ov1 <ov2 ... ov32>
```

All plot statements must begin with the directive "`.plot`" and all print statements must begin with the directive "`.print`." The `.probe` statement saves output variables into the interface and graph data files. The parameter `analysis_type` specifies the type of analysis (`dc` or `trans`) to be displayed. `ov1` to `ov32` are output variables. At least one output variable must be specified. In HSPICE, the statement `.option probe` needs to be used when only the output variables are to be saved; otherwise, all voltages and supply currents in addition to output variables are saved.

2.5.1.4 .alter Statement

The `.alter` statement specifies that a simulation is to be rerun one more time using different circuit topologies, models, library components, elements, and parameter values, among others.

In the following example, the `.alter` statement is used to study the *I-V* characteristics of an nMOS transistor of different design corners.

■ **Example 2-14: (A study of design corners.)** In this example, the current-voltage characteristics of an nMOS transistor of different design corners are studied. To this end, we first print out the *I-V* characteristics with a typical library. Then, we use `.alter` statements to switch to slow-n and slow-p (SS) and fast-n and fast-p (FF) libraries, respectively, and carry out the DC analysis again.

```
nMOS I–V Characteristics —— 0.25 um process
******** Parameters and model *********
.lib '..\cmos25.txt' TT
.opt scale = 0.125u
******** Circuit description **********
MN1 2 1 0 0 nmos L = 2 W = 4  VD1 2 0 dc
VG1 1 0 dc
******** analysis statement *******
.dc VD1 0 2.5 0.001 sweep VG1 0.0 2.5 0.2
******** Output statements *******
.print id1 = i(mn1)
.option post
.alter * slow-n slow-p
.lib '..\cmos25.txt' SS
.alter * fast-n fast-p
.lib '..\cmos25.txt' FF
.end
```

■

2.5.1.5 .subckt Statement The `.subckt` statement is used to define a circuit that will be invoked several times with the same or different parameter values. The syntax of the `.subckt` statement is as follows:

```
.subckt subname n1 <n2 n3...> <param = val...>
  ..... *** circuit description
.ends <subname>
```

Node `n1` is the node for external reference; it cannot be the ground node (0). Any element nodes appearing between `.subckt` and `.ends` statements but not included in this list `<n2 n3...>` are strictly local, with the following exceptions: ground node (0), nodes assigned using the `.global` statement, and nodes assigned using `BULK = node` in MOS transistor or BJT models.

Parameter `param` can only be used within the `.subckt` statements or `.subckt` calls; it is overridden by an assignment in a `.subckt` call or by values set in the `.param` statement.

A circuit defined with the `.subckt` statement can be invoked at any time by issuing a `.subckt` call using the following form:

```
Xyyyy n1 <n2 n3...> subname <param = val...> <M=val>
```

where `Xyyyy` is the instance name, `<n2 n3...>` is the node list, `subname` is the subcircuit name, `<param = val...>` is the subcircuit parameters, and `<M = val>` is the multiplier.

An illustration of how to define and invoke a subcircuit is demonstrated in the following example.

■ **Example 2-15: (A simple subcircuit example.)** In this example, we first define an inverter subcircuit and then invoke it twice with different multipliers, one with multiplier 1 and the other with multiplier 4. Finally, we carry out a DC analysis to study their voltage-transfer characteristics.

```
A subcircuit example
```

```
********** Parameters and model **********
.lib '..\cmos18.txt' cmos
.param SUPPLY = 1.8V
.param T = 4
.option scale = 0.09u
********** Subcircuit definition **********
.global vdd gnd
.subckt inv x f N = 2*M P = 4*M
Mn f x gnd gnd NMOS L = 2  W = 'N'
Mp f x vdd vdd PMOS L = 2  W = 'P'
.ends
********* Circuit description ***********
Vdd vdd gnd 'SUPPLY'
Vin a   gnd
X1  a b inv       * first stage
X2  a c inv M = 'T' * output stage
********* Analysis statement **********
.dc Vin 0   'Supply' 'Supply/100'
********* Output statements *************
.print V(b) V(c)
.option post
.end
```

■

2.5.1.6 .measure (or .meas for short) Statement The `.measure` statement in HSPICE is a useful tool that can be employed to measure rise and fall times as well as propagation delays. To use this statement, we need to set both trigger and target points. The trigger point is the reference point while the target point is the desired point to be measured relative to the reference point. The syntax of the `.measure` statement is as follows:

```
.measure <dc|ac|tran> result_var
+ trig trig_var val=trig_value
+ <td=delay><cross=#crsooings><rise=#rises><fall=#falls>
+ targ targ_var val=targ_value
+ <td=delay><cross=#crsooings><rise=#rises><fall=#falls|last>
```

where `result_var` denotes the variable name to be measured, `trig` and `targ` separately identify the beginning of trigger and target specifications, `trig_var` and `trig_value` specify the reference variable and value, `targ_var` and `targ_value` specify target variable and value to be measured, and `#crsooings`, `#rises` as well as `#falls`, indicate the number of occurrences of a `cross`, `fall`, and `rise` event causing a measurement to be performed.

■ Example 2-16: (The use of .measure statement.) In this example, we use `.measure` statements to measure both propagation delays, t_{pdr} (t_{pLH}) and t_{pdf} (t_{pHL}), and their average value t_{pd}. For measuring t_{pd}, the `trig_value` of both `trig` substatements are set to $V_{DD}/2$. The number of rising and falling times, `rise` and `fall`, is set to 1.

```
The .measure statement study —— 0.18-um process
```

```
******** Parameters and model *********
.lib   '..\tsmc18.txt' cmos
.param Supply=1.8V   * Set value of Vdd
.opt   scale=0.09u
******** Circuit description **********
MN  Vout Vin Gnd Gnd nmos L=2 W=2 AD=10 AS=10 PD=14 PS=14
MP  Vout Vin Vdd Vdd pmos L=2 W=4 AD=20 AS=20 PD=18 PS=18
Vdd Vdd  Gnd 'Supply'
C   Vout Gnd 1pF
Vin Vin  Gnd pulse 0 'Supply' 0ps 100ps 100ps 100ns 200ns
******** analysis statement *******
.tran 10ps 400ns
******** Output statements *******
.measure tpdr  * rising propagation delay
+ trig V(Vin)  val='Supply/2' fall=1
+ targ V(Vout) val='Supply/2' rise=1
.measure tpdf  * falling propagation delay
+ trig V(Vin)  val='Supply/2' rise=1
+ targ V(Vout) val='Supply/2' fall=1
.measure tpd param='(tpdr+tpdf)/2'
.probe V(Vin) V(Vout)
.option post
.end
```

The outputs of the `measure` statement are given in the following:

```
tpdr = 8.9684E-09  targ = 1.0912E-07  trig = 1.0015E-07
tpdf = 5.3277E-09  targ = 5.3777E-09  trig = 5.0000E-11
tpd  = 7.1480E-09
```

■

2.5.2 Diode Model

Recall that the depletion region of a pn diode functions as a capacitor with the capacitance expressed as the following equation:

$$C_j = C_{j0}\left[1 - \frac{V_D}{\phi_0}\right]^{-m} \tag{2.102}$$

Combining this equation with the ideal-diode equation

$$I_D = I_s\left[\exp\left(\frac{V_D}{nV_t}\right) - 1\right] \tag{2.103}$$

the charge in the depletion region of a diode can then be represented as

$$Q_D = \tau I_s\left[\exp\left(\frac{V_D}{nV_t}\right) - 1\right] + C_{j0}\int_0^{V_D}\left(1 - \frac{V_D}{\phi_0}\right)^{-m} dV \tag{2.104}$$

where the first term is the charge induced under the forward bias while the second term is the charge induced under the reverse bias. The parameter τ is the *carrier transit time* (life time).

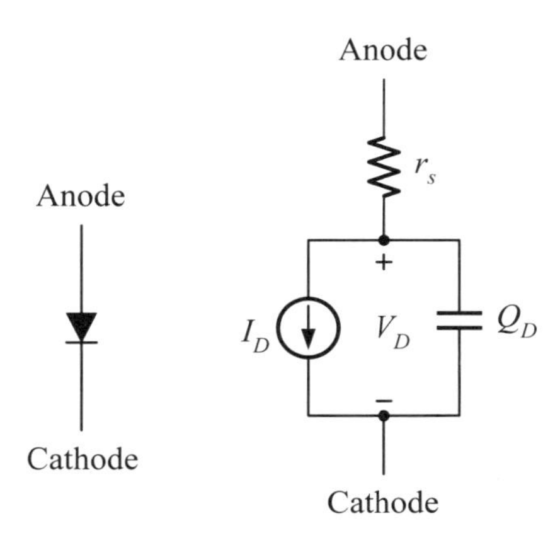

Figure 2.32: SPICE model of a typical *pn*-junction diode.

Table 2.9: SPICE model parameters of a typical *pn*-junction diode.

Name	SPICE	Meaning	Name	SPICE	Meaning
I_s	IS	Saturation current	C_{j0}	CJ0	Zero-bias junction capacitance
r_s	RS	Series resistance	ϕ_0	VJ	Built-in potential
n	N	Emission coefficient	m	M	Grading coefficient
V_{bd}	BV	Breakdown voltage	τ	TT	Carrier transit time
I_{bd}	IBV	Current flowing during V_{bd}			

The capacitance is given by taking the derivative of Q_D with respect to V_D and is as follows:

$$C_D = \frac{dQ_D}{dV_D} = \frac{\tau I_s}{nV_t} \exp\left(\frac{V_D}{nV_t}\right) + C_{j0}\frac{1}{(1 - V_D/\phi_0)^m} \tag{2.105}$$

where the first term is the diffusion capacitance, contributed by the storage charge caused by minority carriers of both the n- and p-side of the junction under forward bias. The second term is the junction capacitance under the reverse bias.

The complete SPICE model of a typical diode is depicted in Figure 2.32. The series resistance r_s results from the finite resistance of the semiconductor used to fabricate the diode and the contact resistance between the metal terminals and semiconductor. The SPICE diode model parameters are listed in Table 2.9.

2.5.3 MOS Transistor Models

We first take a look at basic SPICE models for MOS transistors. Then, we briefly examine the more advanced BSIM (Berkeley Short-Channel IGFET Model) model.

2.5.3.1 Basic MOS Transistor Models Recall that SPICE is a simulation tool but not a design tool of a circuit. It is only accurate as the model for the transistor. Hence, different models are often required for different processes. Most versions of SPICE support three basic models. These are named as levels 1 to 3. We briefly describe each of these.

1. Level-1 (Schichman-Hodges model) model is based on the GCA model and is precisely down to 4 μm. It is only for hand calculation and is described by two quadratic current-voltage equations.
2. Level-2 (Grove-Frohman model) model includes the subthreshold current and is a detailed analytical model.
3. Level-3 (Empirical model) model is a semiempirical model and is precisely down to 1 μm. It is usually used in PSPICE for modeling discrete devices. It was also widely used in 0.8-μm processes.

The level-1 model equations. The level-1 model is the simplest current-voltage description derived from the GCA model. The current-voltage equation for an nMOS transistor being operated in the linear region can be expressed as

$$I_{DS} = \frac{W_{eff}}{L_{eff}} \frac{\text{KP}}{2} \left[2(V_{GS} - V_T)V_{DS} - V_{DS}^2\right](1 + \text{LAMBDA} \times V_{DS})$$
$$\text{for } V_{GS} \geq V_T, V_{DS} \leq V_{GS} - V_T \tag{2.106}$$

The current-voltage equation for an nMOS transistor being operated in the saturation region is as follows:

$$I_{DS} = \frac{W_{eff}}{L_{eff}} \frac{\text{KP}}{2} (V_{GS} - V_T)^2 (1 + \text{LAMBDA} \times V_{DS})$$
$$\text{for } V_{GS} \geq V_T, V_{DS} \geq V_{GS} - V_T \tag{2.107}$$

where the LAMBDA is the channel-length modulation coefficient and KP is the process transconductance. If the KP is not defined in the `.model` statement, it is computed by using the following equation:

$$\text{KP} = \text{UO} \times C_{ox} \tag{2.108}$$

The parameter UO is the surface mobility and C_{ox} is the gate capacitance per unit area. The C_{ox} is defined to be

$$C_{ox} = \frac{\varepsilon_{ox}}{\text{TOX}} \tag{2.109}$$

where ε_{ox} is the permittivity of silicon dioxide and TOX is the thin-oxide thickness.

The threshold voltage of an nMOS transistor is modeled as follows:

$$V_T = \text{VTO} + \text{GAMMA}(\sqrt{\text{PHI} - V_{BS}} - \sqrt{\text{PHI}}) \tag{2.110}$$

where the built-in potential PHI is defined to be

$$\text{PHI} = 2 \times \frac{kT}{e} \ln\left(\frac{\text{NSUB}}{n_i}\right) \tag{2.111}$$

The body-effect coefficient GAMMA is given as

$$\text{GAMMA} = \frac{\sqrt{2e\varepsilon_{si} \times \text{NSUB}}}{C_{ox}} \tag{2.112}$$

where e is the electron charge, ε_{si} is the permittivity of silicon, and NSUB is the substrate impurity concentration.

For the level-1 model, five parameters, KP, VTO, GAMMA, PHI, and LAMBDA, completely characterize the transistor. These parameters can be specified directly in the `.model` statement. In general, one should use the values supplied by the device or process vendors rather than try to derive them from some other parameters.

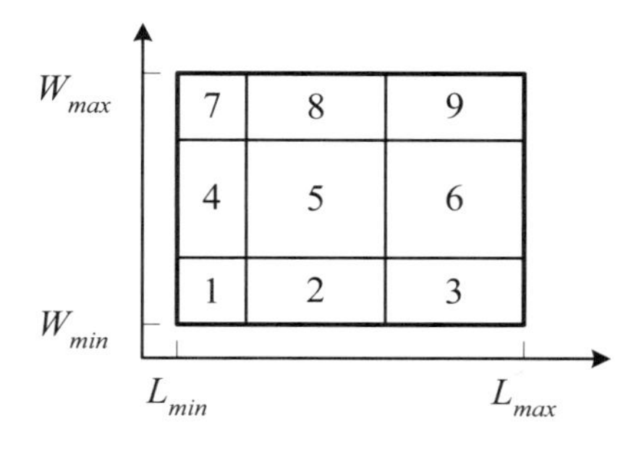

Figure 2.33: The concept of a binning process.

2.5.3.2 BSIM BSIM is the acronym of the *Berkeley Short-Channel IGFET Model* and was developed at UC Berkeley in the 1980s and 1990s. It was first targeted to the submicron process and widely used in the processes of feature sizes of 0.5 μm and 0.6 μm. BSIM has also evolved with the reduction of feature sizes. To precisely capture the key features of the DC and AC behaviors of deep submicron MOS transistors, the concept of binning process is introduced in BSIM3 to subdivide the model parameters into several bins, with each containing only a small area corresponding to a small range of channel length and width. The actual number of bins varies from foundry to foundry, even from process to process. Figure 2.33 shows a typical *binning process* with nine bins, with each having minimum and maximum channel lengths and widths. In the IC industry, many widely used models originating from BSIM are as follows:

1. Level 13 (BSIM) model is the first Berkeley short-channel IGFET model.
2. Level 28 (Modified BSIM2) model was used in 0.5-μm and 0.6-μm processes.
3. Level 49 (BSIM3) model is capable of modeling the features size down to 0.1 μm. It is typically used in 0.35-μm, 0.25-μm, and 0.18-μm processes.
4. Level 54 (BSIM4) model is capable of modeling the features size below 0.13 μm.

Since the BSIM model is too complex to present in this book, in what follows we only briefly describe the threshold voltage and mobility models. The interested reader can consult the BSIM3/4 manual for details.

Threshold voltage. BSIM3 models the subthreshold voltage by using the following equation:

$$V_T = \text{VTHO} + \text{K1}(\sqrt{\text{PHI} - V_{BS}} - \sqrt{\text{PHI}}) - \text{K2} \times V_{BS} \tag{2.113}$$

where PHI is defined as follows:

$$\text{PHI} = 2 \times \frac{kT}{e} \ln\left(\frac{\text{NCH}}{n_i}\right) \tag{2.114}$$

The parameter NCH denotes the surface doping concentration due to the threshold implant. Constants K1 and K2 are a function of PHI and two other parameters related to NCH and NSUB, respectively. They can be expressed as follows:

$$\text{K1} = f(\gamma_2, \text{PHI}) \qquad \text{and} \qquad \text{K2} = f(\gamma_1, \gamma_2, \text{PHI}) \tag{2.115}$$

The two constants, γ_2 and γ_2, are separately defined as

$$\gamma_1 = \frac{\sqrt{2e\varepsilon_{si} \times \text{NCH}}}{C_{ox}} \tag{2.116}$$

Table 2.10: The design parameters for typical 0.18- and 0.13-μm processes.

Parameter	Symbol	0.18 μm nMOS	0.18 μm pMOS	0.13 μm nMOS	0.13 μm pMOS	Unit
Supply voltage	V_{DD}	1.8	1.8	1.2	1.2	V
Oxide thickness	t_{ox}	4.1	4.1	3.1	3.1	nm
Oxide capacitance	C_{ox}	1.33	1.33	1.08	1.08	μF/cm^2
Threshold voltage	V_{T0}	0.37	-0.39	0.35	-0.35	V
Body-effect coefficient	γ	0.3	0.3	0.2	0.2	V$^{0.5}$
Fermi potential	$2\lvert\phi_F\rvert$	0.84	0.84	0.88	0.88	V
Junction capacitance	C_{j0}	0.95	1.17	0.95	1.15	fF/μm^2
Built-in potential	ϕ_B	0.8	0.86	0.98	0.8	V
Grading coefficient	m	0.37	0.42	0.40	0.44	—
Nominal mobility	μ_0	292	112	430	100	cm^2/V-s
Effective mobility	μ_{eff}	287	88	298	97	cm^2/V-s
Saturation electric field	E_{sat}	1.2×10^5	2.5×10^5	9.5×10^4	4.2×10^5	V/cm
Equivalent on-resistance	R_{eq}	8	22	18	37	kΩ/□
Saturation velocity	v_{sat}	8×10^6	8×10^6	1.5×10^7	1.5×10^7	cm/s

$$\gamma_2 = \frac{\sqrt{2e\varepsilon_{si} \times \text{NSUB}}}{C_{ox}} \tag{2.117}$$

For deep submicron devices, the subthreshold voltage is also affected by both L_{eff} and W_{eff}. The values of L_{eff} and W_{eff} are in turn determined by many effects, including short-channel effect (SCE), reverse short-channel effect (RSCE), and narrow-channel effect (NCE). Recall that the SCE results in the reduction of the threshold voltage while the RSCE causes the increase of the threshold voltage. The total short-channel effect is the sum of both SCE and RSCE and is called ΔV_{SCE}. The NCE results in the increase of the threshold voltage and is accounted for by a factor of ΔV_{NCE}. The DIBL effect on lowering the barrier for current flow by the drain-source voltage is accounted for as ηV_{DS}. By combining the above three factors with Equation (2.113), the threshold voltage equation becomes as

$$\begin{aligned} V_T &= \text{VTHO} + \text{K1}(\sqrt{\text{PHI} - V_{BS}} - \sqrt{\text{PHI}}) - \text{K2} \times V_{BS} \\ &\quad -\eta V_{DS} - \Delta V_{SCE} + \Delta V_{NCE} \end{aligned} \tag{2.118}$$

Mobility model. In addition to the nominal mobility UO, three parameters are used to account for the effect of vertical field. These three parameters are UA, UB, and UC. The resulting mobility model is as follows:

$$\mu_v = \frac{\text{UO}}{1 + (\text{UA} + \text{UC} + V_{BS})\left(\dfrac{V_{GS} - V_T}{t_{ox}}\right) + \text{UB}\left(\dfrac{V_{GS} - V_T}{t_{ox}}\right)^2} \tag{2.119}$$

The mobility μ_v is related to the saturation velocity VSAT by a proportional factor called the *saturation electric field* E_{sat} by the following equation:

$$E_{sat} = \frac{2\text{VSAT}}{\mu_v} \tag{2.120}$$

Table 2.10 lists useful design parameters of typical 0.18- and 0.13-μm processes.

2.6 Summary

In this chapter, we dealt with semiconductor fundamentals, the *pn*-junctions, MOS systems, MOS transistors, as well as SPICE models for *pn* junction and MOS transistors. The fundamental features of semiconductors were addressed. The intrinsic and extrinsic semiconductors coupled with the Fermi-energy level, work function, and the relationships between impurity concentrations and Fermi-energy levels of both *p*-type and *n*-type materials were also introduced briefly.

A MOS transistor, regardless of the n or p type, is composed of three major kinds of components: *pn*-junctions, metal-semiconductor junctions, and a metal-oxide-semiconductor (MOS) system. There are two pn^+/p^+n junctions for each nMOS/pMOS transistor and an extra *pn* junction exists between the *n*-well and *p*-type substrate. To connect an nMOS or a pMOS transistor with others, metal wires are contacted with n^+ or p^+ diffusions regions. Hence, metal-semiconductor junctions are also formed parts of a MOS transistor.

The heart of a MOS transistor is the MOS system. Based on the MOS system, a MOS transistor is constructed by adding two n^+/p^+ regions on each side of a MOS system to serve as source and drain regions. The basic operations of MOS transistors (MOSFET) and their current-voltage (I-V) characteristics were examined in detail. After this, the scaling theory of CMOS processes was explored briefly.

The advanced features of MOS transistors, including nonideal features, threshold voltage effects, and leakage currents, were dealt with in great detail. The temperature effects on the drain currents of both nMOS and pMOS transistors and limitations of MOS transistors followed.

We concluded this chapter with the introduction of SPICE and looked at how to model diodes and MOS transistors along with some examples.

References

1. K. Chen, C. Hu, P. Fang, M. R. Lin, and D. L. Wollesen, "Predicting CMOS speed with gate oxide and voltage scaling and interconnect loading effects," *IEEE Trans. on Electron Devices,* Vol. 44, No. 11, pp., 1951–1957, November 1997.
2. R. Dennard et al.,"Design of ion-implanted MOSFET's with very small physical dimensions," *IEEE J. of Solid State Circuits,* Vol. 9, No. 5, pp. 256–268, October 1974.
3. Y. Goto, "A triple gate oxide CMOS technology using fluorine implant for system-on-a-chip," *The 2000 Symposium VLSI Technology Digest of Technical Papers,* pp. 148–149, 2000.
4. P. R. Gray, P. J. Hurst, S. H. Lewis, and R. G. Meyer, *Analysis and Design of Analog Integrated Circuits,* 4th ed. New York: John Wiley & Sons, 2001.
5. S. O. Kasap, *Principles of Electronic Materials and Devices,* 2nd ed. New York: McGraw-Hill Books, 2002.
6. W. Liu, *MOSFET Models for SPICE Simulation: Including BSIM3v3 and BSIM4.* New York: John Wiley & Sons, 2001.
7. W. Liu et al., *BSIM3v3.3 MOSFET Model Users' Manual.* Berkeley: University of California, 2005.
8. G. S. May and C. J. Spanos, *Fundamentals of Semiconductor Manufacturing and Process Control.* New York: John Wiley & Sons, 2006.

9. R. McGowen et al., "Power and temperature control on a 90-nm Itanium family processor," *IEEE J. of Solid-State Circuits,* Vol. 41, No. 1, pp. 229–237, January 2006.

10. T. H. Morshed et al., *BSIM4.6.4 MOSFET Model Users' Manual.* Berkeley: University of California, 2009.

11. R. S. Muller and T. I. Kamins, *Device Electronics for Integrated Circuits,* 3rd ed. New York: John Wiley & Sons, 2003.

12. B. Murphy, "Unified field-effect transistor theory including velocity saturation," *IEEE J. of Solid-State Circuits,* Vol. 15, No. 3, pp. 325–328, June 1980.

13. D. A. Neamen, *Semiconductor Physics and Devices: Basic Principles,* 3rd ed. New York: McGraw-Hill Books, 2003.

14. R. F. Pierret, *Semiconductor Device Fundamentals.* Reading, MA: Addison-Wesley, 1996.

15. K. Roy, S. Mukhopadhyay, and H. Mah moodi-Meimand, "Leakage current mechanisms and leakage reduction techniques in deep-submicrometer CMOS circuits," *Proc. IEEE,* Vol. 91, No. 2, pp. 305–327, February 2003.

16. T. Sakurai, K. Nogami, M. Kakumu, and T. Iizuka, "Hot-carrier generation in submicrometer VLSI environment," *IEEE J. of Solid-State Circuits,* Vol. 21, No. 1, pp. 187–192, February 1986.

17. T. Sakurai and R. Newton, "Alpha-power law MOSFET model and its applications to CMOS inverter delay and other formulas," *IEEE J. of Solid-State Circuits,* Vol. 25, No. 2, pp. 584–594, April 1990.

18. B. Sheu, D. Scharfetter, P. Ko, and M. Jeng, "BSIM: Berkeley short-channel IGFET model for MOS transistors," *IEEE J. of Solid-State Circuits,* Vol. 22, No. 4, pp. 558–566, August 1987.

19. H. Shichman and D. Hodges, "Modeling and simulation of insulated-gate field-effect transistor switching circuits," *IEEE J. of Solid-State Circuits,* Vol. 3, No. 3, pp. 285–289, September 1968.

20. C. G. Sodini, P. K. Ko, and J. L. Moll, "The effect of high fields on MOS device and circuit performance," *IEEE Trans. on Electronic Devices,* Vol. 31, No. 10, pp. 1386–1393, October 1984.

21. H. Soeleman, K. Roy, and B. Paul, "Robust subthreshold logic for ultra-low power operation," *IEEE Trans. on VLSI Systems,* Vol. 9, No. 1, pp. 90–99, February 2001.

22. Y. Taur and T. H. Ning, *Fundamentals of Modern VLSI Devices.* New York: Cambridge University Press, 1998.

23. L. Wei, Z. Chen, M. Johnson, K. Roy, and V. De, "Design and optimization of low voltage high performance dual threshold CMOS circuits," *Proceedings in Design Automation Conf.,* pp. 489–494, 1998.

Problems

2-1. Calculate the intrinsic carrier concentration of silicon at (a) $T = 350$ K and (b) $T = 650$ K.

2-2. Consider a p-type silicon doped with boron impurity and assume that the concentration of boron impurity is 8.5×10^{15} cm^{-3}.

(a) Calculate the Fermi potential of the p-type material.

(b) Calculate the thermal-equilibrium electron concentration at $T = 300$ K.

2-3. Consider an n-type silicon doped with antimony impurity and assume that at $T = 300$ K. The Fermi-energy level is 0.35 eV below the conduction band.

(a) Calculate the concentration of antimony impurity, assuming that all impurity atoms are ionized.

(b) Calculate the thermal-equilibrium hole concentration.

2-4. Assume that a p-type material with an impurity concentration of $N_a = 5 \times 10^{18}/\text{cm}^3$ and an n-type material with an impurity concentration of $N_d = 8.4 \times 10^{15}/\text{cm}^3$ are combined into a pn junction. Calculate the built-in potential of the pn junction at $T = 300$ K.

2-5. Consider a pn junction with acceptor and donor concentrations of $N_a = 1.2 \times 10^{18}/\text{cm}^3$ and $N_d = 2.6 \times 10^{15}/\text{cm}^3$, respectively. Calculate the depletion-region width at $T = 300$ K.

2-6. Represent the junction capacitance versus the applied diode voltage V. The diode is formed between the n-well and p-type substrate. The area and depth of the n-well are 80×60 μm^2 and 2 μm, respectively, as shown in Figure 2.8(a). Suppose that $N_a = 4.5 \times 10^{16}$ cm^{-3} and $N_d = 1.6 \times 10^{17}$ cm^{-3}. The measured zero-bias junction capacitance is 85 aF/μm^2 and $m = 0.45$.

2-7. For an abrupt n^+p junction diode with $N_d = 4.5 \times 10^{18}$ cm^{-3} and $N_a = 5.6 \times 10^{15}$ cm^{-3}, assume that the junction area is 50×50 μm^2.

(a) Find ϕ_0 and C_{j0} at $T = 300$ K.

(b) Find C_j for $V = -1.8$ V.

(c) Find C_{eq} for $V_1 = -1.8$ V and $V_2 = 0$ V.

2-8. Supposing that the p-type substrate has an impurity concentration of $N_a = 6.5 \times 10^{15}$ cm^{-3} and using $n_i = 1.45 \times 10^{10}$ cm^{-3}, find the maximum depletion-layer thickness $x_{d(max)}$ at $T = 300$ K.

2-9. Assume that the substrate is an n-type with donor concentration of $N_d = 2 \times 10^{16}$ cm^{-3} and the temperature is 300 K. ϕ_G and ϕ_{Si} are the work function (Fermi-energy level) of a gate and silicon substrate, respectively. Referring to Figure 2.1, complete Table 2.11 if the electron affinity is 4.05 eV and bandgap energy $E_g = 1.12$ eV.

Table 2.11: The Table for Problem 2-9.

	Aluminum	p^+ polysilicon	n^+ polysilicon
ϕ_G (V)	4.10	5.17	4.05
ϕ_{Si} (V)			
V_{GS} (V)			

2-10. For an aluminum-gate MOS system, the gate-oxide thickness t_{ox} is 6.5 nm and $\phi_{GS} = -0.88$ V. Assume that acceptor concentration N_a is 6.5×10^{15} cm^{-3} and Q_{ss} is 6.5×10^{10} cm^{-2}.

(a) Find the threshold voltage V_{Tn} at $T = 300$ K.

(b) Assuming that delta function is used, find the impurity density required to adjust the threshold voltage to 0.55 V.

2-11. Assume that the gate-oxide thickness t_{ox} is 6.5 nm and $\phi_{GS} = -0.88$ V. The impurity concentration of the substrate is $N_a = 6.5 \times 10^{15}$ cm^{-3}. If the body bias V_{SB} is 1.8 V, find the body-effect coefficient γ and the threshold voltage V_{Tn} under this body bias, assuming that the threshold voltage without body bias V_{T0n} is 0.55 V.

2-12. Assume that $t_{ox} = 3.2$ nm in a 0.13-μm process.

(a) Calculate the effective electron mobility as $V_{GS} = 1.2$ V and $V_{Tn} = 0.35$ V.

(b) Calculate the effective hole mobility as $V_{GS} = 1.2$ V and $V_{Tp} = -0.35$ V.

2-13. Using the parameters given in Table 2.12 and assuming that $2\phi_f = -0.8$ V, determine the device parameters, V_{T0}, k, γ, and λ.

Table 2.12: Parameters for Problem 2.13.

V_{GS}(V)	V_{DS}(V)	V_{SB}(V)	I_D(μA)
0.8	0.8	0	37
0.8	1.2	0	40
1.2	1.2	0	130
1.2	1.2	0.3	44
1.8	1.8	0.3	178

2-14. Using the parameters given in Table 2.13 and assuming that $2\phi_f = -0.8$ V, determine the device parameters, V_{T0}, k, γ, and λ.

Table 2.13: Parameters for Problem 2.14.

V_{GS}(V)	V_{DS}(V)	V_{SB}(V)	I_D(μA)
0.8	1.2	0	32
1.2	1.2	0	116
1.2	1.2	0.4	17
1.2	1.8	0	120

2-15. Show that the temperature dependence of the threshold voltage can be expressed as follows:

$$\frac{dV_{Tn}}{dT} = -\frac{1}{T}\left(\frac{E_g}{2e} - |\phi_{fp}|\right)\left(2 + \frac{\gamma}{\sqrt{2|\phi_{fp}|}}\right)$$

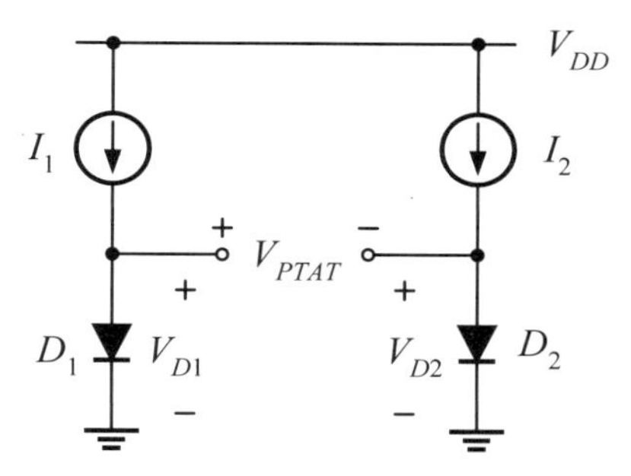

Figure 2.34: The circuit of PTAT.

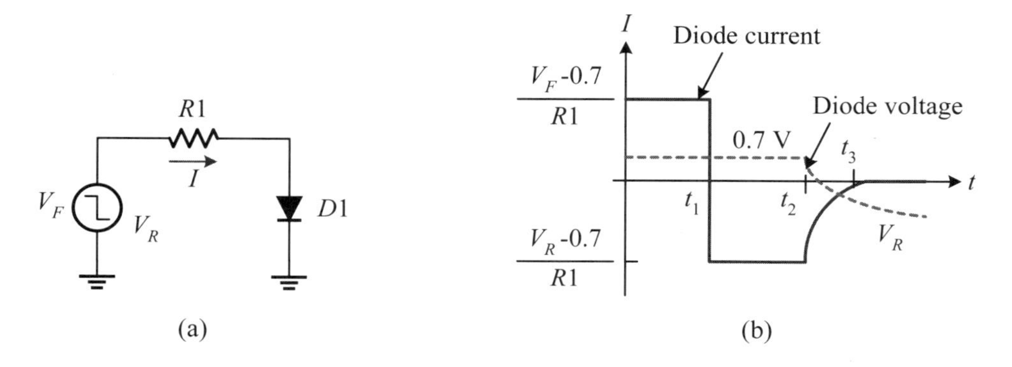

Figure 2.35: The (a) circuit of a *pn*-junction diode for studying (b) transition characteristic.

2-16. The well-defined temperature dependence of diode voltage can be used to develop a voltage directly proportional to absolute temperature (PTAT), referred to as the PTAT voltage or V_{PTAT}.

(a) Referring to Figure 2.34, show that

$$V_{PTAT} = V_t \ln\left(\frac{I_{D1}}{I_{D2}}\right)$$

(b) Calculate V_{PTAT} if I_1 is set to 10 times I_2.

2-17. In this problem, we would like to study the transient characteristics of a *pn*-junction diode. Referring to the circuit shown in Figure 2.35(a), use SPICE to verify the transition characteristic given in Figure 2.35(b). Assume that the diode parameters are as follows:

```
.model Diode D IS=1.0E−15 TT=10E−9 CJO=1E−12
        VJ=0.7 M=0.33
```

2-18. Referring to Figure 1.34(a), design a two-stage buffer. The first stage is a standard $1X$ inverter and the second stage is a $4X$ inverter. Assume that $L = 2\lambda$ and $W = 3\lambda$.

2-19. Referring to Figure 1.34(b), write a SPICE data file to verify the function of the two-input NAND gate. Assume that $L = 2\lambda$ and $W = 3\lambda$.

2-20. Referring to Figure 1.34(c), write a SPICE data file to verify the function of the two-input NOR gate. Assume that $L = 2\lambda$ and $W = 3\lambda$.

2-21. Referring to Figure 1.36(a), write a SPICE data file to verify the function of the 2-to-1 multiplexer. Assume that $L = 2\lambda$ and $W = 3\lambda$ are used for all transistors.

2-22. Referring to Figure 1.36(b), define a subcircuit of TG and write a SPICE data file to verify the function of the 2-to-1 multiplexer. Assume that $L = 2\lambda$ and $W = 3\lambda$ are used for all transistors.

2-23. Referring to Figure 1.37(a), write a SPICE data file to verify the function of the 1-to-2 demultiplexer. Assume that $L = 2\lambda$ and $W = 3\lambda$ are used for all transistors.

2-24. Referring to Figure 1.37(b), define a subcircuit of TG and write a SPICE data file to verify the function of the 1-to-2 demultiplexer. Assume that $L = 2\lambda$ and $W = 3\lambda$ are used for all transistors.

2-25. Referring to Figure 1.38(a), define a subcircuit of TG and write a SPICE data file to verify the function of the positive D-type latch. Assume that $L = 2\lambda$ and $W = 3\lambda$ are used for all transistors.

2-26. Referring to Figure 1.38(a), define a subcircuit of TG and write a SPICE data file to verify the function of the negative D-type latch. Assume that $L = 2\lambda$ and $W = 3\lambda$ are used for all transistors.

2-27. Referring to Figure 1.39, define a subcircuit of the D-type latch and write a SPICE data file to verify the function of the master-slave positive edge-triggered D-type flip-flop. Assume that $L = 2\lambda$ and $W = 3\lambda$ are used for all transistors.

2-28. Referring to Figure 1.39, define a subcircuit of the D-type latch and write a SPICE data file to verify the function of the master-slave negative edge-triggered D-type flip-flop. Assume that $L = 2\lambda$ and $W = 3\lambda$ are used for all transistors.

3

Fabrication of CMOS ICs

The essential feature of a semiconductor is that its types and resistivity or conductivity of designated areas can be modified by adding proper types and concentrations of dopant impurities. This means that the concentrations of dopant impurities—where to be added—and how to add are the major concerns when a semiconductor device is manufactured.

The fabricating process of a semiconductor device is basically around the following five basic unit processes: thermal oxidation, doping, photolithography, thin-film removal, and thin-film deposition. By combining basic unit processes with appropriate materials, a complementary metal-oxide-semiconductor (CMOS) integrated circuit can then be fabricated. A typical fabricating process may take as many as four to eight weeks and involve several hundred steps to complete.

After a fabricating process is completed, a series of postfabrication processes are carried out. These processes include a wafer test (or a *wafer probe*), die separation, packaging, and a final test as well as a burn-in test. To facilitate a complex system, multiple chips are often packed into a single package. The most common techniques include the multichip module (MCM), system-in-package (SiP), system-on-package (SoP), and system-on-wafer (SoW). The last three techniques are also referred to as 3-D packaging and the resulting integrated circuit (IC) is called a 3-D IC.

3.1 Basic Processes

The fabricating (or manufacturing) process of semiconductor devices is basically around the following five basic unit processes: *thermal oxidation*, *doping process*, *photolithography*, *thin-film removal*, and *thin-film deposition*. We deal with each of these in great detail.

3.1.1 Thermal Oxidation

Thermal oxidation (oxidation for short) is the process by which a layer of *silicon dioxide* (SiO_2) is formed on the surface of a silicon wafer with a specific thickness. Two widely used thermal oxidation processes are *dry oxidation* and *wet oxidation*. The major difference between these two processes is that the former uses pure oxygen (O_2) while the latter uses water vapor (H_2O) as the oxygen source. We address each of these briefly.

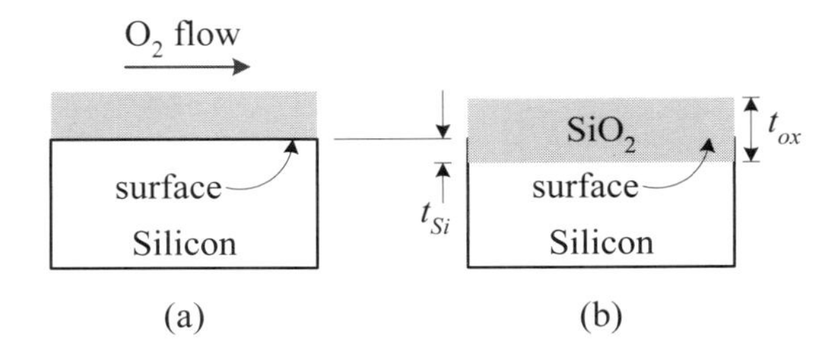

Figure 3.1: An illustration of the oxidation process: (a) before oxidation; (b) after oxidation.

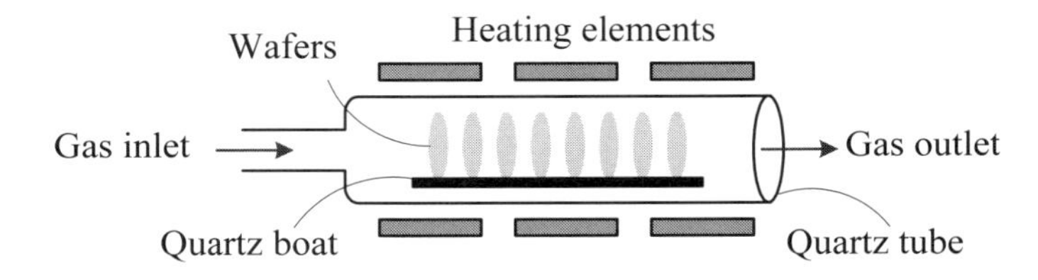

Figure 3.2: An illustration of the oxidation process using a tube furnace.

3.1.1.1 Dry Oxidation Dry oxidation is so named because the oxidizing atmosphere contains pure oxygen during the course of oxidation. In dry oxidation, the working temperature is between 900°C and 1200°C to achieve an acceptable growth rate. The reaction equation is as follows:

$$Si + O_2 \longrightarrow SiO_2 \tag{3.1}$$

3.1.1.2 Wet Oxidation Wet oxidation is so called because the water vapor is used as the oxidizing atmosphere during the oxidation process. In wet oxidation, the required temperature is usually less than that of dry oxidation and is between 900°C and 1000°C. The reaction equation is given as follows:

$$Si + 2H_2O \longrightarrow SiO_2 + 2H_2 \tag{3.2}$$

3.1.1.3 General Considerations The reaction between oxygen and silicon regardless of dry or wet oxidation is a recessed process. About 44% of the thickness of final silicon dioxide is contributed by the silicon surface, as depicted in Figure 3.1.

Although the silicon dioxide film obtained using wet oxidation is about 5 to 10 times thicker than that using dry oxidation at a given temperature and time, the oxidation process is a slow reaction; a thick oxide takes several hours to grow. As a result, it is often to process wafers in a batch, with a large number of wafers at a time.

An oxidation process can proceed by using either a *tube furnace* or a *rapid thermal processing* (RTP) tool. A tube furnace is generally composed of a quartz tube surrounded by many heating elements, as portrayed in Figure 3.2. Because the oxidation process is slow, the tube furnace is designed as a batch tool to process hundreds of wafers at the same time to provide an acceptable throughput.

The RTP tool is similar to the tube furnace with an exception that the thermal source is heating lamps, which are usually tungsten-halogen lamps, as shown in Figure 3.3. An RTP tool consists of a quartz tube surrounded by tungsten-halogen lamps.

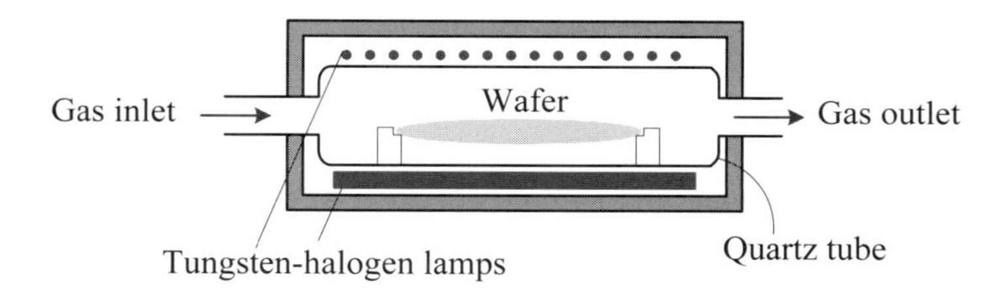

Figure 3.3: An illustration of the oxidation process using an RTP tool.

The wafer to be processed is placed within the quartz tube. The lamps are arranged in a way that allows the wafer to be heated uniformly by the infrared (IR) radiation. A useful feature of IR radiation is that it can heat up or cool down the chamber at a temperature rate between 100°C/sec to 200°C/sec. Because of this feature, it is named as an RTP tool.

The feature of an RTP tool is that it can heat the chamber up to the required temperature or lower the temperature to its original value in a few seconds. Consequently, the thermal budget of an RTP system is much smaller than that of a tube furnace, which needs several hours to process a batch of wafers. Because of this property, an RTP tool usually processes a single wafer at a time; that is, it is a single-wafer process in contrast to the batch process using a tube furnace. RTP is a system favorable for ultra-thin silicon dioxide layer growth, postimplantation annealing, and silicide annealing.

■ Review Questions

Q3-1. What are the two thermal oxidation processes?

Q3-2. What is the distinction between a tube furnace and an RTP tool?

Q3-3. What is the intention of the thermal oxidation process?

3.1.2 Doping Process

The astonishing feature of a semiconductor is that its conductivity can be controlled by doping proper impurities into it. A doping process is a procedure by which specific-type impurity atoms can be added into the selective area of the surface of a semiconductor material. In this subsection, we will deal with the two most widely used doping approaches in CMOS IC fabricating: *diffusion* and *ion implantation.*

3.1.2.1 Diffusion. Diffusion is a natural process in which materials move from high-concentration into low-concentration regions driven by the thermal motion of molecules. In semiconductor fabrications, the wafer is often placed into a high-temperature environment to speed up the diffusion process. The diffusion process usually proceeds in two phases: *predeposition* and *drive-in*, as shown in Figure 3.4(a). During the predeposition phase, a high-concentration dopant is deposited at the surface of a material; during the drive-in phase, the dopant deposited at the surface is driven into the bulk of the semiconductor through a high-temperature diffusion process. The most commonly used temperature is about 900 to 1200°C.

The needed drive-in time is determined by the required *junction depth.* The junction depth is defined as the point where the diffused dopant concentration is equal to the substrate concentration, as shown in Figure 3.4(b). Because both concentration

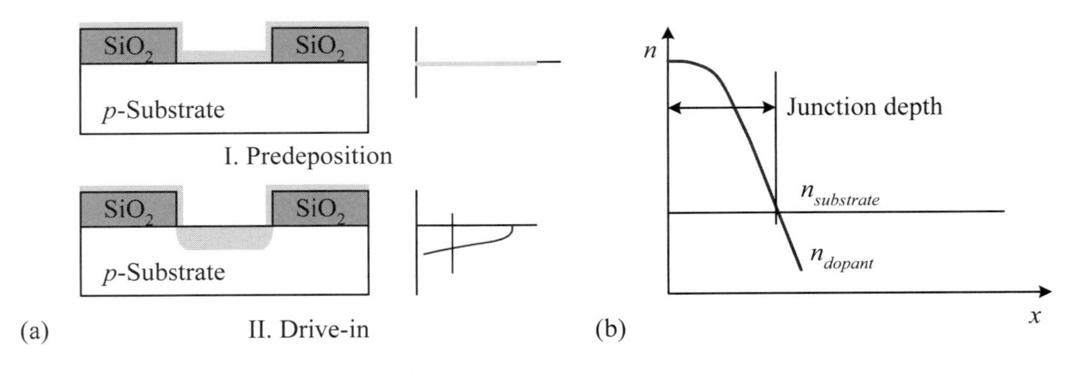

Figure 3.4: The concepts of the diffusion process: (a) diffusion process; (b) junction depth.

and junction depth are dependent on temperature and time when using a diffusion process, they cannot be controlled independently. As a result, in modern CMOS processes, the diffusion process is commonly replaced by the ion implantation.

Review Questions

Q3-4. What are the two doping approaches in CMOS IC fabrication?

Q3-5. What are the two basic steps of diffusion?

Q3-6. What is the junction depth?

3.1.2.2 Ion Implantation. Ion implantation is the process by which ions of a particular dopant (impurity) are accelerated by an electric field to a high velocity to bombard the wafer surface and physically lodge within the silicon material. The ion implantation has been widely used in CMOS processes to create wells, adjust threshold voltage of metal-oxide-semiconductor (MOS) transistors, and create source/drain n^+ and p^+ regions for MOS transistors, because of its better control on both dopant concentration and junction depth. Dopant concentration can be controlled by the combination of ion beam current and implantation time; junction depth can be controlled by the ion energy.

A conceptual view of an ion implantation system is shown in Figure 3.5. The entire system must be operated in a high vacuum environment to minimize collisions between the energetic ions and neutral gas molecules along the ion travel way. The ion source is responsible for generating dopant ions required for the implanting process. The ion source can be generated by one of the following three types: *hot-filament*, *radio-frequency* (RF), and *microwave*.

The analyzing magnet of the implantation system exactly selects the ion species required for implantation and generates a pure ion beam. This ion beam passes through a resolving aperture and is accelerated by a high electric field of 0 to 500 kV. The accelerated ion beam is further purified by a neutral beam gate and then deflected by X and Y scan systems prior to reaching the wafer surface.

The ion implantation generally proceeds in two major steps: *ion implantation* and *annealing* (that is, drive-in). The high-energy dopant ions are first implanted into the wafer surface using an ion implantation system. Because of the *bombardment effect* during ion implantation, the wafer surface lattice is damaged. To repair such a

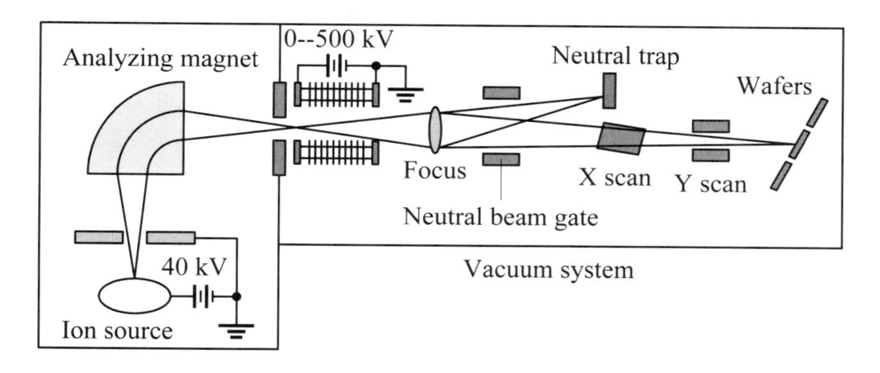

Figure 3.5: A conceptual view of an ion implantation system.

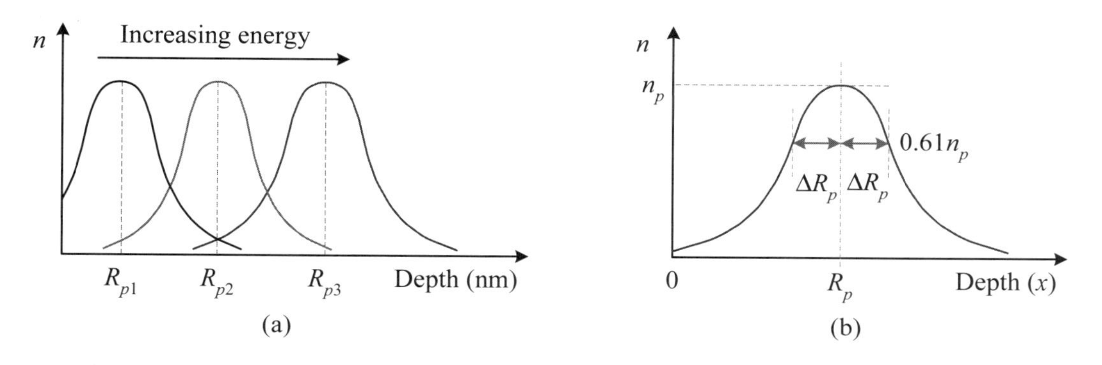

Figure 3.6: (a) Ion implantation profiles. (b) Ideal implant profile.

damaged lattice, an annealing process is needed to restore the lattice structure back to its single-crystal structure and activate the dopants. Annealing is a process that anneals the wafer at the appropriate combination of time and temperature to drive ions into the semiconductor and restore the mobility, as well as other material parameters, of the damaged region caused by ion implantation.

In the ion implantation, the average distance that ions can travel is called the *range of the ions*. The projection of this distance along the axis of incidence is called *projected range* R_p of the ions. The values of R_p range from about 0.1 μm to 5 μm, dependent on the incident energy, the species, and the crystal orientation. The incident energy of ion beams may be from a few keV for the ultra-shallow junction (junction depth is about 0.1 μm) to a few MeV for well implementation (well depth is about 2 to 4 μm). Higher-energy ion beams can penetrate deeper into the substrate and hence have a longer projected range. Figure 3.6(a) gives the profiles of three different ion beams with different levels of energy.

The distribution of ions implanted into the silicon substrate can be approximated to the first order using the Gaussian distribution

$$n(x) = n_p \exp\left[\frac{-(x - R_p)^2}{2\Delta R_P^2}\right] \tag{3.3}$$

where n_p is the peak concentration in a unit of cm^{-3}, R_P is the projected range, and the standard deviation is called the *straggle* ΔR_p. It is of interest to note that the peak

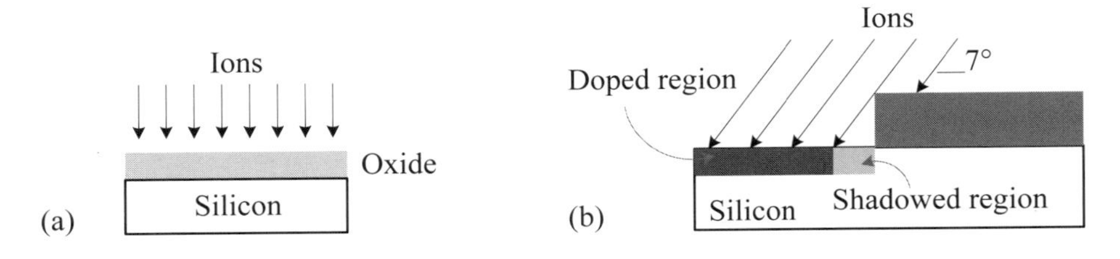

Figure 3.7: Two basic approaches to avoiding the ion-channeling effect: (a) screen oxide; (b) tilted wafer.

concentration n_p occurs at $x = R_p$ rather than at the surface $x = 0$, the situation of the diffusion process.

The implant dose in a unit of cm^{-2} is related to the distribution of implanted ions as in the following equation.

$$D_I = \int_{all\ x} n(x)dx \tag{3.4}$$

The implant dose can be precisely measured using charge counters.

Ion-channeling effect. During ion implantation, if an ion with the right implantation angle enters the channel, it can travel a long distance without colliding with the lattice. Such an effect of an incident ion aligning with a major crystallographic direction so that the range of channeled ion can be significantly larger than it would be in normal conditions is known as the *ion-channeling effect*. The ion-channeling effect may cause the trouble about the control of junction depth and affect device performance. It is an undesired dopant profile and should be minimized.

There are two basic approaches that can be used to minimize the ion-channeling effect. As shown in Figure 3.7(a), the first approach is to use a thermally grown *screen oxide layer*, which is also called a *sacrificial oxide layer* and is an amorphous material. The ions passing through it collide with oxide atoms and are scattered before they enter the single-crystal silicon substrate. The incident angles of ions are then randomized into a wide range, thereby significantly reducing the chance of the ion-channeling effect.

Figures 3.7(b) shows another approach that the wafer is tilted at an angle, typically 7°, so that the incident ions impact with the wafer at an angle and cannot reach the channel. Most ion implantation systems use this approach combined with the screen oxide to avoid the ion-channeling effect. It is worth noting that the wafer could cause shadowed regions by the photoresist when the tilted wafer approach is used, as shown in Figure 3.7(b). However, this *shadowing effect* can be easily removed by rotating the wafer. In addition, a small amount of dopant diffusion during the annealing process can also minimize this shadowing effect.

Review Questions

Q3-7. What are the two basic steps of ion implantation?

Q3-8. What are the major uses of ion implantation in CMOS processes?

Q3-9. What is the ion-channeling effect?

Q3-10. How would you minimize the ion-channeling effect?

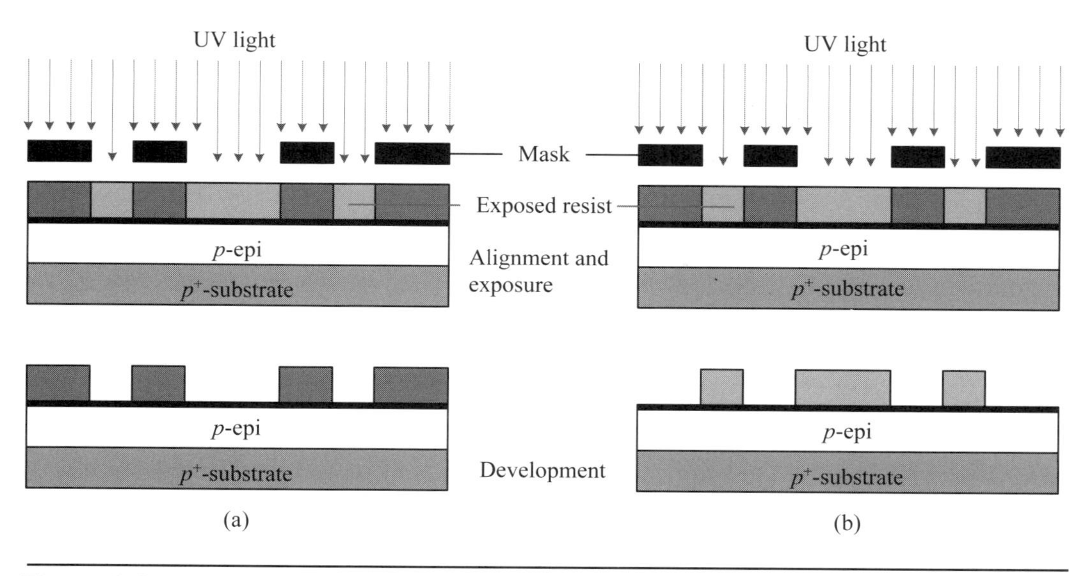

Figure 3.8: An illustration of (a) positive and (b) negative photoresists.

3.1.3 Photolithography

Photolithography (also called *lithography* or *photo* for short) is the process that transfers the image of a circuit pattern on a *mask* (also called a *photomask*) onto the *photoresist* covering the wafer surface to define various regions, such as active regions, contact windows, and bonding-pad areas, in an IC. The essential components in a photolithography process are photoresist (or *resist* for short) and masks. We first describe photoresist and then address the related issues, including exposure methods, mask making, pattern transfer, limitations, wet photolithography, as well as a clean room.

3.1.3.1 Photoresist A photoresist is a radiation-sensitive hydrocarbon-based material whose properties can be changed when it is exposed to light. After being exposed to light, the photoresist material changes its properties by chemically restructuring itself into a new stable form. Depending on how they respond to radiation, photoresists can be either *positive* or *negative*. The positive photoresist responds to light by becoming more soluble in the developer solution while the negative photoresist responds to light by becoming less soluble.

Figure 3.8 shows the difference between positive and negative photoresists. When a positive photoresist is used, the photoresist of those areas exposed to ultraviolet (UV) light are removed after development, as depicted in Figure 3.8(a); thereby, the same patterns are transferred on the photoresist from the mask. In contrast, when a negative photoresist is used, the photoresist of those areas not exposed to UV light are removed after development, as illustrated in Figure 3.8(b); thereby, the inverted patterns are transferred on the photoresist from the mask. Nowadays, positive photoresists are the primary materials used in the semiconductor industry since they have better resolution than negative photoresists.

Three important parameters determine the usefulness of a particular photoresist. These are *sensitivity*, *resolution*, and *etch resistance*. Sensitivity is a measure of how much light energy is required to expose the photoresist. High sensitivity is generally desired because it can shorten the exposure time of the photoresist and therefore increase the throughout of the photolithography process. However, extreme sensitivity is not usually desired because this tends to make the photoresist material unstable and very sensitive to ambient temperatures. Resolution is the smallest feature size that can be printed on a wafer; that is, it is the ability to differentiate between two closely spaced patterned features on the wafer surface. Etch resistance refers to the property that it must maintain its adhesion and protect the substrate surface from the subsequent etching processes or ion implantation.

Review Questions

Q3-11. What are the two types of photoresists?

Q3-12. Why positive photoresists are the primary materials used in the semiconductor industry?

Q3-13. What are the three important parameters that determine the usefulness of a particular photoresist?

3.1.3.2 Exposure Methods The most critical step of the photolithography process is alignment and exposure. This step is essential for the whole IC processing because it determines whether the IC design patterns on the mask can be successfully transferred onto the photoresist and then on the wafer surface. Because many mask steps are needed to process an IC, precise alignment of each mask with respect to the previous ones is vital.

The four exposure methods are: *contact printing*, *proximity printing*, *projection printing*, and *scanning projection printing*. The contact and proximity printing processes are widely used for the alignment and exposure process in the early years of the semiconductor industry. In the contact printing process, the mask is made directly to contact with the photoresist on the wafer surface, as shown in Figure 3.9(a). Even though the contact printing process can produce good image resolution on the wafer surface due to the mask pattern being closely contacted with the wafer surface, it has two vital problems. First, it is an operator-dependent process and often causes problems with control and repeatability. Second, it is prone to particle contamination, which will damage the photoresist, the mask, or both, and needs to be replaced every 10 to 20 operations. Because of these reasons, this method is only used in the process with feature sizes of about 5 μm and above.

The second exposure method is proximity printing. This printing method attempts to alleviate the contamination problem associated with the contact printing process. Instead of making direct contact with the wafer surface, the mask used in the proximity printing process is kept away from the wafer surface a distance of about 5 to 20 μm, as depicted in Figure 3.9(b). However, this process will reduce performance because a gap between the mask and wafer surface, may cause the effects of *edge diffraction* and *surface reflectivity* when UV-light rays pass through the transparent regions of the mask and the air gap between the mask and wafer surface.

The third exposure method is projection printing. The mask and wafer surface are far apart and thus no particle contamination problem exists. The projection printing process takes advantage of optical systems to transfer a 1:1 image onto the wafer surface. The concept of projection printing is shown in Figure 3.9(c). The UV-light source passes through a lens system to convert the incident light into parallel light rays. These parallel light rays then project onto the mask and expose the photoresist coated on the wafer surface.

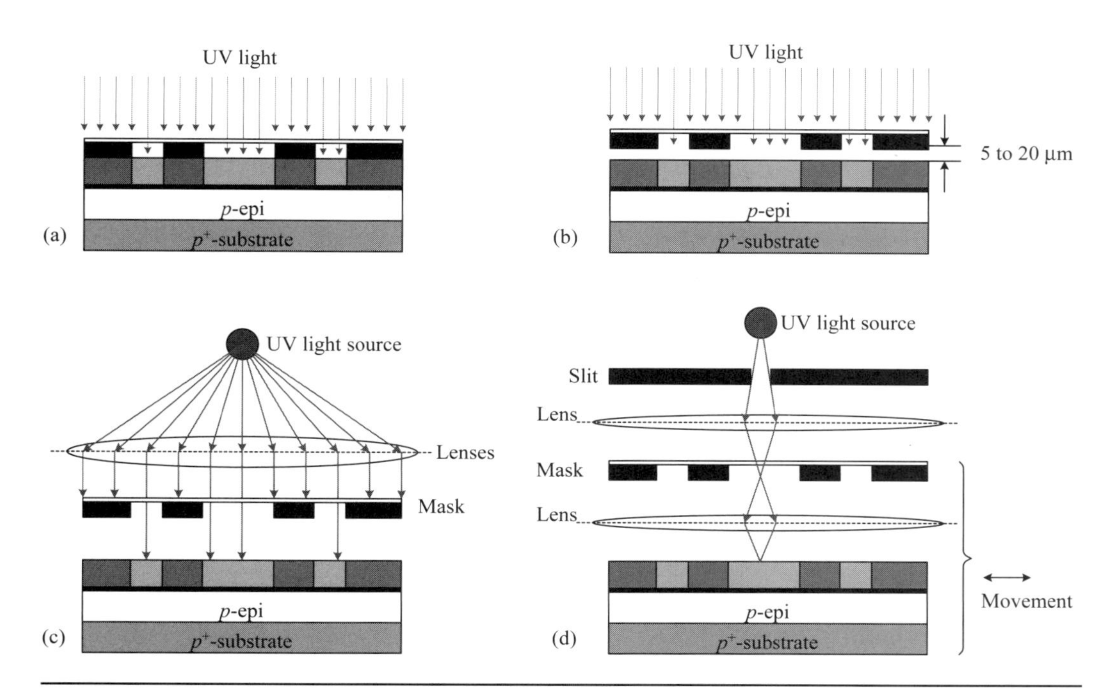

Figure 3.9: An illustration of various exposure methods: (a) contact printing; (b) proximity printing; (c) projection printing; (d) scanning projection printing.

Nowadays, the widely used projection printing in industry is the one shown in Figure 3.9(d), called the scanning projection printing. The scanning projection printing uses a slit to block a large amount of UV light and to produce a uniform light source to reduce light scattering and improve the exposure resolution. However, this results in only a small area of wafer surface that can be exposed at a time. To make the entire wafer exposed, a scanning mechanism is needed. As shown in Figure 3.9(d), the mask and wafer are mounted on a scanning carriage and moved in synchronism to allow the UV light scanning across the mask to refocus on the wafer surface.

One challenge of scanning projection printing is to make a cost-effective 1X mask that contains all chips on the wafer; in particular, as the feature size comes into the submicron era, it is very difficult to make a defect-free mask in a cost-effective way. One way to solve this problem is to make only a part of the mask. Such a mask containing only one or more chips is called a *reticle*. To make the entire wafer surface exposed with a reticle, a new printing approach known as *step-and-repeat printing* is developed. The resulting exposure system is often called a *stepper*.

The conceptual view of step-and-repeat printing is shown in Figure 3.10. In this approach, a mechanism, somewhat like scanning projection printing, is used to make the entire wafer surface exposed. As illustrated in Figure 3.10, a carriage holding the wafer can be moved in both X and Y directions so that the entire wafer surface can be exposed after a sequence of steps.

Because step-and-repeat printing shrinks the image from the reticle, the feature size on the reticle may be larger than the actual image to be patterned on the wafer surface. Hence, it is much easier to make a reticle than a 1:1 mask. The typical reduction factors are $4\times$, $5\times$, and $10\times$.

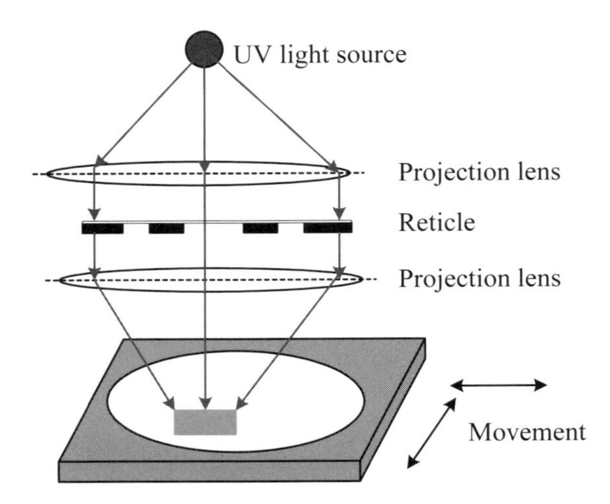

Figure 3.10: The conceptual view of step-and-repeat printing.

3.1.3.3 Mask Making As stated, masks or reticles are another important component in the photolithography process. Both mask and reticle are a *chromium* (Cr) glass image. The difference between them is that a mask covers the entire wafer, whereas a reticle only covers a part of the wafer.

A mask normally transfers the image to the wafer surface in a 1:1 ratio. The highest resolution for a mask is about 1.5 μm. Exposure systems such as contact printing, proximity printing, and projection printing systems use masks. A reticle has a larger image and feature size than the image actually projected on the wafer surface. The reduction ratios are typically 4:1 (4×), 5:1 (5×), or 10:1 (10×). When using reticles, exposure systems need to expose several times to cover the whole wafer.

The mask or reticle is made by printing the layout image generated by an *electronic design automation* (EDA) software on a piece of quartz glass coated with a layer of chromium, as shown in Figure 3.11(a). To make a mask or reticle, the following major steps are carried out. First, a layer of chromium is coated on the surface of a piece of quartz glass. Second, a photoresist layer is spun on top of the chromium layer. Third, a computer-controlled laser beam projects the layout image onto the photoresist-coated chromium glass surface. Fourth, the photoresist is developed to dissolve the exposed regions. Fifth, the chromium is at a location where no photoresist protection is etched away. Finally, a pellicle film is mounted a short distance from the chromium glass surface to keep the surface of the mask or reticle from being contaminated.

Phase-shift mask. As the feature sizes are continually shrunk, the light diffraction and interference may distort the features when a number of small features are closely packed together. An illustration for this is given in Figure 3.12(a), where the light passing through two adjacent openings produces a constructive interference and results in an undesired pattern. To solve this problem, a natural way is to use a light source of shorter wavelength. However, this means that more expensive photolithography equipment is required. Another way is to make use of the feature of the destructive interference of light. This results in the invention of the phase-shift mask (PSM).

The idea of PSM is to pattern a dielectric layer at every other opening on the mask, as shown in Figure 3.11(b). The purpose of the dielectric layer is to shift the incident light to a phase of 180° so that the light passing through two adjacent openings will destructively interfere with one another and result in a sharper image in the densely

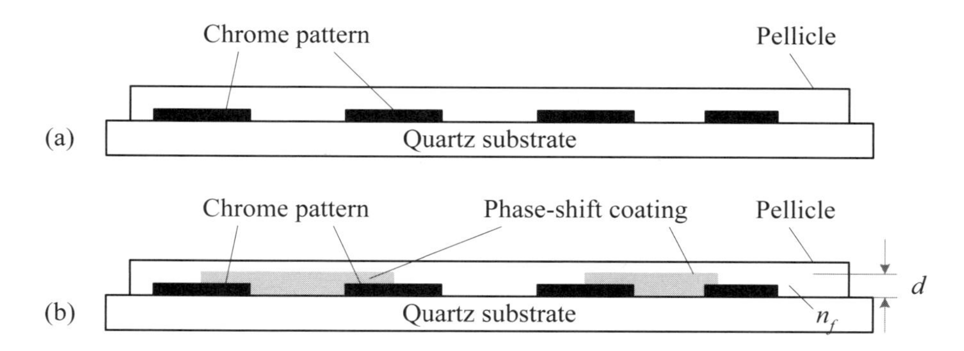

Figure 3.11: The structures of (a) general and (b) phase-shift masks.

packed area, as shown in Figure 3.12(b).

By using a phase-shift mask, the minimum feature size can be as small as one-fifth of the exposure wavelength. The resulting process is called *subwavelength photolithography*. The thickness d of the dielectric and the dielectric constant n_f are related to the exposure light wavelength λ by the following equation:

$$d(n_f - 1) = \lambda/2 \tag{3.5}$$

The widely used light sources have wavelengths of 365 nm (i-line) and 193 nm deep ultraviolet (DUV), respectively. Refer to Figure 3.16 for more details.

Review Questions

Q3-14. What is the most critical step of the photolithography process?

Q3-15. What are the four exposure methods?

Q3-16. What is a photomask?

Q3-17. Distinguish a reticle from a mask?

Q3-18. What is a phase-shift mask?

3.1.3.4 Pattern Transfer The goal of pattern transfer is to transfer the image on the mask to the wafer surface. To achieve this, two phases are often performed when fabricating microelectronic devices. In the first phase, a photolithography process is used to transfer the pattern on the mask or reticle onto the photoresist; in the second phase, the thin-film removal is employed to copy the image on the photoresist onto the wafer surface. In what follows, we describe the first phase in more detail. The details of thin-film removal are deferred to the next section.

The major steps required for pattern transfer in the photolithography process are shown in Figure 3.13. We take a look at these major steps along with some minor enhancement steps.

1. **Wafer clean and prime.** The first step of the photolithography is to clean the wafer because it might be contaminated during the previous step, as shown in Figure 3.13(a). Then, a priming process is utilized to deposit a thin primer layer to wet the wafer surface and enhance the adhesion between the photoresist and the wafer surface.
2. **Photoresist coating.** The wafer is coated with a liquid photoresist by a spin coating method, as shown in Figure 3.13(b). The spin speed and the viscosity of

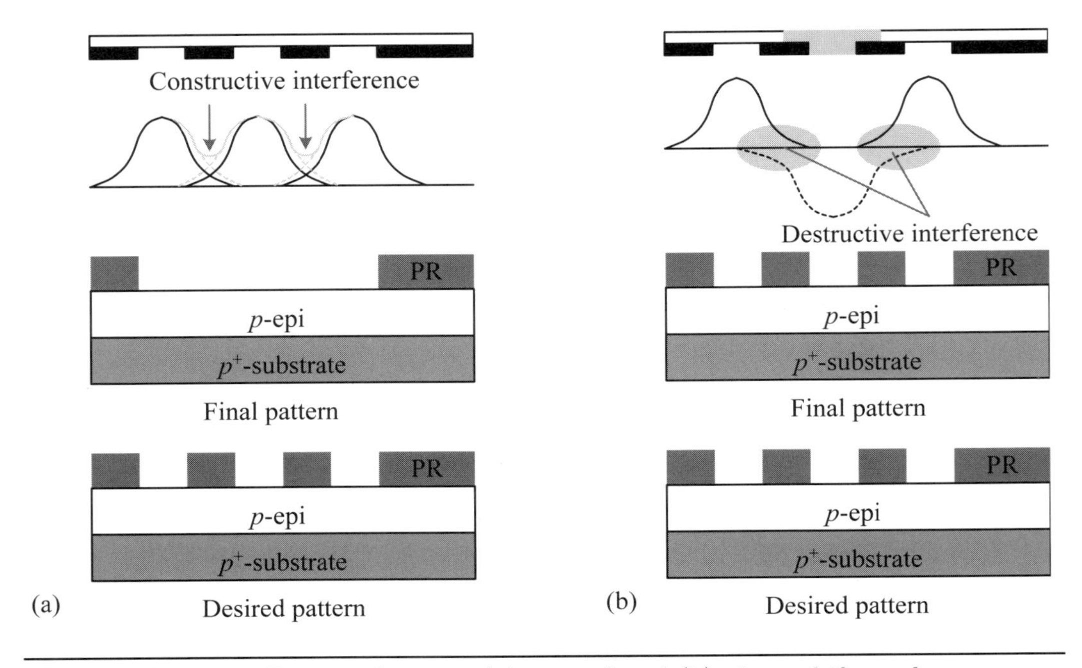

Figure 3.12: The differences between (a) general and (b) phase-shift masks.

photoresist material determines the final photoresist thickness, ranging from 0.6 to 1 μm.

(a). **Soft bake.** Before going to the alignment and exposure step, a soft bake process is required to drive off most of the solvent in the photoresist material. The soft bake process is to place the wafer on a hot plate at a temperature of 90° to 100°C for about 30 seconds.

3. **Alignment and exposure.** The mask is aligned to the correct location of the photoresist-coated wafer and then both the mask (or reticle) and wafer are exposed to a controlled UV light to transfer the mask image onto the photoresist on the wafer surface, as shown in Figure 3.13(c).

   (a). **Postexposure bake.** The postexposure bake is intended to minimize striations of overexposed and underexposed areas through the photoresist caused by the standing-wave effect that might be occurring from the interference between the incident light and the light reflected from the photoresist-substrate interface. In modern processes, a thin *antireflective coating* (ARC) layer is often used to help reduce the amount of reflective light.

4. **Development.** Development is the critical step for creating the pattern in photoresist on the wafer surface. In this step, the soluble regions are removed by developer chemicals, as shown in Figure 3.13(d). After development, the following two steps are often carried out.

   (a). **Hard bake.** After development, the wafer needs to be baked again on a hot plate at a temperature of 100° to 130°C for about 1 to 2 minutes to drive out the remaining solvent in the photoresist material. This step improves not only the strength and adhesion but also the etch and ion implantation resistance of the photoresist.

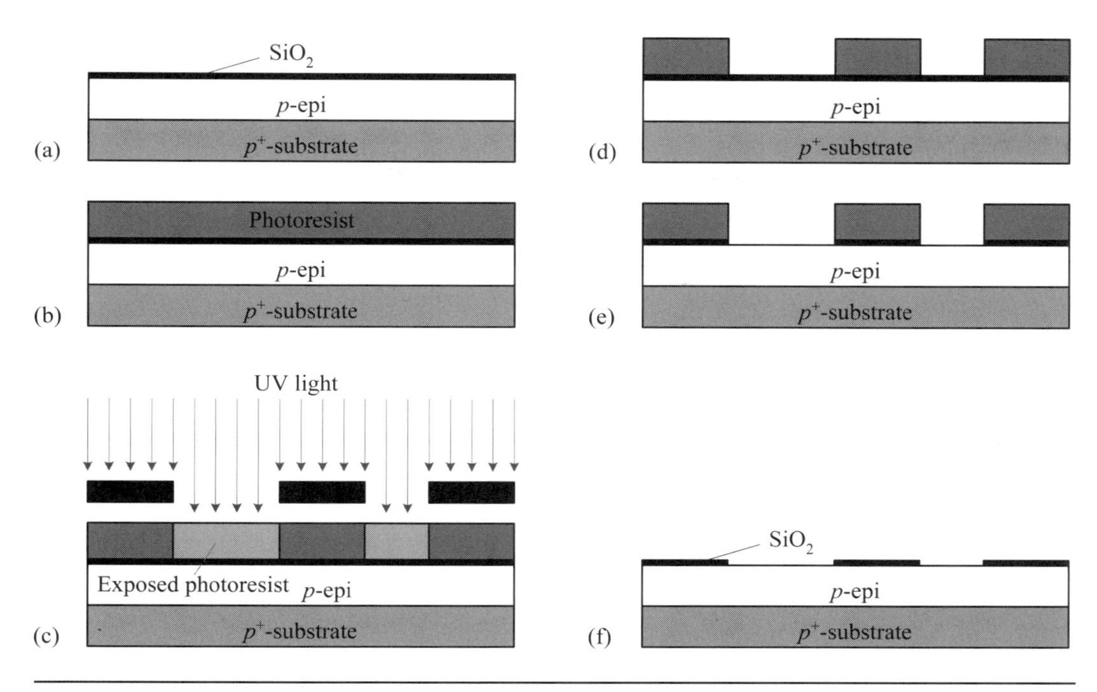

Figure 3.13: The major steps for pattern transfer: (a) wafer clean and prime; (b) photoresist coating; (c) alignment and exposure; (d) development; (e) silicon dioxide etch; (f) photoresist strip.

(b). **Pattern inspection.** The final step of the pattern transfer process is pattern inspection, which checks whether the image on the mask or reticle is correctly transferred onto the photoresist. If the wafer fails to pass inspection, the photoresist needs to be stripped and the whole pattern transfer process is repeated again.

The remaining part of the pattern transfer of the running example is completed by etching away the silicon dioxide exposed and then removing the photoresist. The illustrations of these two steps are depicted in Figures 3.13(e) and (f), respectively.

3.1.3.5 Limitations of Photolithography. Since steppers are widely used in modern very-large-scale integration (VLSI) processes, in what follows, we will look at some limitations of this method. In general, there are three important parameters associated with a stepper: *resolution*, *depth of focus* (DoF), and *registration error*.

Resolution. Resolution R is defined as the minimum feature size that can be printed on the wafer surface. Resolution can be related to the UV light wavelength (λ) and the numerical aperture (NA) of the projection lens by the following equation:

$$R = \frac{k \cdot \lambda}{NA} \tag{3.6}$$

where k is a process constant with values ranging from 0.5 to 0.8.

The numerical aperture (NA) of a lens is the capability to collect diffracted light. It can be expressed as

$$NA = n \cdot \sin\theta \tag{3.7}$$

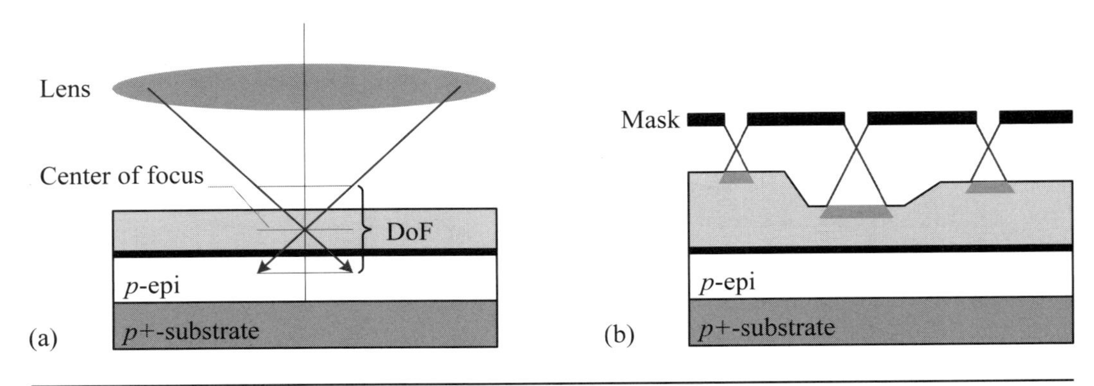

Figure 3.14: (a) Definition of depth of focus of lens. (b) DoF effects on a nonplanar surface.

where n is the index of refraction of the space between the wafer and lens. The angle θ is the acute angle between the focus on the wafer surface and the edge of the lens radii. The NA is proportional to the diameter of a lens. Therefore, to obtain a high NA, a larger diameter lens is required.

From Equation (3.6), it is apparent that resolution is determined by the wavelength of light source and the NA of the projection lens. To obtain a better resolution, a smaller wavelength of light source and/or a lens with high NA have to be used. Another way to obtain a better resolution without virtually using a shorter wavelength of light source and a higher NA lens is to use PSM, as described before.

Although PSM can relieve a little bit of the limitations of light wavelength and improve the resolution of photolithography, the rectangle pattern—such as contact windows—still cannot be properly printed on the photoresist on the wafer surface due to the *corner rounding effect*. One approach to solving this effect without compromising the wavelength of light source is the *optical proximity correction* (OPC). The OPC method uses computer algorithms to compensate for the corner rounding effect by introducing selective image size biases (that is, the precompensated image) into the reticle pattern. It has been widely used in modern deep submicron CMOS processes, referring to Section 4.1.3 for more details.

Depth of focus. The depth of focus, also known as *depth of field* (DoF), is the range around the focal point over which the image is continuously on focus, as shown in Figure 3.14(a). The DoF of a projection lens is described by the following equation:

$$DoF = \frac{\lambda}{2(NA)^2} \tag{3.8}$$

where λ is the wavelength of UV light source and NA is the numerical aperture of the projection lens.

The DoF of the projection lens limits the ability to pattern features at different heights, as shown in Figure 3.14(b). As we can see from the figure, three different exposed areas are produced for the same feature on a nonplanar wafer surface. This situation is more severe as the feature size is further reduced. Hence, it is very important to keep the wafer surface as planar as possible. For this purpose, in modern VLSI industry, an important technique called *chemical mechanical planarization* (CMP) is widely used to planarize the wafer surface. The CMP will be introduced later in this section.

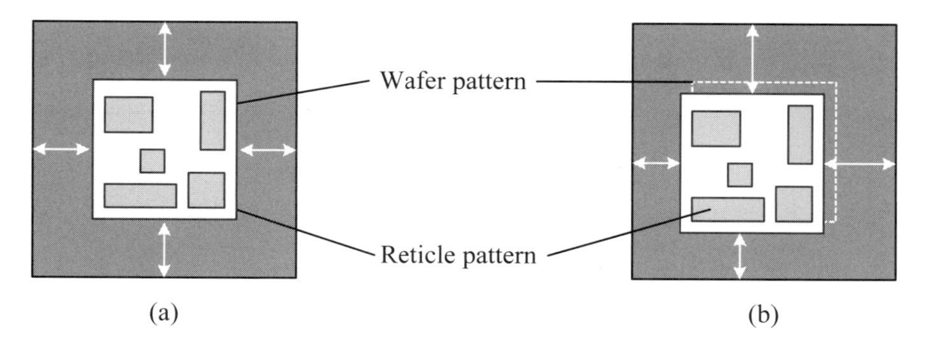

Figure 3.15: Some registration issues: (a) no registration error; (b) shift in registration.

Registration error. Recall that an IC is manufactured as a sequence of layers, with each layer being defined by an appropriate mask or reticle. Each projected image from a mask or reticle must be placed in correct relation with respect to patterns already on the wafer surface. To achieve this, an alignment process is used to determine the position, orientation, and distortion of patterns already on the wafer surface, and place them in correct relation to the projected image from the mask or reticle. The result of the alignment process is known as *registration*. In other words, the registration is a measure of how accurately a mask is aligned with respect to its previous patterns on the wafer; that is, it is a measure of the layer-to-layer alignment error. Some registration cases are shown in Figure 3.15. Figure 3.15(a) shows the case of perfect overlay accuracy, namely, no registration error, while Figure 3.15(b) gives the situation of a shift in registration.

To help reduce registration error to a minimum degree, some visible patterns called *alignment marks* are placed on the mask or reticle and hence on the wafer to determine their position and orientation. Alignment marks, also known as *targets* or *fiducial marks*, are either one or more lines on the mask or reticle. These alignment marks become trenches when printed on the wafer surface, thereby serving as the alignment marks for the subsequent layer.

3.1.3.6 Wet Photolithography The wavelengths of the light source used in various generations of CMOS processes accompanied with the feature sizes are plotted in Figure 3.16. Before the 0.25-μm process, the wavelength of the light source used in a photolithography process is smaller than the feature size. After that, the feature size is smaller than the wavelength of the light source. The feature size is continuously shrunk, but the wavelength of the light source starts to be freezed at 193 nm with the era of a 90-nm feature size.

Because of the difficulty of finding suitable materials for making lenses and masks at a wavelength of less than 193 nm, a technique called *wet photolithography* or *immersion photolithography* is used. In this technique, the objective lens is placed very closely to the wafer waiting to be exposed, and the gap between the lens and the wafer is filled with water; indeed, the entire system is immersed in the water. The resulting wavelength used to expose the wafer is then reduced by a factor of the refraction index of water, 1.43. Other materials are also used to replace the water to obtain the shorter effective wavelength.

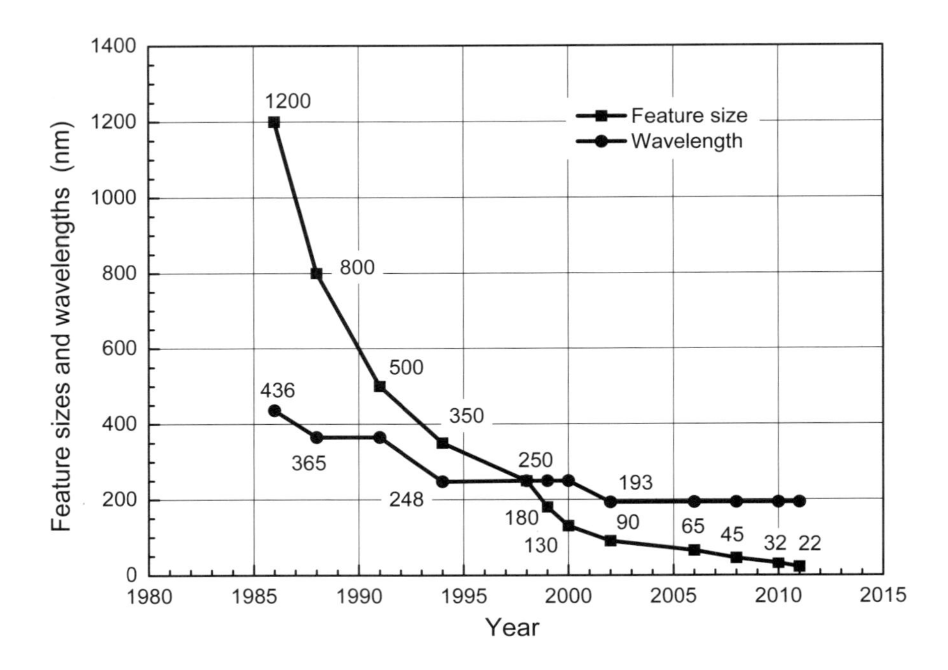

Figure 3.16: Wavelengths used in various generations of CMOS processes.

3.1.3.7 Clean Room Since photolithography is intended to print the image on the mask or reticle on a wafer surface, any dust particles from the air settled on a semiconductor wafer and a lithographic mask might be deemed as a part of the image to be printed. This may result in defects and cause circuit failures. Hence, the photolithography process must be carried out in a clean room, where the total number of dust particles per unit volume, together with the temperature and humidity, is tightly controlled.

A clean room is regulated and categorized according to how many particles with a specific diameter are allowed in a specific volume. A class-X clean room means that there may exist at most X particles with a diameter larger than 0.5 μm per cubic foot. A foundry that manufactures microelectronic devices with a feature size of 0.25 μm needs a class-1 clean room to achieve an acceptable yield. The topmost-level clean room is class $M - 1$, which allows up to 10 particles with a diameter larger than 0.5 μm per cubic meter, corresponding to 0.28 particles per cubic foot.

Review Questions

Q3-19. What is the intention of pattern transfer?

Q3-20. What are the factors that limit the use of photolithography?

Q3-21. Explain the meaning of optical proximity correction (OPC).

Q3-22. Explain the registration error.

Q3-23. What is the intention of alignment marks?

Q3-24. What is wet photolithography?

Q3-25. What is the meaning of a class-X clean room?

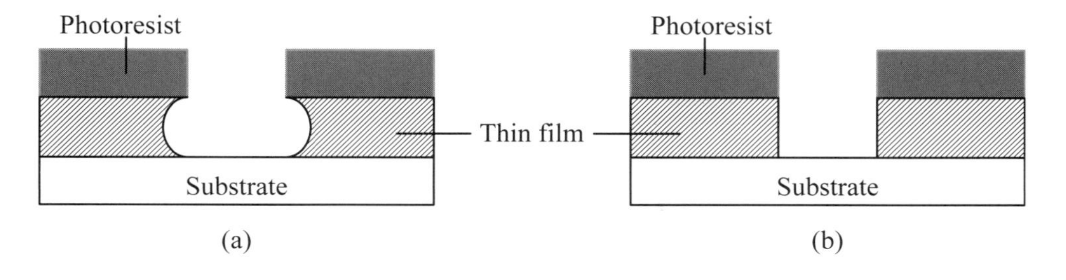

Figure 3.17: The etching profiles: (a) isotropic etch; (b) anisotropic etch.

3.1.4 Thin-Film Removal

Almost all principal layers of an IC have to be processed by a sequence of thin-film deposition, photolithography, and thin-film removal. Here, the thin film may be any one of the following: silicon dioxide (SiO_2), silicon nitride (Si_3N_4), polysilicon (poly-Si), aluminum-copper alloy, borophosphosilicate glass (BPSG), and others. Thin-film removal, also known as *thin-film etch* or just *etch* for short, is the process of selectively removing the unneeded (unprotected) material by a chemical (wet) or physical (dry) means. Its objective is to reproduce the features defined on the mask or reticle on the photoresist-coated wafer surface. The thin-film removal process is called *wet etching* when chemical solvents are used and *dry etching* when some kind of physical means is used. We first describe some important parameters related to the etch process and then briefly describe the etching processes widely used in today's IC fabricating.

3.1.4.1 Etch Parameters The most important parameters that determine the use of which etching process, wet etching or dry etching, is appropriate when removing a specific thin film are *etch rate*, *etch profile*, *selectivity*, *uniformity*, and *degree of anisotropy*.

Etch rate. Etch rate is a measure of the thickness removed per unit of time. It is usually a strong function of solution concentration and etching temperature. High etch rate is generally favorable due to a higher throughput.

Etch profile. Etch profile refers to the fraction of sidewall of the etched feature removed during an etching process. There are two basic etch profiles, as shown in Figure 3.17. In an *isotropic etch* profile, all directions are etched at the same rate, leading to an undercut of the etched material under the mask, as shown in Figure 3.17(a). This results in the reduction of the actual width of a line such as a polysilicon or a metal wire. The other etch profile is called an *anisotropic etch* profile. In this profile, the etch rate is in only one direction perpendicular to the wafer surface, as shown in Figure 3.17(b).

Selectivity. Selectivity means how much faster one material is etched than another under the same condition. Selectivity S is defined as the etch rate ratio of one material to another and is given by

$$S = \frac{R_1}{R_2} \tag{3.9}$$

where R_1 is the etch rate of the material intended to be removed and R_2 is the etch rate of the material not intended to be removed.

Uniformity. Uniformity is a measure of the capability of the etching process to etch evenly across the entire wafer surface.

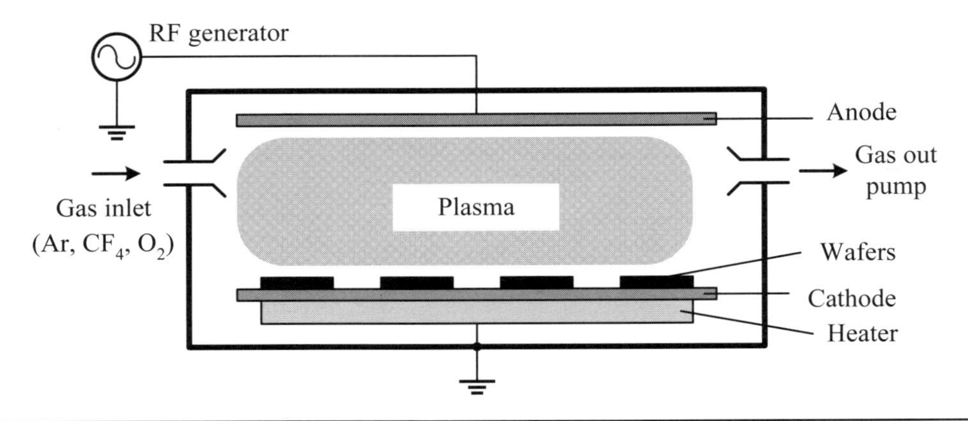

Figure 3.18: A conceptual view of plasma etching system.

Degree of anisotropy. Degree of anisotropy A_f is a measure of how rapidly an etchant removes material in different directions and can be given by

$$A_f = 1 - \frac{R_l}{R_v} \tag{3.10}$$

where R_l is the lateral etch rate, whereas R_v is the vertical etch rate. For isotropic etch, R_l is equal to R_v, and therefore $A_f = 0$; for anisotropic etch, R_l is equal to 0, and therefore $A_f = 1$.

3.1.4.2 Wet Etching Process The wet etching process is the earliest used etching process. It uses the chemical reaction between thin film and solvent to remove the thin film unprotected by photoresist. The wet etching process is commonly used to etch silicon dioxide, single-crystal silicon, silicon nitride, and metal. The wet etching process has high throughput compared with the dry etching process and is usually an isotropic process even though it can also be anisotropic. Consequently, in modern deep submicron processes, it is not suitable for defining the line features. However, because of high selectivity, it still plays an important role in cleaning the wafer surface and thin-film removal, such as silicon dioxide cleaning, residue removal, and stripping surface layers, such as the blanket thin film.

3.1.4.3 Dry Etching Process The dry etching (also called *plasma etching*) process has gradually replaced the wet etching process for all patterned etching processes since the feature size reached 3 μm in the late 1980s. Nowadays, because it has an excellent anisotropic profile and can generate very reactive chemical species, the dry etching process has become the primary etching approach in semiconductor fabricating. The dry etching process is commonly used to etch dielectric, single-crystal silicon, polysilicon, and metal, as well as to strip photoresist.

Plasma is an ionized gas composed of ions, electrons, and neutral atoms or molecules with an equal amount of positive and negative charge. To achieve the etching action, plasma provides energetic positive ions that are accelerated toward the wafer surface by a high electric field. These ions physically bombard the unprotected wafer surface material, causing material to be ejected off the wafer surface. A conceptual view of the plasma etching system is shown in Figure 3.18.

The two main types of species involved in plasma etching are: the reactive neutral chemical species and the ions. The reactive neutral chemical species are mainly responsible for the chemical component in the plasma etching process, while the ions are for the physical component. According to whether these two components work together or independently, the plasma etching process can be differentiated into three processes: the *plasma etching process*, the *sputter etching process*, and the *reactive ion etch (RIE) process*. All of these are the most important dry etching processes used in the IC industry nowadays.

Plasma etching process. The plasma etching process is also referred to as the *plasma chemical etching* process since, in such a system, the material to be etched is commonly removed by a chemical reaction between it and the reactive chemical species. The reactive chemical species are free radicals, which are neutral species having incomplete bonding. Because of their incomplete bonding structures, free radicals are very highly reactive chemical species. The species used in the plasma etching process need to have the property that it will generate a volatile by-product when reacting with the material to be etched. Thereby, the by-product can be easily removed and the surface is made to expose for more material to be etched. Because of the chemical reaction involved in this process, it has high selectivity but poor direction, that is, an isotropic profile.

Sputter etching process. Another plasma etching process makes use of the ions; that is, ions are responsible for the etching operation. In this etching process, ions accelerated by a high electric field bombard the atoms on the wafer surface to physically dislodge them. This bombardment results in more physical components of etching, that is, sputtered surface materials. Because it is fundamentally a physical reaction, it has poor selectivity but high direction, that is, an anisotropic profile.

Reactive ion etching (RIE) process. The third approach is that both free radicals and ions work together in a synergistic manner to etch the material. In other words, the etching process involves both ion sputtering and radicals reacting with the wafer surface. The result not only has a high degree of selectivity but also achieves a very anisotropic etch profile. In this etching process, the actual etch profile is between isotropic and anisotropic and can be controlled by adjusting the plasma conditions and gas composition.

3.1.4.4 CMP Process As described previously, because of the DoF effect, the global planarization is increasingly important in deep submicron processes in which a very high density of components is required. *Planarization* is a process that removes the surface topologies and smoothes as well as flattens the surface. The chemical mechanical polishing (CMP) process is a global planarization process widely used in industry today. The CMP process combines both *chemical etching* and *mechanical sanding* to produce planar surfaces on silicon wafers.

A conceptual view of the CMP system is given in Figure 3.19(a), where the major components are shown. The wafer is held face down against the polishing pad in between which a colloidal silica slurry is kept. The polishing table and the wafer carrier are rotated in opposite directions to reduce polishing time. An application of the CMP process to planarize the surface is shown in Figure 3.19(b).

Some advantages of the CMP process are that it allows more uniform thin-film deposition, minimizes defects, and improves yield, as well as widens the IC chip design parameters.

■ Review Questions

Q3-26. What is the distinction between wet etching and dry etching?

Q3-27. Why is the plasma etching process also called plasma chemical etching?

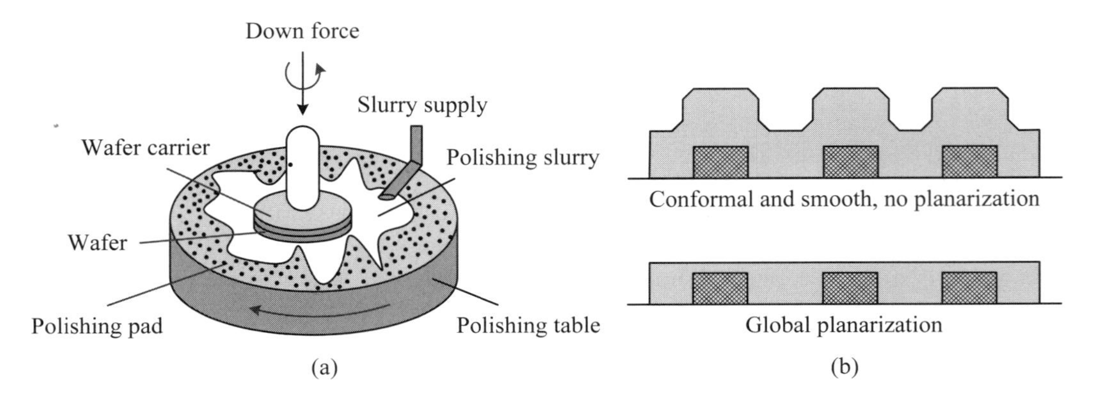

Figure 3.19: The illustration of using CMP to smooth a surface globally: (a) CMP operation; (b) CMP application.

Q3-28. Describe the sputter etching process.
Q3-29. Describe the reactive ion etching (RIE) process.
Q3-30. Why is global planarization so important in deep submicron processes?

3.1.5 Thin-Film Deposition

Thin-film deposition is the means by which thin films of various materials, including insulators, semiconductors, and conductors, can be deposited on the silicon wafer. Like the thin-film removal, thin-film deposition can also be done through either a physical or a chemical process. The former employs a physical process to deposit a thin film on the wafer surface while the latter relies on a chemical reaction between gases of a gas mixture to produce the desired material, which in turn accumulates on the wafer surface. We first discuss the thin-film formation and some important parameters associated with thin-film deposition. Then, we briefly address a variety of widely used thin-film deposition processes.

3.1.5.1 Thin-Film Formation A thin film is a thin, solid layer of a material created on a substrate. A thin film is formed through the following three major stages. The first stage is called *nucleation*, in which clusters of stable nuclei are formed. As the first few atoms or molecules of the reactants coming down to the wafer surface, they may be absorbed to the wafer surface and become adatoms. These adatoms then combine to form isolated patches of film.

The second stage is known as *grain growth* (also *island growth* or *nuclei coalescence*). As the stable nuclei are formed, the nuclei coalesce into clusters in random orientations. The required atoms are no longer limited to the adatoms; they may be from gaseous atoms or molecules in the chamber. The grain growth is based on the surface mobility and the density of the cluster.

The third stage is *continuous film*. The island clusters continue to grow and eventually develop into a continuous film. These island clusters are then further condensed into a solid sheet spreading across the substrate surface.

3.1.5.2 Parameters The important parameters associated with thin-film deposition are *step coverage* and *aspect ratio*. We look at these parameters in more detail.

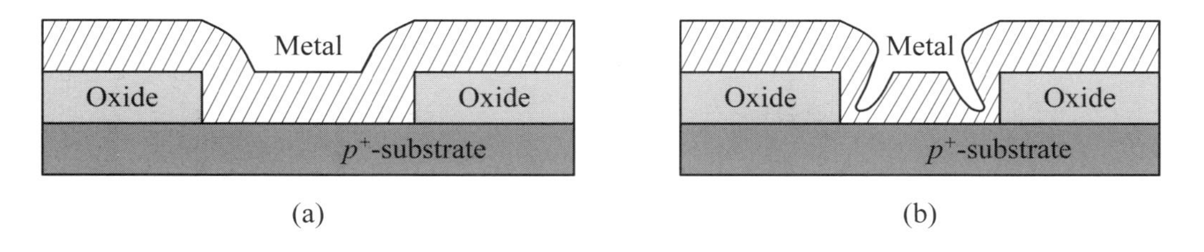

Figure 3.20: An illustration of thin-film step coverage: (a) good coverage; (b) poor coverage.

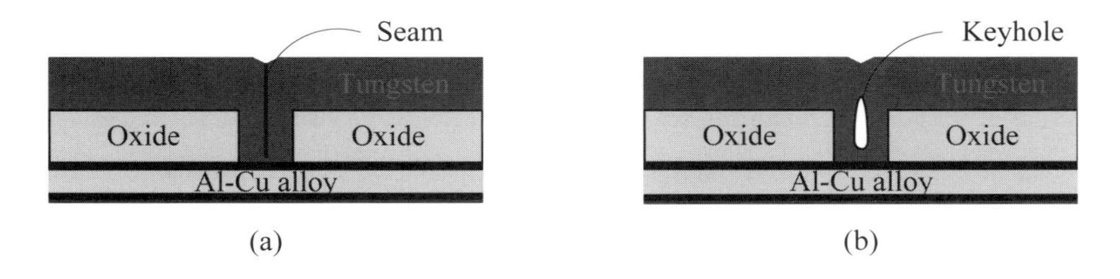

Figure 3.21: The profiles of gap fill: (a) good gap fill; (b) poor gap fill.

Step coverage. It is desirable to form a thin film uniformly over surface features, as shown in Figure 3.20(a), where thin-film thickness is uniform over all surfaces of the trench. Because of the surface topography created by patterned features, the thickness of a thin-film will not be conformal, especially at the steps of trenches. To measure the uniformity of thin-film thickness, step coverage is used. Figure 3.20(b) shows the case of poor step coverage. For good step coverage, it is desirable to carefully control the uniformity of thin-film thickness to a high degree across the wafer diameter. The poor step coverage may cause high stress. Thin films with minimum stress are important because stress may lead to substrate deformation in a convex or concave shape.

Aspect ratio. Aspect ratio is defined as the ratio of its depth to width; it is often denoted as a ratio such as 3:1, meaning the depth of a trench is three times the width. In modern deep submicron CMOS processes, gaps or holes, such as vias, often result in a very high aspect ratio. Thus, the ability to fill such gaps and holes becomes a very important fabricating feature for devices. High aspect ratio gaps or holes make it difficult to uniformly deposit a thin film, thereby leading to a void or keyholes, as shown in Figure 3.21(b). A good gap fill should be the one like that shown in Figure 3.21(a).

3.1.5.3 Physical Vapor Deposition (PVD)

Physical vapor deposition uses a physical process to produce the constituent atoms (or molecules) required for thin-film formation. The material to be deposited is first dismantled into gas-phase atoms or molecules by a physical process in a very low-pressure chamber. These atoms or molecules are then condensed on the surface of the substrate. The two most commonly used PVD processes are *evaporation* and *sputter deposition*. In what follows, we describe each of these in greater detail. One of the disadvantages of PVD is that the resulting thin films often have poor step coverage.

Evaporation. A conceptual view of the evaporation system is shown in Figure 3.22. The source material to be deposited on the wafer surface is heated past its melting point in a crucible within a low-pressure chamber. The vapor form of the

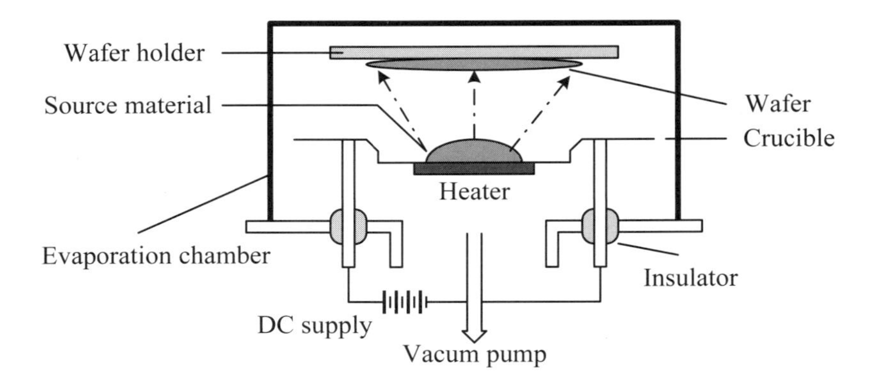

Figure 3.22: A conceptual view of the evaporation system.

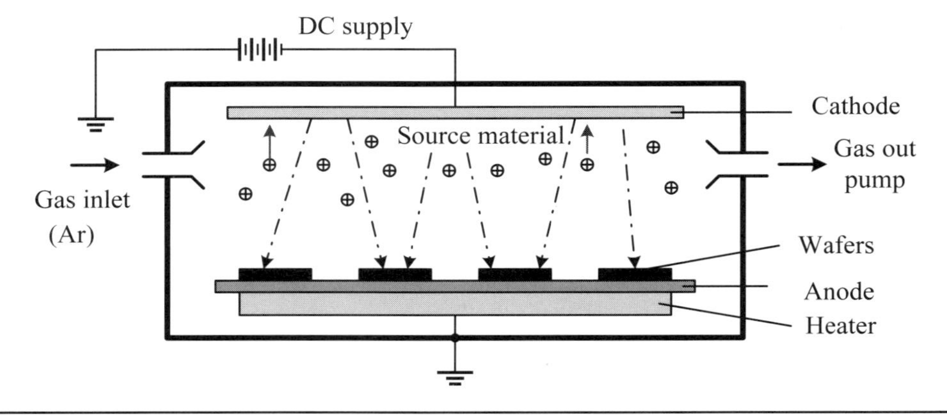

Figure 3.23: A conceptual view of the sputter system.

material coats all exposed wafer surface. The heating system used in the evaporation system can be a heating filament or focused electron beam.

Sputter deposition. In a sputter system shown in Figure 3.23, an inert gas such as argon (Ar) is ionized into positive charged ions in a low-pressure chamber by a high electric field. These positive charged ions are attracted toward the cathode and bombard the source material to be deposited there. The sputtered (ejected) atoms or molecules then reside on the wafer surface, forming the desired thin film.

It is interesting to note the similarity between sputter etching and sputter deposition. In sputter etching, the wafer serves as the cathode so that the accelerated positive ions bombard the unprotected areas of wafer surface and dislodge the atoms or molecules; in the sputter deposition, the wafers serve as the anode and the source material to be deposited serves as the cathode so that the accelerated positive ions bombard the source material to sputter atoms or molecules, which then accumulate on the wafer surface, thereby forming the desired thin film.

3.1.5.4 Chemical Vapor Deposition (CVD) Process

In a *chemical vapor deposition* (CVD) process, a gas mixture is introduced into a reaction chamber where chemical reactions between gases in the gas mixture at the surface of wafers produce a desired thin film. To speed up the reaction rate, the wafer surface is often heated to provide additional energy to the reaction process. In comparison with PVD, CVD often results

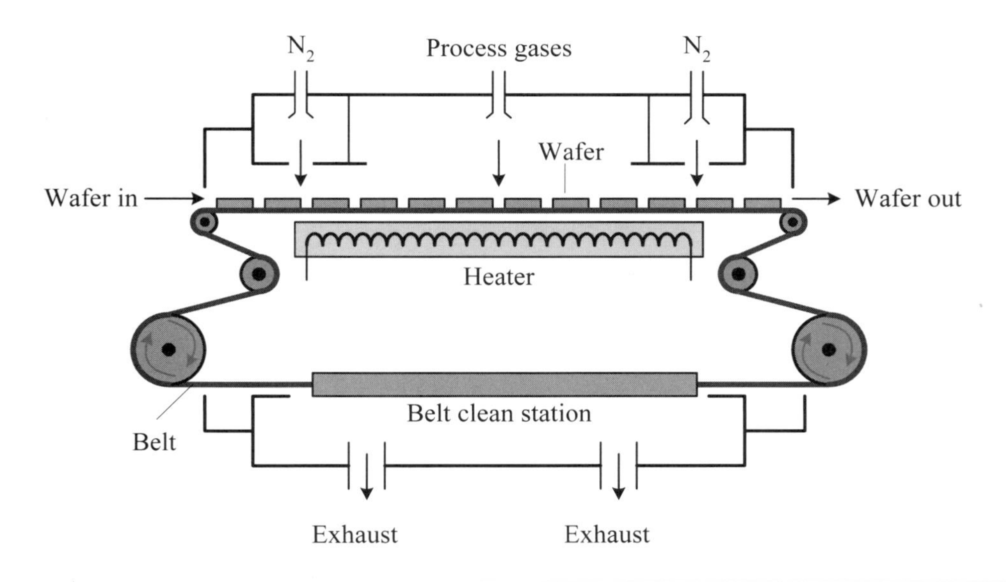

Figure 3.24: A conceptual view of the APCVD system.

in good step coverage of the resulting thin films. The common CVD processes include *atmospheric-pressure CVD* (APCVD), *low-pressure CVD* (LPCVD), *plasma-enhanced CVD* (PECVD), and *high-density plasma CVD* (HDPCVD). We briefly address each of these.

Atmospheric-pressure CVD (APCVD) process. A conceptual view of a typical atmospheric-pressure CVD system is illustrated in Figure 3.24. The APCVD process is operated at a pressure of about 1 atm (760 Torr at sea level at 0°C). One major advantage of the APCVD process is that it has a very fast deposition rate of about 600 to 1000 nm/min. To make use of this feature, a transport system, such as a belt, as depicted in Figure 3.24, is used to convey the wafers through the CVD system. To speed up the reaction process, a heater is used to heat the wafer surface. At the two ends, two nitrogen (N_2) buffers are used to prevent the process gases from leaking into the atmosphere.

Low-pressure CVD (LPCVD) process. A conceptual view of the low-pressure CVD system is shown in Figure 3.25. A gas mixture is introduced into a low-pressure chamber where chemical reactions between gases of the gas mixture at the surface of a substrate occur and yield a desired thin film. Compared to the APCVD process, the LPCVD process can produce highly conformal thin films but at the expense of higher deposition temperature. The typical temperature is in the range of about 450 to 800°C.

Plasma-enhanced CVD (PECVD) process. The basic idea of the PECVD process is the use of plasma to supply the additional energy to the reactant gases so that the reactions needed for thin-film deposition can occur at temperatures much lower than those when only thermal energy is provided. The PECVD process usually operates at a temperature of less than 450°C. A conceptual view of the PECVD system is shown in Figure 3.26, where anode and cathode plates (electrodes) are placed a few inches apart within a vacuum reaction chamber at 1 to 10 Torr. The RF power is applied to the top plate while the wafers are placed on the bottom plate. A plasma develops when the mixture of gases is introduced into the chamber. This plasma is

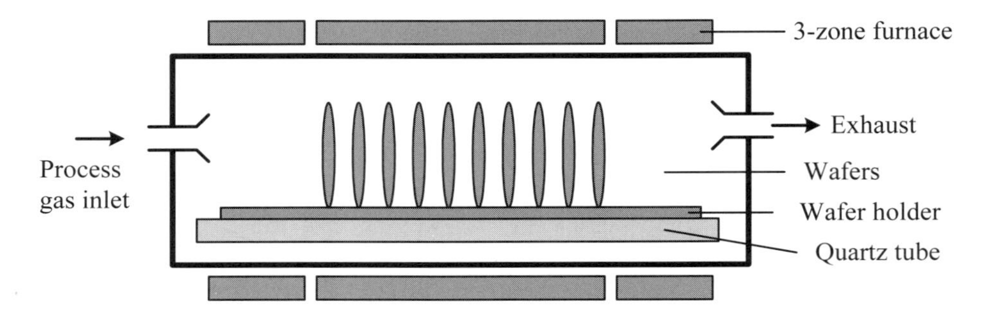

Figure 3.25: A conceptual view of the LPCVD system.

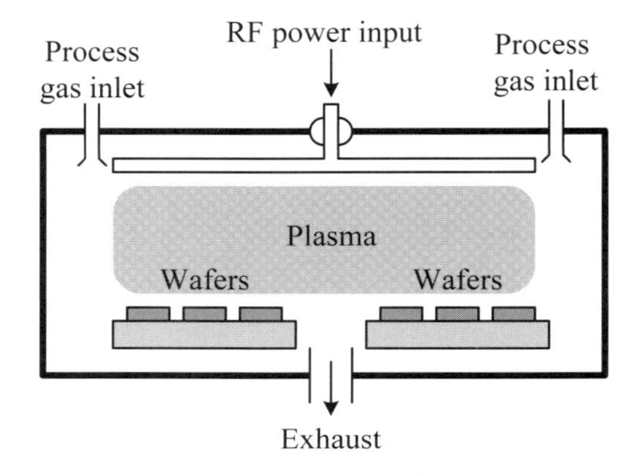

Figure 3.26: A conceptual view of the PECVD system.

toward the wafer surface in the reaction chamber. Exhaust gases are pumped out from the bottom of the chamber.

High-density plasma CVD (HDPCVD) process. The HDPCVD process is developed for depositing intermetal dielectrics (IMDs), shallow trench isolations (STIs), etch-stop layers, and low-k dielectrics; it combines the PECVD deposition with sputtering to obtain very good filling of high aspect ratio gaps with a deposition temperature of 150° to 400°C.

The HDPCVD reaction involves a chemical reaction between two or more gas precursors, which are intermediate reactions in gas-phase reactions. These precursors form a gas species that does not contain the original gas components and are transported to the wafer surface for adsorption and reaction. Adsorption is the chemical binding that occurs during deposition, causing the gaseous atoms and molecules to chemically attach to the solid wafer surface.

The high-density plasma can be generated by a variety of sources, including *inductively coupled plasma* (ICP) and *electron cyclotron resonance* (ECR). In ICP, a source RF power is used to generate a changing electric field that accelerates electrons and causes ionization collisions. The main feature of ICP is that the induced electric field is in the angular direction and hence accelerates the electrons in the same direction. This allows electrons to have longer mean free paths. Thus, a high-density plasma can be generated at low pressure. In ECR, a microwave power is used to resonate electrons in a harmonic way and a magnetic field is used to determine the resonance position. Electrons acquire energy from the microwaves and collide with other atoms or molecules. The ionization collisions yield more electrons. These newly created electrons resonate again with the microwave frequency and repeat the above process. Therefore, it is able to generate high-density plasma at low pressure.

Both ICP and ECR systems have a bias RF power to control the ion bombardment energy (and hence deposit rate) and a cooling system to control the wafer temperature. One major advantage of ECR is that the resonance position can be controlled by adjusting the magnetic current, which means that the uniformity of thin films can be optimized by controlling the magnetic current.

3.1.5.5 Spin-on-Glass Process The *spin-on-glass* (SoG) process was widely used to fill gaps and do the planarization prior to the wide acceptance of the CMP process. The SoG process proceeds in a way much like photoresist coating and baking processes. The major steps of SoG are as follows. First, a barrier layer of *undoped silicate glass* (USG) is deposited using a PECVD process. Next, liquid SoG is uniformly spun on the wafer surface to form a thin film with a thickness of a few hundred nanometers. Then, the wafer is baked and cured in two steps. It is first annealed on a hot plate at a temperature of 150° to 250°C and then placed into a furnace at a temperature of 350° to 450°C to drive out solvent from the SoG. The resulting SoG becomes solid silicate glass. The SoG may or may not then be etched back to remove most SoG from the surface, depending on specific applications. Finally, a cap layer of silicon dioxide is deposited using the PECVD process to cover the SoG to prevent it from moisture absorption.

■ Review Questions

Q3-31. Describe the three major steps of forming a thin film.

Q3-32. What is the distinction between PVD and CVD processes?

Q3-33. Describe the two important parameters of thin-film deposition.

Q3-34. Describe the two types of PVD process.

Q3-35. How would you distinguish sputter etching from sputter deposition?

3.2 Materials and Their Applications

In CMOS processes, insulators, semiconductors, and conductors are used. In this section, we take a look at some important materials widely used in modern CMOS processes.

3.2.1 Insulators

The most commonly used insulators (dielectrics) in the semiconductor industry are silicon-based compounds, such as silicon dioxide (SiO_2), silicon nitride (Si_3N_4), and doped glasses, such as borophosphosilicate glass (BPSG). These insulators are used for gate dielectrics, device isolation, metal1-to-substrate isolation, intermetal isolation, etch masks, implantation masks, diffusion barriers, sidewall spacers, and passivation. Both silicon dioxide and silicon nitride are good electrical insulators with very high dielectric strength, that is, breakdown voltage. We will deal with these two insulators in more detail.

3.2.1.1 Silicon Dioxide Silicon dioxide (SiO_2) is also called *quartz glass* or simply glass or oxide. It is used for the gate oxide in MOS transistors because it is an excellent electrical insulator, adheres very well to most materials, and can be grown on a silicon wafer or be deposited on top of the wafer. Silicon dioxide has a dielectric constant k of 3.9.

When silicon is exposed to an oxygen or ambient environment, a thin silicon dioxide will be formed at room temperature with a growth rate of about 1.5 nm per hour up to a maximum thickness of about 4.0 nm. This kind of silicon dioxide is often called a *native oxide* and is considered an undesirable contaminant. A silicon dioxide layer can be grown on top of a wafer using a thermal oxidation or CVD process. When it is created by a thermal oxidation process, the wafer surface is exposed to an oxygenated atmosphere; when it is created by a CVD process, a chemical reaction process using silane (SiH_4) is used to produce the SiO_2 layer above the wafer.

Silicon dioxide can be used in a variety of places during fabricating a CMOS integrated circuit. These mainly include gate oxide, field oxide, dopant barrier, sacrificial oxide, and pad oxide.

Gate oxide. Gate oxide serves as the dielectric between the gate and substrate of MOS transistors. The common thickness ranges from about 1.0 nm to a few tens of nanometers, depending on the specific process. To obtain a silicon dioxide of stable and good quality, dry thermal oxidation is the preferred growth method.

Field oxide. Field oxide serves as the isolation barrier between neighboring MOS transistors. Its thickness ranges from 250 nm to $1,500$ nm. Field oxide can be formed by using local oxidation of silicon (LOCOS) or shallow trench isolation (STI) process. Wet thermal oxidation is the preferred growth method for LOCOS, whereas the CVD process is often used to fill STI trenches.

Dopant barrier. Silicon dioxide can also be used as a mask to selectively introduce dopants into the silicon surface. Its thickness lies in the range from 40 nm to 120 nm. The silicon dioxide protects the silicon surface without openings being diffused into impurity during the doping process, thereby allowing the impurity to be doped into a selective area into the wafer surface.

Sacrificial oxide. Sacrificial oxide is widely used to minimize the ion-channeling effect in ion implantation and to help prevent silicon contamination by blocking the sputtered photoresist. The sacrificial oxide, also called *screen oxide*, is usually grown thermally; its thickness is about 10 to 20 nm.

Pad oxide. In both LOCOS and STI processes, silicon nitride (Si_3N_4) is usually used as the mask for depositing thick field oxide thermally by a wet oxidation process or the CVD oxide process. To relieve the stress of silicon nitride, a thermally grown pad oxide with a thickness of about 10 to 30 nm is deposited prior to the deposition of silicon nitride. Other uses of pad oxide are to serve as a protective layer between metal layers and as an etch-stop layer.

3.2.1.2 Silicon Nitride Silicon nitride is an ideal material to keep the sensitive silicon circuits from contamination. It has a relatively high dielectric constant, $k = 7.8$. Silicon nitride finds its widespread use in CMOS processes, especially in the following: oxidation mask in LOCOS and STI processes, sidewall spacers and an etch-stop layer of lightly doped drain (LDD) extensions, dopant diffusion barrier layer for premetal dielectric (PMD), the intermetal dielectric (IMD) seal layer, as well as the etch-stop layer in the dual-damascene copper process.

Silicon nitride (Si_3N_4) is also called *nitride* for short. It is a dense material and often deposited with silane (SiH_4) and ammonia (NH_3) chemicals at about 900°C. The reaction equation is as follows:

$$3\text{SiH}_4 \text{ (gas)} + 4\text{NH}_3 \text{ (gas)} \longrightarrow \text{Si}_3\text{N}_4 \text{ (solid)} + 12\text{H}_2 \text{ (gas)} \tag{3.11}$$

Silicon nitride can also be deposited using the LPCVD process with dichlorosilane (SiH_2Cl_2) and ammonia (NH_3) chemicals at a temperature of 700° to 800°C. Another way to deposit silicon nitride is the PECVD process at a temperature of 400°C. Most IMD silicon nitride depositions use such a PECVD process.

■ Review Questions

Q3-36. What is the native oxide?
Q3-37. Describe the meaning of dopant barrier.
Q3-38. Describe the meaning of sacrificial oxide.
Q3-39. Describe the meaning of pad oxide.

3.2.2 Semiconductors

Silicon crystalline can exist in three forms: *amorphous silicon*, *polycrystalline silicon* (also called *polysilicon* for short), and *single-crystalline silicon*. Amorphous silicon has an order within a few atomic dimensions and no crystalline regions formed; it is formed when a silicon layer is deposited using a CVD process at a temperature below about 600°C. Polysilicon has a high degree of order over many atomic dimensions and is formed by raising the CVD temperature to about 600° to 700°C. Single-crystalline silicon has a high degree of order throughout the entire volume of the silicon material and is formed at temperatures between 850° and 1200°C.

In modern VLSI processes, the following two types of silicon layer are widely used: *epitaxial silicon* and *polycrystalline silicon* (polysilicon). The polycrystalline silicon is often used as the gate electrodes and local area interconnect. We describe each of them briefly in the following sections.

3.2.2.1 Epitaxial Silicon The deposition of a thin layer of an ordered crystalline on top of a single crystalline is known as *epitaxy*. Epitaxy is a kind of interface between a thin film and a substrate. Epitaxy may be *homoepitaxy* or *heteroepitaxy*. Homoepitaxy means a crystalline film is grown on a thin film or a substrate of the same material while heteroepitaxy is a crystalline film grown on a thin film or a substrate of different materials. Homoepitaxy is often used to either grow a more purified thin film than the substrate or fabricate layers with different doping levels (hence different resistance layers). Heteroepitaxy is employed to fabricate integrated crystalline layers of different materials and to grow crystalline films of materials of which a single crystal cannot be formed. Homoepitaxy is widely used in CMOS processes to enhance the performance of dynamic random access memory (DRAM) and CMOS ICs or in BiCMOS processes to enhance the performance of bipolar junction transistors (BJTs) through the insertion of a high-doping layer between an epitaxial layer and the substrate. Heteroepitaxy finds its widespread use in constructing optical devices, such as aluminum gallium indium phosphide (AlGaInP) on gallium arsenide (GaAs) and gallium nitride (GaN) on sapphire.

The epitaxial layer growth is often the most expensive process in CMOS IC fabrication. At present, the silicon epitaxy is usually grown with a CVD process called a *vapor-phase epitaxy* (VPE) process, where the silane (SiH_4) is often used as the silicon source. The typical reaction temperature is about 1150° to 1200°C. Other silicon sources are also possible. Refer to Sze [52] for more details.

3.2.2.2 Polycrystalline Silicon Another semiconductor thin film commonly used in CMOS processes is polycrystalline silicon, usually called *polysilicon* or simply *poly*. Polysilicon is employed as the gate material of MOS transistors or local interconnect. Polysilicon is usually deposited with a silane LPCVD process at a temperature of about 575° to 650°C with the following reaction equation:

$$SiH_4 \longrightarrow Si + 2H_2 \qquad (3.12)$$

Undoped polysilicon is a high-resistance material. Polysilicon can be doped by diffusion, ion implantation, or in situ doping through the addition of doping gas during the CVD process. However, ion implantation is the most commonly used approach because of its lower processing temperature. To be used as gate materials, polysilicon is often heavily doped with an n- or a p-type impurity to reduce its sheet resistance to the level of about 200 Ω/square. Since this value is still much larger than that of metal wire, polysilicon is only used for a local interconnect between two or a few adjacent transistors.

■ Review Questions

Q3-40. What are the three forms in which silicon crystalline may exist?

Q3-41. What is epitaxy?

Q3-42. How would you distinguish homoepitaxy from heteroepitaxy?

3.2.3 Conductors

Metals are the major conductor materials used for local interconnect, contacts, vias, diffusion barriers, global interconnect, and bond pads. In this subsection, we look at metal materials widely used in CMOS processes. These metal materials include *aluminum* (Al), *tantalum* (Ta), *titanium* (Ti), *tungsten* (W), and *copper* (Cu).

Aluminum (Al). Al is the most common metal used for interconnect in CMOS processes above 0.18 μm. Aluminum metal has an electrical resistivity of 2.75 μΩ·cm and is the fourth best electrical conducting metal only after silver (1.62 μΩ·cm), copper (1.69 μΩ·cm), and gold (2.28 μΩ·cm).

Both PVD and CVD processes can be used to deposit aluminum. When aluminum makes direct contact with silicon, it will dissolve silicon and form the aluminum spikes. Hence, it is necessary to add a barrier layer before such a contact is made. Another feature of aluminum material is the electromigration caused by the electron stream that constantly bombards the monocrystalline grains composed of the metallic aluminum. The electromigration can cause the reliability problem since it may cause open loops because of an aluminum wire broken after a period of work. The quantitative description of electromigration can be referred to in Section 14.1.1.2.

The electromigration is often alleviated by alloying a small amount (about 1%) of copper with aluminum to form a Cu·Au alloy at the cost of increasing sheet resistance, from 2.75 μΩ·cm to 3.3 μΩ·cm, in practical applications.

Titanium (Ti). Titanium has an electrical resistivity of 40 μΩ·cm and can be deposited by a sputter process or a CVD process. Titanium is often used as a welding layer for tungsten and aluminum alloy to reduce contact resistance. Except for forming titanium silicide ($TiSi_2$) with polysilicon and silicon materials, titanium can also be used to form titanium nitride (TiN). Titanium nitride finds widespread use for barrier layers, adhesion layers, and antireflective coating (ARC) layers.

Tungsten (W). Tungsten has an electrical resistivity of 5.25 $\mu\Omega\cdot$cm and is normally deposited with a CVD process. Tungsten is mainly used for filling contacts to wires between the metal layer and silicon or via holes (plugs) to wires between two different metal layers. To improve reliability, titanium nitride is often used as the adhesion layer between silicon dioxide and tungsten metal.

Tantalum (Ta). Tantalum has an electrical resistivity of 12.45 $\mu\Omega\cdot$cm and is normally deposited by a sputter process. It is mainly used as the barrier layer for copper deposition to prevent copper from diffusing across silicon dioxide into silicon substrate and causing possible device damage.

Copper (Cu). Copper has an electrical resistivity of 1.69 $\mu\Omega\cdot$cm and can be deposited by a sputter process, a CVD process, or an electrochemical plating deposition (EPD) process. Copper is the metal used to replace Al for interconnect in process of 0.18 μm and has been widely used in processes below 0.13 μm. The reason why copper does not find its application in early processes is that copper material is not easy to be etched by the dry etching process since the generated by-product is not volatile so that it cannot be easily removed.

3.2.3.1 Silicide and Salicide When the device dimension shrinks, the resistance of the polysilicon local interconnect increases, leading to more power dissipation and longer RC delay. To reduce the sheet resistance of contacts and vias, a refractory metal is usually used to react with silicon to form a metal compound called a *silicide* and with polysilicon to form a *polycide*. It is generally used to name both silicide and polycide as silicide because polysilicon is also silicon. The most common silicide is *titanium silicide* ($TiSi_2$), which is formed by making titanium (Ti) metal react with the silicon or polysilicon to form a metal compound. Another two common silicide compounds are *tungsten silicide* (WSi_2) and *cobalt silicide* ($CoSi_2$), which are formed by reacting with the refractory metals tungsten and cobalt with silicon or polysilicon, respectively.

A self-aligned silicide process is often named as *salicide*. The salicide process relies on the fact that metals often do not react with dielectrics. Hence, the dielectrics can be used as a mask when forming silicide without the need of an additional photolithography process and mask. Based on this idea, the salicide can be formed as follows. A refractory metal is first deposited over the entire wafer surface, and then the wafer is annealed at a high temperature chamber. The metal reacts with the exposed silicon and polysilicon to form silicide. Finally, the unwanted part of the metal is removed by a wet etching process. More details about salicide will be described in the next section.

■ Review Questions

Q3-43. What is the meaning of a barrier layer?

Q3-44. What is electromigration?

Q3-45. What are silicide and salicide?

3.3 Process Integration

We have examined the five basic unit processes, thermal oxidation, doping process, photolithography, thin-film removal, thin-film deposition, and a variety of materials commonly used in CMOS processes. In this section, we address how to combine basic unit processes with appropriate materials to fabricate a CMOS integrated circuit. Such

Table 3.1: The masks used in a generic CMOS process.

Layer name	Aligns to level	Times used	Description
1 (n-well)	Notch	1	Defines n-well
2 (p-well)	1	1	Defines p-well
3 (active)	1	2	Defines FOX region and active area
4 (polysilicon)	1	1	Defines polysilicon and local interconnect
5 (n-select)	1	3	Defines V_{Tn} implants, nLDD, and n^+
6 (p-select)	1	3	Defines V_{Tp} implants, pLDD, and p^+
7 (contact)	4	1	Defines contacts
8 (metal1)	7	1	Defines metal1
9 (via1)	8	1	Defines via1
10 (metal2)	9	1	Defines metal2
Passivation	Top-level metal	1	Defines bonding pad openings

a well-defined collection of unit processes accompanied with proper materials required to fabricate CMOS integrated circuits starting from virgin silicon wafers is known as a *process integration*.

The CMOS process integration is often divided into two major parts: the *front end of line* (FEOL) and *back end of line* (BEOL). The FEOL includes the following modules.

1. Twin-well process
2. Shallow trench isolation (STI) process
3. Threshold voltage adjustment
4. Polysilicon gate structure process
5. Lightly doped drain (LDD) extension implant process
6. Source/drain (S/D) implant process

In some modern CMOS processes, the STI process is performed prior to the twin-well process. However, in this book, we will follow the sequence described above as most typical CMOS processes do.

The BEOL includes the following modules.

1. Contact (silicide and salicide) formation
2. Metal1 interconnect formation
3. Via1 and plug1 formation
4. Metal2 interconnect formation
5. Via2 and plug2 formation
6. Metal3 interconnect formation
7. Other metal layers
8. Passivation and bonding pad

Table 3.1 summarizes the required masks in a generic CMOS process along with the sequence applied. Some masks are only used once while others may be used many times.

After a wafer is fabricated, there are many processes that still need to be processed. These processes are known as back-end processes and include the wafer test and die separation and packaging, as well as the final test and burn-in test.

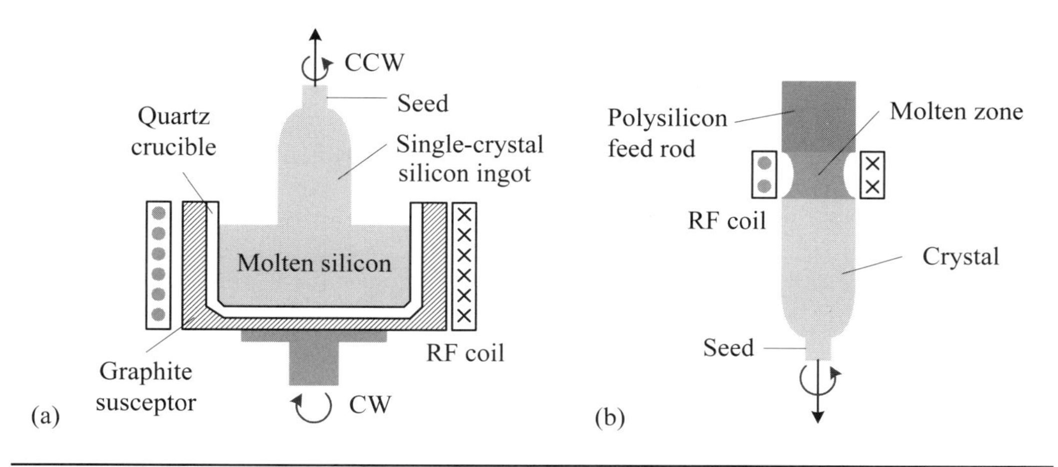

Figure 3.27: Two approaches for preparing wafers: (a) CZ method; (b) FZ method.

3.3.1 FEOL

The FEOL refers to all processes required to fabricate the isolated and fully functioned MOS transistors. It mainly encompasses all processes from wafer preparation until the source/drain regions of MOS transistors are formed.

3.3.1.1 Wafer Preparation There are many approaches that may be used to make silicon wafers. However, only the *Czochralski* (CZ) method and the *floating zone* (FZ) method depicted in Figure 3.27 are widely used in industry. The CZ method is the only approach that can make wafers with diameters larger than 200 mm (8 in). Hence, the 300-mm (12-in) wafers widely used nowadays are made by this method. The largest wafer diameter made with the FZ method is 150 mm (6 in).

CZ method. In the CZ method, the high-purity polycrystalline silicon is melted in a quartz crucible at 1415°C, just above the melting point 1414°C of silicon. As shown in Figure 3.27(a), the crucible is heated by radio-frequency (RF) power or resistive heating coils. A single-crystal seed mounted on the crystal puller and rotation mechanism is slowly lowered into the molten silicon. As the seed submerges in the molten silicon, it starts to melt. The temperature of the seed is precisely controlled at just below the silicon melting point. The seed crystal is then slowly pulled up, dragging some molten silicon to condense again around it with the same crystal orientation. The ingot growth rate is about a few millimeters per minute. The diameter of the ingot in the CZ method can be controlled by the temperature and the pulling rate.

Advantages of the CZ method are as follows. First, it is relatively low cost because a single piece of crystal and polysilicon can be used. Second, dopant materials may be added to the molten silicon during the ingot growth to obtain the desired electrical resistivity.

FZ method. The FZ method is another approach to making single-crystal ingots. As shown in Figure 3.27(b), the method starts with a polysilicon bar placed vertically in a furnace chamber. The RF heating coil melts the lower end of the polysilicon bar. A seed crystal fuses into the molten zone and rotates counterclockwise slowly by a rotation mechanism. The molten silicon at the seed end starts to freeze and solidify, forming the same crystal orientation as the seed crystal. The diameter of the grown ingot is determined by the relative motion between the top polysilicon rod and the bottom crystal section.

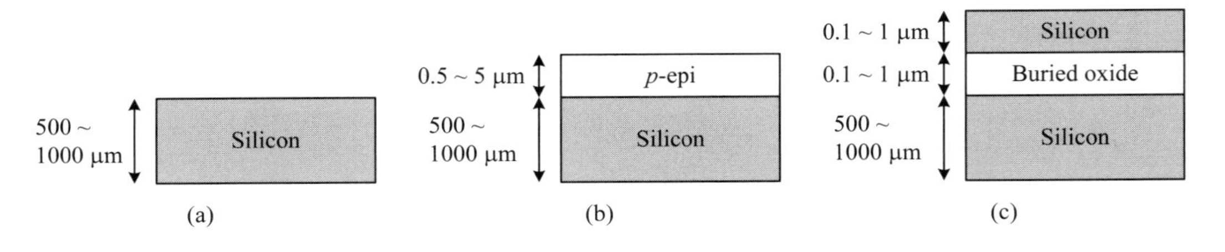

Figure 3.28: The three general types of silicon wafers used for CMOS fabricating: (a) bulk-silicon wafer; (b) epitaxial-silicon wafer; (c) silicon-on-insulator wafer.

One major advantage of the FZ method is that it produces the high-purity silicon without the use of a crucible. However, as mentioned previously, the maximum diameter produced by the FZ method is limited to 150 mm. As a result, it is only used in some special applications in which a high-purity silicon crystal is needed.

Types of silicon wafers. At present, there are three general types of silicon wafers used for CMOS fabricating. These are *bulk-silicon wafer*, *epitaxial-silicon wafer*, and *silicon-on-insulator* (SoI) *wafer*, as shown in Figure 3.28. The thickness of a typical wafer is about 500 to 1000 μm. The actual MOS transistors are fabricated in the top 1 μm or less of wafer, whereas the remaining hundreds of microns of wafer are used only for mechanical support during device fabricating.

The bulk-silicon wafer is least costly but may not be the optimal choice in high performance applications. The epitaxial-silicon wafer is formed by depositing a lightly doped epitaxial silicon layer with a thickness of about 0.5 to 5 μm on top of a bulk-silicon wafer. The goals of the epitaxial layer are to improve device performance and provide good immunity of the latch-up problem inherent in CMOS devices. The SoI wafer is formed by depositing a thick oxide with a thickness of about 0.1 to 1 μm on top of a bulk-silicon wafer; then an epitaxial silicon layer with a thickness of about 0.1 to 1 μm is deposited on top of the oxide. This type of wafer is most costly to implement but has the best performance and can completely eliminate the latch-up problem.

In the following discussion, we assume that an epitaxial-silicon wafer is used. The same discussion can be applied equally well to a bulk-silicon wafer.

3.3.1.2 Twin-Well Process The first step in modern CMOS processes is to define the active areas where MOS transistors are to be made. A twin-well (twin-tub) approach is often used to define the active regions, where the n-well defines pMOS transistors while the p-well defines nMOS transistors.

The major steps for the formation of both the n-well and p-well are shown in Figure 3.29. The wafer is first rinsed and dried to remove contamination and native oxide. Then, the wafer is grown with a screen (that is, sacrificial) oxide with a thickness of about 10 to 15 nm by placing the wafer in a high temperature (about 1000°C) furnace chamber, as shown in Figure 3.29(a). The goals of the screen oxide are as follows. First, it helps control the depth of dopants and prevents the silicon from being excessively damaged (namely, the ion-channeling effect) during ion implantation. Second, it protects the epitaxial layer from contamination.

The n-well mask is applied to define the n-well region, as depicted in Figure 3.29(a). As mentioned earlier, a photolithography process is required to transfer the n-well pattern onto the wafer surface. After the required pattern is transferred, phosphorus ions are implanted onto the areas of wafer surface unprotected by photoresist. The

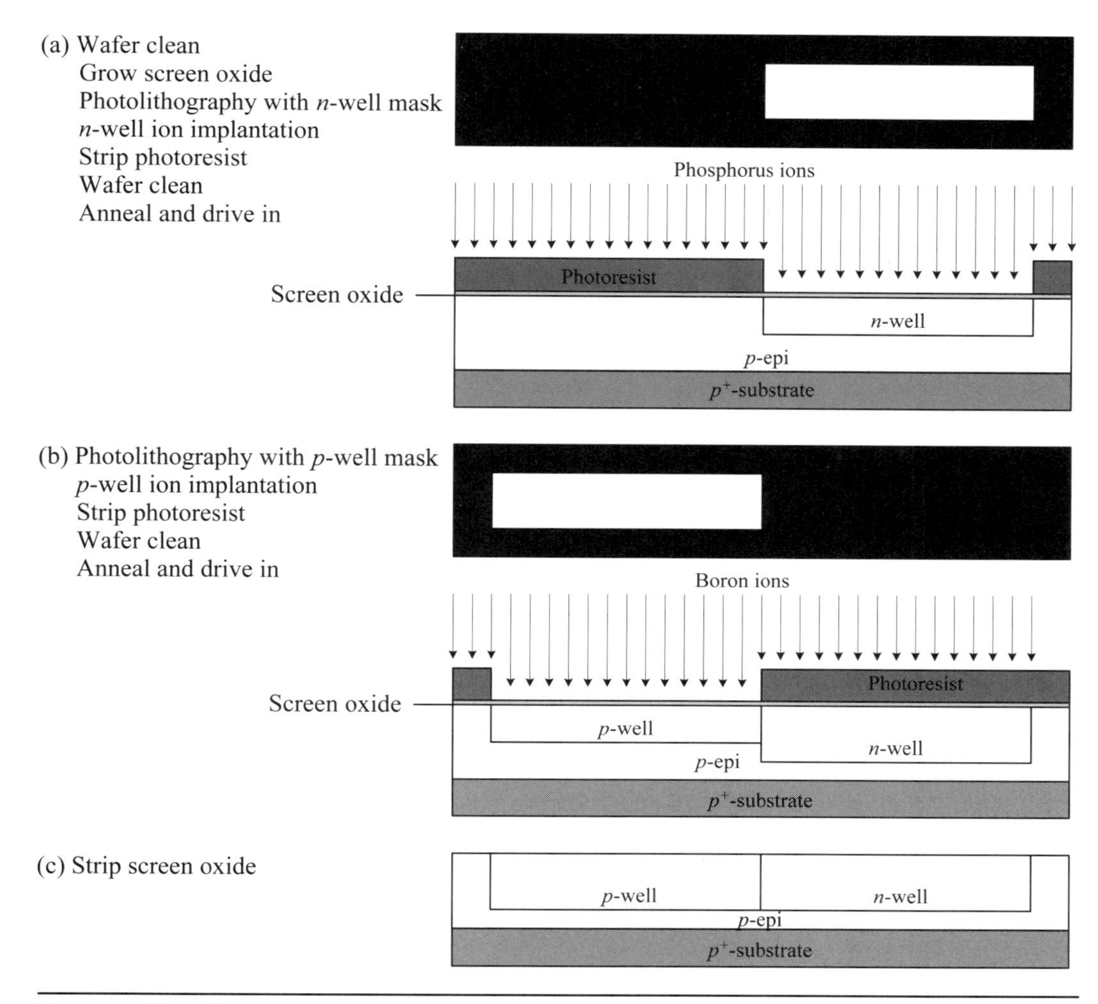

Figure 3.29: The major steps in n-well and p-well formation.

following two steps are to strip the photoresist off and to clean the wafer to remove residual photoresist and polymers created by the plasma process.

The next step is to place the wafer into an anneal furnace to drive the dopants further into the silicon (a diffusion process) and restore the damaged lattice structure back to its single-crystal structure.

The p-well formation proceeds in the same steps as the n-well formation, except that the p-well mask and boron ions are used instead of the n-well mask and phosphorus ions. The details are portrayed in Figures 3.29(b). The last step is to remove the screen oxide, as shown in Figures 3.29(c).

3.3.1.3 Isolation Methods In an IC, transistors must be electrically isolated from one to another to reduce leakage currents between transistors of the same type or different types. The isolation can be done either by making a reverse-biased pn junction between transistors or placing a thick dielectric between transistors. This thick dielectric is often called *field oxide* (FOX). The isolation method based on field oxide can be further subdivided into two methods: *local oxidation of silicon* (LOCOS) and

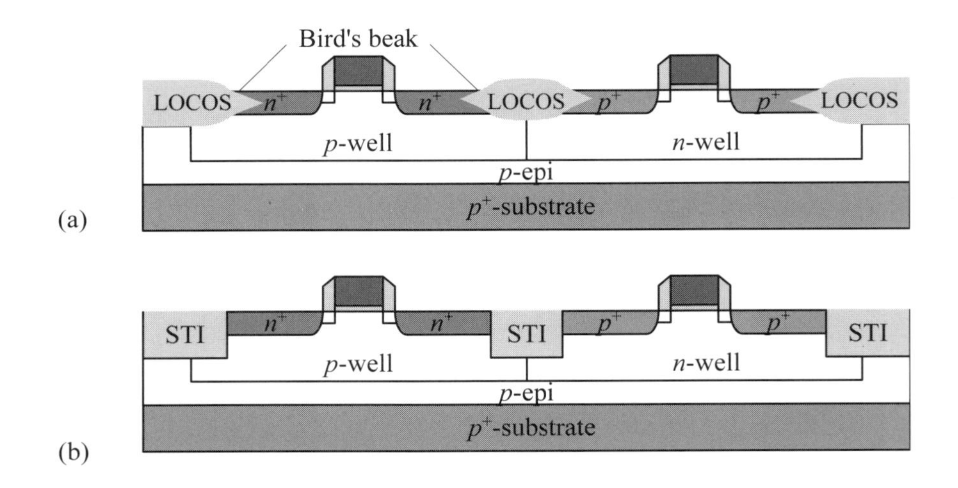

Figure 3.30: The two widely used isolation methods based on field oxide: (a) LOCOS; (b) STI.

shallow trench isolation (STI). These two isolation methods are compared and shown in Figure 3.30.

In the LOCOS method, a thick dielectric (such as silicon dioxide) with a thickness of 500 to 1,000 nm is formed in all regions where no transistors are defined. As shown in Figure 3.30(a), because of the isotropic diffusion of oxygen inside the silicon dioxide during the thermal oxidation process, both sides of LOCOS are encroached into source/drain regions of MOS transistors. This encroachment is known as "bird's beak" due to its bird-beak shape. The bird's beak encroachment almost has the same dimension of the oxide thickness. As a result, it severely limits the transistor packaging density.

In deep submicron technology, the STI method is widely used instead of the LOCOS method as the isolation between neighboring transistors because it has a more accurate vertical profile, as shown in Figure 3.30(b). The STI is so named because its depth is generally less than 1 μm, in contrast to deep trenches used in DRAM devices, with a depth of about 4 to 5 μm. Compared with the LOCOS method, the STI method not only can avoid the bird's beak problem but also has a more planarized surface topology. Consequently, it is widely used in CMOS processes below 0.25 μm.

Local oxidation of silicon (LOCOS). The major steps of the LOCOS process is portrayed as Figure 3.31. First, the wafer is cleaned and a pad oxide is grown on the surface. The pad oxide layer is necessary to reduce the tensile stress during the subsequent deposition of silicon nitride using an LPCVD process. After the deposition of pad oxide, a silicon nitride layer is deposited using an LPCVD process, as shown in Figure 3.31(a).

Next, the active mask (negative mask) is applied to reserve the active regions of the wafer surface using a photolithography process. After this, the unprotected regions of both silicon nitride and pad oxide are etched away, thereby leaving openings for depositing field oxide. However, prior to this oxidation process, two channel-stop implantation processes with boron (p^+) and phosphorous (n^+) are usually carried out separately. The *channel stop* is used to increase the threshold voltage of the parasitic MOS transistors possibly created by the conductor above the field oxide, the field oxide itself, and the substrate. The channel stop is generally an impurity with the same type

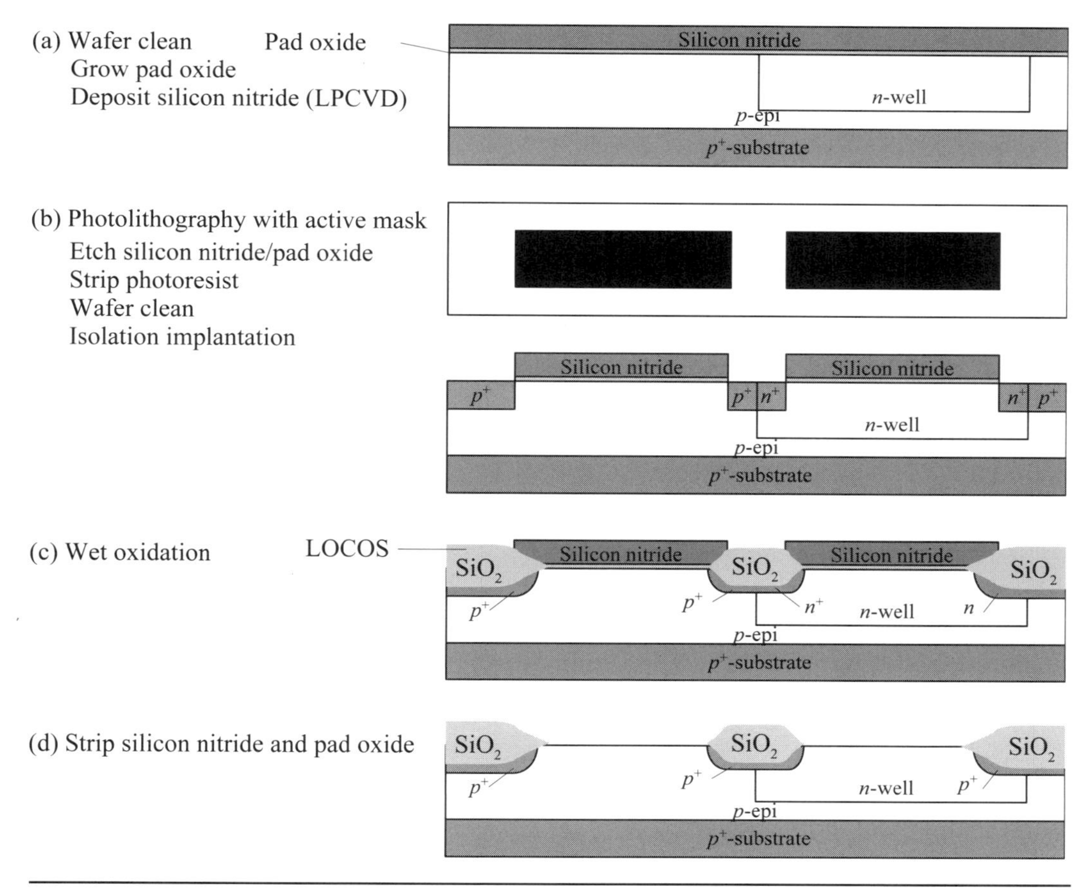

Figure 3.31: The major steps of local oxidation of the silicon (LOCOS) process.

as its substrate but opposite to that of the channel of MOS transistors. The result is shown in Figure 3.31(b).

Then, the field oxide is thermally grown by a wet oxidation process using silicon nitride as a mask. The result is shown in Figure 3.31(c). Finally, both silicon nitride and pad oxide are removed. The side view of the resultant wafer is shown in Figure 3.31(d).

Shallow trench isolation (STI). The major steps for forming STIs are portrayed in Figure 3.32. The first step is to clear the wafer surface and grow a new pad oxide with a thickness of about 10 to 15 nm. The pad oxide is used to protect active regions from chemical contamination and reduce the tensile stress during the silicon nitride strip. A silicon nitride layer is deposited on the wafer surface. The objectives of this silicon nitride layer are to serve as a mask to protect active regions during the trench oxide deposition process and to serve as the polish-stop material during the CMP process. The result is shown in Figure 3.32(a).

The second step is to form shallow trenches. To do this, the active mask (negative mask) is applied to reserve the active regions on the wafer surface using a photolithography process. A dry etching process is then applied to etch away the unprotected regions of silicon nitride and pad oxide, as shown in Figures 3.32(b) and (c). Finally, the photoresist is stripped and the wafer is cleaned to remove contamination. The sili-

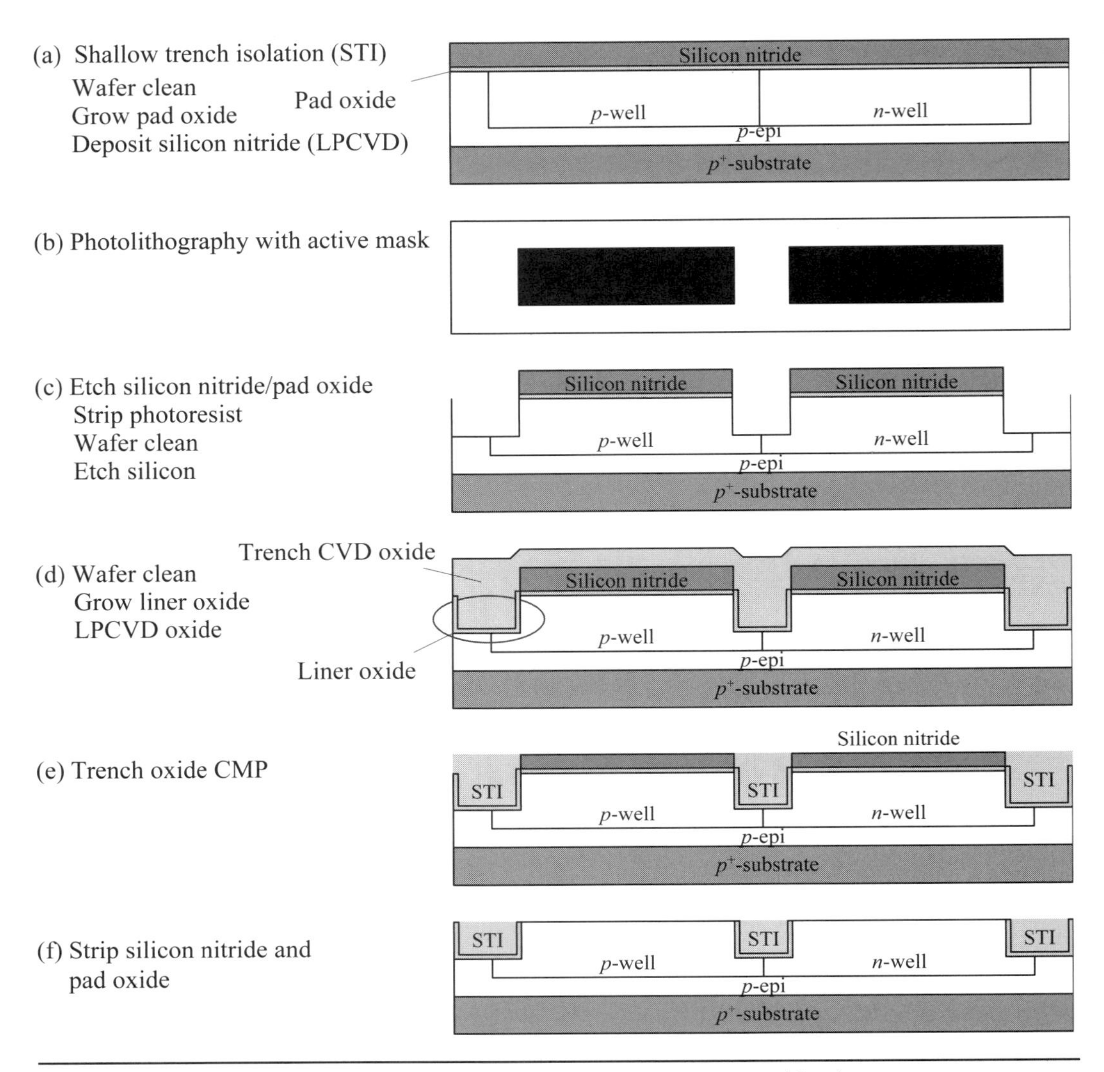

Figure 3.32: The major steps of the shallow trench isolation (STI) process.

con regions without being protected by silicon nitride are etched away to form shallow trenches.

The third step is to fill shallow trenches. As shown in Figure 3.32(d), the wafer is first cleaned to remove contamination and native oxide. An oxide layer with a thickness of about 15 nm is then grown on the walls of shallow trenches. This oxide layer is called the *liner oxide* and is used to improve the interface quality, such as reducing the leakage current between silicon and the trench oxide to be filled. After the deposition of liner oxide, the shallow trenches are filled with oxide by an LPCVD process.

The final step is to polish the wafer surface by applying a CMP process, as depicted in Figure 3.32(e). The silicon nitride is a harder material than oxide and serves as the polish-stop material; that is, the CMP process continues to remove the trench oxide until it reaches the silicon nitride. After the CMP process, the silicon nitride and pad oxide are removed by a wet etching process. The final result is shown in Figure 3.32(f).

3.3.1.4 Threshold Voltage Adjustment As discussed in the previous chapter, it is desirable to add dopants to adjust the threshold voltage of MOS transistors to form an enhancement device regardless of whether it is an n-type or a p-type. Hence, in CMOS processes, the threshold voltage adjustment is a very important module. Since the threshold voltage is positive for an nMOS transistor and negative for a pMOS transistor, the adjustment needs to be separately performed on each type of transistor.

The threshold voltage adjustment is usually done by a low-energy, low-current ion implantation and performed prior to the gate formation. Figure 3.33(a) shows the case of nMOS transistors. The wafer is first cleaned to remove contamination and native oxide, and then grown a sacrificial oxide with a thickness of about 10 to 20 nm. After this, the n-channel V_{Tn} (n-select) mask is applied to the transfer pattern onto the wafer surface by a photolithography process, and boron ions with a specific dose and energy are implanted into active regions. The result is depicted in Figure 3.33(b).

The threshold voltage adjustment for pMOS transistors is the same as that for nMOS transistors except that here the p-select mask and phosphorus ions are used instead of the n-select mask and boron ions. The result is revealed in Figure 3.33(c). After this ion implantation is done, the wafer is thermally annealed by an RTP process. Finally, the sacrificial oxide is stripped off. The result is shown in Figure 3.33(d).

3.3.1.5 Polysilicon Gate Formation The main steps in fabricating the polysilicon gate structure are shown in Figure 3.34. This process defines the gate oxide (or called thin oxide), polysilicon gate, and local interconnect of polysilicon. It is a critical step for modern CMOS processes because it is one of the smallest physical structures in the entire IC process.

The process proceeds as follows. First, the wafer is cleaned to remove contamination and native oxide. A *gate oxide* with a thickness ranging from about 1 nm to 5 nm is grown. The gate oxide serves as the dielectric between the polysilicon gate and substrate. After this, the wafer is deposited on a polysilicon layer with 500 nm in thickness by an LPCVD process. The polysilicon may also be doped into an n-type or a p-type to reduce its electrical resistivity. Some processes may call for doping the polysilicon immediately following the deposition step. The result is shown in Figure 3.34(a). For a deep submicron process in which deep UV photolithography processes are used, an antireflective coating (ARC) is commonly applied between the polysilicon and photoresist to reduce undesirable reflections.

Next, a gate mask is applied to a transfer pattern onto the wafer surface by a photolithography process. After this, an anisotropic plasma etching process is used to remove the unprotected polysilicon. Then, photoresist and ARC are stripped, as shown in Figure 3.34(b). Finally, the wafer is cleaned and placed in a high temperature (about 900°C) furnace chamber to reoxidize the polysilicon gate. The resulting oxide is thicker on the polysilicon than on the active region. This thermal oxidation process not only activates the implanted dopants on the polysilicon but also grows a buffer pad oxide for silicon nitride in the subsequent sidewall spacer deposition if silicon nitride is used. The result is depicted in Figure 3.34(c). It is worth noting that we will not show the threshold-voltage implantation regions in figures again hereafter to make these figures focus more on their main subjects.

3.3.1.6 LDD Extension Formation As mentioned earlier, the LDD structure not only suppresses the hot-carrier injection into the gate but also reduces short-channel effects of deep submicron MOS devices. The approach to making LDD extensions is as follows. First, n^- and p^- implants are introduced into the source/drain regions of nMOS and pMOS transistors, respectively. Second, a thin-oxide layer is deposited

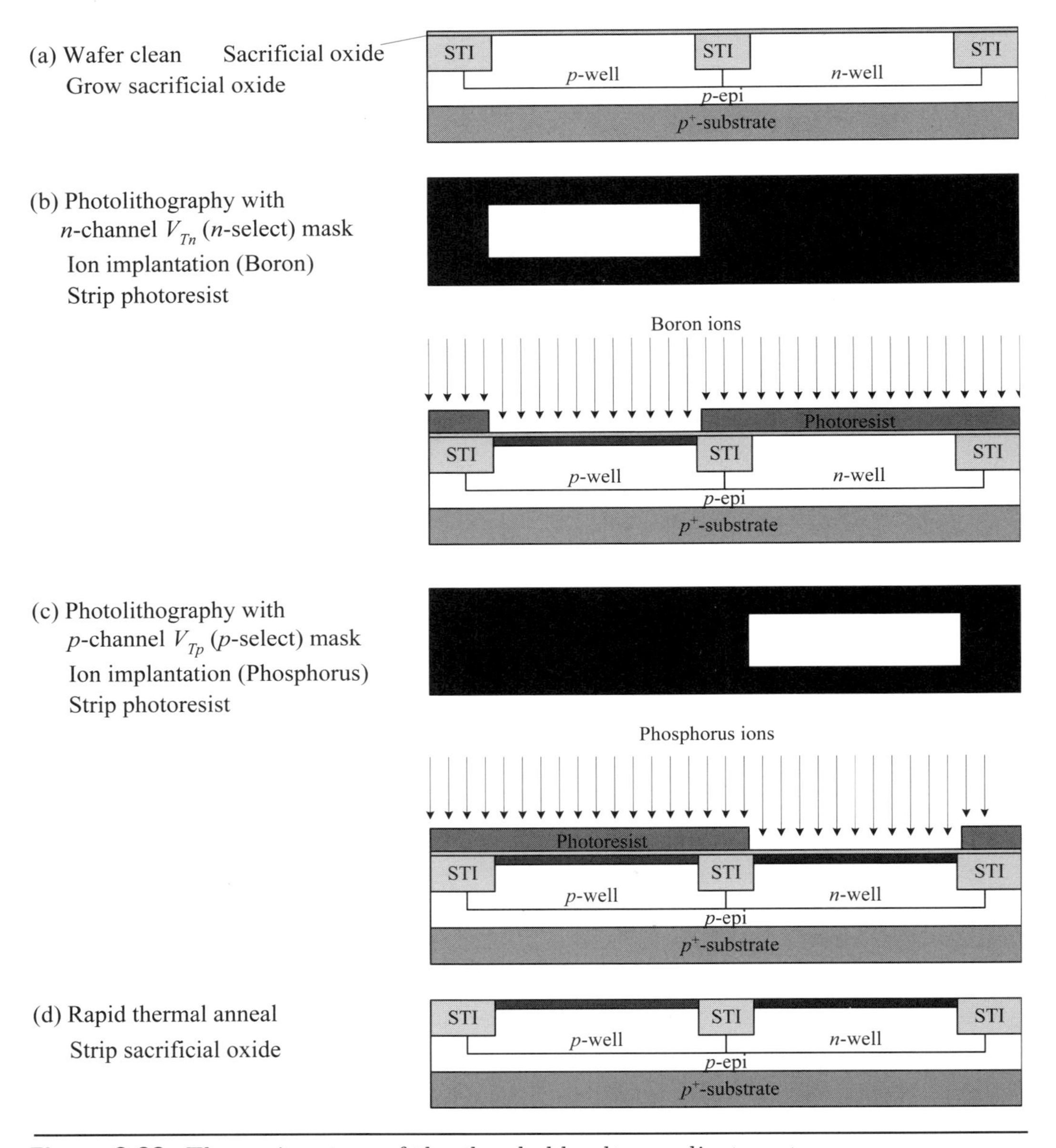

Figure 3.33: The major steps of the threshold voltage adjustment process.

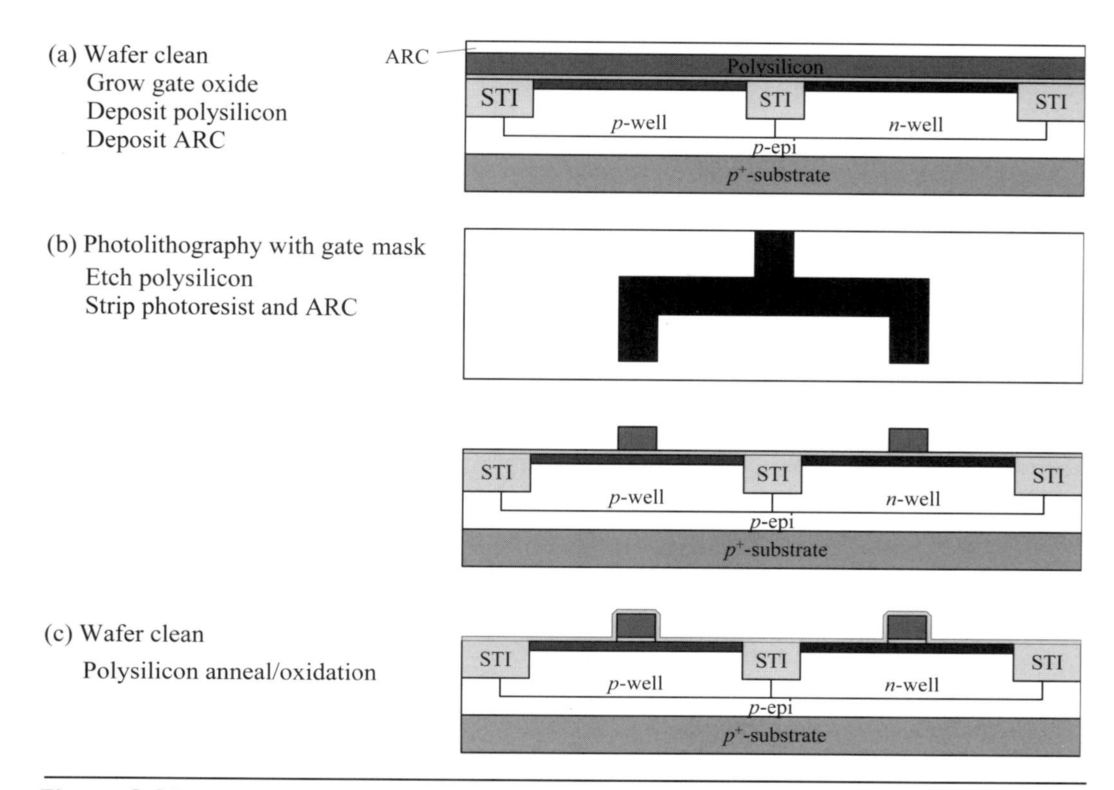

Figure 3.34: The major steps of the polysilicon gate formation process.

along the edges of the polysilicon gates. This thin-oxide layer is called a *sidewall spacer layer*.

The major steps for making LDD extensions are illustrated in Figure 3.35. The first two steps separately apply an n-channel LDD (n-select) mask and a p-channel LDD (p-select) mask to implant light concentrations of n-type and p-type impurity materials. The results are depicted in Figures 3.35 (a) and (b).

The last two steps are to form the sidewall spacers along the edges of the polysilicon gates of both nMOS and pMOS transistors. The sidewall spacers serve two functions: a mask to the source/drain implants, and a barrier to the subsequent silicide formation. As depicted in Figure 3.35(c), a thickness of 100 nm of silicon dioxide is deposited using a CVD process. The silicon dioxide is then etched back by an anisotropic plasma etching process, leaving behind the thicker silicon dioxide on the sidewalls of polysilicon gates. The sidewall spacers can also use silicon nitride instead of silicon dioxide.

3.3.1.7 Source/Drain (S/D) Formation As shown in Figure 3.36, to form the source/drain regions of nMOS transistors, the n-select mask coupled with the combination of polysilicon gate and sidewall spacers is applied as a mask, and a low-energy, high-dose arsenic ion implantation is performed to form the n^+ regions.

The source/drain regions of pMOS transistors are formed in a similar way. At this point, the p-select mask, coupled with the combination of polysilicon gate and sidewall spacers, is used as a mask, and a low-energy, high-dose boron ion implantation is performed to form the p^+ regions. The implants penetrate the silicon lightly beyond the LDD junction depth. The source/drain implantation process concludes with a high

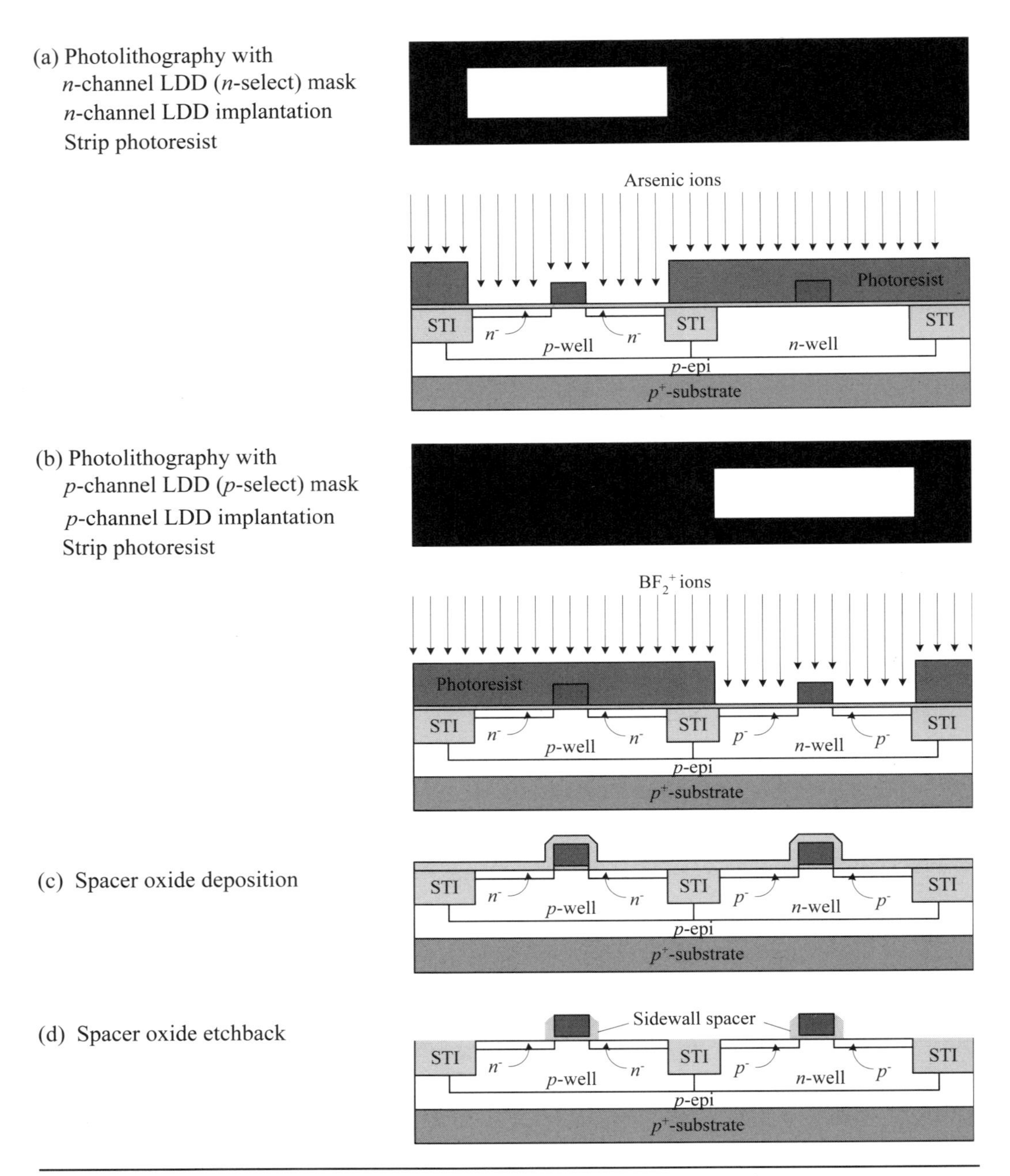

Figure 3.35: The major steps of the lightly doped drain (LDD) extension implant process.

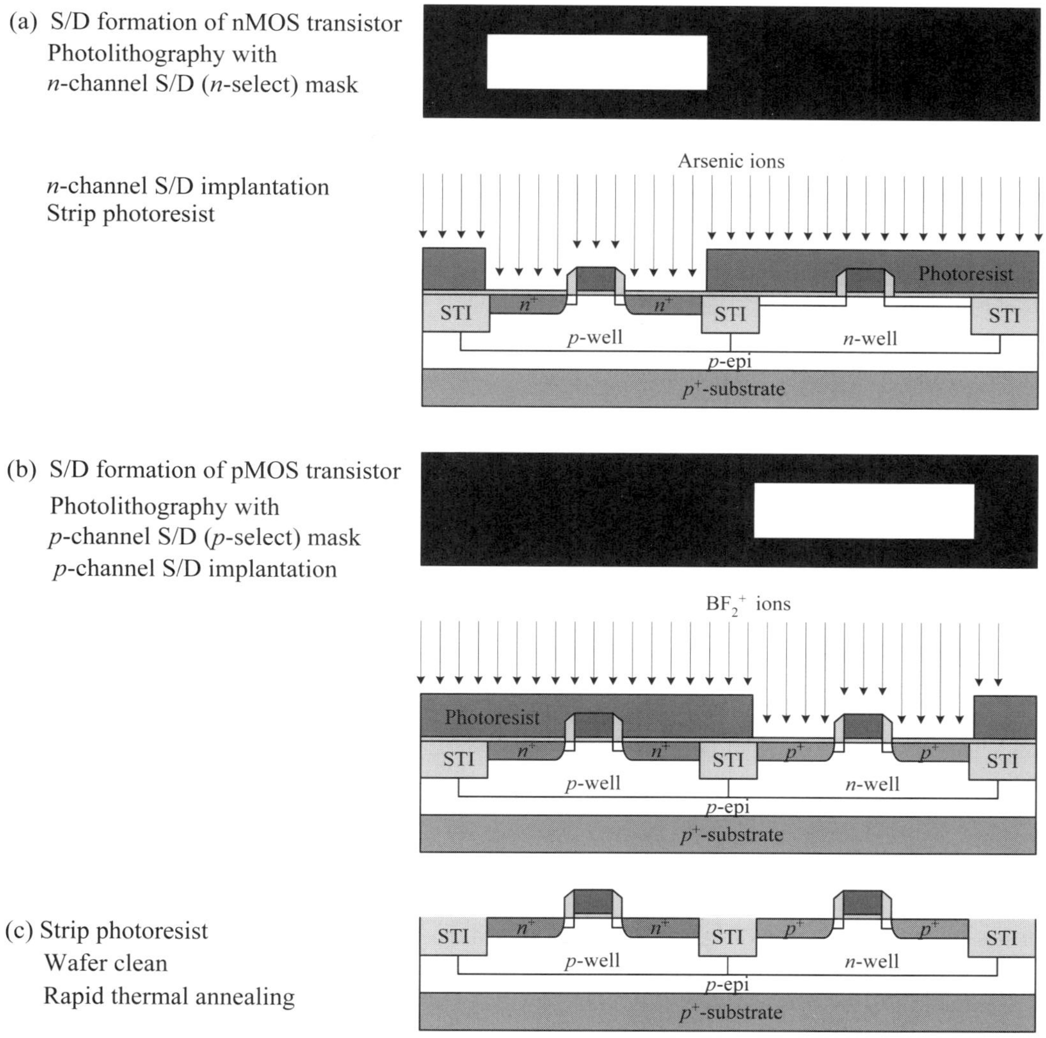

(a) S/D formation of nMOS transistor
Photolithography with
n-channel S/D (n-select) mask
n-channel S/D implantation
Strip photoresist
Arsenic ions
Photoresist
STI
n+
p-well
n-well
p-epi
p+-substrate
(b) S/D formation of pMOS transistor
Photolithography with
p-channel S/D (p-select) mask
p-channel S/D implantation
BF2+ ions
p+
(c) Strip photoresist
Wafer clean
Rapid thermal annealing

Figure 3.36: The major steps of the source/drain implant process.

temperature annealing, an RTP at 1000°C for several seconds, to remedy the damaged silicon structure and activate the implants.

■ Review Questions

Q3-46. What are the differences between LOCOS and STI?

Q3-47. What is the intention of channel-stop implantation?

Q3-48. What is the function of antireflective coating (ARC)?

Q3-49. What is the purpose of sidewall spacers?

Q3-50. How would you make LDD extensions?

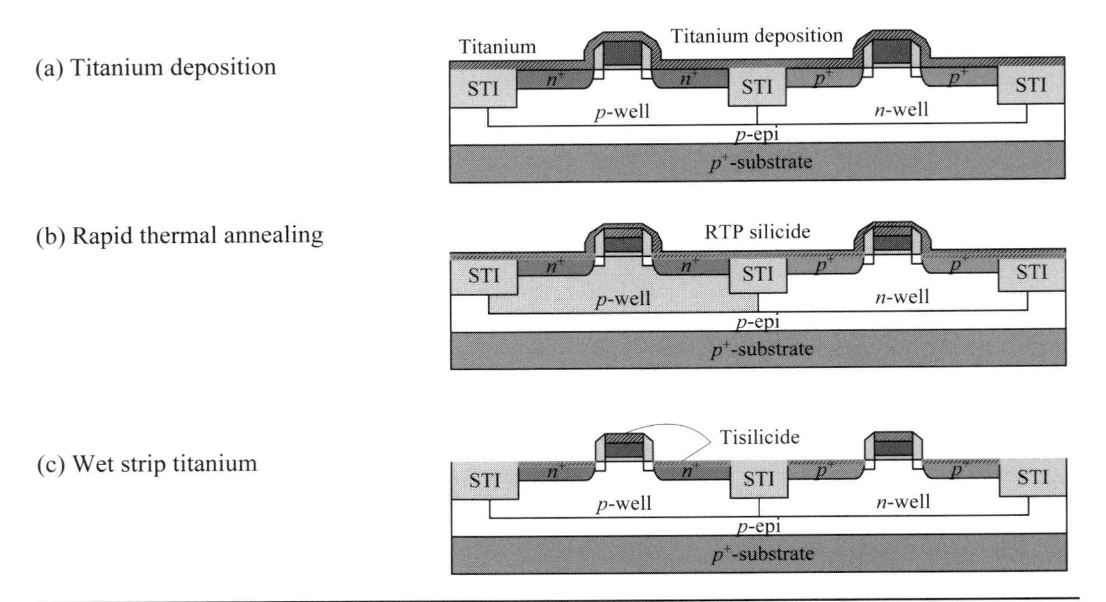

Figure 3.37: The major steps in titanium contact formation.

3.3.2 BEOL

BEOL refers to all processes required to interconnect all isolated MOS transistors into a complete circuit or module and to I/O pads. We describe each of these processes in more detail.

3.3.2.1 Silicide Formation The silicide formation process is the interface between FEOL and BEOL. Its functions are to form metal contacts on all active and polysilicon areas to improve the adhesion between the silicon/polysilicon and metal conductor material. The two most common contact materials are titanium and cobalt. The interesting properties of titanium material are that it has low resistivity and is easy to react with silicon but not with silicon dioxide. As the temperature is raised above 700°C, titanium material bonds with silicon and forms a titanium compound called a *titanium silicide* ($TiSi_2$), called *tisilicide* for short.

The major steps in titanium contact formation are shown in Figure 3.37. The first step is to clean the wafer surface to thoroughly remove contamination and native oxide. Next, a thin film of titanium (Ti) is deposited on the wafer surface using a PVD process, as shown in Figure 3.37(a). Then, the wafer is annealed in an RTP tool at a temperature above 700°C. As stated above, the titanium reacts with the silicon and forms a tisilicide ($TiSi_2$), as shown in Figure 3.37(b). Finally, using a proper wet etching process etches away the unreacted titanium, leaving behind the tisilicide over the active silicon and polysilicon regions. The result is shown in Figure 3.37(c).

3.3.2.2 Contact Formation The purposes of contact formation are to open contact holes for metal1. As shown in Figure 3.38, the first step is to deposit a barrier layer of silicon nitride or silicon dioxide using a CVD process. The premetal dielectric (PMD) formed from SiO_2 deposited by a CVD process follows. The PMD provides the isolation between metal1 layer and silicon/polysilicon. To improve its dielectric qualities, the PMD layer is then lightly doped with phosphorus or boron. An additional RTP step

(a) Silicon nitride deposition
Oxidation deposition
Lightly doped oxide CVD
PMD deposition by CVD
PMD polish by CMP

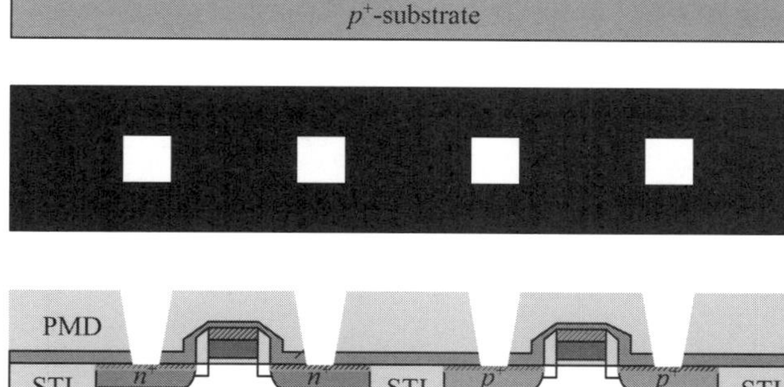

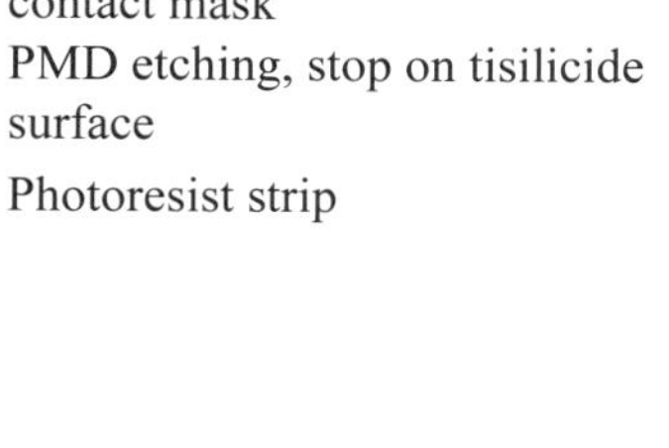
(b) Photolithography with contact mask
PMD etching, stop on tisilicide surface
Photoresist strip

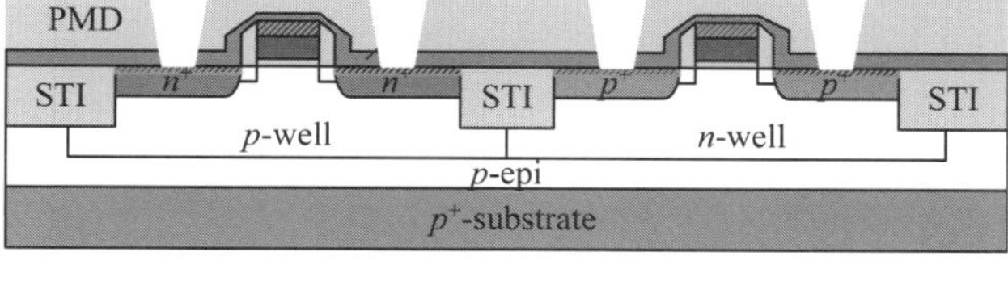

(c) Titanium deposition by PVD
Titanium nitride deposition
Tungsten desposition by CVD
Tungsten polish by CMP

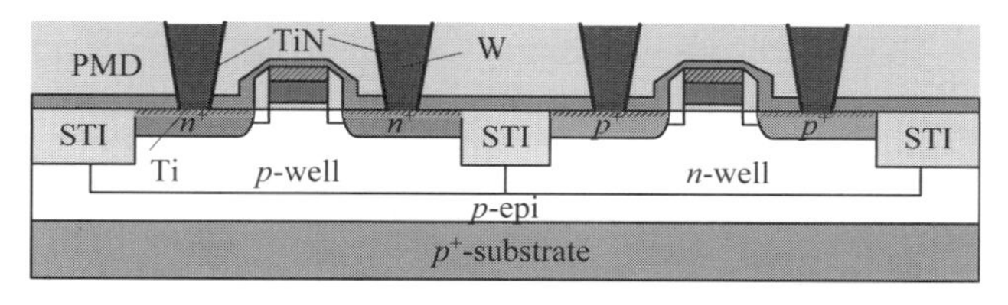

Figure 3.38: The major steps of contact formation.

may be used to smooth the surface. A CMP process is used to planarize the surface and the resulting thickness of the PMD is about 800 nm, as depicted in Figure 3.38(a).

The wafer is first patterned with the contact mask, as shown in Figure 3.38(b) and then the PMD is etched to create the narrow trenches that will serve as the contact holes for defining the paths between the metal1 layer and silicon/polysilicon. The PMD etching process is stopped on the tisilicide surface.

After the trenches are created in the PMD layer, the plugs are formed in the following steps. First, a thin barrier of titanium (Ti) is deposited by a PVD process on the bottom and inside of trenches to improve the adhesion between tungsten (W) and SiO_2. Then, a titanium nitride (TiN) layer is deposited immediately over the titanium to serve as a diffusion barrier to fluorine, which readily etches silicon, used in the subsequent tungsten deposition. Finally, the contact openings are filled with tungsten using a WF_6 CVD process. The resulting wafer is etched back to the top of planarized PMD using a CMP process, as shown in Figure 3.38(c).

3.3.2.3 Formation of Metal1 Interconnect The metal1 is used to connect contact to contact and contact to via1. It is formed by a stack of titanium, titanium nitride, aluminum-copper alloy, and titanium nitride. As depicted in Figure 3.39, a titanium layer is first deposited by a PVD process followed by a thin layer of titanium nitride (TiN). The titanium is the first metal to be deposited on the entire wafer. It not only serves as the glue between the tungsten plugs and the next metal, aluminum and

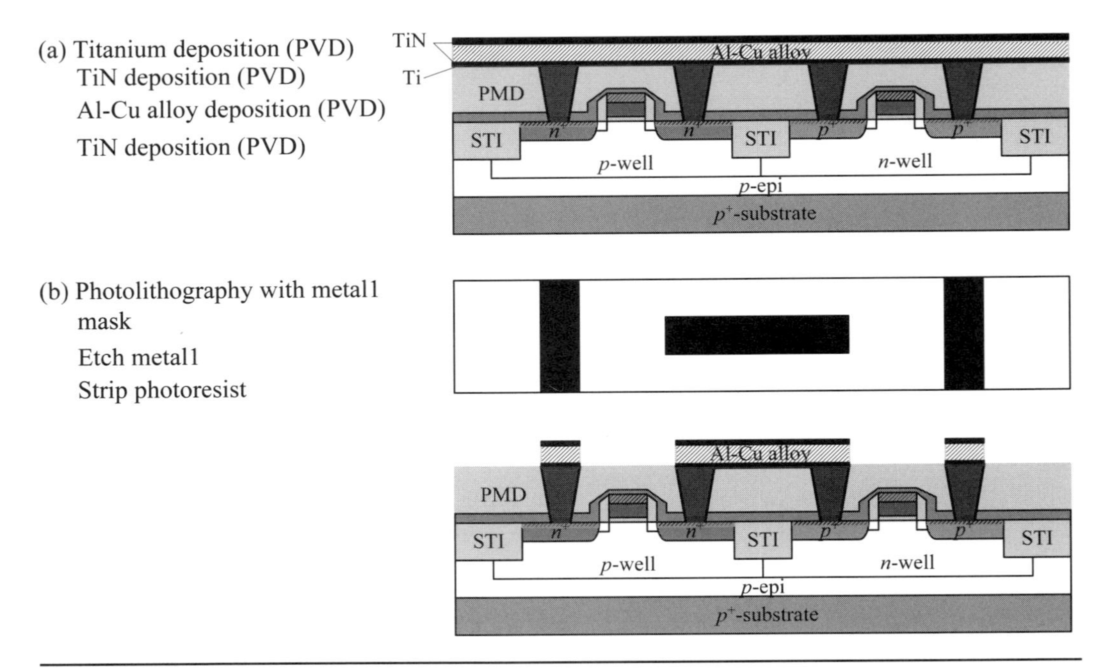

Figure 3.39: The major steps of metal1 interconnect formation.

copper alloy, but also adheres well to the PMD layer material to improve the reliability of the metal stack.

After the deposition of titanium nitride, a layer of aluminum-copper alloy is deposited using a PVD process. The use of copper (about 1%) is to alleviate the electromigration effect of the aluminum material alone. Next, a thin layer of titanium nitride (TiN) is deposited using a PVD process on top of the aluminum-copper alloy to serve as an antireflective coating (ARC) layer for the next photolithography process. The resulting structure is shown in Figure 3.39(a).

A photolithography process with the metal1 mask is then applied to etch away the undesired part of the metal stack, leaving behind the desired part protected by the metal1 mask, as shown in Figure 3.39(b).

3.3.2.4 Via1 and Plug1 Formation Metal1 and metal2 are electrically contacted by via1 and electrically isolated by an intermetal dielectric (IMD). It is of interest to note that in the IC industry the connections between active regions or polysilicon and metal1 are called *contacts* and the connections between any two adjacent metal layers are called *vias*. Via1 is the connection between metal1 and metal2, via2 is the connection between metal2 and metal3, and so on. Similarly, the dielectric material deposited between metal1 and metal2 is called IMD1, the dielectric material deposited between metal2 and metal3 is called IMD2, and so forth.

As shown in Figure 3.40(a), the IMDi layer is usually deposited using a HDPCVD process and then followed by a CMP process to globally planarize the IMDi surface topography. The next few steps follow the same paradigm of contact formation. That is, the via1 mask is applied to define where the via1 and plug1 are needed, and then the IMD1 is etched away and the photoresist is stripped off. At this point, the via trenches are formed as shown in Figure 3.40(b).

(a) IMD1 oxide deposition (CVD)
IMD1 oxide polish (CMP)

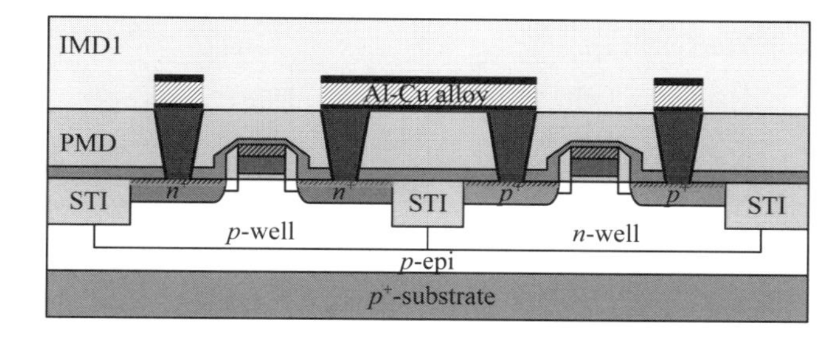

(b) Photolithography with
via1 mask
Etch IMD1
Strip photoresist

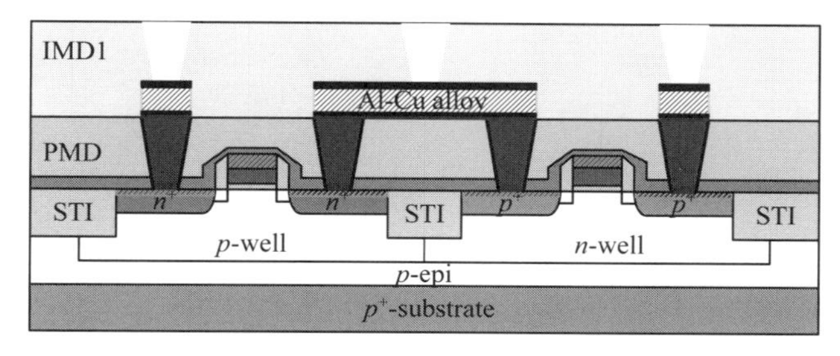

(c) Titanium deposition (PVD)
TiN deposition (CVD)
Tungsten deposition (CVD)
Tungsten polish

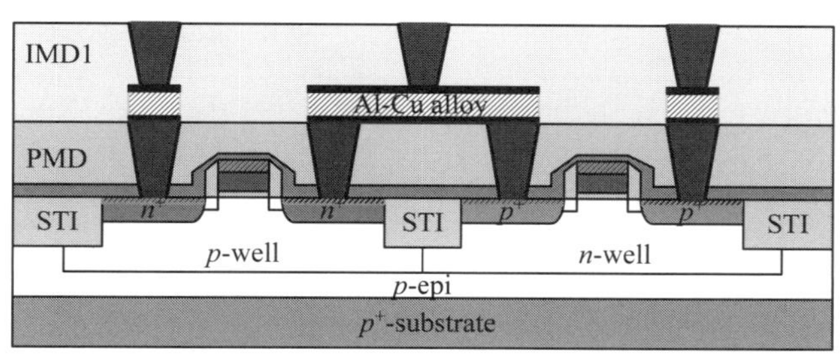

Figure 3.40: The major steps of via1 and plug1 formation.

After the trenches are created in the IMD1 layer, the plugs are formed as in the following steps. First, a thin barrier of titanium (Ti) is deposited by a PVD process on the bottom and inside of the trenches to improve the adhesion between the tungsten (W) and the SiO_2. Then, a titanium nitride (TiN) layer is deposited immediately over the titanium to serve as a diffusion barrier to fluorine, which readily etches silicon, used in the subsequent tungsten deposition. Finally, the contact openings are filled with tungsten using a WF_6 CVD process. The resulting wafer is etched back to the top of the planarized IMD1 using a CMP process, as shown in Figure 3.40(c).

3.3.2.5 Formation of Metal2 Interconnect The metal2 is used to connect via1 to via1 and processed in a similar way as the metal1. It is also formed by a stack of titanium, titanium nitride, aluminum-copper alloy, and titanium nitride. The major steps of metal2 interconnect formation is shown in Figure 3.41. Because of its similarity to the metal1 process, we would like to omit its details here.

For a multilayer-metal process, additional tiers of dielectric/metal layers can be formed by replicating the above-mentioned via-and-plug process followed by a metal

(a) Titanium deposition (PVD)
TiN deposition (PVD)
Al-Cu alloy deposition (PVD)
TiN deposition (PVD)

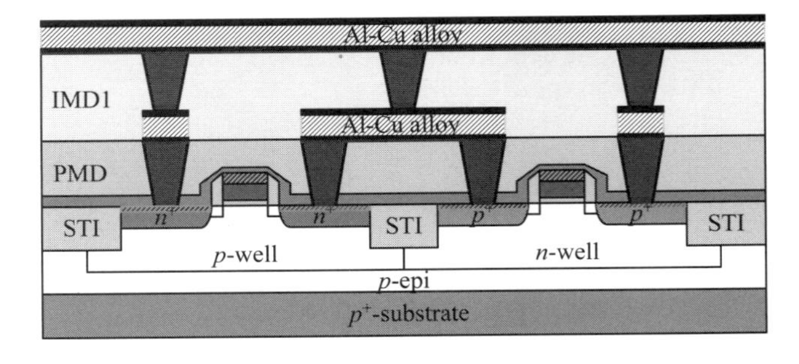

(b) Photolithography with metal2 mask
Etch metal2
Strip photoresist

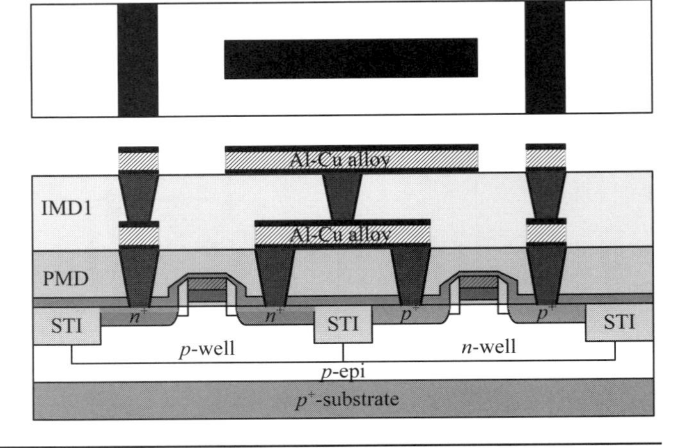

Figure 3.41: The major steps of metal2 interconnect formation.

formation process. In modern deep submicron CMOS processes, there may be up to nine, even more, metal layers.

3.3.2.6 Passivation and Bonding Pad The final step of BEOL is to deposit a silicon nitride or doped glass layer of a thickness of about 200 nm on the wafer surface, as shown in Figure 3.42(a). This layer is referred to as the *passivation layer*. Its purposes are to protect the wafer from scratches and moisture during wafer probe and packaging, and to provide a barrier to prevent the wafer from contamination. Since silicon nitride does not adhere to aluminum very well due to stress mismatch, an oxide layer (IMD4) is deposited prior to the formation of thick silicon-nitride passivation to buffer the stress and help silicon-nitride adhesion.

A *bonding pad mask*, also known as a *passivation mask*, is then applied to open holes for bonding pads, as illustrated in Figure 3.42(b). At this point, the entire wafer is virtually processed. A final polyamide coating is added to protect the wafer from mechanically scratching during wafer shipping and the microelectronic devices from being damaged by alpha particles. The result is shown in Figure 3.42(c). The wafers are then shipped to test and package.

3.3.2.7 Dual-Damascene Copper Process In deep submicron CMOS processes, copper has been widely used as an interconnect due to its lower sheet resistivity and higher electromigration resistance than aluminum-copper alloy. Although copper can be deposited using either a CVD or PVD process, it is very difficult to be etched. Fortunately, the copper can be polished by using the CMP process. The major challenges

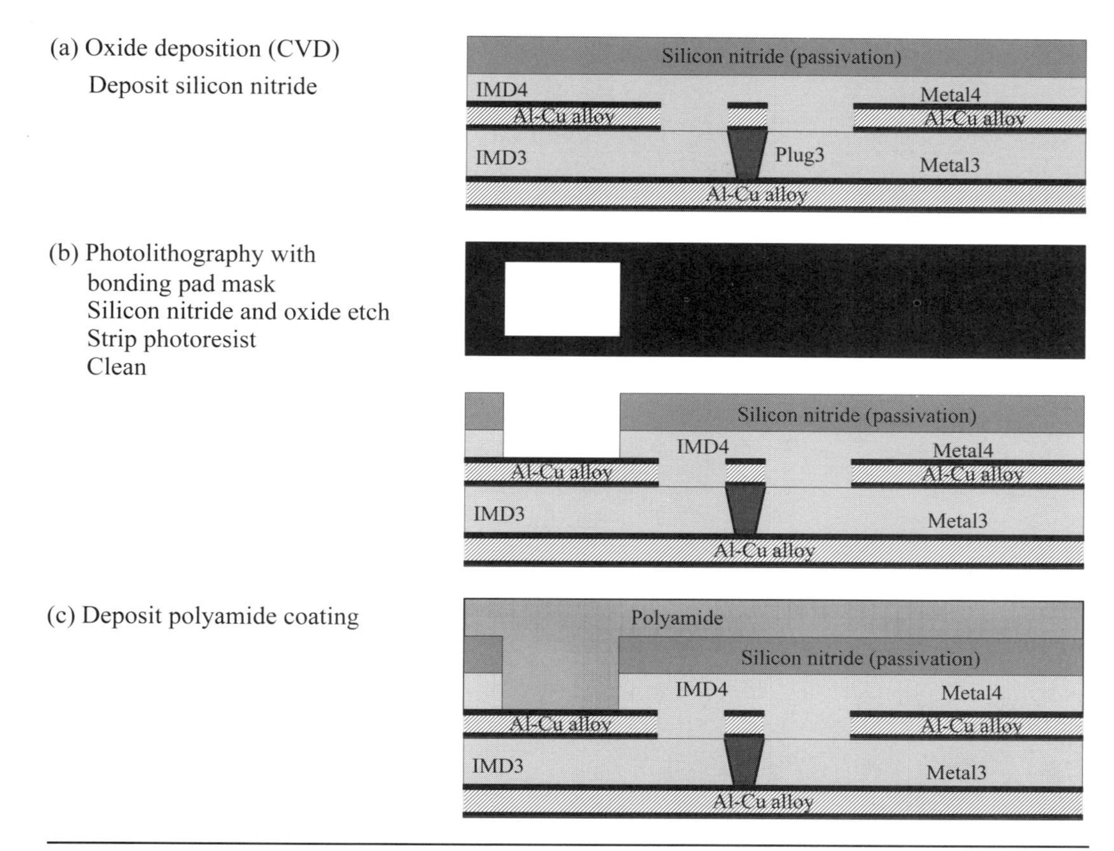

Figure 3.42: The major steps of passivation layer and bonding pad formation.

of using copper as an interconnect are dielectric etch, metal deposition, and metal CMP processes.

The copper interconnect can be done by either of the following methods: *damascene* or *dual damascene*. In the damascene method, the copper pattern is first etched into a silicon dioxide layer and then copper is deposited on the surface. A CMP process follows to planarize the surface and remove copper not in silicon dioxide trenches. The detailed steps are depicted in Figure 3.43. In the dual-damascene method, the major steps are the same as the damascene method except that two silicon dioxide etch steps are used to create two stacked holes of different sizes. And then copper is deposited into these two stacked holes. After the planarization of the dual-damascene copper process, the result is shown in Figure 3.44. The difference between the dual-damascene copper process and the conventional interconnect process is that the former requires two dielectric etches and no metal etch while the latter needs one dielectric etch and one metal etch.

The major steps of the dual-damascene copper process are shown in Figure 3.45. As shown in Figure 3.45(a), silicon dioxide (PMD) and silicon nitride are deposited on the wafer surface in sequence. Then, a contact (via) mask is applied to the wafer surface to etch away the unprotected silicon nitride. The photoresist is stripped, leaving the result depicted in Figure 3.45(b).

Unlike the metalization process based on aluminum-copper alloy, the metal1 mask is applied to the wafer immediately following the contact mask after depositing an

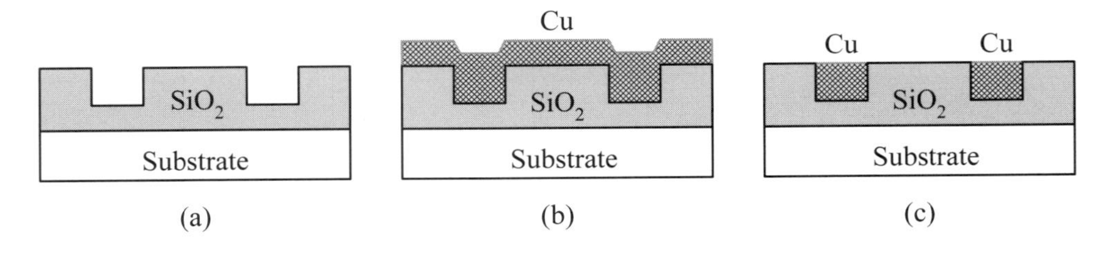

Figure 3.43: The major steps of the damascene copper process: (a) oxide patterning; (b) copper deposition; (c) planarization.

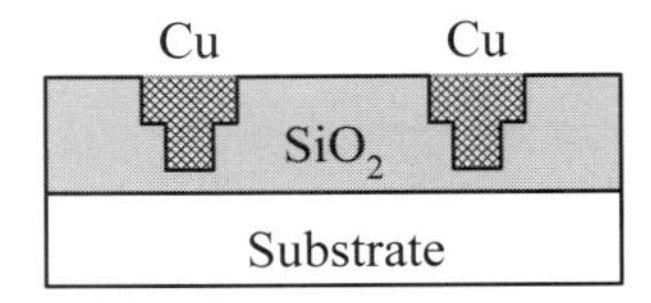

Figure 3.44: The result after planarization of the dual-damascene copper process.

IMD layer. The trenches corresponding to the metal1 interconnect are formed. At the same time, the contact holes formed on the top dielectric layer are etched through to the bottom of the second dielectric layer and stop on the silicon nitride and tisilicide. The details are shown in Figures 3.45(c) and (d).

The copper layer is deposited in the openings of both contacts and the interconnect trenches. However, prior to doing this, the wafer is cleaned first to remove native oxide and possible polymer depositions from dielectric etch on the surface of the bottom of contact or via openings. Then, a diffusion barrier layer of tantalum (Ta) is deposited to prevent copper from diffusing into the silicon substrate and damaging the microelectronic devices. A copper seed layer follows. The copper seed layer provides a nucleation site for the bulk copper grain and film formation so that the copper can adhere to the wafer surface very well. The result is portrayed in Figure 3.45(e).

After the copper seed layer, the copper layer is deposited by an *electrochemical plating* (ECP) process and a CMP process is employed to planarize the wafer surface. Finally, the resulting wafer is deposited on a thin layer of silicon nitride using a CVD process to prepare for the next via/metal layer deposition. The result is shown in Figure 3.45(f).

Review Questions

Q3-51. What does BEOL refer to?

Q3-52. What is the purpose of the passivation layer?

Q3-53. What is the intention of the passivation mask?

Q3-54. Describe the damascene and dual-damascene methods.

Q3-55. Why is a polyamide coating added to the wafer?

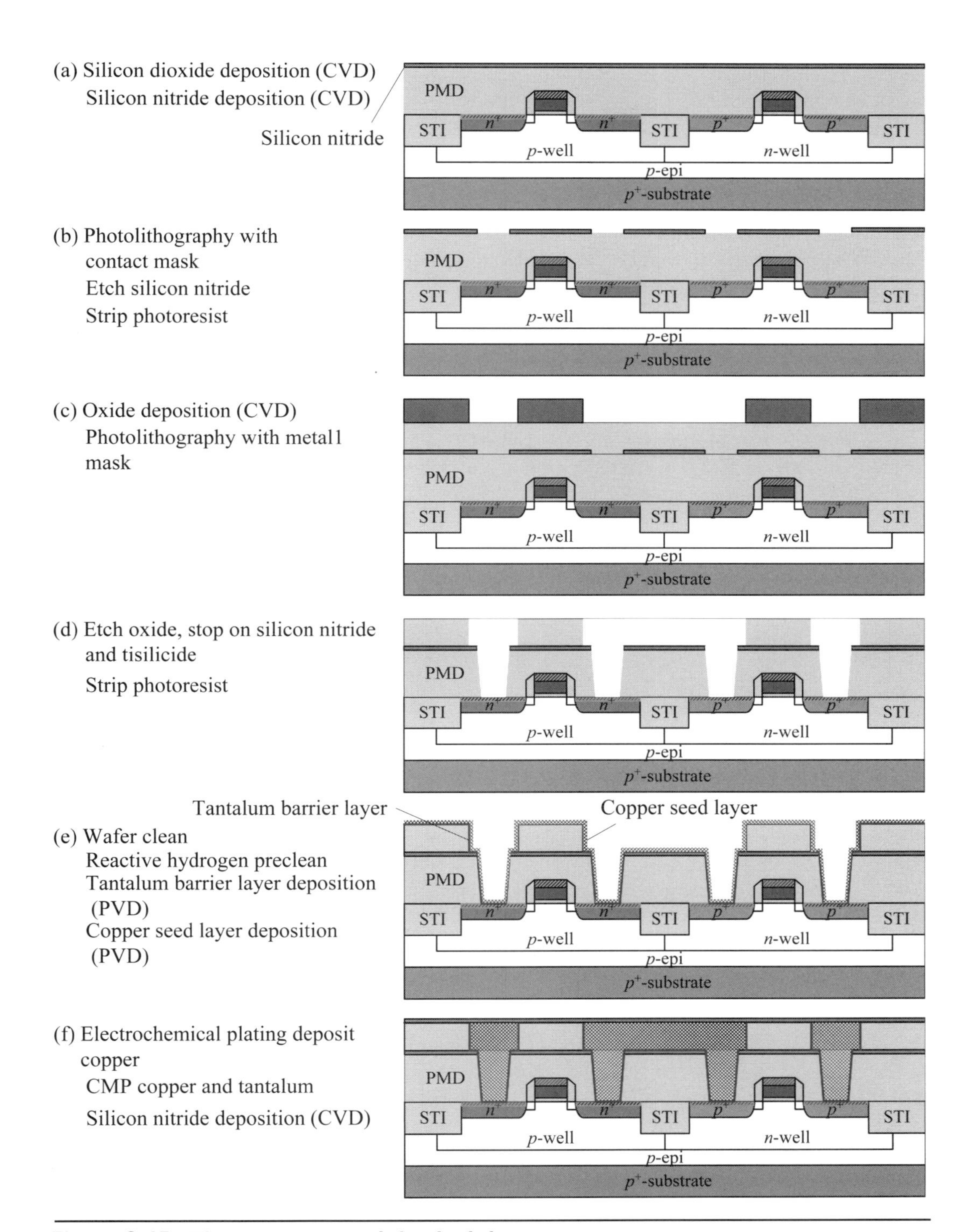

Figure 3.45: The major steps of the dual-damascene copper process.

3.3.3 Back-End Processes

After the completion of the final passivation layer of a wafer, a series of postfabrication processes, called *back-end processes*, are carried out. These back-end processes include the *wafer test* (or called the *wafer probe*), *die separation*, *packaging*, and *final test* as well as *burn-in test*. More details about the wafer test, final test, and burn-in test can be found in Section 16.1.

3.3.3.1 Wafer Test The wafer test is performed while the wafers are being processed and after the wafers are removed from the IC fab to check device parameters, functionality, and performance. The wafer test is generally further subdivided into two tests: *in-line parameter test* and *wafer sort*. The in-line parameter test, also known as *wafer electrical test* (WET), is an electrical test performed on some special test structures, referred to as *wafer test structures*. These test structures are commonly arranged in the *scribe line* region between the individual dies on the wafer. The scribe line region is like a street between two blocks in a city and surrounds a complete chip where it is cut with a diamond saw. Each scribe line has a width between 100 to 150 μm. The wafer test structures are also referred to as *process control monitors* (PCMs). Some examples of wafer test structures are as follows.

- *Discrete transistors* are used to measure the characteristics of transistors, including leakage current, breakdown voltage, threshold voltage, and effective channel length as well.
- *Various line-widths* are used to measure critical dimensions.
- *Contact or via chain* are used to measure contact resistance and connection.

Wafer sort is also known as *wafer probe* since it is accomplished by using sophisticated equipment to probe individual dies and apply test vectors to test circuit behavior. Inevitably, some dies do not pass the test and are considered as failures. These dies are marked to indicate that they need not be further packaged.

3.3.3.2 Die Separation After the wafer test is completed and the failed dies are marked, the dies are cut from the wafer by a diamond-blade dicing saw, which is about 25 μm in thickness and is rotated up to 20,000 rpm. It is apparent that the die separation process may damage some dies and reduce the yield.

3.3.3.3 Packaging After good dies are cut from the wafer, they are attached in a lead frame or a substrate in the appropriate package type and then the bonding pads are wired to the leads of the package. Except for protecting the chip from outside mechanical damage, the other functions of packaging include delivering power to the chip, to remove heat away from the chip, and to transfer information into and out of the chip to the printed circuit board (PCB) as well.

A chip can be physically attached to the lead frame using one of the following techniques: *epoxy attach*, *eutectic attach*, and *glass frit attach*. The lead frame packaging requires wire bond attachment and is inexpensive. It is suitable for small IC with low-pin count (300 pins max), such as low-capacity memory devices and ASICs. Nowadays, about 75% of ICs are in lead-frame packages.

The commonly used wire bonding techniques include *thermocompression bonding*, *ultrasonic bonding*, and *thermosonic ball bonding*. Regardless of which bonding technique, a bonding mechanism called *capillary tip* is used to position the gold wire on a chip bonding pad or a lead of the package. In thermocompression bonding, thermal energy and pressure are used to form the wire bond of a gold wire to the chip bonding

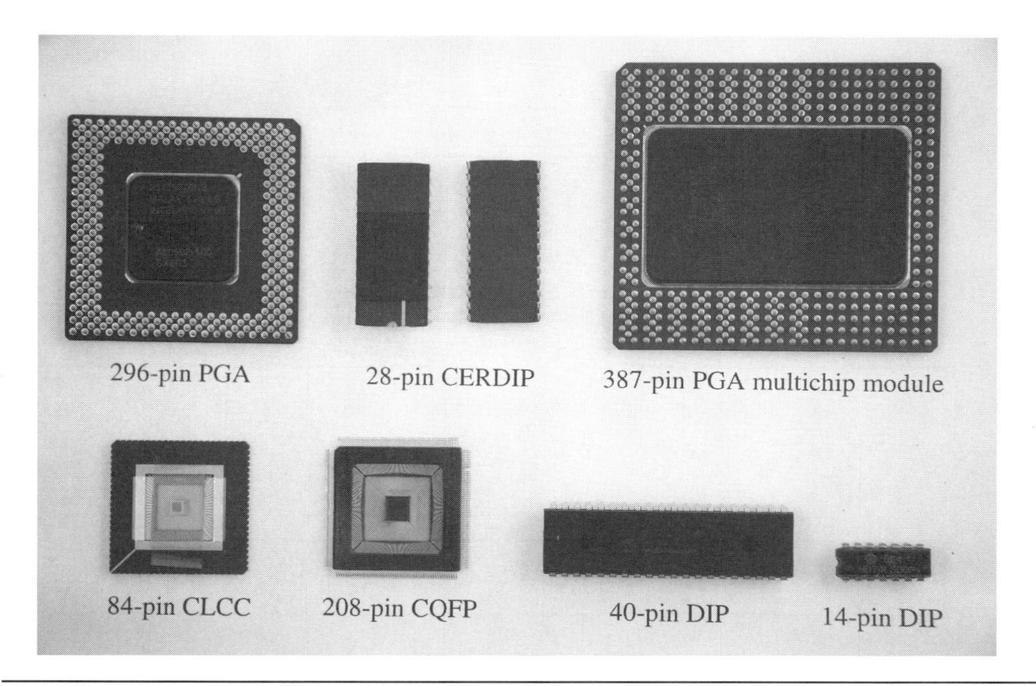

Figure 3.46: Some widespread use of integrated-circuit packages.

pad and the lead frame lead. In ultrasonic bonding, ultrasonic energy and pressure are used. In thermosonic ball bonding, a combination of ultrasonic and thermal energy, as well as pressure is used to form a bond, referred to as a ball bond.

After a chip is attached on a lead frame and its bonding pads are appropriately wired to the leads of the lead frame, the resulting set still needs to be housed in a package to protect itself from contamination and moisture. The two most widely used types of traditional IC packaging materials are *plastic packaging* and *ceramic packaging*. The choice of which type of material is suitable for a particular chip depends on the power dissipation and performance (namely, operating frequency) of the chip.

There are many different types of plastic package. The most common ones include *dual-in-line package* (DIP), *single in-line package* (SIP), *thin small outline package* (TSOP), *plastic leaded chip carrier* (PLCC), and *leadless chip carrier* (LCC). Ceramic packages are used in state-of-the-art applications that require either high-power or maximum reliability. Ceramic packaging mainly includes *pin grid array* (PGA) and *ceramic DIP* (CERDIP). Some widely used integrated-circuit packaging are shown in Figure 3.46.

Recently, many new packages are applied to provide more reliable, faster, and higher-density circuits at low cost. Two of them are introduced in the following: *ball grid array* (BGA) and *chip on board* (COB). Before introducing these package types, we first briefly describe the flip-chip technique. *Flip chip* is a packaging technique that places the bumped chip on a routable substrate in an upside-down fashion relative to the wire-bonding packaging approach. It results in the shortest paths from the chip devices to the substrate, thereby reducing the packaging inductance and resistance and being suitable for high-speed signals. Combining the flip-chip technique with the most common solder bump process, referred to as C4 (controlled collapse chip carrier), results in a packaging technique known as the *flip-chip C4 technique*.

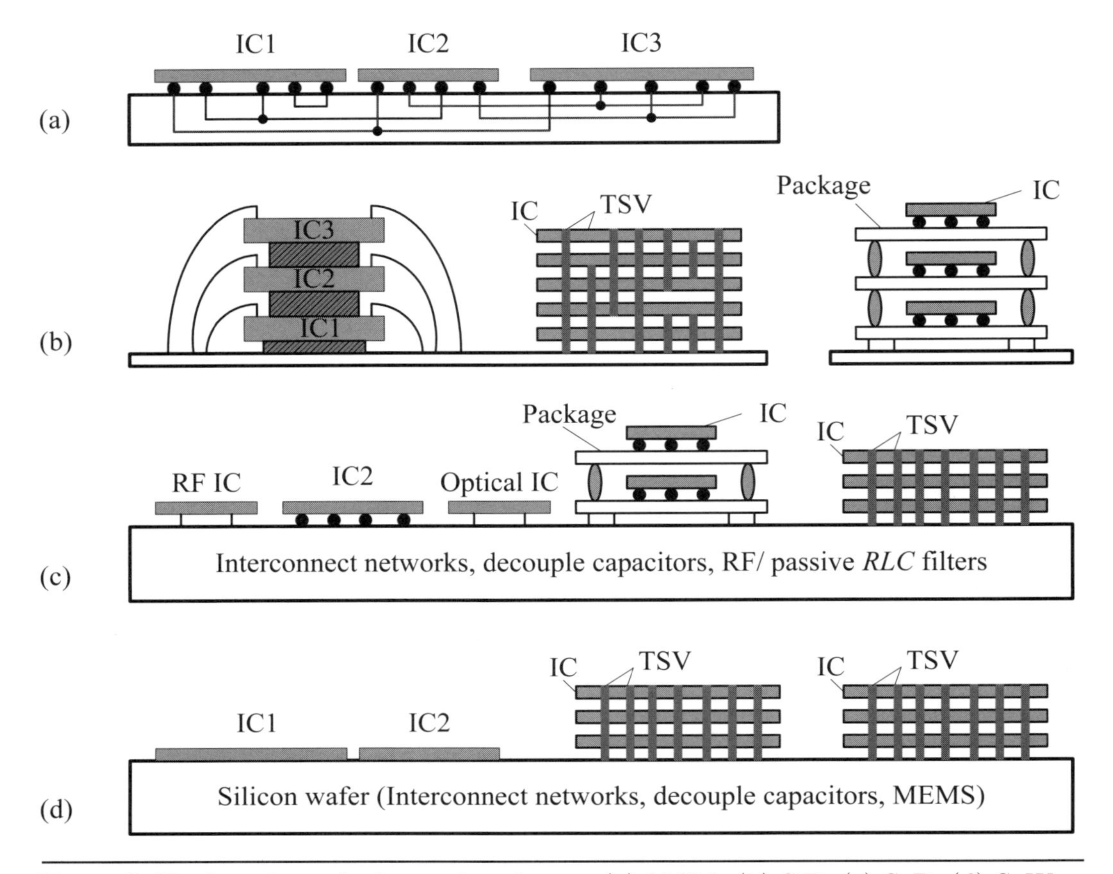

Figure 3.47: A variety of advanced packages: (a) MCM; (b) SiP; (c) SoP; (d) SoW.

In the BGA package, a routable multilayer substrate with organic and ceramic materials is used. The BGA package is similar to PGA but instead of using pin connections, it uses an array of *solder balls* to connect the substrate to the circuit board. The silicon chip may be attached to the top of a substrate using wire bond or the flip-chip C4 technique. Since the solder ball pitch is much less than pin pitch, BGA has a higher lead count than PGA.

The COB process, also called *direct chip attach* (DCA), is a process that mounts IC chips directly onto the substrate accompanied with other *surface-mount technology* (SMT) components. Because of no package around the silicon chip, the COB approach significantly reduces the package size and hence is very suitable for the applications where size and cost are important.

3.3.3.4 Advanced Packaging All of the aforementioned packages are used to package a single chip. In modern complex systems, it is often to pack multiple chips in a single package to reduce power dissipation and increase performance. To achieve this, the following techniques are the most common: *multichip module* (MCM), *system-in-package* (SiP), *system-on-package* (SoP), and *system-on-wafer* (SoW), as illustrated in Figure 3.47. The last three techniques belong to 3-*D* packaging. When packaged with one of these three techniques, the IC is often called a 3-D IC.

With the MCM technique, as shown in Figure 3.47(a), many chips are placed on a routable substrate and interconnected through the interconnect mechanism facilitated by the substrate to serve their use as a single IC. Depending on the system complexity and design philosophy, MCM techniques come in a variety of forms, ranging from mounting prepackaged ICs on a small PCB to fully integrating many chips on a high-density interconnect substrate.

An SiP is a system that is constructed as the vertical stacking of similar or dissimilar components in the same package rather than the horizontal nature of SoC. An illustration is given in Figure 3.47(b), where an SiP may be constructed as a combination of 3-D stacking of similar ICs, such as DRAM modules, the stacking of dissimilar ICs, such as microprocessors, DRAMs, or Flash memory, or the stacking of packaged ICs with embedding of discrete active and passive components.

With the SoP technique, an SoP itself is a system in the sense that it integrates multiple system functions into one compact packaged system, leading to higher performance, smaller size, and more reliable full-system modules at lower cost. As shown in Figure 3.47(c), the SoP includes both active and passive components in a thin-film form rather than in a discrete or thick-film form. It may also include digital modules, RF circuits, and even optical components as well in a miniaturized package.

In the SoW technique, a silicon wafer is used for both components and substrates, as shown in Figure 3.47(d). As a result, the standard process technique may be used for the fabrication of dies and substrates, and to directly integrate active and/or passive functions into the silicon substrate. The advantage of SoW is that it may perform heterogeneous integration, in particular, including *microelectromechanical system* (MEMS) devices.

The *through-silicon via* (TSV), also referred to as *silicon through via* (STV), is a very promising technology for the replacement of wire bonding, in SiP, SoP, or SoW packaging techniques as well as chip-stacking technologies. The TSV technique allows the stacked silicon chips to be interconnected through direct contact. The use of TSV technology has the following advantages: size reduction, function complexity increasing, performance improvements, and decrease of the cost of connections.

The formation of TSVs is as follows. First, the target wafer, about 500 to 1000 μm in thickness, is attached to a handle wafer by adhesive film and then thinned down to about 100 μm (even 20 μm) in thickness by back-grinding and polishing processes. Second, a back-side photolithography process, along with deep-Si etching and SiO_2 etching processes, is applied to form the desired TSVs, each having a rectangle hole of about 60×60 μm^2. The following step is to make the sidewall insulation and to fill the TSV holes with conductive material. After forming TSVs, the handle wafer is removed and the target wafer is continued to process back-side interconnect and bumps, as well as diced and packaged.

3.3.3.5 Final Test and Burn-In Test All packaged ICs undergo a final electrical and burn-in test to ensure their quality. The electrical test is carried out for each IC to guarantee the functionality and performance. After the electrical test, a burn-in test at extreme temperatures and voltages is performed to wear out the infant failures. Of course, both tests may reduce the yield a little bit.

Review Questions

Q3-56. What do back-end processes mean?

Q3-57. What does the wafer test mean?

Q3-58. What are the goals of wafer test structures?

Q3-59. What is the meaning of 3-*D* packaging and 3-*D* IC?

Q3-60. What is the through-silicon via (TSV) technique?

3.4 Enhancements of CMOS Processes and Devices

Once we have understood the basic issues of CMOS processes, we continue to examine in this section some advanced CMOS-process devices and CMOS enhancement processes.

3.4.1 Advanced CMOS-Process Devices

In what follows, we would like to look at many advanced CMOS-related devices. These include *dual-gate CMOS devices*, *high-k MOS transistors*, *finFETs*, *plastic transistors*, *silicon-on-insulator CMOS transistors*, *silicon-germanium-base (SiGe-base) npn transistors*, *high-mobility transistors*, and *high-voltage transistors*.

3.4.1.1 Dual-Gate CMOS Transistors Because of high sheet resistance of pure polysilicon, it is often to dope the polysilicon used for gate electrodes and local interconnect with a high concentration n-type material, as shown in Figure 3.48(a). However, the resulting pMOS transistors have a threshold voltage V_{T0p} ranging from -0.5 V to -1.0 V. To make the threshold voltage of pMOS transistors fall into the desired range between -0.35 V and -0.5 V, boron ion implantation is often used, thereby yielding buried-channel pMOS transistors. The buried-channel pMOS transistors suffer short-channel effects severely as the transistor size shrinks to 0.25 μm and below.

One way to alleviate this problem is to dope the polysilicon gate of pMOS transistors with p-type instead of n-type material. In other words, the n^+-polysilicon is used for the gate of nMOS transistors and the p^+-polysilicon is used for the gate of pMOS transistors, as shown in Figure 3.48(b). The resulting CMOS transistors are called *dual-gate CMOS transistors* and both n- and p-type transistors are surface-channel transistors.

It is feasible to fabricate dual-gate CMOS transistors using the same implants to dope the polysilicon gates and to form shallow n^+ and p^+ source/drain regions through a careful control of implant dose and thermal cycle. Hence, it does not increase the number of masks and the fabrication cost.

3.4.1.2 High-k MOS Transistors Traditional or standard MOS transistors use silicon dioxide as the gate dielectric and have the structure as shown in Figure 3.49(a). The n^+ and p^+ polysilicon materials are used as the gates for nMOS and pMOS transistors, respectively. As the thickness of the gate dielectric is scaled down, the gate leakage current due to barrier tunneling through the gate dielectric increases profoundly, leading to the significant increase of static power dissipation.

The structure of a high-k MOS transistor is illustrated in Figure 3.49(b). The polysilicon gate in the traditional transistor is replaced by a metal and the gate dielectric is replaced by a hafnium-based material, such as *hafnium oxide* (HfQ_2), which has a much higher k value of about 20. Other proposed high-k dielectrics include zirconium oxide (ZrO_2) with $k = 23$, tantalum oxide (Fa_2O_5) with $k = 20$ to 30, silicon nitride (Si_3N_4) with $k = 6.5$ to 7.5, among others. Different metals are needed for pMOS and nMOS transistors. When a high-V_T MOS transistor is desired, the silicon dioxide

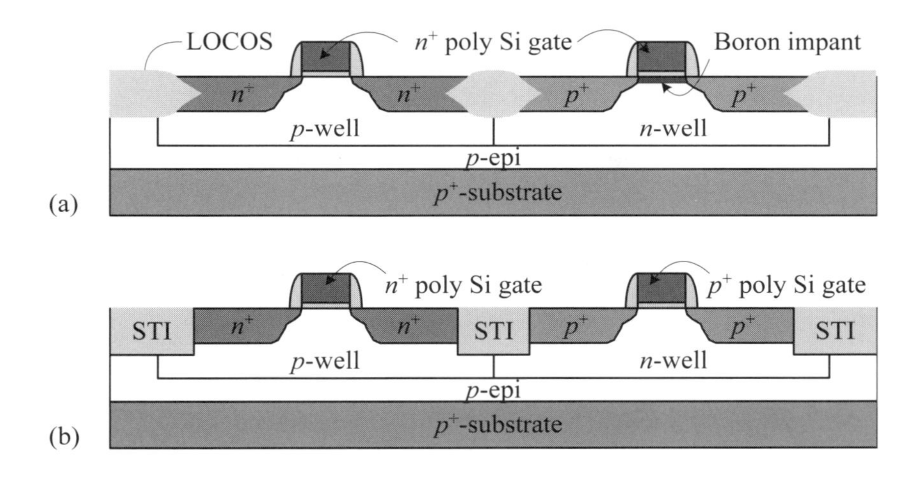

Figure 3.48: Comparison of CMOS structures between conventional (a) single polysilicon gate and advanced (b) double polysilicon gates.

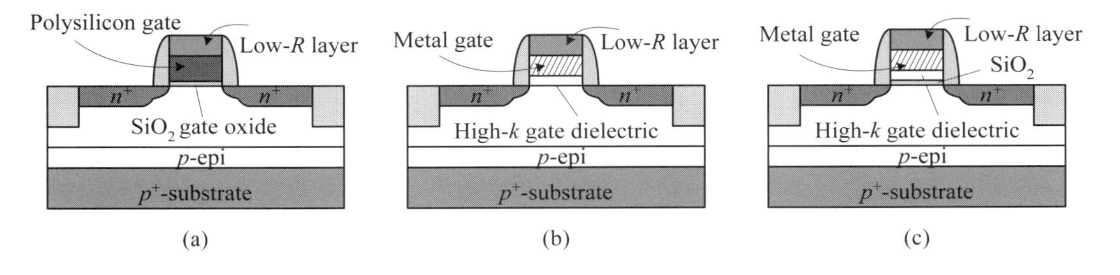

Figure 3.49: Cross-sectional views of various nMOS transistors: (a) traditional standard transistor; (b) high-k transistor; (c) high-k, high-V_T transistor.

can be added below the high-k dielectric, resulting in a high-k, high-V_T structure as depicted in Figure 3.49(c).

The rationale behind the use of high-k dielectric in building MOS transistors to improve performance and reduce static power dissipation is as follows. As the thickness of the gate dielectric is increasingly reduced from generation to generation, the gate leakage current due to barrier tunneling through the gate dielectric increases significantly. This can be illustrated by the tunneling probability from quantum mechanics. The tunneling probability is given as follows:

$$P \approx \exp\left[\frac{-2(2m_n^*\phi_{eff})^{1/2}t_{ox}}{\hbar}\right] = \exp(\alpha t_{ox}) \tag{3.13}$$

where α is defined to be

$$\alpha = \frac{-2(2m_n^*\phi_{eff})^{1/2}}{\hbar} \tag{3.14}$$

The ϕ_{eff} is defined to be the *effective work function* and is given as the difference between the barrier potential U and the Fermi energy of carriers E_F; that is, $U - E_F$. The m_n^* is the electron effective mass and $\hbar$ is $\frac{h}{2\pi}$, where $h = 4.135 \times 10^{-15}$ eV-s is Planck's constant.

The following gives an example to illustrate how the thickness of the gate oxide severely affects the gate tunneling current and hence the gate leakage current, which in turn constitutes the static power dissipation.

■ Example 3-1: (Gate tunneling current.) Suppose that the $\phi_{eff} = 4.16$ eV and $m_n^* = 1.06\ m_0$. Find the tunneling probability when the thickness of the gate dielectric is specified as follows:

1. 1.2 nm
2. 3.0 nm

Solution: We first calculate α as follows:

$$\begin{aligned}\alpha &= \frac{-2\left[2(1.05 \times 9.11 \times 10^{-31}) \times (4.16)(1.6 \times 10^{-19})\right]^{1/2}}{1.05 \times 10^{-34}} \\ &= -7.12 \times 10^{9}\ \text{m}^{-1}\end{aligned}$$

1. The tunneling probability for $t_{ox} = 1.2$ nm is
   $$P = \exp(-7.12 \times 10^9 \times 1.2 \times 10^{-9}) = 1.95 \times 10^{-4}$$
2. The tunneling probability for $t_{ox} = 3.0$ nm is
   $$P = \exp(-7.12 \times 10^9 \times 3.0 \times 10^{-9}) = 5.29 \times 10^{-10}$$

Hence, the tunneling probability is a very strong function of the thickness of the gate dielectric.

■

Hence, to control the gate leakage current under an acceptable level, the thickness of the gate dielectric must be greater than some critical value, say, 3 nm. However, if the thickness of the gate dielectric is not scaled down by the same factor as the gate voltage, the underlying MOS transistors would not be turned on under the nominal operating voltage since now the gate voltage is unable to induce enough charge required for forming the strong inversion layer (that is, channel) at the surface of the substrate. One way to solve this problem is to increase the gate capacitance to keep the total charge as it should be. By the parallel-plate capacitor model, the capacitance is given as

$$C = k\varepsilon_0 \frac{A}{t_{ox}} \tag{3.15}$$

where A is the area of the plate and t_{ox} is the distance between two plates forming the capacitor. Constants k and ε_0 are relative permittivity and free-space permittivity, respectively. Consequently, a high-k dielectric is required to compensate for the loss of t_{ox}. For convenience, we define an *equivalent oxide thickness* ($t_{ox(equiv)}$) as follows:

$$t_{ox(equiv)} = \frac{k_{ox}}{k_{high-k}} t_{high-k} \tag{3.16}$$

For instance, when the nominal capacitance using silicon dioxide as the gate insulator and $t_{ox} = 1.2$ nm is required, the desired k_{high-k} as the thickness of the gate insulator (t_{high-k}) is 3.0 nm is $3.9 \times 3.0/1.2 = 9.75$, where $k_{ox} = 3.9$.

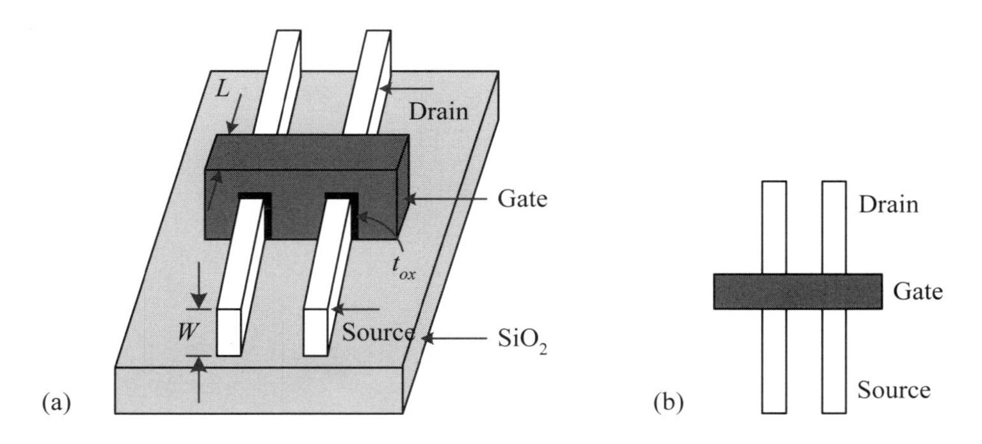

Figure 3.50: The (a) 3-D view and (b) top view of a finFET.

3.4.1.3 FinFETs In addition to the gate leakage current, another factor that causes undesired static power dissipation is the subthreshold current from the drain to source due to the inability of the gate to turn off the transistor. A possible approach to alleviating this problem is to change the gate structure of MOS transistors, such as the one shown in Figure 3.50(a). This new type of nMOS transistor is indeed a 3-*D* structure, which places the gate on three sides of the channel. Such a device is called a "finFET," because, as depicted in Figure 3.50(a), its source/drain region forms fins in the silicon surface.

The channel width of a finFET is defined by the height of the "fin" and the channel length is determined by the length of the gate electrode. To get a larger device (that is, channel width) to provide a larger drive current, many "fins" can be made in parallel. For example, the one shown in Figure 3.50(a) is a two-fin structure. Figure 3.50(b) is the top view of the two-fin MOS transistor.

3.4.1.4 Plastic Transistors MOS transistors can also be made with organic chemical materials. These polymer-based *thin-film transistors* (TFTs), also known as *plastic transistors*, have found their widespread use in active-matrix display and flexible *electronic paper* because of the good mechanical properties of polymer materials.

The structure of a plastic transistor is shown in Figure 3.51. The device is built in an upside-down manner. First, a gold gate and interconnect pattern are built on the plastic substrate. Next, an organic insulator, such as polymer or silicon nitride (Si_3N_4), is laid down. Then, the gold source and drain connections are made on top of the insulator. Finally, solution-processed pentacene is used as the semiconductor, which is a *p*-type and operating in the accumulation mode. Although the carrier mobility in the resulting plastic pMOS transistor is quite small, only about 0.02 cm^2/Vs, it is sufficient for some applications, such as a small-scale active-matrix display. The typical pitches of channel length (L) and width (W) of a plastic pMOS transistor are 5 μm and 400 μm, respectively.

The use of organic materials has a number of important advantages over conventional silicon-based materials. First, they can be processed from an inexpensive chemical solution, thereby simplifying the device fabricating process and reducing the fabricating cost. Second, the low processing temperature makes the possibility of using plastic materials instead of glass as substrates. As a result, a flexible device can be made.

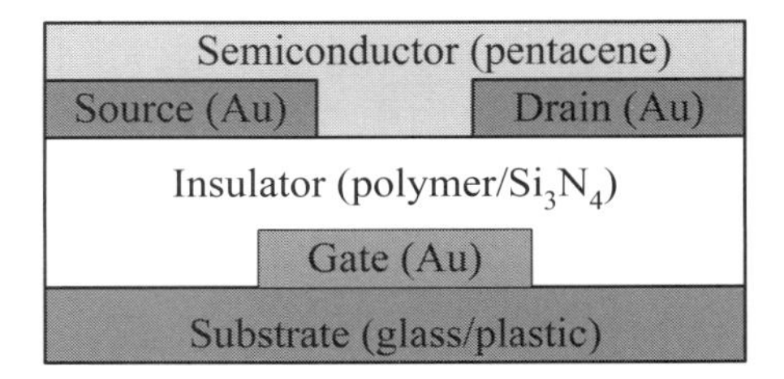

Figure 3.51: The cross-sectional view of a plastic transistor.

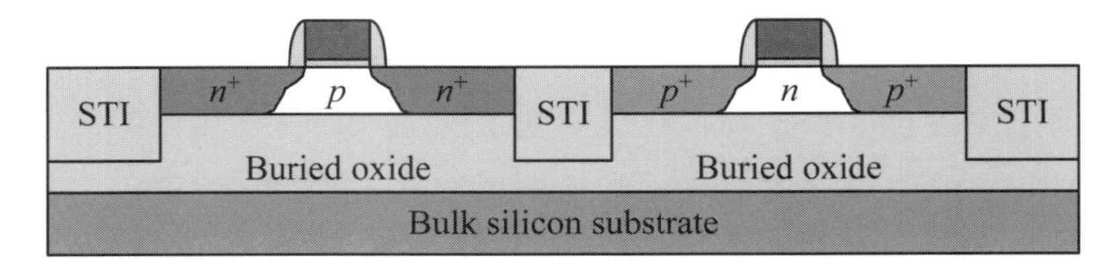

Figure 3.52: The cross-sectional view of an SoI CMOS device.

3.4.1.5 Silicon-on-Insulator CMOS Transistors

Silicon-on-insulator (SoI) CMOS transistors built on silicon dioxide have been proposed for use in many applications, including microprocessors, SRAMs, as well as other high-performance and low-power devices. A typical SoI CMOS device is depicted in Figure 3.52, where both types of MOS device are built on the silicon dioxide on top of a silicon substrate. Other insulation materials used to make SoI CMOS devices include sapphire, spinel, and nitride.

Compared with the CMOS device built on a bulk-silicon substrate, namely, bulk CMOS, an SoI CMOS device is simpler and does not need a complicated well structure, thereby increasing the device density.

The major features of SoI CMOS are as follows. First, it eliminates bottom-area junction capacitance, thereby significantly reducing the source and drain parasitic capacitance. Second, it lacks the reverse body effect in stacked circuits. Third, the SoI body is slightly forward biased under most operating conditions. Fourth, the body of the device is floating; therefore, V_{BS} is never less than zero but indeed is greater than 0 V under most operating conditions. Fifth, the SoI CMOS devices are free of latch-up problems inherent in the bulk CMOS. Sixth, because of the small volume of silicon available for electron-hole pair generation by radiation, it has better radiation-immunity capability than the bulk CMOS.

The SoI CMOS devices can be cast into *partially depleted* (PD) and *fully depleted* (FD) types according to the thickness of the silicon channel layer, as shown in Figure 3.52. When the silicon channel layer is thicker than the maximum depletion-layer thickness before the channel is formed, the device is called a PD-SoI; otherwise, the device is called a FD-SoI. Circuit design and performance of PD-SoI is similar to that of bulk CMOS. FD-SoI is very attractive for low-power applications since it can be operated at a lower electric field. However, FD-SoI is very sensitive to the thickness variation of the silicon channel layer; an FD-SoI circuit may be unstable if it is built on a nonuniform thickness of the silicon channel layer.

The floating body of SoI CMOS devices gives rise to a number of unique effects, referred to as *floating body effects*. The most important ones include the *self-heating effect*, *pass-gate leakage*, and *history effect*. Since oxide is a good thermal insulator,

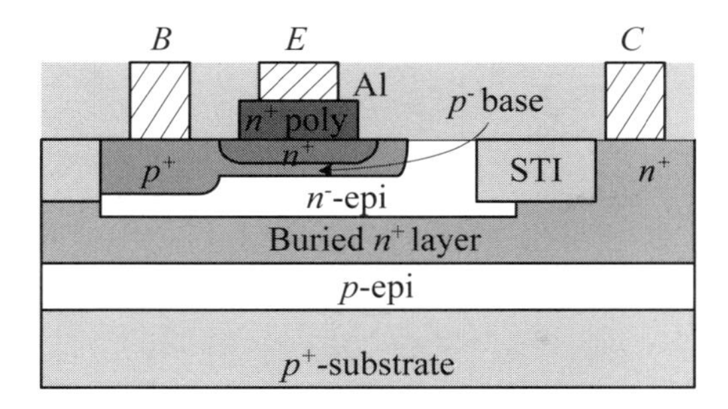

Figure 3.53: The cross-sectional view of a SiGe-base *npn* transistor.

the buried oxide layer prevents the heat generated by the switching transistor from being removed through the bulk silicon substrate into the environment. Thereby, the switching transistor dissipating much more power will be warmer than the average temperature of the wafer. This phenomenon is called the *self-heating effect*.

As shown in Figure 3.52, a parasitic transistor, composed of source, drain, and body, exists. If the SoI MOS transistor is off and both the source and drain of it are at high-level voltages and the gate is at a low-level voltage, the body/base will be floated into a high-level voltage due to leakage current of diodes existing between source/emitter and body/base as well as drain/collector and body/base. Now, if the source is pulled down to a low-level voltage due to some reason, the body/base and source/emitter junction will be forward-biased, thereby turning on the parasitic transistor and causing a current to flow between the drain and source. This effect is referred to as *pass-gate leakage*. The pass-gate leakage can cause malfunctions in dynamic logic circuits and latches but has less impact on static logic circuits.

The history effect means that the propagation delay of a logic gate depends on the switching history. This is due to the following two reasons. First, the body voltage is dependent on whether the logic gate has been idle or switching. Second, any change of the body voltage modulates the threshold voltage. The increased body voltage reduces the threshold voltage, thereby leading to a faster logic gate. However, this uncertainty makes the circuit design more difficult and challenging.

3.4.1.6 SiGe-Base Transistors The cross-sectional view of a modern *npn* transistor with a polysilicon emitter is shown in Figure 3.53. The inherent high-bandwidth feature makes the bipolar transistors very useful in microwave or high-frequency transceiver applications. Recently, a silicon-germanium-base *npn* heterojunction bipolar transistor is proposed to be used in high-frequency realms. Such a transistor is known as a *SiGe-base transistor*. At present, CMOS-compatible SiGe-base transistors are widely used for 5 to 15 GHz applications.

In the SiGe-base transistor with a linear concentration profile into the silicon base, the largest amount of Ge profile is introduced near the base-collector junction, while the least amount of Ge profile is introduced near the base-emitter junction to reduce the bandgap energy near the base-collector junction. This results in a transistor that has almost the same base current as a pure-Si transistor but has the increasing collector current. This implies that the current gain is obtained. To see this, we first note that the adding of Ge into silicon effectively reduces the bandgap energy E_g of silicon, which in turn causes the collector current and current gain to be increased. For example,

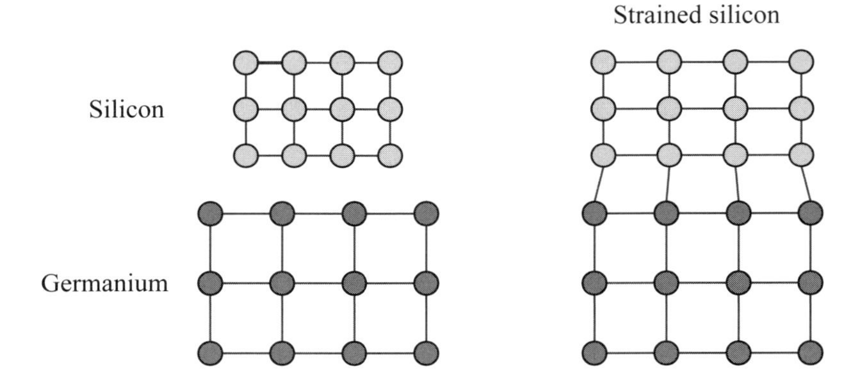

Figure 3.54: A conceptual illustration of strained silicon.

the collector current and current gain will be increased by a factor of about four if the bandgap energy is narrowed by a 0.1 eV.

3.4.1.7 High-Mobility Transistors High mobility means high-speed and high-drive current of MOS transistors. One way to improve the mobility of MOS transistors is to use *strained silicon*. Strained silicon is a layer of silicon where the silicon atoms are stretched beyond their normal interatomic distance. This is because as silicon is deposited on top of a substrate such as silicon germanium (SiGe) with atoms spaced farther apart, the atoms in silicon stretch to line up with the atoms beneath. A conceptual illustration of this is revealed in Figure 3.54. Moving these silicon atoms farther apart reduces the atomic forces that interfere with the movement of electrons through the transistors and thus has better mobility, thereby resulting in better performance in speed. In the strained silicon on top of silicon germanium (SiGe), electrons can move 70% faster and hence strained silicon transistors can switch 35% faster, without having to shrink the size of the transistors.

3.4.1.8 High-Voltage Transistors Because incorporating high-power and DC-to-DC conversion circuits into designs are increasingly popular in communication and consumer chips, especially, in SoC designs, the needs of high-voltage MOS transistors will be increased profoundly. To provide a high breakdown voltage needed in these MOS transistors, the gate oxide of such devices must be much thicker than that of the minimum feature-size MOS transistors; the channel length must be also much longer. Hence, to incorporate these two different types of devices onto the same chip, a more complicated and challenging process is required.

■ Review Questions

Q3-61. What do dual-gate CMOS devices mean?

Q3-62. What is a plastic transistor?

Q3-63. What is the rationale behind the SiGe-base transistor?

Q3-64. Why are high-k MOS transistors preferred in deep submicron processes?

Q3-65. What are the PD transistors in SoI CMOS technology?

Q3-66. What is strained silicon?

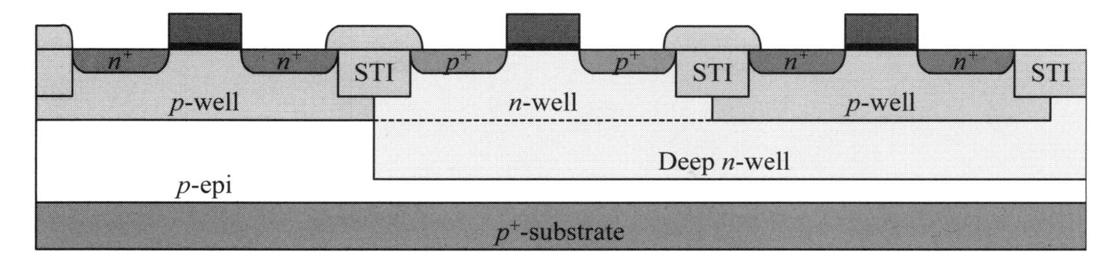

Figure 3.55: The cross-sectional view of a triple-well structure.

3.4.2 Enhancements of CMOS Processes

With the maturity of the CMOS VLSI process, three variants of the CMOS process are widely used in the IC industry today. These are the *triple-well process*, *bipolar CMOS* (BiCMOS) process, and microelectromechanical system (MEMS) process. In this subsection, we briefly address the basic features of these CMOS-related processes.

3.4.2.1 Triple-Well Processes A triple-well structure contains at least one p-well in a p-type substrate, a number of deep n-wells, and a p-well formed within one of the deep n-wells, as depicted in Figure 3.55. The triple-well structure for semiconductor IC devices allows for improved data storage stability and improved immunity capability against interference from I/O bouncing of device and alpha particles.

Triple-well structure can be formed using either diffusion, called the *diffused triple well* or high-energy (a few MeVs) ion implantation, called the *retrograde triple well*. Because of many advantages over the diffused triple well, the retrograde triple-well structure has dominated and been widely used in modern CMOS devices, including memory (SRAM, Flash, and DRAM) and embedded memory and logic on the same chip.

The advantages of the retrograde triple well include a negligible thermal budget (less than 950°C), simpler process steps, and hence reduced fabricating cost, greater packaging density, and an improved device performance due to retrograde dopant profile of the deep n-well structure.

The process for fabricating the triple-well structure begins to form deep n-wells with a high-energy (1 to 3 MeV) ion implantation in the p-type substrate at a depth of 2 to 4 μm. The n-well mask is subsequently used to form n-wells at a depth of about 1 μm where p-type MOS transistors are to be fabricated. The threshold voltage adjustment of pMOS transistors is also done in this step. Then, the p-well mask is employed to form p-wells where n-type MOS transistors are to be fabricated; these p-wells are at about the same depth as n-wells. The threshold voltage adjustment of nMOS transistors is also done in this step. Finally, a high temperature drive-in process is applied to form p-wells and n-wells.

3.4.2.2 BiCMOS Processes As we know, CMOS technology has the features of low-power dissipation, high noise margin, and high integration density, while bipolar technology has the features of high driving current, high operating frequency, and better performance in the analog circuit. BiCMOS is a technology that takes full advantage from both technologies and contains both CMOS and bipolar devices on the same chip; that is, it makes use of the high integration density associated with CMOS technology for implementing logic circuits and the capability of high driving current from bipolar technology for driving heavy capacitive loading. Therefore, only

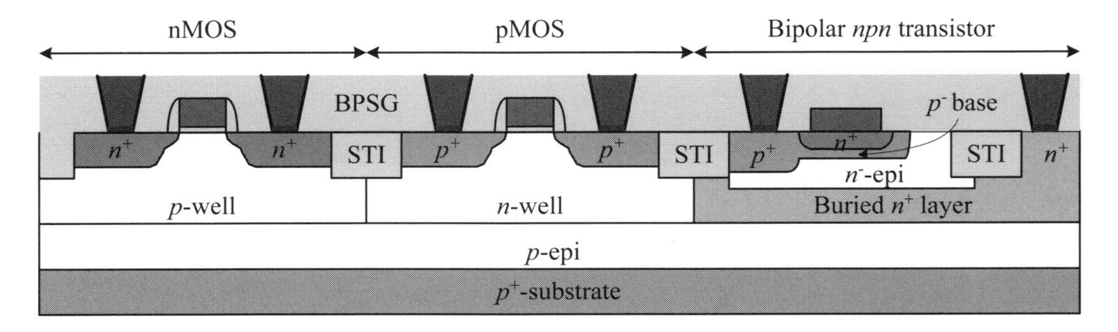

Figure 3.56: The cross-sectional view of a BiCMOS structure.

a very small part of the chip is occupied by bipolar transistors; most parts of the chip are taken up by CMOS devices.

The cross-sectional view of a BiCMOS structure is shown in Figure 3.56, which contains both nMOS and pMOS transistors and a bipolar *npn* transistor. A BiCMOS process is an extension of the CMOS process with some extra masking steps. These extra steps include the buried n^+ mask, the collector deep-n^+ mask, the base-p^- mask, and the polyemitter mask. A buried n^+ layer is implanted to reduce the collector's resistance. The p^+-region for base contact can be formed together with the source/drain formation of pMOS transistors; the n^+-region for emitter contact can be formed along with the source/drain formation of nMOS transistors.

3.4.2.3 MEMS Processes Generally, a MEMS, also called a *micromachine*, is a miniaturized system composed of both mechanical and electronic devices. These miniaturized devices are called *micromechanical devices* and *microelectronic devices*, respectively. The study of micromechanical devices is known as *micromechanics*, including microsensors and microactuators. Microsensors sense or monitor their environments and microactuators modify their environments under the control of microelectronic circuits.

A fabricating process used to fabricate a MEMS is called a MEMS process. A MEMS process is similar to a CMOS process with some modified steps. At present, most MEMS devices are manufactured using *silicon bulk micromachining*, *silicon surface micromachining*, and *LIGA* (a German acronym for lithographie, galvanoformung, abformung) processes.

Silicon bulk micromachining is the most mature of the two silicon micromachining technologies. It allows the selective removal of a large amount of silicon substrate to form various structures such as trenches, holes, or other structures. Instead of shaping the bulk silicon, silicon surface micromachining builds structures on the surface of silicon by depositing thin films, including both sacrificial layers as well as structural layers. The sacrificial layers are eventually removed to form the desired mechanical structures. The LIGA process uses the advanced X-ray lithography to pattern desired structures and includes three basic steps: *lithography*, *electroplating*, and *molding*. The LIGA process can create 3-D structures as the silicon bulk micromachining process while retaining the design flexibility as the silicon surface micromachining process.

In summary, a MEMS is a device composed of extremely small parts. A MEMS process uses the following five major types of materials: metals, semiconductors, ceramics, polymers, and composites, to make the desired devices. Nowadays, MEMS devices have found their widespread use in many areas, including various RF-MEMS devices, such as inductors, capacitors, filters, as well as transmission lines, air-bag sensors, microsensors, microactuators, micromotors, and so on.

■ Review Questions

Q3-67. What is a diffused triple well? What is a retrograde triple well?

Q3-68. Describe the extra masking steps needed in a BiCMOS process in addition to the standard CMOS process.

Q3-69. Describe the MEMS process.

3.5 Summary

A typical wafer fabrication process may take as many as four to eight weeks and involve several hundred steps to complete the full process flow. The entire fabricating process is basically around the following five basic unit processes: thermal oxidation, doping process, photolithography, thin-film removal, and thin-film deposition. Thermal oxidation produces a layer of silicon dioxide on the wafer surface with a specific thickness. The doping process adds the needed dopant impurities into the selective areas of the silicon wafer. The photolithography process copies the image on the mask (also photomask) onto the photoresist on the wafer surface, which is then transferred into the wafer surface through using some thin-film removal process. The thin-film deposition process deposits thin films, including insulators, semiconductors, and conductors, on the wafer surface.

By combining basic unit processes with appropriate materials, a CMOS integrated circuit can be fabricated. Such a well-defined collection of unit processes required to fabricate CMOS integrated circuits starting from virgin silicon wafers is known as a *process integration*. A process integration is often partitioned into two parts: front end of line (FEOL) and back end of line (BEOL). The FEOL includes all processes needed to fabricate isolated and fully functioned MOS transistors and the BEOL covers all processes required to interconnect together all related transistors and passive components into workable circuits.

After a fabricating process is completed, a series of postfabrication processes are carried out. These processes include the wafer test (or called the wafer probe), die separation, packaging, and the final test as well as the burn-in test. To facilitate a complex system, multiple chips are often packed into a single package. The most common techniques include the multichip module (MCM), system-in-package (SiP), system-on-package (SoP), and system-on-wafer (SoW). The last three techniques are also referred to as 3-D packaging, and the resulting IC is called a 3-D IC.

Finally, we examine some advanced CMOS-process devices and CMOS enhancement processes. The advanced CMOS-related devices include dual-gate CMOS devices, high-k MOS transistors, finFETs, plastic transistors, silicon-on-insulator CMOS transistors, silicon-germanium-base (SiGe-base) npn transistors, high-mobility transistors, and high-voltage transistors. The CMOS enhancement processes cover the triple-well process, bipolar CMOS (BiCMOS) process, and microelectromechanical system (MEMS) process.

References

1. M. Agostinelli, M. Alioto, D. Esseni, and L. Selmi, "Leakage-delay tradeoff in finFET logic circuits: a comparative analysis with bulk technology," *IEEE Trans.*

on Very Large Scale Integration (VLSI) Systems, Vol. 18, No. 2, pp. 232–245, February 2010.

2. M. Armacost et al., "Plasma-etching processes for ULSI semiconductor circuits," *IBM J. of Research and Development,* Vol. 43, No. 1–2, pp. 39–72, January-March 1999.

3. R. Jacob Baker, *CMOS Circuit Design, Layout, and Simulation,* 2nd ed. New York: John Wiley & Sons, 2005.

4. J. Baliga, "Depositing diffusion barriers," *Semiconductor International,* Vol. 20, No. 3, pp. 76–80, March 1997.

5. K. E. Bean, "Anisotropic etching of silicon," *IEEE Trans. on Electronic Devices,* Vol. 25, No. 10, pp. 1185–1193, October 1978.

6. C. Bencher et al., "Dielectric antireflective coatings for DUV lithography," *Solid State Technology,* Vol. 40, No. 3, pp. 109–112, March 1997.

7. J. O. Borland, "Triple well applications profit from MeV implant technology," *Semiconductor International,* Vol. 21, No. 4, pp. 69–72, April 1998.

8. A. E. Braun, "Copper electroplating enters mainstream processing," *Semiconductor International,* Vol. 22, No. 4, pp. 58–63, April 1999.

9. D. M. Brown, M. Ghezzo, and J. M. Pimbley, "Trends in advanced process technology–submicrometer CMOS device design and process requirements," *Proc. IEEE,* Vol. 74, No. 12, pp. 1678–1702, December 1986.

10. J. M. Bustillo, R. T. Howe, and R. S. Muller, "Surface micromachining for microelectromechanical systems," *Proc. IEEE,* Vol. 86, No. 8, pp. 1552–1574, August 1998.

11. S. A. Campbell, *The Science and Engineering of Microelectronic Fabrication,* 2nd ed., New York: Oxford University Press, 2001.

12. C. Y. Chang and S. M. Sze, *ULSI Technology.* New York: McGraw-Hill Books, 1996.

13. J. P. Colinge, "Thin-film SOI technology: the solution to many submicron CMOS problems," *IEEE Int'l Electron Devices Meeting (IEDM) Technical Digest,* pp. 817–820, 1989.

14. F. H. Dill, "Optical lithography," *IEEE Trans. on Electronic Devices,* Vol. 22, No. 7, pp. 440–444, July 1975.

15. F. H. Dill et al., "Characterization of positive photoresist," *IEEE Trans. on Electronic Devices,* Vol. 22, No. 7, pp. 445–452, July 1975.

16. D. Edelstein et al., "Full copper wiring in a sub-0.25 μm CMOS ULSI technology," *IEEE Int'l Electron Devices Meeting (IEDM) Technical Digest,* pp. 773–776, December 1997.

17. R. B. Fair, "Challenges in manufacturing submicron, ultra-large scale integrated circuits," *Proc. IEEE,* Vol. 78, No. 11, pp. 1687–1705, November 1990.

18. H. Fujita, "Microactuators and micromachines," *Proc. IEEE,* Vol. 86, No. 8, pp. 1721–1732, August 1998.

19. M. A. Fury, "CMP processing with low-k dielectric," *Solid State Technology,* Vol. 42, No. 7, pp. 87–92, July 1999.

20. J. W. Gardner, V. K. Varadan, and O. O. Awadelkarim, *Microsensors MEMS and Smart Devices.* New York: John Wiley & Sons, 2001.

21. H. Guckel, "High-speed-ratio micromachining via deep X-ray lithography," *Proc. IEEE,* Vol. 86, No. 8, pp. 1586–1593, August 1998.

22. D. L. Harame et al., "Si/SiGe epitaxial-base transistors—part I: materials, physics, and circuits," *IEEE Trans. on Electron Devices,* Vol. 42, No. 3, pp. 455–468, March 1995.

23. D. L. Harame et al., "Si/SiGe epitaxial-base transistors—part II: process integration and analog applications," *IEEE Trans. on Electron Devices,* Vol. 42, No. 3, pp. 469–482, March 1995.

24. C. C. Hu, *Modern Semiconductor Devices for Integrated Circuits.* Upper Saddle River, NJ: Prentice-Hall, 2010.

25. E. Huitema et al.,"Plastic transistors in active-matrix displays," *Proc. of IEEE Int'l Solid-State Circuits Conf.,* pp. 380-381, February 2003.

26. H. Ito, "Deep-UV resists: evolution and status," *Solid State Technology,* Vol. 39, No. 7, pp. 164–168, July 1996.

27. J. L. de Jong et al., "Single polysilicon layer advanced super high-speed BiCMOS technology," *Proc. of the IEEE Bipolar Circuits and Technology Meeting,* pp. 182–185, September 1989.

28. C. O. Jung et al., "Advanced plasma technology in microelectronics," *Thin Solid Films,* Vol. 341, pp. 112–119, 1999.

29. J. Kedzierski et al.,"High-performance symmetric gate and CMOS compatible V_t asymmetric gate FinFET devices," *IEEE Int'l Electron Devices Meeting* $(IEDM)$ *Technical Digest,* pp. 437–440, 2001.

30. J. A. Kittl et al., "Salicides and alternative technologies for future ICs: part 1," *Solid State Technology,* Vol. 42, No. 6, pp. 81–87, June 1999.

31. J. A. Kittl et al., "Salicides and alternative technologies for future ICs: part 2," *Solid State Technology,* Vol. 42, No. 8, pp. 55–59, August 1999.

32. E. Korczynski, "Low-k dielectric costs for dual-damascene integration," *Solid State Technology,* Vol. 42, No. 5, pp. 43–47, May 1999.

33. G. T. A. Kovacs, N. I. Maluf, and K. E. Petersen, "Bulk micromachining of silicon," *Proc. IEEE,* Vol. 86, No. 8, pp. 1536–1551, August 1998.

34. D. H. Lee and J. W. Mayer, "Ion-implanted semiconductor devices," *Proc. IEEE,* Vol 62, No. 9, pp. 1241–1255, September 1974.

35. M. D. Levenson, N. S. Viswanathan, and R. A. Simpson, "Improving resolution in photolithography with a phase-shifting mask," *IEEE Trans. on Electronic Devices,* Vol. 29, No. 12, pp. 1812–1846, December 1982.

36. G. P. Li et al.,"An advanced high-performance trench-isolated self-aligned bipolar technology," *IEEE Trans. on Electronic Devices,* Vol. 34, No. 11, pp. 2246–2254, November 1987.

37. H. C. Lin, J. C. Ho, R. R. Iyer, and K. Kwong, "Complementary MOS-bipolar transistor structure," *IEEE Trans. on Electronic Devices,* Vol. 16, No. 11, pp. 945–951, September 1969.

38. X. W. Lin and D. Pramanik, "Future interconnection technologies and copper metalization," *Solid State Technology,* Vol. 41, No. 10, pp. 63–69, October 1998.

39. C. M. Melliar-Smith, D. E. Haggan, and W. W. Troutman, "Key steps to the integrated circuit," *Bell Labs Technical Journal,* pp. 15–28, Autumn 1997.

40. G. Moore, "No exponential is forever: but 'forever' can be delayed!" *Proc. of IEEE Int'l Solid-State Circuits Conf.*, pp. 1–19, 2003.

41. M. Motoyoshi, "Through-silicon via (TSV)," *Proc. IEEE*, Vol. 97, No. 1, pp. 43–48, January 2009.

42. J. D. Plummer, M. D. Deal, and P. B. Griffin, *Silicon VLSI Technology: Fundamentals, Practice and Modeling.* Upper Saddle River, NJ: Prentice-Hall, 2000.

43. G. Poupon et al., "System on wafer: a new silicon concept in SiP," *Proc. IEEE*, Vol. 97, No. 1, pp. 60–69, January 2009.

44. M. Quirk and J. Serda, *Semiconductor Manufacturing Technology.* Upper Saddle River, NJ: Prentice-Hall, 2001.

45. G. M. Rebeiz, *RF MEMS Theory, Design, and Technology.* New York: John Wiley & Sons, 2003.

46. L. Rubin and W. Morris, "High-energy ion implanters and applications take off," *Semiconductor International,* Vol. 20, No. 4, pp. 77–83, April 1997.

47. R. D. Rung, H. Momose and Y. Nagakubo, "Deep trench isolation CMOS devices," *IEEE Int'l Electron Devices Meeting (IEDM) Technical Digest,* pp. 237–243, 1982.

48. C. Ryu et al., "Barriers for copper interconnections," *Solid State Technology,* Vol. 42, No. 4, pp. 53–55, April 1999.

49. G. Shahidi,"SOI technology for the GHz era," *IBM J. of Research and Development,* Vol. 46, No. 2/3, pp. 121–131, March/May 2002.

50. R. Sharangpani, R. P. S. Thakur, N. Shah, and S. P. Tay, "Steam-based RTP for advanced processes," *Solid State Technology,* Vol. 41, No. 10, pp. 91–95, October 1998.

51. V. Sundaram et al., "System-on-a-package (SOP) substrate and module with digital, RF and optical integration," *Proc. of the IEEE 54th Electronic Components and Technology Conference,* Vol. 1, pp. 17–23, June 2004.

52. S. M. Sze, *Semiconductor Devices: Physics and Technology.* 2nd ed., New York: John Wiley & Sons, 2002.

53. G. K. Teal, "Single crystals of germanium and silicon—basic to the transistor and integrated circuit," *IEEE Trans. on Electronic Devices,* Vol. 23, No. 7, pp. 621–639, July 1976.

54. S. E. Thompson et al., "A logic nanotechnology featuring strained-silicon," *IEEE Electron Device Letters,* Vol. 25, No. 4, pp. 191–193, April 2004.

55. R. R. Tummala, "SOP: what is it and why? A new microsystem-integration technology paradigm—Moore's law for system integration of miniaturized convergent systems of the next decade," *IEEE Trans. on Advanced Packaging*, Vol. 27, No. 2, pp. 241–249, May 2004.

56. J. P. Uyemura, *Introduction to VLSI Circuits and Systems.* New York: John Wiley & Sons, 2002.

57. V. K. Varadan, K. J. Vinoy, and K. A. Jose, *RF MEMS and Their Applications.* New York: John Wiley & Sons, 2003.

58. E. J. Walker, "Reduction of photoresist standing-wave effects by postexposure bake," *IEEE Trans. on Electronic Devices,* Vol. 22, No. 7, pp. 464–466, July 1975.

59. P. Wambacq et al., "The potential of finFETs for analog and RF circuit applications," *IEEE Trans. on Circuits and Systems–I: Regular papers,* Vol. 54, No. 11, pp. 2541–2551, November 2007.

60. C. Y. Wong et al., "Doping of n^+ and p^+ of polysilicon in a dual-gate CMOS process," *IEEE Int'l Electron Devices Meeting (IEDM) Technical Digest,* pp. 238–241, 1988.

61. H. Wong, "Beyond the conventional transistor," *IBM J. of Research and Development,* Vol. 46, No. 2/3, pp. 133–168, March/May 2002.

62. H. Xiao, *Introduction to Semiconductor Manufacturing Technology.* Upper Saddle River, NJ: Prentice-Hall, 2001.

Problems

3-1. Assume that two parallel wires are separated by a polyamide dielectric layer with a thickness of 0.5 μm. The polyamide has a relative permittivity of 2.7. Both wires have a cross-sectional area of 0.5×0.6 μm^2, length of 1 cm, and resistivity of 2.7 $\mu\Omega$-cm.

(a) Calculate the resistance of each of the wires.

(b) Calculate the capacitance between two wires.

3-2. Assume that a DRAM cell uses silicon dioxide as its dielectric and has a capacitance of 35 fF and an area of 1.22 μm^2. If the silicon dioxide is replaced by tantalum oxide (Ta_2O_5) without changing the thickness, calculate the cell area required. The relative permittivity of tantalum oxide is 25.

3-3. In modern CMOS processes, the Al-alloyed wire is replaced by Cu wire and some low-k dielectric ($k = 2.5$) is used instead of silicon dioxide ($k = 3.9$). The resistivity of Al is 2.7 $\mu\Omega$-cm and the resistivity of Cu is 1.7 $\mu\Omega$-cm. Calculate the percentage improvement of the use of Cu in terms of the intrinsic RC delay.

3-4. Assume that a 300-mm wafer is exposed to an air stream under a laminar-flow condition at 25 m/min for 5 minutes.

(a) How many particles will land on the wafer in a class 100 clean room?

(b) How many particles will land on the wafer in a class 10 clean room?

3-5. The resistivity of a thin film is equal to the product of the sheet resistance and thin-film thickness. Assume that a sheet resistance of 0.5 $\Omega/\Box$ is required.

(a) Calculate the thickness of titanium silicide. The resistivity of titanium silicide is 25 $\mu\Omega$-cm.

(b) Calculate the thickness of cobalt silicide. The resistivity of cobalt silicide is 15 $\mu\Omega$-cm.

3-6. Suppose that the $\phi_{eff} = 4.16$ eV and $m_n^* = 1.06\ m_0$. Find the tunneling probability when the thickness of the gate dielectric is (a) 1.5 nm and (b) 4.0 nm.

4

Layout Designs

Once we have understood the logic design and fabrication process, we are in a position to look at the interface between the fabrication process and circuit design. As mentioned earlier, the layout is another view of a circuit and denotes the set of mask (photomask) patterns needed to fabricate the circuit.

To design the layout of a circuit, it is necessary to follow a set of layout design rules specifying the geometrical information, such as minimum width (*feature size*) and separations (spacings) for each layer as well as overlaps for different layers. These layout design rules are used to ensure that the fabricated circuits are functional and reliable, and have reasonable yields as well. Indeed, layout design rules are the interface between a fabricating process and the design of circuits.

The layout of a circuit not only closely relates to the signal integrity, manufacturability, and reliability of the circuit but also is in close connection with the latch-up problem inherent in complementary metal-oxide-semiconductor (CMOS) processes. To avoid the latch-up problem, care must be taken and some basic guidelines must be followed in designing the layouts of critical circuits. This chapter is concluded with the Euler path approach, a widely used approach to carrying out a regular layout design of a CMOS logic circuit.

4.1 Layout Design Rules

In this section, we first consider the basic concepts of *layout designs* and typical layout design rules. Then, we deal with layout designs of some basic structures. Finally, we briefly address some layout considerations about signal integrity, manufacturability, and reliability.

4.1.1 Basic Concepts of Layout Designs

A layout design, also called a *physical layout design*, *physical design*, *physical layout*, or *layout* for short, of a circuit means to define and draw a set of mask patterns. These mask patterns are in turn employed to make related photomasks required for fabricating the circuit. During the layout design phase, a circuit is translated into a set of geometry patterns, consisting of rectangles, squares, polygons, or other allowed shapes. Hence, a layout design defines the geometrical shapes and creates a set of corresponding mask patterns that define the final integrated circuit.

For a twin-well process, the geometrical patterns at least include the following

- n-well (needed to layout explicitly)
- p-well (can be generated automatically by CAD tools)
- Active region
- Polysilicon (or poly, polygate)
- p-select (or p-implant, pimp, pplus) and n-select (or n-implant, nimp, nplus)
- Active contact
- Polysilicon (poly) contact
- Metal1
- Via1
- Metal2
- Passivation (or overglass)

The layout design rules are a set of specifications for the mask patterns used in the layout and provide geometrical information, such as minimum width (feature size) and separations (spacings) for each layer as well as overlaps for different layers. Layout design rules are unique to each CMOS process. They are guidelines for constructing photomasks and must be followed in designing the layout of a circuit to ensure that the fabricated circuit is functional and reliable, and has a reasonable yield as well.

4.1.1.1 Specifications of Layout Design Rules Layout design rules are usually specified in one of the following two ways: lambda (λ) rules and micron (μ) rules. The λ rules are also called *scalable layout design rules*.

Lambda (λ) rules. Lambda rules represent all rules as the function of a single parameter, called λ, proposed by Mead and Conway. The minimum wire width of a process is typically set to 2λ. For instance, for a process with a feature size of 0.18 μm, 1 λ is set equal to 0.09 μm, or sometimes rounded to 0.1 μm. When using lambda rules, the scaling of the minimum dimension is done by simply changing the λ value. The λ rules are chosen in a way such that a design can be easily portable over different processes. Lambda rules are very conservative as compared with micron rules since the rule specifications are usually rounded to integer multiples of λ. For example, a feature with a value of 0.25 μm needs to be set into 3λ in a 0.18-μm process.

Micron (μ) rules. Micron rules define all rules in absolute dimensions and can exploit the features of a given process to its maximum extent. Hence, micron rules are more aggressive as compared with lambda rules. Nevertheless, they tend to differ from company to company, even from process to process. The layout is more compact using micron rules than using lambda rules.

4.1.1.2 Types of Layout Design Rules Recall that an integrated circuit is fabricated layer by layer and each layer needs a mask pattern to define its geometrical shape. Some mask patterns are only confined to the same layer but others are extended to two closely adjacent layers. For the features on the same layer, minimum width and minimum spacing are specified to tolerate the isotropic etch profile inherent in etching processes. For the features staying on two different layers, surround and minimum extension are used to overcome the possible registration error originated from steppers and the isotropic etch profile encountered in the most etching processes.

Layout design rules can generally be categorized into the following five main types: *minimum width*, *minimum spacing*, *surround*, *minimum extension*, and *exact size*, as shown in Figure 4.1, regardless of whether lambda or micron rules are used.

Minimum width. The minimum width takes into consideration is the wire-width limitation of an imaging system and an etching process. In an imaging system, the reticle shadow projected to the photoresist surface does not have a sharp edge due to

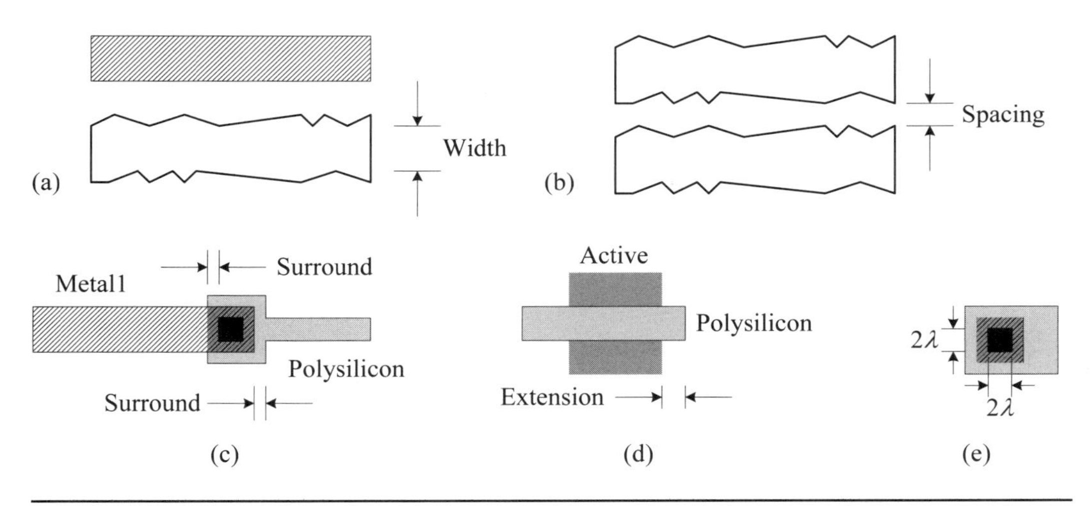

Figure 4.1: The types of layout design rules: (a) minimum width rule; (b) minimum spacing rule; (c) surround rule; (d) minimum extension rule; (e) exact rule.

optical diffraction, even if the postbake process is applied. In an etching process, the lateral etching limits the reachable resolution. As a result, the widths of the geometries defined on a mask must exceed a minimum value; otherwise, the features would not be reliably made. Figure 4.1(a) illustrates the requirement of the design rule concerning the minimum width.

Minimum spacing. With the same reasons as minimum width rules, the geometries built on the same mask must be separated by a minimum spacing to fabricate a reliable feature. Figure 4.1(b) illustrates the requirement of the design rule concerning the minimum spacing.

Surround. When a geometrical feature must be placed inside an existing geometry, it must be surrounded with a sufficient margin to guarantee the feature contained by the existing geometry. Figure 4.1(c) examines the requirement of the design rule concerning the surround. Because metal1 and polysilicon belong to two different layers and connect by a contact, insufficient surround will cause metal1, polysilicon, or both to shift over the active contact, resulting in a high-contact resistance, or even an open circuit.

Minimum extension. Some geometrical features must extend beyond the edge of the others by a minimum value. This design rule is also based on the misalignment problem caused by steppers. An illustration of this is depicted in Figure 4.1(d), where an extension rule of the polysilicon layer is used to ensure that a desired metal-oxide-semiconductor (MOS) transistor can be correctly fabricated. Since in the self-aligned process, the polysilicon layer is used as a dopant mask for n^+/p^+-type ion implants that defines source and drain regions of a MOS transistor, insufficient extension will result in an imperfect MOS transistor, leading to a short circuit between source and drain regions.

Exact size. An exact size means that the feature can only have the dimensions specified in the layout design rules. Other sizes are not allowed. For instance, the contacts and vias for many 0.35-μm processes are set to fixed sizes, 0.4 μm × 0.4 μm.

In addition to layout design rules associated with geometrical features, *electrical design rules* are also provided by CMOS processes to specify some basic values when

Table 4.1: Some differences between SCMOS, SUBM, and DEEP rules.

Rule	SCMOS	SUBM	DEEP
Well width	10	12	12
Source/drain active to well edge	5	6	6
Poly spacing over filed/active	2	3	3/4
Active extension of polysilicon	3	3	4
Spacing to contact	2	3	4
Spacing to Metal1	2	3	3

certain electrical conditions occur. One such example is to specify the allowed maximum current density for a given wire width to avoid the electromigration effect.

4.1.1.3 Scalable CMOS Layout Design Rules. In the scalable CMOS (SCMOS) rules, circuitry geometries are specified in terms of λ. The λ value can be easily scaled to different fabrication processes as semiconductor technology advances. At present, the rules break down into three types: generic SCMOS, submicron (SUBM), and deep submicron (DEEP). The generic SCMOS rules can be applied to feature sizes larger than about 1 μm; SUBM rules typically are employed for about 0.8 μm to 0.35 μm, while DEEP rules are often used below 0.35 μm. An example of SCMOS rules is listed in Figure 4.2. Some differences between SCMOS, SUBM, and DEEP rules from MOSIS are listed in Table 4.1. The full set of SCMOS rules can be found from MOSIS Web site.[1] *You have to check the layout design rules before doing the actual layout if you want to submit a chip design to an IC foundry.*

Before leaving this subsection, we would like to summarize some key points closely related to layout design. These are listed as follows:

1. Any layout must conform to layout design rules.
2. Layout design rules may be λ-based or μ-based.
3. The grid in layout editor helps with design-rule conformity.
4. The *design rule checker* (DRC) should be frequently used to check the layout of a design.
5. A layout plan has to be created and followed in a design.
6. Layouts of cells should have standard locations for inputs, outputs, power, and ground.

■ Review Questions

Q4-1. What are the intentions of layout design rules?

Q4-2. How would you specify layout design rules?

Q4-3. What are the types of layout design rules?

Q4-4. What is the surround rule?

Q4-5. Why is the minimum extension rule required?

4.1.2 Layouts of Basic Structures

In this subsection, we are concerned with layouts of basic structures. Recall that a CMOS circuit is fabricated in a predefined masking sequence, as shown in Table 3.1.

[1] The MOSIS Web site is at http://www.mosis.com/Technical/Designrules/scmos/scmos-main.html.

1. Well rules

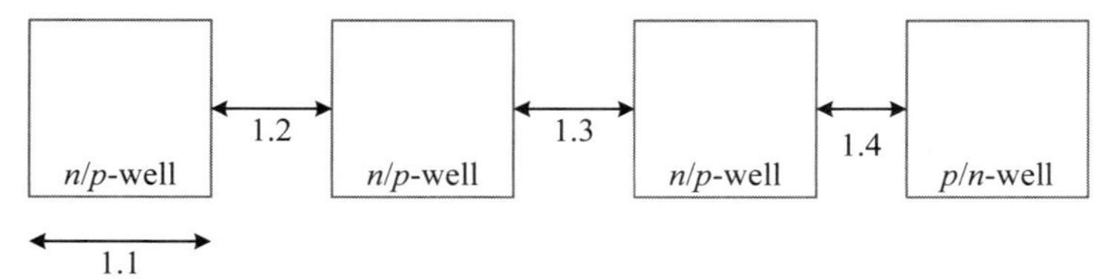

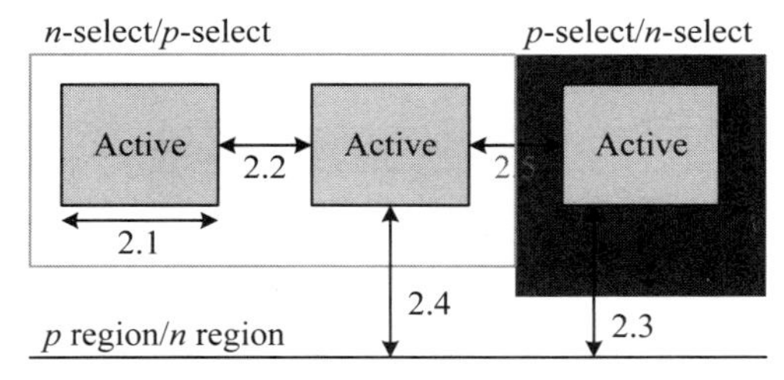

Description	λ
1.1: Width	12
1.2: Spacing to well at different potential	18
1.3: Spacing to well at same potential	6
1.4: Spacing to well of different types	0

2. Active area (transistor) rules

Description	λ
2.1: Width	3
2.2: Spacing to active	3
2.3: Substrate/well contact active to well edge	3
2.4: Source/drain active to well edge	6
2.5: Spacing to active of different type	4

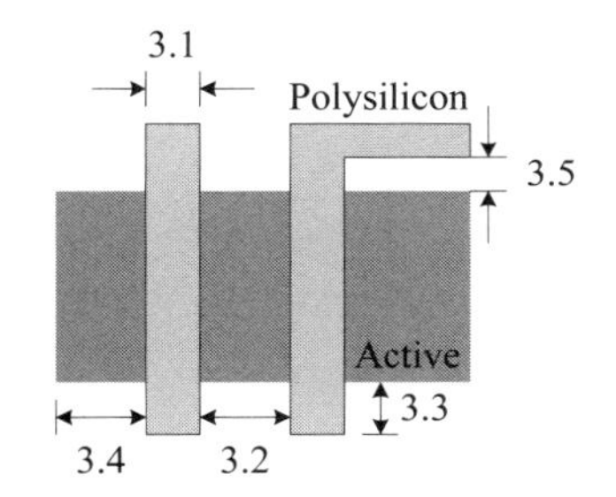

3. Polysilicon design rules

Description	λ
3.1: Width	2
3.2: Spacing over field/active	3
3.3: Gate extension of active	2
3.4: Active extension of polysilicon	3
3.5: Spacing of polysilicon to active	1

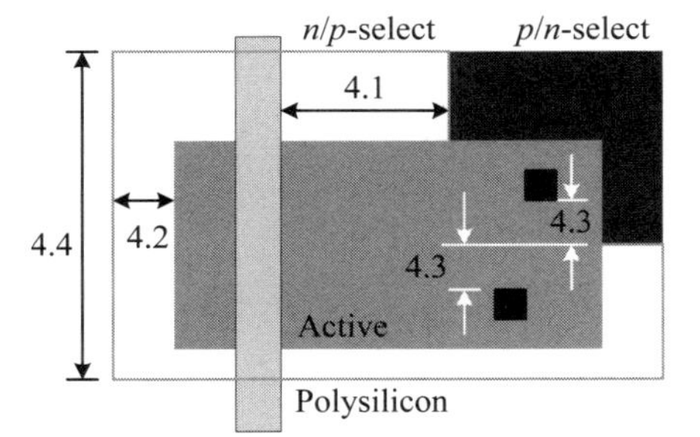

4. Select design rules

Description	λ
4.1: Spacing to channel	3
4.2: Overlap of active	2
4.3: Overlap of contact	1
4.4: Width and spacing to select	2

5. Polysilicon contact design rules

Description	λ
5.1: Width (exact)	2×2
5.2: Polysilicon overlap	1
5.3: Spacing to contact	3
5.4: Spacing to gate	2
5.5b: Spacing to polysilicon	4

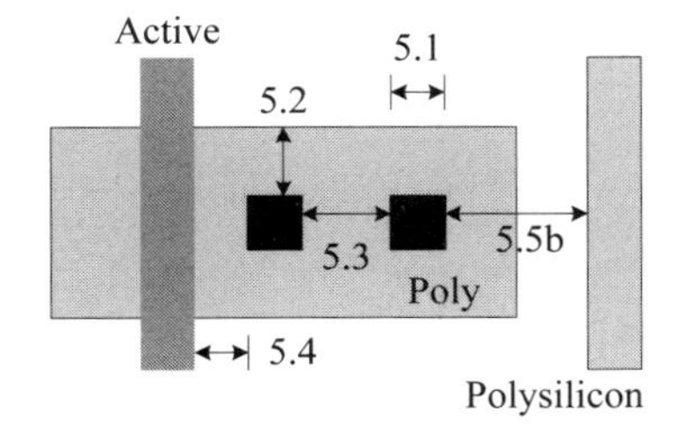

Figure 4.2: An example of SCMOS layout design rules.

6. Active contact design rules

Description	λ
6.1: Width (exact)	2×2
6.2: Active overlap	1
6.3: Spacing to contact	3
6.4: Spacing to gate	2
6.5b: Spacing to active	5

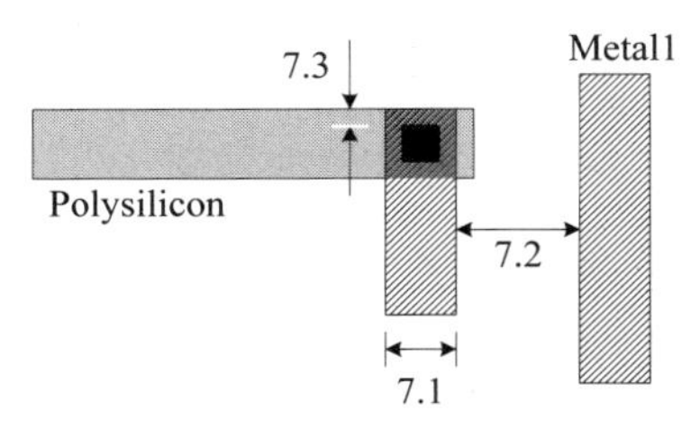

7. Metal1 design rules

Description	λ
7.1: Width	3
7.2: Spacing to metal1	3
7.3: Overlap of any contact	1
7.4: Spacing to metal1 (wider than 10 λ)	6

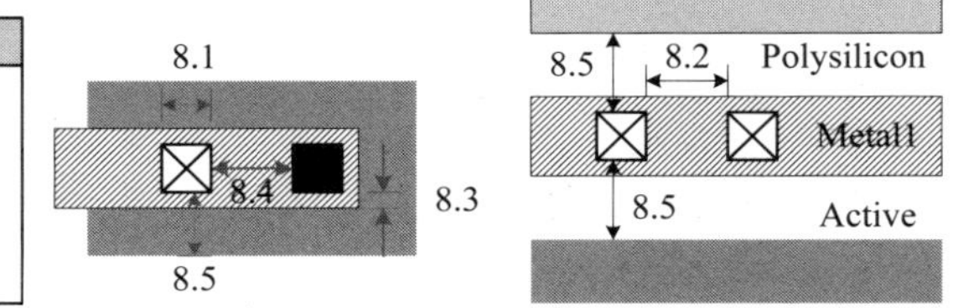

8. Via design rules

Description	λ
8.1: Width (exact)	2
8.2: Spacing to via1	3
8.3: Overlap by metal1	1
8.4: Spacing to contact	3
8.5: Spacing to polysilicon or active	2

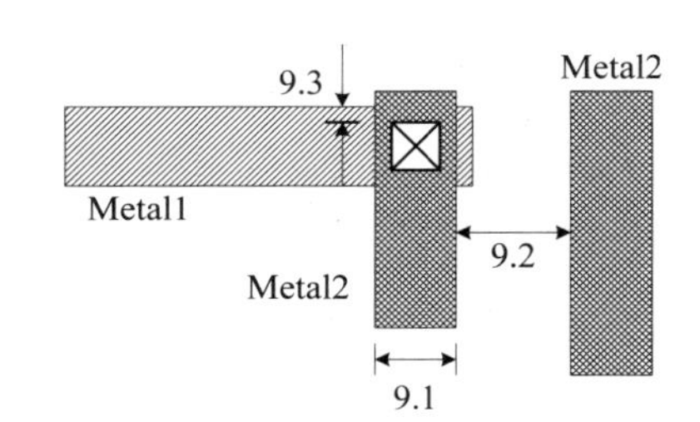

9. Metals 2 to 4 design rules

Description	Metal2	Metal3	Metal4
	λ	λ	λ
9.1: Width	3	6	6
9.2: Spacing to metal$x$	3	4	6
9.3: Overlap via(i-1)	1	2	2
9.4: Spacing to metal$x$ (wider than 10 λ)	6	8	12

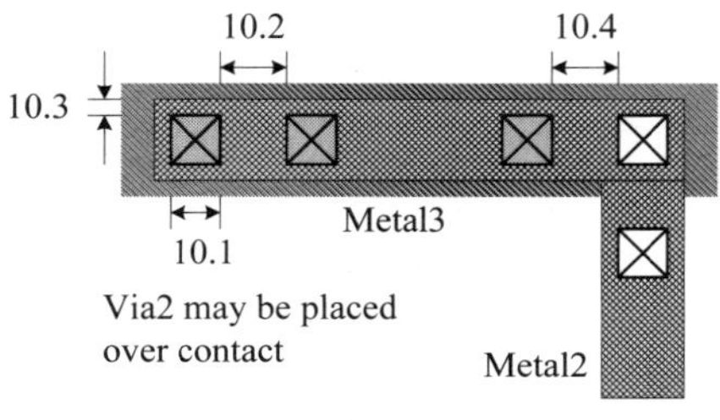

10. Via2 design rules

Description	λ
10.1: Width (exact)	2 × 2
10.2: Spacing to via2	3
10.3: Overlap by metal2	1
10.4: Spacing to via1 (for not allowed stack)	2

10.2 10.4 10.3 10.1 Metal3

Via2 may be placed over contact

Metal2

11. Via3 design rules

Description	λ
11.1: Width	2 × 2
11.2: Spacing to via3	3
11.3: Overlap by metal3	1

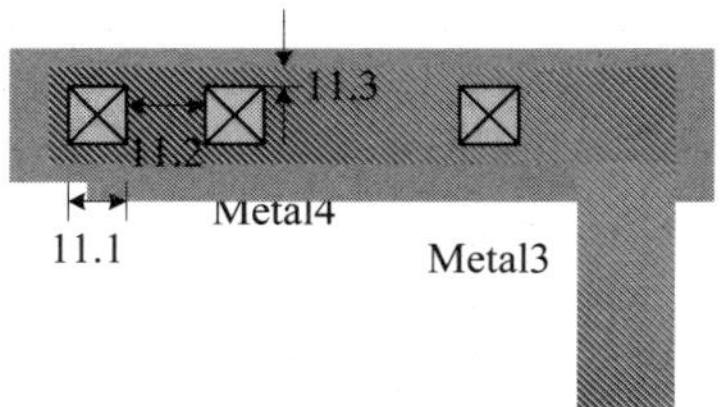

Figure 4.2: (Continued.)

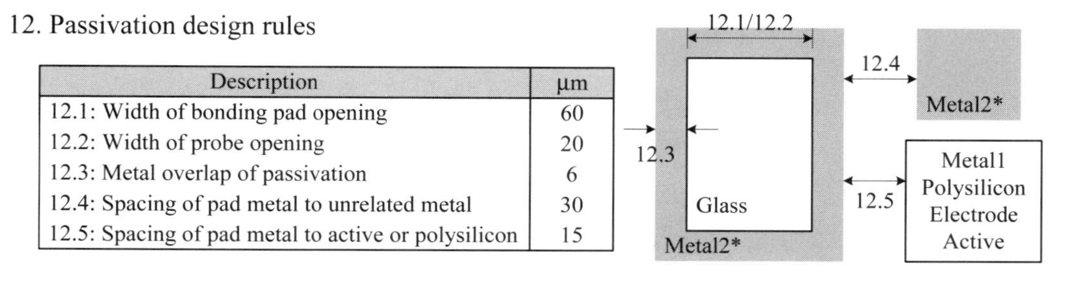

12. Passivation design rules

Description	µm
12.1: Width of bonding pad opening	60
12.2: Width of probe opening	20
12.3: Metal overlap of passivation	6
12.4: Spacing of pad metal to unrelated metal	30
12.5: Spacing of pad metal to active or polysilicon	15

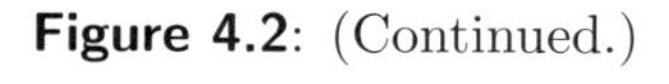

Figure 4.2: (Continued.)

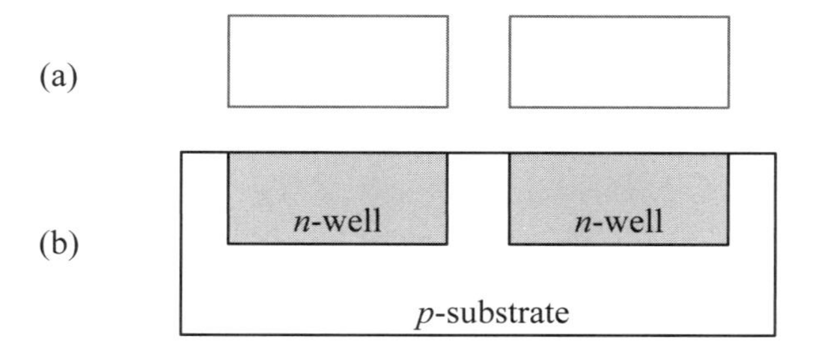

Figure 4.3: An example of two n-wells: (a) n-well mask; (b) cross-sectional view.

Hence, in what follows, we describe each of these mask patterns in order. The related layout design rules are illustrated in Figure 4.2.

4.1.2.1 The n/p-Wells An n-well is required at every location where a p-type MOS (pMOS) transistor is to be made; a p-well is required at every location where an n-type MOS (nMOS) transistor is desired. It is often possible to merge together adjacent wells with the same type. In order to avoid the occurrence of latch-up, an n-well must have a connection to the power supply V_{DD} when used as the substrate for pMOS transistors, whereas a p-well must have a connection to the ground V_{SS} when used as the substrate for nMOS transistors.

An example of an n-well is shown in Figure 4.3, where two n-wells are created on the p-substrate. By and large, an n-well is defined using a closed polygon to indicate where the n-well is to be placed. Figure 4.3(a) shows the n-well mask set and Figure 4.3(b) is the cross-sectional view of the two n-wells created.

In a twin-well process, both n-wells and p-wells are defined and created whenever they are necessary. Two wells with the same type must be separated with a definite distance but with different types may be placed adjacently. In addition, wells may be at different potentials. The separation between two wells of the same type but at different potentials is usually larger than they are at the same potential. The n/p-well layout design rules are shown in Figure 4.2.

4.1.2.2 Active Regions Active regions of the substrate, including the p-type substrate and n-type well, are used to build silicon devices. Each active device is defined by a closed polygon on an active mask. The active mask usually contains as many closed polygons as the active devices needed. Figure 4.4(a) shows the active mask set and Figure 4.4(b) is the cross-sectional view of two active regions with different sizes.

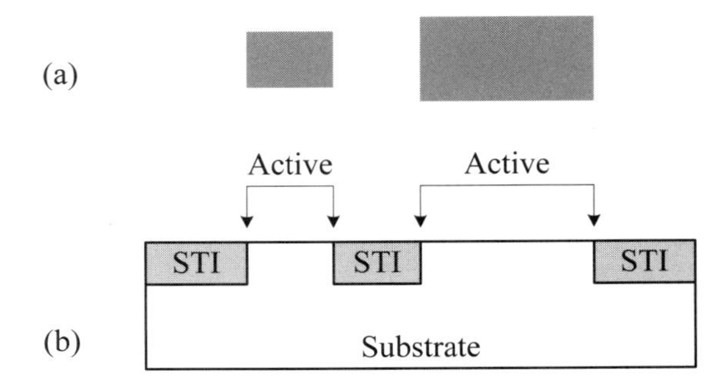

Figure 4.4: An example of active regions: (a) Active mask; (b) cross-sectional view.

To electrically isolate these active devices, the regions on the wafer surface without active devices are filled with field oxide (FOX), formed by either the local oxidation of silicon (LOCOS) or shallow trench isolation (STI) process. The field oxide regions can be easily derived from the active mask by the following relation.

$$\text{FOX (STI/LOCOS)} = \text{NOT (active regions)} \tag{4.1}$$

That is, any region on the wafer surface is a FOX region if it is not an active region. The wafer surface is the union of FOX and active regions.

$$\text{Wafer surface} = \text{FOX (STI/LOCOS)} \cup \text{active regions} \tag{4.2}$$

4.1.2.3 Doped Silicon Regions The n^+ and p^+ regions are also called *n-diffusion* and *p-diffusion*, respectively, or ndiff and pdiff for short, since before the ion implantation found its widespread use in IC fabrication, these two regions were formed by the diffusion process. To create the n^+/p^+ regions, both the n-select/p-select and Active masks are needed. An n^+ region is created by implanting arsenic or phosphorous ions into the p-substrate areas defined by the n-select mask and Active mask. In other words,

$$n^+ = n\text{-select} \cap \text{Active} \tag{4.3}$$

Similarly, a p^+ region is created by implanting boron ions into the n-well areas defined by the p-select mask and Active mask. That is,

$$p^+ = p\text{-select} \cap \text{Active} \cap n\text{-well} \tag{4.4}$$

Figure 4.5 shows the examples of n^+ and p^+ regions along with their associated mask sets. Figure 4.5(a) is an n-type doped silicon region, whereas Figure 4.5(b) depicts a p-type doped silicon region. An n^+ region is formed whenever an active region is surrounded by an n-select region, while a p^+ region is formed whenever an active region is surrounded by a p-select region.

An n^+ or a p^+ region alone cannot form an nMOS or a pMOS transistor. However, n^+ and p^+ regions are employed to connect the n-type well and p-type substrate to V_{DD} and ground, respectively, through metal wires. Schottky diodes may be obtained if metal wires are simply connected to an n-type well and p-type substrate without using n^+ and p^+ regions. To form an nMOS or a pMOS transistor, a polysilicon layer is also needed, which will be described next.

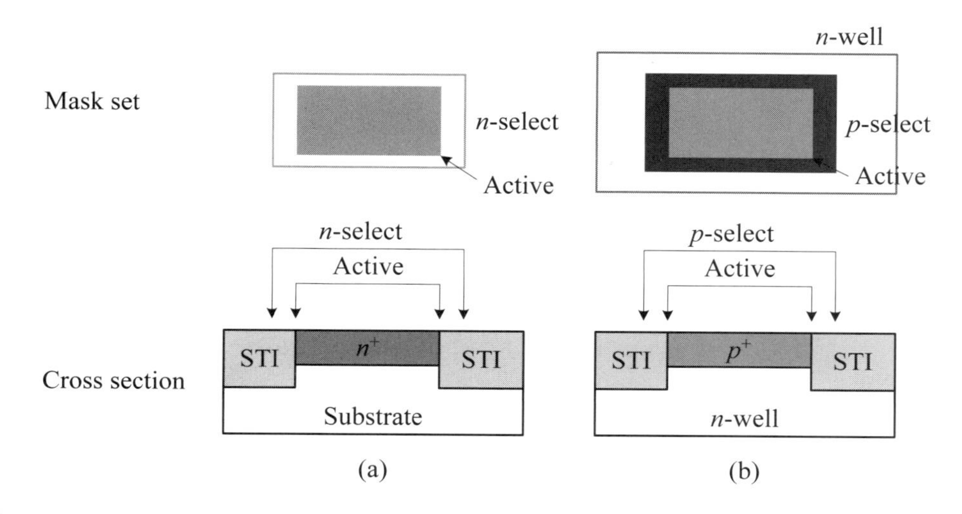

Figure 4.5: Examples of doped n^+ and p^+ silicon regions: (a) n-type; (b) p-type.

4.1.2.4 MOS Transistors A self-aligned MOS transistor is properly created whenever a polysilicon layer completely crosses an n^+ or a p^+ region. An nMOS transistor is formed whenever a polysilicon area cuts an n^+ region into two separate segments. Therefore, an nMOS transistor is created by the intersection of three masks: n-select, Active, and Polysilicon masks.

$$\text{nMOS} = n\text{-select} \cap \text{Active} \cap \text{Polysilicon} \tag{4.5}$$

The source and drain regions of the nMOS transistor are defined by the following relation.

$$n^+ = n\text{-select} \cap \text{Active} \cap \text{NOT (Polysilicon)} \tag{4.6}$$

The cross-sectional view, layout view, and mask set of an nMOS transistor are shown in Figure 4.6(a).

Similarly, a pMOS transistor is formed whenever a polysilicon area cuts a p^+ region into two separate segments. Therefore, a pMOS transistor is created by the intersection of the following mask set: p-select, Active, Polysilicon, and n-well masks. In other words,

$$\text{pMOS} = p\text{-select} \cap \text{Active} \cap \text{Polysilicon} \cap n\text{-well} \tag{4.7}$$

The source and drain regions of the pMOS transistor are defined by the following

$$p^+ = p\text{-select} \cap \text{Active} \cap n\text{-well} \cap \text{NOT (Polysilicon)} \tag{4.8}$$

The cross-sectional view, layout view, and mask set of a pMOS transistor are shown in Figure 4.6(b).

4.1.2.5 Active and Polysilicon Contacts An active contact is a cut in the oxide (premetal dielectric, PMD) that allows the first metal layer to contact an active n^+ or p^+ region, as shown in Figure 4.7(a). An active contact has an exact area size of $2\lambda \times 2\lambda$ and is subject to the surround layout design rule. That is, it has to include one λ space on each side of the contact.

Similarly, a polysilicon contact is a cut in the oxide that allows the first metal layer to contact a polysilicon region or local interconnect, as shown in Figure 4.7(b). Like active contacts, polysilicon contacts are also subject to the surround layout design rule.

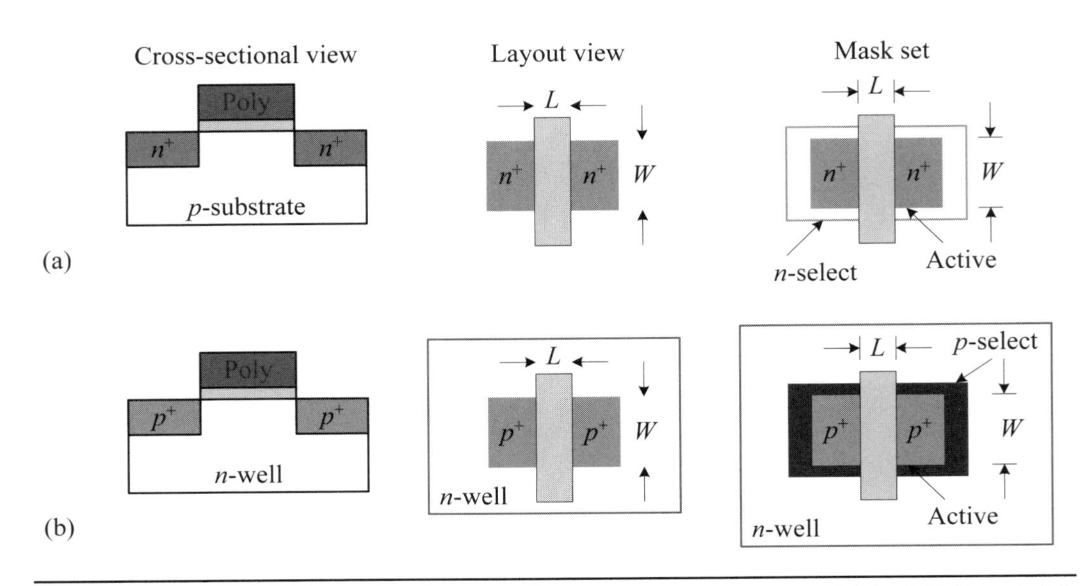

Figure 4.6: The cross-sectional view, layout view, and mask set of (a) nMOS and (b) pMOS transistors.

4.1.2.6 Metal1 Interconnect Metal1 is used as an interconnect for signals and for a power distribution network within a cell. Metal1 may connect to n^+/p^+ regions or polysilicon layer through contacts and to Metal2 through via1. Every contact or via has a *contact resistance* R_c (Ω). As described above, each contact has an exact area of $2\lambda \times 2\lambda$. A larger connection is done by an array of contacts to avoid the current crowding at the periphery and to reduce the effective contact resistance because these contact resistances are effectively connected in parallel. Examples of using metal1 to connect n^+ and p^+ regions through contacts are exhibited in Figure 4.8.

4.1.2.7 Vias Recall that the fabrication of an IC is a layered process; namely, each feature has a definite sequence in the process and cannot be exchanged with other features. The sequence of fabricating different metal layers in a multiple metal process is as in the following.

$$\text{Metal1} \longrightarrow \text{Metal2} \longrightarrow \text{Metal3} \longrightarrow \text{other metal layers (if any)}$$

Along the progress of the fabricating process, a proper dielectric (interlayer dielectric, ILD) is deposited between two metal layers by a chemical vapor deposition (CVD) process to become electrically insulated from one to another. The desired connections between two adjacent metal layers are then done through vias defined by a via mask. An illustration of the connection between two metal wires through a via is shown in Figure 4.9. The connection of metal1 to an active region or to a polysilicon layer is usually called a *contact*, and the connection between two different metal layers is called a *via*. Examples of contact and via are shown in Figures 4.8 and 4.9, respectively.

4.1.2.8 Metal2 to Metal4 Rules Modern deep submicron processes usually use multiple metal layers to increase the integration density of an integrated circuit (IC). The basic rules of these higher metal layers are the same as those of metal1. Nevertheless, the minimum width and spacing of a higher metal layer are usually wider than those

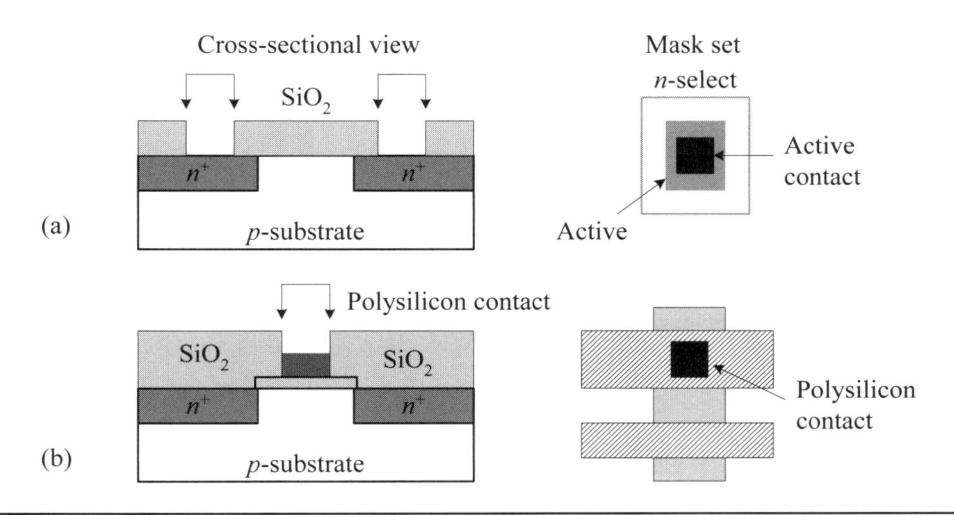

Figure 4.7: An example of (a) active and (b) polysilicon contacts.

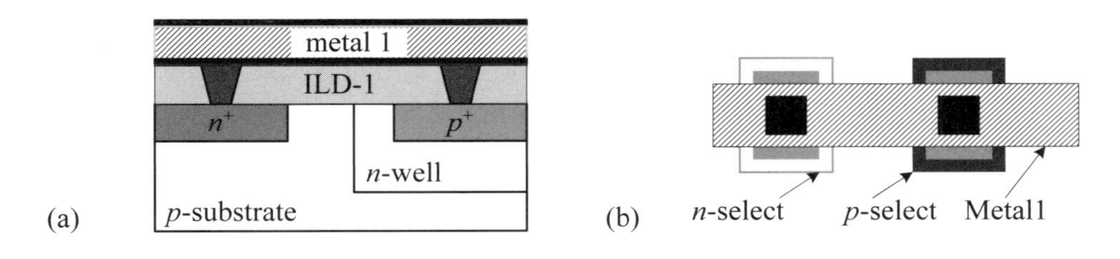

Figure 4.8: The connection of metal1 to n^+ and p^+ regions through contacts: (a) cross-sectional view; (b) mask set.

of metal1. In addition, some processes allow the stack of contact and vias on top of one another to further increase integration density.

4.1.2.9 Passivation After fabrication, the die is vulnerable to dirt and other foreign matters. As a consequence, a *passivation layer* of glass material is often deposited on the wafer surface to protect the die from contamination. This layer is also known as an *overglass layer*. To allow the die to be accessible from outside, a passivation mask is employed to open holes in the passivation layer. These holes are called *bonding pads*, which are in the highest metal layer and define the entrance and exit ports of the die. The size of bonding pads ranges from 100 μm× 100 μm to 60 μm× 60 μm in some more advanced processes.

Before leaving this subsection, some basic layout examples are given in the following.

■ **Example 4-1: (Basic layout examples.)** Figure 4.10 gives three layout examples. Figure 4.10(a) is a single transistor. Figure 4.10(b) shows two transistors connected in parallel. These two transistors share their drain/source area to reduce the capacitance of an internal node. Figure 4.10(c) gives two transistors connected in series. Once again these two transistors share their drain/source area, leading to an internal node of smaller area and less capacitance.

■

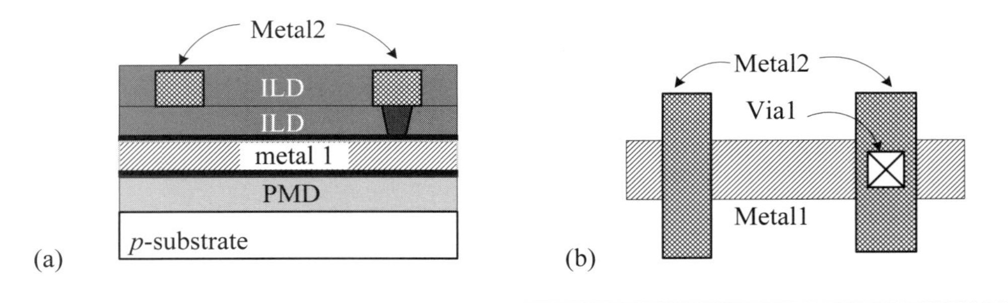

Figure 4.9: The connection between two metal wires through a via: (a) cross-sectional view; (b) mask set.

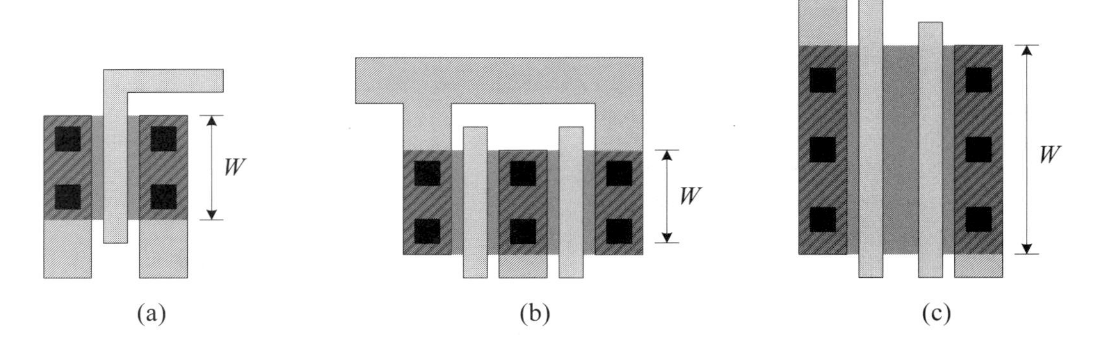

Figure 4.10: Three basic layout examples of transistors: (a) single; (b) parallel-connected; (c) series-connected.

Review Questions

Q4-6. How would you create an n^+ region?
Q4-7. What is the mask set required to define an nMOS transistor?
Q4-8. What is the mask set required to define a pMOS transistor?
Q4-9. What is the distinction between contacts and vias?
Q4-10. What is the goal of an overglass layer?

4.1.3 Advanced Layout Considerations

Modern nanometer processes face severe challenges in signal integrity, manufacturability, and reliability. In this subsection, we are concerned with layout problems related to these issues.

4.1.3.1 Routing for Signal Integrity From the circuit viewpoint, the term *signal integrity problem* addresses two issues: the timing and the signal quality. In other words, the signal integrity problem concerns whether a signal reaches its destination at the right time in good condition. With the advance of fabrication processes, the feature sizes continue to decrease, the clock rates keep increasing, and the devices and wires are placed in closer proximity to reduce the routing area and propagation delays, thereby leading to the increase of coupling capacitance between two adjacent long wires. The coupling capacitance gives rise to crosstalk, which in turn affects signal

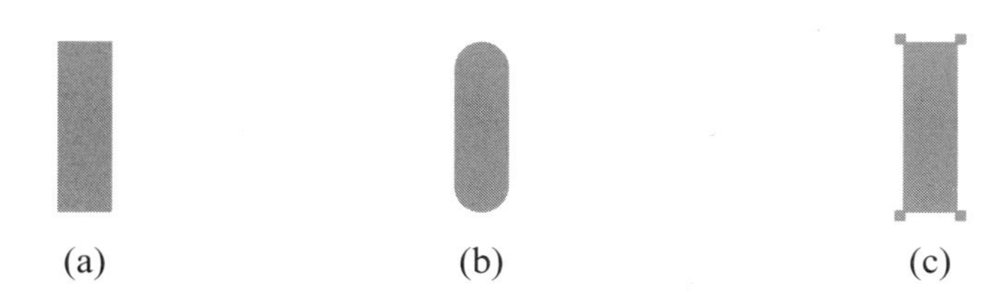

Figure 4.11: The notching effect in nanometer scale photolithography: (a) mask pattern; (b) wafer pattern; (c) corrected mask pattern.

integrity. The magnitude of crosstalk is proportional to the coupling capacitance. The coupling capacitance between two orthogonal wires is negligible compared with that between two adjacent parallel wires. The two most common approaches to reducing the crosstalk caused by coupling capacitance are to increase the separation between and to twist two parallel wires. Refer to Section 13.3.2 for more details.

4.1.3.2 Routing for Manufacturability During manufacturing, optical proximity correction (OPC) and chemical-mechanical polishing (CMP) are two important issues needed to be considered. The OPC uses precompensated feature patterns on a mask to improve the printability of the mask pattern and the CMP improves layout uniformity and chip planarization to achieve higher manufacturing yield.

Optical Proximity Correction. In deep submicron processes, especially below 0.18 μm, the minimum feature sizes of circuit patterns become significantly smaller than the photolithographic wavelengths, as shown in Figure 3.16. As a consequence, it is very difficult to obtain the exact image that we desire on the wafer. As shown in Figure 4.11(a), the desired pattern is a rectangle but the transferred pattern on the wafer is a distorted one, as depicted in Figure 4.11(b). This means that some kinds of resolution enhancement techniques are needed to achieve acceptable process accuracy. Among these, the optical proximity correction (OPC) is the most popular one in industry. The OPC is a process that modifies the layout patterns on the mask or reticle provided by the designers to compensate for the nonideal features of the photolithography process. The corrected mask pattern is given in Figure 4.11(c).

Chemical-Mechanical Polishing. As stated, the chemical-mechanical polishing (CMP) process has gained its widespread use in copper (Cu) interconnect process in the back end of line (BEOL) of deep submicron processes, especially, below 130 nm. It is shown that the Cu and oxide thickness after the damascene process is not uniform across the whole chip. Instead, systematic Cu and oxide thickness variations are observed. These systematic variations are found to be layout dependent. Consequently, to improve the CMP quality, modern IC foundries often impose some CMP-related layout design rules, even density gradient rules on each layer. For instance, a metal layer might have to have 30% minimum fill. Dummy features are also filled into layouts to reduce the variations on each layer. However, the fill may introduce unexpected parasitic capacitance to nearby wires. Designers must be aware of this.

4.1.3.3 Routing for Reliability With the advent of the nanometer era, manufacturing reliability and yield become crucial challenges. To achieve higher reliability and yield, the antenna effect caused in the plasma process and *via-open defects* are the two most important issues.

Antenna Effects. An antenna is a device that collects electrons. Recall that during a chip fabricating process metal1 is initially deposited over the entire chip and

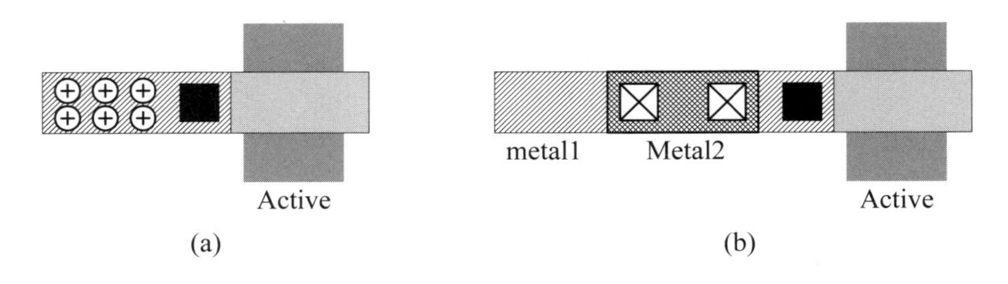

Figure 4.12: (a) The antenna effect. (b) A prevention method.

then the undesired portions are etched away using a plasma process. The exposed area of metal1 collects charged ions from plasma and raises the potential over its surface. Hence, the metal1 acts like an antenna, as shown in Figure 4.12(a). If the amount of accumulated charge exceeds a threshold, the *Fowler-Nordheim tunneling* current will discharge through the gate oxide and may cause the gate-oxide to be damaged. Similarly, the antenna effect may also occur for any large piece of conductive material tied to the gate, including polysilicon itself.

The antenna effect can be reduced in one of the following three methods:

- *Jumper insertion*: In this approach, signal wires are broken and routed to upper levels, thereby leading to the reduction of the amount of charge accumulated on each wire during fabricating.
- *Embedded protection diode*: In this method, protection (that is, clamped) diodes are added on every input port for every standard cell. Nevertheless, this approach will consume unnecessary area and increase fabricating cost because these diodes are embedded and consume the cell area.
- *Dynamically inserting diode after placement and route*: Protection diodes are only added to those wires with antenna violation. During wafer fabricating, all of the inserted diodes are floating (or ground); during operation mode, these diodes act like reverse diodes. One diode can be used to protect all input ports that are connected to the same output ports. However, this approach can only be applied to the case where there is enough room for diode insertion.

At present, the jumper insertion method is the most popular way to reduce the antenna effect. To use this, the total area of conductor directly connected to the gate is limited to some threshold, which is defined as the ratio of the conductor area to gate area. When the ratio exceeds the threshold, the desired large conductor (metal1) is broken into two pieces and connected using metal2 as a bridge, as shown in Figure 4.12(b). This set of rules is popularized as *antenna rules* or *process-induced damage rules*.

Via-Open Defects. With the progress of very-large-scale integration (VLSI) fabrication into the nanometer era, via-open defects are one of dominant failures due to the *copper cladding process*. To improve via yield and reliability, *redundant-via insertion* is a recommended technique proposed by IC foundries. If one via fails, the redundant via can serve as a fault-tolerant substitute for the failing one. Double-via insertion may be performed at the postlayout optimization by computer-aided design (CAD) tools. Nevertheless, because of the increasing design complexity, very limited space is left for postlayout optimization. As a consequence, the double-via insertion is better considered at both the routing and postrouting stages.

4.1.4 Related CAD Tools

The CAD tools related to layout design include the following:

1. *Design rule checker* (DRC) uses the layout database and checks every occurrence of the design rule list on the layout.
2. *Parameter extraction routine* translates the polygon patterns and layers into an equivalent electrical network, usually in a SPICE (simulation program with integrated circuit emphasis) data file.
3. *Layout versus schematic* (LVS) checks the layout against the schematic diagram. It is based on a graph isomorphism algorithm to check whether two graphs are identical.
4. *Electrical rule checker* (ERC) checks the electrical continuity in the layout.

■ Review Questions

Q4-11. What are the two general approaches to reducing the crosstalk caused by two adjacent wires?

Q4-12. What are the two routing issues related to manufacturability?

Q4-13. What are the two routing issues related to reliability?

Q4-14. Describe the antenna effect.

Q4-15. How would you avoid the antenna effect?

Q4-16. What is the meaning of redundant-via insertion?

4.2 CMOS Latch-Up and Prevention

In this section, we are concerned with the latch-up problem inherent in basic CMOS processes. We begin with an introduction to *I-V* characteristic of a typical *pnpn* device. Then, we discuss the origin of the parasitic bipolar junction transistors (BJTs) in a basic CMOS inverter and how they form a *pnpn* device. Finally, we take a look at the issues of the turn-on condition of the parasitic *pnpn* device and the approaches to avoiding it.

4.2.1 CMOS Latch-Up

Before considering the latch-up problem encountered in basic CMOS processes, we first consider the structure and operation of a *pnpn* device, as shown in Figure 4.13. As illustrated in Figure 4.13(a), a *pnpn* device is a four-layer structure comprising three *pn* junctions, J_1, J_2, and J_3, in series. Conceptually, an *pnpn* device can be thought of as a combination of a *pnp* and an *npn* BJTs with the equivalent circuit shown in Figure 4.13(b); physically, when two BJTs are connected in such a way, the resulting circuit indeed reveals the same behavior as that of a *pnpn* device.

The typical *pnpn* device has an *I-V* characteristic as displayed in Figure 4.13(c). When the applied voltage V_{AK} is small, the current I_A is also very small; however, when the applied voltage V_{AK} increases to a value in excess of the *forward-breakover voltage* ($V_{BR(F)}$), the device enters the forward conduction region. At the onset of this region, the decrease of voltage V_{AK} will cause an increase of current I_A until the applied voltage reaches a point, called the *hold state*, denoted by (V_H, I_H), beyond

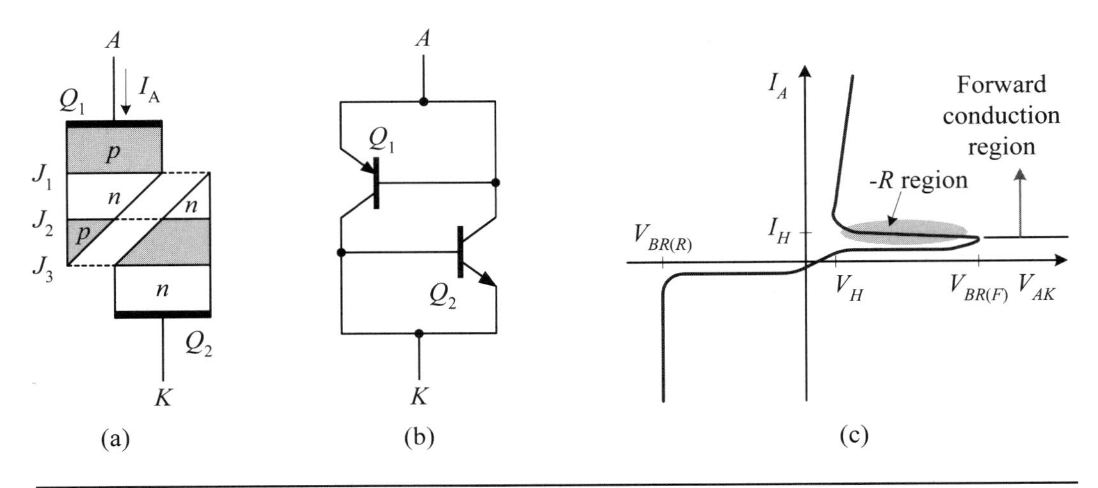

Figure 4.13: The (a) physical structure, (b) equivalent circuit, and (c) I-V characteristic of a typical *pnpn* device.

which an increase of small voltage will cause the current I_A rapidly to grow in an almost unlimited way. The small voltage across the device at the hold state is called the *hold voltage*, V_H, and the corresponding current is called the *hold current*, I_H. In order to avoid the damage of the device, an external resistor, usually the load, with an appropriate resistance must be connected in series with the device to limit the current to a safe range.

A salient feature of *pnpn* devices is that in the region between forward-breakover voltage ($V_{BR(F)}$) and V_H in the forward conduction region, a negative incremental resistance exists, that is, $\Delta R = \Delta V/\Delta I < 0$, as shown in Figure 4.13(c). This region is called a *negative resistance region* ($-R$ region). It is worth noting that once a *pnpn* device is turned on and enters its forward conduction region, it cannot be turned off unless the applied voltage across it is removed. Consequently, the turn-on condition is also referred to as the *latch-up condition* in terms of CMOS process terminology.

4.2.1.1 Parasitic pnpn Device The cross-sectional view of a CMOS inverter accompanied with parasitic components is shown in Figure 4.14. The *pn* junction between the output and the n-well forms the capacitor C_1; the *pn* junction between the output and the p-substrate forms the capacitor C_2. The *pnp* transistor is formed from the p^+ region connected to the V_{DD}, n-well, and p-substrate; the *npn* transistor is formed from the n^+ region connected to ground, p-substrate, and n-well. Both n-well and p-substrate have relative high resistance and their corresponding equivalent resistances are called R_{W1}, R_{W2}, R_{S1}, and R_{S2}, respectively, as shown in the figure. Among these four resistances, R_{W1} and R_{S2} are the two most important ones, which are strongly related to the latch-up condition of the parasitic *pnpn* device, as we will see.

To illustrate how the parasitic BJTs form a *pnpn* device and how the latch-up condition is reached—that is, the condition that both parasitic BJTs are turned on—the parasitic BJT circuit is redrawn as an equivalent circuit schematically shown in Figure 4.15. Since resistances R_{S1} and R_{W2} do not matter with the latch-up condition, they are ignored in the circuit. In what follows, we will use this equivalent circuit to derive the latch-up condition.

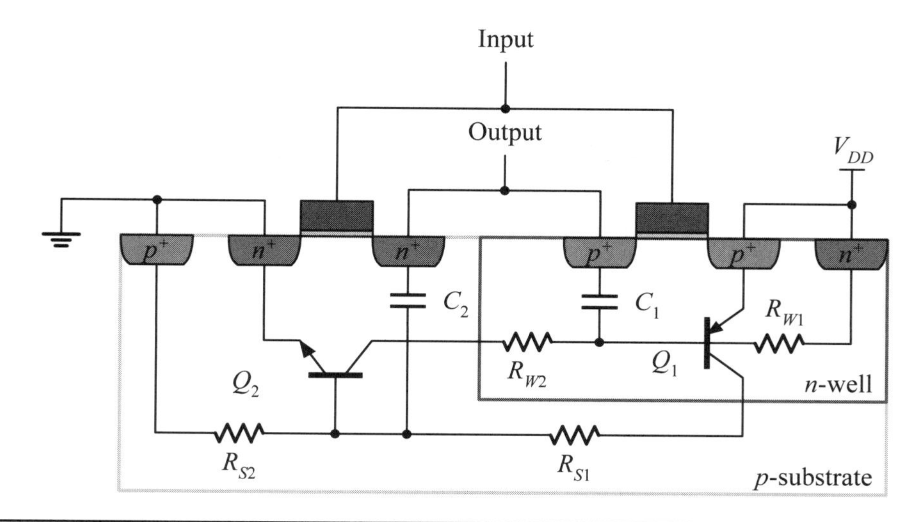

Figure 4.14: The parasitic BJTs in a basic CMOS inverter.

4.2.1.2 Latch-Up Condition As mentioned above, the condition for latch-up is that both parasitic BJTs have to be turned on and remain at this state; that is, the parasitic *pnpn* device enters and stays at the hold state. To be more quantitative, let us consider the equivalent circuit depicted in Figure 4.15. In order to maintain both BJTs at their turn-on states, the current gain of the loop formed by both BJTs, as indicated in the figure, must be greater than 1. In other words, the following current relationship must hold

$$(I_{bn}\beta_n - I_{well})\beta_p - I_{sub} > I_{bn} \tag{4.9}$$

which implies

$$I_{bn}(\beta_n\beta_p - 1) > I_{sub} + \beta_p I_{well} \tag{4.10}$$

In addition, from the ground node, we can obtain

$$I_{DD} = I_{sub} + I_{bn}(\beta_n + 1) \tag{4.11}$$

Rearranging the terms, the above equation can be expressed as

$$I_{bn} = (I_{DD} - I_{sub})/(\beta_n + 1) \tag{4.12}$$

Substituting Equation (4.12) into Equation (4.10) and after rearranging some terms, we have

$$\beta_n\beta_p > 1 + \frac{(\beta_n + 1)(I_{sub} + \beta_p I_{well})}{I_{DD} - I_{sub}} \tag{4.13}$$

Consequently, to turn on both transistors, the product of their current gains, β_n and β_p, must be greater than 1 plus a term determined by the current gains of both BJTs and the currents in the n-well and p-substrate, I_{well} and I_{sub}. Because the turn-on voltage of both BJTs are almost fixed with a constant value of 0.6 to 0.7 V, the currents in both the n-well and p-substrate, I_{well} and I_{sub}, are inversely proportional to the resistance of the n-well and p-substrate, namely, R_{W1} and R_{S2}, respectively.

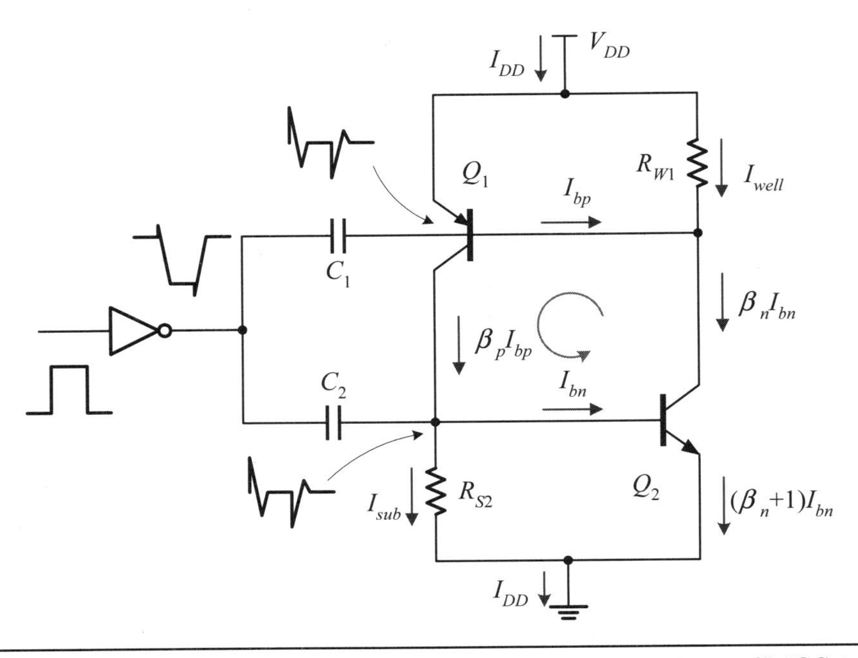

Figure 4.15: The equivalent circuit of the parasitic BJTs in a basic CMOS inverter.

4.2.2 Latch-Up Prevention

In practice, there are two basic techniques that can be used to eliminate or alleviate the possibility of latch-up occurrence. These are the use of a latch-up resistant process and the use of layout techniques. A popular latch-up resistant process is the *SoI process* in which both nMOS and pMOS transistors are completely isolated so that there exists no parasitic *pnpn* device at all. A brief description of a silicon-based SoI process can be referred to in Chapter 3. In what follows, we are concerned with the avoidance of the latch-up problem through the use of layout techniques in conventional CMOS processes.

4.2.2.1 Core Logic Circuits From Equation (4.13), we can see that two basic approaches can be used to alleviate or eliminate the latch-up problem in conventional CMOS processes. The first approach is of course to reduce the current gains of parasitic BJTs. This can be done through a carefully designed fabricating process. Another way is to increase both currents, I_{well} and I_{sub}, corresponding to reducing the resistances of resistors R_{W1} and R_{S2} because $I_{well} = V_{BEon(pnp)}/R_{W1}$ and $I_{sub} = V_{BEon(npn)}/R_{S2}$. In most BJTs, the turn-on voltage V_{BEon} is almost fixed to a constant value, ranging from 0.6 to 0.7 V.

The R_{W1} denotes the sum of both contact and well resistances and R_{S2} is the sum of both contact and substrate resistances. The values of R_{W1} and R_{S2} are a strong function of the distance between the source connections of MOS transistors and well/substrate contacts. Consequently, the key technique to reduce the latch-up occurrence is to make use of substrate and well contacts to reduce both resistance ingredients. Based on this, the following general guidelines immediately follow in designing the layout of a core circuit:

- Every well and the substrate must be properly connected to a power-supply rail through metal wires directly.

- Both substrate and well contacts should be placed as close as possible to the source connection of MOS transistors, thereby reducing the resistances of both resistors R_{S2} and R_{W1}.
- A conservative rule is to place a substrate or well contact for about every 5 to 10 MOS transistors. The actual rule should be consulted with the layout design rules from the IC foundry.
- The nMOS transistors and pMOS transistors should be separately placed toward V_{SS} and V_{DD}. Avoid placing nMOS transistors with pMOS transistors in a convoluted way.

4.2.2.2 I/O Circuits For I/O circuits, since large signal swings must be exercised and heavier capacitive loading has to be driven, much larger signal transients may occur in these circuits compared with core circuits. This means that the I/O circuits are much more inclined to encounter latch-up problem than core circuits. Therefore, care must be taken in designing I/O circuits to reduce the possibility of latch-up occurrence.

The most common approach to preventing the I/O latch-up problem is based on the use of *guard rings*. Conceptually, the guard-ring technique is an extension of substrate or well contacts. In this technique, instead of using a single contact with a small area, a p^+ or an n^+ ring is used to surround the entire nMOS or pMOS transistor, as shown in Figure 4.16. The function of guard rings can be thought of as *dummy collectors*, which separately collect holes and electrons leaking from nMOS and pMOS transistors, respectively, before they reach the bases of the parasitic BJTs, thereby significantly reducing the conduction possibility of both parasitic BJTs and hence the likelihood of latch-up occurrence. For practical applications, both n^+- and p^+-type guard rings should be densely connected to the V_{DD} and ground through metal wires, respectively.

Some basic guidelines for designing the layouts of typical I/O circuits are as follows:

- It should include p^+ guard rings connected to the V_{SS} to surround nMOS transistors and n^+ guard rings connected to the V_{DD} to surround pMOS transistors.
- Source fingers of large MOS transistors should be perpendicular rather than parallel to the dominant direction of current flow.
- It should keep both distances between the substrate contacts and the source connections of nMOS transistors, as well as the well contacts and the source connections of pMOS transistors to a minimum, hence reducing both substrate and well resistances, R_{S2} and R_{W1}, to their minimum values.

Review Questions

Q4-17. Describe the features of a *pnpn* device.

Q4-18. Describe the parasitic *pnpn* device in a CMOS inverter.

Q4-19. What is the latch-up condition in a CMOS inverter?

Q4-20. Why are I/O circuits more prone to a latch-up problem than core circuits?

Q4-21. What is the distinction between a well contact and a guard ring?

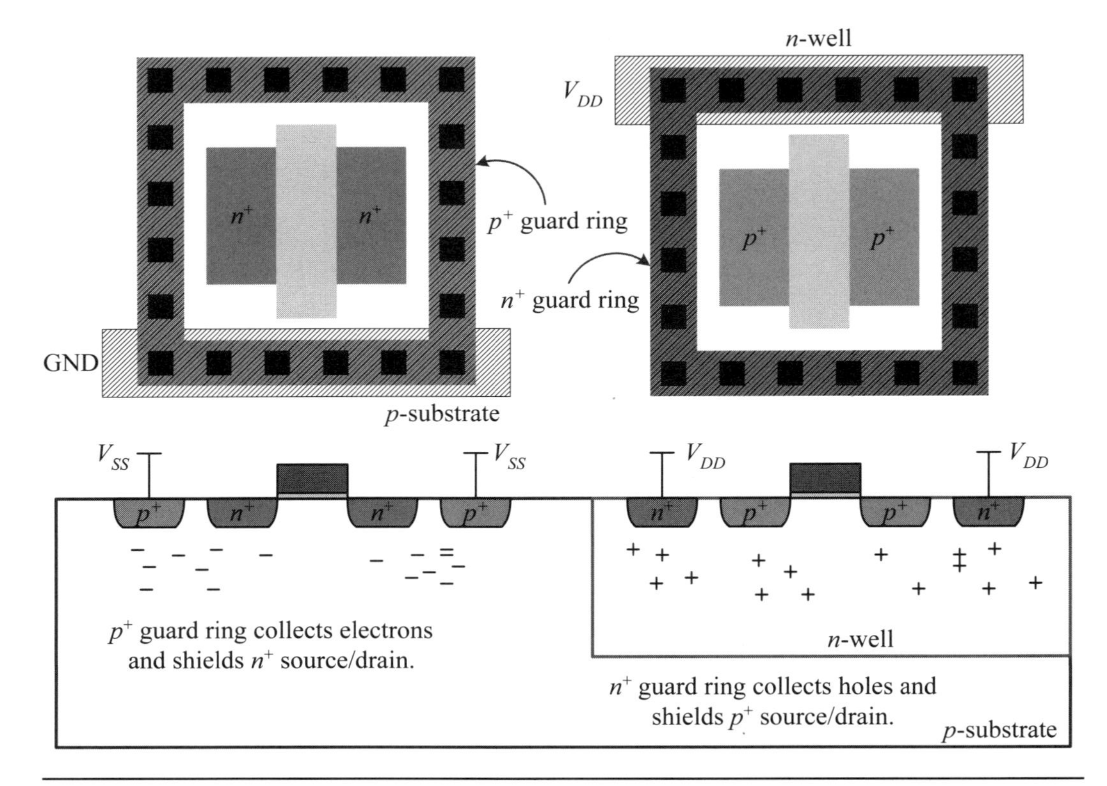

Figure 4.16: The use of the guard-ring technique to reduce the possibility of latch-up in the I/O circuits.

4.3 Layout Designs

The (physical) layout design of a circuit means to design and draw the mask patterns needed for fabricating the circuit. In this section, we first introduce cell concepts and then address layout designs of a number of MOS logic circuits.

4.3.1 Cell Concepts

Modern VLSI chips are based on the concept of hierarchical design in which a design is subdivided into many smaller modules. These modules are further partitioned into even smaller modules until the modules can be easily handled. One benefit of hierarchical design is that it can reuse some basic modules as many times as needed.

The basic building blocks in layout designs are usually referred to as *cells*. A cell may be as simple as a single MOS transistor, a primitive gate, an arithmetic logic unit (ALU), or as complex as a entral processing unit (CPU) module. Every cell may be used as a component to create a larger and complex logic module. Each cell has input and output terminals called *ports*. There are two basic types of ports: *signal ports* and *power ports*. Signal ports are used to communicate with other cells and can be further subdivided into *input port*, *output port*, and *bidirectional port*. Power ports, also known as *power-supply ports*, are used to provide the cell with power and ground connections and include V_{DD} and V_{SS} (that is, ground). Power ports are often placed

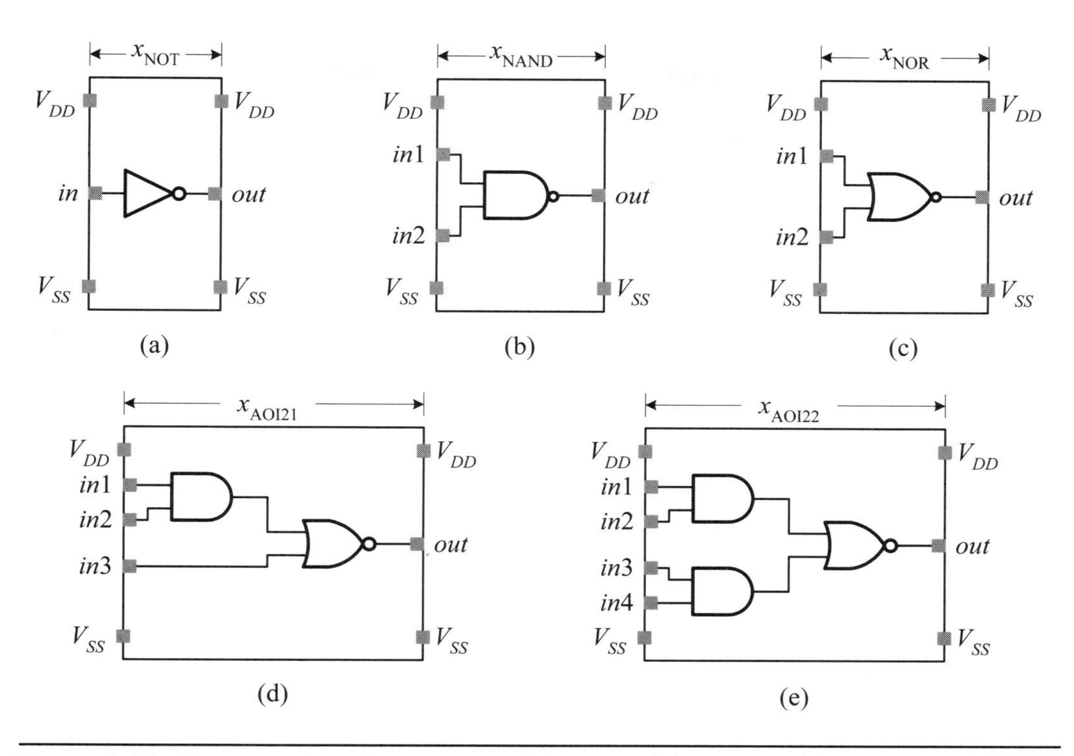

Figure 4.17: Five samples of basic cells: (a) NOT; (b) NAND; (c) NOR; (d) AOI21; (e) AOI22.

at the same location for every cell in order for the abutted connection between two adjacent cells when they are placed side by side.

■ Example 4-2: (Basic cells.) Five samples of basic cells: NOT, NAND, NOR, AOI21, and AOI22 gates are shown in Figure 4.17. These cells have the same height but different widths. Their power ports are located at the same position to allow the abutted connection when two such basic cells are placed side by side to function as a more complicated logic circuit.

■

A complex logic cell can be created using basic cells. When a complex cell based on basic cells or other cells is designed, metal1 wires are used to connect together the ports of these cells as needed. Two such examples are illustrated in Figure 4.18, where two new cells, AND and OR, are separately built from a NAND cell and a NOT cell, as well as a NOR cell and a NOT cell. Because the three basic cells, NOT, NAND, and NOR, have the same height and port positions, the resulting AND and OR cells still have the same height but have the cell width equal to the sum of the cell widths of the two basic cells used to construct them.

Using the hierarchical design approach allows us to create and construct a complex logic circuit on the basis of other simple or smaller modules or cells. The hierarchical design takes advantage of reusing lower-level modules or cells. To get more insight into this concept, consider the following example that illustrates how to create a two-input

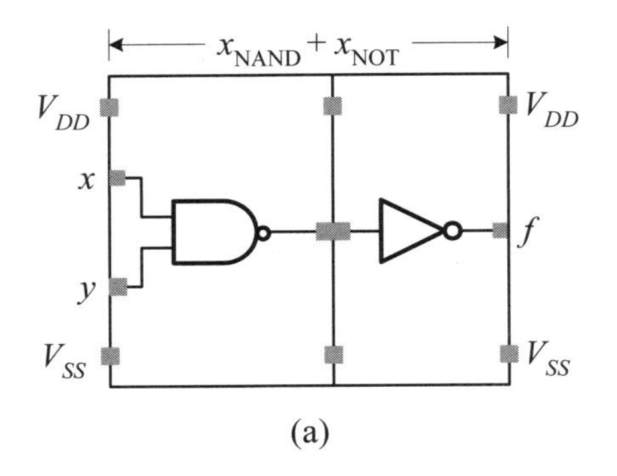

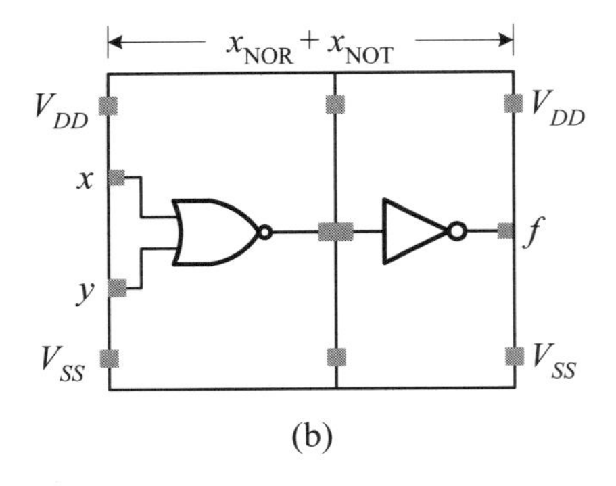

Figure 4.18: Examples of (a) AND and (b) OR cells constructed from the three basic cells: NOT, NOR, and NAND.

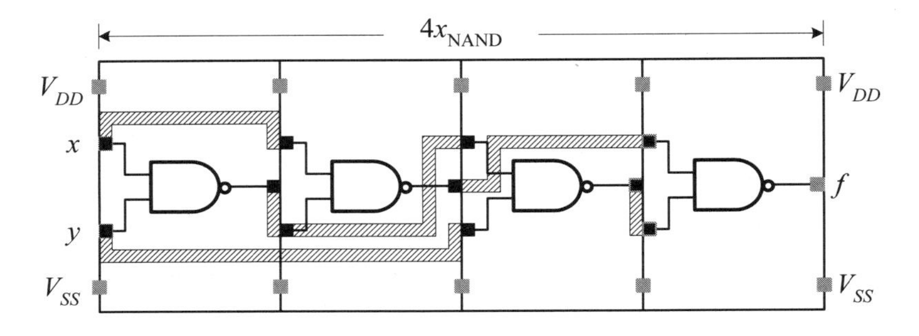

Figure 4.19: A new XOR cell constructed from four two-input NAND cells.

XOR cell from four two-input NAND cells.

■ Example 4-3: (A NAND-based XOR cell.) Figure 4.19 shows an XOR cell constructed from four two-input NAND cells. Intuitively, we can construct this cell in a way like that in constructing AND and OR cells from the three basic cells: AND, OR, and NOT. That is, we first place these four two-input NAND cells side by side as shown in Figure 4.19 and then properly connect their ports through metal wires. The resulting XOR cell has the same height but with a width equal to four times that of the NAND cell. ■

4.3.1.1 Power-Supply Wires From the viewpoint of cell layout, we can arbitrarily allocate power-supply wires in any place within cells. However, as we described previously, to alleviate or avoid the latch-up problem, it is better to place the n-well region being used for constructing pMOS transistors adjacent to the V_{DD} power-supply rail and place the p-substrate region being used for building nMOS transistors adjacent to the V_{SS} power-rail (ground). Hence, the resulting layout is like the one shown in Figure 4.20, where an n-well is placed beside the V_{DD} wire and a p-substrate is placed along the V_{SS} to reduce the resistances between the power-supply rail and an n-well as well as the power-supply rail and the substrate.

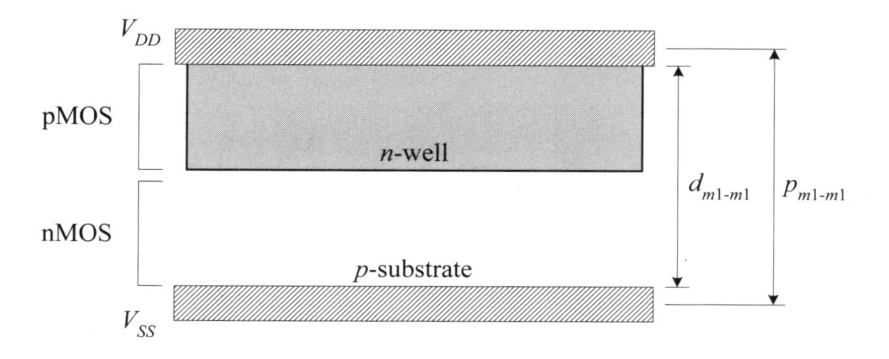

Figure 4.20: A layout example of V_{DD} and V_{SS} (ground) wires in a typical cell.

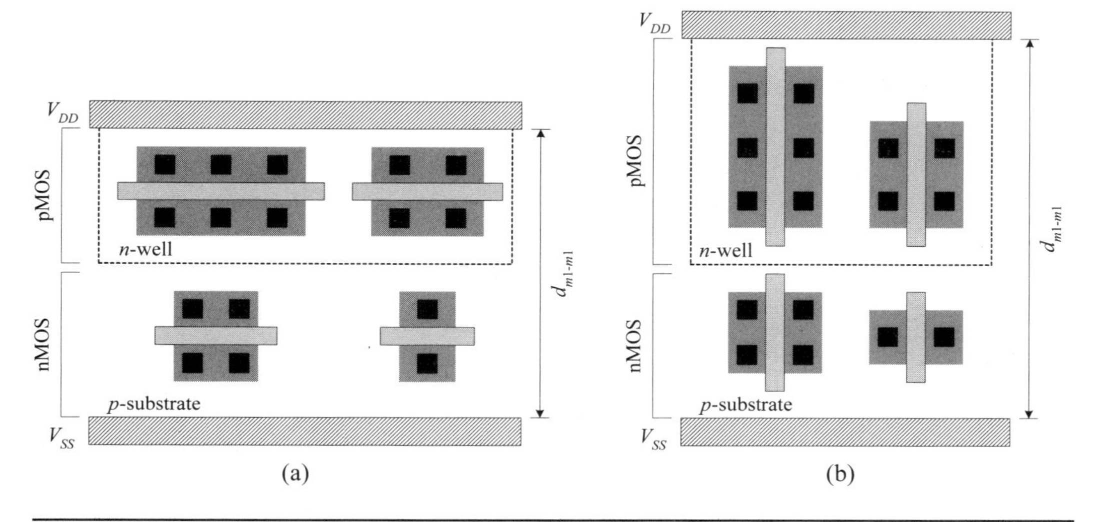

Figure 4.21: The orientation effects of MOS transistors within cells: (a) vertical orientation; (b) horizontal orientation.

4.3.1.2 MOS Transistor Orientations and Their Effects One fundamental feature of cell designs is to keep the same height for all or a group of cells. Hence, as shown in Figure 4.20, what we need is to choose a constant that can be used for every cell in the group under consideration. There are two parameters that can be used. The distance $d_{m1\text{-}m1}$ measures the spacing between the inside edge of power-supply wires V_{DD} and V_{SS} and the *pitch* $p_{m1\text{-}m1}$ is the distance between the centers of power-supply wires V_{DD} and V_{SS}. The former is more convenient for layout design and the latter is widely used in industry. Both parameters are determined by the cell orientation, vertical or horizontal. As a consequence, we should be aware of the effects of transistor orientations on the cell dimensions. Different orientations of transistors will result in different aspects of cells. Figure 4.21 shows the orientation effects of MOS transistors within cells. Of course, both transistor orientations are feasible, dependent on which cell layout style is preferred.

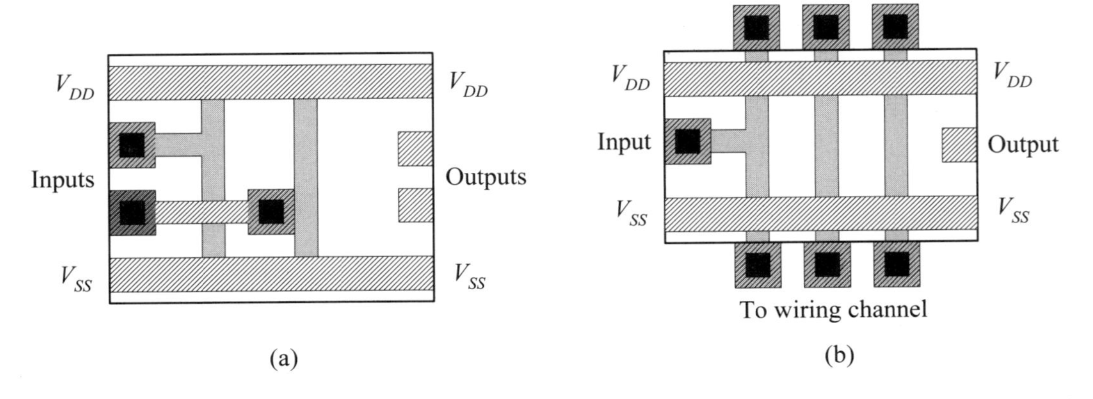

Figure 4.22: An illustration of the port placement used in a basic layout module: (a) horizontal style; (b) a combination of horizontal and vertical styles.

■ Review Questions

Q4-22. Describe the features of layout editors.

Q4-23. Which layers can be used as conducting wires in CMOS processes?

Q4-24. What are the two types of ports?

Q4-25. What are the three types of signal ports?

Q4-26. Describe the concept of hierarchical design.

4.3.1.3 Port Placement In designing a cell, it is also important to take into account the port placement because the port positions will affect the ease of using the cell or not. Ports should be located at convenient places to facilitate interconnect wiring. The input ports are usually supposed to be the gates of MOS transistors. Hence, the input polysilicon wires should provide a polysilicon contact to facilitate the capability of interconnecting with the output port from other cells. The output port is at the metal layer, thereby allowing cells to be connected with the same metal layer.

Figure 4.22(a) presents a horizontal style in which all ports are located in a horizontal way. Input ports are located at the left end while output ports are located at the right end. Figure 4.22(b) gives another style, where, in addition to locating both input and output ports horizontally, polysilicon wires are also used to provide contacts to wiring channels vertically for facilitating routing connections.

4.3.1.4 Wiring Channels In a complex digital system, basic cells are repeatedly used to construct other complex cells and these complex cells are in turn used to build other even more complicated cells or modules. This process is repeated until the final desired system is obtained. It is worth noting that a cell, regardless of how simple, is indeed a module. Hence, the terms cell and module are often used interchangeably.

In modern CMOS processes, there are many metal layers that can be utilized to route a cell. However, for simple cells, the interconnect within them is often accomplished by the lowest one or two metal layers along with the polysilicon layer. The other higher-layer metals are reserved for interconnecting different cells or modules and routing global signals as well as the power supply.

Figure 4.23 shows two possible and feasible approaches to constructing a complex cell or module. The first one assumes that a single metal layer is available for routing

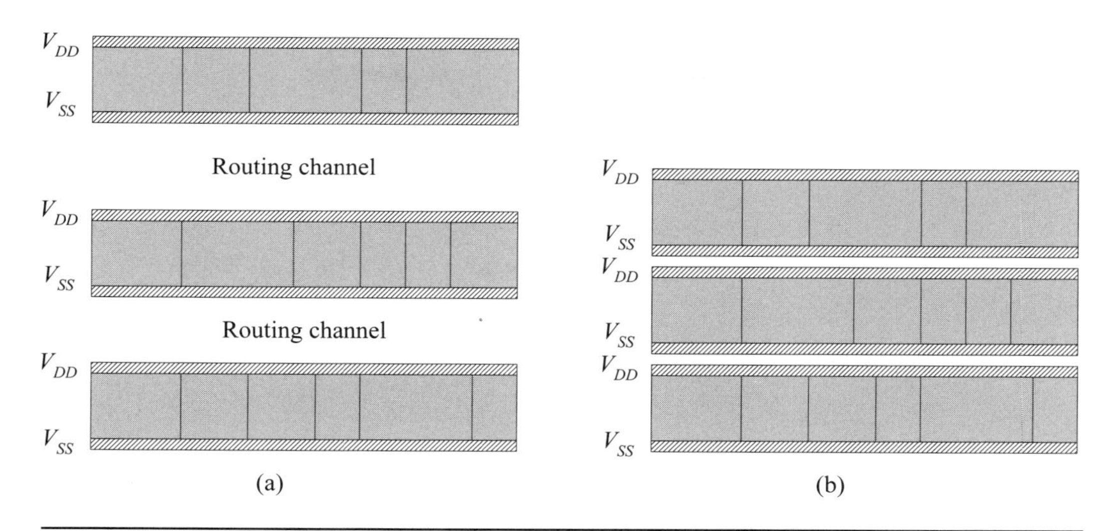

Figure 4.23: An illustration of two layout styles of complex cells or modules: (a) single-metal layout style; (b) multiple-metal layout style.

requirements whereas the second one supposes that two or more metal layers are ready to be used for the interconnect. When only a single metal layer is available, a space must be reserved between two rows of cells to complete the required interconnect among cells, as shown in Figure 4.23(a). These spaces between two rows of cells are called *routing channels*, or *wiring channels*. The width of a routing channel is dynamically adjustable to accommodate the actual spacing needed for completing the desired routing.

When multilayer metals are available for routing the required connections between cells, there is no need of a routing channel, as illustrated in Figure 4.23(b). Hence, the area required for a module is minimal compared with the situation that only one metal layer is available. Virtually, the area in this situation is only determined by the cells without the need of accounting for routing channels.

4.3.1.5 Weinberger Arrays In Weinberger arrays, V_{DD}- and V_{SS}-supply wires are alternated and shared by the logic rows above and below, as shown in Figure 4.24. As a consequence, a high-density layout can be reached because a V_{SS}-supply wire along with the required separation is removed. The Weinberger array layout style finds its widespread use in various types of memory design, such as read-only memory (ROM). One disadvantage of this layout style is that the connections between two rows must be accomplished by using the metal2 layer because metal1 is already used for the power-supply wires.

■ Review Questions

Q4-27. Why is it so important to plan the port placement of cells?

Q4-28. What do routing channels mean?

Q4-29. Describe the features of Weinberger arrays.

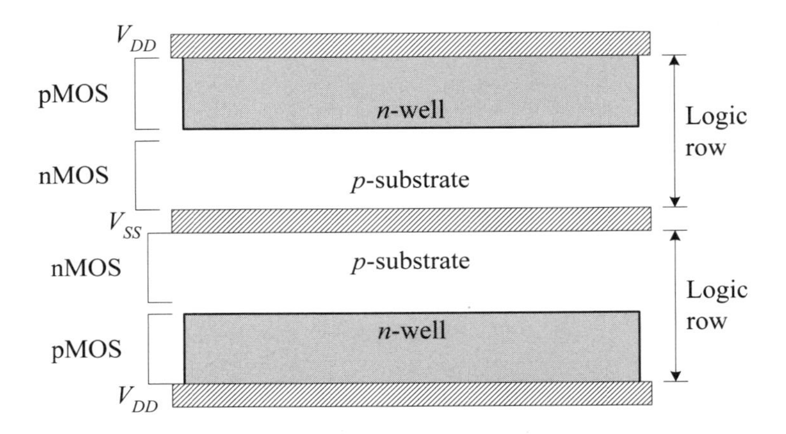

Figure 4.24: An illustration of the Weinberger array used in a basic layout module.

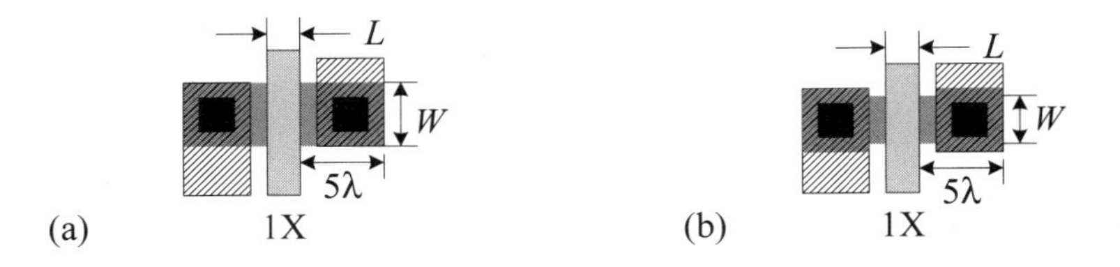

Figure 4.25: Two different layouts of an nMOS transistor: (a) with an active feature; (b) with a minimum width.

4.3.2 Basic Layout Designs

In this section, we begin with the introduction of basic geometry of MOS transistors and then look at the basic layouts of some primitive gates.

Recall that the feature size is the minimum wire width or the spacing between two wires. For a minimum-size MOS transistor, the length is set to feature size but the width is often a little wider, usually set to 3λ. Two different layouts of an nMOS transistor are shown in Figure 4.25. Figure 4.25(a) is a layout with an active feature, whereas Figure 4.25(b) shows the layout with a minimum size. In Figure 4.25(a), the nMOS transistor has the size of $W \times L$, which is $4\lambda \times 2\lambda$, not a feature-size transistor. In Figure 4.25(b), the nMOS transistor has the size of $3\lambda \times 2\lambda$ and is a feature-size transistor.

In practical applications, it is useful to design a set of transistors with different sizes to provide a variety of different driving currents. As described in Chapter 2, the current of a MOS transistor is proportional to its channel width. Therefore, a larger driving current can be achieved simply by widening the channel width of the transistor under consideration. Several different geometries of nMOS transistors are exhibited in Figure 4.26. Figure 4.26(a) is the feature-size transistor of area $3\lambda \times 2\lambda$, which is often denoted as a $1X$ transistor. Figures 4.26(b) and (c) show a $2X$ and a $4X$ transistors with a channel width equal to two times and four times that of the $1X$ transistor, respectively. In summary, an nX transistor means that its channel width and current driving capability are nX times those of the $1X$ transistor.

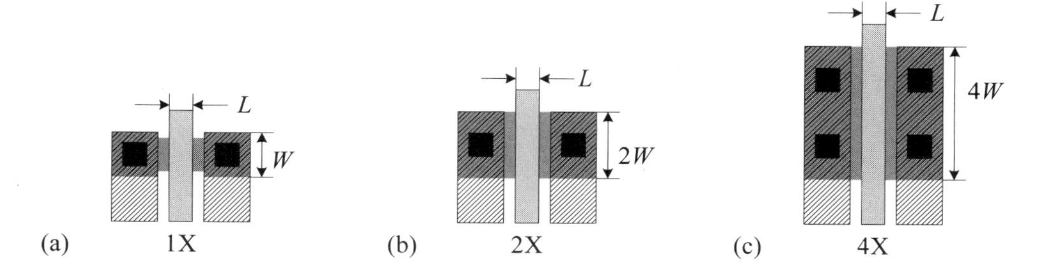

Figure 4.26: An illustration of the basic geometries of nMOS transistors: (a) basic size; (b) double width; (c) quadruple width.

4.3.2.1 Basic Steps for Layout Design Designing a layout of a logic circuit usually involves the following basic steps:

1. The CMOS or MOS logic circuit is designed and drawn schematically.
2. Using the resulting logic circuit creates a routing (namely, stick) diagram where only wiring paths and leaves are important, namely, creating a layout outline for the logic circuit. An example of a stick diagram can be referred to in Figure 1.32.
3. The routing (stick) diagram is modified and adjusted to include proper sizes for all features and to adhere to the layout design rules.

To illustrate these basic steps, we present in the following several examples to explain the above layout design procedure. The layouts of a NOT gate in which only two transistors are involved is first considered and then two more complex logic circuits, two-input NAND and NOR gates, are dealt with in detail.

■ **Example 4-4: (A NOT gate.)** Two basic layout styles with a minimum-size channel width of a NOT gate are shown in Figure 4.27. In Figure 4.27(a), both transistors, nMOS and pMOS, are horizontally placed so that the input and output port signals are aligned in parallel with the power-supply wires and transistors. In Figure 4.27(b), both transistors, nMOS and pMOS, are vertically placed. Thereby, the input and output port signals are orthogonal to transistors but still aligned with the direction of power-supply wires.

■

In practice, it often needs to balance both pMOS and nMOS current-driving capability to a load. To achieve this, the channel width of the pMOS transistor is usually set to two times that of an nMOS transistor to compensate for the loss of hole mobility, which is about two to three times less than electron mobility. The following example describes the basic 1X and 2X layouts of a NOT gate.

■ **Example 4-5: (Another NOT gate.)** An illustration of the basic 1X and 2X layouts of a NOT gate is depicted in Figure 4.28. In these two layouts, pMOS transistors always have their channel widths twice nMOS transistors. In addition, both nMOS and pMOS transistors are placed vertically, and both input and output port signals are orthogonal to transistors and aligned with the direction of power-supply wires.

■

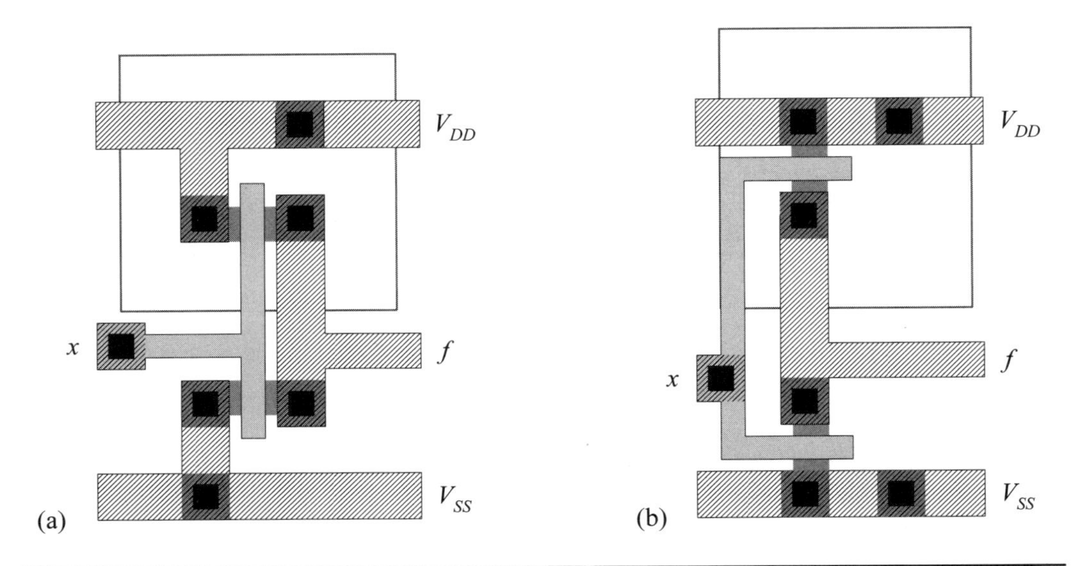

Figure 4.27: An illustration of two basic layouts of a NOT gate: (a) horizontal orientation; (b) vertical orientation.

In practical applications, we usually denote the above 1X NOT gate as a basic NOT gate; namely, a basic NOT gate is the one with its pMOS transistor having a channel width two times an nMOS transistor. In the rest of this subsection, we consider the layouts of a two-input NAND gate and a two-input NOR gate.

■ Example 4-6: (A two-input NAND gate.) A two-input NAND gate circuit has two nMOS transistors connected in series and two pMOS transistors connected in parallel. The basic layout of a two-input NAND gate is illustrated in Figure 4.29. As in the case of NOT gate, vertical orientation is used to carry out the layout of the NAND gate circuit. Since both nMOS transistors are connected in series, their drain and source regions are shared to reduce area, capacitance, and power dissipation. The channel widths of both nMOS transistors are doubled to make the resulting nMOS stack have the same resistance as that of the basic NOT gate. Because two pMOS transistors are connected in parallel, they still have the same channel width as that of the pMOS transistor in the basic NOT gate to obtain the same resistance in the worst case.

■

Basically, a two-input NOR gate is dual to the two-input NAND gate in the sense that the series connection and parallel connection are interchanged. The following example illustrates this duality.

■ Example 4-7: (A two-input NOR gate.) A two-input NOR gate circuit has two nMOS transistors connected in parallel and two pMOS transistors connected in series. A basic layout of a two-input NOR gate is illustrated in Figure 4.30. As in the case of the NOT gate, vertical orientation is used to carry out the layout of the NOR gate circuit. Because both pMOS transistors are connected in series, their drain and source regions are shared to reduce area, capacitance, and power dissipation. The

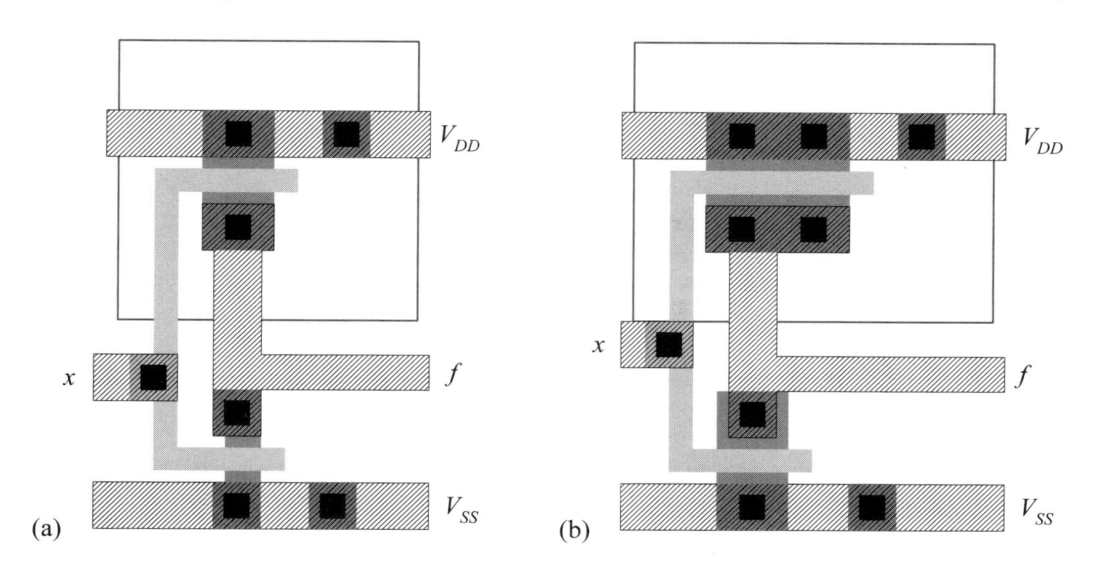

Figure 4.28: An illustration of the basic layouts of a NOT gate: (a) 1X cell; (b) 2X cell.

channel width of both pMOS transistors are doubled to make the resulting pMOS stack have the same resistance as that of the basic NOT gate. Since two nMOS transistors are connected in parallel, they still have the same channel width as that of the nMOS transistor in the basic NOT gate to obtain the same resistance in the worst case.

■

4.3.2.2 NAND versus NOR Gates From the above two examples, we may conclude that when n nMOS or pMOS transistors are connected in series, their channel widths need to be widened by a factor of n, whereas when n nMOS or pMOS transistors are in parallel connection, their channel widths remain unchanged.

To compare the area needed for an n-input NAND and NOR gates, respectively, let us consider for simplicity only the active regions and contact areas of both nMOS and pMOS transistors and ignore all other parts. Based on this, the area of an n-input NAND gate can be calculated as follows:

$$\begin{aligned} A_{NANDn} &= 2(3n \times 4) + 3n[(1+2+1) + (n-1)(3+2)] = 15n^2 + 21n\ (\lambda^2) \\ A_{NANDp} &= n[2 \times (6 \times 4)] + n(2 \times 3 \times 4) = 72n\ (\lambda^2) \end{aligned} \qquad (4.14)$$

where A_{NANDn} and A_{NANDp} are the areas of nMOS and pMOS transistors, respectively. In both equations, the first term is the contact area and the second term is the basic transistor area. The total area needed for an n-input NAND gate is equal to $15n^2 + 93n$.

Similarly, the area of an n-input NOR gate is calculated as follows.

$$\begin{aligned} A_{NORn} &= n[2(4 \times 4)] + n(3 \times 4) = 44n\ (\lambda^2) \\ A_{NORp} &= 2(6n \times 4) + 6n[(1+2+1) + (n-1)(3+2)] = 42n + 30n^2\ (\lambda^2) \end{aligned} \qquad (4.15)$$

where A_{NORn} and A_{NORp} are the areas of nMOS and pMOS transistors of the n-input NOR gate, respectively. In both equations, the first term is the contact area and the

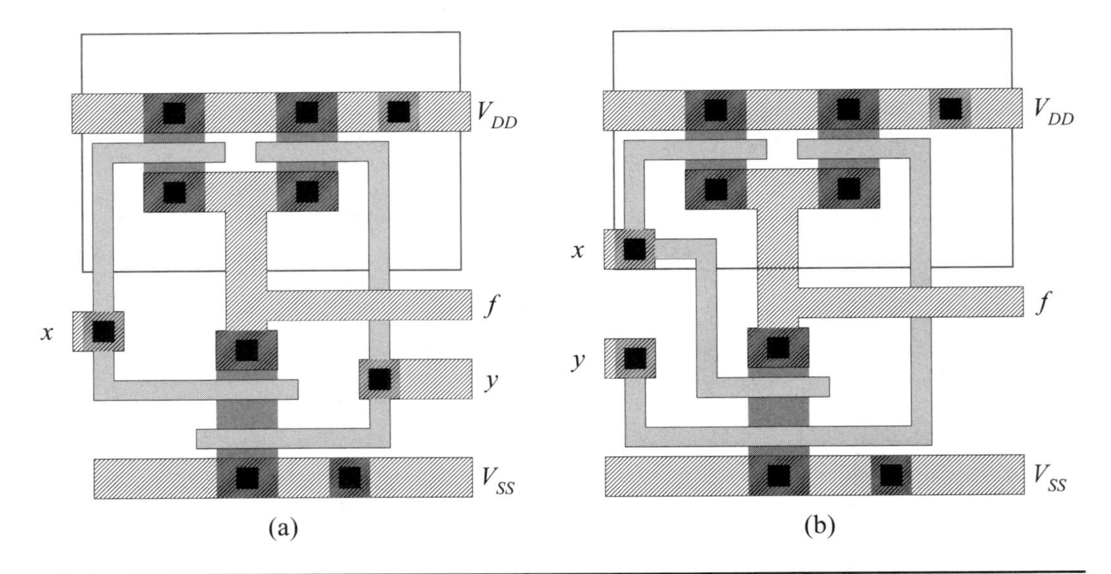

Figure 4.29: An illustration of the basic layout of a two-input NAND gate whose input ports are (a) at both sides; (b) at the same side.

second term is the basic transistor area. The total area required for an n-input NAND gate is equal to $30n^2 + 86n$.

■ Example 4-8: (Comparison of n-input NAND and NOR gates.) According to the above equations, the areas required for NAND gates with the number of inputs equal to 2, 3, and 4, are 246, 414, and 612 λ^2, respectively. The areas required for NOR gates with the number of inputs equal to 2, 3, and 4, are 292, 528, and 824 λ^2, respectively. Consequently, NAND gates generally need less area than NOR gates, in particular, when the number of inputs is large.

■

■ Review Questions

Q4-30. Give why the size of a pMOS transistor is two times that of an nMOS transistor in the basic NOT gate.

Q4-31. Explain the meanings of 1X, 2X, and 4X gates.

Q4-32. Give the reasons why NAND gates are generally preferred to NOR gates in CMOS processes.

4.4 Layout Methods for Complex Logic Gates

One systematic way to construct a layout for an irregular switching function is to use the *Euler path approach*. In this section, we first introduce the importance of source/drain sharing in the sense of layout area, performance, and power dissipation, and then describe the Euler path approach along with a number of examples.

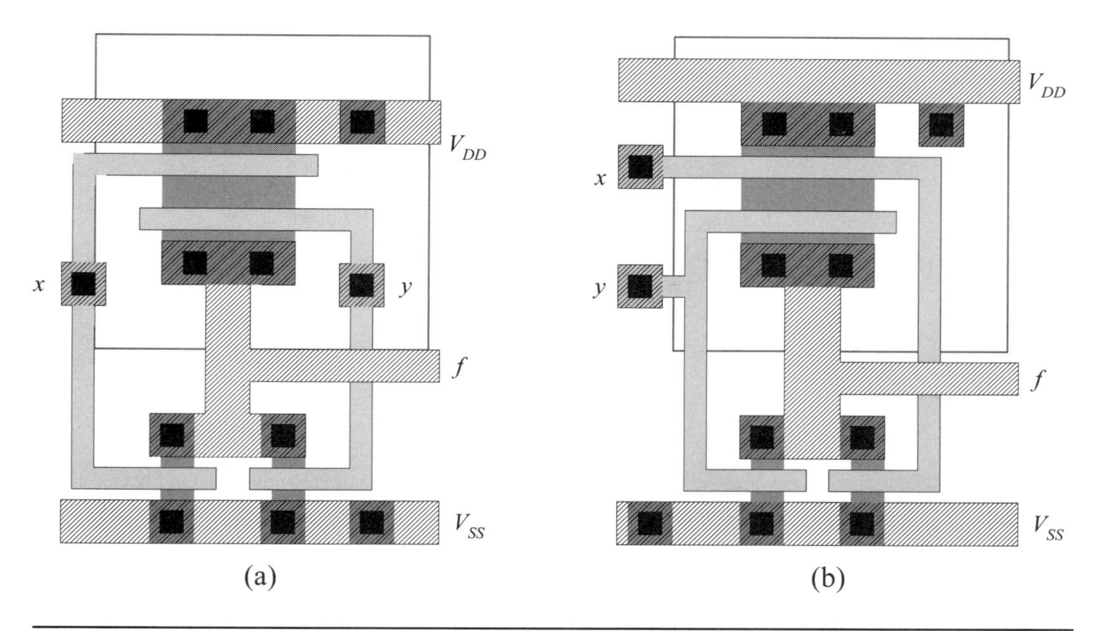

Figure 4.30: An illustration of the basic layout of a two-input NOR gate whose input ports are (a) at both sides; (b) at the same side..

4.4.1 Source/Drain Sharing

The area of a cell is calculated as the product of width and height of the cell. For standard cells, the height of a group of cells is usually assumed to be a constant. Hence, the simplest way to reduce the area of a cell is to reduce its width. In addition, when two transistors are connected in series or in parallel, a better way to save the cell area is by sharing their source and drain regions as much as they can. The following two examples concretely illustrate the benefits from sharing source and drain regions when two transistors are in serial or in parallel connection.

■ **Example 4-9: (Two MOS transistors in serial connection.)** When two MOS transistors need to be connected in series, at least two ways can be used to do this. The first method is to carry out the layouts of two individual MOS transistors and then connect them together through a metal wire, as shown in Figure 4.31(a). The resulting layout needs to account for both layout design rules associated with contacts and the separation between two active regions. Hence, it may occupy an area larger than what would be expected. The second method is to eliminate the contacts and merge together the source and drain regions of two MOS transistors. Consequently, the layout design rules associated with the contacts and adjacent active regions are simply replaced by the separation rule between two polysilicon wires over active regions. The resulting layout has less area compared with the first method, as shown in Figure 4.31(b).

■

Whenever possible one should make use of source/ drain sharing to its maximum extent when designing a layout because the resulting layout not only occupies less area

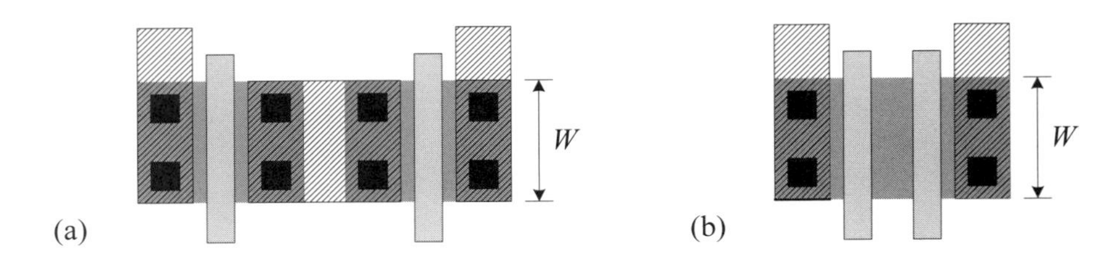

Figure 4.31: The series connection of two MOS transistors: (a) without source/ drain sharing; (b) with source/drain sharing.

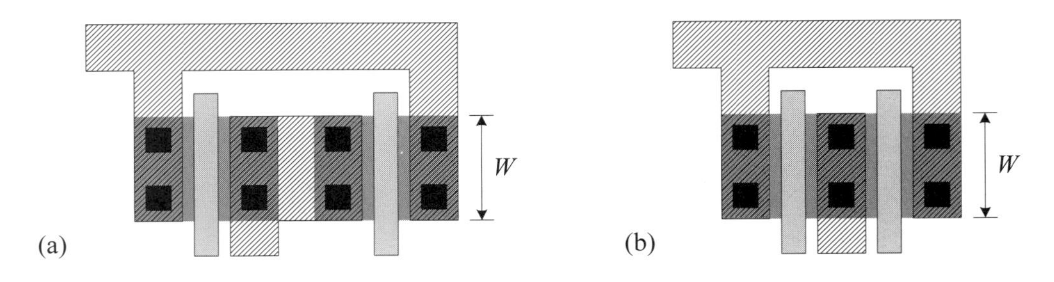

Figure 4.32: The parallel connection of two MOS transistors: (a) without source/ drain sharing; (b) with source/drain sharing.

but also has better performance due to less internal capacitance associated with the sharing node.

■ **Example 4-10: (Two MOS transistors in parallel connection.)** When two MOS transistors need to be connected in parallel, we can proceed with it in a similar way as the situation of serial connection. Of course, at least two ways can be used to do this. One way is to do the layouts of two individual MOS transistors and then connect them together through the use of a metal wire, as shown in Figure 4.32(a). The resulting layout has to consider both layout design rules associated with contacts and the separation between two active regions. Hence, it may occupy an area exceeding what would be expected. The other way is to eliminate one set of contacts and merge together the source and drain regions of two MOS transistors, as shown in Figure 4.32(b). This results in a layout that has less area compared with the first method.

■

From the previous discussion, we can recognize that the sharing of source and drain regions of two transistors may reduce the resulting layout area. As a consequence, we will introduce in the following a systematic approach to maximizing the degree of source/drain sharing between two transistors of a logic circuit. This approach is known as the *Euler path approach.* However, prior to introducing this approach, let us consider the following simple example.

■ **Example 4-11: (A simple layout example.)** Figure 4.33(a) shows a logic circuit implementing the switching function: $f(x, y, z) = \overline{x + yz}$. From the logic circuit, we are able to carry out the layout of the circuit by using separate transistor segments

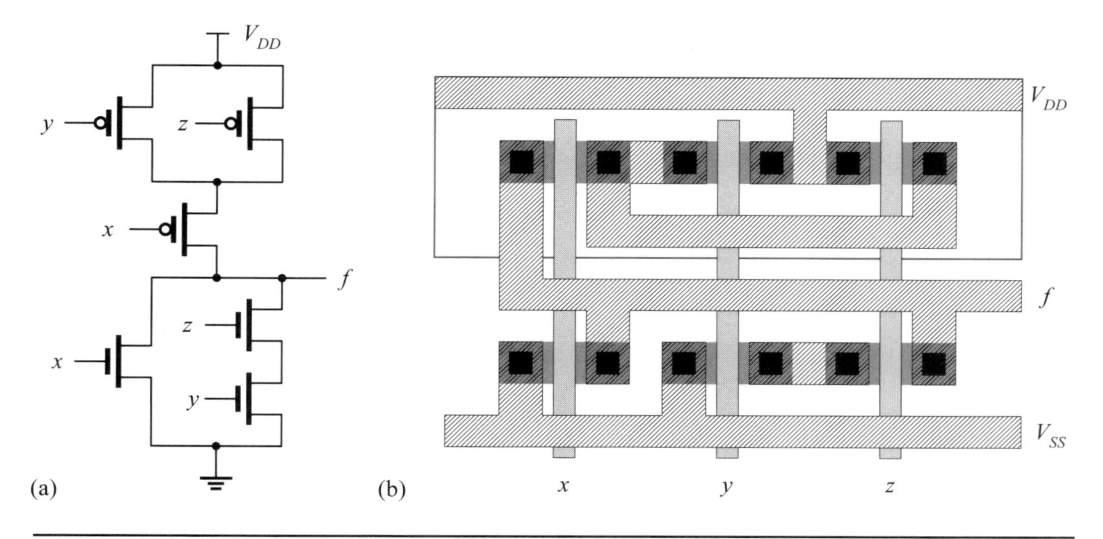

Figure 4.33: The (a) logic circuit and (b) a straightforward layout of switching function: $f(x, y, z) = \overline{x + yz}$.

and then connecting them through metal wires. Based on this idea, the resulting layout is shown in Figure 4.33(b), which includes one separation. By taking a closer look at the resultant layout, we can remove the contacts connected by metal wires between two adjacent transistors and merge their source/drain regions, namely, removing away from the separation.

■

4.4.2 Euler Path Approach

The essence of the Euler path approach is to rearrange the transistor order in such a way that the degree of source/drain sharing is maximized. In other words, the number of separations between sources/drains is minimized. The transistor order is found by using an undirected graph property called the *Euler path*. A Euler path for an undirected graph is defined as an *open trail* that visits each edge exactly once.

Before further exploring how the Euler path can be used to carry out the desired layout for a logic circuit, let us consider how to convert a CMOS logic circuit into an equivalent undirected graph, denoted as a *logic graph*. A logic graph $G = (V, E)$ is an undirected graph such that the vertex set V consists of the source/drain connections and the edge set E consists of the transistors in the logic circuit. Each edge of G is labeled with the gate signal name. For convenience, we define a *label order* as a sequence of labels on each edge, namely, a sequence of gate signal names.

Having both definitions of logic graph and label order, the Euler path approach can be described as follows.

Algorithm: Euler path approach

1. Form the pMOS and nMOS logic graphs separately from the CMOS logic circuit under consideration.
2. Find all Euler paths that separately cover nMOS and pMOS logic graphs.

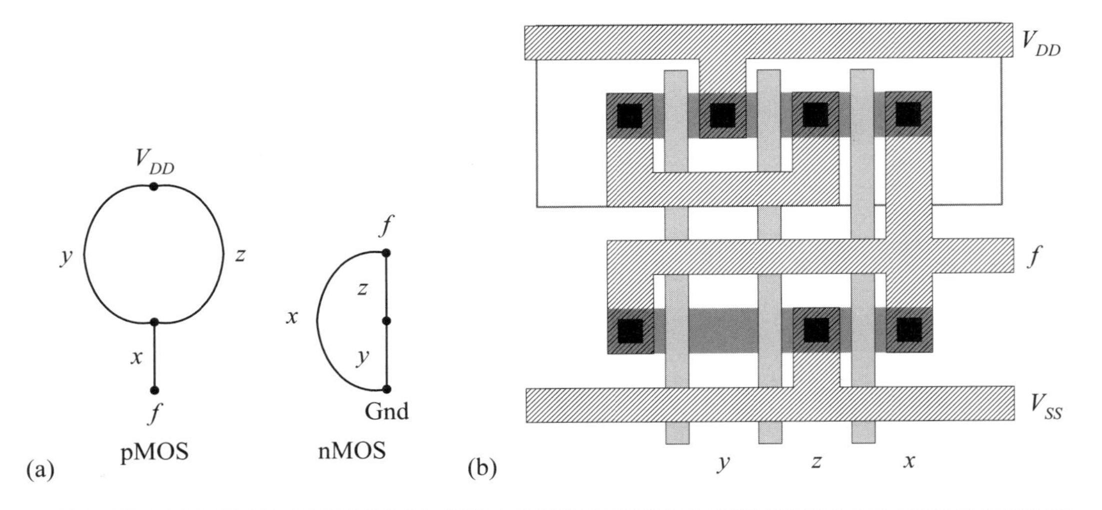

Figure 4.34: The (a) logic graphs and (b) resulting layout of the switching function: $f(x, y, z) = \overline{x + yz}$ using the Euler path approach.

3. Find n- and p-Euler paths that have an identical label order. If such a path cannot be found, then break the logic graphs in the minimum number of places to obtain such an Euler path by using the separated Euler paths.
4. Each Euler path corresponds to a diffusion segment with source/drain sharing.
5. Use metal wire segments to connect or configure the diffusion segments so that transistors are structured as its original CMOS logic circuit.

■

The Euler path approach used to perform the layout of a CMOS logic circuit is also called a *diffusion line rule*, or *diffusion segment rule*. The following example illustrates how to use the Euler path approach to redesign the layout of the preceding example.

■ Example 4-12: (Euler path approach.) From the logic circuit shown in Figure 4.33(a), the logic graphs for pMOS and nMOS logic circuits can be obtained, as shown in Figure 4.34(a). The Euler paths for the pMOS logic graph are as follows.

$$y \longrightarrow z \longrightarrow x$$
$$z \longrightarrow y \longrightarrow x$$
$$x \longrightarrow y \longrightarrow z$$
$$x \longrightarrow z \longrightarrow y$$

The Euler paths for the nMOS logic graph are as follows:

$$x \longrightarrow z \longrightarrow y$$
$$x \longrightarrow y \longrightarrow z$$
$$y \longrightarrow z \longrightarrow x$$
$$z \longrightarrow y \longrightarrow x$$

By examining the above Euler paths, we find there exist four identical label orders. Supposing that the label order: $y \longrightarrow z \longrightarrow x$ is chosen, the resulting layout is displayed in Figure 4.34(b). The other three label orders and their associated layouts are left to the reader as exercises.

■

The following example illustrates the situation that seems to be impossible to use only one diffusion segment for nMOS and pMOS transistors. However, by examining the logic graphs for both nMOS and pMOS transistors more carefully, we are indeed able to find a layout with only one diffusion segment.

■ Example 4-13: (Euler path approach.) Find all Euler paths that cover the pMOS and nMOS logic graphs of switching function $f(w, x, y, z) = \overline{\overline{wx} + (y + z)}$. Carry out the layout using the resulting Euler path.

Solution: It is easy to realize the switching function with CMOS logic circuitry, as shown in Figure 4.35(a). Both pMOS and nMOS logic graphs are readily derived from the logic circuit shown in Figure 4.35(a) and are displayed in Figure 4.35(b) and (c), respectively.

From both pMOS and nMOS logic graphs, one possible Euler path with the same label order is as follows:

$$w \longrightarrow x \longrightarrow y \longrightarrow z \longrightarrow g$$

where the signal g is only used internally. The resulting layout is depicted in Figure 4.35(d).

■

The following example illustrates the use of the Euler path approach to design the layout of the 1-bit full adder circuit given in Section 1.3.2. In this example, we attempt to illustrate the case of a layout that cannot be done with only one diffusion segment.

■ Example 4-14: (The logic graph and layout of a full adder.) Consider the full adder circuit shown in Figure 1.41. It is easy to derive its pMOS and nMOS logic graphs, as shown in Figure 4.36. For convenience, we append different numbers to an input signal in different circuit parts to differentiate the use of the input signal in these circuits, such as x_1, x_2, and so on. They are indeed the same input signal and need to be connected together in the layout. From both pMOS and nMOS logic graphs, we can obtain the following Euler path with an identical label order:

$$\begin{aligned}\bar{S} \quad &\longrightarrow \quad x_4 \longrightarrow y_4 \longrightarrow C_{in3} \longrightarrow \bar{C}_{out} \longrightarrow C_{in2} \longrightarrow x_3 \longrightarrow y_3 \\ &\longrightarrow \quad x_2 \longrightarrow y_2 \longrightarrow C_{in1} \longrightarrow x_1 \longrightarrow y_1\end{aligned}$$

and

$$\bar{C}_{out}$$

Therefore, the resulting layout has two segments and is presented in Figure 4.37.

■

Of course, the break of the logic graphs of the previous example is arbitrary. For instance, the following is another possible Euler path with an identical label order:

$$\bar{S} \longrightarrow x_4 \longrightarrow y_4 \longrightarrow C_{in3} \longrightarrow \bar{C}_{out} \longrightarrow C_{in2} \longrightarrow x_3 \longrightarrow y_3$$

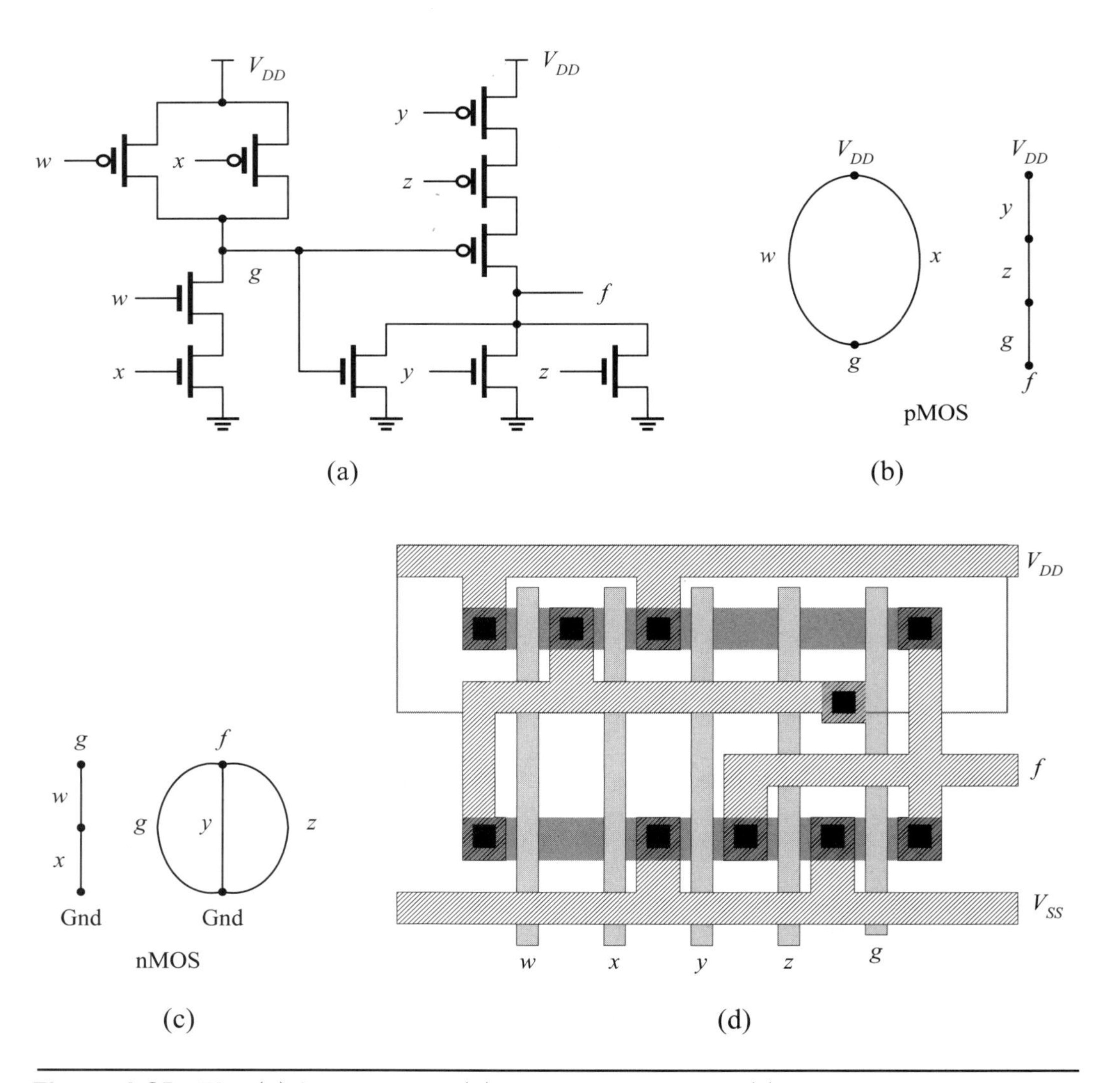

Figure 4.35: The (a) logic circuit, (b) pMOS logic graph, (c) nMOS logic graph, and the resulting layout of the switching function: $f(w, x, y, z) = \overline{\overline{wx} + (y + z)}$ using the Euler path approach.

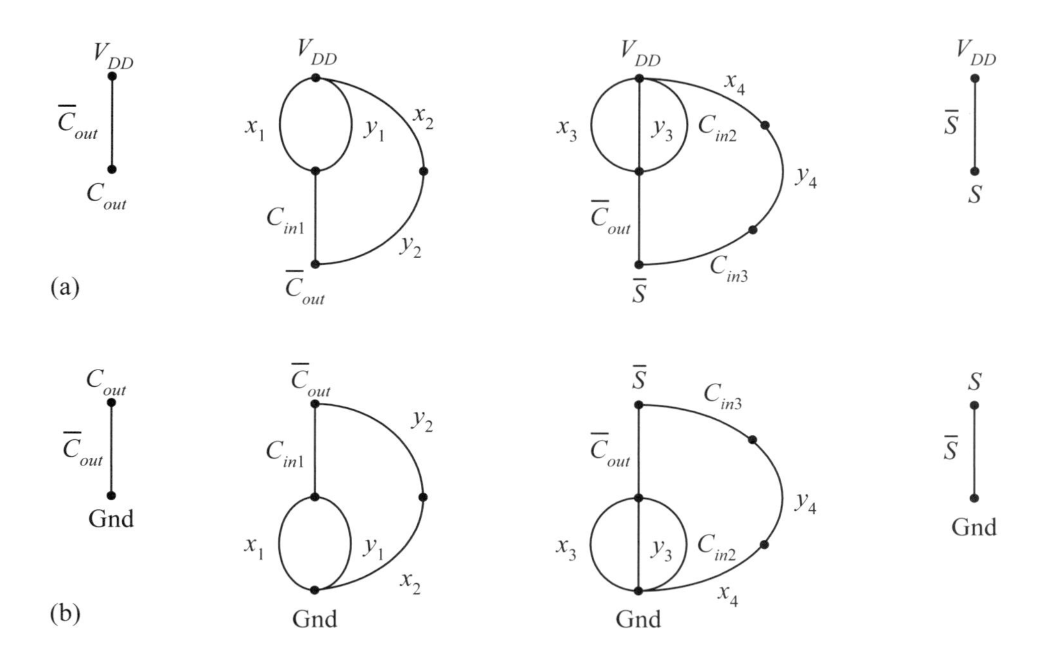

Figure 4.36: The (a) pMOS and (b) nMOS logic graphs of the full adder circuit shown in Figure 1.41. (All nodes represented by V_{DD} denote the same node and all nodes named Gnd are also the same node.)

and

$$C_{in1} \longrightarrow y_2 \longrightarrow x_2 \longrightarrow x_1 \longrightarrow y_1 \longrightarrow \bar{C}_{out}$$

Many other breaks are also possible and are left to the interested reader as an exercise.

■ Review Questions

Q4-33. Define the Euler paths of an undirected graph.

Q4-34. Describe the rationale behind the Euler path approach.

Q4-35. Define both the logic graph and label order used in the Euler path approach.

Q4-36. Describe the Euler path approach.

4.4.3 Summary in Layout Designs

Before leaving this section, we summarize some general guidelines for the layout designs of a CMOS logic cell as follows.

1. Run V_{DD} and V_{SS} in metal wires at the top and bottom of the cell.
2. Run a vertical polysilicon wire for each gate input.
3. Order the polysilicon gate signals to allow the maximum source/drain sharing.
4. Place nMOS segments close to V_{SS} and pMOS segments close to V_{DD}.
5. Complete the connections of the logic gate in polysilicon, metal, or where appropriate in diffusion.
6. Keep the capacitance on internal nodes to a minimum.

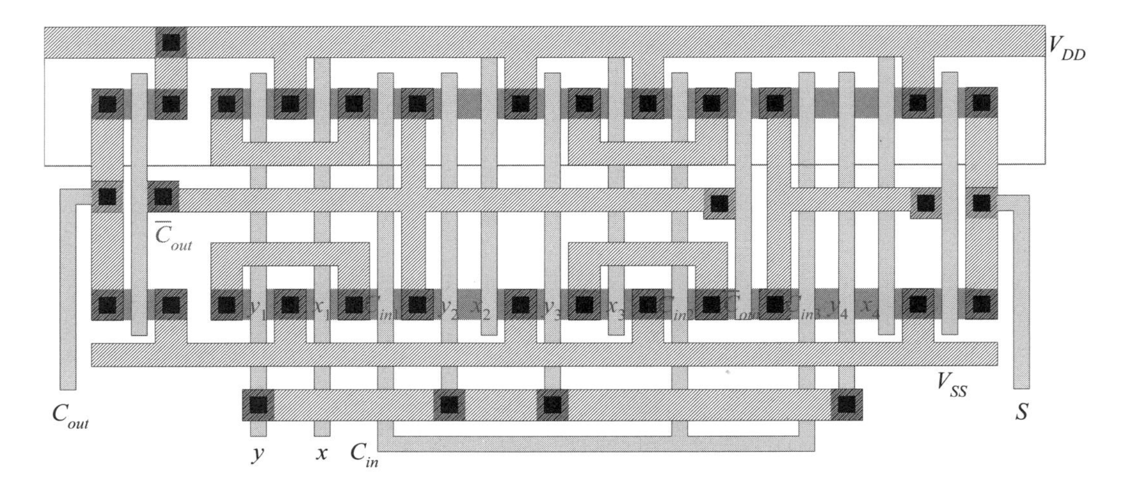

Figure 4.37: The resulting layout of the full adder circuit shown in Figure 1.41 when leaving the $\overline{C_{out}}$ edge as a separate diffusion segment.

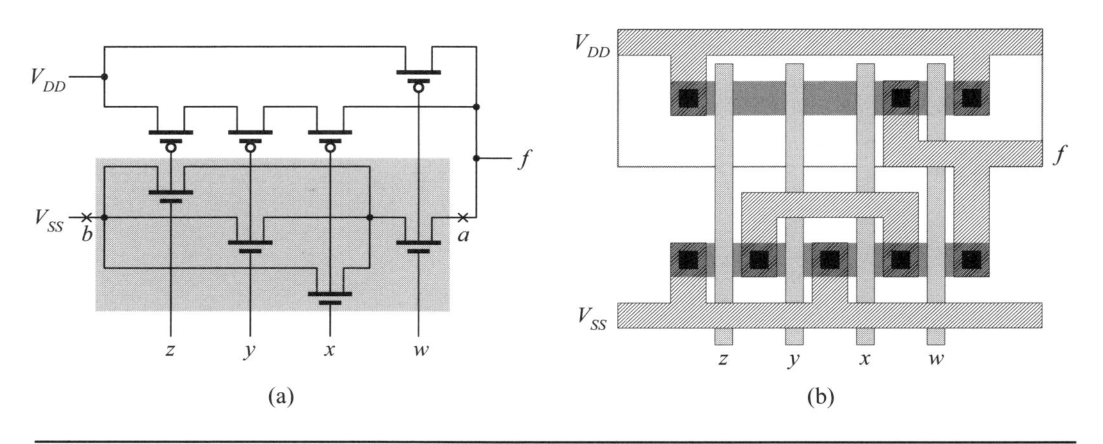

Figure 4.38: An example of optimized layout: (a) logic circuit; (b) layout.

■ Example 4-15: (An optimized layout example.) Figure 4.38 shows a logic circuit and its layout. The stack of nMOS transistors consists of three transistors connected in parallel and then in series with another nMOS transistor. Thus, it has two options to connect it to the output f. As shown in Figure 4.38(a), the node of the single nMOS transistor is used as the output node of the nMOS stack. This arrangement leads to a better performance due to less output capacitance, as shown in Figure 4.38(b).

■

■ Example 4-16: (A nonoptimized layout example.) Figure 4.39(a) shows another arrangement of the circuit shown in Figure 4.38(a), where nodes a and b are swapped. This results in a large amount of capacitance at the output node, which can be easily seen from the layout given in Figure 4.39(b).

■

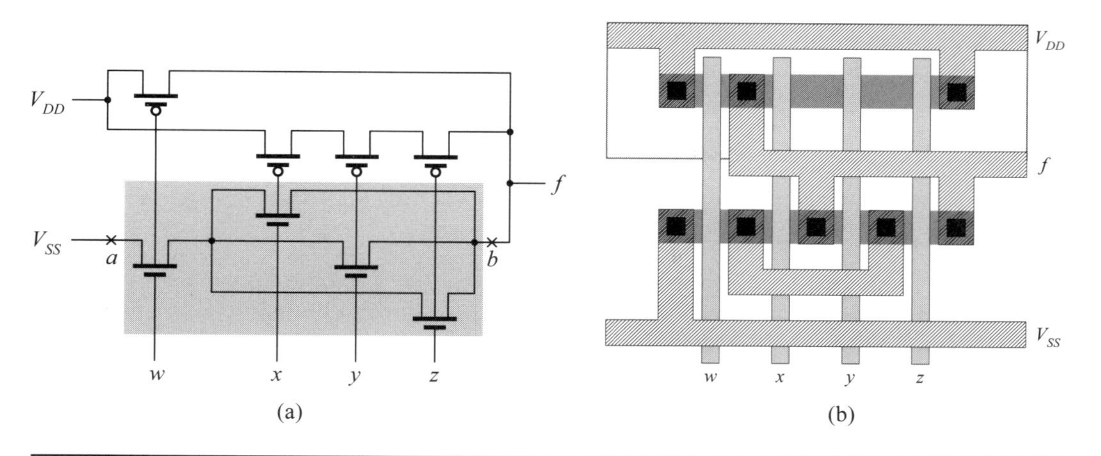

Figure 4.39: The resulting (a) logic circuit and (b) layout when nodes a and b are swapped.

In summary, when designing the layout of a logic circuit, we should make use of source/drain sharing to reduce the needed area and the capacitance of internal nodes. Moreover, unless it is necessary, the output capacitance of a logic circuit should be made as less as possible by properly arranging the transistor order of the logic circuit. The reduction of internal-node and output capacitances means that resulting logic circuit not only consumes less power but also has a smaller propagation delay.

■ Review Questions

Q4-37. What are the advantages of source/drain sharing?

Q4-38. What are the implications of the reduction of internal-node capacitance of a logic circuit?

4.5 Summary

Layout design rules are an interface between the fabrication process and circuit design, and in general can be described by one of two types: lambda (λ) rules and micron (μ) rules. The λ rules are also called scalable layout design rules. Lambda rules represent all rules as the function of a single parameter λ, and micron rules define all rules in absolute dimensions.

After the introduction of layout design rules, we dealt with the challenges of layouts. These include signal integrity, manufacturability, and reliability. Signal integrity means the capability of maintaining the timing and the quality of the signal. Crosstalk is the most important issue related to signal integrity. The issues of manufacturability include optical proximity correction (OPC) and chemical-mechanical polishing (CMP). The OPC uses precompensated feature patterns on a mask to improve the printability of the mask pattern, and the CMP process improves layout uniformity and chip planarization to achieve higher manufacturing yield. To achieve higher reliability and yield, the antenna effect caused in the plasma process and via-open defects are the two most important issues.

The latch-up problem inherent in CMOS processes and its prevention methods were dealt with in detail. After this, the layout designs of basic CMOS logic circuits were addressed. In addition, the Euler path approach was introduced along with examples for realizing irregular switching functions as a regular layout structure.

References

1. R. J. Baker, *CMOS Circuit Design, Layout, and Simulation.* 2nd ed., New York: John Wiley & Sons, 2005.
2. P. H. Chen et al., "Fixing antenna problem by dynamic diode dropping and jumper insertion," *Proc. of IEEE Int'l Symposium on Quality Electronic Design,* (ISQED 2000), pp. 275–282, March 2000.
3. T. C. Chen and Y. W. Chang, "Multilevel full-chip gridless routing with applications to optical-proximity correction," *IEEE Trans. on Computer-Aided Design of Integrated Circuits and Systems,* Vol. 26, No. 6, pp. 1041–1053, June 2007.
4. H. Y. Chen et al., "Full-chip routing considering double-via insertion," *IEEE Trans. on Computer-Aided Design of Integrated Circuits and Systems,* Vol. 27, No. 5, pp. 844–857, May 2008.
5. W. M. Coughran, M. R. Pinto, and R. K. Smith, "Computation of steady-state CMOS latchup characteristics," *IEEE Trans. on Computer-Aided Design,* Vol. 7, No. 2, pp. 307–323, February 1988.
6. D. B. Estreich and R. W. Dutton, "Modeling latch-up in CMOS integrated circuits and systems," *IEEE Trans. on Computer-Aided Design,* Vol. 1, No. 4, pp. 347–354, October 1982.
7. J. Luo, Q. Su, C. Chiang and J. Kawa, "A layout dependent full-chip copper electroplating topography model," *IEEE/ACM Int'l Conf. on Computer-Aided Design* (ICCAD-2005), pp. 133–140, November, 2005.
8. S. M. Sait and H. Youssef, *VLSI Physical Design Automation Theory and Practice.* Piscataway, NJ: IEEE Press, 1995.
9. M. Sarrafzadeh and C. K. Wong, *An Introduction to VLSI Physical Design.* New York: McGraw-Hill Books, 1996.
10. H. W. Trickey, "Flamel: a high level hardware compiler," *IEEE Trans. on Computer-Aided Design,* Vol. 6, No. 2, pp. 259–269, March 1987.
11. R. R. Troutman, *Latch-Up in CMOS Technology: The Problem and Its Cure.* Boston: Kluwer Academic Publishers, 1986.
12. T. Uehara and W. van Cleemput, "Optimal layout of CMOS functional arrays," *IEEE Trans. on Computers,* Vol. 30, No. 5, pp. 305–312, May 1981.
13. J. P. Uyemura, *Introduction to VLSI Circuits and Systems.* New York: John Wiley & Sons, 2002.
14. L. T. Wang, Y. W. Chang, and K. T. (TIM) Cheng, *Electronic Design Automation: Synthesis, Verification, and Test.* Boston: Morgan Kaufmann, 2009.
15. A. Weinberger, "Large scale integration of MOS complex logic: a layout method," *IEEE J. of Solid State Circuits,* Vol. 2, No. 4, pp. 182–190, 1967.
16. A. Weinberger, "4-2 carry-save adder module," *IBM Technical Disclosure Bulletin,* Vol. 23, No. 8, pp. 3811–3814, January 1981.

Problems

4-1. Consider the nMOS 2-to-1 multiplexer shown in Figure 1.36(a).

(a) Draw a stick diagram.
(b) Design and construct a layout.

4-2. Consider the TG 2-to-1 multiplexer shown in Figure 1.36(b).

(a) Draw a stick diagram.
(b) Design and construct a layout.

4-3. Consider the positive D-type latch shown in Figure 1.38(a).

(a) Draw a stick diagram.
(b) Design and construct a layout.

4-4. Consider the negative D-type latch shown in Figure 1.38(b).

(a) Draw a stick diagram.
(b) Design and construct a layout.

4-5. Consider the master-slave positive edge-triggered D-type flip-flop shown in Figure 1.39.

(a) Draw a stick diagram.
(b) Design and construct a layout.

4-6. Consider the NOR-type SR latch shown in Figure 9.9(a).

(a) Realize the SR latch using a CMOS logic circuit.
(b) Draw a stick diagram.
(c) Design and construct a layout.

4-7. Consider the NAND-type SR latch shown in Figure 9.9(b).

(a) Realize the SR latch using a CMOS logic circuit.
(b) Draw a stick diagram.
(c) Design and construct a layout.

4-8. Consider the gated-SR latch shown in Figure 9.11.

(a) Realize the SR latch using a CMOS logic circuit.
(b) Draw a stick diagram.
(c) Design and construct a layout.

4-9. Consider the two-way arbiter circuit given in Figure 9.8(b).

(a) Realize the two-way arbiter using a CMOS logic circuit.
(b) Draw a stick diagram.
(c) Design and construct a layout.

4-10. Consider the simple D-type latch circuit depicted in Figure 9.10(b).

(a) Draw a stick diagram.

(b) Design and construct a layout.

4-11. Consider the simple gated D-type latch circuit depicted in Figure 9.13(b).

(a) Draw a stick diagram.

(b) Design and construct a layout.

4-12. Consider the switching function of XOR gate: $f(x, y) = \bar{x}y + x\bar{y}$.

(a) Prove that $f(x, y) = \overline{\overline{x + y} + xy}$.

(b) Realize it using a CMOS logic circuit.

(c) Draw the pMOS and nMOS logic graphs and find the Euler paths.

(d) Design and construct a layout.

4-13. Consider the switching function of the XNOR gate: $f(x, y) = xy + \bar{x}\bar{y}$.

(a) Prove that $f(x, y) = \overline{\overline{xy} \cdot (x + y)}$.

(b) Realize it using a CMOS logic circuit.

(c) Draw the pMOS and nMOS logic graphs and find the Euler paths.

(d) Design and construct a layout.

4-14. Consider the switching function: $f(w, x, y, z) = \overline{wx + (y + z)}$.

(a) Find all Euler paths that cover the pMOS and nMOS logic graphs of the switching function.

(b) Carry out the layout using the resulting Euler path.

4-15. A full subtractor is a device that accepts three inputs, x, y, and b_{in}, and produces two outputs, b_{out} and d, according to the truth table shown in Table 4.2.

Table 4.2: The truth table of a full subtractor.

x	y	b_{in}	b_{out}	d
0	0	0	0	0
0	0	1	1	1
0	1	0	1	1
0	1	1	1	0
1	0	0	0	1
1	0	1	0	0
1	1	0	0	0
1	1	1	1	1

(a) Derive the b_{out} and d switching functions.

(b) Realize it using CMOS logic circuits.

(c) Draw the pMOS and nMOS logic graphs and find the Euler paths.

(d) Design and construct a layout.

4-16. A full adder/subtractor is a device that accepts four inputs, x, y, c_{in}, and $mode$, and produces two outputs, s_{out} and d. The operation is addition if the $mode$ is 0 and subtraction if the $mode$ is 1.

(a) Derive the truth table for this adder/subtractor.

(b) From the truth table, derive the s_{out} and d switching functions.

(c) Realize it using CMOS logic circuits.

(d) Draw the pMOS and nMOS logic graphs and find the Euler paths.

(e) Design and construct a layout.

4-17. Suppose that we want to design and implement a 4-bit two's complement adder circuit. For simplicity, this adder is composed of four 1-bit two's complement adders. Assume that the inputs of a 1-bit two's complement adder are x, y, and c_{in} and the outputs are sum and c_{out}.

(a) Using the CMOS approach, implement the 1-bit two's complement adder. Use SPICE to verify the function of the resulting circuit with the process parameters you have.

(b) Cascade four 1-bit two's complement adders into the desired 4-bit two's complement adder. Verify the resulting circuit using SPICE.

(c) Carry out the layout design of the resulting 4-bit two's complement adder and verify the resulting layout using SPICE.

5

Delay Models and Path-Delay Optimization

Once that we have studied the features and layouts of metal-oxide-semiconductor (MOS) transistors, in this chapter we continually explore the parasitic resistance and capacitances of MOS transistors. The resistance alone can cause the MOS transistor to consume power. Combining resistance with capacitance produces RC delay of desired signal passing through the MOS transistor.

To be more precise, we first describe the resistance and capacitance models of MOS transistors and then address three common approaches to calculating the propagation delays: t_{pHL} and t_{pLH}. These are the average-current approach, equivalent-resistance approach, and differential-current approach as well. In addition, cell delay and Elmore delay models are discussed in detail. The cell delay model is widely used in cell libraries to model the propagation delays of a cell.

Finally, we deal with the path-delay optimization problem. The path-delay optimization problem concerns how to optimize the MOS transistor sizes of a logic chain to minimize the total propagation delay when such a logic chain drives a large capacitor. The path-delay optimization problems of general logic chains, including inverter chains (super-buffers) and logic chains of generic logic gates, are dealt with in detail.

5.1 Resistance and Capacitance of MOS Transistors

In this section, we deal with the resistance and capacitance of MOS transistors. The effect of resistance yields power dissipation and the interaction between resistance and capacitance results in RC delays.

5.1.1 Resistances of MOS Transistors

The MOS transistor can also be thought of as a resistor with definite resistance. Table 5.1 lists the equivalent on-resistances, R_{eqn}/R_{eqp}, with an aspect ratio of $2\lambda/2\lambda$ of a variety of processes. To calculate both rise and fall times, the values at $\frac{1}{2}V_{DD}$ are preferred, while to calculate both propagation delays, t_{PHL} and t_{pLH}, the values at $\frac{3}{4}V_{DD}$ are used. In either case, the on-resistance of a MOS transistor other than

Table 5.1: The equivalent on-resistances of nMOS and pMOS transistors in various CMOS processes.

	0.35 μm	0.25 μm	0.18 μm	0.13 μm	90 nm	65 nm	45 nm	32 nm
R_{eqn} (kΩ)								
$\frac{1}{2}V_{DD}$	8	7	5	13	11	12	17	21
$\frac{3}{4}V_{DD}$	12	10	8	18	16	17	22	28
R_{eqp} (kΩ)								
$\frac{1}{2}V_{DD}$	24	20	16	28	30	36	55	71
$\frac{3}{4}V_{DD}$	32	26	22	37	40	46	68	82

feature size can be calculated as follows:

$$R_n = R_{eqn} \times \left(\frac{L_n}{W_n}\right) \tag{5.1}$$

$$R_p = R_{eqp} \times \left(\frac{L_p}{W_p}\right) \tag{5.2}$$

Example 5-1: (An example of a MOS transistor.) Referring to the layout shown in Figure 5.1(a), the transistor uses the feature size, namely, 2λ, for its channel width and length. Assuming that the 0.18-μm process is under consideration, calculate the on-resistance of this MOS transistor, which can be n-type MOS (nMOS) or p-type MOS (pMOS).

Solution: The on-resistance of this MOS transistor is estimated as follows:

$$R_n = R_{eqn} \times \left(\frac{L_n}{W_n}\right) = 8\text{ k}\Omega \times \left(\frac{2\lambda}{2\lambda}\right) = 8\text{ k}\Omega$$

or

$$R_p = R_{eqp} \times \left(\frac{L_p}{W_p}\right) = 22\text{ k}\Omega \times \left(\frac{2\lambda}{2\lambda}\right) = 22\text{ k}\Omega$$

where the equivalent on-resistances R_{eqn} and R_{eqp} at $(3/4)V_{DD}$ are used.

The minimum channel width of an nMOS or a pMOS transistor in most processes is usually a little bigger than the minimum channel length, that is, the feature size. As a consequence, when using lambda rules, the minimum channel width of a MOS transistor is set to 3λ.

Example 5-2: (Another example of a MOS transistor.) Referring to the layout depicted in Figure 5.1(b), the transistor has a channel width of 4λ and a channel length of 2λ. Assuming that a 0.18-μm process is under consideration, calculate the on-resistance of this MOS transistor, which can be an nMOS or a pMOS.

Solution: The on-resistance of this MOS transistor can be estimated as follows:

$$R_n = R_{eqn} \times \left(\frac{L_n}{W_n}\right) = 8\text{ k}\Omega \times \left(\frac{2\lambda}{4\lambda}\right) = 4\text{ k}\Omega$$

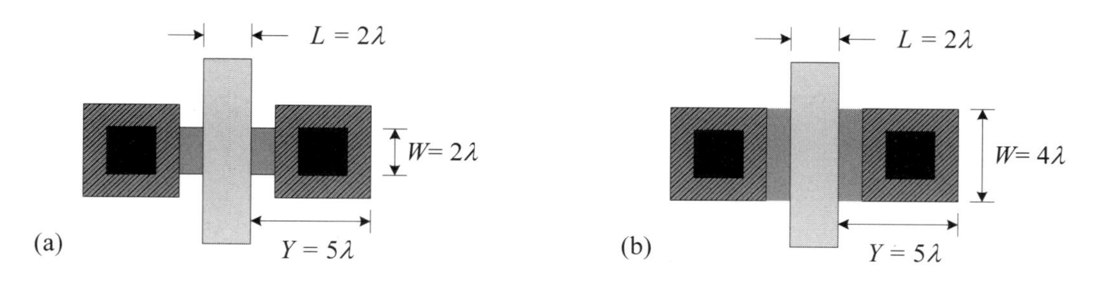

Figure 5.1: Two layout examples of MOS transistors: (a) W = feature size; (b) $W = 4\lambda$.

Table 5.2: The equivalent on-resistances (in Ω) of nMOS and pMOS transistors in different supply voltages.

	1.8 V	1.5 V	1.2 V	0.9 V	0.6 V	0.3 V
R_{eqn} (nMOS)	10 k	11 k	12 k	16 k	38 k	1.16 M
R_{eqp} (pMOS)	28 k	37 k	58 k	120 k	669 k	139 M

or

$$R_p = R_{eqp} \times \left(\frac{L_p}{W_p}\right) = 22 \text{ k}\Omega \times \left(\frac{2\lambda}{4\lambda}\right) = 11 \text{ k}\Omega$$

where the equivalent on-resistances R_{eqn} and R_{eqp} at $(3/4)V_{DD}$ are used.

The equivalent on-resistance R_{eqn}/R_{eqp} is also strongly dependent on the operating voltages. In general, the on-resistance of a MOS transistor is inversely proportional to the operating voltage. Table 5.2 shows the equivalent on-resistances of nMOS and pMOS transistors of a 0.18-μm process under a variety of supply voltages. We can see that the pMOS transistor is more sensitive to the variations of supply voltage than the nMOS transistor.

Review Questions

Q5-1. What is the resistance effect?

Q5-2. What is the combining effect of resistance and capacitance?

Q5-3. What is the relationship between the on-resistance and operating voltage of a MOS transistor?

5.1.2 Capacitances of MOS Transistors

The parasitic capacitances associated with MOS transistors can be categorized into the following two types: *oxide-related capacitance* and *junction capacitance*. Oxide-related capacitances include overlap capacitances C_{GDO} and C_{GSO}, gate-bulk (body or substrate) capacitance C_{gb}, gate-source capacitance C_{gs}, and gate-drain capacitance C_{gd}. Junction capacitances contain source-bulk junction capacitance C_{sb} and drain-bulk junction capacitance C_{db}.

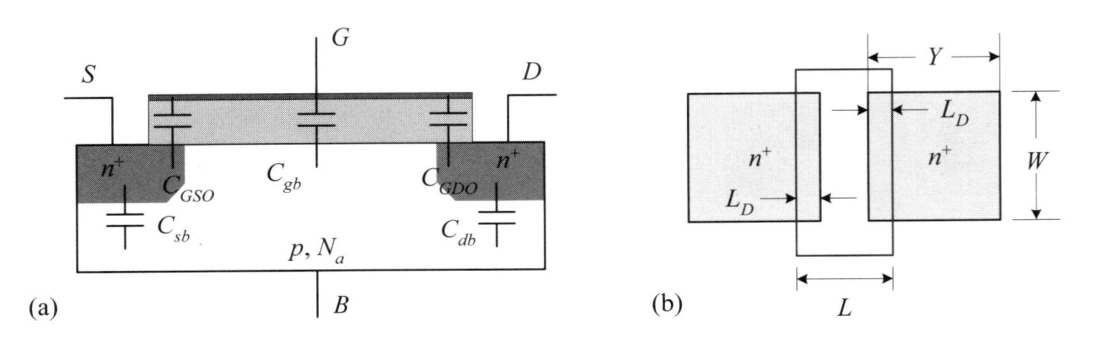

Figure 5.2: (a) The gate capacitance model of MOS transistors. (b) Actual and mask channel lengths.

5.1.2.1 Oxide-Related Capacitances Except for the overlap capacitances, all the other oxide-related capacitances are a function of gate-to-source voltage V_{GS}. In other words, their capacitances depend on the operating modes of a MOS transistor. To quantify the values of oxide-related capacitances, consider the gate capacitance model of the MOS transistor shown in Figure 5.2. Because of the drive-in process following the ion implantation, both source and drain regions are overlapped with the gate with a length of L_D. This results in the reduction of actual channel length, L_{actual}, by an amount of $2L_D$.

$$L_{actual} = L - 2L_D \tag{5.3}$$

where L is the mask channel length, that is, the drawing channel length.

Because of the symmetric structure of a MOS transistor, both overlap capacitances C_{GSO} and C_{GDO} have the same value and can be expressed as follows:

$$C_{GSO} = C_{GDO} = C_{ox} W L_D \tag{5.4}$$

The C_{ox} is the gate-oxide capacitance in units of $\mu\text{F/cm}^2$ and is equal to $\varepsilon_r\varepsilon_0/t_{ox}$, where ε_r is the relative permittivity of gate oxide, ε_0 is the permittivity of free space, and t_{ox} is the gate-oxide thickness.

Both overlap capacitances have a constant value, independent of the operating modes of a MOS transistor. In what follows, we consider the other oxide-related capacitances at various operating modes of MOS transistors, that is, at cutoff, linear, and saturation modes, in sequence.

When a MOS transistor operates at the cutoff mode, the gate-to-source voltage V_{GS} is less than its threshold voltage V_T, and hence, there is no induced channel under the gate on the surface of substrate. In this situation, as shown in Figure 5.3, the values of both components C_{gs} and C_{gd} of gate-related capacitors are zero. The oxide-related capacitance is simply the gate-to-substrate capacitance C_{gb}. That is,

$$\begin{aligned} C_{gs} &= C_{gd} = 0 \\ C_{gb} &= C_{ox} W L_{actual} \end{aligned} \tag{5.5}$$

When a MOS transistor operates at the linear mode, the gate-to-source voltage V_{GS} is a little greater than its threshold voltage V_T and the drain-to-source voltage V_{DS} is a small positive value. In this case, there exists a channel under the gate at the surface of the substrate. As shown in Figure 5.4, both components C_{gs} and C_{gd} are not zero since the channel serves as a conducting plate to shield the substrate bulk.

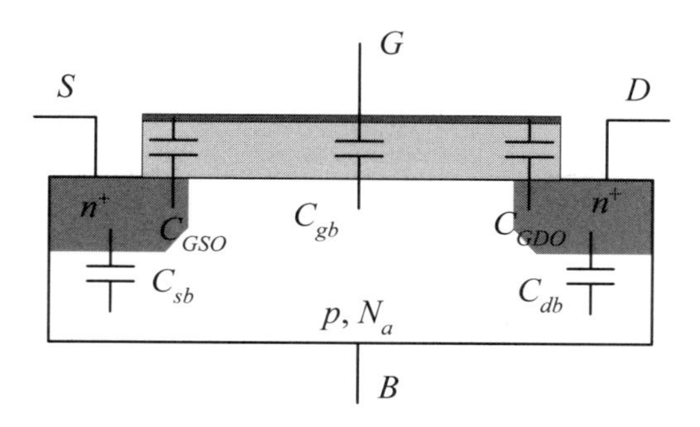

Figure 5.3: The capacitance model of a MOS transistor in cutoff mode.

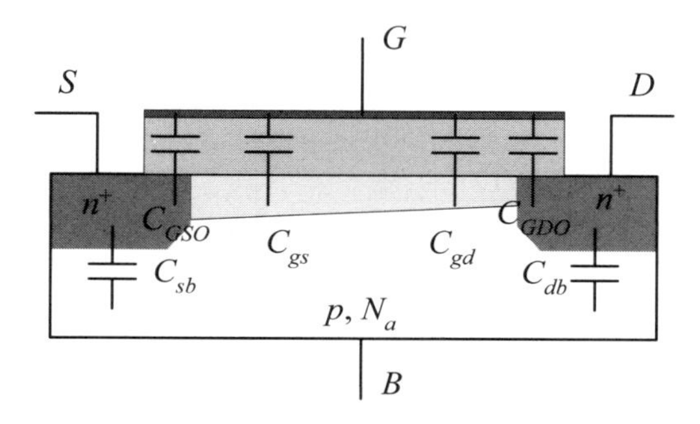

Figure 5.4: The capacitance model of a MOS transistor in linear mode.

This results in a zero-value component C_{gb} and the capacitance between the gate and channel is conventionally partitioned into two equal parts and represented as follows:

$$\begin{aligned} C_{gs} &= C_{gd} = \frac{1}{2} C_{ox} W L_{actual} \\ C_{gb} &= 0 \end{aligned} \tag{5.6}$$

When a MOS transistor operates at the saturation mode, the gate-to-source voltage V_{GS} is larger than its threshold voltage V_T. A channel exists under the gate on the surface of the substrate. Nevertheless, because of the large drain-to-source voltage V_{DS}, the channel near the drain end is effectively pinched off and the depletion region around the drain end is pushed toward the source end. In this situation, as shown in Figure 5.5, the component C_{gd} is virtually zero and C_{gs} is accounted for as $\frac{2}{3} C_{ox} W L_{actual}$. That is,

$$\begin{aligned} C_{gs} &= \frac{2}{3} C_{ox} W L_{actual} \\ C_{gd} &= C_{gb} = 0 \end{aligned} \tag{5.7}$$

The gate capacitance C_G is the sum of all oxide-related capacitances. The results are summarized in Table 5.3. From the table, we can see that, when the MOS transistor operates at both cutoff and linear modes, its gate capacitance is equal to

$$\begin{aligned} C_G &= C_{ox} W L_{actual} + C_{GSO} + C_{GDO} \\ &= C_{ox} W (L_{actual} + L_D + L_D) \\ &= C_{ox} W L \end{aligned} \tag{5.8}$$

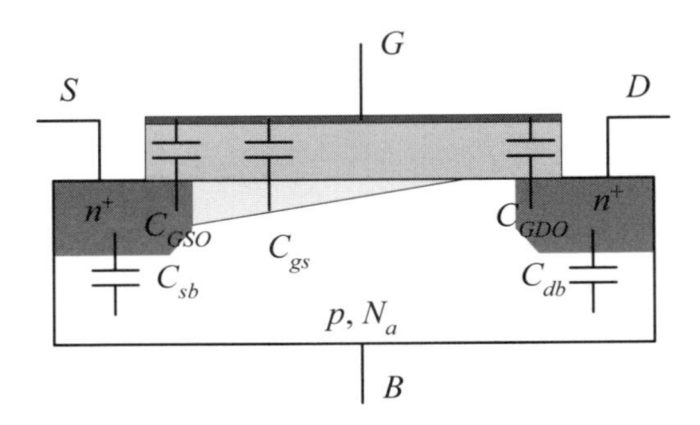

Figure 5.5: The capacitance model of a MOS transistor in saturation mode.

Table 5.3: The total oxide-related capacitances of MOS transistors.

Capacitance	Cutoff	Linear	Saturation
$C_{gb(total)}$	$C_{ox}WL_{actual}$	0	0
$C_{gs(total)}$	C_{GSO}	$\frac{1}{2}C_{ox}WL_{actual} + C_{GSO}$	$\frac{2}{3}C_{ox}WL_{actual} + C_{GSO}$
$C_{gd(total)}$	C_{GDO}	$\frac{1}{2}C_{ox}WL_{actual} + C_{GDO}$	C_{GDO}

and when the MOS transistor operates at the saturation mode, its gate capacitance can be expressed as follows:

$$\begin{aligned} C_G &= \frac{2}{3}C_{ox}WL_{actual} + C_{GSO} + C_{GDO} \\ &\leq C_{ox}WL \end{aligned} \tag{5.9}$$

Because in digital applications, any logic circuit can sweep across all of the three operating modes when it changes state, it is useful to consider the maximum capacitance, which determines the worst-case timing. Consequently, for hand calculation, it is safe to use WLC_{ox} for gate capacitance C_G regardless of which operation mode the MOS transistor operates in.

■ Example 5-3: (MOS transistor capacitance.) Referring to Figure 5.1(b) and assuming that the gate-oxide thickness is 7.7 nm in a 0.35-μm process and is 4.1 nm in a 0.18-μm process, calculate the gate capacitance of the MOS transistor in each process.

Solution: We first calculate the gate-oxide capacitance C_{ox} of the 0.35-μm process as follows:

$$\begin{aligned} C_{ox} &= \frac{\varepsilon_{SiO_2}\varepsilon_0}{t_{ox}} = \frac{3.9 \times 8.854 \times 10^{-14}}{7.7 \times 10^{-7}} \\ &= 4.5 \times 10^{-7} \text{ F/cm}^2 = 0.45\ \mu\text{F/cm}^2 \end{aligned}$$

In addition, in the 0.35 μm process, $2\lambda = 0.35\ \mu$m. Hence, the gate capacitance is

$$\begin{aligned} C_G &= C_{ox}A \\ &= 0.45 \times (0.35 \times 10^{-4} \times 2 \times 0.35 \times^{-4}) \\ &= 0.11 \times 10^{-8}\ \mu\text{F} = 1.1 \text{ fF} \end{aligned}$$

Table 5.4: The gate-oxide capacitances of a variety of CMOS processes.

	0.35 μm	0.25 μm	0.18 μm	0.13 μm	90 nm	65 nm	45 nm	32 nm	Unit
t_{ox}	7.7	5.7	4.1	3.1	2.5	1.85	1.75	1.65	nm
C_{ox}	0.32	0.84	1.33	1.08	1.62	1.34	1.56	0.97	μF/cm^2
$C_{ox}L$	1.13	2.1	2.4	1.4	1.46	0.87	0.70	0.31	fF/μm

Next, we calculate the gate-oxide capacitance C_{ox} of the 0.18-μm process as follows:

$$\begin{aligned} C_{ox} &= \frac{\varepsilon_{SiO_2}\varepsilon_0}{t_{ox}} = \frac{3.9 \times 8.854 \times 10^{-14}}{4.1 \times 10^{-7}} \\ &= 8.4 \times 10^{-7}\ \text{F/cm}^2 = 0.84\ \mu\text{F/cm}^2 \end{aligned}$$

In the 0.18 μm process, $2\lambda = 0.18$ μm. As a consequence, the gate capacitance is equal to

$$\begin{aligned} C_G &= C_{ox}A \\ &= 0.84 \times (0.18 \times 10^{-4} \times 2 \times 0.18 \times^{-4}) \\ &= 0.54 \times 10^{-9}\ \mu\text{F} = 0.54\ \text{fF} \end{aligned}$$

■

Table 5.4 lists the gate-oxide thickness and capacitances of a variety of processes. The gate-oxide capacitance in a real-world process may be quite different from the one calculated above. The actual capacitance is usually extracted from the test structures or using circuit simulations with the parameters from the IC foundry.

5.1.2.2 Junction Capacitances As illustrated in Figure 5.6, each MOS transistor has two junctions, a heavily doped source and drain regions with the substrate/well. The n^+p junctions are for nMOS transistors and p^+n junctions are for pMOS transistors. Under normal operation, both junctions are in their reverse-bias conditions. Hence, they exhibit reverse-bias junction capacitances. In general, the junction capacitance is composed of two components: bottom-area capacitance and sidewall capacitance. The bottom-area capacitance is determined by the product of length Y and width W. The sidewall capacitance is determined by the product of periphery $2(Y+W)$ and junction depth x_j. Since the junction depth x_j is a process parameter, not controlled by the designer, the effect of junction depth is included into sheet capacitance, C_{jp}. As a result, the junction capacitance can be expressed as follows:

$$C_{D/S} = C_{ja} \times (YW) + C_{jp} \times (2Y + 2W) \tag{5.10}$$

where C_{ja} is the bottom-area capacitance in F/μm^2 and C_{jp} is the periphery capacitance in F/μm; parameters Y and W are the width and length of the diffusion region, respectively. Both Y and W have units of micrometers. Both bottom-area and periphery capacitances are the junction capacitance under the reverse-bias condition and can be expressed as

$$C_{j(a/p)} = C_{j0}\left(1 - \frac{V_j}{\phi_0}\right)^{-m} \tag{5.11}$$

where m is the grading coefficient, C_{j0} is the zero-bias junction capacitance of the n^+p or p^+n junction, V_j is the applied voltage, and ϕ_0 is the built-in potential of the junction.

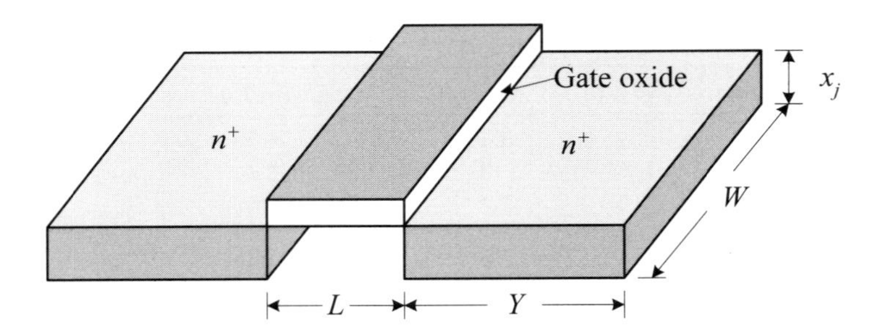

Figure 5.6: The basic structure showing the junction capacitance of a MOS transistor.

In digital applications, we often set the channel length equal to the feature size and use the channel width as a design parameter to adjust the driving current. As a consequence, the zero-bias junction capacitance is often represented as a function of channel width; namely, the junction capacitance $C_{D/S}$ can be expressed as follows:

$$C_{D/s} = C_j W \tag{5.12}$$

where C_j is in units of fF/μm. The zero-bias junction capacitances of a variety of typical CMOS processes are listed in Table 5.5.

■ Example 5-4: (An example of the junction capacitance.) Referring to Figure 5.1 again and using the zero-bias junction capacitance of the 0.18-μm process given in Table 5.5, calculate the junction capacitance of the MOS transistors shown in Figures 5.1(a) and (b).

Solution: Both MOS transistors shown in Figures 5.1(b) and (b) have the same junction capacitance.

$$C_{D/S} = C_j W = 1.66 \times 0.18 = 0.30 \text{ fF}$$

The capacitance obtained is comparable to gate capacitance. Thus, the junction capacitance of MOS transistors in deep submicron processes cannot be simply ignored. It must be taken into account when considering the timing of the logic circuit in most practical applications.

■

5.1.2.3 Gate Capacitance Model in SPICE. As stated, there are three capacitors, C_{gs}, C_{gd}, and C_{gb}, and their capacitances must be computed according to the operation modes of a MOS transistor. In addition, there are three constant capacitors, C_{GSO}, C_{GDO}, and C_{GBO}, representing gate-source, gate-drain, and gate-bulk overlap capacitors, respectively. C_{GBO} is due to the gate extension into the field region and is relatively small. These three capacitances need to be considered when modeling a MOS transistor. In addition to the overlap (that is, lateral diffusion) capacitance, in deep submicron devices, both C_{GSO} and C_{GDO} also take into account the fringing capacitance from the edge of polysilicon to the surface of silicon. In SPICE, the above-mentioned three capacitances, C_{GSO}, C_{GDO}, and C_{GBO}, are denoted as CGSO, CGDO, and CGBO, respectively.

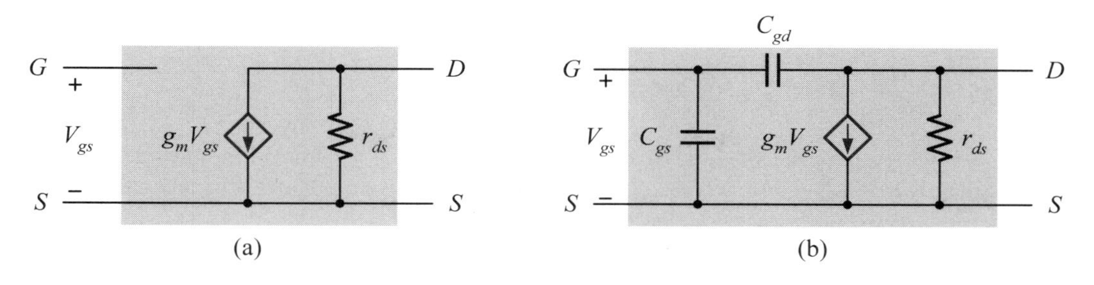

Figure 5.7: The small-signal equivalent circuits of a MOS transistor at (a) low frequencies; (b) high frequencies.

The drain-bulk junction capacitance can be modeled in SPICE by the following equation:

$$C_{DB} = \frac{\mathrm{CJ} \times \mathrm{AD}}{\left(1 - \dfrac{V_J}{\mathrm{PB}}\right)^{\mathrm{MJ}}} + \frac{\mathrm{CJSW} \times \mathrm{PD}}{\left(1 - \dfrac{V_J}{\mathrm{PB}}\right)^{\mathrm{MJSW}}} \tag{5.13}$$

and the source-bulk junction capacitance can be modeled as

$$C_{SB} = \frac{\mathrm{CJ} \times \mathrm{AS}}{\left(1 - \dfrac{V_J}{\mathrm{PB}}\right)^{\mathrm{MJ}}} + \frac{\mathrm{CJSW} \times \mathrm{PS}}{\left(1 - \dfrac{V_J}{\mathrm{PB}}\right)^{\mathrm{MJSW}}} \tag{5.14}$$

where CJ and CJSW represent zero-bias junction capacitances at the bottom and sidewall areas, respectively; AD and AS denote the areas of drain and source, respectively; PD and PS denote the perimeters of drain and source, respectively; VJ is the voltage applied to the junction and PB is the built-in potential of the junction; MJ and MJSW are the grading coefficients of the bottom and sidewall junctions, respectively.

Review Questions

Q5-4. What are two types of capacitances of MOS transistors?

Q5-5. Describe the oxide-related capacitances of MOS transistors.

Q5-6. Describe the junction capacitances of MOS transistors.

Q5-7. Give why an overlap capacitance is created between the gate and drain/source of MOS transistors.

5.1.2.4 Small-Signal Circuit Model When a MOS transistor is used as an amplifier instead of a switch, it is more convenient to represent the MOS transistor as an equivalent circuit, referred to as a *small-signal model*, as shown in Figure 5.7. The small-signal model exhibits the relationship between i_D, v_{GS}, and v_{DS}. At low frequencies, both capacitors C_{gs} and C_{gd} are unimportant and can be ignored, as indicated in Figure 5.7(a). However, at high frequencies, these two capacitors can no longer be neglected; they should be incorporated into calculation when analyzing the frequency response of the circuit, as shown in Figure 5.7(b).

The parameter that relates the change of I_D to the change of V_{GS} is referred to as the *MOS-transistor transconductance* g_m. For a long-channel MOS transistor in the linear region of operation, the transconductance g_m can be expressed as

$$g_m = \frac{\partial I_D}{\partial V_{GS}} = \mu C_{ox}\left(\frac{W}{L}\right) V_{DS} \tag{5.15}$$

which is linearly proportional to drain-to-source voltage V_{DS} but independent of gate-to-source voltage V_{GS}. For a long-channel MOS transistor operating in the saturation region, the transconductance g_m is equal to

$$g_{m(sat)} = \frac{\partial I_D}{\partial V_{GS}} = \mu C_{ox}\left(\frac{W}{L}\right)(V_{GS} - V_T) = \sqrt{2kI_{Dsat}} \tag{5.16}$$

In this region of operation, the transconductance g_m is independent of drain-to-source voltage V_{DS} but depends linearly on gate-to-source voltage V_{GS}.

For a short-channel MOS transistor, when it operates in the saturation region, the transconductance $g_{m(sat)}$ becomes

$$\begin{aligned} g_{m(sat)} &= \frac{\partial I_D}{\partial V_{GS}} = \frac{\partial}{\partial V_{GS}} W C_{ox}(V_{GS} - V_T - V_{Dsat}) v_{sat} \\ &= v_{sat} W C_{ox}\left(1 - \frac{\partial V_{DSsat}}{\partial V_{GS}}\right) \end{aligned}$$

Combining with Equation (2.94), we obtain

$$g_{m(sat)} = v_{sat} W C_{ox} \frac{(V_{GS} - V_T)(V_{GS} - V_T + 2E_{sat}L)}{(V_{GS} - V_T + E_{sat}L)^2} \tag{5.17}$$

Because of the effect of channel-length modulation, the drain current is affected by the drain-to-source voltage V_{DS} in the saturation region. To quantify this effect in the small-signal model, an *output resistance* r_{ds}, also denoted as r_o, is used. The output resistance r_{ds} can be expressed as

$$r_{ds} = \frac{1}{\lambda I_D} \tag{5.18}$$

where λ denotes the channel-length modulation coefficient and I_D is the drain current without channel-length modulation.

Speed of MOS transistors. There are two basic frequency limitation factors in MOS transistors: *channel transit time* and *unity-gain frequency*. The channel transit time is the time for carriers to pass from the source to drain. To quantify the channel transit time, let the carrier drift velocity be v_d and the channel length be L, then the channel transit time can be represented as

$$\tau_t = L/v_d \qquad \text{and} \qquad f = 1/\tau_t \tag{5.19}$$

This usually results in an overestimation of the highest frequency that a MOS transistor can respond.

The *unity-gain frequency* denotes the frequency at which the magnitude of the current gain of MOS transistor is unity. For an ideal MOS transistor, the overlap or parasitic capacitance is zero. When the MOS transistor is operated in the saturation region, C_{gd} is zero and $C_{gs} = \frac{2}{3}(C_{ox}WL)$. For a long-channel MOS transistor, the unity-gain frequency f_T can be expressed as follows:

$$f_T = \frac{g_m}{2\pi C_G} = \frac{3\mu(V_{GS} - V_T)}{4\pi L^2} \tag{5.20}$$

The unity-gain frequency is inversely proportional to the square of channel length. To get a high speed, we need to minimize the channel length.

For a short-channel MOS transistor, the unity-gain frequency can be expressed by the following equation:

$$f_T = \frac{3v_{sat}(V_{GS} - V_T)(V_{GS} - V_T + 2E_{sat}L)}{4\pi L(V_{GS} - V_T + E_{sat}L)^2} \approx \frac{3v_{sat}}{4\pi L} \tag{5.21}$$

where the last term is obtained by assuming that $E_{sat}L \ll V_{GS} - V_T$. The highest frequency that a MOS transistor can respond to is determined by the unity-gain frequency.

Example 5-5: (The unity-gain frequency.) Calculate the unity-gain frequency of each process specified in the following.

1. Use 0.18-μm parameters and set $V_{GS} = 1.8$ V and $L = 0.18$ μm.
2. Use 0.13-μm parameters and set $V_{GS} = 1.2$ V and $L = 0.13$ μm.

Solution: By using Equation (5.21), the unity-gain frequency of each process can be calculated as follows:

1. For the 0.18-μm process, the unity-gain frequency is equal to

$$f_T = \frac{3 \times 8 \times 10^6(1.8 - 0.4)(1.8 - 0.4 + 2 \times 2.16)}{4\pi \times 0.18 \times 10^{-4}(1.8 - 0.4 + 2.16)^2} = 67.04 \text{ GHz}$$

2. For the 0.13-μm process, the unity-gain frequency is equal to

$$f_T = \frac{3 \times 1.5 \times 10^7(1.2 - 0.35)(1.2 - 0.35 + 2 \times 1.24)}{4\pi \times 0.13 \times 10^{-4}(1.2 - 0.35 + 1.24)^2} = 178.49 \text{ GHz}$$

■

Review Questions

Q5-8. What is the transconductance g_m?

Q5-9. What is the output resistance r_{ds}?

Q5-10. What is the channel transit time?

Q5-11. What is the unity-gain frequency?

5.2 Propagation Delays and Delay Models

For a logic circuit, noise margins, power dissipation, and propagation delays are closely related. In this section, we first address voltage levels and noise margins. Then, we deal with the approaches to calculating the propagation delay of a logic circuit. Finally, we describe the cell delay model and Elmore delay model. The cell delay model is widely used in cell libraries to model the propagation delays of a cell.

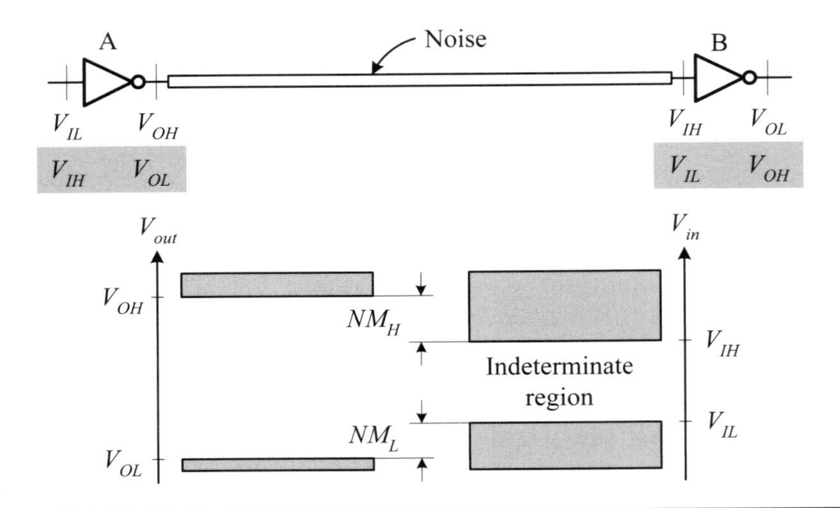

Figure 5.8: The definition of noise margins.

5.2.1 Voltage Levels and Noise Margins

Figure 5.8 shows a circuit with two inverters cascaded together through a wire. The wire may induce some unwanted signals, known as *noise*. The inverter A outputs a logic 1 (a voltage greater than V_{OH}) when its input voltage is less than V_{IL}. This logic-1 output voltage is then passed through the wire to serve as the input voltage to the inverter B, which in turn outputs a logic 0 at the output.

To allow two logic circuits of the same or different families to be cascaded and to tolerate *process, voltage, and temperature* (PVT) variations, four voltage levels are defined to specify the input and output voltage values. The two input voltage levels, V_{IL} and V_{IH}, are used to specify the input logic 0 and 1, respectively. The two output voltage levels, V_{OL} and V_{OH}, are separately used to specify the output logic 0 and 1.[1] More precisely, these four voltage levels are defined as follows:

- V_{IH} (high-level input voltage) is the minimum high input voltage to be recognized as a logic 1.
- V_{IL} (low-level input voltage) is the maximum low input voltage to be recognized as logic 0.
- V_{OH} (high-level output voltage) is the minimum high output voltage when the output is logic 1.
- V_{OL} (low-level output voltage) is the maximum low output voltage when the output is logic 0.

As indicated in Figure 5.8, to make the inverter B properly recognize the input signal as a high-level input voltage, V_{IH}, the output voltage V_{OH} of inverter A must be greater than the V_{IH} of inverter B. Similarly, to make the inverter B properly recognize the input signal as a low-level input voltage, V_{IL}, the output voltage V_{OL} of inverter A must be less than the V_{IL} of inverter B. The difference between V_{OH} and V_{IH} is defined as a *high-noise margin*, NM_H, and the difference between V_{IL} and V_{OL} is defined as a *low-noise margin*, NM_L. That is, both NM_H and NM_L can be

[1] When a CMOS logic circuit is interfaced with a transistor-transistor logic (TTL) circuit, the four current levels (I_{IL}, I_{IH}, I_{OL}, and I_{OH}) also need to be taken into account. Refer to Lin [11] for more detail.

expressed as follows:

$$\begin{aligned} NM_H &= V_{OH} - V_{IH} \\ NM_L &= V_{IL} - V_{OL} \end{aligned} \tag{5.22}$$

Both noise margins must be positive for a logic circuit to function correctly. In practice, noise margins should be as large as possible to tolerate the potential noises that may occur in a weird environment.

The region between V_{IL} and V_{IH} is an *indeterminate region*, or referred to as a *transition region*, in which a voltage value may be interpreted as either logic 0 or 1 by the inverter B in a somewhat random manner. For instance, some inverters will recognize the voltage value as logic 1 but the others as logic 0, even when the inverters with the same circuit type of the same family are used. As a consequence, to ensure the proper operation of a cascaded logic circuit, one should follow the definition of the aforementioned voltage levels when designing a logic circuit so that the resulting logic circuit can function correctly, even interfacing with other logic circuits of the same or different families.

■ Example 5-6: (Noise margins.) For a commercial 5-V CMOS logic family, the specification states that $V_{IL} = 0.8$ V, $V_{IH} = 2.0$ V, $V_{OL} = 0.1$ V, and $V_{OH} = 4.4$ V. Calculate the high- and low-noise margins.

Solution: From Equation(5.22), we obtain

$$\begin{aligned} NM_H &= V_{OH} - V_{IH} = 4.4 - 2.0 = 2.4 \text{ V} \\ NM_L &= V_{IL} - V_{OL} = 0.8 - 0.1 = 0.7 \text{ V} \end{aligned}$$

where the high-noise margin is much higher than the low-noise margin. This gives a clear reason to explain why the active-low signals are usually used in a high-noise operating environment for electronic equipment.

■

In general, noise margins tell us about what amount of noise that a logic circuit can tolerate. Beyond this, the logic circuit would not work reliably, even properly. Hence, in principle, it is better to make noise margins of a logic circuit as large as possible. Nevertheless, high-noise margins mean large voltage excursions, leading to much more power dissipation and longer propagation delays. Therefore, a trade-off between noise margins and power dissipation as well as propagation delays exists.

■ Review Questions

Q5-12. Define the voltage levels: V_{IL} and V_{IH}.

Q5-13. Define the voltage levels: V_{OL} and V_{OH}.

Q5-14. Explain the meaning of the indeterminate region between V_{IL} and V_{IH}.

Q5-15. Define high-noise margin and low-noise margin.

Q5-16. Explain the relationship between noise margin and power dissipation, and propagation delays.

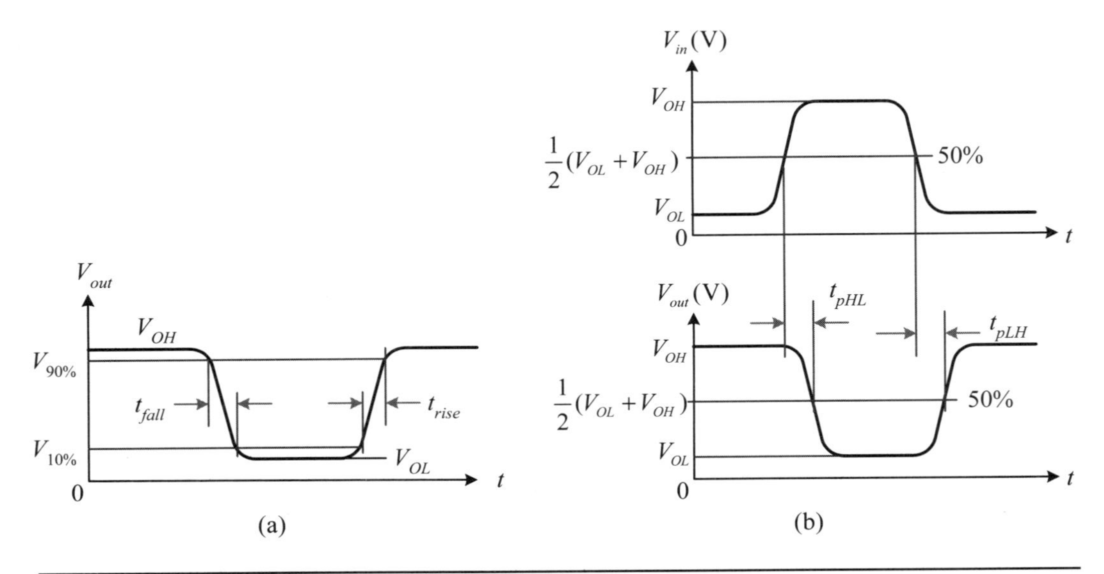

Figure 5.9: The definitions of (a) rise and fall times as well as (b) propagation delays.

5.2.2 Basic Timing-Related Terminology

Before the timing model and the timing analysis of digital circuits are defined, it is necessary to define some basic timing-related terms. These include *rise time*, *fall time*, and *propagation delay*.

As illustrated in Figure 5.9(a), the rise time t_{rise} (t_r) of a signal is defined as the time for a signal to rise from 10% to 90% of its maximum excursion. The fall time t_{fall} (t_f) of a signal is defined as the time for a signal to fall from 90% to 10% of its maximum excursion.

More formally, rise and fall times, t_{rise} and t_{fall}, can be defined in terms of the two output voltage levels, V_{OL} and V_{OH}, as follows:

- *Rise time* (t_r, t_{rise}) is the time required for the output voltage to rise from 10% of $(V_{OH} - V_{OL})$ to 90% of $(V_{OH} - V_{OL})$.
- *Fall time* (t_f, t_{fall}) is the time required for the output voltage to drop from 90% of $(V_{OH} - V_{OL})$ to 10% of $(V_{OH} - V_{OL})$.

Referring to Figure 5.9(a), both voltages $V_{10\%}$ and $V_{90\%}$ can be expressed as

$$V_{10\%} = V_{OL} + 0.1(V_{OH} - V_{OL}) \tag{5.23}$$

$$V_{90\%} = V_{OL} + 0.9(V_{OH} - V_{OL}) \tag{5.24}$$

The propagation delay denotes the time interval required for a logic circuit to respond to its input signal and outputs a stable signal at its output. The propagation delay of a logic circuit is usually defined as the time interval between 50% of the maximum input excursion and 50% of the maximum output excursion. Since the output signal can go from either high to low or low to high, two propagation delays, t_{pHL} and t_{pLH}, are defined accordingly. The detailed illustration of these two propagation delays are shown in Figure 5.9(b).

Like rise and fall times, propagation delays, t_{pLH} and t_{pHL}, can be described in terms of two output voltage levels, V_{OL} and V_{OH}, more formally as in the following:

- *Low-to-high propagation delay* (t_{pLH}) is the time required for the output voltage to rise from V_{OL} to $V_{50\%}$ of $(V_{OH} - V_{OL})$ with respect to the $V_{50\%}$ of $(V_{OH} - V_{OL})$ of the input signal.
- *High-to-low propagation delay* (t_{pHL}) is the time required for the output voltage to drop from V_{OH} to $V_{50\%}$ of $(V_{OH} - V_{OL})$ with respect to the $V_{50\%}$ of $(V_{OH} - V_{OL})$ of the input signal.

As shown in Figure 5.9(b), $V_{50\%}$ can be alternatively represented as

$$V_{50\%} = V_{OL} + \frac{1}{2}(V_{OH} - V_{OL}) = \frac{1}{2}(V_{OH} + V_{OL}) \tag{5.25}$$

Review Questions

Q5-17. Define rise and fall times.

Q5-18. Define propagation delays: t_{pLH} and t_{pHL}.

Q5-19. How would you represent $V_{10\%}$ in terms of V_{OH} and V_{OL}?

Q5-20. How would you represent $V_{90\%}$ in terms of V_{OH} and V_{OL}?

Q5-21. How would you represent $V_{50\%}$ in terms of V_{OH} and V_{OL}?

5.2.3 Propagation Delays

Once we have defined various timing-related terms, we are now in a position to consider the methods to estimate the propagation delays of a logic circuit. Generally, there are three common approaches that can be used to calculate the propagation delays: t_{pHL} and t_{pLH}. These are

1. Average-current approach
2. Equivalent-RC approach
3. Differential-current approach

All approaches are based on the use of $Q = CV = It$. Their differences are on the relationship between V and I. In the average-current approach, the voltage V is constant and an average value is used for current I. In the equivalent-RC approach, the ratio of V and I is considered as a constant, that is, an equivalent resistance R_{eqv}. In the differential approach, the voltage I is a function of V, thereby leading to a much more complicated situation than the other two approaches.

5.2.3.1 Average-Current Approach Like the case of finding large-signal equivalent capacitance, both propagation delays of a logic circuit can also be estimated by using the average current of two extreme points associated with the circuit. This results in the so-called *average-current approach*.

As we know, the output node of a CMOS logic circuit is a capacitive load. Hence, both propagation delays are equivalent to the charge and discharge times of the load capacitor, respectively. From the simple relation $Q = CV = It$, both propagation delays can be expressed as follows:

$$t_{pHL} = \frac{C_L \cdot \Delta V_{HL}}{I_{HL(avg)}} = \frac{C_L \cdot (V_{OH} - V_{50\%})}{I_{HL(avg)}} \tag{5.26}$$

$$t_{pLH} = \frac{C_L \cdot \Delta V_{LH}}{I_{LH(avg)}} = \frac{C_L \cdot (V_{50\%} - V_{OL})}{I_{HL(avg)}} \tag{5.27}$$

where the average current $I_{HL(avg)}$ and $I_{LH(avg)}$ can be calculated, respectively, as follows:

$$I_{HL(avg)} = \frac{1}{2}\left[i_C(V_{in} = V_{OH}, V_{out} = V_{OH}) + i_C(V_{in} = V_{OH}, V_{out} = V_{50\%})\right] \quad (5.28)$$

$$I_{LH(avg)} = \frac{1}{2}\left[i_C(V_{in} = V_{OL}, V_{out} = V_{OL}) + i_C(V_{in} = V_{OL}, V_{out} = V_{50\%})\right] \quad (5.29)$$

The input signal is assumed to be a step function with zero rise and fall times.

An example is given in the following to illustrate how to use the average-current approach to calculate the propagation delays, t_{pHL} and t_{pLH}, of an inverter.

■ Example 5-7: (An example of the average-current approach.) Assume that $V_{DD} = 1.8$ V, load capacitance $C_L = 0.1$ pF, $L_n = L_p = 2\lambda$, $W_n = 18\lambda$, and $W_p = 42\lambda$. Using 0.18-μm parameters, $k'_n = 96\ \mu\text{A/V}^2$, $k'_p = 48\ \mu\text{A/V}^2$, and $V_{T0n} = |V_{T0p}| = 0.4$ V, find t_{pHL} and t_{pLH}.

Solution: We begin with the calculation of k_n and k_p as follows:

$$\begin{aligned} k_n &= k'_n\left(\frac{W_n}{L_n}\right) = 96 \times \left(\frac{18}{2}\right) = 0.864 \text{ mA}/V^2 \\ k_p &= k'_p\left(\frac{W_p}{L_p}\right) = 48 \times \left(\frac{42}{2}\right) = 1.008 \text{ mA}/V^2 \end{aligned}$$

Using $V_{OH} = V_{DD} = 1.8$ V and $V_{50\%} = \frac{1}{2}V_{DD} = 0.9$ V, according to Equation (5.28), the $I_{HL(avg)}$ can be calculated as follows:

$$\begin{aligned} I_{HL(avg)} &= \frac{1}{2}\left[i_C(V_{in} = V_{OH}, V_{out} = V_{OH}) + i_C(V_{in} = V_{OH}, V_{out} = V_{50\%})\right] \\ &= \frac{1}{2}\left[\frac{1}{2}k_n\left(V_{OH} - V_{T0n}\right)^2 + \frac{1}{2}k_n\left[2\left(V_{OH} - V_{T0n}\right)V_{50\%} - V_{50\%}^2\right]\right] \\ &= \frac{1}{4}0.864\left[(1.8 - 0.4)^2 + 2(1.8 - 0.4)0.9 - 0.9^2\right] \\ &= 0.788 \text{ mA} \end{aligned}$$

Similarly, assuming $V_{OL} = 0$ V and using Equation (5.29), the $I_{LH(avg)}$ can be calculated as follows:

$$\begin{aligned} I_{LH(avg)} &= \frac{1}{2}\left[i_C(V_{in} = V_{OL}, V_{out} = V_{OL}) + i_C(V_{in} = V_{OL}, V_{out} = V_{50\%})\right] \\ &= \frac{1}{2}\left[\frac{1}{2}k_p\left(V_{DD} - |V_{T0p}|\right)^2 + \frac{1}{2}k_p\left[2\left(V_{DD} - |V_{T0p}|\right)V_{50\%} - V_{50\%}^2\right]\right] \\ &= \frac{1}{4}1.008\left[(1.8 - 0.4)^2 + 2(1.8 - 0.4)0.9 - 0.9^2\right] \\ &= 0.92 \text{ mA} \end{aligned}$$

Using Equations (5.39) and (5.40), the propagation delays t_{pHL} and t_{pLH} can be computed as follows:

$$\begin{aligned} t_{pHL} &= \frac{C_L \cdot \Delta V_{HL}}{I_{HL(avg)}} = \frac{C_L \cdot (V_{OH} - V_{50\%})}{I_{HL(avg)}} \\ &= \frac{0.1 \times 10^{-12} \cdot (1.8 - 0.9)}{0.788} = 114.2 \text{ ps} \end{aligned}$$

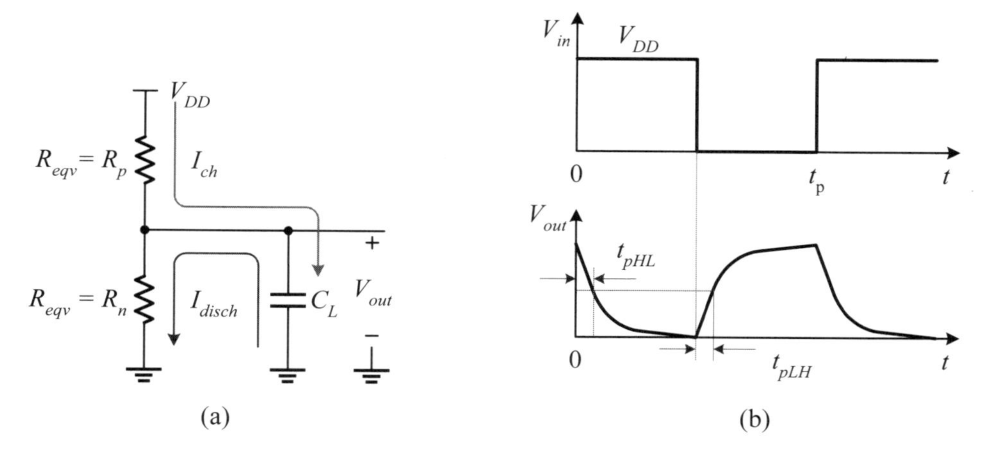

Figure 5.10: (a) The RC-equivalent delay model. (b) Timing diagram.

$$\begin{aligned} t_{pLH} &= \frac{C_L \cdot \Delta V_{LH}}{I_{LH(avg)}} = \frac{C_L \cdot (V_{50\%} - V_{OL})}{I_{HL(avg)}} \\ &= \frac{0.1 \times 10^{-12} \cdot (0.9 - 0)}{0.92} = 97.8 \text{ ps} \end{aligned}$$

The simulation results are 115 and 110 ps, respectively.

5.2.3.2 Equivalent-RC Approach In this approach, both pMOS and nMOS transistors of an inverter are regarded as a simple resistor with *equivalent on-resistance* R_{eqv} and the output of the inverter is a load capacitor C_L, as shown in Figure 5.10. Based on this, we are able to calculate the propagation delays as well as the rise and fall times of the inverter circuit.

Propagation delays. When the input voltage is low, the pMOS transistor is turned on and the load capacitor C_L is charged to V_{DD}; when the input voltage is high, the nMOS transistor is turned on and the load capacitor C_L is discharged to the ground level. For simplicity, we assume here that both pMOS and nMOS transistors cannot be turned on at the same time. For most logic circuits, both pMOS and nMOS transistors usually have a minimum channel length to minimize the area, propagation delays, and power dissipation. As a consequence, the propagation delays of a logic circuit are solely determined by the widths of the transistors in the circuit and the load capacitance needed to be driven.

From the basic circuit theory, the voltage across the capacitor of a *single time constant* (STC) circuit can be expressed as follows:

$$V_{out} = V_{DD}\left[1 - \exp\left(\frac{-t}{R_p C_L}\right)\right] \tag{5.30}$$

when it is charging, and

$$V_{out} = V_{DD} \exp\left(\frac{-t}{R_n C_L}\right) \tag{5.31}$$

when it is discharging.

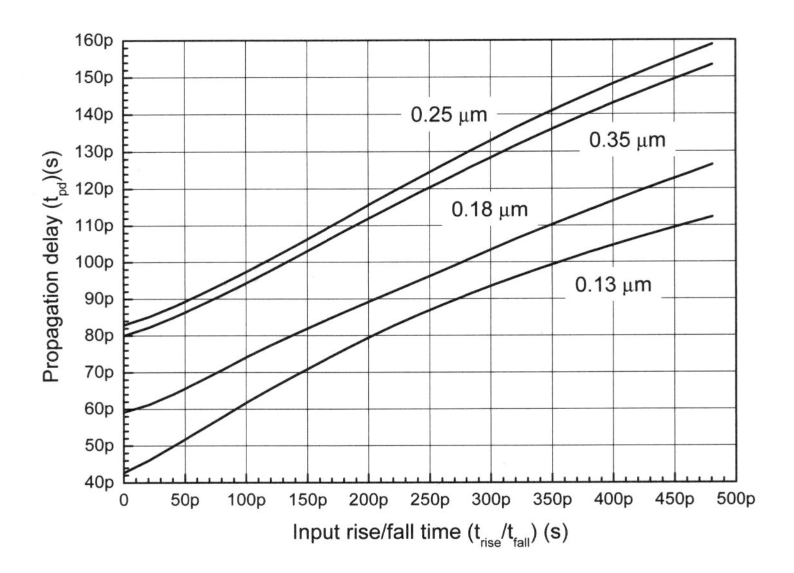

Figure 5.11: The propagation delay as a function of input rise/fall time in a variety of processes.

Therefore, the propagation delay defined at the 50% point of the voltage excursion occurs at

$$t_{pd} = 0.69 R_{eqv} C_L = t_{pLH} = t_{pHL} \tag{5.32}$$

when a step input is applied to the inverter. The equivalent on-resistance R_{eqv} can be R_n or R_p, depending on whether the t_{pHL} or t_{pLH} is under consideration. R_n and R_p denote the equivalent on-resistances of nMOS and pMOS transistors, respectively, and are a function of the ratio L/W of the corresponding transistor, referring to Section 5.1.1 and Table 5.1 for more details.

Effects of input-signal slopes. The aforementioned RC delay model assumes that the input is an ideal step signal, namely, with a zero rise/fall time. Nevertheless, in most practical applications, an input signal cannot have a zero rise/fall time. This nonzero rise/fall time of input signals results in a propagation delay a little longer than the one calculated from the use of an ideal step input. Figure 5.11 shows how the input rise/fall time impacts on the propagation delay. Roughly speaking, the increased amount of propagation delay of a logic circuit is proportional to the rise/fall time of an input signal.

Although it is not easy to quantify the slope effects of the input signal on the propagation delays, the propagation delay with a ramp input can be approximated by the following equation:

$$t_{pLH} = t_{pHL} \approx \frac{1}{2} t_{ramp} + t_{step} \approx R_{eqv} C_L \tag{5.33}$$

where t_{ramp} is the rise/fall time of the ramp input and the t_{step} is the propagation delay of the STC circuit under the ideal step input. Some other approximations are also possible. For instance, the following model is also widely used in practice:

$$t_{pLH} = t_{pHL} \approx \sqrt{t_{ramp}^2 + t_{step}^2} \tag{5.34}$$

Based on the above discussion, we will omit the constant 0.69 when designing or analyzing an inverter in practice. The following example illustrates how to design an inverter using the simple RC-equivalent delay model.

The following example is an illustration of using the equivalent-RC approach to design an inverter.

■ Example 5-8: (An example of the inverter design.) Assume that $V_{DD} =$ 1.8 V and load capacitance $C_L = 0.1$ pF. If $L_n = L_p = 2\lambda$, use 0.18-μm parameters to determine both W_n and W_p such that t_{pHL} and t_{pLH} are both equal to 100 ps.

Solution: According to Equation (5.33),

$$t_{pHL} = t_{pLH} = R_{eqv}C_L$$

we find the equivalent on-resistance R_n of the pull-down nMOS transistor as

$$R_n = \frac{t_{pHL}}{C_L} = \frac{100 \text{ ps}}{0.1 \text{ pF}} = 1.0 \text{ k}\Omega$$

From this value, we can find the aspect ratio of the nMOS transistor as

$$\begin{aligned} R_n &= R_{eqn} \times \left(\frac{L_n}{W_n}\right) \Rightarrow 1.0 = 8 \times \left(\frac{L_n}{W_n}\right) \\ &\Rightarrow \left(\frac{W_n}{L_n}\right) = 8 \Rightarrow \left(\frac{W_n}{L_n}\right) = \frac{16\lambda}{2\lambda} \end{aligned}$$

Similarly, the aspect ratio of the pull-up pMOS transistor can be found by assuming that the inverter is symmetric and hence has the same equivalent on-resistance as that of an nMOS transistor. The result is as follows:

$$\begin{aligned} R_p &= R_{eqp} \times \left(\frac{L_p}{W_p}\right) \Rightarrow 1.0 = 22 \times \left(\frac{L_p}{W_p}\right) \\ &\Rightarrow \left(\frac{W_p}{L_p}\right) = 22 \Rightarrow \left(\frac{W_p}{L_p}\right) = \frac{44\lambda}{2\lambda} \end{aligned}$$

The simulation results are as follows: $t_{pHL} = 128$ ps and $t_{pLH} = 105$ ps. Modifying the aspect ratio of an nMOS transistor into $20\lambda/2\lambda$, the resulting propagation delays are both 106 ps.

■

Rise and fall times. A simple expression can also be derived for both rise and fall times. For this purpose, the following equation is obtained by rearranging Equation (5.30).

$$t_{ch} = R_p C_L \ln \frac{V_{DD}}{V_{DD} - V_{out}} \tag{5.35}$$

Consequently, the rise time, t_{rise}, can be calculated as follows:

$$\begin{aligned} t_{rise} &= t(V_{90\%}) - t(V_{10\%}) \\ &= 2.2 R_p C_L \end{aligned}$$

Similarly, from Equation (5.31), we obtain

$$t_{disch} = R_n C_L \ln \frac{V_{DD}}{V_{out}} \tag{5.36}$$

The fall time, t_{fall}, can then be computed as follows:

$$\begin{aligned} t_{fall} &= t(V_{10\%}) - t(V_{90\%}) \\ &= 2.2R_nC_L \end{aligned} \tag{5.37}$$

As a result, both rise and fall times can be expressed as the following simple equation:

$$t_{rise} = t_{fall} = 2.2R_{eqv}C_L \tag{5.38}$$

where R_{eqv} is R_p when computing t_{rise} and is R_n when computing t_{fall}. The R_{eqv} for a feature-size nMOS transistor of different processes can be referred to in Table 5.1. When designing or analyzing the rise/fall time of an inverter, we should use the equivalent on-resistance at $1/2V_{DD}$ because it is roughly the average value between the two extreme points of interest. For example, in the 0.18-μm process, the equivalent on-resistance is 5 kΩ for a feature-size nMOS transistor and is 16 kΩ for a feature-size pMOS transistor. For those transistors other than feature-size, their equivalent on-resistances can be scaled by using Equations (5.1) and (5.2), respectively.

Example 5-9: (Rise and fall times.) Consider the circuit in Figure 5.10(a) again. Calculate the rise and fall times of the output voltage if the load capacitance C_L is 1 pF and $(W_p/L_p) = 2(W_n/L_n)$, where $L_p = L_n = 2\lambda$ and $W_n = 4\lambda$.

Solution: The equivalent on-resistances of both MOS transistors can be calculated as follows:

$$\begin{aligned} R_n &= R_{eqn} \times \left(\frac{L_n}{W_n}\right) = 5\text{ k}\Omega \times \left(\frac{2\lambda}{4\lambda}\right) = 2.5\text{ k}\Omega \\ R_p &= R_{eqp} \times \left(\frac{L_p}{W_p}\right) = 16\text{ k}\Omega \times \left(\frac{2\lambda}{2 \times 4\lambda}\right) = 4\text{ k}\Omega \end{aligned}$$

Consequently, the rise time is equal to

$$\begin{aligned} t_{rise} &= 2.2R_pC_L \\ &= 2.2 \times 4\text{k}\Omega \times 1\text{ pF} = 8.8\text{ ns} \end{aligned}$$

The fall time is

$$\begin{aligned} t_{fall} &= 2.2R_nC_L \\ &= 2.2 \times 2.5\text{ k}\Omega \times 1\text{ pF} = 5.5\text{ ns} \end{aligned}$$

The simulation results are 10 and 6.2 ns, respectively.

5.2.3.3 Differential-Current Approach In this approach, the required propagation delays are calculated by solving the current equations from the circuit under consideration. To reach this, we need to consider the operating regions of the MOS transistor of interest and use the appropriate current equation accordingly. We divide the analysis into two parts: discharging phase and charging phase.

Discharging phase. During discharging phase, the nMOS transistor is turned on to remove the charge previously stored on the load capacitor through it to ground, as depicted in Figure 5.12(a). Hence, the output voltage V_{out} will decay from its initial value V_{OH} toward V_{OL}. The nMOS transistor initially operates in its saturation region

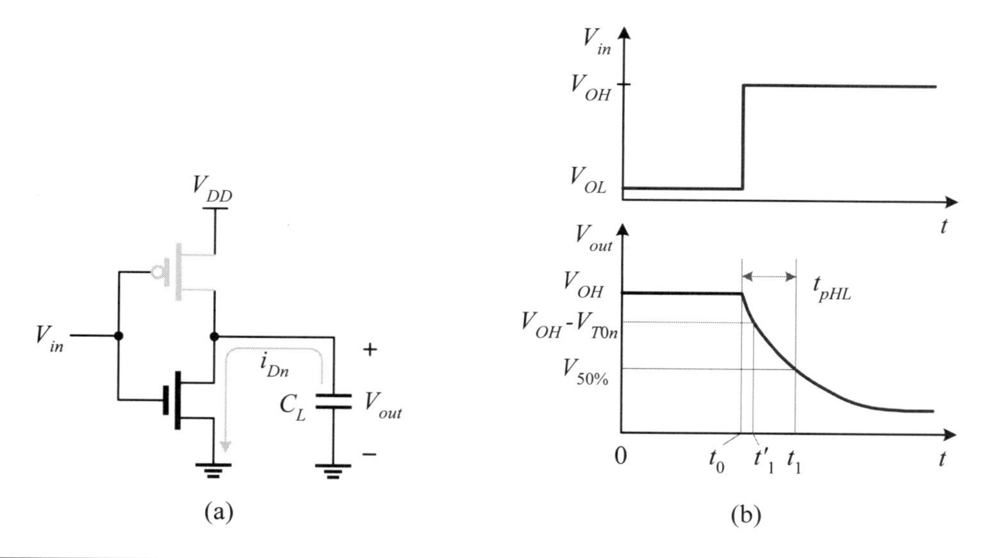

Figure 5.12: The discharging-phase of typical CMOS inverter: (a) equivalent circuit; (b) timing diagram.

and then in its linear region as the output voltage V_{out} drops below V_{OH} to an amount of V_{T0n}.

The high-to-low propagation delay t_{pHL} is the sum of two time intervals, $(t_1' - t_0)$ and $(t_1 - t_1')$, as illustrated in Figure 5.12(b), and can be expressed as

$$t_{pHL} = \frac{C_L}{k_n}\frac{1}{(V_{DD} - V_{T0n})}\left[\frac{2V_{T0n}}{V_{DD} - V_{T0n}} + \ln\left(\frac{4(V_{DD} - V_{T0n})}{V_{DD}} - 1\right)\right] \quad (5.39)$$

where we assume that $V_{OH} = V_{DD}$ and $V_{OL} = 0$.

Charging phase. The charging phase is virtually an image of the discharging phase in which the roles of both nMOS and pMOS transistors are interchanged, as depicted in Figure 5.13(a). During the charging phase, the pMOS transistor starts its operation in the saturation region due to the low-output voltage V_{OL} usually being less than the absolute value of the pMOS threshold voltage $|V_{T0p}|$. This condition sustains until the output voltage rises to $V_{OL} + |V_{T0p}|$ in which the pMOS transistor transits into its linear region of operation and stays there during the rest of the charging phase of interest.

The low-to-high propagation delay t_{pLH} is the sum of two time intervals, $(t_1' - t_0)$ and $(t_1 - t_1')$, as illustrated in Figure 5.13(b), and can be expressed as

$$t_{pLH} = \frac{C_L}{k_p}\frac{1}{(V_{DD} - |V_{T0p}|)}\left[\frac{2|V_{T0p}|}{V_{DD} - |V_{T0p}|} + \ln\left(\frac{4(V_{DD} - |V_{T0p}|)}{V_{DD}} - 1\right)\right] \quad (5.40)$$

where $V_{OH} = V_{DD}$ and $V_{OL} = 0$ are assumed.

To get more insights into the relationship between propagation delay and supply voltage, the normalized propagation delay is plotted in Figure 5.14 as a function of supply voltages using threshold voltages as parameters. From this figure, we can see the following two implications: First, for a specific process, the propagation delays (t_{pHL} and t_{pLH}) are inversely proportional to the supply voltages with a specific threshold voltage. The reduction of supply voltage may cause the propagation delay to increase

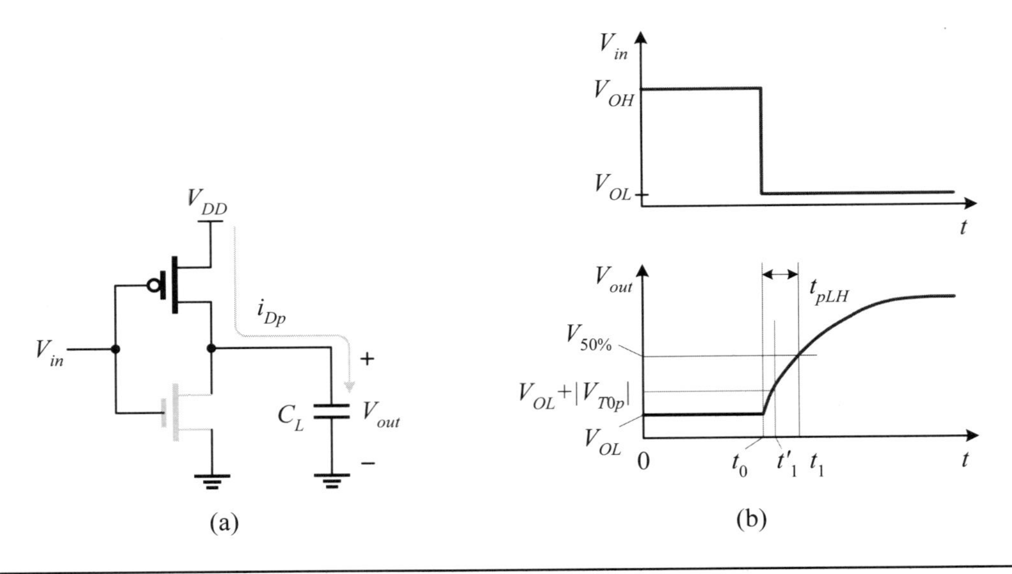

Figure 5.13: The charging-phase of typical CMOS inverter: (a) equivalent circuit; (b) timing diagram.

significantly, especially as the supply voltage is small. Second, for small supply voltages, the increase of the propagation delay due to the reduction of the supply voltage may be compensated for by lowering the threshold voltage. Unfortunately, the reduction of the threshold voltage will impact on the noise margins. Hence, a trade-off exists between the reduction of the threshold voltage and the needed noise margins.

■ Example 5-10: (Differential current approach.) Assume that $V_{DD} = 1.8$ V, load capacitance $C_L = 0.1$ pF, $L_n = L_p = 2\lambda$, $W_n = 18\lambda$, and $W_p = 42\lambda$. Using 0.18-μm parameters, $k'_n = 96\ \mu\text{A/V}^2$, $k'_p = 48\ \mu\text{A/V}^2$, and $V_{T0n} = |V_{T0p}| = 0.4$ V, find t_{pHL} and t_{pLH}.

Solution: From Equation (5.39), we obtain

$$\begin{aligned} t_{pHL} &= \frac{0.1 \times 10^{-12}}{0.864 \times 10^{-3}} \frac{1}{(1.8-0.4)} \left[\frac{2 \times 0.4}{1.8 - 0.4} + \ln\left(\frac{4(1.8-0.4)}{1.8} - 1 \right) \right] \\ &= 109 \text{ ps} \end{aligned}$$

Similarly, from Equation (5.40), we have

$$\begin{aligned} t_{pLH} &= \frac{0.1 \times 10^{-12}}{1.008 \times 10^{-3}} \frac{1}{(1.8-0.4)} \left[\frac{2 \times 0.4}{1.8 - 0.4} + \ln\left(\frac{4(1.8-0.4)}{1.8} - 1 \right) \right] \\ &= 94.2 \text{ ps} \end{aligned}$$

■

■ Review Questions

Q5-22. What is the philosophy behind the average-current approach?

Q5-23. What is the philosophy behind the equivalent-resistance approach?

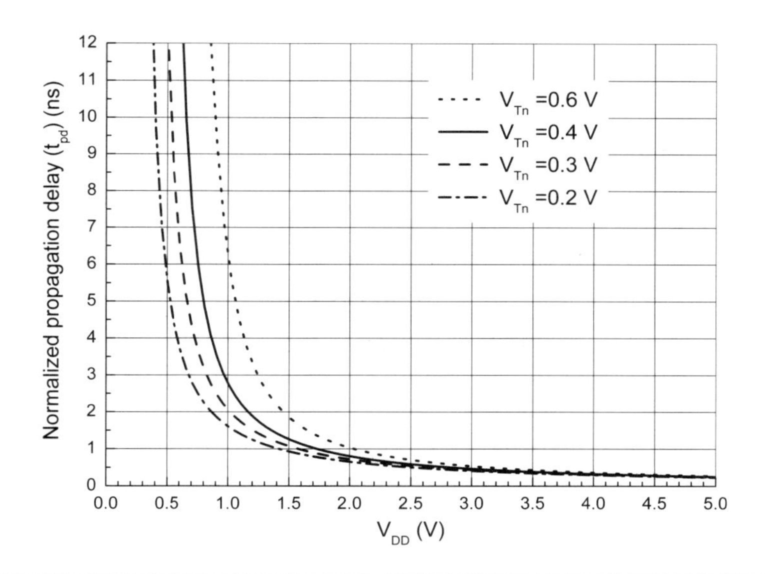

Figure 5.14: Normalized propagation delay versus supply voltage of typical CMOS inverter using threshold voltages as parameters.

Q5-24. What is the philosophy behind the differential-current approach?

Q5-25. Describe the relationship between propagation delay and supply voltage.

5.2.4 Cell Delay Model

For the ease of calculating propagation delays of a module, each cell within the module must have a simple delay model. In this subsection, we are first concerned with the components of loading capacitance. Then, we consider the cell delay model, consisting of the intrinsic delay and logical effort delay. Finally, we define the fan-out and fan-out-of-4 (FO4) delays of a logic circuit.

5.2.4.1 Components of Loading Capacitance Recall that each MOS transistor has two types of parasitic capacitance: gate (that is, oxide-related) capacitance and junction capacitance. Consequently, the loading capacitance of a logic circuit is composed of three components: the junction capacitance of the driving logic circuit, the wire capacitance, and the gate capacitance of the driven logic circuit, as depicted in Figure 5.15.

- *Junction capacitance* ($C_{D/S}$): This is the parasitic capacitance inherent in a MOS transistor itself and is often referred to as the *self-loading capacitance* ($C_{self-load}$) or *diffusion capacitance* ($C_{D/S}$).
- *Interconnect capacitance* (C_{wire}): This part comprises the capacitance contributed by all wires connected to the output node of the driving logic circuit.
- *Gate capacitance* ($C_{G(load)}$): It is composed of all input capacitances, including the gate capacitance and the Miller capacitance, of all driven logic circuits.

The propagation delays can then be expressed as in the following:

$$t_{pLH} = t_{pHL} = R_{eqv}(C_{self-load} + C_{wire} + C_{G(load)})$$

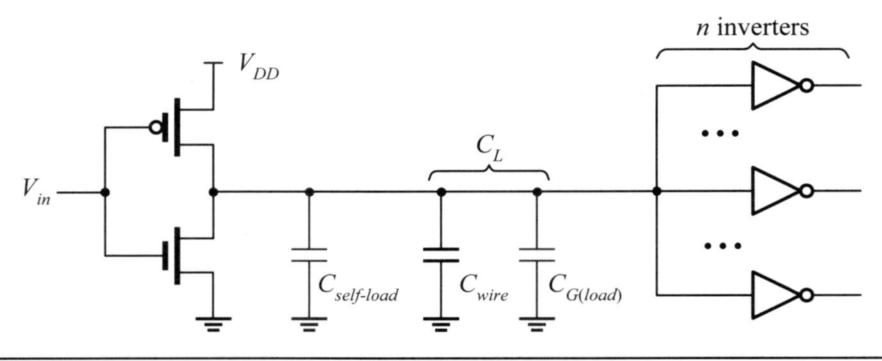

Figure 5.15: The components of loading capacitance.

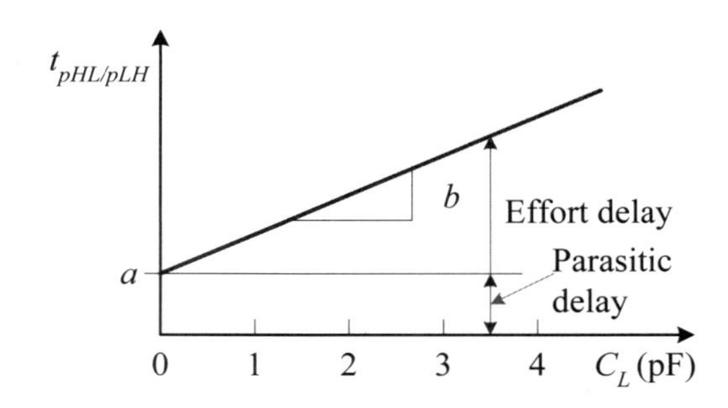

Figure 5.16: The cell delay model showing the propagation delay as a function of load capacitance.

$$
\begin{aligned}
&= R_{eqv} C_{self-load} + R_{eqv}(C_{wire} + C_{G(load)}) \\
&= a + bC_L \qquad (5.41)
\end{aligned}
$$

where $C_L = C_{wire} + C_{G(load)}$. The first term a denotes the *parasitic delay*, and the second term bC_L represents the *effort delay*. The relationship of the propagation delay versus load capacitance is depicted in Figure 5.16. The intercept at the y-axis is the parasitic delay a and the slope of the delay function is b. This linear delay model is widely used in cell libraries due to its simplicity. Hence, it is often referred to as the *cell delay model*. The cell delay model is also used in the method of logical effort, to be introduced in the next section. Next, we deal with each component in more detail.

Junction capacitance. As stated, only the 50%-point transitions of input and output signals of a logic circuit are of interest when both propagation delays t_{pHL} and t_{pLH} are considered. Hence, we may assume that the transistors are in saturation or cutoff mode of operation and ignore C_{gd} totally.

As shown in Figure 5.17(a), the self-loading capacitance at the output node of the driving inverter can be expressed as follows:

$$
\begin{aligned}
C_{self-load} &= C_{DBn} + 2C_{OL} + C_{DBp} + 2C_{OL} \\
&= C_{jn}W_n + 2C_{ol}W_n + C_{jp}W_p + 2C_{ol}W_p \\
&= (C_j + 2C_{ol})(W_n + W_p) \\
&= C_{para}(W_n + W_p) \qquad (5.42)
\end{aligned}
$$

where C_{jn} and C_{jp} are assumed to be the same and equal to C_j, and $C_{para} = C_j + 2C_{ol}$. The Miller effect as shown in Figure 5.17(b) has been used to calculate the overlap

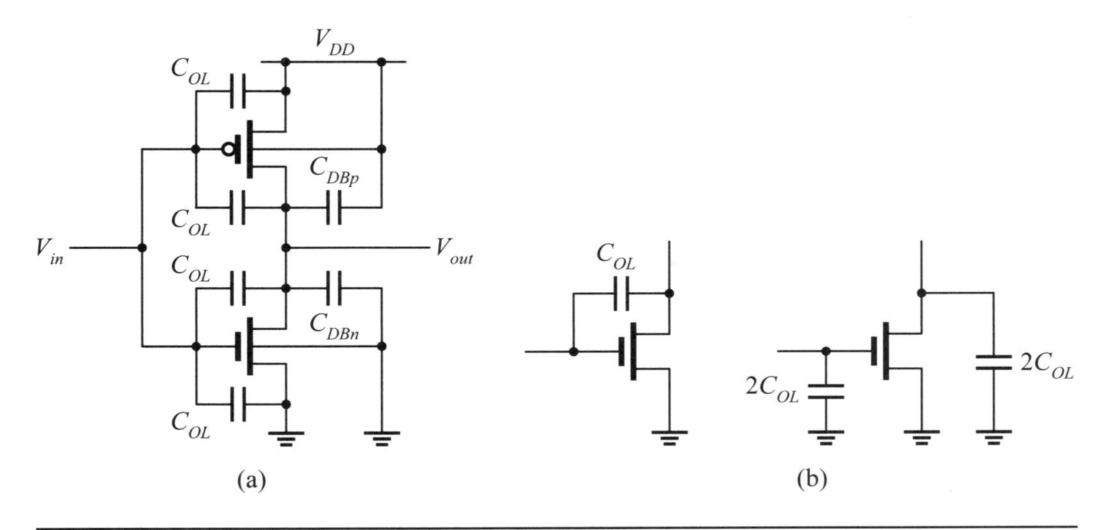

Figure 5.17: (a) The calculation of self-loading capacitance in CMOS inverters. (b) Miller effect.

Table 5.5: The oxide-related and junction capacitances of a variety of CMOS processes.

Process	0.35 μm	0.25 μm	0.18 μm	0.13 μm	90 nm	65 nm	45 nm	32 nm	Unit
C_{ol}	0.36	0.80	1.01	0.45	0.63	0.86	0.89	0.88	fF/μm
C_j	2.11	2.88	1.66	0.98	1.42	2.63	2.76	2.35	fF/μm
$C_{ox}L$	1.13	2.1	2.4	1.4	1.46	0.87	0.70	0.31	fF/μm
C_{ox}	0.32	0.84	1.33	1.08	1.62	1.34	1.56	0.97	μF/cm^2
C_g	1.49	2.90	3.41	1.85	2.09	1.73	1.59	1.19	fF/μm
C_{para}	2.83	4.48	3.68	1.88	2.68	4.35	4.54	4.11	fF/μm
$\frac{C_{para}}{C_g}$	1.90	1.54	1.08	1.02	1.28	2.51	2.86	3.45	

capacitance C_{OL}. The typical values of C_{para} in a variety of processes are listed in Table 5.5.

Interconnect capacitance. The capacitance effect of a wire can be classified into the following three cases according to the wire length in question:

- *Short wire*: For a wire with a length less than a few microns, we may ignore the wire capacitance.
- *Mediate-length wire*: For a wire with a length greater than a few microns, we include the lumped wire capacitance.
- *Long wire*: For a very long wire, we have to deal with both distributed-RC and capacitive-coupling effects.

An approximate rule of thumb for estimating the wire capacitance independent of processes is to use

$$C_{wire} = C_{int}l \tag{5.43}$$

where $C_{int} = 0.1$ fF/μm for widely spaced wires and $C_{int} = 0.15$ fF/μm for closely spaced wires, assuming that the minimum width of metals is used. Refer to Section 13.1.2 for more details.

Gate capacitance. As we described in Section 5.1.2, each transistor consists of one gate-oxide capacitance and two overlap capacitances, as depicted in Figure 5.18.

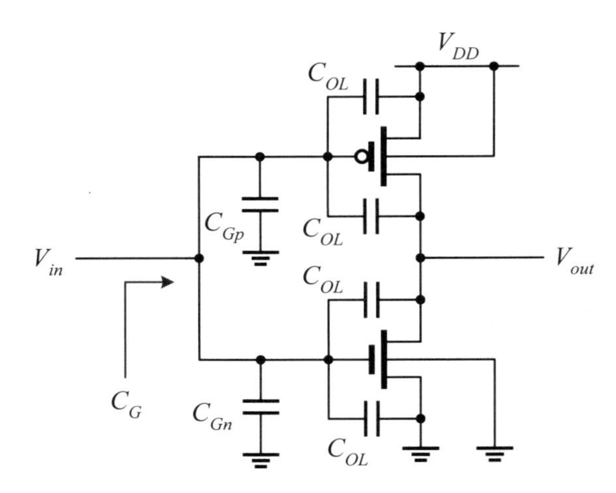

Figure 5.18: The calculation of the gate capacitance of a CMOS inverter.

The total gate capacitance looking into an inverter is thus equal to

$$\begin{aligned} C_G &= C_{Gn} + 3C_{OL} + C_{Gp} + 3C_{OL} \\ &= C_{ox}L_{actual}W_n + 3C_{ol}W_n + C_{ox}L_{actual}W_p + 3C_{ol}W_p \\ &= (C_{ox}L_{actual} + 3C_{ol})(W_n + W_p) \\ &= C_g(W_n + W_p) \end{aligned} \tag{5.44}$$

where $C_g = C_{ox}L_{actual} + 3C_{ol} = C_{ox}L + C_{ol}$. The Miller effect has been applied to account for the overlap capacitance C_{OL} between the gate and drain/source. The total gate capacitance $C_{G(load)}$ of n driven inverters is equal to nC_G, as illustrated in Figure 5.15.

The values of C_g for a variety of typical processes are listed in Table 5.5. Using these values, we may quickly estimate the gate capacitance without resorting to a complicated computation process.

5.2.4.2 Intrinsic Delay of Inverters For a typical inverter, we usually make the size of a pMOS transistor twice that of an nMOS transistor, namely, $W_p = 2W_n$. Hence, without considering the self-loading and wire capacitances, an inverter has a propagation delay equal to

$$\begin{aligned} \tau_{inv} &= R_{eqv}C_{in} = R_{eqv}C_g(W_n + W_p) = R_{eqv}C_g(3W_n) \\ &= R_{eqn}\left(\frac{L_n}{W_n}\right)C_g(3W_n) = 3R_{eqn}C_gL_n \end{aligned} \tag{5.45}$$

when driving another inverter with the same size. The resulting propagation delay is defined as the *intrinsic delay* of an inverter. It is independent of the channel width of an inverter.

Now we consider the situation when an inverter is driving a loading capacitance C_{out} other than an inverter with the same size. Referring to Figure 5.19 and assuming that the wire capacitance is combined into the output capacitance, that is, $C_{out} = C_L$, the propagation delay of the inverter can then be represented as a function of intrinsic delay by normalizing both capacitances with respect to the input capacitance of the

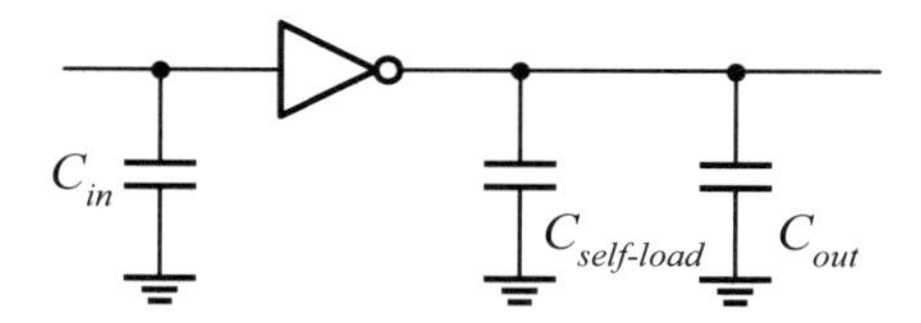

Figure 5.19: An illustration of effort and parasitic delays of an inverter.

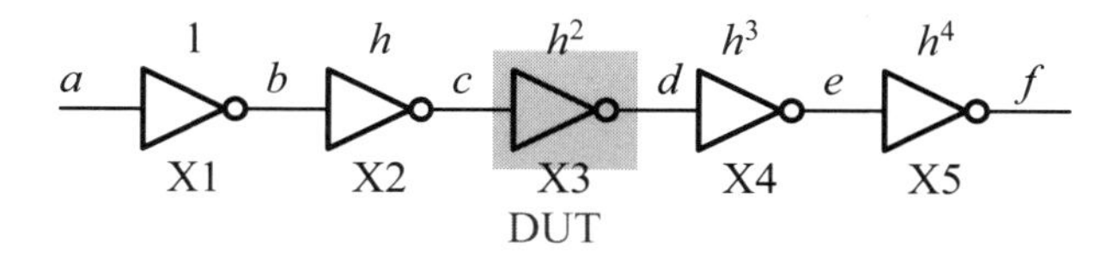

Figure 5.20: A circuit used to explore propagation delays versus the logical effort of a CMOS inverter.

inverter and can be expressed as follows:

$$\begin{aligned} t_{pd} &= R_{eqv}(C_{out} + C_{self-load}) \\ &= R_{eqv}C_{in}\left(\frac{C_{out}}{C_{in}} + \frac{C_{self-load}}{C_{in}}\right) \\ &= \tau_{inv}\left(\frac{C_{out}}{C_{in}} + \gamma_{inv}\right) \\ &= t_{pd(effort)} + t_{pd(para)} \end{aligned} \tag{5.46}$$

where γ_{inv} is the ratio of the self-loading capacitance to the input capacitance of the inverter, $\gamma_{inv} = C_{self-load}/C_{in} = C_{para}/C_g$. Its value is highly dependent on the layout of the inverter since $C_{self-load}$ is a strong function of its layout, as we have discussed. The propagation delay of an inverter comprises two parts: *(logical) effort delay* $t_{pd(effort)}$ and *parasitic delay* $t_{pd(para)}$. The effort delay involves the feature of loading capacitance while the parasitic delay only concerns the properties of the inverter itself.

5.2.4.3 Parasitic Capacitances The parasitic capacitance of a MOS transistor can be extracted by using a simple circuit, as shown in Figure 5.20. Since from Equation (5.46), the propagation delay t_{pd} can be represented as

$$\begin{aligned} t_{pd} &= \tau_{inv}\left(h + \frac{C_j + 2C_{ol}}{C_g}\right) \\ &= \tau_{inv}\left(\frac{C_j + 2C_{ol}}{C_g}\right) + \tau_{inv}h \\ &= a + b \cdot h \end{aligned} \tag{5.47}$$

where h is defined as C_{out}/C_{in}, $C_{in} = C_g(W_n + W_p)$, $C_{self-load} = (C_j + 2C_{ol})(W_n + W_p)$, and b is the intrinsic delay τ_{inv}, we may extract both C_{ol} and C_j of an inverter in various processes by simulating the circuit shown in Figure 5.20 without and with junction capacitance. The simulation results are shown in Figure 5.21, where constants a and b are listed in Table 5.6.

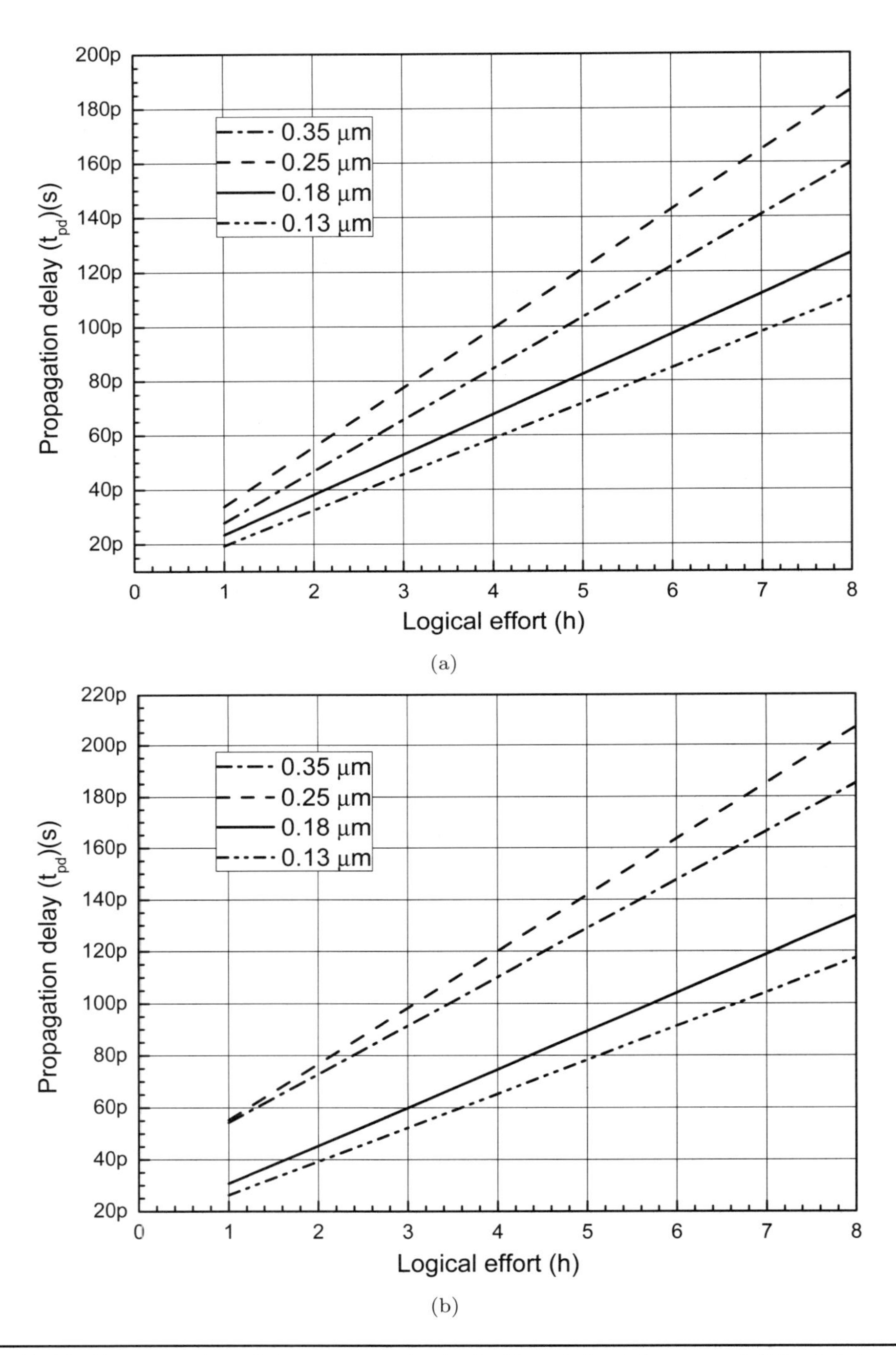

(a)

(b)

Figure 5.21: Propagation delays versus logical effort (a) without and (b) with junction capacitance.

Table 5.6: The values of a and b in the relation: $t_{pd} = a + b \cdot h$.

Process	0.35 μm	0.25 μm	0.18 μm	0.13 μm	90 nm	65 nm	45 nm	32 nm	Unit
Without junction capacitance									
a	9.03	12.03	8.76	6.34	5.43	5.67	5.26	4.71	ps
b	18.87	21.77	14.74	13.06	9.07	5.73	4.72	3.20	ps
With junction capacitance									
a	35.63	33.63	15.95	13.23	11.56	14.37	13.44	11.04	ps
b	18.67	21.67	14.75	12.97	9.04	5.73	4.76	3.26	ps

■ Example 5-11: (Calculation of C_j, C_{ol}, and C_{para}.) Using the values given in Table 5.6, calculate C_j, C_{ol}, and C_{para} of the 0.18-μm process.

Solution: From Equation (5.45) and the equivalent on-resistance from Table 5.1, we obtain the value of C_g as

$$C_g = \frac{\tau_{inv}}{3R_{eqn}L_n} = \frac{14.75}{3 \times 8 \times 0.18} = 3.41 \text{ fF}/\mu\text{m}$$

Using the value of a without the junction capacitance, we may find the value of C_{ol} as follows:

$$C_{ol} = \frac{aC_g}{2\tau_{inv}} = \frac{8.76 \times 3.41}{2 \times 14.75} = 1.01 \text{ fF}/\mu\text{m}$$

The junction capacitance can be calculated by noting that the propagation delay caused by this part is equal to the difference of both values of a. Hence,

$$C_j = \frac{(15.95 - 8.76)C_g}{\tau_{inv}} = \frac{7.19 \times 3.41}{14.75} = 1.66 \text{ fF}/\mu\text{m}$$

As stated, the parasitic capacitance C_{para} is equal to $C_j + 2C_{ol} = 3.68$ fF/μm. ■

The capacitances of other processes can be computed in the same way and their results are also given in Table 5.5.

5.2.4.4 Fan-Out and FO4 Definition Before proceeding to define the fan-out of a logic circuit, consider the logic circuit shown in Figure 5.22, where an inverter drives a NAND gate in two different ways. In Figure 5.22(a), the output of an inverter is connected to one of the two inputs of the NAND gate, and the second input of the NAND gate is connected to V_{DD}, while in Figure 5.22(b) the output of an inverter is connected to both inputs of the NAND gate. Both logic circuits have the same logic function, but the circuit of Figure 5.22(a) is faster than that of Figure 5.22(b) because the former has only the half-loading capacitance of the latter.

In order to quantify the loading capacitance of a logic gate, the concept of *fan-out* is introduced. Informally, the fan-out of a logic gate is the number of gates that it drives. For instance, each inverter in Figure 5.22 drives a single NAND gate. Hence, they have the same fan-out. Nevertheless, both inverters indeed have different loading capacitances. Consequently, a more precise definition is needed. Recall that one major feature of CMOS gates is that each input of any CMOS logic gate is connected to one nMOS and one pMOS transistor, exactly the same as an inverter. Because of this reason, the fan-out of a logic gate is usually defined as the number of inverters of the

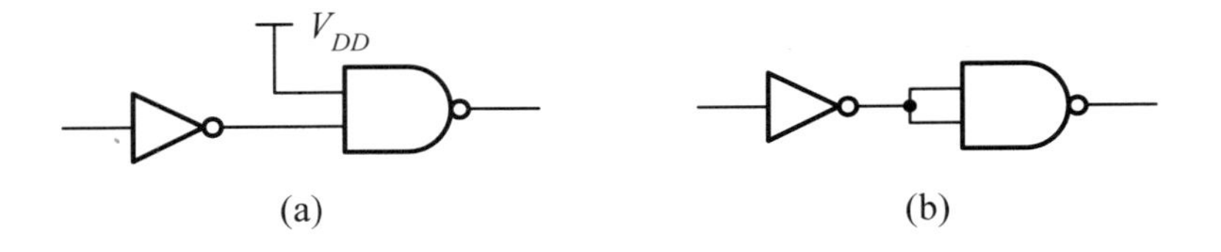

Figure 5.22: An illustration of fan-out concepts: (a) fan-out = 4/3; (b) fan-out = 8/3.

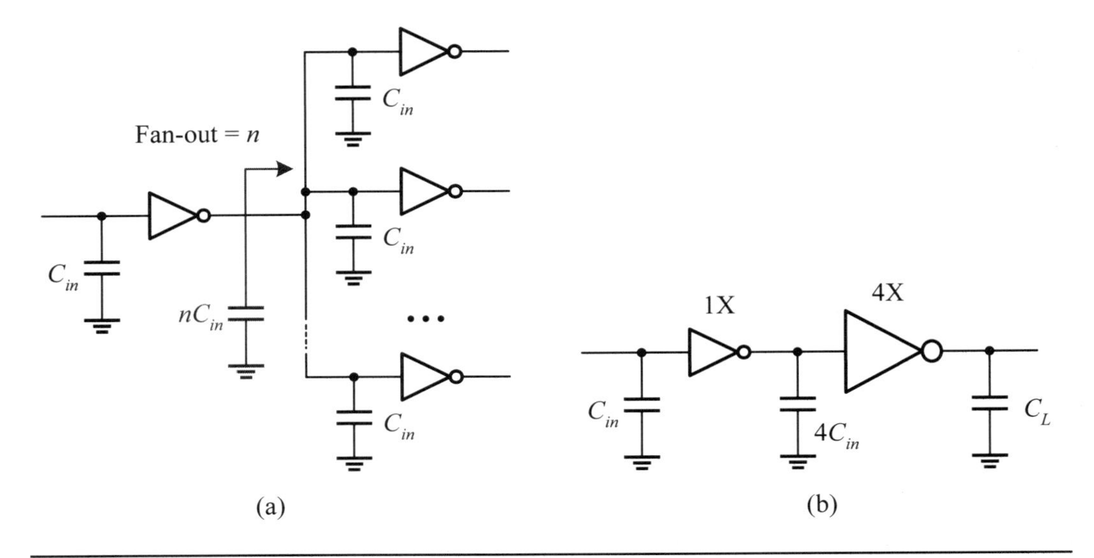

Figure 5.23: (a) Definitions of fan-out of n. (b) An example of fan-out of 4.

same size that it drives. In other words, the fan-out of a logic gate can be stated as the ratio of the loading capacitance[2] to the input capacitance. Based on this definition, the fan-out of the inverter in Figure 5.22(a) is 4/3 while in Figure 5.22(b) it is 8/3.

Figure 5.23(a) illustrates the general meaning of fan-out of n. In this case, an inverter drives n inverters with the same size. Without considering the wire capacitance, the loading capacitance of the inverter is n times its input capacitance. Hence, the fan-out of the driving inverter is n. When n is set to 4, the fan-out is referred to as the fan-out of 4 (FO4) and the corresponding propagation delay is denoted as the *FO4 delay*.

Figure 5.23(b) shows another situation that the size of the driven inverter is four times that of the driving inverter. This means that the input capacitance of the driven inverter is four times that of the driving inverter. Thereby, the fan-out of the driving inverter is 4. The simulation results of FO4 delays with different processes are summarized in Table 5.7. In the table, we also list the intrinsic delays of inverters for reference. For processes scaled beyond 90 nm, strained silicon is widely used to improve the performance. When this is the situation, the propagation delays listed in the table would be reduced by an amount of about 30%.

[2]Here, the loading capacitance includes interconnect capacitance (C_{wire}) and gate capacitance ($C_{G(load)}$) of all driven logic circuits. It does not take into account the junction capacitance ($C_{self-load}$).

Table 5.7: FO4 and intrinsic delays of CMOS inverters in a variety of processes.

	0.35 μm	0.25 μm	0.18 μm	0.13 μm	90 nm	65 nm	45 nm	32 nm	Unit
τ_{inv}	19	22	15	13	9	5.7	4.7	3.2	ps
t_{FO4}	110	120	75	65	48	37	32	24	ps

The FO4 delay is often used as a metric to compare different designs in the same or different processes. Full-custom designs generally offer a 3 to 5 times frequency advantage over cell-based designs. Full-custom designs typically have cycle times of 10 to 25 FO4 inverter delays with sequencing (flip-flop) overhead of about 2 to 3 FO4 delays and a clock skew overhead of about 1 to 2 FO4 delays, roughly 10% of a cycle time. Cell-based designs typically have cycle times of 50 to 100 FO4 inverter delays with a sequential overhead of about 4 to 6 FO4 delays and a clock skew overhead of about 5 to 10 FO4 delays. High-performance cell-based designs can have cycle times of 25 to 50 FO4 inverter delays.

■ Review Questions

Q5-26. What are the three components of the loading capacitance of a logic circuit?

Q5-27. What is the intrinsic delay of an inverter?

Q5-28. What is the effort delay of an inverter?

Q5-29. Define the fan-out of an inverter.

Q5-30. Explain why the fan-out of Figure 5.22(a) is 4/3.

Q5-31. What is the meaning of FO4?

5.2.5 Elmore Delay Model

In CMOS logic circuits, there often exists an RC-ladder circuit, such as the one shown in Figure 5.24(a). However, to obtain an STC-equivalent circuit, as shown in Figure 5.24(b), for it is not an easy task, the propagation delay of such a network can be estimated by the following observation. For a source voltage to reach the output node, it must pass through all capacitors associated with the RC-ladder network. Therefore, the source voltage must charge or discharge all capacitors along the path toward the output node to some extent. Based on this, the Elmore delay model is the result.

5.2.5.1 Elmore Delay Model For an RC-ladder network shown in Figure 5.24(a), the propagation delay at the output node can be described as a simple expression known as the *Elmore delay model*, which is stated as follows:

$$t_{Elmore} = \sum_{i=1}^{n}\left(\sum_{j=1}^{i} R_j\right) C_i = \sum_{i=1}^{n}\left(\sum_{j=i}^{n} C_j\right) R_i \quad (5.48)$$

where both R_i and R_j denote the segment resistances and both C_i and C_j denote the segment capacitances. We may restate the above relation as follows. The propagation delay of an RC-ladder network is the sum of all segment time constants, where each segment time constant is the product of the segment capacitance and the total resistance along the path from the segment to the source.

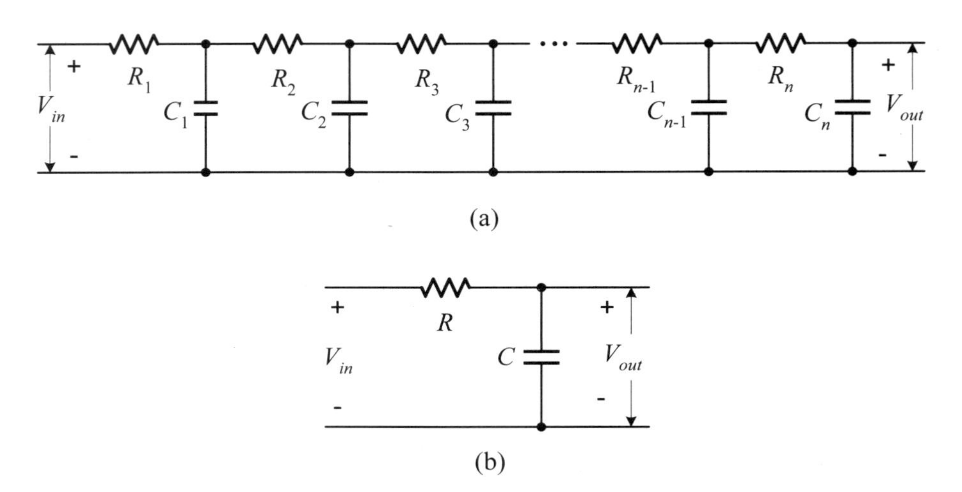

Figure 5.24: Elmore delay model: (a) an n-stage RC-ladder circuit; (b) an STC circuit.

Taking the RC-ladder network depicted in Figure 5.24(a) as an example and applying the Elmore delay model to it, we obtain the propagation delay at the output node as follows:

$$\begin{aligned} t_{Elmore} &= nRC + (n-1)RC + \cdots + 2RC + RC \\ &= \frac{1}{2}n(n+1)RC \propto n^2RC \end{aligned} \tag{5.49}$$

Consequently, the propagation delay is a quadratical function of the number of segments of the RC ladder circuit. Based on this result, we can see that, to keep the propagation delay at a tolerable level, the number of RC segments should be limited. In general, n is set to 4 or 5 at most.

As an illustration of applying the Elmore delay model to a real-world CMOS logic gate, consider the four-input NAND gate depicted in Figure 5.25. The details are described in the following example.

■ Example 5-12: (Propagation delay of a four-input NAND gate.) Apply the Elmore delay model to the four-input NAND gate shown in Figure 5.25(a) and calculate the worst-case propagation delay at the output.

Solution: Assume that the capacitance C_{out} is the sum of all parasitic capacitances of nMOS w and the four pMOS transistors. The other three capacitances are the junction capacitance of nMOS transistors. The worst-case propagation delay occurs when the output voltage is discharged to the ground level. In this case, the equivalent RC-ladder network is shown in Figure 5.25(b), where $V_{in} = 0$. Applying the Elmore delay model to the equivalent RC-ladder network, the propagation delay at the output node is

$$\begin{aligned} t_{pd(out)} &= C_yR_z + C_x(R_y + R_z) + C_w(R_x + R_y + R_z) + \\ &\quad C_{out}(R_w + R_x + R_y + R_z) \end{aligned}$$

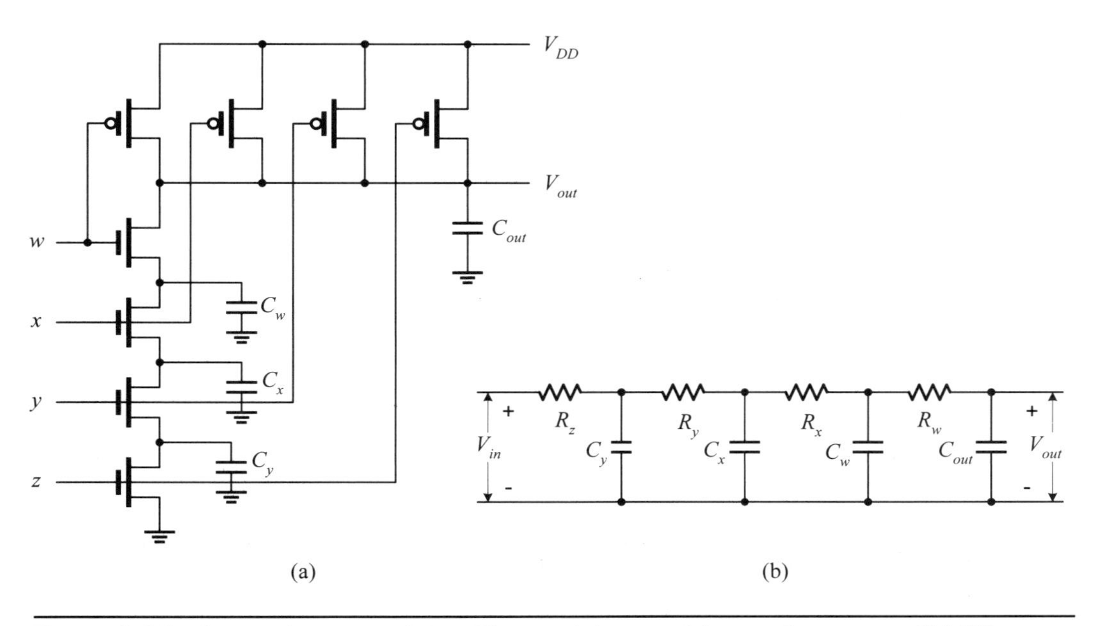

Figure 5.25: An application of the Elmore delay model to a four-input NAND gate: (a) logic circuit; (b) equivalent circuit.

From this example, we reconfirm the importance of the reduction of the parasitic capacitance of MOS transistors because the parasitic capacitance might affect the propagation delay severely.

■

5.2.5.2 RC-Tree Delay Model For a complex logic gate, the equivalent-RC network is usually not a simple ladder network. Instead, it may be an RC-tree network, such as the one shown in Figure 5.26. In such a situation, the Elmore delay model can be no longer applied.

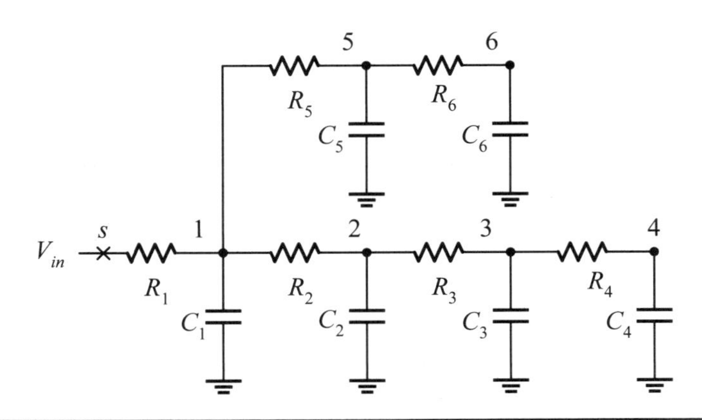

Figure 5.26: The RC-tree delay model generalized from the Elmore delay model.

Nevertheless, based on the same concept as we described, the propagation delay at any node i can be described as in the following modified Elmore delay model.

$$t_{pdi} = \sum_{k=1}^{n} C_k R_{i,k} \tag{5.50}$$

where $R_{i,k}$ is defined as follows:

$$R_{i,k} = \sum R_j \Rightarrow R_j \in [\text{path}(i \to s) \cap \text{path}(k \to s)] \tag{5.51}$$

which means that the resistance $R_{i,k}$ is the sum of resistances along the common path segment from node i to source node s and from node k to source node s. Node i is the one of interest and node k, where $1 \leq k \leq n$, are nodes to which capacitors connect. The propagation delay at node i is the sum of all time constants formed by each capacitor and its associated resistance, $R_{i,k}$.

The following example shows how to apply this modified Elmore delay model to the RC-tree network shown in Figure 5.26.

■ Example 5-13: (An RC-tree example.) Considering the RC-tree network in Figure 5.26, calculate the propagation delays at nodes 3 and 5, respectively.

Solution: Applying the modified Elmore delay model to the RC-tree network, the propagation delay at node 3 is

$$\begin{aligned} t_{pd3} &= R_1C_1 + (R_1 + R_2)C_2 + (R_1 + R_2 + R_3)C_3 \\ &\quad + (R_1 + R_2 + R_3)C_4 + R_1C_5 + R_1C_6 \end{aligned}$$

The propagation delay at node 5 is

$$t_{pd5} = R_1C_1 + R_1C_2 + R_1C_3 + R_1C_4 + (R_1 + R_5)C_5 + (R_1 + R_5)C_6$$

■

The modified Elmore delay model is reduced to the Elmore delay model when the RC-tree network is degenerated to an RC-ladder network. In other words, the Elmore delay model is a special case of modified Elmore delay model.

■ Review Questions

Q5-32. Describe the Elmore delay model.
Q5-33. Describe the modified Elmore delay model.
Q5-34. Show the equality of Equation (5.48).

5.3 Path-Delay Optimization

To make a logic circuit useful, it is usually necessary to cascade many different types of logic gates to form a *logic chain* or a logic network. When such a logic chain needs to drive a large loading capacitance, a problem about how to optimize the logic chain to minimize the total propagation delay is arisen. This problem is known as a *path-delay optimization problem* and is the focus of this section. We first address the principle of path-delay optimization, including inverter chains (super-buffers) and logic chains of generic logic gates. Then, we define *logical effort* and apply it to the path-delay optimization problem.

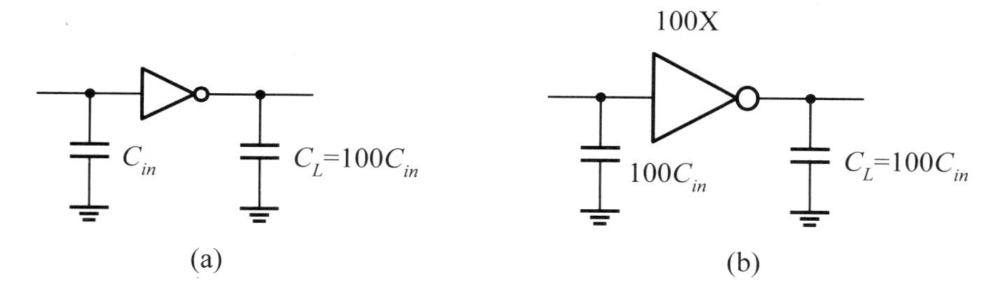

Figure 5.27: An example of driving a large capacitive load: (a) basic inverter; (b) enlarged inverter.

5.3.1 Driving Large Capacitive Loads

Before proceeding, let us consider the case that a basic inverter needs to drive a large loading capacitor, say, $C_L = 100C_{in}$, as shown in Figure 5.27(a). Probably, one straightforward method is just to enlarge the inverter 100 times, as shown in Figure 5.27(b). Of course, this approach solves the problem at the output node because the resistance of the inverter is reduced to $R/100$, and hence, the propagation delay is equal to RC_L. However, it simply moves the tough problem to the input node of the inverter since now the input capacitance of the inverter is scaled up to $100C_{in}$, which in turn must be driven by its previous stage. Consequently, the problem is still not solved but only moves backward.

From the above discussion, we may conclude that to drive a large loading capacitor, the driver must at least contain two stages; thereby, an inverter chain is named. The first stage interfaces the input, while the last stage drives the loading capacitor. The sizes of all stages are successively increased to optimize the total propagation delay. This inverter chain is historically called a *super-buffer*, and the related design technique is known as a *super-buffer design*. For now, the inverter chain has been generalized into a logic chain that may allow the use of mixed types of logic gates. The resulting logic chain along with its design technique is known as a path-delay optimization problem. We first consider an inverter chain and then extend the technique to the general logic chain with generic logic gates. It is worth noting that an inverter chain is only composed of inverters, while a logic chain may comprise logic gates of the same or different types.

A path-delay optimization problem may be stated more formally as follows: *Assume that a logic chain consisting of many logic-gate stages of increasing sizes is used to drive a large loading capacitor. Determine the number of logic-gate stages needed in the chain and their corresponding sizes to minimize the path delay.* Of course, to solve this problem, two questions arise in nature. First, what is the optimal number of stages? Second, what are the optimal sizes of the gates in the logic chain? We explore and answer these two questions in the following.

■ Review Questions

Q5-35. What is a super-buffer?

Q5-36. Distinguish between an inverter chain and a logic chain.

Q5-37. Define the path-delay optimization problem.

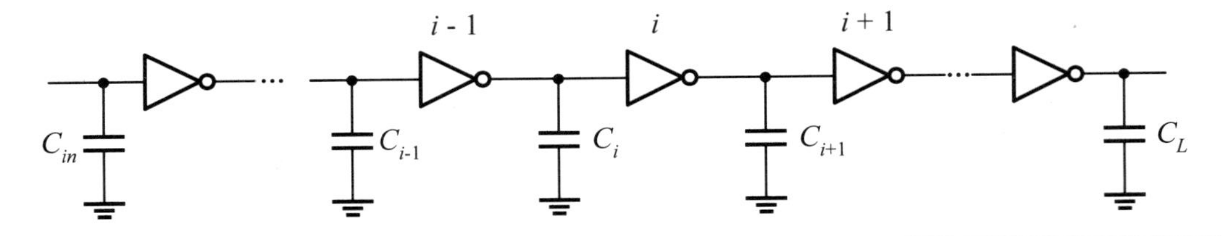

Figure 5.28: An illustration of path-delay optimization.

5.3.2 Path-Delay Optimization

We first illustrate the principle of path-delay optimization using the design of a super-buffer (an inverter chain) as an example and then generalize the results into a logic chain of generic logic gates.

5.3.2.1 Super-Buffer Designs Figure 5.28 shows an inverter chain of n inverters, where all MOS transistors use the same channel length, and each individual inverter has its own size of S_i multiples of the basic inverter. The path delay of the inverter chain is the sum of the propagation delay of each stage in the chain. To minimize the path delay, the propagation delay of each stage or the sum of propagation delays of two or a few stages must be minimized because the propagation delay of each stage is greater than zero. Hence, we first consider the optimal propagation delay of two successive stages.

Remember that the propagation delay of an inverter can be expressed as follows:

$$t_{pd} = \tau_{inv}\left(\frac{C_{out}}{C_{in}} + \gamma_{inv}\right) \tag{5.52}$$

where τ_{inv} is the intrinsic delay of the inverter, C_{out} and C_{in} are the output and input capacitances of the inverter, respectively, and γ_{inv} is the ratio of the self-loading capacitance to the input capacitance of the inverter.

The propagation delay of two successive stages, namely, the $(i-1)$th and ith stages, can then be expressed as

$$t_{two-stage} = \tau_{inv}\left(\frac{C_i}{C_{i-1}} + \gamma_{inv}\right) + \tau_{inv}\left(\frac{C_{i+1}}{C_i} + \gamma_{inv}\right) \tag{5.53}$$

To find the optimal propagation delay, taking the derivative of $t_{two-stage}$ with respect to C_i, we obtain

$$\frac{\partial t_{two-stage}}{\partial C_i} = 0 \Rightarrow \tau_{inv}\left(\frac{1}{C_{i-1}} - \frac{C_{i+1}}{C_i^2}\right) = 0 \tag{5.54}$$

Hence, $\tau_{inv}C_i/C_{i-1} = \tau_{inv}C_{i+1}/C_i$, which means that both inverters must have the same effort delay. This implies that $FO_{i-1} = FO_i$, where FO_{i-1} and FO_i are the fan-outs of the $(i-1)$th and ith stages, respectively. Since the input capacitance of an inverter is proportional to the size of the inverter, the above result also means that the size of an inverter is the geometric mean of the sizes of its previous and succeeding inverters. In other words, the size ratio between two successive stages is a constant, that is, the geometric ratio, denoted a. Based on this, we can show that all stages have the same propagation delay and are equal to

$$t_{pd} = \tau_{inv}\left(\frac{C_i}{C_{i-1}} + \gamma_{inv}\right) = \tau_{inv}\left(a + \gamma_{inv}\right) \tag{5.55}$$

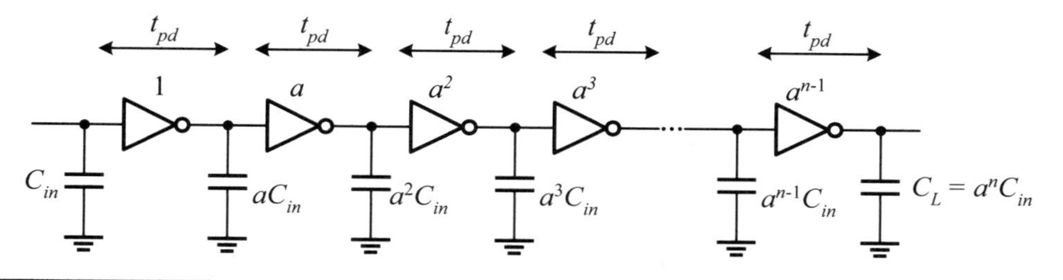

Figure 5.29: The principle of path-delay optimization.

where $a = C_i/C_{i-1}$ and C_i and C_{i-1} are input capacitances of the ith and $(i-1)$th inverters, respectively.

Based on the above result, an inverter chain can be constructed by cascading n inverters in the way shown in Figure 5.29, where each successive inverter has the size a times its previous stage. The input capacitance of the first stage is C_{in} and the nth stage is equal to $a^nC_{in} = C_L$. The nth inverter does not actually exist. To complete the design of an inverter chain, we need to determine the required number of stages, n, and the size of each stage, a.

By solving equation $a^nC_{in} = C_L$, the n can be determined and expressed as

$$n = \frac{\ln(C_L/C_{in})}{\ln a} \tag{5.56}$$

Because all stages have the same propagation delay, the total propagation delay of the inverter chain is equal to nt_{pd}, where t_{pd} is the propagation delay of each stage. By combining Equation (5.55) with Equation (5.56), we obtain the total propagation delay, $t_{pd(total)}$, as follows:

$$\begin{aligned} t_{pd(total)} &= n \times \tau_{inv}\,(a + \gamma_{inv}) \\ &= \frac{\ln(C_L/C_{in})}{\ln a} \times \tau_{inv}\,(a + \gamma_{inv}) \end{aligned} \tag{5.57}$$

Taking the partial derivative of $t_{pd(total)}$ with respect to a and setting the result to equal 0, we obtain

$$a(\ln a - 1) = \gamma_{inv} \tag{5.58}$$

The optimal value of a is e when γ_{inv} is set to 0. Nevertheless, there is no closed-form expression for the optimal a when γ_{inv} is not equal to 0. To get more insight into the relationship between the propagation delay and fan-out (a), the normalized propagation delay versus fan-out (a) for different γ_{inv} values is plotted in Figure 5.30. From this figure, we can see that the optimal values of a with a variety of γ_{inv} values fall in the range from 2.5 to 5. As a consequence, in practical applications, the fan-out of each gate is usually set to 4, and hence, FO4 (the fan-out of 4 delay) delay is named.

■ Example 5-14: (An example of the super-buffer design.)

1. Consider the single-stage buffer circuit shown in Figure 5.31(a). Calculate the propagation delay of the single-stage buffer when it drives a load capacitance of 100 fF. Assume that $R_{eqn} = 8$ kΩ and $R_{eqp} = 22$ kΩ.
2. Applying the FO4 rule to the inverter chain shown in Figure 5.31(b), calculate the size of each inverter in the inverter chain when the load capacitance of the last stage is 100 fF. Calculate the propagation delay.

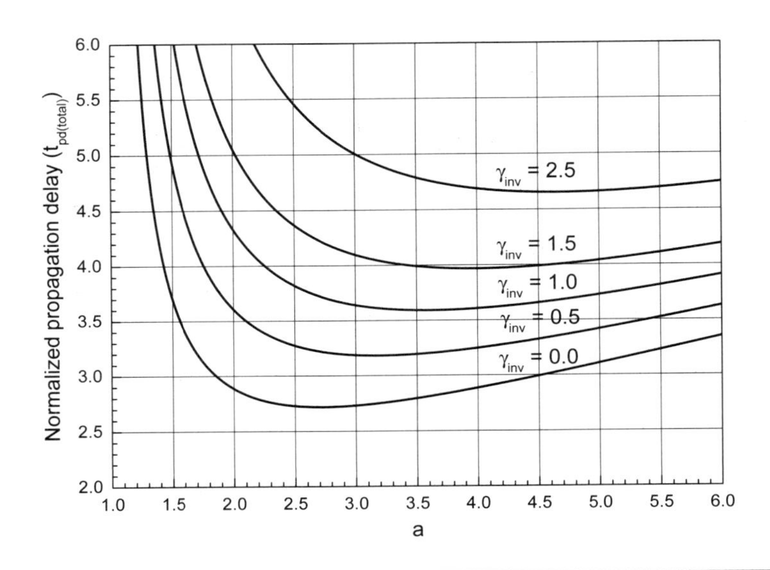

Figure 5.30: Normalized propagation delay ($t_{pd(total)}$) versus fan-out (a) for different γ_{inv} values.

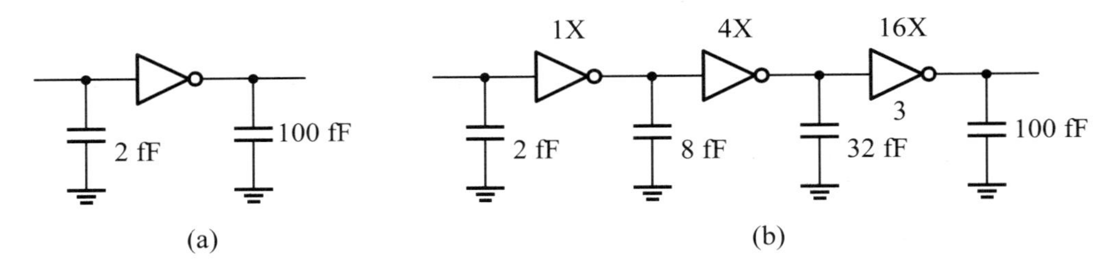

Figure 5.31: Examples of (a) single-stage buffer and (b) a three-stage buffer using the FO4 rule.

Solution:

1. The propagation delay of the single-stage buffer is calculated as follows:

$$\begin{aligned} t_{pd} &= \frac{1}{2}(t_{pLH} + t_{pHL}) = \frac{1}{2}(R_p C_L + R_n C_L) \\ &= \frac{1}{2}(22\ \text{k}\Omega \cdot \frac{2}{6} + 8\ \text{k}\Omega \cdot \frac{2}{3}) \cdot 100\ \text{fF} = 633\ \text{ps} \end{aligned}$$

2. The size of each inverter in the inverter chain is calculated as in the following:

$$C_L = 4^n C_{in}$$

Hence,

$$n = \log_4 \frac{C_L}{C_{in}} = \log_4 \frac{100}{2} = 2.82$$

So we take the whole number, that is, 3. The sizes of each buffer are

$$\begin{aligned} L_{p/n1} &= L_{p/n2} = L_{p/n3} = L_{min} \\ W_{p1} &= 2W_{n1} = 3L_{min} \end{aligned}$$

$$\begin{aligned} W_{p2} &= 4W_{p1} \\ W_{n2} &= 4W_{n1} \\ W_{p3} &= 4W_{p2} \\ W_{n3} &= 4W_{n2} \end{aligned}$$

The propagation delay of the inverter chain can be calculated as in the following:

$$\begin{aligned} t_{pd} &\leq 3 \cdot \frac{1}{2}(t_{pLH} + t_{pHL}) = 3 \cdot \frac{1}{2}(R_p + R_n) \cdot 4C_{in} \\ &= 3 \cdot \frac{1}{2}(22 \text{ k}\Omega \cdot \frac{2}{6} + 8 \text{ k}\Omega \cdot \frac{2}{3}) \cdot 8 \text{ fF} = 152 \text{ ps} \end{aligned}$$

which is much shorter than the propagation delay of the single-stage buffer. Of course, the shorter propagation delay is obtained at the expense of more silicon area needed for the super-buffer, an example of trading off between area and performance (namely, timing).

■

5.3.2.2 Path-Delay Optimization After studying the optimization problem of an inverter chain, we are in a position to generalize the inverter chain to a logic chain composed of n generic logic gates. To this end, we first consider in the following the intrinsic delays of both NAND and NOR gates and then cope with the general path-delay optimization problem.

Intrinsic delays τ_{nor} and τ_{nand} of NAND and NOR gates can be defined in the same way as that of inverters we have studied previously. Referring to Figure 1.34, both intrinsic propagation delays can be expressed as follows:

$$\tau_{nand} = R_{eqv}C_{in} = R_{eqn}\left(\frac{L_n}{W_n}\right)(4W_nC_g) = 4R_{eqn}C_gL_n \tag{5.59}$$

and

$$\tau_{nor} = R_{eqv}C_{in} = R_{eqn}\left(\frac{L_n}{W_n}\right)(5W_nC_g) = 5R_{eqn}C_gL_n \tag{5.60}$$

where C_{in} is the input capacitance of the gate under consideration. The value of C_{in} is $4W_nC_g$ for a two-input NAND gate and is $5W_nC_g$ for a two-input NOR gate.

By generalizing Equation (5.52) to logic gates other than inverters, the total propagation delay of the logic chain shown in Figure 5.32 can be expressed as the sum of the propagation delay of each individual gate. That is,

$$t_{pd(total)} = \sum_{i}^{n} \tau_i \left(\frac{C_{i+1}}{C_i} + \gamma_i\right) \tag{5.61}$$

where τ_i denotes the intrinsic delay of gate i and γ_i is the ratio of self-loading capacitance to input capacitance of gate i.

Now, we are ready to deal with the general path-delay optimization problem. From the discussion of the inverter chain, we have realized that to optimize the propagation delay of the inverter chain all stages in the inverter chain must have the same effort delay. To explore how this result can also be applied to a logic chain, refer to the logic chain shown in Figure 5.32 and consider the two middle stages, the ith and $(i+1)$th stages. The total propagation delay of these two stages can be expressed as follows:

$$t_{i \to i+1} = \tau_{nor}\left(\frac{C_{i+1}}{C_i} + \gamma_{nor}\right) + \tau_{nand}\left(\frac{C_{i+2}}{C_{i+1}} + \gamma_{nand}\right) \tag{5.62}$$

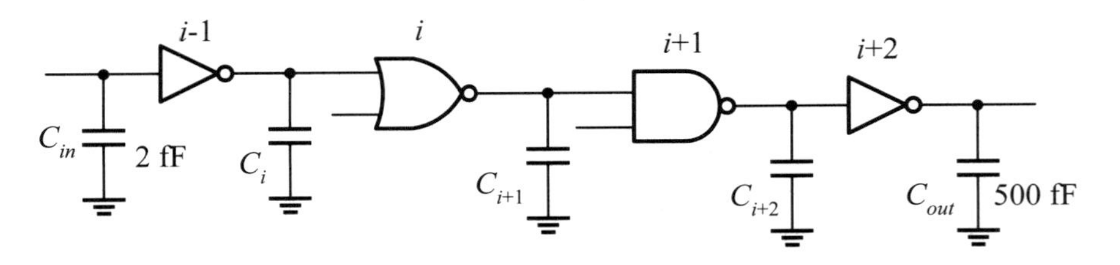

Figure 5.32: An example of path-delay optimization with generic logic gates.

To find the optimal propagation delay, taking the derivative of $t_{i\rightarrow i+1}$ relative to C_{i+1}, we obtain

$$\frac{\partial t_{i\rightarrow i+1}}{\partial C_{i+1}} = \tau_{nor}\left(\frac{1}{C_i}\right) - \tau_{nand}\left(\frac{C_{i+2}}{C_{i+1}^2}\right) = 0 \tag{5.63}$$

As a consequence, we have the following relation:

$$\tau_{nor}\left(\frac{C_{i+1}}{C_i}\right) = \tau_{nand}\left(\frac{C_{i+2}}{C_{i+1}}\right) \tag{5.64}$$

which implies that the effort delay of these two stages must be the same; that is,

$$\tau_{nor}FO_i = \tau_{nand}FO_{i+1} \tag{5.65}$$

where FO_i and FO_{i+1} are the fan-outs of the ith and $(i+1)$th stages, respectively. One more time, we reconfirm that, to optimize the path delay, all stages along the logic path must have the same effort delay. The following example illustrates how to use this result to solve the path-delay optimization problem.

■ **Example 5-15: (An example of the path-delay optimization.)** Referring to Figure 5.32, find the sizes of all logic gates when the optimal path delay is required.

Solution: To optimize the path delay, all stages along the path must have the same effort delay. Hence, we obtain the following relation:

$$\tau_{inv}\left(\frac{C_i}{C_{in}}\right) = \tau_{nor}\left(\frac{C_{i+1}}{C_i}\right) = \tau_{nand}\left(\frac{C_{i+2}}{C_{i+1}}\right) = \tau_{inv}\left(\frac{C_{out}}{C_{i+2}}\right)$$

The optimal stage effort delay is then equal to their geometrical mean.

$$t_{fan-out} = \sqrt[4]{\tau_{inv}\left(\frac{C_i}{C_{in}}\right) \times \tau_{nor}\left(\frac{C_{i+1}}{C_i}\right) \times \tau_{nand}\left(\frac{C_{i+2}}{C_{i+1}}\right) \times \tau_{inv}\left(\frac{C_{out}}{C_{i+2}}\right)}$$

By using $\tau_{inv} = 3R_{eqn}C_gL_n$, $\tau_{nand} = 4R_{eqn}C_gL_n$, and $\tau_{nor} = 5R_{eqn}C_gL_n$, we have the optimal stage effort delay as follows:

$$\begin{aligned} t_{fan-out} &= \sqrt[4]{\tau_{inv} \times \tau_{nor} \times \tau_{nand} \times \tau_{inv} \times \left(\frac{C_{out}}{C_{in}}\right)} \\ &= \sqrt[4]{3 \times 5 \times 4 \times 3 \times \left(\frac{500}{2}\right)} R_{eqn}C_gL_n \\ &= 14.56R_{eqn}C_gL_n \end{aligned}$$

Once we obtain the optimal stage effort delay, we may determine the size of each stage by equating the delay of that stage to the optimal stage effort delay. To do this, we proceed from the last stage backward to the first one.

$$\tau_{inv}\left(\frac{C_{out}}{C_{i+2}}\right) = 3R_{eqn}C_gL_n\left(\frac{500}{C_{i+2}}\right) = 14.56R_{eqn}C_gL_n$$

Hence, the input capacitance of the last inverter is $C_{i+2} = 103$ fF.

The input capacitance of the NAND gate can then be determined by the C_{i+2} and the optimal stage effort delay.

$$\tau_{nand}\left(\frac{C_{i+2}}{C_{i+1}}\right) = 4R_{eqn}C_gL_n\left(\frac{103}{C_{i+1}}\right) = 14.56R_{eqn}C_gL_n$$

Hence, the input capacitance of the NAND gate is $C_{i+1} = 28.30$ fF.

Similarly, the input capacitance of the NOR gate is calculated as follows:

$$\tau_{nor}\left(\frac{C_{i+1}}{C_i}\right) = 5R_{eqn}C_gL_n\left(\frac{28.30}{C_i}\right) = 14.56R_{eqn}C_gL_n$$

The result is $C_i = 9.72$ fF.

Finally, the input capacitance of the first inverter is found to be

$$\tau_{inv}\left(\frac{C_i}{C_{in}}\right) = 3R_{eqn}C_gL_n\left(\frac{9.72}{C_{in}}\right) = 14.56R_{eqn}C_gL_n$$

which is $C_{in} = 2$ fF, exactly as expected.

The actual sizes of each stage can then be determined by its input capacitance using the following equation:

$$C_{in} = C_G = C_g\left(W_p + W_n\right)$$

Using the value listed in Table 5.5, the channel widths of all MOS transistors of each gate can be easily determined. The details are left to the reader as an exercise. ■

■ Review Questions

Q5-38. What role does the effort delay play in the path-delay optimization?

Q5-39. What is the condition for an optimized inverter chain?

Q5-40. What is the ratio of the sizes of two successive stages in a super-buffer?

Q5-41. What is the condition for an optimized logic chain?

5.3.3 Logical Effort and Path-Delay Optimization

The method of *logical effort* (LE) has been proposed as an alternative to solve the path-delay optimization problem. It has the following features. First, it simplifies the propagation delay along a logic path by normalizing it relative to the intrinsic delay of inverter, τ_{inv}. Second, it provides more insights into the key factors in the path delay of a logic chain. Third, it allows designers to quickly determine the optimal propagation delay of a given logic path without knowing in advance the actual sizes of logic gates in the path. Fourth, it provides a method to compare the performance of the design of a logic structure across different CMOS processes.

Table 5.8: Logical effort and parasitic delay of gates in various processes.

	0.35 μm	0.25 μm	0.18 μm	0.13 μm	90 nm	65 nm	45 nm	32 nm
				Logical effort				
Inverter	1.00	1.00	1.00	1.00	1.00	1.00	1.00	1.00
NAND2	1.33	1.21	1.33	1.18	1.19	1.19	1.22	1.26
NAND3	1.64	1.43	1.65	1.37	1.41	1.41	1.45	1.53
NAND4	1.95	1.66	1.96	1.59	1.62	1.61	1.68	1.79
NOR2	1.74	1.65	1.60	1.64	1.55	1.75	1.82	2.01
NOR3	2.49	2.31	2.20	2.26	2.10	2.46	2.65	2.99
NOR4	3.08	2.92	2.76	2.75	2.62	3.09	3.35	3.80
				Parasitic delay				
Inverter	1.88	1.53	1.06	1.02	1.28	2.52	2.86	3.45
NAND2	3.19	2.51	1.90	1.63	2.01	3.89	4.42	5.21
NAND3	4.41	3.48	2.74	2.23	2.74	5.24	5.96	6.97
NAND4	5.63	4.39	3.56	2.84	3.48	6.56	7.47	8.71
NOR2	4.68	3.76	2.70	2.83	3.04	6.73	7.77	9.33
NOR3	8.51	8.69	4.80	5.25	5.42	11.7	13.45	16.38
NOR4	9.86	7.90	5.50	5.64	6.16	13.54	15.78	19.33

5.3.3.1 Definition of Logical Effort The essential idea of the method of logical effort is that it normalizes the propagation delay of a gate relative to that of the basic inverter gate. More specifically, the logical effort of a gate is defined as the ratio of the intrinsic propagation delay of the gate to that of the basic inverter. Based on this, by referring to Figure 1.34, the logical effort values of the three basic gates can be calculated as follows:

$$g_{inv} = \frac{\tau_{inv}}{\tau_{inv}} = \frac{3R_{eqn}C_gL_n}{3R_{eqn}C_gL_n} = 1 \tag{5.66}$$

$$g_{nand} = \frac{\tau_{nand}}{\tau_{inv}} = \frac{4R_{eqn}C_gL_n}{3R_{eqn}C_gL_n} = \frac{4}{3} \tag{5.67}$$

$$g_{nor} = \frac{\tau_{nor}}{\tau_{inv}} = \frac{5R_{eqn}C_gL_n}{3R_{eqn}C_gL_n} = \frac{5}{3} \tag{5.68}$$

Therefore, the NOR gate has a higher logical effort than the NAND gate. By and large, NAND gates have a better performance than NOR gates in terms of the output driving capability and the input loading. It should be noted that the above logical effort values are specific to the circuits given in Figure 1.34. Different configurations of a logic circuit may have different logical effort values. Average logical effort and parasitic delay of the three basic gates in a variety of processes, from 0.35 μm to 32 nm, are compared and summarized in Table 5.8.

5.3.3.2 Path-Delay Optimization To illustrate how the method of logical effort can be applied to solve the path-delay optimization problem, let us normalize Equation (5.61) with respect to the intrinsic delay of the inverter τ_{inv}. This results in the following equation:

$$\frac{t_{pd(total)}}{\tau_{inv}} = t_D = \sum_i^n \frac{\tau_i}{\tau_{inv}} \left(\frac{C_{i+1}}{C_i} + \gamma_i \right) \tag{5.69}$$

The normalized path delay t_D is found to be

$$t_D = \sum_{i}^{n}(g_i \times h_i + p_i) = t_{Deffort} + t_{Dpara} \tag{5.70}$$

where g_i is the logical effort of gate i and is defined as τ_i/τ_{inv}, h_i is the *electrical effort* of gate i and is defined as C_{i+1}/C_i, and p_i is the parasitic term and is defined as $g_i \times \gamma_i$. The terms $t_{Deffort}$ and t_{Dpara} are called the *normalized path effort delay* and *normalized path parasitic delay*, respectively. The logical effort g captures the properties of the logic gate itself while the electrical effort h captures the features of the load. The electrical effort h_i is distinguished from fan-out in that it only takes into account the on-path output capacitance (C_{i+1}) of gate i while fan-out (FO_i) also incorporates the off-path capacitance.

From Equation (5.69), the parasitic term p_i of a specific gate can be expressed as

$$p_i = g_i \times \gamma_i = g_i \times \frac{C_{self-load(i)}}{C_{in(i)}} \tag{5.71}$$

The parasitic term (p_i) strongly depends on process technology, gate type, and layout as well. Hence, for different processes, this term should have to be recomputed for all gates. The value of p_i depends on the junction and gate capacitances. Maybe the best way to accurately compute the value is through circuit simulations. See Table 5.8 for more details.

Recall that to optimize the path delay of a logic chain, the effort delay $t_{Deffort}$ $(g_i \times h_i)$ of each stage along the logic chain must be the same. Based on this, the normalized stage effort delay t_{Dstage} of a logic chain with n gates can be expressed as the geometric mean of effort delays of all stages. That is,

$$\begin{aligned} t_{Dstage} &= \sqrt[n]{\Pi_{i=1}^{n} g_i h_i} \\ &= \sqrt[n]{(\Pi_{i=1}^{n} g_i)(\Pi_{i=1}^{n} h_i)} = \sqrt[n]{GH} \\ &= \sqrt[n]{G\left(\frac{C_{out}}{C_{in}}\right)} \end{aligned} \tag{5.72}$$

where G is the product of logical effort g_i of all logic gates along the logic path of interest and is called the *path logical effort*. H is called the *path electrical effort* and is defined to be the product of electrical effort h_i of all logic gates along the logic path. The product of the path logical effort and path electrical effort is referred to as the *path effort*, often denoted as $F = GH$.

We give an example to illustrate how this principle can be applied to a practical logic circuit.

■ Example 5-16: (Path-delay optimization using the method of logical effort.) Design the optimal size of each gate in the logic chain shown in Figure 5.33 when the loading capacitance C_{out} is 200 fF. The input capacitance of the inverter is assumed to be 2 fF. The parasitic terms p_i of inverter, NAND, and NOR gates are assumed to be 1.0, 1.6, and 2.8, respectively.

Solution: Since there are three gates in the logic chain, the normalized stage effort delay is computed as follows:

$$t_{Dstage} = \sqrt[3]{g_{inv} \times g_{nor} \times g_{nand} \times \left(\frac{C_L}{C_{in}}\right)}$$

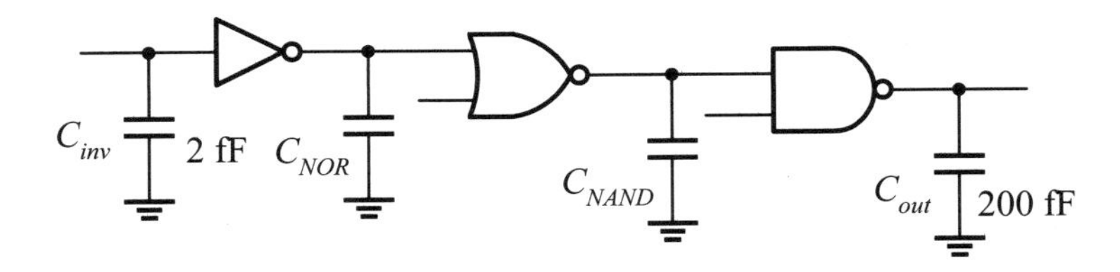

Figure 5.33: An example of path-delay optimization using the method of logical effort.

$$= \sqrt[3]{1 \times \frac{5}{3} \times \frac{4}{3} \times \frac{200}{2}} = 6.06$$

The normalized path delay t_D is three times this value plus the parasitic term for each gate.

$$\begin{aligned} t_D &= n \times t_{Dstage} + (p_{inv} + p_{nor} + p_{nand}) \\ &= 3 \times 6.06 + (1.0 + 1.6 + 2.8) = 23.58 \end{aligned}$$

The actual path delay can then be found by incorporating the intrinsic delay of the inverter into it. For example, $\tau_{inv} = 13$ ps in a 0.13-μm process, hence the actual path delay is

$$t_d = \tau_{inv} \times t_D = 13 \text{ ps} \times 23.58 = 306.54 \text{ ps}$$

The gate size of each stage in terms of capacitance can be computed using the normalized stage effort delay. To do this, we proceed from the last stage backward to the first one. The input capacitance of the NAND gate can be calculated as follows:

$$g_{nand}\left(\frac{C_L}{C_{nand}}\right) = 6.06 \Rightarrow C_{nand} = g_{nand}\left(\frac{C_L}{6.06}\right) = \frac{4}{3}\left(\frac{200 \text{ fF}}{6.06}\right) = 44.00 \text{ fF}$$

This capacitance is then employed to find the input capacitance of the NOR gate and is equal to

$$g_{nor}\left(\frac{C_{nand}}{C_{nor}}\right) = 6.06 \Rightarrow C_{nor} = g_{nor}\left(\frac{C_{nand}}{6.06}\right) = \frac{5}{3}\left(\frac{44.00 \text{ fF}}{6.06}\right) = 12.10 \text{ fF}$$

This capacitance is used to find the input capacitance of the inverter and is found to be

$$g_{inv}\left(\frac{C_{nor}}{C_{in}}\right) = 6.06 \Rightarrow C_{in} = g_{inv}\left(\frac{C_{nor}}{6.06}\right) = 1 \times \left(\frac{12.10 \text{ fF}}{6.06}\right) = 2.0 \text{ fF}$$

As before, once the input capacitance of a gate is known, the gate size can then be easily determined.

■

The above example does not give an optimal path delay because, when one extra inverter is added, the normalized stage effort reduces to 3.86 and the normalized path delay is 21.84, rather than 23.58. To obtain the best number of stages in a logic chain to minimize the total path delay, the FO4 rule can be applied. This means that the normalized stage effort delay should be chosen to be around 4. Based on this, the optimal number of stages in a specific logic chain can then be determined as follows:

$$n = \lceil \log_4 (\Pi_{i=1}^{n} g_i h_i) \rceil = \lceil \log_4 GH \rceil = \lceil \log_4 F \rceil \tag{5.73}$$

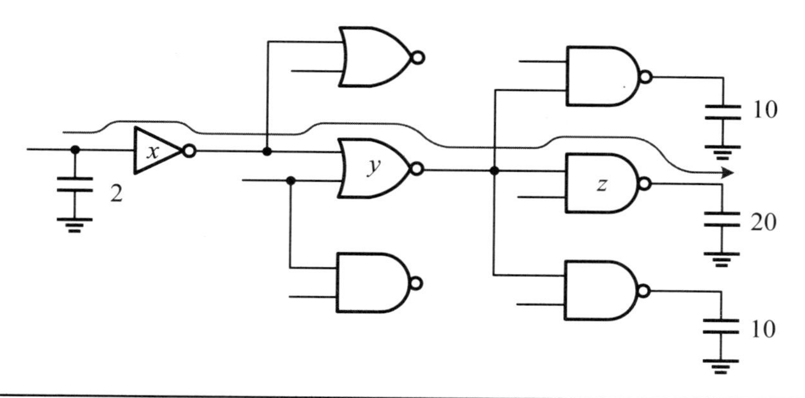

Figure 5.34: An example of illustrating the branching effort.

When the optimal number of stages is greater than that of an existing specific logic chain, a proper number of buffer stages need to be added to minimize the total path delay.

From the preceding example, we can see that the optimal path delay can be determined before sizing the gates by using the method of logical effort. Once the target delay is achieved, gate sizing can then be carried out in a specific process, namely, by first finding the input capacitance of a gate and then determining the size of the gate.

5.3.3.3 Branching Effort Up to now, we are concerned with the situation where each gate only drives another gate. In practice, it is very common that the output of a gate drives many gates at the same time. An example is illustrated in Figure 5.34, where the first inverter x drives two NOR gates and the NOR gate y drives three NAND gates in turn. The gates not on the logic path of interest but need to be driven also affect the path delay. The effect of these off-path gates on the path delay is referred to as the *branching effect*.

In order to capture the branching effects on the path effort, a new factor called a *branching effort* (b) of a node is defined. The branching effort (b) at the output of a logic gate is defined as

$$b = \frac{C_{on-path} + C_{off-path}}{C_{on-path}} = \frac{C_{total}}{C_{useful}} \tag{5.74}$$

where $C_{on-path}$ is the load capacitance along the path of interest and $C_{off-path}$ is the capacitance of connections that lead off the path. From the definition of b, the branching effort b of a gate is 1 if its output does not branch.

By incorporating into branching effort b, the normalized path effort delay $t_{Deffort}$ can be generalized as follows:

$$t_{Deffort} = \sum_{i}^{n}(g_i \times h_i \times b_i) \tag{5.75}$$

which is the sum of a normalized effort delay of each stage along the logic path of interest. Similarly, the normalized stage effort delay is generalized as in the following:

$$\begin{aligned} t_{Dstage} &= \sqrt[n]{\Pi_{i=1}^{n} g_i h_i b_i} \\ &= \sqrt[n]{GHB} \end{aligned}$$

$$= \sqrt[n]{GB\left(\frac{C_{out}}{C_{in}}\right)} \tag{5.76}$$

where B is the product of branching effort b_i of all logic gates along the logic path and is called the *path branching effort*. By incorporating path branching effort B, the path effort becomes $F = GHB$.

It is interesting to note that the product of electrical effort h_i and branching effort b_{i+1} at the output of gate i is equal to the fan-out of the gate i; that is,

$$\begin{aligned} h_i \cdot b_{i+1} &= \frac{C_{i+1}}{C_i}\frac{C_{i+1} + C_{off-path}}{C_{i+1}} \\ &= \frac{C_{outi}}{C_i} = FO_i \end{aligned} \tag{5.77}$$

where C_{outi} is the loading capacitance of gate i and equals the sum of the on-path and off-path capacitances, that is, $C_{i+1} + C_{off-path}$.

■ **Example 5-17: (An example of the branching effect.)** Considering the circuit shown in Figure 5.34 and using $C_g = 2$ fF/μm, design the sizes of gates y and z to minimize the propagation delay in the highlighted path.

Solution: The branching effort b of the inverter is 2 and of the NOR gate is 3. Hence, the path branching effort is 6. The path logical effort is $1 \times 5/3 \times 4/3 = 20/9$, and the $C_{out}/C_{in} = 20/2 = 10$. Hence, from Equation (5.76), the normalized stage effort delay is

$$\begin{aligned} t_{Dstage} &= \sqrt[n]{GB\left(\frac{C_{out}}{C_{in}}\right)} \\ &= \sqrt[3]{\frac{20}{9} \times 6 \times 10} = 5.11 \end{aligned}$$

The normalized path delay is

$$t_D = n \times t_{Dstage} + \sum P = 3 \times 5.11 + (1.0 + 1.6 + 2.8) = 20.73$$

The gate sizes of y and z can then be calculated by equating the normalized effort delay of the gate to the normalized stage effort delay. We begin with the consideration of gate z.

$$g_{nand} \times b_{nand} \times \frac{C_L}{C_{nand}} = t_{Dstage} \Rightarrow \frac{4}{3} \times 1 \times \frac{20}{C_{nand}} = 5.11$$

Hence,

$$C_{nand} = 5.22 \text{ fF}$$

and the size of the NAND gate is found to be

$$W_{nand} = \frac{C_{nand}}{4 \times C_g} = \frac{5.22}{4 \times 2} = 0.65 \ \mu\text{m}$$

From this, we may obtain $W_p = W_n = 2W_{nand} = 1.3$ μm.

Next, we deal with the size of NOR gate y. Using the same equation and the normalized effort delay, we have

$$g_{nor} \times b_{nor} \times \frac{C_{nand}}{C_{nor}} = t_{Dstage} \Rightarrow \frac{5}{3} \times 3 \times \frac{5.22}{C_{nor}} = 5.11$$

where the branching effort of the NOR gate is 3. The input capacitance of the NOR gate equals

$$C_{nor} = 5.11 \text{ fF}$$

Consequently, the size of the NOR gate is

$$W_{nor} = \frac{C_{nor}}{5 \times C_g} = \frac{5.11}{5 \times 2} = 0.51 \ \mu\text{m}$$

From this, we may obtain $W_p = 4W_{nor} = 2.04$ μm and $W_n = W_{nor} = 0.51$ μm. ■

■ Review Questions

Q5-42. What are the features of the method of logical effort?

Q5-43. Define the logical effort of a logic gate.

Q5-44. How would you relate the fan-out, electrical effort, and branching effort?

Q5-45. How would you determine the optimal number of stages in a logic chain?

Q5-46. How would you determine the optimal normalized stage effort delay in a logic chain?

5.4 Summary

In this chapter, we introduced the parasitic resistance and capacitances of MOS transistors and their impact on the performance of digital circuits. The resistance alone can cause the MOS transistor to consume power. Combining resistance with capacitance produces the RC delay of desired signal passing through the MOS transistor.

To calculate the propagation delay of a logic circuit, three widely used approaches are average-current approach, equivalent-RC approach, and differential-current approach. For the ease of calculating propagation delays of a module, a linear delay model, referred to as a cell delay model, is often used. In the cell delay model, the propagation delay of a cell comprises the parasitic delay and (logical) effort delay. The parasitic delay concerns only the properties of the gate itself, while the effort delay involves the feature of loading capacitance. In addition, the Elmore delay model is also widely used to calculate the propagation delay of an RC-ladder and an RC-tree network.

The path-delay optimization problem states how to optimize a logic chain to minimize the total propagation delay along the logic chain. For this purpose, we first addressed the principle of path-delay optimization, including inverter chains (superbuffers) and logic chains of generic logic gates. Then, we defined logical effort and used it to optimize path delays.

References

1. A. I. Abou-Seido, B. Nowak, and C. Chu, "Fitted Elmore delay: a simple and accurate interconnect delay model," *IEEE Trans. on Very Large Scale Integration (VLSI) Systems,* Vol. 12, No. 7, pp. 691–696, July 2004.
2. R. J. Baker, *CMOS Circuit Design, Layout, and Simulation.* New York: John Wiley & Sons, Inc., 2005.
3. H. Bakoglu, *Circuits, Interconnections, and Packaging for VLSI.* Reading, MA: Addison-Wesley, 1990.
4. W. Elmore, "The transient response of damped linear networks with particular regard to wideband amplifiers," *Journal of Applied Physics,* Vol. 19, No. 1, pp. 55–63, January 1948.
5. P. R. Gray, P. J. Hurst, S. H. Lewis, and R. G. Meyer, *Analysis and Design of Analog Integrated Circuits,* 4th ed. New York: John Wiley & Sons, 2001.
6. R. Ho, K. Mai, and M. Horowitz, "The future of wires," *Proc. IEEE,* Vol. 89, No. 4, pp. 490–504, April 2001.
7. D. A. Hodges, H. G. Jackson, and R. A. Saleh, *Analysis and Design of Digital Integrated Circuits: In Deep Submicron Technology,* 3rd ed. New York: McGraw-Hill Books, 2004.
8. M. Hrishikesh et al., "The optimal logic depth per pipeline stage is 6 to 8 FO4 inverter delays," *Proc. of the 29th Annual Int'l Symposium on Computer Architecture (ISCA.02),* pp. 14–24, 2002.
9. Y. I. Ismail, E. G. Friedman, and J. L. Neves, "Equivalent Elmore delay for *RLC* trees," *IEEE Trans. on Computer-Aided Design of Integrated Circuits and systems,* Vol. 19, No. 1, pp. 83–97, January 2000.
10. B. E. Keiser, *Principles of Electromagnetic Compatibility.* Dedham, MA: Artech House, 1979.
11. M. B. Lin, *Digital System Design: Principles, Practices, and Applications,* 4th ed. Taipei, Taiwan: Chuan Hwa Book Ltd., 2010.
12. J. Lohstroh, "Static and dynamic noise margins of logic circuits," *IEEE J. of Solid-State Circuits,* Vol. 14, No. 3, pp. 591–598, June 1979.
13. J. Lohstroh, E. Seevinck, and J. De Groot "Worst-case static noise margin criteria for logic circuits and their mathematical equivalence," *IEEE J. of Solid-State Circuits,* Vol. 18, No. 6, pp. 803–807, December 1983.
14. A. M. Mohsen and C. A. Mead, "Delay-time optimization for driving and sensing of signals on high-capacitance paths of VLSI systems," *IEEE J. of Solid-State Circuits,* Vol. 14, No. 2, pp. 462–470, April 1979.
15. Predictive Technology Model (PTM) Web site: http://ptm.asu.edu/.
16. J. M. Rabaey, A. Chandrakasan, and B. Nikolić, *Digital Integrated Circuits: A Design Perspective,* 2nd ed. Upper Saddle River, NJ: Pearson Education, 2003.
17. J. Rubinstein, P. Penfield, Jr., and M. Horowitz, "Signal delay in *RC* tree networks," *IEEE Trans. on Computer-Aided Design,* Vol. 2, No. 3, pp. 202–211, July 1983.
18. I. Sutherland, B. Sproull, and D. Harris, *Logical Effort: Designing Fast CMOS Circuits.* San Francisco, CA: Morgan Kaufmann, 1999.

19. The MOSIS Service Web site: http://www.mosis.com/.

20. J. P. Uyemura, *Introduction to VLSI Circuits and Systems.* New York: John Wiley & Sons, 2002.

21. Q. Zhou and K. Mohanram, "Elmore model for energy estimation in *RC* trees," *2006 ACM/IEEE Design Automation Conf.*, pp. 965–970, July 2006.

Problems

5-1. Referring to Figure 5.1(b) and assuming that the gate-oxide thickness is 2.5 nm in a 0.13-μm process, calculate the gate capacitance of the MOS transistor.

5-2. Show that the gate-to-source capacitance C_{gs} of a MOS transistor in the saturation mode of operation can be expressed as

$$C_{gs} = \frac{2}{3} W L C_{ox}$$

5-3. Calculate the unity-gain frequency of each process specified in the following:

(a) Use 0.18-μm parameters and set $V_{GS} = 1.2$ V and $L = 0.4$ μm.

(b) Use 0.13-μm parameters and set $V_{GS} = 0.8$ V and $L = 0.2$ μm.

5-4. Assume that $V_{DD} = 1.8$ V, load capacitance $C_L = 0.2$ pF, $L_n = L_p = 2\lambda$, $W_n = 10\lambda$, and $W_p = 22\lambda$. Using the average-current approach and 0.18-μm parameters, $k'_n = 96$ μA/V^2, $k'_p = 48$ μA/V^2, and $V_{T0n} = |V_{T0p}| = 0.4$ V, find t_{pHL} and t_{pLH}.

5-5. Assume that $V_{DD} = 1.8$ V and load capacitance $C_L = 0.5$ pF. If $L_n = L_p = 2\lambda$, use 0.18-μm parameters and equivalent-RC approach to determine both W_n and W_p such that t_{pHL} and t_{pLH} are both equal to 200 ps.

5-6. Consider the circuit shown in Figure 5.10(a). Calculate the rise and fall times of the output voltage if the load capacitance is 5 pF and $(W_p/L_p) = 2(W_n/L_n)$, where $L_p = L_n = 2\lambda$ and $W_n = 4\lambda$.

5-7. Show that the Equation (5.39) is valid.

5-8. Show that the Equation (5.40) is valid.

5-9. Assume that $V_{DD} = 1.8$ V, load capacitance $C_L = 0.5$ pF, $L_n = L_p = 2\lambda$, $W_n = 20\lambda$, and $W_p = 48\lambda$. Using the differential-current approach and 0.18-μm parameters, $k'_n = 96$ μA/V^2, $k'_p = 48$ μA/V^2, and $V_{T0n} = |V_{T0p}| = 0.4$ V, find t_{pHL} and t_{pLH}.

5-10. Referring to the circuit shown in Figure 5.35, show that the effects of C_{DBp} and C_{DBn} is equivalent to two capacitors connected in parallel.

5-11. Consider the two networks shown in Figure 5.36:

(a) Show that $Z_1 = Z/(1 - A)$ and $Z_2 = Z/(1 - 1/A)$, where $A = V_2/V_1$, if Figure 5.36(a) can be converted into Figure 5.36(b).

(b) Show that $C_1 = C_2 = 2C$ if a capacitor with capacitance C is used as the component Z and $A = -1$.

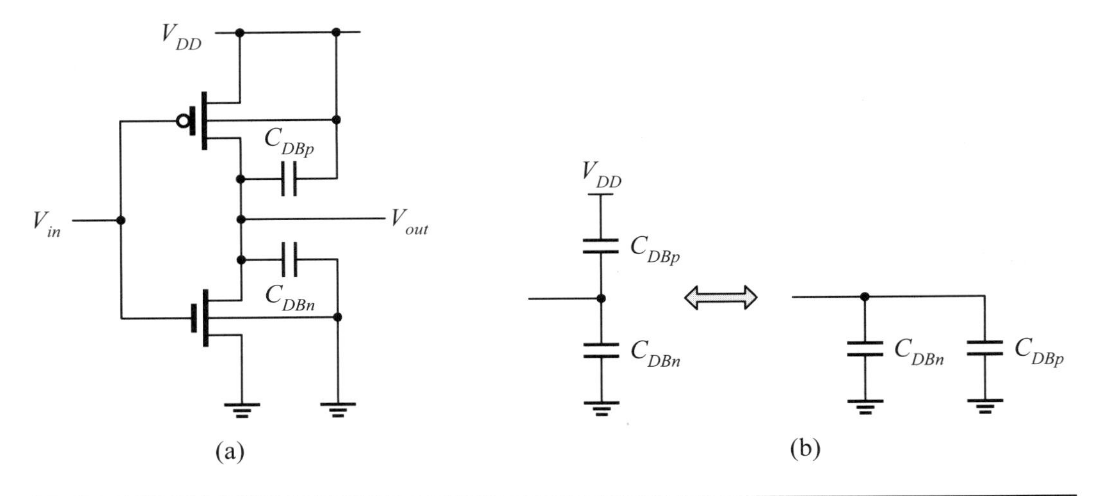

Figure 5.35: An illustration of the self-loading capacitance in CMOS inverters: (a) logic circuit showing C_{DBp} and C_{DBn}; (b) C_{DBp} and C_{DBn}.

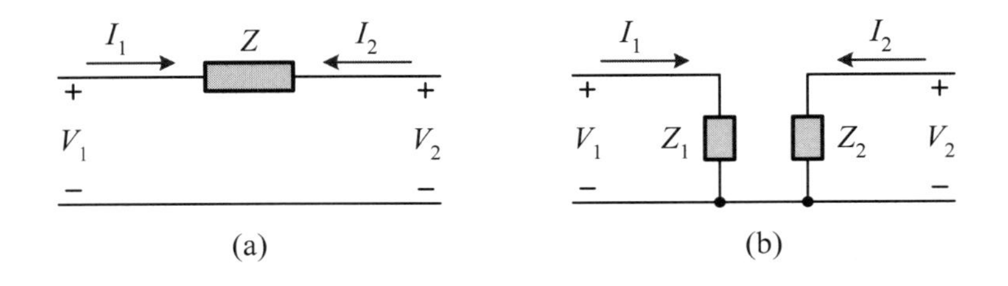

Figure 5.36: Miller effect: (a) original circuit; (b) equivalent circuit after applying the Miller effect.

5-12. Using the parameter values of the 0.13-μm process given in Table 5.6, calculate C_j, C_{ol}, and C_{para}.

5-13. Using the data of the 0.18-μm process listed in Table 5.7, calculate the allowable operating frequencies of the design based on full-custom, cell-based, and high-performance cell-based approaches.

5-14. Using the data of the 0.13-μm process listed in Table 5.7, calculate the allowable operating frequencies of the design based on full-custom, cell-based, and high-performance cell-based approaches.

5-15. Consider the logic circuit shown in Figure 5.37:

(a) What is the function of nMOS transistors, M_{n1} and M_{n2}?

(b) Supposing that both inverters are basic inverters and both nMOS transistors, M_{n1} and M_{n2}, are feature-size transistors, use HSPICE to estimate the propagation delay of the circuit.

(c) What will be the results if we double the channel widths of both nMOS transistors, M_{n1} and M_{n2}.

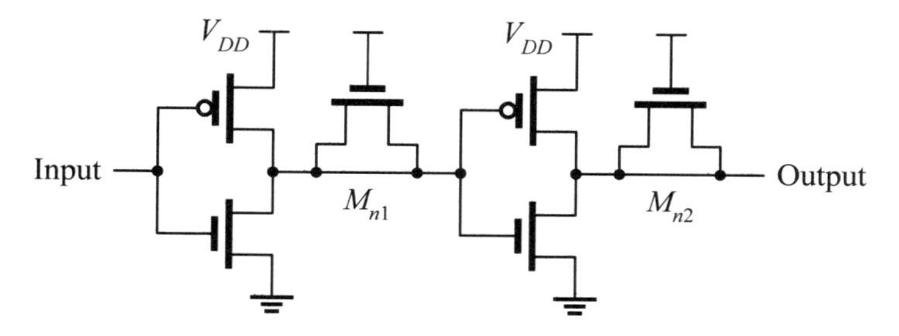

Figure 5.37: A buffer delay element.

5-16. Show that, in the inverter chain, all stages have the same propagation delay if each successive inverter has its size a times its previous stage.

5-17. Assume that a super-buffer is used to drive a load capacitor with capacitance of 2 pF. Using that $R_{eqn} = 8$ kΩ, $R_{eqp} = 22$ kΩ, and $C_{in} = 2$ fF, design the super-buffer to optimize the propagation delay.

(a) Determine the optimal number of stages.

(b) Determine the size of each stage.

(c) Calculate the buffered propagation delay.

5-18. Assume that a 5-pF loading capacitor needs to be driven. The propagation delay of the output stage should be less than 0.5 ns. Using that $R_{eqn} =$ 8 kΩ, $R_{eqp} = 22$ kΩ, and $C_{in} = 2$ fF, design a super-buffer to optimize the propagation delay.

5-19. Referring to Figure 5.32, replace the final-stage inverter with a NOR gate and the loading capacitance with 1 pF. Find the sizes in terms of input capacitance of all logic gates when the optimal path delay is required.

5-20. Referring to Figure 5.32 and using $C_g = 2$ fF/μm, calculate the sizes of all gates.

5-21. Using the method of logical effort, design the optimal size of each gate in the logic chain shown in Figure 5.32 when the loading capacitance C_{out} is 600 fF. The input capacitance of the inverter is assumed to be 2 fF.

5-22. (A study of inverter circuits)

(a) Design an inverter with the following circuit parameters: $L_p = L_n = L_{min}$ and $W_p = W_n = \frac{3}{2}L_{min}$. Construct a layout.

(b) Cascade three copies of the layout of the above inverter and then extract the RC parameters from the resulting layout.

(c) Apply an ideal square wave (that is, both rise and fall times are 0 ns) with the 50% duty cycle of an appropriate frequency to the input of the first stage and measure the output of the second stage.

(d) Repeat above steps with the following circuit parameters: $W_n = \frac{3}{2}L_{min}$ and $W_p = 3L_{min}$.

(e) Compare both results and give your comments.

5-23. (A study of path delay)

(a) Draw the schematic diagram of a three-stage logic circuit with a load capacitor of 0.2 pF. The output of the two-input NOR gate is connected to an inverter and the output of the inverter is in turn connected to one input of a two-input NAND gate. The output of the NAND gate drives the load capacitor. Suppose that all transistors use L_{min} and $W_p = 3 \times L_{min}$, $W_n = \frac{3}{2} \times L_{min}$. The unused inputs of the NAND and NOR gates are connected to V_{DD} and ground, respectively.

(b) Apply an ideal square wave (that is, both rise and fall times are 0 ns) of an appropriate frequency to the input of the first stage, and measure both t_{pLH} and t_{pHL} of the output of the last stage.

(c) (FO4 rule) Scale up the W of both pMOS and nMOS transistors for the second and last stages by factors of 4 and 16, respectively, and repeat the above step.

(d) Compare both results and give your comments.

6

Power Dissipation and Low-Power Designs

The power dissipation of complementary metal-oxide-semiconductor (CMOS) logic circuits comprises static and dynamic components. Static power dissipation is due to the leakage current of reverse-biased junctions, gate leakage current, and subthreshold current or other currents continuously drawn from the power supply. Dynamic power dissipation is caused by a short circuit as well as the currents that charge and discharge loading capacitors.

To limit and reduce the power dissipation of a system, low-power logic design and power management are two important design issues to consider in designing a system. Low-power logic design means that some techniques are applied to reduce power dissipation of a logic circuit, while, at the same time, keeping the performance of the logic circuit unchanged. The power management technique removes unnecessary power dissipation of various modules in a system during operation. A useful power management technique needs to take into account the overall hardware and software management overhead to achieve true power reduction and meanwhile satisfy the performance requirement of applications.

6.1 Power Dissipation

In this section, we deal with the power dissipation of CMOS logic circuits. The I^2R_{on} generated on metal-oxide-semiconductor (MOS) transistors mainly contributes to power dissipation due to the activities of charging and discharging capacitors. Recall that the basic feature of a CMOS logic circuit is that each input of it corresponds to an inverter, composed of an nMOS and a pMOS transistor. Hence, we will focus our attention on the individual components of power dissipation of a static CMOS inverter.

6.1.1 Components of Power Dissipation

The cross-sectional view of a CMOS inverter is shown in Figure 6.1. In a CMOS inverter, there are four reverse-biased diodes and two MOS transistors. All of these components contribute an amount of power dissipation to the inverter. For conve-

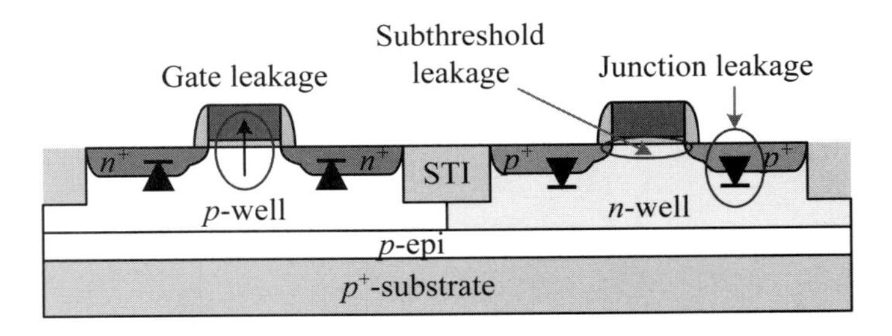

Figure 6.1: The sources of static power dissipation in a CMOS inverter.

nience, we may partition the power dissipation of an inverter into the following two major components:

- *Static power dissipation* (P_s): Static power dissipation is due to the leakage current of reverse-biased junctions, gate leakage current, and subthreshold current. Refer to Section 2.4.3 for more details about leakage currents. Based on this, the total static power dissipation can then be expressed as the product of the sum of all leakage currents and the supply voltage.

$$P_s = \sum I_{leakage} \times V_{DD} \tag{6.1}$$

- *Dynamic power dissipation* (P_d): Dynamic power dissipation is yielded by the *short-circuit* (also called the *switching transient*) *current* as well as by the currents that charge and discharge loading capacitors.

In general, the power dissipation of a CMOS inverter can be formulated as follows:

$$P_{total} = P_s + P_d = P_s + P_{d(cd)} + P_{d(sc)} \tag{6.2}$$

where P_d is the sum of $P_{d(cd)}$ and $P_{d(sc)}$. P_s, P_d, $P_{d(cd)}$, and $P_{d(sc)}$ represent static, dynamic, charging and discharging, and short-circuit power dissipation, respectively.

6.1.2 Dynamic Power Dissipation

The dynamic power dissipation includes two components: the power consumed by short-circuit current flowing through the series-connected p-type MOS (pMOS) and n-type MOS (nMOS) transistors as the inverter switches states and the power consumed by the pMOS and nMOS transistors during charging and discharging loading capacitors.

6.1.2.1 Charging/Discharging Power Dissipation In practice, the dynamic power dissipation is dominated by the power consumed by charging and discharging the loading capacitor. Figure 6.2(a) shows the circuit model used to calculate the dynamic power dissipation during charging and discharging a loading capacitor. The related timing diagram is shown in Figure 6.2(b).

Assume that the input signal is a square wave with voltage V_{in} and has a repetition frequency of $f_p = 1/T$, as shown in Figure 6.2(b). Then, the average dynamic power dissipation $P_{d(cd)}$ at both nMOS and pMOS transistors can be calculated as follows:

$$P_{d(cd)} = \frac{1}{T}\int_0^{T/2} i_n(t)V_{out}dt + \frac{1}{T}\int_{T/2}^{T} i_p(t)(V_{DD} - V_{out})dt$$

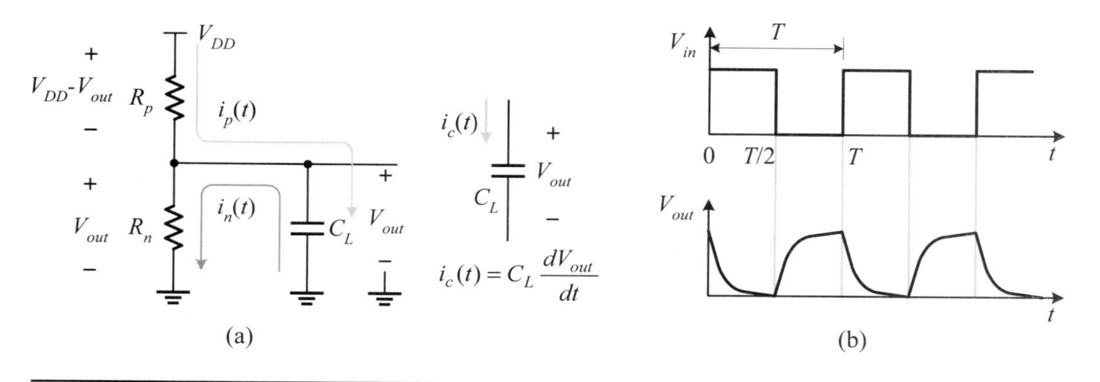

Figure 6.2: The (a) circuit model and (b) timing diagram for estimating the dynamic power dissipation of a CMOS inverter.

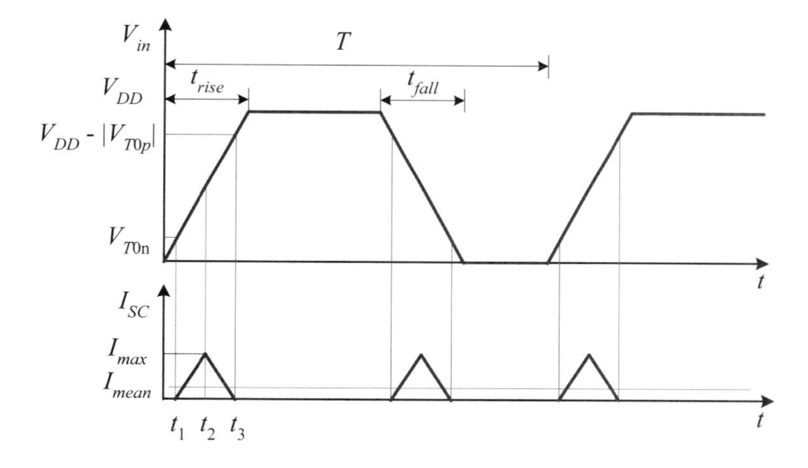

Figure 6.3: The timing diagram for estimating the short-circuit power dissipation of a CMOS inverter.

$$
\begin{aligned}
&= \frac{C_L}{T}\int_0^{V_{DD}} V_{out} dV_{out} + \frac{C_L}{T}\int_0^{V_{DD}} (V_{DD} - V_{out}) d(V_{DD} - V_{out}) \\
&= \frac{C_L}{T} V_{DD}^2 = C_L V_{DD}^2 f_p \qquad (6.3)
\end{aligned}
$$

where $i_n(t) = -C_L dV_{out}/dt$ and $i_p(t) = C_L dV_{out}/dt$. As a consequence, the power dissipation is proportional to the square of supply voltage and linearly proportional to both loading capacitance and operating frequency.

6.1.2.2 Short-Circuit Power Dissipation Referring to Figure 6.3, the following assumptions are made for simplicity: First, both the rising and falling edges of an input signal are a ramp function and have the same slope. Second, the switching transient current is also a ramp function at both edges of the input signal. Third, both rise and fall times are calculated between 0 and V_{DD} instead of between 10% and 90% of V_{DD} to simplify the calculation. Based on these, the switching transient current I_{mean} can then be approximately estimated as follows:

$$I_{mean} = 2 \times \left[\frac{1}{T}\int_{t_1}^{t_2} I(t)dt + \frac{1}{T}\int_{t_2}^{t_3} I(t)dt\right] \tag{6.4}$$

If $V_{T0n} = |V_{T0p}| = V_T$ and $k_n = k_p = k$, then

$$I_{mean} = 2 \times \left[\frac{2}{T}\int_{t_1}^{t_2} \frac{1}{2}k(V_{in} - V_T)^2 dt\right] \tag{6.5}$$

Using the facts: $V_{in} = (t/t_{rise})V_{DD}$, $t_1 = (V_T/V_{DD})t_{rise}$, and $t_2 = t_{rise}/2$, the short-circuit power dissipation $P_{d(sc)}$ is

$$\begin{aligned} P_{d(sc)} &= I_{mean} \times V_{DD} = \frac{2}{T}k \times \int_{t_1}^{t_2} \left(\frac{V_{DD}}{t_{rise}}t - V_T\right)^2 dt \cdot V_{DD} \\ &= \frac{2}{T}k \times \frac{1}{3}\frac{t_{rise}}{V_{DD}}V_{DD}\left[\left(\frac{V_{DD}}{t_{rise}}t - V_T\right)^3\right]_{t_1}^{t_2} \\ &= \frac{k}{12}(V_{DD} - 2V_T)^3 \frac{t_{rise}}{T} \end{aligned} \tag{6.6}$$

Consequently, the short-circuit power dissipation is linearly proportional to the rise and fall times of the input signal. To reduce the short-circuit power dissipation, both rise and fall times of the input signal should be made as small as possible. Before leaving this subsection, it is valuable to note that the V_{DD} factor within the dynamic power dissipation equations, $P_{d(cd)}$ and $P_{d(sc)}$, is indeed the output voltage swing, ΔV_{swing}.

6.1.2.3 Power and Delay Trade-offs Since the hardware cost is increasingly reduced, the primary trade-off in most very-large-scale integration (VLSI) designs at present is between power and delay. Hence, a metric is needed to evaluate various possible designs. For this purpose, the *power-delay product* (PDP) was proposed and has been used for a long time. The PDP represents the energy required for a gate to perform a specific operation, for instance, from low to high or vice versa. To specify the PDP more precisely, assume that the gate is switched at its maximum possible rate of $f_{max} = 1/(2t_{pd})$, where t_{pd} is the propagation delay of the gate. Then, the PDP can be expressed as follows:

$$PDP = P_{avg} \cdot t_{pd} = CV_{DD}^2 f \cdot \frac{1}{2f} = \frac{1}{2}CV_{DD}^2 \tag{6.7}$$

Hence, the PDP may be lowered by reducing the capacitance C, the voltage swing ΔV_{swing}, or the supply voltage V_{DD}. A disadvantage of the PDP metric is that it loses timing information associated with the design of the gate.

To incorporate the timing information, another widely used metric is the *energy-delay product* (EDP), which represents the energy and delay required for a gate to perform a specific switching operation. By definition, the EDP is the product of PDP and the propagation delay t_{pd} of the gate. When long-channel devices are used, the EDP of a gate can be expressed as

$$EDP = PDP \cdot t_{pd} = \frac{1}{2}CV_{DD}^2 \cdot \frac{C\Delta V}{I_{sat}}$$

$$= \frac{1}{2}CV_{DD}^2 \cdot \frac{CV_{DD}}{\frac{1}{2}k\left(\frac{W}{L}\right)(V_{GS}-V_T)^2}$$

$$= \frac{C^2V_{DD}^3}{k\left(\frac{W}{L}\right)(V_{GS}-V_T)^2} \quad (6.8)$$

As a consequence, both supply voltage and device size are captured by the *EDP*.

Review Questions

Q6-1. What are the components of power dissipation of a CMOS inverter?

Q6-2. Describe the static power dissipation of a CMOS inverter.

Q6-3. Describe the dynamic power dissipation of a CMOS inverter.

Q6-4. Describe the rise and fall times of the input signal impact on the power dissipation of an inverter.

Q6-5. Define the terms: PDP and EDP.

6.1.3 Design Margins

The performance of a logic circuit will be subject to change after it has been designed because of the process, voltage, and temperature (PVT) variations. The goal of a circuit design is to ensure that the circuit will reliably operate over all extremes of the PVT variations. We briefly explain the meanings of PVT variations.

Process variations. The variations in device performance may be caused by variations in various processes, such as doping densities, implant doses, the widths and thicknesses of active diffusion, oxide layers, and passive conductors. To reflect these variations, the parameters of transistors usually have the following three boundary values: slow, nominal (typical), and fast. The designers need to combine these values with appropriate supply voltages and temperatures to explore the performance of the device that they designed to make them reliably operate at various realistic environments.

Supply voltage. The tolerance of supply voltage is ±5% for commercial products and ±10% for military products. Hence, for commercial products, the operating voltage ranges from 4.75 V to 5.25 V for a nominal 5-volt power supply, from 3.14 V to 3.44 V for a nominal 3.3-volt power supply, from 2.38 V to 2.63 V for a nominal 2.5-volt power supply, and from 1.71 V to 1.89 V for a nominal 1.8-volt power supply.

In designing analog circuits, care must be taken with the voltage coefficient of each related component, including the transistor, resistor, and capacitor, because these circuits are very sensitive to small changes in voltage.

Temperature variation. The temperature dependence of the drain current of MOS transistors is found to be proportional to $T^{-1.5}$ in normal operation. That is, the performance of MOS transistors is inversely proportional to the operating temperature. The nominal ambient temperature ranges from 0°C to 70°C for commercial parts, from −40°C to 85°C for industrial parts, and from −55°C to 125°C for military parts.

The junction temperature of a device (or chip) is related to ambient temperature and power dissipation of the device by the following expression:

$$T_j = T_a + \theta_{ja} \times P_d \quad (6.9)$$

where T_j is the junction (in the chip itself) temperature in °C, T_a is the ambient temperature in °C, θ_{ja} is the package thermal impedance in °C/W, and P_d is the power dissipation of the device.

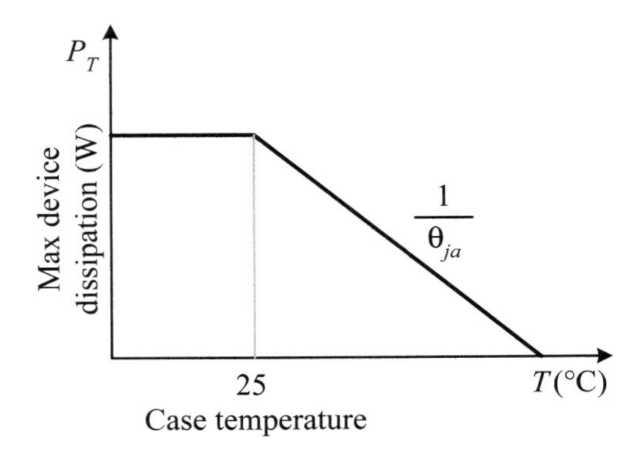

Figure 6.4: The power-derating curve of a typical device.

The maximum power dissipation of a device with respect to the case temperature is usually plotted as a *power-derating curve*, as shown in Figure 6.4. The power-derating curve indicates the maximum allowable power dissipation of the device for a given case (ambient) temperature. The junction temperature determines the maximum power that the device can dissipate. Therefore, the power rating of a device must be decreased as the case temperature rises to keep the junction temperature within a safe limit.

Example 6-1: (Power-derating curve.) Consider a device with a maximum power dissipation of 100 W. If the allowable junction temperature is 120°C and the package thermal impedance θ_{ja} is 2.4°C/W, what is the maximum power dissipation allowed at the ambient temperature of 50°C?

Solution: From Equation (6.9), we find the maximum allowable power dissipation is

$$P_D = \frac{T_{j(max)} - T_{a(max)}}{\theta_{ja}} = \frac{120 - 50}{2.4} = 29.16 \text{ W}$$

The 100-W device can dissipate only 29.16 W because of the high-ambient temperature requirement and the high package thermal impedance. To increase the allowable power dissipation, one may use a package with lower thermal impedance.

■

6.1.3.1 Design Corners. The designers usually need to combine process variations with appropriate supply voltages and temperatures to explore the performance of the device that they designed to make the device reliably operate at various real-world environments. Each combination is called a *design corner*. The following exhibits the two most common design corners:

- The *worst-power or highest-speed corner* combines the lowest temperature and the highest operating voltage. This corner uses fast-n/fast-p parameters and sets the lowest allowable operating temperature ($T = 0$°C) and maximum allowable supply voltage ($V_{DD} = 5.25$ V (3.44 V, 2.63 V, 1.89 V)). It is employed to test (DC) power dissipation, clock races, and hold time constraints of the circuit.
- The *best-power* or *worst-speed corner* combines the highest temperature and the lowest operating voltage. This corner uses slow-n/slow-p parameters and sets the

highest allowable operating temperature ($T = 125°C$ (85°C, 70°C)) and minimum allowable supply voltage ($V_{DD} = 4.75$ V (3.14 V, 2.38 V, 1.71 V)). It is used to test circuit speed and setup time constraints.

Of course, other combinations are also possible and are employed to test some other desired features and constraints of devices.

Review Questions

Q6-6. What are the PVT variations?

Q6-7. Explain the meaning of the power-derating curve of a device.

Q6-8. Explain the meaning of design corners.

Q6-9. Why does the worst-power corner combine the lowest temperature and the highest operating voltage?

Q6-10. Why does the worst-speed corner combine the highest temperature and the lowest operating voltage?

6.1.4 Sizing Wires

Wires are indispensable in any VLSI circuit. They serve as the interconnect for supply voltages, clocks, and general signals. Although the copper wires used in deep submicron processes are more resistant to electromigration, which is a function of temperature and crystal structure, than aluminum-copper wires, copper wires still have their definite current density limit, usually ranging from 0.5 to 1.0 mA/μm. Thus, when doing the layout of a circuit, one should always keep in mind that wires must be properly sized to prevent the electromigration of metal wires from occurring.

The other reasons for sizing the wire widths are as follows. First, clock transitions give rise to current spikes, thereby yielding power bounce and ground bounce, namely, power-supply noise. This may cause the *power-supply integrity* problem. Second, because of the resistance and capacitance of wires, RC delays may introduce into the signals passing through the wires, thus, resulting in the *signal integrity* problem. Sizing a wire properly may effectively reduce the RC delay caused by the wire.

Example 6-2: (Sizing power-supply wires.) Assume that the electromigration rule is $J_{AL} = 1.0$ mA/μm and $V_{DD} = 1.8$ V.

1. Find the required widths of power and ground wires connected to a 500-MHz clock buffer that drives 5-pF on-chip capacitance.
2. Suppose that the clock buffer is 400 μm away from both the power and ground pads. The rise and fall times of the clock are the same and equal to 200 ps. What is the ground bounce with the chosen size of wire?

Solution:

1. From Equation (6.3), we have the power dissipation of the clock buffer

$$\begin{aligned} P &= C_L V_{DD}^2 f \\ &= 5 \times 10^{-12} \times 1.8^2 \times 500 \times 10^6 \\ &= 8.1 \text{ mW} \end{aligned}$$

The current from the power-supply pad is found to be

$I = 8.1 \text{ mW}/1.8 \text{ V} = 4.5 \text{ mA}$

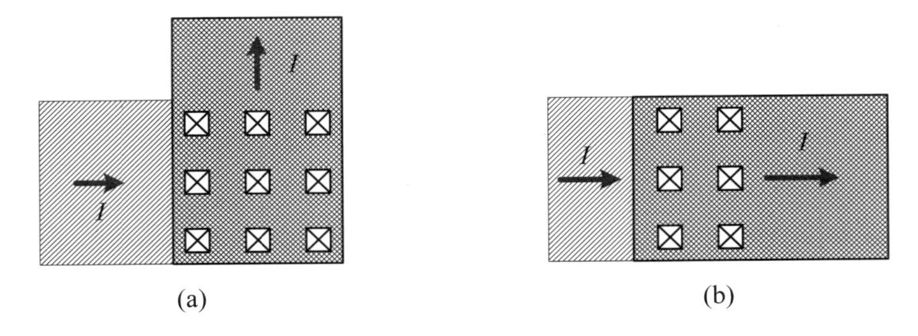

Figure 6.5: Examples of two metal wires being connected together: (a) at a right angle; (b) in the same direction.

Hence, the width of power-supply wires for the clock buffer should be at least 5 μm.

2. Let the wire resistance be 57 m$\Omega/\square$. The resistance of power-supply wires is

$$R = \frac{400}{5} \times 57 \times 10^{-3} = 4.56\ \Omega$$

The IR drop on the power supply-wire is

$$\Delta V = IR = C_L \frac{dV}{dt} \times R = 5 \times 10^{-12} \times \frac{1.8}{0.2 \times 10^{-9}} \times 4.56$$
$$= 205 \text{ mV}$$

which may be intolerable in some applications. To reduce the ground bounce, the power-supply wire should be widened to reduce its resistance.

■

Before leaving this section, let us consider another problem related to wires. Recall that two adjacent-layer wires can only be connected together through the use of contacts or vias. Because the current-crowding phenomenon tends to occur at the periphery of contacts and vias, multiple contacts or vias rather than single large contacts or vias are preferred, in particular, when the wire width is large. For this reason, most modern CMOS processes set fixed-size contacts and vias as their design rules. The resistance of each contact or via is about 2 to 15 Ω, depending on the materials and size of the contact or via.

Examples of two wires being connected together are illustrated in Figure 6.5. To accommodate sufficient current density, it is necessary to incorporate enough numbers of contacts or vias in the interface between two adjacent-layer metal wires. In addition, a square array of contacts or vias is generally needed when both wires are at a right angle, and fewer contacts or vias may be used when both wires are at the same direction. The use of arrays of contacts or vias effectively reduces the total resistance of contacts or vias since they are connected in parallel.

■ Review Questions

Q6-11. What are the functions of wires?

Q6-12. Describe the power bounce and ground bounce.

Q6-13. Give the reasons why the wire width needs to be sized.

Q6-14. In what situation is a square array of contacts or vias needed?

Q6-15. What is the resistance of a typical contact or via?

6.2 Principles of Low-Power Logic Designs

Low-power logic design means that some techniques are applied to reduce power dissipation of a logic circuit while keeping the performance of the logic circuit unchanged at the same time. In this section, we first discuss the basic principle of low-power logic design and then describe the issues related to low-power design in greater detail.

6.2.1 Basic Principles

CMOS logic style has been prevailing as the technology for implementing low-power digital systems due to its reduction of switching energy per device caused by the continually shrinking feature sizes and a negligible static power dissipation compared to the dynamic (switching, transient) power dissipation. Nevertheless, for large systems containing many million, even billion, transistors, it is necessary to further reduce the dynamic power dissipation because system-level issues such as battery life, weight, and size are directly affected by the power dissipation.

Recall that the power dissipation of a CMOS circuit consists of charging and discharging power dissipation, as well as short-circuit and leakage current power dissipation. It can be expressed as follows:

$$P_d = V_{DD}^2 \cdot f_{clk} \cdot \sum_n \alpha_n \cdot C_n + V_{DD} \cdot \sum_m (I_{sc} + I_{leakage}) \tag{6.10}$$

where α_n denotes the node switching activities, C_n represents the node capacitance, I_{sc} is the node short-circuit current and $I_{leakage}$ is the leakage current, n represents the number of nodes, and m is the number of devices causing leakage currents. From this equation, it can be seen that, to reduce power dissipation, we may address the following issues:

- Reduction of voltage swing
- Reduction of switching activity
- Reduction of switched capacitance
- Reduction of short-circuit and leakage currents

Except for the last one, each of these will be considered in more detail in the following. The V_{DD} factor within the dynamic power dissipation equation is indeed the output voltage swing, ΔV_{swing}.

6.2.2 Reduction of Voltage Swing

Because of the square factor associated with the voltage, the most effective way to reduce the power dissipation is by scaling down the swing voltages. For this purpose, reduction of supply voltage and reduction of voltage swing are commonly used. The former directly reduces the supply voltage for a specific process technology and hence may cause performance degradation, whereas the latter limits the voltage swing but keeps the supply voltage unchanged.

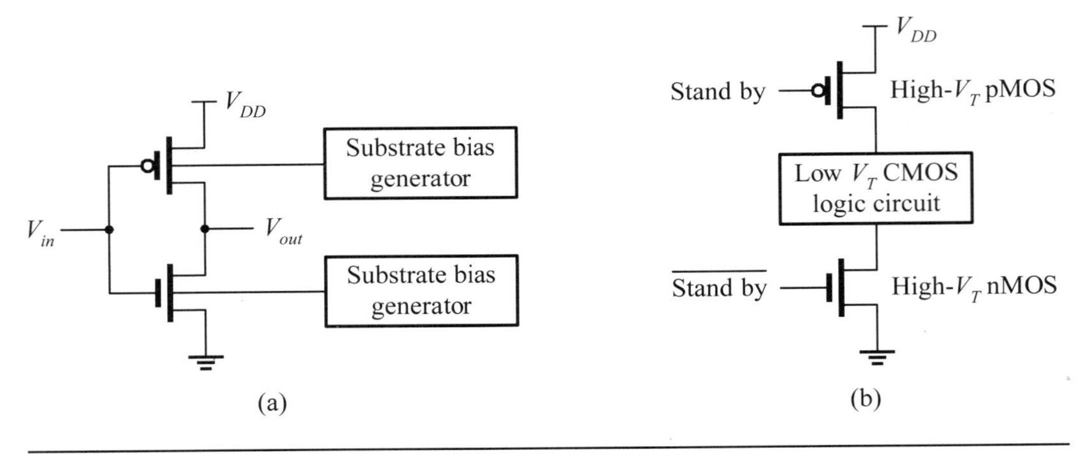

Figure 6.6: Concepts of (a) VTCMOS and (b) MTCMOS logic circuits.

6.2.2.1 Reduction of Supply Voltage Recall that the propagation delay of a logic circuit would increase significantly if we reduce the supply voltage while keeping the threshold voltage unchanged (see Figure 5.14). To compensate for this, either the use of low threshold voltage transistors or the increase of the concurrency of the system can be used. We cope with the use of low threshold voltage transistors. The techniques of increasing concurrency of the system will be detailed in Section 6.3.

From Equations (5.39) and (5.40), we can see that to compensate for the performance degradation due to the reduction of supply voltage is to scale down the threshold voltage by the same factor such that the amount of $V_{DD} - V_T$ remains the same. Nevertheless, as the threshold voltage decreases, the noise margins will be reduced accordingly and subthreshold currents will be increased significantly. To reduce the sleep power dissipation caused by the subthreshold current, *variable-threshold CMOS* (VTCMOS) and *multiple-threshold voltage CMOS* (MTCMOS) techniques and transistors are proposed. We describe each of these two approaches briefly.

VTCMOS technique. Figure 6.6(a) exhibits the general paradigm of VTCMOS circuits. As stated, the threshold voltage of an nMOS or a pMOS transistor may be affected by the substrate (body) bias. Therefore, by properly setting the substrate bias, we are able to control the threshold voltage of an nMOS or a pMOS transistor. For example, assume that the threshold voltages of nMOS/pMOS transistors are set to 0.3/−0.3 V during the active mode and set to 0.7/−0.7 V in the sleep mode to reduce the subthreshold current and hence the leakage power dissipation. To accomplish this, a substrate bias generator is needed to generate a negative bias for nMOS transistors and a positive bias for pMOS transistors.

Another application of the VTCMOS technique, referred to as a *self-adjusting threshold-voltage scheme* (SATS), is to automatically control the threshold voltage to reduce leakage currents and compensate for the process variations.

Even though the VTCMOS technique is an effective way to control the subthreshold currents and threshold voltages in low-V_{DD} and low-V_T applications, it needs a twin-well or triple-well CMOS process to independently control the threshold voltage in different parts of a chip. In addition, a substrate bias generator is needed to generate the required substrate bias.

MTCMOS technique. Another way for reducing sleep power dissipation due to subthreshold currents in low-voltage circuits is by using transistors with two or more different threshold voltages. Such a process is referred to as a multiple-threshold voltage CMOS (MTCMOS). As illustrated in Figure 6.6(b), low-V_T transistors are used to design the desired logic while high-V_T transistors are used to isolate the desired logic during the sleep mode. In the active mode, the high-V_T transistors are turned on and the desired logic can operate with low switching power dissipation; in the sleep mode, high-V_T transistors are turned off to cut off the subthreshold currents caused by the low-V_T logic circuits. Actually, this is a kind of power gating to be described later.

Compared with the VTCMOS technique, the MTCMOS technique does not need the complicated twin-well or triple-well process and the sophisticated substrate bias generator. Nonetheless, it needs a process capable of providing transistors with two different threshold voltages. Fortunately, two or more threshold voltages are widely provided in modern deep submicron processes; refer to Table 2.4 for more details. Moreover, because of the presence of series-connected high-V_T transistors, the resulting circuit needs a larger area and has a larger parasitic capacitance compared with the VTCMOS circuit.

The above-mentioned two techniques are feasible only when the underlying CMOS process is able to support the required features. It is infeasible when a process can only provide two types of transistors with a single threshold voltage. In such a case, the pipelining and parallel processing techniques can be used instead. The essence of these two techniques is the same, that is, to increase the "concurrency" of the system. Refer to Section 6.3 for the details of these two techniques.

6.2.2.2 Reduction of Voltage Swing In those systems implemented with deep-submicron processes, the power dissipation associated with interconnect wires, such as bus, clock, and timing signals, and their driver and receiver circuits, is found to be a large fraction of the total power dissipation. A study showed that the power dissipation of interconnect wires and clocks may be up to 40% and 50% of the total on-chip power dissipation for cell-based and gate-array-based designs, respectively. For traditional FPGA devices, this fraction may be even more than 90% of the total power dissipation.

One way to reduce the power dissipation of interconnect wires and their related driver and receiver circuits is to limit the voltage swing. The rationale behind this is because the voltage factor in the dynamic power dissipation is indeed the voltage swing (ΔV_{swing}) on the charged/discharged capacitor rather than the fixed supply voltage. For instance, using an nMOS transistor to pull up the output will limit the swing to $V_{DD} - V_{Tn}$, instead of V_{DD}. The power consumed for a 0 to $(V_{DD} - V_{Tn})$ transition will be $C_L(V_{DD} - V_{Tn})^2$, rather than $C_L V_{DD}^2$. Nonetheless, this approach has two consequences: First, the high-noise margin (N_{MH}) for the output is reduced by an amount of V_{Tn}. Second, because the output does not rise to V_{DD}, a static gate following the output will consume static power for a high-level output voltage, thereby increasing the effective energy per transition. As a result, to effectively reduce the power dissipation by the shrinking voltage swing, special circuit designs are often needed to compensate for the loss of noise margins, and to eliminate short-circuit currents in the stage that follows. In addition, some sort of voltage-level converters (called *level shifters*) may be needed to interface between two different voltage levels. The combination of the bit-line voltage swing with the associated voltage sense amplifiers used in most semiconductor memories is such an example.

■ Example 6-3: (An example of the low-swing voltage signaling technique.) An example of the low-swing voltage signaling technique is illustrated

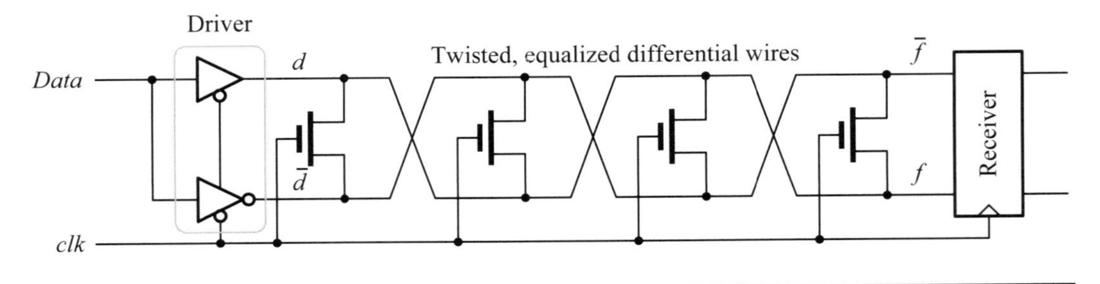

Figure 6.7: An example of the low-voltage swing signaling technique.

in Figure 6.7. The signaling system consists of three components: a driver consisting of one tristate buffer and one tristate inverter, a pair of twisted, equalized differential wires, and a receiver based on the sense-amplifier-based D-type flip-flop (SAFF) (refer to Figure 9.27 for more details about SAFF). Both the tristate buffer and inverter are $\overline{clk}$ enabled. Their details can be referred to in Section 15.3.2. As the clock clk goes high, the driver is at a high impedance and the differential wires are equalized to the same voltage to prevent interference from the previous data transmitted. As the clock clk goes low, the driver turns on. At the end of the half cycle, the receiver senses the differential voltage and amplifies to the full-swing voltage level.

■

6.2.3 Reduction of Switching Activity

The switching activity also plays an important role in the dynamic power dissipation. Hence, we consider in the following the concepts of switching activity and give some approaches for reducing it.

6.2.3.1 The Concept of Switching Activity Switching activity means that an output signal switches from 0 to 1 or vice versa. Since dynamic power dissipation depends on node switching activity along with other factors, the reduction of node switching activity can effectively lower the power dissipation. The switching activity is a function of the logic circuit, logic family, and input data profile. Different logic functions may yield different static transition probabilities and hence switching activities.

To illustrate the concept of switching activity, consider a two-input NAND gate. Assume that only one input transition is possible during a clock cycle, and the inputs to the NAND gate have a uniform input distribution of high and low logic levels. That is, the four possible states (00, 01, 10, 11) for inputs x and y are equally likely. Let P_0 and P_1 denote the probabilities of having 0 and 1 at the output, respectively. It should be noted that $P_0 + P_1 = 1$. For a two-input NAND gate, the probability P_0 is 3/4 and P_1 is 1/4. The probability that power-dissipation transition occurs at the output is given by the probability that the output is in the 0 state multiplied by the probability whose next state is 1. Consequently, the probability of transition from 0 to 1 is

$$P_{0\to 1} = P_0 \cdot P_1 = \frac{3}{4} \cdot \frac{1}{4} = \frac{3}{16} \tag{6.11}$$

The transition probabilities can be exhibited on a transition diagram, which is composed of two possible output states and possible transitions between them. The

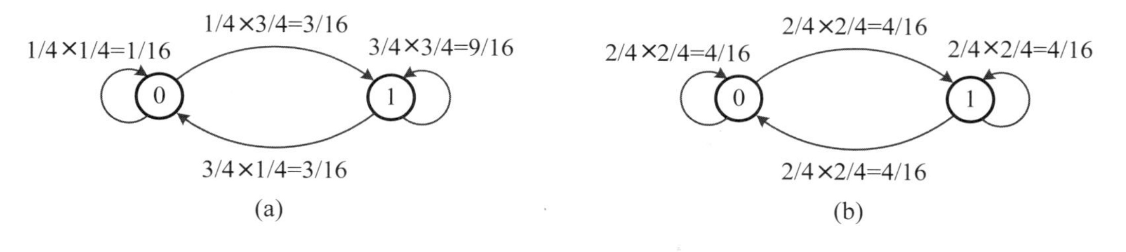

Figure 6.8: The concept of switching activity: (a) two-input NAND gate; (b) two-input XOR gate.

transition diagrams of two-input NAND and XOR gates are shown in Figures 6.8(a) and (b), respectively.

For a general n-input CMOS logic gate, the probability of a power-dissipation output transition can be expressed as a function of n_0, where n_0 is the number of zeros out of the 2^n possible input combinations.

$$P_{0\to1} = P_0 \cdot P_1 = \left(\frac{n_0}{2^n}\right) \cdot \left(\frac{2^n - n_0}{2^n}\right) \tag{6.12}$$

6.2.3.2 Reduction of Switching Activity Reduction of switching activity can be done at various abstraction levels: from the system to the gate and to the circuit. Except for the clock-gating method that is systematic and can be applied in a variety of different applications, most of the approaches used to reduce switching activity are application dependent and need to be considered on a case-by-case basis. We briefly introduce some fundamental concepts and examples. The details of clock gating are deferred to Section 6.4.1.

At the system level, some sorts of algorithmic and architecture optimization are often employed to exploit the statistics of switching activity and then appropriate coding schemes, such as work-zone encoding, bus-invert coding, Gray coding, even sign-magnitude instead of two's complement representation, among other strategies, are used to accomplish the reduction of switching activity. In general, the effectiveness of reduction of switching activity strongly depends on the applications; there are no general schemes for all systems.

Finite-state machine encoding. Recall that one-hot and Gray encoding schemes consume less power in comparison with a binary encoding scheme. This is because one-hot and Gray encoding schemes have only a single-bit change while making transitions from one state to another. We give an example that uses minimum *Hamming distance* codes to reduce the switching activity. Here, the Hamming distance is defined as the number of bit positions that corresponding bits differ. For instance, the Hamming distance between 10010101 and 11000010 is 5.

■ **Example 6-4: (A BCD to excess-3 code converter.)** Figure 6.9 gives the state diagram of a binary-coded decimal (BCD) to an excess-3 code converter. Suppose that the transition probabilities of the dark edges are greater than those of the lighter edges. To make less switching activities, the minimum Hamming distance codes are applied to the pairs of states that have large interstate transition probabilities. For example, the following coding $S_0 = 000$, $S_1 = 001$, $S_2 = 011$, $S_3 = 010$, $S_4 = 100$, $S_5 = 101$, and $S_6 = 111$ may satisfy this requirement.

■

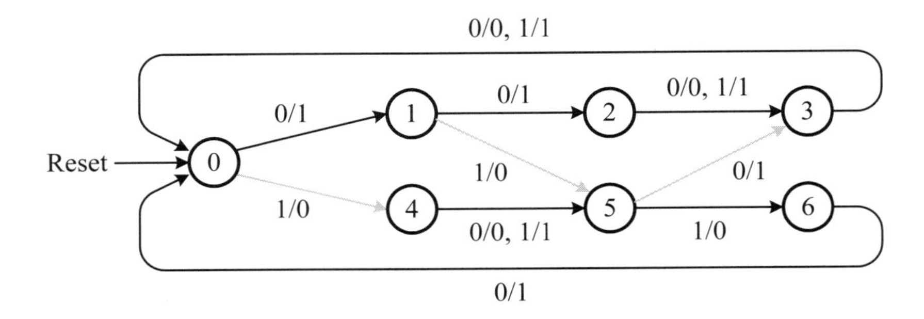

Figure 6.9: The state diagram of BCD to the excess-3 code converter.

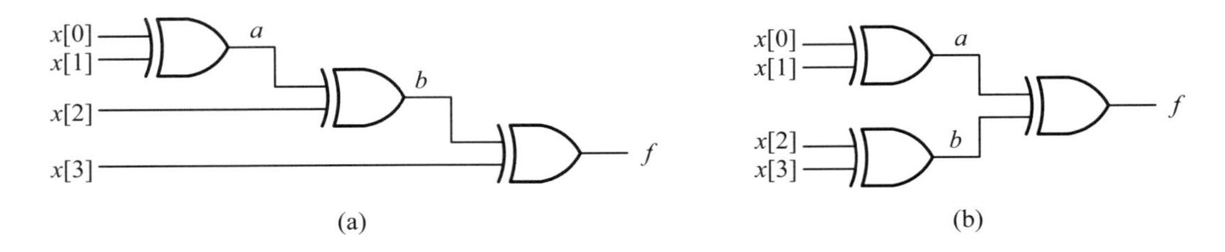

Figure 6.10: Balancing signal paths by logic restructuring: (a) linear structure; (b) binary-tree structure.

Avoid unnecessary signal transitions. Unnecessary signal transitions are often referred to as *glitches* and may be caused by a variety of factors, such as *imbalance signal paths* and *unnecessary signal transitions* in logic circuits.

The imbalance signal paths in a logic circuit may be removed by either restructuring the logic circuit or adding buffers to the shorter signal paths. To get more insights into this, consider the 4-bit parity generator shown in Figure 6.10(a), which is implemented with a linear structure and results in imbalanced signal paths. To balance signal paths and hence remove unwanted glitches, the binary-tree structure shown in Figure 6.10(b) may be used instead. The binary-tree structure not only has balanced signal paths but also has the optimized propagation delay. Unfortunately, not all logic circuits with imbalanced signal paths can be ameliorated by using *logic restructuring*. When this is the case, buffer insertion may be used instead. The idea behind buffer insertion is to balance signal paths by adding buffers in the shorter signal paths so that the propagation delays of these signal paths is increased and compatible to others.

The removal of unnecessary signal transitions in logic circuits can be illustrated by the circuit in Figure 6.11. The arithmetic unit is only enabled when the input 0 of the multiplexer is selected, as shown in Figure 6.11(a). Hence, by gating the inputs to the arithmetic unit appropriately as depicted in Figure 6.11(b), the unwanted signal transitions can be reduced profoundly.

Using Gray encoding for addressing memories. The main feature of the Gray code is also proved useful to significantly reduce the power dissipation in the access of memories because there are fewer number of transitions that the address counter performs.

Using bus-invert coding for I/Os or long data paths. *Bus-invert coding* is a technique that inverts the bits to be transmitted on the bus if the *Hamming distance* (d) between the current data and the previous one is more that $n/2$, where n is the

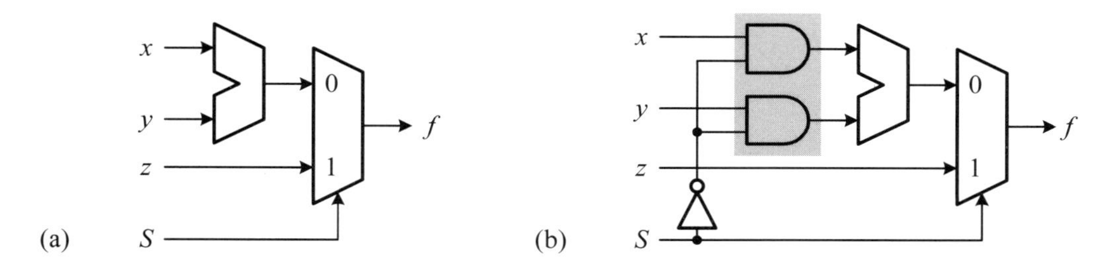

Figure 6.11: The removal of unnecessary signal transitions in logic circuits: (a) original circuit; (b) gated-input circuit.

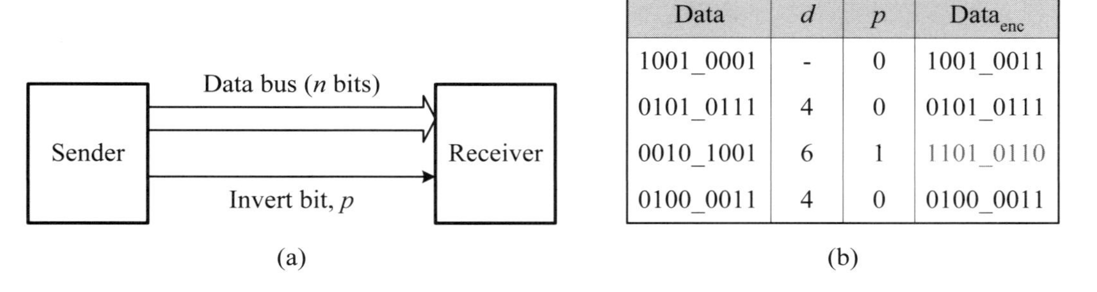

Data	d	p	$Data_{enc}$
1001_0001	-	0	1001_0011
0101_0111	4	0	0101_0111
0010_1001	6	1	1101_0110
0100_0011	4	0	0100_0011

Figure 6.12: An conceptual illustration of bus-invert coding approach: (a) block diagram; (b) examples.

bus width. Based on this, the number of transitions on the bus can be minimized. Of course, a control bit must come along with the data to indicate the receiver whether the data is inverted when using this technique. A conceptual illustration is shown in Figure 6.12.

6.2.4 Reduction of Switched Capacitance

Another factor of the dynamic power dissipation in CMOS logic circuits is switched capacitance. The switched capacitance is the amount of capacitance that needs to be charged/discharged when necessary. The reduction of switched capacitance can be done at various levels: from the system to the gate, and to the circuit. A typical example to illustrate the reduction of switched capacitance is the column multiplexer described in Section 11.2.4. When it is implemented with a uniform structure, each input must drive the capacitance combined from the output capacitances of all nMOS switches. When it is implemented with a multilevel-tree (binary or heterogeneous) structure, each input only needs to drive a fraction of the output capacitances. Hence, the switched capacitance is significantly reduced.

■ Review Questions

Q6-16. What design issues may be addressed when considering a low-power logic design?

Q6-17. How would you reduce power dissipation by scaling the supply voltage?

Q6-18. Give examples of the reduction of switching activity.

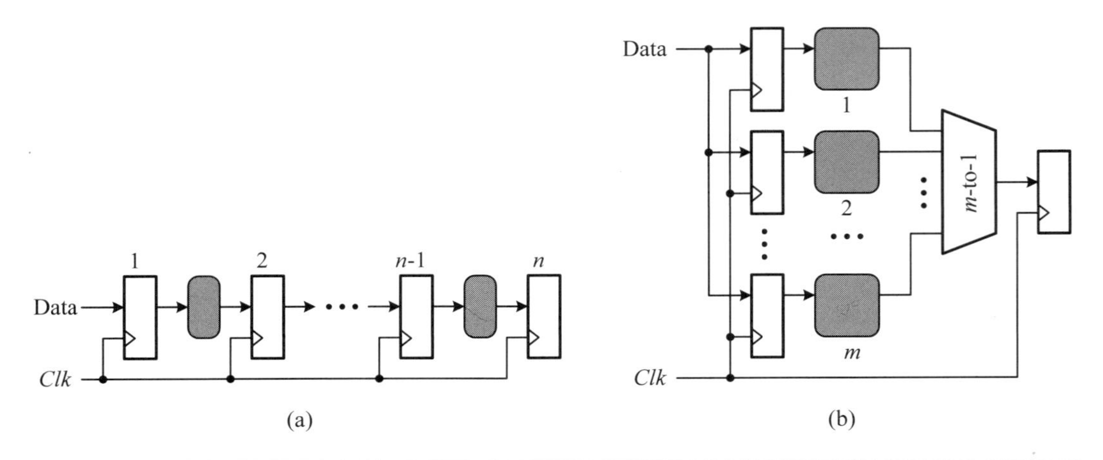

Figure 6.13: Concepts of (a) pipelining technique and (b) parallel processing for the reduction of power dissipation.

Q6-19. Give examples of the reduction of switched capacitance.

Q6-20. Describe the bus-invert coding approach.

6.3 Low-Power Logic Architectures

To compensate for the performance loss due to the reduction of the supply voltage with an unchanged threshold voltage, the pipelining technique and parallel processing are the two widely used approaches for increasing the concurrency and hence the performance of the system. We introduce these two techniques in the context of low-power logic design. The details of the pipelining technique in general can be referred to in Section 9.4.2.

6.3.1 Pipelining Technique

The pipelining technique is a means by which a data stream can be continuously applied to a computational circuit in a regular way to yield results in a sequence. The pipelining technique can be used to save power by allowing the reduction of the supply voltage. To see this, assume that P_{seq} is the power dissipation of the original system and is equal to $C_{total}V^2f$. Now consider an n-stage pipelined structure with a reduced critical path of $1/n$ of its original length, as shown in Figure 6.13(a). The capacitance to be charged/discharged in a single clock period is reduced to C_{total}/n without taking into account the capacitance of pipelining registers. This implies that if the same clock speed is maintained, the supply voltage can be reduced to αV, where α is a positive constant less than 1. As a result, without considering the capacitances of pipeline registers, the power dissipation of the pipelined structure is equal to

$$P_{pipeline} = n \times \frac{C_{total}}{n} \times f \times \alpha^2 V^2 = \alpha^2 P_{seq} \tag{6.13}$$

Therefore, the power dissipation of the pipelined structure is reduced by a factor of α^2 compared to the original structure. The constant α can be determined by noting that the propagation delay of both structures are assumed to be equal. Based on this, the following relationship must be valid.

$$n(\alpha V - V_T)^2 = \alpha(V - V_T)^2 \tag{6.14}$$

where the capacitors are assumed to be charged/discharged by transistors being operated in saturation regions, for simplicity.

For practical systems, the capacitances introduced by pipeline registers may not be neglected. In this case, we need to take these capacitances into account when considering the power dissipation of the pipelined structure.

■ Example 6-5: (A pipelining example.) Supposing that the capacitance of each pipeline register is C_{reg}, calculate the power savings when the supply voltage is reduced from 1.8 to 1.2 V, namely, $\alpha = 1.2/1.8 = 0.67$.

Solution: The power dissipation of the n-stage pipelined structure taking into account C_{reg} is

$$P_{pipeline} = [C_{total} + (n-1)C_{reg}] \times f \times \alpha^2 V^2$$

The power reduction factor achieved in an n-stage pipelined structure is then equal to

$$\frac{P_{pipeline}}{P_{seq}} = \left[\left(1 + \frac{(n-1)C_{reg}}{C_{total}}\right) \times \alpha^2\right]$$

For instance, if $(n-1)C_{reg} = 0.1C_{total}$, the power reduction factor is 0.49. As a result, the pipelined structure only consumes about half the power of the original structure. The power savings are about 50%.

■

In a practical pipelined structure, the allowable voltage scale factor is also a function of the capacitance of pipeline registers. Hence, to determine the voltage scale factor α, one should also take these capacitances into consideration.

6.3.2 Parallel Processing

Parallel processing means that many copies of the same circuit are used to perform the same computation on the different data sets in parallel. Like the pipelining technique, parallel processing can also be used to lower the power dissipation by allowing the reduction of the supply voltage. To see this, consider an m-parallel system with m copies of function units, as shown in Figure 6.13(b). The total capacitance is increased m times. To maintain the same performance, the clock frequency is reduced by a factor of m. Thereby, there is more time to charge/discharge the m capacitors (mC_{total}), and hence, the supply voltage can be reduced by a factor of α. Without considering the capacitances of parallel-processing registers, the power dissipation of the m-parallel system can be expressed as

$$P_{parallel} = (mC_{total}) \times \frac{f}{m} \times \alpha^2 V^2 = \alpha^2 P_{seq} \tag{6.15}$$

Therefore, the power dissipation of the m-parallel system is reduced by a factor of α^2 compared with the original system. The constant α can be determined by noting that the propagation delay of m-parallel system is m times that of the original system. Based on this, the following relationship must be valid.

$$m(\alpha V - V_T)^2 = \alpha(V - V_T)^2 \tag{6.16}$$

where the capacitors are assumed to be charged/discharged by transistors being operated in saturation regions, for simplicity.

For practical systems, the capacitances introduced by parallel-processing registers may not be neglected. In this situation, we need to take into account these capacitances when considering the power dissipation of the parallel-processing system.

■ Example 6-6: (An example of parallel processing.) Assuming that the capacitance of each register is C_{reg} and the input capacitance of the m-to-1 multiplexer is C_{mux}, calculate the power savings, when the supply voltage is reduced from 1.8 to 1.2 V, namely, $\alpha = 1.2/1.8 = 0.67$.

Solution: The power dissipation of the m-parallel system taking into account C_{reg} is

$$P_{parallel} = (mC_{total} + mC_{reg} + C_{mux}) \times \frac{f}{m} \times \alpha^2 V^2$$

The power reduction factor achieved in an m-parallel system is then equal to

$$\frac{P_{parallel}}{P_{seq}} = \left[\left(1 + \frac{C_{reg} + C_{mux}/m}{C_{total}}\right) \times \alpha^2\right]$$

For instance, if $C_{reg} + C_{mux}/m = 0.2C_{total}$, the power reduction factor is 0.54. As a result, the m-parallel system only consumes about half the power of the original structure. The power savings are 46%.

■

■ Review Questions

Q6-21. What is the philosophy behind the use of the pipelining technique to save power?

Q6-22. What is the rationale behind the use of parallel processing to save power?

Q6-23. Is it possible to mix the use of pipelining and parallel techniques?

6.4 Power Management

The power management approach broadly spans a variety of levels, from the circuit and logic level to architecture, software, and the system level. In this section, we begin to discuss the basic power management techniques and then address the dynamic power management.

6.4.1 Basic Techniques

We are concerned in the following with a variety of techniques that can be employed to design power-manageable components, which are in turn used to configure the desired system. These techniques include *clock gating*, *power gating*, multiple supply voltages, and *dynamic voltage and frequency scaling* (DVFS).

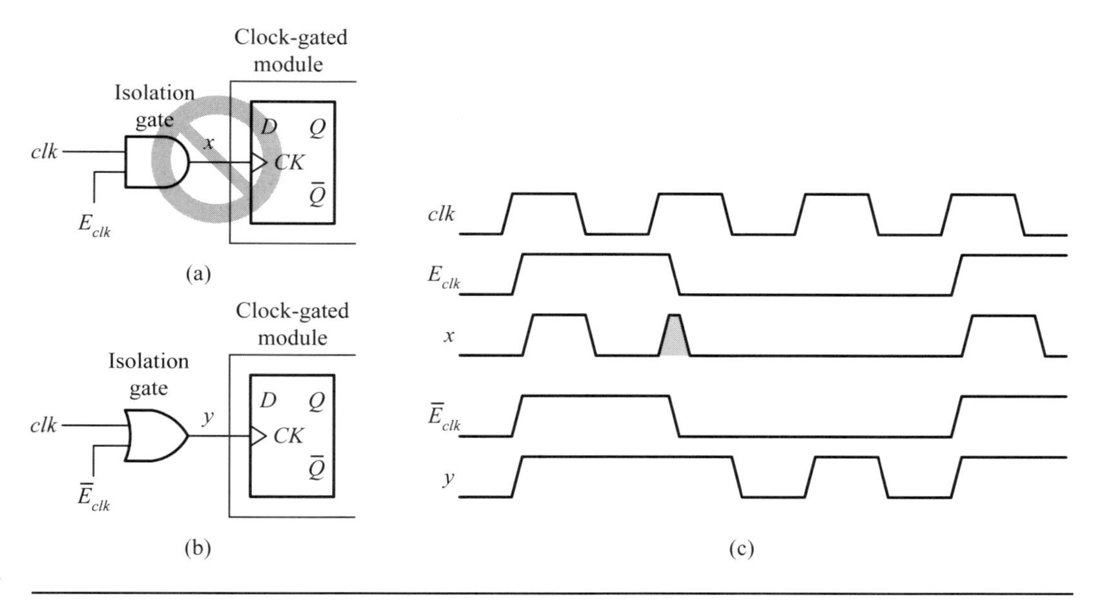

Figure 6.14: The concept of clock gating: (a) bad circuit; (b) good circuit; (c) timing diagram.

6.4.1.1 Clock Gating Clock gating is a widely used approach to reducing power dissipation in a digital system by disabling the clocks entering those circuits that are not being used. The rationale behind the success of this technique is that not all circuits need to be active at all times. Activations of individual circuits vary in a broad way across applications and time. Consequently, there are many opportunities to use the clock-gating technique.

Conceptually, the clock-gating technique can be illustrated by the circuit shown in Figure 6.14(a), where an AND gate is used to gate the clock being routed to the storage element, a D-type flip-flop. The clock is blocked as the clock enable E_{clk} is set to 0 and is routed to the D-type flip-flop as the clock enable E_{clk} is set to 1.

The problems associated with the use of the AND gate to control the clock are as follows. First, if E_{clk} is the output from a finite-state machine or a register clocked by the clock clk, the E_{clk} signal changes some time after the clock clk has risen to high. This results in a glitch in the clock x, as shown in Figure 6.14(c). Second, the AND-gate propagation delay gives excessive clock skew. Hence, the use of the AND gate to control the clock is considered as a bad circuit style. A better approach is shown in Figure 6.14(b) to use an OR gate in place of the AND gate. In this approach, the clock clk is disabled when the $\overline{E}_{clk}$ is 1 and is enabled when the $\overline{E}_{clk}$ is 0. As indicated in Figure 6.14(c), there is no "glitch" (a small duty-cycle clock pulse) occurring.

The features of clock gating are as follows. First, to make the clock gating effective, it is very important to decide the proper locations to gate the clock. Second, the extra gates used in the clock-gating circuit may add skew to the clock. Consequently, the gates must be inserted in a balanced way if it is possible. Third, clock gating increases the complexity of timing verification and the difficulty of reusability of a block. Fourth, the logic circuit needed for the enable signal of clock gating may affect system performance and consume a significant amount of power. Hence, care must be taken in trying to use clock gating to reduce power dissipation of the target system.

Although clock gating can effectively reduce the power dissipation of unused clocked

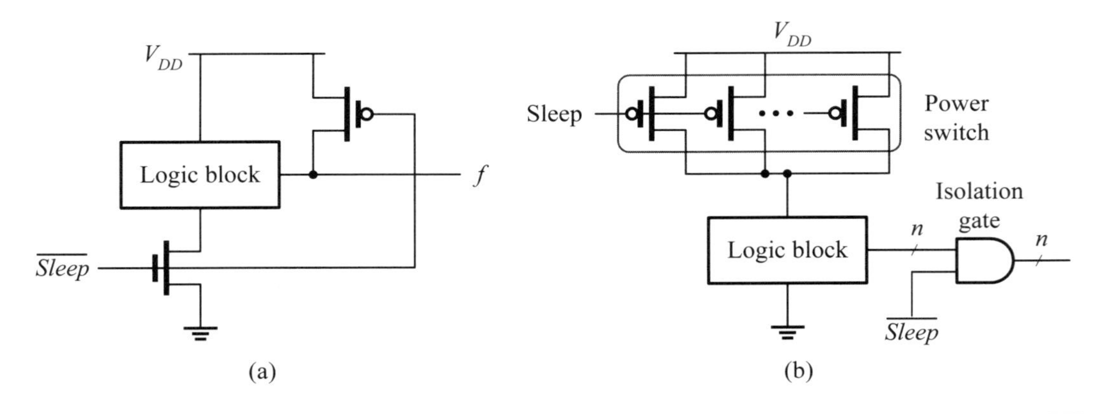

Figure 6.15: The concept of power gating: (a) fine grain; (b) coarse grain.

circuits, including sequential and dynamic circuits, it is of no use for static circuits and the reduction of leakage currents. Fortunately, this difficulty may be solved by the power-gating technique.

6.4.1.2 Power Gating Power gating is the most effective method for reducing leakage power dissipation during sleep (standby) mode. Fundamentally, it is an extension of MTCMOS shown in Figure 6.6(b) in which *power switches* are used as sleep transistors to shut off power supplies to the logic block in sleep mode. A power switch is referred to either a pMOS or an nMOS low-leakage transistor that connects the permanent power supply (V_{DD}/V_{SS}) to the circuit power supply, called the "virtual power supply" (V_{VDD}/V_{VSS}).

Most practical systems cannot tolerate the significant IR drop on the two high-V_T power switches (pMOS and nMOS transistors) connected in series. Hence, in practice, only one of the power rails is switched. This results in gated-V_{SS} and gated-V_{DD} schemes, as illustrated in Figures 6.15(a) and (b), respectively. The nMOS transistor is used in the gated-V_{SS} scheme, whereas a pMOS transistor is used in the gated-V_{DD} scheme as the power switch. Depending on the amount of current sunk in the logic block, the power switch can be as simple as a single pMOS or an nMOS transistor, or needs to be composed of a number of MOS transistors connected in parallel.

Granularity of power gating. As illustrated in Figure 6.15, power gating can be done at either a fine-grained or coarse-grained level. When implemented in the fine-grained level, the power switch is a part of the cell circuit. The power switch must be properly sized not to slow down the normal operation of the underlying cell circuit. When implemented in the coarse-grained level, the power switch is a part of the *power-switching network*. In such a situation, an appropriate management strategy is needed to schedule the power switches so that only a controlled number of power switches and, hence, modules can be switched on and off at a time to avoid overwhelming the power-switching network by the demanded current.

■ **Example 6-7: (A power-gating example.)** An example of fine-grained power gating is illustrated in Figure 6.15(a), where the logic block can be any CMOS logic circuit. As the logic block is active, the foot nMOS transistor turns on, while the pull-up pMOS transistor turns off to leave the logic block to operate in its normal operation mode. As the logic block goes to sleep, the foot nMOS transistor turns off, whereas the pull-up pMOS transistor turns on to pull up the output node to a high-level voltage.

Figure 6.15(b) shows an example of coarse-grained power gating, where a power switch is used to connect the power supply V_{DD} to a logic block. As the logic block is active, the power switch is turned on, connecting the power supply V_{DD} to the logic block. As the logic block goes to sleep mode, the power switch turns off and disconnects the power supply from the logic block. All outputs are gated to 0s as the logic block is in sleep mode.

■

Isolation circuits. *Isolation circuits* need to be inserted between powered-down and powered-up modules to ensure that there are no floating inputs to the powered-up modules and all inputs of powered-up modules are connected to appropriate voltage levels. Isolation circuits can be attached to the outputs of powered-down modules or the inputs to the powered-up modules.

An example of output isolation is illustrated in Figure 6.15(b). Isolation circuits can be done with the AND gate or OR gate to clamp the output signals to logic 0 or logic 1. Sometimes latches are used to clamp the output signals to the last values. The detailed CMOS implementation of the AND gate and OR gate can be referred to in Section 1.3.1.

State retention power gating. A number of design issues arise in the power-gating technique. During the active mode, the power switch should add minimal delay and during the sleep mode, the power switch should have a low leakage current. In addition, power gating is a time-consuming process. Both power up and down processes may take up a large number of clock cycles in high-speed applications. To reduce this overhead, the *state retention power gating* (SRPG) technique can be used instead. SRPG is a technique for retaining states while avoiding unnecessary leakage of power consumption. To implement such a technique, two power supplies are used: continuous power supply (V_{DDC}) and switchable power supply (V_{DD}). Only a very small number of registers used for retaining the state of the system are connected to the continuous supply voltage, and thus, significant leakage power savings can be achieved in idle or sleep mode. Details of retention registers and memory cells can be referred to in Sections 9.2.5 and 11.2.1, respectively.

Power-switching networks. Power-switching networks can be implemented in a variety of styles, such as a ring or a mesh style. An illustration in Figure 6.16(a) is a ring style in which a ring of power switches connects V_{DD} to a mesh of virtual power supply (V_{VDD}). This style is the only way that can be used to control the power of an existing hardware block.

In the mesh style, power switches are distributed throughout the region that needs to be power-gated. They are connected between the permanent power supply mesh and the virtual power supply networks, as illustrated in Figure 6.16(b). The main advantage of mesh style is the ability to share the charging or discharging current among the power switches. Consequently, it is less sensitive to PVT variation and introduces smaller IR drop variations than the ring style. The power switch sharing also reduces the area overhead significantly.

■ Review Questions

Q6-24. What is the philosophy behind the clock-gating technique?

Q6-25. What are the limitations of the clock-gating technique?

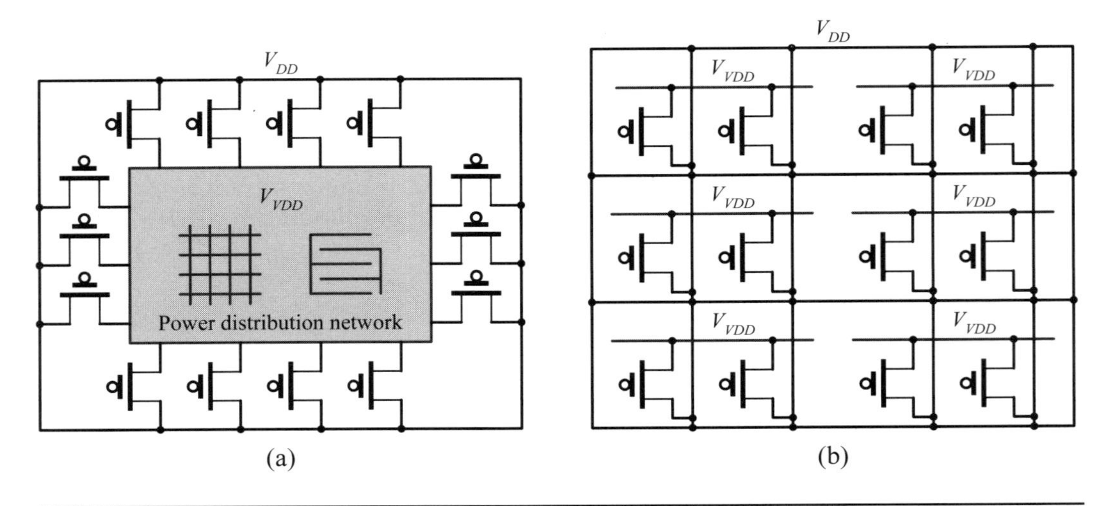

Figure 6.16: Examples of the power-switching network: (a) ring style; (b) mesh style.

Q6-26. What is the rationale behind the power gating?

Q6-27. What is the objective of the power-switching network?

Q6-28. What are the features of the ring-style power-switching network?

Q6-29. What are the features of the mesh-style power-switching network?

Q6-30. Give the reasons why the isolation circuits are needed in between power-gated and power-up circuits.

Q6-31. What is the rationale behind the use of state retention power gating?

6.4.1.3 Multiple Supply Voltages For those components that are not idle but can be operated in lower frequencies, the reduction of power dissipation can be achieved by lowering the supply voltage. The most intuitive approach to providing such a scheme is to partition the system into many voltage domains in a static power-directed way, with each voltage domain powered by its own supply voltage. Such a scheme is often referred to as *static voltage scaling* (SVS). Remember that, as the supply voltage is reduced, the maximum operating frequency also decreases accordingly due to the increase of the propagation delay of logic circuits (see Figure 5.14). Hence, the net effect is that both supply voltage and operating frequency are scaled down at the same time, leading to the high efficiency in the reduction of power dissipation.

Currently, most systems with static voltage scaling use two or three supply-voltage levels and employ level shifters at the border of voltage domains running on different supply voltages to convert the required signal levels. Of course, the use of supply voltages of more than three levels is also possible. However, this will result in a more complicated system design because more voltage domains need to be handled. In a system with static voltage scaling, each module operates in its own fixed voltage and operating frequency at all times.

■ **Example 6-8: (Level shifters.)** Figure 6.17 gives two level shifters. Figure 6.17(a) is a level shifter based on cascode voltage switch logic (CVSL), which will be discussed in Section 7.3.1. The arriving input x is up-converted to the desired

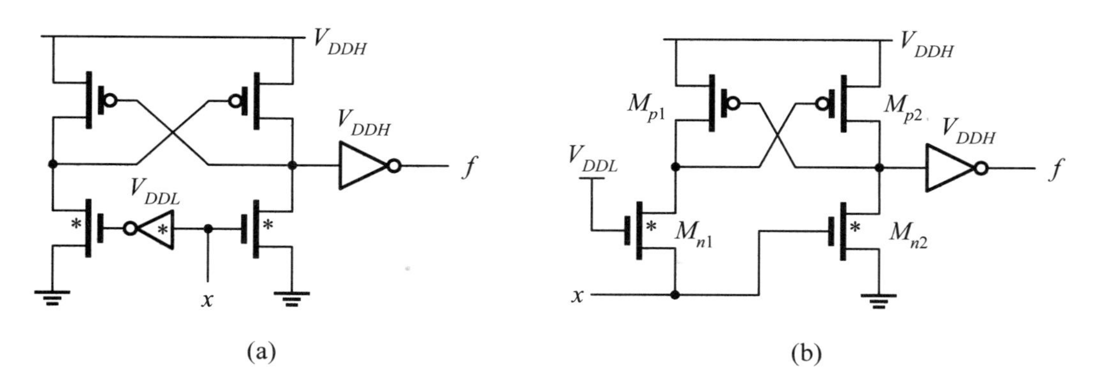

Figure 6.17: Two examples of level shifters: (a) CVSL level converter; (b) pass-gate level converter (Transistors labeled with (*) indicate low-V_T devices.)

voltage level through the cross-coupled pMOS device pair. Nevertheless, this converter consumes significant power due to the contention at the connection points of the cross-coupled pair and the pull-down nMOS transistors.

Figure 6.17(b) shows a level shifter based on a weak feedback pull-up pMOS transistor (M_{p1}) and an nMOS transistor (M_{n1}). Through the use of an nMOS transistor (M_{n1}) to isolate the input of a pMOS transistor M_{p2} from the previous stage, the feedback pMOS transistor (M_{p1}) can then pull up the internal node to V_{DDH} without being affected by its previous stage. Because of fewer transistors being used and reduced contention, this level shifter consumes less power than the CVSL level converter. ■

■ Example 6-9: (An example of multiple supply voltages.) As illustrated in Figure 6.18, suppose that a system with a switching capacitance of 500 pF, operating at a frequency of 500 MHz, and a supply voltage of 1.8 V can be partitioned into two modules operating at the following conditions:

- Module one operates at a frequency of 500 MHz and a supply voltage of 1.8 V. The switching capacitance is 300 pF.
- Module two operates at a frequency of 200 MHz and a supply voltage of 1.2 V. The switching capacitance is 200 pF.

Calculate the power saving of the partitioned system.

Solution: The power dissipation of the original system is

$$\begin{aligned} P_D &= C_T f V_{DD}^2 \\ &= 500 \times 10^{-12} \times 500 \times 10^6 \times 1.8^2 = 0.81 \text{ W} \end{aligned}$$

The power dissipation of module one is found to be

$$P_{D1} = 300 \times 10^{-12} \times 500 \times 10^6 \times 1.8^2 = 0.486 \text{ W}$$

The power dissipation of module two is calculated as follows:

$$P_{D2} = 200 \times 10^{-12} \times 200 \times 10^6 \times 1.2^2 = 0.058 \text{ W}$$

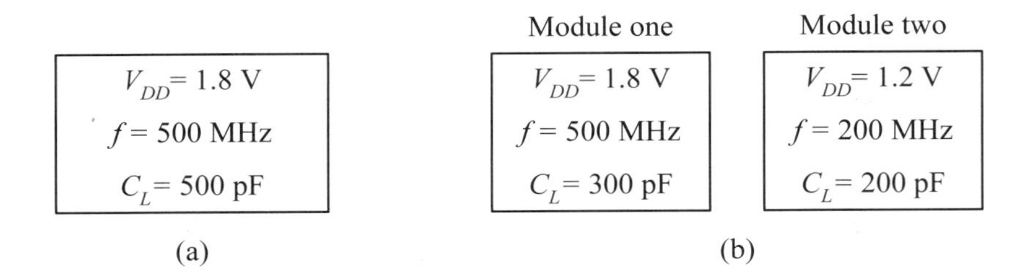

Figure 6.18: An example of multiple supply voltages: (a) original system; (b) partitioned system.

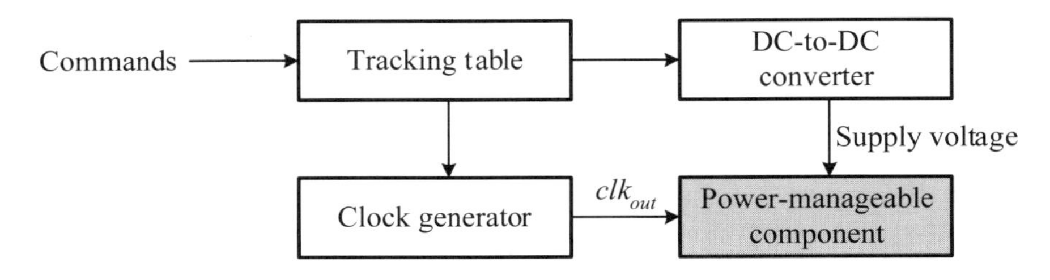

Figure 6.19: The conceptual block diagram of the tracking-table DVFS approach.

As a consequence, the power saving is equal to

$$\text{Power saving (\%)} = \frac{0.81 - (0.486 + 0.058)}{0.81} \times 100\ \% = 33.3\ \%$$

■

To further increase the possible power saving, each module may be allowed to operate in different voltage/frequency pairs at different times when necessary. This results in *multilevel voltage scaling* (MVS). The MVS is an extension of static voltage scaling in which each voltage/frequency domain can be switched between two or more voltage/frequency pairs dynamically, leading to a technique called *dynamic voltage and frequency scaling* (DVFS).

6.4.1.4 Dynamic Voltage and Frequency Scaling

There are two approaches widely used to implement the DVFS technique: the *tracking-table approach* and the *adaptive DVFS approach*. The former is better than the latter in respect of stability. Nevertheless, the latter can more efficiently manage the power of the load due to continuous tracking of a minimum energy point. Unfortunately, the adaptive DVFS approach is much more complicated and difficult to implement than the tracking-table approach. Hence, most practical systems adopt the tracking-table approach currently.

Tracking-table approach. In the tracking-table approach, a set of discrete workable voltage/frequency pairs for a power-manageable component are obtained by simulation and summarized in a static look-up table, known as a *tracking table*, which controls the DC-DC converter and clock generator to generate the desired supply voltage and clock signal for the power-manageable component, respectively. A conceptual block diagram of the tracking-table DVFS approach is illustrated in Figure 6.19.

When using the tracking-table approach, the following scenario is applied to control the power-manageable component. First, the operating frequency is lowered to reduce

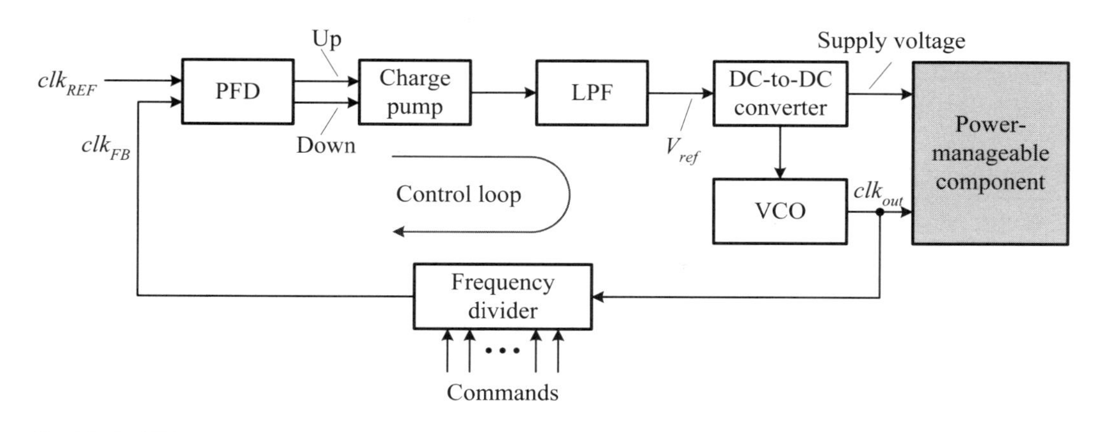

Figure 6.20: The conceptual block diagram of the adaptive DVFS approach.

the power dissipation provided that the power-manageable component still meets the required performance of applications. Second, the supply voltage is scaled down to further reduce the power dissipation. To restore the high performance, the supply voltage is first recovered and then the operating frequency.

Adaptive DVFS approach. In the adaptive DVFS approach, a feedback control loop is used to track both supply voltage and operating frequency of a set of power-manageable components at the specific operating performance. For the time being, most implementations use a frequency-to-voltage converter to generate the required supply voltages according to the desired operating frequencies. An example is illustrated in Figure 6.20, where a DC-to-DC converter is embedded into the *phase-locked loop* (PLL). The frequency divider accepts control signals, indicating the performance requirement of the target system, from the power-management unit to set the desired operating frequency and cause the *voltage-controlled oscillator* (VCO) to track this requirement and generate the needed supply voltage.

Example 6-10: (A DVFS example.) Figure 6.20 is a conceptual block diagram of the adaptive DVFS approach. Basically, it is a modified phase-locked loop (PLL) (see Section 14.3) and comprises a phase frequency detector (PFD), charge pump (CP), low pass filter (LPF), DC-to-DC converter, and voltage-controlled oscillator (VCO). The PFD detects phase error between the reference clock (clk_{REF}) and feedback clock (clk_{FB}), and generates Up, Down control signals. The LPF produces the reference voltage (V_{ref}) for the DC-to-DC converter. The VCO tracks the load condition and finds the minimum energy point for the target system. The frequency divider accepts control signals from the power-management unit and generates the feedback clock (clk_{FB}) to PFD to form a control loop. Because of the same reference voltage being used to generate the supply voltage via the DC-to-DC converter and as the VCO control voltage, the supply voltage and VCO output clock frequency (clk_{out}) are dynamically locked at the same time.

■

Two performance parameters closely related to DVFS are *transition time* and *transition energy*. Transition time is defined as the duration required to alter the operating frequency and supply voltage. Transition energy is the extra energy required to change the supply voltage of the power-manageable component. The transition time impacts both interrupt latency and wake-up latency when the system is in its lowest-energy sleep state.

Review Questions

Q6-32. What is the SVS technique?

Q6-33. Give the reasons why the level shifters are needed in the interface between two different voltage domains from low to high.

Q6-34. What is the tracking-table DVFS approach?

Q6-35. What is the adaptive DVFS approach?

Q6-36. How would you apply the DVFS technique to a system design?

6.4.2 Dynamic Power Management

With the rapid advance of hardware and software techniques, building a very complicated digital system becomes feasible and popular. However, because of many millions, even billions, of transistors contained on a chip, the management of power dissipation of various modules in the system to remove unnecessary power dissipation or to operate with limited battery charge becomes an important design issue of digital systems. A useful power-management technique needs to take into account the overall hardware and software management overhead to achieve true power reduction and meanwhile satisfy the performance requirement of applications.

6.4.2.1 Dynamic Power Management *Dynamic power management* (DPM) is a design approach for dynamically reconfiguring a VLSI (digital) system according to activity profiles to provide the demanded services and performance levels within a minimum number of active components or a minimum load on such components. DPM includes a set of techniques to achieve energy-efficient computation by selectively turning off or reducing the performance of system components as they are idle or partially functioning. A conceptual block diagram of DPM is illustrated in Figure 6.21(a) and an example of an activity profile is given in Figure 6.21(b). The DPM comprises an observer and a controller; the system consists of a set of power-manageable components. The observer receives observations from the system and generates workload information for the controller to generate the required commands for the system to adjust the system performance and hence power dissipation. For a small system, it is sufficient for the controller to use a simple finite-state machine, referred to as a *power-state machine* (PSM), to carry out the policy. For a complex system, the controller may need a microprocessor system to execute the complex policy, namely, a complex PSM.

The rationale behind the applicability of DPM is that systems (and their components) experience nonuniform workloads during the operation of interest. In addition, such workloads are possible to be predicted with a certain degree of confidence. To make the DPM feasible, both workload observation and prediction should not consume significant energy. By and large, a DPM unit implements a control process (that is, policy) based on some observations and/or assumptions on the workloads of system components. For example, a widely used policy is the use of the timeout scheme, which shuts down a system component after a fixed inactivity time. This is based on a simple-minded assumption that it is highly likely that a component remains idle if it has been idle for a period of expected time. Nevertheless, such a simple policy may turn out

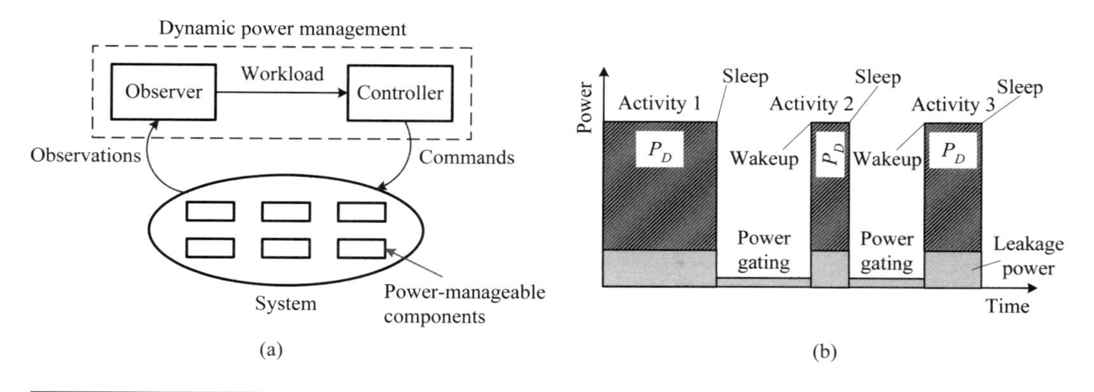

Figure 6.21: The dynamic power management: (a) a conceptual block diagram; (b) an example of activity profile.

to be inefficient in practice. Consequently, a more complex policy accompanied with power-controlled techniques needs to be used for a practical system.

6.4.2.2 Implementations of Dynamic Power Management As stated, a power-managed system is composed of a set of interacting power-manageable components controlled by a *power management unit*, as shown in Figure 6.21(a). For now, most practical systems simply combine clock gating, power gating, and a set of discrete voltage/frequency pairs to manage the power dissipation of the underlying system. To achieve this, power-manageable components in the target systems are partitioned and classified into many operating modes according to application profiles. For each mode, a set of power-manageable components are clock gating, power gating, operating with lower or higher supply voltage, and/or with different frequencies. The transitions between different modes are scheduled by a power-state machine (PSM). For a simple system, this PSM can be implemented by a simple logic circuit; for a complex system, the PSM might need a microprocessor system.

A simple example of power management is given in the following. In this system, only three operation states (modes) are provided and only clock gating and power gating are used to control the power dissipation.

■ **Example 6-11: (A simple DPM unit.)** Figure 6.22 shows an example of PSM for a microprocessor system. The system has three states of operation: run, idle, and sleep. The run state is the normal operating mode in which every module is functional. The system enters the run state after successful power-up and reset. The idle state allows an application to stop the central processing unit (CPU) through clock gating as it is not in use, while continuing to monitor interrupt requests on or off system. In the idle state, the CPU can return to the run state by enabling the clock signal whenever an interrupt occurs. The transitions between run and idle states are quick (it only takes 10 μs). The sleep state offers the greatest power saving and hence the lowest level of available functionality. In the transition from run or idle to the sleep state, the system performs an orderly shutdown of system activity through power gating. In a transition from the sleep to run state, the system goes through a rather complex wake-up sequence before it can resume normal activity. Hence, a quite long transition time (150 ms) is needed.

■

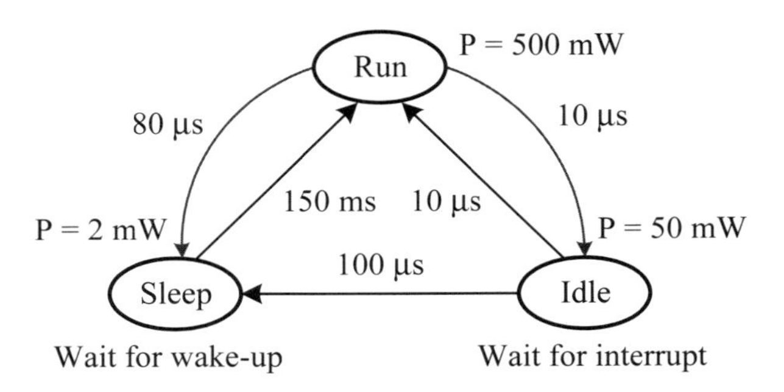

Figure 6.22: A simple example of the power-state machine.

Table 6.1: The dynamic power management modes of an Intel Atom processor.

	C0 HFM	C0 LHM	C1/C2	C4	C6
Core voltage	Full	$\longrightarrow$ decreasing			
Core clock	On	On	Off	Off	Off
PLL	On	On	On	Off	Off
L1 caches	On	On	Flushed	Flushed	Off
L2 caches	On	On	On	Partially flushed	Off
Wakeup time	Active	Active	$< 1\ \mu s$	$< 30\ \mu s$	$< 100\ \mu s$
Power	Full	$\longrightarrow$ decreasing			

A more complicated example of power management is illustrated in the following. In this system, there are six (actually five) operation states (modes). The combination of clock gating, power gating, multiple supply voltages, and multiple operating frequencies is used to obtain the minimum power dissipation of the target system.

■ Example 6-12: (An example of power management.) Table 6.1 gives the dynamic power management states of an Intel Atom processor. The peak frequency is 2 GHz at a 1-V supply voltage, consuming 2 W. At a lower frequency mode, the core frequency drops down to 600 MHz at a 0.75-V supply voltage. The power states (C-states) are shown in Table 6.1, where lower power is achieved as more modules are turned off from the C-0 state to the C-6 state. In the C-0 state, the processor may operate at its high frequency mode (HFM), and its low frequency mode (LFM), respectively. In the C-1 state, the core clock is gated off and the L1 caches flushed. The exit latency is under 1 μs. In the C-4 state, the PLLs are shut down and the L1 caches are flushed. The exit latencies are in the order of 30 μs. Finally, in the C-6 state, the state of the machine is stored in an on-chip static random access memory (SRAM) and the core supply voltage is shut down, resulting in the lowest power dissipation. The exit latencies are in the order of 100 μs.

■

■ Review Questions

Q6-37. What is the philosophy behind the power-management technique?

Q6-38. Describe the basic principle of the power-management technique.

Q6-39. What is the power-state machine?

6.5 Summary

The power dissipation of a CMOS logic circuit is composed of static power dissipation and dynamic power dissipation. Static power dissipation is due to the leakage current of reverse-biased junctions, gate leakage current, and subthreshold current or other currents continuously drawn from the power supply. Dynamic dissipation (P_d) may be caused by the short-circuit (also called the switching transient) current between power-supply rails as well as charge and discharge currents to and from load capacitors.

To limit and reduce the power dissipation of a system, low-power logic design and power management are the two important design issues that need to be taken into account in designing a system. Low-power logic design means that some techniques are applied to reduce power dissipation of a logic circuit while keeping the performance of the logic circuit unchanged at the same time. A variety of basic principles of low-power logic design and related issues were dealt with in great detail.

As for power management, we first addressed a variety of techniques suitable for designing power-manageable components and then introduced the principle of dynamic power management along with practical examples. The techniques include clock gating, power gating, multiple supply voltages, and dynamic voltage and frequency scaling (DVFS).

References

1. A. Agarwal et al., "Leakage power analysis and reduction for nanoscale circuits," *IEEE Micro*, pp. 68–80, March–April 2006.
2. L. Benini, A. Bogliolo, and G. De Micheli, "A survey of design techniques for system-level dynamic power management," *IEEE Trans. on Very Large Scale Integration (VLSI) Systems*, Vol. 8, No. 3, pp. 299–316, June 2000.
3. T. D. Burd et al., "A dynamic voltage scaled microprocessor system," *IEEE J. of Solid State Circuits*, Vol. 35, No. 11, pp. 1571–1580, November 2000.
4. B. H. Calhoun et al., "Ultra-dynamic voltage scaling using subthreshold operation and local voltage dithering," *IEEE J. of Solid State Circuits*, Vol. 41, pp. 238–245, January 2006.
5. A. P. Chandrakasan and R. W. Brodersen, "Minimizing power consumption in digital CMOS circuits," *Proc. IEEE*, Vol. 83, No. 4, pp. 498–523, April 1995.
6. L. Clark, S. Demmons, N. Deutscher, and F. Ricci, "Standby power management for a 0.18-μm microprocessor," *Proc. of Int'l Symposium on Low Power Electronics and Design*, pp. 7–12, August 2002.
7. G. Gerosa et al., "A sub-2 W low power IA processor for mobile Internet devices in 45 nm high-k metal gate CMOS," *IEEE J. of Solid State Circuits*, Vol. 44, No. 1, pp. 73–82, January 2009.
8. V. Gutnik and A. P. Chandrakasan, "Embedded power supply for low-power DSP," *IEEE Trans. on Very Large Scale Integration (VLSI) Systems*, Vol. 5, No. 4, pp. 425–435, December 1997.
9. R. Ho, K. Mai, and M. Horowitz, "Efficient on-chip global interconnects," *2003 Symposium on VLSl Circuits Digest of Technical Papers*, pp. 271–274, 2003.

10. Y. S. Hwang, S. K. Ku, C. M. Jung, and K. S. Chung, "Predictive power aware management for embedded mobile devices," *2008 Asia and South Pacific Design Automation Conf.* (ASPDAC 2008), pp. 36–41, March 2008.

11. S. M. Kang and Y. Leblebici, *CMOS Digital Integrated Circuits: Analysis and Design,* 3rd ed. New York: McGraw-Hill Books, 2003.

12. M. Keating et al., *Low Power Methodology Manual: For System-on-Chip Design.* Boston: Springer, 2007.

13. S. H. Kulkarni and D. Sylvester, "High performance level conversion for dual V_{DD} design," *IEEE Trans. on Very Large Scale Integration (VLSI) Systems,* Vol. 12, No. 9, pp. 926–936, September 2004.

14. T. Kuroda et al., "A 0.9-V, 150-MHz, 10-mW, 4 mm^2, 2-D discrete cosine transform core processor with variable threshold-voltage (V_T) scheme," *IEEE J. of Solid-State Circuits,* Vol. 31, No. 11, pp. 1770–1779, November 1996.

15. T. Kuroda, "Optimization and control of V_{DD} and V_{TH} for low-power high-speed CMOS design," *IEEE/ACM Int'l Conf. on Computer Aided Design,* pp. 28–34, 2002.

16. J. Lee, B. G. Nam, and H. J. Yoo, "Dynamic voltage and frequency scaling (DVFS) scheme for multi-domains power management," *IEEE Asian Solid-State Circuits Conf.,* pp. 360-363, Jeju, Korea, November 12–14, 2007.

17. J. Lee, B. G. Nam, S. J. Song, N. Cho, and H. J. Yoo, "A power management unit with continuous co-locking of clock frequency and supply voltage for dynamic voltage and frequency scaling," *IEEE Int'l Symposium on Circuits and Systems,* (ISCAS 2007), pp. 2112-2115, 27–30 May, 2007.

18. J. D. Meindl, "Low power microelectronics: retrospect and prospect," *Proc. IEEE,* Vol. 83, No. 4, pp. 619–635, April 1995.

19. S. Mutoh, "1-V power supply high-speed digital circuit technology with multi-threshold-voltage CMOS," *IEEE J. of Solid-State Circuits,* Vol. 30, No. 8, pp. 847–854, August 1995.

20. M. Pedram and A. Abdollahi, "Low-power RTL synthesis techniques: a tutorial," *IEE Proceedings-Computers and Digital Techniques,* Vol. 152 , No. 3, pp. 333–343, May 2005.

21. J. M. Rabaey, A. Chandrakasan, and B. Nikolic, *Digital Integrated Circuits: A Design Perspective,* 2nd ed. Upper Saddle River, NJ: Prentice-Hall, 2003.

22. K. Roy and S. C. Prasad, *Low-Power CMOS VLSI Circuit Design.* New York: John Wiley & Sons, 2000.

23. S. Rusu et al., "A 45 nm 8-core enterprise Xeon processor," *IEEE J. of Solid State Circuits,* Vol. 45, No. 1, pp. 7–14, January 2010.

24. K. Shi, Z. Lin, and Y. M. Jiang, "A power network synthesis method for industrial power gating designs," *The 8th Int'l Symposium on Quality Electronic Design* (ISQED), pp. 362–367, 2007.

25. M. R. Stan and W. P. Burleson, "Bus-invert coding for low-power I/O," *IEEE Trans. on Very Large Scale Integration (VLSI) Systems,* Vol. 3, No. 1, pp. 49–58, March 1995.

26. K. Usami et al., "Automated low-power technique exploiting multiple supply voltages applied to a media processor," *IEEE J. of Solid-State Circuits,* Vol. 33, No. 3, pp. 463–471, March 1998.

27. H. Veendrick, "Short-circuit dissipation of static CMOS circuitry and its impact on the design of buffer circuits," *IEEE J. of Solid-State Circuits,* Vol. 19, No. 4, pp. 468–473, August 1984.

28. N. H. E. Weste and D. Harris, *CMOS VLSI Design: A Circuit and Systems Perspective,* 4th ed. Boston: Addison-Wesley, 2011.

Problems

6-1. Consider a device with a maximum power dissipation of 120 W. If the allowable junction temperature is 120°C and package thermal impedance θ_{ja} is 1.5°C/W, what is the maximum power dissipation allowed at the ambient temperature of 60°C?

6-2. Assume that the metal-migration rule is $J_{AL} = 1.0$ mA/μm and $V_{DD} = 1.8$ V.

(a) Find the required widths of power and ground wires connected to a 1200-MHz clock buffer that drives 5-pF on-chip load capacitance.

(b) Suppose that the clock buffer is 500 μm away from both power and ground pads. The rise and fall times of the clock are the same and equal to 100 ps. What is the ground bounce with the chosen wire size?

6-3. Assuming that all transistors operate in their saturation regions and using the relation $Q = It = CV$, prove the following two equations:

(a) Equation (6.14).

(b) Equation (6.16).

6-4. Supposing that the capacitance of each pipeline register is C_{reg} and $(n-1)C_{reg} = 0.3C_{total}$, calculate the power savings, when the supply voltage is reduced from 1.8 to 1.5 V, namely, $\alpha = 1.5/1.8 = 0.83$.

6-5. Assuming that the capacitance of each register is C_{reg}, the input capacitance of the m-to-1 multiplexer is C_{mux}, and $C_{reg} + C_{mux}/m = 0.2C_{total}$, calculate the power savings when the supply voltage is reduced from 1.8 to 1.5 V, namely, $\alpha = 1.5/1.8 = 0.83$.

6-6. Assume that a system with a switching capacitance of 1 nF, operating at a frequency of 600 MHz, and a supply voltage of 1.2 V can be partitioned into three modules operating at the following conditions:

- Module one operates at a frequency of 600 MHz and a supply voltage of 1.2 V. The switching capacitance is 700 pF.
- Module two operates at a frequency of 400 MHz and a supply voltage of 1.0 V. The switching capacitance is 200 pF.
- Module three operates at a frequency of 100 MHz and a supply voltage of 0.8 V. The switching capacitance is 100 pF.

Calculate the power savings of the partitioned system.

7

Static Logic Circuits

Complementary metal-oxide-semiconductor (CMOS) logic circuits can be roughly categorized into two broad categories: *static logic circuits* and *dynamic logic circuits*, depending on how a logic value is retained at the output node of a given logic circuit. In a static logic circuit, a logic value is retained by using the circuit states, while in a dynamic logic circuit, a logic value is stored in the form of charge. In this chapter, we first consider the static logic circuits. Dynamic logic circuits are discussed in the next chapter. To make use of a static logic circuit, it is necessary to understand its features thoroughly. To achieve this, we discuss in detail some basic static logic circuits, including CMOS inverters, and NAND and NOR gates.

Static logic circuits can be further subdivided into single-rail and dual-rail logic circuits according to the signal system used. The single-rail logic circuits receive as its inputs only either true or complementary signals and produce either true or complementary signals at their outputs. The conventional CMOS logic, transmission-gate-based (TG-based) logic, and pseudo n-type metal-oxide-semiconductor (pseudo-nMOS) logic are such examples. The dual-rail logic circuits receive as its inputs both true and complementary signals and yield both true and complementary signals at their outputs. Because of this, dual-rail logic is also called differential logic. The major types of dual-rail logic circuits include the following: cascode voltage switch logic (CVSL), complementary pass-transistor logic (CPL), a combination of both, differential cascode voltage switch with pass-gate (DCVSPG) logic, and double pass-transistor logic (DPL).

7.1 Basic Static Logic Circuits

In this section, we first introduce the concepts of single-rail as well as dual-rail signal systems and their corresponding logic circuit families. Next, we study the features of an inverter to a great extent. Then, we cope with the detailed features of both NAND and NOR gates. Finally, we address the problems about the sizes of logic circuits.

7.1.1 Types of Static Logic Circuits

When designing a digital system, we often need to define the signal system used throughout the entire or a part of the system. Two signal systems widely used in digital systems are *single-rail* and *dual-rail* signal systems, as shown in Figure 7.1.

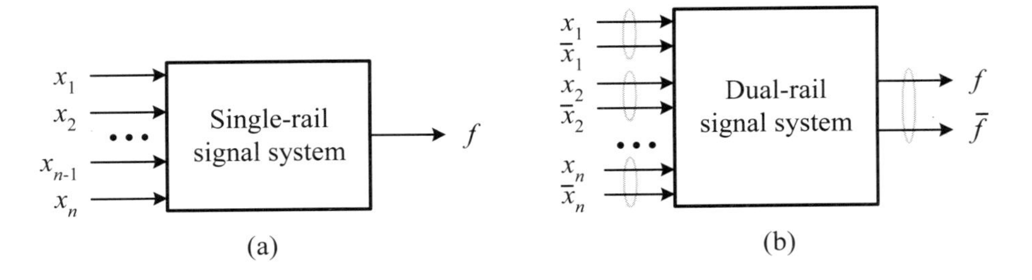

Figure 7.1: The (a) single-rail and (b) dual-rail signal systems.

In the single-rail signal system, as shown in Figure 7.1(a), both signals input to and output from the system are represented in either true or complementary form but not both. The most popular single-rail CMOS static logic circuits include the following:

- CMOS (or called *fully CMOS*)
- TG-based logic (including pass-transistor-based logic)
- Ratioed logic
  1. Pseudo-nMOS (pseudo p-type metal-oxide-semiconductor (pseudo-pMOS) logic is also possible)
  2. Nonthreshold logic (NTL)
  3. Ganged CMOS logic

In the dual-rail signal system, as indicated in Figure 7.1(b), both signals input to and output from the system are represented in both true and complementary forms at the same time. The most popular dual-rail CMOS static logic circuits include the following:

- CVSL (cascode voltage switch logic)
- CPL (complementary pass-transistor logic)
- DCVSPG (differential cascode voltage switch with pass-gate logic)
- DPL (double-pass-transistor logic)

In the rest of this chapter, we will take a look at both single-rail and dual-rail signal systems and their related logic circuits in more detail. The interested reader is encouraged to refer to the references listed at the end of this chapter to get more insights into these logic circuits.

7.1.2 CMOS Inverters

A CMOS inverter (also called a *NOT gate*) consists of a pMOS and an nMOS transistor shown in Figure 7.2(a). Its logic symbol is shown in Figure 7.2(b). In the inverter, the source of an nMOS transistor is connected to ground (sometimes called V_{SS}), while the source of a pMOS transistor is connected to V_{DD}. The drains of both transistors are connected together to serve as the output. In addition, the substrates of nMOS and pMOS transistors are directly routed to ground and V_{DD}, respectively. Consequently, both transistors in a CMOS inverter do not encounter the body effect.

The inverter circuit works as follows. As the input voltage is below the low-level input voltage (V_{IL}), the pMOS transistor is turned on, whereas the nMOS transistor is turned off or is only slightly on. The output voltage rises to V_{DD}. On the other

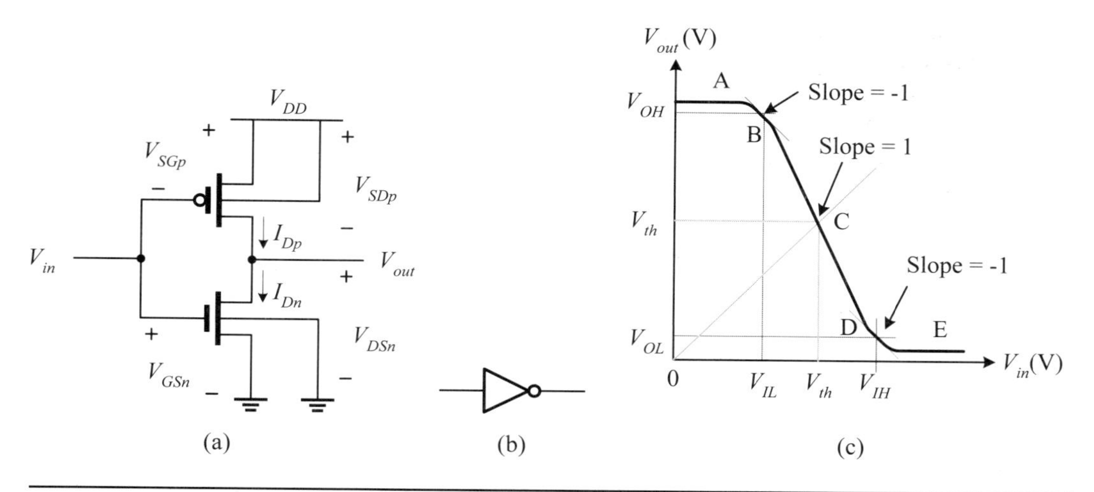

Figure 7.2: The (a) circuit, (b) logic symbol, and (c) voltage-transfer characteristic (VTC) of a CMOS inverter.

Table 7.1: The operating regions of nMOS and pMOS transistors in the inverter.

Region	V_{in}	V_{out}	nMOS	pMOS
A	$< V_{T0n}$	V_{OH}	Cutoff	Linear
B	V_{IL}	V_{OH}	Saturation	Linear
C	V_{th}	V_{th}	Saturation	Saturation
D	V_{IH}	V_{OL}	Linear	Saturation
E	$> V_{DD} - \lvert V_{T0p} \rvert$	V_{OL}	Linear	Cutoff

hand, as the input voltage rises above the high-level input voltage (V_{IH}), the pMOS transistor is turned off while the nMOS transistor is turned on. The output voltage falls to 0 V. As a result, it functions as an inverter.

7.1.2.1 Voltage-Transfer Characteristic The circuit behavior of a CMOS inverter can be characterized by a *voltage-transfer characteristic* (VTC), which expresses the output voltage as a function of the input voltage, ranging from 0 V to V_{DD}. The VTC of the inverter shown in Figure 7.2(a) is plotted in Figure 7.2(c). It can be partitioned into five regions, as indicated in the figure with capital letters: A, B, C, D, and E. The operating regions of nMOS and pMOS transistors in the inverter are summarized in Table 7.1.

The nMOS transistor is off while the pMOS transistor is on when the input voltage V_{in} is smaller than the threshold voltage V_{T0n} of the nMOS transistor. Hence, in region A, the output is virtually at V_{DD}. The nMOS transistor starts to conduct and lower the output voltage V_{out} a little bit as the input voltage increases above V_{T0n}. This results in the region B.

When the input voltage V_{in} further increases a little more, both nMOS and pMOS transistors go into their saturation regions of operation such that there exists a low-resistance path between the power supply and ground. This region is denoted as region C. It is also called the *transition region* of the inverter and is useful when the inverter is used as an amplifier. The voltage gain of this region is about −10 to −20 (V/V) for most CMOS inverters.

The other two regions are in parallel to regions B and A. When the input voltage V_{in} approaches $V_{DD} - |V_{T0p}|$, the pMOS transistor is no longer on, and instead, it is turned off. However, the nMOS transistor is still on, and hence, the output voltage V_{out} is continually discharged toward ground. Finally, the output voltage stays in region E.

From the VTC shown in Figure 7.2(c), we can see that two points on the VTC curve have slopes of minus 1. These two points define the input voltages associated with them as the low-level input voltage V_{IL} and high-level input voltage V_{IH}, respectively. The corresponding output voltages are separately defined as high-level output voltage V_{OH} and low-level output voltage V_{OL}. From these four voltages, we can calculate the low-noise and high-noise margins of the inverter. Refer to Section 5.2.1 for more details.

The VTC of an inverter is characterized by an *inverter threshold voltage*,[1] V_{th}, which is defined as the input voltage when it is equal to the output voltage, namely, $V_{in} = V_{out} = V_{th}$. In this case, both nMOS and pMOS transistors operate in their saturation regions. This can be easily checked by using the condition: $V_{DS} \geq V_{GS} - V_{T0n}$ and $V_{SD} \geq V_{SG} - |V_{T0p}|$. As a result, because both nMOS and pMOS transistors are in series, and hence, their currents must be equal, the inverter threshold voltage V_{th} can be readily obtained as follows:

$$V_{th} = \frac{V_{DD} - |V_{T0p}| + \sqrt{k_R}V_{T0n}}{1 + \sqrt{k_R}} \tag{7.1}$$

where the subscript "0" in both threshold voltages, V_{T0n} and V_{T0p}, is used to emphasize that both nMOS and pMOS transistors encounter no body effect. The term k_R is known as the *ratio of device transconductances* of nMOS and pMOS transistors and is defined as

$$k_R = \frac{k_n}{k_p} = \frac{\mu_n C_{ox}\left(\dfrac{W_n}{L_n}\right)}{\mu_p C_{ox}\left(\dfrac{W_p}{L_p}\right)} = \frac{\mu_n}{\mu_p}\left(\frac{W_n}{L_n}\right)\left(\frac{L_p}{W_p}\right) \tag{7.2}$$

where μ_n and μ_p are electron and hole mobilities, respectively, and in general $\mu_n = 2 \sim 3\mu_p$. The inverter threshold voltage V_{th} is at $\frac{1}{2}V_{DD}$ if k_R is 1 and V_{T0n} equals $|V_{T0p}|$.

The following example further illustrates some more insights into the inverter threshold voltage of a typical inverter circuit.

■ **Example 7-1: (The inverter threshold voltage.)** Consider the inverter shown in Figure 7.2(a). Assume that $V_{DD} = 3.3$ V and $V_{T0n} = |V_{T0p}| = 0.5$ V. The electron and hole mobilities are 540 cm^2/V-s and 180 cm^2/V-s, respectively.

1. Calculate the inverter threshold voltage, V_{th}, if both nMOS and pMOS transistors have the same size.
2. Determine the sizes of both nMOS and pMOS transistors if the inverter threshold voltage is set to $\frac{1}{2}V_{DD}$.

Solution: The detailed calculations follow:

[1]Do not confuse with the threshold voltage of MOS transistors.

1. We first calculate the k_R using Equation (7.2).

$$k_R = \frac{k_n}{k_p} = \frac{\mu_n}{\mu_p}\left(\frac{W_n}{L_n}\right)\left(\frac{L_p}{W_p}\right) = \frac{540}{180}\left(\frac{W_n}{L_n}\right)\left(\frac{L_p}{W_p}\right) = 3$$

Hence, by using Equation (7.1), the inverter threshold voltage is equal to

$$\begin{aligned} V_{th} &= \frac{V_{DD} - |V_{T0p}| + \sqrt{k_R}V_{T0n}}{1+\sqrt{k_R}} \\ &= \frac{3.3 - 0.5 + \sqrt{3}\times 0.5}{1+\sqrt{3}} = 1.34 \text{ V} \end{aligned}$$

2. From the above discussion, we know that to set the inverter threshold voltage to $\frac{1}{2}V_{DD}$, k_R must be set to 1. That is,

$$k_R = \frac{k_n}{k_p} = \frac{\mu_n}{\mu_p}\left(\frac{W_n}{L_n}\right)\left(\frac{L_p}{W_p}\right) = 1$$

Consequently, if we set the channel lengths of both nMOS and pMOS transistors to the minimum length L_{min} allowed by the process at hand, the channel width ratio of nMOS to pMOS transistor is

$$\frac{W_n}{W_p} = \frac{\mu_p}{\mu_n} = \frac{180}{540} = \frac{1}{3}$$

As a consequence, the channel width of the pMOS transistor is three times that of the nMOS transistor. That is, $W_p = 3W_n$.

■

7.1.2.2 The k_R-Ratio Effects In general, all CMOS inverters have the same VTC shape, just like the one shown in Figure 7.2(c). The specific curve is identified by the inverter threshold voltage, which is in turn determined by the k_R value. Three curves with different k_R values are depicted in Figure 7.3(a). It is easy to see that curves are moved leftward as the k_R values increase and moved rightward as the k_R values decrease. The effects of k_R on inverter threshold voltages (V_{th}) are plotted in Figure 7.3(b).

One immediate consequence of Figure 7.3(a) is that we can easily modify an inverter into a circuit with hysteresis if we can design the inverter circuit to have different k_R values as its input voltage goes forward to V_{DD} from 0 V and backward to 0 V from V_{DD}. Such a circuit with hysteresis is known as a *Schmitt circuit*, which will be discussed later in this book (see Chapter 15).

■ Review Questions

Q7-1. Define the single-rail and dual-rail signal systems.

Q7-2. How would you characterize a CMOS inverter?

Q7-3. Define the inverter threshold voltage of a CMOS inverter.

Q7-4. How would you define the two voltages: V_{OL} and V_{OH} from a VTC of an inverter?

Q7-5. How would you define the two voltages: V_{IL} and V_{IH} from a VTC of an inverter?

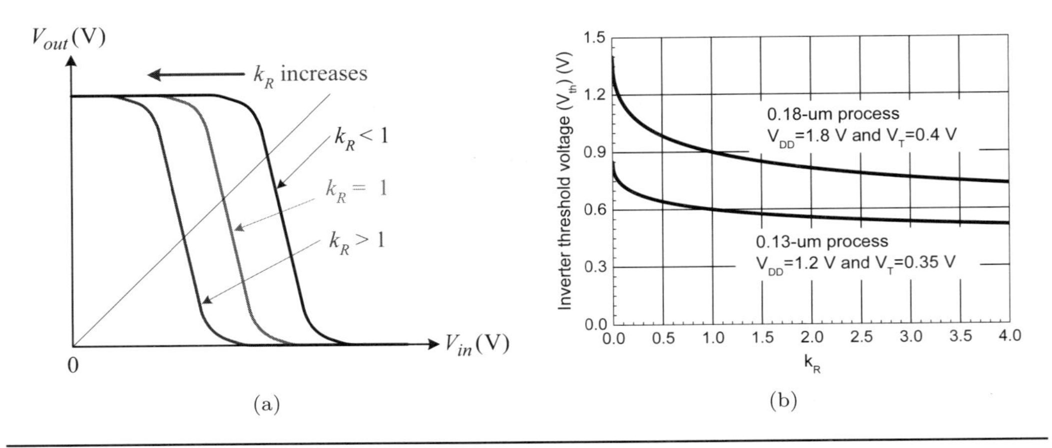

Figure 7.3: The effects of k_R ratios of an inverter: (a) on VTC; (b) on inverter threshold voltages.

7.1.2.3 A Mathematical Analysis We are concerned with the mathematical analysis of the voltage-transfer characteristic (VTC) of a static inverter. Referring to Figure 7.2 and Table 7.1, we can see that V_{IL}, V_{IH}, and V_{th} are the three important parameters determining the VTC of an inverter. Hence, we will focus on these three voltages in the following.

Calculation of $\mathbf{V_{IL}}$. At this point, the nMOS transistor is in the saturation region of operation and the pMOS transistor operates in the linear region. Because of an in-series connection of both MOS transistors, we have the following expression:

$$\frac{1}{2}k_n(V_{GSn} - V_{T0n})^2 = \frac{1}{2}k_p[2(V_{SGp} - |V_{T0p}|)V_{SDp} - V_{SDp}^2] \tag{7.3}$$

By substituting $V_{GSn} = V_{in}$, $V_{SGp} = V_{DD} - V_{in}$, and $V_{SDp} = V_{DD} - V_{out}$ into the above equation, we obtain

$$\frac{1}{2}k_n(V_{in} - V_{T0n})^2 = \frac{1}{2}k_p[2(V_{DD} - V_{in} - |V_{T0p}|)(V_{DD} - V_{out}) - (V_{DD} - V_{out})^2] \tag{7.4}$$

Using the definition of V_{IL}, that is, $dV_{out}/dV_{in} = -1$, and setting $V_{in} = V_{IL}$, the following representation is the result.

$$k_n(V_{IL} - V_{T0n}) = k_p(2V_{out} - V_{IL} - |V_{T0p}| - V_{DD}) \tag{7.5}$$

Solving it for V_{IL}, we get

$$V_{IL} = \frac{2V_{out} - |V_{T0p}| - V_{DD} + k_R V_{T0n}}{1 + k_R} \tag{7.6}$$

where $k_R = k_n/k_p$. Because the V_{IL} is still a function of output voltage V_{out}, to solve it we need another equation. By solving both Equations (7.4) and (7.6) simultaneously and setting $V_{out} = V_{OH}$ and $V_{in} = V_{IL}$, V_{IL} and V_{OH} can be obtained through complicated algebraic manipulations and separately expressed as follows:

$$V_{IL} = \frac{2\sqrt{k_R}(V_{DD} - V_{T0n} - |V_{T0p}|)}{(k_R - 1)\sqrt{k_R + 3}} - \frac{V_{DD} - k_R V_{T0n} - |V_{T0p}|}{k_R - 1} \tag{7.7}$$

and

$$V_{OH} = \frac{(1 + k_R)V_{IL} + V_{DD} - k_R V_{T0n} + |V_{T0p}|}{2} \tag{7.8}$$

Calculation of $\mathbf{V_{IH}}$. The calculation of V_{IH} is quite similar to that of V_{IL}. When the input signal reaches the operating region D, the nMOS transistor is in the linear region of operation and the pMOS transistor operates in the saturation region. As a consequence, we have the following current equation:

$$\frac{1}{2}k_p(V_{SGp} - |V_{T0p}|)^2 = \frac{1}{2}k_n[2(V_{SGn} - V_{T0n})V_{DSn} - V_{DSn}^2] \tag{7.9}$$

By substituting $V_{GSn} = V_{in}$, $V_{SGp} = V_{DD} - V_{in}$, and $V_{DSn} = V_{out}$ into the above equation, we have

$$\frac{1}{2}k_p(V_{DD} - V_{in} - |V_{T0p}|)^2 = \frac{1}{2}k_n[2(V_{in} - V_{T0n})V_{out} - V_{out}^2] \tag{7.10}$$

Using the definition of V_{IH}, that is, $dV_{out}/dV_{in} = -1$, and setting $V_{in} = V_{IH}$, we obtain

$$k_p(V_{IH} - V_{DD} + |V_{T0p}|) = k_n(2V_{out} - V_{IH} + V_{T0n}) \tag{7.11}$$

Solving it for V_{IH}, the result is as follows:

$$V_{IH} = \frac{V_{DD} - |V_{T0p}| + k_R(2V_{out} + V_{T0n})}{1 + k_R} \tag{7.12}$$

By solving both Equations (7.10) and (7.12) simultaneously and setting $V_{out} = V_{OL}$ and $V_{in} = V_{IH}$, V_{IH} and V_{OL} are obtained through complicated algebraic manipulations and can be separately represented as follows:

$$V_{IH} = \frac{2k_R(V_{DD} - V_{T0n} - |V_{T0p}|)}{(k_R - 1)\sqrt{3k_R + 1}} - \frac{V_{DD} - k_R V_{T0n} - |V_{T0p}|}{k_R - 1} \tag{7.13}$$

and

$$V_{OL} = \frac{(1 + k_R)V_{IH} - V_{DD} - k_R V_{T0n} + |V_{T0p}|}{2k_R} \tag{7.14}$$

Calculation of $\mathbf{V_{th}}$. Finally, we proceed to find the inverter threshold voltage V_{th}. The inverter threshold voltage V_{th} is defined as the point where $V_{in} = V_{out} = V_{th}$. At this point, both nMOS and pMOS transistors are in the saturation regions of operation. Hence, we have the following current expression:

$$\frac{1}{2}k_p(V_{GSn} - V_{T0n})^2 = \frac{1}{2}k_p(V_{SGp} - |V_{T0p}|)^2 \tag{7.15}$$

By substituting $V_{GSn} = V_{in} = V_{th}$ and $V_{SGp} = V_{DD} - V_{in} = V_{DD} - V_{th}$ into the above expression, the following equation is the result:

$$\frac{1}{2}k_p(V_{th} - V_{T0n})^2 = \frac{1}{2}k_p(V_{DD} - V_{th} - |V_{T0p}|)^2 \tag{7.16}$$

Solving the above equation for V_{th}, a closed-form expression is obtained as follows:

$$V_{th} = \frac{V_{DD} - |V_{T0p}| + \sqrt{k_R}V_{T0n}}{1 + \sqrt{k_R}} \tag{7.17}$$

For an ideal inverter, the inverter threshold voltage is set to half of the supply voltage, that is,

$$V_{th} = \frac{1}{2}V_{DD} \tag{7.18}$$

Substituting it into Equation (7.17) and solving for the k_R ratio, we have

$$k_R = \frac{k_n}{k_p} = \left(\frac{V_{DD} - |V_{T0p}| - V_{th}}{V_{th} - V_{T0n}}\right)^2 = \left(\frac{0.5V_{DD} - |V_{T0p}|}{0.5V_{DD} - V_{T0n}}\right)^2 \tag{7.19}$$

If $V_{T0} = V_{T0n} = |V_{T0p}|$, then $k_R = 1$ and

$$k_R = 1 = \frac{k_n}{k_p} = \frac{\mu_n}{\mu_p}\frac{(W_n/L_n)}{(W_p/L_p)} \tag{7.20}$$

Therefore, the ratio of the aspect ratios of the nMOS to the pMOS transistor is equal to the ratio of the hole to the electron mobility.

$$\frac{(W_n/L_n)}{(W_p/L_p)} = \frac{\mu_p}{\mu_n} \tag{7.21}$$

When designing or analyzing an inverter in 0.18-μm or below processes, the effective electron and hole mobilities should be used to obtain more accurate results. An example is used to illustrate this in the following.

■ Example 7-2: (An analysis example of an inverter.) Assume that $V_{DD} =$ 1.8 V, $W_n = 4\lambda$, $W_p = 20\lambda$, and $L_n = L_p = 2\lambda$. Using 0.18-μm parameters: $\mu_{effn} =$ 287 cm^2/V-s, $\mu_{effp} = 88$ cm^2/V-s, and $V_{Tn} = |V_{Tp}| = 0.4$ V, calculate V_{OL}, V_{OH}, V_{IL}, V_{IH}, and V_{th}.

Solution: The k_R ratio is found to be

$$k_R = \frac{k_n}{k_p} = \frac{\mu_{effn}W_n}{\mu_{effp}W_p} = \frac{287 \times 4}{88 \times 20} = 0.65$$

Using Equation (7.17), the inverter threshold voltage V_{th} is equal to

$$\begin{aligned} V_{th} &= \frac{V_{DD} - |V_{T0p}| + \sqrt{k_R} \times V_{T0n}}{1 + \sqrt{k_R}} \\ &= \frac{1.8 - 0.4 + \sqrt{0.65} \times 0.4}{1 + \sqrt{0.65}} = 0.95 \text{ V} \end{aligned}$$

By using Equation (7.7), V_{IL} can be calculated as follows:

$$\begin{aligned} V_{IL} &= \frac{2\sqrt{0.65}(1.8 - 0.4 - 0.4)}{(0.65 - 1)\sqrt{0.65 + 3}} - \frac{1.8 - 0.65 \times 0.4 - 0.4}{0.65 - 1} \\ &= 0.85 \text{ V} \end{aligned}$$

The corresponding V_{OH} can be obtained by using Equation (7.8) and is

$$\begin{aligned} V_{OH} &= \frac{(1 + 0.65) \times 0.85 + 1.8 + 0.4 - 0.65 \times 0.4}{2} \\ &= 1.67 \text{ V} \end{aligned}$$

From Equation (7.13), we have V_{IH} as follows:

$$\begin{aligned} V_{IH} &= \frac{2 \times 0.65(1.8 - 0.4 - 0.4)}{(0.65 - 1)\sqrt{1 + 3 \times 0.65}} - \frac{1.8 - 0.65 \times 0.4 - 0.4}{0.65 - 1} \\ &= 1.1 \text{ V} \end{aligned}$$

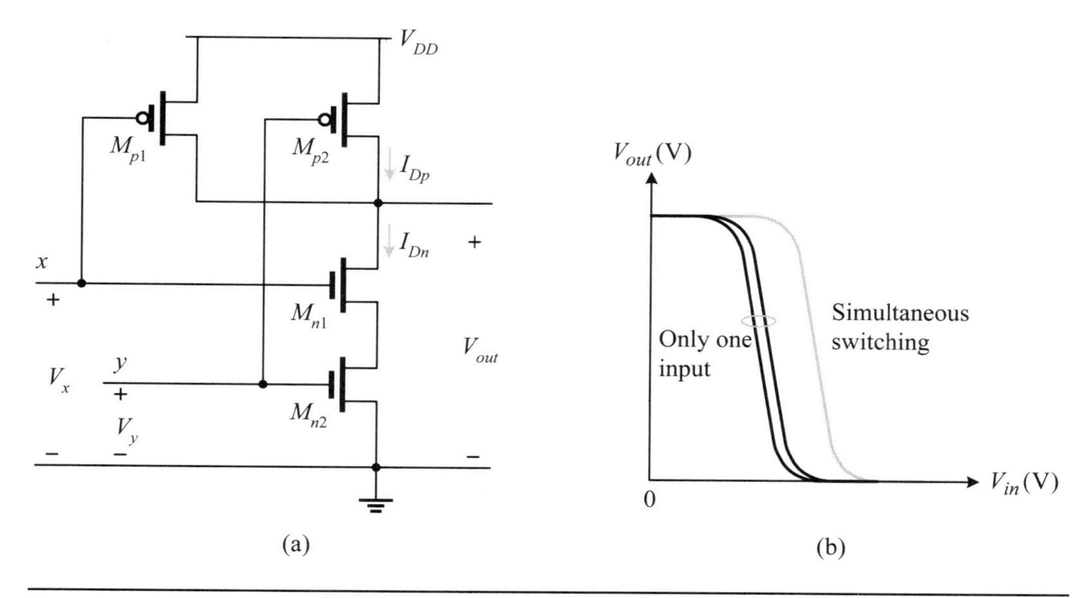

Figure 7.4: The (a) logic circuit and (b) VTC of a two-input NAND gate.

By using Equation (7.14), the corresponding V_{OL} is obtained.

$$\begin{aligned} V_{OL} &= \frac{(1+0.65)\times 1.1 - 1.8 + 0.4 - 0.65 \times 0.4}{2 \times 0.65} \\ &= 0.12 \text{ V} \end{aligned}$$

■

The important issue of designing a CMOS inverter is to properly size the aspect ratio W/L values of both pMOS and nMOS transistors so that they may satisfy a set of design specifications. The sizing process is generally a trade-off between timing, power dissipation, area, and noise margins.

7.1.3 NAND Gates

Like inverters, an n-input NAND gate is also the basic gate in the CMOS logic circuit family. Recall that an n-input NAND gate has n nMOS transistors connected in series and n pMOS transistors connected in parallel. To see this, consider the two-input NAND gate circuit shown in Figure 7.4(a). The NAND gate circuit has two nMOS and two pMOS transistors, where each input is connected to one pair of nMOS and pMOS transistors. Therefore, each input of a two-input NAND gate is much like that of an inverter.

Like an inverter, a two-input NAND gate is also characterized by its VTC, as shown in Figure 7.4(b). In a two-input NAND gate, the two nMOS transistors are connected in series and the two pMOS transistors are connected in parallel. Because of a common substrate shared by two nMOS transistors, the one adjacent to the output node encounters the body effect but the other does not. Nevertheless, none of the pMOS transistors have the body effect. Why?

As can be seen from Figure 7.4(b), there are two different VTCs whenever one input is switching while the other input is held at V_{DD}. This is because the nMOS transistor M_{n1} encounters the body effect, whereas the nMOS transistor M_{n2} does not. Therefore, they have different threshold voltages and hence lead to different VTCs. The third VTC is the one when both inputs are tied together and hence are switching simultaneously.

7.1.3.1 Equivalent-NOT Gate The detailed analysis of an n-input NAND gate is generally quite complicated (see Problem 7-5). Hence, in practice, it is typical to use the concept of an equivalent-NOT gate; namely, we first reduce the NAND gate in question into an equivalent-NOT gate and then find the threshold voltage of the equivalent-NOT gate. In what follows, we will detail this concept.

Before proceeding, we need to consider the combined effects on the device transconductance k of n MOS transistors. It is easy to show that the equivalent device transconductance k_{eqv} is the sum of the device transconductances of individual transistors when they are connected in parallel. That is,

$$k_{eqv} = k_1 + k_2 + \cdots + k_n \qquad \text{(in parallel)} \tag{7.22}$$

The equivalent device transconductance k_{eqv} for n transistors connected in series is found to be the reciprocal of the sum of the reciprocal device transconductances of individual transistors. It can be expressed as follows:

$$\frac{1}{k_{eqv}} = \frac{1}{k_1} + \frac{1}{k_2} + \cdots + \frac{1}{k_n} \qquad \text{(in series)} \tag{7.23}$$

Based on these two results, the equivalent device transconductance k_{eqvp} of the equivalent pMOS transistor for a two-input NAND gate is equal to

$$k_{eqvp} = k_p + k_p = 2k_p \tag{7.24}$$

and the equivalent device transconductance k_{eqvn} of the equivalent nMOS transistor is found to be

$$\frac{1}{k_{eqvn}} = \frac{1}{k_n} + \frac{1}{k_n} \Longrightarrow k_{eqvn} = \frac{1}{2}k_n \tag{7.25}$$

The resulting equivalent inverter is shown in Figure 7.5(b) with the indication of equivalent device transconductances of nMOS and pMOS transistors.

Substituting the equivalent device transconductances of nMOS and pMOS transistors into Equation (7.1), we obtain

$$\begin{aligned} V_{th} &= \frac{V_{DD} - |V_{T0p}| + \sqrt{k_{R(NAND)}}V_{T0n}}{1 + \sqrt{k_{R(NAND)}}} \\ &= \frac{V_{DD} - |V_{T0p}| + \frac{1}{2}\sqrt{k_R}V_{T0n}}{1 + \frac{1}{2}\sqrt{k_R}} \end{aligned} \tag{7.26}$$

Thus, the threshold voltage of the two-input NAND gate is not at $\frac{1}{2}V_{DD}$ when the k_R is set to 1. Some more insights into this are explored in the following example.

Example 7-3: (A two-input NAND gate.) Assume that V_{DD} is 3.3 V and $V_{T0n} = |V_{T0p}| = 0.5$ V. The electron and hole mobilities are 540 cm^2/V-s and 180 cm^2/V-s, respectively. Calculate the threshold voltage V_{th} of a two-input NAND gate under the following conditions:

1. If all nMOS and pMOS transistors have the same size.

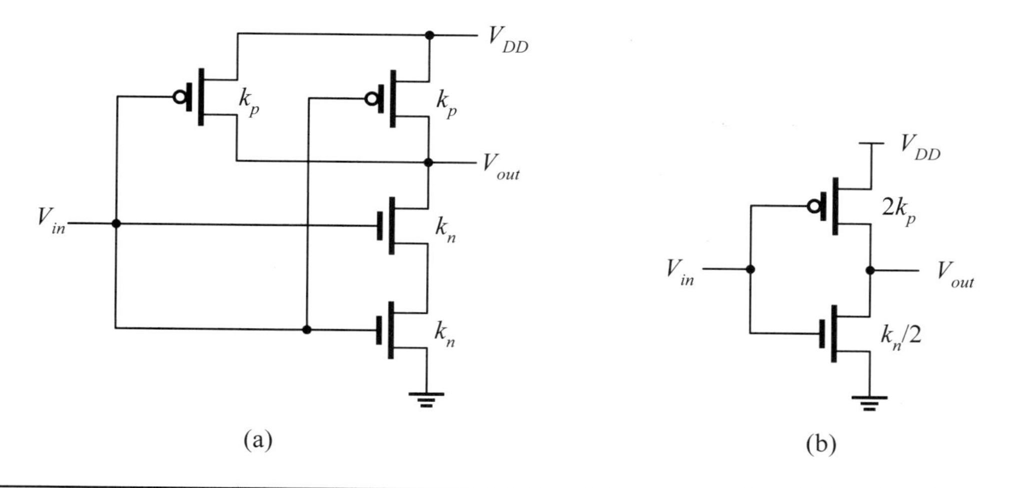

Figure 7.5: The (a) logic circuit and (b) its equivalent-NOT gate of a two-input NAND gate.

2. If all pMOS transistors have twice the size of nMOS transistors and all nMOS transistors have the same size.

Solution: The detailed calculations are as follows:

1. Since all nMOS and pMOS transistors have the same size, the k_R can be calculated as follows:

$$k_R = \frac{k_n}{k_p} = \frac{\mu_n}{\mu_p}\left(\frac{W_n}{L_n}\right)\left(\frac{L_p}{W_p}\right) = \frac{540}{180}\left(\frac{W_n}{L_n}\right)\left(\frac{L_n}{W_n}\right) = 3$$

Hence, by using Equation (7.26), the threshold voltage of the NAND gate is found to be

$$\begin{aligned} V_{th} &= \frac{V_{DD} - |V_{T0p}| + \frac{1}{2}\sqrt{k_R}V_{T0n}}{1 + \frac{1}{2}\sqrt{k_R}} \\ &= \frac{3.3 - 0.5 + \frac{1}{2}\sqrt{3}\cdot 0.5}{1 + \frac{1}{2}\sqrt{3}} = 1.73 \text{ V} \end{aligned}$$

2. Now, we recalculate the equivalent k_R using Equation (7.2). Because all pMOS transistors have twice the size of nMOS transistors, the k_R is changed to the following value:

$$\begin{aligned} k_R &= \frac{k_n}{k_p} = \frac{\mu_n}{\mu_p}\left(\frac{W_n}{L_n}\right)\left(\frac{L_p}{W_p}\right) \\ &= \frac{540}{180}\left(\frac{W_n}{L_n}\right)\left(\frac{L_n}{2W_n}\right) = 1.5 \end{aligned}$$

By using Equation (7.26), the threshold voltage of the NAND gate becomes

$$\begin{aligned} V_{th} &= \frac{V_{DD} - |V_{T0p}| + \frac{1}{2}\sqrt{k_R}V_{T0n}}{1 + \frac{1}{2}\sqrt{k_R}} \\ &= \frac{3.3 - 0.5 + \frac{1}{2}\sqrt{1.5}\cdot 0.5}{1 + \frac{1}{2}\sqrt{1.5}} = 1.93 \text{ V} \end{aligned}$$

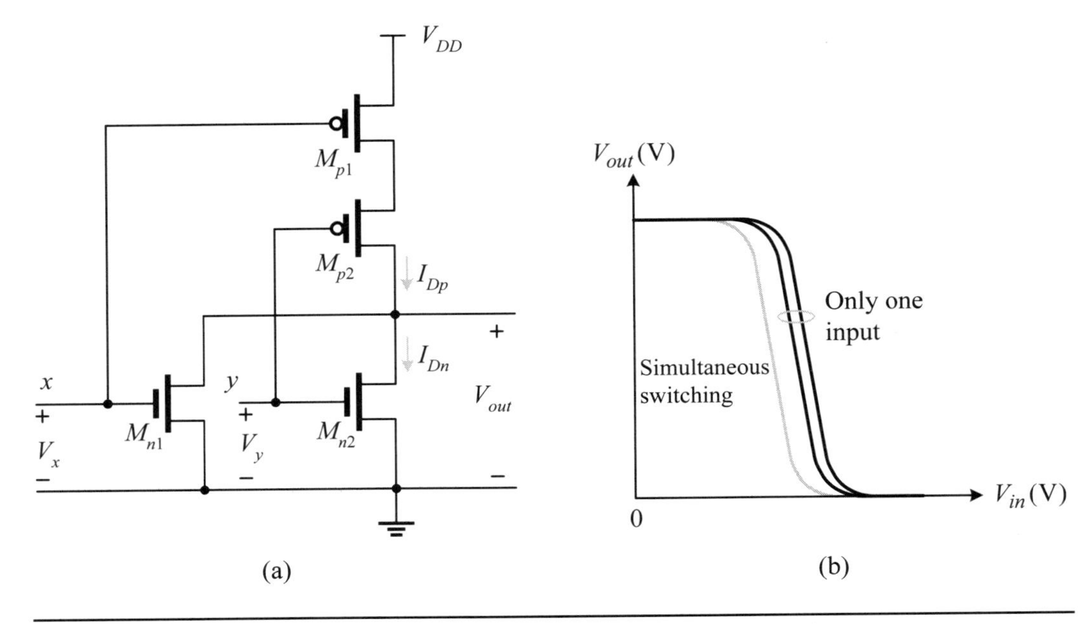

Figure 7.6: The (a) logic circuit and (b) VTC of a two-input NOR gate.

Hence, it is shifted to a more positive voltage, a situation much like the case of the inverter.

■

■ Review Questions

Q7-6. Show that Equation (7.22) is valid.

Q7-7. Show that Equation (7.23) is valid.

Q7-8. Explain why the nMOS transistor M_{n1} in Figure 7.4 encounters the body effect.

Q7-9. Describe the threshold voltage of a 3-input NAND gate.

7.1.4 NOR Gates

In addition to inverters and NAND gates, an n-input NOR gate is also the basic gate in the CMOS logic circuit family. Remember that an n-input NOR gate has n nMOS transistors connected in parallel and n pMOS transistors connected in series. To see this, consider the two-input NOR gate circuit depicted in Figure 7.6(a). The NOR gate circuit has two nMOS and two pMOS transistors, where each input is connected to one pair of nMOS and pMOS transistors. Thus, each input of a NOR gate resembles that of an inverter.

Like an inverter, a two-input NOR gate is also characterized by its VTC, as shown in Figure 7.6(b). In a two-input NOR gate, the two nMOS transistors are connected in parallel while the two pMOS transistors are connected in series. Because of a common substrate (n-well) that is shared by two pMOS transistors, the one adjacent to the

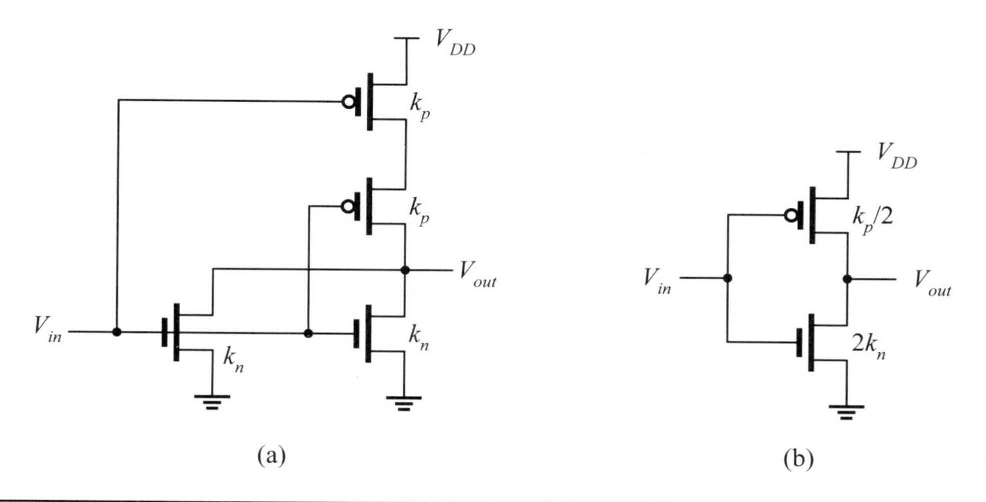

Figure 7.7: The (a) logic circuit and (b) its equivalent-NOT gate of a two-input NOR gate.

output node has the body effect but the other does not. Nevertheless, none of the nMOS transistors encounter the body effect. Why?

As can be seen from Figure 7.6(b), there are two different VTCs whenever one input is switching while the other input is held at the ground level (0 V). This is because the pMOS transistor M_{p2} encounters the body effect, whereas the pMOS transistor M_{p1} does not. Therefore, they have different threshold voltages and hence lead to different VTCs. The third VTC is the one where both inputs are tied together and thus are switching simultaneously.

7.1.4.1 Equivalent-NOT Gate Like NAND gates, the detailed analysis of an n-input NOR gate is generally quite complicated (see Problem **7-7**). Hence, the concept of an equivalent-NOT gate is usually used in practice to avoid the complicated, detailed analysis of an n-input NOR gate. Based on this, we first reduce the NOR gate at hand into an equivalent-NOT gate and then find the threshold voltage of the equivalent-NOT gate. In what follows, we will apply this concept to an n-input NOR gate as we have done in the case of the NAND gate.

As described, the equivalent device transconductance k_{eqvp} of the equivalent pMOS transistor for a two-input NOR gate is equal to

$$\frac{1}{k_{eqvp}} = \frac{1}{k_p} + \frac{1}{k_p} \Longrightarrow k_{eqvp} = \frac{1}{2}k_p \tag{7.27}$$

and the equivalent device transconductance k_{eqvn} of the nMOS transistor is

$$k_{eqvn} = k_n + k_n = 2k_n \tag{7.28}$$

The resulting equivalent-NOT gate is shown in Figure 7.7(b).

Substituting the equivalent device transconductances of nMOS and pMOS transistors into Equation (7.1), we obtain

$$V_{th} = \frac{V_{DD} - |V_{T0p}| + \sqrt{k_{R(NOR)}}V_{T0n}}{1 + \sqrt{k_{R(NOR)}}}$$

$$= \frac{V_{DD} - |V_{T0p}| + 2\sqrt{k_R} V_{T0n}}{1 + 2\sqrt{k_R}} \tag{7.29}$$

Hence, the threshold voltage of the two-input NOR gate is not at $\frac{1}{2}V_{DD}$ when the k_R is set to 1. Some more insight into this is given in the following example.

Example 7-4: (A two-input NOR gate.) Assume that the V_{DD} is 3.3 V and $V_{T0n} = |V_{T0p}| = 0.5$ V. The electron and hole mobilities are 540 cm^2/V-s and 180 cm^2/V-s, respectively. Calculate the threshold voltage V_{th} of a two-input NOR gate under the following conditions:

1. If all nMOS and pMOS transistors have the same size.
2. If all pMOS transistors have four times the size of nMOS transistors and all nMOS transistors have the same size.

Solution: The detailed calculations are as follows:

1. We first calculate the equivalent k_R using Equation (7.2). Since all nMOS and pMOS transistors have the same size, the k_R is as follows:
$$k_R = \frac{k_n}{k_p} = \frac{\mu_n}{\mu_p}\left(\frac{W_n}{L_n}\right)\left(\frac{L_p}{W_p}\right) = \frac{540}{180}\left(\frac{W_n}{L_n}\right)\left(\frac{L_n}{W_n}\right) = 3$$
Hence, by using Equation (7.29), the threshold voltage of the NOR gate is
$$V_{th} = \frac{V_{DD} - |V_{T0p}| + 2\sqrt{k_R} V_{T0n}}{1 + 2\sqrt{k_R}}$$
$$= \frac{3.3 - 0.5 + 2\sqrt{3} \cdot 0.5}{1 + 2\sqrt{3}} = 1.02 \text{ V}$$
2. Now we recalculate the equivalent k_R using Equation (7.2). Because all pMOS transistors have four times the size of nMOS transistors, the k_R is changed to the following value:
$$k_R = \frac{k_n}{k_p} = \frac{\mu_n}{\mu_p}\left(\frac{W_n}{L_n}\right)\left(\frac{L_p}{W_p}\right) = \frac{540}{180}\left(\frac{W_n}{L_n}\right)\left(\frac{L_n}{4W_n}\right) = 0.75$$
By using Equation (7.29), the threshold voltage of NOR gate is found to be
$$V_{th} = \frac{V_{DD} - |V_{T0p}| + 2\sqrt{k_R} V_{T0n}}{1 + 2\sqrt{k_R}}$$
$$= \frac{3.3 - 0.5 + 2\sqrt{0.75} \cdot 0.5}{1 + 2\sqrt{0.75}} = 1.34 \text{ V}$$
Hence, it is shifted to a more positive voltage, a situation much like the case of the inverter and NAND gate.

Review Questions

Q7-10. Explain why the pMOS transistor M_{p2} in Figure 7.6 encounters the body effect.

Q7-11. Describe the threshold voltage of a 3-input NOR gate.

7.1.5 Sizing Basic Gates

Generally, the transistors in a logic gate need to be sized according to different criteria on the basis of different applications. The most widely accepted criteria are propagation delays, area, and power dissipation. In this section, we are concerned with these and classify the gates into the following three types: *symmetric gates*, *asymmetric gates*, and *skewed logic gates*.

7.1.5.1 Symmetric Gates Recall that the on-resistance of the pMOS transistor in an inverter determines the rise time, t_{rise}, and the low-to-high propagation delay, t_{pLH}, of the inverter; the on-resistance of the nMOS transistor in an inverter determines the fall time, t_{fall}, and the high-to-low propagation delay, t_{pHL}, of the inverter.

Because electron mobility is usually two to three times larger than hole mobility in a silicon semiconductor, the current in a pMOS transistor is only half to one-third of that in an nMOS transistor under the same size and operating voltage. In other words, the on-resistance of a pMOS transistor is two to three times larger than that of an nMOS transistor. As a result, the rise time and low-to-high propagation delay of an inverter are larger than the fall time and high-to-low propagation delay by a factor of two to three.

To balance the rise and fall times and low-to-high and high-to-low propagation delays, in practice we often widen the channel width of pMOS transistors to compensate for the effect of smaller hole mobility inherently associated with pMOS transistors. The rationale behind this is based on the following observation. From the current-voltage relationship of both nMOS and pMOS transistors, the current is proportional to device transconductance k, which is in turn determined by the transistor size and carrier mobility. Recall that the device transconductance k can be expressed as follows:

$$k = \mu C_{ox}\left(\frac{W}{L}\right) \tag{7.30}$$

To maintain a definite value of device transconductance k, we have to adjust the channel width to compensate for the loss of carrier mobility. The gate-oxide capacitance $C_{ox} = \epsilon_{ox}/t_{ox}$ is a process parameter, determined by process, not by circuit designers.

Based on the above discussion, we usually make the (channel) width of the pMOS transistor twice that of the nMOS transistor in a basic inverter, as shown in Figure 1.34(a). The channel lengths of both transistors are set to the feature size of a given process at hand. For most digital logic circuits, we usually set the channel length of a transistor to the feature size and leave the channel width as a design parameter for optimizing the circuit performance. Hence, the channel width and size of a MOS transistor are often used synonymously.

For a two-input NAND gate, as shown in Figure 1.34(b), two nMOS transistors are connected in series while two pMOS transistors are connected in parallel. The on-resistance of the nMOS stack is the sum of the on-resistance of each nMOS transistor. To maintain the serial on-resistance with the same unit-resistance R as that of an nMOS transistor of a basic inverter, the channel width (that is, size) of each nMOS transistor is doubled, namely, $2W$. As for the two pMOS transistors, because they are connected in parallel, their channel width is not scaled up and still uses $2W$, the basic size as in the basic inverter. We may generalize this rule to an n-input NAND gate in which an n-nMOS stack is formed. In this case, the channel width of each nMOS transistor has to be widened up to n times that used in a basic inverter, that is, nW.

For a two-input NOR gate, as depicted in Figure 1.34(c), two nMOS transistors are connected in parallel and two pMOS transistors are connected in series. Hence, the

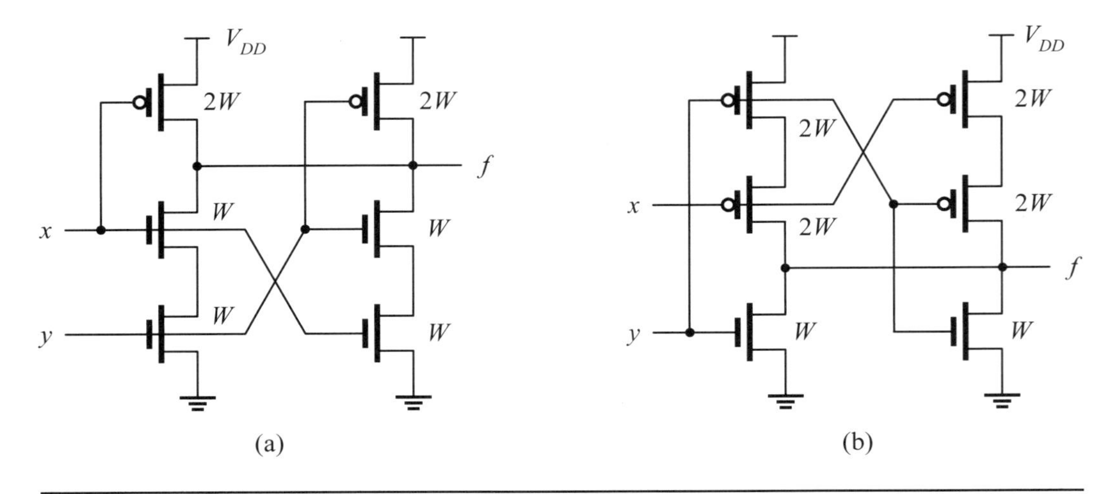

Figure 7.8: Perfectly symmetric two-input (a) NAND and (b) NOR gates.

channel width of each pMOS transistor is doubled to $4W$ to maintain the worst-case on-resistance of the pMOS stack with the same unit-resistance R as that of a pMOS transistor in the basic inverter. As for the two nMOS transistors, because they are an in-parallel connection, their channel width is not scaled up and still uses the basic size W as in the basic inverter. We may generalize this rule to an n-input NOR gate in which an n-pMOS stack is formed. In this situation, the channel width of each pMOS transistor has to be widened up to n times that used in a basic inverter, that is, $2nW$.

Delay effects of input ordering. We generally define the *inner input* of a logic gate as the input closer to the output node and the *outer input* as the input closer to the power-supply rail, either V_{DD} or ground. For an nMOS or a pMOS stack, the inner input usually encounters the body effect and thus has a higher threshold voltage to turn on. To reduce the body effect on this input, it should be activated as late as possible so that all intermediate nodes have been discharged. In summary, if we have known that one signal always arrives at the input later than the others, the logic gate would have a short propagation delay when that signal is applied to the inner input.

■ Example 7-5: (A perfectly symmetric two-input NAND gate.) As we described previously, the two voltage-transfer characteristics of a two-input NAND gate are not exactly matched when only one input is switching, as shown in Figure 7.4(b). This is because the inner input of an nMOS stack has a higher threshold voltage and a larger parasitic capacitance. One way to make both inputs perfectly matched is to split the nMOS stack into two halves and connect their gates together as illustrated in Figure 7.8(a). This results in a perfectly symmetric gate with the same propagation delay from any input to its output.

■

■ Example 7-6: (A perfectly symmetric two-input NOR gate.) Like the case of the NAND gate, the two voltage-transfer characteristics of a two-input NOR gate are not exactly matched when only one input is switching, as shown in Figure 7.6(b). The same manner of what we have done on the two-input NAND

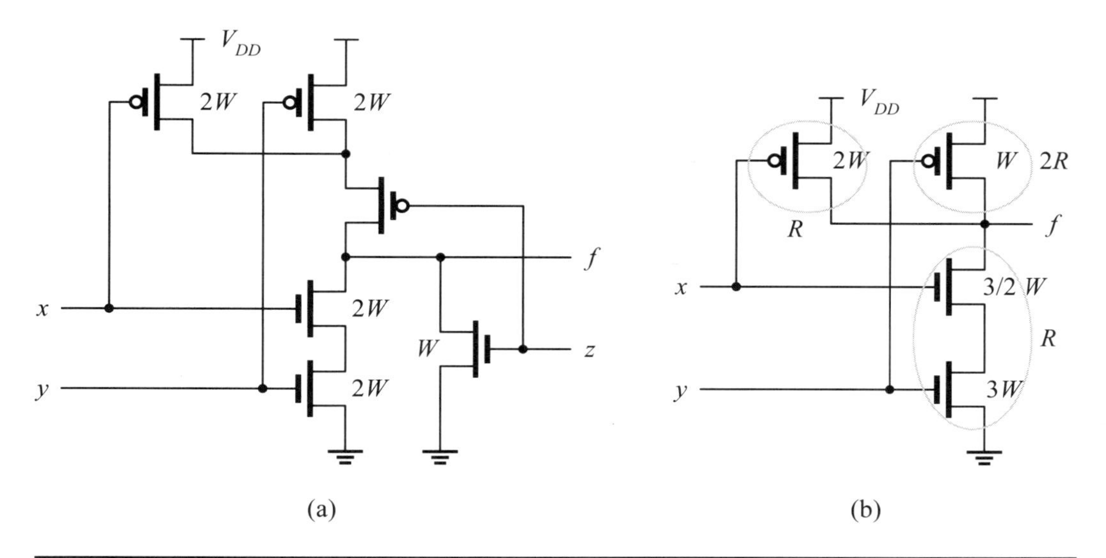

Figure 7.9: Examples of asymmetric logic gates: (a) an asymmetric gate; (b) an intended asymmetric gate.

gate can be applied to compensate for this mismatched characteristic, as illustrated in Figure 7.8(b). The resulting gate is perfectly symmetric and has the same propagation delay from any input to its output.

■

The above approach can be applied to NAND and NOR gates with any number of inputs. The extensions to the cases of three-input NAND and NOR gates are left to the reader as exercises.

pMOS/nMOS size ratio. The theoretical size ratio of pMOS to nMOS for the fastest average propagation delay is $\sqrt{2} \approx 1.4$. However, the practical size ratio of pMOS to nMOS is determined on the basis of area, power dissipation, and reliability, not on the fastest average propagation delay. The most common value of the size ratio of pMOS to nMOS is in the range of 1.5 to 2. Hence, in this book, we will set the ratio to be 2.

7.1.5.2 Asymmetric Gates For most applications, every input of a logic gate should be as symmetric as possible. A logic gate is said to be *symmetric* if every input of it has the same switching characteristics, including the switching point, input capacitance, parasitic capacitance, and driving capability. Otherwise, the logic gate is said to be *asymmetric*. The normal static CMOS logic gates are nominally symmetric, but the input and parasitic capacitances of different inputs are slightly different. Some other logic gates, such as the one shown in Figure 7.9(a), are inherently asymmetric.

In some applications, such as arbiters, the use of symmetric logic gates is definitely desirable since an arbiter must not favor one input over the others. In some other applications, using asymmetric gates to emphasize some desired features is needed. In such cases, it is often necessary to design or modify a symmetric gate into an asymmetric one to optimize the required performance. For example, in Figure 7.9(b), the two-input NAND gate is intentionally designed as an asymmetric gate to favor the input x over y. The stacked nMOS transistors are sized into a unit-resistance R in total,

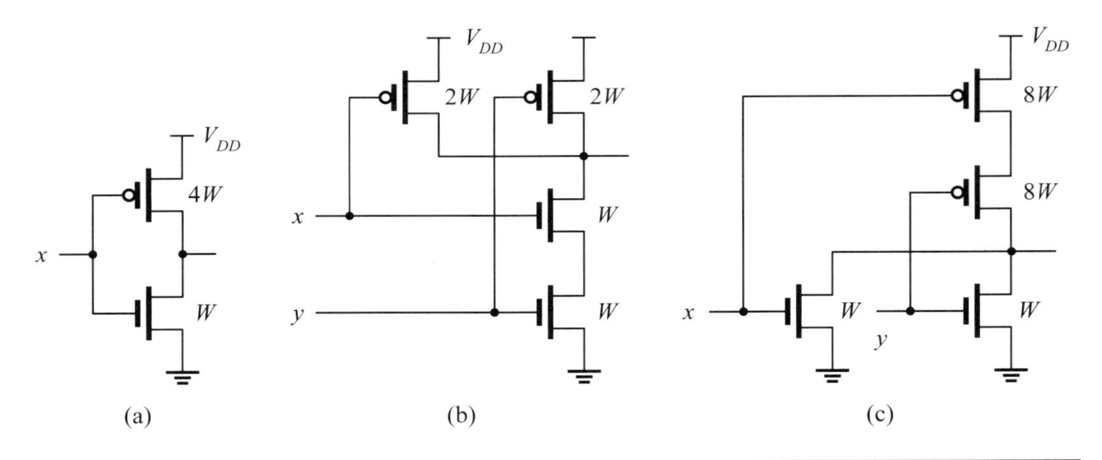

Figure 7.10: Examples of high-skewed logic gates: (a) NOT; (b) NAND; (c) NOR.

which is the same as the pull-up resistance associated with the input x. Consequently, the input x still works in a symmetric way in terms of both rise and fall times.

7.1.5.3 Skewed Logic Gates Skewed logic circuits are static logic in which the sizes of pMOS and nMOS transistors are intentionally made in a way such that one of two transitions is faster than the other. The word "skewed" means that the driving capabilities of pMOS and nMOS blocks are changed to favor one transition direction over another. In general, skewed logic has a performance comparable to that of dynamic logic and has better noise immunity due to no floating nodes associated with it. Furthermore, skewed logic allows a trade-off between the propagation delay of the logic and its noise margin. Because of higher noise immunity, skewed logic is better than domino logic for high-performance, low-voltage, and low-power applications.

Skewed logic can be either high-skewed or low-skewed, depending on whether the low-to-high (rising) or high-to-low (falling) transition needs to be speeded up. When a faster low-to-high transition is required, the channel widths of pMOS transistors of the logic gate are widened and the channel widths of nMOS transistors are reduced. This results in a *high-skewed logic gate*. Some examples of high-skewed logic gates are shown in Figure 7.10.

On the other hand, when a faster high-to-low transition is required, the channel widths of nMOS transistors of the logic gate are widened and the channel widths of pMOS transistors are reduced. The resulting logic circuit is called a *low-skewed logic gate*. Some examples of low-skewed logic gates are given in Figure 7.11.

The term "skew" is often used to denote the speedup degree of a skewed logic gate. For high-skewed logic gates, the *skew* is defined as the ratio of the worst-case driving capability of pMOS (pull-up) to nMOS (pull-down) block, while for low-skewed logic gates, the skew is defined as the ratio of the worst-case driving capability of the nMOS (pull-down) to the pMOS (pull-up) block. The actual amount of skew of a logic gate strongly depends on the actual requirement from a specific application.

■ **Example 7-7: (Logical effort values of high-skewed logic gates.)** Considering the high-skewed logic gates shown in Figure 7.10, calculate their skews and logical effort values.

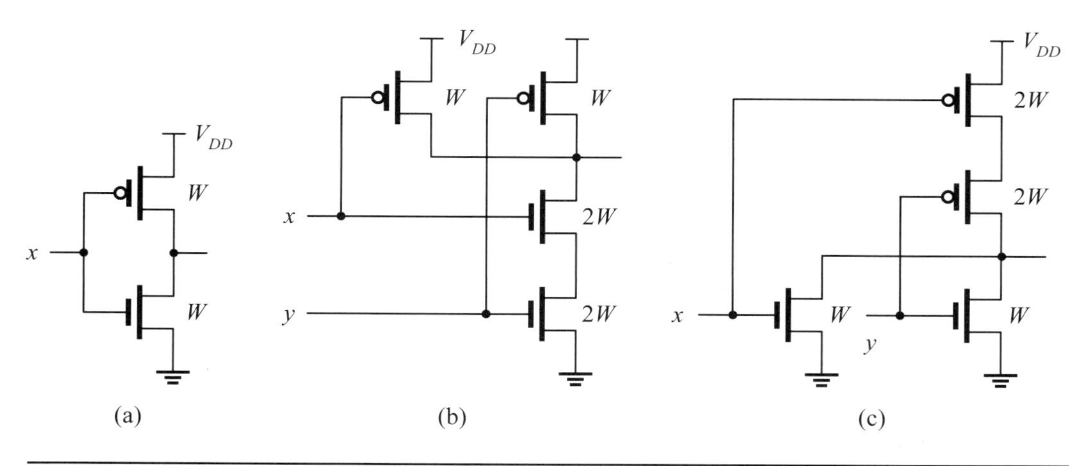

Figure 7.11: Examples of low-skewed logic gates: (a) NOT; (b) NAND; (c) NOR.

Solution: Based on the definition of high-skewed logic gates, the skews of all three logic gates shown in Figure 7.10 are 2. The logical effort values of NOT, NAND, and NOR gates are $\frac{5}{3}$, 1, and 3, respectively.

■

■ **Example 7-8: (Logical effort values of low-skewed logic gates.)** Considering the low-skewed logic gates shown in Figure 7.11, calculate their skews and logical effort values.

Solution: By the definition of low-skewed logic gates, the skews of all three logic gates shown in Figure 7.11 are 2. The logical effort values of NOT, NAND, and NOR gates are $\frac{2}{3}$, 1, and 1, respectively.

■

■ Review Questions

Q7-12. Define symmetric and asymmetric logic gates.

Q7-13. Calculate the logical effort of the input x and y of the circuit shown in Figure 7.9(b).

Q7-14. Explain why the input ordering of a logic gate may affect propagation delays.

Q7-15. Why is a symmetric logic gate needed in an arbiter?

Q7-16. What is a high-skewed logic gate?

Q7-17. What is a low-skewed logic gate?

Q7-18. What are the salient features of skewed logic gates?

Q7-19. Define "skewed" in terms of low-skewed and high-skewed logic gates.

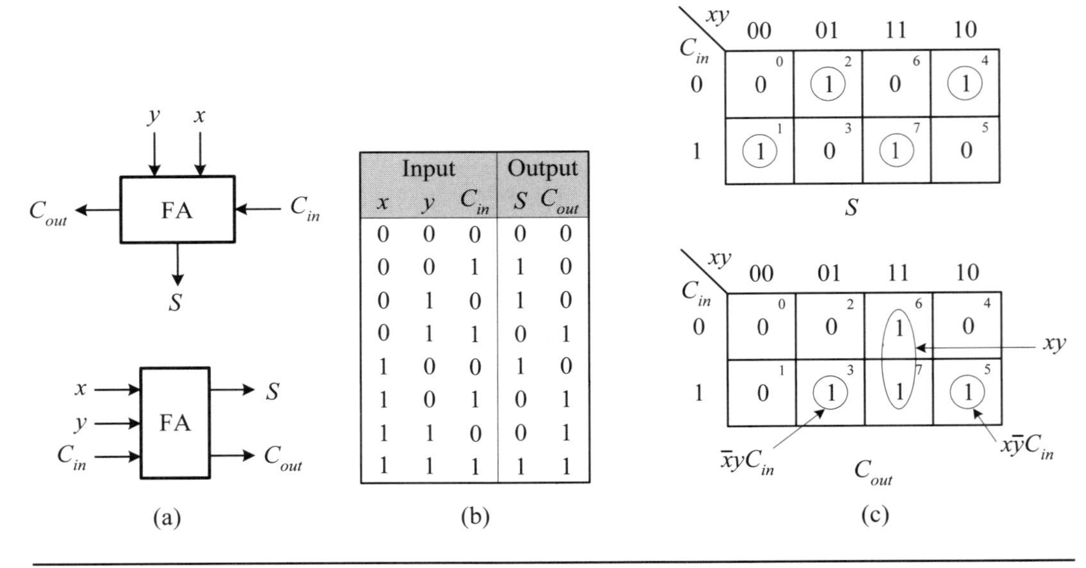

Input			Output	
x	y	C_{in}	S	C_{out}
0	0	0	0	0
0	0	1	1	0
0	1	0	1	0
0	1	1	0	1
1	0	0	1	0
1	0	1	0	1
1	1	0	0	1
1	1	1	1	1

Figure 7.12: The (a) block diagrams, (b) truth table, and (c) Karnaugh maps for a full adder.

7.2 Single-Rail Logic Circuits

Recall that CMOS logic circuits can be either single rail or dual rail. We first address the single-rail logic circuits in this section and defer the dual-rail logic circuits to the next section. As stated, the single-rail logic circuits cover CMOS logic circuits, TG-based logic circuits, and the ratioed logic circuits. The ratioed logic circuits include pseudo-nMOS logic circuits, nonthreshold logic (NTL) circuits, and ganged CMOS logic circuits. Some design examples are given to illustrate the related concepts along with the discussion.

7.2.1 CMOS Logic Circuits

We have described how to design a CMOS logic circuit using the serial-parallel connection of CMOS and nMOS/pMOS switches in Chapter 1. We review the design technique using a more complex example.

A full adder is the most basic arithmetic circuit for any digital system. It is a circuit that outputs the sum of its three one-bit inputs in a two-bit signal. The block diagrams, truth table, and related Karnaugh maps, of a full adder are exhibited in Figure 7.12.

From the Karnaugh maps shown in Figure 7.12(c), we can obtain the following two switching expressions for both carry out C_{out} and sum S.

$$C_{out} = x \cdot y + y \cdot C_{in} + x \cdot C_{in} = x \cdot y + C_{in}(x \oplus y) \quad (7.31)$$

$$S = x \oplus y \oplus C_{in} \quad (7.32)$$

The resulting logic circuit is shown in Figure 1.31, where we intend to present the full adder as two interconnected subcircuits called *half adders*, as the shaded circuits depicted in the figure. An example of CMOS implementation can be referred to Figure 1.41 and the layout of the circuit is exhibited in Figure 4.37. Many other possible

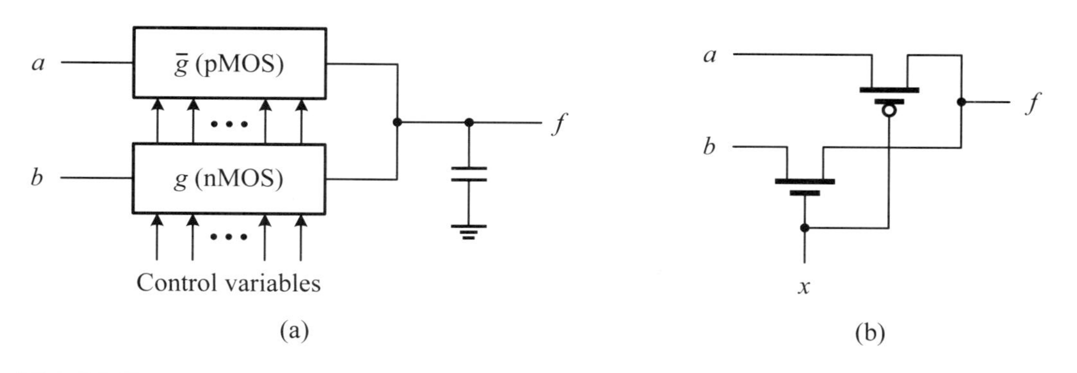

Figure 7.13: The (a) general block diagram and (b) an example of GDI logic.

implementations of a full adder can also be derived. We will return to this issue later in this book.

7.2.1.1 Gate-Diffusion Input Logic *Gate-diffusion input* (GDI) logic is a $f/\bar{f}$ network in which each control variable exactly connects to a pair of pMOS and nMOS switches and the residue of each input can only be 0 V, V_{DD}, or a true literal. GDI logic has the general block diagram as depicted in Figure 7.13(a), where inputs a and b can be a constant 0 V or V_{DD}, or a true literal. Generally, a GDI logic circuit implements the following switching function:

$$f = a \cdot \bar{g} + b \cdot g \tag{7.33}$$

where $\bar{g}$ is implemented by pMOS switches while g is implemented by nMOS switches. An illustration of GDI logic is exploited in the following example:

■ Example 7-9: (An example of GDI logic.) Suppose that we want to implement a two-input AND logic gate, namely, $f = xy$, with GDI logic style. By applying Shannon's expansion theorem to decompose f with respect to variable x, we have

$$f = xy = \bar{x} \cdot 0 + x \cdot y$$

where g is x. Hence, by setting input a to 0 and b to y, the logic circuit shown in Figure 7.13(a) is the desired AND gate. It is worth noting that the inversion of x is canceled out by the inverting feature of the gate input of the pMOS switch. In addition, the logic-0 signal at the input a will be deteriorated by the pMOS switch, as described in Chapter 1.

■

Some other two-input logic gates can also be implemented by using the basic GDI paradigm shown in Figure 7.13(a), in particular, OR and NOT gates. Hence, the two-input GDI logic circuit is a universal logic circuit.

To implement a complex switching function with GDI logic style, the uniform-tree network can be employed with a minor modification that each 2-to-1 multiplexer is implemented with a pair of pMOS and MOS switches. The resulting logic circuit is referred to as *multiplexer-tree* (MUX) logic. It is worth noting that each control variable in MUX logic exactly connects to a pair of pMOS and nMOS transistors, thereby

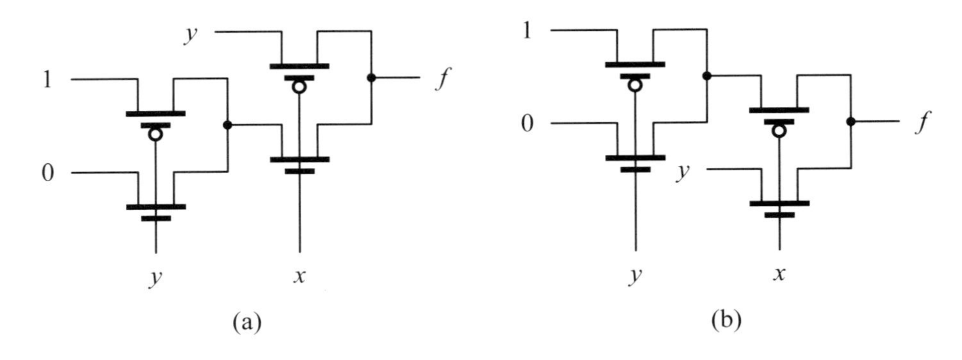

Figure 7.14: Two-input (a) XOR and (b) XNOR gates implemented with GDI logic.

saving the inverter needed in the case of being realized with nMOS switches alone. However, the input signals may be deteriorated when passing through the multiplexer tree to the output due to the inherent features of pMOS and nMOS switches.

To use a uniform-tree network with GDI logic style to implement a complex switching function, Shannon's expansion theorem is applied to decompose the switching function repeatedly until the residue of each input is only left as a 0, 1, or true literal. To illustrate this, consider the following example.

■ Example 7-10: (Another example of GDI logic.) Suppose that we would like to implement a two-input XOR gate with GDI logic style, namely, $f = \bar{x}y + x\bar{y}$. By using Shannon's expansion theorem, we obtain

$$\begin{aligned} f &= \bar{x}y + x\bar{y} \\ &= \bar{x}y + x(\bar{y}\cdot 1 + y\cdot 0) \end{aligned}$$

Hence, the resulting two-stage GDI XOR gate is as depicted in Figure 7.14(a). ■

The two-input XNOR gate can be implemented with GDI logic in a similar way. The resulting logic circuit is shown in Figure 7.14(b). However, we would like to leave the details to the reader as an exercise.

7.2.2 TG-Based Logic Circuits

Because of the poor performance of pMOS transistors, for those applications needing switches, nMOS switches or TG switches are usually used. In addition, even though each TG switch contains two MOS transistors, it has a better performance than an nMOS switch alone in terms of voltage swing. Hence, in the following, we will focus on TG-based logic circuits.

The basic structure of nMOS-based logic with an output inverter (buffer) and a pMOS pull-up transistor is shown in Figure 7.15(a). This structure is also known as *lean integration with pass-transistors* (LEAP). The output inverter can be low-skewed to favor the asymmetric response of the nMOS network. Since the output voltage of the nMOS network is deteriorated by a threshold voltage V_{Tn}, the outputs can only be pulled up to $V_{DD} - V_{Tn}$. Hence, a pMOS feedback transistor is used to pull up the internal node fully high to V_{DD} to avoid the power dissipation in the output inverter.

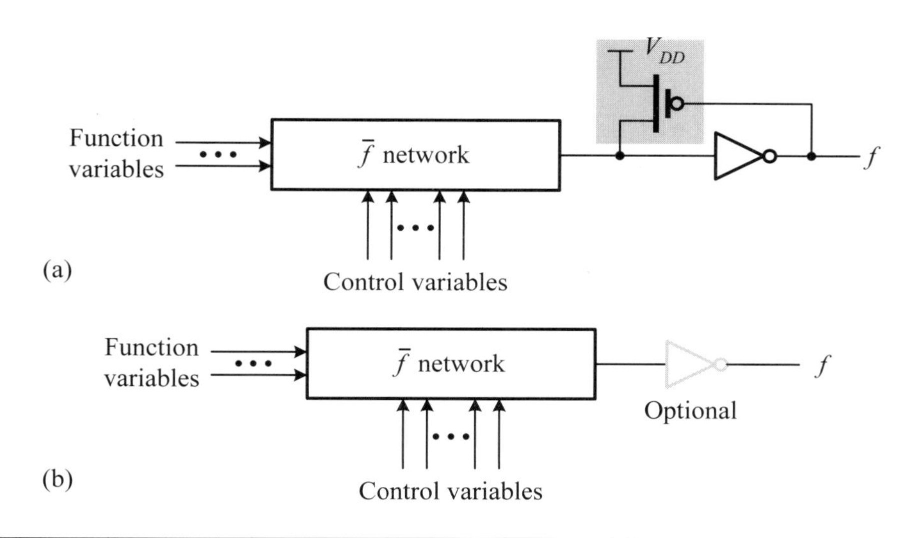

Figure 7.15: The structures of (a) nMOS-based and (b) TG-based logic.

The nMOS network can be designed with Shannon's expansion theorem introduced in Section 1.2.5. The reader may refer to that section for more details.

The general structure of TG-based logic is shown in Figure 7.15(b). The output inverter (buffer) is optional. The TG network implements the complementary output switching function if the output inverter is used and the output switching function otherwise. Like the nMOS-based structure, the TG network is also designed with Shannon's expansion theorem.

We give some TG-based examples. Although these are not easy to be derived by using the pedagogical approach we introduced, they are quite easy to be analyzed by using the concepts of controlled gates.

■ **Example 7-11: (A TG-based XOR gate.)** Figure 7.16 shows two TG-based XOR gate logic circuits. The first one in Figure 7.16(a) has a 2-to-1 multiplexer made up of two TG switches with the input x as the source selection line. Hence, the output is the true input y when the source selection line x is 0 and is the complement of input y when the source selection line x is 1. That is, $f = \bar{x} \cdot y + x \cdot \bar{y} = x \oplus y$, an XOR function.

The second one in Figure 7.16(b) uses a controlled inverter as the output stage with its two power-supply rails being provided and controlled by the input y through an inverter. A TG switch is also controlled by the input y. The output f is the true input x when the input y is 0 and is the complement of input x when the input y is 1. Hence, $f = \bar{x} \cdot y + x \cdot \bar{y} = x \oplus y$, an XOR function.

■

The following example illustrates how to construct TG-based XNOR gate logic circuits. Actually, these two logic circuits have the same structures as those of XOR gates with only some minor modifications.

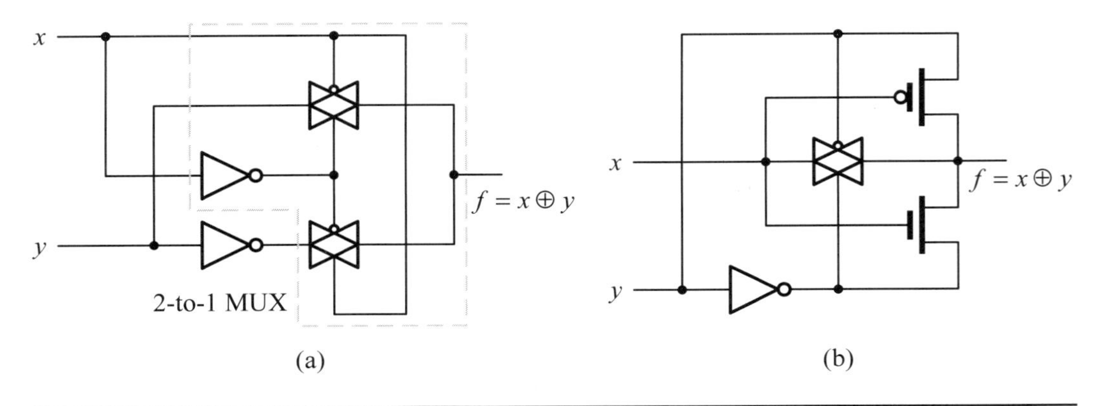

Figure 7.16: Two TG-based XOR gate logic circuits: (a) with 2-to-1 multiplexer; (b) a heuristic circuit.

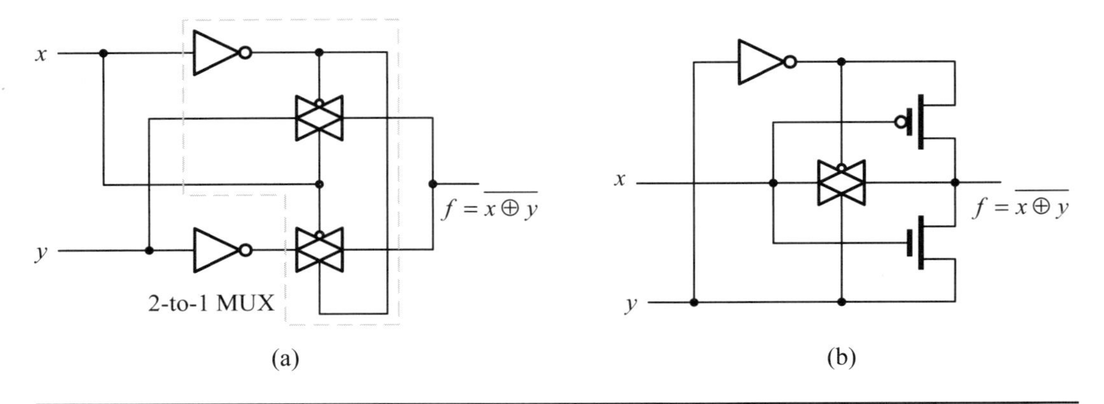

Figure 7.17: Two TG-based XNOR gate logic circuits: (a) with 2-to-1 multiplexer; (b) a heuristic circuit.

■ Example 7-12: (A TG-based XNOR gate.) Figure 7.17 shows two TG-based XNOR gate logic circuits. The first one in Figure 7.17(a) has a 2-to-1 multiplexer made up of two TG switches with the input x as the source selection line. Hence, the output is the complement of input y when the source selection line x is 0 and is the true input y when the source selection line x is 1. That is, $f = \bar{x} \cdot \bar{y} + x \cdot y = x \odot y$, an XNOR function.

The second one in Figure 7.17(b) uses a controlled inverter as the output stage with its two power-suuply rails being provided and controlled by the input y through an inverter. A TG switch is also controlled by the input y. The output f is the true input x when the input y is 1 and is the complement of input x when the input y is 0. Hence, $f = \bar{x} \cdot \bar{y} + x \cdot y = x \odot y$, an XNOR function.

■

The above two TG-based gates, XOR and XNOR gates, can be combined with some other TG switches to construct a full adder. An illustration is exhibited in the following example.

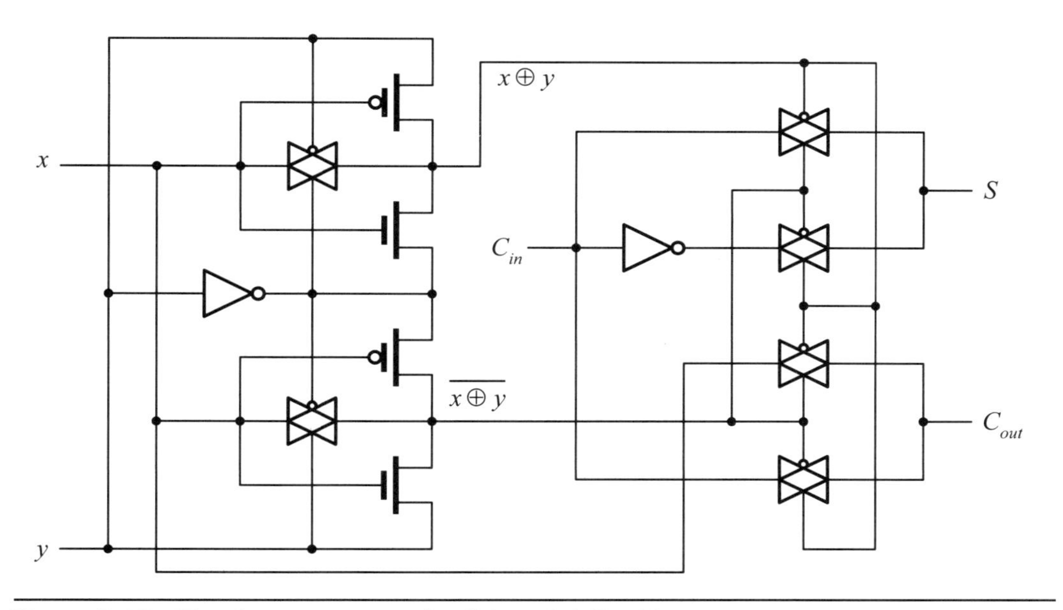

Figure 7.18: The first structure of TG-based full adder.

■ Example 7-13: (A TG-based full adder.) As shown in Figure 7.18, both XOR and XNOR gates are constructed by the second forms of XOR and XNOR gates, respectively. The two TG switches at the output node S comprise a 2-to-1 multiplexer controlled by $x \oplus y$ and $\overline{x \oplus y}$. Consequently, the output S is true C_{in} when $x \oplus y$ is 0 and is the complement of C_{in} when $x \oplus y$ is 1. The resulting output S is then equal to $x \oplus y \oplus C_{in}$, the sum of a full adder.

The carry out C_{out} is derived as follows. The two TG switches at the output node C_{out} also comprise a 2-to-1 multiplexer controlled by $x \oplus y$ and $\overline{x \oplus y}$. The output C_{out} is the input x when the selection signal $x \oplus y$ is 0 and is the C_{in} when the selection signal $x \oplus y$ is 1. As a result, the C_{out} is equal to

$$C_{out} = x \cdot (\overline{x \oplus y}) + C_{in} \cdot (x \oplus y)$$

which can be easily shown to be equal to the following expression,

$$C_{out} = xy + C_{in} \cdot (x \oplus y)$$

and thus performs the carry-out function of a full adder.

■

Of course, some transistors in the TG-based full adder shown in Figure 7.18 can be removed by realizing that the XNOR function is exactly the complement of the XOR function. The resulting structure is given in the following example:

■ Example 7-14: (An area-optimzed TG-based full adder.) Another structure of a TG-based full adder is shown in Figure 7.19. In this structure, the circuit of the XNOR gate is removed from the full adder and replaced with an inverter to provide the XNOR function by complementing the output of the XOR gate. The other part of the structure remains the same as that of the preceding example.

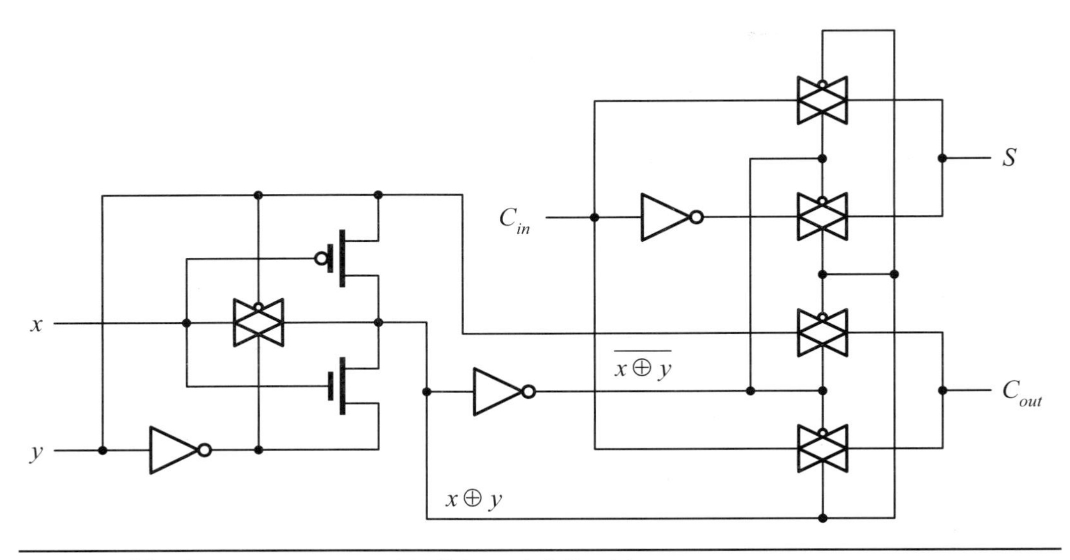

Figure 7.19: An area-optimized TG-based full adder.

The area required in this structure is smaller than the one shown in Figure 7.18. This structure only needs 18 transistors rather than the 20 needed in the one shown in Figure 7.18, and 28 needed in the CMOS implementation shown in Figure 1.41.

■ Review Questions

Q7-20. What is GDI logic? Describe its features.

Q7-21. Explain the operation of the TG-based XOR gate shown in Figure 7.16(a).

Q7-22. Explain the operation of the TG-based XNOR gate shown in Figure 7.17(a).

Q7-23. Show that $C_{out} = x \cdot (\overline{x \oplus y}) + C_{in} \cdot (x \oplus y)$ is equivalent to $C_{out} = xy + C_{in} \cdot (x \oplus y)$.

7.2.3 Ratioed Logic Circuits

In this section, we introduce three ratioed logic circuits: *pseudo-nMOS logic*, *nonthreshold logic* (NTL), and *ganged CMOS logic*. Recall that a ratioed logic circuit is the one that its low-level output voltage is determined by the relative sizes of both nMOS and pMOS transistors. Hence, it is essential to properly size pMOS and nMOS transistors in designing a logic circuit using any one of these three logic styles.

7.2.3.1 Pseudo-nMOS Logic One essential feature of a CMOS logic circuit is that it employs an equal number of transistors in both nMOS and pMOS types. In addition, to balance the rise and fall times of a logic circuit, the size of pMOS transistors generally needs to be larger than that of nMOS transistors. In some applications, it may be required to reduce the area of logic circuits to fit into the specific area budget.

Because both nMOS and pMOS blocks in a CMOS logic gate perform dual logic functions, only one part is indeed needed to carry out the desired logic function.

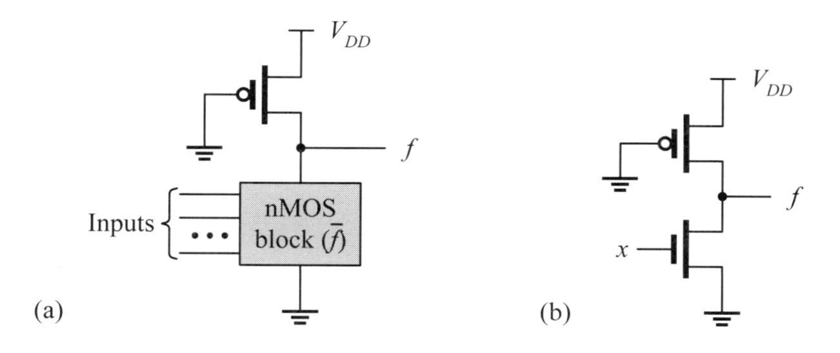

Figure 7.20: The (a) basic block diagram and (b) an inverter of pseudo-nMOS logic circuit.

Because of the better performance of nMOS transistors, we usually keep the nMOS block and remove the pMOS block from a CMOS logic circuit. Nevertheless, once we removed the pMOS block, the resulting logic circuit will be no longer satisfied with the node-value rule; that is, the output node will not function correctly whenever the nMOS block does not pull it down to ground. Of course, there are many ways that can be used to solve this problem. Probably, the simplest way is to connect the output node to V_{DD} with a pull-up resistor. Unfortunately, the needed area of the resistor makes this approach intolerable in VLSI applications. To save the area, the most widely used approach is to replace the pull-up resistor with an always-on pMOS transistor, called a *load* or a *pull-up pMOS transistor*. Because of the use of a pMOS transistor, the resulting logic circuit is referred to as a *pseudo-nMOS* logic circuit.

The general logic block diagram of pseudo-nMOS logic is shown in Figure 7.20(a). A pseudo-nMOS logic circuit consists of a pMOS load transistor and an nMOS block to realize the desired function. The gate of the pMOS transistor is permanently grounded so that it is always on. Sometimes, a bias network is used instead of the grounded connection. A simple pseudo-nMOS inverter is given in Figure 7.20(b). It is also possible to construct a pseudo-pMOS logic circuit by using a pMOS block and an always-on pull-down nMOS transistor although it is uncommonly used in practice due to the poor performance of pMOS transistors.

Pseudo-nMOS logic is a ratioed logic circuit because it violates the node-conflict-free rule, which means that there exist two logic paths with different logic values at the same time. The pull-up pMOS transistor always pulls up the output node to V_{DD} while the nMOS block attempts to pull down the output node to the ground level when it is turned on. As a result, the output voltage is determined by the resistances of both logic paths. The sizes of the pMOS transistor and those nMOS transistors in the nMOS block need to be considered with care to provide a correct logic level at the output node.

To design a pseudo-nMOS logic circuit, an analytical approach using current-voltage equations is often needed. However, in the following, we simply employ the concept of equivalent resistance to illustrate how to determine the sizes of pMOS and nMOS transistors of a pseudo-nMOS inverter circuit to provide a desired low-level output voltage (V_{OL}). The analytical approach will be described later in this section.

■ **Example 7-15: (A design example of a pseudo-nMOS inverter.)** A simple equivalent-resistance rule can be used to design a pseudo-nMOS inverter circuit

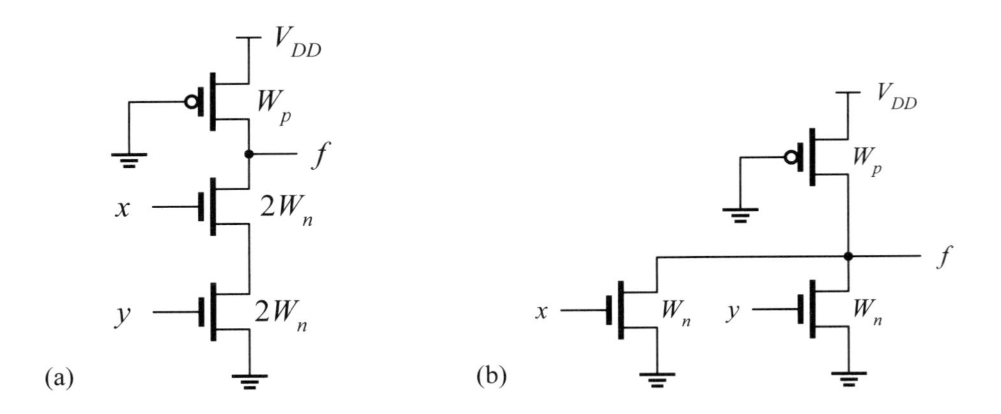

Figure 7.21: Examples of pseudo-nMOS logic gates: (a) two-input NAND; (b) two-input NOR.

in practice. Let R_n be the on-resistance of an nMOS transistor. The low-level output voltage is V_{OL} when the input voltage is at high-level input voltage, V_{IH}. Both nMOS and pMOS transistors are turned on as the input voltage is at V_{IH}. By setting V_{OL} to $0.1V_{DD}$, the on-resistance R_p of a pMOS transistor can then be determined from the following relationship:

$$V_{OL} = V_{DD} \times \frac{R_n}{R_n + R_p} \leq 0.1V_{DD} \tag{7.34}$$

As a result, $9R_n \leq R_p$. Hence, the aspect ratio of a pMOS transistor can then be determined. That is, $W_n = 3W_p = 3W$, if we use $\mu_n = 3\mu_p$ and $L_p = L_n = L$, where W and L are the minimum channel width and length allowed by the process at hand. ■

The low-level output voltage V_{OL} of a pseudo-nMOS logic gate should be less than the low-level input voltage V_{IL} of its driven gate so that the resulting cascaded logic chain can work correctly. Some more pseudo-nMOS logic circuits are given in the following examples.

■ Example 7-16: (Examples of pseudo-nMOS logic gates.) Two more examples of pseudo-nMOS logic gates are shown in Figure 7.21. Figure 7.21(a) shows a two-input NAND gate and Figure 7.21(b) gives a two-input NOR gate. As mentioned, to balance both the pull-up and pull-down timing at the output node, the size of nMOS transistors is set to two times that of the basic inverter in the NAND gate. However, the size of nMOS transistors does not need to be scaled up in the case of the NOR gate because the two nMOS transistors are connected in parallel. ■

As we can see from Figures 7.20 and 7.21, pseudo-nMOS logic circuits generally need less area than the corresponding CMOS logic circuits because they do not need the pMOS block. The use of only an nMOS block in pseudo-nMOS logic circuits also reduces the loading capacitance for the driving circuits. Hence, in theory they should be much faster than the corresponding CMOS logic circuits. Nevertheless, because of

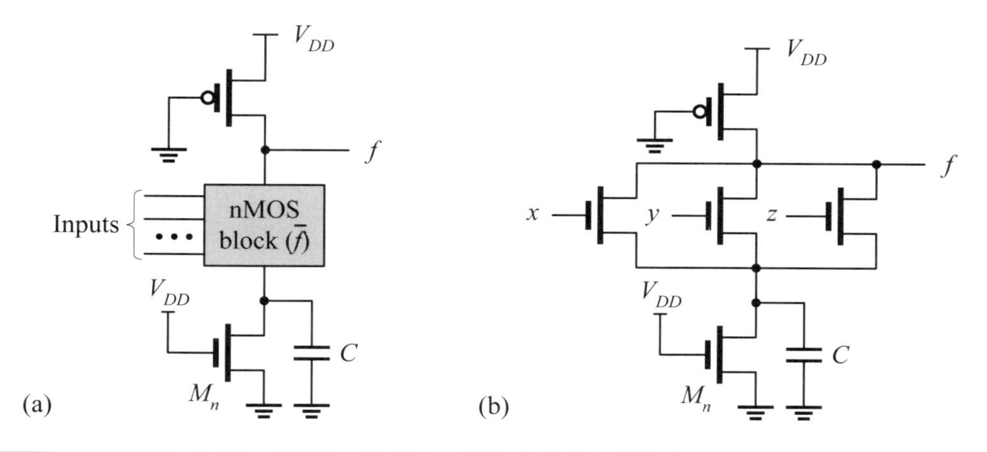

Figure 7.22: The (a) general block diagram and (b) a three-input NOR gate of NTL.

the existence of direct current between power supply V_{DD} and ground when the nMOS block is turned on, the switching speed of a pseudo-nMOS logic circuit is quite slow compared to its CMOS counterpart. In addition, this direct current also causes much static power dissipation and inhibits the use of pseudo-nMOS logic circuits in those power-limited applications. As a consequence, the pseudo-nMOS logic circuits only find their applications in which large fan-in is required. However, in these applications, care must be taken to deal with the problem of static power dissipation caused by direct current.

7.2.3.2 Nonthreshold Logic A variation of pseudo-nMOS logic that uses the concept of a source degenerate resistor to reduce the output voltage swing is known as nonthreshold logic (NTL). The NTL circuit has an improvement in speed at the cost of direct-current (DC) power dissipation. The basic circuit structure of NTL is shown in Figure 7.22(a), where an nMOS transistor is added below the nMOS block together with a bypass (or shunting) capacitor C. Both the pMOS load and the nMOS transistor split the supply voltage to reduce the output voltage swing. The nMOS transistor in the negative feedback path results in a nonthreshold-like DC transfer characteristic. The bypass capacitor C is used to reduce the negative feedback effect caused by the nMOS transistor. In practice, this capacitor is often implemented by using the gate of an nMOS transistor. NTL consumes static power and is usually slower than pseudo-nMOS logic.

Figure 7.22(b) shows an example of NTL in which a three-input NOR gate is constructed. Other logic circuits can be designed in a similar way by using the nMOS block to implement the desired functions.

7.2.3.3 Ganged CMOS Logic A ganged CMOS logic circuit is also a ratioed logic that employs CMOS inverters with their outputs ganged together to form a desired logic circuit, as shown in Figures 7.23. Since all pMOS transistors are connected in parallel like the case of nMOS transistors, this type of logic belongs to symmetric logic. In general, a ganged logic circuit contains n parallel pMOS pull-up transistors and n parallel nMOS pull-down transistors. DC power is consumed except when the inputs are all 0s or all 1s. Ganged logic gates may achieve higher speed and lower input capacitance at the expense of high static power dissipation.

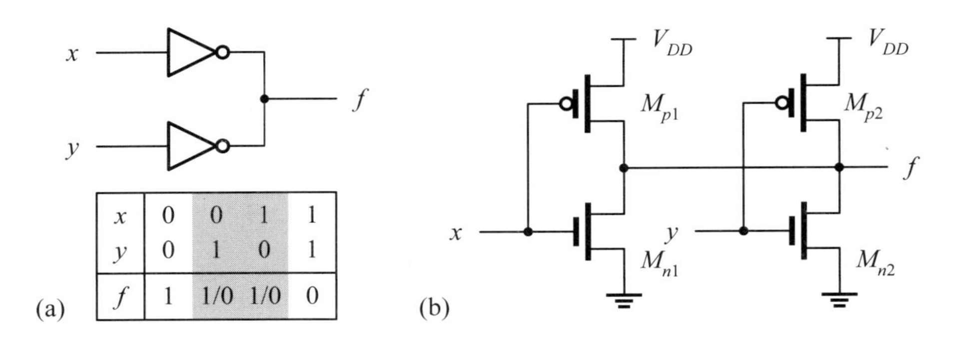

Figure 7.23: The concepts of ganged logic: (a) basic structure; (b) logic circuit.

An example of ganged CMOS logic circuit is shown in Figure 7.23, in which two inverters are ganged together. The output f is 1 when both inputs, x and y, are 0s and is 0 when both inputs, x and y, are 1s. The other two combinations of inputs virtually determine the desired logic type. The resulting logic is a two-input NOR gate if they are set to 0s and is a two-input NAND gate if they are set to 1s. However, because in these two combinations, one pMOS and one nMOS transistors are on, a direct current between power supply and ground exists. To force the output voltage to have the desired logic levels, the size ratio of pMOS and nMOS transistors must be set to a proper value.

We give an example to illustrate how to determine the sizes of pMOS and nMOS transistors of inverters so that the ganged logic circuit can function as desired.

■ Example 7-17: (A two-input NOR and NAND gates.) Determine the sizes of pMOS and nMOS transistors so that the ganged logic circuit shown in Figure 7.23(b) can function as

1. A two-input NOR gate
2. A two-input NAND gate

Solution:

1. From Figure 7.23(a), to make the ganged logic circuit function as a two-input NOR gate, the output f should output a low-level output voltage, V_{OL}, when both combinations of inputs, x and y, are 01 and 10. In general, the V_{OL} is set to $0.2V_{DD}$ in CMOS technology. Hence,

$$V_{out} = V_{OL} = \frac{1}{5}V_{DD} = \frac{R_n}{R_n + R_p}V_{DD}$$

where $R_n = R_{eqn}(L_n/W_n)$ and $R_p = R_{eqp}(L_p/W_p)$. Solving the above equation, we obtain $R_p = 4R_n$. Hence, $W_n = 2W = 2W_p$ if $L_n = L_p = L$ and $R_{eqp} = 2R_{eqn}$. The resulting logic circuit is shown in Figure 7.24(a).

2. From Figure 7.23(a), to make the ganged logic circuit function as a two-input NAND gate, the output f should output a high-level output voltage, V_{OH}, when the combinations of inputs, x and y, are 01 and 10. In general, the V_{OH} is set to $0.8V_{DD}$ in CMOS technology. Hence,

$$V_{out} = V_{OL} = \frac{4}{5}V_{DD} = \frac{R_n}{R_n + R_p}V_{DD}$$

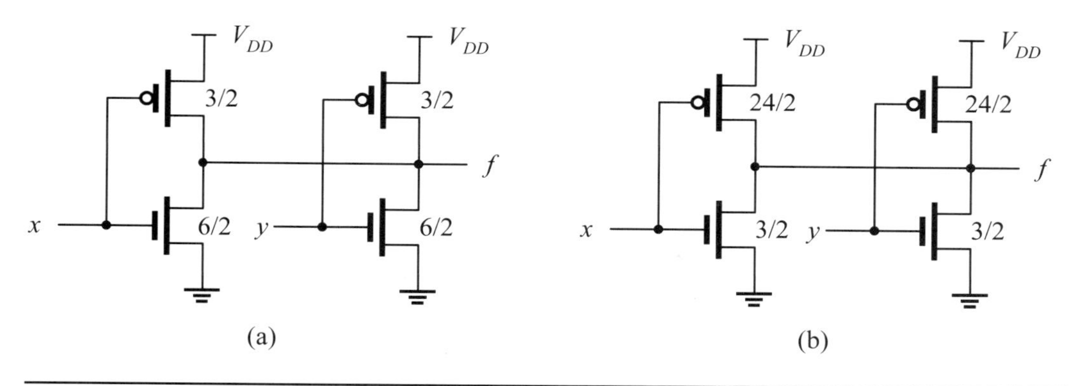

Figure 7.24: Two examples of ganged CMOS logic: (a) NOR gate; (b) NAND gate.

Solving the above equation, we have $R_n = 4R_p$. Hence, $W_n = W$ and $W_p = 8W$ if $L_n = L_p = L$ and $R_{eqp} = 2R_{eqn}$. The resulting logic circuit is shown in Figure 7.24(b).

■

In summary, different logic functions in ganged logic can be achieved with identical topology through properly designing the size ratios between pMOS and nMOS transistors.

■ Review Questions

Q7-24. Draw a block diagram showing the basic structure of a pseudo-pMOS logic circuit.

Q7-25. What is ganged logic?

Q7-26. Calculate the logical effort values of ganged NOR and NAND gates shown in Figure 7.24.

7.2.3.4 Mathematical Analysis of Pseudo-nMOS Inverters

In a pseudo-nMOS inverter, an always-on pMOS transistor is used as the load of the nMOS block, which is used to implement the desired logic function. Referring to Figure 7.25, we may informally set the high-level output voltage V_{OH} of the pseudo-nMOS inverter to V_{DD} because the nMOS transistor is turned off in this situation.

To calculate the low-level output voltage V_{OL}, referring to Figure 7.25(a), the nMOS transistor turns on as its gate receives a high-level input voltage, V_{IH}. Because the low-level output voltage V_{OL} should be far below the threshold voltage V_{Tn}, the pMOS transistor is in the saturation region of operation and the nMOS transistor operates in the linear region. In addition, both transistors have the same amount of current, namely, $I_{Dp(sat)} = I_{Dn(linear)}$, because they are connected in series. Consequently, we have

$$I_{Dp(sat)} = \frac{\mu_n C_{ox}}{2}\left(\frac{W_n}{L_n}\right)[2(V_{GSn} - V_{Tn})V_{OL} - V_{OL}^2] \qquad (7.35)$$

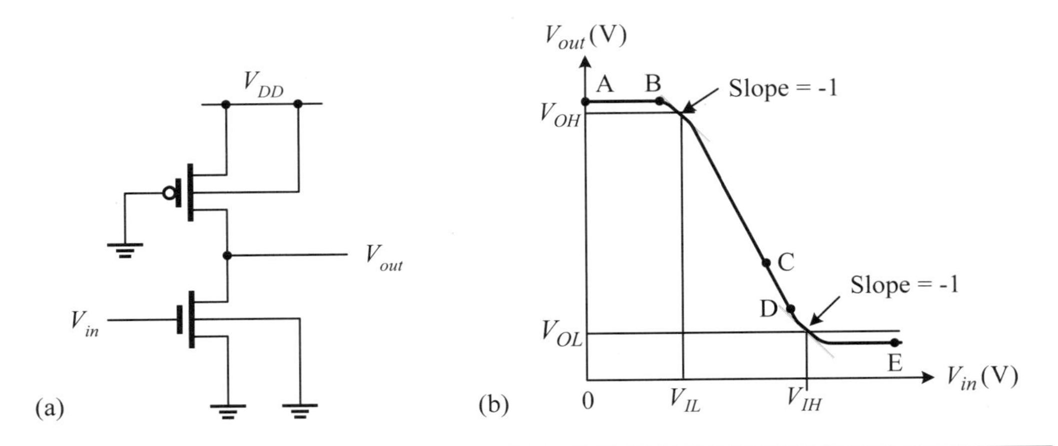

Figure 7.25: The (a) circuit and (b) VTC of a pseudo-nMOS inverter.

Since V_{OL} is expected to be very small, the squared term of V_{OL} may be neglected. Solving the above equation, the low-level output voltage V_{OL} is found to be

$$V_{OL} = \frac{I_{Dp(sat)}}{k_n(V_{DD} - V_{Tn})} \tag{7.36}$$

Short-channel analysis. For deep submicron processes, the short-channel current equation is often used to obtain a more accurate result.

$$\frac{W_p v_{sat} C_{ox}(V_{SGp} - |V_{Tp}|)^2}{(V_{SGp} - |V_{Tp}|) + E_{satp}L_p} = \frac{1}{2}\frac{W_n}{L_n}\frac{\mu_n C_{ox}}{(1 + V_{OL}/E_{satn}L_n)}[2(V_{GSn} - V_{Tn})V_{OL} - V_{OL}^2] \tag{7.37}$$

■ Example 7-18: (Pseudo-nMOS inverter design.) Using 0.18-μm parameters: $\mu_p = 112$ cm^2/V-s, $\mu_n = 292$ cm^2/V-s, $\mu_{effn} = 287$ cm^2/V-s, and $V_{T0n} = |V_{T0p}| = 0.4$ V, design a pseudo-nMOS inverter that provides $V_{OL} = 0.1$ V. Assume that $V_{DD} = 1.8$ V, $v_{sat} = 8 \times 10^6$ cm/s, $E_{satn} = 12 \times 10^4$ V/cm, and $E_{satp} = 25 \times 10^4$ V/cm.

Solution: Because the pMOS transistor is in the saturation region of operation and the nMOS transistor operates in the linear region as the low-level output voltage V_{OL} is to be found, the current relation is as follows:

$$\frac{\mu_p C_{ox}}{2}\left(\frac{W_p}{L_p}\right)(V_{SGp} - |V_{T0p}|)^2 = \frac{\mu_n C_{ox}}{2}\left(\frac{W_n}{L_n}\right)[2(V_{GSn} - V_{T0n})V_{OL} - V_{OL}^2]$$

Substituting appropriate parameters and supply voltage into the above equation, we have

$$112\left(\frac{W_p}{L_p}\right)(1.8 - 0.4)^2 = 292\left(\frac{W_n}{L_n}\right)[2(1.8 - 0.4)0.1 - 0.1^2]$$

since the V_{DSn} is only the low-level output voltage V_{OL}, the nominal mobility value is used for the nMOS transistor. Supposing that $L_n = L_p = L$ and solving it, the ratio of W_n/W_p is found to be

$$\frac{W_n}{W_p} = 2.78 \Rightarrow \frac{W_n}{W_p} = 3 \qquad \text{(Overdesign)}$$

Using the short-channel current equation, that is, Equation (7.37), we have

$$\frac{W_p(8\times 10^6)(1.8-0.4)^2}{(1.8-0.4)+4.5} = \frac{1}{2}\frac{W_n}{0.18\times 10^{-4}}\frac{287}{(1+0.1/2.16)}[2(1.8-0.4)0.1-0.1^2]$$

Solving the above equation, the ratio of W_n/W_p is obtained as follows:

$$\frac{W_n}{W_p} = 1.30 \Rightarrow \frac{W_n}{W_p} = 2$$

This result is consistent with SPICE simulation, as shown in Figure 7.26.

■ Example 7-19: (Pseudo-nMOS inverter design.) Using 0.13 μm parameters, design a pseudo-nMOS inverter that provides $V_{OL} = 0.1$ V. Assume that $V_{DD} = 1.2$ V, $V_{T0n} = |V_{T0p}| = 0.35$ V, $\mu_n = 298$ cm^2/V-s, and $v_{sat} = 1.5\times 10^7$ cm/s.

Solution: Because the pMOS transistor is in the saturation region of operation and the nMOS transistor operates in the linear region as the low-level output voltage V_{OL} is to be found, using the short-channel current relation, that is, Equation (7.37), we have

$$\frac{W_p(1.5\times 10^7)(1.2-0.35)^2}{(1.2-0.35)+5.46} = \frac{1}{2}\frac{W_n}{0.13\times 10^{-4}}\frac{298}{(1+0.1/1.24)}[2(1.2-0.35)0.1-0.1^2]$$

Solving the above equation, the ratio of W_n/W_p is found to be

$$\frac{W_n}{W_p} = 1.01 \Rightarrow \frac{W_n}{W_p} = 2$$

This result is consistent with SPICE simulation, as shown in Figure 7.26.

7.3 Dual-Rail Logic Circuits

Dual-rail logic is also called *differential logic* because it handles both true and complementary signals at any time. The major types of dual-rail logic circuits include the following: cascode voltage switch logic (CVSL), complementary pass-transistor logic (CPL), differential cascode voltage switch with pass-gate (DCVSPG) logic, and double pass-transistor logic (DPL). In this section, we address each of these in detail along with some examples.

7.3.1 Cascode Voltage Switch Logic (CVSL)

Cascode voltage switch logic (CVSL) is also called DCVSL (differential cascode voltage switch logic) or DCVS logic. It consists of two parts, a cross-coupled pMOS latch and two nMOS blocks (or referred to as a logic tree), as shown in Figure 7.27. The cross-coupled pMOS latch provides both true and complementary outputs, f and $\bar{f}$, at the same time, and the two nMOS blocks (the logic tree) realize the desired logic function in both true and complementary forms.

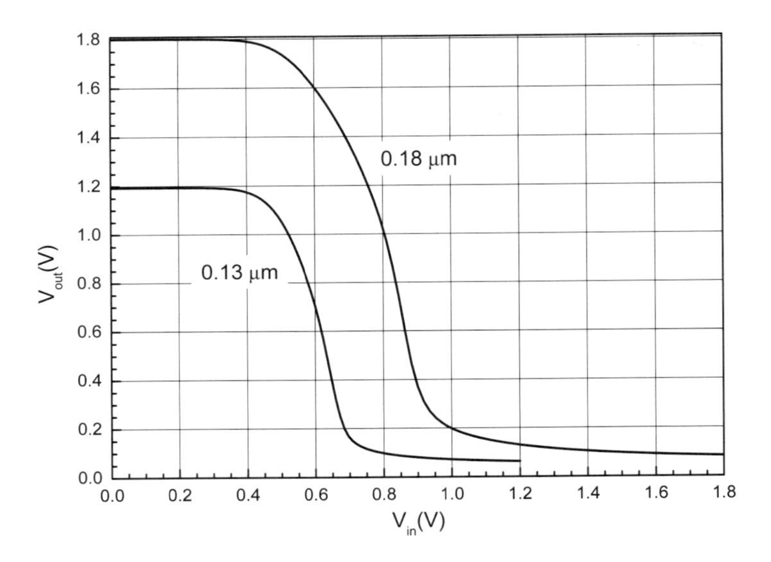

Figure 7.26: The simulation results of the two pseudo-nMOS inverters.

The operation of CVSL is as follows. Since both nMOS blocks in CVSL are complementary to each other, their outputs are mutually exclusive; namely, when one of them conducts, the other is turned off. Hence, both true and complementary switching functions are carried out at the same time. To get more insights into the operation of CVSL logic, consider the block diagram shown in Figure 7.27(a). Assume that output nodes, f and $\bar{f}$, are initially at high and low voltages, respectively. Hence, pMOS transistor M_{p1} is on, whereas M_{p2} is off. Now, for some combination of inputs, nMOS block f is turned off while nMOS block $\bar{f}$ is turned on, thereby causing pMOS transistor M_{p2} to turn on and the output node $\bar{f}$ to be pulled up. This rising voltage at the output node $\bar{f}$ decreases the conduction of pMOS transistor M_{p1}, causing the output node f to further discharge all the way to ground. Finally, pMOS transistor M_{p1} is turned off while M_{p2} is turned on. As a result, both output nodes, f and $\bar{f}$, exchange their states and are at low- and high-level voltages, respectively.

CVSL circuits are fundamentally a ratioed logic circuit because during the transition the in-series connected pMOS transistor and nMOS blocks are turned on simultaneously. The nMOS block must be strong enough to pull the gate voltage of the opposite-side pMOS transistor below $V_{DD} - |V_{Tp}|$ to turn on the pMOS transistor. Fortunately, this can be satisfied easily. Moreover, the resulting CVSL circuit exhibits a rail-to-rail swing, and there is no static power dissipation in the steady state. Nevertheless, there is a power dissipation during the transition owing to the short-circuit (also called *crossover*) current, which is caused by the pMOS transistor and the nMOS block being turned on at the same time.

7.3.1.1 Logic Design Methodology The logic designs of CVSL can be done in either of the following two ways: the *two-nMOS block approach* and one *logic-tree approach*. We introduce each of these coupled with examples.

Two-nMOS block approach. The two-nMOS block approach is to implement separately both switching functions, f and $\bar{f}$, as two independent nMOS blocks. As shown in Figure 7.27(a), it is simply proceeded in the same way as that used in de-

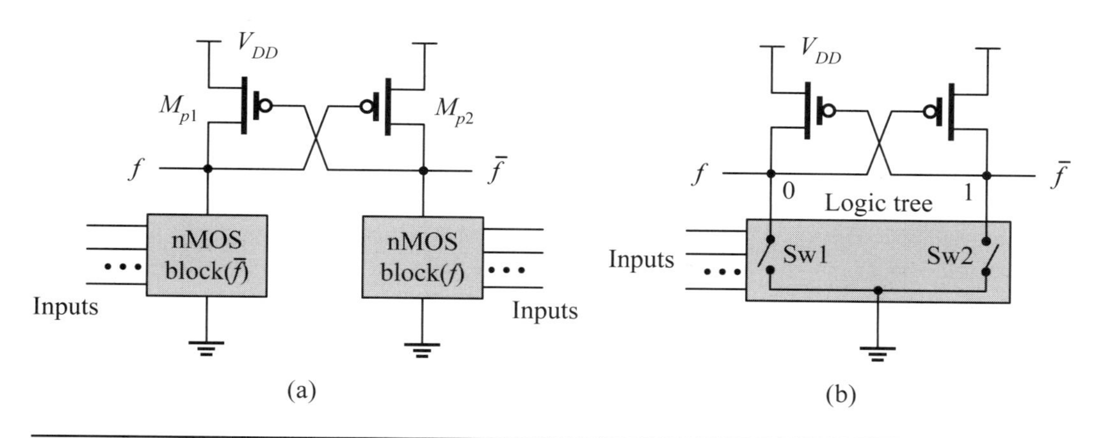

Figure 7.27: The general block diagrams of CVSL logic: (a) two separate nMOS blocks; (b) one logic tree.

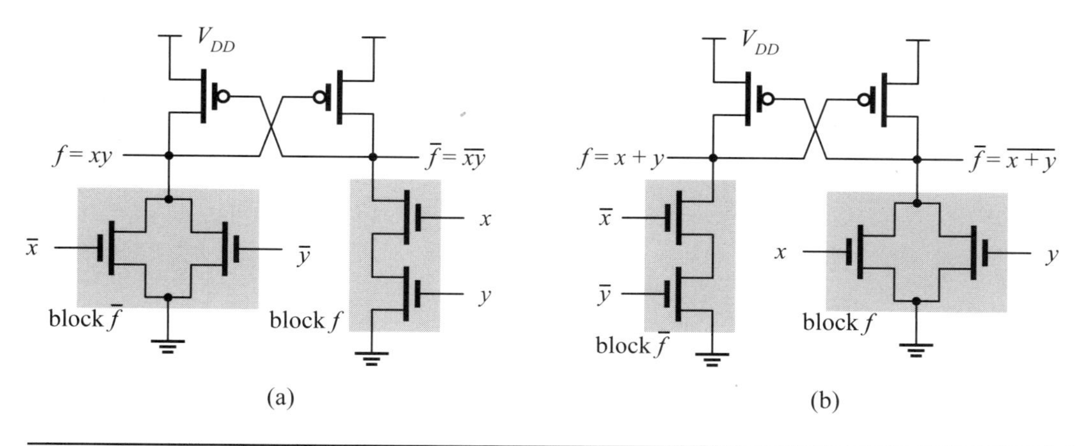

Figure 7.28: The CVSL (a) AND/NAND and (b) OR/NOR gates.

signing the nMOS block in a CMOS logic circuit. We use AND/NAND and OR/NOR gates as examples to illustrate this approach.

■ Example 7-20: (Two examples of the CVSL logic gate.) An example of the two-input AND/NAND gate realized in CVSL style is shown in Figure 7.28(a). The nMOS block $\bar{f}$ is composed of two nMOS transistors connected in parallel. These two nMOS transistors are separately controlled by the complementary signals $\bar{x}$ and $\bar{y}$. The nMOS block f comprises two nMOS transistors connected in series, which are controlled by the signals x and y, respectively. The output of nMOS block $\bar{f}$ is the function of the AND gate, whereas the output of the nMOS block f is the function of the NAND gate.

Figure 7.28(b) shows an example of the two-input OR/NOR gate realized in CVSL style. The nMOS block $\bar{f}$ is composed of two nMOS transistors connected in series. These two nMOS transistors are separately controlled by the complementary signals $\bar{x}$ and $\bar{y}$. The nMOS block f comprises two nMOS transistors connected in parallel, which are controlled by the signals x and y, respectively. The output of nMOS block $\bar{f}$

is the function of the OR gate, whereas the output of the nMOS block f is the function of the NOR gate.

■

One logic-tree approach. This approach is to treat both switching functions, f and $\bar{f}$, as a whole and to realize both switching functions in a (unified) logic tree. It is indeed a variation of the 0/1-tree network introduced in Chapter 1 and is able to explore the transistor sharing between both true and complementary switching functions. Hence, it can achieve a more area-efficient design. However, it is more difficult to handle for a complex switching function.

To implement a logic tree, the 0/1-tree network needs the following minor modification. Because of the use of a cross-coupled pMOS latch, the output node of a 0/1-tree network is grounded and the input node labeled with V_{DD} (or "1") is connected to the output node $\bar{f}$, while the input node labeled with 0 V (or "0") is connected to the output node f, as indicated in Figure 7.27(b).

An example illustrates how to use one logic-tree approach to design the two-input AND/NAND gate shown in Figure 7.28(a) is exhibited in the following.

■ **Example 7-21: (A two-input AND/NAND gate.)** The switching function of a two-input AND gate, $f = x \cdot y$, can be expanded with respect to the variables x and then y as follows:

$$\begin{aligned} f &= x \cdot y \\ &= \bar{x} \cdot \bar{y} \cdot 0 + x \cdot \bar{y} \cdot 0 + \bar{x} \cdot y \cdot 0 + x \cdot y \cdot 1 \\ &= (\bar{x} \cdot \bar{y} \cdot 0 + x \cdot \bar{y} \cdot 0 + \bar{x} \cdot y \cdot 0) + (x \cdot y \cdot 1) \\ &= (\bar{x} + \bar{y}) \cdot 0 + (x \cdot y) \cdot 1 \end{aligned}$$

which means that the 0 tree is $\bar{x} + \bar{y}$ while the 1 tree is $x \cdot y$. The resulting logic circuit is the same as that shown in Figure 7.28(a), exactly the same as that obtained with the use of a two-nMOS block approach.

■

The following example gives another example in which some transistors are shared by both switching functions: f, and $\bar{f}$.

■ **Example 7-22: (A two-input XOR/XNOR gate.)** The switching function of a two-input XOR gate, $f = x \oplus y$, can be expanded with respect to the variables x and then y as follows:

$$\begin{aligned} f &= x \oplus y \\ &= \bar{x} \cdot y + x \cdot \bar{y} \\ &= \bar{x} \cdot (y) + x \cdot (\bar{y}) \\ &= \bar{x} \cdot (\bar{y} \cdot 0 + y \cdot 1) + x \cdot (\bar{y} \cdot 1 + y \cdot 0) \end{aligned}$$

which means that the 0 tree is $\bar{x} \cdot \bar{y} + x \cdot y$, while the 1 tree is $\bar{x} \cdot y + x \cdot \bar{y}$. The resulting logic circuit is shown in Figure 7.29(b). In this case, the two transistors, labeled as x and $\bar{x}$, are common to two minterms and, hence, are shared by these two minterms.

■

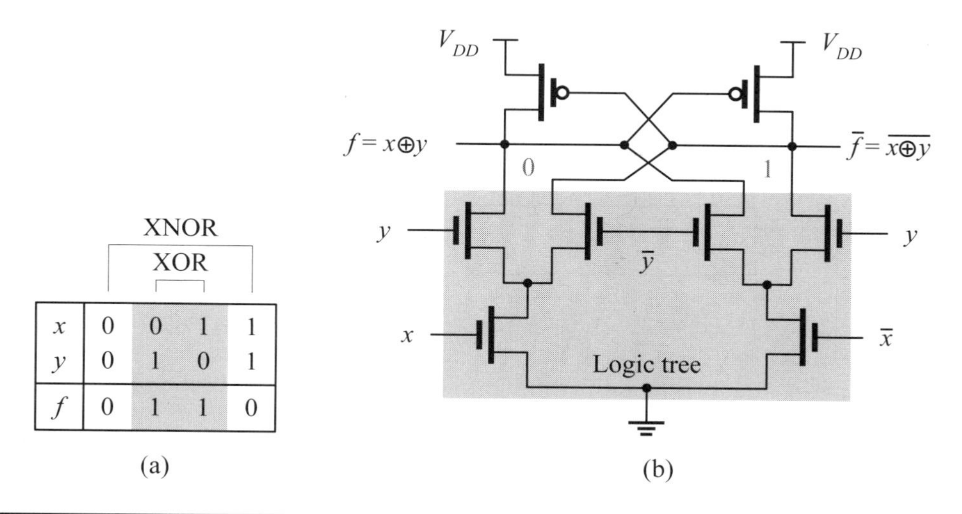

Figure 7.29: An example of using one logic tree to implement a two-input XOR/XNOR gate: (a) truth table; (b) one logic tree circuit.

The logic tree can also be obtained from the truth table of the switching function to be realized. An example is shown in Figure 7.29(a). Nevertheless, when using the truth table, we often encounter difficulty in finding the shared transistors. In the rest of this book, we may only show the topology with two-nMOS blocks for all CVLS-related logic. However, the reader should be aware that the logic tree is also implied in all logic as well. Keep in mind when realizing a complex switching function, that the one logic-tree approach should be taken into account for saving the area from the possible transistor sharing between 0 and 1 trees.

Review Questions

Q7-27. What is CVSL?

Q7-28. What are the features of CVSL?

Q7-29. How would you implement a switching function using CVSL?

7.3.2 Complementary Pass-Transistor Logic (CPL)

Although CVSL uses a pMOS cross-coupled latch as the load, it cannot work as fast as expected because the pMOS cross-coupled latch cannot be easily inverted due to the regenerative property of the latch. This drawback can be overcome by the complementary pass-transistor logic (CPL).

The basic structure of CPL is shown in Figure 7.30. It consists of complementary inputs and outputs, two nMOS networks for realizing both truth and complementary functions, f and $\bar{f}$, and a pair of output buffers. The two nMOS networks function as pull-up and pull-down devices; hence, the pMOS cross-coupled latch can be removed. The output buffers are used to restore the signals to the desired voltage levels to compensate for the V_{Tn} loss due to pass transistors. A pMOS latch can also be added to CPL, as indicated in Figure 7.30 with lighter lines, to reduce static power dissipation caused by the output buffer when receiving a high-level, but not a fully high, input

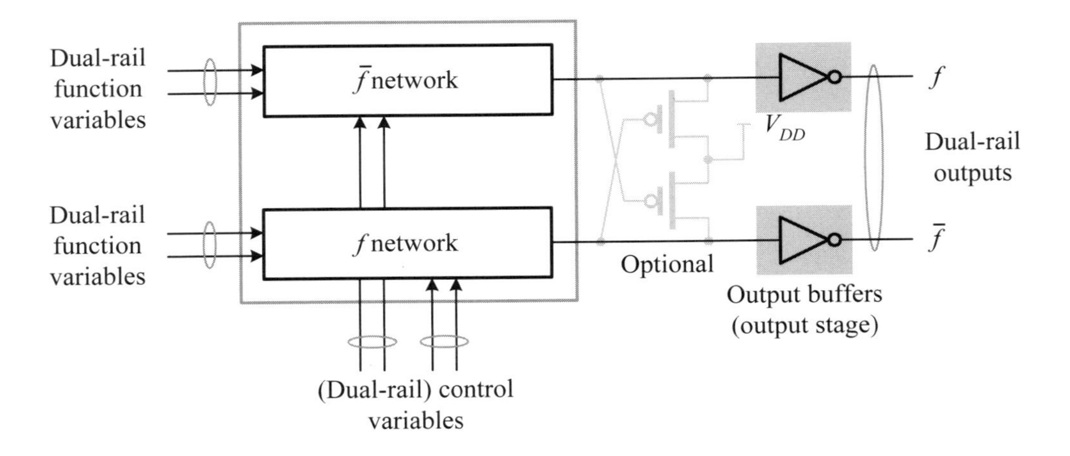

Figure 7.30: The general block diagram of CPL.

voltage. When a CPL circuit is used as an intermediate stage of a logic circuit, the output buffers may be omitted.

The CPL function networks can be realized by the $0/1$-$x/\bar{x}$-tree network paradigm with the following minor modification. The constants 1 and 0 appearing at the dual-rail inputs have to be replaced by a variable using the properties: $x \cdot \bar{x} = x \cdot 0 = 0 \cdot \bar{x}$ and $x = x \cdot x = 1 \cdot x$, from switching algebra, if the resulting logic circuit needs to be cascadable.

■ Example 7-23: (A two-input CPL AND/NAND gate.) Using CPL style, realize a two-input AND/NAND logic function.

Solution: Based on the block diagram depicted in Figure 7.30, the $\bar{f}$ network of the two-input AND/NAND gate is $\bar{f} = \overline{xy}$, which can be expanded with respect to variable y into the following:

$$\bar{f} = \overline{xy} = \bar{x} + \bar{y} = \bar{y} \cdot 1 + y \cdot \bar{x} = \bar{y} \cdot \bar{y} + y \cdot \bar{x}$$

where the rule $x \cdot 1 = x \cdot x$ is used.

Similarly, the f network of the two-input AND/NAND gate is $f = xy$, which can be expanded with respect to variable y into the following:

$$f = xy = \bar{y} \cdot 0 + y \cdot x = \bar{y} \cdot y + y \cdot x$$

where the rule $x \cdot 0 = x \cdot \bar{x}$ is used. The resulting logic circuit is shown in Figure 7.31(a). ■

■ Example 7-24: (A two-input CPL XOR/XNOR gate.) Using CPL style, realize a two-input XOR/XNOR logic function.

Solution: Based on the block diagram depicted in Figure 7.30, the $\bar{f}$ network of the two-input XOR/XNOR gate is $\bar{f} = \overline{x \oplus y}$, which can be expanded with respect to variable y into the following:

$$\bar{f} = \overline{x \oplus y} = \bar{y} \cdot \bar{x} + y \cdot x$$

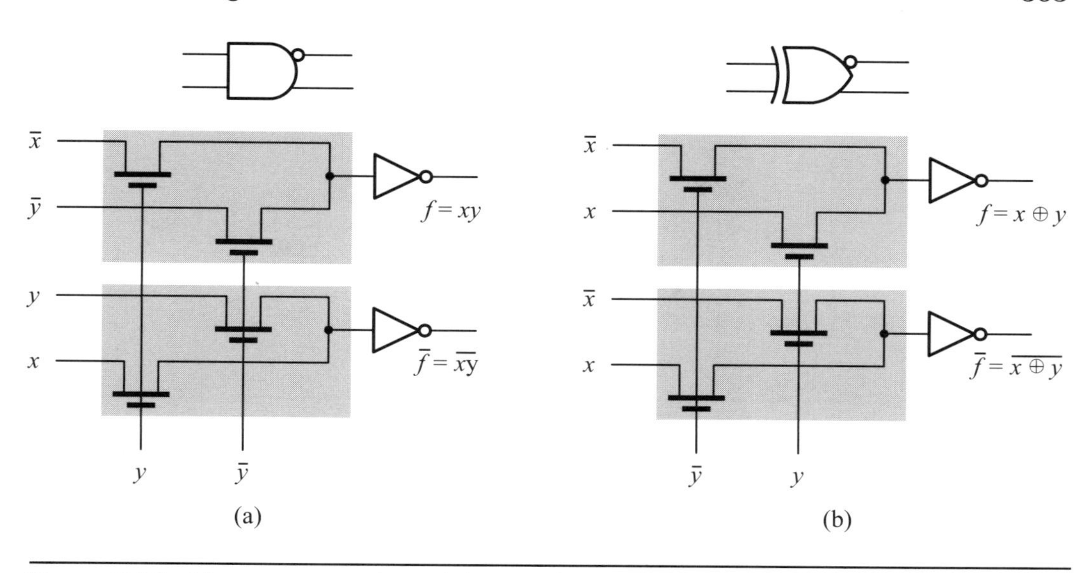

Figure 7.31: The two-input CPL (a) AND/NAND and (b) XOR/XNOR gates.

Similarly, the f network of the two-input XOR/XNOR gate is $f = x \oplus y$, which can be expanded with respect to variable y into the following:

$$f = x \oplus y = \bar{y} \cdot x + y \cdot \bar{x}$$

The resulting logic circuit is shown in Figure 7.31(b).

■

A complex logic circuit can be implemented by using CPL in either a single network or a cascaded network. In the following example, we use a three-input XOR/XNOR gate as an example to illustrate how to design a cascaded CPL network. As shown in Figure 7.32, the first stage comprises a standard two-input CPL XOR/XNOR gate without output buffers while the second stage is an output-buffered CPL XOR/XNOR gate. The first stage combines the inputs x and y into an intermediate result, whereas the second stage combines the result from the first stage and the input z into the desired result, that is, to realize the XOR/XNOR function of three inputs, x, y, and z.

■ **Example 7-25: (A three-input CPL XOR/XNOR gate.)** Show that the CPL circuit shown in Figure 7.32 is a three-input XOR/XNOR gate.

Solution: Referring to Figure 7.32, the h and $\bar{h}$ function can be expressed as $h = \bar{x} \cdot y + x \cdot \bar{y} = x \oplus y$ and $\bar{h} = x \cdot y + \bar{x} \cdot \bar{y} = \overline{x \oplus y}$, respectively. Similarly, g and $\bar{g}$ can be obtained and separately represented as $g = \bar{z} \cdot h + z \cdot \bar{h} = z \oplus h = z \oplus x \oplus y$ and $\bar{g} = z \cdot h + \bar{z} \cdot \bar{h} = \overline{z \oplus h} = \overline{z \oplus x \oplus y}$. As a consequence, the CPL circuit is a three-input XOR/XNOR gate.

■

■ Review Questions

Q7-30. What is CPL?

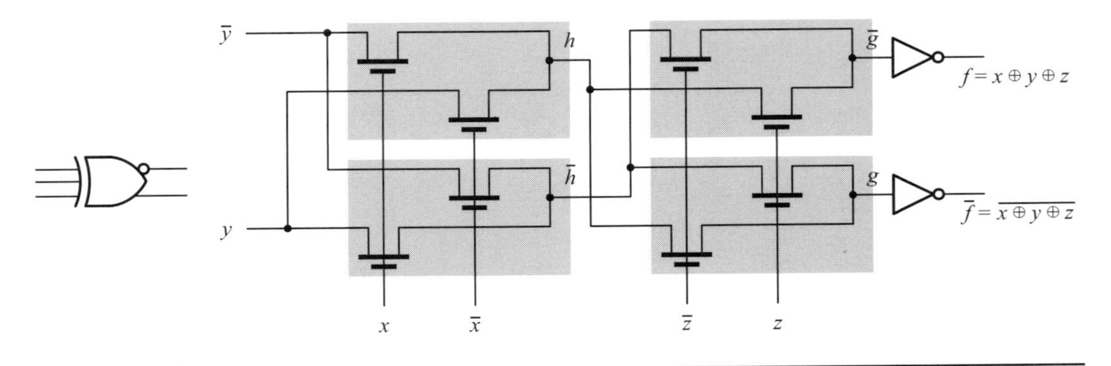

Figure 7.32: A three-input CPL XOR/XNOR gate.

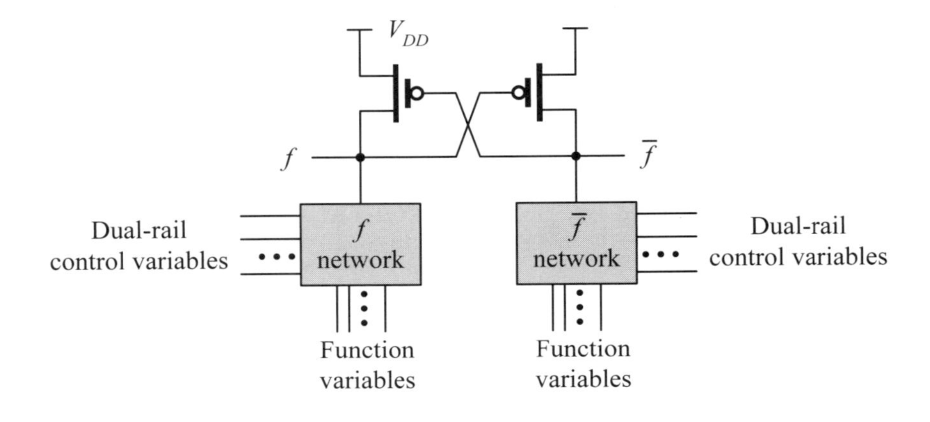

Figure 7.33: The general block diagram of DCVSPG logic.

Q7-31. What are the features of CPL?
Q7-32. How would you implement a switching function using CPL style?

7.3.3 DCVSPG

Differential cascode voltage switch with pass-gate (DCVSPG) logic is a combination of CVSL and CPL design styles; that is, the DCVSPG logic uses the output stage of CVSL as its output stage and uses the logic networks of CPL to implement the desired logic function. However, the DCVSPG logic circuit is generally a stand-alone circuit, not intentionally to be cascadable like CPL. Because of this, the DCVSPG logic function can be designed using the 0/1-$x/\bar{x}$-tree network paradigm directly. The general block diagram of DCVSPG is shown in Figure 7.33. In the following example, we illustrate this.

■ Example 7-26: (A two-input DCVSPG AND/NAND gate.) Using DCVSPG style, realize a two-input AND/NAND logic gate.

Solution: Based on the block diagram depicted in Figure 7.33, the $\bar{f}$ network of the two-input AND/NAND gate is $\bar{f} = \overline{x \cdot y}$, which can be expanded with respect to

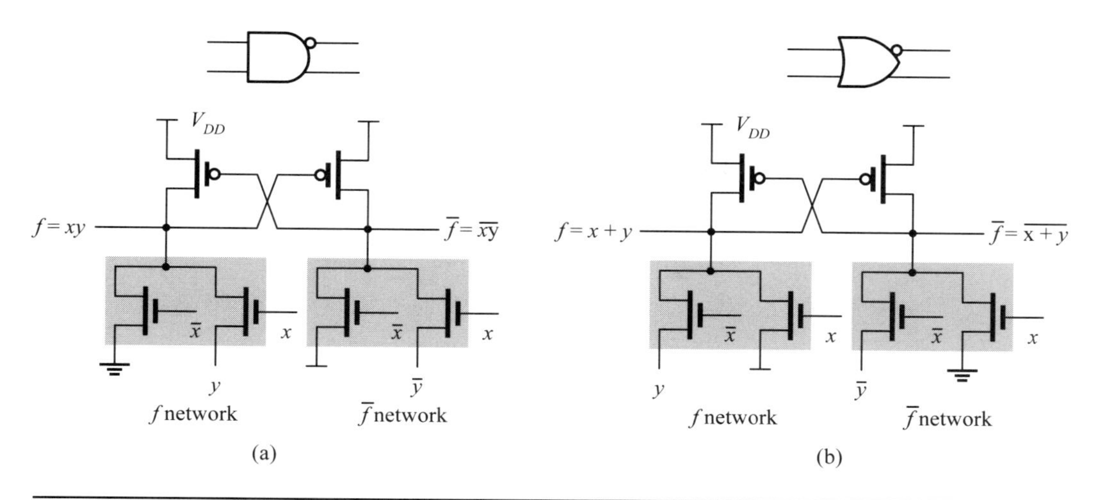

Figure 7.34: Two-input DCVSPG (a) AND/NAND and (b) OR/NOR gates.

variable x into the following:

$$\bar{f} = \overline{x \cdot y} = \bar{x} + \bar{y} = \bar{x} \cdot 1 + x \cdot \bar{y}$$

Similarly, the f network of the two-input AND/NAND gate is $f = x \cdot y$, which can be expanded with respect to variable x into the following:

$$f = x \cdot y = \bar{x} \cdot 0 + x \cdot y$$

The resulting logic circuit is shown in Figure 7.34(a).

■

The two-input DCVSPG OR/NOR gate can be designed in the same way as the two-input DCVSPG AND/NAND gate. The resulting logic circuit is shown in Figure 7.34(b). Because of its intuitive simplicity, the details are left to the reader as an exercise. In the following example, we give some more complex examples of DCVSPG logic circuits.

■ **Example 7-27: (A three-input DCVSPG XOR/XNOR gate.)** Using the DCVSPG style, realize a three-input XOR/XNOR logic gate.

Solution: Based on the block diagram shown in Figure 7.33, the f network of the three-input XOR/XNOR gate, $f = x \oplus y \oplus z$, can be expanded first with respect to variable x and then with respect to variable y as follows:

$$\begin{aligned} f &= x \oplus y \oplus z = x \cdot \bar{y} \cdot \bar{z} + \bar{x} \cdot y \cdot \bar{z} + \bar{x} \cdot \bar{y} \cdot z + x \cdot y \cdot z \\ &= \bar{x} \cdot (\bar{y} \cdot z + y \cdot \bar{z}) + x \cdot (\bar{y} \cdot \bar{z} + y \cdot z) \end{aligned}$$

Similarly, the $\bar{f}$ network of the three-input XOR/XNOR gate, $\bar{f} = \overline{x \oplus y \oplus z}$, can be expanded first with respect to variable x and then with respect to variable y as in the following:

$$\begin{aligned} \bar{f} &= \overline{x \oplus y \oplus z} = \bar{x} \cdot \bar{y} \cdot \bar{z} + \bar{x} \cdot y \cdot z + x \cdot \bar{y} \cdot z + x \cdot y \cdot \bar{z} \\ &= \bar{x} \cdot (\bar{y} \cdot \bar{z} + y \cdot z) + x \cdot (\bar{y} \cdot z + y \cdot \bar{z}) \end{aligned}$$

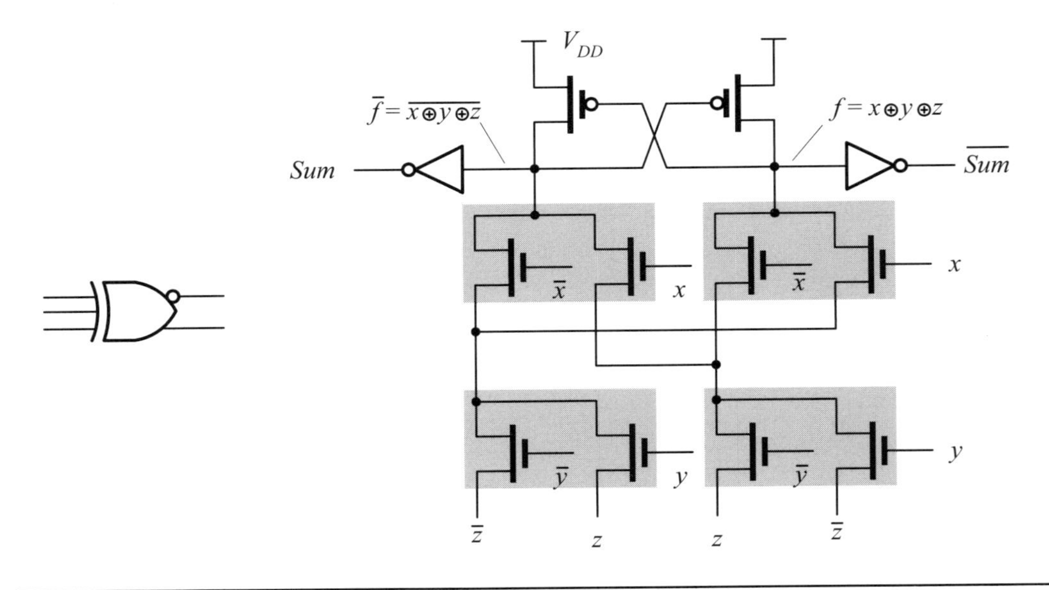

Figure 7.35: A three-input DCVSPG XOR/XNOR gate.

The resulting logic circuit is shown in Figure 7.35. It is instructive to note the parts being shared by the two functions: $\bar{f}$ and f.

■

■ Example 7-28: (The carry function of a full adder.) Using the DCVSPG style, realize the carry function, namely, the three-input majority function, of a full adder.

Solution: Based on the block diagram shown in Figure 7.33, the f network of the carry function, $Carry = x \cdot y + y \cdot z + x \cdot z$, can be expanded first with respect to variable z and then with respect to variable x as in the following:

$$\begin{aligned} Carry &= x \cdot y + y \cdot z + x \cdot z \\ &= \bar{z} \cdot (x \cdot y) + z \cdot (y + x) = \bar{z} \cdot (\bar{x} \cdot 0 + x \cdot y) + z \cdot (\bar{x} \cdot y + x \cdot 1) \end{aligned}$$

Similarly, the $\bar{f}$ network of the carry function, $\overline{Carry} = \overline{x \cdot y + y \cdot z + x \cdot z}$, can also be first expanded with respect to variable z and then with respect to variable x as follows:

$$\begin{aligned} \overline{Carry} &= \overline{x \cdot y + y \cdot z + x \cdot z} = (\bar{x} + \bar{y})(\bar{y} + \bar{z})(\bar{x} + \bar{z}) \\ &= \bar{z} \cdot (\bar{x} + \bar{y}) + z \cdot (x \cdot y) = \bar{z} \cdot (\bar{x} \cdot 1 + x \cdot \bar{y}) + z \cdot (\bar{x} \cdot \bar{y} + x \cdot 0) \end{aligned}$$

The resulting logic is shown in Figure 7.36.

■

■ Review Questions

Q7-33. What is DCVSPG logic?

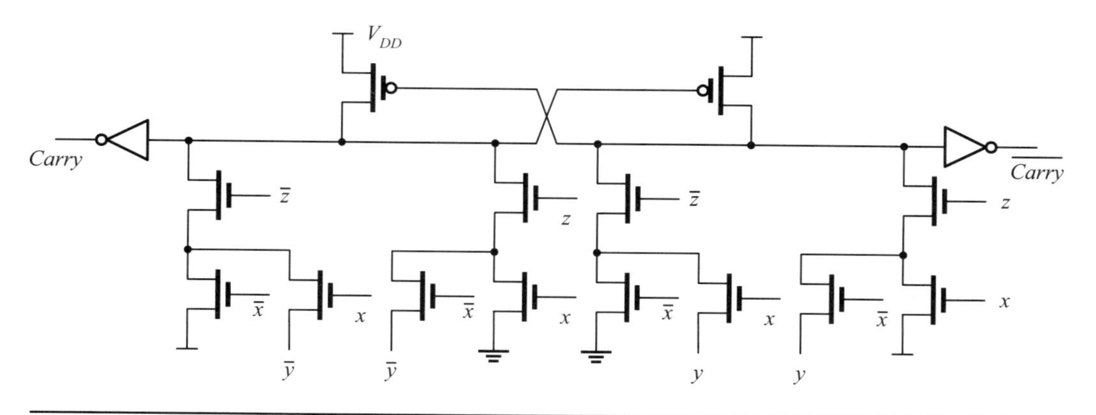

Figure 7.36: The carry function of a full adder realized by the DCVSPG style.

Q7-34. What is the distinction between CPL with a pMOS latch but without output buffers and DCVSPG logic?

Q7-35. How would you implement a switching function with the DCVSPG style?

7.3.4 Double Pass-Transistor Logic (DPL)

In contrast to CPL circuits, *double pass-transistor logic* (DPL) uses both nMOS and pMOS switches to realize the desired functions. Like the case of CPL, a DPL gate consists of both true and complementary inputs/outputs and is thus also a dual-rail logic circuit. The major features of DPL circuits are as follows. First, the DPL is a symmetrical arrangement in the sense that each input is connected to an equal number of gates and sources of MOS transistors. This results in a balanced input capacitance and reduces the dependence of the propagation delay on the data profile. Second, it has double-transmission characteristics due to the use of both nMOS and pMOS transistors. Hence, it does not need the output buffers to restore the voltage levels as in the case of CPL.

Like CPL, DPL also uses the concept of modular design to construct a complex logic circuit. To do this, basic two-input logic gates, including AND/NAND, OR/NOR, XOR/XNOR, 2-to-1 multiplexer, and so on, are constructed by applying Shannon's expansion theorem to decompose the underlying switching function with respect to each input variable. The result associated with each variable is then implemented by a pair of nMOS and pMOS switches.

In order to maintain the equal amount of loading capacitance for each input, the constants 1 and 0 appearing at the dual-rail inputs are replaced by proper literals using the following properties: $x \cdot \bar{x} = x \cdot 0 = 0 \cdot \bar{x}$, $\bar{x} = \bar{x} \cdot \bar{x} = 1 \cdot \bar{x}$, and $x = x \cdot x = 1 \cdot x$, from switching algebra. Of course, these constants can also be directly connected to V_{DD} and ground if the balanced input capacitance is not needed. In the following examples, we assume that the balanced input capacitance is needed.

■ **Example 7-29: (Two-input DPL AND and NAND gates.)** Using the DPL style, realize a two-input AND and a two-input NAND logic gate.

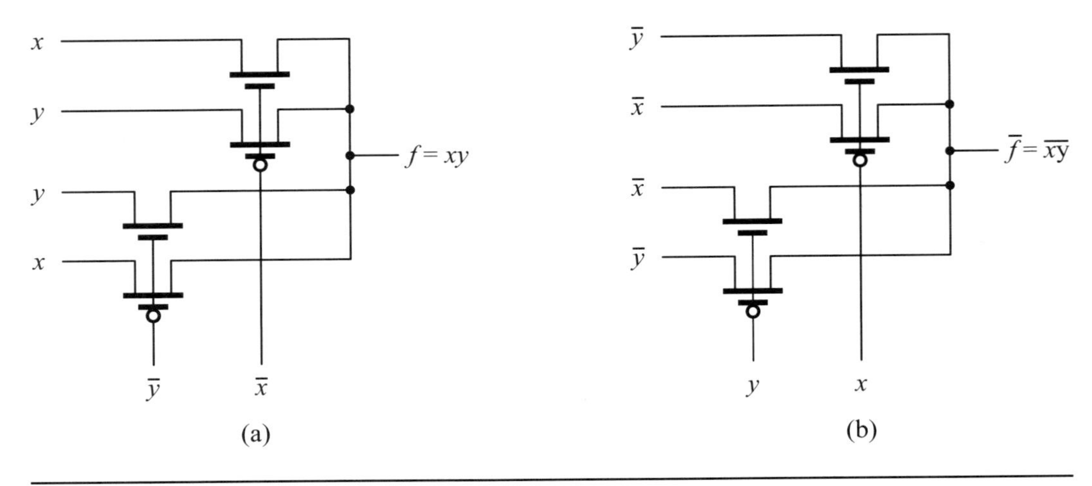

Figure 7.37: Two-input DPL (a) AND and (b) NAND gates.

Solution: Based on Shannon's expansion theorem, the switching function of a two-input AND logic gate, $f = x \cdot y$, can be expanded with respect to both variables x and y into the following:

$$\begin{aligned} f &= x \cdot y = [\bar{x} \cdot 0 + x \cdot y] + [\bar{y} \cdot 0 + y \cdot x] \\ &= [\bar{x} \cdot x + x \cdot y] + [\bar{y} \cdot y + y \cdot x] \end{aligned}$$

where the rule $\bar{x} \cdot 0 = \bar{x} \cdot x$ is used. This results in the DPL circuit shown in Figure 7.37(a), where each expanded result is realized by a pair of nMOS and pMOS transistors. The "circle" at the gate of a pMOS transistor implies an inverting operation.

Similarly, the switching function of a two-input NAND gate, $f = \overline{x \cdot y}$, can be expanded with respect to both variables x and y as follows:

$$\begin{aligned} f &= \overline{x \cdot y} = \bar{x} + \bar{y} = [\bar{x} \cdot 1 + x \cdot \bar{y}] + [\bar{y} \cdot 1 + y \cdot \bar{x}] \\ &= [\bar{x} \cdot \bar{x} + x \cdot \bar{y}] + [\bar{y} \cdot \bar{y} + y \cdot \bar{x}] \end{aligned}$$

where the rule $\bar{x} \cdot 1 = \bar{x} \cdot \bar{x}$ is used. The resulting logic circuit is shown in Figure 7.37(b). ■

In summary, it is not easy to directly implement an arbitrary switching function using DPL. Hence, a complex switching function, such as a full adder, is usually designed and implemented in DPL by combining the basic logic gates described above, thereby probably leading to a long chain of MOS transistors. This would result in a long propagation delay for the input signal to pass through the transistor chain to the output and may be intolerable in some applications.

■ Review Questions

Q7-36. What is DPL? What are its features?

Q7-37. How would you implement a switching function using the DPL style?

Q7-38. What is the distinction between CPL and DPL?

7.4 Summary

CMOS logic circuits can be roughly categorized into two broad categories: *static logic* and *dynamic logic*, depending on how a logic value is maintained at the output node of a given logic circuit. In a static logic circuit, a logic value is retained with the circuit states while in a dynamic logic circuit, a logic value is stored in the form of a charge.

To make use of a static logic circuit, it is necessary to understand its features thoroughly. To this end, the features and VTCs of basic static logic circuits, including CMOS inverters, NAND, and NOR gates were explored. In addition, the transistors in a logic gate may need to be sized according to different requirements for different applications. The most widely accepted requirements are propagation delay, area, noise margins, and power dissipation. Based on this, the static logic gates can be classified into three types: symmetric gates, asymmetric gates, and skewed logic gates.

The two major types of static logic circuits are single-rail and dual-rail logic circuits. The single-rail logic circuits receive as its inputs only either true or complementary signals and produce either true or complementary signals at their outputs. The conventional CMOS logic, TG-based logic, and pseudo-nMOS logic are such circuits. The dual-rail logic circuits receive as its inputs both true and complementary signals and yield both true and complementary signals at their outputs. Because of this, dual-rail logic is also called differential logic. The major types of dual-rail logic circuits include the following: cascode voltage switch logic (CVSL), complementary pass-transistor logic (CPL), differential cascode voltage switch with pass-gate (DCVSPG) logic, and double pass-transistor logic (DPL).

References

1. K. Chu and D. Pulfrey, "Design procedures for diffential cascode voltage switch circuits," *IEEE J. of Solid State Circuits*, Vol. 21, No. 6, pp. 1082–1087, December 1986.
2. K. Chu and D. Pulfrey, "A comparison of CMOS circuit techniques: differential cascode voltage switch logic versus conventional logic," *IEEE J. of Solid State Circuits*, Vol. 22, No. 4, pp. 528–532, August 1987.
3. L. Heller, W. Griffin, J. Davis, and N. Thoma, "Cascode voltage switch logic: a differential CMOS logic family," *Proc. of IEEE Int'l Solid-State Circuits Conf.*, pp. 16–17, 1984.
4. D. A. Hodges, H. G. Jackson, and R. A. Saleh, *Analysis and Design of Digital Integrated Circuits: In Deep Submicron Technology*, 3rd ed. New York: McGraw-Hill Books, 2004.
5. N. Jha, "Testing of differential cascode voltage switch one-count generators," *IEEE J. of Solid State Circuits*, Vol. 25, No. 1, pp. 246–253, February 1990.
6. M. Johnson, "A symmetric CMOS NOR gate for high-speed applications," *IEEE J. of Solid State Circuits*, Vol. 23, No. 5, pp. 1233–1236, October 1988.
7. S. M. Kang and Y. Leblebici, *CMOS Digital Integrated Circuits: Analysis and Design*, 3rd ed. New York: McGraw-Hill Books, 2003.
8. F. Lai and W. Hwang, "Design and implementation of differential cascode voltage switch with pass-gate (DCVSPG) logic for high-performance digital systems," *IEEE J. of Solid State Circuits*, Vol. 32, No. 4, pp. 563–573, April 1997.

9. A. Morgenshtein, A. Fish, and I. A. Wagner, "Gate-diffusion input (GDI): a power-efficient method for digital combinatorial circuits," *IEEE Trans. on Very Large Scale Integration (VLSI) Systems*, Vol. 10, No. 5, pp. 566–581, October 2002.
10. V. Oklobdzija and R. Montoye, "Design-performance tradeoffs in CMOS-domino logic," *IEEE J. of Solid State Circuits*, Vol. 21, No. 2, pp. 304–309, April 1986.
11. J. Pasternak, A. Shubat, and C. Salama, "CMOS differential pass-transistor logic design," *IEEE J. of Solid State Circuits*, Vol. 22, No. 2, pp. 216–222, April 1987.
12. J. Pasternak and C. Salama, "Design of submicrometer CMOS differential pass-transistor logic circuits," *IEEE J. of Solid State Circuits*, Vol. 26, No. 9, pp. 1249–1258, September 1991.
13. K. Schultz, R. Francis, and K. Smith, "Ganged CMOS: trading standby power for speed," *IEEE J. of Solid State Circuits*, Vol. 25, No. 3, pp. 870–873, June 1990.
14. A. Solomatnikov, D. Somasekhar, K. Roy, and C. K. Koh, "Skewed CMOS: noise-immune high-performance low-power static circuit family," *Proc. of IEEE Int'l Conf. on Computer Design*, 17–20 Sept. 2000, pp. 241–246.
15. M. Suzuki et al., "A 1.5-ns 32-b CMOS ALU in double pass-transistor logic," *IEEE J. of Solid State Circuits*, Vol. 28, No. 11, pp. 1145–1151, November 1993.
16. G. Tharakan and S. Kang, "A new design of a fast barrel switch network," *IEEE J. of Solid State Circuits*, Vol. 27, No. 2, pp. 217–221, February 1992.
17. J. P. Uyemura, *CMOS Logic Circuit Design*. Boston: Kluwer Academic Publishers, 1999.
18. J. P. Uyemura, *Introduction to VLSI Circuits and Systems*. New York: John Wiley & Sons, 2002.
19. J. Wang, S. Fang, and W. Feng, "New efficient designs for XOR and XNOR functions on the transistor level," *IEEE J. of Solid State Circuits*, Vol. 29, No. 7, pp. 780–786, July 1994.
20. N. H. E. Weste and D. Harris, *CMOS VLSI Design: A Circuit and Systems Perspective*, 3rd ed. Boston: Addison-Wesley, 2005.
21. K. Yano, T. Yamanaka, T. Nishida, M. Saito, K. Shimohigashi, and A. Shimizu, "A 3.8-ns 16 × 16-b multiplier using complementary pass-transistor logic," *IEEE J. of Solid State Circuits*, Vol. 25, No. 2, pp. 388–395, April 1990.
22. K. Yano, Y. Sasaki, K. Rikino, and K. Seki, "Top-down pass-transistor logic design," *IEEE J. of Solid State Circuits*, Vol. 31, No. 6, pp. 792–803, June 1996.
23. N. Zhuang and H. Wu, "A new design of the CMOS full adder," *IEEE J. of Solid State Circuits*, Vol. 27, No. 5, pp. 840–844, May 1992.
24. R. Zimmermann and W. Fichtner, "Low-power logic styles: CMOS versus pass-transistor logic," *IEEE J. of Solid State Circuits*, Vol. 32, No. 7, pp. 1079–1090, July 1997.

Problems

7-1. Referring to the inverter shown in Figure 7.2(a), answer the following questions. Assume that V_{DD} is 1.8 V and $V_{T0n} = |V_{T0p}| = 0.4$ V. The electron and hole mobilities are 292 cm^2/V-s and 112 cm^2/V-s, respectively.

(a) Calculate the inverter threshold voltage V_{th} if both nMOS and pMOS transistors have the same size.

(b) Determine the sizes of both nMOS and pMOS transistors if the inverter threshold voltage is set to $\frac{1}{2}V_{DD}$.

7-2. Assume that $V_{DD} = 1.2$ V, $W_n = 4\lambda$, $W_p = 12\lambda$, $L_n = L_p = 2\lambda$. Using 0.13-μm parameters: $\mu_{effn} = 298$ cm^2/V-s, $\mu_{effp} = 97$ cm^2/V-s, and $V_{T0n} = |V_{T0p}| = 0.35$ V, calculate V_{OL}, V_{OH}, V_{IL}, V_{IH}, and V_{th} of a CMOS inverter.

7-3. For a CMOS inverter, if $k_R = 1$ and $V_{T0n} = |V_{T0p}| = V_T$, then show that

(a) $V_{IL} = \frac{3}{8}V_{DD} + \frac{1}{4}V_T$

(b) $V_{IH} = \frac{5}{8}V_{DD} - \frac{1}{4}V_T$

(c) $NM_L = NM_H = \frac{3}{8}V_{DD} + \frac{1}{4}V_T$

7-4. Assume that V_{DD} is 1.8 V and $V_{T0n} = |V_{T0p}| = 0.4$ V. The electron and hole mobilities are 292 cm^2/V-s and 112 cm^2/V-s, respectively. Calculate the threshold voltage V_{th} of a two-input NAND gate under each of the following conditions.

(a) If all nMOS and pMOS transistors have the same size.

(b) If all pMOS transistors have twice the size of nMOS transistors and all nMOS transistors have the same size.

7-5. Referring to the two-input NAND gate shown in Figure 7.4(a), consider the situation of simultaneous switching; namely, both inputs x and y are connected together. Show analytically that

(a) The threshold voltage V_{th} can be expressed as

$$V_{th} = \frac{V_{DD} - |V_{T0p}| + \frac{1}{2}\sqrt{k_R}V_{T0n}}{1 + \frac{1}{2}\sqrt{k_R}}$$

where $k_R = k_n/k_p$.

(b) If $k_n = k_p$ and $V_{T0n} = |V_{T0p}| = V_{T0}$, then the threshold voltage is equal to

$$V_{th} = \frac{2V_{DD} - V_{T0}}{3}$$

7-6. Assume that V_{DD} is 1.2 V and $V_{T0n} = |V_{T0p}| = 0.35$ V. The electron and hole mobilities are 298 cm^2/V-s and 97 cm^2/V-s, respectively. Calculate the threshold voltage V_{th} of a two-input NOR gate under each of the following conditions.

(a) If all nMOS and pMOS transistors have the same size.

(b) If all pMOS transistors have four times the size of nMOS transistors and all nMOS transistors have the same size.

7-7. Referring to the two-input NOR gate shown in Figure 7.6(a), consider the situation of simultaneous switching; namely, both inputs x and y are connected together. Show analytically that

(a) The threshold voltage V_{th} can be expressed as

$$V_{th} = \frac{V_{DD} - |V_{T0p}| + 2\sqrt{k_R}V_{T0n}}{1 + 2\sqrt{k_R}}$$

where $k_R = k_n/k_p$.

(b) If $k_n = k_p$ and $V_{T0n} = |V_{T0p}| = V_{T0}$, then the threshold voltage is equal to

$$V_{th} = \frac{V_{DD} + V_{T0}}{3}$$

7-8. Using equivalent-NOT gate, derive the threshold voltage of an n-input NAND gate.

7-9. Using equivalent-NOT gate, derive the threshold voltage of an n-input NOR gate.

7-10. Referring to Figure 7.8(a), draw a circuit schematically to show the perfectly symmetric 3-input NAND gate. Give a general rule to an n-input NAND gate.

7-11. Referring to Figure 7.8(b), draw a circuit schematically to show the perfectly symmetric 3-input NOR gate. Give a general rule to an n-input NOR gate.

7-12. Considering the circuit shown in Figure 7.13(b), answer the following questions:

(a) Show that the circuit is a universal logic module.

(b) Construct a two-input NAND and a two-input NOR gates using the circuit.

(c) Construct a 2-to-1 multiplexer using the circuit.

7-13. Implement the following logic circuits with GDI logic style:

(a) Three-input AND gate

(b) Three-input OR gate

7-14. Using 0.35-μm parameters, $\mu_n = 540$ cm^2/V-s, $\mu_p = 180$ cm^2/V-s, and $V_{T0n} = |V_{T0p}| = 0.5$ V, design a pseudo-nMOS inverter that provides $V_{OL} = 0.1$ V. Assume that $V_{DD} = 3.3$ V.

7-15. Referring to the pseudo-nMOS inverter in Figure 7.25(a), show analytically that

(a) The low-level input voltage V_{IL} can be expressed as follows:

$$V_{IL} = V_{T0n} + \frac{V_{DD} - |V_{T0p}|}{\sqrt{k_R(k_R + 1)}}$$

(b) The high-level input voltage V_{IH} can be expressed as follows:

$$V_{IH} = V_{T0n} + \frac{2(V_{DD} - |V_{T0p}|)}{\sqrt{3k_R}}$$

(c) If $V_{T0n} = |V_{T0p}| = V_{T0}$, the threshold voltage is equal to

$$V_{th} = V_{T0} + \frac{V_{DD} - V_{T0}}{\sqrt{1 + k_R}}$$

7-16. Using the CVSL paradigm, realize the following switching expression:

$$f = vw + x(y + z)$$

7-17. Using the one logic-tree approach, realize the sum and carry functions of a CVSL full adder.

7-18. Using the CPL style, realize a two-input OR/NOR logic function.

7-19. Using the CPL style, realize a three-input AND/NAND gate in a way specified in the following:

(a) Directly implement the AND/NAND gate as a complex logic circuit.

(b) Implement the AND/NAND gate by cascading two-input basic CPL gates.

7-20. Using the CPL style, realize a three-input OR/NOR gate in a way specified in the following:

(a) Directly implement the OR/NOR gate as a complex logic circuit.

(b) Implement the OR/NOR gate by cascading two-input basic CPL gates.

7-21. Using the CPL style, realize a three-input majority logic circuit in a way specified in the following:

(a) Directly implement the majority logic circuit as a complex logic circuit.

(b) Implement the majority logic circuit by cascading two-input basic CPL gates.

7-22. Using the DCVSPG style, realize a two-input OR/NOR gate.

7-23. Using the DCVSPG style, realize a two-input XOR/XNOR logic gate.

7-24. Using the DCVSPG style, realize a three-input AND/NAND gate.

7-25. Using the DCVSPG style, realize a three-input OR/NOR gate.

7-26. Using the DPL style, realize a two-input OR and a two-input NOR logic gates.

7-27. Using the DPL style, realize a two-input XOR and a two-input XNOR logic gates.

7-28. Design and implement a 2-to-1 multiplexer using the DPL style.

7-29. (A study of NAND gate)

(a) Design a two-input NAND gate with the following circuit parameters: $L_p = L_n = L_{min}$ and $W_p = W_n = W_{min}$. Construct a layout of the gate and then extract the parameters from the layout. The output of the gate is connected to a load capacitance of 50 fF. For convenience, the two inputs are labeled as x and y, respectively.

(b) Apply an ideal square wave (that is, both rise and fall times are 0 ns) of an appropriate frequency to input x, and input y is connected to V_{DD}. Measure the propagation delays t_{pLH} and t_{pLH}.

(c) Apply an ideal square wave of an appropriate frequency to both inputs x and y. Measure the propagation delay s t_{pLH} and t_{pLH}.

(d) Repeat above three steps with $W_p = W_n = 2W_{min}$.

7-30. (A study of NOR gate)

(a) Design a two-input NOR gate with the following circuit parameters: $L_p = L_n = L_{min}$ and $W_p = W_n = W_{min}$. Construct a layout of the gate and then extract the parameters from the layout. The output of the gate is connected to a load capacitance of 50 fF. For convenience, the two inputs are labeled as x and y, respectively.

(b) Apply an ideal square wave (that is, both rise and fall times are 0 ns) of an appropriate frequency to input x, and input y is grounded. Measure the propagation delays t_{pLH} and t_{pLH}.

(c) Apply an ideal square wave of an appropriate frequency to both inputs x and y. Measure the propagation delays t_{pLH} and t_{pLH}.

(d) Repeat step (c) with $W_p = 4W_{min}$.

8

Dynamic Logic Circuits

Both n-type metal-oxide-semiconductor (nMOS) and p-type MOS (pMOS) blocks are required in complementary metal-oxide-semiconductor (CMOS) logic to implement a switching function. Although it results in a ratioless logic circuit, this type of logic circuit contains redundant information. For each nMOS transistor, there is a pMOS counterpart. In other words, the information of a switching function is provided by the nMOS transistors and repeated by the pMOS counterpart again. Hence, substantial amounts of silicon and energy are wasted, especially for complex switching functions.

The information redundancy of CMOS logic can be removed by using pseudo-nMOS logic. However, the pseudo-nMOS logic not only becomes ratioed but also consumes power continually as the nMOS block is turned on. A better way probably is dynamic logic, which is also called precharge-evaluate (PE) logic since its operation comprises two phases: precharge and evaluate. The dynamic logic is ratioless and consumes no static power. However, because of the use of a charge to represent information, there are many nonideal effects that severely affect the signal integrity of such a logic circuit. These include leakage current, charge injection, clock (capacitive) feedthrough, back-gate coupling, charge-loss (or charge-leakage) effect, charge-sharing effect, and power-supply noise.

Dynamic logic can be further classified into three classes: single-rail dynamic logic, dual-rail dynamic logic, and clocked CMOS logic. Remember that single-rail logic receives single-rail input signals and also generates single-rail signals at its outputs; dual-rail logic receives dual-rail input signals and generates dual-rail output signals. Clocked CMOS logic means that its output node is controlled by a clock clk. Clocked CMOS logic can also be further cast into single-rail logic and dual-rail logic.

8.1 Introduction to Dynamic Logic

In a dynamic circuit, a logic value is stored in a capacitor made intentionally or a capacitive node being composed of parasitic capacitance inherently existing in the circuit. In this section, we begin to consider the dynamic features of MOS transistors when they are used as switches and then deal with (PE) dynamic logic circuits and explore their features. Finally, we address the hazards and types of dynamic logic circuits.

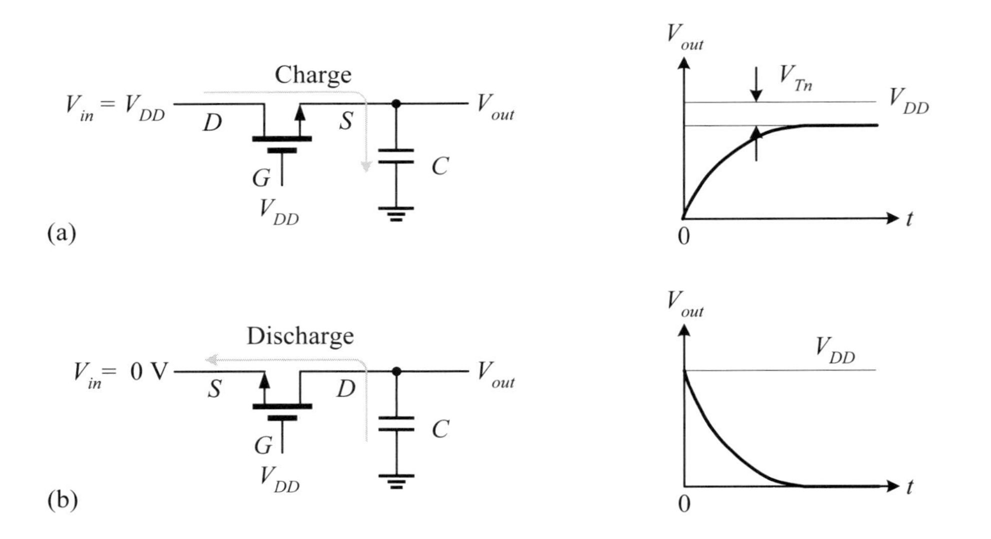

Figure 8.1: The operations of an nMOS switch: (a) $V_{in} = V_{DD}$; (b) $V_{in} = 0$ V.

8.1.1 MOS Transistors as Switches

We describe in the following the dynamic features of nMOS, pMOS, and transmission gate (TG) switches. For both nMOS and pMOS switches, we use them as a switch to control the operations of charging and discharging a capacitor to study how the threshold voltage affects the capacitor voltage. Finally, we deal with the features and delay model of TG switches.

8.1.1.1 nMOS Switches In digital logic circuits, an nMOS transistor can be regarded as a simple switch. The nMOS switch is turned on when the gate-to-source voltage V_{GS} is greater than or equal to the threshold voltage V_{Tn} and turned off otherwise. Remember that the roles of source and drain of an nMOS transistor are determined dynamically rather than statically. This will manifest itself in the following discussion.

Figure 8.1 shows two circuits to explain the operations of an nMOS transistor when it is used as a switch. In both circuits, we suppose that the gate voltage is V_{DD}, which is usually greater than the threshold voltage V_{Tn}, and hence both nMOS transistors are turned on initially. In Figure 8.1(a), the input voltage is V_{DD} and the output voltage is 0 V initially; that is, the output capacitor is completely discharged. Hence, the output capacitor will be charged on its way to V_{DD} once the voltage V_{DD} is applied to the gate. The source and drain nodes of the nMOS transistor are indicated in the figure. The one with more positive voltage is the drain and the other is the source. The output voltage V_{out} is exponentially increased with time until the point when $V_{out} = V_{DD} - V_{Tn}$ is reached. After that, the nMOS transistor is turned off since the V_{GS} is no longer greater than threshold voltage V_{Tn}. Therefore, the output voltage remains at its maximum value of $V_{DD} - V_{Tn}$, assuming that the leakage current is ignored.

Figure 8.1(b) considers the case that the output capacitor is being discharged; that is, the input voltage is 0 V, whereas the output voltage is V_{DD} initially. At this time, the roles of source and drain are interchanged in contrast to the situation of Figure 8.1(a). Consequently, during the discharging process the nMOS transistor is

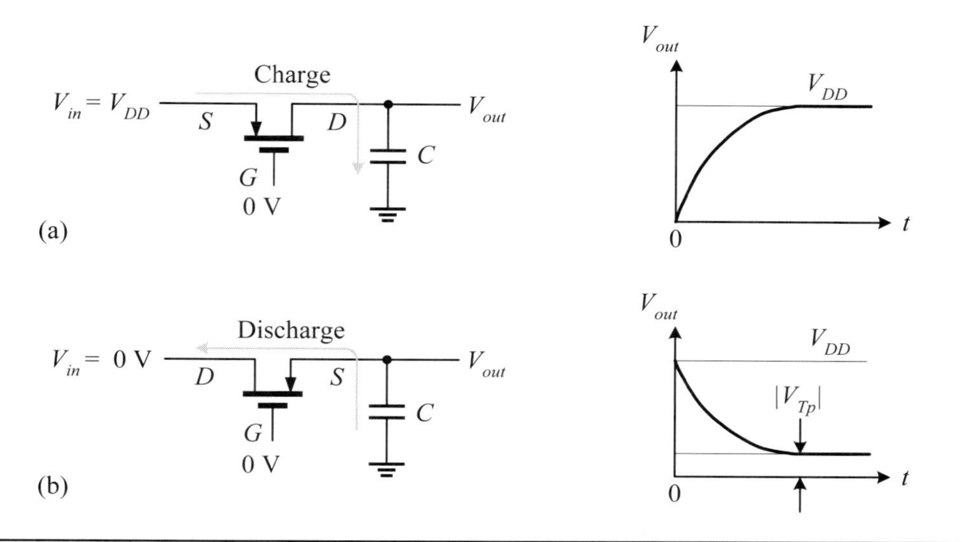

Figure 8.2: The operations of a pMOS switch: (a) $V_{in} = V_{DD}$; (b) $V_{in} = 0$ V.

always on because the voltage V_{GS} is always greater than the threshold voltage V_{Tn}. The output capacitor can completely discharge to 0 V.

8.1.1.2 pMOS Switches Like an nMOS transistor, a pMOS transistor can also be thought of as a simple switch in digital logic circuits. The pMOS switch is turned on when the source-to-gate voltage V_{SG} is greater than or equal to the absolute value of threshold voltage $|V_{Tp}|$ and turned off otherwise. Like the nMOS transistor, the roles of source and drain of a pMOS transistor are determined dynamically rather than statically.

Figure 8.2 shows two circuits to explain the operations of a pMOS transistor when it is used as a switch. In both circuits, we suppose that the gate voltage is 0 V, which may turn on or off the pMOS transistor, depending on whether the source voltage is greater than the threshold voltage $|V_{Tp}|$. Figure 8.2(a) shows the case when the input voltage is V_{DD} while the output voltage is 0 V initially; that is, the output capacitor is completely discharged. Hence, the output capacitor is to be charged toward V_{DD} once a 0-V voltage is applied to the gate. The source and drain of the pMOS transistor are indicated in the figure. The one with more positive voltage is the source and the other is the drain. The output voltage V_{out} is exponentially increased with time until its value reaches the input voltage V_{in} since the pMOS transistor is always on due to its $V_{SG} = V_{DD} - 0 \geq |V_{Tp}|$ in this case.

Figure 8.2(b) considers the situation that the output capacitor is being discharged; that is, the input voltage is 0 V and the output voltage is V_{DD} initially. At this time, the roles of source and drain are interchanged in contrast to the case of Figure 8.2(a). Consequently, during the discharging process, the pMOS transistor is turned on and the output voltage V_{out} is exponentially decreased with time until its value reaches $|V_{Tp}|$. After that, the pMOS transistor is turned off since the V_{SG} is no longer greater than the threshold voltage $|V_{Tp}|$. Hence, the output voltage remains at $|V_{Tp}|$.

8.1.1.3 TG Switches Recall that a TG switch is a device formed by connecting an nMOS and a pMOS switch in parallel, thereby removing the imperfect features

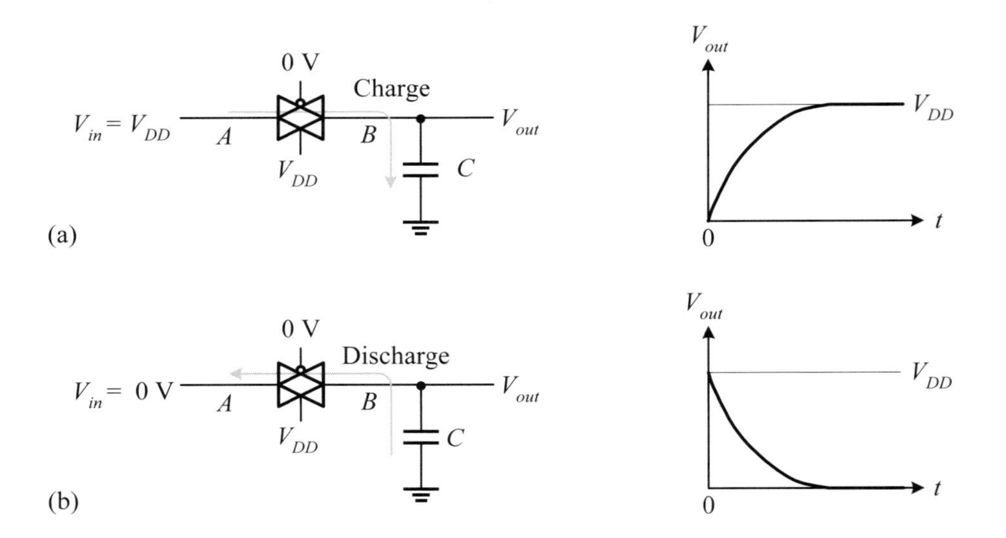

Figure 8.3: The operations of a CMOS switch: (a) $V_{in} = V_{DD}$; (b) $V_{in} = 0$ V.

of nMOS and pMOS switches when they are used alone. We consider the dynamic features of TG switches.

Figure 8.3 shows two circuits to explain the operations of a TG switch when it is used to control the charging and discharging operations of a capacitor. In both circuits, we suppose that the gate voltage of the pMOS transistor is set to 0 V and the gate voltage of the nMOS is set to V_{DD} when we wish to turn on the TG switch. Figure 8.3(a) shows the case when the input voltage is V_{DD} while the output voltage is 0 V initially; that is, the output capacitor is completely discharged. Hence, the output capacitor will be charged toward V_{DD} when the TG switch is turned on. Recall that the nMOS transistor will be turned off as the output voltage V_{out} reaches $V_{DD} - V_{Tn}$. However, the pMOS transistor is still on and hence the output voltage will eventually reach the input voltage V_{DD}. In other words, the TG switch can pass the logic-1 signal without degradation.

Figure 8.3(b) considers the situation when the output capacitor is being discharged; that is, the input voltage is 0 V while the output voltage is V_{DD} initially. At this time, the output voltage V_{out} is decreased toward 0 V when the TG switch is turned on. The pMOS transistor is turned off as the output voltage reaches $|V_{Tp}|$ since its V_{SG} is no longer greater than the threshold voltage $|V_{Tp}|$. However, the nMOS transistor is still on, causing the output capacitor to discharge to 0 V eventually. In other words, the TG switch can pass the logic-0 signal without degradation.

8.1.1.4 Delay Model of TGs A TG switch is a parallel connection of an nMOS and a pMOS transistors to take the good features of both devices, as shown in Figure 8.4(a). Using the equivalent-resistance model, a TG switch can be represented as a Π-shape RC circuit consisting of two capacitors at both ends and an equivalent resistor composed of two resistors connected in parallel, as shown in Figure 8.4(b). The input capacitance C_{in} and output capacitance C_{out} include overlapping capacitance C_{OL} and diffusion capacitance C_{DB} of both MOS transistors. They can be formally expressed as follows:

$$C_{in} = C_{out} = C_{para}(W_n + W_p) = 2C_{para}W \tag{8.1}$$

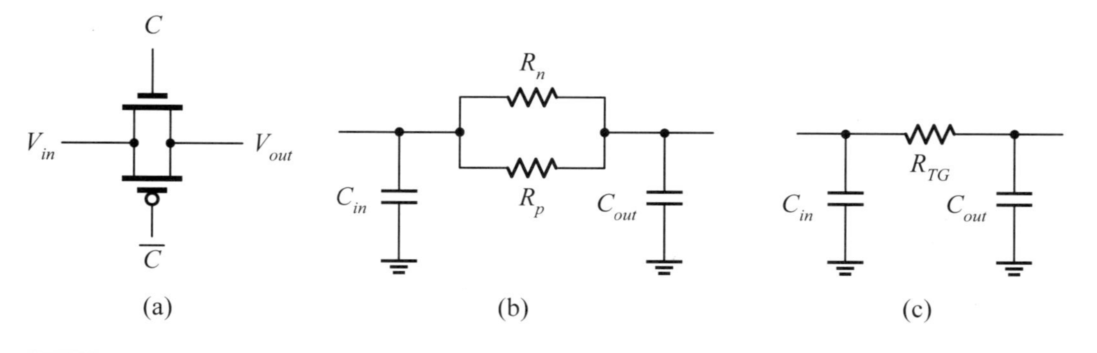

Figure 8.4: The TG switch and its delay model: (a) logic circuit; (b) equivalent circuit; (c) equivalent-Π circuit.

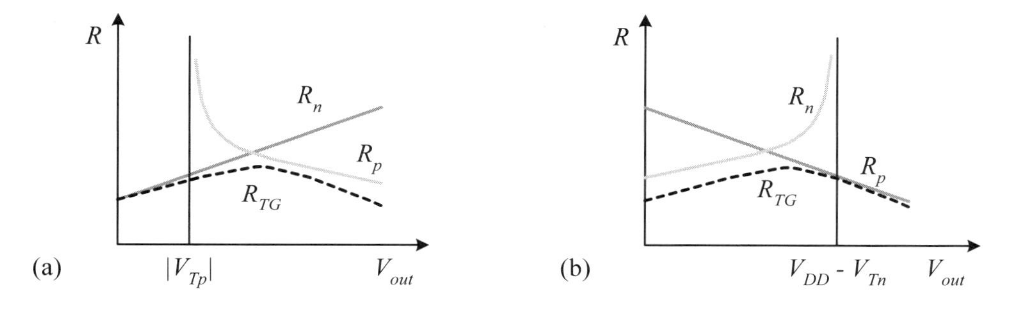

Figure 8.5: The on-resistance of a TG switch when: (a) $V_{in} = 0$ V; (b) $V_{in} = V_{DD}$.

where W_n is assumed to be equal to W_p, a usual case for the TG switch. Since R_n and R_p are connected in parallel, Figure 8.4(b) may be further reduced into Figure 8.4(c), a Π-network, as its shape implied.

The on-resistance of a TG switch is a parallel combination of the on-resistances of both nMOS and pMOS transistors. As shown in Figure 8.5, the on-resistance depends on whether a logic 0 or a logic 1 is being propagated through the switch. When $V_{in} = 0$ V, and V_{out} changes from V_{DD} to 0 V, the on-resistance of the pMOS transistor increases rapidly while the on-resistance of the nMOS transistor drops linearly over the range, as shown in Figure 8.5(a). When $V_{in} = V_{DD}$, and V_{out} changes from 0 V to V_{DD}, the on-resistance of the pMOS transistor drops linearly over the range while the on-resistance of the nMOS transistor increases rapidly, as shown in Figure 8.5(b).

For the purpose of timing calculation, the on-resistance of a TG switch can be approximated by an R_{TG} regardless of whether the V_{in} is at 0 V or V_{DD}. Namely, the equivalent on-resistance R_{TG} of a TG switch can be represented as follows:

$$R_{TG} = R_{eqn}\left(\frac{L}{W}\right) \tag{8.2}$$

where $L_n = L_p = L$ and $W_n = W_p = W$.

■ Example 8-1: (An application of the TG delay model.) Consider the circuit given in Figure 8.6(a), where a TG switch is sandwiched between two inverters. Draw an *RC*-equivalent circuit and apply the Elmore delay model to calculate the delay at the output.

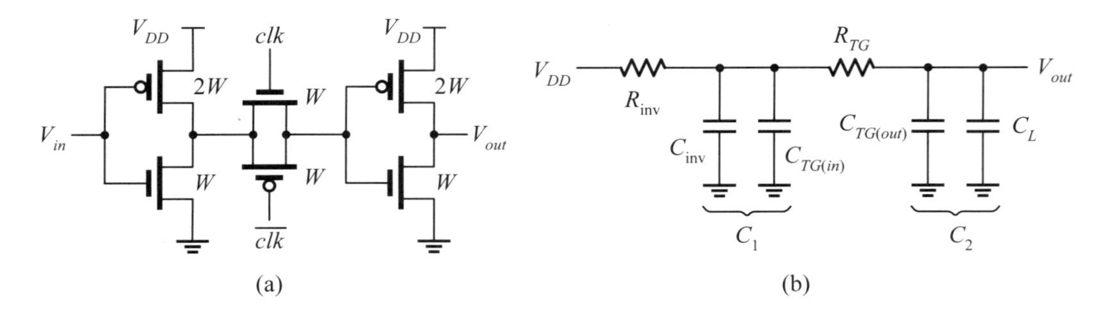

Figure 8.6: A TG switch is sandwiched between two inverters: (a) logic circuit; (b) equivalent circuit.

Solution: The RC-equivalent circuit is exhibited in Figure 8.6(b). We begin to calculate the input and output capacitances of the TG switch. When the TG switch is off, the input and output capacitances can be calculated as follows:

$$C_{TG(in)} = C_{TG(out)} = C_{para}(W_n + W_p) = 2C_{para}W \tag{8.3}$$

When the TG switch is on, assuming that both devices are in their linear regions, the input and output capacitances can be expressed as

$$\begin{aligned} C_{TG(in)} &= C_{TG(out)} = C_{para}(W_n + W_p) + \frac{1}{2}C_g(W_n + W_p) \\ &= 2C_{para}W + C_gW \end{aligned} \tag{8.4}$$

Therefore, the propagation delay can be calculated using the Elmore delay model as follows:

$$\begin{aligned} t_{Elmore} &= R_{inv}(C_{inv} + C_{TG(in)}) + (R_{inv} + R_{TG})(C_{TG(out)} + C_L) \\ &= R_{inv}[(2C_{para}W + C_gW) + C_{TG(in)}] + \\ &\quad (R_{inv} + R_{TG})[(2C_{para}W + C_gW) + C_L] \end{aligned} \tag{8.5}$$

■

■ Review Questions

Q8-1. How would you distinguish the source and drain of an nMOS transistor?

Q8-2. How would you distinguish the source and drain of a pMOS transistor?

Q8-3. Describe the features of a TG switch.

Q8-4. How would you calculate the equivalent on-resistance of a TG switch?

8.1.2 Basic Dynamic Logic

In this section, the evolution of dynamic logic, the principle of dynamic logic, and logical effort of dynamic logic, and footless dynamic logic are addressed in order.

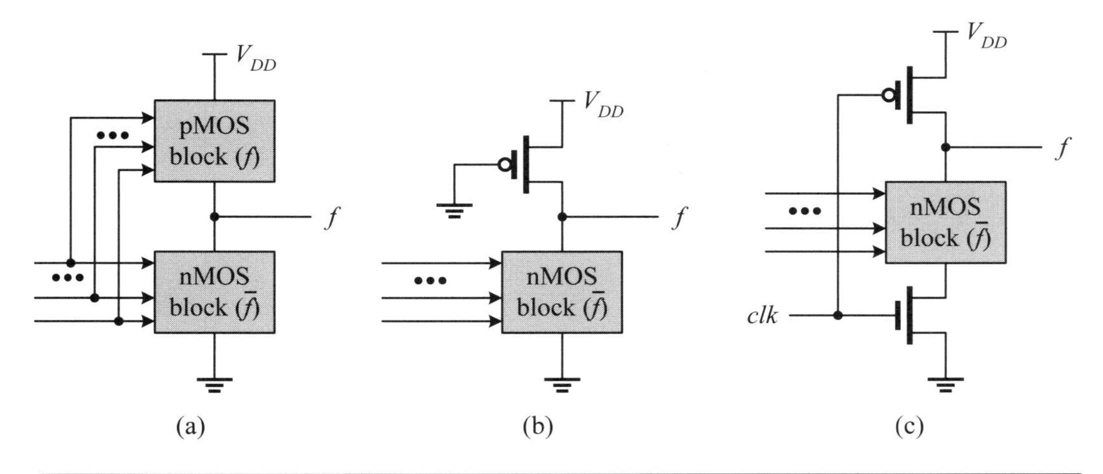

Figure 8.7: The evolution of CMOS dynamic logic from (a) FCMOS logic, (b) pseudo-nMOS logic, to (c) dynamic logic.

8.1.2.1 Evolution of Dynamic Logic Before proceeding, recall that CMOS logic requires two function blocks, pMOS and nMOS, as shown in Figure 8.7(a). Because of redundant functions realized by both blocks, the pseudo-nMOS logic is proposed to reduce the area and the logical effort, as shown in Figure 8.7(b). In this scheme, an always-on pMOS transistor is used as the load of the nMOS block to pull up the output when the nMOS block is off to enforce the node-value rule. However, it violates the node-conflict-free rule and becomes a ratioed logic circuit. In addition, it consumes a large amount of power because of a direct current flowing between power-supply rails through it when the nMOS block is on. To circumvent this flaw, an alternative is to connect the gate of the pMOS transistor to a clock instead of being permanently connected to ground, as shown in Figure 8.7(c). This results in a logic circuit known as a dynamic logic circuit and is ratioless logic because it satisfies both node-value and node-conflict-free rules. The node-value rule is satisfied by using a capacitive node as a meta node to keep the precharged logic value when the nMOS block is off. The node-conflict-free rule is enforced by multiplexing both pull-up and pull-down paths in a time-division manner and by adding a foot nMOS transistor with its gate controlled by the same clock as the pMOS transistor.

8.1.2.2 Principles of Dynamic Logic There are two basic types of basic dynamic logic circuits. One uses an nMOS block and is called the *nMOS dynamic logic*, and the other uses a pMOS block and is denoted as the *pMOS dynamic logic*. Because of better performance, nMOS dynamic logic is preferred in most applications.

nMOS dynamic logic. The nMOS dynamic logic circuit has the block diagram shown in Figure 8.8(a), where an nMOS block, one pMOS pull-up transistor, and a foot nMOS transistor compose the logic circuit. Like pseudo-nMOS logic, the nMOS block is used to carry out the desired logic function. Nevertheless, unlike the pseudo-nMOS logic circuits, where the pMOS transistor is always turned on, the pMOS transistor in the nMOS dynamic logic circuit is only turned on regularly to precharge the output node to V_{DD}. To further guarantee the node-conflict-free rule, a foot nMOS transistor is generally needed to isolate the nMOS block from the ground as the pMOS transistor is turned on. In some special applications, this foot nMOS transistor may be omitted, and hence, the performance of these logic circuits can be improved.

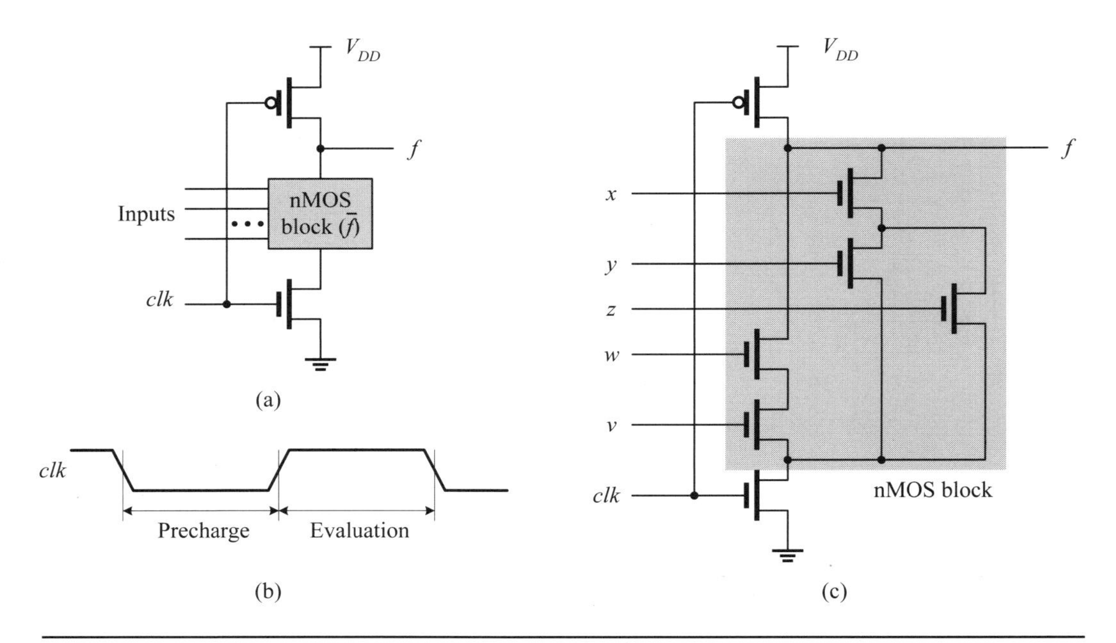

Figure 8.8: The (a) general logic diagram, (b) timing, and (c) an example of nMOS dynamic logic.

The operation of an nMOS dynamic logic circuit is subdivided into two phases: *precharge* and *evaluation*, as shown in Figure 8.8(b). During the precharge phase, the clock is low, thereby turning on the pMOS transistor and causing the output node f to precharge toward V_{DD}. Meanwhile, the foot nMOS transistor is turned off by the low-level clock to prevent the nMOS block from conducting. During the evaluation phase, the clock rises high, thereby turning on the foot nMOS transistor to enable the nMOS block. At the same time, the pMOS transistor is turned off to prevent the output node from continuing to charge toward V_{DD}. The output node f starts to settle down to its final value according to the input value and the function realized by the nMOS block. Because of this two-phase scenario of operation, dynamic logic is often referred to as *precharge-evaluation* (PE) logic.

Example 8-2: (An example of nMOS dynamic logic.) Figure 8.8(c) shows an example of an nMOS dynamic logic circuit. This logic circuit implements the following switching function:

$$f = \overline{x \cdot (y + z) + w \cdot v}$$

The reader should easily verify this, referring to Chapter 1 for more details about the implementation of switching functions with CMOS logic circuits.

pMOS dynamic logic. Although pMOS transistors are not often used in designing logic circuits because of poor performance compared to nMOS transistors, it is also possible to use the pMOS block to form basic dynamic logic for implementing the desired switching function. The general block diagram of pMOS dynamic logic is shown

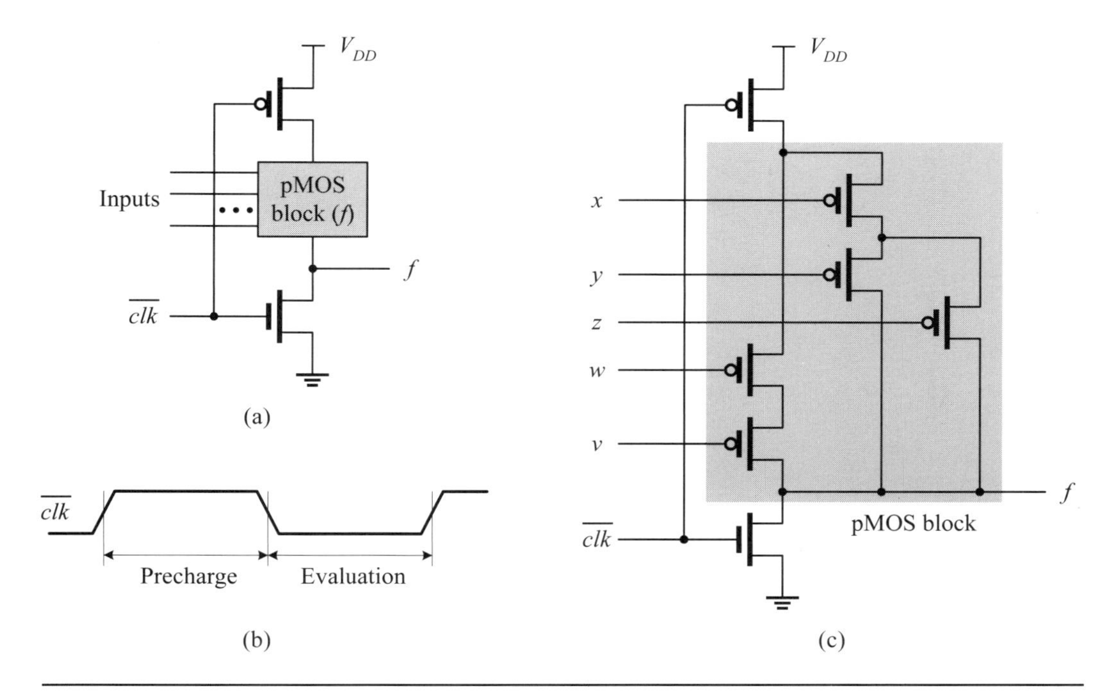

Figure 8.9: The (a) general logic diagram, (b) timing, and (c) an example of pMOS dynamic logic.

in Figure 8.9(a). It has a similar basic structure as the nMOS dynamic logic except that the pMOS block is used instead along with some other minor modifications.

The operation of pMOS dynamic logic is similar to that of nMOS dynamic logic except that both roles of nMOS and pMOS transistors are exchanged, the output node is taken from the connection between the nMOS precharge transistor and pMOS block, and precharged to 0 V rather than V_{DD}. During the precharge phase, the nMOS transistor is on while the pMOS transistor is off to prevent the pMOS block from conducting. The output node f is precharged to the ground level, that is, 0 V. During the evaluation phase, the nMOS transistor is turned off while the pMOS transistor is turned on, thereby allowing the output node f to settle down to its final value according to the logic function realized by the pMOS block and the input value. The detailed timing is exhibited in Figure 8.9(b).

■ **Example 8-3: (An example of pMOS dynamic logic.)** Figure 8.9(c) shows an example of pMOS dynamic logic. This logic circuit implements the following switching function:

$$f = \overline{(x + y \cdot z) \cdot (w + v)}$$

The reader should easily verify this.

■

8.1.2.3 Logical Effort of Dynamic Logic Dynamic logic inherently has smaller logical effort values than static logic since it only contains the nMOS block without the pMOS block. To compare the logical effort between static and dynamic logic, let us

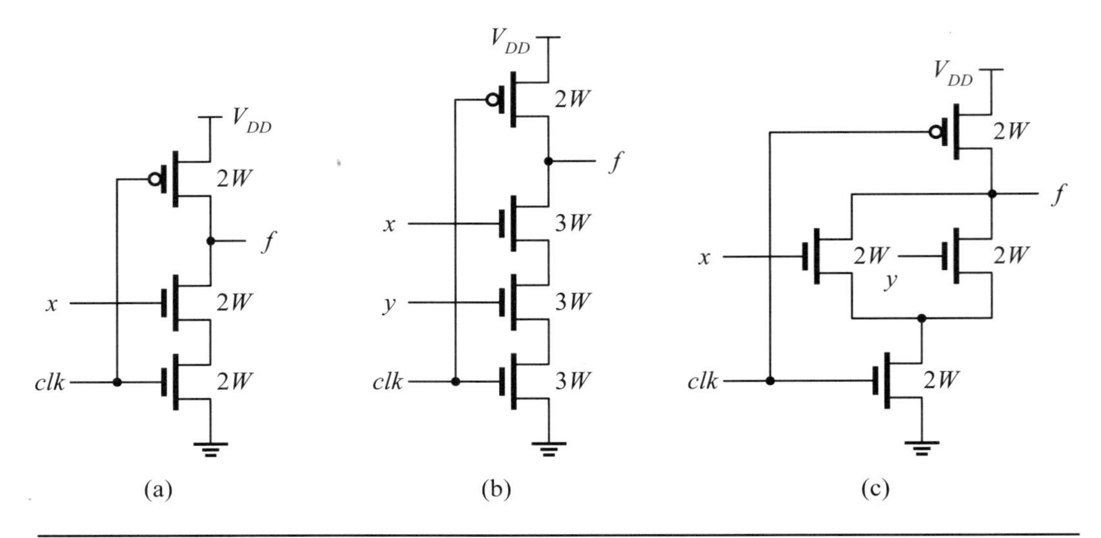

Figure 8.10: The three basic dynamic gates: (a) NOT; (b) NAND; (c) NOR.

make the pull-down strengths of both circuits equal to one another and assume that they drive the same output capacitive load.

An illustration of how to calculate the logical effort of dynamic logic is explored in the following.

■ Example 8-4: (Logical effort of dynamic logic.) Assume that all transistors use the minimum channel length. The logical effort of the three basic dynamic gates depicted in Figure 8.10 can be calculated as follows. For an inverter, the size of each nMOS transistor is $2W$ due to two such transistors being connected in series. Hence, the logical effort is equal to

$$g_{inv} = \frac{\tau_{inv}}{\tau_{inv}} = \frac{2R_{eqn}C_gL_n}{3R_{eqn}C_gL_n} = \frac{2}{3}$$

For a two-input NAND gate, the nMOS stack is three transistors. Thereby, the channel width of each nMOS transistor is $3W$. The logical effort of this gate is

$$g_{nand} = \frac{\tau_{nand}}{\tau_{inv}} = \frac{3R_{eqn}C_gL_n}{3R_{eqn}C_gL_n} = \frac{3}{3} = 1$$

For a two-input NOR gate, since two nMOS transistors in the nMOS block are connected in parallel, the nMOS stack is two transistors. Thereby, the channel width of each nMOS transistor is $2W$. The logical effort of this gate is found to be

$$g_{nor} = \frac{\tau_{nor}}{\tau_{inv}} = \frac{2R_{eqn}C_gL_n}{3R_{eqn}C_gL_n} = \frac{2}{3}$$

It is apparent that the logical effort values of these three basic dynamic logic gates are smaller than those of static gates, which are 1, 4/3, and 5/3.

■

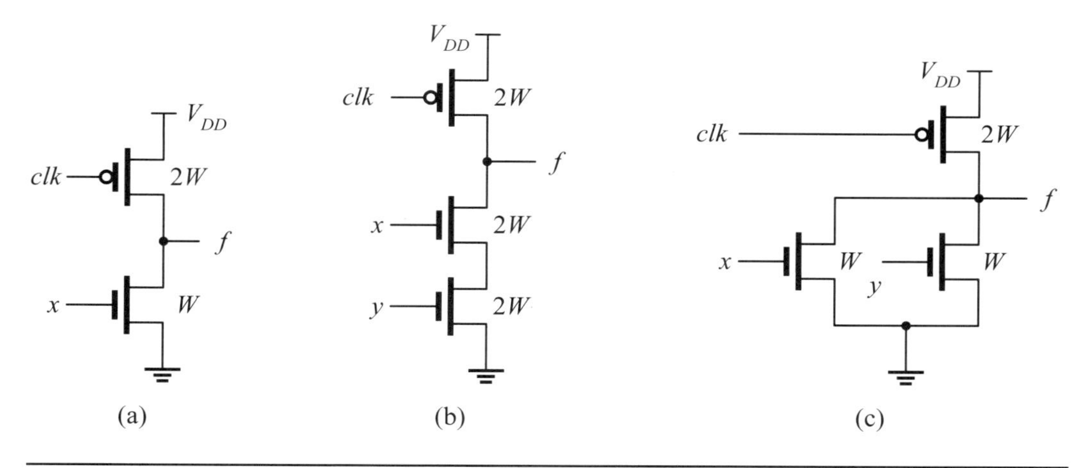

Figure 8.11: Footless basic dynamic logic gates: (a) NOT; (b) NAND; (c) NOR.

8.1.2.4 Footless Dynamic Logic In nMOS dynamic logic, an nMOS transistor is generally needed at the foot of the nMOS block to guarantee the ratioless property, namely, to guarantee that there is no direct current flowing between power-supply rails through the logic circuit. However, this foot nMOS transistor contributes a resistance to the nMOS stack, thereby resulting in a logic circuit with a larger logical effort value.

In some special applications, the ratioless property is inherently guaranteed since all transistors in the nMOS block are off during the precharge phase. In such situations, the foot nMOS transistor can be removed to enhance performance. Next, we give an example to compare the logical effort between footed and footless dynamic logic circuits.

■ **Example 8-5: (Comparison of footed and footless dynamic logic.)** Figure 8.11 shows footless basic dynamic gates. Calculate their logical effort values and compare with those of basic dynamic gates.

Solution: As shown in Figure 8.11, the logical effort values of footless basic dynamic gates are 1/3, 2/3, and 1/3, which are much less than those of basic dynamic gates. Hence, the footless logic has better performance. The precharge pMOS transistor in a dynamic circuit usually uses the smallest size. This is because the precharge phase is often not critical and hence the pMOS transistor is made as small as possible to alleviate the capacitive loading on the clock and save area.

■

■ Review Questions

Q8-5. Explain why the dynamic logic is ratioless.

Q8-6. Draw the nMOS dynamic logic circuits for a three-input NAND gate.

Q8-7. Draw the nMOS dynamic logic circuits for a three-input NOR gate.

Q8-8. Show that the foot nMOS transistor is generally needed in dynamic logic.

Q8-9. Give why dynamic logic has generally less logical effort than static logic.

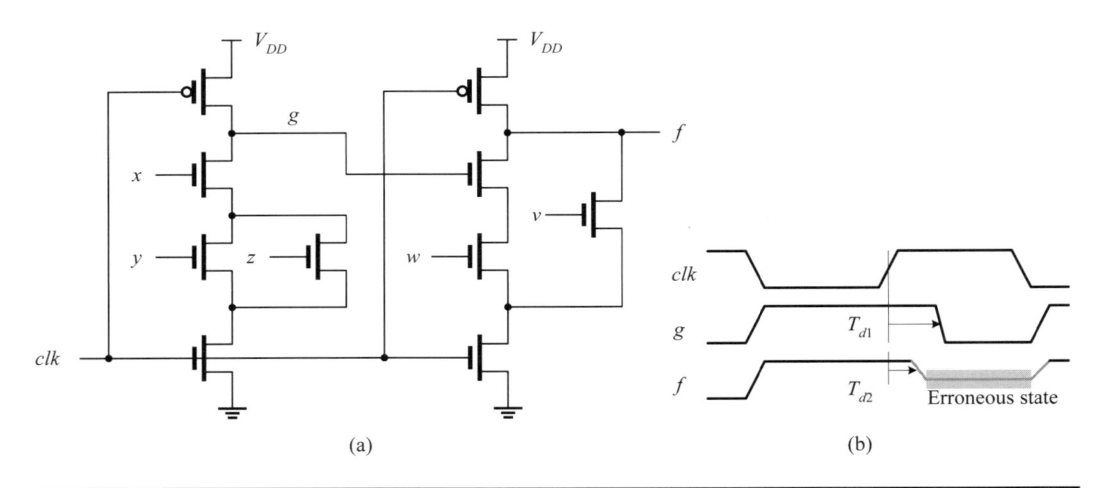

Figure 8.12: The partially discharged hazard of cascaded dynamic logic: (a) circuit; (b) timing.

8.1.3 Partially Discharged Hazards

As stated in Chapter 5, the number of transistors in a stack is often limited to four or five to bound the propagation delay. Hence, it is necessary to cascade many dynamic logic stages into a dynamic circuit to realize a complex switching function. Suppose that we have a dynamic logic circuit composed of two stages. The first stage implements the switching function, $g = \overline{x \cdot (y + z)}$ and the second stage implements the switching function, $f = \overline{g \cdot w + v}$, as shown in Figure 8.12(a). According to the operations of dynamic logic, both nodes g and f are precharged to V_{DD} when $clk = 0$. Node f may begin to discharge toward ground before node g completes its evaluation to the final value when $clk = 1$, as shown in Figure 8.12(b). This race phenomenon of the discharging operations between two successive stages is called a *partially discharged hazard.*

The partially discharged hazard of a dynamic node may reduce the noise margin or even cause the malfunction of a logic circuit. To see this, consider the circuit shown in Figure 8.12(a) again. Assume that $w = 1$ and $v = 0$. During the precharge phase, both nodes g and f are charged to V_{DD}. During the evaluation phase, since both nodes w and g are equal to 1, node f starts to discharge all the way to the ground level. As a result, node f goes faster to its final value if node g is determined to be 1 eventually. However, node f would already go to a value below where it should be, namely, logic 1, if the final value of node g is determined to be 0. Depending on how fast node f discharges, node f may have an erroneous voltage level below the threshold voltage of its driven stage and causes the driven gate to change its state accordingly or have a deteriorated high-level voltage yet still higher than the threshold voltage of its driven stage.

Based on the above discussion, two basic dynamic logic circuits of the same type cannot be simply cascaded to avoid the partially discharged hazard. Nevertheless, cascading two or more such logic circuits to carry out a complex logic function is indeed necessary in many applications. Therefore, we have to find out some way to solve the partially discharged hazard of dynamic logic circuits.

8.1.3.1 Avoidance of Hazards As we reexamine the hazards caused by the two cascaded dynamic logic circuits of the same type shown in Figure 8.12 more carefully, we can find out that the partially discharged hazard caused by the circuit is due to the same polarity of both precharging node g and the turn-on condition of the driven nMOS transistor. That is to say, node g is precharged to a high-level voltage, which in turn causes the nMOS transistor connected to it to be turned on. Consequently, the ways to solve this partially discharged hazard are simply to break this condition. Based on this idea, the major approaches include

1. *Domino logic*: The domino logic is a logic circuit that an inverter is added to the output of the basic dynamic logic circuit to make the output of the resulting logic circuit precharged to a low-level voltage; thereby, it is unable to turn on the driven nMOS transistor.
2. *np-domino logic*: By interlacing nMOS dynamic logic with pMOS dynamic logic, the precharged high-level or low-level voltage cannot turn on the driven pMOS or nMOS transistor. Hence, there exists no partially discharged hazard.
3. *Special timing design*: By designing a special clock timing, the partially discharged hazard may be eliminated. The two most popular schemes are as follows:
   - *Two-phase nonoverlapping clocking scheme*: Using a two-phase nonoverlapping clock to isolate the precharge and evaluation phase of two adjacent stages, the partially discharged hazard can be eliminated.
   - *Self-timed clock*: The clock of the driven stage is delayed by an amount equal to the propagation delay of the function block of the driver stage. As a consequence, the driven stage can start its evaluation only after the driver stage has completed its computation.

We will deal with each of these in detail later in the context of single-rail dynamic logic.

8.1.4 Types of Dynamic Logic Circuits

Dynamic logic circuits can be classified into the following three classes: single-rail logic, dual-rail logic, and clocked CMOS logic. Each class contains a variety of logic circuits. We briefly summarize these logic circuits in terms of these classes.

Single-rail logic receives single-rail input signals and also generates single-rail signals at its outputs. The popular single-rail logic circuits include the following types:

- Domino logic
  1. Basic domino logic
  2. Multiple-output domino logic (MODL)
  3. Compound domino logic (CDL)
- *np*-domino (NORA) logic
- Two-phase nonoverlapping clocking scheme
- Clock-delayed domino logic

Dual-rail logic receives dual-rail input signals and generates dual-rail output signals. Dual-rail logic can be further partitioned into three major types as follows:

- Dual-rail domino logic
- Dynamic CVSL (dynamic cascode voltage switch logic)
- Sense-amplifier-based (SA-based) dynamic logic

1. Dynamic SSDL (dynamic sample-set differential logic)
2. SODS (switched output differential structure)

Clocked CMOS logic means that its output is controlled by a clock *clk*. The output follows the value calculated by its logic function when it is enabled by the clock and is floated when it is disabled. Clocked CMOS logic can also be cast into two types in terms of the signals being processed: single-rail logic and dual-rail logic. The clocked single-rail logic includes

- Basic clocked CMOS logic (C^2MOS)
- NORA (no-race) clocked dynamic logic)
- *np*-domino clocked dynamic logic
- True single-phase clock (TSPC) logic
- All-*n* dynamic logic

and the clocked dual-rail logic includes

- Differential NORA dynamic logic

Review Questions

Q8-10. Explain why two basic dynamic logic circuits of the same type cannot be simply cascaded together.

Q8-11. Define the partially discharged hazard.

Q8-12. Explain why domino logic can solve the partially discharged hazard of the basic dynamic logic circuit.

Q8-13. Explain why *np*-domino logic can solve the partially discharged hazard of the basic dynamic logic circuit.

8.2 Nonideal Effects of Dynamic Logic

Although a dynamic logic circuit has less logical effort value than its static counterpart, there are many nonideal effects that severely affect the signal integrity of a dynamic logic circuit. *Signal integrity* means how well a signal can transfer from one place to another along a signal path, which may be a logic chain, a wire, or both. There are many factors that can affect the signal integrity. These include leakage current, charge injection, clock (capacitive) feedthrough, back-gate coupling, charge-loss (or charge-leakage) effect, charge-sharing effect, and power-supply noise. These effects are often named as *nonideal effects* of dynamic logic.

8.2.1 Leakage Current of Switches

Recall that an amount of off current exists in a MOS transistor. In long-channel transistors, the off current is dominated by the reverse-bias current of drain-well and well-substrate *pn* junctions. For short-channel transistors, the off current is made up of many ingredients, including *pn* reverse-bias current, subthreshold current, current caused by drain-induced barrier-lowering (DIBL) effect, gate-induced drain leakage (GIDL), punchthrough current, current caused by narrow-width effect, gate-oxide tunneling current, and hot-carrier injection current. As a consequence, in deep submicron processes, the better way to determine the off current of a specific nMOS or pMOS transistor is through simulation.

The off current is a function of drain-to-source voltage V_{DS} and the size of the transistor. For a specific process, the off current is proportional to the drain-to-source voltage V_{DS} (V_{SD}) and the channel width in both nMOS and pMOS transistors, as shown in Figure 8.13. The off current of MOS transistors plays an important role in dynamic logic circuits.

8.2.2 Charge Injection and Capacitive Coupling

At first glance, charge injection and capacitive coupling seem to be two different effects; they indeed have a close relationship in terms of charge. The former concerns with the charge induced in the channel and the latter deals with the charge induced in the capacitance between gate and drain/source diffusion region.

8.2.2.1 Charge Injection An illustration of charge injection is portrayed in Figure 8.14. The charge under the gate oxide resulting from the inverted channel is Q_{ch} when the nMOS switch is on and V_{DS} is small. Half of this charge is injected onto the capacitor C_{out} and half into V_{in} when the nMOS switch turns off. Note that we also have charge injection but in the opposite direction when we turn on the nMOS switch.

Charge injection will change the voltage across capacitor C_{out}, but it does not affect the input voltage V_{in} due to the low-impedance feature of the input voltage source. From Equation (2.53), the channel charge can be expressed as follows:

$$Q_{ch} = WLC_{ox}(V_{GS} - V_{Tn}) \tag{8.6}$$

Consequently, the change of output voltage caused by the charge injection is equal to

$$\Delta V_L = -\frac{WLC_{ox}(V_{GS} - V_{Tn})}{2C_{out}} = -\frac{WLC_{ox}(V_{DD} - V_{in} - V_{Tn})}{2C_{out}} \tag{8.7}$$

supposing that half of the charge is injected onto the capacitor C_{out}.

8.2.2.2 Capacitive Coupling As we know, one salient feature of a capacitor is that the voltage across its two plates cannot be changed instantaneously; namely, it needs to take a finite time to balance its charge across its two plates. As a consequence, any disturbance at one end of a capacitor will cause an equal voltage to be developed at the other end momentarily.

In deep submicron CMOS technologies, wires are placed very closely together. Therefore, a wire over or next to a dynamic node may couple capacitively and destroy the charge stored in the dynamic node. The details of capacitive coupling between wires are deferred to Chapter 13. We are concerned with *clock feedthrough* and *back-gate coupling*.

Clock feedthrough. Clock feedthrough (also called *capacitive feedthrough*) is a special case of capacitive coupling in which the coupling is through the capacitances between the gate (clock input) and drain, and between the gate and source (dynamic node) of MOS transistors. The coupling capacitance transfers the charge from the gate to both drain and source and, hence, yields undesirable signals. Clock feedthrough is an undesirable and unavoidable effect, in particular, when a MOS transistor is employed as a switch under the control of a high-speed clock.

To quantify the clock-feedthrough effect, let C_{ft} be the capacitance between the gate and the output of a MOS switch. The clock can be fed to the output through C_{ft}, as shown in Figure 8.15(a). The amount of clock fed through to the output can then

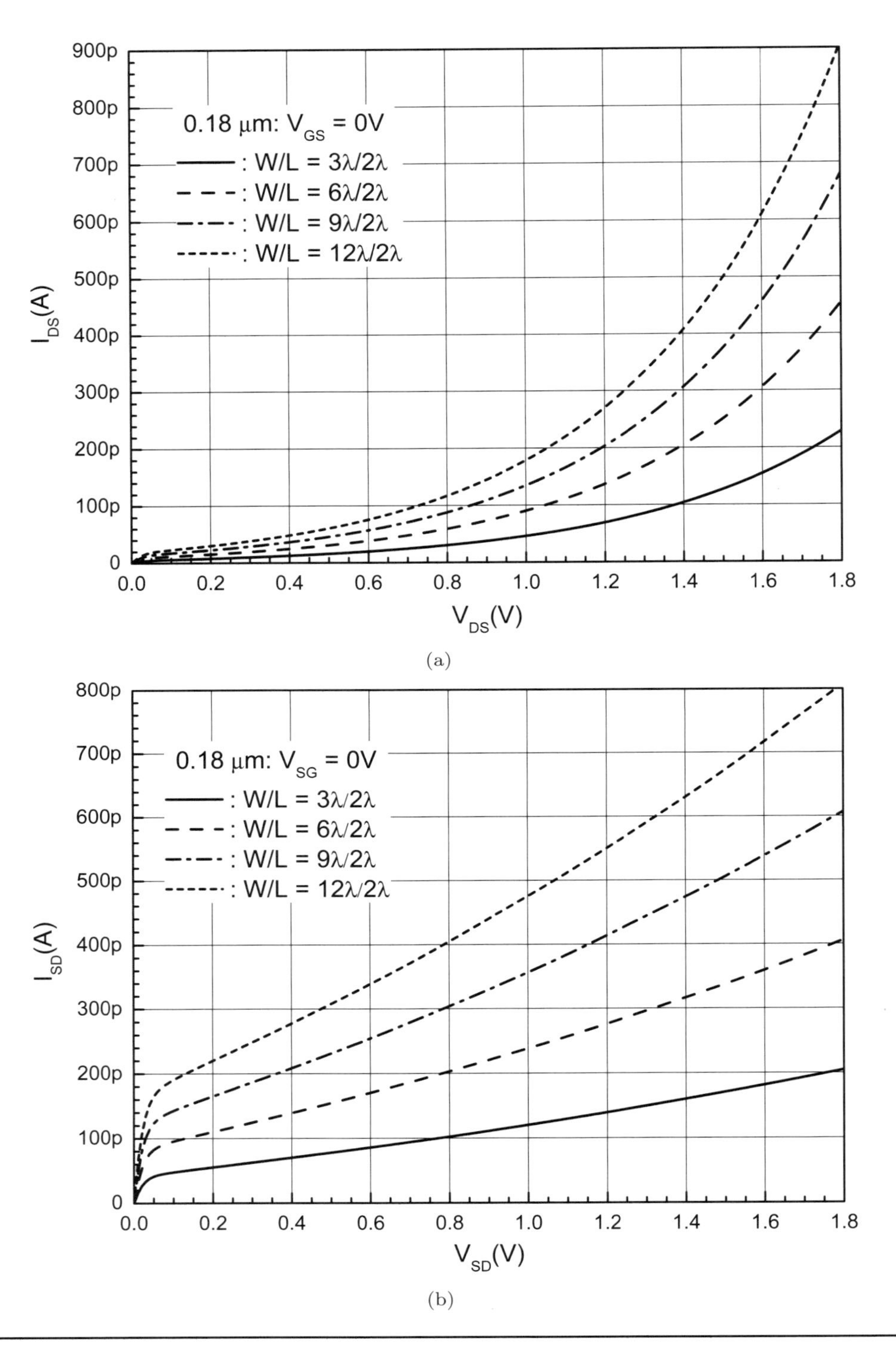

Figure 8.13: The leakage currents of (a) nMOS and (b) pMOS transistors when $V_{GS/SG} = 0$ V.

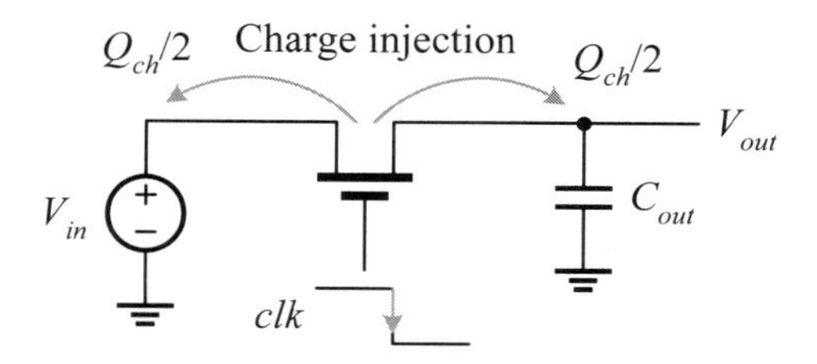

Figure 8.14: An illustration of the charge injection effect.

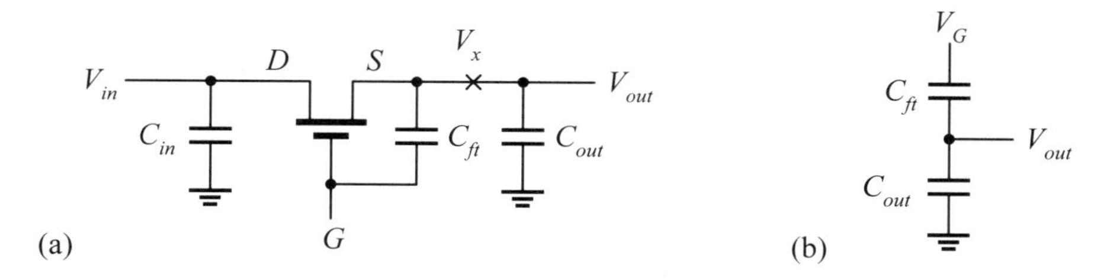

Figure 8.15: An illustration of clock-feedthrough effect: (a) circuit; (b) equivalent circuit.

be determined by the capacitance C_{ft} and output capacitance C_{out}. More specifically, the amount of clock fed through to the output can be accounted for by notifying that

$$C_{ft}(V_G - V_{out}) = C_{out}V_{out} \tag{8.8}$$

Therefore, the fraction of output voltage caused by the clock feedthrough is

$$V_{out} = \frac{C_{ft}}{C_{ft} + C_{out}} V_G \tag{8.9}$$

If there is an abrupt change in voltage at the gate G of the switch, then there is a corresponding change at the output (that is, source S).

■ **Example 8-6: (Clock feedthrough.)** Consider the circuit shown in Figure 8.15(a).

1. Assuming that $V_{in} = 1.8$ V, what is the initial value of output voltage when the clock is at 1.8 V? Estimate the final value of output voltage after the clock goes to low.
2. Assuming that $V_{in} = 0$ V, what is the initial value of the output voltage when the clock is at 1.8 V? Estimate the final value of output voltage after the clock goes to low.

Solution:

1. First, we calculate the stable voltage at the output (that is, source S) by taking into account the body effect.

$$\begin{aligned} V_{out} &= V_{DD} - V_{Tn} \\ &= 1.8 - [0.4 + 0.3(\sqrt{0.84 + V_{out}} - \sqrt{0.84})] \end{aligned}$$

Hence, $V_{out} = 1.24$ V. When the clock switches from high to low

$$V_x = 1.24 - \Delta V = 1.24 - \frac{C_{ft}}{C_{ft} + C_{out}} \cdot 1.8 = 0.57 \text{ V}$$

where C_{ft} and C_{out} are calculated by noting that the transistor is in the cutoff region of operation in this case and are as follows:

$$C_{ft} = C_{OL} = C_{ol}W = 1.01 \text{ fF}/\mu\text{m} \times 0.27\ \mu\text{m} = 0.27 \text{ fF}$$

$$C_{out} = C_j W = 1.66 \text{ fF}/\mu\text{m} \times 0.27\ \mu\text{m} = 0.45 \text{ fF}$$

2. Now consider the situation of $V_{in} = 0$ V. When the clock switches from high to low, the transistor is in its linear region of operation. Hence, C_{ft} can be calculated as follows:

$$\begin{aligned} C_{ft} &= \frac{1}{2}(C_g - 3C_{ol})W + C_{ol}W \\ &= \frac{1}{2}(3.41 - 3 \times 1.01) \times 0.27\ \mu\text{m} + 1.01 \text{ fF}/\mu\text{m} \times 0.27\ \mu\text{m} \\ &= 0.32 \text{ fF} \end{aligned}$$

where we use $C_g = C_{ox}L_{actual} + 3C_{ol}$. The output voltage is

$$V_x = 0 - \Delta V = 0 - \frac{C_{ft}}{C_{ft} + C_{out}} \cdot 1.8 = -0.75 \text{ V}$$

■

Back-gate coupling. Another important form of capacitive coupling is the effect of *back-gate coupling*, which states that the dynamic nodes are affected by the capacitive coupling between the output of a static gate and the output of a dynamic gate. To illustrate this, consider the circuit shown in Figure 8.16, where a dynamic logic gate drives a static two-input NAND gate. Because of the existence of capacitances between gate and drain and between gate and source, the input signal transition at the input x may cause the dynamic node a to go low. If this voltage droop is large enough, it may cause the dynamic logic circuit to evaluate incorrectly and cause the malfunction of the static NAND gate. To overcome this problem, the dynamic node should be connected to the input closer to the power-supply rail, namely, input x.

■ Review Questions

Q8-14. What is the meaning of signal integrity?

Q8-15. What are the ingredients of the off current?

Q8-16. Explain the charge injection and clock feedthrough.

Q8-17. Explain the meaning of back-gate coupling.

8.2.3 Charge-Loss Effects

As stated, a dynamic circuit stores its information on a capacitor or a *capacitive node*, which is also called a *dynamic node*, *soft node*, or *critical node* since its charge will leak away over time. Ideally, a dynamic node should retain its charge for a long time, at least during the evaluation phase. However, in any practical dynamic circuit a dynamic node often gradually leaks its charge away due to charge loss (or charge leakage) and eventually this will result in a malfunction of the logic circuit. The potential sources

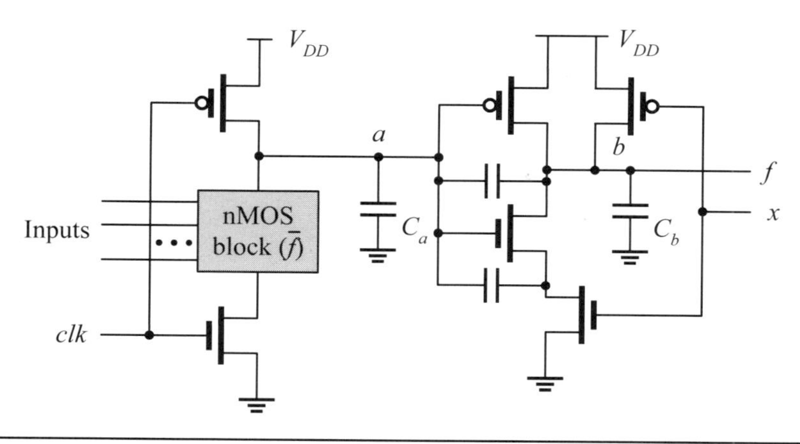

Figure 8.16: An illustration of the effect of back-gate coupling.

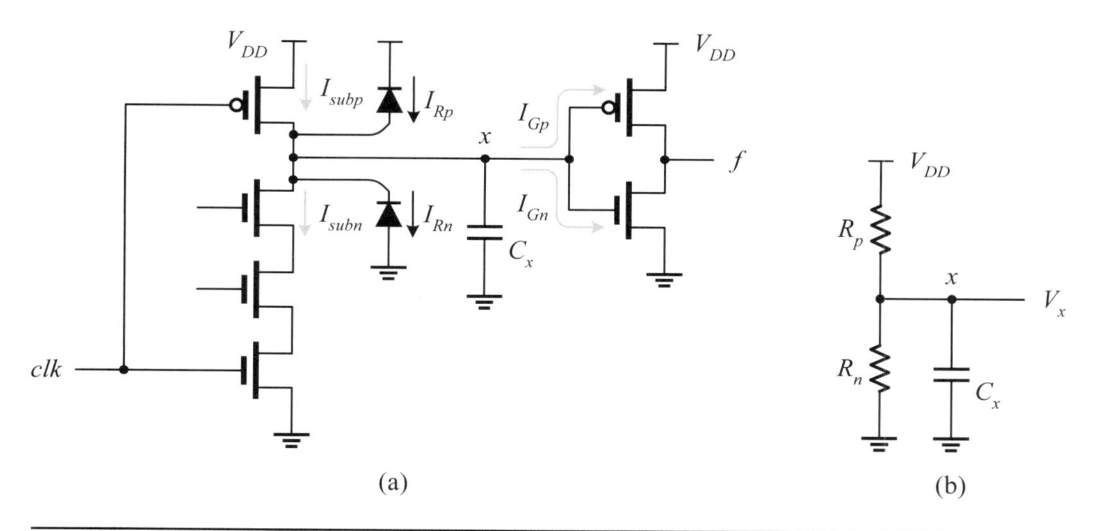

Figure 8.17: The (a) leakage-current sources and (b) equivalent circuit of node x of a dynamic logic circuit.

of charge loss include the following: the reverse-bias leakage current from the drain junction, the noise injection from neighboring wires, subthreshold current, gate leakage current, and others. We ignore the noise effects from neighboring wires.

8.2.3.1 A Dynamic Logic Example An example is shown in Figure 8.17(a), where the leakage current comprises three major sources, including junction leakage currents from both reverse-biased diodes, subthreshold currents of pMOS and nMOS transistors, and gate leakage currents of pMOS and nMOS transistors. The two reverse-biased diodes are the *pn* junctions existing between the drain and substrate of the nMOS transistor and between the drain and *n*-well of the pMOS transistor, respectively. The subthreshold current of nominally off nMOS and pMOS transistors also combine into the leakage current of the dynamic node. The gate leakage current is due to the barrier tunneling from gate oxide. The impact of gate leakage current on the leakage current of a dynamic node will be more severe as the technology is further scaled down.

Both pull-up and pull-down paths are virtually turned on in terms of leakage current and in some sense the resulting logic circuit is a ratioed logic circuit. Hence, the maximum voltage of dynamic node x is set by the resistive voltage divider composed of the pull-up and pull-down paths, as shown in Figure 8.17(b), where the equivalent resistances of both paths are called R_p and R_n, respectively.

In summary, the amount of leakage current determines the minimum clock rate needed in the dynamic logic circuit. The clock period is composed of two parts: precharge time and evaluation time. The precharge time must be long enough for the precharge pMOS transistors of the dynamic circuits to charge their soft nodes toward power supply, V_{DD}, but may be at $V_{x(max)}$. The required time is in turn determined by the sizes of the precharge pMOS transistors. The evaluation time must be long enough for all cascaded combinational logic circuits to evaluate completely their functions. It also needs to be short enough to prevent their stored information from being lost by the leakage current of dynamic nodes. The maximum allowable evaluation time is determined by the maximum hold time, $t_{hold(\max)}$, of the underlying soft node, which is calculated as follows:

$$t_{hold(\max)} = \frac{\Delta Q_{crit(\min)}}{I_{leak(\max)}} \tag{8.10}$$

where $\Delta Q_{crit(\min)} = C_{x(\min)}(V_{x(\max)} - V_{th})$, $V_{x(\max)}$ is the maximum voltage of soft node x, the V_{th} is the threshold voltage of loading inverter, and $I_{leak(\max)}$ is the maximum leakage current of soft node x.

Example 8-7: (The charge-loss effect.) Considering the circuit shown in Figure 8.17, calculate the maximum hold time $t_{hold(\max)}$. The related parameter values of node x are as follows: $V_{x(max)} = 1.08$ V, $V_{th} = 0.84$ V, $I_{leak(\max)} = 70$ pA, and $C_{x(min)} = 3.35$ fF.

Solution: The maximum hold time $t_{hold(\max)}$ before causing the incorrect operation of the loading inverter is equal to

$$t_{hold(\max)} = \frac{\Delta Q_{crit(\min)}}{I_{leak(\max)}} = \frac{3.35 \text{ fF} \cdot (1.08 - 0.84)}{70 \text{ pA}} = 11.5\ \mu\text{s}$$

■

The above example is only used to illustrate how to estimate the maximum hold time of a soft node. This value is strongly dependent on the process technology. Different processes have their own parasitic and gate capacitances, as well as leakage current. All of these parameters are critical for a soft node to keep its stored charge. Hence, different hold times will result from different processes.

8.2.3.2 Charge Keepers The charge keepers used to compensate for the charge-sharing effect, to be introduced later, may also be used for this purpose to provide sufficient current to overcome the leakage current under all process, voltage, and temperature (PVT) variations.

Review Questions

Q8-18. What does a soft node mean?

Q8-19. Explain the charge-loss effect.

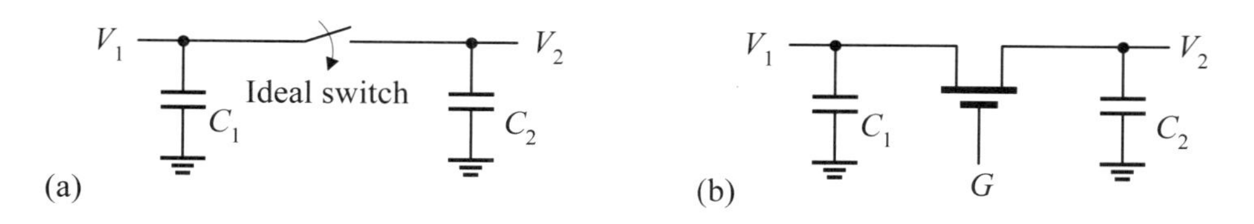

Figure 8.18: Illustrations of charge sharing with: (a) an ideal switch; (b) an nMOS transistor as a switch.

Q8-20. How would you determine the minimum and maximum clock rates of a dynamic logic circuit?

8.2.4 Charge-Sharing Effects

Charge sharing means that two or more dynamic nodes in a circuit share their charge when they are connected together. The charge-sharing effect may cause the voltage across a dynamic node to be changed. The rationale behind this stems from the conservative of the charge. To get more insight into the charge-sharing effect, consider two dynamic nodes with different voltages isolated by an ideal switch, as illustrated in Figure 8.18(a). As the switch is turned on, the charge associated with the two nodes will be redistributed until the voltage levels at these two nodes are equal because they are connected in parallel. That is,

$$Q_{total} = C_1V_1 + C_2V_2 = (C_1 + C_2)V$$

where C_1 and C_2 are the capacitances of the two nodes of interest. Solving the above equation, the final voltage V is found to be

$$V = \frac{C_1V_1 + C_2V_2}{C_1 + C_2} \tag{8.11}$$

Care must be taken in applying the above equation to the circuit in Figure 8.18(b), where an nMOS switch is used in place of the ideal switch. Because of the feature of the nMOS switch, the final voltage V will not be changed if both voltages on capacitors C_1 and C_2 are greater than $V_G - V_{Tn}$ but not necessary to be equal before the nMOS switch is on. If one voltage is less than $V_G - V_{Tn}$ and the other is greater than $V_G - V_{Tn}$, then the node (say, C_1) with lower voltage could only have a maximum final voltage of $V_G - V_{Tn}$ and the other node (C_2) will drop its voltage to a level of $[Q_{total} - C_1(V_G - V_{Tn})]/C_2$ after charge sharing. If both voltages are less than $V_G - V_{Tn}$, then Equation (8.11) can be applied to find the final voltage V after charge sharing.

■ **Example 8-8: (Charge-sharing effect.)** Consider the circuit shown in Figure 8.18(b). Assume that $C_1 = 10$ fF and $C_2 = 20$ fF. The voltages on both capacitors before the switch is turned on are $V_1 = 0$ V and $V_2 = 1.8$ V, respectively. Determine the voltage on both capacitors after the nMOS switch is turned on, ignoring the on-resistance of the nMOS switch.

Solution: Since the total charge remains unchanged before and after the nMOS switch is turned on, from Equation (8.11), the voltage on both capacitors after the nMOS switch is closed is equal to

$$V = \frac{C_1V_1 + C_2V_2}{C_1 + C_2} = \frac{10 \cdot 0 + 20 \cdot 1.8}{10 + 20} = 1.2 \text{ V}$$

which is above $V_G - V_{Tn} = 1.8 - 0.7 = 1.1$ V. Hence, the final voltages of V_{C1} is 1.1 V and V_{C2} is indeed equal to

$$V_{C2} = \frac{Q_{total} - (C_1)(V_G - V_{Tn})}{C_2} = \frac{20 \times 1.8 - 10(1.8 - 0.7)}{20} = 1.25 \text{ V}$$

As a result, the voltage on capacitor C_2 is dropped to 1.25 V after the charge sharing with capacitor C_1. This may not be tolerable in some applications.

■

The larger amount of capacitance of the critical node, the less the charge-sharing effect. Hence, for this purpose, it is not uncommon to add some extra amount of capacitance to those critical nodes intentionally. Next, we describe some common approaches to alleviating the charge-sharing effects.

8.2.4.1 Reduction of Charge-Sharing Effects The common approaches to reducing the charge-sharing effect are as follows:

1. Increasing the capacitance of the critical node
2. Using the charge keeper
3. Precharging the internal nodes (e.g., multiple-output domino logic)

The use of a charge keeper to lessen the charge-sharing effect will be described. As for the approach of precharging internal nodes, we only address it briefly here. The more details of it will be coped with later along with multiple-output domino logic (MODL) in Section 8.3.1.

Charge keepers. In the charge keeper approach, a weak pMOS transistor is used to pull up the critical (or sensitive) node to V_{DD} to provide a charging path from the V_{DD} to the critical node permanently or in a control manner. This pMOS transistor is usually referred to as a charge keeper. An illustration of using the charge keeper to alleviate the charge-sharing effect is given in the following example.

■ Example 8-9: (Examples of charge keepers.) Figure 8.19(a) shows a *simple charge keeper* being applied to domino logic. A gate-grounded pMOS transistor is simply connected across the critical node, that is, precharge node, and the V_{DD} node to provide a permanent charge path. As a consequence, the charge loss from the critical node can be compensated for by the charge keeper if its size is designed properly. One disadvantage of this simple charge keeper is that it slows down the operations of the domino logic circuit because the charge keeper always provides a charging path to the output node. Consequently, the discharge current of the critical node is reduced by an amount of the charging current contributed by the charge keeper.

Figure 8.19(b) shows another type of charge keeper, referred to as *feedback charge keeper*, where the gate of the pMOS transistor is connected to the output of the inverter of domino logic rather than grounded permanently. Hence, the charge keeper is not always conducted but instead only turned on when the output of the inverter is low. This type of charge keeper is often found in a situation where the output of a logic circuit is inverted. Figure 8.19(b) also shows a practical implementation, in which a gate-grounded pMOS transistor is serially connected between the pMOS transistor and V_{DD}. This results in smaller capacitive loading on the output of the inverter.

■

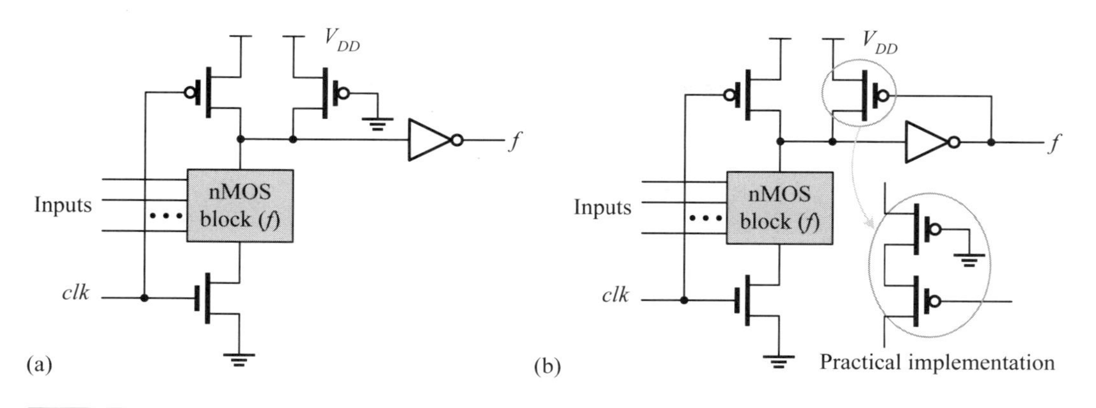

Figure 8.19: Alleviating the charge-sharing effect with charge keepers: (a) a simple charge keeper; (b) a feedback charge keeper.

Even though the use of a charge keeper can reduce the charge-sharing effect, it slows down the operations of the dynamic logic circuit. In addition, it also causes the resulting dynamic logic circuit to become a ratioed logic circuit. Therefore, it is necessary to properly size the transistors of the charge-keeper circuit and the pull-down stack to provide a correct low-level output voltage. In general, the transistor of a charge keeper is on the order of 1/10 strength of the pull-down stack; namely, the on-resistance of the charge keeper is 10 times that of the pull-down stack. An example to illustrate how to design a charge keeper based on this idea is explored in the following.

■ **Example 8-10: (A design example of a charge keeper.)** Consider the dynamic logic circuit shown in Figure 8.20, where a gate-grounded pMOS transistor is used as a charge keeper. To keep the low-level output voltage, V_{OL}, at an acceptable value, say, $0.1V_{DD}$, the on-resistance of the pMOS charge keeper can be found from the following expression:

$$V_{OL} = \frac{R_n}{R_n + R_p} V_{DD} \leq \frac{1}{10} V_{DD}$$

As a result, $R_p \geq 9R_n$; namely, the aspect ratio of the nMOS transistor is $3W/L$ and of the pMOS charge keeper is $W/4.5L$ if $R_{eqp} = 2R_{eqn}$. The aspect ratio of the precharge pMOS transistor is set to W/L.

■

Precharge internal nodes. The third way to reduce the charge-sharing effect associated with dynamic logic circuits is to precharge their internal nodes, which is illustrated in Figure 8.21(a). The nMOS block is partitioned into several smaller blocks with each being precharged by a precharge pMOS transistor. A simple, trivial example to illustrate this idea is depicted in Figure 8.21(b), where two additional pMOS precharge transistors are added to precharge the two internal nodes. We will return to this topic in greater detail later.

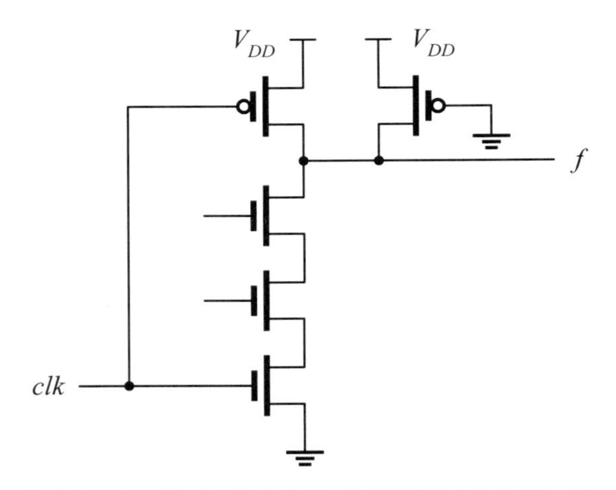

Figure 8.20: A design example of a charge keeper.

8.2.5 Power-Supply Noise

In designing a digital system, especially a high-frequency circuit, a trade-off between performance and noise margins, is often needed to be made. Noises may be caused by DC (direct current) and AC (alternating current) sources. The DC source is mainly the IR drop arising from the resistance of power-supply wires and the AC source is the Ldi/dt noise caused by the inductance of power-supply wires. The DC noise margins are used to characterize the effects caused by the power-supply rails on the logic circuit. The AC noise margins characterize the behavior of the logic gate when it is subjected to pulses of varying duration and amplitude at the inputs.

Noise margins can be improved either by increasing the size of the pMOS charge keeper or by decreasing the size ratio of the pMOS to nMOS transistor of the output inverter. However, either way will increase the propagation delay of the gate. Thus, a trade-off between performance and noise-margin needs to be made.

■ Review Questions

Q8-21. What is the charge-sharing effect?

Q8-22. How does the charge sharing affect a dynamic logic circuit?

Q8-23. How would you reduce the charge-sharing effects?

Q8-24. Explain why the capacitance of internal nodes needs to be kept as small as possible.

Q8-25. What is the charge keeper?

Q8-26. What are the drawbacks of using charge keepers?

8.3 Single-Rail Dynamic Logic

As stated, the partially discharged hazard can be removed by the use of domino logic, np-domino logic, or a special clock-timing scheme. These approaches result in a variety of single-rail dynamic logic circuits, including domino logic, np domino (NORA) logic,

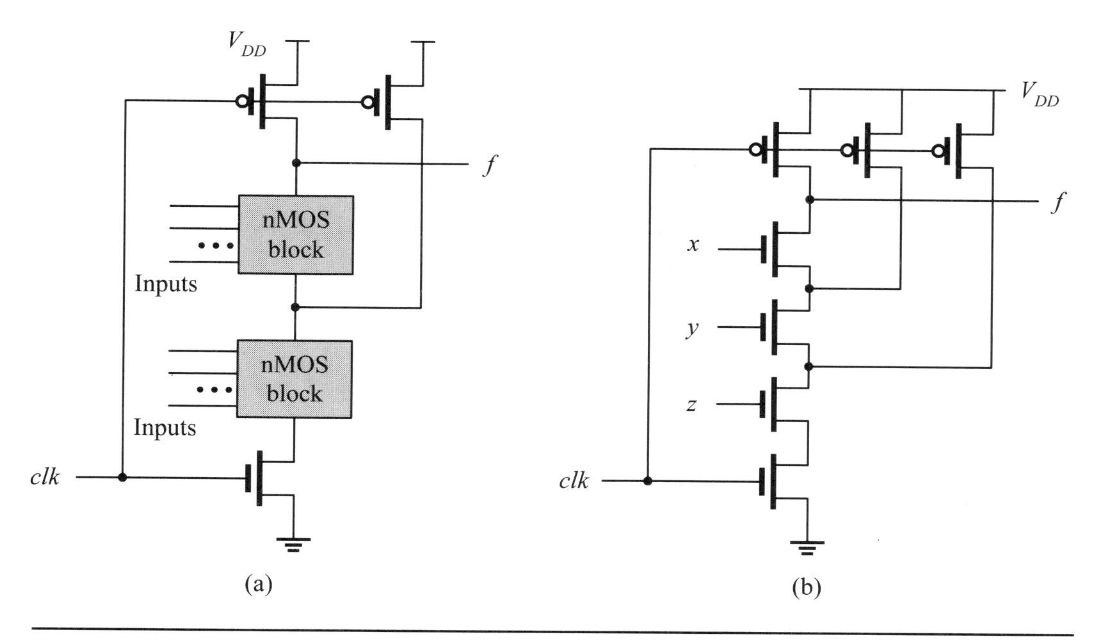

Figure 8.21: The method of precharging internal nodes: (a) block diagram; (b) an example.

two-phase nonoverlapping clocking scheme, and clock-delayed domino logic. All of these are variations of the basic dynamic logic and will be discussed in the following.

8.3.1 Domino Logic

A domino logic circuit is constructed by appending a static inverter at the output node of a basic dynamic logic circuit, as shown in Figure 8.22(a). This means that the output node is precharged to a low-level rather than a high-level voltage so that it never causes the driven nMOS transistor to be turned on before the function computed by the logic circuit is done. Consequently, there exists no partially discharged hazard when two or more domino logic circuits are cascaded. However, because of the static inverter appearing at the output node, a domino logic circuit can only realize a positive switching function. That is, it cannot implement inverting switching functions, such as a NOT gate and an XOR gate. Despite this, domino logic has been widely used in arithmetic logic and other logic circuits involving complex gates with high fan-in and fan-out, in particular, the datapaths of modern high-speed microprocessors.

The general block diagrams of domino logic are shown in Figure 8.22. Like the basic dynamic logic, a domino logic circuit can be constructed using either an nMOS block or a pMOS block, as depicted in Figures 8.22(a) and (b). The output f of nMOS domino logic is precharged to a low-level voltage while the output f of pMOS domino logic is precharged to a high-level voltage. In practical applications, nMOS domino logic is more popular than pMOS domino logic because of the better performance of nMOS transistors. Hence, the term domino logic usually means nMOS domino logic unless stated otherwise.

Basically, the design approach of domino logic is exactly the same as that of the basic dynamic logic circuit except that the nMOS block in domino logic performs the

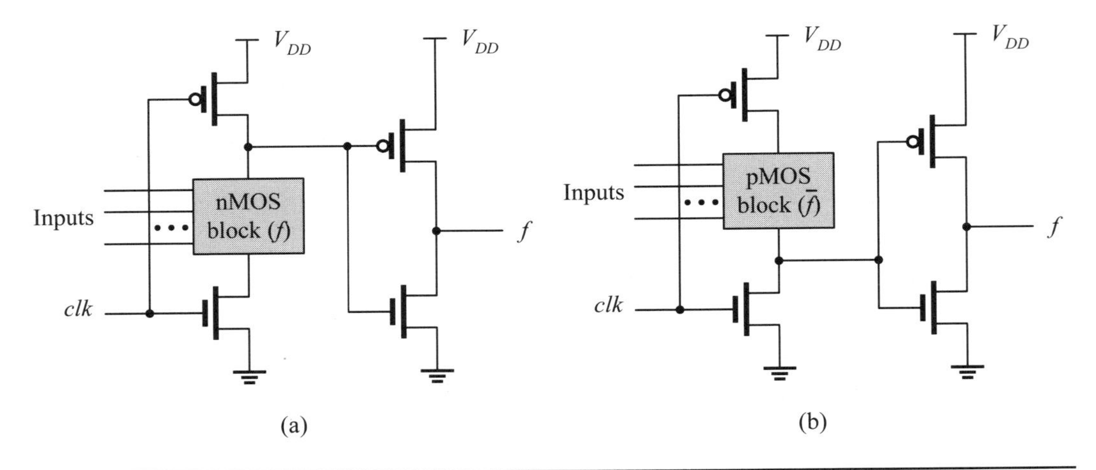

Figure 8.22: The general block diagrams of domino logic: (a) nMOS block; (b) pMOS block.

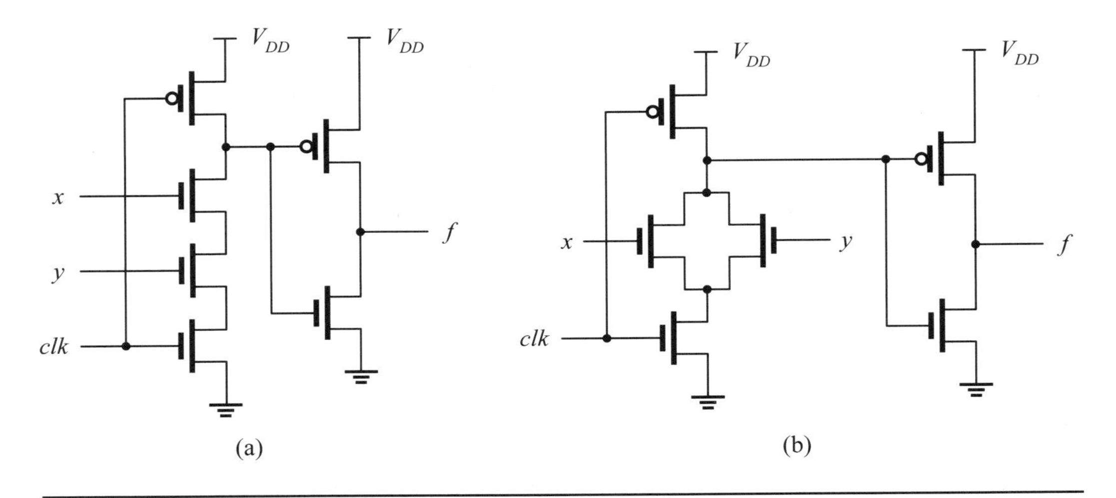

Figure 8.23: The two-input domino (a) AND and (b) OR logic gates.

true output function because of the inverter at the output. We give an example to illustrate how to realize desired switching functions with domino logic.

Example 8-11: (Two-input domino AND and OR gates.) Figure 8.23(a) shows a two-input AND gate and Figure 8.23(b) shows a two-input OR gate. Both gates are realized by using domino logic. Because of the output inverter, the noninverting (true) switching functions are implemented by the nMOS block.

Even though any complex switching function can be realized with a single domino logic, in practice we often partition a complex switching function into many smaller subfunctions, and then implement each with a small domino logic circuit to optimize the propagation delay. Finally, these domino logic circuits are cascaded together to

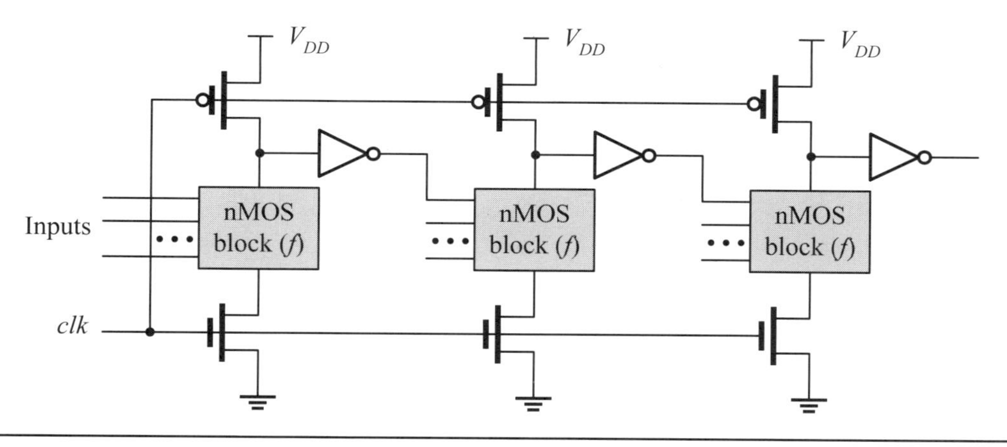

Figure 8.24: The block diagram of a cascaded domino logic circuit.

carry out the original function in a single clock period. An example of a cascaded domino logic circuit is exhibited in Figure 8.24. Assume that each domino logic circuit has a propagation delay t_{pdi}, and there are n domino logic circuits cascaded together. Then, the evaluation time $t_{evaluation}$ must be greater than the sum of t_{pdi}, for all $1 \leq i \leq n$. That is,

$$t_{evaluation} \geq \sum_{1}^{n} t_{pdi}. \tag{8.12}$$

The major limitation of domino logic is that it can only implement a noninverting switching function. This feature also limits it to becoming a universal logic circuit. Even though it is not always possible, in many situations, a logic circuit can be realized by using domino logic to restructure the logic circuit. An illustration of this is given next.

■ **Example 8-12: (Restructuring logic to fit into domino logic.)** Figure 8.25 shows a logic circuit that may be restructured to be implemented by using domino logic circuitry. Figure 8.25(a) is the logic circuit to be realized and Figure 8.25(b) shows the transformed result. It can be realized by one or two domino logic circuits.

■

8.3.1.1 Variations of Domino Logic Because domino logic is widely used in various arithmetic and complex circuits, a variety of variations of it have been proposed to fit into special requirements associated with particular applications. We introduce two of these: *multiple-output domino logic* (MODL) and *compound domino logic* (CDL).

Multiple-output domino logic (MODL). In domino logic and other CMOS logic styles, there is only one output available from the specific logic gate. Hence, when multiple switching functions are implemented in the nMOS block with one being a subfunction of the other, each subfunction has to be realized as an additional logic gate. For example, when a switching function is $f = f_1 \cdot f_2$ and both f and f_1 need to be used as separate output signals, both f and f_1 need to be implemented as two

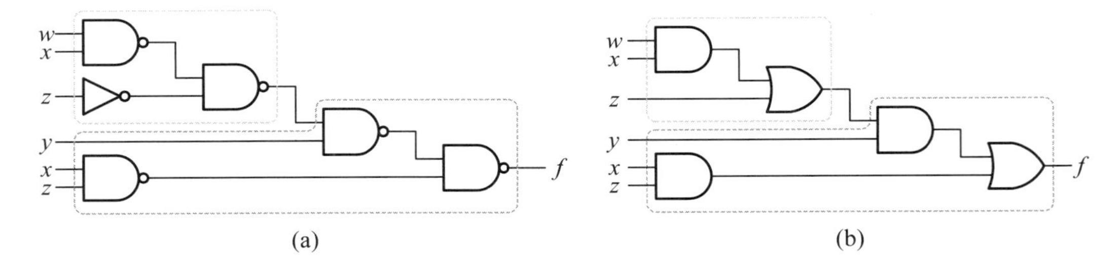

Figure 8.25: Restructuring logic to enable implementation by using domino logic circuits: (a) original logic circuit; (b) resulting logic circuit.

separate domino logic circuits because each domino logic circuit can only provide one output signal.

Multiple-output domino logic (MODL) is a modification of domino logic in a way such that the available subfunctions within the domino logic can also be output as signals by adding precharge devices and static inverters at the corresponding nodes. The technique of precharging internal nodes alone can also be used in domino logic or other dynamic logic to help alleviate the charge-sharing and charge-loss effects, as described in Section 8.2.4. MODL is often used when a switching function to be implemented contains a subfunction of another logic output. In such a case, the main switching function together with its subfunctions can be built in a single domino logic tree with multiple outputs to save duplicate circuits.

As an illustration of MODL, we give an example to illustrate how MODL can be used to design a four-bit carry-lookahead (CLA) generator necessary in the four-bit CLA adder. Details of the CLA generator and adder are deferred to Section 10.4.2.5.

■ Example 8-13: (An example of MODL.) Figure 8.26 shows an example of using MODL to design a carry-lookahead generator. The carries, c_i, for all $i = 1, \ldots, 4$, are represented in terms of g_i and p_i as follows:

$$\begin{aligned} c_1 &= g_1 + p_1 \cdot g_0 \\ c_2 &= g_2 + p_2 \cdot (g_1 + p_1 \cdot g_0) \\ c_3 &= g_3 + p_3 \cdot (g_2 + p_2 \cdot (g_1 + p_1 \cdot g_0)) \\ c_4 &= g_4 + p_4 \cdot (g_3 + p_3 \cdot (g_2 + p_2 \cdot (g_1 + p_1 \cdot g_0))) \end{aligned}$$

where g_0 is the input carry c_{in}, or called c_0, and g_i and p_i are known as *carry generate* and *carry propagate* functions, respectively, and are defined as

$$\begin{aligned} g_i &= a_i \cdot b_i \\ p_i &= a_i \oplus b_i \end{aligned}$$

where a_i and b_i are the inputs to the ith bit adder.

■

Compound domino logic (CDL). Recall that the maximum number of transistors in an nMOS stack is often limited to four or five to control the propagation delay to an acceptable level. Consequently, it is often necessary to decompose a complex positive switching function into many simpler ones and then combine together with another domino logic gate or a static logic gate. However, when these simpler switching

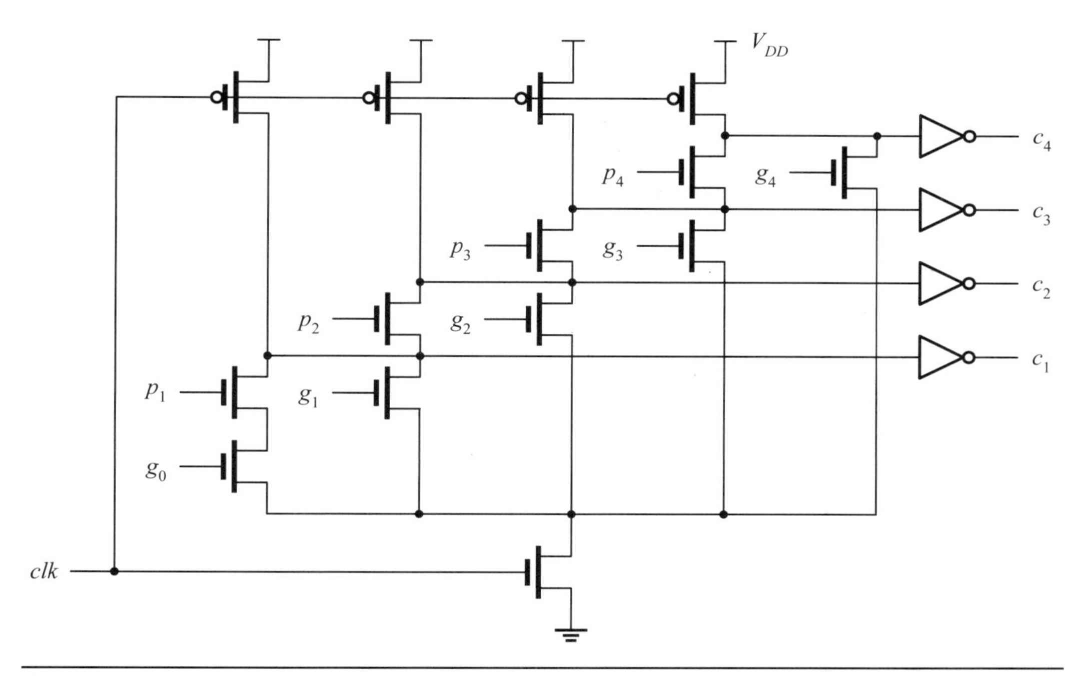

Figure 8.26: The use of MODL to design a four-bit carry-lookahead generator.

functions are implemented in separate domino logic gates, the overall overhead from each domino logic circuit may not be tolerable. One way to solve this is by using a compound domino logic, a generalized form of MODL. To explore this idea, consider the following example.

■ **Example 8-14: (An example of compound domino logic.)** Realize the following switching function on the condition that each nMOS stack can only have three transistors at most, excluding the evaluate (that is, foot) nMOS transistor.

$$f = tuvwxyz + stw$$

Solution: Since each nMOS stack can have three transistors at most, by using DeMorgan's law, the switching function f may be decomposed into the following:

$$\begin{aligned} f &= \overline{(\overline{tuv} + \overline{wxy} + \bar{z}) \cdot \overline{stw}} \\ &= \overline{(g + h + i) \cdot k}, \end{aligned}$$

where $g = \overline{tuv}$, $h = \overline{wxy}$, $i = \bar{z}$, and $k = \overline{stw}$. Consequently, a three-input OR gate and a two-input NAND gate (or an OAI31 gate) are required to combine the final result. The resulting compound domino logic is given in Figure 8.27. It is instructive to compare the result with the one implemented by using standard domino logic.

■

Self-resetting (oostcharge) domino logic. The concept of self-resetting (or postcharge) (SRCMOS) domino logic is to control the prcharge operation based on the output of the domino circuit. Because of this, the self-resetting domino logic is also called *postcharge domino logic*.

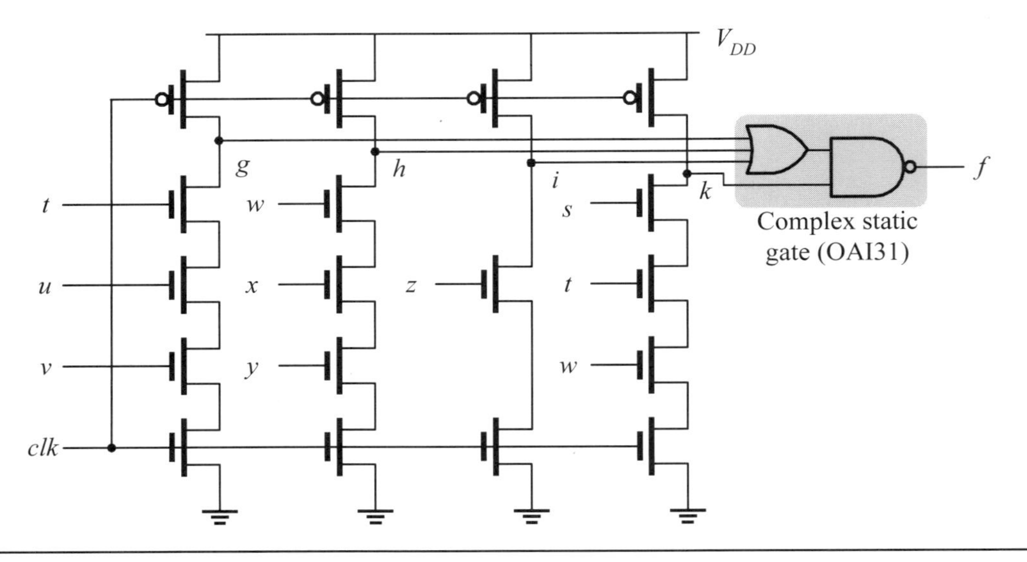

Figure 8.27: A compound domino logic circuit for implementing the switching function: $f = tuvwxyz + stw$.

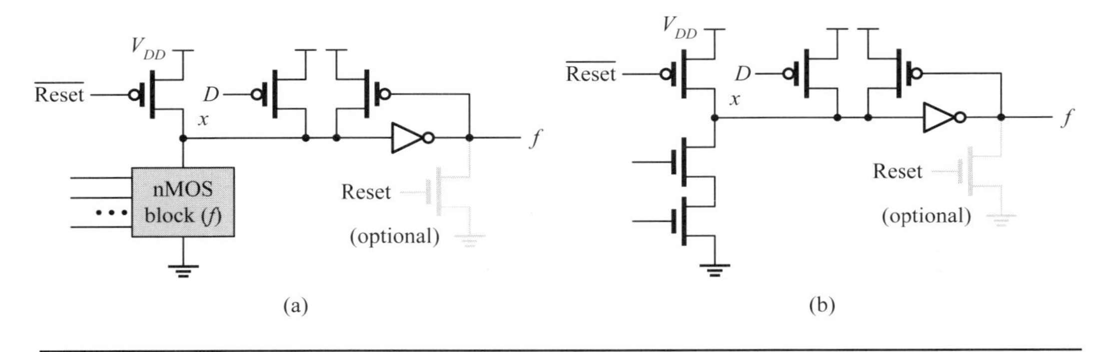

Figure 8.28: The (a) general block diagram and (b) a two-input AND gate of SRCMOS logic.

The general block diagram of SRCMOS logic circuits is shown in Figure 8.28(a) and an example of a two-input AND gate is given in Figure 8.28(b). A SRCMOS logic circuit is basically a *footless domino logic* circuit with a charge keeper but operates in a different way. As usual, the SRCMOS logic conditionally discharges the dynamic node (say, x) to evaluate the desired logic function, but it then resets the dynamic node back to its original charged state via a local feedback timing chain instead of a global clock.

The pMOS reset transistor is controlled by the reset signal $\overline{Reset}$. An optional nMOS reset transistor controlled with the complement of $\overline{Reset}$ for pulling the inverter output to ground may also be used. One more feature of this logic style is the addition of a diagnostic pMOS transistor with its gate connected to the global testing signal D. By activating the global testing signal D, the SRCMOS gate is converted from a precharged logic style into a static ratioed logic circuit, thereby enabling testability.

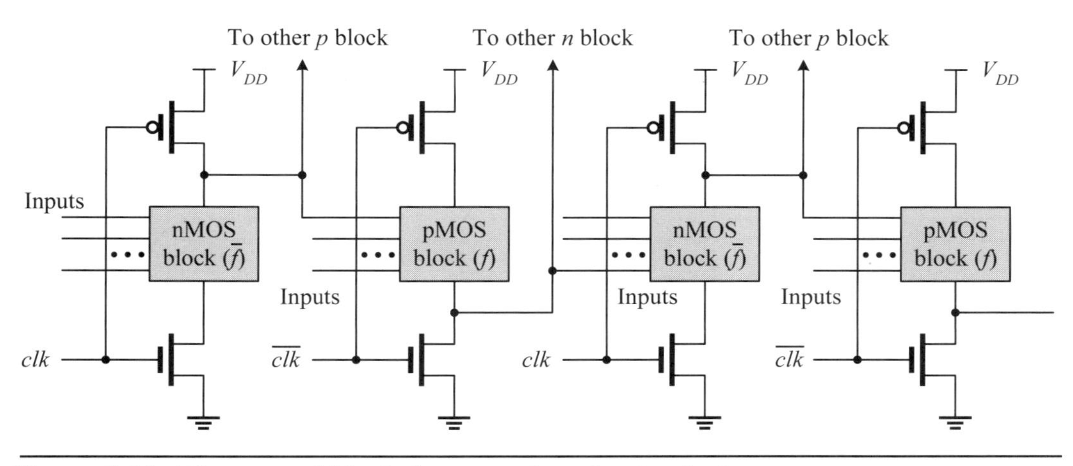

Figure 8.29: The general block diagram of *np*-domino logic.

■ Review Questions

Q8-27. Describe the features of domino logic.

Q8-28. What would happen when a pMOS domino logic follows an nMOS domino logic?

Q8-29. Can any switching function be implemented by domino logic?

Q8-30. Describe the features of multiple-output domino logic (MODL).

Q8-31. What are the features of compound domino logic (CDL)?

Q8-32. What are the features of SRCMOS logic?

8.3.2 np-Domino Logic

Another approach to overcoming the partially discharged hazard encountered in cascading basic dynamic logic is to interleave pMOS blocks with nMOS blocks, as shown in Figure 8.29. Such a new dynamic logic is called *np-domino logic* because each nMOS block is followed by a pMOS block and vice versa. The essence of *np*-domino logic is to explore the duality between nMOS-block logic and pMOS-block logic so that the partially discharged hazard in cascading dynamic logic can be removed. Both nMOS and pMOS blocks are controlled by a pair of clocks, clk and $\overline{clk}$. If the nMOS-block logic is controlled by the true clock clk, then the pMOS-block logic is controlled by the complementary clock $\overline{clk}$, and vice versa.

The general block diagram of *np*-domino logic is shown in Figure 8.29, where the nMOS block uses the clock clk, whereas the pMOS block uses the complementary clock $\overline{clk}$. During the precharge phase, the clock clk is low. The outputs of nMOS blocks are precharged to a high-level voltage and may conditionally make a transition from 1 to 0 during the evaluation phase, depending on the inputs of the nMOS blocks. In contrast, the outputs of pMOS blocks are precharged to a low-level voltage and may conditionally make a transition from 0 to 1 during the evaluation phase, depending on the inputs of the pMOS blocks. The precharged high-level voltage at the outputs of nMOS blocks cannot turn on pMOS transistors and the precharged low-level voltage at the outputs of pMOS blocks cannot turn on the nMOS transistors. As a consequence, there exists no partially discharged hazard. Note that the outputs of both nMOS and pMOS blocks are taken from different places.

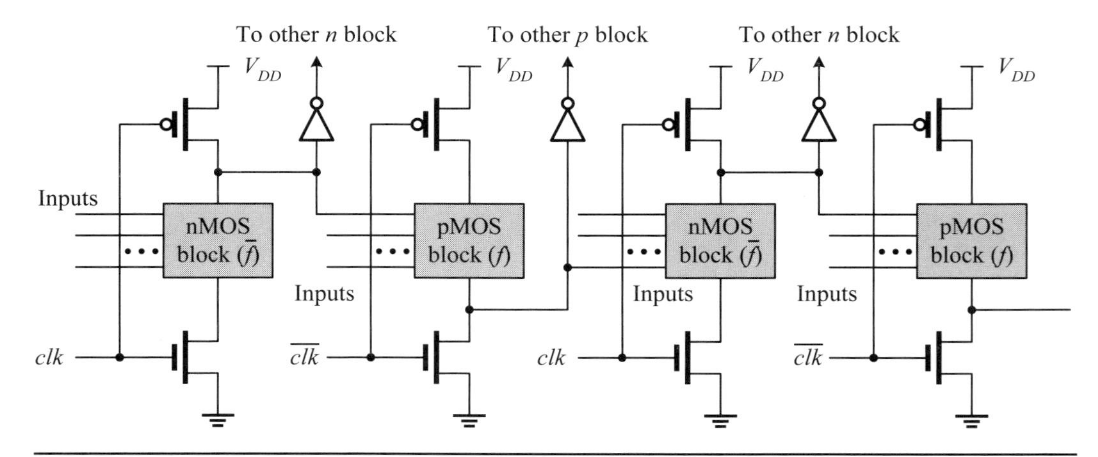

Figure 8.30: The general block diagram of *np*-domino logic connecting to other dynamic logic or static logic blocks.

The *np*-domino logic has the general block diagram as shown in Figure 8.30 when connecting to other dynamic logic blocks of the same type or static logic. An inverter is necessary when the output of the dynamic block is connected to other dynamic blocks of the same type or a static logic circuit. Because of the inherent inferior performance of pMOS transistors, *np*-domino logic has poorer performance than domino logic.

Review Questions

Q8-33. Describe the features of *np*-domino logic.

Q8-34. How would you connect an *np*-domino logic circuit to a static logic circuit?

Q8-35. Why does *np*-domino logic have poorer performance than domino logic?

8.3.3 Two-Phase Nonoverlapping Clocking Scheme

The partially discharged hazard encountered in the cascading basic dynamic logic of the same type can also be removed by using an isolated technique to electronically separate the consecutive stages. To do this, nMOS switches or TG switches are employed to schedule the signal connections between two adjacent stages. Each stage consists of a basic dynamic logic circuit and input switches. It is often called a ϕ_i *logic block*, where $i \in \{1, 2, \ldots, k\}$ and the most common values of k are 2 and 4, even though k can be any positive integer greater than 1. The clocking structure is called a *two-phase nonoverlapping clocking scheme* if k is set to 2 and a *four-phase nonoverlapping clocking scheme* if k is set to 4. In order to make the resulting logic circuit work properly, all clock phases applied to the logic circuit must not be overlapped; namely, at any time

$$\phi_1(t) \cdot \phi_2(t) \cdot \ldots \cdot \phi_k(t) = 0. \tag{8.13}$$

We only deal with the two-phase nonoverlapping clocking scheme.

The basic dynamic logic structure using a two-phase nonoverlapping clocking scheme is depicted in Figure 8.31(a). The ϕ_1 logic block consists of a ϕ_1-enabled TG switch and a ϕ_1-precharge dynamic logic stage, and the ϕ_2 logic block consists of a ϕ_2-enabled TG switch and a ϕ_2-precharge dynamic logic stage. As ϕ_1 is high, the ϕ_1 logic block

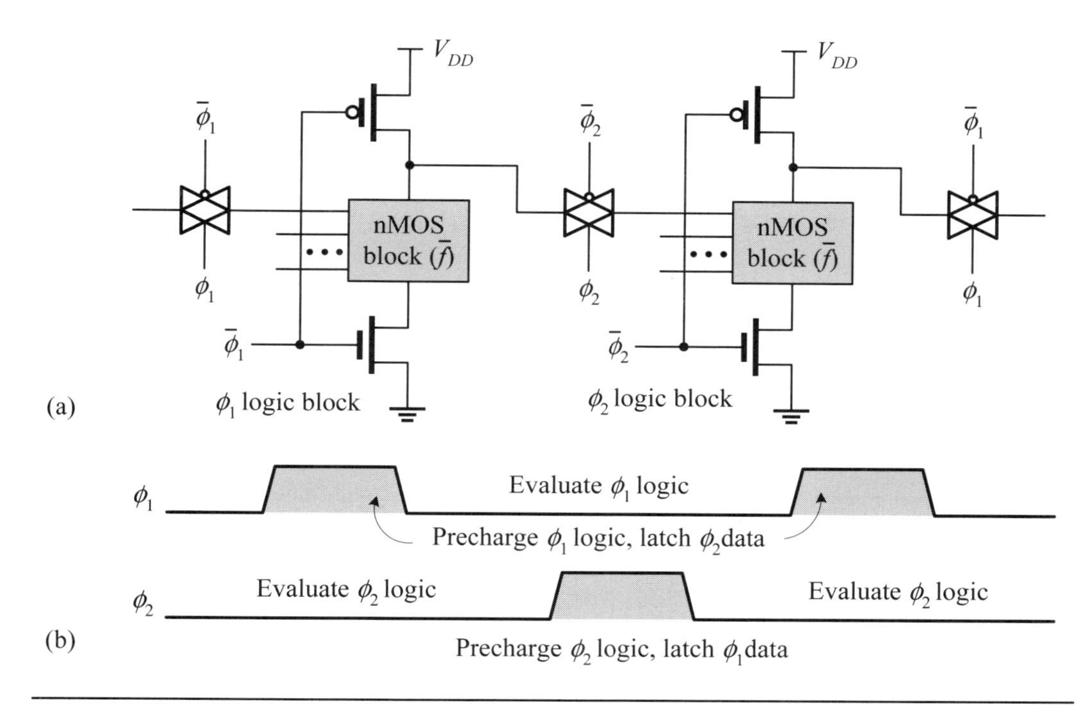

Figure 8.31: The (a) general block and (b) timing diagrams of a two-phase nonoverlapping clocking scheme.

is precharged to V_{DD} and the logic value from ϕ_2 logic block is latched; as ϕ_2 is high, the ϕ_2 logic block is precharged to V_{DD} and the logic value from the ϕ_1 logic block is latched. The operating timing is illustrated in Figure 8.31(b) in detail.

As mentioned above, to maintain the correct operation of the cascading dynamic logic, the two clock phases, $\phi_1(t)$ and $\phi_2(t)$, must not be overlapped. That is, the two-phase clocks, $\phi_1(t)$ and $\phi_2(t)$, must satisfy the following condition at any time.

$$\phi_1(t) \cdot \phi_2(t) = 0. \tag{8.14}$$

8.3.3.1 Two-Phase Nonoverlapping Clock Generators The two-phase nonoverlapping clock can be generated by a number of simple logic circuits. To get more insights into this, we give two examples in the following.

■ Example 8-15: (Two-phase nonoverlapping clock generators.) The simplest two-phase clock generator is made possible by using an inverter to invert the input clock and then output it along with the input clock as the desired two-phase clocks, ϕ_1 and ϕ_2, as shown in Figure 8.32(a). However, although this scheme is simple and costs only one extra inverter, the generated two-phase clocks are indeed overlapped due to the propagation delay of the inverter, as the timing diagram shown in Figure 8.32(a). An improved version is given in Figure 8.32(b), in which an always-on TG switch is employed to balance the propagation delay of the inverter. Despite the difficulty of balancing both propagation delays of the inverter and TG switch, this two-phase clock generator may work much better than the previous one if we properly size both nMOS and pMOS transistors of the TG switch.

■

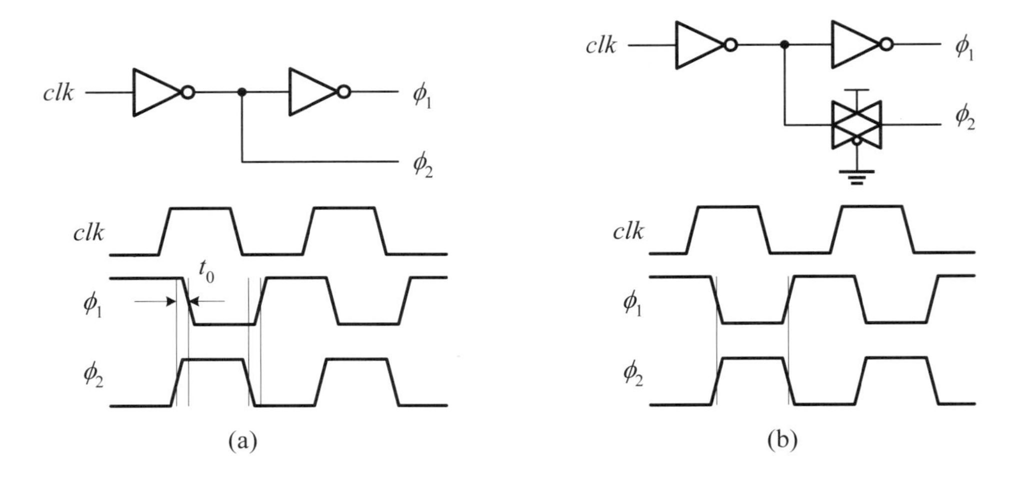

Figure 8.32: Two simple circuits for generating two-phase nonoverlapping clocks: (a) a simple circuit; (b) an improved circuit.

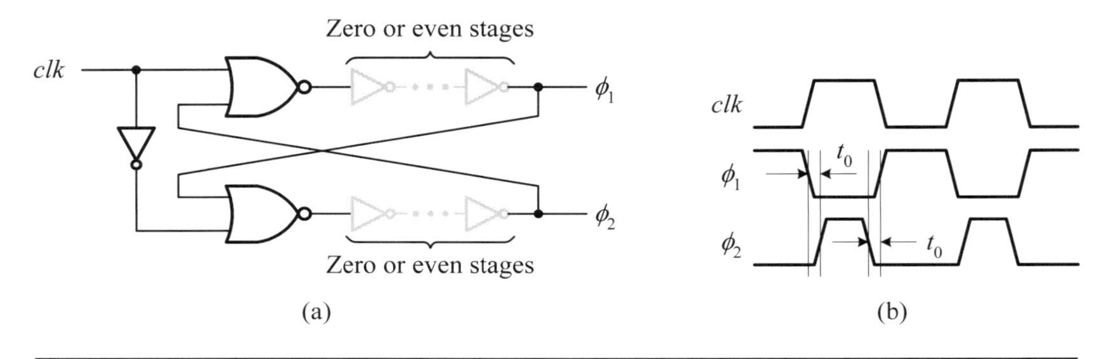

Figure 8.33: A (a) two-phase nonoverlapping clock generator and its (b) timing based on a NOR-SR latch.

Another more complex two-phase nonoverlapping clock generator is explored in the following example. The essential element of this clock generator is a bistable latch, which will be described in the next chapter.

■ Example 8-16: (Another two-phase nonoverlapping clock generator.) Figure 8.33(a) shows a simple but practical two-phase nonoverlapping clock generator. A NOR-SR latch is used to construct a D-type latch. Since a latch always has a pair of complementary outputs, Q and $\bar{Q}$, these two outputs can be used as the desired two-phase clocks, ϕ_1 and ϕ_2, as shown in Figure 8.33(a). Figure 8.33(b) gives its timing diagram. The nonoverlapping duration, t_0, can be adjusted to the desired value by adding pairs of inverters at the output nodes of both NOR gates, as depicted in lighter lines of Figure 8.33(a).

■

Because a two-phase clock generator is usually used to drive a clock distribution network, the skew caused by the unbalanced propagation delays from its generator

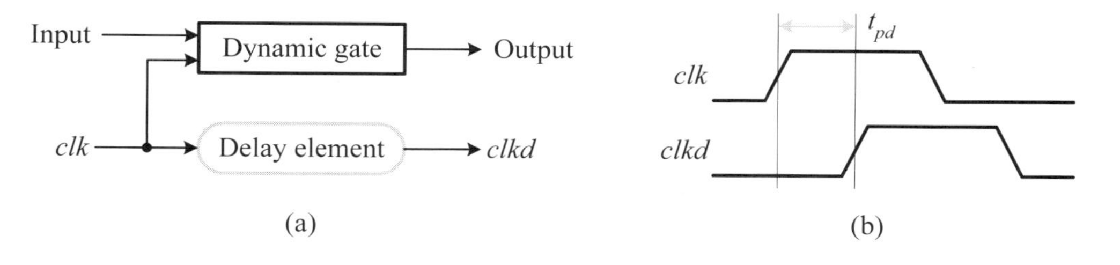

Figure 8.34: The (a) basic structure and (b) related timing of clock-delayed (CD) domino logic.

circuit may be compensated by the propagation delays from the clock distribution network. Hence, one should also consider the capacitance of the clock distribution network when trying to reduce the clock skew between the two-phase clocks.

In practice, it is hard to distribute a pair of two-phase nonoverlapping clocks in a large system, especially, in a high-speed system, because of the *RLC* parasitics associated with wires, unbalancing capacitive loading on the two clock lines, and capacitive-coupling effects between adjacent wires. As a consequence, it is more popular to globally distribute a single-phase clock over the system and then to locally generate the desired two-phase clocks.

Review Questions

Q8-36. Explain the principle of a two-phase nonoverlapping clocking scheme.

Q8-37. How would you generate two-phase nonoverlapping clocks?

Q8-38. How would you control the nonoverlapping duration of two-phase nonoverlapping clocks?

Q8-39. Explain the difficulties of distributing two-phase nonoverlapping clocks over a large system.

8.3.4 Clock-Delayed Domino Logic

The *clock-delayed domino logic* (*CD domino logic*) uses self-timed clocks to remove the partially discharged hazard associated with the cascading dynamic logic of the same type. By using the clock-delayed technique, the next stage starts its evaluation only after its previous stage has completed its evaluation. We first introduce the principle of clock-delayed domino logic and then deal with a variety of delay elements.

8.3.4.1 Clock-Delayed Domino Logic Clock-delayed (CD) domino logic is dynamic logic that can provide single-rail inverting or noninverting signal outputs; namely, it is complete logic and hence can be used to realize any switching function. CD domino logic uses self-timed delay signals as the precharge and evaluate clocks. Each CD domino logic circuit consists of a dynamic logic circuit, such as basic dynamic logic or domino logic, and a delay element, if necessary, as shown in Figure 8.34. The self-timed delay signal tells the next-stage logic circuit when the data output is ready. As a consequence, by appropriately controlling the propagation delays of delay elements, the partially discharged hazard encountered in cascading dynamic logic of the same type can be avoided.

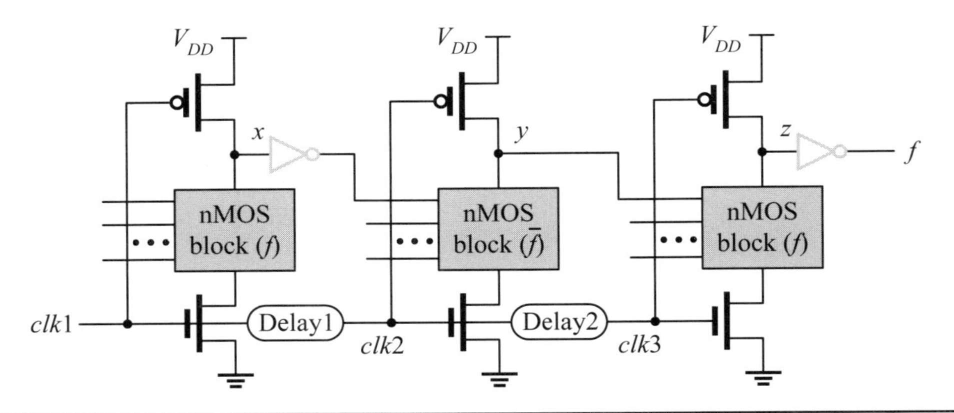

Figure 8.35: A clock-delayed (CD) domino logic.

The essence of designing a clock-delayed domino logic circuit is that the rising edge of the self-timed signal should occur after the output of the logic circuit has switched and never before, as shown in Figure 8.34(b). In other words, we have to control the delay of each delay element associated with a dynamic logic circuit in the CD domino logic very carefully. One feature of CD domino logic is that the delay elements will always be the critical path because the delay set by each delay element is always greater than the worst-case delay of the dynamic logic circuit plus some margin.

Figure 8.35 shows an example of clock-delayed domino logic. Here, three dynamic logic circuits are cascaded, in which two are domino logic and one is basic dynamic logic. In practical applications, a CD domino logic circuit may consist of any type of dynamic logic circuits. The CD domino logic circuit works as follows. The delay element associated with each stage tells its next stage when the data output is ready. As a consequence, the second stage starts to evaluate after the first stage has completed its evaluation. The same scenario is applied to the third stage. Of course, to make the CD domino logic work correctly, one stage should not enter precharge phase before its next stage consumes its output results.

8.3.4.2 Delay Elements Three delay elements are widely used in CMOS processes. These include the *current-starved inverter*, *buffer chain*, and *switched capacitor*, as shown in Figure 8.36. The current-starved inverter uses an nMOS transistor as a *voltage-controlled resistor* to control the driving and sinking currents of loading capacitance. An illustration is shown in Figure 8.36(a). The buffer chain used as a delay element is simply to delay its input signal an amount of the propagation delay associated with it. Two such examples are depicted in Figure 8.36(b). The first simply cascades two inverters while the second inserts an extra delay-controlled stage composed of a pair of pMOS and nMOS transistors connected in parallel between two inverters. The switched-capacitor method employs a voltage-controlled resistor connected in series with a MOS capacitor to serve as an *RC* delay component. By properly controlling the resistance of the voltage-controlled resistor, the desired signal delay can be obtained. In order to isolate the *RC* delay component, two inverters are used as buffers. An example is depicted in Figure 8.36(c).

Regardless of which approach is used, the propagation delay of delay element used in CD logic should be matched with its corresponding dynamic logic circuit. The propagation delay of the delay element consists of four components: the gate delay, the wiring delay of the output net, the fan-out delay of the gate, and a margin. The

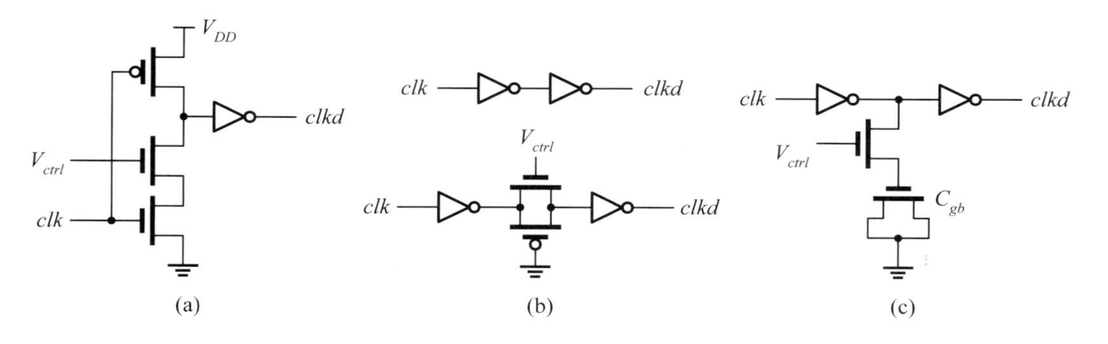

Figure 8.36: Three popular delay elements in CMOS processes: (a) current-starved inverter; (b) buffer chain; (c) switched capacitor.

gate delay is set equal to the worst-case pull-down delay of the corresponding dynamic logic circuit during the evaluation phase. To match the delay of the dynamic logic circuit, a dummy domino logic circuit may be used. A margin is added to account for the setup times to the next gate, the PVT variations between the delay element and its corresponding dynamic logic circuit, and the differences in the signal delay due to output wiring, fan-out loading, and coupling parasitic capacitances. A margin of at least 10 to 20% of the delay of the dynamic logic circuit should be added to the delay elements to ensure proper precharge and evaluate operations of the CD domino logic circuits.

■ Review Questions

Q8-40. What are the features of clock-delayed (CD) domino logic?

Q8-41. Explain the operation of the switched capacitor as a delay element.

Q8-42. Explain the operation of the current-starved inverter as a delay element.

Q8-43. How would you design a delay element for working with CD domino logic?

8.3.5 Conditional Charge Keepers

Even though a domino logic circuit is faster than its static logic counterpart, it is very sensitive to noise sources, such as crosstalk, leakage current, charge sharing, power-supply bump, and ground bounce, because its dynamic nodes cannot be recovered after its stored data are corrupted by these noise sources. To overcome this drawback, a pMOS pull-up transistor, namely, a charge keeper, is often used to compensate for the charge loss at dynamic nodes due to various sources. In addition to simple charge keepers and feedback charge keepers introduced in Section 8.2.4, many other types of charge keepers are also used in various specific applications. We deal with two such examples: the *delay charge keeper* and *burn-in charge keeper*.

8.3.5.1 Delay Charge Keeper For small leakage current, standard weak charge keepers are sufficient to maintain the voltage levels of precharged nodes without significant impact on the performance of dynamic logic circuits. However, in deep submicron (DSM) processes, the leakage currents of dynamic nodes are gradually increased with the reduction of feature sizes due to the following reasons. First, wires are packed together closer and closer and their thicknesses are increased to reduce their sheet

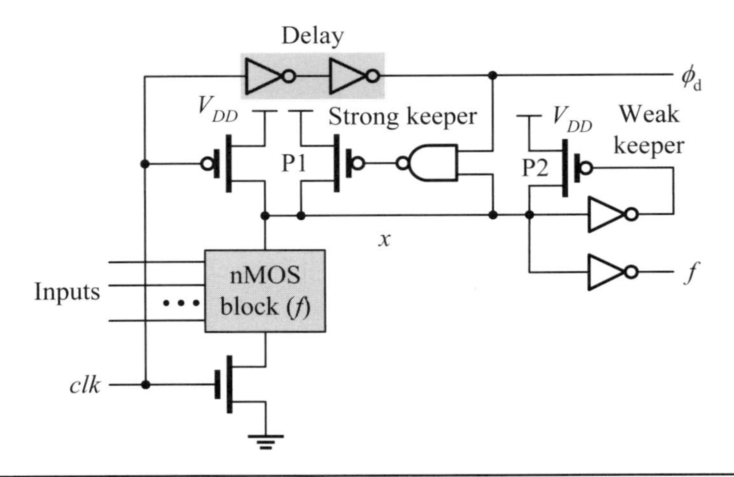

Figure 8.37: The general block diagram of a delay charge keeper.

resistances. This increases the cross-coupling capacitance between two adjacent wires. Second, the leakage (subthreshold) currents of DSM, low-threshold voltage devices are increased to the point that may significantly degrade the performance of dynamic logic circuits. As a consequence, the charge keepers must be sized to compensate for the cumulative dynamic node leakage currents and subthreshold currents due to the worst-case noise on the inputs of pull-down devices lest they degrade the performance of dynamic logic circuits profoundly.

In conventional dynamic logic circuits, the standard charge keeper has to be sized to cover the entire evaluation time, thereby degrading the performance of the logic circuit because it is turned on unconditionally as soon as at the start of the evaluation phase. To alleviate the performance degradation, the charge keeper should be weak enough during the output transition window and strong enough for the rest of the evaluation time if the dynamic output should remain high. One way to achieve this is illustrated in Figure 8.37 by using two charge keepers: a *weak charge keeper* and a *strong charge keeper*. The weak charge keeper results in reduced contention between itself and the nMOS block during the transition window and a faster output transition, while the strong charge keeper results in good robustness to leakage currents and noises during the rest of evaluation time. This kind of charge keeper is also referred to as a *delay charge keeper*.

8.3.5.2 Burn-in Charge Keeper The *burn-in* or *stress test* is used to test the reliability of devices in which the extreme operating conditions, where both temperature and supply voltage may exceed the upper bound of their target operating range (by up to 40%), are used. This burn-in test uses elevated temperature and voltage to accelerate the time to early failures, which are dominantly due to manufacturing defects. Since each part must pass the test (between 2 to 24 hours), the burn-in is a time-consuming and relatively expensive process. Therefore, to reduce the burn-in time and cost, it is highly desirable to provide the maximum burn-in temperature and voltage levels, which are eventually bounded only by process reliability considerations. This results in the increase of leakage currents. In addition, during the burn-in test, the clock frequency of the circuit can be relatively low, while the functionality still needs to be maintained. As a consequence, it is desirable to have a charge keeper that can meet these two goals in addition to the normal operation.

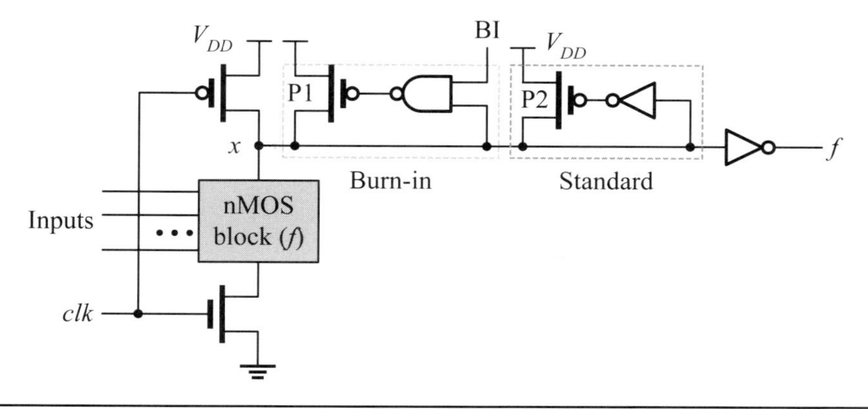

Figure 8.38: The general block diagram of a burn-in charge keeper.

A conditional charge keeper that can be used to compensate for the large leakage currents during the burn-in test is shown in Figure 8.38. The charge keeper consisting of the pMOS transistor $P2$ and its associated inverter is the standard charge keeper, used in normal operation and burn-in test. The charge keeper consisting of the pMOS transistor $P1$ and its associated NAND gate is the burn-in charge keeper, which is a strong charge keeper but is only turned on during the burn-in test. During the burn-in test, the NAND gate acts as an inverter and the effective channel width of the charge keeper is the sum of both pMOS transistors, $P1$ and $P2$. During normal mode, only the standard charge keeper is activated.

Review Questions

Q8-44. What is the philosophy of using delay charge keepers?

Q8-45. What is the burn-in test?

Q8-46. What is the philosophy of using burn-in charge keepers?

8.4 Dual-Rail Dynamic Logic

The three major dual-rail dynamic logic styles are *dual-rail domino logic*, *dynamic CVSL*, and *sense-amplifier-based* (SA) dynamic logic. The dual-rail domino logic is an extension of domino logic by adding the complementary part and hence overcoming the difficulty in implementing inverting functions. The dynamic CVSL is a dynamic version of CVSL. The SA dynamic logic uses the features of sense amplifiers to detect a small differential voltage and amplify it to a full-rail output. It is of interest to note that all dual-rail dynamic logic circuits are stemmed from the CVSL circuit. Consequently, the logic design methodology used in CVSL can be applied equally well to all of dual-rail dynamic logic.

8.4.1 Dual-Rail Domino Logic

Domino logic circuits are widely used in modern high-performance microprocessors because of their better performance over static CMOS logic circuits in terms of area and

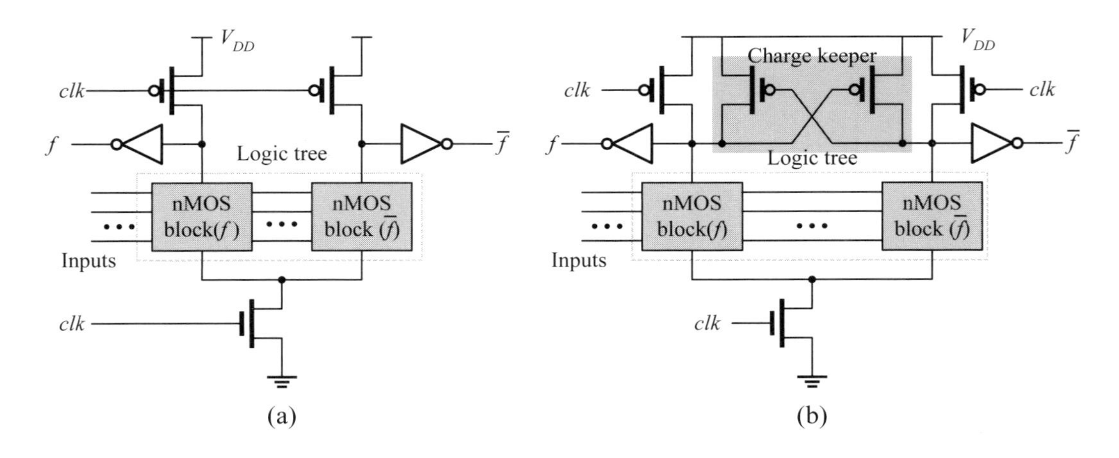

Figure 8.39: The general block diagrams (a) without and (b) with charge keepers of dual-rail domino logic.

propagation delay. The increased performance is owed to the reduced input capacitance, lower switching thresholds, and circuit implementations that typically use fewer levels of logic because of the use of efficient and wide complex gates. It is shown that domino logic can achieve about 60% average speed improvements over static CMOS for random logic blocks. Nevertheless, domino logic can only implement noninverting switching functions. To overcome this difficulty, a redundant part is used to realize the dual function of the original domino logic. This results in new dynamic domino logic, known as *dual-rail domino logic*.

The general block diagrams of dual-rail domino logic are shown in Figure 8.39(a), where an nMOS block ($\bar{f}$) is added to implement the complementary part of function f. Like domino logic, charge keepers can also be added to the outputs to increase noise margins and avoid the charge-sharing effect. The resulting logic is shown in Figure 8.39(b). The design of dual-rail domino logic circuits is the same as that of CVSL, described in Section 7.3.1. As we will see later, dual-rail domino logic with charge keepers is topologically the same as dynamic CVSL.

The following is an example of using the dual-rail logic style to design and implement a two-input XOR/XNOR gate. This example also illustrates some transistors that can be shared by two nMOS blocks, f and $\bar{f}$. Consequently, the resulting circuit needs fewer transistors.

■ **Example 8-17: (A dual-rail domino XOR/XNOR gate.)** Figure 8.40 shows the two-input XOR/XNOR logic circuit implemented by the dual-rail domino logic style. The logic tree is exactly the same as that exhibited in Figure 7.29, where two nMOS transistors are shared by both blocks. Of course, as mentioned before, both nMOS blocks can also be realized separately.

■

8.4.2 Dynamic CVSL

As stated previously, one disadvantage of the domino logic is that it can only implement a noninverting function. This limits the logic flexibility and implies that logic inversion

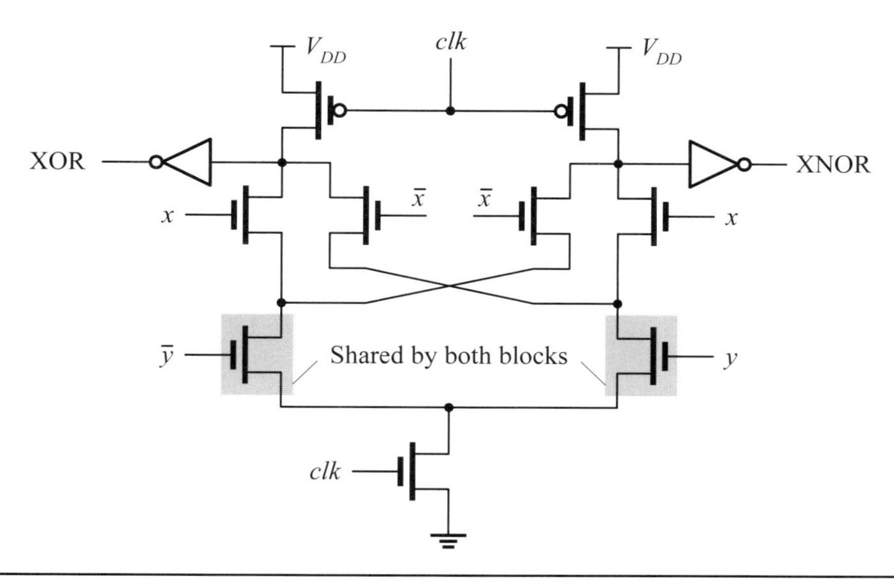

Figure 8.40: The XOR gate built from the dual-rail domino logic style.

must be performed at the inputs or outputs of domino logic blocks. To remove this limitation, the dynamic CVSL is proposed as an alternative in addition to dual-rail domino logic.

Dynamic CVSL, also called *clocked CVSL*, can be considered as either a dynamic version of CVSL or a basic nMOS dynamic logic with the complementary counterpart. Like static CVSL, dynamic CVSL also receives dual-rail inputs and produces dual-rail outputs. Dynamic CVSL has the basic structure shown in Figure 8.41(a). It consists of a logic tree for implementing the required logic function, a foot nMOS transistor, and two pMOS precharge transistors. Sometimes feedback charge keepers are added to the basic dynamic CVSL, as shown in Figures 8.41(b), to improve noise margins and reduce the charge-sharing effect.

Dynamic CVSL has been widely used in various applications with excellent performance and showed a good trade-off between occupied area, propagation delay, and power dissipation. The use of dynamic CVSL for implementing switching functions proceed in the same way as its static version, CVSL. Hence, we omit it here.

■ Review Questions

Q8-47. Describe the features of dual-rail domino logic.

Q8-48. Explain why dual-rail domino logic is not an effective way to implement an n-input NOR gate.

Q8-49. Describe the features of dynamic CVSL.

Q8-50. Design the carry function with the dynamic CVSL style using the two-nMOS block approach.

Q8-51. Design the sum function with the dynamic CVSL style using the two-nMOS block approach.

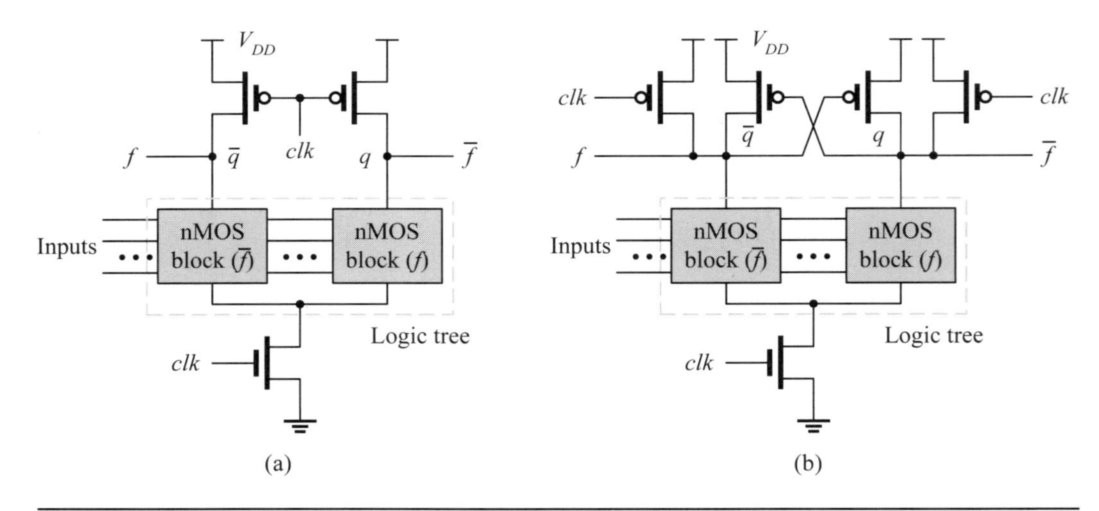

Figure 8.41: The general block diagrams (a) without and (b) with charge keepers of dynamic CVSL logic.

8.4.3 Sense-Amplifier-Based Dynamic Logic

A *sense amplifier* (SA) is a positive feedback circuit capable of sensing the small difference of two input signals and generating a full-rail output. Sense amplifiers are extensively used in memory cores and in low-swing bus drivers to improve performance or reduce power dissipation. We introduce some dynamic logic circuits that incorporate some sort of sense amplifier in their circuits to reduce the propagation delays.

The general block diagram of SA-based logic is shown in Figure 8.42. It is fundamentally the same as dual-rail domino logic depicted in Figure 8.39(a) except that a sense amplifier is added to speed up the transitions between the two states of output nodes. A variety of variations have been proposed during the past decades. We only describe sample set differential logic (SSDL) and switched output differential structure (SODS).

8.4.3.1 Sample Set Differential Logic (SSDL) Sample set differential logic (SSDL) modifies the dual-rail domino logic by adding a clocked latching sense amplifier, as shown in Figure 8.43(a). The operation of SSDL uses the concepts of sample and set rather than precharge and evaluate. During the sample phase, the clock clk is low and both precharge pMOS and evaluate nMOS transistors are turned on. The latching sense amplifier is off. A path exists from either internal nodes to ground through either $\bar{f}$ or f nMOS block. The result is that one internal node will be at the voltage V_{DD} and the other at a voltage less than V_{DD}.

During the set phase, the clock clk is high. Both precharge pMOS and evaluate nMOS transistors are off and the latching sense amplifier is turned on through a clock-controlled nMOS transistor. The internal node at the lower voltage is discharged quickly to ground due to the large driving capability of the sense amplifier, which only consists of two nMOS transistors connected in series. Consequently, without being pulled down through a series connection of many transistors, SSDL circuits have a speed advantage over the dual-rail domino logic and dynamic CVSL circuits.

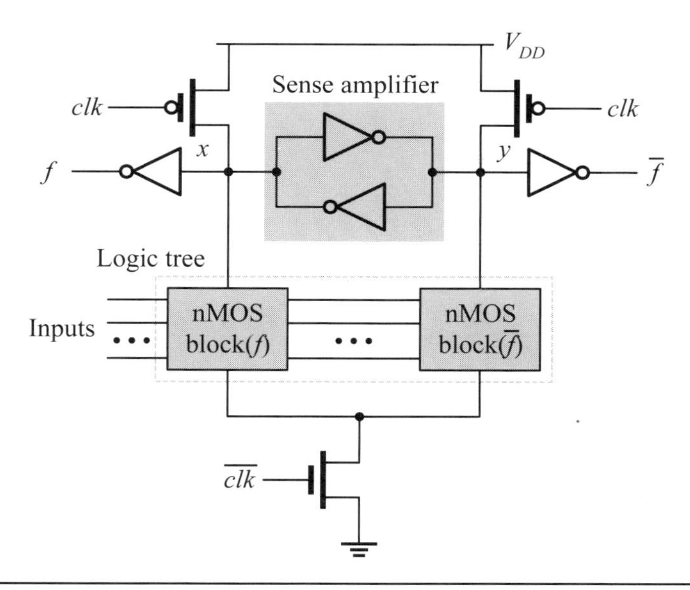

Figure 8.42: The general block diagram of SA-based dynamic logic.

One obvious problem of SSDL is that it consumes the static power during the sample phase because all precharge pMOS and evaluate nMOS transistors are turned on. A simple way to solve this problem is shown in Figure 8.43(b) to insert an nMOS transistor between the precharge pMOS transistor and nMOS block on each side. However, since in such a situation these two nMOS transistors are not turned on even though the evaluate nMOS is turned on when the clock *clk* is low, that is, the sample phase, both internal nodes, x and y, are precharged to the V_{DD}. When the clock *clk* is high, that is, the set phase, both nMOS transistors are turned on along with the latching sense amplifier and both internal nodes, x and y, are evaluated to their final values with the help of the positive feedback effect of the latching sense amplifier.

8.4.3.2 Switched Output Differential Structure (SODS) The switched output differential structure (SODS) modifies the load circuitry of previously introduced SA-based dynamic logic structures to gain the saving of transistor count and area. The general block diagram is illustrated in Figure 8.44. The load circuit is fundamentally a latch formed by two pMOS transistors and three switchs, including two nMOS and one pMOS transistors, controlled by a clock *clk*.

The operation of SODS is as follows. During the precharge phase, the clock *clk* goes low. The pMOS switch is on while both nMOS switches are off. One of the nMOS blocks is on and turns on one of the two load pMOS transistors in turn. Since both two nMOS switches are off, there is no path connecting the V_{DD} and ground and hence no power dissipation. During the evaluation phase, the clock *clk* rises high. The pMOS switch is off while both nMOS switches are on, therefore, passing the generated function values to the outputs. Since the pMOS switch turns off, data remain stable owing to the action of the latch formed by the two load pMOS transistors.

Review Questions

Q8-52. What is the function of sense amplifiers?

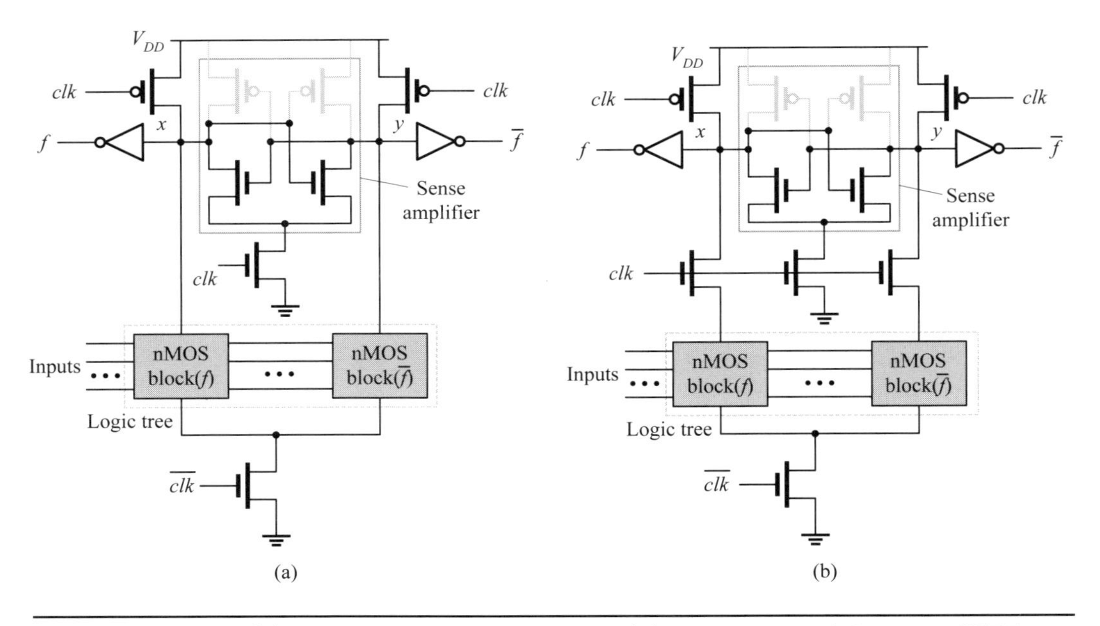

Figure 8.43: The (a) general block diagram and (b) a variation of dynamic SSDL.

Q8-53. Explain why sense amplifiers can be used in low-swing bus drivers.

Q8-54. What is the advantage of using sense amplifiers in dual-rail dynamic logic?

Q8-55. What is the distinction between dual-rail domino logic and sample set differential logic (SSDL)?

Q8-56. Describe the operation of the switched output differential structure (SODS).

8.5 Clocked CMOS Logic

Clocked CMOS logic means that the output node of a logic circuit is controlled by a clock *clk*. The clocked CMOS logic has its output node follow the value being calculated by the logic function when it is enabled by the clock and float when it is disabled. A clocked CMOS logic circuit usually uses a C^2MOS inverter or a modified C^2MOS inverter called a single-phase clock (SPC) latch or a true single-phase-clock (TSPC) latch as its output latch or latches. In this section, we explore many examples of this type of dynamic logic. For consistency, we also partition these into two types: single-rail logic and dual-rail logic.

8.5.1 Clocked Single-Rail Logic

The clocked single-rail logic that we will consider in this subsection includes basic *clocked CMOS logic* (C^2MOS), NORA (no-race) *clocked dynamic logic*, *np-domino clocked dynamic logic*, *true single-phase-clock* (TSPC) *logic*, and *all-n dynamic logic*.

8.5.1.1 Basic Clocked CMOS Logic Clocked-CMOS logic (C^2MOS) is dynamic logic that combines a static logic circuit with a clocked logic circuit. The general block diagram of C^2MOS logic is shown in Figure 8.45(a), where two clocked transistors,

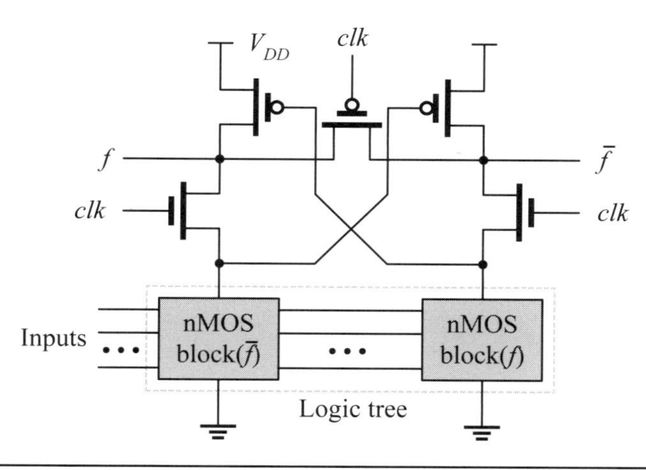

Figure 8.44: The general block diagram of dynamic SODS.

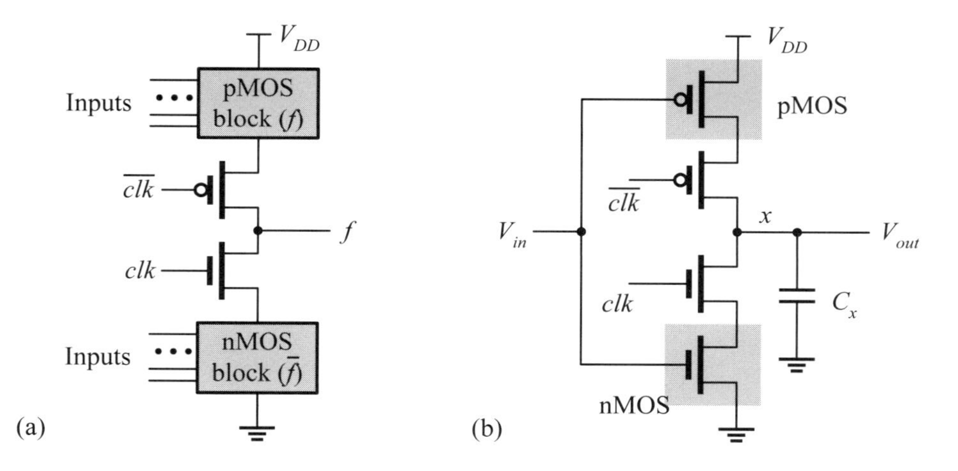

Figure 8.45: The (a) block diagram and (b) an example of clocked CMOS (C^2MOS) logic.

one nMOS and one pMOS, are inserted between the logic blocks and the output node. Both pMOS and nMOS blocks perform the same function as in CMOS logic. The same logic design methodology used in CMOS logic circuits can be applied to C^2MOS logic. A simple example is portrayed in Figure 8.45(b), where an inverter is implemented. To avoid the charge-sharing effect between the output node and internal nodes of both pMOS and nMOS blocks, the pair of clocked nMOS and pMOS transistors should be placed closely to the output node.

The operation of C^2MOS logic is as follows. When the clock *clk* is high, both clocked MOS transistors are turned on and the C^2MOS logic circuit works as a CMOS logic circuit. The output logic function is computed by both nMOS and pMOS blocks. When the clock *clk* is low, both clocked MOS transistors are turned off. The output node is floated and retains its state on the capacitor. Hence, the output node is a dynamic node just like the other dynamic logic circuits.

The drawback of clocked CMOS logic is that it needs both pMOS and nMOS blocks,

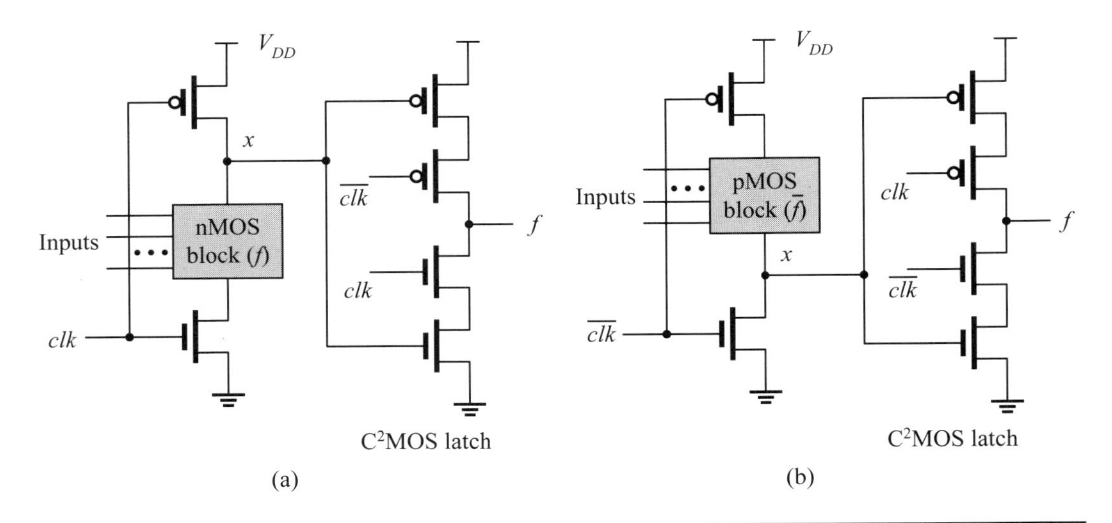

Figure 8.46: The NORA clocked dynamic logic: (a) precharge-high stage; (b) precharge-low stage.

exactly the same as CMOS logic, and requires more area than the other dynamic logic. Hence, it is seldom used today except as the clocked inverter, shown in Figure 8.45(b). The clocked inverter is indeed a dynamic latch, which may be used to keep the previous value of the output node during the period when the clock *clk* is deactivated. This clocked inverter is often combined with the basic dynamic circuit and *np*-domino logic to form the other two types of logic circuits: NORA clocked dynamic circuit and *np*-domino clocked dynamic logic. The clocked inverter is usually called a *C^2MOS latch.*

8.5.1.2 NORA Clocked Dynamic Logic

The basic structure of NORA (no-race) clocked dynamic logic consists of a basic dynamic logic circuit followed by a C^2MOS latch, as shown in Figure 8.46. Like the basic dynamic logic, two different types are possible. Figure 8.46(a) shows the one using an nMOS block to perform its logic function. This dynamic logic is composed of an nMOS dynamic logic followed by a C^2MOS latch and is called a *precharge-high stage*. The other type is shown in Figure 8.46(b), where a pMOS block is used to perform its logic function. This dynamic logic comprises a pMOS dynamic logic followed by a C^2MOS latch and is referred to as a *precharge-low stage*. The precharge-high stage is evaluated as the clock *clk* is high, whereas the precharge-low stage is evaluated as the clock *clk* is low. Because C^2MOS latches are included at the output nodes, both nMOS and pMOS blocks implement f and $\bar{f}$ functions, respectively.

To illustrate the operation of NORA clocked dynamic logic, we use the precharge-high stage as an example. When the clock *clk* is low, the output of the nMOS dynamic logic is precharged to high and the C^2MOS latch is disabled, forcing its output to be floated and thereby retaining its state on the capacitor. When the clock *clk* is high, the nMOS dynamic logic enters its evaluation phase and the C^2MOS latch is enabled, forcing its output to follow the inverted value of the output from the nMOS dynamic logic. As a consequence, the logic function performed by the nMOS dynamic logic is noninverting. The precharge-low stage works in a complementary way. Hence, we omit here.

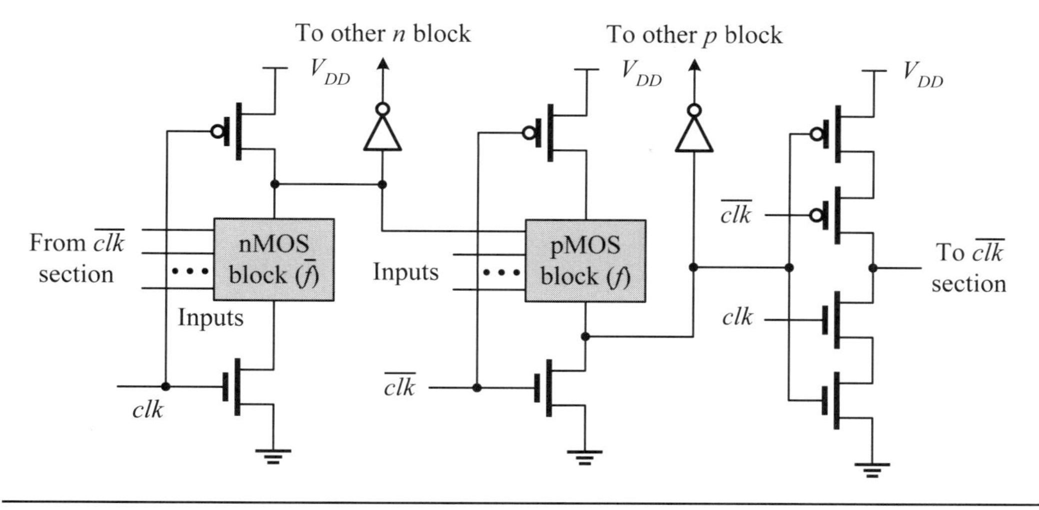

Figure 8.47: The *np*-domino clocked dynamic logic.

8.5.1.3 np-Domino Clocked Dynamic Logic The basic structure of *np*-domino clocked dynamic logic consists of an *np*-domino logic circuit and a C^2MOS latch, as shown in Figure 8.47. Like the *np*-domino logic, it also consists of nMOS and pMOS dynamic logic circuits but is followed by a C^2MOS latch. The desired logic function is carried out by both nMOS and pMOS blocks. Basically, this logic circuit operates in the same way as *np*-domino logic except that here a C^2MOS latch is added at the output node.

The operation of the *np*-domino clocked dynamic logic is as follows. When the clock *clk* is low, the *np*-domino logic enters its precharge phase and both nMOS and pMOS dynamic logic circuits precharge their outputs to high-level and low-level voltages, respectively. The C^2MOS latch is disabled, forcing its output to be floated. Hence, the output node retains its state on the capacitor. When the clock *clk* is high, the *np*-domino logic enters the evaluation phase and both nMOS and pMOS dynamic logic circuits evaluate their logic function values. The C^2MOS latch is enabled, forcing its output to follow the inverted value of the output from the pMOS dynamic logic.

Review Questions

Q8-57. What is a clocked CMOS logic circuit?

Q8-58. What is the C^2MOS latch?

Q8-59. Explain the operation of NORA clocked dynamic logic.

Q8-60. Explain the operation of *np*-domino clocked dynamic logic.

8.5.1.4 True Single-Phase Clock Logic All clocked CMOS logic circuits discussed so far need both true and complementary clocks, *clk* and $\overline{clk}$. We deal with another clocked CMOS logic, which only needs a true clock *clk*. Therefore, no clock skew exists except the clock delay problem, and an even higher clock frequency can be reached. This new type of clocked CMOS logic is known as *true single-phase-clock* (TSPC) logic.

Like NORA (no-race) clocked dynamic logic, the basic structures of TSPC logic can also be classified into two types: precharge-high and precharge-low, as shown in

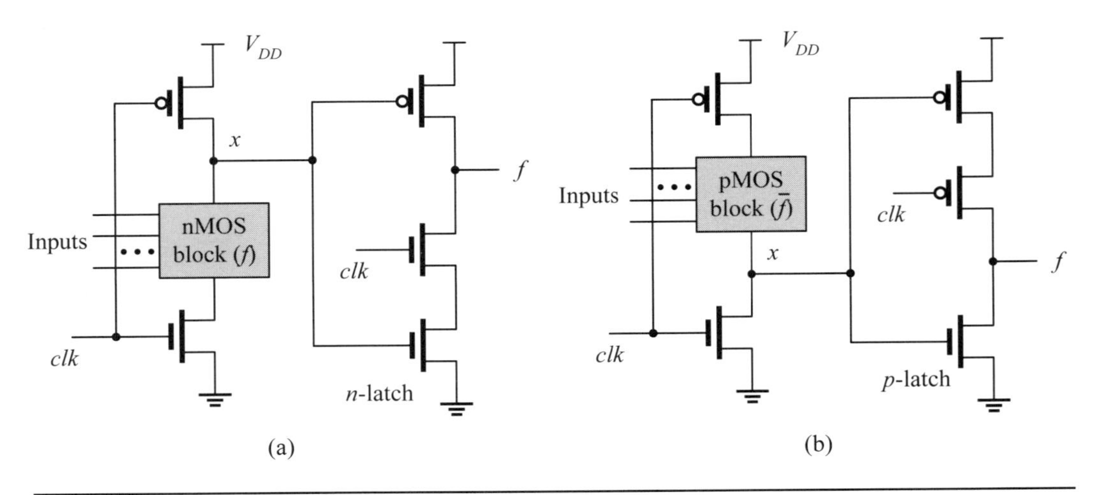

Figure 8.48: The general block diagrams of TSPC logic: (a) precharge-high stage; (b) precharge-low stage.

Figure 8.48. The TSPC logic is virtually evolved from the NORA (no-race) clocked dynamic logic by removing the complementary clock-controlled MOS transistor from each C^2MOS latch. In other words, the pMOS transistor is removed from the precharge-high stage while the nMOS transistor is removed from the precharge-low stage. The resulting modified C^2MOS latches are called *n-latch* and *p-latch*, respectively, as shown in Figure 8.48. Because clock-controlled inverters are included at the output nodes, both nMOS and pMOS blocks implement f and $\bar{f}$ functions, respectively.

The precharge-high stage is used as an example to explain the operation of the TSPC logic. When the clock *clk* is low, the TSPC logic enters the precharge phase and node x is precharged to a high-level voltage. The output latch is off because the clocked nMOS transistor is turned off. The output is therefore stable and retains its previous value. When the clock *clk* is high, the TSPC logic enters the evaluation phase. The nMOS block evaluates its logic function and reflects its final value on its output. An inverted value is then output to the output node through the output n-latch because the clocked nMOS transistor of the latch is turned on now. The precharge-low stage works in a complementary way. Hence, we omit it here.

8.5.1.5 All-n Dynamic Logic Even though domino logic and np-domino logic remove the partially discharged hazard of cascading basic dynamic logic circuits of the same type, both logic styles have their own shortcomings. The domino logic can only implement a positive switching function while np-domino logic requires two different types of logic blocks, nMOS and pMOS blocks. In all-n dynamic logic, only nMOS blocks are required and can realize any switching function by cascading together two different types of logic blocks.

The circuit structures of all-n dynamic logic are shown in Figure 8.49. There are two different function blocks, noninverting and inverting. The noninverting block is a modified version of the TSPC precharge-high stage, shown in Figure 8.48(a), by adding a pMOS transistor M_{p2} and replacing the ground connection of the n-latch with the connection to the drain of the foot nMOS transistor. The inverting block shown in Figure 8.49(b) is a modification of the TSPC precharge-low stage in which an nMOS block is used in place of the pMOS block and two transistors, M_{p2} and M_{n2}, are added.

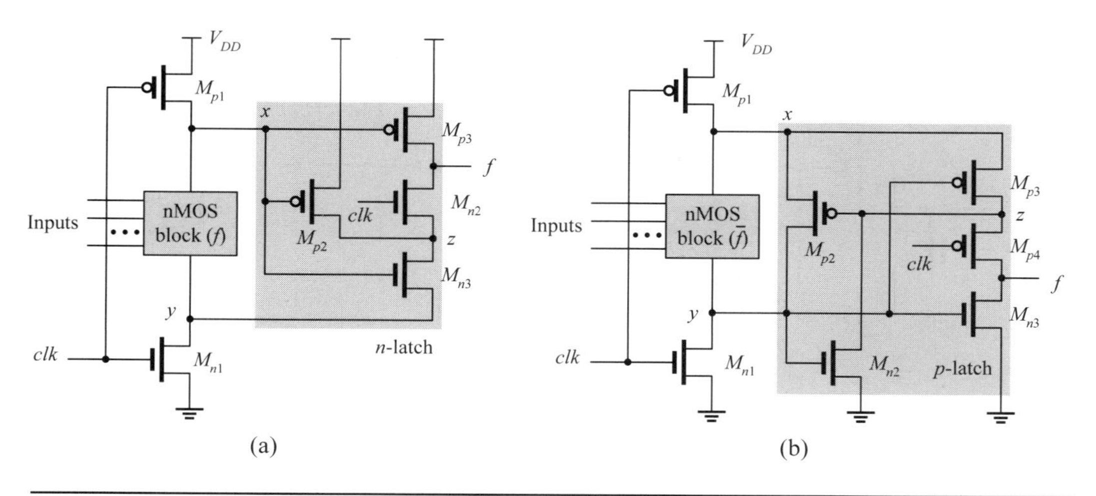

Figure 8.49: The general block diagrams of all-n dynamic logic: (a) noninverting block; (b) inverting block.

In addition, the V_{DD} connection of the p-latch is replaced by the connection to the drain of the precharge pMOS transistor.

The operation of the noninverting block is as follows. As depicted in Figure 8.49(a), when the clock *clk* is low, the noninverting block enters the precharge phase. The precharge pMOS transistor M_{p1} is on and charges node x to V_{DD}. The pMOS transistor M_{p3} is off. Because the nMOS transistor M_{n2} is off, the output node f floats and holds its previous value. When the clock *clk* is high, the noninverting block enters the evaluation phase. Both nMOS transistors M_{n1} and M_{n2} are turned on, and node y is discharged to ground, thereby enabling the n-latch. If the nMOS block evaluates a logic 1, node x is also discharged to ground and the output node f rises high. If the nMOS block evaluates a logic 0, node x remains high and causes the output node f to fall to ground. Hence, a noninverting function is performed.

The use of pMOS transistor M_{p2} is to avoid the charge redistribution between the output node f and node z when the output node f is logic 1 and the clock switches, by charging node z to V_{DD} in advance as the nMOS-block logic is logic 1, so that both output node f and node z are at the same voltage level.

The operating principle of the inverting block is as follows. As shown in Figure 8.49(b), when the clock *clk* is high, the inverting block enters the precharge phase. The nMOS precharge transistor M_{n1} is on to discharge node y to a low-level voltage, which in turn causes the nMOS transistor M_{n2} to be off. The output node f floats and holds its previous value because pMOS transistor M_{p4} is off. When the clock *clk* is low, the inverting block enters the evaluation phase. Both pMOS transistors M_{p1} and M_{p4} are turned on, and node x is charged to a high-level voltage. The p-latch is enabled. If the nMOS block evaluates a logic 1, node y is charged to a high-level voltage and the output node f is discharged to ground. If the nMOS block evaluates a logic 0, node y remains low and causes the output node f to charge to a high-level voltage. Hence, an inverting function is carried out. The use of pMOS transistor M_{p2} is to make node y quickly rise to V_{DD} if necessary. Another feature of pMOS transistor M_{p2} is to improve the noise margin by supplying enough current to cancel the effect of noise in node y during the evaluation phase.

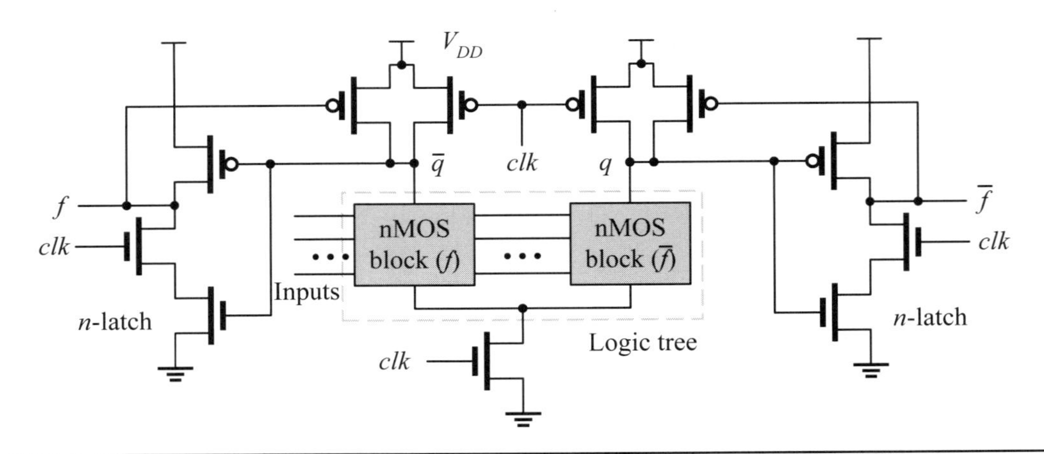

Figure 8.50: The general block diagram of the differential NORA dynamic logic circuit.

8.5.2 Clocked Dual-Rail Logic

Next, we look at clocked dual-rail logic: differential NORA dynamic logic. The basic structure of the differential NORA dynamic logic circuit is shown in Figure 8.50. Basically, its structure is the same as dual-rail domino logic with feedback charge keepers except that n-latches are used to replace the output inverters. The differential NORA dynamic logic works as follows. During the precharge phase, both outputs of the logic tree are precharged to V_{DD}. As a consequence, the clock-controlled pMOS transistor in the C^{2}MOS latch is not necessary so that TSPC n-latches can be used instead. During the precharge phase, both outputs, f and $\bar{f}$, holds their previous values. During the evaluation phase, the two n-latches work as inverters, and the logic circuit is identical to dual-rail domino logic.

Review Questions

Q8-61. What are the n-latch and p-latch?

Q8-62. What are the features of TSPC logic?

Q8-63. Describe the features of all-n dynamic logic.

Q8-64. Describe the features of differential NORA dynamic logic.

8.6 Summary

In a dynamic circuit, the logic value is stored in a capacitor made intentionally or a capacitive node, which is composed of the parasitic capacitance inherently existing in the circuit. Dynamic circuits can be roughly cast into two classes: sample-and-hold circuits and dynamic logic circuits. A sample-and-hold circuit is basically composed of a switch and a capacitor or a capacitive node. It does not perform any logic function but simply stores logic values. Dynamic logic is also called precharge-evaluate (PE) logic because its operation is composed of two phases: precharge and evaluate.

The dynamic features of both nMOS and pMOS transistors when used as switches were studied extensively. The transmission gate (TG) or TG switch combines nMOS with pMOS transistors into one switch. The delay model of a TG switch was described. After this, dynamic logic circuits accompanied with their general features and nonideal effects were discussed in detail.

When two basic dynamic logic circuits of the same type are cascaded together, a race problem of the discharging operations between these two circuits may exist. This problem is referred to as the partially discharged hazard. The partially discharged hazard can be removed by domino logic, *np*-domino logic, or a special clock-timing scheme. Generally, the dynamic logic circuits can be classified into three classes: single-rail dynamic logic, dual-rail domino logic, and clocked CMOS logic.

Single-rail dynamic logic is a variation of basic dynamic logic and can be further subdivided into domino logic, *np*-domino (NORA) logic, two-phase nonoverlapping clocking scheme, and clock-delayed domino logic. The variations of domino logic, including multiple-output domino logic (MODL) and compound domino logic (CDL) were also discussed.

There are three major dual-rail dynamic logic styles: dual-rail domino logic, dynamic CVSL, and SA-based dynamic logic. The dual-rail domino logic is an extension of domino logic by adding the complementary counterpart and hence overcomes the difficulty of implementing inverting functions. The dynamic CVSL is a dynamic version of CVSL. The SA-based dynamic logic uses the features of sense amplifiers to detect a small differential voltage and amplify it to a full-rail output. SA-based dynamic logic has many variations, including SSDL and SODS.

Clocked CMOS logic means that its output node is controlled by a clock *clk*. A clocked CMOS logic has its output node follow the value being calculated by its logic function when it is enabled by the clock and float when it is disabled. Clocked CMOS logic can also be partitioned into single-rail logic and dual-rail logic. The clocked single-rail logic includes basic clocked CMOS logic (C^2MOS), NORA (no-race) clocked dynamic logic, *np*-domino clocked dynamic logic, true single-phase-clock (TSPC) logic, and all-*n* dynamic logic. The differential NORA dynamic logic belongs to the clocked dual-rail logic.

References

1. A. J. Acosta et al., "SODS: a new CMOS differential-type structure," *IEEE J. of Solid-State Circuits,* Vol. 30, No. 7, pp. 835–838, July 1995.

2. M. Allam, M. Anis, and M. Elmasry, "High-speed dynamic logic styles for scaled-down CMOS and MTCMOS technologies," *Proc. of Int'l Symposium on Low Power Electronics and Design,* pp. 155–160, 2000.

3. A. Alvandpour et al.,"A sub-130-nm conditional keeper technique," *IEEE J. of Solid State Circuits,* Vol. 37, No. 5, pp. 633–638, May 2002.

4. R. J. Baker, *CMOS Circuit Design, Layout, and Simulation.* New York: John Wiley & Sons, 2005.

5. A. Chandrakasan, W. Bowhill, and F. Fox, ed., *Design of High-Performance Microprocessor Circuits.* Piscataway, NJ: IEEE Press, 2001.

6. V. Friedman and S. Liu,"Dynamic logic CMOS circuits," *IEEE J. of Solid-State Circuits,* Vol. 19, No. 2, pp. 263–266, April 1984.

7. N. Gonclaves and H. DeMan, "NORA: a racefree dynamic CMOS technique for pipelined logic structures," *IEEE J. of Solid-State Circuits,* Vol. 18, No. 3, pp. 261–266, June 1983.

8. T. Grotjohn and B. Hoefflinger, "Sample-set differential logic (SSDL) for complex high-speed VLSI," *IEEE J. of Solid-State Circuits,* Vol. 21, No. 2, pp. 367–369, April 1986.

9. R. X. Gu, and M. I. Elmasry, "All-n-logic high-speed true-single-phase dynamic CMOS logic," *IEEE J. of Solid-State Circuits,* Vol. 31, No. 2, pp. 221–229, February 1996.

10. D. A. Hodges, Horace G. Jackson, and Resve A. Saleh, *Analysis and Design of Digital Integrated Circuits: In Deep Submicron Technology.* 3rd ed., New York: McGraw-Hill Books, 2004.

11. I. Hwang and A. Fisher, "Ultrafast compact 32-bit CMOS adders in multiple-output domino logic," *IEEE J. of Solid-State Circuits,* Vol. 24, No.. 2, pp. 358–369, April 1989.

12. Y. Ji-ren, I. Karlsson, and C. Svensson, "A true single-phase-clock dynamic CMOS circuit technique," *IEEE J. of Solid-State Circuits,* Vol. 22, No. 5, pp. 899–901, October 1987.

13. R. H. Krambeck, C. M. Lee, and H.-F. S. Law, "High-speed compact circuits with CMOS," *IEEE J. of Solid-State Circuits,* Vol. 17, No. 3, pp. 614–619, June 1982.

14. P. Larsson and C. Svensson, "Impact of clock slope on true single phase clocked (TSPC) CMOS circuits," *IEEE J. of Solid-State Circuits,* Vol. 29, No. 6, pp. 723–726, June 1994.

15. F. Murabayashi et al., "2.5 V CMOS circuit techniques for a 200 MHz superscalar RISC processor," *IEEE J. of Solid-State Circuits,* Vol. 31, No. 7, pp. 972–980, July 1996.

16. S. Naffziger et al., "The implementation of the Itanium 2 microprocessor," *IEEE J. of Solid-State Circuits,* Vol. 37, No. 11, pp. 1448–1460, November 2002.

17. P. Ng, P. Balsara, and D. Steiss, "Performance of CMOS differential circuits," *IEEE J. of Solid-State Circuits,* Vol. 31, No. 6, pp. 841–846, June 1996.

18. B. Nikolic et al., "Improved sense-amplifier-based flip-flop: design and measurements," *IEEE J. of Solid-State Circuits,* Vol. 35, No. 6, pp. 876–884, June 2000.

19. K. Nowka and T. Galambos, "Circuit design techniques for a gigahertz integer microprocessor," *Proc. of Int'l Conf. on Computer Design,* pp. 11–16, 1998.

20. R. Pereira, J. A. Michell, and J. M. Solana, "Fully pipelined TSPC barrel shifter for high-speed applications," *IEEE J. of Solid-State Circuits,* Vol. 30, No. 6, pp. 686–690, June 1995.

21. J. A. Pretorius, A. S. Shunat, and C. A. T. Salama, "Latched domino CMOS logic," *IEEE J. of Solid-State Circuits,* Vol. 21, No. 4, pp. 514–522, August, 1986.

22. Y. Suzuki, K. Odagawa, and T. Abe, "Clocked CMOS calculator circuitry," *IEEE J. of Solid-State Circuits,* Vol. 8, No. 6, pp. 462–469, June 1973.

23. T. Thorp, G. Yee, and C. Sechen, "Design and synthesis of monotonic circuits," *Proc. of IEEE Int'l Conf. on Computer Design,* pp. 569–572, 1999.

24. John P. Uyemura, *CMOS Logic Circuit Design.* Boston: Kluwer Academic Publishers, 1999.

25. John P. Uyemura, *Introduction to VLSI Circuits and Systems.* New York: John Wiley & Sons, 2002.

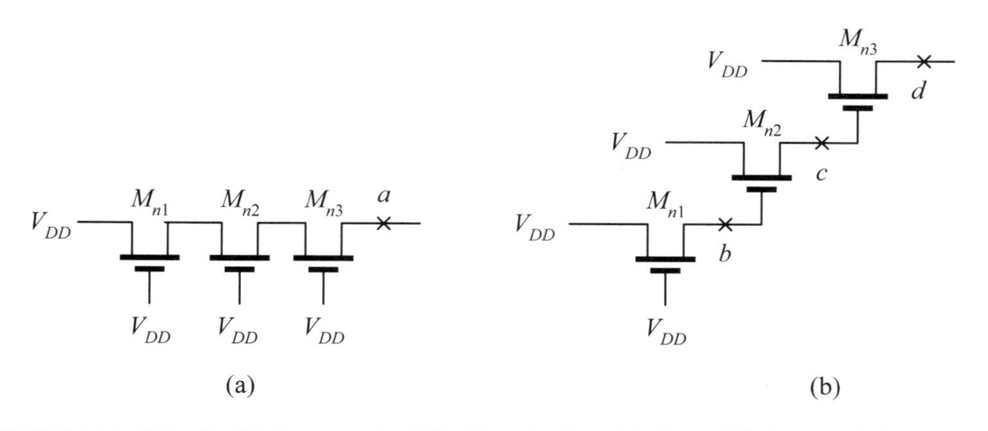

Figure 8.51: The threshold-voltage effects on two different combinations of nMOS switches: (a) series connection; (b) driving-driven connection.

26. Z. Wang et al., "Fast adders using enhanced multiple-output domino logic," *IEEE J. of Solid-State Circuits,* Vol. 32, No. 2, pp. 206–214, February 1997.

27. N. H. E. Weste and D. Harris, *CMOS VLSI Design: A Circuit and Systems Perspective.* 3rd ed. Boston: Addison-Wesley, 2005.

28. G. Yee and C. Sechen, "Clock-delayed domino for dynamic circuit design," *IEEE Trans. on Very Large Scale Integration (VLSI) Systems,* Vol. 8, No. 4, pp. 425–430, August 2000.

29. J. Yuan and C. Svensson, "High-speed CMOS circuit technique," *IEEE J. of Solid-State Circuits,* Vol. 24, No. 1, pp. 62–70, February 1989.

Problems

8-1. Referring to Figure 8.51, answer the following questions.

(a) What is the voltage of node a?

(b) What are the voltages of nodes b, c, and d?

8-2. Referring to Figure 8.52, answer the following questions.

(a) What is the voltage of node a?

(b) What are the voltages of nodes b, c, and d?

8-3. Referring to Figure 8.51, assume that all nMOS transistors have feature sizes.

(a) Using SPICE, simulate the circuit shown in Figure 8.51(a) and observe the results. What is the output voltage?

(b) Using SPICE, simulate the circuit shown in Figure 8.51(b) and observe the results. What are the output voltages of the three nMOS transistors, M_{n1}, M_{n2}, and M_{n3}?

8-4. Referring to Figure 8.52, assume that all pMOS transistors have feature sizes.

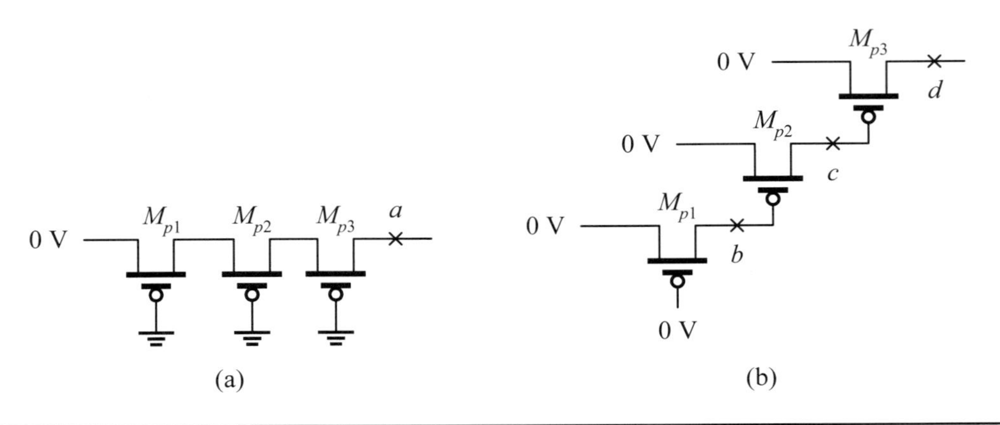

Figure 8.52: The threshold-voltage effects on two different combinations of pMOS switches: (a) series connection; (b) driving-driven connection.

(a) Using SPICE, simulate the circuit shown in Figure 8.52(a) and observe the results. What is the output voltage?

(b) Using SPICE, simulate the circuit shown in Figure 8.52(b) and observe the results. What are the output voltages of the three pMOS transistors, M_{p1}, M_{p2}, and M_{p3}?

8-5. Using the Elmore delay model, calculate the propagation delay of the logic circuit shown in Figure 8.6. Assume that the 0.18-μm parameters are used. The loading capacitance is 500 fF.

8-6. Considering the domino logic circuit shown in Figure 8.53, let capacitor C_1 be initially charged to V_{DD} and capacitors C_2 to C_4 be precharged to 0. Assume that $x = y = 1$ and $z = 0$. Determine the voltage on capacitor C_1 when

(a) All capacitors have the same capacitance C.

(b) Capacitor C_1 has capacitance $6C$ and capacitors C_2, C_3, and C_4 have the same capacitance C.

8-7. Considering the dynamic logic circuit shown in Figure 8.53, let capacitor C_1 be initially charged to V_{DD} and capacitors C_2 to C_4 be precharged to 0. Assume that $x = y = 1$ and $z = 0$. Determine the voltage on capacitor C_1 under each of the following specified conditions:

(a) Capacitor C_1 has capacitance of $5C$ and capacitors C_2, C_3, and C_4 have the same capacitance C.

(b) Capacitor C_1 has capacitance of $10C$ and capacitors C_2, C_3, and C_4 have the same capacitance C.

8-8. Find the logical effort values of the input and the clock input of each logic circuit shown in Figure 8.54.

8-9. Referring to Figure 8.20, design the charge keeper.

(a) Determine the size of the charge keeper.

(b) Use SPICE to simulate and verify the resulting circuit.

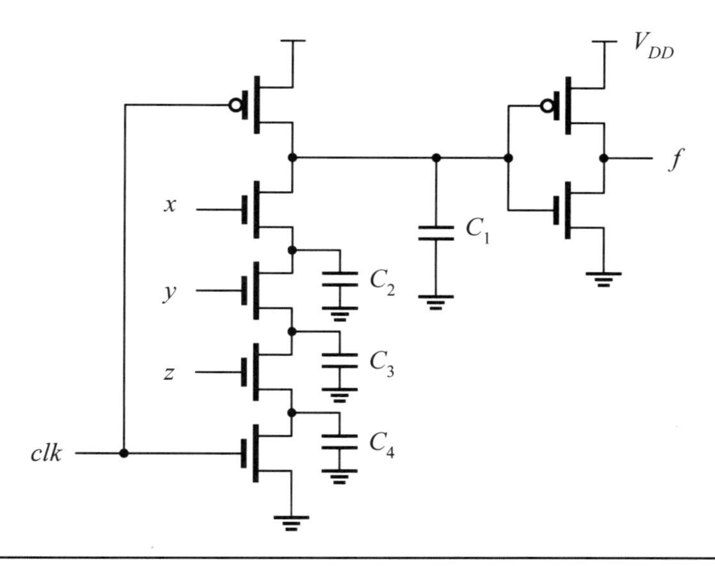

Figure 8.53: The charge-sharing effect in a domino logic circuit.

8-10. Realize the following switching function using the domino logic style.

$$f = v \cdot w + x \cdot (y + z)$$

8-11. Referring to the SRCMOS block in Figure 8.28(a), answer the following short questions:

(a) Design a two-input OR gate and draw its logic circuit.
(b) Calculate the logical effort of the resulting circuit.
(c) Compare the logical effort with the counterpart in domino logic.
(d) Compare the logical effort with the CMOS counterpart.

8-12. Referring to the SRCMOS block in Figure 8.28(a), answer the following short questions:

(a) Design a three-input AND gate and draw its logic circuit.
(b) Calculate the logical effort of the resulting circuit.
(c) Compare the logical effort with the counterpart in domino logic.
(d) Compare the logical effort with the CMOS counterpart.

8-13. Referring to the SRCMOS block in Figure 8.28(a), answer the following short questions:

(a) Design a three-input OR gate and draw its logic circuit.
(b) Calculate the logical effort of the resulting circuit.
(c) Compare the logical effort with the counterpart in domino logic.
(d) Compare the logical effort with the CMOS counterpart.

8-14. This problem is to compare the performance between static and dynamic logic circuits.

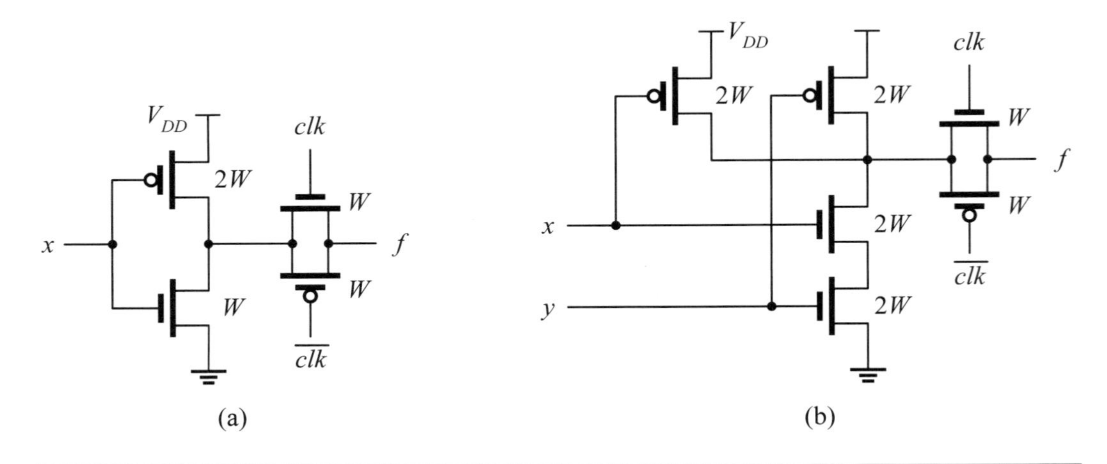

Figure 8.54: The logical effort of TGs associated with gates: (a) TG with an inverter; (b) TG with a NAND gate.

(a) Design a two-input AND gate with CMOS static logic and domino logic, respectively. Draw the resulting logic circuits.

(b) Use SPICE to explore the propagation delays of both circuits.

(c) Give comments about the results you obtain.

8-15. A two-phase nonoverlapping clock generator based on a NOR-SR latch is shown in Figure 8.33.

(a) Design an experiment with SPICE to measure the output clocks.

(b) Change the number of inverters and measure the output clocks again.

(c) Give comments about the results you obtain.

8-16. Referring to Figure 8.12, answer the following questions.

(a) Design an experiment with SPICE to study the partially discharged hazard of the circuit.

(b) Using clock-delayed domino logic (Figure 8.35), redesign the logic function. Draw the logic circuit.

(c) Design an experiment with SPICE to verify the result. Can the partially discharged hazard be removed with the clock-delayed domino logic style?

8-17. Referring to Figure 8.36, study the three popular delay elements.

(a) Using standard sizes for all pMOS and nMOS transistors, design an experiment with SPICE to measure the relationship between the propagation delay and control voltage of each circuit.

(b) Give comments of the results that you obtain.

8-18. Realize a two-input AND/NAND and OR/NOR gates using the dual-rail domino logic style.

8-19. Referring to the annihilation gate in Figure 8.55, answer the following short questions.

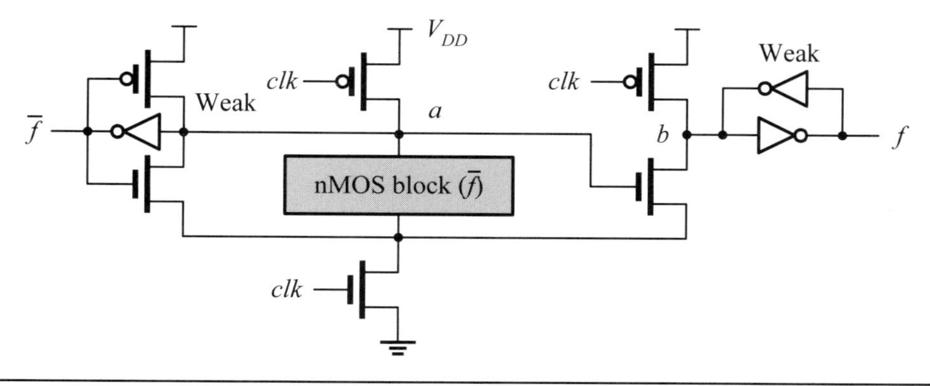

Figure 8.55: An annihilation gate.

(a) Explain the operation of this gate.

(b) Realize a five-input NOR gate with this annihilation gate.

8-20. Referring to Figure 8.55, calculate the logical effort of the annihilation gate that realizes a five-input NOR gate and compare it with the one realized using the dual-rail domino logic circuit.

8-21. Using the dynamic CVSL logic style, realize the two-input AND/NAND and OR/NOR gates.

8-22. Using the dynamic CVSL logic style, realize the carry function of a full adder.

8-23. Using the dynamic CVSL logic style, realize the sum function of a full adder.

8-24. Using the dynamic CVSL logic style, design a complex gate to implement the following switching function.

$$f(v, w, x, y, z) = vw + x(y + z)$$

8-25. Using the TSPC logic style, realize two-input AND and OR gates.

8-26. Using the SA-based dynamic logic style shown in Figure 8.42, design a two-input AND/NAND gate.

(a) Design an experiment with SPICE to measure the propagation delay.

(b) Compare the result with the dual-rail domino logic obtained from Problem 8-18.

(c) Give comments about the results that you obtain.

8-27. Using the dynamic SODS logic style, design a two-input AND/NAND gate.

(a) Draw the logic circuit.

(b) Measure the propagation delay of this circuit with SPICE.

(c) Compare the results with that of dual-rail domino logic.

9

Sequential Logic Designs

After introducing both static and dynamic logic circuits for building combinational logic in the previous chapters, in this chapter we explore how these logic circuits can also be used to implement sequential logic circuitry. The output of combinational logic is determined solely by the present input. In contrast, the output of sequential logic is determined not only by the present input but also by its preceding inputs, thereby the previous outputs. In other words, the sequential logic has a memory that memorizes the history of its previous outputs.

We begin to introduce in this chapter the fundamentals of sequential logic circuits, including the sequential logic models, basic bistable devices, metastable states and hazards, and arbiters. After this, we consider a variety of static and dynamic memory elements that are widely used in complementary metal-oxide-semiconductor (CMOS) technology. These include latches, flip-flops, and pulsed latches.

Both combinational and sequential logic circuits are then combined together with clocking circuits to perform the desired logic functions. The timing issues of systems using flip-flops, latches, and pulsed latches are dealt with in detail. For effectively processing streams of data, parallelism and pipelining are the two major techniques. With parallelism, many copies of the same circuit are used to perform the same computation on the different data sets in parallel. With pipelining, a data stream can be continuously applied to a computational circuit in a regular way to yield results in sequence. This chapter concludes with the introduction of pipeline systems.

9.1 Sequential Logic Fundamentals

Combinational logic is a kind of logic circuit whose output is determined solely by the present input. On the contrary, sequential logic is a type of circuit whose output is determined not only by the present input but also by its preceding inputs. In other words, the sequential logic has a memory that memorizes the history of its previous outputs. We address the sequential logic model referred to as *Huffman's model*, *basic memory devices*, *metastable states* and *hazards*, and *arbiters* in order.

9.1.1 Huffman's Model

Any sequential logic circuit can be modeled by a mathematical model, known as the *finite-state machine* (FSM). An FSM $\mathcal{M}$ is a quintuple $\mathcal{M} = (\mathcal{I}, \mathcal{O}, \mathcal{S}, \delta, \lambda)$, where $\mathcal{I}$, $\mathcal{O}$, and $\mathcal{S}$ are finite, nonempty sets of input, output, and state symbols, respectively.

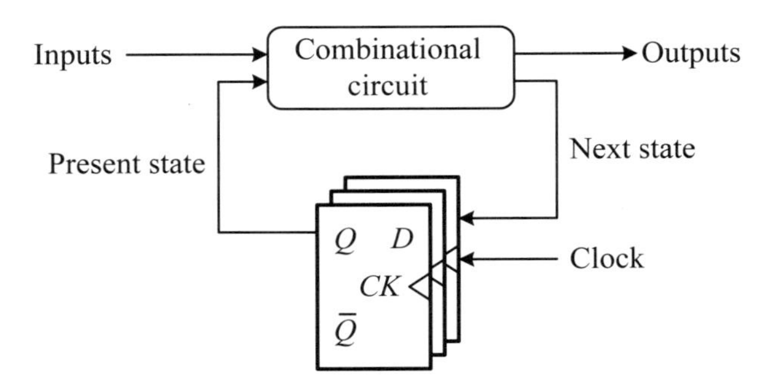

Figure 9.1: The block diagram of Huffman's model.

Here, a symbol in an individual set is a combination of the values of its variables. δ and λ are the state transition and output functions, respectively.

The state transition function δ is defined as follows:

$$\delta : \mathcal{I} \times \mathcal{S} \rightarrow \mathcal{S} \tag{9.1}$$

which means that the δ function is determined by both input and state symbols.

The output function λ is defined as follows:

$$\lambda : \mathcal{I} \times \mathcal{S} \rightarrow \mathcal{O} \quad \text{(Mealy machine)} \tag{9.2}$$
$$\lambda : \mathcal{S} \rightarrow \mathcal{O} \quad \text{(Moore machine)} \tag{9.3}$$

where, in a *Mealy machine*, the output function is determined by both present input and present state, whereas, in a *Moore machine*, the output function is solely determined by the present state.

The general circuit model for sequential logic circuitry is referred to as *Huffman's model*, as shown in Figure 9.1. It consists of a combinational logic circuit and one or more flip-flops. The combinational logic circuit performs all the required computations of both output and next-state functions from both the present input and present state. An illustration of using Huffman's model to model a pipeline system is given in the following.

■ Example 9-1: (A simple pipeline system.) Figure 9.2 shows an example of a simple pipeline scheme. Compared to the basic Huffman's model, the outputs of all flip-flops compose the present state and all outputs from combinational logic determine the next state. The outputs are simply to be taken from the last flip-flops. Note that in this example all data are assumed to be n bits.

■

In practice, we often further categorize the sequential logic circuits into two broad types, Moore and Mealy machines, in terms of the dependence of the outputs on their inputs. As shown in Figure 9.3, in a Moore machine, the output is only determined by its present state, while in a Mealy machine, the output is determined by both the present input and present state. Although both machines have the same computational power, subtle differences indeed exist between them in practical applications. For a Moore machine, the output can only be changed with the state of the machine. However, in a Mealy machine, the output can be changed as either the state of the

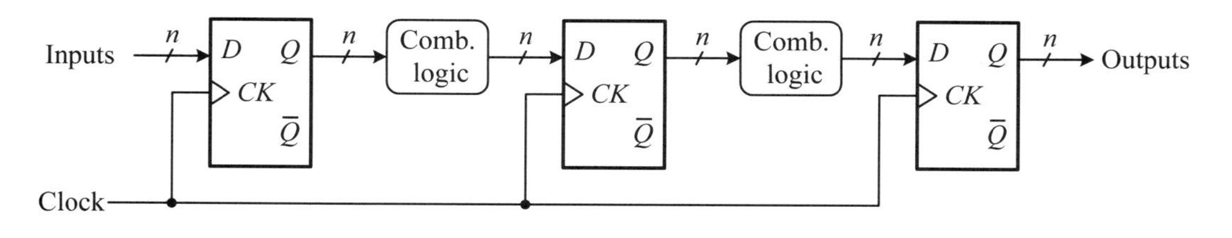

Figure 9.2: A simple pipeline structure.

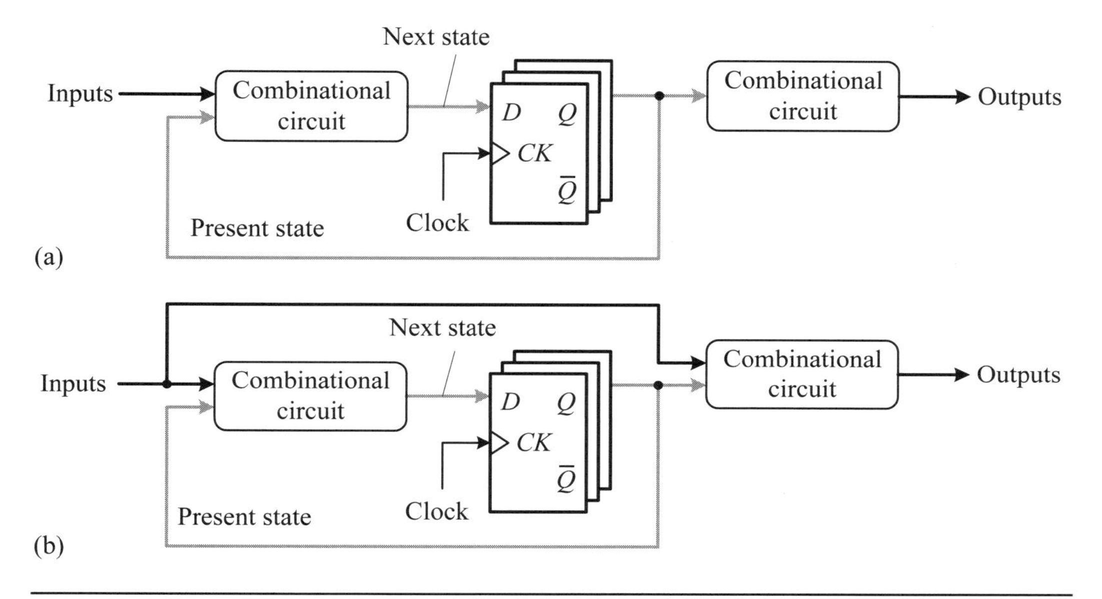

Figure 9.3: The block diagrams of (a) Moore and (b) Mealy machines.

machine or the input is changed. In addition, Mealy machines usually use less number of states than Moore machines.

In summary, the design issues of a sequential logic circuit are to determine the output and next-state functions from both the present input and present state. To do this, we first need to derive a state diagram or table and an output table from the relationship between inputs and desired outputs. Then, we follow a deterministic procedure to obtain both output and next-state functions. Finally, we implement the output and next-state functions in any available implementation options. Details of Moore and Mealy machines and how they are applied to design sequential logic circuits can be referred to any related textbooks in digital logic, such as Kohavi and Niraj [13] and Lin [14, 15].

Review Questions

Q9-1. Define an FSM.

Q9-2. What is Huffman's model?

Q9-3. What are the differences between the Mealy machine and Moore machine?

Q9-4. How would you use Huffnan's mode to describe a pipeline system?

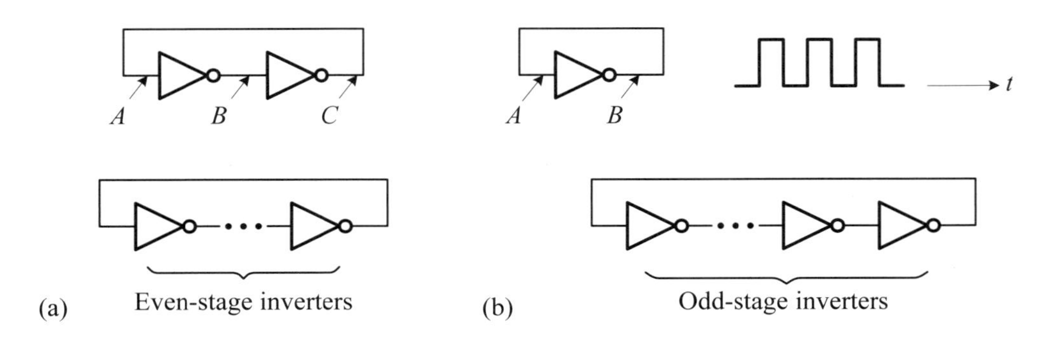

Figure 9.4: An illustration of (a) bistable circuits and (b) oscillators.

Q9-5. How would you use Huffnan's mode to describe a parallel system?

9.1.2 Basic Memory Devices

In digital systems, we often need to distinguish the following three memory elements: *latch*, *flip-flop*, and *register*. All of these three devices are constructed from *bistable devices*. A latch is a level-sensitive bistable device, whereas a flip-flop is an edge-triggered bistable device. Both latches and flip-flops can be used as memory (storage) elements. From an information viewpoint, a latch or flip-flop is a device capable of storing 1-bit information. The word "register" often means memory elements in digital systems. A register is a device that consists of a specific number of flip-flops. To be more specific, an n-bit register contains n flip-flops. Hence, a flip-flop is a single bit register.

As the name implies, a bistable device is a circuit having two stable states. That is, it can remain at one state until it is forced to change or transfer to another state by an external way. One simple way to build such a bistable-state device is to connect two inverters (NOT gates) in a cross-coupled manner. In theory, an even number of inverters connected in a ring structure is a bistable device, as illustrated in Figure 9.4(a).

The operation of the bistable device built on a two-stage inverter chain is illustrated as follows. As shown in Figure 9.4(a), the signal A is inverted when passing through the first inverter and becomes signal B, which is inverted again when passing through the second inverter and becomes signal C. It is then fed back to the input of the first inverter. Because of the same polarity of both signals A and C, the circuit is said to be stable. Since signal A can be 0 or 1, this circuit is a bistable device.

An interesting circuit that is also constructed by connecting a number of inverters but in an odd number of stages is shown in Figure 9.4(b). This logic circuit is known as an *oscillator* or *astable device* because it has no stable state. As we know, an oscillator is a device that can generate a signal with a given frequency. The signal generated by this logic circuit is a square wave with a period equal to the propagation delay of the inverter chain.

The operation of a general odd-stage inverter circuit can be illustrated with a single inverter circuit as follows. As shown in Figure 9.4(b), the signal A appears as an inverted copy at the output of the inverter. This signal is then fed back to the input of the inverter and causes the inverter to change its state again. By repeating this process, the output of the inverter will appear as a series of 0 and 1 signals. As a result, the ring-connected inverter circuit works as an oscillator, referred to as a *ring*

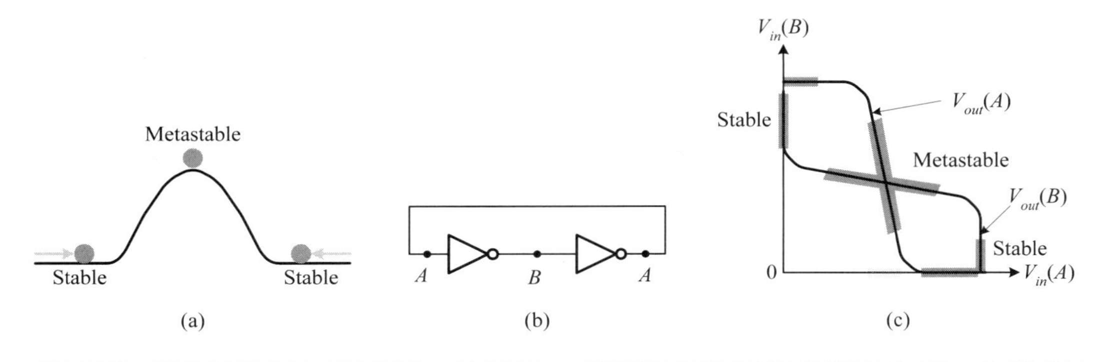

Figure 9.5: An illustration of metastable state of typical latches: (a) basic concept; (b) bistable circuit; (c) VTC.

oscillator. A practical inverter-based oscillator needs at least three stages in CMOS technology.

From the above discussion, we can conclude that a logic circuit is said to be bistable if it has two stable states and to be astable (that is, an oscillator) if it has no stable state. Of course, there exists another type of logic circuit that has only one stable state. This type of logic circuit is often referred to as a *monostable circuit*. Normally, a monostable circuit stays in the stable state and transits to an unstable state for a specific time and then goes back to its stable state as it is triggered by an external signal.

Review Questions

Q9-6. Describe the meanings of bistable, monostable, and astable.

Q9-7. How would you construct a ring oscillator?

Q9-8. How would you distinguish between a latch, a flip-flop, and a register in terms of information storage?

9.1.3 Metastable States and Hazards

To get more insight into the dynamics of both latches and flip-flops, let us consider the concept of the metastable state shown in Figure 9.5(a). The ball has two stable states, one on each side of the hill. When the ball is changed its state by moving from one side of the hill to the other, an external force must be applied to it to overcome the barrier. Depending on the magnitude of external force, the ball may or may not cross over the hill; it may fall down to its original state, stay at the summit, or cross over the hill to reach the other state. When the ball stays at the summit of the hill, it is said to be at a *metastable state*, a state between two stable states. A ball at a metastable state may eventually return to its original stable state or go to the other stable state.

Figure 9.5(b) is a bistable device that consists of two back-to-back connected inverters. Figure 9.5(c) is the voltage transfer characteristic (VTC) of this device. Remember that a bistable device can latch a value and stay at a stable state forever until we change it. To transfer a bistable device from one state to the other, an external signal must be applied to it. Nonetheless, as illustrated in Figure 9.5(c), the device may stay at one of the two stable states or at the metastable state, depending on

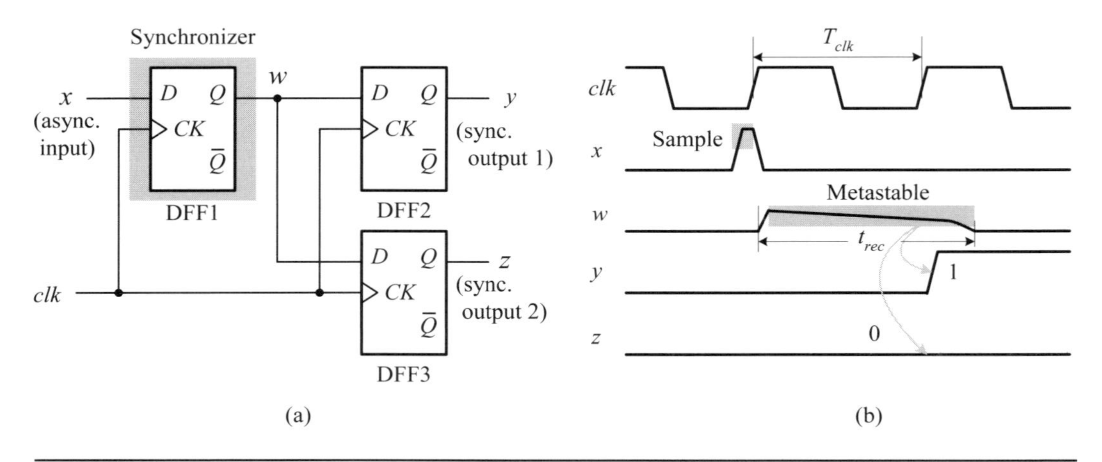

Figure 9.6: The hazards caused by the metastable state of D-type flip-flops: (a) logic circuit; (b) timing.

how strong the external signal is. For example, if the applied external signal is only strong enough to push the first inverter to its metastable state, then its output signal at node B is also very likely to push the second inverter to its metastable state. As a consequence, the device is very likely at the metastable state.

Once a bistable device goes into the metastable state, it can recovery itself from the metastable state back to one of two stable states because of the regenerative property inherent in the bistable device. As we can see from Figure 9.5(c), the metastable state is just located at the linear region with the maximum voltage gain of the VTC. Both inverters work as a cascaded linear amplifier with positive feedback. Because the voltage gain of a general CMOS inverter is about -10, the voltage gain of the linear amplifier is roughly equal to 100. As a consequence, the outputs of both inverters quickly reach one end of the power-supply rails, V_{DD} or ground, namely, a stable state. More details about the bistable devices can be referred to in Section 11.2.5. The amount of time required for a bistable device to leave from its metastable state and to go back to its normal stable state is called the *recovery time*, t_{rec}.

The metastable state of a bistable device may cause hazards in practical applications. To see this, consider the example illustrated in Figure 9.6. The device used to sample an asynchronous signal from the outside of a synchronous system is called a *synchronizer*. A simple synchronizer is just a single D-type flip-flop, such as the one shown in Figure 9.6(a). For a bistable device, a latch or a flip-flop, to properly sample the asynchronous input signal, the input signal must be stable for a specific time before the sample point and remain stable for another specific time after the sample point. If the input signal does not satisfy this requirement, the bistable device may enter the metastable state and output an indeterminate value, between logic 0 and 1. This indeterminate value might be recognized by two different D-type flip-flops as two distinct values, say, logic 1 for DFF2 and logic 0 for DFF3, which would in turn cause an inconsistency for the same external event. The may result in a catastrophic disaster for the system.

From the timing shown in Figure 9.6(b), we can see that the clock period must be at least equal to the sum of the recovery time of the synchronizer and the setup time of the following flip-flop, that is,

$$T_{clk} \geq t_{rec} + t_{setup} \tag{9.4}$$

so that flip-flops DFF2 and DFF3 are allowed to sample stable values, even when the synchronizer has entered the metastable state.

To quantitatively describe how well a synchronizer is, a metric called *mean time between failure* (MTBF) is widely accepted to measure the probability of a given synchronizer entering the metastable state. It is defined as

$$MTBF(t_{rec}) = \frac{1}{T_0 f a} \exp\left(\frac{t_{rec}}{\tau}\right) \tag{9.5}$$

where f is the operating frequency of the synchronizer, a is the number of asynchronous input changes per second, and T_0 and τ are constants, depending on the electrical characteristics of the synchronizer.

Example 9-2: (The MTBF of a synchronizer.) Suppose that $a = 100$ kHz, and a 74ALS74 D-type flip-flop ($t_{setup} = 10$ ns) is used. Calculate the $MTBF(t_{rec})$ at $f = 20$ MHz and 40 MHz, respectively. T_0 and τ of the 74ALS74 are 8.7×10^{-6} sec and 1.0 ns, respectively.

Solution: At $f = 20$ MHz, $t_{clk} = 50$ ns, and $t_{rec} = t_{clk} - t_{setup} = 40$ ns. Hence, the $MTBF$ is computed as follows:

$$\begin{aligned} MTBF(40 \text{ ns}) &= \frac{1}{8.7 \times 10^{-6} \cdot 20 \times 10^{6} \cdot 100 \times 10^{3}} \exp\left(\frac{40 \text{ ns}}{1.0 \text{ ns}}\right) \\ &= 1.35 \times 10^{10} \text{ sec} \approx 429 \text{ years} \end{aligned}$$

As a result, the mean time between two metastable states is approximately 429 years.

When the synchronizer operates at $f = 40$ MHz, $t_{clk} = 25$ ns, and $t_{rec} = t_{clk} - t_{setup} = 15$ ns.

$$\begin{aligned} MTBF(15 \text{ ns}) &= \frac{1}{8.7 \times 10^{-6} \cdot 40 \times 10^{6} \cdot 100 \times 10^{3}} \exp\left(\frac{15 \text{ ns}}{1.0 \text{ ns}}\right) \\ &= 93.94 \times 10^{-3} \text{ sec} \end{aligned}$$

Consequently, it is a high probability for this circuit to enter a metastable state when operating in this environment.

■

As illustrated in Figure 9.7, there are two popular approaches for designing synchronizers. The first is called a *cascaded synchronizer*, as shown in Figure 9.7(a). It is based on the concept that every stage of the synchronizer has the same probability of entering the metastable state. Hence, the probability of an n-stage synchronizer is the product of the individual probability of all stages, which is quite small when a D-type flip-flop with a good $MTBF$ is used as the basic building block. The second is called a *frequency-divided synchronizer*, as shown in Figure 9.7(b). The frequency of the system clock is first divided by N using a frequency divider and then applied to the synchronizer. As a result, the clock period of the synchronizer is increased N times. The available recovery time is then increased to $t_{rec} = N \times T_{clk} - t_{setup}$. Consequently, the $MTBF$ is improved and a flip-flop with a larger setup time can be used to construct the synchronizer.

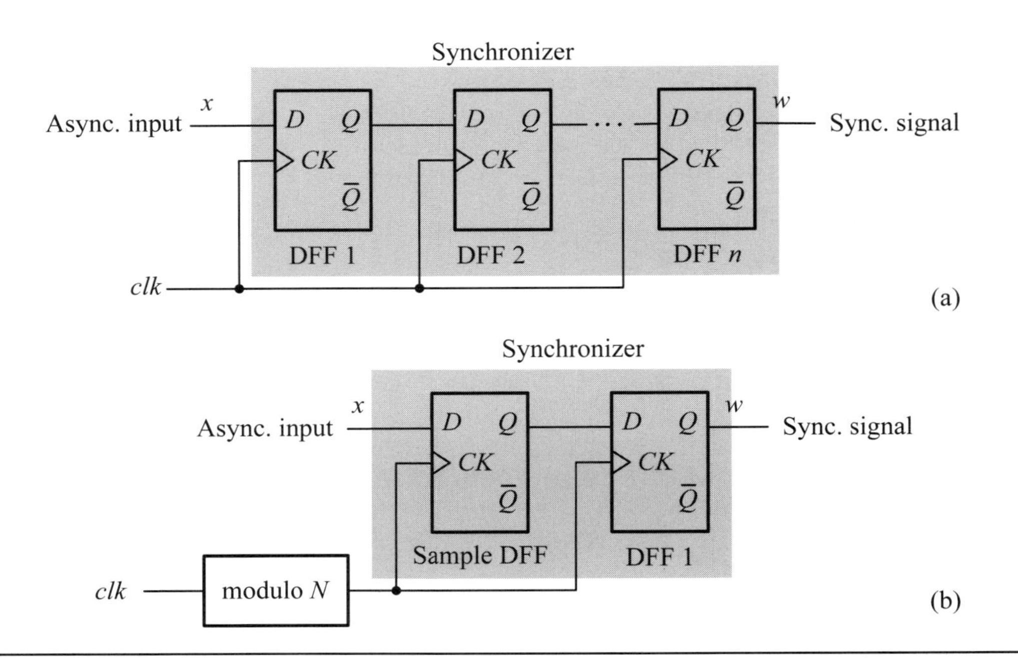

Figure 9.7: Two popular approaches for designing synchronizers: (a) cascaded synchronizer; (b) frequency-divided synchronizer.

9.1.4 Arbiters

In digital systems, there are many situations where two or more devices want to share a resource exclusively. In such a case, an arbiter is needed to arbitrate the use of the shared resource one at a time. The block diagram of a two-way (two-input) arbiter is shown in Figure 9.8(a), where two requests R_0 and R_1 are used to request the use of the shared resource and two grant lines G_0 and G_1 are used to indicate which request is granted and can use the resource.

A circuit for realizing the simple two-way arbiter is shown in Figure 9.8(b), where a set-reset (SR) latch records which device has granted its request. The request lines, R_0 and R_1, are cross-coupled in a way such that when one of them is asserted it disables the other to prevent the SR latch from being changed until the original request line has been deasserted. During the event that the shared resource is in use when a device asserts its request, the corresponding grant signal will not be asserted until the resource is released by the other device. If both requests are asserted in a continuously overlapping manner, their grant signals will be asserted on an alternating basis. Hence, it allows the two devices to share the resource alternatively.

Figure 9.8(c) shows a CMOS implementation of the two-way arbiter. Here, an SR latch is also used to memorize which device has granted its request and two inverters are employed to filter the metastable states and yield the arbitration. The arbiter will respond appropriately whenever one of the requests arrives well before the other. However, when both requests arrive at about the same time, the SR latch enters its metastable state. As we stated, the SR latch will eventually depart from its metastable state and goes back to its normal state; namely, one of the two outputs will be pulled up and the other is guaranteed to remain low. Even though the outputs of this two-way arbiter never enter the metastable state, the time required to make a decision may be unbounded.

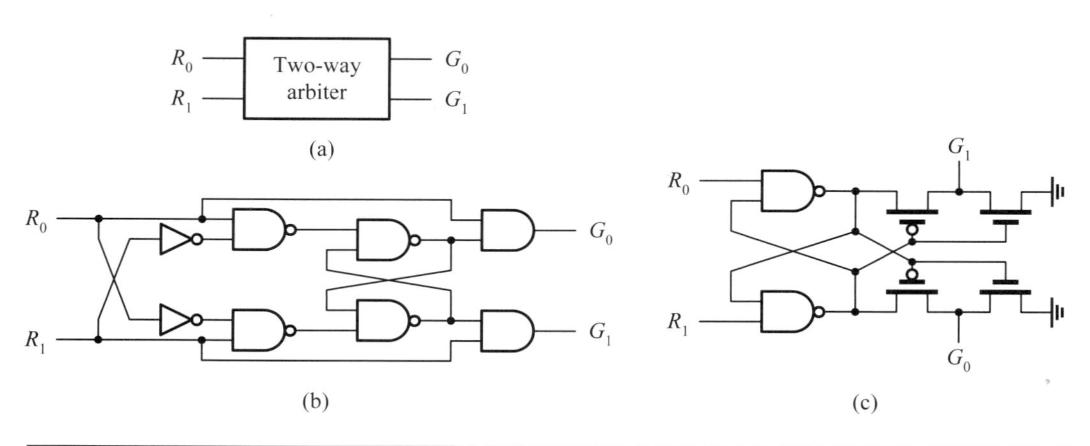

Figure 9.8: The (a) logic symbol and the implementation of two-way arbiters in (b) gates and (c) CMOS logic circuit.

■ Review Questions

Q9-9. Describe the meaning of a metastable state.

Q9-10. What is a synchronizer?

Q9-11. What is the recovery time (t_{rec})?

Q9-12. What is the meaning of $MTBF$ in terms of a synchronizer?

Q9-13. Describe the rationale behind the cascaded synchronizer.

Q9-14. Describe the principle on which the frequency-divided synchronizer is based.

Q9-15. What is the function of an arbiter?

9.2 Memory Elements

Like logic circuits, memory elements used in CMOS technology can also be either static or dynamic, depending on whether the information is stored in circuit states or the form of a charge. In this section, we introduce both fundamental static and dynamic memory elements. These memory elements include latches and flip-flops, regardless of whether the logic style is static or dynamic.

9.2.1 Static Memory Elements

A static memory (storage) element is a device capable of storing information forever if the power supply applied to it is not interrupted. Static memory elements are usually constructed so that the information is stored as circuit states. In this section, we describe in greater detail some common static memory elements, including latches and flip-flops, from the circuit viewpoint.

9.2.1.1 Latches A bistable device is called a *latch* if it can accept external data and change its states accordingly. Of course, the external data must be introduced into a bistable device in a way without violating both node-value and node-conflict-free rules so that the resulting latch is still ratioless. Next, we describe two widely used techniques that can satisfy this requirement. These two techniques are *controlled-gate* and *multiplexer techniques*.

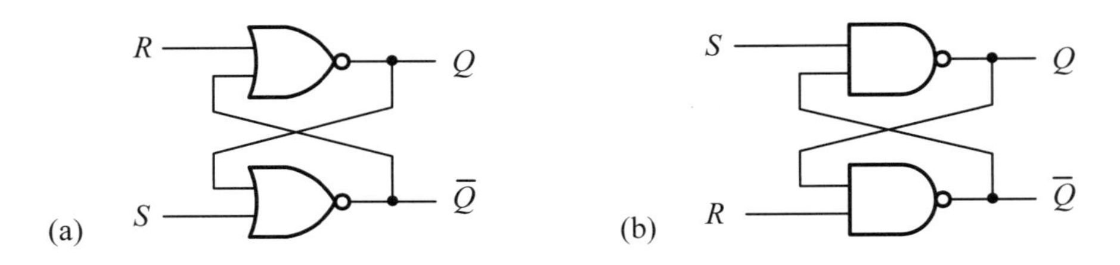

Figure 9.9: Two types of SR latches: (a) NOR-based; (b) NAND-based.

- *Controlled-gate technique*: With this technique, both inverters of the bistable device shown in Figure 9.4(a) are replaced by two-input controlled gates. This results in a circuit, referred to as an SR latch. Depending on whether NOR or NAND gates are used in place of the inverters, SR latches can be further distinguished into two types, the NOR-based SR latch and NAND-based SR latch. For convenience, we will denote such latches as *controlled-gate latches*.
- *Multiplexing technique*: In this technique, a 2-to-1 multiplexer is used to break the feedback loop of the bistable device whenever external data needs to be input to the latch. A latch based on this technique is often referred to as a *multiplexer latch*.

We next describe each of these minutely, from the gate level along with their detailed circuit implementations.

Controlled-gate latches. Figure 9.9(a) illustrates a NOR-based SR latch. Since the output of a NOR gate is 0 if any of its inputs is 1, the input R (reset) needs to be set to 1 to reset the output Q to 0 and the input S (set) needs to be set to 1 to set the output Q to 1. However, both R and S cannot be set to 1 at the same time; otherwise, both outputs Q and $\bar{Q}$ would have the same value, an undesired situation. Therefore, to make the NOR-based SR latch work correctly, the following condition must be held: $S(t) \cdot R(t) = 0$.

Figure 9.9(b) shows a NAND-based SR latch. Since the output of a NAND gate is 1 if any of its inputs is 0, the input R needs to be set to 0 to reset the output Q to 0 and the input S needs to be set to 0 to set the output Q to 1. However, both R and S cannot be set to 0 simultaneously; otherwise, both outputs Q and $\bar{Q}$ would have the same value, an undesired situation. That is, to make the NAND-based SR latch work correctly, the following condition must be held: $\bar{S}(t) \cdot \bar{R}(t) = 0$.

Figure 9.10(a) shows a simple static D-type latch composed of two NOR gates and one inverter. The corresponding CMOS logic circuit is shown in Figure 9.10(b). It is implemented by two NOR gates in a straightforward way. Hence, we will not further discuss it here.

In many practical applications, an SR latch often needs a *gate control* (G) input, or called an *enable control* (E), to enable the latch only at some specific time. This results in a logic circuit, referred to as a *gated SR latch*. Two examples of gated SR-latch circuits are exhibited in Figure 9.11.

■ **Example 9-3: (A gated NOR-based SR-latch circuit.)** Figure 9.11(a) shows a gated NOR-based SR latch built from the NOR-based SR latch. Two AND gates are added to enable or disable the inputs S and R to the NOR-based SR latch. Figure 9.11(b) shows a gated NAND-based SR latch built from the NAND-based SR

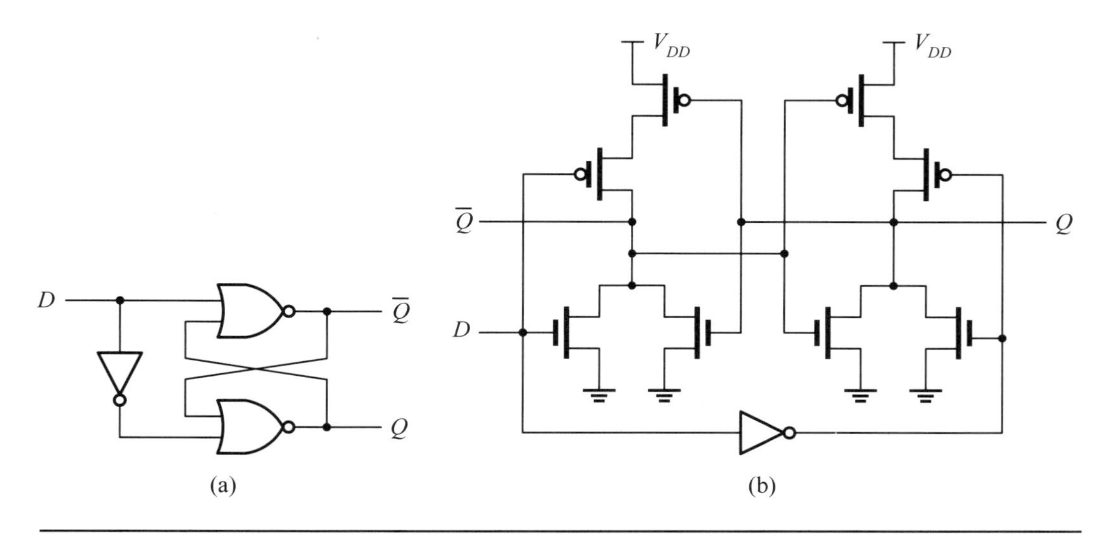

Figure 9.10: The (a) logic and (b) circuit diagrams of a simple controlled-gate D-type latch.

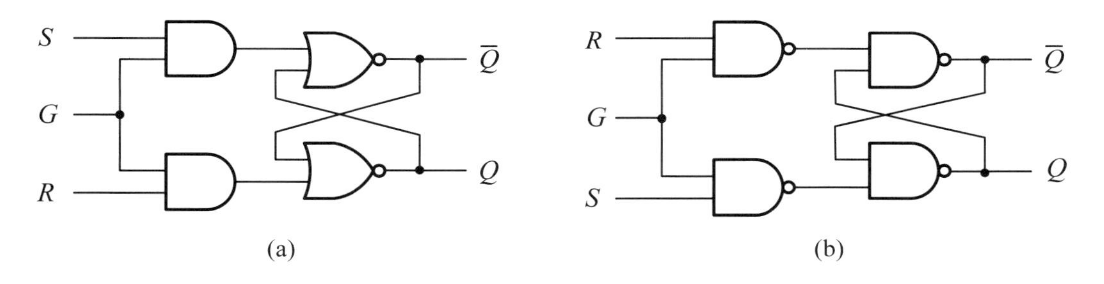

Figure 9.11: Gated SR latches based on (a) NOR-based SR latch and (b) NAND-based SR latch.

latch. Two NAND gates are added to enable or disable the inputs S and R to the NAND-based SR latch.

■

For an SR latch to be useful, it is necessary to further include a mechanism that can enforce the requirement: $S(t) \cdot R(t) = 0$ for a NOR-based SR latch and $\bar{S}(t) \cdot \bar{R}(t) = 0$ for a NAND-based SR latch by the circuit itself implicitly rather than by the user explicitly. The following example illustrates how this idea is applied to a practical latch.

■ Example 9-4: (An SR latch-based D-type latch.) To make the NOR-based SR latch work correctly, the condition $S(t) \cdot R(t) = 0$ must be held at all times. Hence, to reach this, an inverter is placed between the inputs S and R of the gated SR latch depicted in Figure 9.11(a) and results in the SR-latch-based D-type latch shown in Figure 9.12(a). The external data D and its complement $\bar{D}$ are passed to the S and R inputs, respectively, of the SR latch when the gate control G is asserted. The

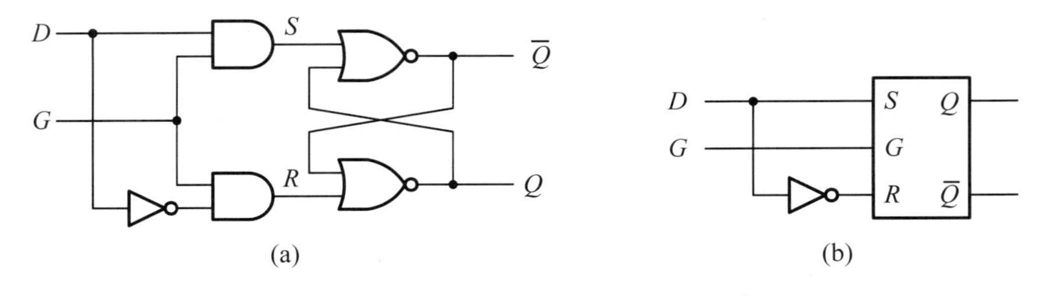

Figure 9.12: An (a) SR-latch-based D-type latch and (b) its logic symbol.

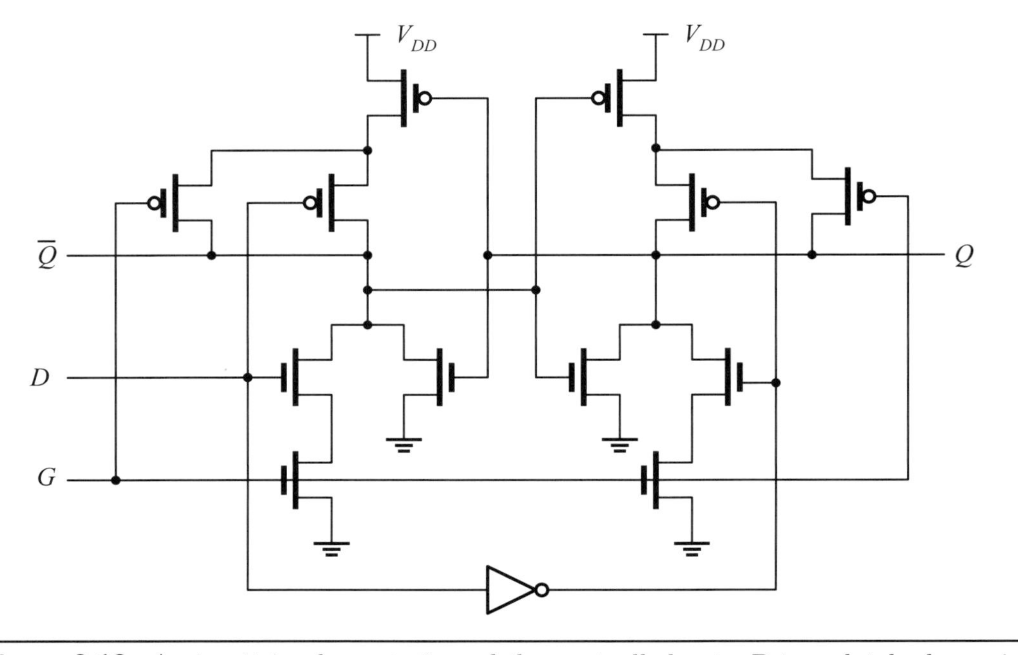

Figure 9.13: A circuit implementation of the controlled-gate D-type latch shown in Figure 9.12(a).

SR latch remains its state unchanged as the gate control is deasserted. Figure 9.12(b) gives its logic diagram.

■

The circuit implementation of the SR-latch-based D-type latch depicted in Figure 9.12(a) is shown in Figure 9.13. In the circuit, both AND gates are incorporated into the NOR gates to form a complex gate, AND-OR-Inverter (AOI21).

Features of latches. From the above discussion, we realize that a latch is a bistable device that changes its output value after a propagation delay from its input through the circuit to the output whenever its input value changes. That is, the output follows its input after the definite propagation delay of the device as the gate control is asserted. This feature is called the *transparent property*.

In addition, the gate control (or enable) signal of a practical latch can be either active-high or active-low. An active-high enable latch is called a *positive latch* and an

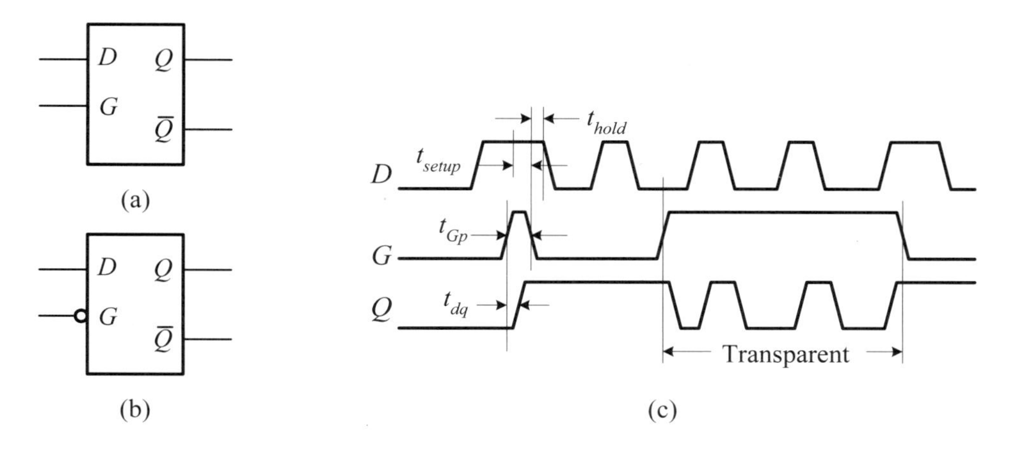

Figure 9.14: The logic symbols for (a) positive and (b) negative D-type latches and (c) timing example of a positive D-type latch.

active-low enable latch is referred to as a *negative latch*. The logic symbols of these two types of latches are displayed in Figures 9.14(a) and (b), respectively. One should be aware of the circle before the gate control signal in Figure 9.14(b).

In the following example, a positive D-type latch is used to further illustrate the meaning of transparent property.

■ Example 9-5: (The timing of the D-type latch.) Figure 9.14(c) shows the timing relationship among the data input D, enable control G, and the output Q. The input D can only be applied to the D-type latch when the gate input G is activated, namely, at a high-level voltage. If the gate input G is activated for a long time, the output Q will follow the input D. In this situation, the D latch is virtually transparent to the input signal; it only provides a definite propagation delay t_{dq} of the input signal. If the gate input pulse t_{Gp} is only long enough for the D-type latch to capture the input data properly, then the output Q will not follow the changes of the input D.

■

Multiplexer latches. The external data D can also be introduced into a bistable device under the control of a transmission gate (TG) switch, as depicted in Figure 9.15(a). As the clock clk is high, the external data D is allowed to reach the node l and make the output node $\bar{Q}$ change its value if the input value has the same value as the current output $\bar{Q}$. Because of the feedback path from the output node through an inverter to node l, the node l may violate the node-conflict-free rule (**Rule 2**), leading to a ratioed logic circuit. Hence, to correctly input the external data into the latch, the transistor sizes of both paths, the feedback inverter and the driver of the input TG switch, should be set appropriately. It is common practice to make the feedback inverter much weaker than the driver of the input TG switch.

From the viewpoint of node l, both feedback and external datapaths are accessed exclusively, that is, not at the same time. Hence, a 2-to-1 multiplexer may be naturally used to connect one of them to node l at a time. Based on this, a multiplexer latch is resulted. In other words, a multiplexer latch is composed of a bistable device and a 2-to-1 multiplexer. The general structures of multiplexer latches are depicted in

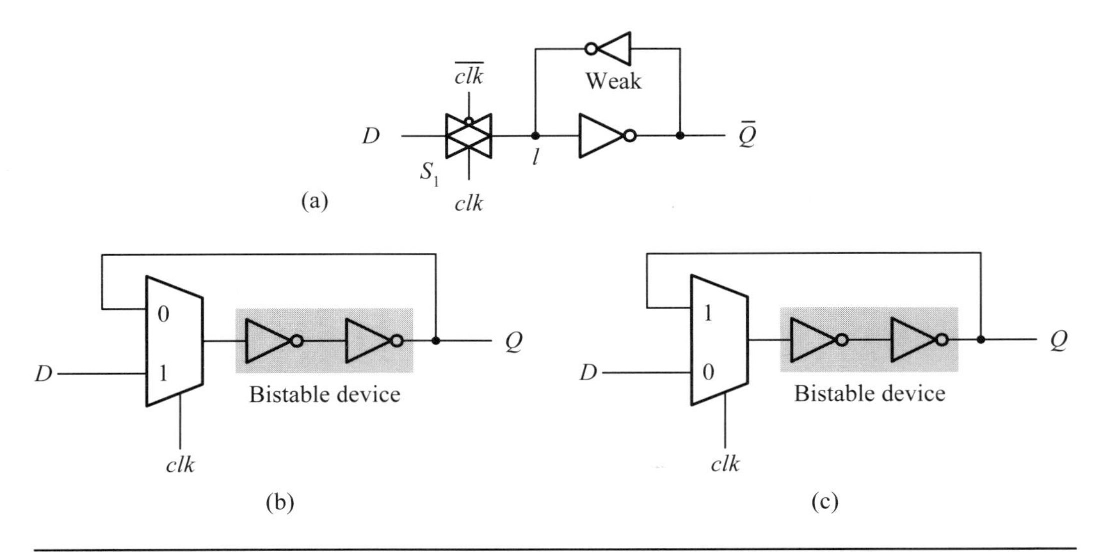

Figure 9.15: General representations of multiplexer D-type latches: (a) basic latch; (b) positive latch; (c) negative latch.

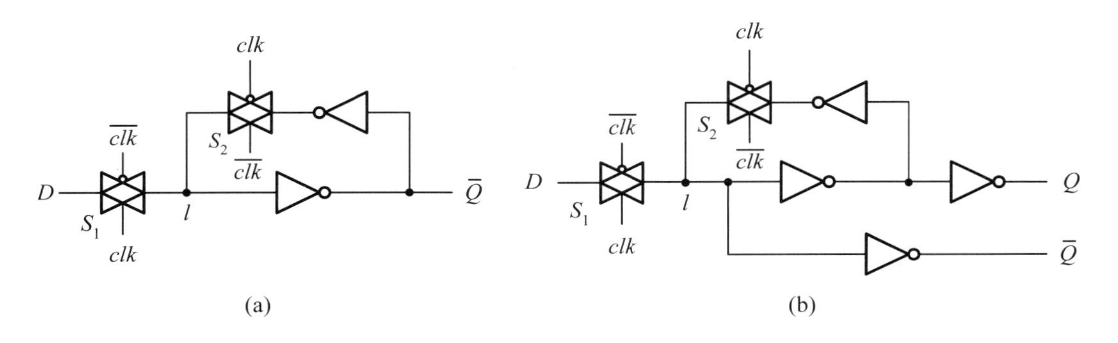

Figure 9.16: Multiplexer D-type latches based on TG switches: (a) basic D-type latch; (b) buffered D-type latch.

Figure 9.15. Like controlled-gate latches, multiplexer latches also have two general types: a positive latch and negative latch. Figure 9.15(b) is a positive latch because it samples the input data as the clock clk is high. Figure 9.15(c) shows a negative latch that samples the input data as the clock clk is low.

In CMOS technology, a multiplexer D-type latch can be simply implemented by using two TG switches and two inverters, as shown Figure 9.16(a). The 2-to-1 multiplexer formed by the two TG switches, S_1 and S_2, uses a clock clk as the selection control. As the clock clk is 1, the input data D is allowed to reach node l without any competing value; as the clock clk is 0, the input data D is isolated from the node l and the two inverters form a loop, thereby retaining its previous data. Since both paths connected to node l are mutually exclusive, the resulting logic circuit is ratioless logic. The node l receives its data from either the input data D or the feedback path but not both at the same time.

A practical latch often needs to be isolated from its loading circuits to keep its stored data value intact. This may be simply accomplished by adding two buffers at

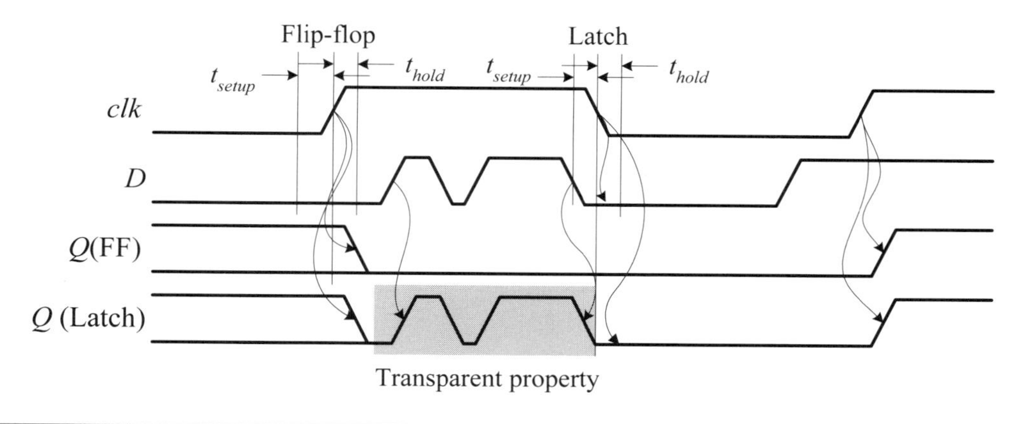

Figure 9.17: Comparison of latches and flip-flops in terms of transparent property.

the output nodes, Q and $\bar{Q}$, as portrayed in Figure 9.16(b). The resulting latch is referred to as a *buffered D-type latch*.

Based on the above discussion, a bistable device capable of being accessed externally is called a *latch* if it has the transparent property. However, in many applications, the transparent property is not desired. Such a bistable device without the transparent property is known as a *flip-flop*. Figure 9.17 shows the difference between latches and flip-flops in terms of transparent property. The flip-flop only captures the input data at the rising edge of the clock *clk* and then holds the data for the rest of clock period. The latch continuously captures and outputs the input data whenever the clock *clk* is high; it captures and outputs the input data at the falling edge of the clock *clk* and holds the data thereafter when the clock *clk* is low.

■ Review Questions

Q9-16. How would you distinguish between a bistable device and a latch?

Q9-17. What are the two basic techniques used to construct a latch?

Q9-18. Define positive and negative latches.

Q9-19. How would you construct a *D*-type latch using the circuit shown in Figure 9.11(b)?

Q9-20. Describe the meaning of transparent property.

Q9-21. How would you distinguish between a latch and a flip-flop?

9.2.1.2 Flip-Flops The essential idea of constructing a flip-flop is on the basis of removing the inherent transparent property of latches. Three approaches result in three different types of flip-flops: *edge-triggered flip-flops*, *master-slave flip-flops*, and *pulsed latches*, respectively. An edge-triggered flip-flop is a circuit that is designed in a special way to facilitate the edge-triggered capability, namely, only sampling the input value at the instant of the clock edge. This type of flip-flop is often found in transistor-transistor logic (TTL) technology. A master-slave flip-flop is a circuit that combines two latches with different polarities: positive latch and negative latch. Such a flip-flop is usually found in CMOS technology. A pulsed latch is a bistable device in which the transparent property is removed by properly controlling the pulse width of the enable control to be only wide enough for the bistable device to sample the input

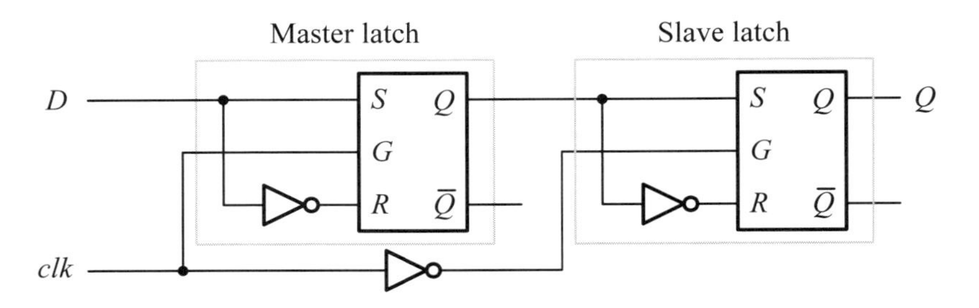

Figure 9.18: A flip-flop constructed from two D-type latches.

data reliably. As a result, the latch can function as a flip-flop. The details of the pulsed latch will be dealt with later in Section 9.3.

We describe many basic static flip-flops, including master-slave flip-flop, D-type flip-flop, CVSL-based D-type flip-flop, RAM-based D-type flip-flop,[1] and dual-edge triggered flip-flop.

Master-slave flip-flop. A master-slave flip-flop is constructed by cascading positive and negative latches. An illustration of this is revealed in the following.

■ Example 9-6: (A D-type flip-flop.) A simple example of a master-slave D-type flip-flop is given in Figure 9.18. At the positive edge of the clock *clk*, the master latch is transparent and the slave latch is opaque. The data input D appears at the output of the master latch, and the output of the slave latch remains unchanged. At the negative edge of the clock *clk*, the master latch is opaque, and the slave latch is transparent. The master latch retains the data appearing at its input before the clock *clk* changes to the negative edge. These data also appear at the output of the slave latch. As as result, the transparent property is eliminated. This type of flip-flop is often called a *negative edge-triggered flip-flop* because it latches the data just before the negative edge of the clock.

■

■ Review Questions

Q9-22. What is the master-slave flip-flop?

Q9-23. What is a pulsed latch?

Q9-24. Referring to Figure 9.18, construct a positive edge-triggered flip-flop.

D-type flip-flops. In CMOS technology, the most commonly used flip-flops are D-type flip-flops. The D-type flip-flops have the symbol as shown in Figure 9.19(a). A D-type flip-flop can also be constructed from an SR flip-flop, as depicted in Figure 9.19(b). In a practical D-type flip-flop, two additional inputs, *preset* and *clear*, are also provided in addition to the data and clock inputs to override the output values of the flip-flop. Whenever the preset input is asserted, the output of flip-flop is set to high immediately regardless of the current values of both data and clock inputs. Similarly, whenever the clear input is activated, the output of the flip-flop is set to low immediately regardless

[1]The acronym CVSL means cascode voltage switch logic and RAM means random access memory.

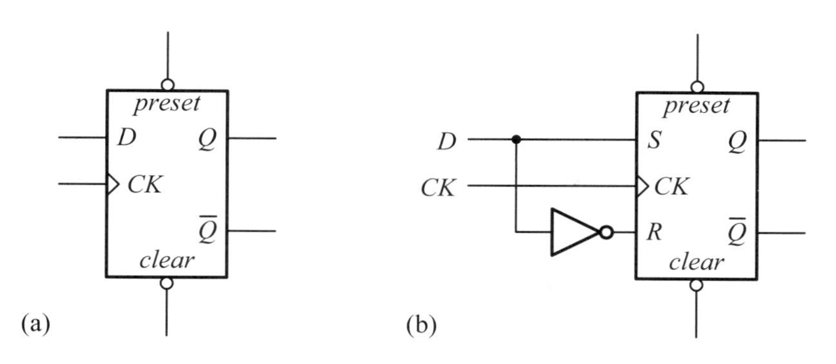

Figure 9.19: The (a) logic symbol and (b) SR-FF realization of a D-type flip-flop.

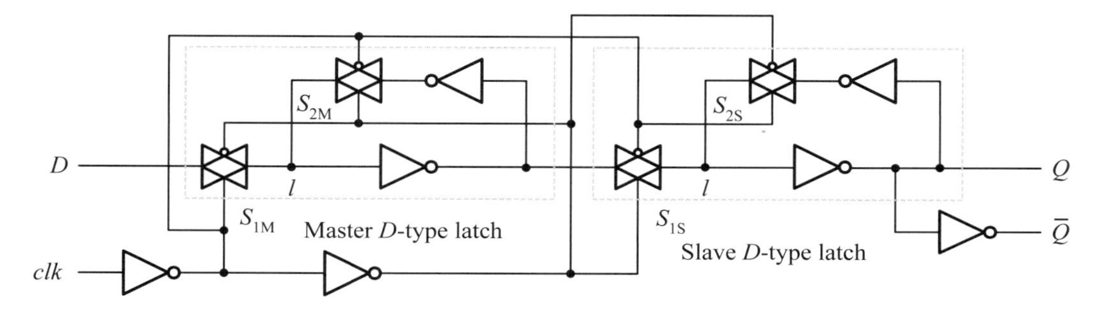

Figure 9.20: A D-type master-slave flip-flop consisting of two D-type latches.

of the current values of both data and clock inputs. Both preset and clear inputs are generally active-low and cannot be asserted at the same time.

A D-type flip-flop is usually composed of two D-type latches, the *master D-type latch* and *slave D-type latch*. Depending on the ways of combining a positive D-type latch with a negative D-type latch, there are two types of D-type flip-flops. A positive edge-triggered D-type flip-flop is formed when the master latch is a negative D-type latch and the slave latch is a positive D-type latch. A negative edge-triggered D-type flip-flop is the result of exchanging the polarities of both master and slave latches. Figure 9.20 shows a positive edge-triggered D-type flip-flop.

The operation of the master-slave D-type flip-flop is as follows. As shown in Figure 9.20, when the clock clk is low, the switch S_{1M} of the master D-type latch and the switch S_{2S} of the slave D-type latch are turned on; the other two switches are turned off. The equivalent circuit is shown in Figure 9.21(a). The master D-type latch samples the input data, and the slave D-type latch holds its previous stored value. Now, if the clock clk rises high, both switches S_{1M} and S_{2S} of the master D-type latch and slave D-type latch, respectively, are turned off and the switches S_{2M} and S_{1S} of the master D-type latch and slave D-type latch, respectively, are turned on. The equivalent circuit is shown in Figure 9.21(b). The master D-type latch stores the sampled value at the positive edge of the clock clk. Meanwhile, this value is passed to the output node through the slave D-type latch. As a consequence, the master-slave flip-flop behaves as a positive edge-triggered flip-flop.

As stated, a practical D-type flip-flop often needs to have the capability of setting its output value asynchronously. A master-slave D-type flip-flop with the asynchronous preset is illustrated in Figure 9.22. Here, two NAND gates are used to replace the two

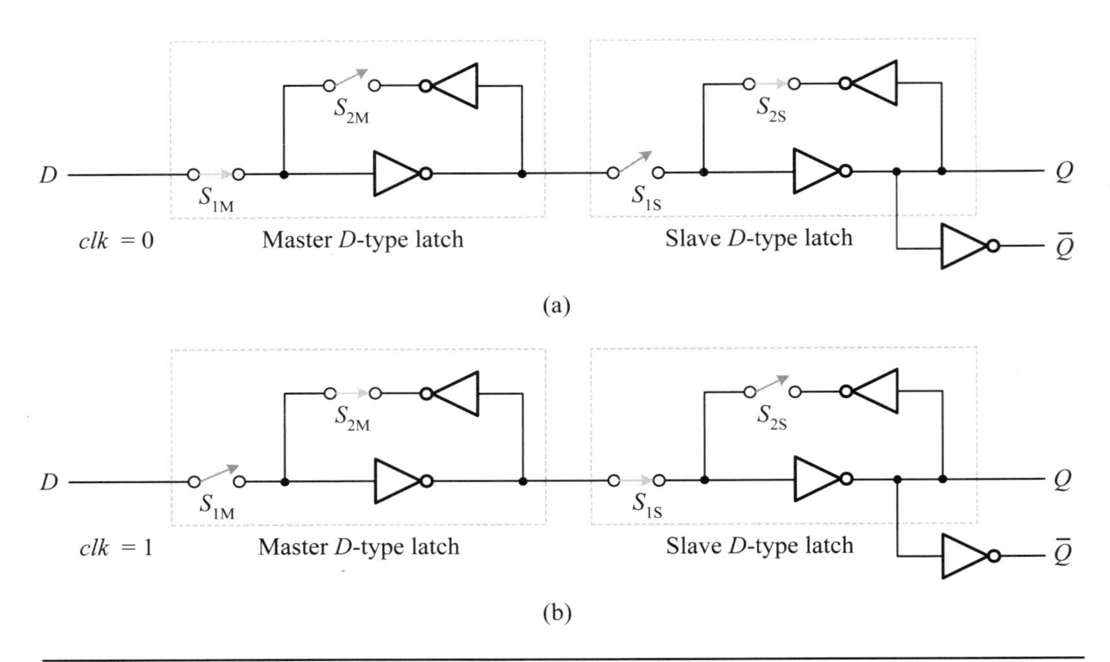

Figure 9.21: The operation of a D-type master-slave (positive edge-triggered) flip-flop. The switch status when: (a) $clk = 0$; (b) $clk = 1$.

inverters in the flip-flop, one from each latch, because the preset is an active-low control signal. If an active-high preset is required, two NOR gates are used in place of both NAND gates.

The operation of the preset control is as follows. The output Q is set to high by the NAND gate of the slave D-type latch as the preset input $\overline{preset}$ is asserted and set its normal value as the preset input $\overline{preset}$ is inactive. A NAND gate is also needed in the master D-type latch to set its value. To see this, let us first assume that the master D-type latch does not have the NAND gate. Referring to Figure 9.21(b), when the preset signal $\overline{preset}$ with a pulse width less than the duty cycle of the clock clk is applied to the D-type flip-flop after the positive edge of the clock clk, the output will be set to high immediately and holds at this voltage level during the activation of the

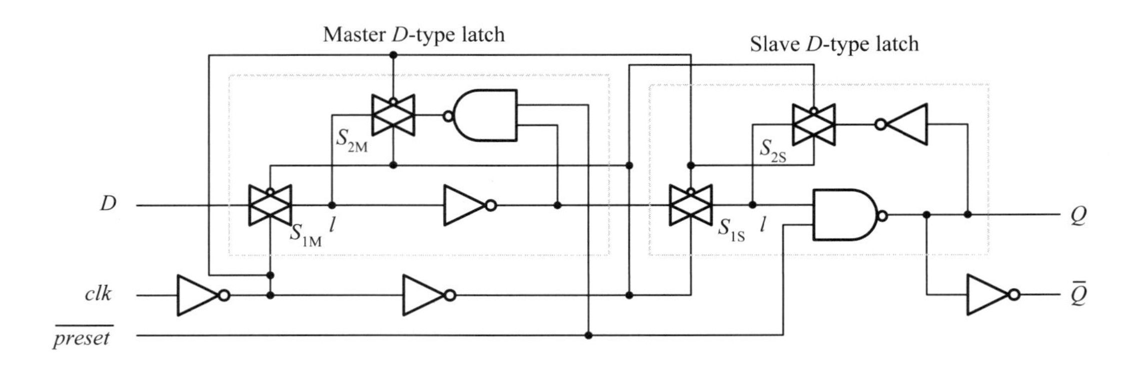

Figure 9.22: A master-slave D-type flip-flop with asynchronous preset.

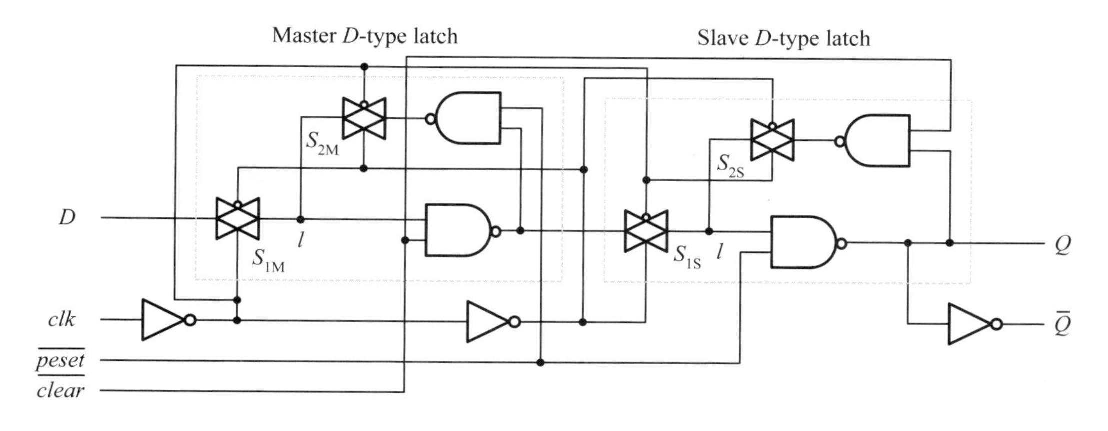

Figure 9.23: A master-slave D-type flip-flop with asynchronous preset and clear.

preset signal. However, the output will be pulled back to its normal value from the master D-type latch after the preset signal is removed. Consequently, a NAND gate is also needed to preset the output value of the master D-type latch.

Both asynchronous preset and clear control inputs can be added into the master-slave D-type flip-flop, as illustrated in Figure 9.23. The output is set to high when the preset input is asserted and is cleared to low when the clear input is asserted. Both preset and clear inputs cannot be activated at the same time and must be maintained at a high-level voltage, namely, at an inactive state for normal operation.

CVSL-based D-type flip-flops. Another type of master-slave D-type flip-flop comprises two CVSL-based latches, as shown in Figure 9.24. The first latch is a positive latch and the second is a negative latch. The operation of this flip-flop is as follows. As the clock clk is high, the positive latch carries out its normal operation. The output of the positive latch follows its data input D. As the clock clk changes from high to low, the positive latch holds its sampled value at the negative edge of the clock clk. At the same time, the output of the negative latch follows its input, namely, the stored value of the positive latch. As a consequence, the resulting flip-flop is a negative edge-triggered type. A positive edge-triggered type can be constructed by interchanging the order of both latches.

RAM-based D-type flip-flops. Flip-flops can also be constructed on the basis of a random-access memory (RAM) cell, or a sense amplifier (SA), which in turn consists of two inverters connected in a back-to-back manner. An example is shown in Figure 9.25. The RAM cell is used to store the input data, and the other four n-type metal-oxide-semiconductor (nMOS) transistors are employed to control the access of the RAM cell. To input an external data D into the RAM cell, the clock clk is set to high. The data D and its complement value are applied to both inputs of the RAM cell, which are also the outputs of the two inverters of the RAM cell, to force the RAM cell to reflect the new data value.

Dual-edge triggered flip-flops. Recall that the power dissipation of a CMOS logic circuit is proportional to the frequency of the clock and the activities of capacitive nodes. In addition, in the traditional flip-flop-based systems, one of the two clock transitions does nothing but only makes some capacitive nodes charged or discharged. As a consequence, one way to reduce the power dissipation in logic circuits is to use dual-edge triggered flip-flops in which both edges carry out some useful operations. This also reduces the system frequency accordingly. Another way to reduce the power

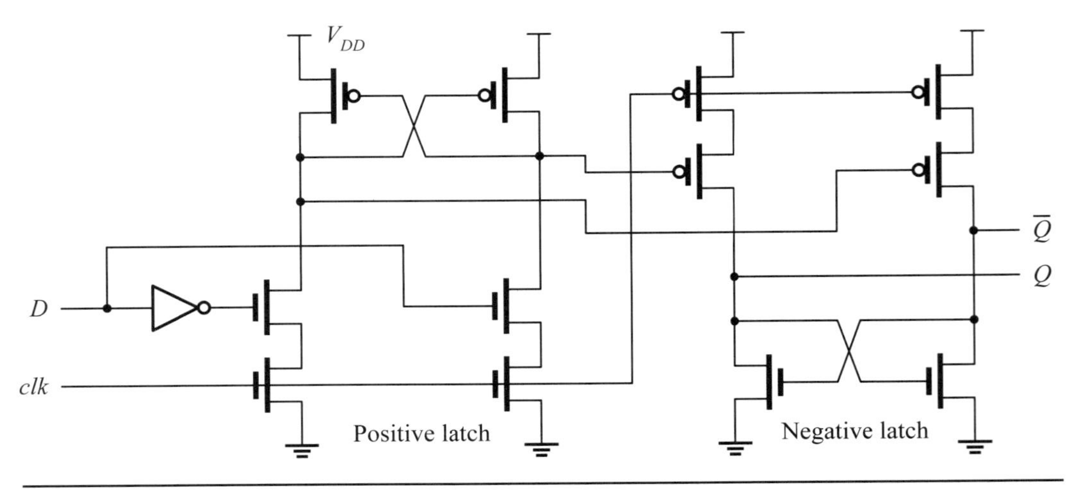

Figure 9.24: A CVSL-based D-type flip-flop.

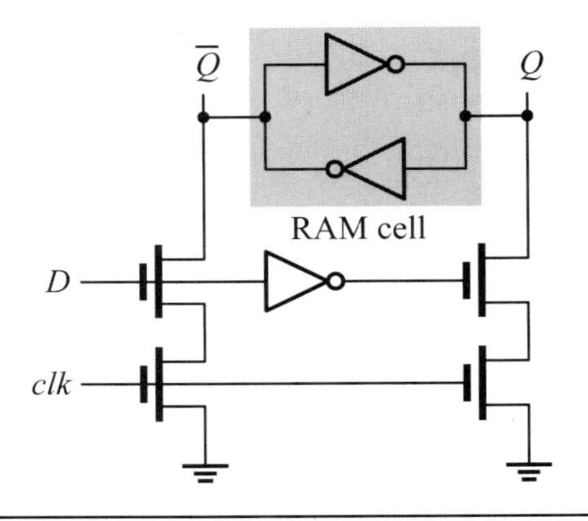

Figure 9.25: The RAM-based D-type flip-flop.

dissipation and speed up the circuit operation is by using pulsed latches, which will be discussed later.

A dual-edge triggered flip-flop based on the cascade of two edge-triggered half circuits is shown in Figure 9.26. Here, both edge-triggered half circuits are constructed by using a footed sense amplifier. As shown in Figures 9.26(a) and (b), the operation of both positive and negative edge-triggered half circuits are controlled by a clock clk. As the clock clk is low, the positive edge-triggered half circuit is disabled, while the negative edge-triggered half circuit is enabled. Both outputs Q_1 and $\bar{Q}_1$ of the positive edge-triggered half circuit are promoted to V_{DD} through the two clock-controlled p-type metal-oxide-semiconductor (pMOS) transistors. As the clock clk is high, the negative edge-triggered half circuit is disabled, whereas the positive edge-triggered half circuit is enabled. Both outputs Q_2 and $\bar{Q}_2$ of the negative edge-triggered half circuit are discharged to the ground level through the two clock-controlled nMOS transistors. The related timing diagram is shown in Figure 9.26(c).

During the rising edge of the clock clk, the positive edge-triggered half circuit is enabled. Depending on the value of the input D, either of the two pMOS transistors is turned on before the clock-controlled pMOS transistor with its source connected to V_{DD} switches off. Hence, either output Q_1 or $\bar{Q}_1$ will remain charged to V_{DD} while the

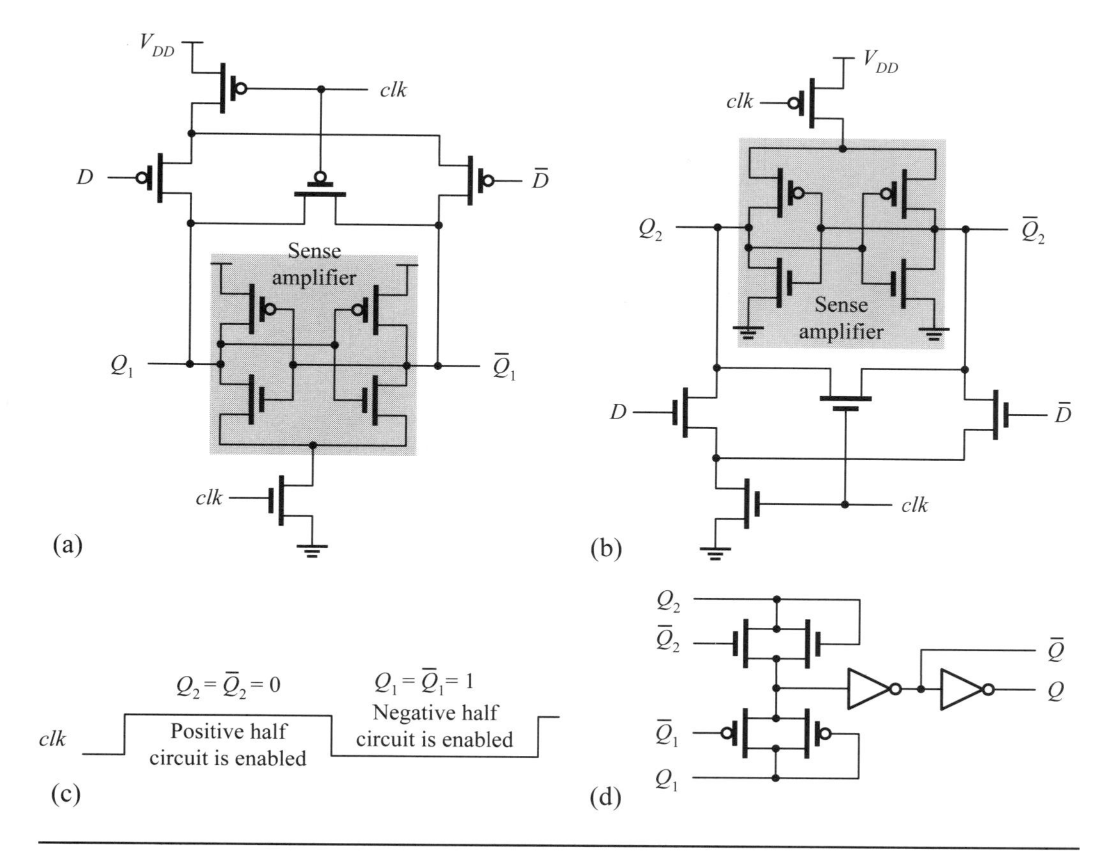

Figure 9.26: A SA-based dual-edge triggered D-type static flip-flop: (a) positive edge-triggered half circuit; (b) negative edge-triggered half circuit; (c) timing diagram; (d) output circuit.

other is discharged to the ground level. Similarly, during the falling edge of the clock clk, the negative edge-triggered half circuit is enabled. Depending on the value of the input D, either of the two nMOS transistors is turned on before the clock-controlled nMOS transistor with its source grounded switches off. Hence, one of output Q_2 and $\bar{Q}_2$ will remain at the ground level while the other is charged to V_{DD}.

From the above operations, when the clock clk is low, both outputs Q_1 and $\bar{Q}_1$ are high. The final value is the value of Q_2 and its complement $\bar{Q}_2$. Similarly, when the clock clk is high, both outputs Q_2 and $\bar{Q}_2$ are low. The final value is the value of Q_1 and its complement $\bar{Q}_1$. By combining these outputs through an appropriate logic circuit, as shown in Figure 9.26(d), the result is a dual-edge triggered flip-flop.

Review Questions

Q9-25. Explain why the preset and clear inputs are usually active-low.

Q9-26. Design a positive edge-triggered CVSL-based D-type flip-flop.

Q9-27. Explain why in Figure 9.22 the reset signal $\overline{preset}$ must also be applied to the master D-type latch.

Q9-28. Explain why the positive/negative half circuit in Figure 9.26(a) is not a latch.

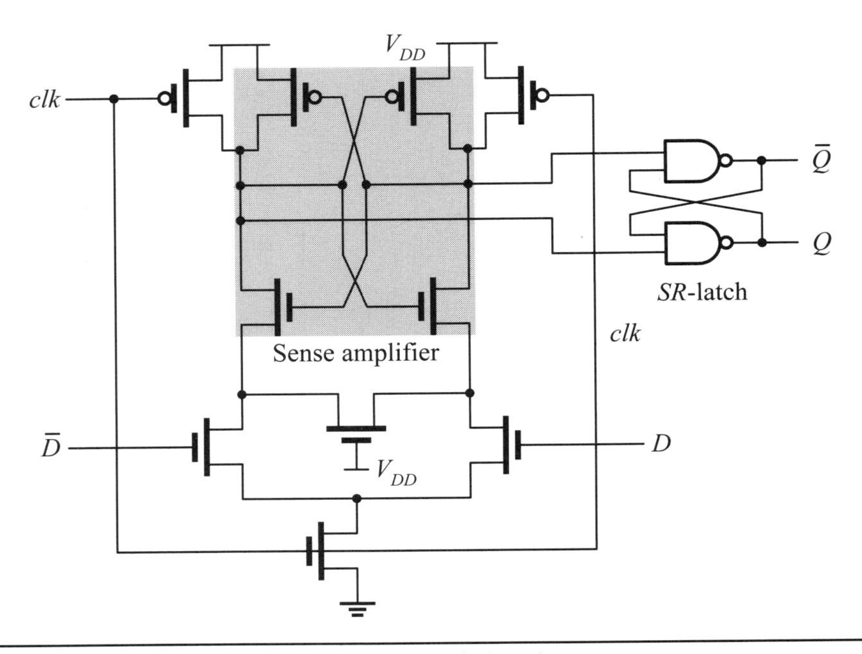

Figure 9.27: The SA-based *D*-type flip-flop (SAFF).

Q9-29. Explain the operation of the output combination circuit in Figure 9.26(c).

9.2.1.3 Differential (Dual-Rail) Flip-Flops. *Differential flip-flops* accept both true and complementary inputs and generate both true and complementary outputs; namely, they are dual-rail storage elements. In this subsection, we introduce two differential flip-flops: one is a simple differential flip-flop and the other is a differential flip-flop with a self-resetting capability. The essential idea of these two flip-flops is that they employ a pulse generator to produce a short pulse to sample the input data to avoid the transparent property encountered in latches. The sampled data are then stored in a slave latch. Each of these are described in more detail.

Sense-amplifier-based D-type flip-flops. A *sense-amplifier-based D-type flip-flop* (SAFF) is shown in Figure 9.27. Like a master-slave flip-flop, the SAFF also comprises two stages: a sense amplifier (SA) and a NAND-based *SR* latch. The SA is a variant of SA-based dynamic logic shown in Figure 8.42 and serves as the pulse generator; the NAND-based *SR* latch is used to hold the sampled value.

The operation of the SAFF is as follows. When the clock *clk* is low, both outputs of SA are charged to V_{DD} and the *SR* latch holds the previous value. During the positive edge of the clock *clk*, the SA is activated and captures the differential value ΔV_{in} between the inputs *D* and $\bar{D}$, passing the stored value in the SA to the complementary outputs *Q* and $\bar{Q}$ of the *SR* latch. Since the SA has the capability of detecting an input voltage difference, ΔV_{in}, as small as 100 mV, the two signal lines, *D* and $\bar{D}$, of the dual-rail pair must be routed adjacent to each other, to decrease the differential-mode noise.

The advantages of SAFF are as follows. First, the SAFF is comparable to the conventional *D*-type flip-flop in terms of area and power dissipation. Second, the SAFF can operate in a true single-phase clock and, hence, requires no additional inverter to generate a local clock with the opposite polarity. However, the major disadvantage of

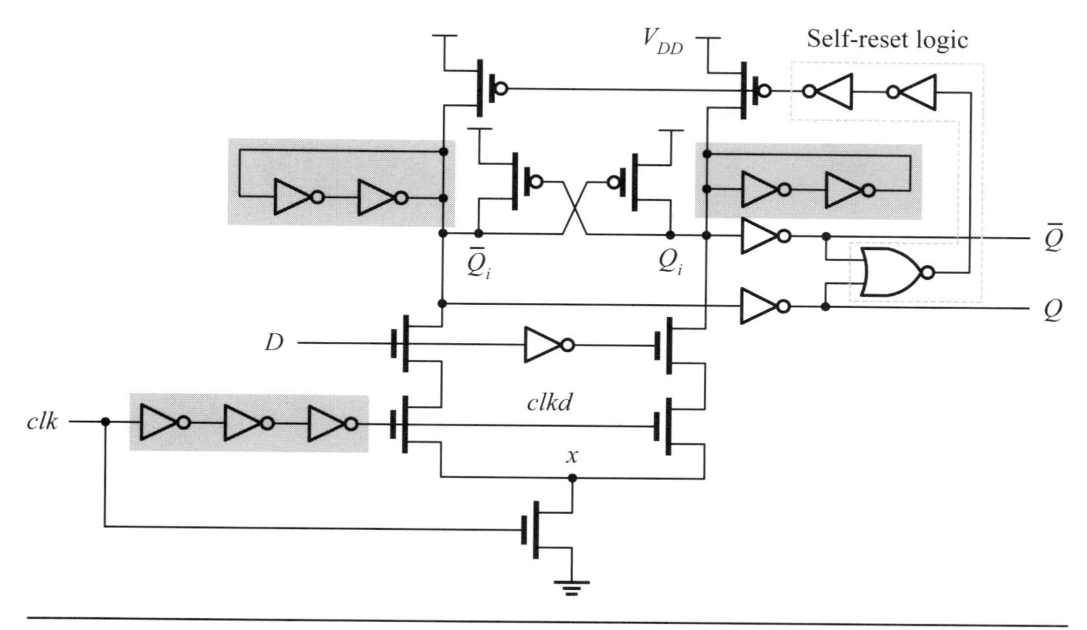

Figure 9.28: A differential D-type flip-flop with self-resetting logic.

the SAFF is the clock-to-Q delay (from the clock clk to output Q) (t_q) is longer than that of the conventional D-type flip-flop due to the sense-amplifier requiring additional time to amplify the low-swing differential input values.

Differential D-type flip-flop with self-resetting logic. Another differential flip-flop has the structure shown in Figure 9.28. This differential flip-flop comprises a *dual-rail pulsed edge-triggered latch* (ETL), two cross-coupled pMOS transistors, two static latches, and a *self-resetting logic*. The dual-rail pulsed ETL employs a precharged node x and a one-shot circuit with a pulse width equal to three-inverter delay to provide a transparent period for sampling the input data D and its complement $\bar{D}$ into the latch. The two cross-coupled pMOS transistors improve the noise immunity, and the two static latches are used to avoid the dynamic features of internal nodes.

The operation of the differential D-type flip-flop with self-resetting logic is as follows. During the period of quiescence, both outputs Q and $\bar{Q}$ are low, namely, $Q = \bar{Q} = 0$. As the clock clk rises from low to high, either of the output nodes, Q or $\bar{Q}$, will go to high, depending on the input value of D. The assertion of one of the outputs activates a delay chain, being composed of a NOR gate and two inverters, that eventually resets the flip-flop to its precharged, quiescent state. As a consequence, the differential flip-flop only produces pulsed outputs.

The differential D-type flip-flop with self-resetting logic may act as a buffer to interface static and dynamic logic. It converts single-ended signals to dual-rail monotonic signals suitable for the applications of dynamic logic. Moreover, it may act as a flip-flop for a brief period determined by the reset path, namely, the delay chain associated with the output. However, in designing such a circuit, care must be taken to ensure that the propagation delay of this reset path is long enough so that the flip-flop data can propagate to and be used by the succeeding stage of dynamic logic before it is reset.

Review Questions

Q9-30. What is the meaning of dual-rail flip-flop?

Q9-31. What is the SAFF?

Q9-32. What is the ETL?

Q9-33. What is the purpose of the three-inverter chain of Figure 9.28?

Q9-34. What is the function of the self-reset logic of Figure 9.28?

9.2.2 Dynamic Memory Elements

Recall that a dynamic memory (storage) element is a device that stores information in the form of a charge rather than in circuit states used in static memory elements. To facilitate this, a capacitor or a capacitive node is usually used. Unfortunately, the information stored in such a capacitive node will get lost with time because of the inevitable leakage currents encountered in the semiconductor. Consequently, to keep the information stored in a dynamic node intact, the dynamic node must be refreshed periodically. In this section, we describe in more detail some common dynamic memory elements, including latches and flip-flops.

9.2.2.1 Dynamic Latches As stated, a latch is a bistable device capable of being accessed externally and has a transparent property; namely, its output follows the input when it is enabled. In CMOS technology, there are many ways that can be used to construct a dynamic latch. Next, we only address a few examples: the TG-based latch, C^2MOS latch,[2] and TSPC latch.

TG-based latches. As indicated in Figure 9.29(a), a TG switch controlled by a clock *clk* is employed to sample the input voltage V_{in}. As the clock *clk* goes high, the TG switch is turned on and the capacitor C_x is charged/discharged to the input voltage V_{in}. This voltage value passes through the inverter and reveals an inverted voltage value V_{out} at the output. As the clock *clk* goes low, the TG switch is turned off. The node x is isolated from the input and holds the voltage value before the TG switch is turned off. In summary, the dynamic latch is in the *transparent mode* if the clock *clk* is high and is in the *opaque (hold) mode* if the clock *clk* is low.

In practice, the TG-based dynamic latch is usually implemented as the circuit depicted in Figure 9.29(b). This circuit is indeed the C^2MOS latch introduced earlier. It should be noted that the clock *clk* and its complement $\overline{clk}$ must be applied to the two inner transistors as illustrated in Figure 9.29(b); otherwise, the charge-sharing effect may be encountered whenever the clock *clk* goes low, as illustrated in the following example.

Example 9-7: (An improper realization of a TG-based dynamic latch.)

Figure 9.29(c) shows an improper realization of a TG-based dynamic latch. This realization encounters the charge-sharing effect when the clock *clk* is low and the input voltage V_{in} changes. The reasons are as follows. As the clock *clk* is high and $V_{in} = 0$, capacitors C_x and C_a charge to V_{DD}. At the same time, the capacitor C_b discharges to 0 V. Now, if the clock *clk* goes low and the input voltage V_{in} is V_{DD}, because of the conduction of the nMOS transistor, capacitors C_x and C_b redistribute their charge to each other, thereby causing the output voltage V_{out} to drop below V_{DD}. ■

[2] C^2MOS is the acronym of clocked CMOS.

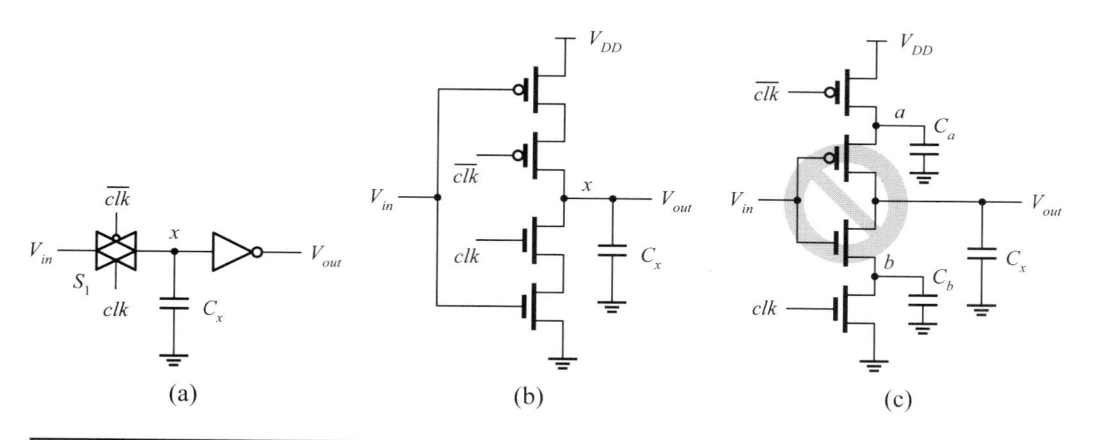

Figure 9.29: A TG-based dynamic latch: (a) logic diagram; (b) a CMOS realization; (c) an improper realization.

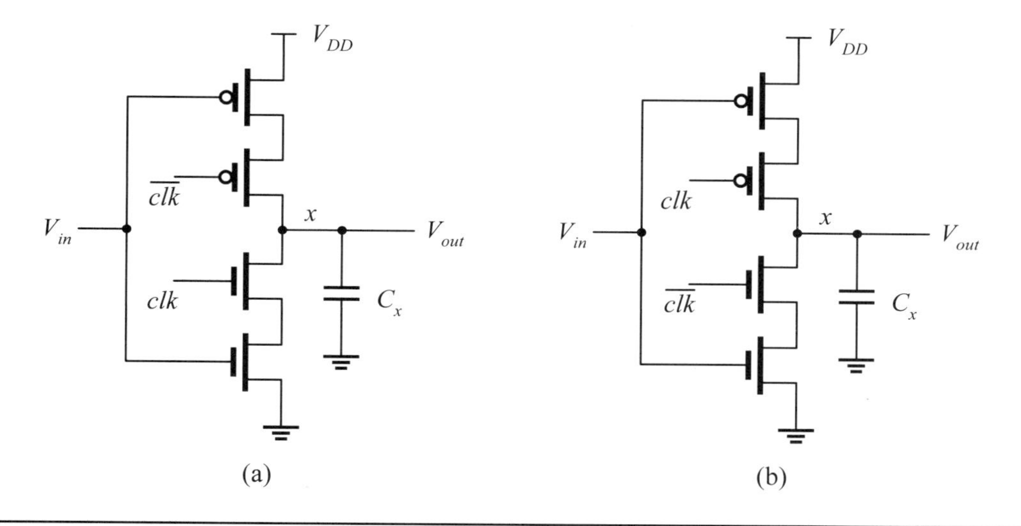

Figure 9.30: (a) Positive and (b) negative C^{2}MOS dynamic latches.

C^{2}MOS latches. As stated, a C^{2}MOS latch is virtually equivalent to the TG-based dynamic latch. Both positive and negative C^{2}MOS latches are exhibited in Figure 9.30. Fundamentally, these two latches have the same circuit structure except for the placement of the clock clk and its complement $\overline{clk}$. In the positive latch, the clock clk is applied to the nMOS transistor and the complementary clock $\overline{clk}$ is applied to the pMOS transistor. The reverse situation is applied to the negative latch. As a result, the positive latch is in the transparent mode, while the negative latch is in the opaque mode when the clock clk is high; the positive latch is in the opaque mode, while the negative latch is in the transparent mode when the clock clk is low.

TSPC latches. In a TSPC logic circuit, only a single-phase clock clk is required. TSPC dynamic latches, including both positive and negative latches, are presented in Figure 9.31. Fundamentally, these two latches have the same circuit structure except that the clock-controlled devices are different. Both latches are constructed by cascading two clock-controlled inverters into a two-stage inverter chain. In the positive

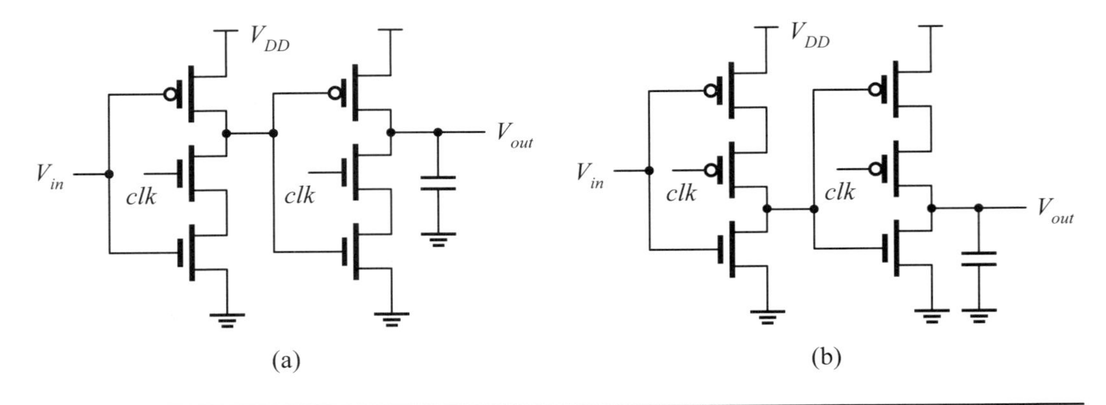

Figure 9.31: (a) Positive and (b) negative TSPC dynamic latches.

latch, the clock-controlled devices are nMOS transistors, while in the negative latch, the clock-controlled devices are pMOS transistors.

The operation of a TSPC dynamic latch can be illustrated by using the positive latch as an example. When the clock *clk* is high, both clock-controlled nMOS transistors are on. The output follows the input; namely, the circuit is in the transparent mode. When the clock *clk* is low, both clock-controlled nMOS transistors are off, and hence, the two-stage inverter chain is disabled. No signal is allowed to propagate from the input to the output of the latch. The output retains its previous value; namely, the circuit is in hold mode.

Compared to the C^2MOS latch, a TSPC latch has a longer propagation delay because the input data pass through more transistors to reach the output. A careful study found that TSPC latches occupy more area, present more clock loading, consume more power, and are slower than traditional dynamic latches.

9.2.2.2 Dynamic Flip-Flops Like a static flip-flop, a dynamic flip-flop is also constructed by cascading two latches: a positive and a negative latch. A few such examples are addressed.

TG-based master-slave flip-flops. Maybe the simplest dynamic flip-flop in CMOS is the TG-based master-slave flip-flop, which is constructed by cascading two TG-based dynamic latches, as shown in Figure 9.32(a). Because a negative dynamic latch is followed by a positive latch, the resulting flip-flop is a positive edge-triggered type.

The detailed operations are illustrated in Figures 9.32(b) and (c). As the clock *clk* is low, TG switches S_1 and S_2 are turned on and off, respectively. The input data are sampled and stored on the storage node x, and the output still retains its previous value; namely, the master latch is in the transparent mode, while the slave latch is in the hold mode. On the rising edge of the clock *clk*, TG switches S_1 and S_2 are turned off and on, respectively. The input value sampled right before the rising edge of the clock *clk* is propagated to the output node through the two inverters and is stored on node y. As a consequence, the master latch is in the hold mode, whereas the slave latch is in the transparent mode.

The correct operations of the TG-based dynamic D-type flip-flop rely on the proper scheduling of both master and slave latches. As can be seen from Figure 9.32, to make the flip-flop work correctly, both clock *clk* and its complement $\overline{clk}$ must not be overlapped. Otherwise, the flip-flop could not work without the transparent property.

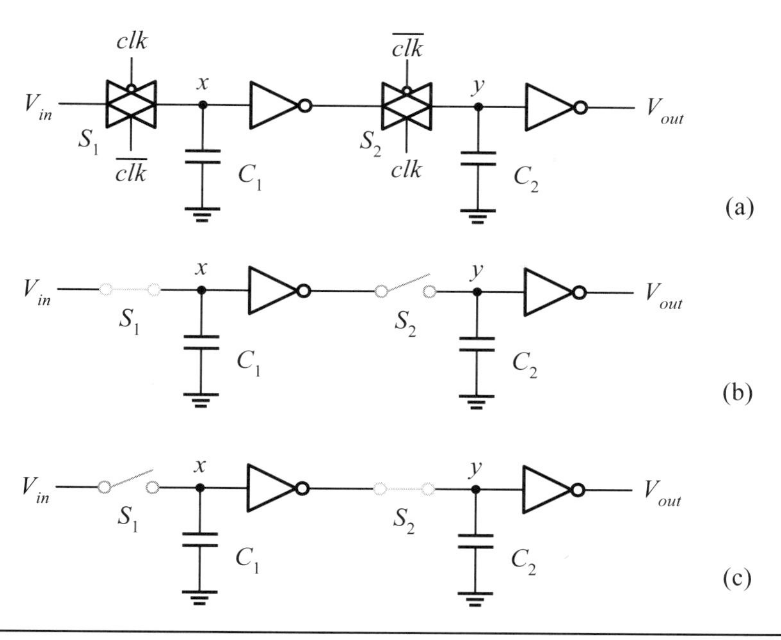

Figure 9.32: The (a) circuit of a TG-based dynamic *D*-type flip-flop and its operations when (b) $clk = 0$ and (c) $clk = 1$.

An illustration is portrayed in Figure 9.33(a), in which both TG switches S_1 and S_2 are turned on due to either of the following two causes: clock skew, as well as *slow-rise and slow-fall times* of the clock clk and its complement $\overline{clk}$. The clock skew is the phenomenon of two clocks arriving at two latches or flip-flops at different times; namely, they have a difference in time. This time difference is defined as the skew t_{skew}. As shown in Figure 9.33(b), the skew between the clock clk and its complement $\overline{clk}$ may cause a nonzero overlapping between them. The slow-rise and slow-fall times of the clock clk and its complement $\overline{clk}$ may also cause a nonzero overlapping between them. During the period of nonzero overlapping, both switches S_1 and S_2 are turned on and the flip-flop is transparent between its output and input. As a consequence, the output follows the input. This is an undesired situation and should be avoided when using flip-flops to design a digital system.

C²MOS master-slave flip-flops. Like other CMOS flip-flops, a C²MOS flip-flop is also built from the cascade of two latches. To form a master-slave positive edge-triggered flip-flop, we need to cascade a negative C²MOS dynamic latch as the master latch with a positive C²MOS dynamic latch as the slave latch, as shown in Figure 9.34.

The operation of a C²MOS dynamic *D*-type flip-flop is as follows. As the clock clk is low, the master latch turns on and serves as an inverter to sample the input data and store the inverted value at the node x. The slave latch at this time is disabled and its output node is floated. Hence, the master latch is in the transparent mode, while the slave latch is in the hold mode. As the clock clk is high, the master latch turns off and makes it output node x floated, thereby retaining its previously sampled value. The slave latch at this time is enabled and serves as an inverter, which in turn presents the value at node x at its output node, inverted one more time. That is, the master latch is in the hold mode, while the slave latch is in the transparent mode.

TSPC edge-triggered flip-flops. Two TSPC latches with different polarities can be cascaded into an edge-triggered flip-flop, namely, a master-slave flip-flop. An

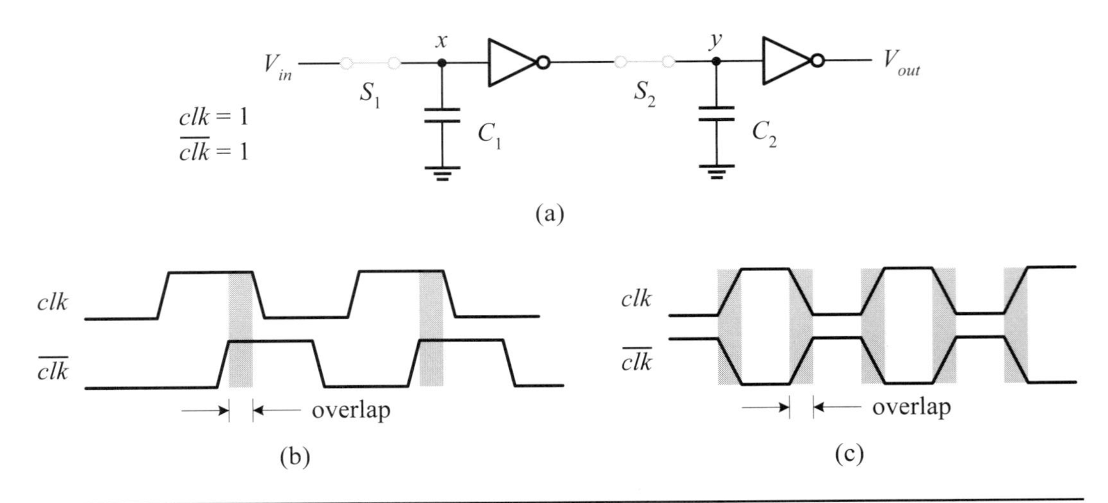

Figure 9.33: The failures of TG-based dynamic D-type flip-flop: (a) circuit showing the failures; (b) clock skew; (c) slow-rise and slow-fall times.

example of a TSPC positive edge-triggered flip-flop is shown in Figure 9.35, where the master latch is a negative latch, while the slave latch is a positive latch. The operation of this flip-flop is as follows. As the clock clk is low, the master (negative) latch is in the transparent mode, while the slave (positive) latch is in the hold mode. Consequently, the master latch samples the input data and stores it at its output node x. At this time, the output of the flip-flop is floated and retains its previously stored value. As the clock clk is high, the master latch is in the hold mode, while the slave latch is in the transparent mode. Hence, the value at node x of the master latch transfers to the output node y of the flip-flop.

Although a TSPC flip-flop only needs a single-phase clock, it requires 12 transistors. An improved version only needing 9 transistors is depicted in Figure 9.36. This flip-flop consists of three instead of four clock-controlled inverters, which are cascaded together. The operation of the TSPC flip-flop is as follows. When the clock clk is low, the input inverter samples the input data D while the second inverter is in the precharge mode. The third inverter is in the hold mode. Hence, the output node Q retains its previously stored value. When the clock clk is high, the first inverter is in the hold mode, the second inverter evaluates, and the third inverter is on and in the transparent mode. The value on node y is passed to the output Q through an inverter.

Overlap-based flip-flops with embedded logic. An overlap-based positive edge-triggered D-type flip-flop is shown in Figure 9.37. It combines a logic stage with an n-latch. The logic stage is an nMOS dynamic logic with the addition of an nMOS transistor controlled by the complementary clock $\overline{clk}$ so that the nMOS dynamic logic can only evaluate its function value during the overlap period between clock clk and its complement $\overline{clk}$. Because of this, this circuit is called an *overlap-based flip-flop*. The clock clk and its complement $\overline{clk}$ are used to latch and hold the data. To make the circuit properly operated, the clk must lead the $\overline{clk}$. The resulting circuit is a D-type flip-flop if the nMOS block is only an nMOS transistor.

The overlap-based flip-flop works as follows. During the evaluation phase, clock clk and its complement $\overline{clk}$ are overlapped, that is, 1-1 overlap. All clock-controlled nMOS transistors are on. The n-latch simply behaves as an inverter, and the input data can pass through an nMOS block and reach the output Q. During the precharge

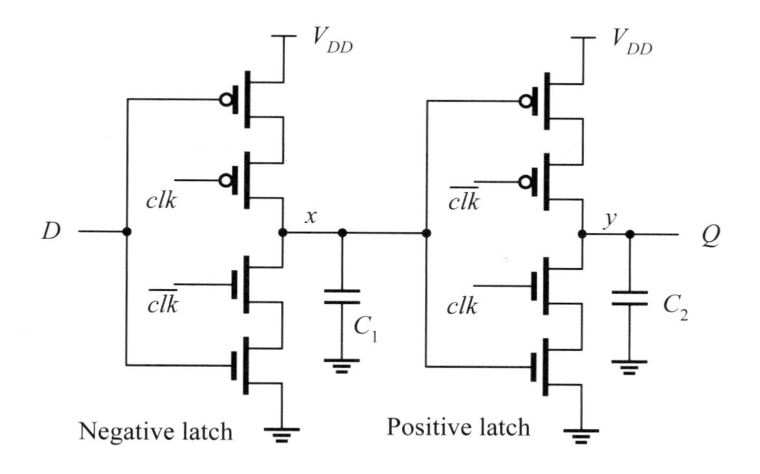

Figure 9.34: A C^2MOS dynamic D-type flip-flop.

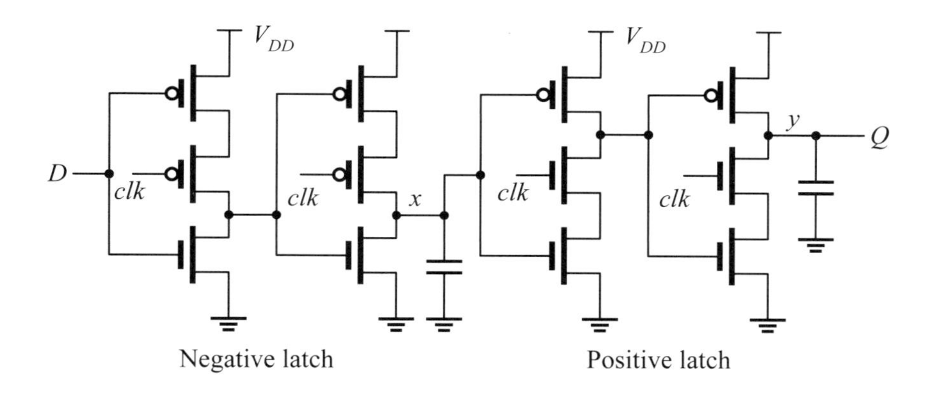

Figure 9.35: A TSPC dynamic D-type flip-flop.

mode, the precharge pMOS transistor is turned on and precharges node x to V_{DD}. The clock clk and its complement $\overline{clk}$ can be either a 1-0 sequence (that is, $clk = 1$ and $\overline{clk} = 0$) or a 0-ϕ sequence (that is, $clk = 0$ and $\overline{clk} = \phi$.) In the former, node x stores the evaluated value, and this value is inverted and passed to the output Q. In the latter, node x is disconnected from the output node and isolated with the nMOS dynamic logic if the complement clock $\overline{clk}$ is also equal to 0.

The major design issue in overlap-based flip-flop is the overlap period between the clock clk and its complement $\overline{clk}$. The overlap period should be long enough for the nMOS dynamic logic to evaluate its function and discharge the parasitic capacitance C_x. It should also be smaller than the minimum time needed to charge the output capacitance C_L so that the input data cannot pass through to the output during the evaluation phase.

Dual-edge triggered flip-flops. In dynamic logic, it is also possible to construct a dual-edge triggered flip-flop. To illustrate this, we use C^2MOS latches as an example.

A C^2MOS-based dual-edge triggered flip-flop is illustrated in Figure 9.38. It is simply to connect two edge-triggered flip-flops in parallel. One is positive edge-triggered and the other is negative edge-triggered. The input stages of both edge-triggered

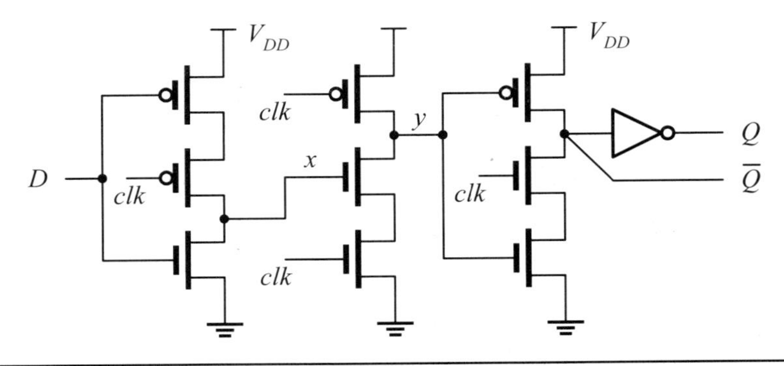

Figure 9.36: The 9-transistor TSPC positive edge-triggered dynamic D-type flip-flop.

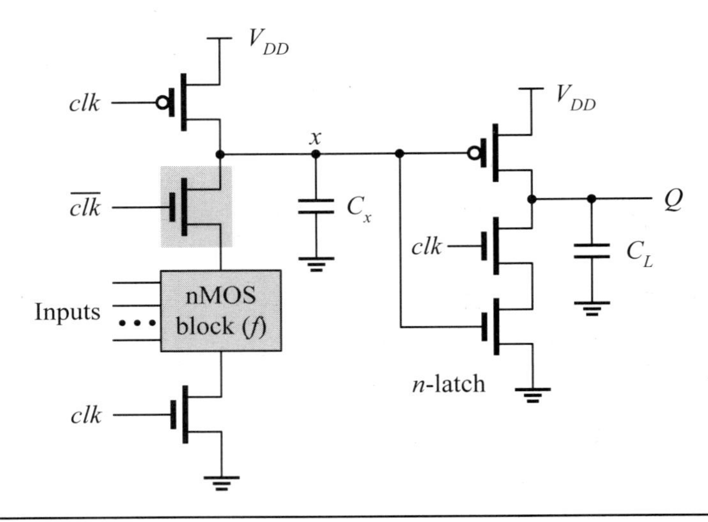

Figure 9.37: The general block diagram of overlap-based flip-flop with embedded logic.

flip-flops are combined into one stage, but the clock-controlled nMOS and pMOS transistors are left as two individual sets. One is conducted when the clock clk is high, while the other is conducted when the clock clk is low. Hence, the resulting circuit is a dual-edge triggered flip-flop.

■ Review Questions

Q9-35. What is the major feature that distinguishes a static memory element from a dynamic one?

Q9-36. Explain why the implementation shown in Figure 9.29(c) is prone to the charge-sharing effect.

Q9-37. How would you construct a TG-based negative edge-triggered dynamic D-type flip-flop?

Q9-38. Explain why clock skew may cause the TG-based dynamic D-type flip-flop to have the transparent property.

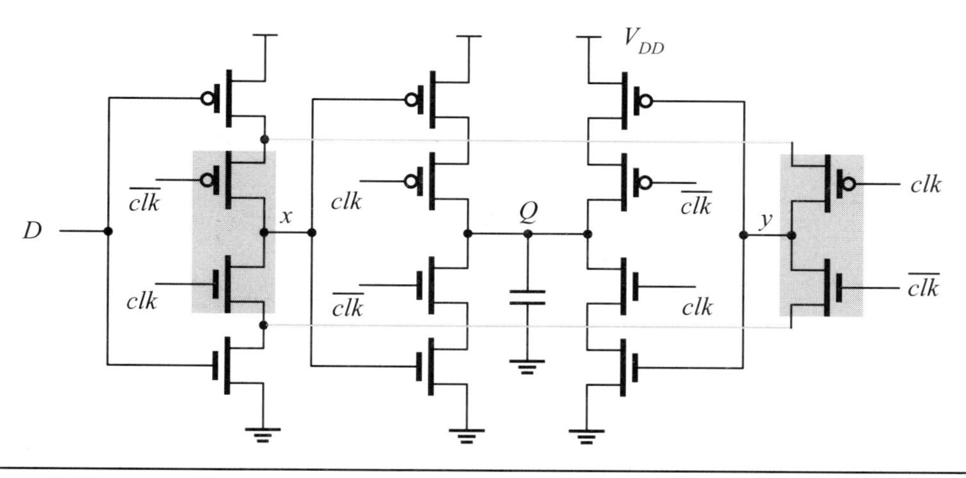

Figure 9.38: The C^{2}MOS dual-edge triggered dynamic D-type flip-flop.

9.2.3 Pulsed Latches

Recall that a latch can also have no transparent property if the pulse width of its enable signal is just wide enough to sample the input data reliably. Based on this idea, a new type of latch is resulted and referred to as a *pulsed latch*. We introduce two representative examples. One is a simple pulsed latch and the other is known as the *Partovi pulsed latch*. Even though these two examples are dynamic, any static latch can also be pulsed.

Basically, a pulsed latch is identical to a transparent latch except that it receives a narrow clock pulse instead of a 50% duty cycle clock. A pulsed latch can be constructed from a C^{2}MOS latch by simply adding a *clock chopper* to reshape the clock into a short pulse, with a width of t_{pw}, as shown in Figure 9.39. This clock pulse should be designed in a way such that it is just wide enough to sample the input data reliably to remove the transparent property. This pulse and its complement are then applied to the C^{2}MOS latch. The pulse width can be controlled by the number of inverters and must be ensured by simulation.

A practical pulsed latch is known as *Partovi pulsed latch* or basic *hybrid-latch flip-flop* (HLFF), as shown in Figure 9.40. The operation of the Partovi pulsed latch is as follows. When the clock clk is low, the delayed clock $clkd$ is initially high and node x is set to high. The output node Q floats its previously stored value. Hence, the latch is in the hold mode. When the clock clk is high, both clock clk and delayed clock $clkd$ are briefly high before the delayed clock $clkd$ falls to low. During this period, the latch is transparent and hence the input data are sampled and stored in the output node Q. The duration of both clock clk and delayed clock $clkd$ at high, denoted as t_{pw}, can be adjusted by increasing or decreasing the number of inverters, just like the case of the simple pulsed latch.

9.2.4 Semidynamic Flip-Flops

The semidynamic flip-flop (SDFF) shown in Figure 9.41 is also called the *Klass semi-dynamic flip-flop*. It combines the features of a dynamic pulsed latch and a static latch. The pulsed latch is used as the input stage to provide the edge-triggered nature

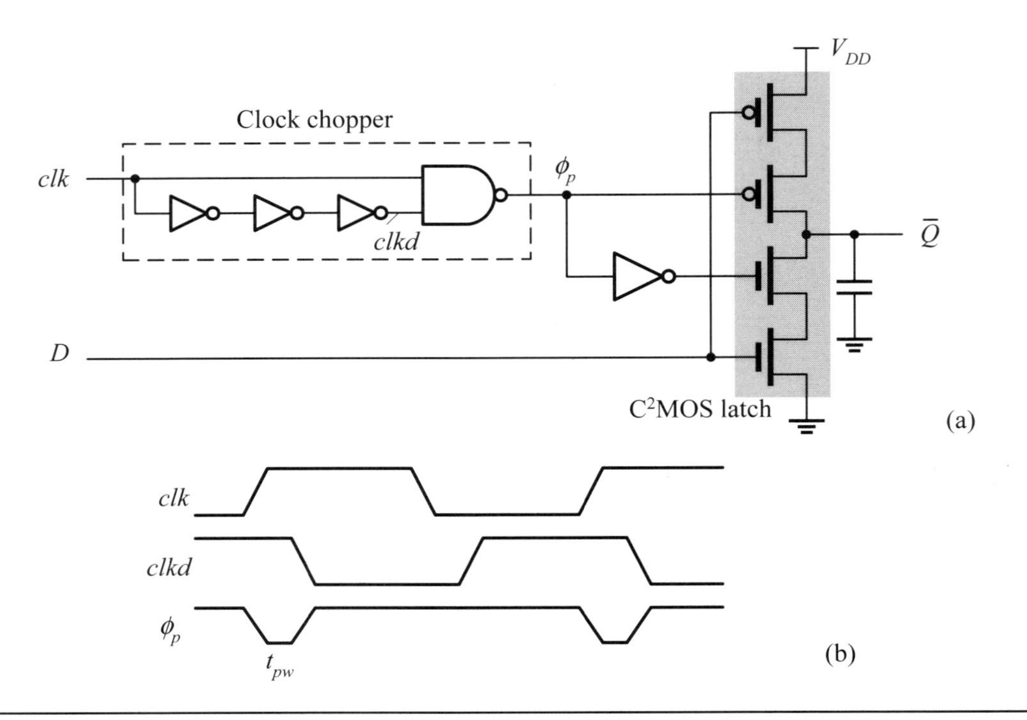

Figure 9.39: The (a) circuit and (b) timing diagram of a simple pulsed latch.

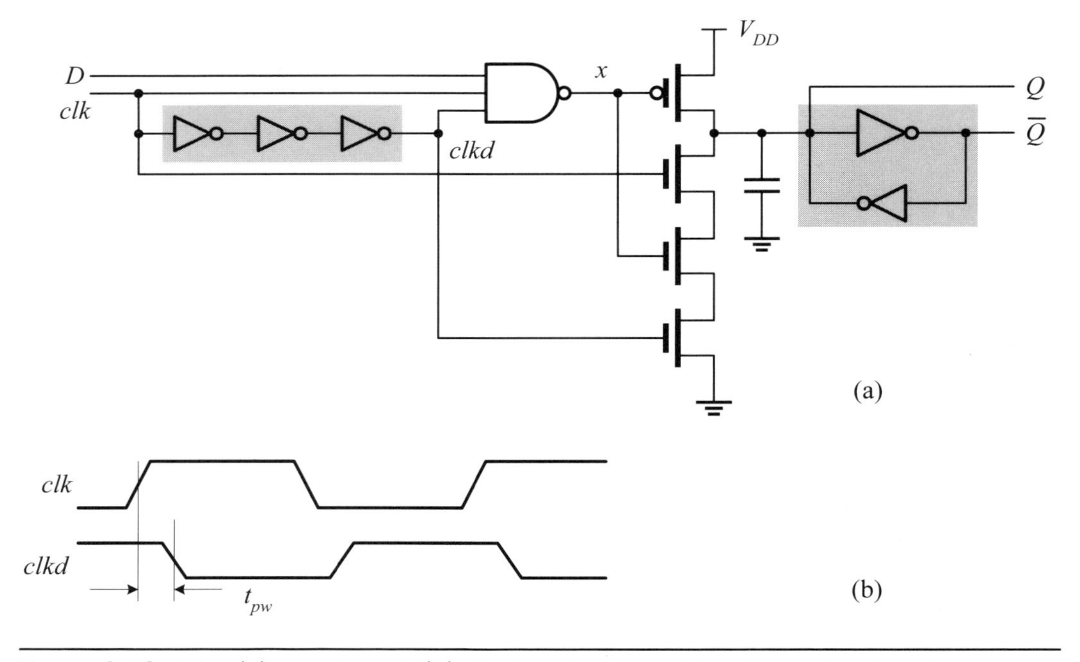

Figure 9.40: The (a) circuit and (b) timing diagram of the Partovi pulsed latch.

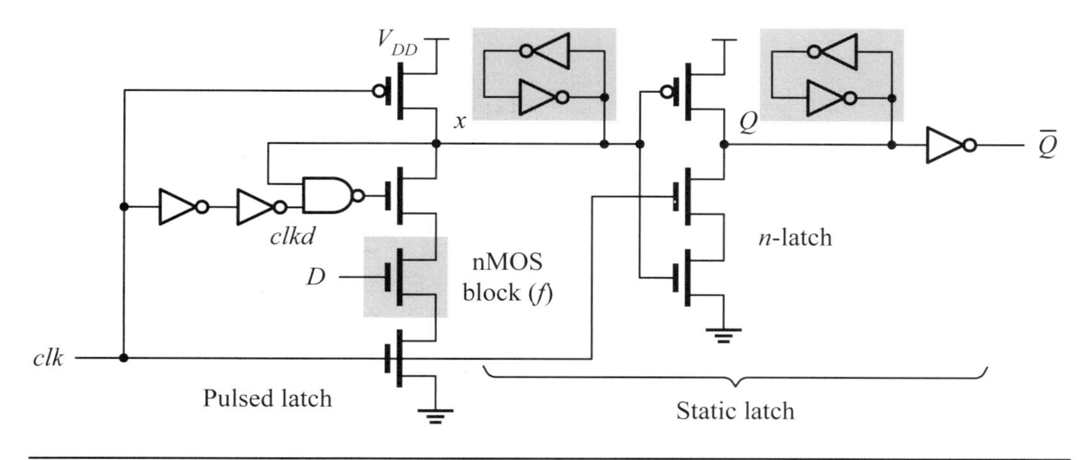

Figure 9.41: The Klass semidynamic D-type flip-flop (SDFF).

and two static latches are employed to stabilize the sampled data to avoid the dynamic nodes at x and Q.

The operation of the SDFF is as follows. When the clock *clk* is low, the flip-flop enters the precharge phase. Node x is precharged to high, and the output static latch holds the previous logic levels of Q and $\bar{Q}$. The top nMOS transistor of the input stage is on because the low-level clock *clk* causes the output of the NAND gate to rise high. When the clock *clk* rises high, the input stage evaluates and the output dynamic inverter is transparent. If the input D is low, node x remains high and the top nMOS transistor turns off. If the input D is high, the top nMOS transistor remains on for three gate delays to complete the transition.

■ Review Questions

Q9-39. Explain the principle of a pulsed latch.

Q9-40. Explain the operation of the Partovi pulsed latch.

Q9-41. Explain the operation of the semidynamic flip-flop (SDFF).

9.2.5 Low-Power Flip-Flops

Remember that for those systems with several different voltage domains, level shifters are generally required to interface the circuits associated with different voltage domains from low to high voltage. In addition, for those systems needing fast resumption of operation after wake-up, the current states of operation need to be saved before going to the sleep mode. For this purpose, we use *retention registers* to save small amounts of information and static random access memory (SRAM) for saving median to large amounts of information. We first deal with a low-power flip-flop with a level shifter and then consider two examples of retention registers. Both of these are based on the conventional D-type flip-flop.

9.2.5.1 Low-Power Flip-Flops Recall that circuits with level shifters are required to interface circuits with different voltage domains from low to high voltage. For such a purpose, level shifters given in Figure 6.17 can be used. Nevertheless, for those circuits with a registered output, it is more effective to combine the level shifter into

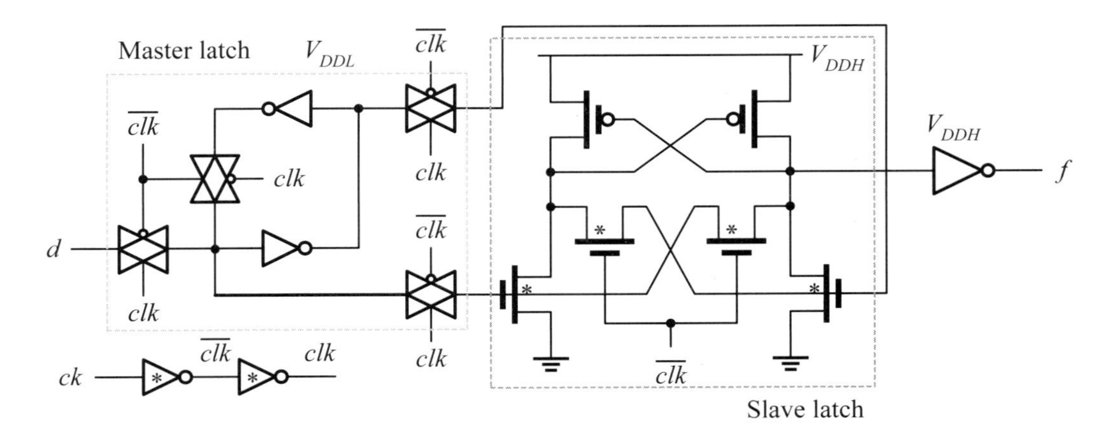

Figure 9.42: A low-power D-type negative edge-triggered flip-flop with level shifter. (Transistors/inverters labeled with (*) indicate low-V_T devices.)

a D-type flip-flop rather than simply cascade the level shifter with the output of a D-type flip-flop. An example is revealed in Figure 9.42. It is composed of a D-type master latch with a CVSL slave latch with a level shifter. Such a circuit is usually referred to as a *slave-latch level-shifting (SLLS) flip-flop.* It is worth noting that the low-swing outputs should only drive nMOS transistors or pass transistors since they cannot fully turn off the pMOS transistors.

The operation of the SLLS flip-flop is as follows. The master latch samples the input data as the clock clk is low. The output of the master latch is then captured by the slave latch as the clock clk becomes low. The cross-coupled nMOS transistors are employed to staticize the slave latch.

9.2.5.2 Retention Registers The simplest retention register is to adapt the conventional D-type flip-flop to provide the low-leakage mode of the slave latch. As illustrated in Figure 9.43, the master latch captures the input data when the clock clk is high. It uses low-V_T MOS transistors to improve the performance. The slave latch captures the output data from the master latch when the clock clk is low. The slave latch uses high-V_T MOS transistors to reduce leakage current during the sleep mode. The output buffer employs low-V_T MOS transistors and is designed with various sizes to provide different drive strengths for different requirements.

A more complicated retention register is depicted in Figure 9.44. It is also built on the basis of a D-type flip-flop. The *retention latch* comprises two high-V_T inverters and a high-V_T TG for the memory circuit, a low-V_T TG switch and a high-V_T TG switch for the switch circuit. The retention (called a *balloon*) latch only changes its states during the transitions between sleep and active modes. Since the retention latch only operates in the sleep mode, it can be designed with the minimum-sized MOS transistors, thereby keeping the area penalty small and the standby power low.

■ Review Questions

Q9-42. Explain the operation of the SLLS flip-flop shown in Figure 9.42.

Q9-43. Explain the operation of the retention register shown in Figure 9.44.

Q9-44. What is the purpose of the retention register?

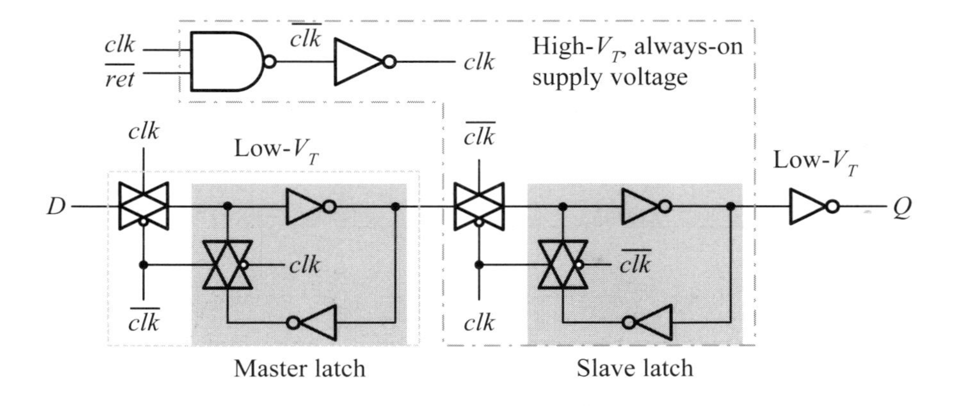

Figure 9.43: A simple D-type negative edge-triggered retention register.

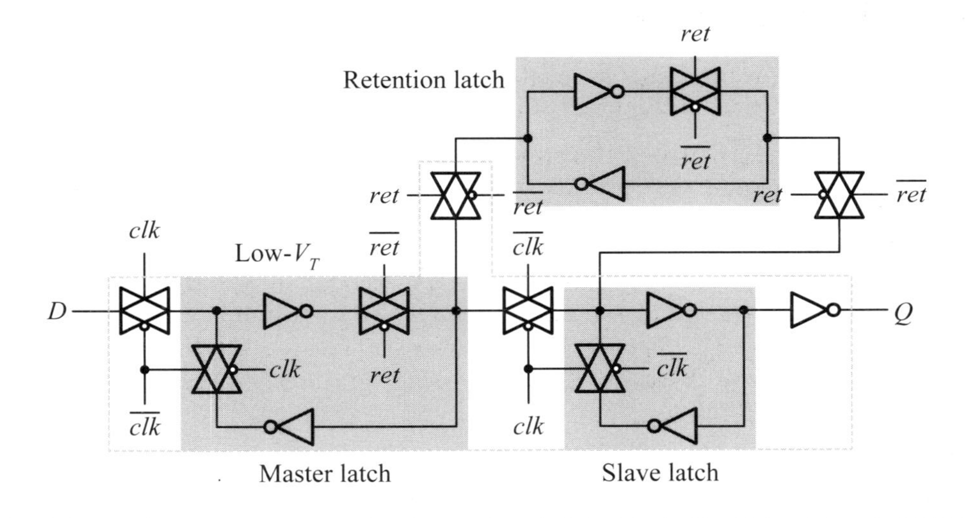

Figure 9.44: A single control balloon register.

Q9-45. What is the retention latch?

9.3 Timing Issues in Clocked Systems

In this section, we deal with the timing issues of flip-flops, latches, and pulsed latches in detail. As stated previously, both flip-flops and pulsed latches are bistable devices without transparent property, while the latches do have the transparent property.

9.3.1 Timing Issues of Flip-Flop Systems

Timing issues of flip-flop-based (FF-based) systems include max-delay and min-delay constraints. The former considers the worst-case delays of both combinational logic and flip-flop, while the latter involves the best-case delays of both combinational logic and flip-flop.

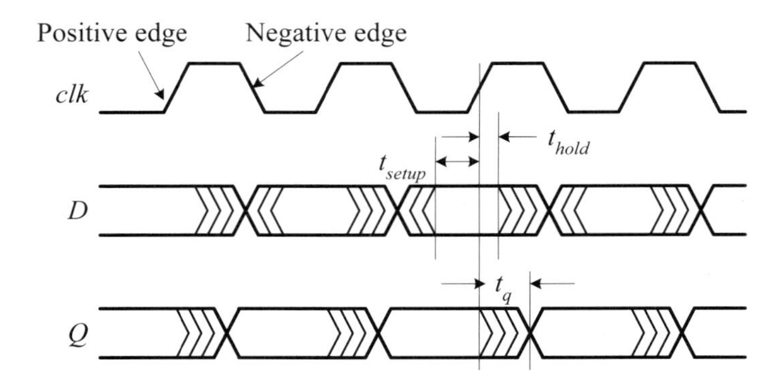

Figure 9.45: The basic timing parameters of flip-flops.

9.3.1.1 Max-Delay Constraint Before proceeding, we have to define some terminology associated with flip-flops when using them to design a digital system. We first define the following three terms associated with a flip-flop, *setup time*, *hold time*, and *clock-to-Q*, as illustrated in Figure 9.45.

- *Setup time* (t_{setup}). The setup time is defined as the amount of time before the clock edge (sampling edge) that the input data (D) of a flip-flop has to be stable. The clock edge may be positive or negative, depending on whether the underlying flip-flop is positive edge-triggered or negative edge-triggered.
- *Hold time* (t_{hold}). The hold time is defined as the amount of time after the clock edge (sampling edge) that the input data (D) of a flip-flop has to remain stable.
- *Clock-to-Q delay* (t_q). The clock-to-Q delay is defined as the maximum propagation delay from the clock edge (sampling edge) to the new value appearing at the output (Q) of a flip-flop.

Both setup time and clock-to-Q delay determine the allowed minimum clock period that has to be used in a system. Figure 9.46(a) shows the max-delay constraint on a path from a flip-flop to the next. The input data pass through the source (first) flip-flop via the combinational logic circuit to reach the input of the destination flip-flop. To quantify the delay constraint on the datapath, assume that the maximum propagation delay of combinational logic is t_{pd}, and the setup time and clock-to-Q delay of D-type flip-flops are t_{setup} and t_q, respectively. Then, according to the timing diagram shown in Figure 9.46(b), the minimum period of the clock can be expressed as follows:

$$T_{clk} \geq t_q + t_{pd} + t_{setup} \tag{9.6}$$

This timing constraint is known as the *max-delay constraint*. When this constraint is violated, the destination flip-flop will miss its setup time and sample the wrong data, even enter the metastable state, as described before. This situation is often called a *setup-time failure* or *max-delay failure*.

The max-delay constraint can be stated in another way. Since both clock-to-Q delay (t_q) and setup time (t_{setup}) are the inherent features of flip-flops and, hence, are the overhead of a system, for a fixed clock period T_{clk} the maximum available time in a clock cycle for the combinational logic to carry out useful computation can be expressed as follows:

$$t_{pd} \leq T_{clk} - (t_q + t_{setup}) \tag{9.7}$$

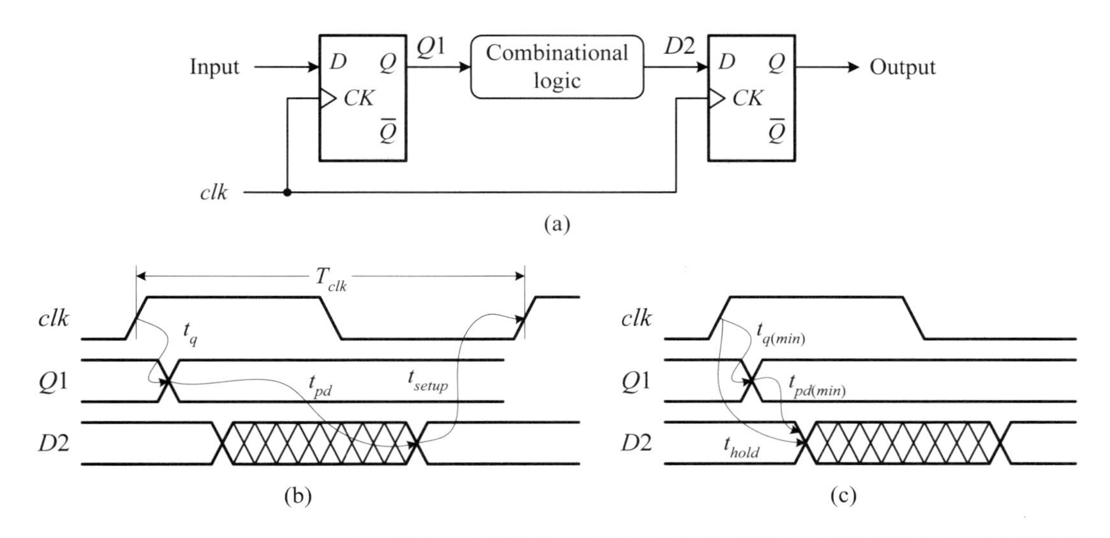

Figure 9.46: The timing constraints of FF-based systems: (a) logic circuit; (b) max-delay constraint; (c) min-delay constraint.

Consequently, the overhead ($t_q + t_{setup}$) of flip-flops must be as small as possible to maximize the available time for the combinational logic to carry out more complicated functions.

9.3.1.2 Min-Delay Constraint The above description is about the max-delay constraint. In practice, in designing a digital system, the *min-delay constraint* must also be considered. When it is violated, the data in the destination flip-flop will be corrupted by the new data, which is supposed to be arrived at later in the current cycle. This situation is called a *race condition*, *hold-time failure*, or *min-delay failure*.

In a practical digital system, a combinational logic circuit and a flip-flop circuit may also exist as the best-case besides the worst-case propagation delay. The reasons are as follows. First, in a general combinational logic circuit, there always exist many different paths with unbalancing propagation delays from the input to the output. Second, the propagation delays of a combinational logic circuit and a flip-flop are determined by both underlying logic gates and input data profiles. Third, because of process, voltage, and temperature (PVT) variations, the propagation delays of a combinational logic circuit and a flip-flop may vary in a broad range.

Based on the above reasons, the best-case propagation delays of both combinational logic and flip-flop are separately defined as the following two terms:

- The *contamination delay* ($t_{pd(min)}$) of a combinational logic circuit is the minimum propagation delay of the combinational logic circuit.
- The *minimum clock-to-Q delay* or *contamination of clock-to-Q delay* ($t_{q(min)}$) of a flip-flop is the minimum propagation delay from the clock edge (sampling edge) to the new value appearing at the output Q of a flip-flop.

The effects of minimum propagation delays of both combinational logic and flip-flops can be illustrated as the timing diagram depicted in Figure 9.46(c). The new data, which is supposed to appear at the later time of the current clock cycle, may corrupt the current data appearing at the destination flip-flop. The destination flip-flop will be contaminated and may enter the metastable state if the sum of contamination delays of the flip-flop and combinational logic is smaller than the hold-time requirement of the destination flip-flop. Because of this reason, the contamination delay is usually given to mean the minimum delay.

In order to ensure that a system is properly worked, the hold time of the destination flip-flop must be shorter than the sum of the minimum clock-to-Q delay of the source flip-flop and the minimum propagation delay of the combinational logic circuit between them. That is, the min-delay constraint is given as

$$t_{hold} \leq t_{q(min)} + t_{pd(min)} \tag{9.8}$$

In summary, the setup-time failure can be eliminated by elongating the clock period, namely, slowing the operating clock or by using flip-flops with shorter setup time and/or clock-to-Q delay. Nevertheless, the hold-time failure can only be fixed by redesigning the logic circuit; it cannot be simply fixed by slowing the operating clock. It is good practice to design a system very conservatively to avoid such failures because redesigning or modifying a system or a chip is very expensive and time consuming.

■ Review Questions

Q9-46. Explain why a max-delay constraint must be enforced in designing a digital system.

Q9-47. Explain the meaning of the min-delay constraint.

Q9-48. Give why the contamination time is so important in designing a digital system.

Q9-49. Give why the hold-time failure cannot be fixed by elongating the clock period.

9.3.2 Clock Skew

The heart of a digital system is the clock whose period determines the rate at which data can be processed. Ideally, a clock should arrive at each storage element at exactly the same time. In reality, because of many uncertainty factors, such as the unbalanced clock paths and differences in the loading of different clock paths, the clock will arrive at different storage elements at different times. This time difference is called the *clock skew*.

The clock skew is generally the variation in arrival times of a clock transition at different storage elements. If node i is the destination, while node j is the source, then the clock skew between two nodes i and j on a circuit is given by $t_{skew} = t_i - t_j$. Depending on whether the directions of clock and data flow are the same, the clock skew can be positive or negative, as shown in Figures 9.47 and 9.48, respectively. Because the clock skew is caused by static mismatches in different clock paths and differences in the loading of clock paths, it is constant from cycle to cycle. The design of a low-skew clock network is essential for modern high-speed digital systems or VLSI chips.

9.3.2.1 Positive Clock Skew Positive clock skew may arise when a clock is routed in the same direction as the data flow through a pipeline structure, as shown in Figure 9.47. The source flip-flop receives its clock earlier than the destination flip-flop. As a result, the positive clock skew occurs, that is, $t_{skew} > 0$. The constraints on the timing of a FF-based system including the clock skew are as follows:

$$T_{clk} \geq t_q + t_{pd} + t_{setup} - t_{skew}$$

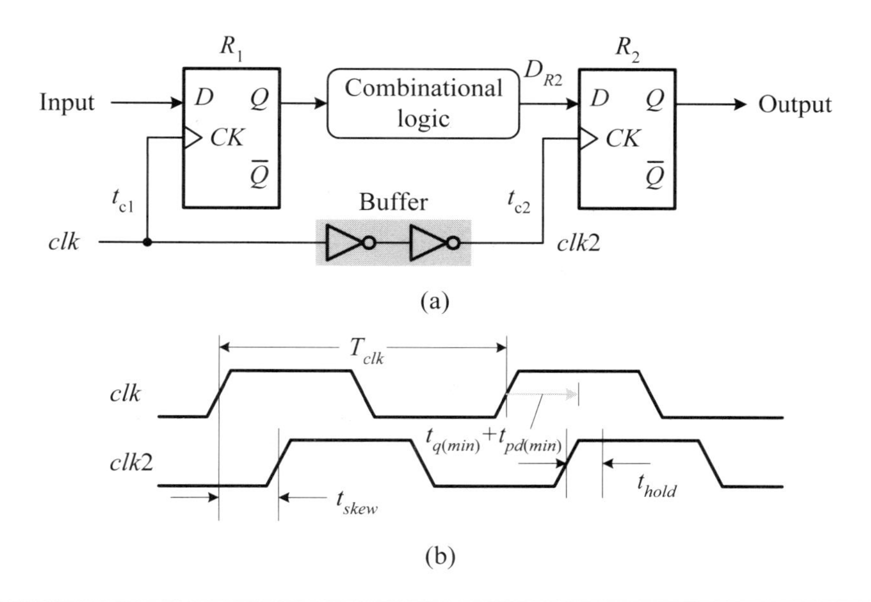

Figure 9.47: The effect of positive clock skew on a FF-based system: (a) logic circuit; (b) timing diagram.

$$\begin{aligned} t_{hold} &\leq t_{q(min)} + t_{pd(min)} - t_{skew} \\ t_{skew} &> 0 \end{aligned} \tag{9.9}$$

This means that the positive clock skew effectively reduces the minimum clock period required for a given system and hence promises to improve the system performance. However, positive clock skew may result in a hold-time constraint violation. For a system to be reliably operated, the hold-time constraint stated above must be satisfied. In addition, if $t_{skew} \geq t_{q(min)} + t_{pd(min)}$, then the destination register R_2 will sample the new data instead of the old one from the source register R_1, as indicated in Figure 9.47(b).

9.3.2.2 Negative Clock Skew Negative clock skew may occur when a clock is routed in the opposite direction as the data flow through a pipeline structure, as shown in Figure 9.48. The source flip-flop receives its clock later than the destination flip-flop. This results in a negative clock skew, namely, $t_{skew} < 0$. The constraints on the timing of a FF-based system can be represented as follows:

$$\begin{aligned} T_{clk} &\geq t_q + t_{pd} + t_{setup} - t_{skew} \\ t_{hold} &\leq t_{q(min)} + t_{pd(min)} - t_{skew} \\ t_{skew} &< 0 \end{aligned} \tag{9.10}$$

This means that the clock skew reduces the time available for actual computation so that the clock period has to be elongated by an amount of $|t_{skew}|$ to maintain the validation of setup time if the other two factors are not changed. However, the hold-time constraint in this situation is unconditionally satisfied.

From the previous discussion, we may conclude that although positive clock skew promises to improve system performance while negative clock skew can eliminate races and make hold-time constraint unconditionally satisfied, they do not always properly

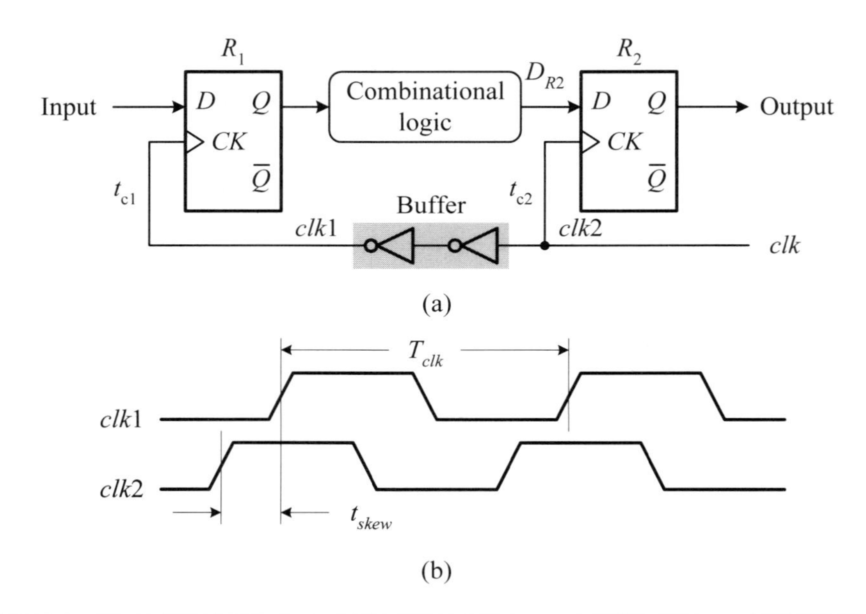

Figure 9.48: The effect of negative clock skew on a FF-based system: (a) logic circuit; (b) timing diagram.

work because a generic logic circuit can generally have data flow in both directions relative to the clock due to feedback paths that may exist in the system, in particular, in the datapath. Hence, it is necessary to control the clock skew regardless of whether it is positive or negative in an acceptable range when designing a system.

■ Review Questions

Q9-50. What is clock skew? Define it.

Q9-51. What is the meaning of positive and negative clock skews?

Q9-52. Explain why the positive clock skew may cause hold-time constraint to be violated.

9.3.3 Timing Issues of Latch Systems

Like flip-flops, latches can also be used as memory elements to design sequential logic circuits. However, the design of a system with latches is much more difficult than with flip-flops due to the transparent property inherently associated with latches. Hence, it is usually recommended to use flip-flops when a digital system only requires a moderate-level performance and is designed with a synthesis flow. For high-performance systems, latches indeed widely find their positions with a full-custom flow, sometimes even with a synthesis flow.

As described, in an edge-triggered FF-based system, the minimum clock period for the entire system is determined by the worst-case combinational logic path between two flip-flops. In a level-sensitive latch-based system, a more flexible timing is provided when one stage is allowed to borrow time from the other stage. As a consequence, a

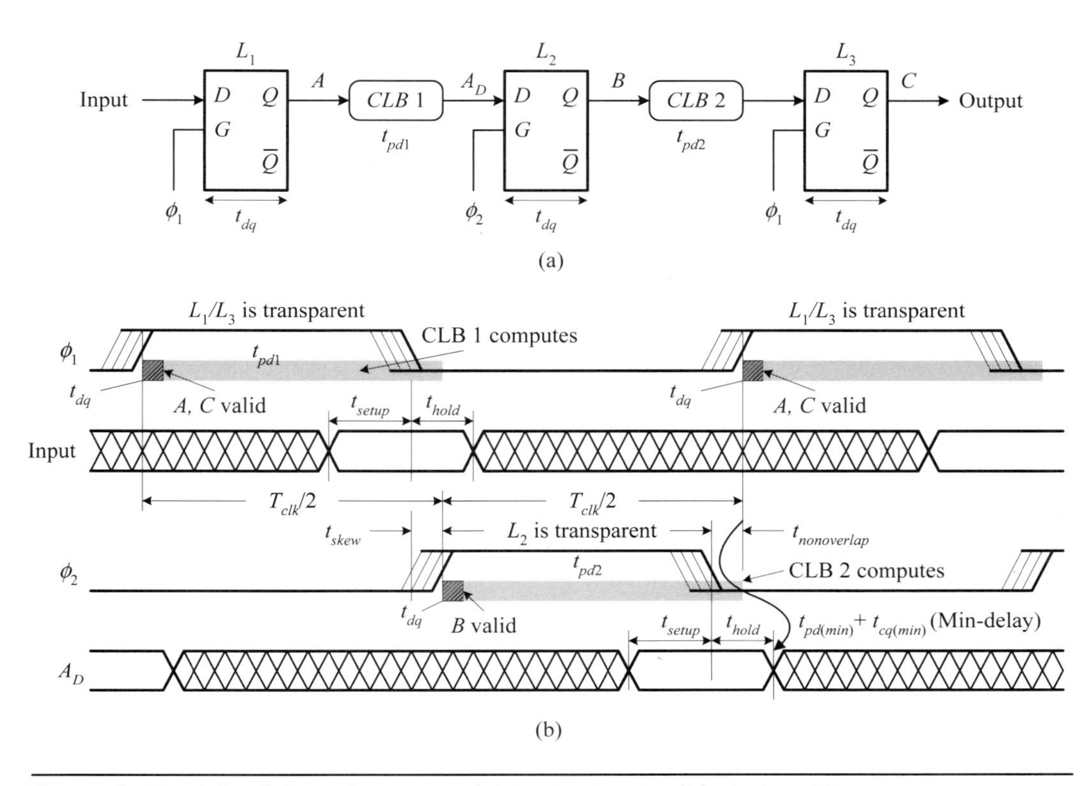

Figure 9.49: A latch-based system: (a) logic circuit; (b) timing diagram.

latch-based system may have better overall performance than an FF-based system. We will illustrate this by way of an example later in the subsection.

9.3.3.1 Max-Delay Constraint Like flip-flops, there are many time constraints that must be satisfied when latches are used to design a digital system. To begin with, we define the following terminology:

- *Setup time* (t_{setup}). Setup time is the amount of time that the input data (D) must be stable before it is sampled.
- *Hold time* (t_{hold}). The hold time is the amount of time that the input data (D) of a latch must remain stable after it is sampled.
- *D-to-Q delay* (t_{dq}). D-to-Q delay is the maximum propagation delay from when the new data arrives at the input D, while the latch is in the transparent mode until the data reaches the output Q.

An example of a latch-based system is shown in Figure 9.49, where a two-phase nonoverlapping clocking scheme is employed. As illustrated in Figure 9.49(b), when clock ϕ_1 rises high, latch L_1 is in the transparent mode and launches data A after a short delay t_{dq}. This data A passes through the combinational logic block CLB_1; thereby, the CLB_1 starts to compute its function. At the falling edge of the clock ϕ_1, the latch L_1 enters hold mode and latches the input. Similarly, latch L_2 enters the transparent mode at the rising edge of clock ϕ_2 and launches data B after a short delay t_{dq}; thereby, the CLB_2 starts to compute its function. These data will be continually computed by the combinational logic block CLB_2 even when L_1 enters its transparent mode again. This means that the computation of CLB_2 may extend from its half-cycle to the next when latch L_3 is transparent. Similarly, the computation of CLB_1 may extend from its half-cycle to the next when latch L_2 is transparent. This feature is referred to as *time borrowing*. Because latch L_2 will latch its data on the falling edge of ϕ_2, the computation extension of CLB_1 cannot go beyond this point. Similarly, the computation extension of CLB_2 cannot go beyond the falling edge of ϕ_1. It is of interest to note that in a latch-based system, latches may be placed at any point in the half-cycle on the condition that there must be one latch at each half cycle.

From the above discussion, we know that the important features of a latch-based system are as follows. First, each latch is transparent when its input arrives and incurs a D-to-Q delay (t_{dq}) rather than a clock-to-Q delay (t_{cq}). Second, because data arrive well before the falling edge of the clock, setup times are trivially satisfied. If the clock period is T_{clk}, then the timing constraint of a latch-based system using two-phase nonoverlapping clocking is as follows:

$$T_{clk} \geq t_{pd} + 2t_{dq} \tag{9.11}$$

where $t_{pd} = t_{pd1} + t_{pd2}$. Both t_{pd1} and t_{pd2} are the propagation delays of combinational logic blocks, CLB_1 and CLB_2, respectively. The maximum useful time t_{pd} for both combinational logic blocks is equal to

$$t_{pd} = T_{clk} - 2t_{dq} \tag{9.12}$$

Because latches can be used in a way such that data arrive more than a setup time before the falling edge of the clock, even with clock skew, neither setup time nor clock skew needs to be considered for the clock period.

9.3.3.2 Min-Delay Constraint To quantitatively describe the min-delay constraint, let us first consider the following related definition.

- The *minimum clock-to-Q delay* or *contamination of clock-to-Q delay* ($t_{cq(min)}$) of a latch is the minimum propagation delay from the clock edge to the new value appearing at the output Q of latch.

The min-delay constraint of a latch-based system can be obtained as follows: For each half cycle, data depart the latch on the rising edge of the clock. Since a latch samples its input data on the falling edge of the clock, the sum of both clock-to-Q delay of latch and combinational logic delay must be greater than a hold time after the falling edge of the previous clock, as illustrated in Figure 9.49(b). The min-delay constraint can then be expressed as

$$t_{hold} \leq t_{pd(min)} + t_{cq(min)} + t_{nonoverlap} - t_{skew} \tag{9.13}$$

where $t_{nonoverlap}$ is the time from the falling edge of one clock to the rising edge of the next and t_{skew} is the clock skew, as illustrated in Figure 9.49(b).

9.3.3.3 Time Borrowing As stated, in a latch-based system with a two-phase nonoverlapping clocking scheme, each latch is nominally assigned a half cycle regardless of whether its logic is fast or slow. However, the slow logic can use more time than its designated half cycle automatically without the need of any explicit design changes. This ability is referred to as *time borrowing*, *slack borrowing*, or *cycle stealing*.

The maximum time borrowing in a two-phase nonoverlapping clocking latch-based system can be derived from the timing diagram shown in Figure 9.50 and can be expressed as follows:

$$t_{borrow(max)} \leq \frac{T_{clk}}{2} - t_{setup} - t_{skew} - t_{nonoverlap} \tag{9.14}$$

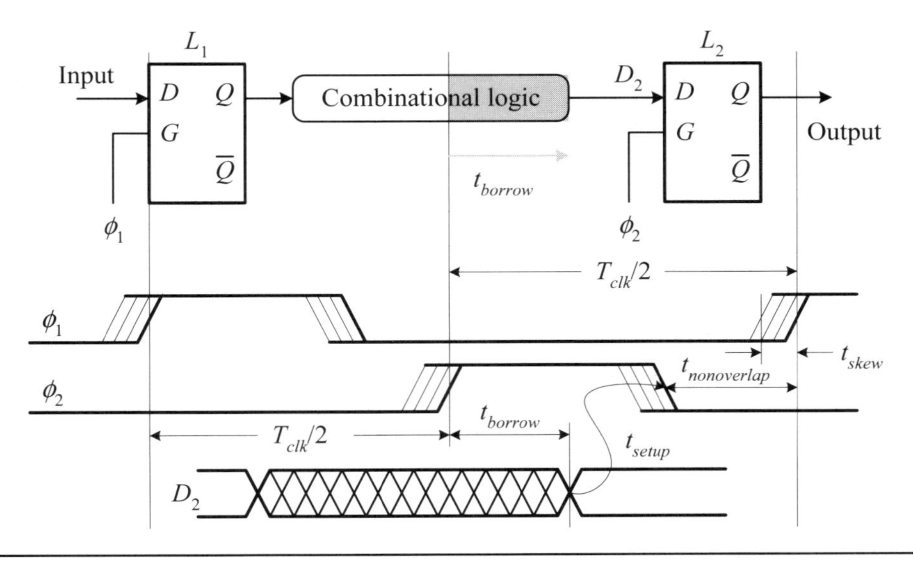

Figure 9.50: The maximum time borrowing in a latch-based system.

In the extreme limit of long cycle time T_{clk}, this approaches a half cycle. For shorter cycle times, the overhead of clock skew, setup time, and nonoverlapping duration reduces the amount of available time to be borrowed.

Because of the feature of automatic time borrowing inherently existing in a level-sensitive latch-based system, a latch-based system may have better overall performance than a FF-based system. We illustrate this in the following by way of an example.

■ Example 9-8: (Time borrowing in a latch-based system.) Suppose that an FF-based pipeline system consists of two stages in which the maximum propagation delays of both combinational logic are of 75 ns and 60 ns, respectively, ignoring clock-to-Q and setup time of the pipeline registers for simplicity. The period of pipeline clock has to be set as 75 ns.

To improve the performance of this system, suppose that the bottleneck combinational logic can be partitioned into two pieces with propagation delays of 50 ns and 25 ns, respectively. The other stage can be partitioned evenly into two pieces with a propagation delay of 30 ns. Furthermore, a latch-based system is used to implement the resulting system, as shown in Figure 9.51(a). The data-to-Q of latches is assumed to be 3 ns. As we can see from the timing diagram depicted in Figure 9.51(b), a 60-ns clock period is adequate for this system. Consequently, a latch-based system in general has better performance than an FF-based system. It is instructive to note the time borrowing between two half cycles.

■

■ Review Questions

Q9-53. What is the meaning of time borrowing?

Q9-54. Explain why the max-delay constraint does not need to take into account the setup time.

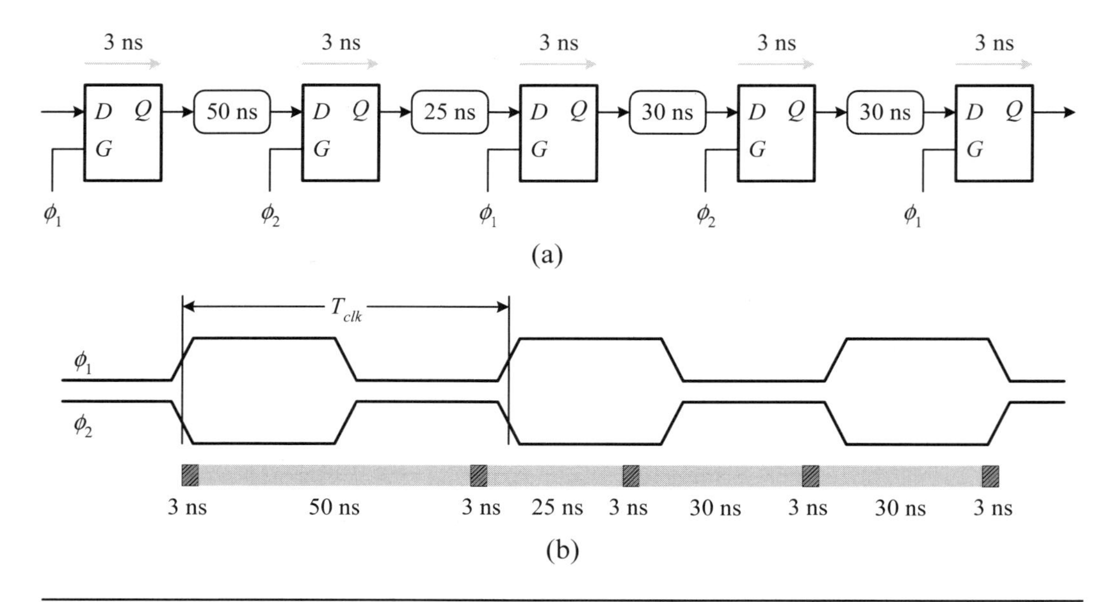

Figure 9.51: An example of time borrowing in a latch-based system: (a) logic circuit; (b) timing diagram.

Q9-55. Explain why a latch-based pipeline system in general has better performance than an FF-based pipeline system.

9.3.4 Timing Issues of Pulsed-Latch Systems

A pulsed latch is also called a *one-phase* or *glitch latch*. Even though it is a latch circuit, a pulsed latch behaves like a flip-flop due to the use of an elaborately designed short-duration pulse to remove the transparent property. As a consequence, like the FF-based system, in the pulsed-latch-based system only one pulsed latch is necessary in each clock cycle. A pulsed-latch-based system is illustrated in Figure 9.52.

9.3.4.1 Max-Delay Constraint Like the latch-based system, data must be set up and held around the falling edge of the pulse in a pulsed-latch-based system. If the pulse is wider than the setup time plus clock skew, then a pulsed latch behaves like a latch and the data can arrive while the pulsed latch is transparent and passes through with only a single latch propagation delay. However, if the pulse is narrow, data must arrive by a setup time and clock skew before the nominal falling edge. Based on these observations, the max-delay constraint of a pulsed-latch-based system can be represented as follows:

$$T_{clk} \geq \max\{t_{pd} + t_{dq}, t_{pd} + t_{cq} + t_{setup} + t_{skew} - t_{pw}\} \quad (9.15)$$

9.3.4.2 Min-Delay Constraint For a pulsed-latch system, data depart the source pulsed latch on the rising edge of the pulse and must not reach the destination pulsed latch until a hold time after the falling edge of the pulse has been satisfied, as shown in Figure 9.52. Consequently, the sum of minimum contamination delay $t_{pd(min)}$ of the combinational logic between two pulsed latches and the minimum clock-to-Q delay

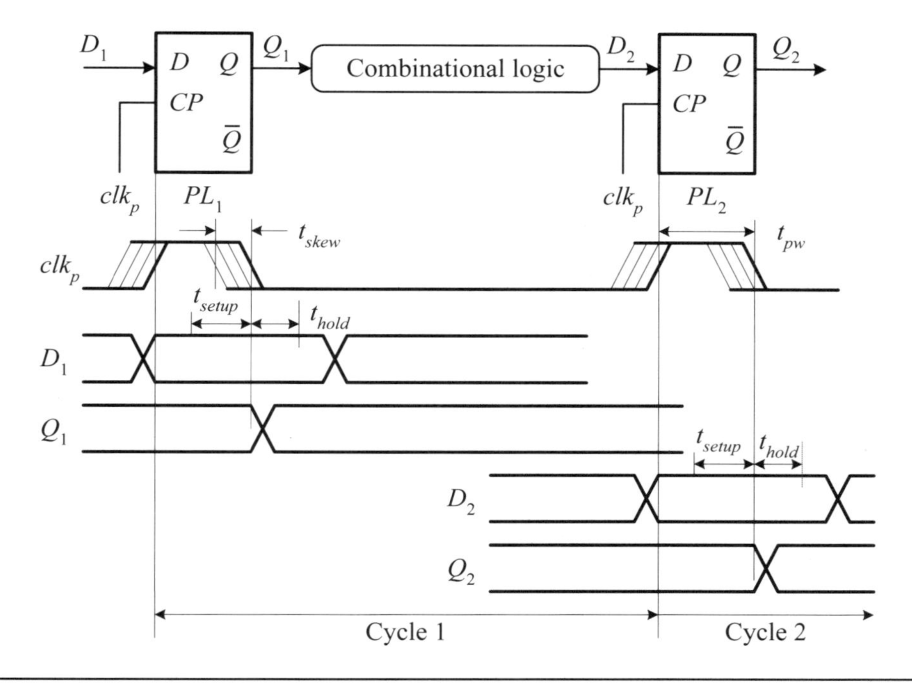

Figure 9.52: A pulsed-latch-based system.

$t_{cq(min)}$ of the pulsed latch must satisfy the following expression

$$t_{pd(min)} + t_{cq(min)} \geq t_{pw} + t_{hold} + t_{skew} \tag{9.16}$$

As a result, the presence of clock skew between pulses and the use of wider pulses increase the minimum amount of delays between latches.

9.3.4.3 Time Borrowing Like a latch-based system, a pulsed-latch-based system can also borrow time from the next cycle because pulsed latches are transparent when their pulses are high. However, the maximum allowable time borrowing is only limited to the duration when the pulse is high and is equal to

$$t_{borrow(max)} = t_{pw} - (t_{setup} + t_{skew}) \tag{9.17}$$

Because of a very narrow pulse, t_{pw}, intended to use in practice, the actual time borrowing in a pulsed-latch-based system is quite small.

■ Review Questions

Q9-56. Explain why a pulsed latch can function as a flip-flop rather than a latch.

Q9-57. Can the time-borrowing technique be applied to a pulsed-latch-based system?

9.4 Pipeline Systems

Pipeline systems can be classified into three types: (synchronous) pipelining, asynchronous pipelining, and wave pipelining. In this section, we begin to describe the rationale behind the classification of pipeline systems and then deal with each type in greater detail.

9.4.1 Types of Pipeline Systems

For a stream of data, it is often needed to increase the *concurrency* of data processing to improve the system performance. There are two approaches widely used to reach this purpose: *parallelism* and *pipelining*. Parallelism means the use of many duplicated circuits or resources such that every circuit has the same function and performs the same computation on the different sets of data at the same time. Pipelining is a means by which a stream of data can be continuously applied to a computational circuit in a uniform way so that output results are yielded regularly in sequence.

Depending on the data-forwarded fashion, namely, when applied to new data, pipelining techniques can be classified into (*synchronous*) *pipelining*, *asynchronous* (or *self-timed*) *pipelining*, and *wave pipelining*. In the (synchronous) pipelining, new data are applied to a computational circuit in intervals determined by the maximum propagation delay of the computational circuit. In the asynchronous pipelining, new data are applied to a computational circuit in intervals determined by the average propagation delay of the computational circuit. In the wave pipelining, new data can be applied to a computational circuit in intervals determined by the difference between maximum and minimum propagation delays of the computational circuit.

Asynchronous logic circuitry offers potential power saving and performance improvements at the cost of design complexity and a small area penalty. The traditional synchronous circuit design features a global clock that drives latches/flip-flops surrounding combinational logic that is used to perform a particular function. The clock speed is determined by the critical path through the system. A synchronous design has periodic power dissipation peaks, thereby leading to *electromagnetic interference* (EMI). Besides, the global clock tree consumes a significant portion of the required power. In contrast, an asynchronous implementation produces very little EMI because there is no periodic power dissipation peaks and it is event-triggered, therefore processing new data using the minimum number of gate transitions.

In practical applications, (synchronous) pipelining can be simply implemented by partitioning the computational circuit into several smaller circuits with roughly the same propagation delay, separated by registers, called *pipeline registers*. These circuits are then synchronized by a global clock with a period greater than the worst-case propagation delay of the circuits and some overhead of the pipeline registers. For this reason, the (synchronous) pipelining is the most widely used approach among the three pipelining techniques. Also because of this, it is often referred to as *pipelining* for short.

9.4.2 Synchronous Pipelining

Pipelining is a system design technique that increases the overall system performance by partitioning a complex combinational logic circuit into many smaller circuits and uses a faster system clock. As shown in Figure 9.53, the computation of data within the combinational logic proceeds in a step-by-step fashion; namely, at time $t = 0$, the first part of the combinational logic performs its computation on the input data. At time $t = t_1$, the second part of the combinational logic performs its computation on the output data from the first part, and so on.

The smallest allowable clock period T_{clk} is determined by the following condition:

$$T_{clk} \geq t_{pd} + t_q + t_{setup} + t_{skew} \tag{9.18}$$

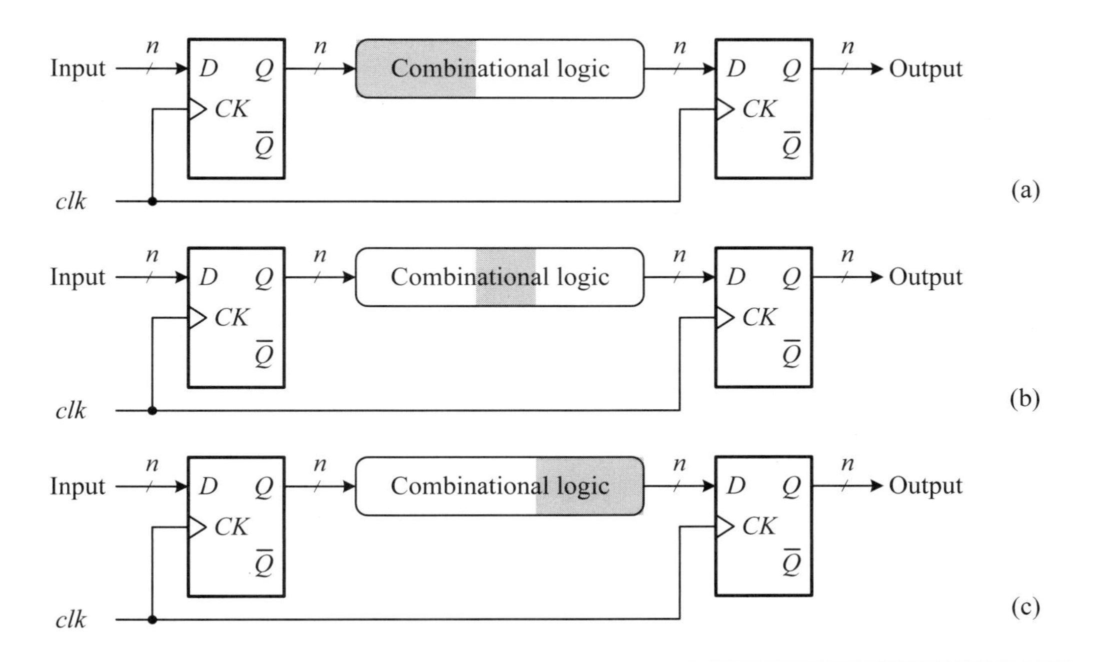

Figure 9.53: An illustration of the pipelining principle: (a) $t = 0$; (b) $t = t_1$; (c) $t = t_2$.

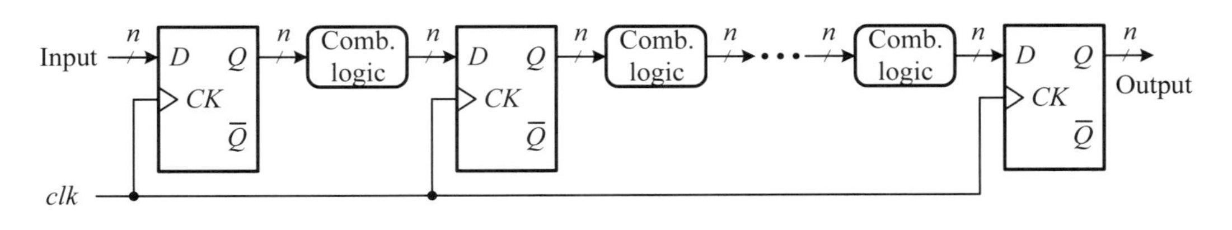

Figure 9.54: A pipeline system.

where t_{pd} is the propagation delay of combinational logic, t_q and t_{setup} are the clock-to-Q delay and setup time of flip-flops, respectively, and t_{skew} is the clock skew. The t_q, t_{setup}, and t_{skew} compose the system overhead.

For processing a stream of data, it is often to partition the combinational logic being used to compute a function into many (say, m) smaller logic pieces and then insert a register between two logic pieces. This results in an m-stage *pipeline system*, as shown in Figure 9.54, where each stage performs a fraction computation of the complete function.

The pipeline clock frequency f_{pipe} is set by the slowest stage and is generally larger than the frequency used in the original nonpipelined system. For the ith stage, the smallest allowable clock period T_i is determined by the following condition:

$$T_i \geq t_{pdi} + t_q + t_{setup} + t_{skew,i} \tag{9.19}$$

where t_q, t_{setup}, and $t_{skew,i}$ compose the *pipeline overhead*. The pipeline clock period for an m-stage pipeline system should then be chosen as

$$T_{pipe} = \max\{T_i | i = 1, 2, \cdots, m\} \tag{9.20}$$

The pipeline clock frequency is equal to the reciprocal of T_{pipe}, namely, $f_{pipe} = 1/T_{pipe}$.

■ **Example 9-9: (Pipelining principles.)** Referring to Figure 9.53, suppose that the combinational logic can be partitioned into three pieces with propagation delays of 15, 25, and 10 ns, respectively. The setup time and clock-to-Q delay of flip-flops are assumed to be 3 ns and 2 ns, respectively. Ignoring the clock skew, determine the pipeline clock frequency.

Solution: According to Equations (9.19) and (9.20), the maximum pipeline clock period is $3 + 2 + 25 = 30$ ns. Hence, the pipeline clock frequency is $1/30$ GHz.

■

It is interesting to note that for a single piece of data, the use of pipelining may even deteriorate the system performance due to the pipeline overhead, including the setup time and clock-to-Q delay, of pipeline registers. For instance, as shown in Figure 9.53, the latency from input to output is 50 ns without pipelining and is $30 \times 3 = 90$ ns when the three-stage pipeline of the preceding example is used. Hence, it needs more time in the pipeline system when only a single piece of data needs to be computed.

The bottleneck stage of a pipeline system can be solved by using the technique of parallelism. An illustration of this is exploited in the following example.

■ **Example 9-10: (The bottleneck of the pipeline system.)** The bottleneck stage of the preceding example is the second stage, which has a propagation delay of 25 ns. By using the technique of parallelism, two duplicates of this stage are used to cut off the propagation delay into half of its original 25 ns. Assume that the sum of propagation delays of both demultiplexer and multiplexer at the beginning and end of the stage does not exceed 2.5 ns, then the pipeline clock period is limited by the first stage and is $3+2+15 = 20$ ns. Consequently, an improvement over the above example is achieved at the cost of extra hardware.

■

■ Review Questions

Q9-58. Distinguish wave pipelining, pipelining, and asynchronous pipelining.

Q9-59. How would you determine the pipeline clock period in a pipeline system?

Q9-60. How would you solve the bottleneck stage of a pipeline system?

9.4.3 Asynchronous Pipelining

In asynchronous pipelining, new data are applied to each stage under the control of the local communication protocol. We first explore the principles of asynchronous pipelining and then introduce a widely used communication protocol, the four-phase handshaking control.

9.4.3.1 Basic Principles An asynchronous design is based on the concept of modular functional blocks intercommunicating using some communication protocols. The general block diagrams of asynchronous pipeline systems are shown in Figure 9.55. Figure 9.55(a) shows a scheme with one-way control, and Figure 9.55(b) shows a scheme

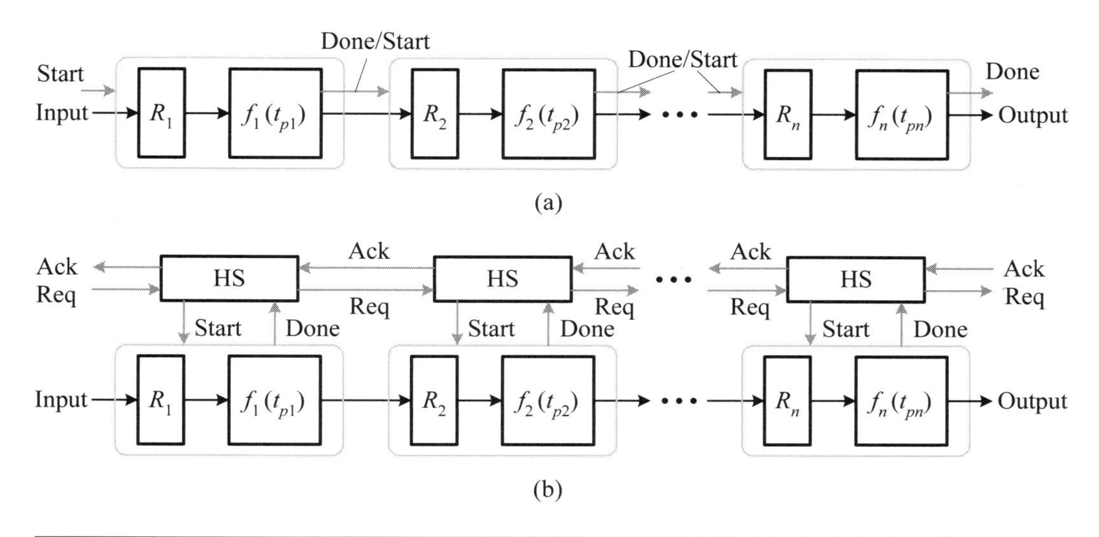

Figure 9.55: Asynchronous pipeline systems with (a) one-way and (b) two-way control schemes.

with two-way control, where HS means handshaking. The one-way control is also called a *strobe control*, and the two-way control scheme is generally referred to as a *handshaking control*. We only consider the details of handshaking control.

Regardless of which scheme is used and whether the underlying system is synchronous or asynchronous, any sequential logic must satisfy two constraints to work properly. These include *physical timing constraint* and *event logical order*. Physical timing constraint guarantees that each function unit and memory element have enough time to complete the specific operations. For example, to guarantee that the output data from a memory element is valid, the physical timing constraints, such as setup time and hold time of latches or flip-flops, must always be satisfied. Event logical order means that events in the system must occur in the logical order set by the designer. In a synchronous system, this is commonly achieved by using a clock to provide a time base for determining what and when to happen. In an asynchronous system, some other schemes, such as handshaking, must be explored to accomplish this.

9.4.3.2 Handshaking Conceptually, *handshaking* is a technique that provides a two-way control scheme for asynchronous data transfer. In this kind of data transfer, each transfer is sequenced by the edges of two control signals: *req* (request or valid) and *ack* (acknowledge), as shown in Figure 9.56. Generally speaking, in a handshaking data transfer, there are four events that proceed in a cyclic order. These events are *ready* (*request*), *data valid*, *data acceptance*, and *acknowledge* in that order. A handshaking control scheme can be initiated by either a source device or destination device.

Source-initiated transfer. The source-initiated data transfer is shown in Figure 9.56(a). The sequence of events is as follows:

1. *Ready*: The destination device deasserts the acknowledge signal and is ready to accept the next data.
2. *Data valid*: The source device places the data onto the data bus and asserts the valid signal to notify the destination device that the data on the data bus are valid.

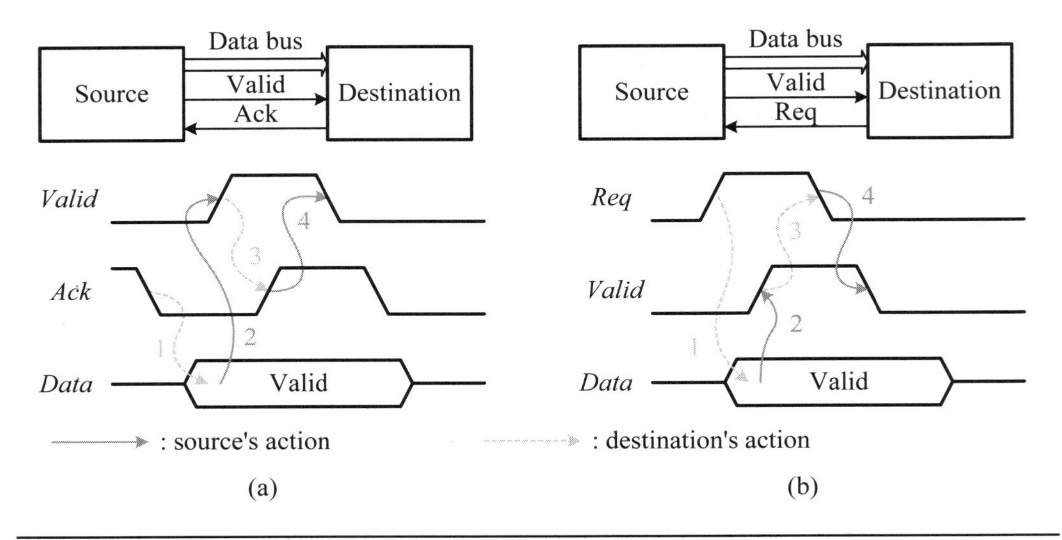

Figure 9.56: Timing diagrams of handshaking data transfers: (a) source-initiated transfer; (b) destination-initiated transfer.

3. *Data acceptance*: The destination device accepts (latches) the data from the data bus and asserts the acknowledge signal.
4. *Acknowledge*: The source device invalidates data on the data bus and deasserts the valid signal.

Because of the inherent back-and-forth operations of the scenario, the asynchronous data transfer described above is called the *handshaking protocol* or *handshaking transfer*.

Destination-initiated transfer. The destination-initiated data transfer is shown in Figure 9.56(b). The sequence of events is as follows:

1. *Request*: The destination device asserts the request signal to request data from the source device.
2. *Data valid*: The source device places the data on the data bus and asserts the valid signal to notify the destination device that the data are valid now.
3. *Data acceptance*: The destination device accepts (latches) the data from the data bus and deasserts the request signal.
4. *Acknowledge*: The source device invalidates data on the data bus and deasserts the valid signal to notify the destination device that it has removed the data from the data bus.

Review Questions

Q9-61. What factors will limit the widespread use of asynchronous pipelining?

Q9-62. What are the features of asynchronous pipelining?

Q9-63. What is the rationale behind asynchronous pipelining?

Q9-64. What is the handshaking mechanism?

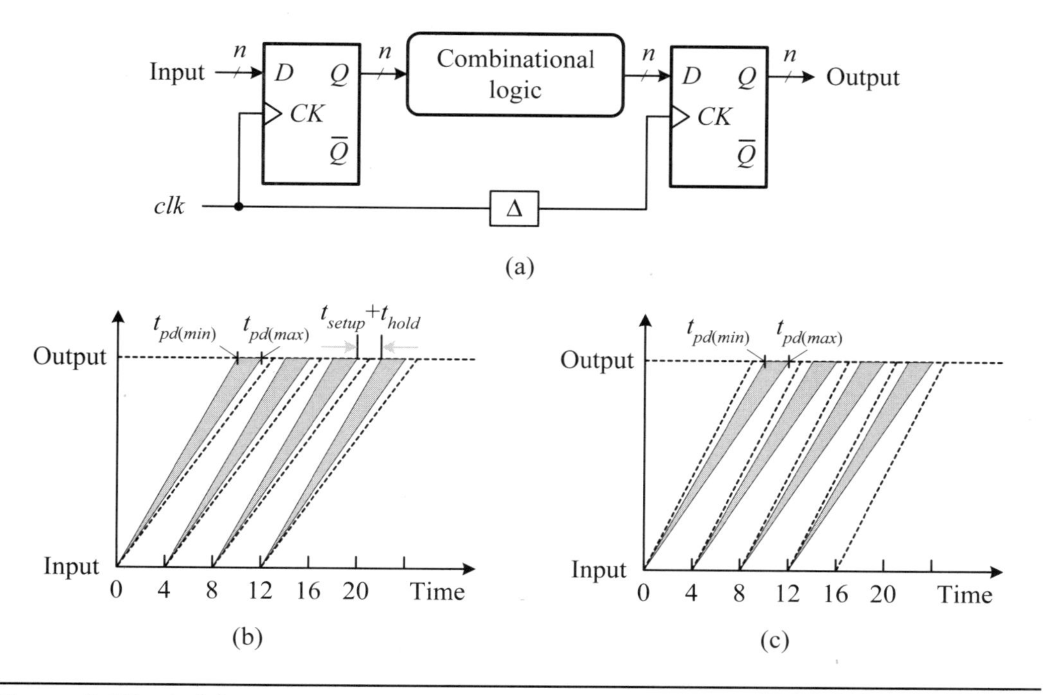

Figure 9.57: A (a) wave-pipeline system and its data flows when (b) $k = 0, \Delta = 13$ and (c) $k = 1, \Delta = 9$.

9.4.4 Wave Pipelining

Wave pipelining is a timing approach allowing new data to be applied to the inputs of a combinational block before the previous outputs are available, thus effectively pipelining the combinational logic and maximizing the utilization of the logic to achieve the maximal-rate operation without inserting registers. The rationale behind wave pipelining is that the throughput of the system does not depend on the maximum propagation delay but instead depends on the difference between the maximum and minimum propagation delays. As a consequence, clock speed can be increased if the idle time of the noncritical paths can be reduced.

Since wave pipelining is a timing method that implements pipelining in combinational logic without the use of intermediate latches or registers, it may significantly reduce the clock loading and associated area, power dissipation, and latency while maintaining the external functionality and timing of a synchronous circuit. The wave pipelining can also be viewed as a virtual pipelining, in which each gate serves as a virtual storage element.

To further illustrate the concept of wave pipelining, consider the circuit shown in Figure 9.57(a). Assume that the maximum and minimum propagation delays of combinational logic are $t_{pd(max)}$ and $t_{pd(min)}$, respectively. Then, it can be shown that the clock period T_{clk} should be greater than the sum of the difference between $t_{pd(max)}$ and $t_{pd(min)}$, the setup time, and hold time of the output register, namely,

$$T_{clk} \geq [t_{pd(max)} - t_{pd(min)}] + t_{setup} + t_{hold} \quad (9.21)$$

supposing that the clock skew is not taken into account. In other words, new data can be launched at a rate of T_{clk}, as illustrated in Figure 9.57(b).

The data launched by the input register at time 0 can be captured by the output register at time $kT_{clk} + \Delta$, where k is the number of clock cycles elapsed since the data is launched. The value of $kT_{clk} + \Delta$ must be greater than $t_{pd(max)} + t_{setup} + t_{hold}$ so that the data through the combinational logic can be latched by the output register. It is of interest to note that the wave-pipeline system shown in Figure 9.57(a) is reduced to the synchronous pipeline system if k is set to one and Δ is set to zero.

■ Example 9-11: (An example of the wave-pipeline system.) Suppose that the maximum and minimum propagation delays of the combinational logic are 12 ns and 10 ns, respectively, and the setup and hold times of the output register can be ignored; namely, t_{setup} and t_{hold} are both zeros. Then, one case with $k = 0$ and $\Delta = 13$ is given in Figure 9.57(b) in which data are captured by a delayed version of the same global clock edge that launched the data. Another case with $k = 1$ and $\Delta = 9$ is shown in Figure 9.57(c) in which data are captured by a delayed version of the global clock edge of the next cycle after the launching edge.

■

To maximize the throughput of wave pipelining, the difference of $[t_{pd(max)} - t_{pd(min)}]$ should be minimized. In other words, the propagation delays through all signal paths have to be balanced. Nonetheless, this is a challenging task for a large logic circuit due to the following reasons. First, propagation delays of paths vary with logic depth in the architecture. Second, the basic building blocks have variable delays. Third, different logic paths may have unequal rise and fall times, as well as PVT variations. Many approaches have been proposed to overcome these problems, ranging from restructuring the logic architecture to inserting buffers in the noncritical paths.

Compared to synchronous pipelining, wave pipelining promises higher throughput and less power dissipation and area, as well as a simpler clock distribution network. However, it needs an accurate SPICE model for the transistors and a careful layout for delay balancing. In addition, the wave-pipelining circuits are more sensitive to PVT variations and clock skew, making the design more challenging. Moreover, the inevitable data-dependent delay fluctuations existing in wave pipelining degrade the performance from the ideal case. These factors limit the widespread use of wave pipelining in general-purpose logic circuits.

■ Review Questions

Q9-65. What factors will limit the widespread use of wave pipelining?

Q9-66. What are the features of wave pipelining?

Q9-67. What is the rationale behind wave pipelining?

9.5 Summary

We have described the fundamentals of sequential logic circuits, including the sequential logic model, basic bistable devices, metastable states and hazards, and arbiters. In addition, a variety of static and dynamic memory elements used in CMOS technology, including latches, flip-flops, and pulsed latches, were introduced.

To perform a desired operation, combinational logic circuits need to be properly combined with sequential logic circuits along with appropriate clocking circuits. The timing issues of systems using flip-flops, latches, and pulsed latches were then dealt with in detail.

For effectively processing streams of data, pipelining and parallelism are the two major speed-up approaches. Parallelism means the use of many duplicated circuits of the same function such that every circuit performs the same computation on different sets of data at the same time. Pipelining is the uniform way that a stream of data can be continuously applied to a computational circuit to yield output results regularly in sequence. Three pipeline systems, including (synchronous) pipelining, asynchronous pipelining, and wave pipelining, were described concisely.

References

1. M. Afghahi and C. Svensson, "A unified single-phase clocking scheme for VLSI systems," *IEEE J. of Solid-State Circuits*, Vol. 25, No. 1, pp. 225–233, February 1990.
2. W. Burleson, M. Ciesielski, F. Klass, and W. Liu, "Wave-pipelining: a tutorial and research survey," *IEEE Trans. on Very Large Scale Integration (VLSI) Systems*, Vol. 6, No. 3, pp. 464–474, September 1998.
3. C. Dike and E. Burton, "Miller and noise effects in a synchronizing flip-flop," *IEEE J. of Solid-State Circuits*, Vol. 34, No. 6, pp. 849–855, June 1999.
4. D. Draper et al., "Circuit techniques in a 266-MHz MMX-enabled processor," *IEEE J. of Solid-State Circuits*, Vol. 32, No. 11, pp. 1650–1664, November 1997.
5. D. Ghosh, and S. K. Nandy, "Design and realization of high-performance wave-pipelined 8 × 8 b multiplier in CMOS technology," *IEEE Trans. on Very Large Scale Integration (VLSI) Systems*, Vol. 3, No. 1, pp. 36–48, March 1995.
6. C. T. Gray, W. Liu, and R. K. Cavin III, "Timing constraints for wave-pipeline systems," *IEEE Trans. on Computer-Aided Design of Integrated Circuits and Systems*, Vol. 13, No. 8, pp. 987–1004, August 1994.
7. M. Hamada et al., "A top-down low power design technique using clustered voltage scaling with variable supply-voltage scheme," in *Proc. of IEEE 1998 Custom Integrated Circuits Conf.*, pp. 495–498, 1998.
8. D. Harris, *Skew-Tolerant Circuit Design.* San Francisco: Morgan Kaufmann Publishers, 2001.
9. J. Horstmann, H. Eichel, and R. Coates, "Metastability behavior of CMOS ASIC flip-flops in theory and test," *IEEE J. of Solid-State Circuits*, Vol. 24, No. 1, pp. 146–157, February 1989.
10. M. Keating et al., *Low Power Methodology Manual: For System-on-Chip Design.* Boston: Springer, 2007.
11. F. Klass et al., "A new family of semidynamic and dynamic flip-flops with embedded logic for high-performance processors," *IEEE J. of Solid-State Circuits*, Vol. 34, No. 5, pp. 712–716, May 1999.
12. U. Ko and P. T. Balsara, "High-performance energy-efficient D-flip-flop circuits," *IEEE Trans. on Very Large Scale Integration (VLSI) Systems*, Vol. 8, No. 1, pp. 94–98, February 2000.
13. Z. Kohavi and K. J. Niraj, *Switching Theory and Finite Automata.* 3rd ed. Cambridge: Cambridge University Press, 2010.

14. M. B. Lin, *Digital System Design: Principles, Practices, and Applications.* 4th ed. Taipei, Taiwan: Chuan Hwa Book Ltd., 2010.

15. M. B. Lin, *Digital Logic Design.* 4th ed. Taipei, Taiwan: Chuan Hwa Book Ltd., 2011.

16. M. B. Lin, *Digital System Designs and Practices: Using Verilog HDL and FPGAs.* Singapore: John Wiley & Sons, 2008.

17. S. Lu and M. Ercegovac, "A novel CMOS implementation of double-edge-triggered flip-flops," *IEEE J. of Solid-State Circuits,* Vol. 25, No. 4, pp. 1008–1010, August 1990.

18. M. Matsui et al., "A 200 MHz 13 mm^2 2-D DCT macrocell using sense-amplifier pipeline flip-flop scheme," *IEEE J. of Solid-State Circuits,* Vol. 29, No. 12, pp. 1482–1490, December 1994.

19. D. Messerschmitt, "Synchronization in digital system design," *IEEE J. of Selected Areas in Communications,* Vol. 8, No. 8, pp. 1404–1419, October 1990.

20. A. Morgenshtein, A. Fish, and I. A. Wagner, "An efficient implementation of D-flip-flop using the GDI technique," in *Proc. of Int'l Symposium on Circuits and Systems* (ISCAS), Vol. 2, pp. 673–676, May 2004.

21. H. Partovi et al., "Flow-through latch and edge-triggered flip-flop hybrid elements," in *Proc. of IEEE Int'l Solid-State Circuits Conf.* (ISSCC), pp. 138–139, 1996.

22. R. C. Pearce, J. A. Field, and W. D. Little, "Asynchronous arbiter module," *IEEE Trans. on Computers,* Vol. 24, No. 9, pp. 931–932, September 1975.

23. J. M. Rabaey, A. Chandrakasan, and B. Nikolic, *Digital Integrated Circuits: A Design Perspective.* 2nd ed. Upper Saddle River, NJ: Prentice-Hall, 2003.

24. O. Sarbishei and M. Maymandi-Nejad, "A novel overlap-based logic cell: an efficient implementation of flip-flops with embedded logic," *IEEE Trans. on Very Large Scale Integration (VLSI) Systems,* Vol. 18, No. 2, pp. 222–231, February 2010.

25. S. Shigematsu et al., "A 1-V high-speed MTCMOS circuit scheme for power-down application circuits," *IEEE J. of Solid-State Circuits,* Vol. 324, No. 6, pp. 861–869, June 1997.

26. V. Stojanovic and V. Oklobdzija, "Comparative analysis of master-slave latches and flip-flops for high-performance and low-power systems," *IEEE J. of Solid-State Circuits,* Vol. 34, No. 4, pp. 536–548, April 1999.

27. S. H. Unger and C.-J. Tan, "Clocking schemes for high-speed digital systems," *IEEE Trans. on Computers,* Vol. 35, No. 10, pp. 880–895, October 1986.

28. C. H. (Kees) van Berkel and C. E. Molnar, "Beware the three-way arbiter," *IEEE J. of Solid-State Circuits,* Vol. 34, No. 6, pp. 840–848, June 1999.

29. H. J. M. Veendrick, "The behavior of flip-flops used as synchronizers and prediction of their failure rate," *IEEE J. of Solid-State Circuits,* Vol. 15, No. 2, pp. 169–176, April 1980.

30. J. S. Wang, P. H. Yang, and D. Sheng, "Design of a 3-V 300-MHz low-power 8-b$\times$8-b pipelined multiplier using pulse-triggered TSPC flip-flops," *IEEE J. of Solid-State Circuits,* Vol. 35, No. 4, pp. 583–592, April 2000.

31. N. H. E. Weste and D. Harris, *CMOS VLSI Design: A Circuit and Systems Perspective.* 3rd ed. Boston: Addison-Wesley, 2005.

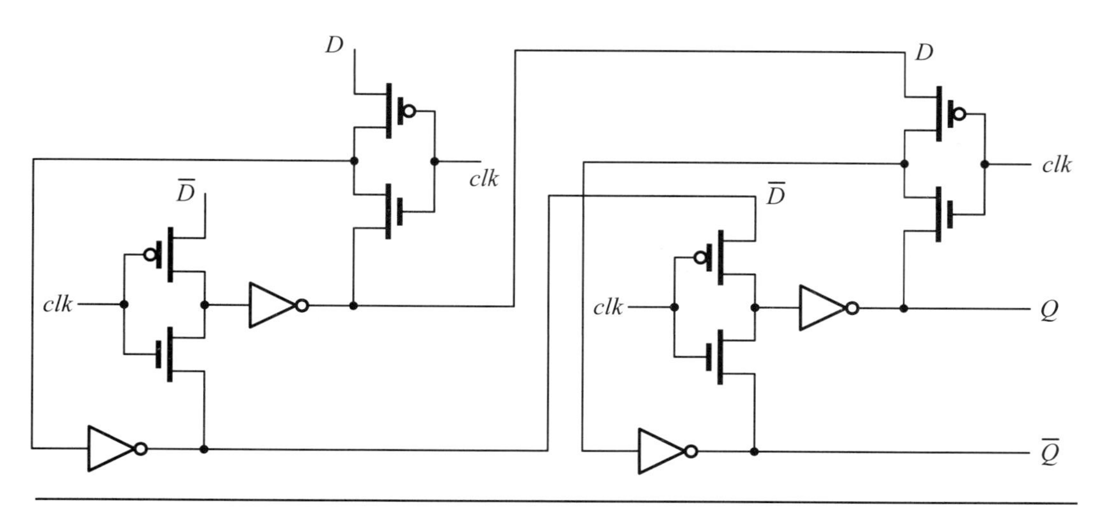

Figure 9.58: A GDI edge-triggered D-type flip-flop.

32. S. H. Yang, C. H. Lee, and K. R. Cho, "A CMOS dual-modulus prescaler based on a new charge sharing free D-flip-flop," in *Proc. of the 14th Annual IEEE Int'l ASIC/SOC Conf.*, pp. 276–280, September 2001.
33. J. R. Yuan, I. Karlsson and C. Svensson, "A true single-phase-clock dynamic CMOS circuit technique," *IEEE J. of Solid-State Circuits*, Vol. 22, No. 5, pp. 899–901, October 1987.
34. J. Yuan and C. Svensson, "New single-clock CMOS latches and flip flops with improved speed and power savings," *IEEE J. of Solid-State Circuits*, Vol. 32, No. 1, pp. 62–69, January 1997.

Problems

9-1. Suppose that $a = 500$ kHz, and a D-type flip-flop with $t_{setup} = 10$ ns is used. Calculate the $MTBF(t_{rec})$ at $f = 25$ MHz and 50 MHz, respectively. T_0 and τ of the flip-flop are 8.7×10^{-6} sec and 1.0 ns, respectively.

9-2. Suppose that $a = 200$ kHz, and a D-type flip-flop with $t_{setup} = 10$ ns is used. T_0 and τ of the flip-flop are 8.7×10^{-6} sec and 1.0 ns, respectively. The system clock frequency is 50 MHz.

(a) Assuming that the system clock is directly applied to the synchronizer, calculate the $MTBF(t_{rec})$.

(b) Assuming that a modulo-4 frequency divider is used to provide the clock to the synchronizer, calculate the $MTBF(t_{rec})$.

9-3. Using the two-way arbiter shown in Figure 9.8(a), design a three-way arbiter. Explain its operation principle.

9-4. Considering the gate-diffusion input (GDI) edge-triggered D-type flip-flop shown in Figure 9.58, answer the following questions:

(a) Explain the operation of the edge-triggered D-type flip-flop. Is it positive or negative edge-triggered?

(b) Using SPICE, simulate the circuit to verify its functionality.
(c) What is the maximum operating frequency that the flip-flop can operate?

9-5. Consider the CVSL-based D-type flip-flop shown in Figure 9.24.

(a) Using SPICE, simulate the circuit to verify its functionality.
(b) What is the maximum operating frequency that the CVSL-based D-type flip-flop can operate?

9-6. Consider the RAM-based D-type flip-flop shown in Figure 9.25.

(a) Using SPICE, simulate the circuit to verify its functionality.
(b) What is the maximum operating frequency that the RAM-based D-type flip-flop can operate?

9-7. Consider the sense-amplifier-based D-type flip-flop (SAFF) shown in Figure 9.27.

(a) Using SPICE, simulate the circuit to verify its functionality.
(b) What is the maximum operating frequency that the SAFF can operate?

9-8. Consider the differential D-type flip-flop with self-resetting logic shown in Figure 9.28.

(a) Using SPICE, simulate the circuit to verify its functionality.
(b) What is the maximum operating frequency that the differential D-type flip-flop with self-resetting logic can operate?
(c) Design an experiment with SPICE to measure the waveforms of clk, $clkd$, node x, and output Q.

9-9. Show that C^2MOS dynamic D-type flip-flops can be immune to the effects of overlapped clocks.

9-10. A 9-transistor TSPC negative edge-triggered D-type dynamic flip-flop is shown in Figure 9.59.

(a) Explain its operation.
(b) Using SPICE, simulate the circuit to verify its functionality.

9-11. Considering the 9-transistor TSPC positive edge-triggered dynamic D-type flip-flop shown in Figure 9.36, design a three-stage ripple-up counter.

(a) Using SPICE, simulate the circuit to verify its functionality.
(b) What is the maximum operating frequency that the three-stage ripple-up counter can operate?

9-12. Considering the 9-transistor TSPC positive edge-triggered dynamic D-type flip-flop shown in Figure 9.60, answer the following questions:

(a) Explain the operation of the positive edge-triggered dynamic D-type flip-flop.
(b) Using SPICE, simulate the circuit to verify its functionality.

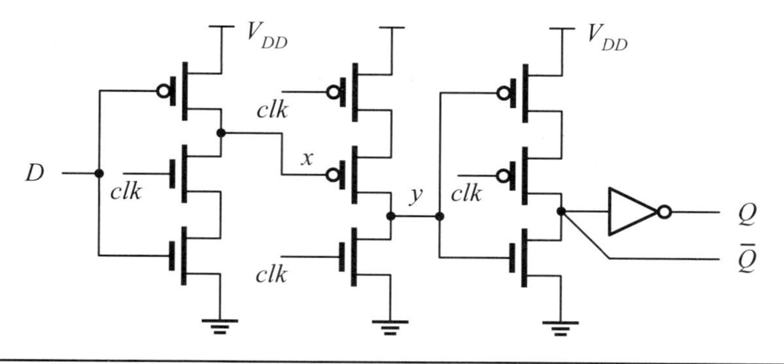

Figure 9.59: The 9-transistor TSPC negative edge-triggered dynamic D-type flip-flop.

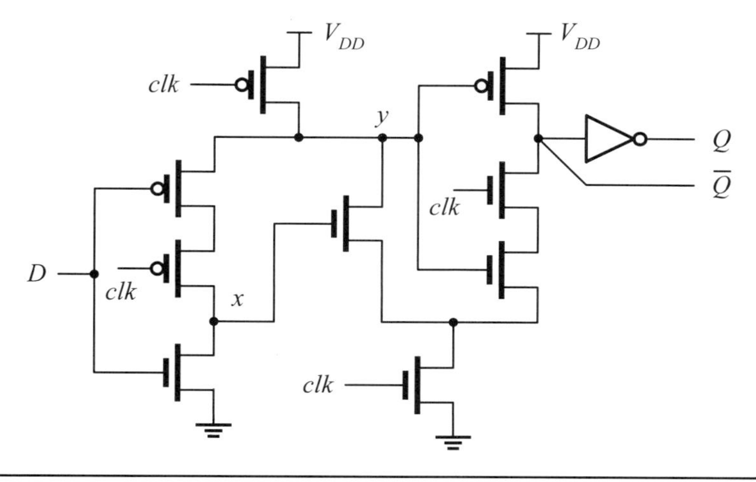

Figure 9.60: A charge-sharing and glitch free positive edge-triggered dynamic D-type flip-flop.

(c) Design an experiment to verify that it is indeed charge-sharing and glitch-free.

(d) What is the maximum operating frequency that the flip-flop can operate?

9-13. Figure 9.61 shows a Klass semidynamic D-type flip-flop (SDFF) that can embed into a logic circuit.

(a) Explain the operation of the Klass SDFF circuit.

(b) Using SPICE, verify the functionality of the Klass SDFF logic circuit. Assume that the embedded logic is a simple two-input AND function.

(c) Design an experiment with SPICE to measure the waveforms of clk, $clkd$, node x, and output Q.

9-14. Consider the pulsed-latch circuit given in Figure 9.39, answer the following questions.

(a) Using SPICE, verify the functionality of the pulsed-latch circuit.

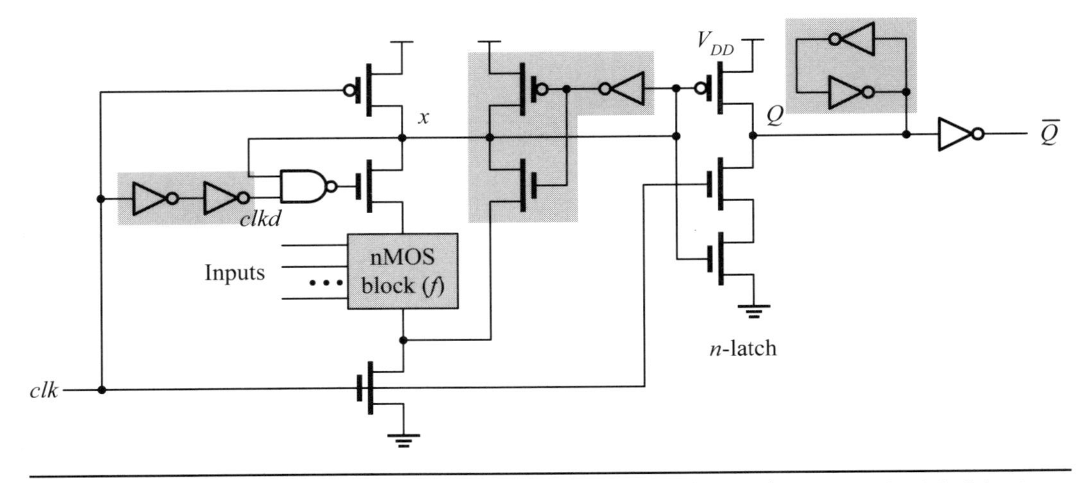

Figure 9.61: The Klass semidynamic D-type flip-flop (SDFF) with embedded logic.

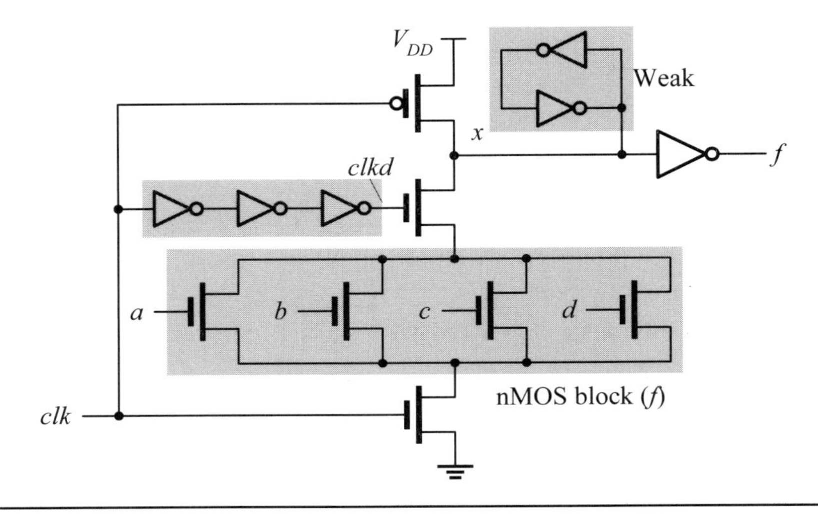

Figure 9.62: A 4-input OR gate implemented with pulsed-domino logic.

(b) Design an experiment with SPICE to measure the waveforms of D, clk, ϕ_p, node x, and output $\bar{Q}$.

9-15. The basic hybrid latch-flip-flop (HLFF) circuit is given in Figure 9.40.

(a) Draw the detailed CMOS logic circuit of the HLFF.

(b) Using SPICE, verify the functionality of the HLFF circuit.

(c) Design an experiment with SPICE to measure the waveforms of D, clk, clk_d, node x, and output Q.

9-16. A 4-input OR gate implemented with pulsed-domino logic is shown in Figure 9.62.

(a) Explain the operation of the pulsed-domino logic circuit.

(b) Using SPICE, verify the functionality of the pulsed-domino logic circuit.

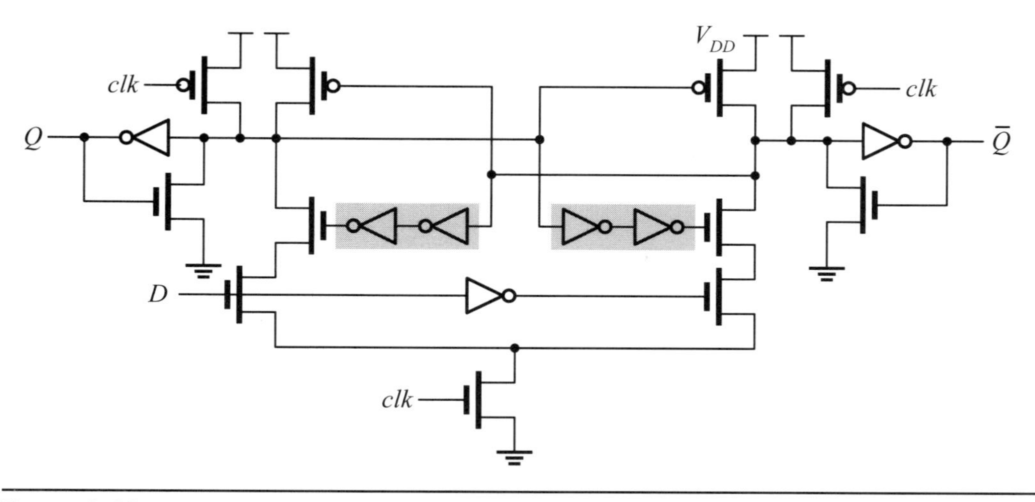

Figure 9.63: The dual-rail dynamic D-type flip-flop.

(c) Design an experiment with SPICE to measure the waveforms of clk, $clkd$, node x, and output f.

9-17. A dual-rail dynamic D-type flip-flop is depicted in Figure 9.63.

(a) Explain the operation of the dual-rail dynamic D-type flip-flop.

(b) Using SPICE, verify the functionality of the dual-rail dynamic D-type flip-flop.

(c) Design an experiment with SPICE to measure the waveforms of D, clk, and output Q.

10

Datapath Subsystem Designs

Many digital systems are often partitioned into two parts: a datapath and a control unit. The datapath performs the required operations to the input data, whereas the control unit schedules the operations being carried out in the datapath in a proper order. Sometimes, a datapath also contains one or many memory modules to store the needed intermediate information in some computations.

In this chapter, we explore the basic components widely used in datapaths. These include basic combinational and sequential components. The former usually comprises decoders, encoders, multiplexers, demultiplexers, magnitude comparators, and shifters, whereas the latter often consists of registers and counters. In addition, arithmetic operations, including addition, subtraction, multiplication, and division, are also dealt with in this chapter in detail. The details of memory module will be discussed in the next chapter.

An arithmetic algorithm can usually be realized by using a multiple-cycle or a single-cycle structure in accordance with the trade-off among area, performance, and power dissipation. The multiple-cycle structure is essentially a sequential logic circuit, while the single-cycle structure is fundamentally a combinational logic circuit. The shift, addition, multiplication, and division algorithms are used as examples to manifest this idea repeatedly. The single-cycle structure naturally evolves into array structures, such as an unsigned array multiplier, a modified Baugh-Wooley signed array multiplier, and a nonrestoring array divider.

10.1 Basic Combinational Components

Code conversion and signal routing circuits are the two most important combinational modules. The code conversion circuits include decoders and encoders, while signal routing circuits contain multiplexers and demultiplexers. In addition, magnitude comparators are also the important components in some applications. In this section, we are concerned with these basic components in detail.

10.1.1 Decoders

A code conversion is a process that transforms one code into another. The circuit used to perform a code conversion is known as a *code converter*. A code conversion is known as an *encoding process* if it transforms an outside world code into the system code; the reverse process, which transforms the system code into the outside world code, is

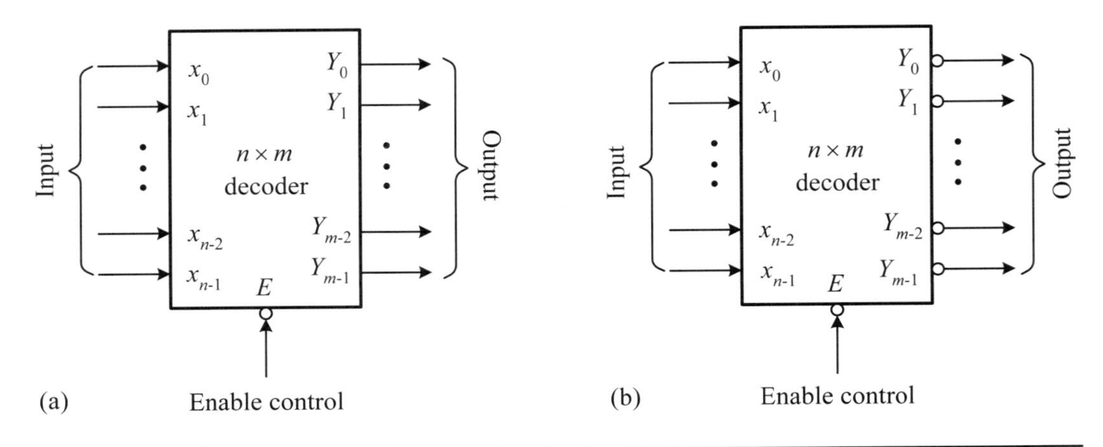

Figure 10.1: The general block diagrams of decoders with: (a) active-high output; (b) active-low output.

called the *decoding process*. The corresponding circuits are referred to as an *encoder* and a *decoder*, respectively. In this book, we assume that the system code is a binary code and the outside world code is a 1-out-of-m code or others.

A standard $n \times m$ decoder has n input lines and m output lines. Each output line Y_i is activated whenever the input combination corresponds to the ith minterm. A minterm is a product term containing every input in either true or complementary form. Decoders can be subdivided into two types: *total decoding* and *partial decoding*. Total decoding means that all input combinations can be recognized exclusively; that is, $m = 2^n$, such as a 3-to-8 decoder. Partial decoding means that only a partial set of input combinations can be recognized exclusively; namely, $m < 2^n$, such as a binary-to-decimal decoder (that is, a type of 4-to-10 decoder). Both types are widely used in digital systems.

Figures 10.1 shows the general block diagrams of decoders. The output of a decoder can be either active-low or active-high, depending on the actual need of applications and the logic circuit designed. For most practical decoders, one or more enable controls may also be embedded into decoders to allow expanding capability. The enable control may be active-low or active-high.

■ Example 10-1: (A 2 × 4 decoder with active-low output and an enable control.) An example of a 2-to-4 decoder with both active-low output and enable control is shown in Figure 10.2. Figure 10.2(a) is the logic symbol of the circuit and Figure 10.2(b) is its function table. From the function table, all of four output switching functions are obtained as follows:

$$\begin{aligned} Y_0 &= \overline{\bar{E} \cdot \bar{x}_1 \cdot \bar{x}_0} \\ Y_1 &= \overline{\bar{E} \cdot \bar{x}_1 \cdot x_0} \\ Y_2 &= \overline{\bar{E} \cdot x_1 \cdot \bar{x}_0} \\ Y_3 &= \overline{\bar{E} \cdot x_1 \cdot x_0} \end{aligned}$$

The resulting decoder is depicted in Figure 10.2(c).

■

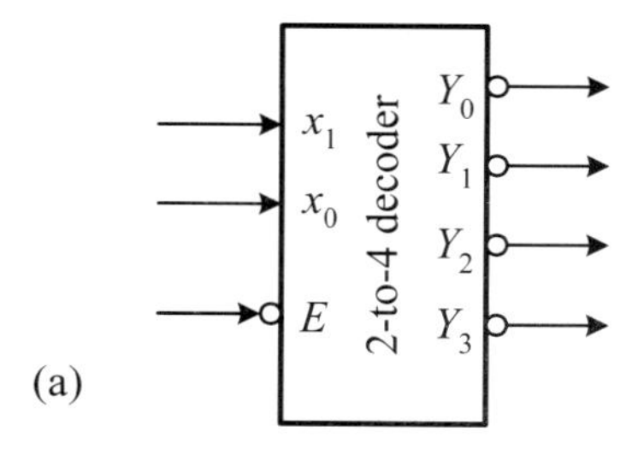

E	x_1	x_0	Y_3	Y_2	Y_1	Y_0
1	ϕ	ϕ	1	1	1	1
0	0	0	1	1	1	0
0	0	1	1	1	0	1
0	1	0	1	0	1	1
0	1	1	0	1	1	1

(b)

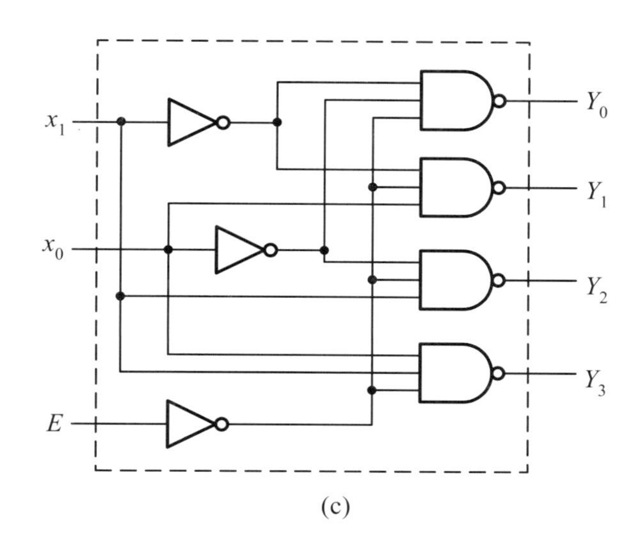

Figure 10.2: A 2×4 decoder active-low output realized by basic gates: (a) logic symbol; (b) function table; (c) logic circuit.

■ Review Questions

Q10-1. Define the encoding process and decoding process.

Q10-2. Define the decoder and encoder.

Q10-3. Define partial decoding and total decoding.

Q10-4. What is the function of the enable control of decoders?

10.1.2 Encoders

The encoding process is the reverse of the decoding process and transforms an outside world code into the system code. The logic circuit used to perform such an encoding process is known as an *encoder*. An encoder has $m = 2^n$ (or fewer) input lines and n output lines. All input lines are arranged and fixed in positions such that all inputs are numbered from 0 up to $m - 1$ in sequence. The output lines generate the binary code corresponding to the position of the activated input line. The general block diagrams of encoders are shown in Figure 10.3, where Figure 10.3(a) is an active-high output while Figure 10.3(b) is an active-low output. Like decoders, an encoder usually associates with one or more optional enable controls, which may be either active-high or active-low, depending on the actual needs.

The purpose of enable control is to control the operations of the encoder. The output of the encoder is a binary code when the enable controls are asserted and is all 1s, all 0s, or even high-impedance, depending on the actual requirement and design, when the enable controls are deactivated. One advantage of using enable control is to make the cascading of two or more encoders to form a bigger one easier.

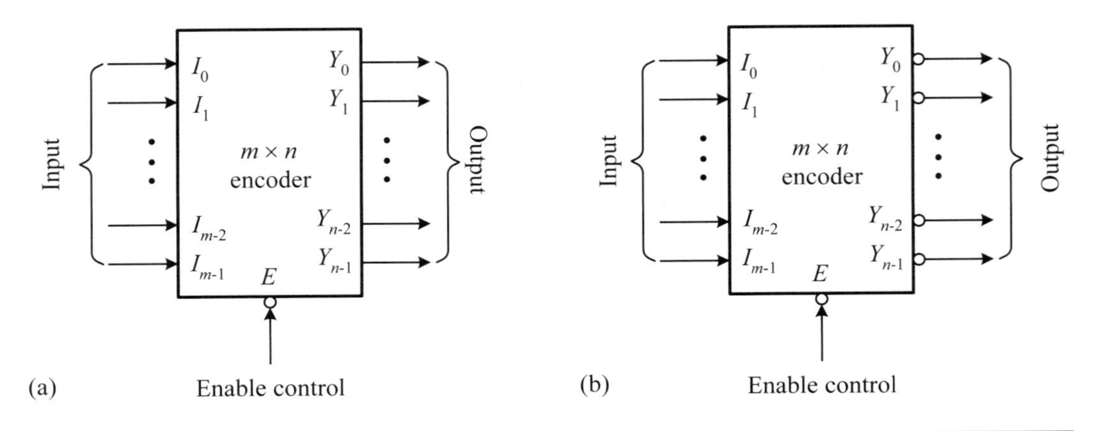

Figure 10.3: The general block diagrams of encoders with: (a) active-high output; (b) active-low output.

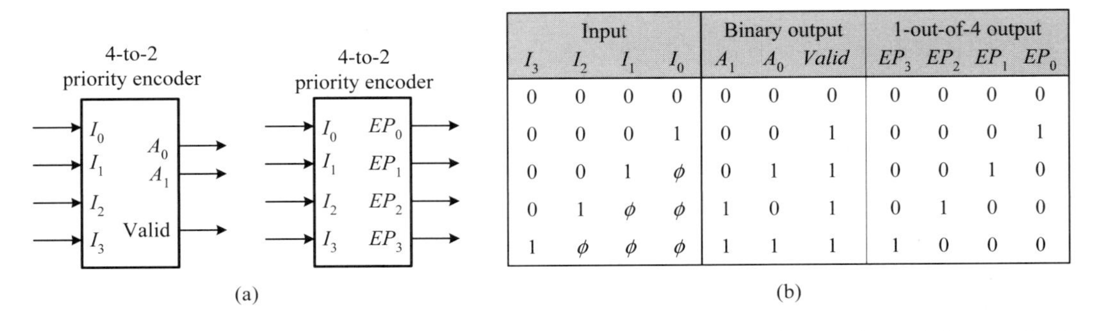

Input				Binary output			1-out-of-4 output			
I_3	I_2	I_1	I_0	A_1	A_0	*Valid*	EP_3	EP_2	EP_1	EP_0
0	0	0	0	0	0	0	0	0	0	0
0	0	0	1	0	0	1	0	0	0	1
0	0	1	ϕ	0	1	1	0	0	1	0
0	1	ϕ	ϕ	1	0	1	0	1	0	0
1	ϕ	ϕ	ϕ	1	1	1	1	0	0	0

Figure 10.4: The (a) block diagrams and (b) function table of a 4×2 priority encoder.

10.1.2.1 Priority Encoders A *priority encoder* is a device that generates an output code corresponding to the activated input with the highest priority. The output of a priority encoder may be a binary code or a 1-out-of-n code, depending on the actual requirements. The conversion from a 1-out-of-n code to the binary code can be simply accomplished by a simple encoder, like the one described previously.

Like simple encoders, a general priority encoder has m inputs and n outputs. The value of m is equal to (or fewer than) 2^n when the output is a binary code and equals n when the output is a 1-out-of-n code. Each input has its own predetermined priority. In addition, an extra output, referred to as *valid*, is often used to indicate that any input is activated. For example, as shown in Figure 10.4, the priority is associated with the index values of the input lines; namely, the priority of the input lines are $I_3 > I_2 > I_1 > I_0$. As a consequence, when the input I_3 is 1, the output Y_1Y_0 is 11 regardless of the values of the other inputs.

■ Example 10-2: (A 4-to-2 priority encoder.) Figure 10.4 shows a 4-to-2 priority encoder. Figure 10.4(a) shows the block diagram and Figure 10.4(b) gives the function table. From the function table, we may obtain the following switching expressions for all outputs, $Valid$, A_0, and A_1:

$$Valid = I_3 + I_2 + I_1 + I_0$$

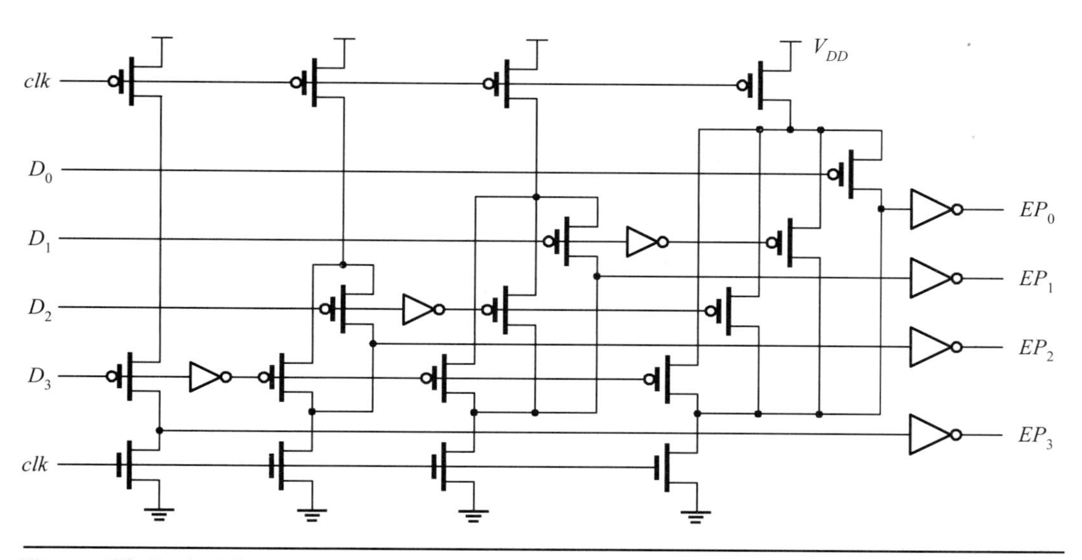

Figure 10.5: A 4-bit priority encoder implemented with domino logic.

$$A_1 = \bar{I}_3 \cdot I_2 + I_3$$
$$A_0 = \bar{I}_3 \bar{I}_2 I_0 + I_3$$

The resulting logic circuit is easily realized using basic logic gates or MOS switches. ■

As we mentioned, a dynamic logic circuit has a faster speed than its static logic counterpart. We give several examples to illustrate how the priority encoders are implemented with dynamic logic circuits. For convenience, we will consider the case that the output code is a 1-out-of-4 code rather than a binary code. We begin with a priority encoder implemented with a domino logic circuit and then give two modular design examples.

10.1.2.2 Implementation with Domino Logic A simple 4-bit priority encoder implemented with domino logic is shown in Figure 10.5. Assume that D_0 to D_3 are the input signals. The D_3 input has the highest priority while the D_0 has the lowest. Each EP_i corresponds to each D_i, for all $0 \leq i \leq 3$. Then, the following four switching expressions describe the priority function of each output EP_i.

$$\begin{aligned} EP_3 &= D_3 \\ EP_2 &= \bar{D}_3 \cdot D_2 \\ EP_1 &= \bar{D}_3 \cdot \bar{D}_2 \cdot D_1 \\ EP_0 &= \bar{D}_3 \cdot \bar{D}_2 \cdot \bar{D}_1 \cdot D_0 \end{aligned} \quad (10.1)$$

To implement the above four switching functions, the p-type domino logic is used because in the p-type domino logic circuit, the$p$-type metal-oxide-semiconductor (pMOS) transistors being used to implement the AND function are connected in parallel rather than in series. Thus, it results in fast operating speed and smaller area.

10.1.2.3 Modular Priority Encoders Another approach to implementing a priority encoder is based on the modular design. A possible basic unit cell of an n-bit priority

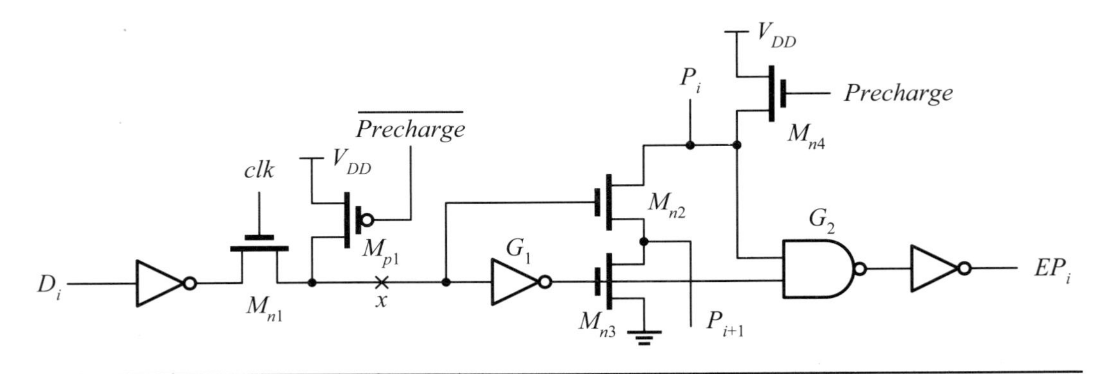

Figure 10.6: The unit cell EP_i of the conventional priority encoder.

encoder is shown in Figure 10.6. It works as follows. During the precharge phase, the signal *precharge* becomes high and n-type metal-oxide-semiconductor (nMOS) transistor M_{n4} turns on such that node P_i is set to a high voltage. The clock clk should be set to 0 to prevent the input D_i from entering the circuit. The pMOS transistor M_{p1} turns on and precharges node x to V_{DD}. This high voltage passes through inverter G_1 and the NAND gate G_2 to cause the output EP_i to become 0. During the evaluation phase, the signal *precharge* becomes 0 and the clock clk rises high. The nMOS transistor M_{n1} turns on such that the input D_i is allowed to pass into the circuit. If the input D_i is 1, nMOS transistor M_{n2} would not be turned on but nMOS transistor M_{n3} is on. Hence, the priority output P_{i+1} is set to 0 to inhibit that EP_{i+1} becomes high. The EP_i outputs a high value. On the other hand, if the input D_i is 0, the nMOS transistor M_{n2} is turned on but the nMOS transistor M_{n3} would not be turned on. Hence, the priority P_i is passed to the output P_{i+1} to enable the next stage. The EP_i remains at 0.

Because nMOS transistor M_{n4} is used as a precharge switch, the priority status P_i can only be precharged to $V_{DD} - V_{Tn}$ rather than V_{DD}. This is beneficial to reduce the time required to discharge P_{i+1} when nMOS transistor M_{n3} is set to 1. Nevertheless, this design will induce some DC power dissipation on the gate following P_i.

In order to reduce the propagation delays of the priority status, a lookahead circuit may be used. As shown in Figure 10.7, a lookahead line (that is, the *LA line*) and a pull-down nMOS transistor M_{n5} whose gate is driven by the request are added to the unit cell shown in Figure 10.6. The feature of this priority lookahead circuit is to provide a fast path for P_i to propagate to other cells through the lookahead line (*LA line*). To see this, let us consider the operation of this circuit. During the precharge phase, the lookahead line is set to 1. When the clock clk goes to 1 and if input D_i is 1, nMOS transistors M_{n3} and M_{n5} are turned on and both nodes *LA line* and P_{i+1} are discharged. The signal on the priority lookahead line (*LA line*) propagates to the lower cells much faster than the priority signal that propagates through the pass transistors (M_{n2}). As a consequence, it effectively reduces the increase of delay in propagating $P_i = 0$ when several basic cells are cascaded, resulting in a chain of pass transistors (M_{n2}).

Review Questions

Q10-5. Explain why a priority encoder is needed.

Q10-6. Give why the p-type domino logic circuit is used in Figure 10.5.

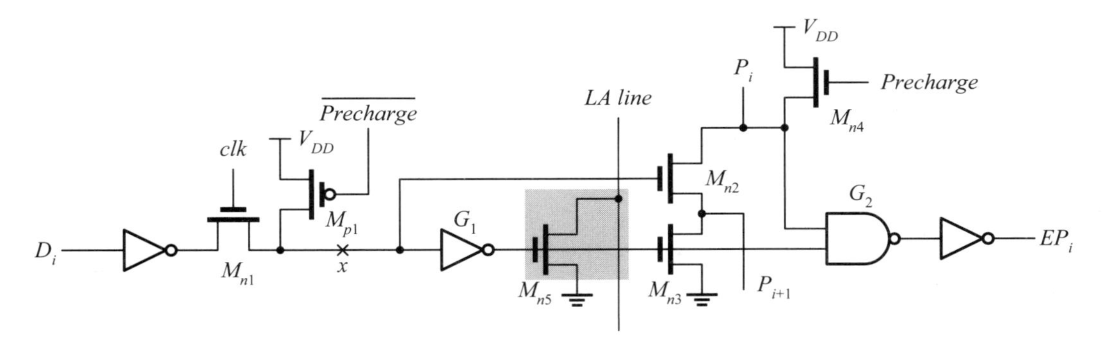

Figure 10.7: The unit cell EP_i of the priority encoder with lookahead.

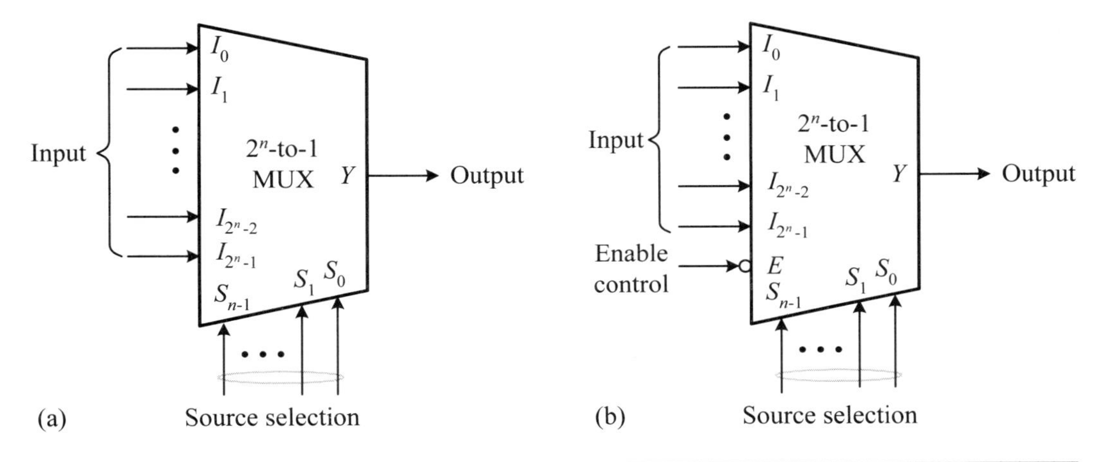

Figure 10.8: The general block diagrams of multiplexers (a) without and (b) with enable control.

Q10-7. Give why the *valid* signal is needed in a priority encoder.

Q10-8. Explain the operation of the unit cell shown in Figure 10.7.

10.1.3 Multiplexers

An m-to-1 ($m \leq 2^n$) multiplexer has m input lines, one output line, and n source-selection lines. The input line I_i, selected by the binary combination of n source-selection lines, is routed to the output line, Y. In other words, the input line I_i is routed to the output Y if the value of the combination of source-selection lines is i. The block diagrams of general multiplexers are shown in Figure 10.8. Figure 10.8(a) does not contain an enable control while Figure 10.8(b) includes an active-low enable control.

Like decoders and encoders, a multiplexer usually associates with one or more optional enable controls, which may be either active-high or active-low, depending on the actual requirement. The objective of enable control is to control the operations of the multiplexer. The output of a multiplexer is the value of a selected input when the

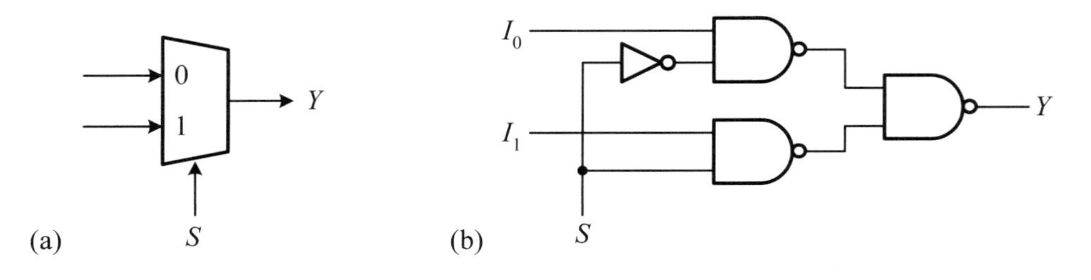

Figure 10.9: The (a) logic symbol and (b) logic circuit of a gate-based 2-to-1 multiplexer.

enable controls are asserted and is 1, 0, or even a high-impedance, depending on the actual need and design, when the enable controls are deactivated.

10.1.3.1 Implementations of Multiplexers Multiplexers may be generally implemented by using either the gate-based style or switch-based style. We are concerned with both styles along with appropriate examples. Also, the differences in features between different implementations are discussed.

Gate-based multiplexers. An example of a 2-to-1 multiplexer without an enable control is shown in Figure 10.9. Figure 10.9(a) is the logic symbol and Figure 10.9(b) is the logic circuit that realizes the following switching expression directly:

$$Y = \bar{S} \cdot I_0 + S \cdot I_1$$

■ Example 10-3: (A gate-based 4-to-1 multiplexer.) An example of a 4-to-1 multiplexer without an enable control is shown in Figure 10.10. Figure 10.10(a) is the logic symbol and Figure 10.10(b) is the function table. From the function table, the output function can be represented as follows:

$$Y = \bar{S}_1 \cdot \bar{S}_0 \cdot I_0 + \bar{S}_1 \cdot S_0 \cdot I_1 + S_1 \cdot \bar{S}_0 \cdot I_2 + S_1 \cdot S_0 \cdot I_3$$

The logic circuit based on basic gates is shown in Figure 10.10(c).

■

Domino logic multiplexers. Instead of implementing with individual logic gates, a 4-to-1 multiplexer may also be implemented with a complex complementary metal-oxide-semiconductor (CMOS) logic circuit. An illustration of using domino logic to implement a 4-to-1 multiplexer is explored in the following example.

■ Example 10-4: (A 4-to-1 multiplexer implemented with domino logic.) Realizing a 4-to-1 multiplexer using domino logic is as simple as the gate-based style, namely, simply replacing each gate with a domino logic circuit. However, a single domino logic circuit may be used to implement the switching expression of the 4-to-1 multiplexer directly whenever the 4-to-1 multiplexer can be expressed as in the following:

$$\begin{aligned} Y &= \bar{S}_1 \cdot \bar{S}_0 \cdot I_0 + \bar{S}_1 \cdot S_0 \cdot I_1 + S_1 \cdot \bar{S}_0 \cdot I_2 + S_1 \cdot S_0 \cdot I_3 \\ &= E_0 \cdot I_0 + E_1 \cdot I_1 + E_2 \cdot I_2 + E_3 \cdot I_3 \end{aligned}$$

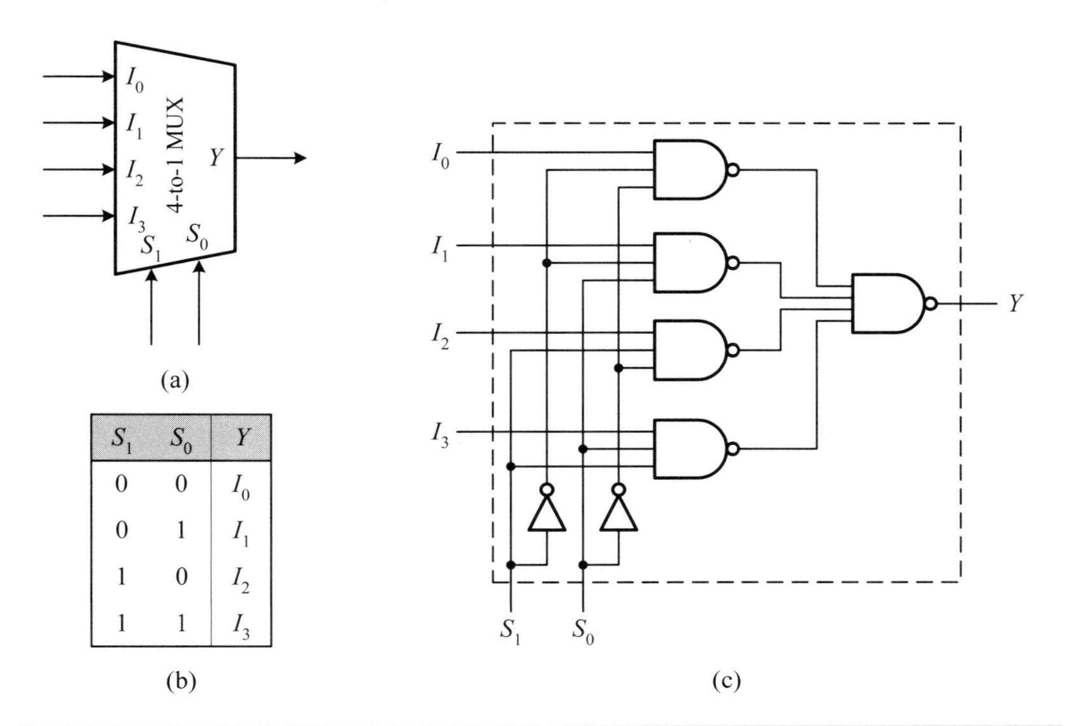

S_1	S_0	Y
0	0	I_0
0	1	I_1
1	0	I_2
1	1	I_3

Figure 10.10: A gate-based 4-to-1 multiplexer: (a) logic symbol; (b) function table; (c) logic circuit.

where $E_0 = \bar{S}_1 \cdot \bar{S}_0$, $E_1 = \bar{S}_1 \cdot S_0$, $E_2 = S_1 \cdot \bar{S}_0$, and $E_3 = S_1 \cdot S_0$ are assumed to be readily available from some other circuits. The resulting domino logic circuit is shown in Figure 10.11.

■

Switch-based multiplexers. A switch-based multiplexer can be implemented in either uniform structure or tree structure. In the uniform structure, only one nMOS or transmission gate (TG) switch is used in each branch to control the datapath from the input to the output. These switches are in turn enabled by the control signals generated from the combinations of control variables by using an n-to-2^n decoder. The following example illustrates how the uniform structure can be used to implement a switch-based 2-to-1 multiplexer. The tree structure will be dealt with in Section 11.2.4.

■ Example 10-5: (A switch-based 2-to-1 multiplexer.) In CMOS technology, it is usual to implement 2-to-1 multiplexers with TG or nMOS switches. From Figure 10.9(a) or (b), we know that the switching expression for the 2-to-1 multiplexer is as follows:

$$Y = \bar{S} \cdot I_0 + S \cdot I_1$$

Hence, by using the selection variable S as the control variable, the resulting circuit is obtained. The TG-switch 2-to-1 multiplexer is shown in Figure 10.12(a) and the nMOS-switch 2-to-1 multiplexer is given in Figure 10.12(b). Note that a buffer formed with two inverters is added to the output to restore the voltage level deteriorated by

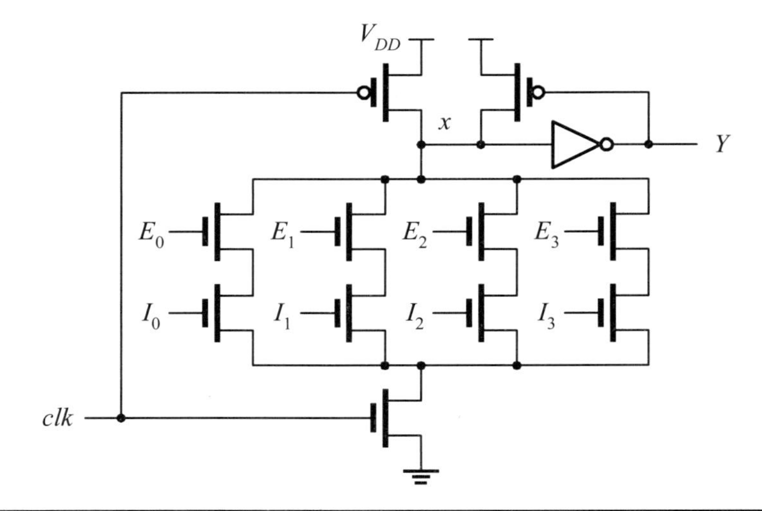

Figure 10.11: A 4-to-1 multiplexer implemented in domino logic style.

the nMOS switches. The other implementations of the 2-to-1 multiplexer with different logic styles are given in Figures 10.12(c) to 10.12(f).

■

■ Review Questions

Q10-9. What are the features of gate-based multiplexers?

Q10-10. What are the features of tristate-buffer multiplexers?

Q10-11. What are the features of switch-based multiplexers?

10.1.4 Demultiplexers

Recall that demultiplexing is the reverse process of multiplexing and routes the input to a selected output line. The logic circuit used to perform demultiplexing is known as a demultiplexer. A 1-to-m ($m \leq 2^n$) demultiplexer has one input line, m output lines, and n destination selection lines. The input line D is routed to the output line Y_i selected by the binary combination of n destination selection lines. In other words, the output line Y_i is set to D if the value of the combination of destination selection lines is i. Like multiplexers, a demultiplexer usually associates with one or more optional enable controls, which may be either active-high or active-low, depending on the actual needs.

The block diagrams of general demultiplexers are shown in Figure 10.13. Figure 10.13(a) does not contain an enable control and Figure 10.13(b) includes an active-low enable control. The purpose of enable controls is to control the operations of the demultiplexer. The output line selected by the destination selection lines has the same value as the input line when the enable controls are asserted and all output lines are 1s, 0s, or even high-impedance, depending on the actual requirement and design, when the enable controls are deactivated.

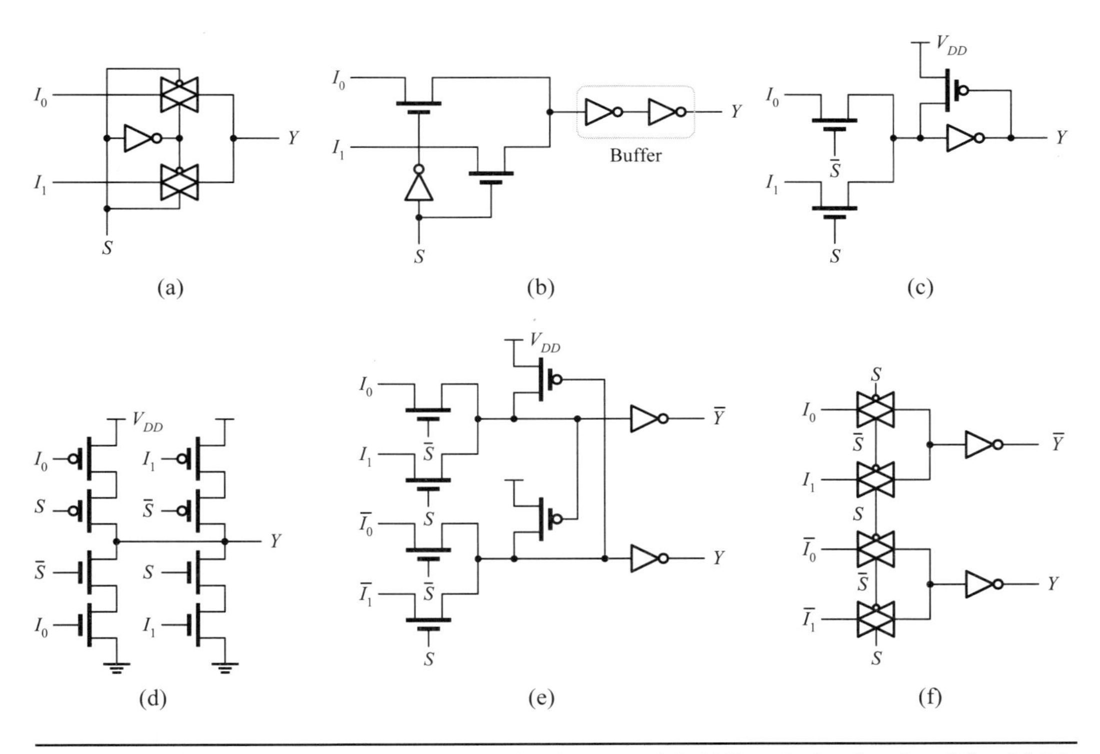

Figure 10.12: Switch-based 2-to-1 multiplexers: (a) TG-switch circuit; (b) nMOS circuit; (c) LEAP circuit; (d) CMOS circuit; (e) CPL circuit; (f) DPL circuit.

10.1.4.1 Implementations of Demultiplexers Like multiplexers, demultiplexers may be generally implemented by using either the gate-based style or switch-based style. We are concerned with both styles by way of appropriate examples.

Gate-based demultiplexers. Like multiplexers, demultiplexers may be implemented by using either individual logic gates or a complex CMOS logic circuit with multiple outputs.

■ Example 10-6: (A gate-based 1-to-4 demultiplexer.) An example of a 1-to-4 demultiplexer with an active-low enable control is shown in Figure 10.14. Figure 10.14(a) is the logic symbol and Figure 10.14(b) is its function table. From the function table, we may obtain the output switching expressions as follows:

$$\begin{aligned}
Y_0 &= \bar{E} \cdot \bar{S}_1 \cdot \bar{S}_0 \cdot D \\
Y_1 &= \bar{E} \cdot \bar{S}_1 \cdot S_0 \cdot D \\
Y_2 &= \bar{E} \cdot S_1 \cdot \bar{S}_0 \cdot D \\
Y_3 &= \bar{E} \cdot S_1 \cdot S_0 \cdot D
\end{aligned}$$

The resulting logic circuit is shown in Figure 10.14(c).

■

Switch-based demultiplexers. Switch-based demultiplexers are indeed the inverse network of the multiplexers. Here, the "inverse" means that when inputs and outputs are exchanged, the network topology remains the same. As an illustration,

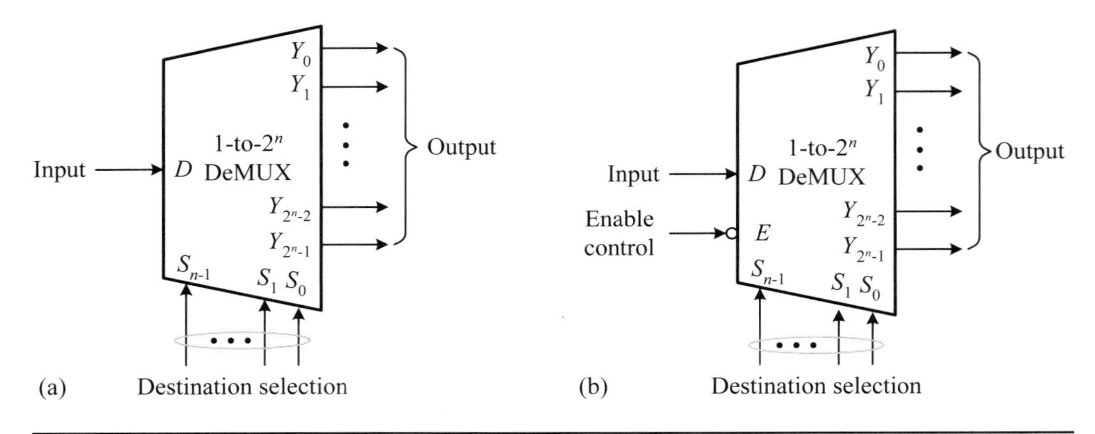

Figure 10.13: The general block diagrams of demultiplexers (a) without and (b) with enable control.

consider the TG-based 2-to-1 multiplexer shown in Figure 10.12(a). As the output Y is set as the input D and both inputs I_0 and I_1 are set as the outputs Y_0 and Y_1, respectively, the result is a 1-to-2 demultiplexer.

An important difference between the gate-based and switch-based demultiplexers is that all outputs of a gated-based demultiplexer always have definite values regardless of whether it is selected, but the unselected outputs of a switch-based demultiplexer have high-impedance values (namely, floating). In many applications, this feature may need to be considered very carefully; otherwise, some unexpected errors or failures may arise.

■ Review Questions

Q10-12. Describe the features of tristate-buffer demultiplexers.

Q10-13. What type of 1-to-2^n demultiplexer can be used as an n-to-2^n decoder?

Q10-14. What type of n-to-2^n decoders can be used as a 1-to-2^n demultiplexer?

10.1.5 Magnitude Comparators

It is often desirable to compare two numbers for equality in digital systems. A circuit that compares two numbers and indicates whether they are equal is called a *comparator*, or *equality detector*. A *magnitude comparator* is a circuit that not only compares two numbers for testing their equality but also indicates their arithmetic relationship. The numbers input to a magnitude comparator can be signed or unsigned generally.

■ Example 10-7: (An n-bit comparator.) A 4-bit equality detector consisting of four XOR gates and an AND gate is shown in Figure 10.15. To construct an n-bit equality detector, at least two approaches may be used. One is first to partition the n-bit inputs into $\lceil \frac{n}{4} \rceil$ groups and then to use the 4-bit equality detector as a basic module, one for each group. Finally, an AND gate tree is employed to combine the comparison results from all groups. This approach needs n XNOR gates and $(n-1)$ two-input AND gates.

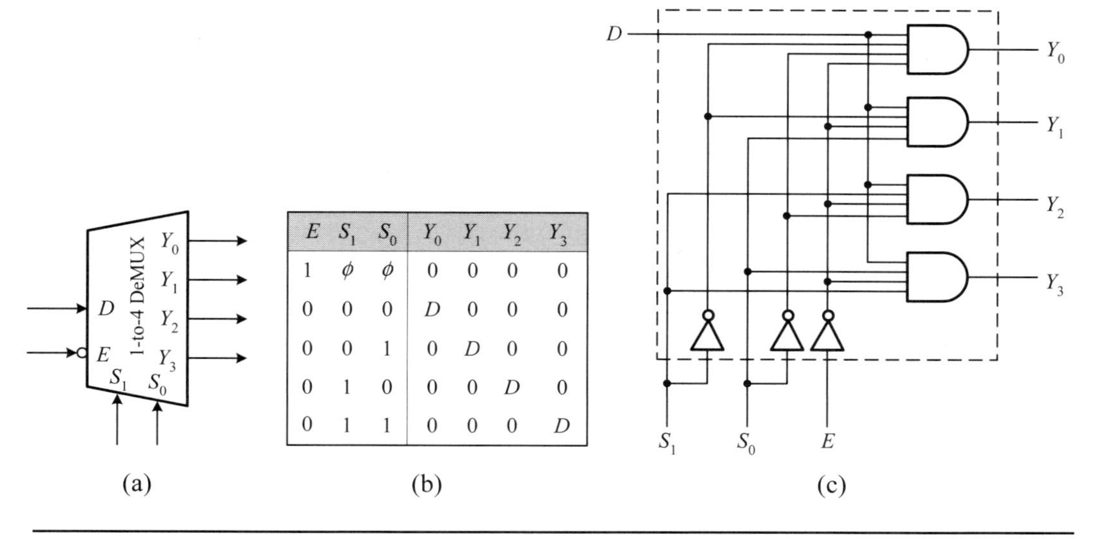

E	S_1	S_0	Y_0	Y_1	Y_2	Y_3
1	ϕ	ϕ	0	0	0	0
0	0	0	D	0	0	0
0	0	1	0	D	0	0
0	1	0	0	0	D	0
0	1	1	0	0	0	D

Figure 10.14: A gate-based 1-to-4 demultiplexer: (a) logic symbol; (b) function table; (c) logic circuit.

Another approach is to use an XNOR tree to obtain the comparison result from n inputs directly. This approach needs $(2n - 1)$ 2-input XNOR gates. Since an XNOR gate is generally more complicated than an AND gate, this approach may not be better than the first one although it requires only one XNOR tree.

■

A magnitude comparator is used to compare two numbers, signed or unsigned, as well as to determine and indicate their arithmetic relationship, namely, less than, equal to, or greater than. By properly combining the input relationship, a magnitude comparator can be readily made into a *cascadable magnitude comparator*.

■ Example 10-8: (A 4-bit magnitude comparator.) Figure 10.16 shows an

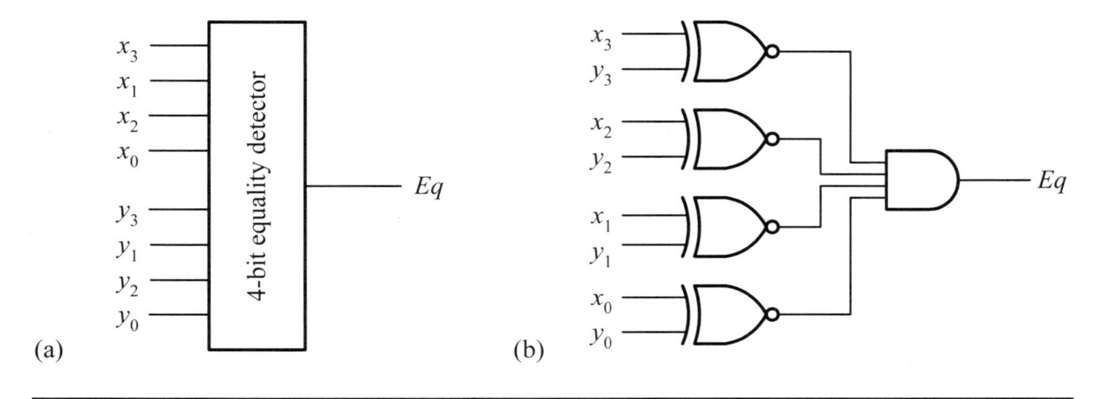

Figure 10.15: A 4-bit equality detector based on logic gates: (a) logic symbol; (b) logic circuit.

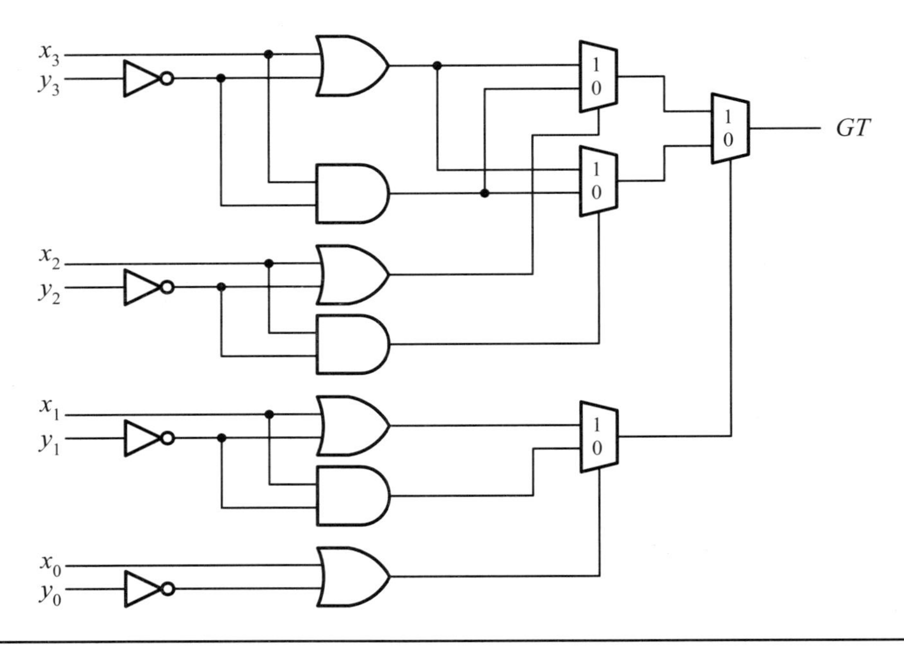

Figure 10.16: A 4-bit magnitude comparator realized at gate level.

unsigned magnitude comparator that is able to determine whether input number X is greater than or equal to Y. The output is set to 1 if $X \geq Y$ and set to 0 otherwise. ■

A magnitude comparator with full functionality can be constructed by combining the two comparators described above. Because of the intuitive simplicity, it is left to the reader as an exercise.

■ Review Questions

Q10-15. Define the comparator and magnitude comparator.

Q10-16. Combine the two comparators described in this subsection into a magnitude comparator.

10.2 Basic Sequential Components

Basic building blocks, such as latches and flip-flops, for constructing sequential components have been described in Chapter 9. In this section, we describe some basic sequential components built with these basic blocks. These components include registers, shift registers, counters, and sequence generators.

10.2.1 Registers

A data register (or register for short) is a set of flip-flops or latches, in which each flip-flop or latch is capable of storing one-bit information. A single-bit register is just a single flip-flop or latch. For most digital systems, flip-flops rather than latches are

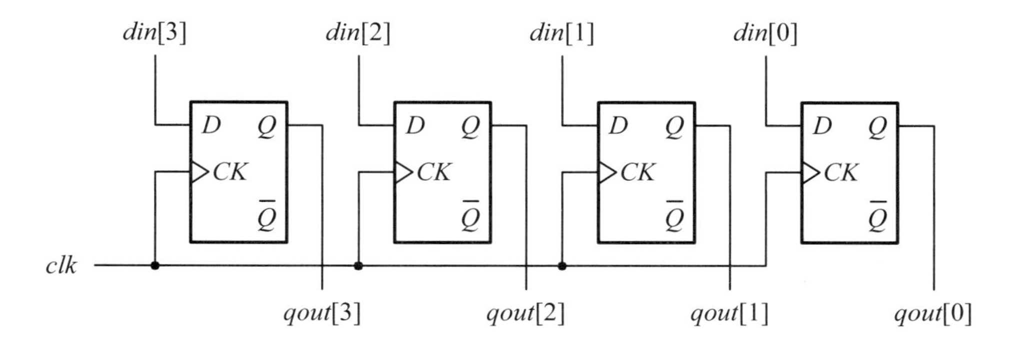

Figure 10.17: The logic diagram of a 4-bit data register.

usually used to construct a register. An n-bit register is a set of n flip-flops placed in parallel with a common clock and a clear (or reset) signals. Registers are usually used to store small amounts of information in digital systems. Figure 10.17 shows a 4-bit data register consisting of four D-type flip-flops with a common clock.

10.2.2 Shift Registers

Another type of register widely used in digital systems is a *shift register*. As its name implies, a shift register may shift its contents left or right a specified number of bit positions. In addition, shift registers are often employed in digital systems to perform various data format conversions. We begin with the introduction of a simple shift register and then deal with the concept of a universal shift register, which can be used to carry out various data format conversions.

The basic structure of a 4-bit shift register is shown in Figure 10.18(a). An n-bit shift register generally consists of n D-type flip-flops connected in series with a common clock. From the timing diagram depicted in Figure 10.18(b), the serial input data are sampled at each positive edge of the clock *clk* and shifted to the next stage per clock cycle.

Shift registers are widely used as data format converters, including serial-to-parallel conversion and vice versa. In such and other applications, a shift register often needs to perform the following operations exclusively:

- Left shift
- Right shift
- Parallel load
- No operation

A shift register having the capability of performing the above operations is known as a *universal shift register*. Refer to Lin [30, 32] for more detail about universal shift registers.

■ Review Questions

Q10-17. Define an n-bit register.

Q10-18. What is a universal shift register?

Q10-19. What are the main functions of shift registers?

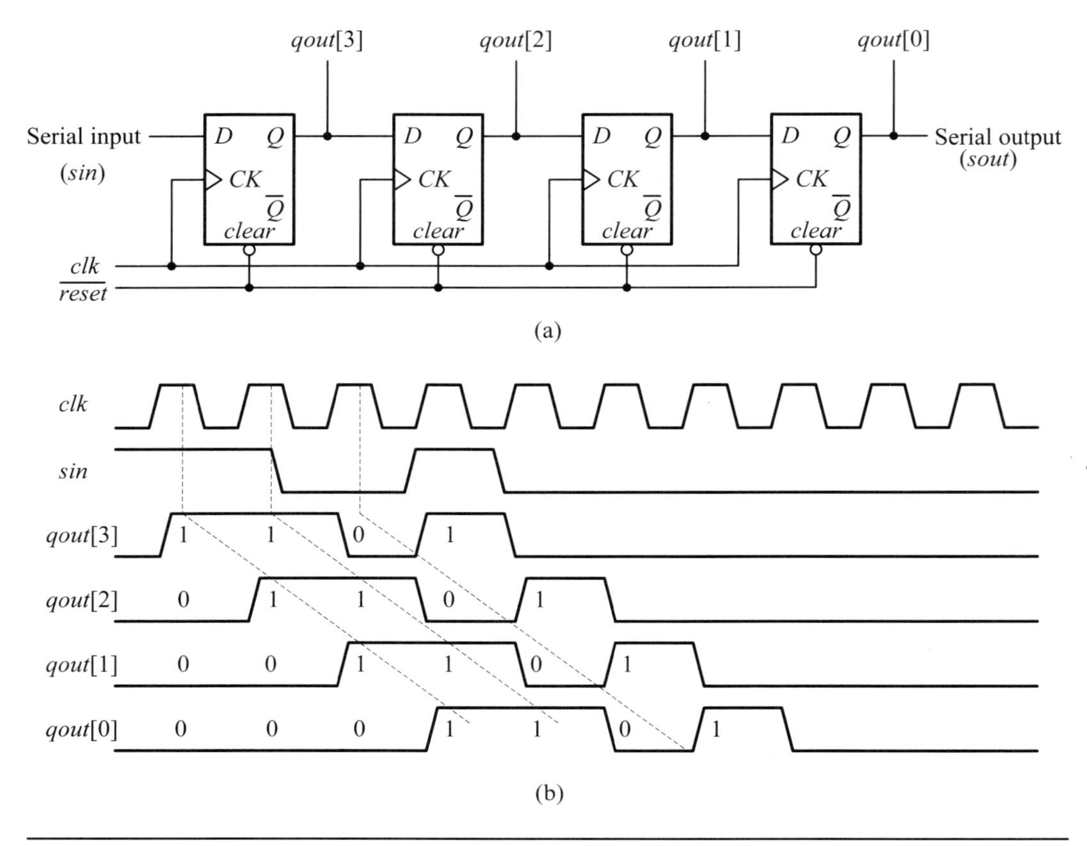

Figure 10.18: The (a) logic and (b) timing diagrams of a 4-bit shift register.

10.2.3 Counters

A counter is a device that counts the input events, such as input pulses or clock pulses. Many kinds of counters are found in a variety of digital systems, depending on the attempted usage. Counters can be classified into asynchronous and synchronous in terms of whether they are synchronous with a clock. In practical applications, the most common asynchronous (ripple) counters are binary counters and the most widely used synchronous counters include binary counters, BCD counters, as well as Gray counters. A counter is called a *timer* when its input or clock source is from a standard or known timing source, such as the system clock.

10.2.3.1 Ripple Counters Probably, the most commonly known type of counter is the ripple counter. An n-bit binary ripple counter consists of n T-type flip-flops (that is, JK flip-flops with their J and K inputs tied together to V_{DD}) connected in series; namely, the output of each flip-flop is connected to the clock input of its succeeding flip-flop. Figure 10.19(a) shows an example of a 3-bit binary ripple counter. As illustrated in the figure, the essential feature of a ripple counter is that each flip-flop is triggered by the output of its preceding stage except the first stage, which is triggered by an external signal, namely, the clock.

10.2.3.2 Synchronous Counters In a synchronous counter, all flip-flops change their states at the same time, usually under the control of a clock. An example of a 3-bit

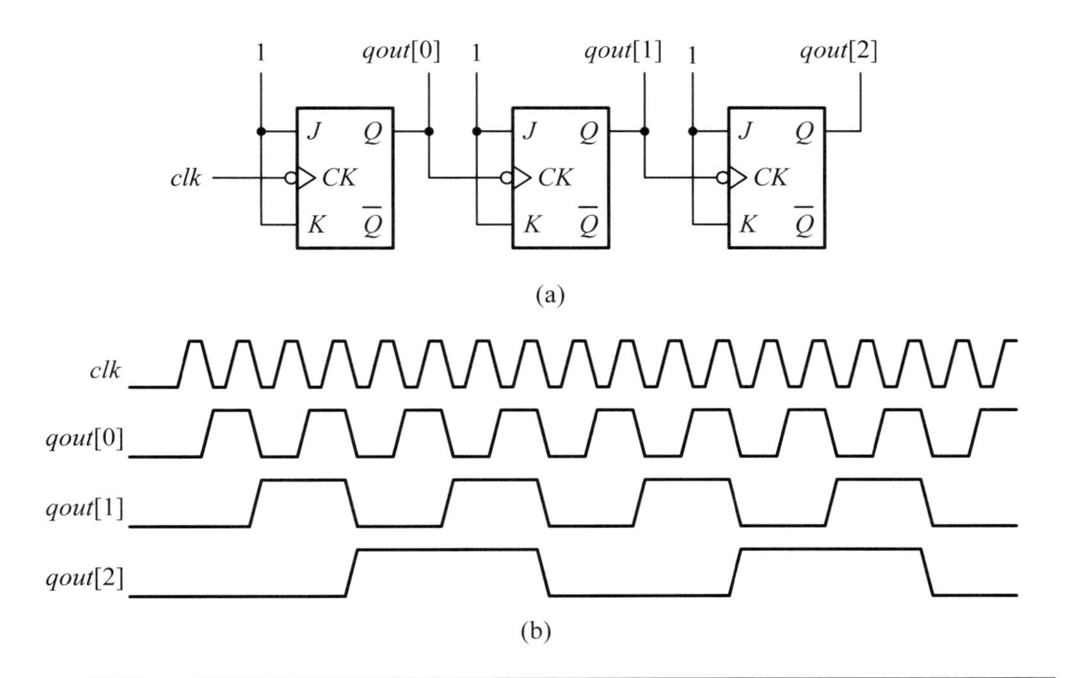

Figure 10.19: The (a) logic and (b) timing diagrams of a 3-bit ripple binary counter.

synchronous binary counter is shown in Figure 10.20. Figure 10.20(a) operates in the parallel-carry mode while Figure 10.20(b) operates in the ripple-carry mode. The maximum operating frequency of the parallel-carry mode is equal to

$$f_{max} = \frac{1}{t_q + t_{setup} + t_{pd}} \tag{10.2}$$

where t_q and t_{setup} are the clock-to-Q and setup time of the flip-flops, and t_{pd} is the propagation delay of the AND gates, assuming that the propagation delay of AND gates is independent of fan-in. However, as we mentioned previously, the propagation delay of a gate is approximately a quadratical function of its fan-in. Hence, in practice, a tree structure consisting of gates with a fan-in of 2 to 4 is often employed to replace the single large fan-in gate to reduce the propagation delay from a quadratical function to a logarithmic function of a fan-in.

The maximum operating frequency of the ripple-carry mode is

$$f_{max} = \frac{1}{t_q + t_{setup} + (n-1)t_{pd}} \tag{10.3}$$

Thus, the maximum operating frequency of it is less than the one being operated in the parallel-carry mode.

10.2.4 Sequence Generators

In addition to being used as data registers and data format converters, shift registers also found their widespread use as sequence generators. In this section, we focus on the following circuits: *ring counters*, the *Johnson counter*, and *pseudo-random sequence* (PR-sequence) *generators*.

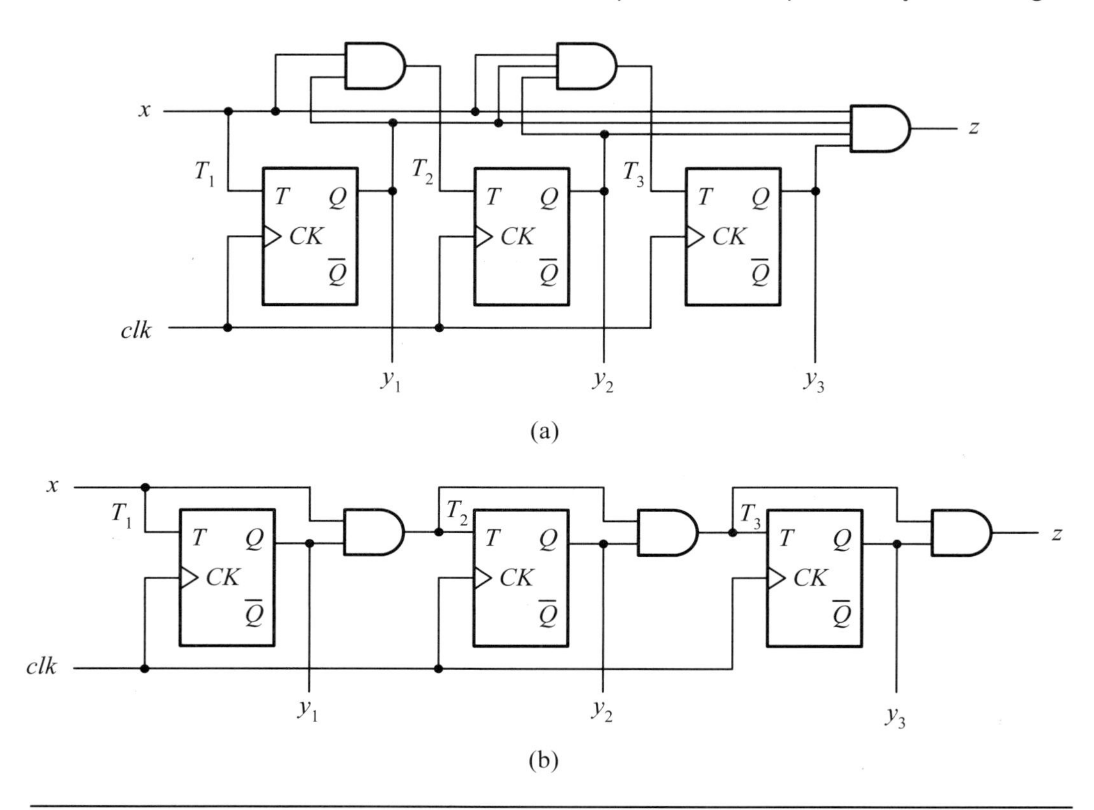

Figure 10.20: The logic diagrams of a 3-bit synchronous binary counter: (a) parallel-carry mode; (b) ripple-carry mode.

The general block diagram of sequence generators based on a shift register is shown in Figure 10.21, where the outputs of all D-type flip-flops of the shift register are combined through a combinational logic circuit, and the result is then fed back into the input of the shift register. The sequence generator is called a *linear feedback shift register* (LFSR) if the combinational logic circuit is a network composed of only XOR gates. An LFSR is called a *pseudo-random sequence generator* (PRSG) if it is capable of generating a maximum-length sequence; namely, the sequence has a period of $2^n - 1$, excluding the case of all 0s, where n is the number of flip-flops. Such a circuit is also known as an *autonomous linear feedback shift register* (ALFSR) and its generated sequence is often called a *pseudo-random sequence* (PR-sequence). PR-sequences are widely used in data networks, communications, and very-large-scale integration (VLSI) testing.

Depending on how the combinational circuit is constructed, Figure 10.21 can also be used to generate sequences other than those that will be discussed in this subsection. However, these applications are beyond the scope of this book. The reader interested in these issues can refer to Lin [30].

10.2.4.1 Ring Counters There are two basic types of ring counters: *standard ring counter* and *Johnson counter*. A modulo-n standard ring counter is an n-bit (or n-stage) shift register with the serial output fed back to the serial input. An essential feature of an n-bit ring counter is that it outputs a 1-out-of-n code directly from the flip-flop outputs.

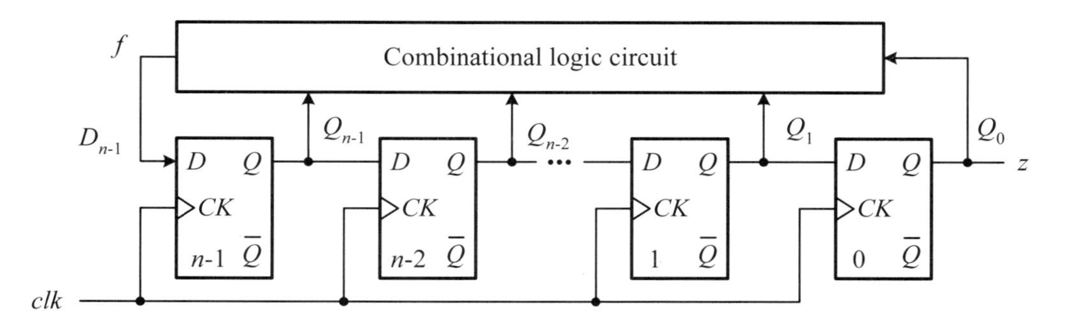

Figure 10.21: The general block diagram of sequence generators.

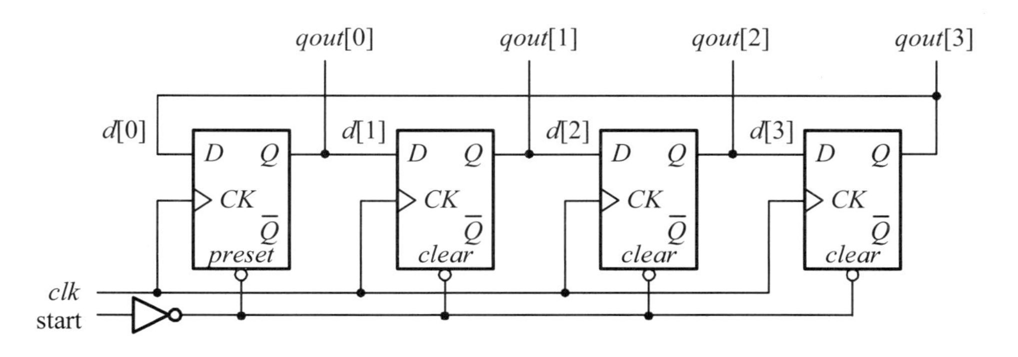

Figure 10.22: An example of a 4-bit ring counter.

An example of a 4-bit ring counter is shown in Figure 10.22, where the *qout*[3] is connected backward to the input *d*[0]. In addition, a start circuit is used to set the initial value of the ring counter to 1000.

10.2.4.2 Johnson Counters A Johnson counter is also known as a *twisted-ring counter*, a *Moebius*, or a *switched-tail counter*. A modulo-$2n$ Johnson counter is an n-bit shift register with the complementary serial output fed back to the serial input. The essential feature of a Johnson counter is that it only requires a half number of flip-flops needed in a ring counter to achieve the same number of counting states.

An example of a 4-bit Johnson counter is shown in Figure 10.23, where the complement of the last-stage output is fed back to the input of the first stage. A start circuit is also employed to clear the counter to set the initial value of the counter.

10.2.4.3 PR-Sequence Generators A PR-sequence is a polynomial code that is based on treating bit strings as a polynomial with coefficients of 0 and 1 only. Let x denote a unit delay, corresponding to a D-type flip-flop, and x^k denote a k-unit delay. Then, an n-stage LFSR can be represented as the following polynomial:

$$f(x) = a_n x^n + a_{n-1}x^{n-1} + \cdots + a_1 x + a_0 \tag{10.4}$$

where coefficients $a_i \in \{0,1\}$, for all $0 \leq i \leq n-1$. Each combination of coefficients a_i of a given n corresponds to a function $f(x)$. However, not all functions $f(x)$ can produce a maximum-length sequence. In fact, only a few of them can achieve this. When a polynomial $f(x)$ generates a *maximum-length sequence*, it is called a *primitive*

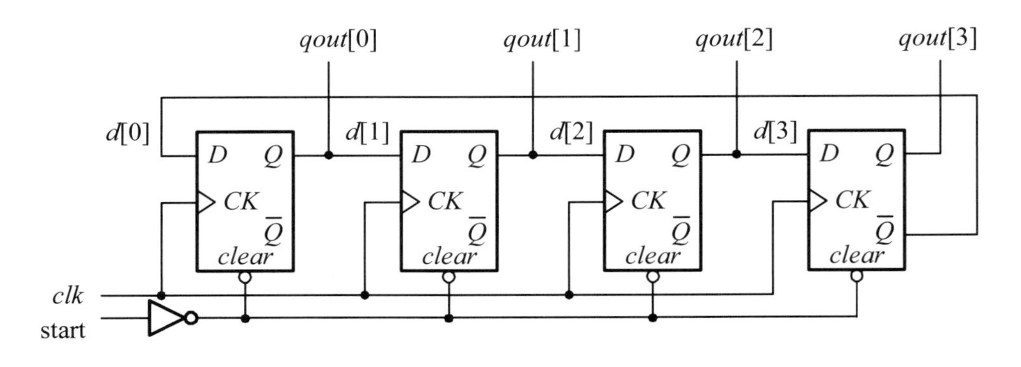

Figure 10.23: An example of a 4-bit Johnson counter.

Table 10.1: Sample primitive polynomials for n ranging from 1 to 60.

n	$f(x)$	n	$f(x)$	n	$F(x)$
1,2,3,4,6	$1+x+x^n$	24	$1+x+x^2+x^7+x^n$	44	$1+x+x^{26}+x^{27}+x^n$
7,15,22		26	$1+x+x^2+x^6+x^n$	50	
60		30	$1+x+x^2+x^{23}+x^n$	45	$1+x+x^3+x^4+x^n$
5,11,21,29	$1+x^2+x^n$	32	$1+x+x^2+x^{22}+x^n$	46	$1+x+x^{20}+x^{21}+x^n$
10,17,20	$1+x^3+x^n$	33	$1+x^{13}+x^n$	48	$1+x+x^{27}+x^{28}+x^n$
25,28,31		34	$1+x+x^{14}+x^{15}+x^n$	48	$1+x+x^{27}+x^{28}+x^n$
41,52		35	$1+x^2+x^n$	49	$1+x^9+x^n$
8	$1+x^2+x^3+x^4+x^n$	36	$1+x^{11}+x^n$	51	$1+x+x^{15}+x^{16}+x^n$
9	$1+x^4+x^n$	37	$1+x^2+x^{10}+x^{12}+x^n$	53	
12	$1+x+x^4+x^6+x^n$	38	$1+x+x^5+x^6+x^n$	54	$1+x+x^{36}+x^{37}+x^n$
13	$1+x+x^3+x^4+x^n$	39	$1+x^4+x^n$	55	$1+x^{24}+x^n$
14,16	$1+x^3+x^4+x^5+x^n$	40	$1+x^2+x^{19}+x^{21}+x^n$	56	$1+x+x^{21}+x^{22}+x^n$
18	$1+x^7+x^n$	42	$1+x+x^{22}+x^{23}+x^n$	59	
19,27	$1+x+x^2+x^5+x^n$	43	$1+x+x^5+x^6+x^n$	57	$1+x^7+x^n$
23,47	$1+x^5+x^n$			58	$1+x^{19}+x^n$

polynomial. In other words, a PR-sequence is generated by a primitive polynomial. Sample primitive polynomials for n ranging from 1 to 60 are listed in Table 10.1.

There are two standard paradigms for implementing primitive polynomials of any given n. These are known as *standard format* and *modular format*, respectively, as shown in Figure 10.24. In the standard format, all D-type flip-flops are connected as a shift register and an XOR gate network is used to compute the primitive function $f(x)$ and feed the result back to the input of the $n-1$ stage, as depicted in Figure 10.24(a). In the modular format, an XOR gate is placed between two D-type flip-flops to compute the primitive function $f(x)$ in place, as depicted in Figure 10.24(b).

■ **Example 10-9: (A 4-bit PR-sequence generator.)** An example of a 4-bit PR-sequence generator having a primitive polynomial, $1+x+x^4$, and being realized in standard form is depicted in Figure 10.25. A start circuit is also utilized to set the initial value to 1000. The reader is encouraged to verify that the PR-sequence generator indeed generates a maximum-length sequence.

▬

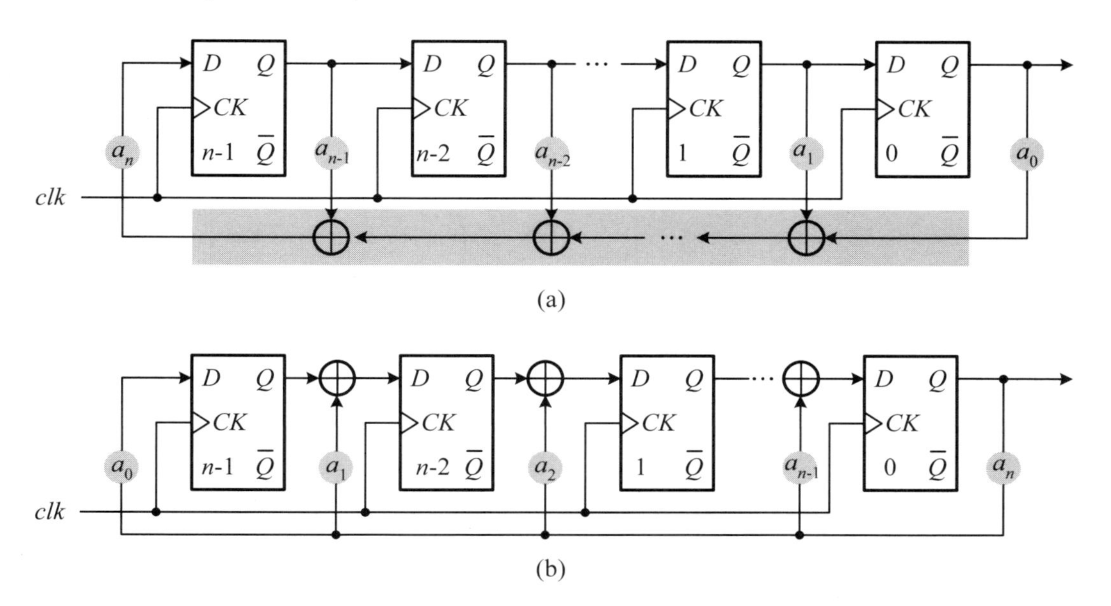

Figure 10.24: The (a) standard and (b) modular formats for implementing primitive polynomials.

■ Review Questions

Q10-20. What is the distinction between a counter and a timer?

Q10-21. What is the distinction between a ring counter and a Johnson counter?

Q10-22. What is the pseudo-random sequence?

Q10-23. What is the primitive polynomial?

Q10-24. How would you generate a PR sequence?

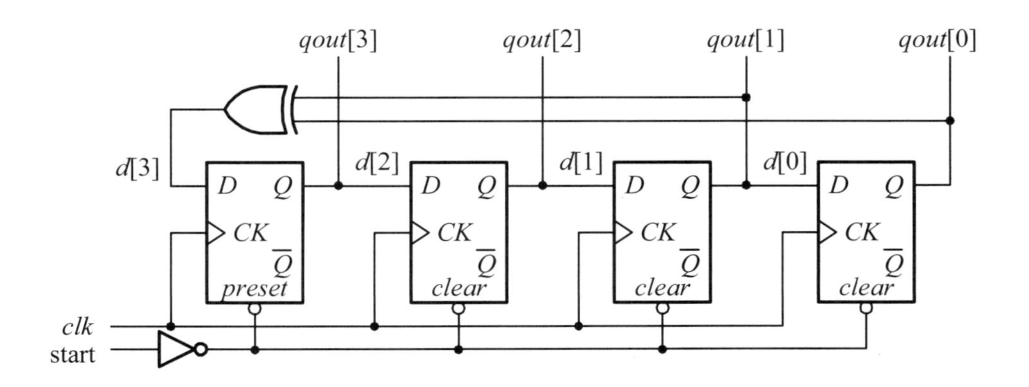

Figure 10.25: A 4-bit PR-sequence generator with a primitive polynomial: $1 + x + x^4$.

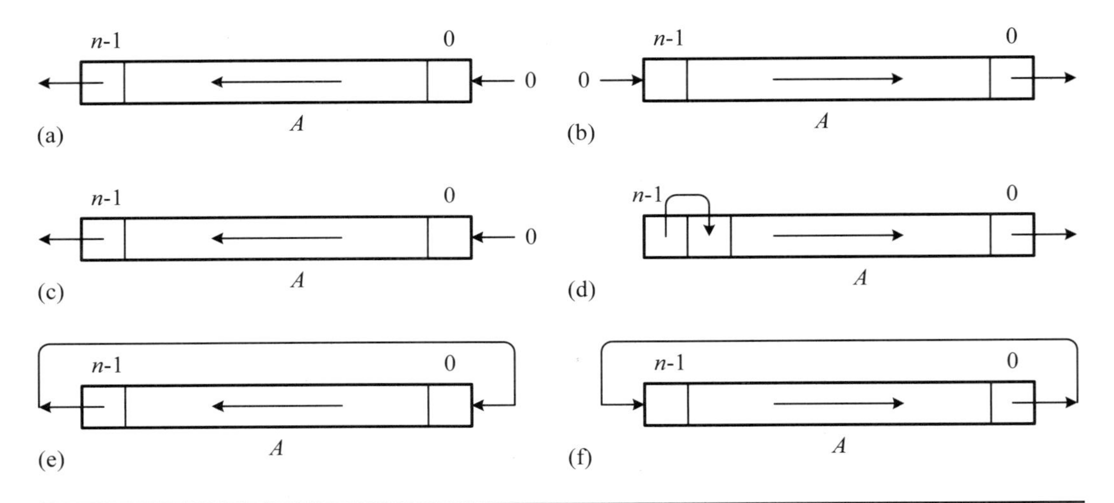

Figure 10.26: The block diagrams of a variety of shifts and rotations: (a) logical left shift; (b) logical right shift; (c) arithmetic left shift; (d) arithmetic right shift; (e) left rotation; (f) right rotation.

10.3 Shifters

Shift operations and shifters play important roles in arithmetic and logical operations. In this section, we begin with the definitions of basic shift operations, and then introduce their implementations with combinational logic circuits.

10.3.1 Basic Shift Operations

A shift operation is to shift an input number left or right a specified number of bit positions with zeros or a sign bit filled in the vacancies left after the shift. Depending on whether the vacancy is filled with a 0 or the sign bit, the shift operation can be cast into *logical shift* and *arithmetic shift*.

Logical shift. In the logical shift operation, the vacant bits are filled with 0s. Depending on the shift direction, it can be further divided into the following two types:

- *Logical left shift:* The input is shifted left a specified number of bit positions, and all vacant bits are filled with 0s, as depicted in Figure 10.26(a).
- *Logical right shift:* The input is shifted right a specified number of bit positions, and all vacant bits are filled with 0s, as depicted in Figure 10.26(b).

Arithmetic shift. The basic feature of an arithmetic shift is that the shifted result is the input number divided by 2 when the operation is right shift and is the input number multiplied by 2 when the operation is left shift. More formally, arithmetic shift can be subdivided into the following two types:

- *Arithmetic left shift:* The input is shifted left a specified number of bit positions and all vacant bits are filled with 0s, as shown in Figure 10.26(c). Indeed, this is exactly the same as logical left shift.
- *Arithmetic right shift:* The input is shifted right a specified number of bit positions and all vacant bits are filled with the sign bit, as shown in Figure 10.26(d).

Rotation. *Rotation*, also called *circular shift*, circulates the bits around the two ends without loss of information. More formally, rotation can be classified into the following two types:

- *Left rotation:* The input is circularly shifted left a specified number of bit positions and all bits shifted out are refilled from the least-significant bit in sequence, as shown in Figure 10.26(e).
- *Right rotation:* The input is circularly shifted right a specified number of bit positions and all bits shifted out are refilled from the most-significant bit in sequence, as shown in Figure 10.26(f).

10.3.2 Implementation Options of Shifters

The device used to perform the shift operation, arithmetic or logical, described above is known as a *shifter*. In general, an n-bit shifter is a device capable of shifting its input number n bits at most. An n-bit shifter may also be implemented with either a multiple-cycle or a single-cycle structure.

The multiple-cycle structure is basically a sequential logic circuit. The most common approach to implementing such a circuit is the use of a shift register, such as the one described in Section 10.2.2. The shift register loads the data to be shifted and performs the desired number of shift operations at the cost of an equal number of clock cycles. Detailed operations of the shift register can be found in the related section again.

On the contrary, the single-cycle structure of a shifter is simply a combinational logic circuit. We introduce a kind of combinational logic circuits that can implement the shifter with an arbitrary number of shifts. Such a logic circuit is referred to as a *barrel shifter*.

10.3.2.1 Barrel Shifters A barrel shifter is a circuit that consists of an input $A = \langle a_{n-1}, a_{n-2}, \cdots, a_1, a_0 \rangle$, an output $B = \langle b_{n-1}, b_{n-2}, \cdots, b_1, b_0 \rangle$, and the number of shifts $S = \langle s_{m-1}, s_{m-2}, \cdots, s_1, s_0 \rangle$, where $m \leq n$. According to the operations illustrated in Figure 10.26, it is easy to design a barrel shifter to carry out the desired operations. We use left- and right-rotation as well as logical/arithmetic-left shifters as examples to explain how such barrel shifters can be constructed.

The relationship between the output and input of the left- and right-rotation barrel shifter can be described as follows:

$$b_i = \begin{cases} a_{(i-s) \bmod n} & \text{(Left rotation)} \\ a_{(i+s) \bmod n} & \text{(Right rotation)} \end{cases} \tag{10.5}$$

where $i = 0, 1, \cdots, n-1$, and $s = 0, 1, \cdots, m-1$. The logical/arithmetic-left barrel shifter can be described as follows:

$$b_i = \begin{cases} 0 & \text{for all } i < s \\ a_{(i-s) \bmod n} & \text{for all } i \geq s \end{cases} \tag{10.6}$$

The other operations can be defined in a similar way.

Barrel shifters can be realized in either a linear or logarithmic number of stages. For convenience, the former will be denoted as a *linear barrel shifter* and the latter as a *logarithmic barrel shifter*. A linear barrel shifter consists of an array of nMOS or TG switches in which the number of rows equals the word length of the data and the number of columns is equal to the maximum amount of shift. An example of

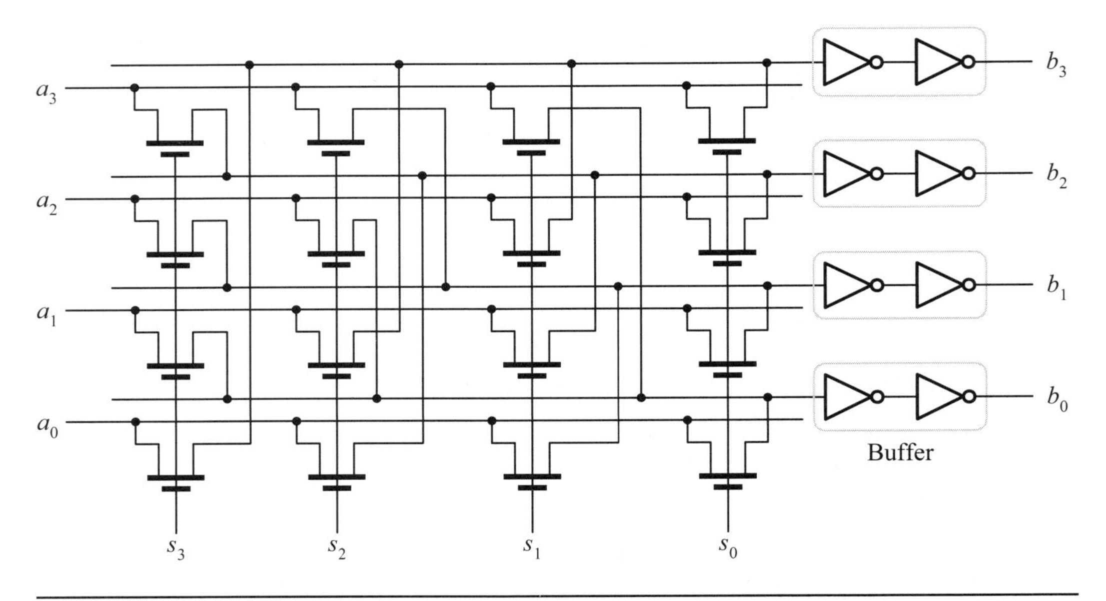

Figure 10.27: An example of a left-rotation linear barrel shifter based on nMOS switches.

using nMOS switches to realize a 4-bit left-rotation linear barrel shifter is given in the following.

■ Example 10-10: (A 4-bit left-rotation linear barrel shifter.) Figure 10.27 shows a 4-bit left-rotation linear barrel shifter implemented by an array of nMOS switches, with each column containing four nMOS switches. Each column is controlled by s_i and circularly shifts left the input of a number of i bit positions. In other words, each stage shifts its input the number of 0 or i bit positions, depending on whether the value of s_i is 0 or 1.

■

A linear barrel shifter can also be implemented with TG switches. An illustration of using TG switches to implement a 4-bit left/right-rotation linear barrel shifter is explored in the following example:

■ Example 10-11: (A 4-bit left/right-rotation linear barrel shifter.) Figure 10.28 shows a 4-bit left/right-rotation linear barrel shifter implemented by an array of TG switches, with each column containing four TG switches. Each column is controlled by s_i and circularly shifts left/right the input of a number of i bit positions. In other words, each stage shifts its input the number of 0 or i bit positions, depending on whether the value of s_i is 0 or 1. The unique feature of this barrel shifter is that it is capable of implementing left rotation and right rotation, depending on the labeling of the inputs and outputs, as indicated in the figure.

■

The major features of barrel shifters with linear number of stages are as follows.

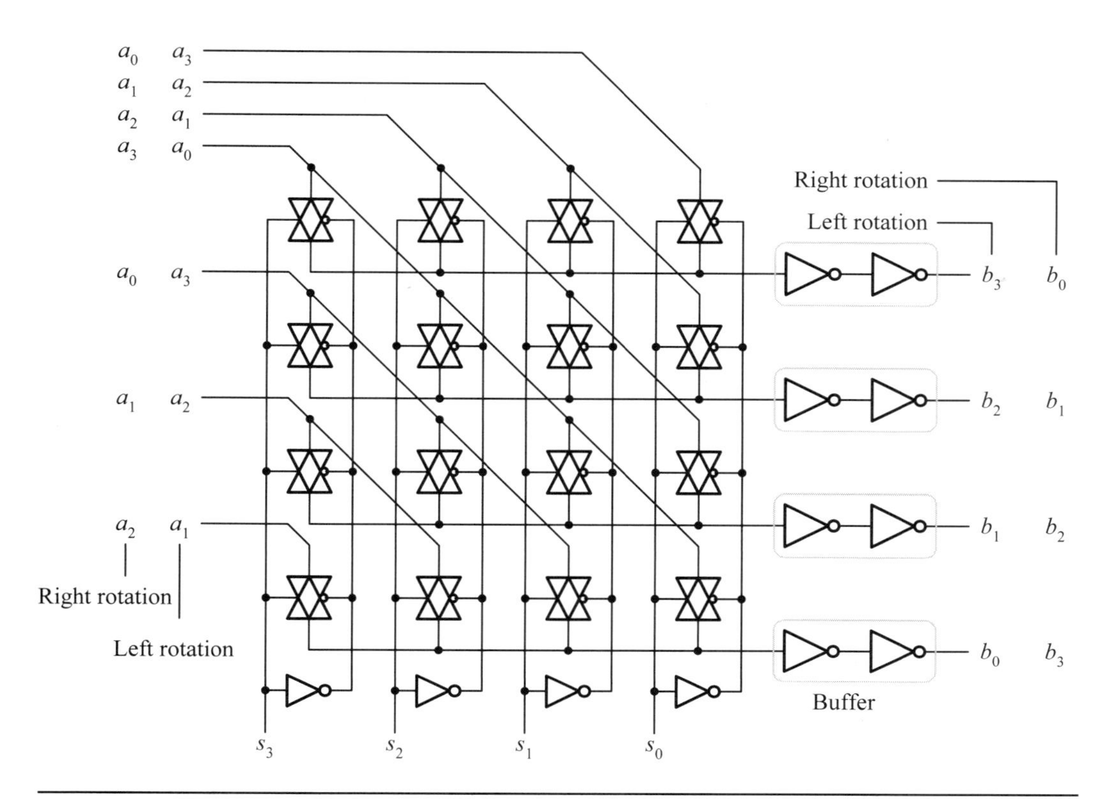

Figure 10.28: An example of a left/right-rotation linear barrel shifter based on TG-switches.

First, the input signal has to pass through at most one nMOS or TG switch, as shown in Figures 10.27 and 10.28. Second, the layout size is not dominated by the active devices, nMOS or TG switches, but instead by the number of wires running through the module. Third, the shift amount of the shifter is selected by individual control signals s_i exclusively. In many applications, the shift amount more likely comes in an encoded binary format. In such a situation, an n-to-2^n decoder is needed to convert the input binary format into the required 1-out-of-2^n code.

Unlike the linear barrel shifter comprising an array of switches, a logarithmic barrel shifter consists of $\log_2 n$ columns, with each containing n 2-to-1 multiplexers. The multiplexers may be constructed with nMOS or TG switches. Each column shifts its input the number of 0 or 2^i bit positions, depending on whether the value of s_i is 0 or 1. The shift amount is represented as a binary code $(s_{\log_2 n-1}, \cdots, s_i, \cdots, s_0)$.

■ **Example 10-12: (An 8-bit logical-left logarithmic barrel shifter.)** An example of an 8-bit logical/arithmetic-left logarithmic barrel shifter is shown in Figure 10.29. It is implemented by three stage multiplexer columns, with each containing eight 2-to-1 multiplexers. Each column of the multiplexers is controlled by s_i and shifts the input a number of $s_i \times 2^i$ bit positions. In other words, each column shifts its input the number of 0 or 2^i bit positions, depending on whether the value of s_i is 0 or 1.

■

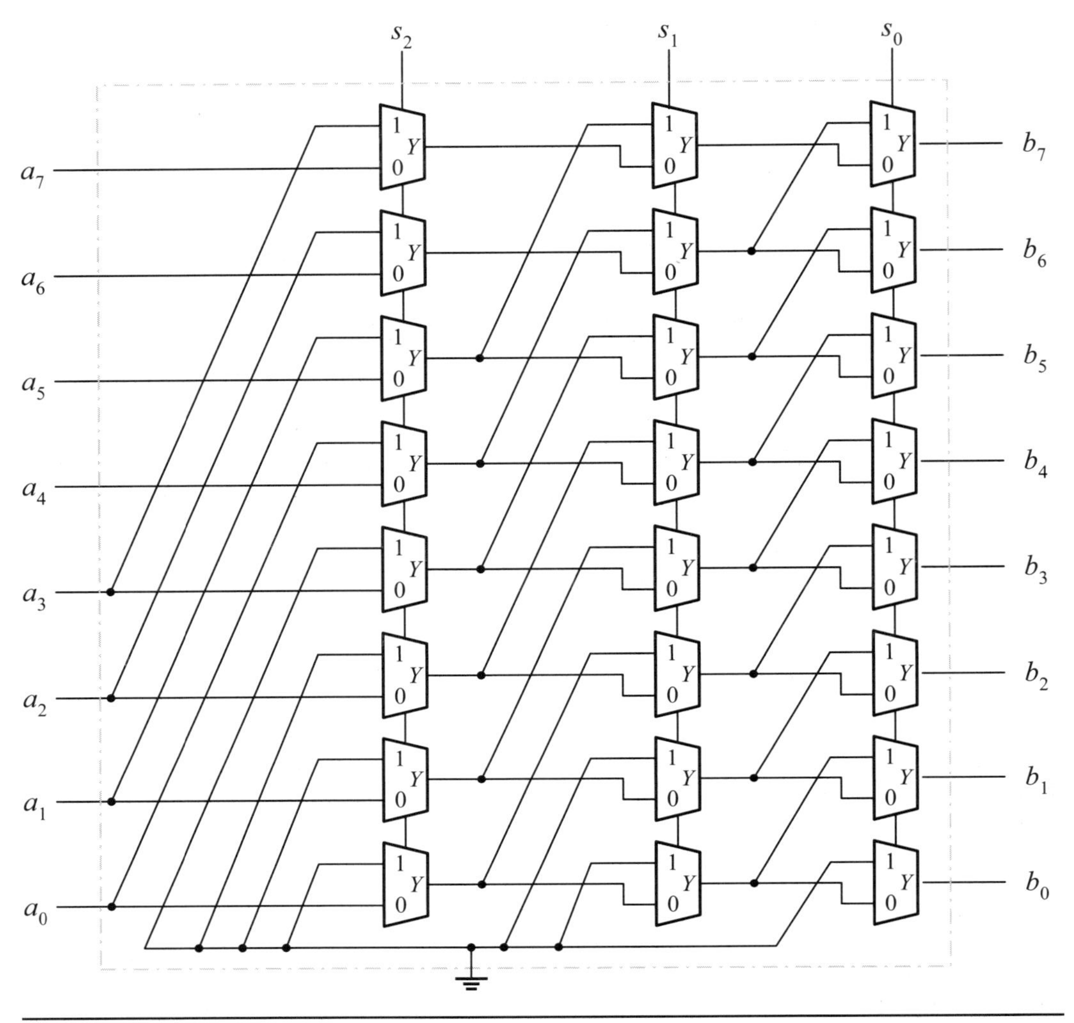

Figure 10.29: An example of a logical/arithmetic-left logarithmic barrel shifter based on 2-to-1 multiplexers.

The logarithmic barrel shifter is generally faster and consumes less area than the linear barrel shifter for large n. In addition, the logarithmic barrel shifter is easier to be parameterized, especially, for the designs using cell-based synthesis flow. Refer to Lin [32] for more about parameterized barrel shifters.

■ Review Questions

Q10-25. Define the barrel shifter.

Q10-26. What are the differences between linear and logarithmic barrel shifters?

Q10-27. How would you distinguish between the logical and arithmetic shifts?

Q10-28. How many 2-to-1 multiplexers are needed in an n-bit logical-left logarithmic barrel shifter?

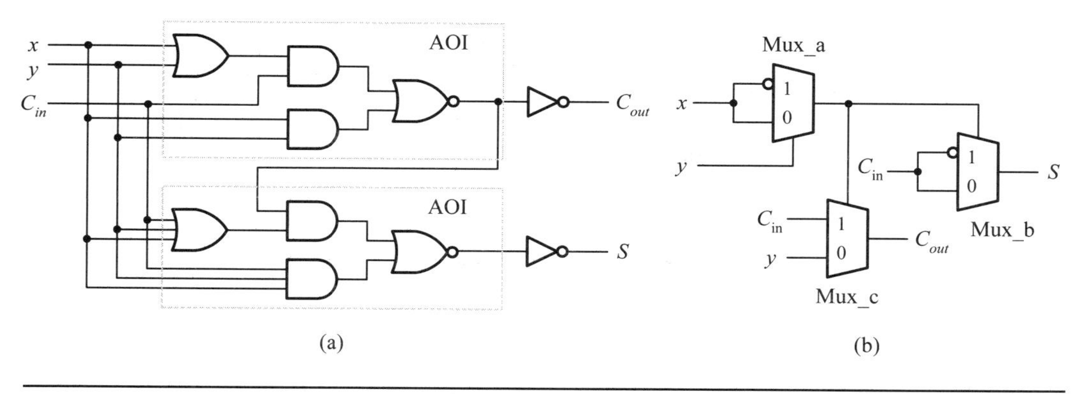

Figure 10.30: (a) An AOI-based full adder and (b) a multiplexer-based full adder.

10.4 Addition/Subtraction

Addition plays very important roles in arithmetic operations, including subtraction, multiplication, and division. By using two's complement technique, addition can also carry out subtraction. Therefore, addition usually means both. In this section, we are concerned with a variety of addition circuits.

10.4.1 Basic Full Adders

A variety of implementations for the full adder have been considered in the previous chapters. In this section, we are only concerned with two additional basic circuits. One is based on AND-OR-Inverter (AOI) gates and the other is based on 2-to-1 multiplexers.

10.4.1.1 Gate-Based Full Adder From Section 7.2.1, the sum and carry of a full adder can be represented as follows:

$$s = (x + y + c_{in})\bar{c}_{out} + xyc_{in} \quad (10.7)$$
$$c = (x + y)c_{in} + xy \quad (10.8)$$

Consequently, both functions can be easily implemented with two AOI gates and two inverters, as shown in Figure 10.30(a).

10.4.1.2 Multiplexer-Based Full Adder The basic full adder can also be realized by using 2-to-1 multiplexers. To see this, we notice that the sum and carry of a full adder may be rewritten as in the following:

$$s = (x \oplus y) \oplus c_{in} \quad (10.9)$$
$$c = (x \oplus y)c_{in} + xy = (x \oplus y)c_{in} + y\overline{(x \oplus y)} \quad (10.10)$$

Therefore, the sum s may be implemented with two 2-to-1 multiplexers, Mux_a and Mux_b, with each being accompanied with an inverter to function as an XOR gate, as shown in Figure 10.30(b). The carry c can be implemented with multiplexers Mux_a and Mux_c, as shown in Figure 10.30(b). Mux_a is shared by both sum s and carry c.

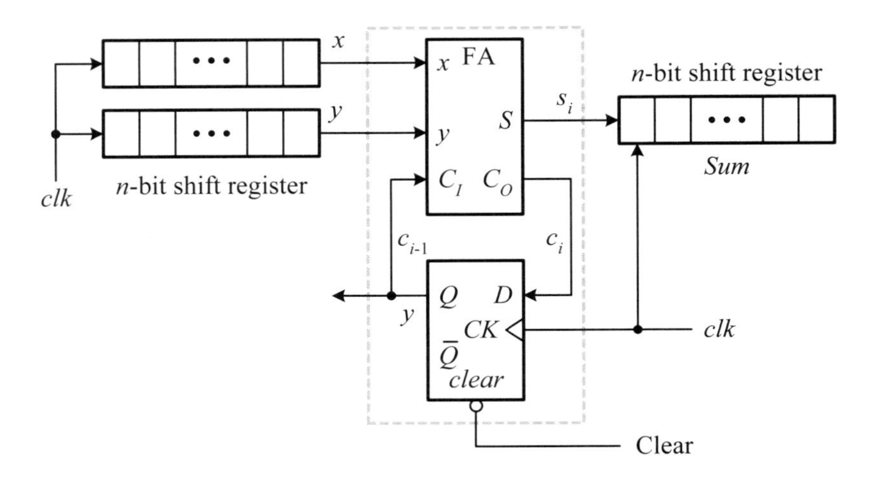

Figure 10.31: An n-bit serial adder.

10.4.2 n-Bit Adders/Subtracters

Once we make out the implementations of basic circuits for full adders, we are now in a position to combine basic full adders to form wider adders. In this subsection, we are concerned with some well-known conventional n-bit adders.

10.4.2.1 n-Bit Serial Adder In many applications, such as communication systems, the data are usually coming in and out as a bit-serial stream. Hence, in such situations, it is natural to take advantage of bit-serial arithmetic operations. We begin with the introduction of n-bit serial addition and then explore how to realize it in a variety of ways and present the features of these different implementation options.

n-bit serial addition is rather easy to follow and can be described as in the following algorithm.

Algorithm: n-bit serial addition

Input: Two numbers to be added.
Output: The sum of the input numbers.
Begin
  $c[-1] = 0;$
  for $(i = 0;\ i < n;\ i = i + 1)$
    $\{s[i], c[i]\} = x[i] + y[i] + c[i-1];$
End ■

An algorithm can be generally implemented with either a multiple-cycle (sequential) or a single-cycle hardware structure. A multiple-cycle structure for implementing the above addition algorithm is shown in Figure 10.31, where only a single-bit full adder is required to carry out the addition of two bits at a time. To hold the carry status of the current addition for the next bit, a D-type flip-flop is used. As the algorithm stated, the D-type flip-flop is reset at the beginning. Then, the output c_{i-1} of the D-type flip-flop is added with the two input operands to yield the sum s_i and carry c_i at each clock cycle. The addition is finished after n clock cycles. A controller is needed to control the operation and to provide the desired reset signal along with the finish signal after n clock cycles.

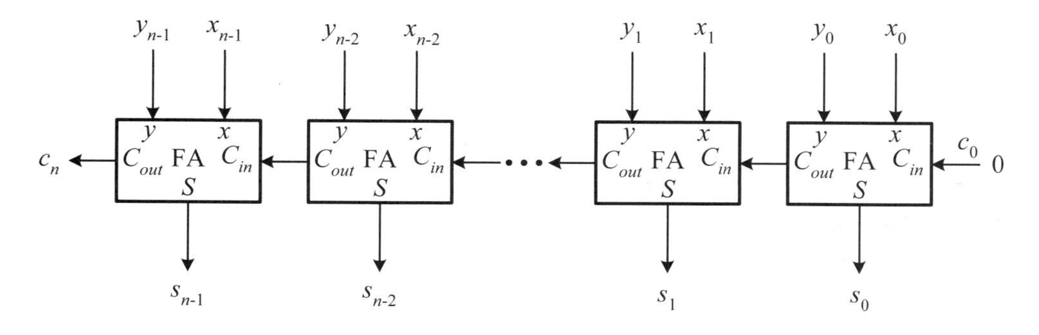

Figure 10.32: An n-bit ripple-carry adder.

To complete n-bit addition, the serial adder needs a time of $n(t_q + t_{setup} + t_{FA})$, where t_q and t_{setup} are the clock-to-Q and setup time of the D-type flip-flop, respectively, and t_{FA} is the propagation delay of the full adder.

10.4.2.2 Ripple-Carry Adder From the n-bit serial addition algorithm, we know that an addition is performed at each time step. Consequently, to speed up the n-bit addition, we may duplicate the full adder n times and cascade them together in a way shown in Figure 10.32. The resulting adder is referred to as a *ripple-carry adder* (RCA, also called a *carry-ripple adder*) because the carry signals are propagated to the next stage ripple.

In an RCA, the propagation delay is proportional to the number of stages. Because no flip-flops are involved, the propagation delay of an n-bit ripple-carry adder is only nt_{FA}. In practical applications, it is preferred to rewrite this propagation delay as two parts, t_{carry} and t_{sum}, to emphasize the critical part contributed by the carry chain. Based on this, the propagation delay of an n-bit ripple-carry adder is equal to

$$t_{RCA} = (n-1)t_{carry} + t_{sum} \tag{10.11}$$

where t_{carry} and t_{sum} are the propagation delays from C_{in} (c_0) to C_{out} and from inputs x_i and y_i to S of each full adder, respectively. In other words, the critical (timing) path of the n-bit ripple-carry adder is on the carry chain. As a consequence, to speed up the addition, it is necessary to reduce the propagation delay of the carry chain.

In addition to addition, subtraction is also fundamentally important. In digital systems, subtraction is usually carried out by two's complement addition. In other words, the subtractor is first converted into the two's complement form and then added to the subtrahend. To facilitate this, n XOR gates are employed as an n-bit true/complement circuit to provide the true and one's complement form of the subtractor. Accompanying the carry input at the least-significant bit, the two's complement form of subtractor is resulted. Hence, as shown in Figure 10.33, the addition is performed when $sub/add = 0$ and subtraction is carried out when $sub/add = 1$.

10.4.2.3 Carry-Select Adder As described above, the critical path of an n-bit ripple-carry adder is on the carry chain. One way to speed up the addition is to partition the n-bit addition into several smaller groups, with each having k bits, say, 4-bit groups, and then for each group, two additions are performed in parallel, one assuming that the carry-in is 0 and the other assuming that the carry-in is 1. When the carry-in is eventually known, the correct sum is simply selected through a k-bit 2-to-1 multiplexer. The adder based on this approach is known as a *carry-select adder*.

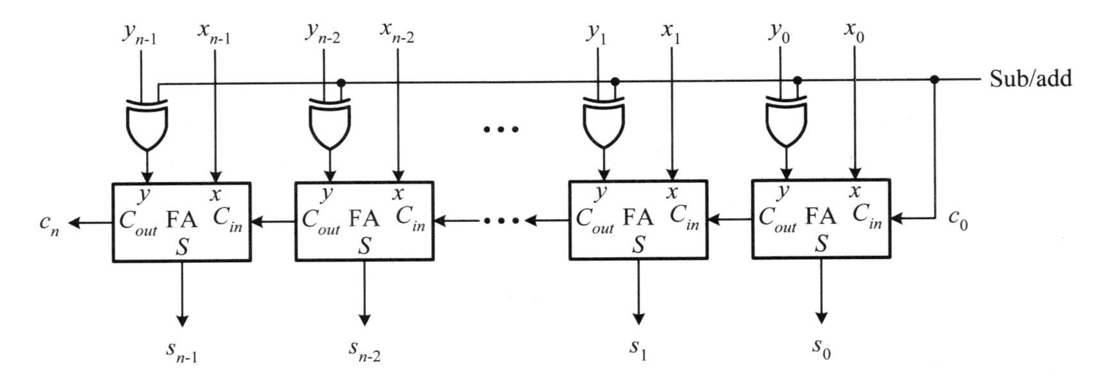

Figure 10.33: An n-bit ripple-carry adder/subtracter (Sub/add = 1: subtraction; Sub/add = 0: addition).

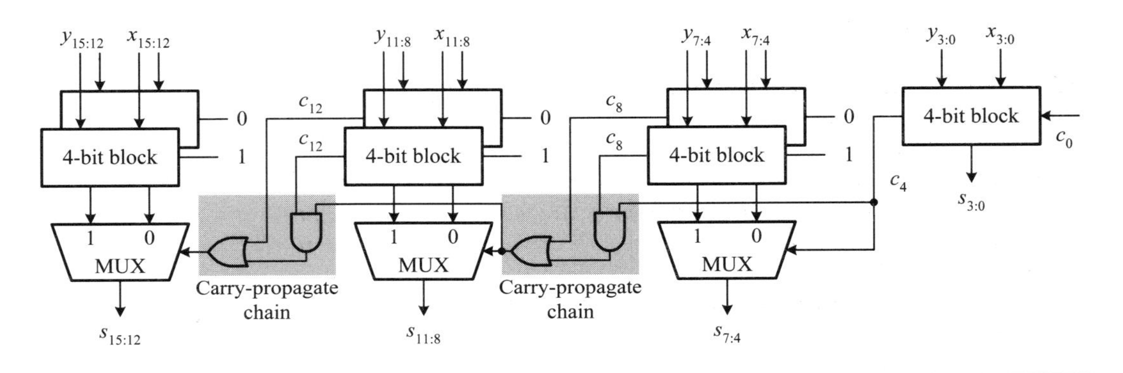

Figure 10.34: A 16-bit carry-select adder.

An example of 16-bit carry-select adder is depicted in Figure 10.34, where each group is 4 bits and contains two 4-bit adders except the first one, which only needs one 4-bit adder. The output sum from each 4-bit adders in each group is selected through a 4-bit 2-to-1 multiplexer by a fast carry-propagate chain, as indicated by the shadow regions. It is instructive to note that a 2-to-1 multiplexer can be used to replace the carry-propagate chain of each group.

The propagation delay of an n-bit carry-select adder is proportional to the length of the carry-propagate chain. From the figure, we can see that the propagation delay of an n-bit carry-select adder is equal to

$$t_{CSA} = t_{kb-adder} + (g-2)(t_{AND} + t_{OR}) + t_{MUX} \tag{10.12}$$

where $t_{kb-adder}$ is the propagation delay of the k-bit block adder and $g = \lceil n/k \rceil$ is the number of groups, with each containing k bits. t_{AND} and t_{OR} are the propagation delays of two-input AND and OR gates, respectively. t_{MUX} is the propagation delay of the 2-to-1 multiplexer.

10.4.2.4 Conditional-Sum Adder Conditional-sum adder is a special application of the carry-select adder such that the carry-select addition starts from single bit and *recursively doubling* to $n/2$ bits. An 8-bit conditional-sum adder is exhibited in

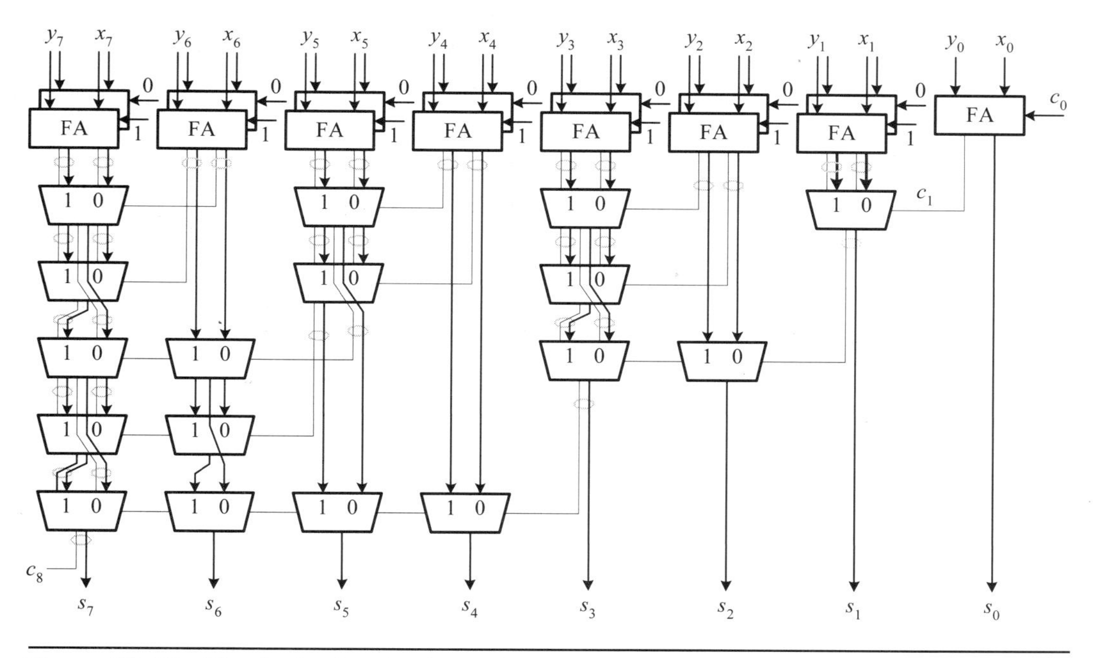

Figure 10.35: The logic circuit of an 8-bit conditional-sum adder.

Figure 10.35. The first two rows consist of full adders and compute the sum and carry-out assuming that the carry-in bits are 0 and 1, respectively. The rest of the adder is the sum-selection tree that selects the proper sum bit. The propagation delay of an n-bit conditional-sum adder is $\log_2 n$. An n-bit conditional-sum adder requires $2n - 1$ full adders and an approximate number of $n(\log_2 n - 1) + 1$ multiplexers.

■ Review Questions

Q10-29. Describe the structures of carry-select adder and conditional-sum adder.

Q10-30. Describe the principle of n-bit ripple-carry adder/subtracter depicted in Figure 10.33.

Q10-31. Replace the carry-propagate chain shown in Figure 10.34 with a 2-to-1 multiplexer.

10.4.2.5 Carry-Lookahead Adder As mentioned, the performance bottleneck of an n-bit carry-ripple adder is on the generation of carriers needed in all stages. In order to further explore the problem of carry generation, let us consider the full adder shown in Figure 10.36, which is implemented by using two half adders that we have introduced in Section 7.2.1.

Let two input numbers be $x = (x_{n-1} \cdots x_1 x_0)$ and $y = (y_{n-1} \cdots y_1 y_0)$, and sum be $s = (s_{n-1} \cdots s_1 s_0)$. At the ith stage of an n-bit ripple-carry adder as depicted in Figure 10.36, the output carry c_{i+1} is generated if both inputs x_i and y_i are 1, regardless of what value of the input carry c_i is. The input carry c_i is propagated to the output if either input of x_i and y_i is 1. Consequently, we can define two new functions: *carry generate* (g_i) and *carry propagate* (p_i) in terms of inputs x_i and y_i as

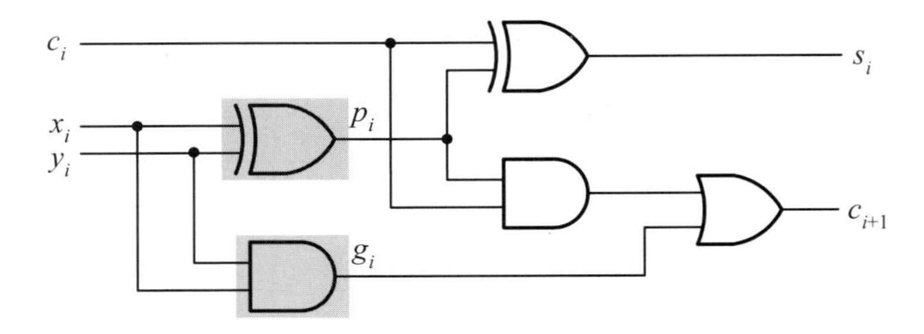

Figure 10.36: The logic circuit used to define carry generator and propagate functions.

follows:

$$g_i = x_i \cdot y_i \tag{10.13}$$
$$p_i = x_i \oplus y_i \tag{10.14}$$

Based on these two functions, the sum and carry-out can then be separately represented as a function of both carry generate (g_i) and carry propagate (p_i), and are as in the following:

$$s_i = p_i \oplus c_i \tag{10.15}$$
$$c_{i+1} = g_i + p_i \cdot c_i \tag{10.16}$$

As a result, the carry-in of the $(i+1)$th-stage full adder can be generated by using the recursive equation of the c_{i+1}. For example, the first four carry signals are

$$\begin{aligned}
c_1 &= g_0 + p_0 \cdot c_0 \\
c_2 &= g_1 + p_1 \cdot c_1 \\
&= g_1 + p_1(g_0 + p_0 \cdot c_0) = g_1 + p_1 \cdot g_0 + p_1 \cdot p_0 \cdot c_0 \\
c_3 &= g_2 + p_2 \cdot c_2 \\
&= g_2 + p_2(g_1 + p_1 \cdot g_0 + p_1 \cdot p_0 \cdot c_0) \\
&= g_2 + p_2 \cdot g_1 + p_2 \cdot p_1 \cdot g_0 + p_2 \cdot p_1 \cdot p_0 \cdot c_0 \\
c_4 &= g_3 + p_3 \cdot c_3 \\
&= g_3 + p_3 \cdot (g_2 + p_2 \cdot g_1 + p_2 \cdot p_1 \cdot g_0 + p_2 \cdot p_1 \cdot p_0 \cdot c_0) \\
&= g_3 + p_3 \cdot g_2 + p_3 \cdot p_2 \cdot g_1 + p_3 \cdot p_2 \cdot p_1 \cdot g_0 + p_3 \cdot p_2 \cdot p_1 \cdot p_0 \cdot c_0
\end{aligned}$$

The resulting carry signals are only functions of both inputs x and y, and the carry-in (c_0). These carry signals can then be implemented by a two-level logic circuit known as a *carry-lookahead* (CLA) *generator*, as shown in Figure 10.37.

By using a carry-lookahead generator, the four-bit adder may be represented as a function of carry-propagate signals p_i and carry signals c_i. That is, $s_i = p_i \oplus c_i$, where $0 \leq i \leq 3$. The resulting circuit is shown in Figure 10.38, which is composed of three components: *pg* generator, CLA generator, and sum generator.

CMOS realization. The 4-bit CLA generator can be realized with any CMOS logic style that we have described. For high-speed applications, it is usually implemented using the domino logic style, as shown in Figure 8.26.

Manchester carry chain. A Manchester carry chain uses MOS transistors to implement the carry-recurrence relation: $c_{i+1} = g_i + p_i \cdot c_i$. Based on this, many different circuits can be built accordingly. Two of these are given in Figure 10.39.

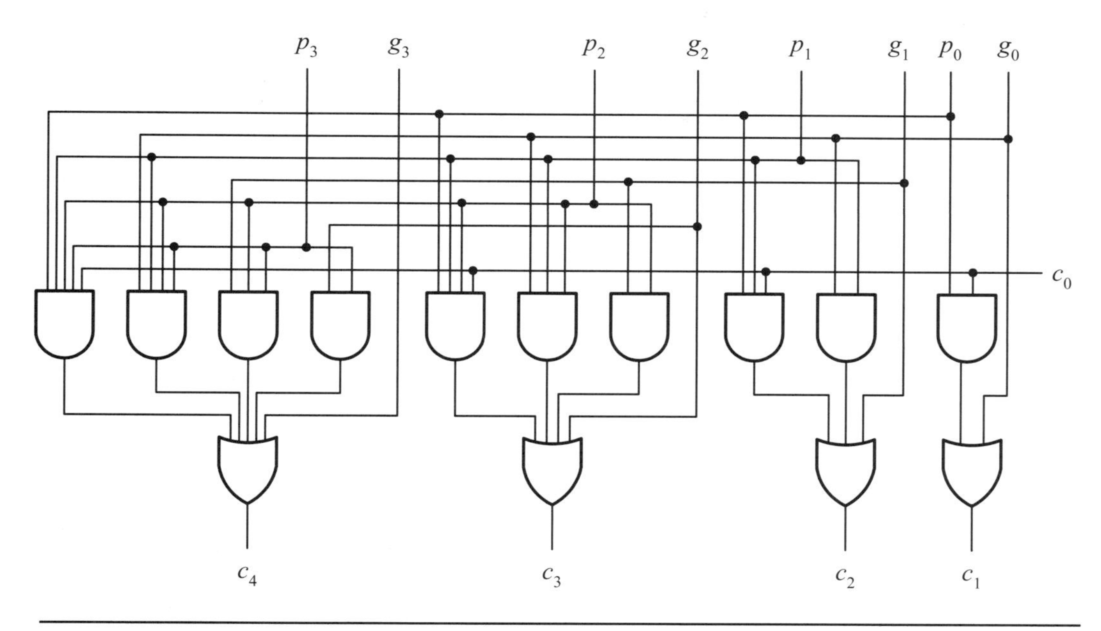

Figure 10.37: The logic circuit of a 4-bit carry-lookahead generator.

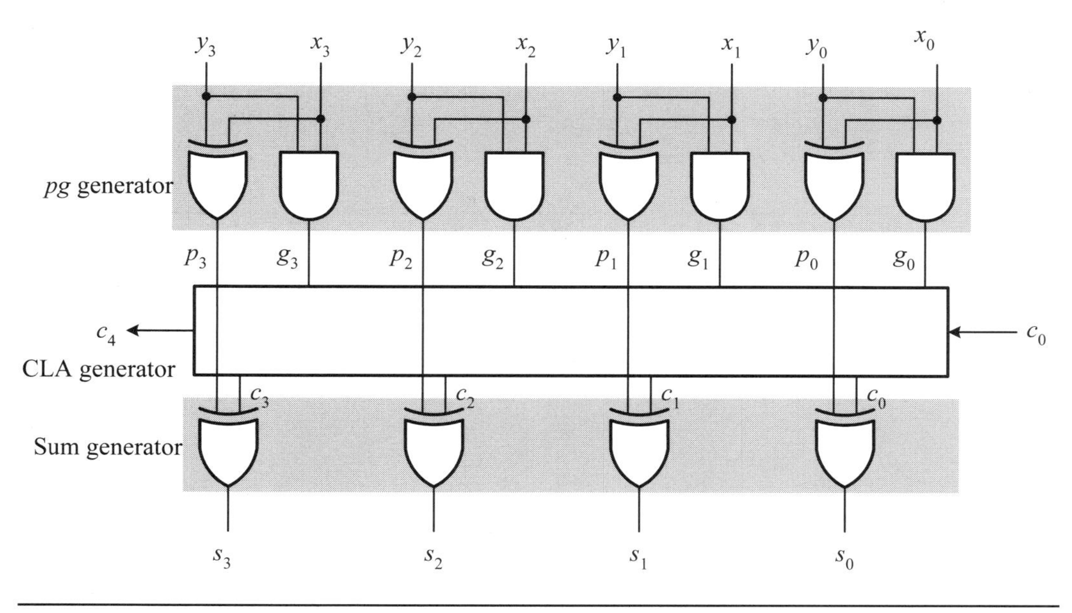

Figure 10.38: A 4-bit carry-lookahead adder.

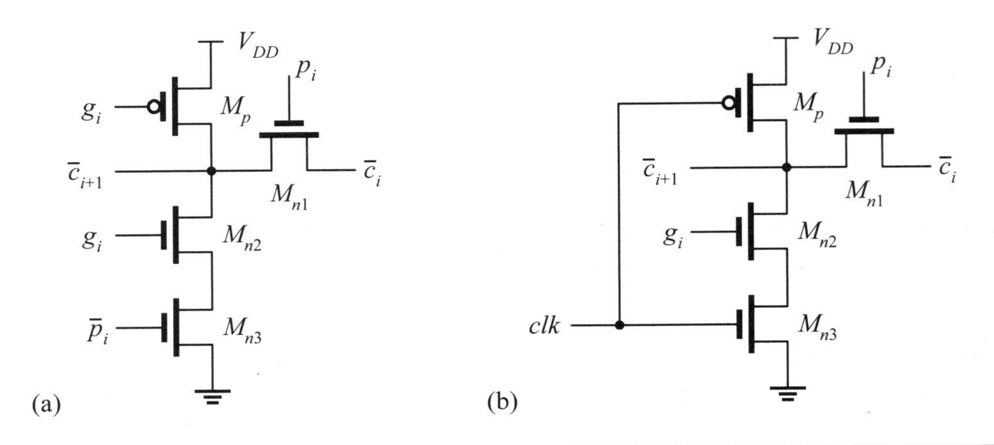

Figure 10.39: The general stages of the Manchester carry style: (a) static circuit; (b) dynamic circuit.

Figure 10.39(a) is a static logic circuit while Figure 10.39(a) is a dynamic logic circuit. Both circuits use $\bar{c}_i$ as an input and generate $\bar{c}_{i+1}$ as the desired output by calculating the expression: $\bar{c}_{i+1} = g_i + p_i \cdot \bar{c}_i$.

The static logic circuit works as follows. As $p_i = 0$, nMOS transistor M_{n1} is off and M_{n3} is on, thereby turning on the carry-generate inverter. The input $\bar{c}_i$ is blocked from propagating through and the carry-generate inverter is enabled. The output carry $\bar{c}_{i+1}$ is equal to the inverted g_i. As $p_i = 1$, nMOS transistor M_{n1} is on and M_{n3} is off, thereby turning off the carry-generate inverter. The input $\bar{c}_i$ is able to propagate through, causing the output carry $\bar{c}_{i+1}$ to be equal to $\bar{c}_i$.

The dynamic logic circuit shown in Figure 10.39(b) is similar to the static logic circuit except that now transistors M_p and M_{n3} are used as precharge and evaluate transistors, respectively, rather than logic transistors. During precharge phase, namely, $clk = 0$, the output node is precharged to V_{DD}. As the clk switches to high, the circuit enters evaluation phase. A carry propagation occurs if $p_i = 1$, whereas the output node discharges to 0 if $g_i = 1$.

Example 10-13: (A 4-bit Manchester carry chain.) An example of a 4-bit Manchester carry chain is shown in Figure 10.40, being formed by cascading four stages of the dynamic logic circuit given in Figure 10.39(b). Some features of this circuit are as follows. First, this circuit is virtually the same as the MODL circuit shown in Figure 8.26. Second, the foot nMOS transistors may be omitted to speed up the computation time of the carries.

10.4.2.6 Multiple-Level Carry-Lookahead Adder Although a wide CLA adder can be simply designed by applying the above design procedure, an alternative approach to designing a wide CLA adder is by using the concept of hierarchical structure, namely, using block CLA generators. To see this, let us reconsider the 4-bit CLA generator and adder. The carry c_4 can be rewritten as follows:

$$c_4 = g_3 + p_3g_2 + p_3p_2g_1 + p_3p_2p_1g_0 + p_3p_2p_1p_0c_0$$

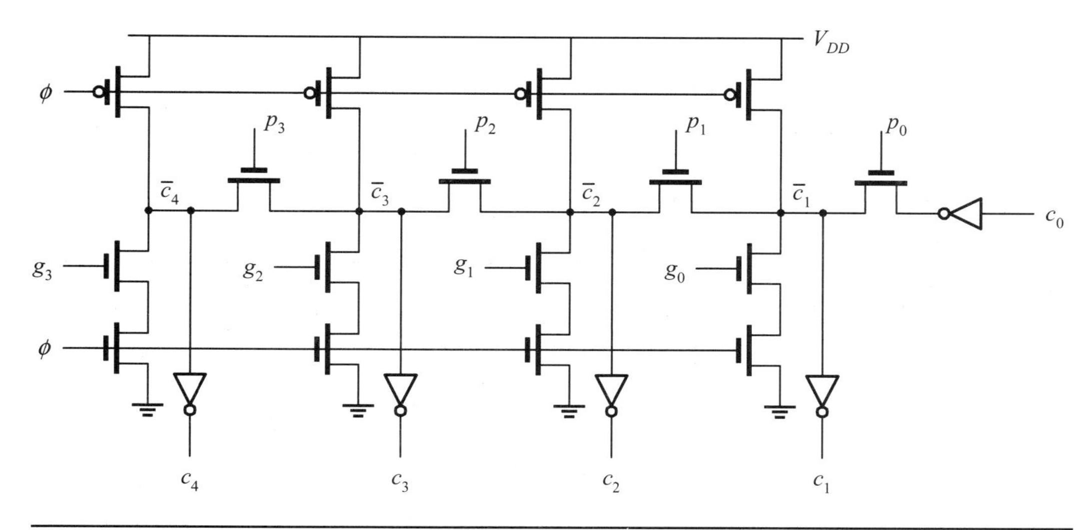

Figure 10.40: The logic circuit of a dynamic Manchester carry chain.

$$= g_{[3:0]} + c_0 p_{[3:0]}$$

where $g_{[3:0]}$ and $p_{[3:0]}$ are called *group-carry generate* and *group-carry propagate*, respectively, and are defined as follows:

$$\begin{aligned} g_{[3:0]} &= g_3 + p_3 g_2 + p_3 p_2 g_1 + p_3 p_2 p_1 g_0 \\ p_{[3:0]} &= p_3 p_2 p_1 p_0 \end{aligned}$$

More generally, both group-carry generate and group-carry propagate can be redefined as in the following:

$$\begin{aligned} g_{[i+3:i]} &= g_{i+3} + p_{i+3} g_{i+2} + p_{i+3} p_{i+2} g_{i+1} + p_{i+3} p_{i+2} p_{i+1} g_i \\ p_{[i+3:i]} &= p_{i+3} p_{i+2} p_{i+1} p_i \end{aligned}$$

Based on this definition, the group carries can then be represented in the same way as the carries generated by the 4-bit CLA generator described before. The carries, c_8, c_{12}, and c_{16} can be represented as in the following:

$$\begin{aligned} c_8 &= g_{[7:4]} + g_{[3:0]} p_{[7:4]} + c_0 p_{[3:0]} p_{[7:4]} \\ c_{12} &= g_{[11:8]} + g_{[7:4]} p_{[11:8]} + g_{[3:0]} p_{[7:4]} p_{[11:8]} + c_0 p_{[3:0]} p_{[7:4]} p_{[11:8]} \\ c_{16} &= g_{[15:12]} + g_{[11:8]} p_{[15:12]} + g_{[7:4]} p_{[11:8]} p_{[15:12]} + \\ &\quad g_{[3:0]} p_{[7:4]} p_{11:8]} p_{[15:12]} + c_0 p_{[3:0]} p_{[7:4]} p_{11:8]} p_{[15:12]} \end{aligned}$$

which can then be implemented by the 4-bit CLA generator with a minor modification. The resulting CLA generator is called a *block CLA generator* and is exhibited in Figure 10.41. Through a proper combination of block CLA generators, a wide adder can be easily implemented. An example of 16-bit CLA adder constructed in such a way is explored next.

■ **Example 10-14: (A 16-bit CLA adder.)** An example of a 16-bit CLA adder constructed with block CLA generators is shown in Figure 10.42. Each 4-bit group generates group-carry generate and group-carry propagate signals, and these signals

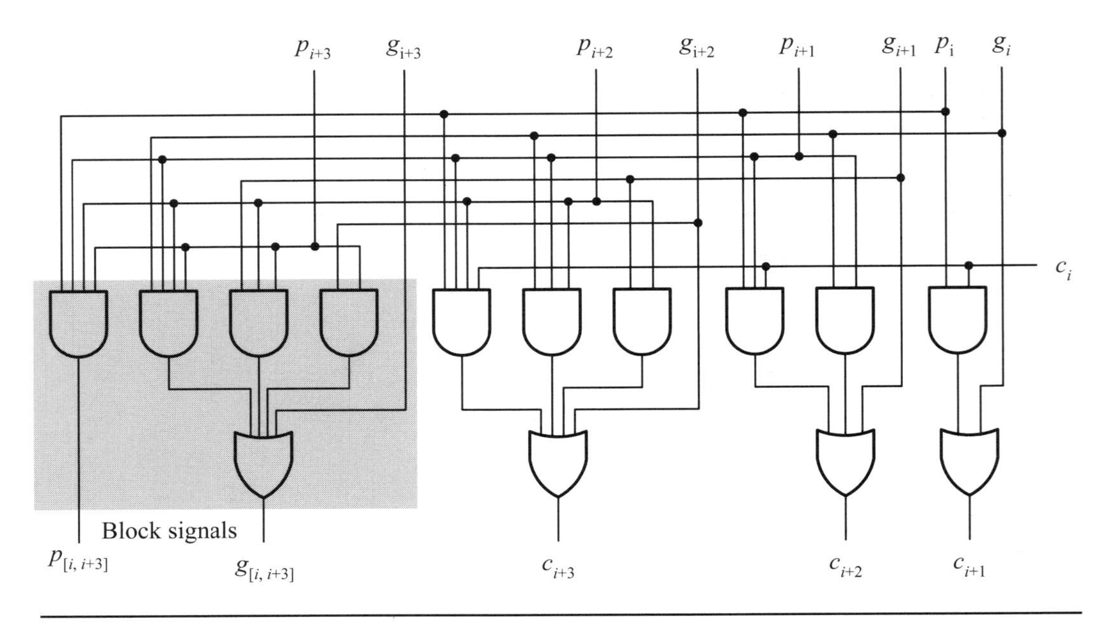

Figure 10.41: The logic circuit of a block CLA generator.

are then combined together by a second-level 4-bit CLA generator to yield the input carries, c_4, c_8, and c_{12}, for all first-level groups to compute their sums. Wider adders can be built in this approach by using more levels of block CLA generators. ■

10.4.2.7 Ling Carry-Lookahead Adder Before proceeding, we rewrite the sum function of CLA adder as follows:

$$s_i = p_i \oplus g_{[i-1:0]} \tag{10.17}$$

where $g_{[i-1:0]}$ is the carry-in c_i of the ith stage and is defined as follows:

$$g_{[i-1:0]} = g_{i-1} + p_{i-1}g_{[i-2:0]} \tag{10.18}$$

Based on this definition, the $g_{[-1:0]}$ is the carry-in c_0 of the least-significant bit.

Ling equations are an alternative to the conventional CLA. Since $p_i g_i = g_i$, the carry generate term $g_{[i:0]}$ can be rewritten as follows:

$$g_{[i:0]} = p_i(g_i + g_{[i-1:0]}) = p_i h_{[i:0]} \tag{10.19}$$

In Ling CLA adder, the pseudo-carry $h_{[i:0]}$ is propagated and combined with the remaining terms in the final sum.

$$\begin{aligned} h_{[i:0]} &= g_i + p_{i-1}h_{[i-1:0]} \\ s_i &= p_i \oplus p_{i-1}h_{[i-1:0]} = p_i \oplus h_{[i:0]} + g_i p_{i-1} h_{[i-1:0]} \end{aligned} \tag{10.20}$$

To see the advantage of Ling CLA adder over the conventional CLA, let us unroll the recursion equations $g_{[i:0]}$ and $h_{[i:0]}$ in the case of 4-bit CLA. The results are as follows:

$$\begin{aligned} g_{[3:0]} &= g_3 + p_3g_2 + p_3p_2g_1 + p_3p_2p_1g_0 \\ h_{[3:0]} &= g_3 + g_2 + p_2g_1 + p_2p_1g_0 \end{aligned} \tag{10.21}$$

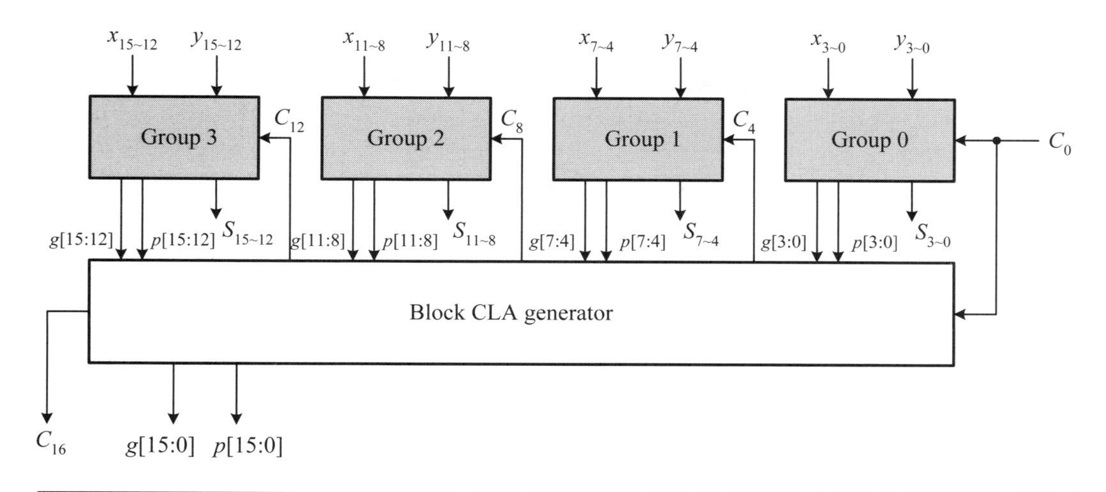

Figure 10.42: A 16-bit CLA adder.

Hence, the $h_{[3:0]}$ term has fewer factors than the $g_{[3:0]}$ term, thereby requiring fewer transistors in CMOS implementation. However, the sum generator in Ling CLA is more complex. To take a look at this, let us compare the sum generators of both adders. For a conventional CLA, the sum generators are

$$\begin{aligned} s_i^0 &= x_i \oplus y_i \\ s_i^1 &= \overline{(x_i \oplus y_i)} \end{aligned} \tag{10.22}$$

where s_i^0 is the sum value with an input carry of 0 and s_i^1 for an input carry of 1. If a Ling CLA is used, s_i^1 changes to

$$s_i^1 = (x_i \oplus y_i) \oplus (x_{i-1} \oplus y_{i-1}) \tag{10.23}$$

which is more complex to implement. From the above discussion, we may conclude that Ling CLA effectively moves complexity from the CLA generator to the sum generator.

10.4.2.8 Carry-Skip Adder Carry-skip (also called *carry-bypass*) adder is the one that speeds up the addition by propagating the carry-in around a group of bits. To reach this, if each block is 4 bits, then a group-carry propagate signal can be expressed as follows:

$$p_{[i+3:i]} = p_{i+3}p_{i+2}p_{i+1}p_i$$

which can be computed from individual propagate signals by a single four-input AND gate. The carry-out of this group is then computed as follows:

$$c_{out} = c_{i+4} + p_{[i+3:i]} \cdot c_i \tag{10.24}$$

where c_i is the carry-in of this group. The carry-out is determined by the carry-in if the group-carry propagate signal $p_{[i+3:i]}$ is 1 and by c_{i+4} otherwise.

The overall performance of a carry-skip adder is determined by the number of bits in each group. Assuming that one stage of full adder (two gate levels) has the same delay as one skip, the worst-case propagation delay in an n-bit carry-skip adder with a fixed block size k can be approximated as follows:

$$t_{p(skip)} \approx 2k + n/k - 3.5 \tag{10.25}$$

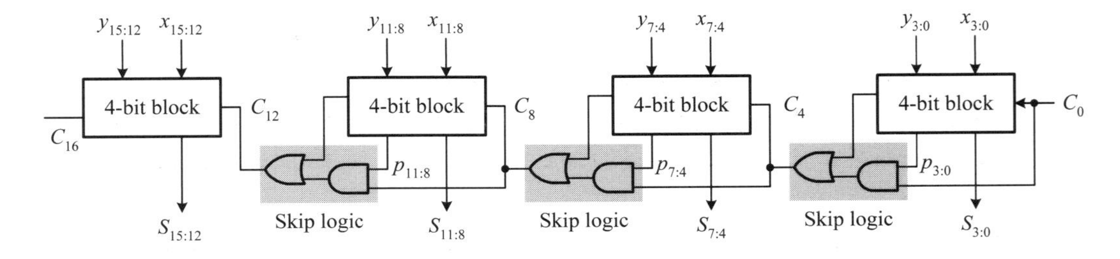

Figure 10.43: A 16-bit carry-skip adder.

Hence, to minimize the propagation delay, the optimal group size k of an n-bit carry-skip adder is approximate to $k = \sqrt{n/2}$. Substituting this result back to $t_{p(skip)}$, the propagation delay of the adder with optimal group size is approximated to

$$t_{opt(skip)} \approx 2\sqrt{2n} - 3.5 \tag{10.26}$$

From the above results, the optimal group size of a 16-bit carry-skip adder is 3 and the propagation delay is 7.81 stages. In practical applications, the group size is often set to 4 bits. Based on this, the 16-bit carry-skip adder shown in Figure 10.43 has the propagation delay of 8.5 stages, a little bigger than the optimal case.

Multiple-level carry-skip adder. Multiple-level carry-skip adder is formed by allowing a carry to skip several groups at a time. Two examples of multiple-level carry-skip adder are shown in Figure 10.44. Figure 10.44(a) shows a 32-bit carry-skip adder with a group size of 8 bits. The second-level carry-skip path is formed by ANDing all four *group-carry propagate signals* (CPS) to form the output carry-propagate signal p_{out}. In this example, we use a 2-to-1 multiplexer to replace the circuit comprising of AND and OR gates. Figure 10.44(b) uses the 32-bit carry-skip adders given in Figure 10.44(a) as basic blocks to construct a 128-bit carry-skip adder.

10.4.2.9 Carry-Save Adder Carry-save adders find their widespread use in the situations whenever multiple operands need to be added. The carry-save adder (CSA)[1] uses the fact that the carry input of a full adder can be regarded as the third regular input. Hence, a single-bit CSA is a device that accepts three 1-bit inputs and generates two 1-bit outputs such that the combination of c_{out} and s represents the number of 1 bits of the inputs. A single-bit CSA is indeed a full adder. It is often referred to as a 3:2 *compressor* because it counts the number of 1s in the three single-bit inputs and indicates the result on the two single-bit outputs. An n-bit CSA is nothing but places n full adders in parallel, as shown in Figure 10.45. An n-bit CSA can accept three n-bit inputs and yield two n-bit outputs, representing the sum and carry portions of the three input operands, respectively. Hence, to obtain the true result, it still needs a *carry-propagate adder* (CPA) to combine both output sum and carry together. In the multiplication section, many examples using CSAs to construct efficient multipliers are concerned.

■ Example 10-15: (An example of CSA operations.) Figure 10.46 shows a numerical example using a 16-bit CSA to find the sum of three 16-bit operands. As described before, the CSA accepts three operands and yields two outputs in carry-save

[1]CSA is usually reserved for the carry-save adder although both carry-select adder and carry-skip adder also have the same acronym.

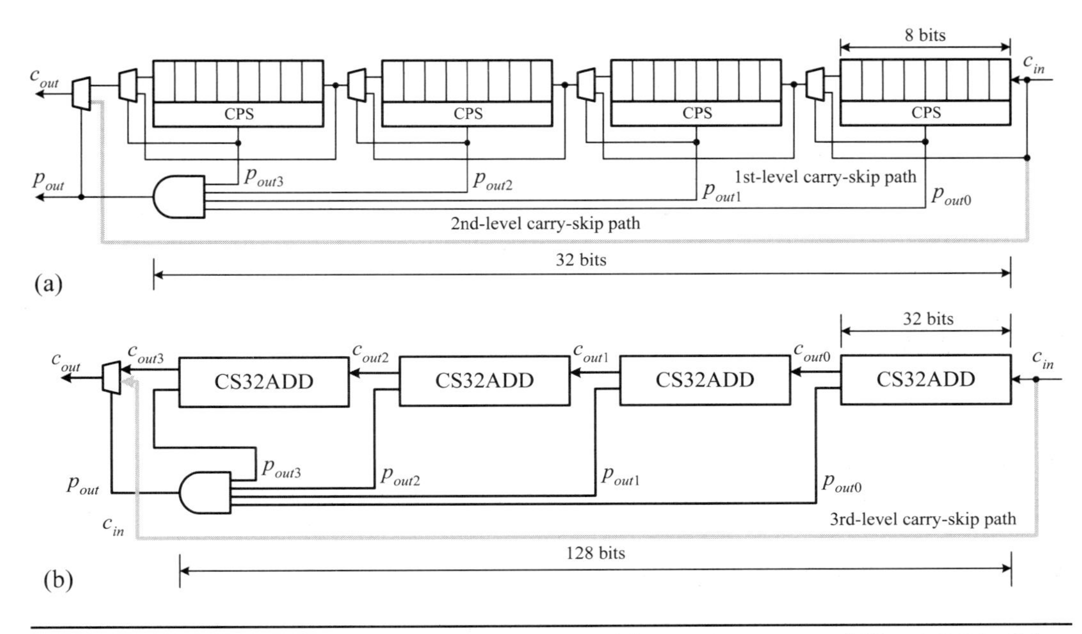

Figure 10.44: Multiple-level carry-skip adders: (a) a 32-bit circuit (CS32ADD); (b) a 128-bit circuit (CS128ADD).

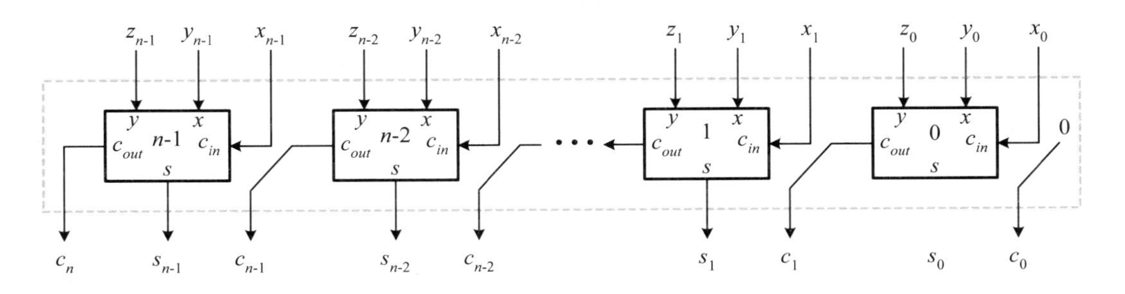

Figure 10.45: An n-bit carry-save adder (CSA).

form. To find the final sum, these two outputs are added up by a 16-bit CPA. The reader is invited to check the result.

■

As compared to CSA, both RCAs and CLA adders as well as other adders introduced earlier except the CSA are often called CPAs because in such adders each carry output from the lower-order bit is added to the adjacent higher-order bit. On the contrary, the carry-out from each CSA is left as another output just like the sum without being added into the higher-order bit.

■ Review Questions

Q10-32. What is the distinction between CLA and block CLA generators?

Q10-33. What is the distinction between CLA and Ling CLA adders?

Q10-34. Describe the philosophy behind carry-skip adders.

		1 0 1 0	1 1 1 0	1 0 0 1	0 0 1 1	= X (CSA)
		0 0 1 1	1 0 1 0	0 0 1 1	1 0 1 1	= Y (CSA)
		1 1 1 0	0 1 0 1	1 1 0 1	1 1 1 1	= Z (CSA)
CPA	0	0 1 1 1	0 0 0 1	0 1 1 1	0 1 1 1	= S
CPA	1	0 1 0 1	1 1 0 1	0 0 1 1	0 1 1 0	= C
	1	1 1 0 0	1 1 1 0	1 0 1 0	1 1 0 1	= Sum

Figure 10.46: A numerical example showing the operations of a 16-bit carry-save adder.

Q10-35. Define an n-bit CSA.

Q10-36. What does a CPA mean?

10.4.3 Parallel-Prefix Adders

As shown in Figure 10.38, a CLA adder comprises three major components: *pg* generator, CLA generator, and sum generator. Depending on the ways of how the CLA generator is constructed, a variety of variations of CLA adders are possible. In this section, we are concerned with a big class in which parallel-prefix computation is used to construct the CLA generator. This class involves the construction of parallel-prefix networks and can be categorized into the following two major types:

- *Full-tree parallel-prefix network*
- *Sparse-tree parallel-prefix network*

In what follows, we begin to introduce how a parallel-prefix network can be used to carry out the generation of carriers and to construct an adder, referred to as a *parallel-prefix adder*. Then, we extend these parallel-prefix networks to perform other datapath operations, such as priority encoding, incrementer/decrementer, and Gray-to-binary code conversion.

10.4.3.1 Parallel-Prefix Computation The prefix sum is defined as follows. Consider a sequence of n elements $\{x_{n-1}, \cdots, x_1, x_0\}$ with an associative binary operator denoted by $\bullet$. The *prefix sums* of this sequence are the n partial sums defined by

$$s_{[i,0]} = x_i \bullet \cdots \bullet x_1 \bullet x_0 \qquad (10.27)$$

where $0 \leq i < n$.

Let $g_{[i,k]}$ and $p_{[i,k]}$ represent the group-carry generate and group-carry propagate from bits i to k, respectively. Both $g_{[i,k]}$ and $p_{[i,k]}$ can then be defined recursively as follows:

$$\begin{aligned} g_{[i,k]} &= g_{[i,j+1]} + p_{[i,j+1]} g_{[j,k]} \\ p_{[i,k]} &= p_{[i,j+1]} p_{[j,k]} \end{aligned} \qquad (10.28)$$

where $0 \leq i < n$, $k \leq j < i$, $0 \leq k < n$, $g_{[i,i]} = x_i \cdot y_i$, and $p_{[i,i]} = x_i \oplus y_i$. Based on this definition, the carry-out $c_{i+1} = g_i + p_i \cdot c_i$ of the ith-bit adder can be written as $g_{[i,0]} = g_{[i,i]} + p_{[i,i]} \cdot g_{[i-1,0]}$. The logic diagram used to implement $g_i = g_{[i,i]} = x_i \cdot y_i$, and $p_i = p_{[i,i]} = x_i \oplus y_i$ is shown in Figure 10.47(a).

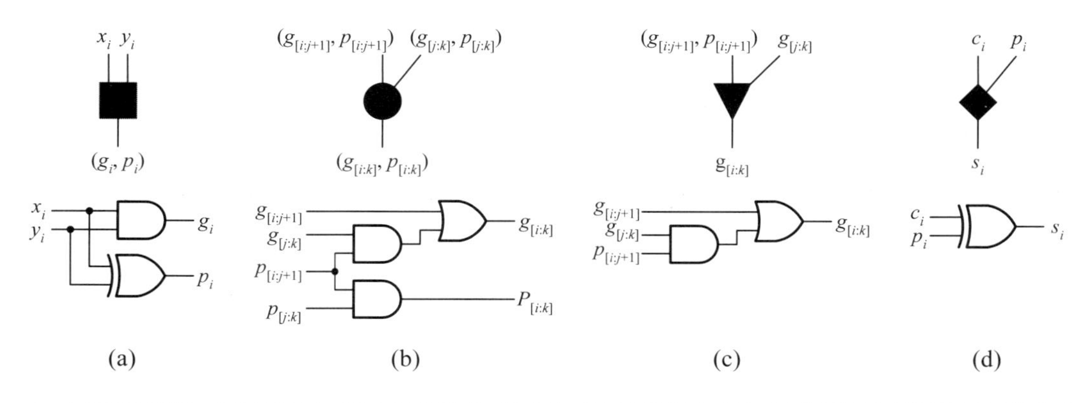

Figure 10.47: Definitions and logic circuits of black computation nodes: (a) *pg* generator; (b) binary operator; (c) simplified binary operator; (d) sum generator.

To parallelize the operations, let us group $g_{[i,j]}$ and $p_{[i,j]}$ together as $w_{[i,j]} = (g_{[i,j]}, p_{[i,j]})$. Then it is ready to show that the binary operator $\bullet$ in the following equation

$$\begin{aligned} w_{[i,k]} &= w_{[i,j+1]} \bullet w_{[j,k]} \\ &= (g_{[i,j+1]}, p_{[i,j+1]}) \bullet (g_{[j,k]}, p_{[j,k]}) \\ &= (g_{[i,j+1]} + p_{[i,j+1]} g_{[j,k]}, p_{[i,j+1]} p_{[j,k]}) \end{aligned} \tag{10.29}$$

is an associative binary operator, where $k \leq j < n$. The binary operator $\bullet$ can be implemented with the logic diagram shown in Figure 10.47(b).

10.4.3.2 Parallel-Prefix Adder In a parallel-prefix adder, a parallel-prefix network is used instead of the CLA generator to generate the carries in parallel. Depending on how to construct the parallel-prefix network, a broad variety of adders can be obtained. In what follows, we deal with three popular parallel-prefix adders: the *Brent-Kung adder*, *Kogge-Stone adder*, and *Ladner-Fisher adder*.

Before proceeding, the common logic cells used in these parallel-prefix networks are introduced. As shown in Figure 10.47, there are four types of black computation nodes (cells), which can be further subdivided into: input logic, group logic, and output logic. The input logic is the *pg* generator, denoted as a black box, as shown in Figure 10.47(a). The group logic carries out actual prefix computation and includes two different types, denoted as a black circle and triangle, as shown in Figures 10.47(b) and (c), respectively. The output logic is the sum generator, denoted as a black diamond, as shown in Figure 10.47(d). It often uses white triangles in the parallel-prefix networks to denote buffers without doing any computation. These buffers are not necessary and may often be omitted in realistic implementations.

Examples of parallel-prefix adders are shown in Figure 10.48. The parallel-prefix networks of these adders are all radix-2 in the sense that the number of groups combined in each black computation node is 2. In general, a radix-k tree means that the maximum number of groups combined in each black computation node is k. We briefly describe each of these parallel-prefix networks.

The Brent-Kung adder uses the recursively doubling approach to compute the prefixes, as shown in Figure 10.48(a). It first computes prefixes for 2-bit groups. These prefixes are then used to compute prefixes for 2^2-bit groups, which in turn are employed to find prefixes for 2^3, and so on. Finally, the prefixes come back to find the carries

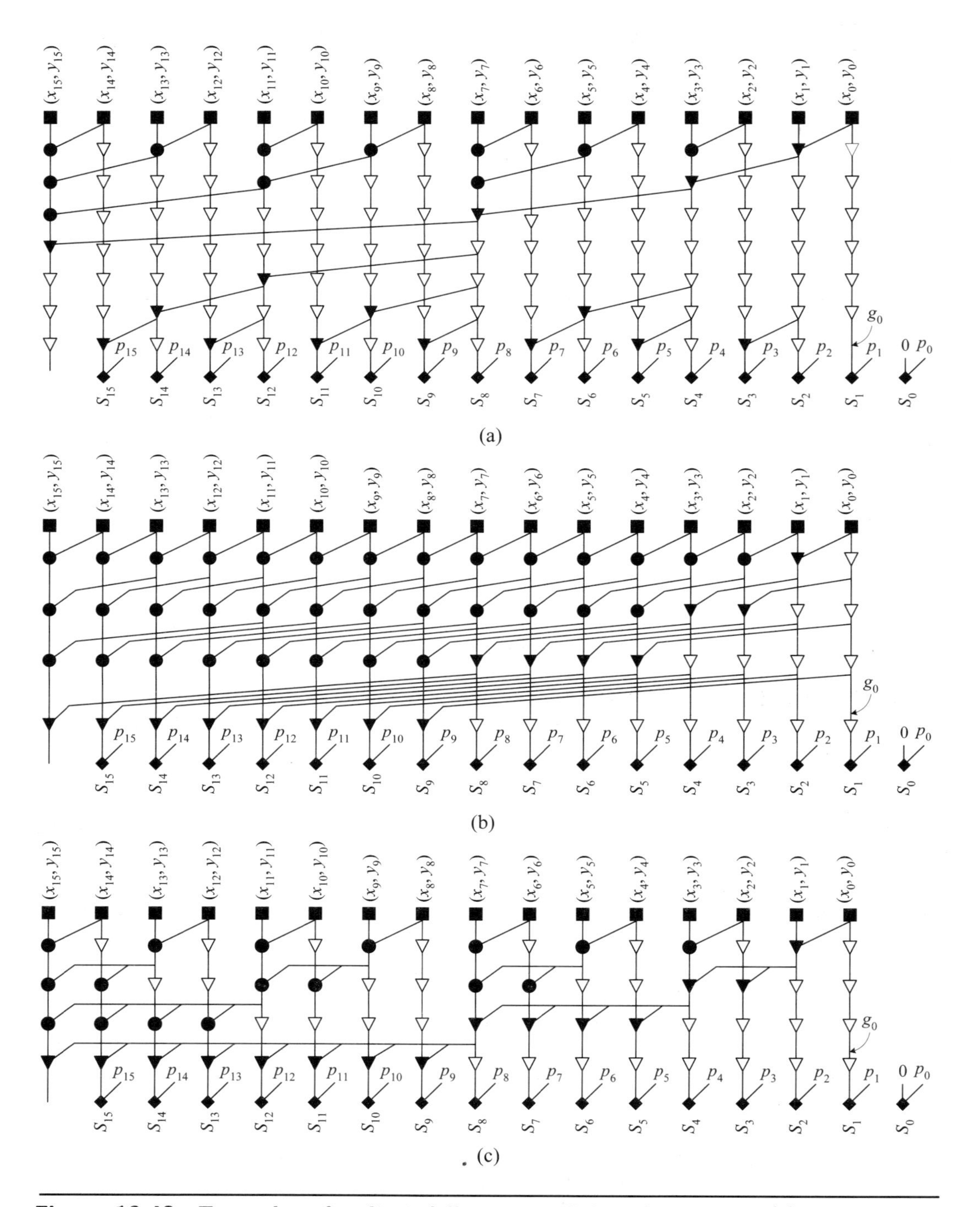

Figure 10.48: Examples of radix-2 full-tree parallel-prefix adders: (a) Brent-Kung adder; (b) Kogge-Stone adder; (c) Ladner-Fisher adder.

to each bit. The features of the Brent-Kung adder are that it uses a parallel-prefix network of $2\log_2 n - 2$ stages and each node, regardless of black computation or buffer node, has a uniform fan-out of 2. The number of group logic nodes (including circles and triangles) of the Brent-Kung parallel-prefix network is $2n - 2 - \log_2 n$.

Compared to the Brent-Kung adder, the Kogge-Stone adder only needs a $\log_2 n$-stage parallel-prefix network with each node still having a fan-out of 2, as shown in Figure 10.48(a). However, the parallel-prefix network used in the Kogge-Stone adder has more black computation nodes and needs more long wires to route between stages, thereby leading to more power dissipation. The number of group logic nodes of the Kogge-Stone parallel-prefix network is $n\log_2 n - n + 1$.

The Ladner-Fisher adder uses a parallel-prefix network to compute the prefixes for the odd-numbered bits and then uses one or more stages to extend these results into even-numbered bits, as shown in Figure 10.48(c). Although the Ladner-Fisher adder also uses a $\log_2 n$-stage parallel-prefix network, the fan-out of each node is not the same. Hence, it is necessary to properly size the high fan-out nodes to achieve a good speed. The number of group logic nodes of the Ladner-Fisher parallel-prefix network is $(1/2)n\log_2 n$.

The parallel-prefix adders with higher radix parallel-prefix networks are also possible. Two radix-4 full-tree parallel-prefix adders are given in Figure 10.49. As stated, in the radix-4 parallel-prefix network, each node can combine together at most four groups. The depth of resulting trees is reduced significantly at the expense of fan-out and fan-in. Hence, to achieve a good speed, it is necessary to appropriately size the nodes. Furthermore, the group logic cells also need to be modified to accommodate the large fan-in.

Carry-select adders or other types of adders can also be merged with a parallel-prefix network to form a hybrid adder. As carry-select adders are combined with a parallel-prefix network to form a hybrid adder, the parallel-prefix network only needs to generate those carries indeed needed by the carry-select adders. Such a parallel-prefix network is called a *sparse-tree*. Some examples are shown in Figure 10.50. In designing such a hybrid adder, care must be taken to balance the propagation delays between carry-select adders and the carry-in generation network by properly selecting the group length of carry-select adders.

10.4.3.3 Other Parallel-Prefix Computations

Many datapath operations can also be reduced into a prefix-sum computation and then computed using a parallel-prefix network. We only introduce three of these.

- Priority encoder
- Incrementer/decrement
- Gray-to-binary code converter

Priority encoder. As stated in Section 10.1.2, assuming that the input is $D = (d_{n-1}d_{n-2}\cdots d_1 d_0)$, an n-input priority encoder can be generally expressed as follows:

$$\begin{aligned}
EP_{n-1} &= d_{n-1}\\
EP_{n-2} &= \bar{d}_{n-1}\cdot d_{n-2}\\
&\cdots\\
EP_1 &= \bar{d}_{n-1}\cdot \bar{d}_{n-2}\cdots \bar{d}_2\cdot d_1\\
EP_0 &= \bar{d}_{n-1}\cdot \bar{d}_{n-2}\cdots \bar{d}_2\cdot \bar{d}_1\cdot d_0
\end{aligned} \tag{10.30}$$

To use a parallel-prefix network to carry out the priority encoding process, we need to define black computation nodes: input logic, group logic, and output logic. These

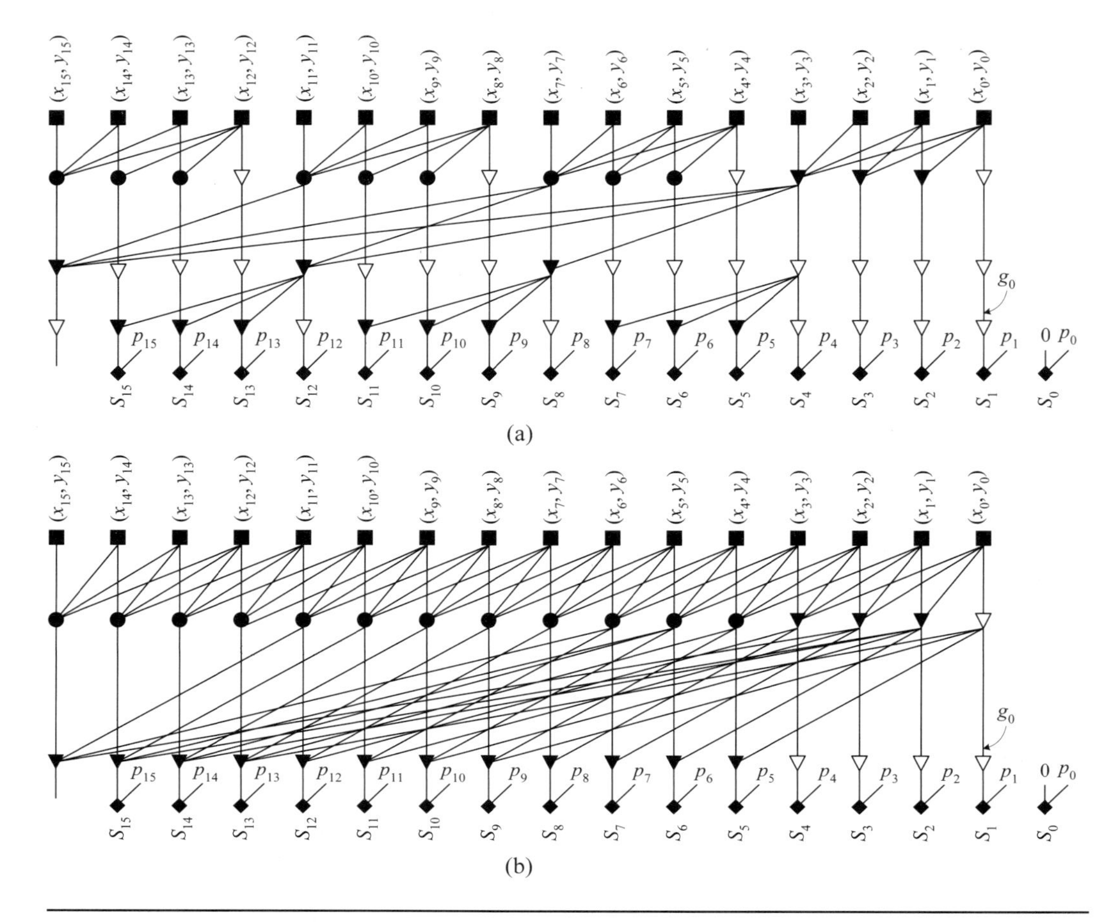

Figure 10.49: Examples of radix-4 full-tree parallel-prefix adders: (a) Brent-Kung adder; (b) Kogge-Stone adder.

nodes are defined as follows:

$$\begin{aligned} x_{[i:i]} &= \bar{d}_i && \text{(input logic)} \\ x_{[i:j]} &= x_{[i:k+1]} \cdot x_{[k:j]} && \text{(group logic)} \\ EP_i &= d_i \cdot x_{[n-1:i+1]} && \text{(output logic)} \end{aligned} \quad (10.31)$$

where $0 \leq i, j \leq n-1$ and $j \leq k < i$. Apparently, any parallel-prefix network can be used to carry out the priority encoding operation.

Incrementer/decrementer. The incrementer/decrementer is essentially used to find the least significant bit with zero value in the input and complement all bits from the least significant bit up to this bit. From this, two operations are implied: search for the least significant bit with zero value in the input and complement all least significant bits up to the bit found. Apparently, the first operation is a priority operation, and the second is only an XOR operation.

Assume that the input data is $D = (d_{n-1}d_{n-2}\cdots d_1 d_0)$. The black computation nodes for parallel-prefix networks used to compute the incrementer/decrementer are as follows:

$$x_{[i:i]} = d_i/\bar{d}_i \qquad \text{(input logic)}$$

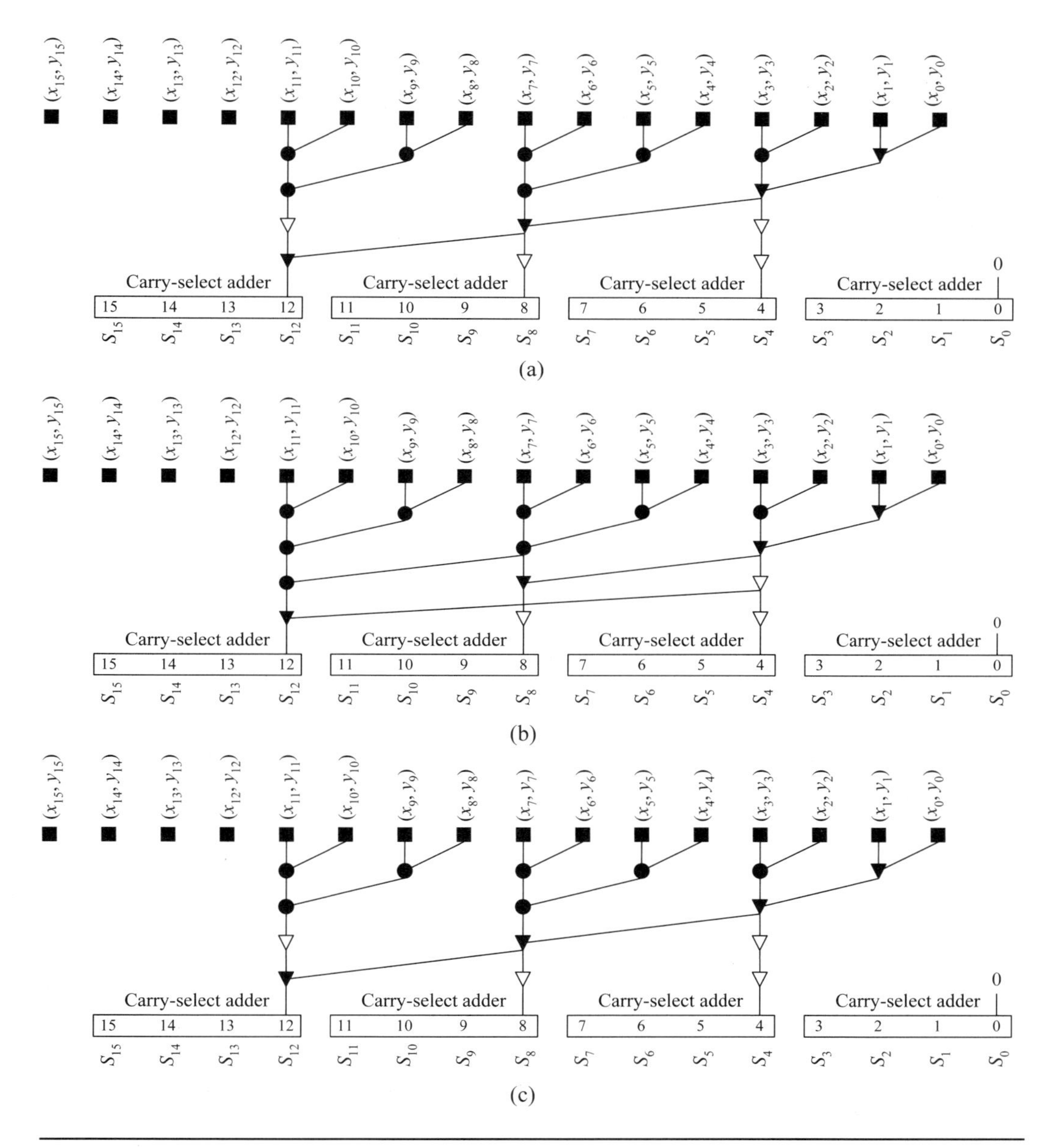

Figure 10.50: Examples of sparse-tree parallel-prefix adders: (a) Brent-Kung adder; (b) Kogge-Stone adder; (c) Ladner-Fisher adder.

$$
\begin{aligned}
x_{[i:j]} &= x_{[i:k+1]} \cdot x_{[k:j]} && \text{(group logic)} \\
y_i &= d_i \oplus x_{[i-1:0]} && \text{(output logic)}
\end{aligned} \tag{10.32}
$$

where $x_{[-1:0]} = 1$, $0 \leq i, j \leq n-1$, and $j \leq k < i$. The $x_{[i:i]}$ is set to d_i for the incrementer and to $\bar{d}_i$ for the decrementer. That is, the input is first complemented for the decrementer operation. Once these nodes are defined, any parallel-prefix network can be utilized to carry out the incrementer/decrementer operation.

Gray-to-binary code converter. Gray code has a salient feature that only one bit is different between two adjacent codes. Hence, it is widely used in communications and low-power systems due to low transitions. Nevertheless, most systems are binary systems; hence, it needs to convert a Gray code into an equivalent binary code. We illustrate how to use a parallel-prefix network to perform such a conversion.

Assume that the input Gray code is $G = (g_{n-1}g_{n-2}\cdots g_1g_0)$, and the output binary code is $B = (b_{n-1}b_{n-2}\cdots b_1b_0)$. The rule of converting a Gray code into an equivalent binary code is that if the number of 1s of the input Gray code, counting from MSB to the current position i is odd, then the ith binary bit is 1; otherwise, the ith binary bit is 0. In other words,

$$
b_i = g_{n-1} \oplus g_{n-2} \oplus \cdots \oplus g_{i+1} \oplus g_i \tag{10.33}
$$

for all $0 \leq i \leq n-1$.

The black computation nodes for parallel-prefix networks used to compute the Gray-to-binary code conversion are as follows:

$$
\begin{aligned}
x_{[i:i]} &= g_i && \text{(input logic)} \\
x_{[i:j]} &= x_{[i:k+1]} \oplus x_{[k:j]} && \text{(group logic)} \\
b_i &= g_i \oplus x_{[n-1:i+1]} && \text{(output logic)}
\end{aligned} \tag{10.34}
$$

where $0 \leq i, j \leq n-1$ and $j \leq k < i$. Once these nodes are defined, any parallel-prefix network can be employed to carry out the Gray-to-binary code conversion.

■ Review Questions

Q10-37. Describe the principle of parallel-prefix networks.

Q10-38. Describe the rationale behind parallel-prefix adders.

Q10-39. Give a numerical example to illustrate the operation of the priority encoder based on a parallel-prefix network.

Q10-40. Give a numerical example to illustrate the operation of the incrementer based on a parallel-prefix network.

Q10-41. Give a numerical example to illustrate the operation of Gray-to-binary code conversion based on a parallel-prefix network.

10.5 Multiplication

Multiplication is one of the essential operations of digital systems. It is usually built on the basis of addition. The basic operation of the multiplication algorithms to be considered in this section is based on the shift-and-add technique and may process unsigned and signed input numbers. A multiplication algorithm may be usually realized by using either a multiple-cycle or a single-cycle structure. The multiple-cycle structure is essentially a sequential logic circuit while the single-cycle structure is fundamentally a combinational logic circuit. The single-cycle structure naturally evolves

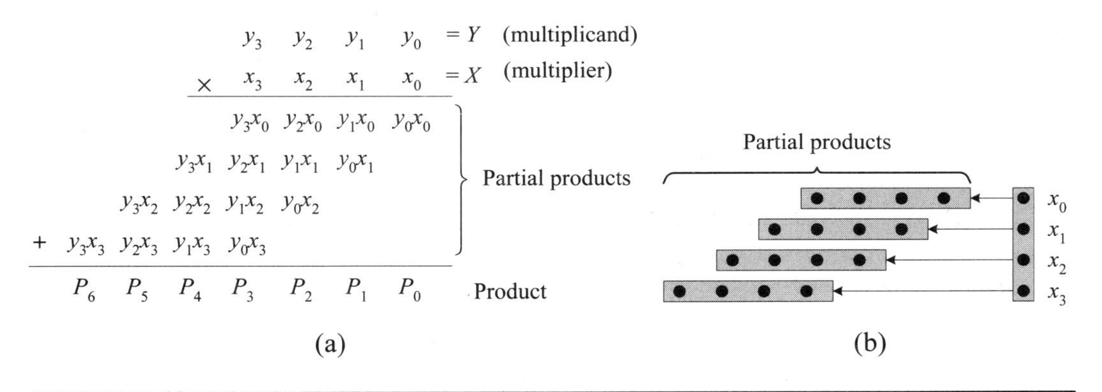

Figure 10.51: The operation of multiplication with two 4-bit operands: (a) partial products; (b) dot diagram.

into array structures, such as an unsigned array multiplier and the modified Baugh-Wooley signed array multiplier.

10.5.1 Unsigned Multipliers

In this section, we first discuss the basic principles of multiplication and then describe how the multiplication can be implemented with multiple-cycle and single-cycle structures.

10.5.1.1 Basic Principles of Multiplication Let multiplier X and multiplicand Y be n- and m-bit numbers, respectively, then the product of X and Y can be expressed as follows:

$$\begin{aligned} P &= X \times Y = \sum_{i=0}^{n-1} x_i \cdot 2^i \cdot \sum_{j=0}^{m-1} y_j \cdot 2^j \\ &= \sum_{i=0}^{n-1}\sum_{j=0}^{m-1} (x_i \cdot y_j) \cdot 2^{i+j} = \sum_{k=0}^{n+m-1} P_k \cdot 2^k \end{aligned} \quad (10.35)$$

where each $x_i \cdot y_j$ can be readily obtained with a two-input AND gate. An illustration of the operation of multiplication with two 4-bit operands is shown in Figure 10.51(a). Here, we generate all partial products, $x_i \cdot y_j$, and arrange them into rows according to the bit order of multiplier. For convenience, a dot diagram is often recommended to simplify the representation of a multiplication algorithm. An example of this is illustrated in Figure 10.51(b). Depending on how to add up the partial products, a variety of algorithms are proposed during the past decades. A few of them will be discussed in the following.

As we can see from Figure 10.51, the basic operation of multiplication in hardware exactly reflects the "paper-and-pencil" method that we use daily. This method is also referred to as the *shift-and-add approach* due to the inherent shift and addition operations associated with it. The rationale behind the shift-and-add approach is based on the following rule when a multiple-bit multiplicand multiplies a single-bit multiplier:

- The partial product is the same as the multiplicand if the multiplier is 1 and is 0 otherwise.

When the multiplier is also a multiple-bit number, the above rule is applied to each individual bit of the multiplier, and then all partial products are added up accordingly to obtain the final product.

The detailed operations of the "paper-and-pencil" method described above can be summarized as the following algorithm.

Algorithm: Unsigned shift-and-add multiplication

Input: An m-bit multiplicand and an n-bit multiplier.
Output: The $(m+n)$-bit product.
Begin
1. Load multiplicand and multiplier into registers M and Q, respectively; clear accumulator A and set the loop count (CNT) equal to n;
2. **repeat**
   2.1 **if** $Q[0] = 1$ **then** $A \leftarrow A + M$;
   2.2 Right shift register pair $A : Q$ one bit;
   2.3 $CNT \leftarrow CNT - 1$;
   until $CNT = 0$

End

This algorithm can be realized in a variety of ways. One extreme is to directly map the algorithm into an n-cycle structure. Each cycle processes a multiplier bit and accumulates the partial product up to this bit. The other extreme generates all partial products at a time and uses a combinational logic circuit, such as an array, to add up all partial products into the final product in one cycle. Of course, a compromise approach is also possible. In this approach, the multiplier is partitioned into (n/k) groups, with each containing k bits. A combinational logic circuit may be used to generate the partial product of the m-bit multiplicand and k-bit multiplier. The resulting (n/k) partial products are then accumulated into the final product.

Two multiple-cycle hardware structures, that is, sequential implementations, are shown in Figure 10.52. Figure 10.52(a) directly maps the above unsigned multiplication algorithm into a sequential structure. After the multiplicand and multiplier are loaded into registers M and Q as well as the accumulator A is cleared, the partial products are generated one by one in sequence and accumulated into the accumulator A. Once a new partial product enters the accumulator A, the accumulator A and the multiplier are shifted right one bit position. This process requires $2n$ clock cycles. To speed up the operation, the two-step accumulator-load and shift-right operations can be combined into one single-step operation, namely, shift-right and load operation. Based on this, the multiplication only needs n clock cycles to finish.

Figure 10.52(b) gives an alternative multiple-cycle structure that needs only $n/2$ clock cycles to finish a multiplication with an n-bit multiplier, when using a single-step shift-right and load operation. In this structure, two multiplier bits are examined each time and hence 0, M, $2M$, or $3M$ is generated. To simplify the hardware structure, two 2-to-1 multiplexers as well as a CSA and CPA are used to accumulate partial products.

10.5.1.2 Bit-Serial Multiplication Another possible implementation of the above algorithm is shown in Figure 10.53, where the multiplier X is input in a bit-serial form while the multiplicand Y is input in parallel. Each time a multiplier bit, with the least-significant bit (LSB) first, is fed into the multiplier circuit and it is ANDed with the multiplicand to yield a partial product. This partial product is then accumulated

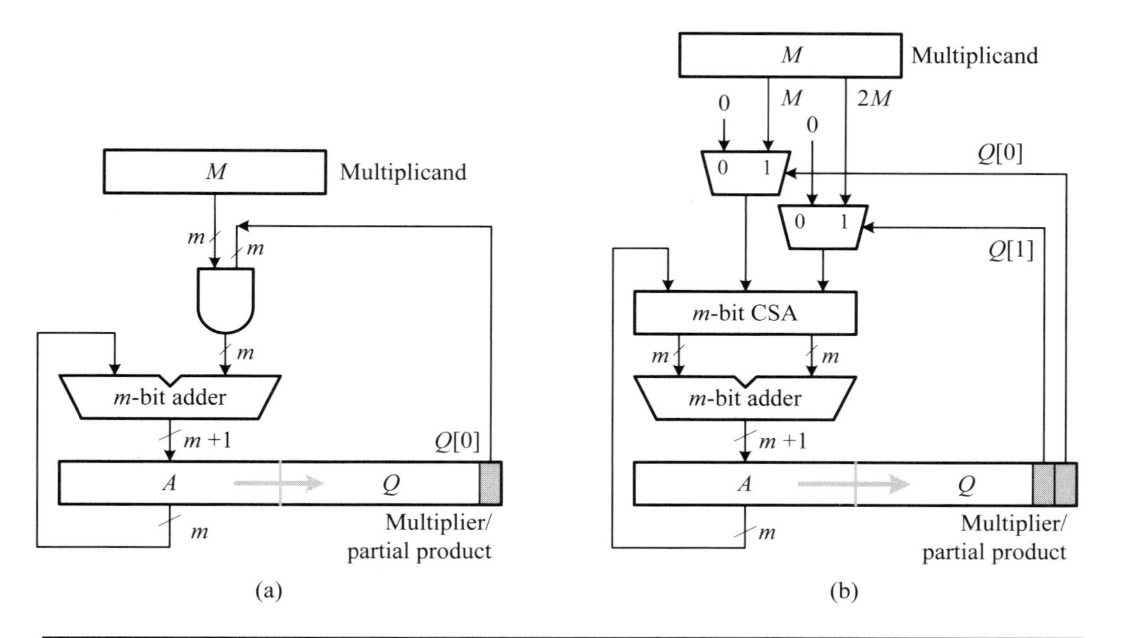

Figure 10.52: Two examples of sequential implementations of unsigned multiplication algorithm: (a) radix-2; (b) radix-4.

to generate one product bit, with the LSB first. After n clock cycles, all multiplier bits are fed into the circuit. The next m clock cycles are used to shift out the product bits from the accumulator. This circuit is referred to as a *bit-serial multiplier*.

10.5.1.3 Unsigned Array Multipliers The single-cycle structure can be regarded as an expansion of the shift-and-add multiplication in the temporal dimension and all partial products are arranged into a proper position without doing any summation. Each partial product is simply generated by ANDing the appropriate bit of the multiplier with the multiplicand. For an $m \times n$ multiplication, this can be done with $m \times n$ two-input AND gates in t_{AND} time, where t_{AND} is the propagation delay of a two-input AND gate. An example of 4×4 multiplication is shown in Figure 10.51(a).

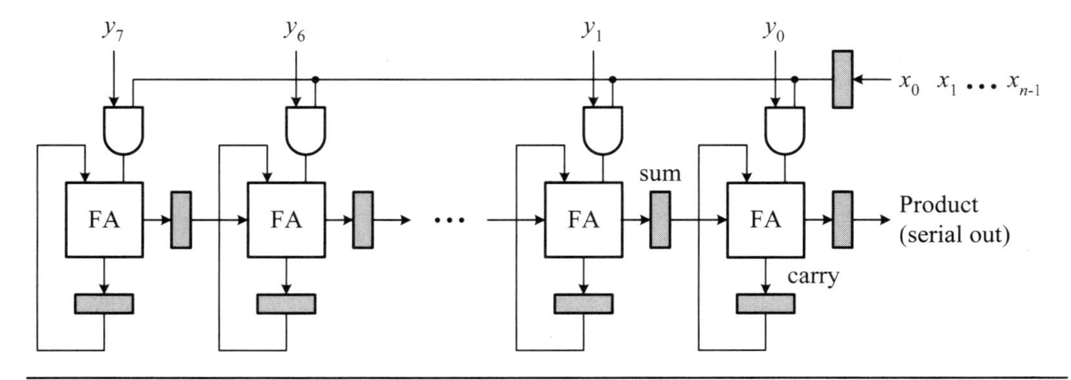

Figure 10.53: An example of an $8 \times n$ bit-serial multiplier.

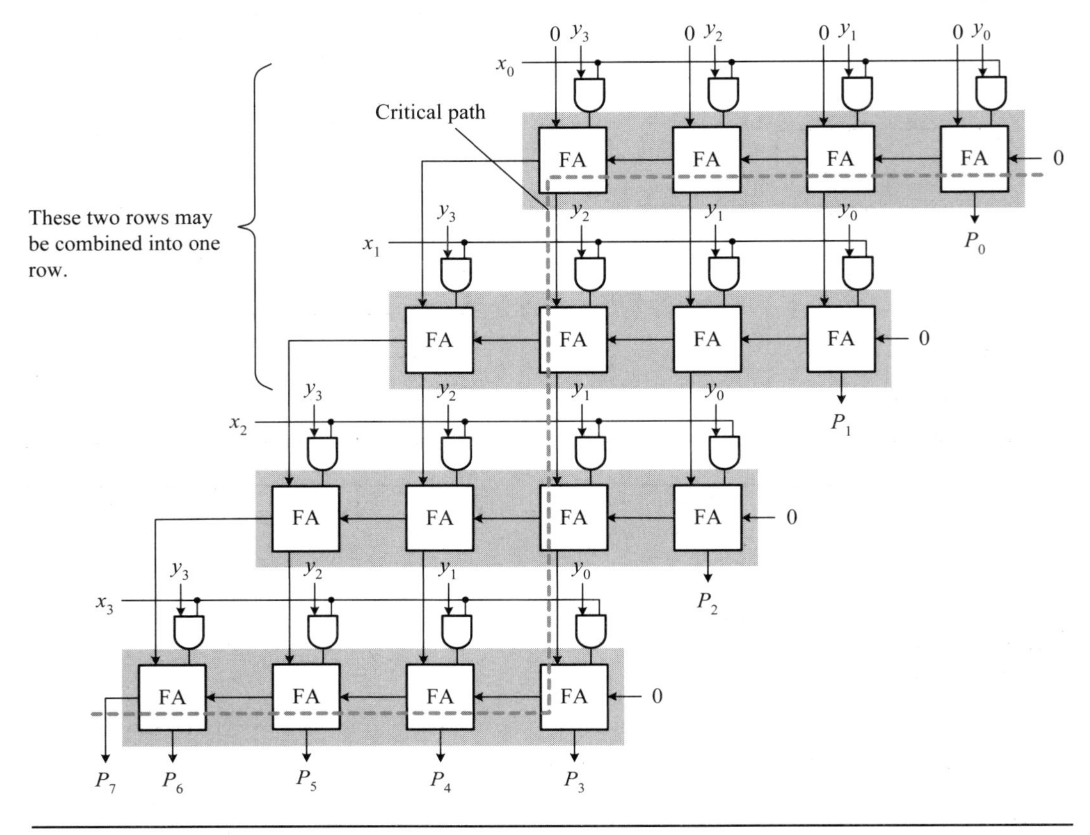

Figure 10.54: An example of an unsigned 4×4 array multiplier.

Once all partial products are generated, the next step is to add them together to obtain the final product. Depending on how to add up the partial products, a variety of approaches have been proposed during the past decades. These approaches yield a variety of array multipliers. We introduce two of these. One uses ripple-carry adders (RCAs), and the other uses carry-save adders (CSAs).

An *array multiplier* based on RCAs accepts an n-bit multiplier and an m-bit multiplicand and uses an array of cells to calculate the bit product $x_i \cdot y_j$ in parallel and then adds them together in a proper way to yield the final product. An example is shown in Figure 10.54. Of course, as we look at the array multiplier more carefully, we can find that the first two rows of the array multiplier may be combined into one row to reduce the number of full adders by a factor of m. The propagation delay of the critical path of this array multiplier is $[2(m-1)+n]t_{FA}$ when the first two rows are not combined and is $[2(m-1)+(n-1)]t_{FA}$ when the first two rows are combined.

■ **Example 10-16: (An unsigned 4×4 RCA array multiplier.)** An example of an unsigned 4×4 array multiplier using RCAs is shown in Figure 10.54. In this array multiplier, 4×4 two-input AND gates and full adders are needed. The propagation delay of the critical path of this array multiplier is $[2(4-1)+4]t_{FA} = 10t_{FA}$ when the first two rows are not combined and is $[2(4-1)+(4-1)]t_{FA} = 9t_{FA}$ when the first two rows are combined. ■

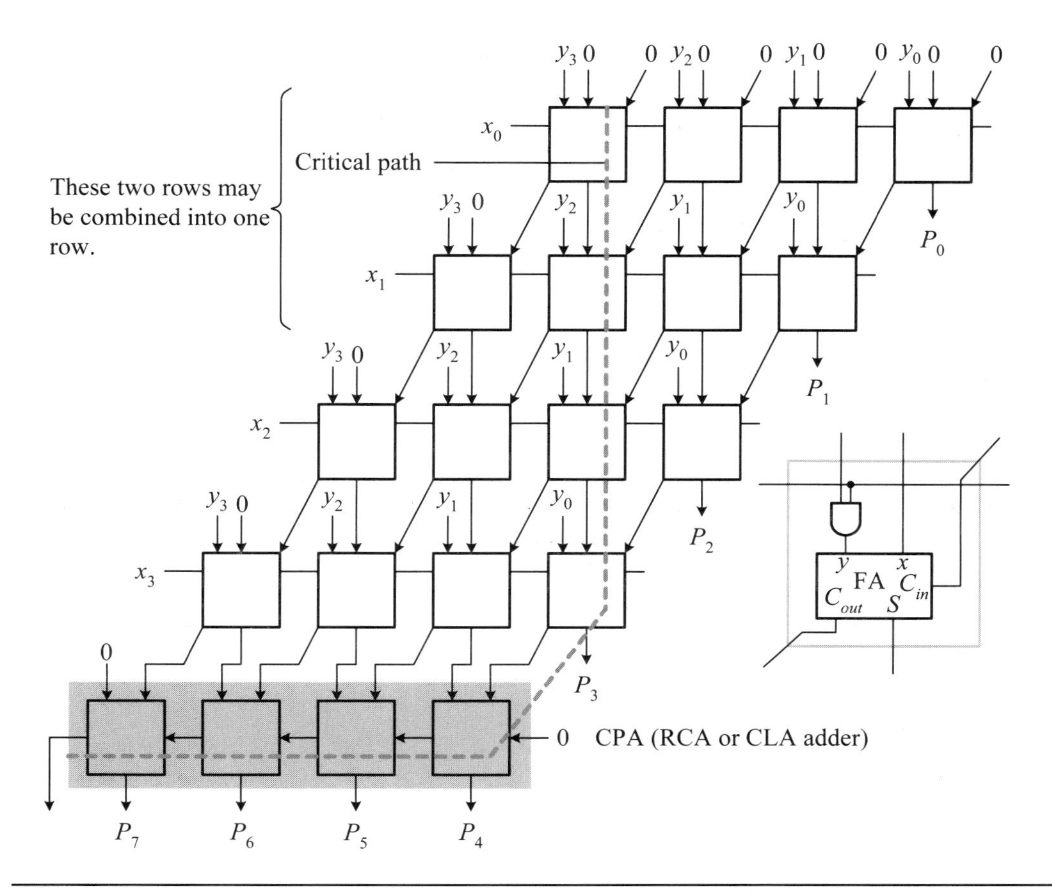

Figure 10.55: An example of an unsigned 4×4 array multiplier using CSAs.

The other array multiplier is based on CSAs, as shown in Figure 10.55. Like the one based on RCAs described above, this array multiplier still needs $m \times n$ AND gates and CSAs, as well as an m-bit CPA, which can be either a RCA or a CLA adder. The propagation delay of the critical path of this array multiplier is $(m+n)t_{FA}$ when the first two rows are not combined and is $[(m+(n-1)]t_{FA}$ when the first two rows are combined. As a consequence, the propagation delay of this array multiplier is reduced by a factor of $(m-2)t_{FA}$ at the cost of an m-bit CPA.

■ Example 10-17: (An unsigned 4×4 CSA array multiplier.) Figure 10.55 shows an example of an unsigned 4×4 array multiplier using CSAs. As we can see from the figure, this array multiplier needs 4×4 AND gates and CSAs, as well as a 4-bit CPA, which can be either a RCA or a CLA adder. The propagation delay of the critical path of this array multiplier is $(4+4)t_{FA} = 8t_{FA}$ when the first two rows are not combined and is $[(4+(4-1)]t_{FA} = 7t_{FA}$ when the first two rows are combined. ■

10.5.1.4 Wallace-Tree Multipliers A k-input Wallace tree accepts k n-bit operands and yields two $(n+\log_2 k-1)$-bit outputs. The basic cells used in the Wallace tree are 3:2 or 4:2 compressors. Because a compressor effectively accounts for the number

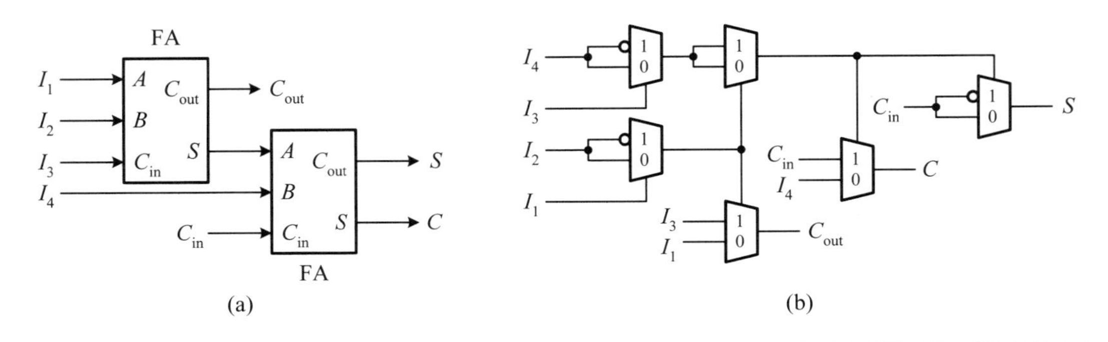

Figure 10.56: Two examples of 4:2 compressor circuits: (a) FA-based; (b) multiplexer-based.

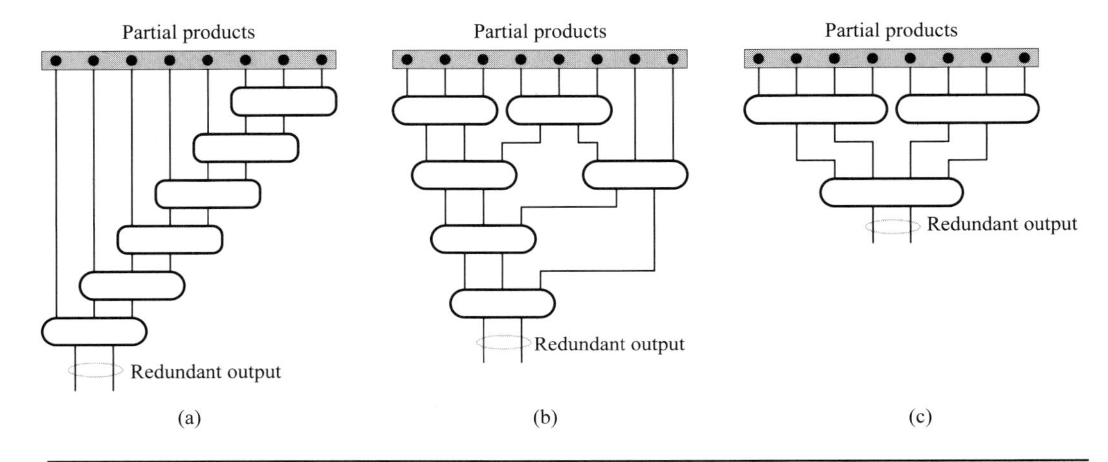

Figure 10.57: Examples of tree multipliers: (a) array multiplier; (b) using 3:2 compressors; (c) using 4:2 compressors.

of 1s in the input and indicates the result as the output, a 3:2 compressor (namely, CSA) is also called a $(3, 2)$ *counter*.

A 4:2 compressor accepts four equal-weight inputs and yields two outputs. Along the way, it also accepts a carry from the previous column and generates an intermediate carry into the next column. Hence, a 4:2 compressor is more appropriate to be named as a $(5, 3)$ counter. Figure 10.56 gives two examples showing how a 4:2 compressor can be built by using either two full adders or six 2-to-1 multiplexers and three inverters. The performance of the multiplexer-based 4:2 compressor is better than that of the FA-based 4:2 compressor.

Figure 10.57 shows three examples of tree multipliers based on 3:2 compressors and 4:2 compressors. All of trees are to add together eight n-bit operands into two $(n+2)$-bit outputs. Figure 10.57(a) is the $n \times 8$ array multiplier, Figure 10.57(b) is the case using 3:2 compressors, and Figure 10.57(c) is the situation using 4:2 compressors. From the figure, we can see that the tree based on the 4:2 compressor has fewer levels and yields more regular layout.

Examples of Wallace- and Dadda-tree multipliers are shown in Figure 10.58. The strategy of the Wallace tree is to add operands as soon as possible and the result is to minimize the overall delay by making the final CPA as short as possible. In contrast,

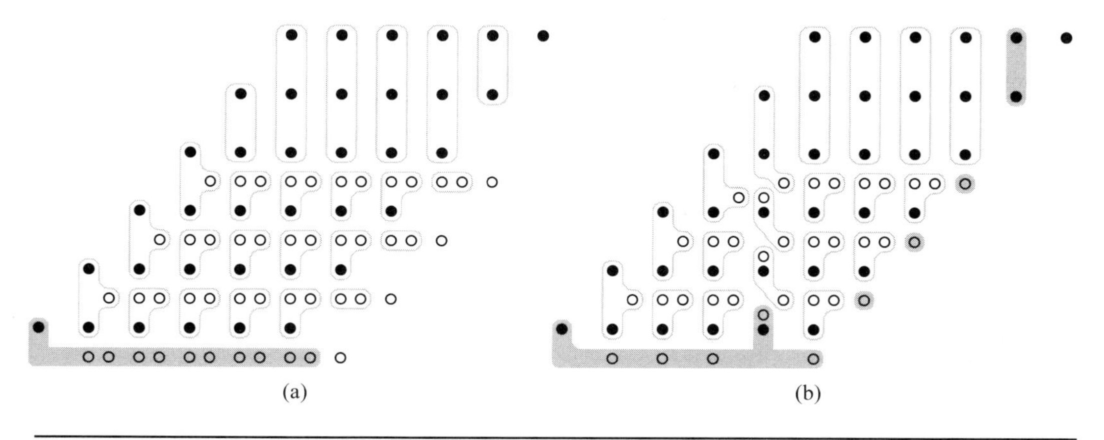

Figure 10.58: Examples of 6×6 (a) Wallace and (b) Dadda multipliers.

the strategy of the Dadda tree is to add operands as late as possible and hence leads to a simple CSA tree and a wider CPA. To reduce k operands into a two carry-save redundant form, the levels of the Wallace tree are equal to

$$\lceil \log_{3/2}(k/2) \rceil \tag{10.36}$$

when using 3:2 compressors and is

$$\lceil \log_2(k/2) \rceil \tag{10.37}$$

when using 4:2 compressors.

In summary, both Wallace and Dadda trees have some features in common. First, both need a logarithmic depth reduction tree and yield an irregular layout, hence making the layout much more difficult. Second, connections and signal paths of varying lengths generate logic hazards and signal skews, leading to performance deterioration and power dissipation. Hence, care must be taken in designing such trees.

■ Review Questions

Q10-42. Show that the ones indicated in Figures 10.54 and 10.55 are indeed the critical paths.

Q10-43. Show that the circuit of Figure 10.56(b) is a 4:2 compressor.

Q10-44. What is the distinction between Wallace and Dadda trees?

10.5.2 Signed Multipliers

Like unsigned multiplication, signed multiplication is also widely used in digital systems. There are many algorithms that can be used to implement signed two's complement multiplication. We only consider the two most common algorithms: the *Booth algorithm* and *modified Baugh-Wooley algorithm*.

10.5.2.1 Booth Multipliers The Booth algorithm accepts an m-bit two's complement multiplicand and an n-bit two's complement multiplier, and generates an $(m+n)$-bit product in two's complement form. To illustrate the operation of the Booth algorithm, let multiplier and multiplicand be $X = (x_{n-1}, \cdots, x_1, x_0)$ and $Y = (y_{m-1}, \cdots, y_1, y_0)$, respectively. Then, the operations of Booth algorithm are as follows. At the

ith step, where $0 \leq i \leq n-1$, one of the following operations is performed according to the values of $x_i x_{i-1}$.

- Add 0 to the partial product P if $x_i x_{i-1} = 00$;
- Add Y to the partial product P if $x_i x_{i-1} = 01$;
- Subtract Y from the partial product P if $x_i x_{i-1} = 10$;
- Add 0 to the partial product P if $x_i x_{i-1} = 11$.

where the initial value of x_{-1} is assumed to be 0.

The above operations can be described as the following algorithm.

Algorithm: Booth algorithm

Input: An m-bit multiplicand and an n-bit multiplier in two's complement form.
Output: The product left in $A : Q[n : 1]$ in two's complement form.
Begin

1. Load multiplicand and multiplier into registers M and $Q[n : 1]$, respectively; clear accumulator A and $Q[0]$, and set the loop count (CNT) equal to n;
2. **repeat**
   2.1 **if** $Q[1 : 0] = 01$ **then** $A \leftarrow A + M$;
   2.2 **if** $Q[1 : 0] = 10$ **then** $A \leftarrow A - M$;
   2.3 Right shift register pair $A : Q$ one bit; $CNT \leftarrow CNT - 1$;
   until $CNT = 0$;

End ∎

A numerical example to illustrate the operations of the Booth algorithm is given in the following.

∎ Example 10-18: (A numerical example of the Booth algorithm.) Suppose that multiplicand and multiplier are 35H and 67H in hexadecimal representation, respectively. Register Q is 9 bits. The multiplier is loaded into register $Q[8 : 1]$ and $Q[0]$ is cleared to 0. The bits $Q[1 : 0]$ is checked to determine the proper operation, addition, subtraction, or shift only, according to the Booth algorithm. After the completion of the algorithm, the final product is left in the most significant 16 bits of register pair $A : Q$, which is 1553H. The details are illustrated in Figure 10.59. ∎

The above Booth algorithm is radix-2. Higher-order radix Booth algorithms are also possible. A radix-4 Booth encoder is given in Table 10.2. Details of the radix-4 Booth algorithm are left to the reader as an exercise (Problem 10-22).

Sequential implementation. The Booth algorithm can be sequentially implemented with the scheme shown in Figure 10.60(a), where registers M and Q are used to record multiplicand and multiplier, respectively. Register Q contains $n+1$ bits with $Q[0]$ to provide the initial value of x_{-1} required by the algorithm. Accumulator A is a register used to hold the partial product up to the ith step. The n-bit adder/subtracter is utilized to perform the addition or subtraction needed by the algorithm and is controlled by the Booth encoder. The detailed coding of the Booth encoder is given in Figure 10.60(b).

Like in the unsigned sequential multiplier, the load and shift operations in the sequential Booth multiplier may be combined into one single step by directly loading the shifted result into the register pair $A : Q$ to speed up the operation.

	A	Q	$Q[0]$	CNT	
Q = 67H	0 0 0 0 0 0 0 0	0 1 1 0 0 1 1 1	0	8	$A \leftarrow A$ - M
M = 35H	- 0 0 1 1 0 1 0 1				
	1 1 0 0 1 0 1 1				
	1 1 1 0 0 1 0 1	1 0 1 1 0 0 1 1	1	7	Right shift A:Q
	1 1 1 1 0 0 1 0	1 1 0 1 1 0 0 1	1	6	Right shift A:Q
	1 1 1 1 1 0 0 1	0 1 1 0 1 1 0 0	1	5	Right shift A:Q
	+ 0 0 1 1 0 1 0 1				$A \leftarrow A + M$
	0 0 1 0 1 1 1 0				
	0 0 0 1 0 1 1 1	0 0 1 1 0 1 1 0	0	4	Right shift A:Q
	0 0 0 0 1 0 1 1	1 0 0 1 1 0 1 1	0	3	Right shift A:Q
	- 0 0 1 1 0 1 0 1				$A \leftarrow A$ - M
	1 1 0 1 0 1 1 0				
	1 1 1 0 1 0 1 1	0 1 0 0 1 1 0 1	1	2	Right shift A:Q
	1 1 1 1 0 1 0 1	1 0 1 0 0 1 1 0	1	1	Right shift A:Q
	+ 0 0 1 1 0 1 0 1				$A \leftarrow A + M$
	0 0 1 0 1 0 1 0				
1553H $\longrightarrow$	0 0 0 1 0 1 0 1	0 1 0 1 0 0 1 1	0	0	Right shift A:Q

Figure 10.59: An illustration of the operation of the Booth algorithm.

Array implementation. Like what we have done before, a single-cycle structure is obtained by expanding the combinational logic part of the sequential structure in the temporal dimension. Based on this, the Booth array multiplier is obtained, as shown in Figure 10.61, where the control ($CTRL$) block is the Booth encoder while each CAS row corresponds to the n-bit adder/subtracter.

■ **Example 10-19: (A 4×4 Booth array multiplier.)** A 4×4 Booth array multiplier is shown in Figure 10.61, which is simply a single-cycle implementation of the Booth algorithm. Basically, each row of the array is a copy of the combinational logic circuit shown in Figure 10.60(a). The shift operation of the partial product is accomplished by skewing the successive rows of the array. The $CTRL$ block is the Booth encoder derived from the encoding table given in Figure 10.61(a). The combination of n CAS cells functions as the n-bit adder/subtracter of Figure 10.60(a). Note how the sign extension is carried out in the array.

■

10.5.2.2 Modified Baugh-Wooley Multipliers To illustrate the operation of modified Baugh-Wooley algorithm, let multiplier and multiplicand be $X = (x_{n-1}, \cdots, x_1, x_0)$ and $Y = (y_{m-1}, \cdots, y_1, y_0)$, respectively. Recall that the most significant bit of a number has a negative weight in the two's complement representation. Based on this

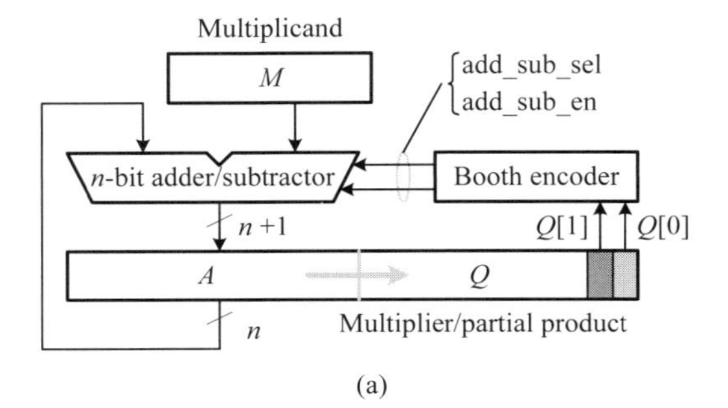

Q[1]	Q[0]	add_sub_sel	add_sub_en	Operation
0	0	ϕ	0	Add 0
0	1	0	1	Add M
1	0	1	1	Subtract M
1	1	ϕ	0	Add 0

(b)

Figure 10.60: An example of the sequential radix-2 Booth multiplier: (a) block diagram; (b) Booth encoder.

property, the two numbers X and Y can be expressed as follows:

$$\begin{aligned} X &= -x_{n-1}2^{n-1} + \sum_{i=0}^{n-2} x_i 2^i \\ Y &= -y_{m-1}2^{m-1} + \sum_{j=0}^{m-2} y_j 2^j \end{aligned}$$

and the product of X and Y can then be represented as

$$\begin{aligned} P &= X \cdot Y \\ &= \left(-x_{n-1}2^{n-1} + \sum_{i=0}^{n-2} x_i 2^i\right)\left(-y_{m-1}2^{m-1} + \sum_{j=0}^{m-2} y_j 2^j\right) \\ &= \sum_{i=0}^{n-2}\sum_{j=0}^{m-2} x_i y_j 2^{i+j} + x_{n-1}y_{m-1}2^{n+m-2} - \\ &\quad \left(\sum_{i=0}^{n-2} x_i y_{m-1} 2^{i+m-1} + \sum_{j=0}^{m-2} x_{n-1} y_j 2^{j+n-1}\right) \end{aligned} \tag{10.38}$$

An illustration of the modified Baugh-Wooley algorithm using a 4×4 example is shown in Figure 10.62.

■ Example 10-20: (A 4×4 Baugh-Wooley array multiplier.) A 4×4 Baugh-Wooley array multiplier is depicted in Figure 10.63. Fundamentally, it is a direct implementation of the scheme shown in Figure 10.62. As can be seen from the figure, two kinds of cell, AND and NAND gates, are required in the array to generate $x_i \cdot y_j$ in parallel. As compared to Figure 10.54, we can see that both unsigned and modified Baugh-Wooley arrays have the same topology. Therefore, a single array may be used for both operations if XOR gates are used to conditionally complement some product terms when signed multiplication is desired. Note the places of 1s in the array. ■

Unlike the unsigned array multiplier and Booth array multiplier, which can accommodate the situation that $m \neq n$, the above discussed modified Baugh-Wooley array

x_i	x_{i-1}	add_sub_sel	add_sub_en	Operation
0	0	ϕ	0	Shift only
0	1	0	1	Add and shift
1	0	1	1	Subtract and shift
1	1	ϕ	0	Shift only

(a)

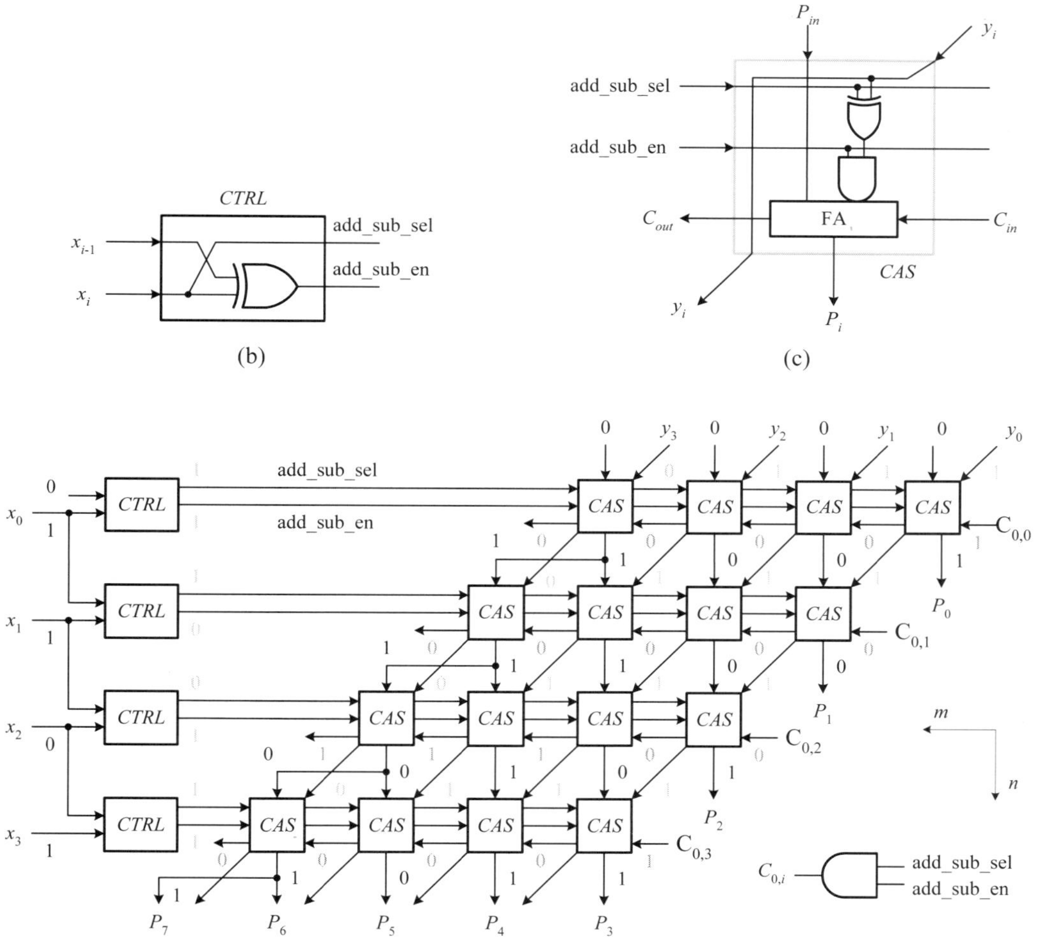

CTRL
x_{i-1}
x_i
add_sub_sel
add_sub_en

(b)

P_{in}
y_i
add_sub_sel
add_sub_en
C_{out}
FA
C_{in}
CAS
y_i
P_i

(c)

0 y_3 0 y_2 0 y_1 0 y_0
add_sub_sel
add_sub_en
x_0 0 1 CTRL CAS CAS CAS CAS $C_{0,0}$
x_1 1 CTRL CAS CAS CAS CAS $C_{0,1}$
x_2 0 CTRL CAS CAS CAS CAS $C_{0,2}$
x_3 1 CTRL CAS CAS CAS CAS $C_{0,3}$
P_0 P_1 P_2
P_7 P_6 P_5 P_4 P_3
m
n
$C_{0,i}$
add_sub_sel
add_sub_en

(d)

Figure 10.61: An example of a 4 × 4 Booth array multiplier: (a) encoding table; (b) CTRL circuit; (c) CAS logic circuit; (d) 4-by-4 Booth array multiplier.

$$
\begin{array}{ccccccccccl}
 & & & & & y_3 & y_2 & y_1 & y_0 & = Y & \text{(multiplicand)} \\
 & & & & \times & x_3 & x_2 & x_1 & x_0 & = X & \text{(multiplier)} \\
\hline
 & & & & 1 & \overline{x_3y_0} & y_2x_0 & y_1x_0 & y_0x_0 & & \\
 & & & & \overline{x_3y_1} & y_2x_1 & y_1x_1 & y_0x_1 & & & \text{Partial product} \\
 & & & \overline{x_3y_2} & y_2x_2 & y_1x_2 & y_0x_2 & & & & \\
+ & 1 & x_3y_3 & \overline{x_2y_3} & \overline{x_1y_3} & \overline{x_0y_3} & & & & & \\
\hline
 & P_7 & P_6 & P_5 & P_4 & P_3 & P_2 & P_1 & P_0 & & \text{Product}
\end{array}
$$

Figure 10.62: An example of a 4×4 modified Baugh-Wooley array multiplier.

multiplier can only work properly when $m = n$. However, the modified Baugh-Wooley algorithm may also work well for any m and n if some proper modifications on the array shown in Figure 10.63 are made. The details of this are left to the reader as an exercise.

10.5.2.3 Mixed Unsigned and Signed Multipliers In many applications, such as general-purpose microprocessors, providing both signed and unsigned multipliers on the same system is often necessary. Depending on how these multipliers are used, many different schemes are possible. If both types of multipliers are only used exclusively, we may then design a multiplier capable of carrying out both signed and unsigned multiplication under the control of a mode selection signal. One possible candidate for such a scheme is the aforementioned scheme, a combination of unsigned array multiplier and Baugh-Wooley array multiplier. Another possible scheme is to use an $[(m+1) \times (n+1)]$-bit Booth multiplier to serve as the desired $(m \times n)$-bit unsigned and signed multiplier. Using this scheme, the unsigned multiplication is carried out simply by setting the highest bit to 0 and the signed multiplication is performed by extending the sign bit to the highest bit. Then the Booth algorithm is performed and the lowest $(m+n)$-bit result is the desired product.

Review Questions

Q10-45. Show that Booth algorithm can correctly carry out the signed multiplication.

Q10-46. Compute the product of $54 \times (-23)$ with the Booth algorithm.

Q10-47. Referring to Figure 10.62, give a numerical example.

Q10-48. Referring to Figure 10.61(b), derive the CTRL logic circuit.

Q10-49. How would you use the Booth algorithm to perform an unsigned multiplication? Give a numerical example to illustrate this.

10.6 Division

The essential operations of multiplication are a sequence of additions. In contrast, the essential operations of division are a sequence of subtractions. Based on this idea, we introduce in this section two basic division algorithms known as *restoring division algorithm* and *nonrestoring division algorithm*, respectively. For simplicity, we only consider the case of unsigned input numbers. The dividend and divisor are assumed

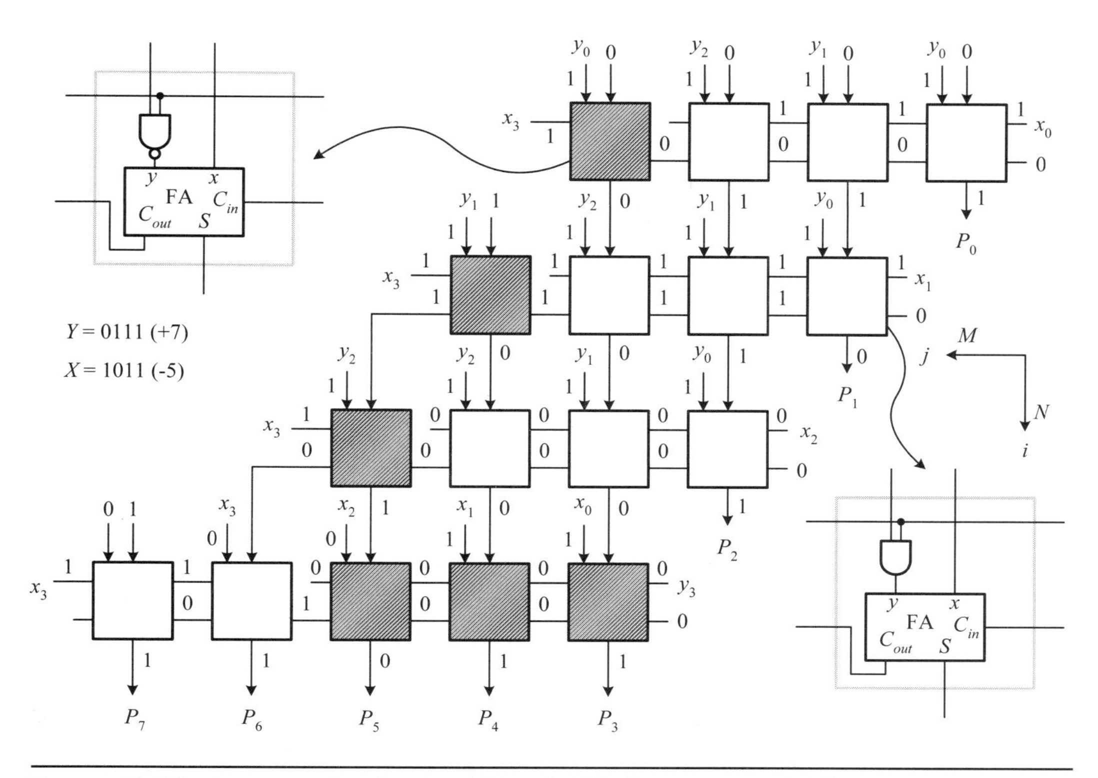

Figure 10.63: An example of a 4×4 Baugh-Wooley array multiplier.

to be m and n bits, respectively. Like multiplication algorithms, a division algorithm may also be realized by using either a multiple-cycle or a single-cycle structure.

10.6.1 Nonrestoring Division

It is often to compare both dividend and divisor in magnitude when a division is carried out. The quotient bit is set to 1 and the divisor is subtracted from the dividend if the dividend is greater than or equal to the divisor; otherwise, the quotient bit is set to 0 and the next bit is proceeded. In digital systems, the comparison is usually done by subtraction directly to simplify the circuit design.

Because of the use of subtraction, the basic design techniques of division circuits can be cast into two types: restoring and nonrestoring methods. In the restoring division method, the quotient bit is set to 1 if the result after subtracting the divisor (M) from the dividend (A) is greater than or equal to 0; otherwise, the quotient bit is set to 0 and the divisor is added back to the dividend to restore the original dividend before proceeding to the next bit. Because of this restoring operation, the restoring division method requires $(3/2)n$ m-bit additions and subtractions on average to complete an n-bit division.

In the nonrestoring division method, the divisor is not added back to the dividend when $A - M < 0$. Instead, the result $A - M$ is shifted left one bit (corresponding to shifting M right one bit) and then added to the divisor M. The result is the same as that the divisor (M) is added back to the dividend (A) and then performs the subtraction. To see this, let $X = A - M$ and assume that $X < 0$. In the restoring division method, the divisor is added back to the X and then the result is shifted left one bit and subtracted by M, namely, $2(X+M)-M=2X+M$, which is equivalent to shifting the X left one bit and then adding with the divisor M. Consequently, both approaches yield the same result. However, the nonrestoring division method needs only n m-bit additions and subtractions. The detailed operations of the nonrestoring division method are summarized as the following algorithm.

Algorithm: Nonrestoring division

Input: An m-bit dividend and an n-bit divisor.
Output: The quotient and remainder.
Begin

1. Load divisor and dividend into registers M and Q, respectively; clear accumulator A and set the loop count (CNT) equal to $n-1$.
2. Left shift register pair $A:Q$ one bit.
3. Compute $A \leftarrow A - M$
4. **repeat**
   4.1 **if** $A < 0$ **then** $\{Q(0) \leftarrow 0$; left shift $A:Q$ one bit; $A \leftarrow A + M;\}$
   else $\{Q(0) \leftarrow 1$; left shift $A:Q$ one bit; $A \leftarrow A - M;\}$
   4.2 $CNT \leftarrow CNT - 1$
   until $CNT = 0$
5. **if** $A < 0$ **then** $\{Q(0) \leftarrow 0$; $A \leftarrow A + M\ \}$
   else $Q(0) \leftarrow 1$

End ∎

A numerical illustration of the operation of the nonrestoring division algorithm is given in Figure 10.64. It is instructive to note that the left shift of $A:Q$ is equivalent to the right shift of M.

■ Example 10-21: (An example of nonrestoring division.) Figure 10.64 gives a numerical example showing the detailed operations of the nonrestoring division algorithm. It is interesting to note that the carry-out for each addition or subtraction is exactly the corresponding quotient bit. The most-significant bit (MSB) of the result after each addition or subtraction indicates that the result is positive or negative; namely, 1 denotes negative whereas 0 denotes positive. ∎

10.6.2 Implementations of Nonrestoring Division

As stated, nonrestoring division can be implemented by using either a multiple-cycle or a single-cycle structure. In what follows, we describe each structure briefly.

10.6.2.1 Sequential Implementation Figure 10.65 shows a sequential implementation of the nonrestoring division algorithm. Three registers are required. Register M stores the divisor, accumulator A is the register used to hold the partial dividend being involved in the ith step, and register Q stores the partial dividend not yet involved in the process and at the same time records the quotient bits up to the ith step. The n-bit adder/subtracter is utilized to perform the addition or subtraction needed by the algorithm and is controlled by the MSB bit of the previous result.

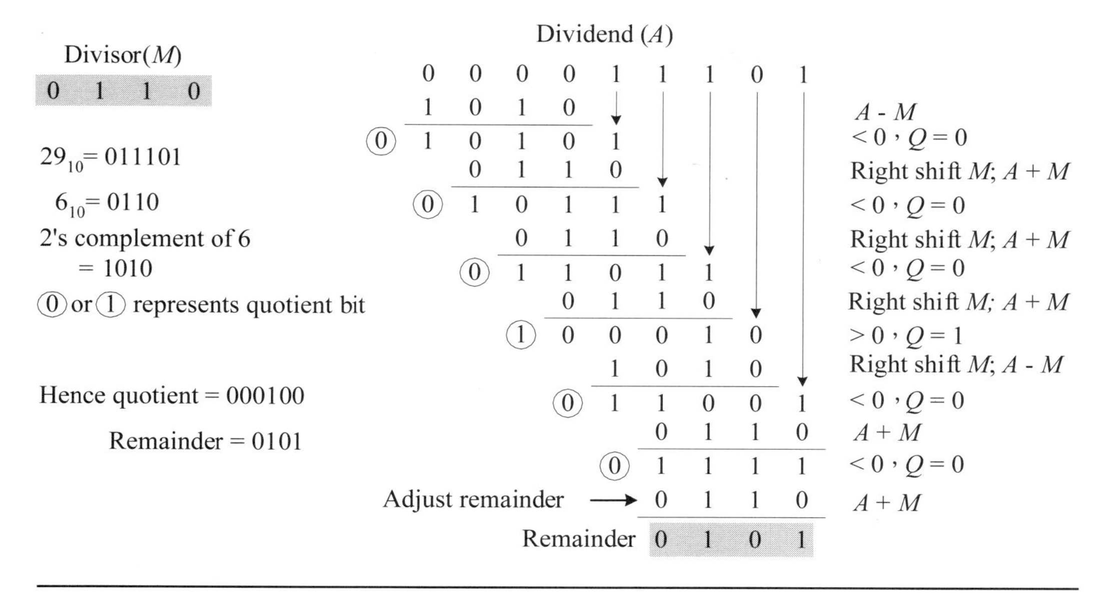

Figure 10.64: An example of the operations of the nonrestoring division algorithm.

After a divisor and a dividend are loaded into the registers M and Q as well as the accumulator A is cleared, the partial dividend is shifted from the register Q into A and then subtracted by the divisor. The remainder is then loaded into the accumulator A. Once this is done, the accumulator A and the dividend (Q) are shifted left one bit position. This process requires $2n$ clock cycles. To speed up the operation, the two-step accumulator-load and shift-left operations can be combined into one single-step operation, namely, shift-left and load operation. Based on this, the division only needs n clock cycles to finish.

10.6.2.2 Array Implementation Like the unsigned array multiplier, an array divider can be thought of as an expansion of the sequential divider in the temporal dimension, and each partial dividend is arranged to the proper position for the next operation. Each partial dividend is simply generated by an n-bit adder/subtracter, as shown in Figure 10.66. For an m-by-n division, this can be done with an $(m-3)$ n-bit *controlled adder and subtracter* (CAS), and n two-input AND gates and full adders. The following is an example of a 7-by-4 array divider.

■ **Example 10-22: (A 7 × 4 nonrestoring array divider.)** Figure 10.66 shows a 7×4 nonrestoring array divider. Each row consisting of four CAS cells and functions as a 4-bit adder/subtractor. The carry-out of the MSB in each row corresponds to the quotient bit, as illustrated in Figure 10.64. The shift operation is carried out by skewing the successive rows of CAS cells. The final row is the remainder adjustment. This array divider needs four 4-bit CASs, and four two-input NAND gates and full adders.

■

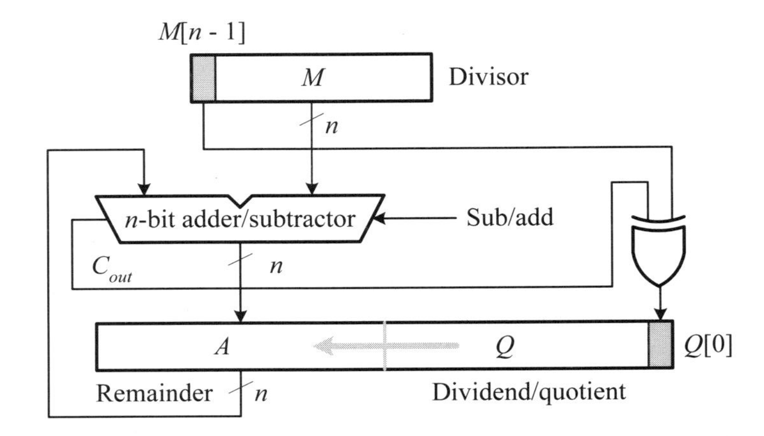

Figure 10.65: The sequential implementation of the nonrestoring division algorithm (Initially sub/add is set to 1 and then controlled by C_{out}.)

Review Questions

Q10-50. What is the main difference between the restoring division and nonrestoring division?

Q10-51. Show that both restoring division and nonrestoring division are functionally equivalent.

Q10-52. Give a numerical example to trace the operations of the array divider shown in Figure 10.66.

10.7 Summary

In combinational logic circuits, code conversion and signal routing circuits are the two most important modules. The code conversion circuits include decoders and encoders while signal routing circuits contain multiplexers and demultiplexers. In addition, magnitude comparators and shifters are also the important components in most digital systems.

Basic sequential components include registers, shift registers, counters, and sequence generators. Registers are usually used to store small amounts of information in digital systems. Shift registers are employed to perform data format conversions or shift their contents left or right a specified number of bit positions. Counters are utilized to count the input events, such as input pulses or clock pulses, and to provide system timing information. Sequence generators are employed to generate pseudo-random sequences (PR-sequence). PR sequences are widely used in data networks, communications, and VLSI testing.

Shift operations play very important roles in arithmetic and logical operations. A shift operation shifts an input number left or right a specified number of bit positions with the vacancies filled with zeros or the sign bit. Shift operations can be subdivided into logical shift and arithmetic shift. Shifters can be implemented with either a multiple-cycle or a single-cycle structure. The basic multiple-cycle structure is a universal shift register while the widely used single-cycle structure is a combinational logic circuit known as a barrel shifter. Barrel shifters can be further subdivided into linear barrel shifters and logarithmic barrel shifters.

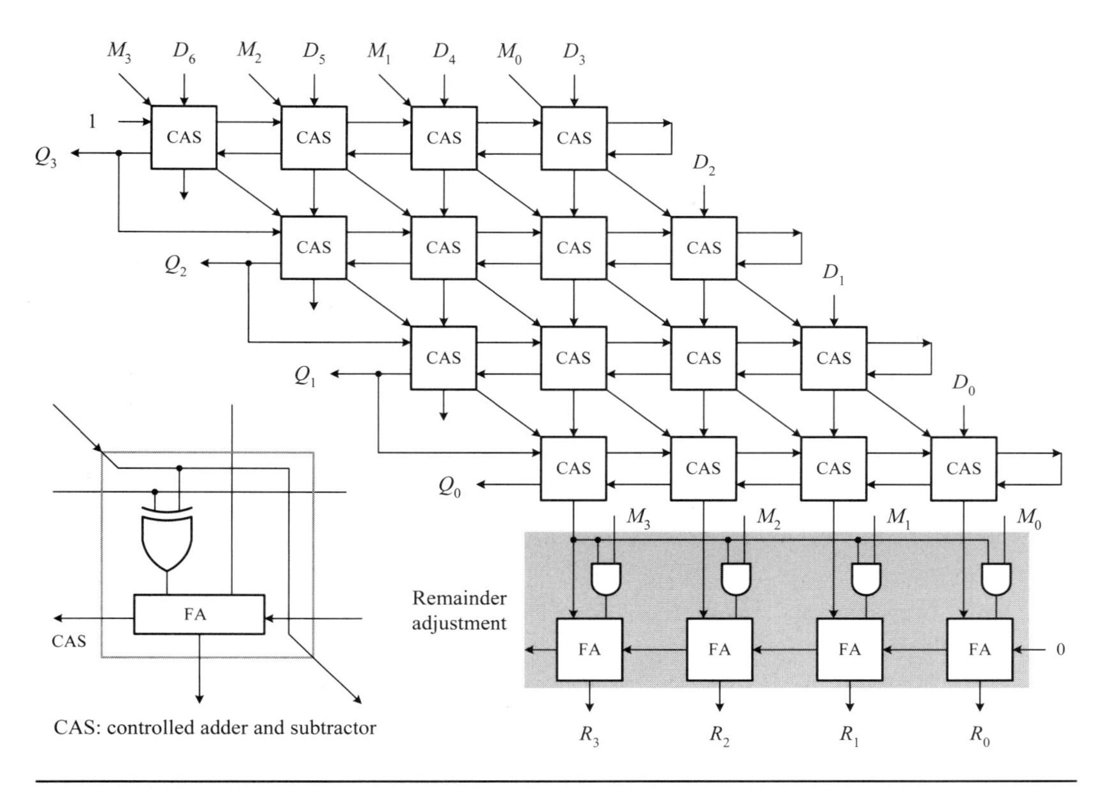

Figure 10.66: An example of a 7 × 4 nonrestoring array divider.

Addition plays very important roles in arithmetic operations, including subtraction, multiplication, and division. The performance bottleneck of an n-bit adder lies on the generation of carriers needed in all stages. To remove the bottleneck, a variety of addition algorithms are proposed, including carry-select adder, conditional-sum adder, carry-skip adder, carry-lookahead adder, Ling carry-lookahead adder, and parallel-prefix adders.

The essential operations of multiplication can be thought of a sequence of additions. A multiplication algorithm may be realized by using either a multiple-cycle or a single-cycle structure. The multiple-cycle structure is essentially a sequential logic circuit while the single-cycle structure is fundamentally a combinational logic circuit. The single-cycle structure naturally evolves into array structures, such as an unsigned array multiplier and the modified Baugh-Wooley signed array multiplier. Likewise, the essential operations of division can be regarded as a sequence of subtractions. Based on this idea, two basic division algorithms, restoring division and nonrestoring division, are obtained. A division algorithm can also be realized by using either a multiple-cycle or a single-cycle structure.

References

1. C. Baugh and B. Wooley, "A two's complement parallel array multiplication algorithm," *IEEE Trans. on Computers*, Vol. 22, No. 12, pp. 1045–1047, December 1973.

2. A. Beaumont-Smith and C. Lim,"Parallel prefix adder design," *Proc. of IEEE Symposium on Computer Arithmetic,* pp. 218–225, 2001.

3. J. Blackburn, L. Arndt, and E. Swartzlander, "Optimization of spanning tree carry lookahead adders," *Proc. of the 30th Asilomar Conf. of Signals, Systems, and Computers,* Vol. 1, pp. 177–181, 1997.

4. A. Booth, "A signed binary multiplication technique," *Quarterly Journal of Mechanics and Applied Mathematics,* Vol. IV, Part 2, pp. 236–240, June 1951.

5. R. P. Brent and H. T. Kung, "A regular layout for parallel adders," *IEEE Trans. on Computers,* Vol. 31, No. 3, pp. 260–264, March 1982.

6. P. Chan and M. Schlag, "Analysis and design of CMOS Manchester adders with variable carry-skip," *IEEE Trans. on Computers,* Vol. 39, No. 8, pp. 983–992, August 1990.

7. S. W. Cheng, "A high-speed magnitude comparator with small transistor count," in *Proc. of the 10th IEEE Int'l Conf. on Electronics, Circuits and Systems (ICECS),* Vol. 3, pp. 1168–1171, 14–17 December, 2003.

8. J. Cortadella and J. Llaberia, "Evaluation of $A + B = K$ conditions without carry propagation," *IEEE Trans. on Computers,* Vol. 41, No. 11, pp. 1484–1488, November 1992.

9. R. L. Davis, "The ILLIAC IV processing element," *IEEE Trans. on Computers,* Vol. 18, No. 9, pp. 800–816, September 1969.

10. J. G. Delgado-Frias, and J. Nyathi, "A VLSI high-performance encoder with priority lookahead," in *Proc. of the 8th IEEE Great Lakes Symposium on VLSI,* pp. 59–64, 1998.

11. J. Delgado-Frias and J. Nyathi, "A high-performance encoder with priority lookahead," *IEEE Trans. on Circuits and Systems–I: Fundamental Theory and Applications,* Vol. 47, No. 9, pp. 1390–1393, September 2000.

12. J. Dobson and G. Blair, "Fast two's complement VLSI adder design," *Electronics Letters,* Vol. 31, No. 20, pp. 1721–1722, September 1995.

13. R. W. Doran, "Variants of an improved carry look-ahead adder," *IEEE Trans. on Computers,* Vol. 37, No. 9, pp. 1110–1113, September 1988.

14. A. Fahim and M. Elmasry, "Low-power high-performance arithmetic circuits and architectures," *IEEE J. of Solid-State Circuits,* Vol. 37, No. 1, pp. 90–94, January 2002.

15. M. Golden et al., "A seventh-generation x86 microprocessor," *IEEE J. of Solid-State Circuits,* Vol. 34, No. 11, pp. 1466–1477, November 1999.

16. A. Guyot and S. Abou-Samra, "Modeling power consumption in arithmetic operators," *Microelectronic Engineering,* Vol. 39, pp. 245–253, 1997.

17. D. Harris, "A taxonomy of prefix networks," *Proc. of the 37th Asilomar Conf. on Signals, Systems, and Computers,* pp. 2213–2217, 2003.

18. R. Hashemian and C. Chen, "A new parallel technique for design of decrement/increment and two's complement circuits," *Proc. of IEEE Midwest Symposium on Circuits and Systems,* Vol. 2, pp. 887–890, 1992.

19. M. Hatamian and G. Cash, "A 70-MHz 8-bit $\times$ 8-bit parallel pipelined multiplier in 2.5-μm CMOS," *IEEE J. of Solid-State Circuits,* Vol. 21, No. 4, pp. 505–513, August 1986.

20. C. Heikes, "A 4.5 mm^2 multiplier array for a 200-MFLOP pipelined coprocessor," *Proc. of IEEE Int'l Solid-State Circuits Conf.*, pp. 290–291, 1994.

21. Z. Huang and M. Ercegovac, "Effect of wire delay on the design of prefix adders in deep-submicron technology," *Proc. of the 34th Asilomar Conf. on Signals, Systems, and Computers,* Vol. 2, pp. 1713–1717, 2000.

22. C. Huang, J. Wang, and Y. Huang, "Design of high-performance CMOS priority encoders and incrementer/decrementers using multilevel lookahead and multilevel folding techniques," *IEEE J. of Solid-State Circuits,* Vol. 37, No. 1, pp. 63–76, January 2002.

23. W. Hwang et al., "Implementation of a self-resetting CMOS 64-bit parallel adder with enhanced testability," *IEEE J. of Solid-State Circuits,* Vol. 34, No. 8, pp. 1108–1117, August 1999.

24. N. Itoh et al., "A 600-MHz 54 $\times$ 54-bit multiplier with rectangular-styled Wallace tree," *IEEE J. of Solid-State Circuits,* Vol. 36, No. 2, pp. 249–257, February 2001.

25. V. Kantabutra, "Designing optimum carry-skip adders," *Proc. of IEEE Symposium on Computer Arithmetic,* pp. 146–153, 1991.

26. V. Kantabutra, "A recursive carry-lookahead/carry-select hybrid adder," *IEEE Trans. on Computers,* Vol. 42, No. 12, pp. 1495–1499, December 1993.

27. S. Knowles, "A family of adders," *Proc. of IEEE Symposium on Computer Arithmetic,* pp. 277–284, 2001.

28. P. Kogge and H. Stone, "A parallel algorithm for the efficient solution of a general class of recurrence equations," *IEEE Trans. on Computers,* Vol. 22, No. 8, pp. 786–793, August 1973.

29. R. Ladner and M. Fischer, "Parallel prefix computation," *Journal of ACM,* Vol. 27, No. 4, pp. 831–838, Oct. 1980.

30. M. B. Lin, *Digital System Design: Principles, Practices, and Applications,* 4th ed. Taipei, Taiwan: Chuan Hwa Book Company, 2010.

31. M. B. Lin and J. C. Chang, "On the design of a two-adder-based RSA encryption/decryption chip," *Journal of the Chinese Institute of Electrical Engineering,* Vol. 9, No. 3, pp. 269–277, August 2002.

32. M. B. Lin, *Digital System Designs and Practices: Using Verilog HDL and FPGAs.* Singapore: John Wiley & Sons, 2008.

33. H. Ling, "High-speed binary adder," *IBM J. of Research and Development,* Vol. 25, No. 3, pp. 156–166, May 1981.

34. T. Lynch and E. Swartzlander, "The redundant cell adder," *Proc. of IEEE Symposium on Computer Arithmetic,* pp. 165–170, 1991.

35. T. Lynch and E. Swartzlander, "A spanning tree carry lookahead adder," *IEEE Trans. on Computers,* Vol. 41, No. 8, pp. 931–939, August 1992.

36. S. Majerski, "On determination of optimal distributions of carry skips in adders," *IEEE Trans. on Electronic Computers,* Vol. 16, No. 1, pp. 45–58, February 1967.

37. S. Mathew et al., "A 4-GHz 130-nm address generation unit with 32-bit sparse-tree adder core," *IEEE J. of Solid-State Circuits,* Vol. 38, No. 5, pp. 689–695, May 2003.

38. N. Ohkubo et al., "A 4.4 ns CMOS 54 × 54-b multiplier using pass-transistor multiplexer," *IEEE J. of Solid-State Circuits,* Vol. 30, No. 3, pp. 251–257, March 1995.

39. J. M. Rabaey, A. Chandrakasan, and B. Nikolic, *Digital Integrated Circuits: A Design Perspective,* 2nd ed. Upper Saddle River, NJ: Prentice-Hall, 2003.

40. J. Sklansky, "Conditional-sum addition logic," *IRE Transactions on Electronic Computers,* Vol. EC-9, No. 3, pp. 226–231, June 1960.

41. J. P. Uyemura, *Introduction to VLSI Circuits and Systems.* New York: John Wiley & Sons, 2002.

42. J. S. Wang and C. S. Huang, "High-speed and low-power CMOS priority encoders," *IEEE J. of Solid-State Circuits,* Vol. 35, No. 10, pp. 1511–1514, October 2000.

43. J. S. Wang and C. S. Huang, "A high-speed single-phase-clocked CMOS priority encoder," *IEEE Int'l Symposium on Circuits and Systems* (ISCAS 2000), May 28-31, pp. V–537–540, 2000.

44. N. H. E. Weste and D. Harris, *CMOS VLSI Design: A Circuit and Systems Perspective,* 3rd ed. Boston: Addison-Wesley, 2005.

45. R. Zlatanovici, S. Kao, and B. Nikolic, "Energy-delay optimization of 64-bit carry-lookahead adders with a 240 ps 90 nm CMOS design example," *IEEE J. of Solid-State Circuits,* Vol. 44, No. 2, pp. 569–583, February 2009.

Problems

10-1. This problem involves the design of a 2-to-4 decoder.

(a) Design a 2-to-4 decoder with active-high output using TG switches.

(b) Design a 2-to-4 decoder with active-low output using NOR gates.

10-2. Answer the following questions.

(a) How would you realize a two-input XOR gate using a 2-to-1 multiplexer?

(b) Show how to realize a 4-to-1 multiplexer with three 2-to-1 multiplexers.

10-3. Show that a 4-to-1 multiplexer is a functionally complete logic module. Here, the functionally complete logic module means that any switching function can be realized by only combining many such logic modules as needed.

10-4. Consider the circuit with n nMOS transistors being connected in series.

(a) Show that the total propagation delay is minimal when all transistors have the same size.

(b) Conduct an experiment using SPICE to verify the above statement.

10-5. Realize a 4-to-1 multiplexer based on the following specified approach.

(a) Using tristate buffers

(b) Using TG switches

10-6. Consider the problem of constructing a 4-to-1 multiplexer.

(a) Design a 4-to-1 multiplexer with TG switches using the uniform-tree networks.

(b) Construct an RC-equivalent circuit for the uniform-tree multiplexer. Then, find the propagation delay using the Elmore delay model.

(c) Design a 4-to-1 multiplexer with TG switches using the uniform structure.

(d) Construct an RC-equivalent circuit for the uniform-structure multiplexer. Then, find the propagation delay using the Elmore delay model.

10-7. Consider the problem of designing a 4-to-1 multiplexer.

(a) Design a 4-to-1 multiplexer using basic dynamic logic.

(b) Conduct a SPICE experiment to measure the propagation delay of the dynamic 4-to-1 multiplexer.

(c) Design a 4-to-1 multiplexer using domino logic.

(d) Conduct a SPICE experiment to measure the propagation delay of the domino 4-to-1 multiplexer.

10-8. Consider the problem of constructing an 8-to-1 multiplexer using smaller multiplexers as building blocks.

(a) Construct an 8-to-1 multiplexer using 4-to-1 and 2-to-1 multiplexers.

(b) Implement the design with a logic style that you prefer.

(c) Assume that the gates are constructed using static CMOS logic circuits. Apply the logical effort technique to design the gates if the loading capacitance of the 8-to-1 multiplexer is 20 times the capacitance of a basic inverter.

10-9. Realize a 1-to-4 demultiplexer based on the following specified approach.

(a) Using tristate buffers

(b) Using TG switches

10-10. Referring to the 4-bit shift register shown in Figure 10.18, answer the following questions:

(a) Construct a positive edge-triggered D-type flip-flop using negative and positive D-type latches. Conduct a SPICE experiment to verify the functionality of the resulting D-type flip-flop.

(b) Cascade four D-type flip-flops that you have designed into a 4-bit shift register. Conduct a SPICE experiment to verify the functionality of the resulting shift register.

10-11. Referring to the 3-bit ripple binary counter shown in Figure 10.19, answer the following questions:

(a) Construct a negative edge-triggered D-type flip-flop using positive and negative D-type latches. Conduct a SPICE experiment to verify the functionality of the resulting D-type flip-flop.

(b) Design a T-type flip-flop using the negative edge-triggered D-type flip-flop that you have designed. Conduct a SPICE experiment to verify the functionality of the resulting T-type flip-flop.

(c) Cascade three T-type flip-flops that you have designed into a 3-bit ripple binary counter. Conduct a SPICE experiment to verify the functionality of the resulting circuit.

10-12. Referring to the 3-bit synchronous binary counter shown in Figure 10.20(a), answer the following questions:

(a) Construct a positive edge-triggered D-type flip-flop using negative and positive D-type latches. Conduct a SPICE experiment to verify the functionality of the resulting flip-flop.

(b) Design a T-type flip-flop using the positive edge-triggered D-type flip-flop that you have designed. Conduct a SPICE experiment to verify the functionality of the resulting T-type flip-flop.

(c) Cascade three T-type flip-flops that you have designed into the 3-bit synchronous binary counter shown in Figure 10.20(a). Conduct a SPICE experiment to verify the functionality of the resulting circuit.

10-13. Referring to the 3-bit synchronous binary counter shown in Figure 10.20(b), answer the following questions:

(a) Construct a positive edge-triggered D-type flip-flop using negative and positive D-type latches. Conduct a SPICE experiment to verify the functionality of the resulting flip-flop.

(b) Design a T-type flip-flop using the positive edge-triggered D-type flip-flop that you have designed. Conduct a SPICE experiment to verify the functionality of the resulting T-type flip-flop.

(c) Cascade three T-type flip-flops that you have designed into the 3-bit synchronous binary counter shown in Figure 10.20(b). Conduct a SPICE experiment to verify the functionality of the resulting circuit.

10-14. Referring to the 4-bit ring counter shown in Figure 10.22, answer the following questions:

(a) Construct a positive edge-triggered D-type flip-flop using negative and positive D-type latches. Conduct a SPICE experiment to verify the functionality of the resulting D-type flip-flop.

(b) Cascade four D-type flip-flops that you have designed into a 4-bit ring counter shown in Figure 10.22. Conduct a SPICE experiment to verify the functionality of the resulting circuit.

10-15. Referring to the 4-bit Johnson counter shown in Figure 10.23, answer the following questions:

(a) Construct a positive edge-triggered D-type flip-flop using negative and positive D-type latches. Conduct a SPICE experiment to verify the functionality of the resulting D-type flip-flop.

(b) Cascade four D-type flip-flops that you have designed into a 4-bit Johnson counter shown in Figure 10.23. Conduct a SPICE experiment to verify the functionality of the resulting circuit.

10-16. Considering the static Manchester carry circuit shown in Figure 10.39(a), answer the following questions:

(a) Examine the size problem for a carry propagate event.

(b) Conduct a SPICE experiment to verify the result.

10-17. Considering the dynamic Manchester carry chain shown in Figure 10.40, answer the following questions:

(a) Construct an RC-equivalent circuit for the carry chain. Then, find the propagation delay using the Elmore delay model.

(b) Conduct a SPICE experiment to verify the result.

(c) How will charge loss affect the operation of the carry chain?

10-18. This is a problem about the design of an 8-input priority encoder using a parallel-prefix network.

(a) Sketch an 8-input priority encoder using the Brent-Kung parallel-prefix network.

(b) Use the logical effort method to estimate the delay of the priority encoder.

10-19. This problem involves the design of an incrementer using a parallel-prefix network.

(a) Sketch an 8-bit incrementer using the Brent-Kung parallel-prefix network.

(b) Use the logical effort method to estimate the delay of the incrementer.

10-20. This problem concerns about the design of a decrementer using a parallel-prefix network.

(a) Sketch an 8-bit decrementer using the Brent-Kung parallel-prefix network.

(b) Use the logical effort method to estimate the delay of the decrementer.

10-21. This is a problem about the design of a Gray-to-binary code converter using a parallel-prefix network.

(a) Sketch an 8-bit Gray-to-binary code converter using the Brent-Kung parallel-prefix network.

(b) Use the logical effort method to estimate the delay of the Gray-to-binary code converter.

10-22. Table 10.2 shows a radix-4 Booth encoder. Based on this table, answer the following questions:

(a) Give a numerical example to illustrate the operation of the radix-4 Booth algorithm.

(b) Construct a sequential radix-4 Booth multiplier. Draw the block diagram and label each component appropriately to indicate its functionality and size.

Table 10.2: The radix-4 Booth encoder.

x_{i+1}	x_i	x_{i-1}	Operation	x_{i+1}	x_i	x_{i-1}	Operation
0	0	0	0	1	0	0	$-2A$
0	0	1	A	1	0	1	$-A$
0	1	0	A	1	1	0	$-A$
0	1	1	$2A$	1	1	1	0

(c) Construct 8×8 radix-4 Booth array multiplier. Draw the block diagram and label each component appropriately to indicate its functionality and size.

10-23. Design a 16-bit CLA adder and then finish it completely using a full-custom approach.

(a) Draw the circuit and label each component appropriately to indicate its functionality.

(b) Carry out the layout, perform a layout versus schematic (LVS) check, and a design rule check (DRC).

(c) Extract the RC parameters, perform a SPICE simulation, and measure the worst-case propagation delay.

(d) Add I/O pads to the circuit core and run off-line DRC.

10-24. Design a 16-bit Brent-Kung adder and finish it completely using a full-custom approach.

(a) Draw the circuit and label each component appropriately to indicate its functionality.

(b) Carry out the layout, perform an LVS check, and a DRC check.

(c) Extract the RC parameters, perform a SPICE simulation, and measure the worst-case propagation delay.

(d) Add I/O pads to the circuit core and run off-line DRC.

10-25. Design a 16-bit Kogge-Stone adder and finish it completely using a full-custom approach.

(a) Draw the circuit and label each component appropriately to indicate its functionality.

(b) Carry out the layout, perform an LVS check, and a DRC check.

(c) Extract the RC parameters, perform a SPICE simulation, and measure the worst-case propagation delay.

(d) Add I/O pads to the circuit core and run off-line DRC.

10-26. Design a 16-bit Ladner-Fisher adder and finish it completely using a full-custom approach.

(a) Draw the circuit and label each component appropriately to indicate its functionality.

(b) Carry out the layout, perform an LVS check, and a DRC check.

(c) Extract the *RC* parameters, perform a SPICE simulation, and measure the worst-case propagation delay.

(d) Add I/O pads to the circuit core and run off-line DRC.

10-27. Design an 8×8 Booth's array multiplier and finish it completely using a full-custom approach.

(a) Draw the circuit and label each component appropriately to indicate its functionality.

(b) Carry out the layout, perform an LVS check, and a DRC check.

(c) Extract the *RC* parameters, perform a SPICE simulation, and measure the worst-case propagation delay.

(d) Add I/O pads to the circuit core and run off-line DRC.

10-28. Design a 16-by-8 nonrestoring array divider and finish it completely using a full-custom approach.

(a) Draw the circuit and label each component appropriately to indicate its functionality.

(b) Carry out the layout, perform an LVS check, and a DRC check.

(c) Extract the *RC* parameters, perform a SPICE simulation, and measure the worst-case propagation delay.

(d) Add I/O pads to the circuit core and run off-line DRC.

11

Memory Subsystems

Roughly speaking, about 30% of the semiconductor business is due to memory chips nowadays. Semiconductor memory capable of storing large amounts of digital information is essential to all digital systems, ranging from 4-bit microprocessor systems to 64-bit computer systems, even cluster computers. In modern digital systems, various memory modules may possibly occupy up to half the area of a digital chip.

The semiconductor memory can be classified in terms of the type of data access and the capability of information retention. According to the type of data access, semiconductor memory can be subdivided into serial access, content addressable, and random access. The data in serial-access memory can only be accessed in a predetermined order and mainly contains shift registers and queues. Content-addressable memory (CAM) is a device capable of parallel search; namely, it behaves like a lookup table in which all entries can be searched in parallel. Random-access memory (RAM) is a memory device where any word can be accessed at random at any time. The random-access memory can be further categorized into read/write memory and read-only memory.

Read/write memory contains two major types: static RAMs (SRAMs) and dynamic RAMs (DRAMs). Read-only memory (ROM), as its name implies, is only allowed to retrieve but not allowed to modify the stored information. The contents of ROM can be committed using a photolithographic mask during being manufactured or programmed in laboratories.

According to the capability of information retention, semiconductor memory may also be cast into volatile and nonvolatile memories. The volatile memory, such as static RAM and dynamic RAM, will lose its information once the power supply is interrupted while the nonvolatile memory, such as the ROM family, ferroelectric RAM (FRAM), and magnetoresistance RAM (MRAM), still retain their information even when the power supply is removed.

11.1 Introduction

In modern digital systems, a large part, probably up to a one-half area, of a digital chip is occupied by various memory modules. These memory modules are used to store and retrieve large amounts of information at high speeds. In addition, memory modules, including ROM and SRAM, with a broad variety of word widths and capacities may be generated by silicon compilers because of their regularity. Nevertheless, DRAM

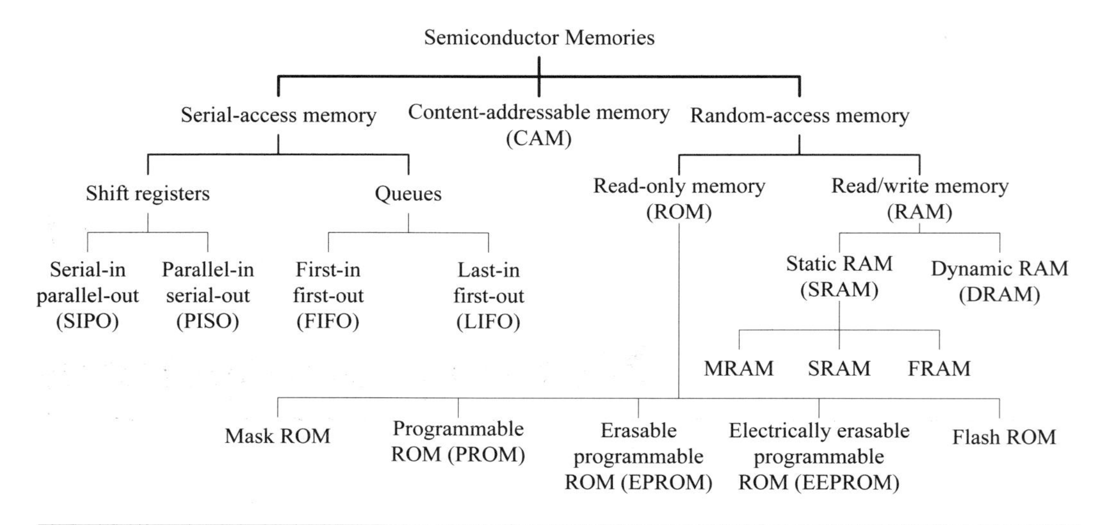

Figure 11.1: Memory classification.

modules are still not easily embedded into the chip design of digital systems, even though some commercial CMOS processes already can do this.

11.1.1 Memory Classification

As stated, about 30% of the semiconductor business is due to memory devices at present. Semiconductor memory capable of storing large amounts of digital information is essential to all digital systems, ranging from 4-bit to 64-bit microcomputer systems, even cluster computers. The semiconductor memory can be classified in accordance with the type of data access and the capability of information retention.

11.1.1.1 Types of Data Access According to the type of data access, semiconductor memory is categorized into serial access, content addressable, and random access, as shown in Figure 11.1. Serial-access memory mainly contains shift registers and queues. The former can be further subdivided into *serial-in parallel-out* (SIPO) and *parallel-in serial-in* (PISO), while the latter contains *first-in first-out* (FIFO) and *first-in last-out* (FILO). A FIFO is generally called a *queue* or a buffer and a FILO is referred to as a *stack*.

Random-access memory (RAM) is a memory device in which any word can be accessed at random in a constant time. The random-access memory can be classified into *read/write memory* and *read-only memory*. Read/write memory can be further subdivided into two types: static RAMs (SRAMs) and dynamic RAMs (DRAMs). SRAM is mainly used as the main memory for those computer systems needing a small memory capacity, and as the cache memory for large computer systems, such as desktop computers and servers. DRAM is used as the main memory for those computer systems requiring a large memory capacity.

The SRAM cell consists of a bistable device accompanied with two access transistors and thus stores its information in a form of circuit state; the DRAM cell comprises an access transistor and a capacitor and retains information in the form of charge. Because of the leakage current of the access transistor, the capacitor (a soft node) in a DRAM cell is gradually losing its charge and eventually its information. To remedy this, a DRAM cell must be read and rewritten periodically, even when the cell is not accessed. This process is known as *refresh*.

Content-addressable memory (CAM) is a device with the capability of parallel search; namely, it behaves like a lookup table in which all entries can be searched in parallel. A CAM cell is fundamentally a SRAM cell to store information, with the addition of some extra circuits to facilitate the operation of parallel interrogation and indicate the result. CAM proceeds the reverse operation of SRAM; namely, it receives as an input data and outputs a matched address when the data matches one in the CAM.

Read-only memory (ROM), as its name implies, is only allowed to retrieve but not allowed to modify the stored information. The contents of ROM are programmed using a photolithographic mask during manufacturing. Hence, this kind of ROM is also called *mask ROM*. Another type of ROM that can be programmed in a laboratory is called a *programmable ROM* (PROM). Programmable ROM can be further categorized into *fuse ROM*, *erasable PROM* (EPROM), *electrically erasable PROM* (EEPROM), and *Flash memory*. Fuse ROM can be only programmed once and therefore belongs to the type of *one-time programmable* (OTP) devices. The contents of an EPROM device can be erased by exposing the device under an ultraviolet light. Such a device is also referred to as an *UV-exposure EPROM* or UV-EPROM to emphasize the erasing feature of using the UV-exposure method. Another feature of UV-EPROM is that the erase operation is done on the entire chip in an off-board fashion, thereby erasing the data of the whole chip. The contents of an EEPROM are erased by a high-voltage pulse in a bundle of bits, say one byte or a block of bytes. Because of the use of voltage pulses, the contents of an EEPROM can be erased partially or entirely on board. Flash memory is similar to EEPROM in which data in the block can be erased by using a high-voltage pulse.

11.1.1.2 Capability of Information Retention Semiconductor memory may also be classified into *volatile* and *nonvolatile* in accordance with the capability of information retention once the power supply is removed. Whenever the power supply is removed, the volatile memories, such as static RAM and dynamic RAM, will lose their information, whereas the nonvolatile memories, such as the ROM family, *ferroelectric RAM* (FRAM), and *magnetoresistance RAM* (MRAM), still retain their information.

Both FRAM and MRAM are nonvolatile, read/write memory in the sense that their read and write operations take the same amount of time. Both FRAM and MRAM employ the hysteresis characteristics to store information. Their cell data can be modified by changing the polarization of ferroelectric material.

■ Review Questions

Q11-1. How would you classify the semiconductor memories?

Q11-2. What are the differences between SRAM cells and DRAM cells?

Q11-3. What does "refresh" mean?

Q11-4. What are the features of CAM?

Q11-5. What does ROM mean?

Q11-6. What is the distinction between volatile and nonvolatile memories?

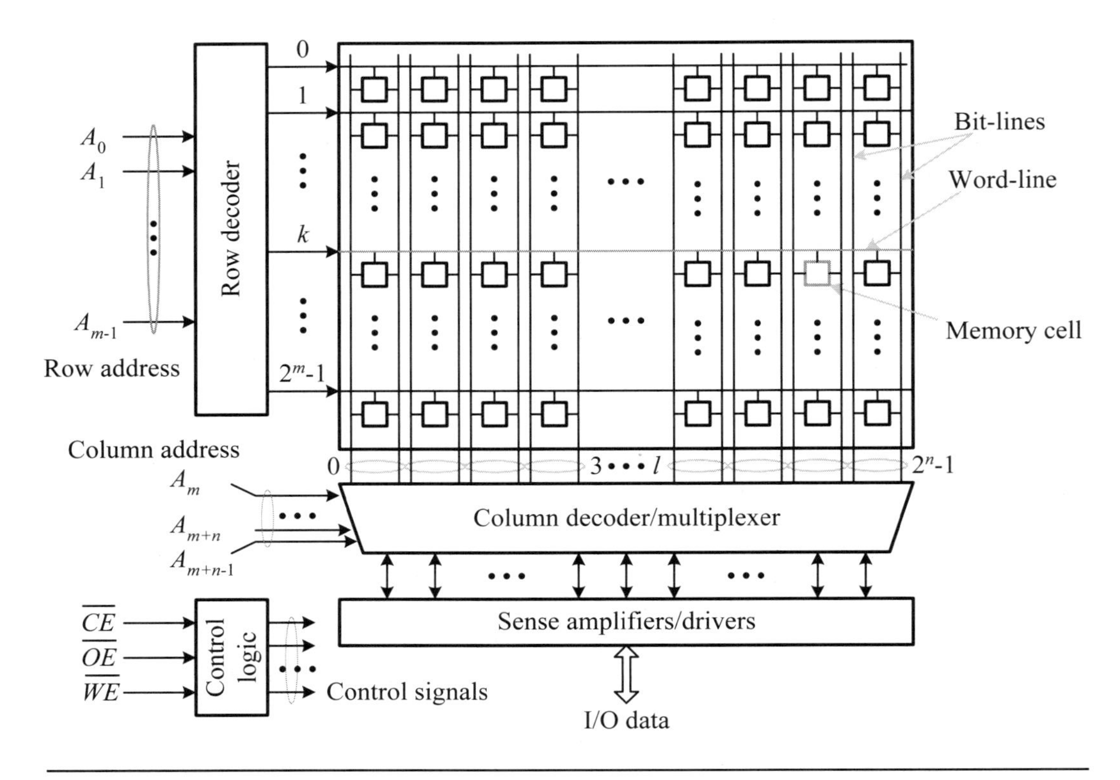

Figure 11.2: The typical memory structure.

11.1.2 Memory Organization

The typical memory organization is illustrated in Figure 11.2. It comprises a *row decoder*, a column decoder/multiplexer, a *memory core*, *sense amplifiers/drivers*, and a piece of control logic. The row decoder takes an m-bit address and produces 2^m word-line enable signals. The column decoder takes an n-bit address and produces 2^n bit-line select signals. The memory core is composed of memory cells. These memory cells are arranged into a two-dimensional array, consisting of horizontal rows and vertical columns. Each individual cell can be accessed in a random order within a constant time, independent of physical location. Each memory cell is connected to a horizontal line called a *word-line*, driven by the row decoder, and two vertical lines, called *bit-line* and *complementary bit-line*, respectively, used to access the data of the cell. Each time, 2^i columns in one row may be accessed, where i is a positive integer or zero, dependent on the organization of the memory. The signals of selected columns are amplified or driven by the sense amplifiers/drivers according to whether the operation is read or write.

The control logic receives three inputs, *chip enable* ($\overline{CE}$), *output enable* ($\overline{OE}$), and *write enable* ($\overline{WE}$), and generates the required internal control signals. The chip select controls the operations of an entire chip. Output enable ($\overline{OE}$) asserts the read operation by enabling the output buffer while write enable ($\overline{WE}$) asserts the write operation. The data bus is in a high-impedance whenever the chip enable is deasserted, regardless of the values of output enable ($\overline{OE}$) and write enable ($\overline{WE}$).

■ Example 11-1: (Memory organization.) As illustrated in Figure 11.2, if the memory capacity is 1 Mb, then it may have 20 address lines at most, depending on how many bits it outputs at a time. For example, if it outputs 8 bits at one time, then it only needs 17 address lines. The memory array may be decomposed into 1024 rows, with each containing 128×8 bits. In other words, each row is decomposed into 8 groups with each having 128 bits. Each time one bit is selected from each group by the column decoder/multiplexer. Hence, seven column address lines are needed to address eight 128-to-1 multiplexers and eight sense amplifiers/drivers are required.

■

11.1.2.1 Advanced Memory Organization Since word-lines are usually built with polysilicon layers to reduce area, their RC delays are significantly large and cannot be neglected, in particular, when the block size is large in which a lot of memory cells are attached. A metal wire may be placed in parallel and contacted to the polysilicon line to reduce the resistance of the word-line and hence the propagation delay. Bit-lines are usually formed in metal so that the resistance is not significant. Nevertheless, a large capacitance may result from the bit-line and access transistors of memory cells connected to it. The large capacitances of bit-lines and word-lines also contribute a large part to the total power dissipation of memory.

To reduce the propagation delay and power dissipation, a number of different partitioning approaches have been used. One simple way is to partition the memory array into many smaller blocks that can be separately accessed. Another way is to restrict the access of the partitioned blocks in an exclusive way, namely, one block at a time. Of course, a hybrid of the above two approaches is also possible.

■ Example 11-2: (Memory array partition.) Figure 11.3 shows an example of partitioning a $2^{k+3} \times 2^m$ array of cells of a (2^{m+k+3})-Mb SRAM for a $4w$-bit access. The array is partitioned into four macro blocks, all of which are accessed at the same time, each providing w bits of the accessed word. Each macro block is further subdivided into four blocks of 2^{m-1} rows and 2^k columns. Each of the blocks provides the w-bit subword. In other words, each row in a block consists of four $(2^k/w)$-bit groups of which w bits may be accessed at a time. For a 4-Mb SRAM with a 16-bit I/O bus, $k = 8$, $m = 11$, and $w = 4$.

■

As illustrated in Figure 11.3, when a memory array is partitioned into several smaller blocks, a *divided word-line* (also called a *hierarchical word-line*) and a *divided bit-line* (also called a *multilevel bit-line*) structure are naturally resulted. An example of divided word-line structure is demonstrated in Figure 11.4. One part of the input address is decoded to activate the *global word-line*, another part of the address is decoded to activate the block group select line, and the remaining address bits activate the *block select line*. The combination of these three parts activates the *local word-line*. The advantages of using divided word-lines are that they use shorter word-lines within blocks and the block address activates only one block. This may significantly reduce power dissipation compared to the traditional memory array structure. It should be noted that metal wires are used for both the global word-line and subglobal word-line to reduce propagation delay.

When the divided bit-line structure is employed, a one-column multiplexer is used in each block. The outputs of column multiplexers from all blocks in the same column

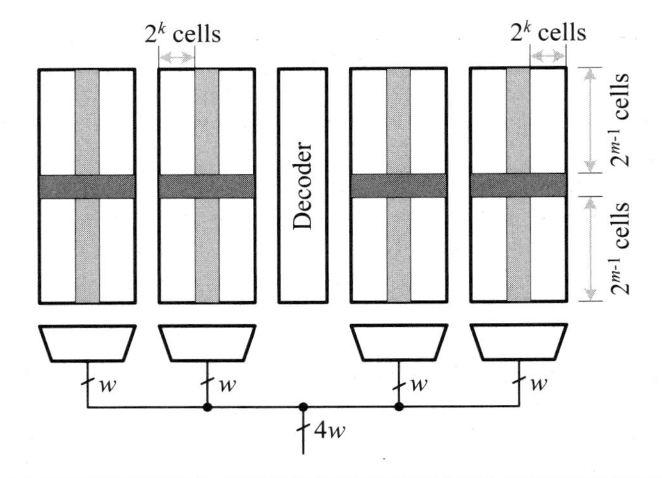

Figure 11.3: An example of the memory array partition.

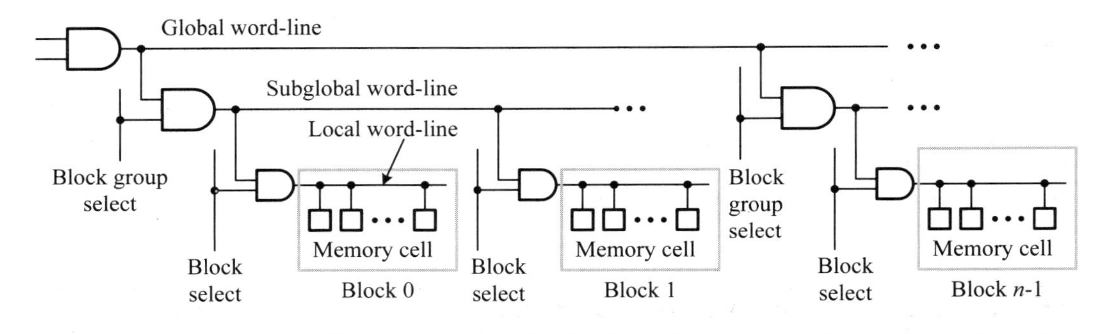

Figure 11.4: A hierarchical word-line for typical high-density SRAM devices.

are then multiplexed again to route the selected bits to the output sense amplifiers. Because of this operation, this structure is often referred to as a multilevel bit-line structure.

■ Review Questions

Q11-7. What components are included in a typical memory?

Q11-8. How would you partition a memory array?

Q11-9. What are the advantages of divided word-line structure?

Q11-10. Describe the divided bit-line structure.

11.1.3 Memory Access Timing

Generally, SRAM devices can be subdivided into asynchronous and synchronous according to whether the access operation is triggered by the external clock signal. For platform-based designs, asynchronous SRAM devices are widely used and for cell-based designs, synchronous SRAM are more popular. We only describe the typical access timing of synchronous SRAM devices briefly.

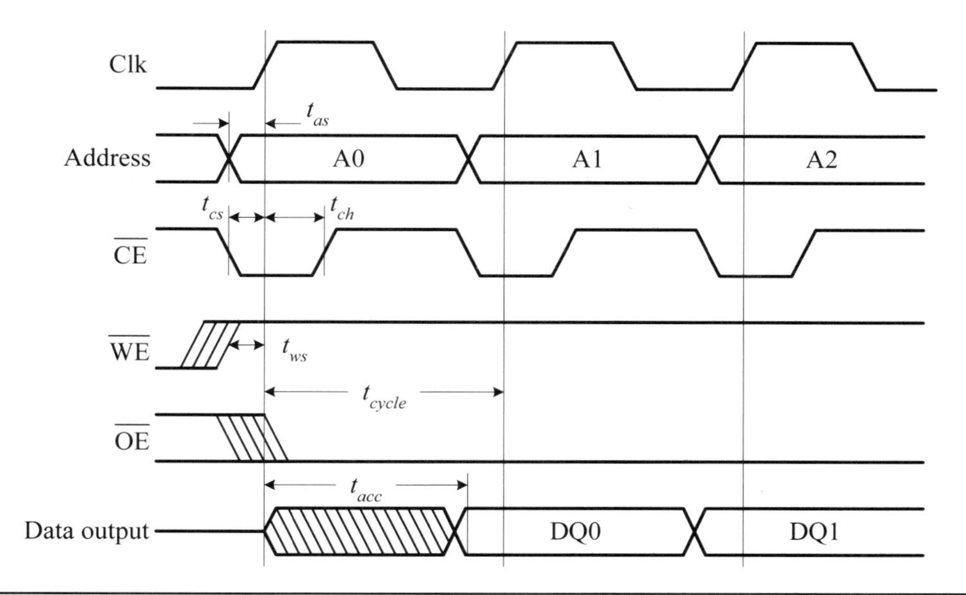

Figure 11.5: Typical synchronous SRAM read-cycle timing.

The read-cycle timing of typical synchronous SRAM is illustrated in Figure 11.5. The read operation is synchronous and triggered by the rising edge of the clock, clk. Address, chip enable ($\overline{CE}$), write enable ($\overline{WE}$), and output enable ($\overline{OE}$) are latched by the rising edge of the clock. To access data from the SRAM device properly, all of these input signals must be subject to individual setup times (t_{as}, t_{cs}, and t_{ws}) and hold time (t_{ch}). For a read operation, the chip enable ($\overline{CE}$) must be set to a low value. Once the chip enable is active, the SRAM device enters the read mode when the value of write enable ($\overline{WE}$) is high. During the read mode, data are read from the memory location specified by the address and appear on the data output bus after a read access time (t_{acc}) if output enable ($\overline{OE}$) is asserted. The read access time is the time counted from the positive edge of clock, clk, to the stable data appearing at the output data bus.

The write-cycle timing of typical synchronous SRAM is illustrated in Figure 11.6. Like in the read mode, the write operation is synchronous and triggered by the rising edge of the clock, clk. Address, chip enable ($\overline{CE}$), write enable ($\overline{WE}$), and input data are latched by the rising edge of the clock. To write data into the SRAM device properly, all of these input signals must be subject to individual setup times (t_{as}, t_{cs}, t_{ws}, and t_{ds}) and hold times (t_{ch}, t_{wh}, and t_{dh}). For a write operation, the chip enable ($\overline{CE}$) and write enable ($\overline{WE}$) must be set to low values. During the write mode, data are written into the memory location specified by the address.

■ Review Questions

Q11-11. How would you define the address setup and hold times?

Q11-12. How would you define the data setup and hold times?

Q11-13. Explain the read operation of synchronous SRAM.

Q11-14. Explain the write operation of synchronous SRAM.

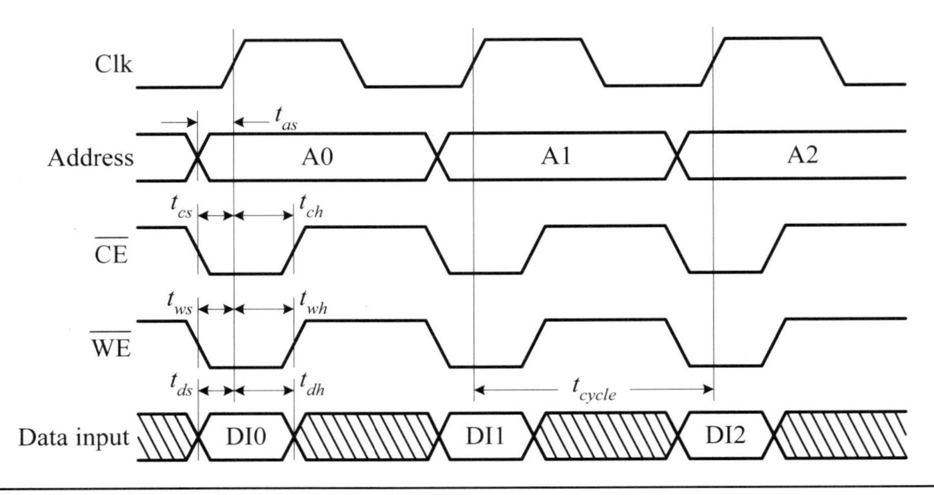

Figure 11.6: Typical synchronous SRAM write-cycle timing.

11.2 Static Random-Access Memory

As stated, the feature of random-access memory is that any memory cell can be accessed at random in a constant time. In this section, we deal with the major components of SRAM, including memory core, operations of SRAM, row decoders, column decoders/multiplexers, and sense amplifiers.

11.2.1 RAM Core Structures

We begin to introduce the basic cell structures and then cope with the read- and write-cycle analysis of standard 6T-SRAM cell. Finally, we discuss the cell stability briefly.

11.2.1.1 Basic Cell Structures Two types of SRAM cells are given in Figure 11.7. They are 4T- and 6T-SRAM cell structures. We can see from the figure that a basic SRAM cell is a bistable device formed by two back-to-back connected inverters along with two access transistors being separately connected to a pair of complementary bit-lines, Bit and $\overline{Bit}$. The 4T-SRAM cell is shown in Figure 11.7(a) and the 6T-SRAM cell is exhibited in Figure 11.7(b). Their structures are roughly the same except for the pull-up devices. The pull-up devices are two resistors in the 4T-SRAM cell and are two pMOS transistors in the 6T-SRAM cell.

The 4T-SRAM cell has a more compact cell size since the resistors (formed from undoped polysilicon) can be stacked on top of the cell, using double-polysilicon technology. Thus, the cell size is reduced from six to four transistors. However, because of the use of passive pull-up resistors, there exists a direct current between power rails since one output of two inverters is at low-level voltage at any time, leading to an amount of power dissipation that cannot be neglected. To reduce the power dissipation, high-value load resistors may be used. Nevertheless, this would result in worse noise margins and slower operation speed. Hence, there is a trade-off between power dissipation and noise margins as well as switching speed.

The 6T-SRAM cell has the lowest static power dissipation and offers superior noise margins and switching speed as well, as compared with 4T-SRAM cell. In addition,

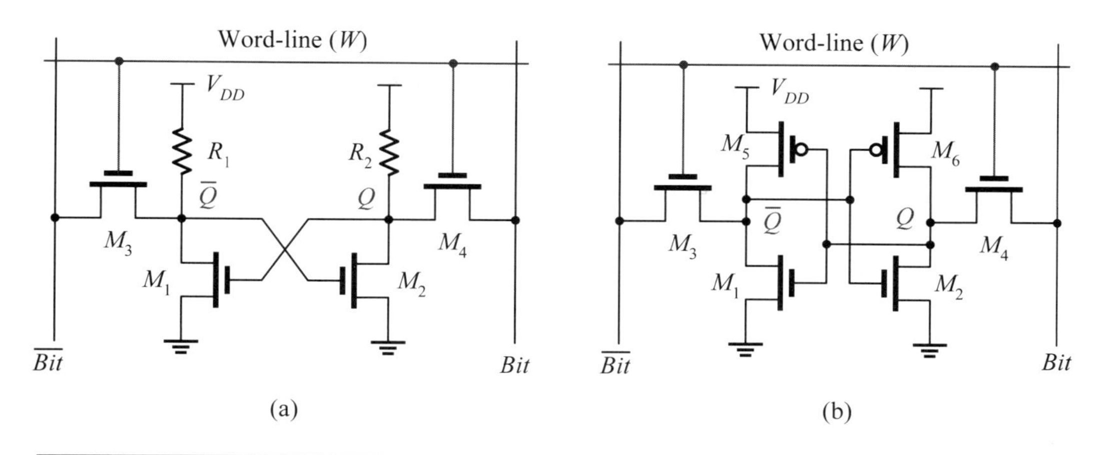

Figure 11.7: (a) 4T- and (b) 6T-SRAM cell structures.

it is compatible with most widely used CMOS logic processes at present. Hence, it finds the widespread use in modern CMOS SRAM modules. Next, we will focus on this type of SRAM cell.

In practical designs, both transistors M_3 and M_4 usually use minimum geometry devices. The sizes of transistors M_1 and M_2, and M_5 and M_6 are determined by the following two basic requirements. First, the data-read operation should not destroy information in the memory cell. Second, the memory cell should be allowed to modify the stored information in the data-write operation. Therefore, the strengths of the three transistors on one side from strongest to weakest are pull-down, access, and pull-up in sequence.

11.2.1.2 Read-Cycle Analysis

For the read operation, both bit-lines are usually precharged to (1/2) V_{DD} or V_{DD}, depending on how the sense amplifier is designed. We assume that both bit-lines are precharged to V_{DD} before the beginning of the read operation. To avoid destroying the information in the memory cell, the $V_{\bar{Q}}$ must be less than the inverter threshold voltage V_{th} after the access transistors are turned on; namely, the following relationship must be held.

$$\frac{R_{M1}}{R_{M1}+R_{M3}}V_{DD} \leq V_{th} = \frac{1}{2}V_{DD} \tag{11.1}$$

where R_{M1} and R_{M3} are the on-resistances of nMOS transistors M_1 and M_3, respectively. $R_{M1} \leq R_{M3} \Rightarrow W_1 \geq W_3$. In practice, we typically set $W_1 = 1.5W_3$. A symmetrical relationship exists for transistors M_2 and M_4.

To analyze the read operation of the 6T-SRAM cell in more detail, assume that the memory cell stores a 1; namely, $v_Q = V_{DD}$ while $v_{\bar{Q}} = 0$ V. The equivalent circuit is given in Figure 11.8. To access the stored value of the memory cell, we raise the voltage of the word-line to V_{DD} to turn on both access transistors, M_3 and M_4, as well as raise the voltages of both bit-lines to V_{DD}.

If the voltage $v_{\bar{B}}$ of $\overline{Bit}$ is still at V_{DD} after the access transistor is turned on, then nMOS transistor M_1 will be in its linear region of operation and nMOS transistor M_3 in the saturation region. From Figure 11.8, we obtain

$$I_{D1} = \mu_n C_{ox}\left(\frac{W_1}{L_1}\right)\left[(V_{DD}-V_{Tn})V_{DS1} - \frac{1}{2}V_{DS1}^2\right] \tag{11.2}$$

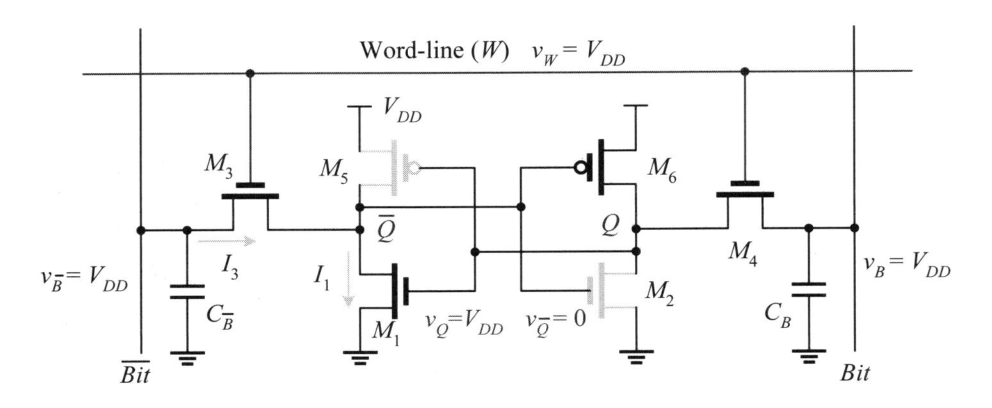

Figure 11.8: The read cycle of a 6T-SRAM cell structure.

and

$$I_{D3} = \frac{1}{2}\mu_n C_{ox}\left(\frac{W_3}{L_3}\right)(V_{DD} - V_{Tn} - V_{DS1})^2 \tag{11.3}$$

The ratio of W_1 to W_3 can be obtained by using the fact that both transistors M_1 and M_3 are in series and hence their currents must be equal. For deep-submicron processes, short-channel current equations are often used instead. An illustration of this is explored in the following example.

Example 11-3: (The ratio of W_1 to W_3.) Using 0.18-μm process parameters: $V_{T0n} = 0.4$ V, $v_{sat} = 8 \times 10^6$ cm/s, $\mu_{effn} = 287$ cm^2/V-s, and $E_{satn} = 12 \times 10^4$ V/cm, determine the ratio of W_1 to W_3. Assume that node $\bar{Q}$ can only tolerate a 0.1 V rise in voltage during the read operation.

Solution: The saturation voltage of transistor M_3 is equal to

$$V_{DSsat} = \frac{(V_{GS} - V_{Tn})E_{satn}L_n}{(V_{GS} - V_{Tn}) + E_{satn}L_n} = \frac{(1.8 - 0.4) \times 2.16}{(1.8 - 0.4) + 2.16} = 0.85 \text{ V}$$

Hence, transistor M_3 operates in its saturation region. Transistor M_1 is in the linear region of operation since its V_{DS} is 0.1 V at most. Hence, we can write the following equation

$$\frac{W_1}{L_1}\frac{\mu_n C_{ox}}{(1 + V_{\bar{Q}}/E_{satn}L_1)}\left[(V_{DD} - V_{Tn1})V_{\bar{Q}} - \frac{V_{\bar{Q}}^2}{2}\right] = \frac{W_3 v_{satn} C_{ox}(V_{DD} - V_{\bar{Q}} - V_{Tn3})^2}{(V_{DD} - V_{\bar{Q}} - V_{Tn3}) + E_{satn}L_3}$$

Eliminating C_{ox} from both sides and substituting parameters into the above equation, we obtain

$$\frac{W_1}{0.18 \times 10^{-4}}\frac{287}{(1 + 0.1/2.16)}\left[(1.4)0.1 - \frac{0.1^2}{2}\right] = \frac{W_3(8 \times 10^6)(1.3)^2}{(1.3) + 2.16}$$

Solving it, the resulting ratio of W_1 to W_3 is approximated to 1.9. Notice that the above calculation does not take into account the body effect. The resulting value will be smaller if the body effect is included. The dependence of the ratio of W_1 to W_3 on $V_{\bar{Q}}$ is plotted in Figure 11.9. The lower ratio of W_1 to W_3, the higher the $V_{\bar{Q}}$.

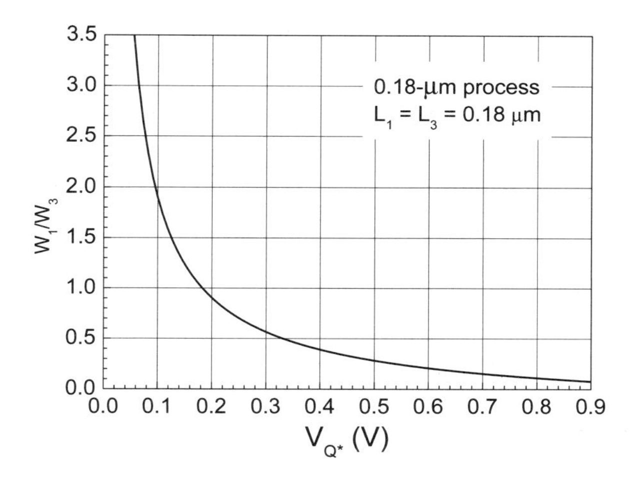

Figure 11.9: The ratio of W_1/W_3 versus $V_{\bar{Q}}$ (Both L_1 and L_3 use minimum length).

The size of transistor M_3 determines the current to discharge the complementary bit-line capacitor $C_{\bar{B}}$. As a result, the actual size is determined by the desired transition time $t_{\Delta\bar{B}}$ for the bit-line voltage $v_{\bar{B}}$ to be changed by an amount of ΔV, about 200 mV, which can be detected by a sense amplifier. The transition time $t_{\Delta\bar{B}}$ can be expressed as

$$t_{\Delta\bar{B}} = \frac{C_{\bar{B}} \cdot \Delta V_{\bar{B}}}{I_{D3}} \tag{11.4}$$

where I_{D3} is the drain current of the access transistor M_3 and $C_{\bar{B}}$ is the capacitance of bit-line $\overline{Bit}$.

Example 11-4: (Read-cycle time.) Assume that the bit-line capacitance is 1 pF. If the desired transition time of the bit-line is 2 ns when its voltage is changed by 200 mV, determine the sizes of transistors M_1 and M_3.

Solution: From Equation (11.4), the drain current of transistor M_3 is found to be

$$I_{D3} = \frac{C_{\bar{B}} \cdot \Delta V_{\bar{B}}}{t_{\Delta\bar{B}}} = \frac{1 \text{ pF} \cdot 0.2}{2 \text{ ns}} = 100\ \mu\text{A}$$

Using a short-channel current equation and substituting parameters into it, we obtain

$$I_{D3} = \frac{W_3(8 \times 10^6)(1.33\ \mu\text{F/cm}^2)(1.8 - 0.1 - 0.4)^2}{(1.8 - 0.1 - 0.4) + 2.16} = 100\ \mu\text{A}$$

By solving it, the channel width W_3 is found to be 0.19 μm. As a consequence, the minimum size 0.27 μm (3 λ) is used. The channel width W_1 is equal to $1.9W_3 = 0.51\ \mu\text{m} \Rightarrow 0.54\ \mu\text{m}$ (6 λ).

11.2.1.3 Write-Cycle Analysis The write operation is accomplished by forcing one bit-line to V_{DD} while keeping the other at about 0 V. For instance, to write a 0 to the memory cell, we set the bit-line voltage v_B to 0 V and the complementary bit-line voltage $v_{\bar{B}}$ to V_{DD}. If Q is 1 and we wish to write a 0 into the memory cell, we

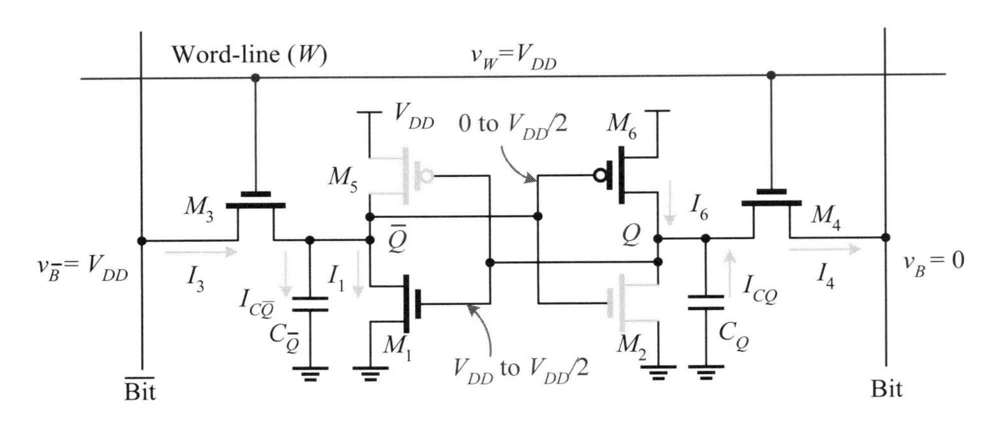

Figure 11.10: The write cycle of a 6T-SRAM cell structure (Write a logic 0 to the cell).

need to pull v_Q below the threshold voltage of the inverter composed of M_5 and M_1 to guarantee the inverter to change its state. As a consequence, the following relationship must be satisfied.

$$\frac{R_{M4}}{R_{M4} + R_{M6}} V_{DD} \leq V_{th} = \frac{1}{2} V_{DD} \tag{11.5}$$

where R_{M4} and R_{M6} are the on-resistances of nMOS and pMOS transistors M_4 and M_6, respectively. From it, we obtain $R_{M4} \leq R_{M6}$. Hence, the maximum ratio of W_6 to W_4 must be less than the ratio of R_{eqp} to R_{eqn} and is equal to 2.75. In practice, we typically set $W_6 = 1.5W_4$. For considering the noise margin, it should pull v_Q below V_{Tn} for transistor M_1 to guarantee that transistor M_1 is turned off.

To analyze the write cycle in more detail, consider the situation that v_Q is equal to V_{DD} and we want to write a 0 into the memory cell. As shown in Figure 11.10, the voltage v_B is forced to 0 V. Once the access transistor M_4 is turned on, C_Q starts to discharge toward the ground level through the access transistor M_4. When $v_Q = V_{th} = \frac{1}{2}V_{DD}$, the inverter composed of transistors M_1 and M_5 starts to switch its state. Meanwhile, transistors M_4 and M_6 operate in their linear regions. From Figure 11.10, we obtain

$$I_{D4} = \mu_n C_{ox} \left(\frac{W_4}{L_4}\right) \left[(V_{DD} - V_{Tn})V_{th} - \frac{1}{2}V_{th}^2\right] \tag{11.6}$$

and

$$I_{D6} = \mu_n C_{ox} \left(\frac{W_6}{L_6}\right) \left[(V_{DD} - |V_{Tp}|)(V_{DD} - V_{th}) - \frac{1}{2}(V_{DD} - V_{th})^2\right] \tag{11.7}$$

The ratio of W_6 to W_4 can be obtained by using the fact that both transistors M_6 and M_4 are in series and hence their currents must be equal. Like the read operation, for deep-submicron processes, short-channel current equations are often used instead.

Example 11-5: (The ratio of W_6 to W_4.) Referring to Figure 11.10 and using 0.18-μm process parameters: $V_{T0n} = |V_{T0p}| = 0.4$ V, $v_{sat} = 8 \times 10^6$ cm/s, $\mu_{effn} = 287$ cm^2/V-s, $\mu_{effp} = 88$ cm^2/V-s, $E_{satn} = 12 \times 10^4$ V/cm, and $E_{satp} = 25 \times 10^4$ V/cm,

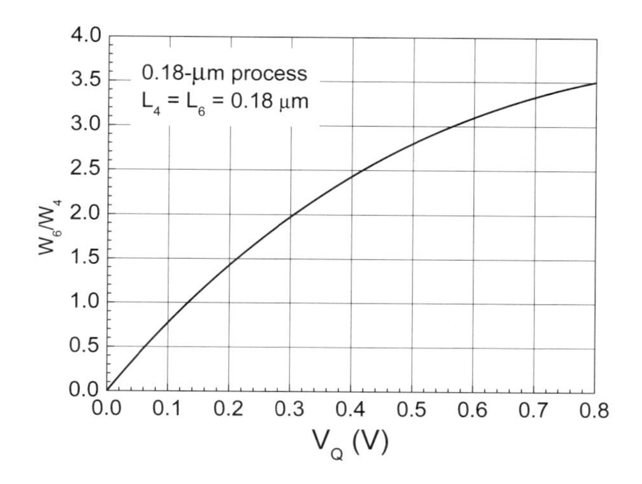

Figure 11.11: The ratio of W_6/W_4 versus V_Q (Both L_4 and L_6 use minimum length).

determine the ratio of W_6 to W_4. Assume that node Q must be pulled down below 0.4 V during the write operation.

Solution: For simplicity, assume that the gate voltage V_{Gp6} of transistor M_6 is 0 V. The saturation voltage V_{DSsat} of it is

$$V_{DSsat} = \frac{(V_{SG} - |V_{Tp}|)E_{satp}L_p}{(V_{SG} - |V_{Tp}|) + E_{satp}L_p} = \frac{(1.8 - 0.4) \times 4.5}{(1.8 - 0.4) + 4.5} = 1.07 \text{ V}$$

The transistor M_6 is at the saturation region of operation as the voltage v_Q drops below 0.73 V. Similarly, the saturation voltage V_{DSsat} of transistor M_4 can be found to be 0.85 V. Hence, transistor M_4 operates at its linear region, and M_6 operates in saturation region as the voltage v_Q drops below 0.73 V. Based on this, the following equation is obtained:

$$\frac{W_4}{L_4}\frac{\mu_n C_{ox}}{(1 + V_{DS4}/E_{satn}L_4)}\left[(V_{GS4} - V_{Tn4})V_{DS4} - \frac{V_{DS4}^2}{2}\right] = \frac{W_6 v_{satp} C_{ox}(V_{DD} - |V_{Tp6}|)^2}{(V_{DD} - |V_{Tp6}|) + E_{satp}L_6}$$

Substituting parameters into the above equation and assuming that both transistors use the minimum channel length, we obtain

$$\frac{287W_4}{(1 + 0.4/2.16)}\left[(1.4)0.4 - \frac{0.4^2}{2}\right] = \frac{W_6 \times 8 \times 10^6 \times 0.18 \times 10^{-4} \times (1.8 - 0.4)^2}{((1.8 - 0.4) + 4.5)}$$

Solving it, the ratio of W_6 to W_4 is equal to 2.43. The dependence of the ratio of W_6 to W_4 on V_Q is plotted in Figure 11.11. The lower the ratio of W_6 to W_4, the lower the V_Q.

■

The size of M_4 determines the current to discharge the capacitor C_Q. As a result, the actual size is determined by the desired transition time $t_{\Delta Q}$ for node voltage v_Q to be changed by an amount of $\Delta v_Q = V_{DD} - V_{th}$.

$$t_{\Delta Q} = \frac{C_Q \cdot \Delta v_Q}{I_{CQ}} = \frac{C_Q(V_{DD} - V_{th})}{I_4 - I_6} \tag{11.8}$$

where I_{D4} is the drain current of the access transistor M_4 and C_Q is the capacitance of node Q.

Example 11-6: (Write-cycle time.) Referring to Figure 11.10, assume that C_Q is 10 fF and the aspect ratios of transistors M_4 and M_6 are $3\lambda/2\lambda$ and $3\lambda/2\lambda$, respectively. Using 0.18 μm-process parameters, determine the transition time for where v_Q drops from V_{DD} to $\frac{1}{2}V_{DD}$.

Solution: As described, for where v_Q drops from V_{DD} to $\frac{1}{2}V_{DD}$, transistor M_4 operates in the saturation region and M_6 is first in cutoff and then in the linear region of operation. Hence, the drain current I_{D4} of transistor M_4 is equal to

$$\begin{aligned} I_{D4}(V_{DS} = V_{DD}) &= \frac{W_4 v_{satn} C_{ox}(V_{DD} - V_{Tn4})^2}{(V_{DD} - V_{Tn4}) + E_{satn} L_4} \\ &= \frac{0.27 \times 10^{-4}(8 \times 10^6)(1.33 \times 10^{-6})(1.8 - 0.4)^2}{(1.8 - 0.4) + 2.16} \\ &= 158\ \mu\text{A} \end{aligned}$$

The average current I_{D6} of pMOS transistor M_6 can be calculated as in the following. The drain current I_{D6} is zero when $V_{SD} = 0$. The pMOS transistor M_6 operates in a linear region when $V_{SD} = \frac{1}{2}V_{DD}$. Thus, the average current of pMOS transistor M_6 is

$$\begin{aligned} I_{D6avg} &= \frac{1}{2}\left(I_{D6}(V_{SD} = 0) + I_{D6}(V_{SD} = \frac{1}{2}V_{DD})\right) \\ &= \frac{1}{2}\left(\frac{W_6}{L_6}\right) \frac{\mu_p C_{ox}}{(1 + V_{SD6}/E_{satp} L_6)} \left[(V_{SG6} - |V_{Tp6}|)V_{SD6} - \frac{V_{SD6}^2}{2}\right] \\ &= \frac{1}{2}\left(\frac{3}{2}\right) \frac{88 \times 1.33 \times 10^{-6}}{(1 + 0.9/4.5)} \left[(1.4)(0.9) - \frac{(0.9)^2}{2}\right] \\ &= 62.5\ \mu\text{A} \end{aligned}$$

Using Equation (11.8), the transition time $t_{\Delta Q}$ can be obtained as follows:

$$t_{\Delta Q} = \frac{C_Q(V_{DD} - V_{th})}{I_4 - I_6} = \frac{10 \times 10^{-15}(1.8 - 0.9)}{(158 - 62.5) \times 10^{-6}} = 94.24 \text{ ps}$$

■

11.2.1.4 Word-Line RC Time Constant

The word-line has a large capacitance C_{word} that must be driven by the row address decoder. The capacitance of each cell is the sum of the gate capacitances of two nMOS access transistors and the wire capacitance of each cell. The capacitance of an entire word-line is the sum of the capacitances of all cells. That is

$$C_{word} = (2C_G + C_{wire}) \times \text{ no. of cells in a row} \tag{11.9}$$

In addition to capacitance, the word-line also has a large resistance R_{word}, which arises from a long polysilicon layer used as the word-line. The total word-line resistance can be expressed as follows:

$$R_{word} = R_{cell} \times \text{ no. of cells in a row} \tag{11.10}$$

Combining word-line resistance (R_{word}) with capacitance (C_{word}) forms an RC time constant, which determines the propagation delay of the word-line. An illustration of this is explored in the following example.

■ Example 11-7: (Word-line resistance and capacitance.) In a 64-kb SRAM, the memory core is composed of 256 × 256 cells. The transistors in each cell have their aspect ratios as follows: M_1 and M_2 are $4\lambda/2\lambda$, and M_3 to M_6 are $3\lambda/2\lambda$. The area of word-line is $26\lambda \times 2\lambda$ and the area of bit-line is $44\lambda \times 3\lambda$. The resistance of an n^+/p^+ polysilicon layer is 5 $\Omega/\square$. Using 0.18-μm process parameters, calculate the capacitance and resistance of each word-line, and the word-line RC time constant as well.

Solution: The gate capacitance is calculated as follows:

$$C_G = C_g \times W_n = 3.41 \times 1.5 \times 0.18 = 0.92 \text{ fF}$$

The wire capacitance of each cell is found to be

$$C_{wire} = 0.15 \text{ fF}/\mu\text{m} \times l_{wire} = 0.15 \times (26 - 6) \times 0.18/2 = 0.27 \text{ fF}$$

where the factor 6 excludes the area of the two access transistors. From Equation (11.9), we have

$$C_{word} = (2C_G + C_{wire}) \times 256 = (2 \times 0.92 + 0.27) \times 256 = 540.16 \text{ fF}$$

The word-line resistance for each cell is

$$R_{cell} = 5 \times (26\lambda/2\lambda) = 65 \ \Omega$$

From Equation (11.10), the word-line resistance is found to be

$$R_{word} = 65 \times 256 = 16.64 \text{ k}\Omega$$

Hence, the word-line RC time constant is

$$\tau = R_{word} \times C_{word} = 16.64 \text{ k}\Omega \times 540.16 \text{ fF} = 8.99 \text{ ns}$$

■

11.2.1.5 Bit-Line RC Time Constant The bit-line also has a large capacitance C_{bit} due to a great number of cells connected to it. The bit-line capacitance consists of three components, source/drain capacitance of the nMOS access transistor, wire capacitance, and source/drain contact capacitance, and can be expressed as follows:

$$C_{bit} = (C_{S/D} + C_{wire} + 2C_{contact}) \times \text{ no. of cells in a column} \tag{11.11}$$

assuming that two contacts are used in each bit cell. Combining the bit-line resistance (R_{bit}) with capacitance (C_{bit}) forms an RC time constant, which determines the propagation delay of the bit-line. An illustration of this is explored in the following example.

■ Example 11-8: (Bit-line resistance and capacitance.) Being continued from the preceding example, assume that each contact on the bit-line has a capacitance of 0.5 fF and the wire capacitance is 0.15 fF/μm. Using 0.18-μm process parameters, calculate the capacitance and resistance of the bit-line, and the bit-line RC time constant as well.

Solution: The source/drain junction capacitance $C_{S/D}$ of the nMOS access transistor is equal to

$$\begin{aligned} C_D &= C_{DBn} + C_{OL} = (C_j + C_{ol}) \times W_n \\ &= 2.67 \text{ fF}/\mu\text{m} \times 1.5 \times 0.18\ \mu\text{m} = 0.72 \text{ fF} \end{aligned}$$

The wire capacitance of each cell is

$$C_{wire} = 0.15 \times l_{wire} = 0.15 \times (44 - 4) \times 0.18/2 = 0.54 \text{ fF}$$

Hence, the total bit-line capacitance is

$$C_{bit} = (0.72 + 0.54 + 2 \times 0.5) \times 256 = 578.56 \text{ fF}$$

Ignoring the contact resistance, about 10 Ω, the bit-line resistance of each cell is calculated as follows:

$$R_{cell} = 54 \text{ m}\Omega \times (40\lambda/3\lambda) = 720 \text{ m}\Omega$$

As a consequence, the total bit-line resistance is found to be

$$R_{bit} = 720 \text{ m}\Omega \times 256 = 184.3\ \Omega$$

The bit-line RC time constant is as follows:

$$\tau = R_{bit} \times C_{bit} = 184.3\ \Omega \times 578.56 \text{ fF} = 106.63 \text{ ps}$$

which may be ignored in most practical applications.

■

11.2.1.6 Cell Stability Recall that the relative strengths of access, pull-up, and pull-down transistors are very important for a memory cell to function correctly. The strengths of these three transistors from strongest to weakest are pull-down, access, and pull-up in sequence. The stability of a 6T-SRAM cell is quantified by the *static noise margin* (SNM). The SNM is defined as the maximum value of v_N that can be tolerated by the flip-flop before changing states. A 6T-SRAM cell should have two stable states in hold and read modes of operation and one stable state in the write mode of operation.

A conceptual setup for modeling SNM of a 6T-SRAM cell is depicted in Figure 11.12(a), where noise sources having value v_N are introduced at each of the internal nodes in the 6T-SRAM cell. As v_N increases, the stability of the 6T-SRAM cell changes. The voltage transfer characteristics (VTC) of both inverters are plotted in Figure 11.12(b). The resulting two-lobed curve is called a "butterfly curve" due to its shape and is used to determine the SNM, which is equal to the side length of the largest square that can be embedded into the lobes of the butterfly curve. If both curves move by more than SNM, then the 6T-SRAM cell is monostable and loses its data.

During the hold mode, the word-line of the 6T-SRAM cell is low, and both access transistors are off and thus they do not affect the behavior of SNM, as shown in Figure 11.12(a). The resulting butterfly diagram is shown in Figure 11.12(b) with lighter lines. A positive noise voltage moves $V_{out}(Q)$ left and $V_{out}(\bar{Q})$ up. At the extreme case, noise voltage eliminates the stable point: $V_{in}(Q) = 0$ and $V_{in}(Q) = V_{DD}$.

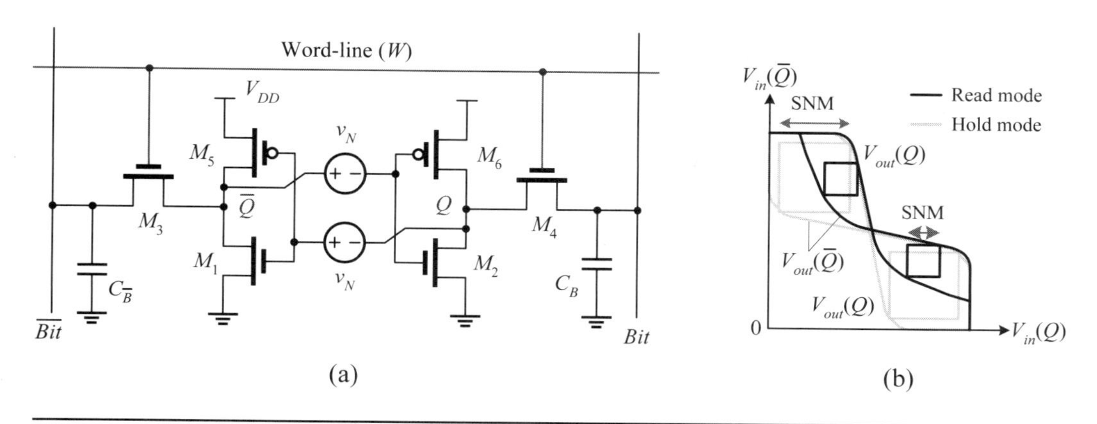

Figure 11.12: A conceptual setup for modeling SNM and the butterfly diagram of 6T-SRAM cell: (a) a 6T-SRAM cell with noise sources; (b) SNM in hold and read modes.

As the 6T-SRAM cell is being read, the word-line rises high and the bit-lines are still precharged to V_{DD} as illustrated in Figure 11.8. The internal node of the 6T-SRAM cell storing a zero gets pulled upward through the access transistor because of the voltage dividing effect across the access and pull-down transistors and hence degrades the SNM, resulting in the butterfly curves shown in Figure 11.12(b) with dark lines.

As the 6T-SRAM cell is being written, the access transistor must overpower the pull-up transistor to force the 6T-SRAM cell to change its state. In other words, the 6T-SRAM cell becomes monostable, thereby forcing the internal voltages to the desired values. If the 6T-SRAM cell remains bistable, then the write does not occur and the SNM is positive on the butterfly plot of 6T-SRAM cell. Therefore, a negative SNM indicates a successful write.

Review Questions

Q11-15. What are the features of a 4T-SRAM cell?

Q11-16. What are the features of a 6T-SRAM cell?

Q11-17. What are the two requirements that determine the sizes of both access transistors of a 6T-SRAM cell?

Q11-18. Define static noise margin.

11.2.1.7 Low-Power SRAM Cells Recall that to reduce the latency time at wake-up process in high performance systems, the on-chip memory contents need to be retained during power gating. For this purpose, many methods are proposed to reduce the leakage current as much as possible during power gating without corrupting the data in the on-chip SRAM. In what follows, we introduce two basic approaches: source biasing and dynamic V_{DD}.

As illustrated in Figure 11.13(a), in the source biasing method the source line voltage is raised during standby mode to generate a negative V_{GS}, thereby leading to the body effect in nMOS transistors and hence lowering the subthreshold and bit-line leakage currents in the SRAM cell. In addition, because raising the source line voltage

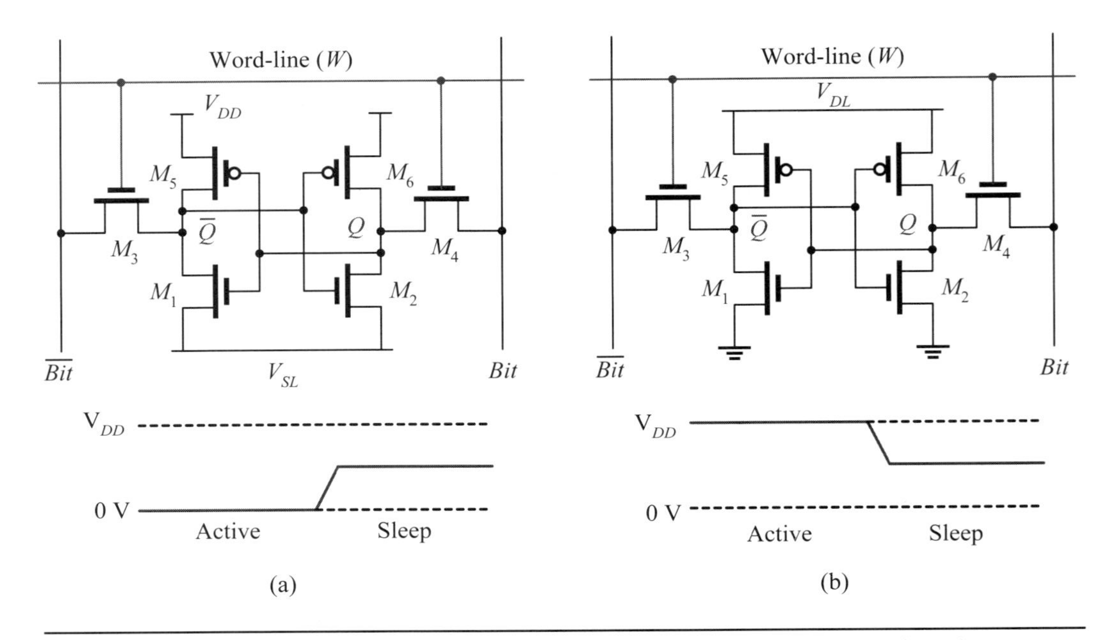

Figure 11.13: (a) Source biasing (V_{SL}) and (b) dynamic supply-voltage (V_{DL}) methods for the reduction of leakage currents.

has a similar effect to the reduction of supply voltage, the gate leakage is also reduced due to the lower voltage stress across the gate and substrate.

One disadvantage of this method is that an nMOS power switch has to be connected in series in the pull-down path to cut off the source line from the actual ground during the sleep mode, thereby increasing propagation delay. Another drawback is the requirement of a separate biasing supply voltage (V_{SL}). Fortunately, because of SRAM only consuming the leakage current in the sleep mode, this biasing supply voltage can be generated on chip through the use of a simple voltage divider.

Figure 11.13(b) shows another method that lowers down the supply voltage to reduce the leakage currents, including subthreshold, gate, and band-to-band tunneling (BTBT) leakage. Nevertheless, the bit-line leakage current cannot be reduced in this scheme since the bias condition in the access transistors does not change. Although there is no impact on propagation delay during the active mode, the large supply-voltage swing between the sleep and active mode imposes a larger transition overhead compared to the source biasing method. Another drawback is the need of a dedicated, switchable supply voltage.

11.2.2 The Operations of SRAM

Two basic operations of a static RAM (SRAM) are read and write. We begin to describe the read operation and then the write operation.

As shown in Figure 11.14, a sense amplifier is used to detect the small voltage difference between true and complementary bit-lines. This also means that sense amplifiers are susceptible to differential noise on the bit-lines. As a consequence, if bit-lines are not precharged long enough, the survival voltages on the bit-lines from previous read operations may cause pattern-dependent failure. To overcome this problem, an equalizing pMOS transistor M_{pc3} is used to guarantee both true and complementary

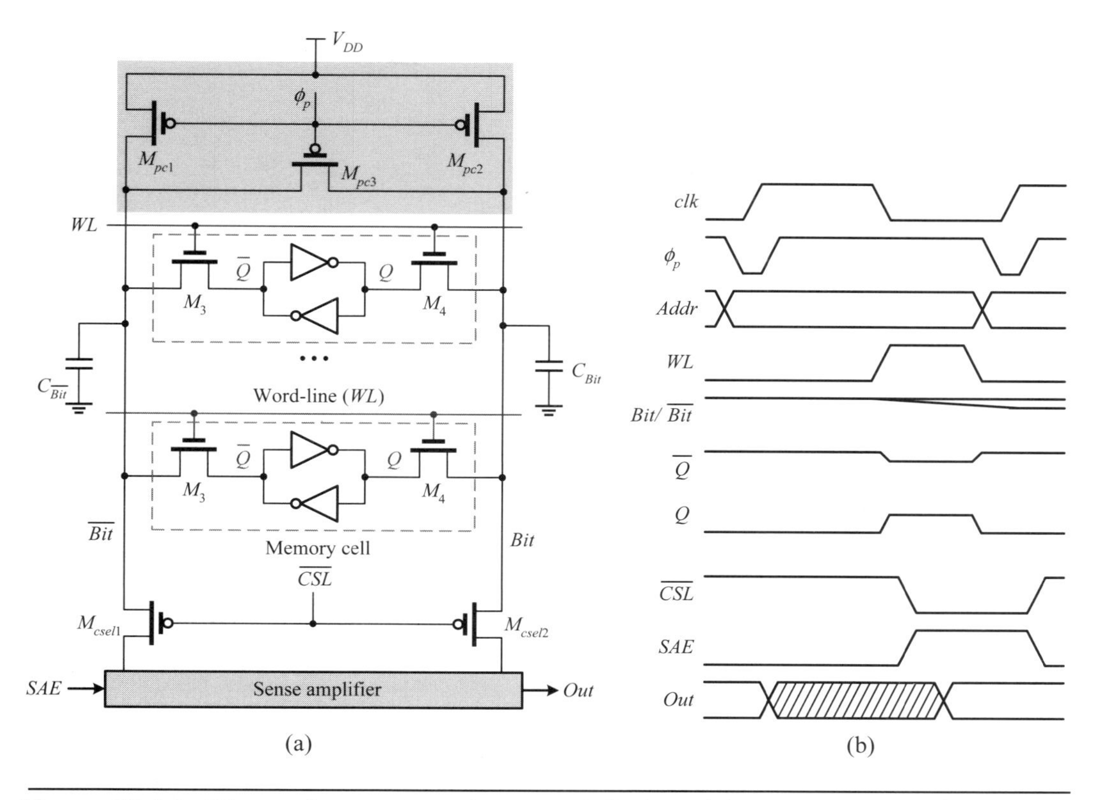

Figure 11.14: The read operation of a typical SRAM: (a) related circuit; (b) timing diagram.

bit-lines to have nearly equal voltage levels even if they are not precharged quite all the way to V_{DD}. Capacitive coupling from neighboring bit-lines may also introduce differential noise. To reduce such a kind of noise, the *twisted-wire* or *twisted-line technique* widely used in data transmission can be utilized to twist the bit-lines. The result is referred to as a *twisted bit-line architecture*.

During the read operation, a precharge signal ϕ_p must be generated to enable the precharge circuit, consisting of three pMOS transistors, M_{pc1}, M_{pc2}, and M_{pc3}, to pull up both bit-lines to V_{DD} (some circuits use $\frac{1}{2}V_{DD}$ or $V_{DD} - V_{Tn}$ by using nMOS transistors to replace pMOS transistors). Many ways can be used for this purpose. We assume that an *address transition detection* (ATD) circuit is used. The details of the ATD circuit can be referred to in Section 11.2.6. After the propagation delay of the row decoder, one word-line rises high to enable the cells associated with it to be read. Depending on whether the stored value is a 1 or 0, either true bit-line or complementary bit-line starts to drop its voltage below V_{DD} through cell transistors. After a small delay relative to the active word-line signal, both column selection ($\overline{CSL}$) and sense amplifier enable (SAE) signals are activated to enable the sense amplifier to capture the voltage change and generate the valid logic value of the stored value. The detailed timing diagram is shown in Figure 11.14(b).

The write operation is as follows. Referring to Figure 11.15, like the read operation, both bit-lines are first precharged to V_{DD} through the precharged circuit. Both address and data are then set up and held stable for a large amount of time before the write enable signal is applied. The address signals are decoded into column selection (CSL)

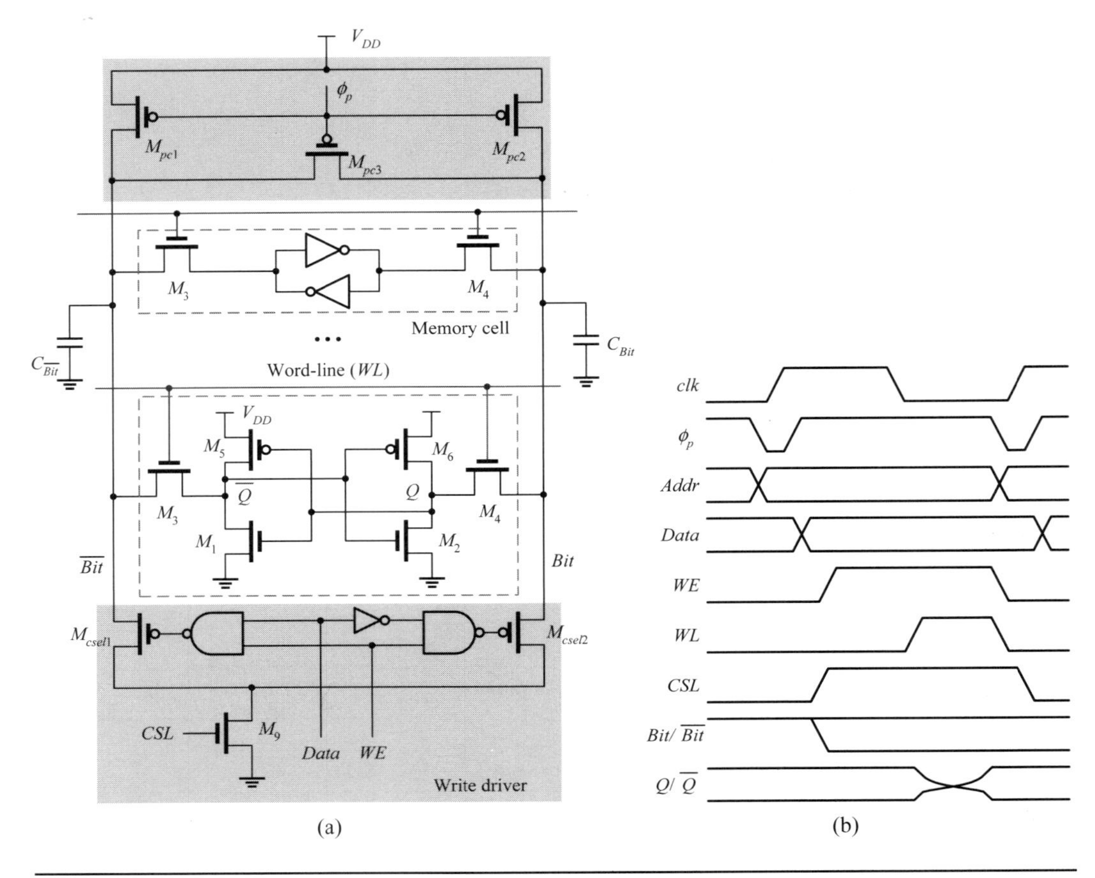

Figure 11.15: The write operation of a typical SRAM: (a) related circuit; (b) timing diagram.

and word-line (WL) enable signals to activate the write driver and the memory cell to be written. Before the word-line (WL) enable signal is activated, data and write enable control signals are applied to bit-lines to pull one of them to 0 V while leaving the other at V_{DD}. As a consequence, the cell may flip its state once the word-line enable signal is enabled. After this, both column selection and word-line enable signals may be removed. A final comment is to note that transistors, M_{csel1}, M_{csel2}, and M_9, need to be properly sized to discharge the bit-lines in a specific amount of time.

■ Review Questions

Q11-19. Describe the function of sense amplifiers in SRAM devices.

Q11-20. What role does equalizing the pMOS transistor M_{pc3} play?

Q11-21. What is the goal of the address transition detection (ATD) circuit?

11.2.3 Row Decoders

The row decoder decodes a portion of address inputs into word-line enable signals. The structures of row decoder can be cast into the following two types: *single-level* and *multilevel*.

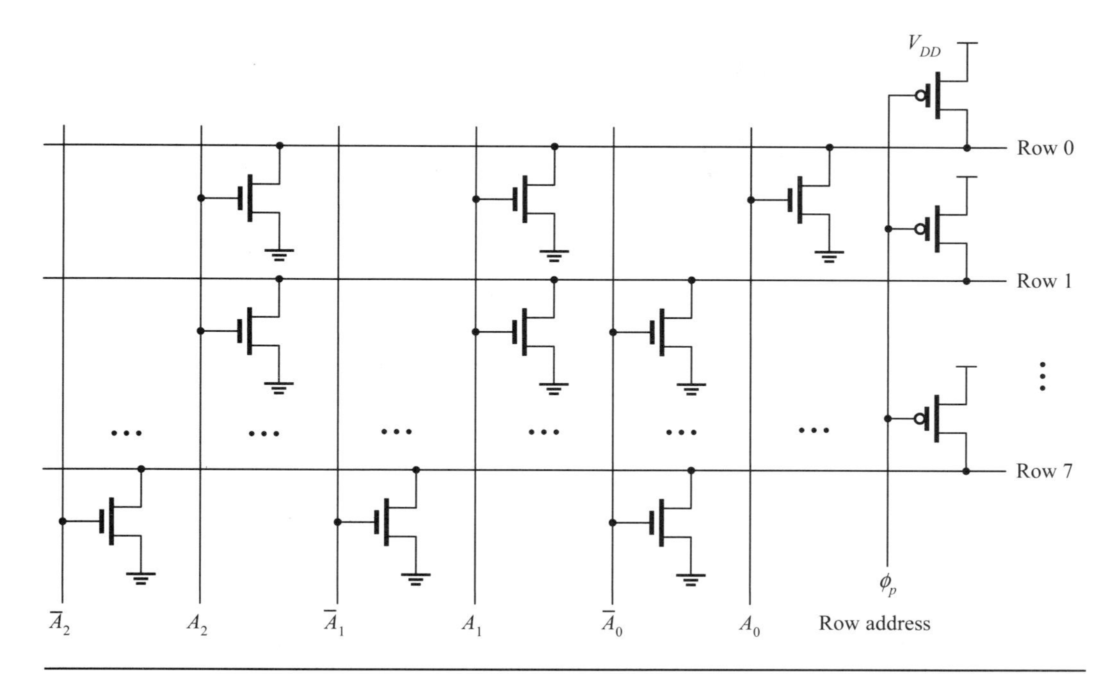

Figure 11.16: An example of a row decoder structure.

11.2.3.1 Single-Level Row Decoders The *single-level row decoder* is an $n \times 2^n$ decoder, which accepts an n-bit input and generates 2^n word-line enable signals. An example is shown in Figure 11.16, where a 3-bit row address is to be decoded into eight word-line enable signals. The row decoder uses footless dynamic NOR gates, with each for one word-line enable signal.

To drive the large loading capacitance on the word-line, a super-buffer is usually used at the output of the decoder, one for each word-line. The design issue of the super-buffer is to properly size the final driver so that it is large enough to provide the word-line voltage with a propagation delay set only by its RC time constant rather than limited by the final driver. An example of a super-buffer word-line driver is shown in Figure 11.17(a). The related timing diagram is depicted in Figure 11.17(b).

Example 11-9: (The maximum current on the word-line.) Consider the situation that $R_{word} = 16.64$ kΩ and $C_{word} = 540.16$ fF, as calculated before. Assuming that the super-buffer approach is used to drive the word-line and $V_{DD} = 1.8$ V, calculate the size of the last stage of the super-buffer.

Solution: Because of the word-line having resistance of 16.64 kΩ, the maximum charging current from the pMOS transistor of the last stage of super-buffer is limited by the word-line resistance R_{word} and is equal to

$$I_{max} = V_{DD}/R_{word} = 1.8 \text{ V}/16.64 \text{ k}\Omega = 108 \ \mu\text{A}$$

At the beginning of the charging phase, the pMOS transistor operates in saturation, hence

$$I_{Dp} = \frac{1}{2}\left(\frac{W}{L}\right)_p \mu_p C_{ox}(V_{SGp} - |V_{Tp}|)^2$$

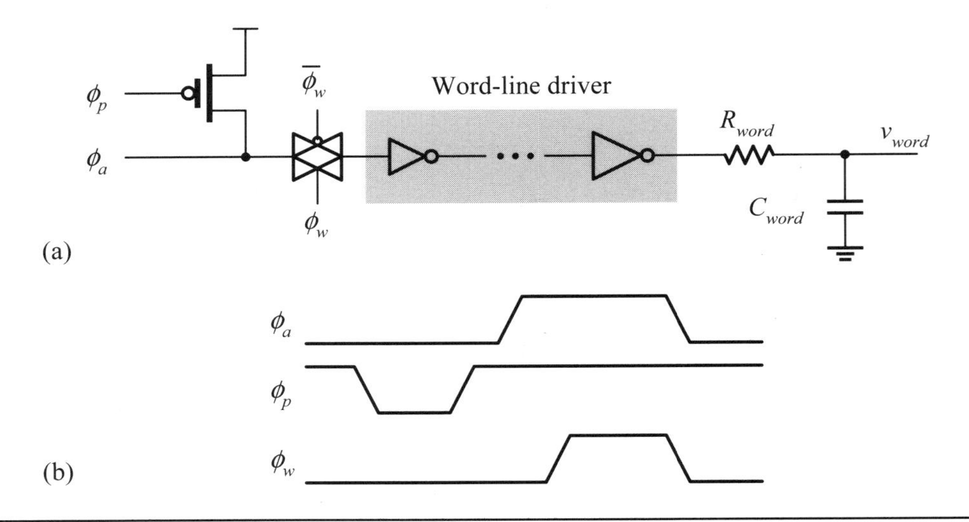

Figure 11.17: An example of a super-buffer word-line driver: (a) circuit; (b) timing diagram.

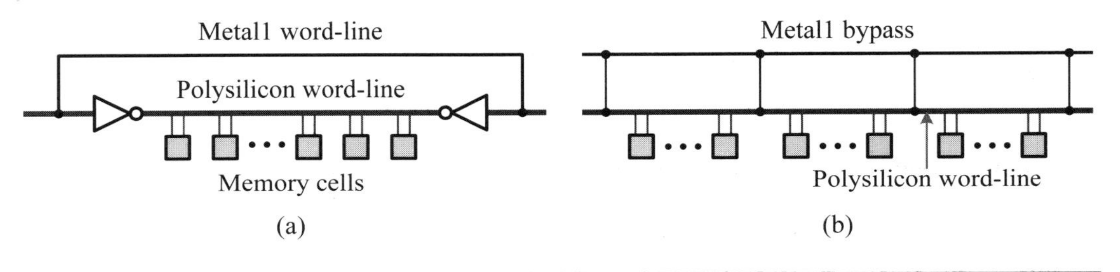

Figure 11.18: Two examples of RC delay reduction by using the metal1 layer: (a) metal word-line; (b) metal bypassed word-line.

Substituting the maximum charging current into the above equation, we obtain

$$108 = \frac{1}{2}\left(\frac{W}{L}\right)_p \cdot 48 \cdot (1.8 - 0.4)^2$$

This results in that $(W_p/L_p) = 2.29$. Hence, the super-buffer cannot work for this case due to the large resistance of the word-line.

■

To reduce the word-line resistance due to the silicided polysilicon layer, the word-lines are routed in both metal1 and polyilicon layers in parallel. Two examples of using metal1 to reduce RC delay of word-lines are illustrated in Figure 11.18. Figure 11.18(a) uses the metal1 layer as a superhighway to feed signals from both ends of the word-line, and Figure 11.18(b) simply uses metal1 as a low-resistance wire to reduce the word-line resistance by tapping it to the polysilicon layer at some specific points. Hence, this results in a low-resistance word-line because both metal1 wire and polysilicon layer are effectively connected in parallel. Nevertheless, the word-line capacitance increases a little bit because of the parasitic capacitance of the metal1 wire.

■ Example 11-10: (Design a word-line super-buffer.) Assuming that the area of a memory cell is $26\lambda \times 44\lambda$ and a metal bypass wire is used to reduce the word-line resistance, design a word-line super-buffer using the 0.18-μm process if 256 cells are connected in the same word-line.

Solution: Because of a metal bypass wire being tapped with the polysilicon word-line, the resulting word-line resistance can be ignored. Hence, a super-buffer is designed to drive a loading capacitance of 540.16 fF. As we described in Chapter 5, the input capacitance of a basic inverter is equal to

$$C_G = C_g(W_n + W_p) = 3.41(3 + 6) \times 0.18 = 5.52 \text{ fF}$$

where the channel widths of nMOS and pMOS are assumed to be 3λ and 6λ, respectively.

Using the FO4 rule, the number of stages needed is

$$n = \left\lceil \log_4 \left(\frac{540.16}{5.52} \right) \right\rceil = 4$$

The resulting super-buffer can then be easily determined, referring to Section 5.3.2 for more details.

■

11.2.3.2 Multilevel Row Decoders As we know, an n-bit decoder requires 2^n logic gates, with each having a fan-in of n. To reduce the propagation delay of the decoder, it is not uncommon to use a cascade of gates instead of n-input gates. The resulting decoder is known as the *multilevel row decoder*, which consists of two or more stages. We introduce a *two-stage decoder* consisting of a predecoder and a final decoder. The predecoder generates intermediate signals for use in the final decoder, which in turn yields the desired word-line selection signals.

■ Example 11-11: (A two-level row decoder.) Figure 11.19 shows a two-level row decoder, where both predecoder and final decoder are composed of two-input NAND gates and inverters. The predecoder comprises two 2-to-4 decoders and separately yields four decoded signals. These signals are then combined into 16 decoded outputs by the final decoder. The salient feature of this decoder is that it only uses 2-input NAND gates rather than 4-input NAND gates in the single-level decoder.

■

The two-level row decoder has the following advantages. First, it significantly reduces the number of transistors required. Second, it effectively shortens the propagation delay of the decoder. Third, by adding a select signal to each predecoder, it is able to save power dissipation profoundly by disabling the predecoders when their associated memory blocks are not accessed.

As in the single-level row decoder, the output of a multilevel row decoder must drive the large loading capacitance on the word-lines. However, in the multilevel row decoder, instead of using the single super-buffer at each output of the word-line enable signal, the approach of path-delay optimization can be employed to optimize the size of decoder to minimize the propagation delay of the decoder.

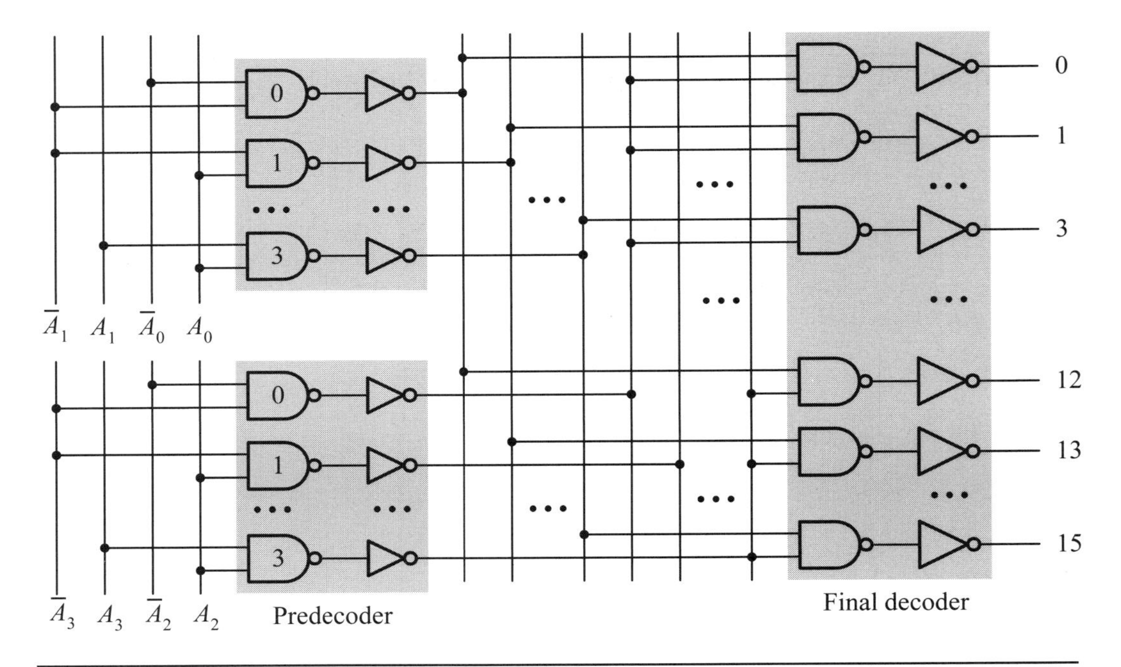

Figure 11.19: An example of a two-level row decoder.

By applying the FO4 rule, the input capacitance C_{in} for each stage can be related to its output capacitance by the following expression:

$$C_{in} = \frac{b \times g \times C_{out}}{4} \tag{11.12}$$

where b and g are the branching effort and logical effort of the gate, respectively.

■ Example 11-12: (Sizing of the two-level row decoder.) Referring to Figure 11.20, the branching effort of each output inverter of the predecoder is 4, as illustrated in Figure 11.19. Hence, the input capacitance of the inverter at the final decoder is $(1/4)C_{word}$ and that of the two-input NAND gate of the final decoder is equal to

$$C_{in(NAND)} = \frac{b \times g \times C_{out}}{4} = \frac{1}{4} \cdot \frac{4}{3} \cdot \frac{1}{4} C_{word}$$

where logical effort g of a two-input NAND gate is 4/3 and the branching effort b is 1. Similarly, the input capacitance of the inverter of the predecoder is equal to

$$C_{in(inv)} = \frac{b \times g \times C_{out}}{4} = 4 \cdot \frac{1}{4} \cdot \frac{1}{4} \cdot \frac{4}{3} \cdot \frac{1}{4} C_{word}$$

and the input capacitance of the NAND gate of the predecoder is found to be

$$C_{in(NAND)} = \frac{b \times g \times C_{out}}{4} = \frac{1}{4} \cdot \frac{4}{3} \cdot 4 \cdot \frac{1}{4} \cdot \frac{1}{4} \cdot \frac{4}{3} \cdot \frac{1}{4} C_{word}$$

■

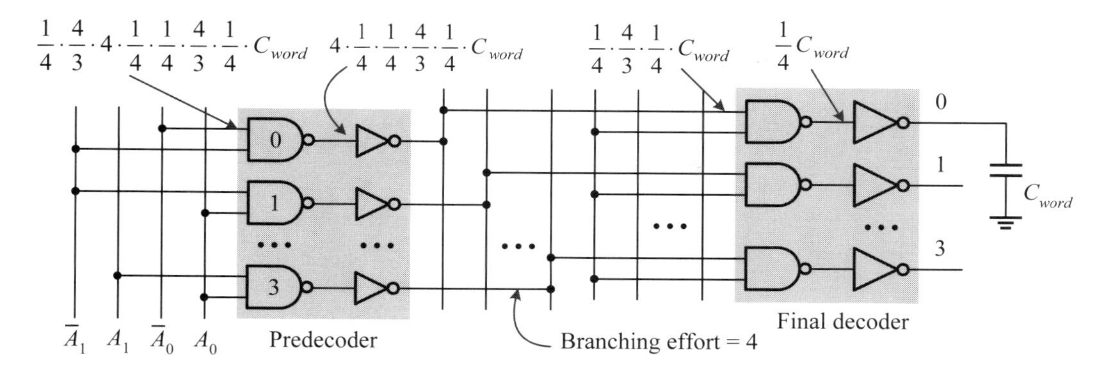

Figure 11.20: An example of sizing a row decoder.

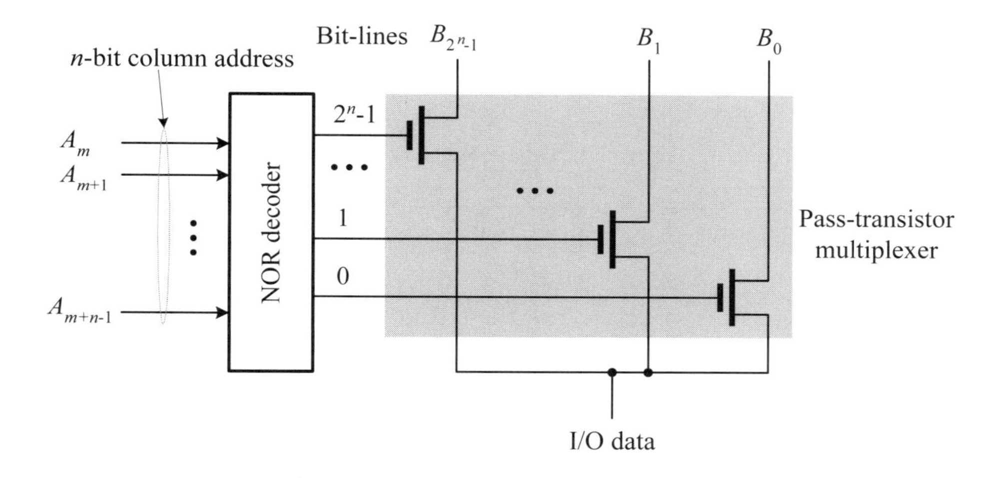

Figure 11.21: The general structure of a single-level column multiplexer.

11.2.4 Column Decoders/Multiplexers

The purpose of a column decoder is to generate the column multiplexer selection signals, which in turn are employed to select the bit-lines for read or write operation. Because of the close relationship between column decoder and column multiplexer, they are generally designed together. The column multiplexer is indeed an n-input multiplexer and can be one of the following three structures: *single-level* (uniform) structure, *binary-tree* structure, or *heterogeneous-tree* structure.

11.2.4.1 Single-Level (Uniform) Structure The single-level (uniform) column multiplexer simply connects together all outputs of n switches controlled by a decoder. An example is illustrated in Figure 11.21, where 2^n nMOS switches are connected as a 2^n-to-1 multiplexer controlled by a NOR address decoder. Generally, the nMOS switches can be replaced with TG switches.

Assuming that nMOS switches are used in the multiplexer and each switch has an output capacitance C and on-resistance R, the propagation delay of the multiplexer is nRC without considering the propagation delay of the address decoder. As a consequence, the propagation delay is linearly proportional to the number of inputs of the

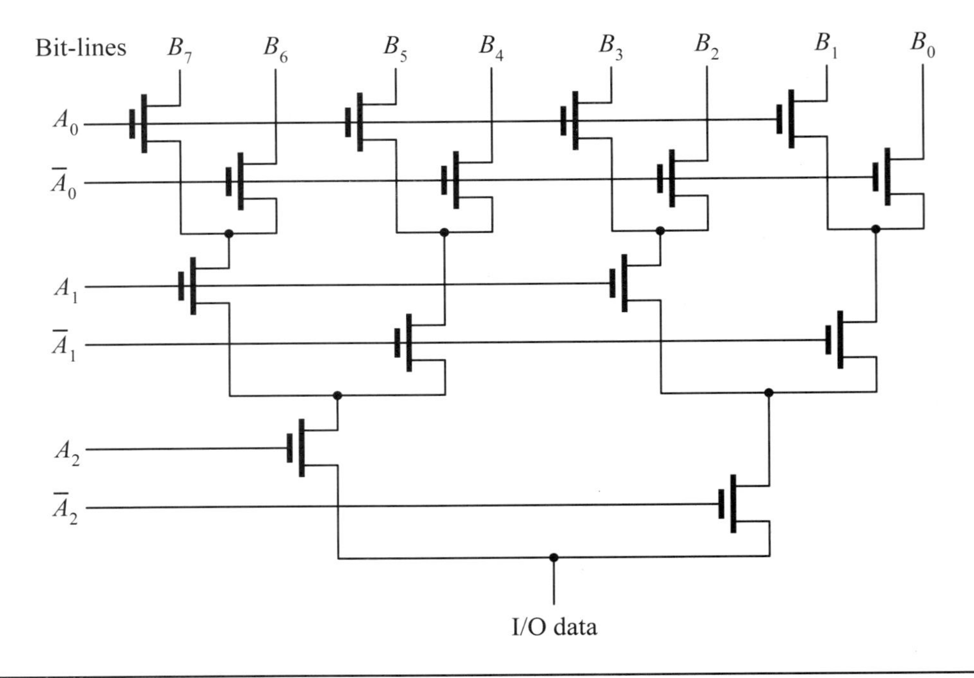

Figure 11.22: An example of an 8-input binary-tree column multiplexer.

multiplexer. However, this structure has the advantage that it only requires n nMOS or TG switches for an n-input multiplexer. Thus, it needs the least area compared to the other two structures.

11.2.4.2 Binary-Tree Structure A binary-tree column multiplexer consists of a binary tree in which each branch is an nMOS or a TG switch. An example of an 8-input binary-tree column multiplexer is given in Figure 11.22. With the binary-tree structure, the column decoder is virtually trivial because each address bit and its complement are directly applied to the nMOS or TG switches. Nevertheless, the maximum fan-out of an address bit can reach $n/2$ for an n-input binary-tree column multiplexer. Hence, an appropriate circuit such as super-buffer may be needed to reduce the propagation delay of the column decoder as n is large.

The binary-tree column multiplexer requires more switches than the single-level column multiplexer but has smaller propagation delay. Using the Elmore delay model and without considering the propagation delay of the column decoder, the propagation delay for an n-input binary-tree column multiplexer is equal to $\frac{1}{2}\log_2 n(1+3\log_2 n)RC$, which is less than nRC when $n \geq 52$. The number of required nMOS or TG switches is $2n - 2$.

11.2.4.3 Heterogenous-Tree Structure A heterogeneous tree is generally a multi-level tree in which each node may contain the same or different number of branches. A two-level column multiplexer with 16 inputs is displayed in Figure 11.23, where each node contains four branches. The propagation delay is 13 RC with the Elmore delay model, less than the $16RC$ in the single-level structure and 15 RC in the binary-tree structure. The total number of required nMOS or TG switches is 20 in this example, greater than 16 in the single-level structure and less than 30 in the binary-tree structure. Another example for the 16-input column multiplexer uses the $(8, 2)$ partition;

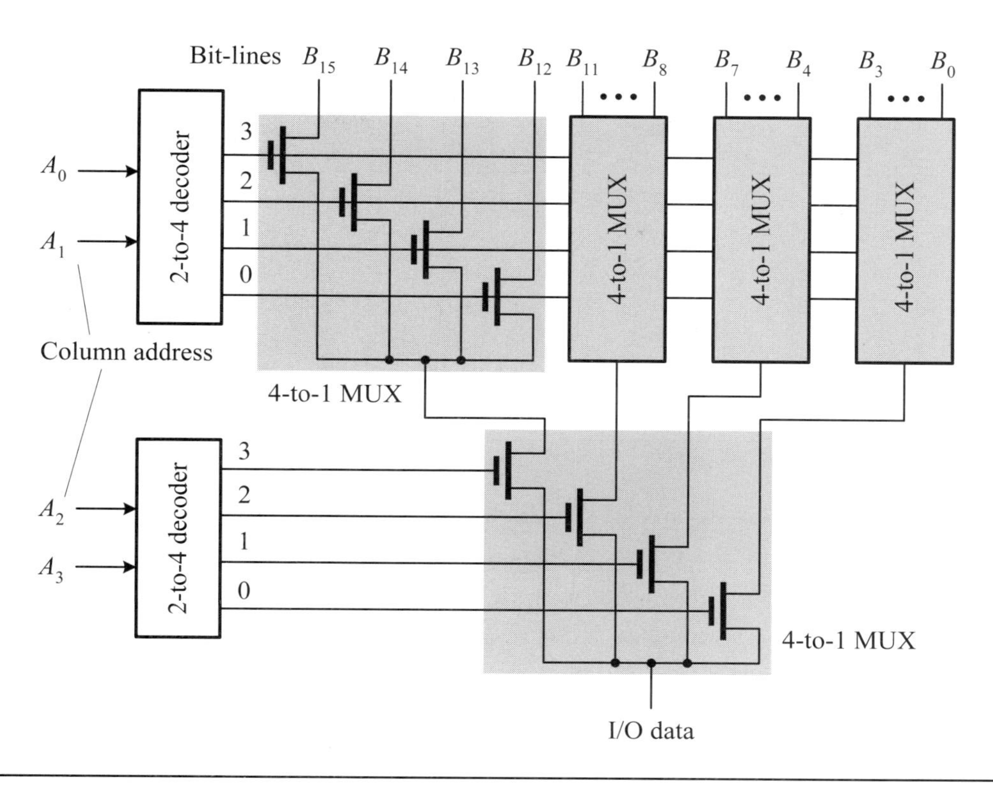

Figure 11.23: An example of a two-level column multiplexer with 16 inputs.

namely, the first stage uses two 8-input multiplexers and the second stage is a 2-to-1 multiplexer. The result needs 18 nMOS or TG switches and takes a 13 RC propagation delay.

11.2.4.4 Comparison of Column Multiplexers

Both aforementioned binary-tree and heterogeneous-tree structures are multilevel-tree structures. They use smaller multiplexers to construct the desired n-input multiplexer in a hierarchical way. The theme difference between these two tree structures is that in the binary-tree structure only 2-to-1 multiplexers are used, while in the heterogeneous-tree structure, a variety of multiplexers of different sizes may be used.

Comparison of three structures in terms of delays and the number of switches with the number of inputs ranging from 4 to 1024 is summarized in Table 11.1. The uniform structure needs fewer switches and requires only a 1-out-of-n decoder with a maximum fan-out of one while the binary-tree structure requires many more switches but only needs $\log_2 n$ 1-out-of-2 decoders with a maximum fan-out of $n/2$, where n is assumed to be an integer power of 2. The hardware cost of heterogeneous-tree structure is between these two structures. More detailed data about the heterogeneous-tree structure can be referred to Lin [26].

■ Review Questions

Q11-22. How would you reduce the word-line resistance?

Q11-23. What are the features of multilevel row decoders?

Table 11.1: Comparison of various column multiplexers.

	Uniform		Binary tree		Heterogenous tree		
n	t_d (RC)	No. of switches	t_d (RC)	No. of switches	t_d (RC)	No. of switches	Partitions
4	4	4	7	6	4	4	(4)
8	8	8	15	14	9	10	(4,2)
16	16	16	26	30	13	18	(8,2)
32	32	32	40	62	17	36	(8,4)
64	64	64	57	126	25	74	(8,4,2)
128	128	128	77	254	31	148	(8,4,4)
256	256	256	100	510	39	292	(8,8,4)
512	512	512	126	1022	50	586	(8,8,4,2)
1024	1024	1024	155	2046	58	1172	(8,8,4,4)

Q11-24. What are the features of the single-level column multiplexer?

Q11-25. What are the three basic structures of column multiplexers?

Q11-26. How would you design an n-input column multiplexer?

Q11-27. Show that the propagation delay of an n-input binary-tree column multiplexer is $\frac{1}{2}\log_2 n(1+3\log_2 n)RC$.

11.2.5 Sense Amplifiers

Sense amplifiers play critical important roles in the functionality, performance, and reliability of memory circuits. In this section, we are concerned with the following three most common sense amplifiers: *differential voltage sense amplifier*, *latch-based sense amplifier*, and *differential current sense amplifier*.

11.2.5.1 Differential Voltage Sense Amplifiers A differential voltage sense amplifier comprises three components: the current mirror, a pair of common-source amplifier, and a biasing current source. As shown in Figure 11.24, pMOS transistors M_3 and M_4 compose the current mirror and serve as the active load of the common-source amplifier, consisting of nMOS transistors M_1 and M_2. The nMOS transistor M_5 is the biasing current source, providing a constant bias current I_{SS} for the circuit.

The current in transistor M_3 is mirrored to transistor M_4; namely, their currents are equal in magnitude if their sizes are identical due to the same V_{GS} voltage. At the steady state, both transistors have the same amount of current $I_{SS}/2$, where I_{SS} is set by the aspect ratio W_5/L_5 of current source M_5, biasing at V_B.

To maintain the differential voltage sense amplifier at normal operation, all transistors of it need to be operated in saturation regions. The transistors M_1 and M_2 constitute the common-source differential amplifier. Their inputs are connected to the bit-lines. Therefore, the precharged voltages at bit-lines must be coped with very carefully because for both input transistors to operate in their saturation regions, their input voltages, V_{in1} and V_{in2}, are only allowed to be up to $V_{DD} - V_{Tn}$ at most. To see this, let us take a closer look at Figure 11.24. To keep both pMOS transistors M_3 and M_4 in their saturation regions of operation, the voltages at nodes D_1 and D_2 are only allowed to be $V_{DD} - |V_{T0p}|$ at most. Hence, if both bit-lines are precharged to V_{DD}, the voltages at nodes D_1 and D_2 would be $V_{DD} - V_{Tn}$, which in turn cause both pMOS transistors M_3 and M_4 at their cutoff edge. On the contrary, if we precharge

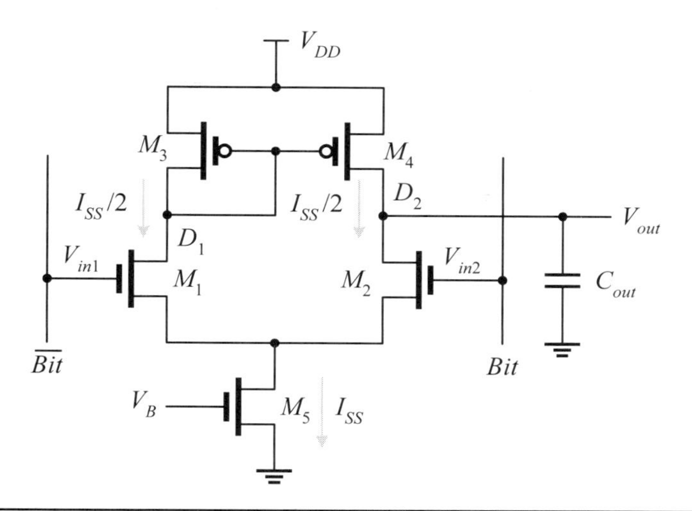

Figure 11.24: An active-load differential voltage sense amplifier.

both bit-lines to $V_{DD} - V_{Tn}$, then

$$V_{D1} = V_{D2} = V_{DD} - 2V_{Tn} \tag{11.13}$$

As a result, there is enough headroom for both pMOS transistors to be operated in saturation regions. Because of this, the three pMOS transistors, M_{pc1}, M_{pc2}, and M_{pc3}, of the precharge circuit shown in Figures 11.14 and 11.15, are replaced by nMOS transistors when differential voltage sense amplifiers are used.

The large-signal operations of the differential voltage sense amplifier are illustrated in Figure 11.25. When the input voltage V_{in1} of nMOS transistor M_1 is dropped below $V_{DD} - V_{Tn}$, the nMOS transistor M_1 is turned off, causing the current of the current mirror M_4 to drop to zero. The total bias current I_{SS} is switched into nMOS transistor M_2, thereby discharging the output node toward the ground level.

Regardless of switching high or low, the rate of change of the output voltage is equal to

$$\frac{dV}{dt} = \frac{I_{SS}}{C_{out}} \tag{11.14}$$

which is named as the *slew rate*, namely, dV/dt at the output. Rearranging the slew rate relationship, we may express the propagation delay through the sense amplifier as

$$t_{sense} = \frac{C_{out}\Delta V_{out}}{I_{SS}} \tag{11.15}$$

To reduce the propagation delay, a large I_{SS} is needed. Nevertheless, it will cause a large power dissipation due to $P_d = I_{SS} \cdot V_{DD}$. As a consequence, a trade-off between propagation delay and power dissipation is needed.

11.2.5.2 Latch-Based Sense Amplifiers The essential feature of a latch-based sense amplifier is a pair of cross-coupled inverters with an enabling transistor, as shown in Figure 11.26(a). As described, the latch-based sense amplifier is fundamentally a dynamic latch circuit, relying on the *regenerative effect* of inverters to generate a high or low valid voltage. It is a low-power circuit in the sense that it is only activated at some interval of proper time.

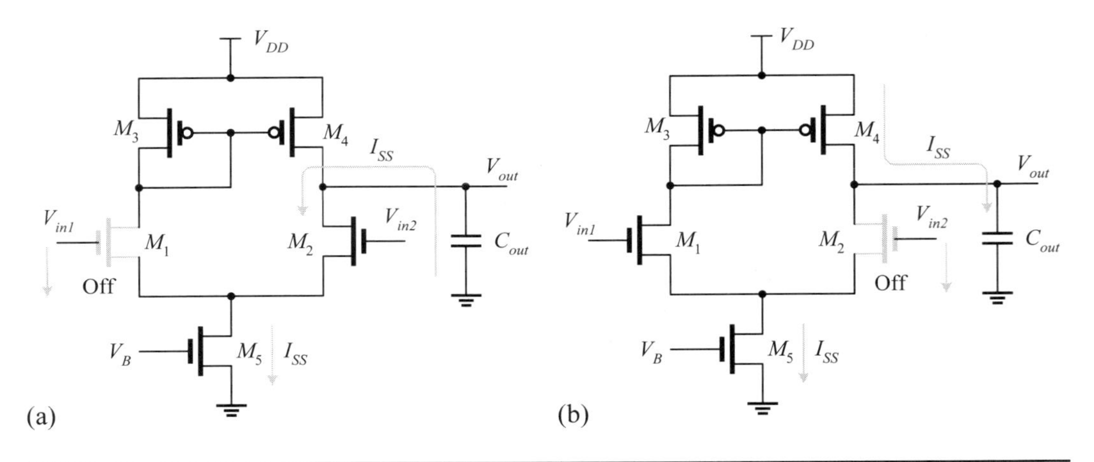

Figure 11.25: The large-signal operations of a differential voltage sense amplifier: (a) discharging output capacitor; (b) charging output capacitor.

The latch-based sense amplifier works as follows. As shown in Figure 11.26(b), after both bit-lines are precharged to $V_{DD}/2$, word-line enable signal (WL) is enabled by the row decoder to select the desired word. Then, the column selection ($\overline{CSL}$) is activated to route the bit-line values to the latch-based sense amplifier. Finally, the sense amplifier is activated to generate the valid logic values. The detailed timing diagram about the voltage change of bit-lines is exhibited in Figure 11.26(c).

The propagation delay t_{sense} of the latch-based sense amplifier can then be written as a function of both input and output voltages and is equal to

$$t_{sense} = \left(\frac{C_{bit}}{g_{mn} + g_{mp}}\right) \ln\left(\frac{\Delta v_{out}(t_{sense})}{\Delta v_{in}}\right) \tag{11.16}$$

■ Example 11-13: (A latch-based sense amplifier.) Referring to Figure 11.26, assume that the bit-line capacitance C_{bit} is 580 fF. Using 0.18-μm-process parameters: $k'_n = 96\ \mu\text{A/V}^2$, $k'_p = 48\ \mu\text{A/V}^2$, and $V_{T0n} = |V_{T0p}| = 0.4$ V, design a latch-based sense amplifier to drive the voltage difference, about 200 mV, in bit-lines to 1.8 V in less than 5 ns.

Solution: From Equation (11.16), assume that $g_{mn} = g_{mp}$, we have

$$\begin{aligned} t_{sense} &= \left(\frac{C_{bit}}{g_{mn} + g_{mp}}\right) \ln\left(\frac{\Delta v_{out}(t_{sense})}{\Delta v_{in}}\right) \\ &= \left(\frac{580 \text{ fF}}{2g_{mn}}\right) \ln\left(\frac{1.8}{0.2}\right) \leq 5 \text{ ns} \end{aligned}$$

Therefore, $g_{mn} \geq 127\ \mu$S. Using the definition of g_{mn} from Equation (5.16), we are able to find the aspect ratio of an nMOS transistor as follows:

$$\begin{aligned} g_{mn} &= \left(\frac{W_n}{L_n}\right) k'_n \left(\frac{V_{DD}}{2} - V_{T0n}\right) \\ &= \left(\frac{W_n}{L_n}\right) 96 \left(\frac{1.8}{2} - 0.4\right) \geq 127 \end{aligned}$$

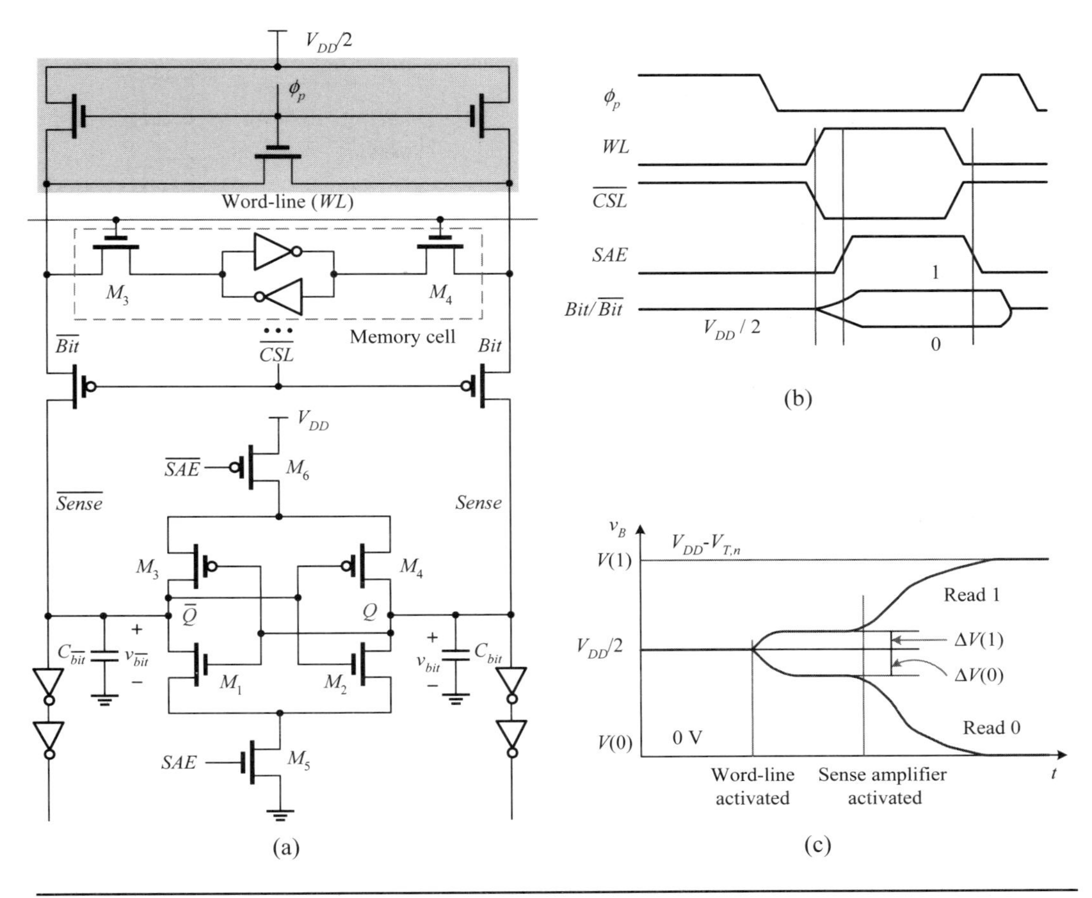

Figure 11.26: A latch-based sense amplifier: (a) circuit; (b) read timing; (c) detailed timing of bit-line.

Solving it, the aspect ratio is equal to

$$\left(\frac{W_n}{L_n}\right) \geq 2.64 \approx 3 \Rightarrow \left(\frac{W_p}{L_p}\right) = 6$$

Next, we determine the aspect ratio of transistor M_5. Since V_{DS5} (say, 0.1 V) is very small compared to $V_{DD} - V_{T0n}$, the squared term of V_{DS5} can be ignored. Therefore,

$$I_{D5} = \left(\frac{W_5}{L_5}\right) \mu_n C_{ox} \left(V_{DD} - V_{T0n}\right) V_{DS5} = 2I_{D1}$$

Because nMOS transistor M_1 operates in its saturation region, its current can be calculated as follows:

$$\begin{aligned} I_{D1} &= \frac{1}{2}\left(\frac{W_1}{L_1}\right) k'_n \left(\frac{V_{DD}}{2} - V_{T0n}\right)^2 \\ &= \frac{1}{2} \cdot 3 \cdot 96 \cdot \left(\frac{1.8}{2} - 0.4\right)^2 = 36\ \mu\text{A} \end{aligned}$$

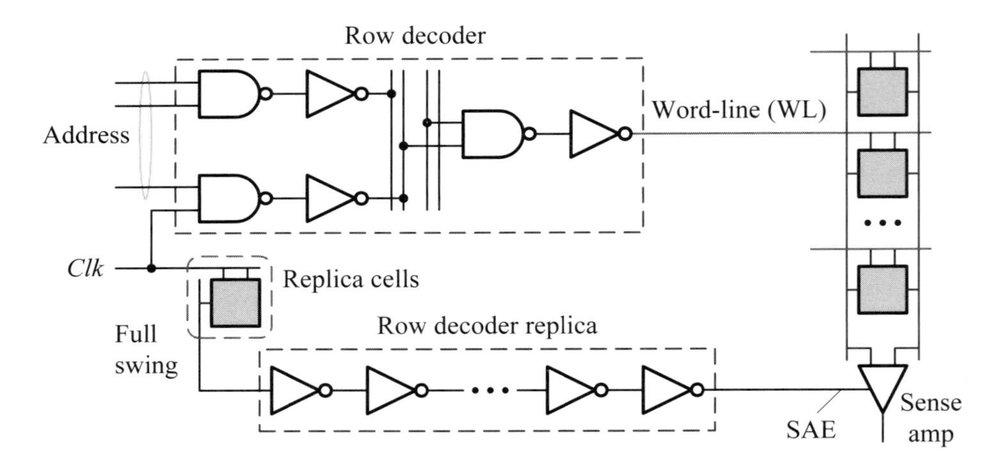

Figure 11.27: Replica circuit for the clock enable of sense amplifier.

Once the drain current of M_1 is found, the aspect ratio of transistor M_5 can be determined as follows:

$$\begin{aligned}\left(\frac{W_5}{L_5}\right) &= \frac{2I_{D1}}{k_n'(V_{DD} - V_{T0n})V_{DS5}} \\ &= \frac{2 \cdot 36}{96(1.8 - 0.4) \cdot 0.1} = 5.4 \Rightarrow 6\end{aligned}$$

Replica circuits. As in the timing shown in Figure 11.26(b), an important design issue of a latch-based sense amplifier is the sense-amplifier enable (SAE) signal, which must come in proper time to ensure a fast and low-power operation. If it is activated too early, the bit-line may not have developed enough voltage difference for the sense amplifier to operate reliably; if it is enabled too late, it adds unnecessary delay to the access time. In addition, process variations virtually determine the timing of the SAE signal. One way to tackle the process variations is by using *replica circuits*, called the *replica delay line*, to mimic the propagation delay of the bit-line path over all conditions to create the SAE signal appropriately.

As illustrated in Figure 11.27, the datapath starts from the row decoder and goes through the word-line driver, memory cell, and bit-line to the input of the sense amplifier. The SAE path often starts from the local block select or some clock phase, and goes through a set of replica cells and a buffer chain (that is, row decoder replica). By properly designing the SAE path in a way such that it can exhibit the same characteristics as the datapath, the SAE can be ensured to arrive at the right time.

The replica cells used in Figure 11.28 are actually a portion of the dummy column, being used to avoid the boundary-dependent etching effect of a memory array. The required number of replica cells is determined by the following criterion. For an m-row memory array, if the supply voltage is V_{DD} and the sensitivity of the sense amplifier is one-tenth of V_{DD}, then the needed number of replica cells is $m/10$ because the full-swing value in the replica cells is required to directly drive the row decoder replica circuit. Here, we assume that the RC-equivalent delay model is used along with the following relationship: the propagation delay is proportional to the capacitance of the

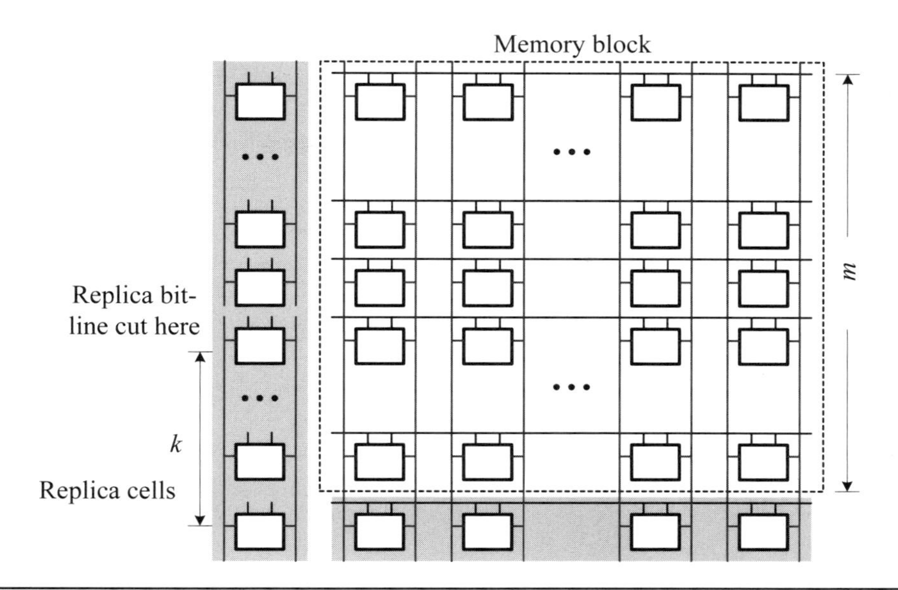

Figure 11.28: The design of replica cells.

bit-line. For instance, if there are 256 cells in a column of a memory array, then the bit-lines in the replica column are cut at a height of $k = 26$ rows (taking a whole number), yielding the capacitance for the replica bit-line which is one-tenth that of the main bit-line.

11.2.5.3 Differential Current Sense Amplifiers

Differential current sense amplifiers are widely used in high-speed SRAM to improve signal sensing speed because they are independent of the bit-line capacitance. The basic structure of a differential current sense amplifier is shown in Figure 11.29. Instead of connecting bit-lines to the gates of a differential voltage sense amplifier, the bit-lines are connected to the sources of latch transistors, M_{SA1} and M_{SA2}. When the sense amplifier enable ($\overline{SAE}$) signal goes to low, the differential current sense amplifier is activated. All transistors of the differential current sense amplifier operate in the saturation region because both bit-lines are maintained at V_{DD} and both data lines are maintained at $|V_{Tp}|$.

The circuit works as follows. As the $\overline{SAE}$ signal goes to low, both bit-lines have the same voltage equal to $V_1 + V_2$. Therefore, both bit-lines have the same load current and both bit-line capacitor currents are also the same. As the memory cell draws current I_{cell}, it follows that the bit-line Bit of the differential current sense amplifier must pass more current than the complementary bit-line $\overline{Bit}$ to keep the same bit-line voltage. The current difference between these two bit-lines is I_{cell}, the memory cell current. The drain currents of M_{SA3} and M_{SA4} are passed to current transporting data lines, DL and $\overline{DL}$.

The main advantages of differential current sense amplifiers are as follows. First, the sensing speed is independent of bit-line capacitance since no differential capacitor discharging is required to sense the cell data. Second, bit-lines need no precharge and equalizing operations during a read access, thereby eliminating the speed and cycle time penalty. Differential current sense amplifiers are widely used in high-speed SRAM and high-density DRAM due to large bit-line capacitance in these circuits.

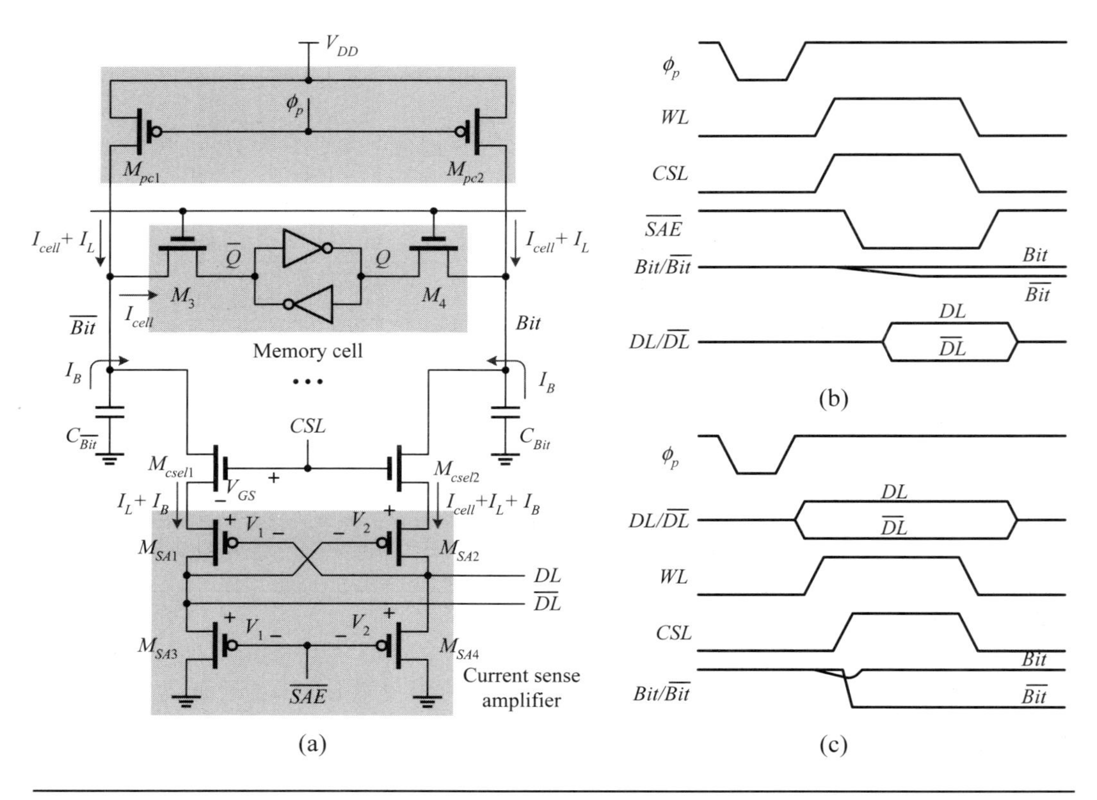

Figure 11.29: A differential current sense amplifier: (a) circuit; (b) read timing; (c) write timing.

11.2.6 ATD Circuit and Timing Generation

Recall that a sequence of actions must be scheduled properly to make a memory device operate correctly. These actions include address latching, word-line decoding, bit-line precharging and equalization, sense-amplifier enabling, and output driving. There are two broad classes of timing control that can be generally used to schedule these actions. The clocked approach is usually used in DRAM and synchronous SRAM circuits, and the self-timed approach is mostly found in asynchronous SRAM circuits. We will focus on a *self-timed approach*, referred to as *address transition detection* (ATD) circuitry.

The address transition detection (ATD) circuitry is the one that detects the address transition and generates related timing signals for harmonizing the read/write operation of SRAM devices. Basically, an ATD circuit is a monostable circuit triggered by any transition on address inputs. As illustrated in Figure 11.30, an ATD circuit is composed of an n-input pseudo-nMOS NOR gate, with each input being fed by an XOR gate. One input of the XOR gate is directly fed by an address input and the other is a delayed version of the address input. The XOR gate will output a short positive pulse when the address input makes any changes. This pulse causes the pseudo-nMOS NOR gate to be turned on and outputs a negative pulse, that is, precharge pulse, ϕ_p. The pulse width is determined by the propagation delay t_{pd} of the delay element at the input of the XOR gate. The propagation delay t_{pd} can be adjusted by modifying the number of inverters, which should be an even number. Other types of delay elements can also be used for this purpose. Refer to Section 8.3.4.2 for more details.

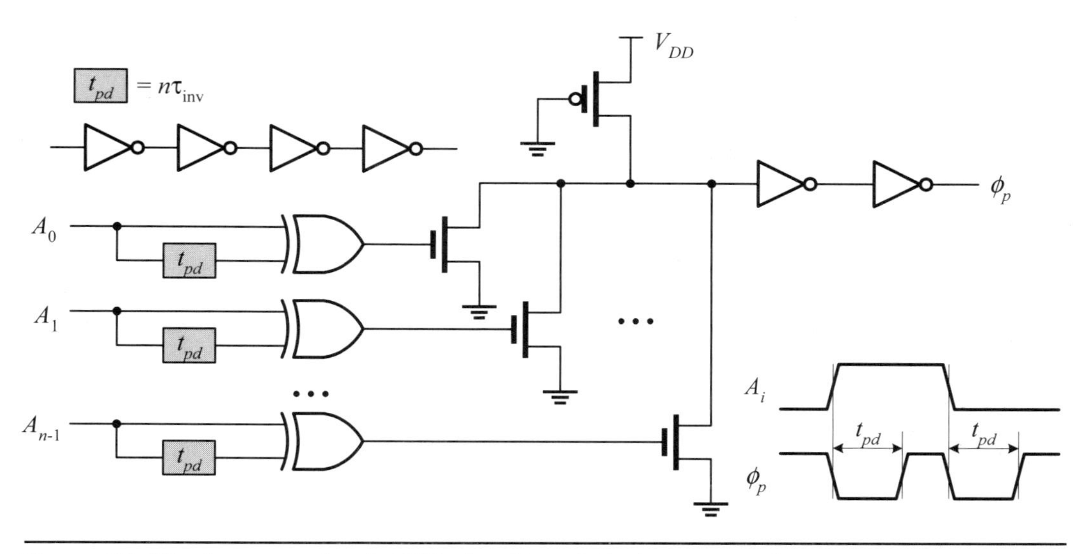

Figure 11.30: An example of ATD circuit and timing generation.

■ Review Questions

Q11-28. Why are the replica cells used associated with latch-based sense amplifiers?
Q11-29. How would you determine the number of replica cells?
Q11-30. Describe the features of latch-based sense amplifiers.
Q11-31. Describe the features of differential current sense amplifiers.
Q11-32. Describe the function of ATD.

11.3 Dynamic Random-Access Memory

Dynamic random-access memories (DRAMs) due to high integration density have been widely used in a wide variety of computer systems over the past decades to form a hierarchical memory system, thereby reducing the cost of memory system profoundly. However, its design is too complex to address in a short section. In what follows, we only describe some basics to see the operation and trade-offs.

11.3.1 Cell Structures

We begin with the introduction of the first widely used dynamic memory cell based on a 3T-circuit structure and then describe two modern 1T-dynamic cell structures.

11.3.1.1 3T-DRAM Cells The first widely used DRAM cell is a three-transistor (3T) structure, as shown in Figure 11.31. The information is stored in the charge form in parasitic capacitor C_S. Transistors M_1 and M_3 act as access transistors during write and read operations, respectively. Transistor M_2 is the storage transistor and is turned on or off, according to the amount of charge stored in capacitor C_S.

Both read and write operations of a 3T-DRAM cell are proceeded in a two-phase process. During the first phase, both bit-lines D_{in} and D_{out} are precharged to $V_{DD} - V_{Tn}$. During the second phase, the actual read or write operation is executed. To

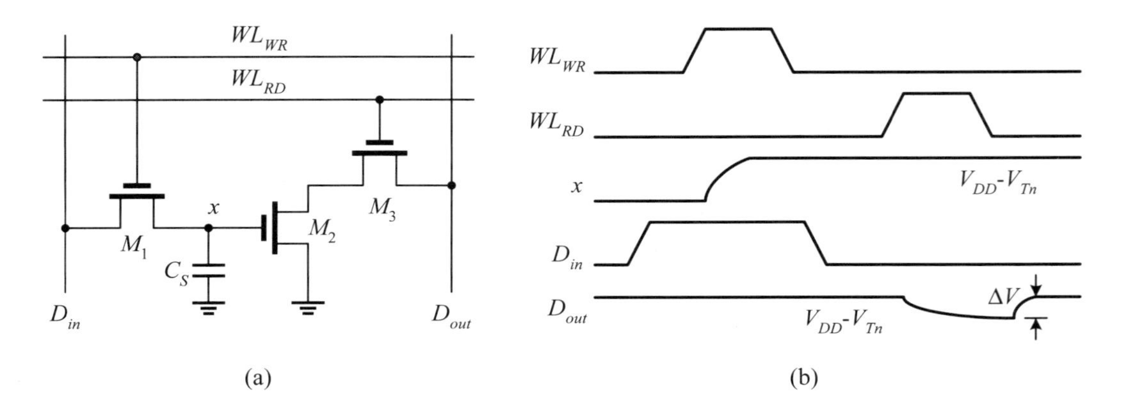

Figure 11.31: The (a) typical circuit structure and (b) timing of a 3T-DRAM cell.

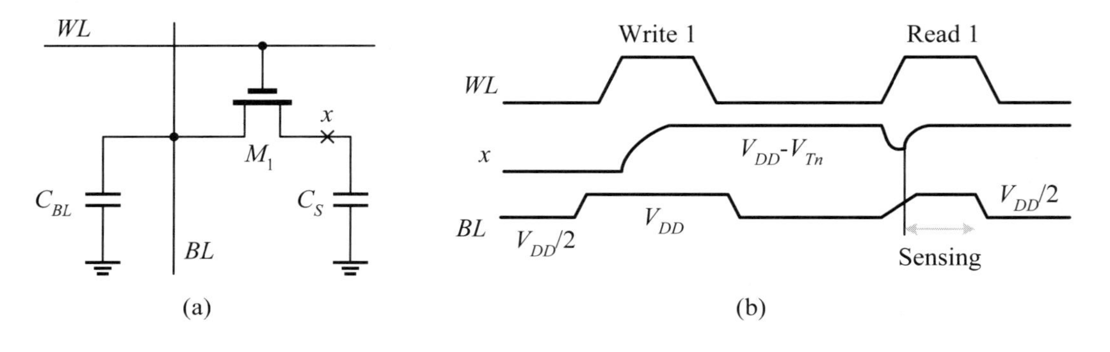

Figure 11.32: The (a) typical circuit structure and (b) timing of a 1T-DRAM cell.

read a 3T-DRAM cell, the read word-line (WL_{RD}) is activated to turn on the read access transistor M_3. The storage transistor M_2 turns on if there is charge stored in the capacitor C_S and remains off otherwise. When the storage transistor M_2 turns on, the voltage of bit-line D_{out} is dropped from $V_{DD} - V_{Tn}$ by an amount of ΔV. This amount of voltage change causes the sense amplifier to restore into a valid logic value. On the other hand, when the storage transistor M_2 stays off, the voltage of bit-line D_{out} would not be dropped. The sense amplifier yields another logic value.

To write a logic value to a 3T-DRAM cell, the write word-line (WL_{WR}) is enabled to turn on the write access transistor M_1. The storage capacitor C_S is then charged by the voltage at bit-line D_{in}. The timing is exhibited in Figure 11.31(b). Because of the body effect of the nMOS access transistor, the maximum voltage at the storage capacitor is only $V_{DD} - V_{Tn}$.

The important features of 3T-DRAM cell are as follows. First, there is no constraint on device ratios. Second, read operations are nondestructive. Third, the voltage value stored at the storage capacitor is at most $V_{DD} - V_{Tn}$. The drawback of the 3T-DRAM cell is that it needs three transistors and one capacitor.

11.3.1.2 1T-DRAM Cells Nowadays, the 1T-DRAM cell is the most widely used storage structure in the DRAM industry because it needs less area than the 3T-DRAM cell. As shown in Figure 11.32, one 1T-DRAM cell only composes one access transistor and one explicit storage capacitor.

Like a 3T-DRAM cell, both read and write operations of a 1T-DRAM cell are also proceeded in a two-phase process. During the first phase, the bit-line (BL) is precharged to $V_{DD}/2$ to improve noise immunity and reduce power dissipation. During the second phase, the actual access operation is carried out. To read the information stored in a 1T-DRAM cell, the word-line (WL) is enabled to turn on the access transistor. This causes charge redistribution to take place between the bit-line parasitic and storage capacitors. Assume that the capacitance of the bit-line is C_{BL}. The voltage at the storage capacitor C_S is $V_{DD} - V_{Tn}$ when it stores a logic 1 and is 0 V when it stores a logic 0. Then, the voltage change of the bit-line can be expressed as follows:

$$\begin{aligned} \Delta V_{bit1} &= V_{after} - V_{BL} \\ &= \left(\frac{C_S}{C_{BL} + C_S}\right)\left(\frac{1}{2}V_{DD} - V_{Tn}\right) \end{aligned} \tag{11.17}$$

when the storage capacitor C_S stores a logic 1 and is

$$\begin{aligned} \Delta V_{bit0} &= V_{after} - V_{BL} \\ &= -\frac{1}{2}V_{DD}\left(\frac{C_S}{C_{BL} + C_S}\right) \end{aligned} \tag{11.18}$$

when the storage capacitor C_S stores a logic 0. The voltage swing of a bit-line is usually small, typically ranging from 20 to 150 mV, depending on the ratio of bit-line capacitance to storage capacitance. The actual required voltage swing is set by the sensitivity of a sense amplifier.

■ Example 11-14: (Bit-line voltage changes of a DRAM cell.) Assume that each DRAM cell has a 30-fF storage capacitor and contributes 2 fF to the total bit-line capacitance. There are 256 bits on each bit-line that result in approximately 512 fF of bit-line capacitance. With the addition of replica cells and other parasitics, assume that the bit-line capacitance is 600 fF. Using 0.18-μm parameters, calculate the bit-line voltage change when a logic 1 and logic 0 are read from a storage capacitor, respectively.

Solution: When the storage capacitor stores a logic 1, its maximum voltage is $V_S = V_{DD} - V_{Tn}$. Because of the body effect of the nMOS access transistor, the threshold voltage increases and can be calculated as follows:

$$\begin{aligned} V_{Tn} &= V_{T0n} + \gamma_n\left(\sqrt{V_{SB} + |2\phi_p|} - \sqrt{|2\phi_p|}\right) \\ &= 0.4 + 0.3\left(\sqrt{V_S + 0.84} - \sqrt{0.84}\right) \end{aligned}$$

Hence, the maximum voltage on the storage capacitor C_S is

$$V_S = 1.8 - \left[0.4 + 0.3\left(\sqrt{V_S + 0.84} - \sqrt{0.84}\right)\right]$$

Solving it iteratively, we obtain $V_S = 1.24$ V and $V_{Tn} = 0.56$ V.

Hence, when the storage capacitor C_S stores a logic 1, the voltage change at the bit-line can be found from Equation (11.17) and is

$$\begin{aligned} \Delta V_{bit1} &= \left(\frac{C_S}{C_{BL} + C_S}\right)\left(\frac{1}{2}V_{DD} - V_{Tn}\right) \\ &= \left(\frac{30}{600 + 30}\right)\left(\frac{1}{2} \cdot 1.8 - 0.56\right) = 16.19 \text{ mV} \end{aligned}$$

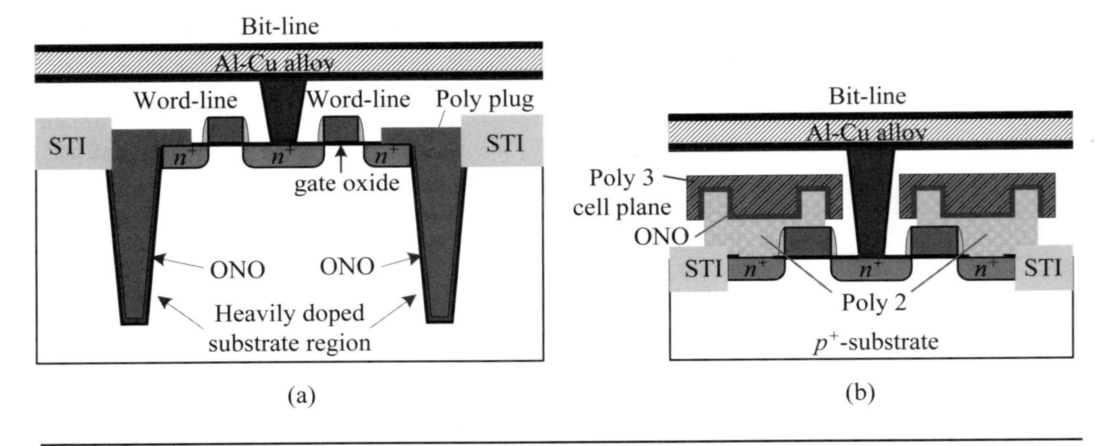

Figure 11.33: Cross-sectional view of (a) trench and (b) stacked capacitor cells.

When the storage capacitor C_S stores a logic 0, the voltage change at the bit-line can be calculated by using Equation (11.18) and is as follows:

$$\begin{aligned}\Delta V_{bit0} &= -\frac{1}{2}V_{DD}\left(\frac{C_S}{C_{BL}+C_S}\right)\\ &= -\frac{1}{2}\cdot 1.8\left(\frac{30}{600+30}\right) = -42.86 \text{ mV}\end{aligned}$$

Thus, the bit-line voltage changes for logic 1 and logic 0 are 16.19 mV and −42.86 mV, respectively.

■

To write data into a 1T-DRAM cell, both word-line *WL* and bit-line *BL* are asserted to charge or discharge C_S, depending on whether a logic 1 or logic 0 is to be written into.

Some important features of a 1T-DRAM cell are as follows: First, 1T DRAM requires a sense amplifier for each bit-line, because of the charge redistribution property associated with the read operation. Second, DRAM cells are single-ended in contrast to SRAM cells, which are dual-ended in nature. Third, the read-out operation of the 1T-DRAM cell is destructive. Hence, a write-back operation is needed after each read for correct operation. Fourth, unlike the 3T-DRAM cell, a 1T-DRAM cell requires the presence of an extra capacitance that must be explicitly included in the design. When writing a logic 1 into a DRAM cell, regardless of a 3T- or 1T-DRAM cell, a threshold voltage V_{Tn} is lost due to the body effect of the nMOS access transistor. Fortunately, this voltage loss can be compensated for by bootstrapping the word-line to a higher value than $V_{DD} + V_{Tn}$ if necessary.

11.3.1.3 Modern 1T-DRAM Cells There are two types of 1T-DRAM cell widely used in the modern DRAM industry. These are *trench capacitor* and *stacked capacitor*, as shown in Figures 11.33(a) and (b), respectively. The trench capacitor is formed by etching a deep trench on the substrate and then filling in *oxide-nitride-oxide* (ONO) dielectric to increase the capacitance. Because of the high aspect ratio of the depth to its diameter of the trench for high-density memory, processing concerns are more complex. Instead of using the trench, a stacked capacitor is constructed by adding

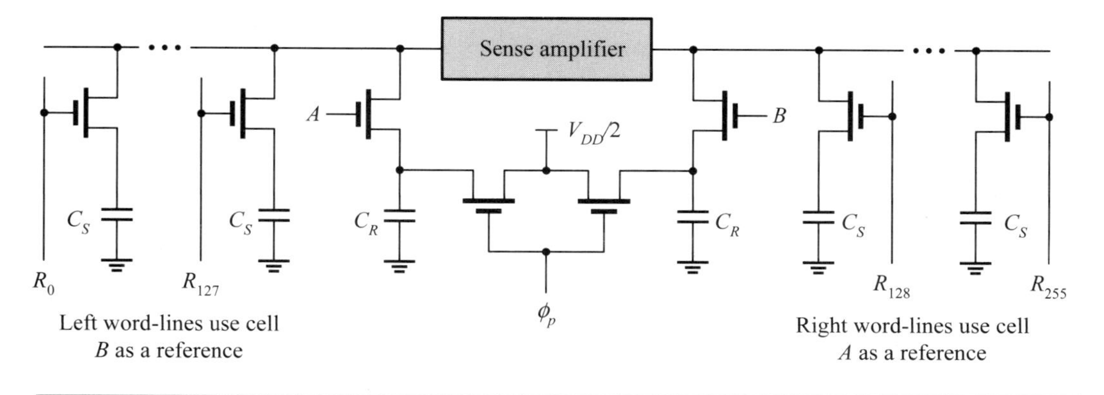

Figure 11.34: The typical open-bit architecture of the 1T DRAM.

two additional polysilicon layers above the nMOS access transistor and below the bit-line with a dielectric sandwiched between them, respectively. Because of simpler processing steps, the stacked capacitor now becomes the preferred approach to realize the cell capacitance. Nonetheless, it has a larger parasitic capacitance as compared to the trench capacitor. Regardless of which approach is used, the capacitance should be in the range of 25 to 40 fF at least to hold the information reliably. Correspondingly, the bit-line capacitance should be limited in the range of 250 to 400 fF, ten times the storage capacitance at most.

11.3.2 Structures of Memory Array

The memory array structure of DRAM is shown in Figure 11.34, where the bit-lines of the storage array are split in half so that equal amount of capacitance is connected to each side of the sense amplifier. This is known as an *open-bit architecture*. Because of the difficulty of detecting a small voltage change in a single-ended sense amplifier, a dummy cell is employed for each side of the sense amplifier to serve as a reference voltage.

The read operation is as follows. The left word-lines use cell B as a reference while the right word-lines use cell A as a reference. The bit-lines and reference cells are precharged to $V_{DD}/2$ by enabling the precharge signal ϕ_p. Now, assume that the word-line R_{127} is activated to read its corresponding cell. To do this, the dummy cell B on the opposite side of the sense amplifier is also enabled by raising the control signal B to balance the common-mode noise arising from the circuit. The resulting small differential voltage difference between two bit-lines on each side of the sense amplifier is then amplified into a valid logic value 0 or 1, as we described previously.

Nowadays, a more popular memory array structure of a 1T DRAM is referred to as a *folded bit-line architecture*, as shown in Figure 11.35(a). The corresponding layout of the cell is given in Figure 11.35(b). In this architecture, each bit-line connects to only a half number of cells and two adjacent bit-lines are grouped as pairs as inputs to the same sense amplifier. When a word-line is activated, one bit-line of a sense amplifier switches while the other serves as the quiet reference. Hence, the dummy cells are no longer needed. Like the open-bit architecture, the quiet bit-line of the sense amplifier functions as the dummy cell and helps reject the common-mode noise likely coupling equally onto the two adjacent bit-lines.

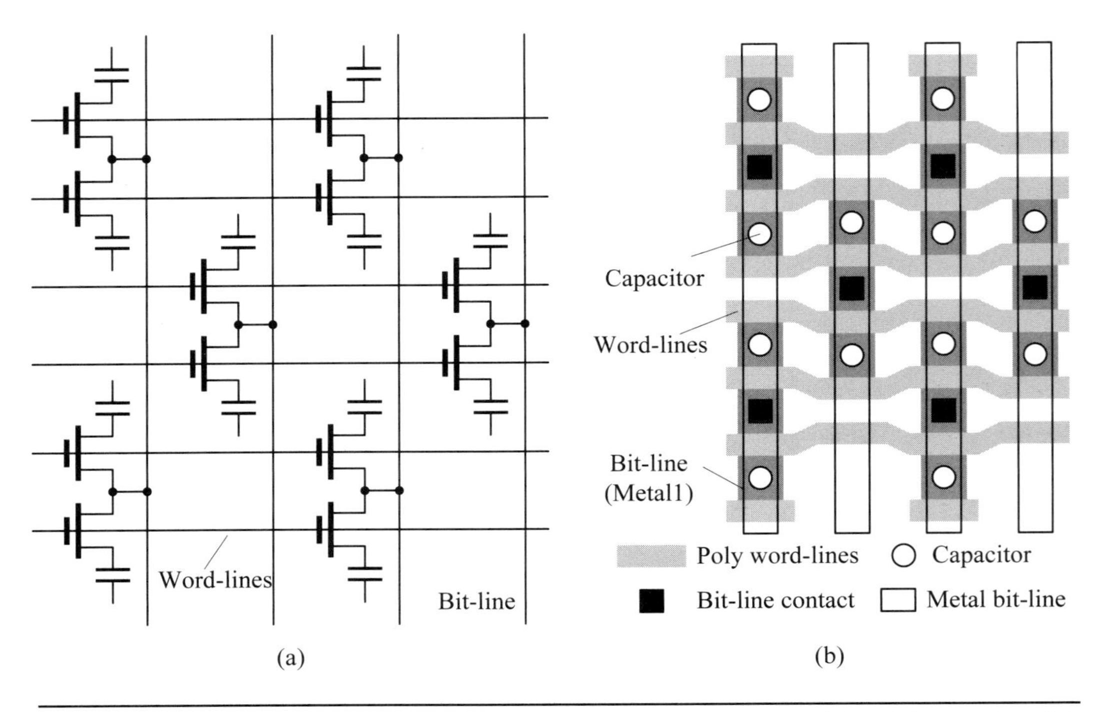

Figure 11.35: The folded bit-line architecture of the 1T DRAM: (a) circuit; (b) layout.

One disadvantage of folded bit-line architecture is that it requires more layout area, about 33%, compared to the open-bit architecture. To further reduce the common-mode noise, the twisted bit-line architecture used in SRAM may be applied.

■ Review Questions

Q11-33. Describe the features of 3T-DRAM cells.

Q11-34. Describe the features of 1T-DRAM cells.

Q11-35. Derive Equations (11.17) and (11.18).

Q11-36. Describe the features of open-bit architecture.

Q11-37. Describe the features of folded bit-line architecture.

11.4 Read-Only Memory

Read-only memory devices are widely used to store programs and permanent data in computers and embedded systems. Basically, a ROM device uses the presence or absence of a transistor to represent the information bit 0 or 1 and can be built based on either a NOR or NAND structure. In the NOR-type structure, each bit-line composes a large pseudo-nMOS NOR gate while in the NAND-type structure, each bit-line is a large pseudo-nMOS NAND gate. We will describe these two structures briefly.

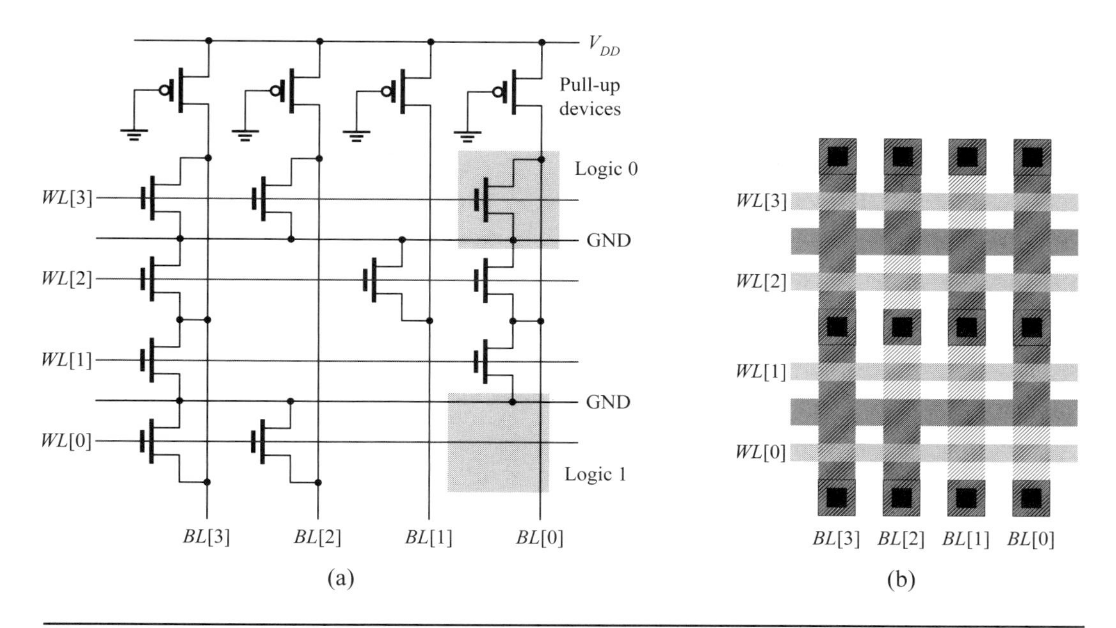

Figure 11.36: A 4 × 4 memory array structure of active-programming NOR-type ROM: (a) circuit; (b) layout.

11.4.1 NOR-Type ROM

In the NOR-type ROM, each bit-line is a pseudo-nMOS NOR gate. The presence of a transistor at the intersection point of a word-line *WL* and a bit-line *BL* means that a logic 0 is stored, whereas the absence of a transistor means that a logic 1 is stored. Based on which layer is used as the programming facility, it can be further subdivided into two types: *active-programming* ROM and *via-programming* ROM. The former only uses the active layer as the programming facility and the latter only uses the contact (or via) layer as the programming facility.

11.4.1.1 Active-Programming ROM Figure 11.36 shows an example of a 4 × 4 memory array structure of active-programming NOR-type ROM. The active layer only appears in the place where there are transistors. The gate-grounded pull-up pMOS transistors can be replaced by precharge transistors to reduce the power dissipation. To keep the memory cell size and the bit-line capacitance small, the sizes of pull-down transistors should be kept as small as possible. The layout of the memory array shown in Figure 11.36(a) is revealed in Figure 11.36(b). Here, we do not show the pull-up devices for simplicity.

From the figure, we can see that each bit-line is a pseudo-nMOS NOR gate. To read a word $WL[i]$ from a NOR-type ROM, we apply a logic 1 to $WL[i]$ and a logic 0 to all other rows. The bit-line will receive a low voltage when its cell stores a logic 0 (that is, the presence of transistor) and a high voltage when its cell stores a logic 1 (that is, the absence of transistor). In practice, because of the large bit-line capacitance, current sense amplifiers are often employed to improve the speed of the read operation.

11.4.1.2 Via-Programming ROM Figure 11.37 shows an example of a 4×4 memory array structure of via-programming NOR-type ROM. The active layer appears uniform under the metal wire regardless of whether the transistor is required. The desired

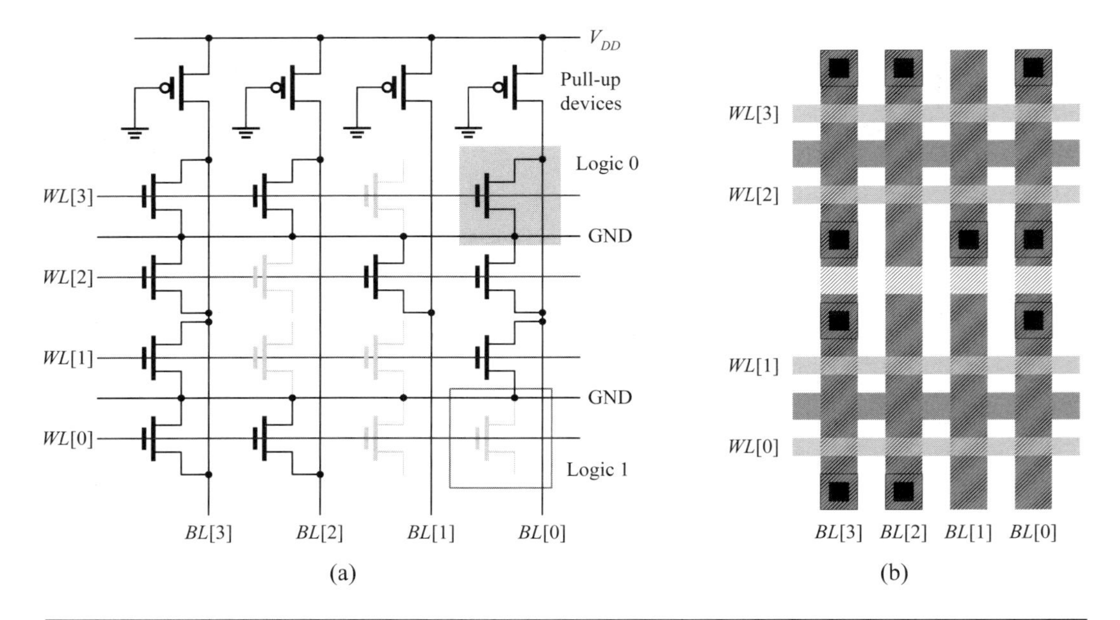

Figure 11.37: A 4×4 memory array structure of via-programming NOR-type ROM: (a) circuit; (b) layout.

transistors are then connected through contact or via connections in higher metal layers in modern embedded ROM modules. The layout of the memory array shown in Figure 11.37(a) is exhibited in Figure 11.37(b). Again, we do not show the pull-up devices for simplicity.

Because via layers are physically higher layers and manufactured later than the active layer, via-programming ROM has shorter turnaround time in manufacturing following ROM-code modification than active-programming ROM. For this reason, via-programming ROM is popular in modern embedded system designs.

11.4.2 NAND-Type ROM

In addition to the pseudo-NOR gate, the NAND gate structure can also be used to implement ROM functions. In the NAND-type ROM, each bit-line is a pseudo-nMOS NAND gate. An example is displayed in Figure 11.38(a), where a 4×4 memory array is assumed. The presence of a transistor between a word-line and a bit-line means that a logic 1 is stored and the absence of a transistor means that a logic 0 is stored. Figure 11.38(b) exhibits its corresponding layout using the via-programming method.

To read the stored value of a word, the word-lines must be operated in a logic mode opposite to the NOR-type ROM; namely, when reading word-line $WL[i]$, we apply a logic 0 to $WL[i]$ and a logic 1 to all other rows. The bit-line will receive a low voltage when its cell stores a logic 0 (absence of transistor) and a high voltage when its cell stores a logic 1 (presence of transistor). For instance, to read word-line $WL[1]$, the word-line enable signals need to be set as $WL[3]WL[2]WL[1]WL[0] = 1101$. The bit-lines outputs a value of 0110 due to a transistor appearing at the intersection of word-line $WL[1]$ and bit-line $BL[1]$ as well as the word-line $WL[1]$ and bit-line $BL[2]$. The transistor is off due to a zero-value voltage being applied to its gate.

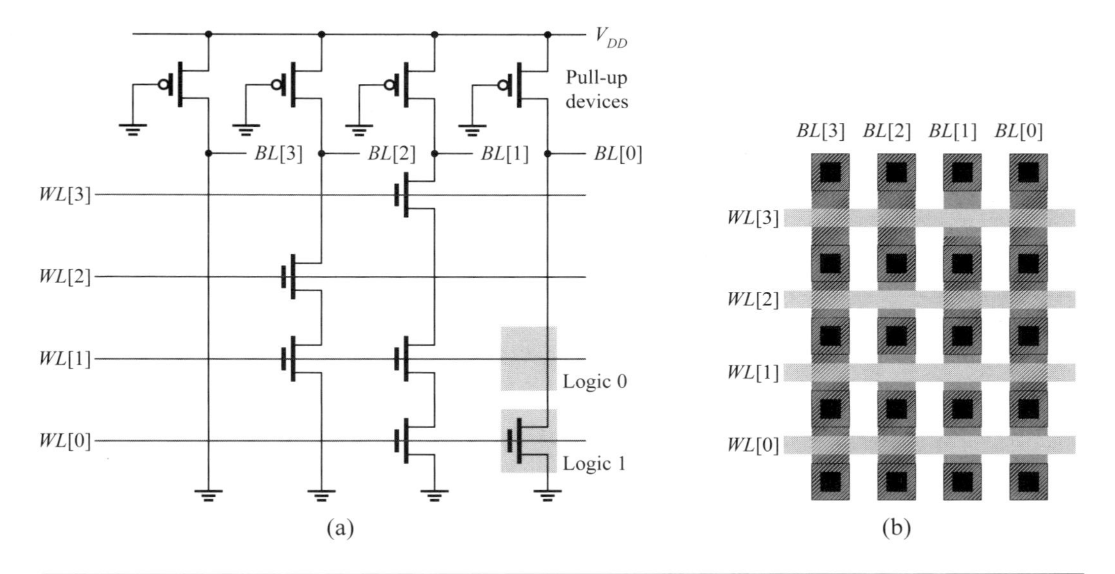

Figure 11.38: A 4×4 memory array structure of NAND-type ROM: (a) circuit; (b) layout.

When using the pseudo-nMOS NAND gate, the resistance of the pull-up devices must be much larger than the pull-down resistance to ensure an adequate low-level voltage. In addition, to keep the cell size and bit-line capacitance small, the pull-down transistors should be kept as small as possible. Hence, it requires a large amount of area for the pull-up devices. To overcome these disadvantages, nowadays most designs use precharge devices in place of the always-on pull-up pMOS transistors to reduce the required area, and use current sense amplifiers instead of voltage sense amplifiers to improve performance.

Compared to the NOR-type ROM, a NAND-type ROM needs less cell area but has inferior performance since many nMOS transistors are connected in series. However, because of the area benefit, the NAND-type structure has been popularized in today's Flash memory.

■ Review Questions

Q11-38. How would you read a word from a NOR-type ROM?

Q11-39. What are the features of NOR-type ROMs?

Q11-40. What is the distinction between active- and via-programming approaches?

Q11-41. How would you read a word from a NAND-type ROM?

Q11-42. What are the features of NAND-type ROMs?

11.5 Nonvolatile Memory

The basic architecture of *nonvolatile memory* (NVM), or also referred to as *nonvolatile read/write memory* (NVRWM), is virtually identical to that of ROM. Both NVM and ROM consist of an array of transistors placed on an intersection point of word-lines

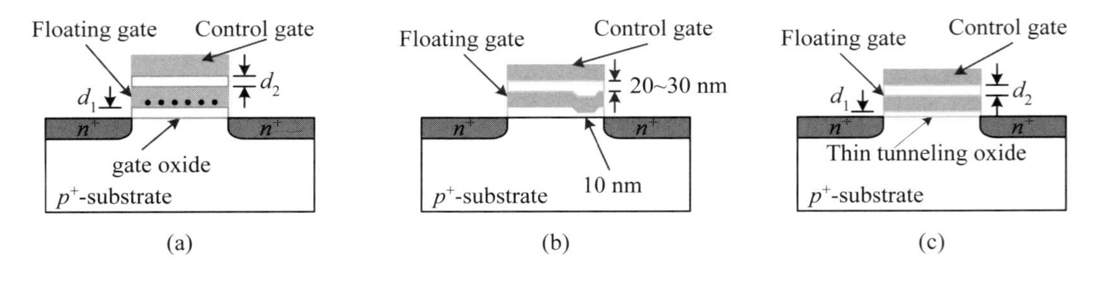

Figure 11.39: The floating-gate structures of NVM memory cells: (a) FAMOS transistor; (b) FLOTOX transistor; (c) ETOX transistor.

and bit-lines. The essential difference between them lies on the fact that ROM devices are mask programmable, whereas NVM devices are field programmable.

Historically, all NVM devices are based on the *floating-gate technique*, as shown in Figure 11.39, and can be classified into the following three types.

- *Erasable PROM* (EPROM): In EPROM, a *floating-gate avalanche-injection MOS* (FAMOS) transistor is used as the basic memory cell. An amount of excess charge is induced into the floating gate to change the threshold voltage of a FAMOS transistor by *hot-electron injection* (HEI), also called *hot-carrier injection* (HCI) or *channel hot-electron* (CHE) injection, and removed by an ultraviolet (UV) procedure.
- *Electrically erasable PROM* (EEPROM): In EEPROM, a *floating-gate tunneling oxide* (FLOTOX) transistor is used as the basic memory cell. The excess charge is induced and removed by a mechanism called *Fowler-Nordheim tunneling* (FNT).
- *Flash memory*: In Flash memory, an *extreme thin-oxide floating-gate* (ETOX) (a trademark of Intel) transistor is used as the basic memory cell. The excess charge is induced by HEI and removed by FNT.

Because both EPROM and EEPROM devices are only for historical interests, we only consider the Flash memory. In addition, two other types of nonvolatile memories, including *magnetoresistive RAM* (MRAM) and *ferroelectric RAM* (FRAM), that are also popular in the market recently, are also introduced briefly.

11.5.1 Flash Memory

We are concerned with the features of the memory cell, programming and erasing operations of Flash memory.

11.5.1.1 Memory Cell An ETOX transistor is essentially a MOS transistor with two gates: a *control gate* and a *floating gate*, as shown in Figure 11.39(c). The control gate is connected to an external circuit while the floating gate has no external connection. The rationale behind the use of a floating gate is due to the difficulty for a trapped electron in the floating gate to escape to its surrounding oxide because it must surmount the oxide energy barrier of 3.1 eV. The loss of excess electrons in the floating gate is very small. Thereby, it can retain a charge permanently. In industry, the word "permanently" is often used to refer to a period of over 10 years.

The threshold voltage of an ETOX transistor can be altered by the amount of charge stayed in the floating gate. Using the equation for threshold voltage described

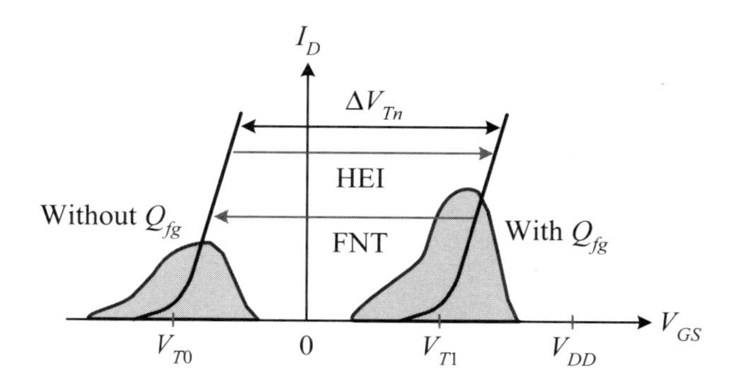

Figure 11.40: The threshold voltage change of the ETOX transistor before and after programming.

in Chapter 2, the new threshold voltage relationship for a FAMOS transistor can be expressed as

$$V_{Tn} = 2|\phi_{fp}| + \phi_{GS} - \frac{Q_{ss}}{C_{ox}} + \frac{|Q_{d(max)}|}{\varepsilon_{ox}}(d_1 + d_2) + \frac{|Q_{fg}|}{\varepsilon_{ox}} d_1 \tag{11.19}$$

where d_1 and d_2 are the dielectric thickness of floating and control gates, respectively, ranging from 30 nm to 10 nm. The amount of electron charge Q_{fg} causes the threshold voltage to be shifted to a more positive position, as shown in Figure 11.40. The change of threshold voltage is as follows:

$$\Delta V_{Tn} = \frac{|Q_{fg}|}{\varepsilon_{ox}} d_1 \tag{11.20}$$

To make the ETOX transistor able to store information, the floating gate is generally programed in a way to make two different threshold voltages, where one is below V_{DD} and the other is higher than V_{DD}. During normal operation, an ETOX transistor can be turned on by the normal V_{DD} if it is not programmed but cannot be turned on if it is programmed. Consequently, the information is stored in an ETOX transistor in the form of these two different threshold voltages. For example, the high threshold voltage represents logic 1 and the low threshold voltage denotes logic 0.

11.5.1.2 Programming Mechanisms To program an ETOX transistor, either an HEI or FNT approach can be used. Nonetheless, most Flash memory devices use an HEI approach to program the devices and FNT approach to erase their contents.

HEI mechanism. To program an ETOX transistor, a high voltage V_{pp}, say, 20 V, is applied to both drain and control gates, as shown in Figure 11.41(a). These two high voltages are used to build up hot electrons and direct the generated hot electrons toward the floating gate through the silicon dioxide. The number of electrons trapped in the floating gate is proportional to the programming time, which can be in turn determined by measuring the threshold voltage at the control gate since the electrons in the floating gate raise the threshold voltage of the control gate. The programming procedure is complete as the threshold voltage of the control gate is far above V_{DD}. After programming, the voltage at the floating gate becomes negative, as shown in Figure 11.41(b), because of the negative charge residing on it.

To read the information from an ETOX transistor, a normal voltage is applied to the control gate, as shown in Figure 11.41(c). As we described before, the ETOX transistor is turned on if it is not programmed but remains off if it is programmed.

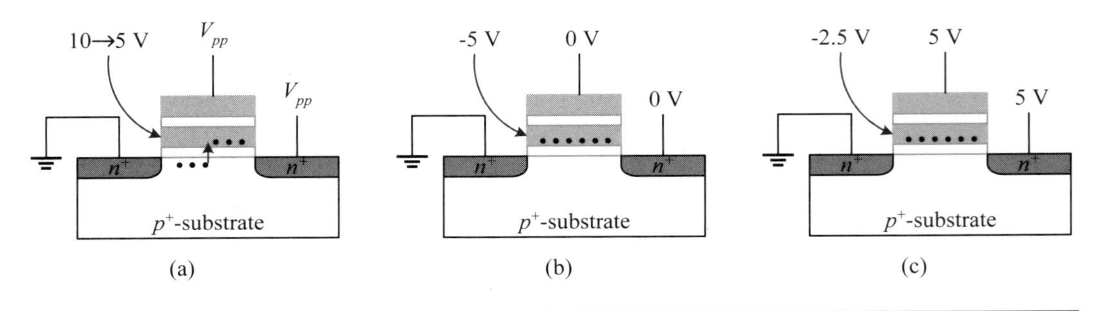

Figure 11.41: The operation of HEI programming: (a) hot-electron injection; (b) after programming; (c) read operation.

FNT mechanism. To program an ETOX transistor with an FNT mechanism, a high programming voltage V_{pp} between 15 V to 20 V, for a oxide thickness of 10 nm, is applied to the control gate of the ETOX transistor, as shown in Figure 11.42(a). Drain as well as both p-well and n-well are grounded; the source is left unconnected, namely, floating. The electrons from the p-well tunnel through the thin oxide via an FNT mechanism and accumulate on the floating gate. This programming process is self-limiting because the amount of tunneling current falls as the electrons accumulate on the floating gate. An ETOX transistor can also not be programmed by raising the drain to a higher voltage above the ground level as its control gate is at the high voltage V_{pp}.

To remove electrons from the floating gate, a positive high voltage V_{pp} is applied to both the p-well and n-well while both the source and drain are left floated; the control gate is grounded, as shown in Figure 11.42(b). The electrons on the floating gate tunnel through the thin oxide to the p-well via an FNT mechanism. Again, this erasing process is self-limiting because as the positive charge accumulates on the floating gate, the amount of tunneling current drops accordingly. However, if the erasing time were too long, a significant amount of positive charge would accumulate on the floating gate, leading to the decrease of threshold voltage. Because of this along with permanently trapped electrons in the SiO_2, repeated programming will gradually cause the shift of threshold voltage, deteriorate the performance of the ETOX transistor, and eventually cause a malfunction or the inability to reprogram the transistor.

Like the other two floating-gate transistors, an ETOX transistor stores the information in the form of different threshold voltages. The programmed cell has a higher threshold voltage, while an unprogrammed one has a lower threshold voltage. It is often the rules to use a programmed cell to represent logic 1 and an unprogrammed one to denote logic 0. The threshold voltage at the end of erasing process strongly depends on the initial threshold voltage of the cell and variations in the oxide thickness. The common solution for this is to first program the cells in the array to make all cells have about the same threshold voltage before applying the erase pulse. Since the access transistor is removed, the erasing process in a Flash memory is performed in bulk for the complete chip or for a subsection of the memory.

■ Review Questions

Q11-43. Define nonvolatile memory (NVM).

Q11-44. What is the rationale behind the floating-gate technique?

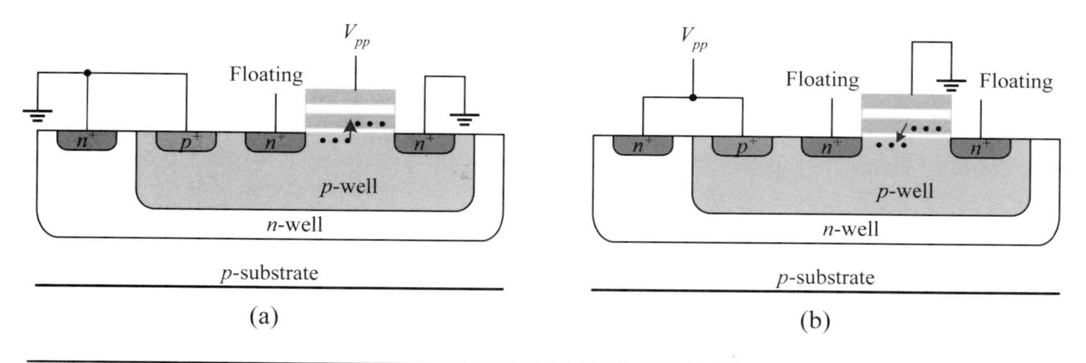

Figure 11.42: The (a) FNT programming and (b) erasing operations of ETOX transistor.

Q11-45. Describe the hot-electron injection (HCI).

Q11-46. Describe Fowler-Nordheim tunneling (FNT).

11.5.1.3 Memory Architectures The architecture of Flash memory can be categorized into the following two types: *NOR-based* and *NAND-based*. The NOR-based Flash memory was proposed by Intel in 1988. It has a faster random access time, but each cell occupies more area than the NAND-based one. NAND-based Flash memory was developed by Toshiba in 1989. Because of the inherent feature of drain/source sharing between adjacent bits, each cell of the NAND-based type takes up less area than that of the NOR-based type. Nonetheless, NAND-based Flash memory has slower random access time compared to the NOR-based Flash memory. Despite this, NAND-based Flash memories have found their widespread use in storing media files, that is, for storing bulk data, such as the MP3 and digital video, and even used as a *solid-state disk* (SSD) for replacing the hard disk (HD) in a number of desktop and laptop computers.

Recently, for high-density NAND Flash memories, a *multilevel* (usually four or eight levels, even more) *cell* (MLC) is used for replacing a *single-level cell* (SLC). MLC uses a multilevel sense amplifier to identify many different charge levels in the cell and an error-correcting code (ECC) algorithm to compensate for the possible bit errors. Because the increase of trapped electrons in SiO_2 after repeated program/erase cycles causes the shifted threshold voltage to be indistinguishable, MLC has less write/erase cycles per cell compared to SLC. In general, an SLC has an endurance level of 100,000 write/erase cycles while an MLC only has an endurance level of 10,000 write/erase cycles, much less than SLC.

NOR Flash memory. NOR-based Flash memory has the same circuit structure as NOR-based ROM except that in Flash memory a Flash cell must be placed at every intersection point of the word-line and bit-line, as depicted in Figure 11.43. Like in ROM, precharge transistors are used as pull-up devices in practical Flash memory devices to save power dissipation.

To write data into a NOR-based Flash memory with an HEI mechanism, as illustrated in Figure 11.43, the source voltage is connected to ground and a programming high voltage V_{pp}, a value much less than V_{DD}, is applied to the word-line. Those bit-lines with their associated bits in the word to be programmed as logic 1 are set to a high value of V_d (for example, V_{pp}), and to be programmed as logic 0 are grounded (0 V). Refer to Figure 11.41 for more details.

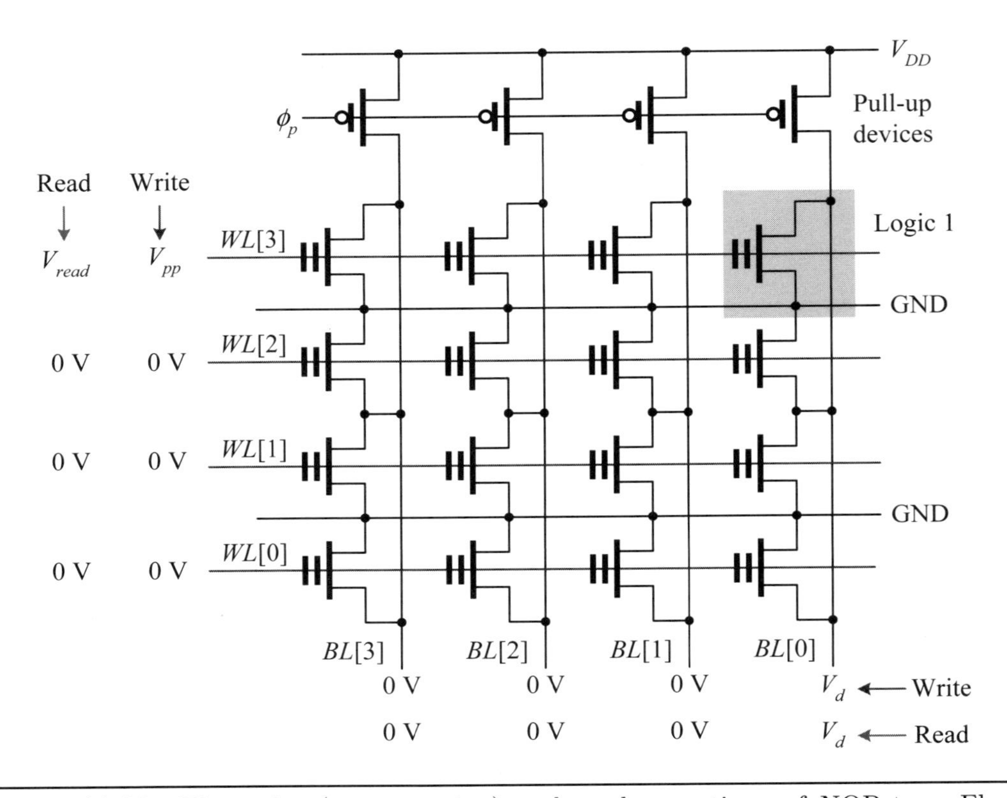

Figure 11.43: The write (programming) and read operations of NOR-type Flash memory.

The read operation of NOR-based Flash memory is performed by grounding the source connection of the selected word, precharging the bit-line to a high voltage of V_d, and activating the word-line with a high voltage V_{read}, a value between two threshold voltages such as V_{DD}, as illustrated in Figure 11.43. If the ETOX transistor is turned on, then a drain current exists, and the bit-line is interpreted as a logic 0; otherwise, the bit-line is referred to as a logic 1.

During the erasing process, all devices connected to the same source line are erased at the same time. As shown in Figure 11.44, to activate the FNT mechanism, a high voltage of V_s is applied to the source and the ground level is applied to both gate and drain. As a consequence, the erasing operation is always carried out in the block mode.

NAND Flash memory. The advantage of NOR-based architecture is that it has a fast random access time. Nonetheless, many applications require large storage density, fast erasing and programming operations, and fast serial access. These requirements lead to the NAND-based architecture, whose cell size is 40% smaller than the NOR cell due to the feature of drain/source sharing. Figure 11.45 shows both layout structure and a cross-sectional view of a cell module of typical NAND-based Flash memory. The oxide-nitride-oxide (ONO) dielectric is often used between floating and control gates of a flash cell to increase Q_{fg}.

To illustrate the operation of a NAND-type Flash memory, assume that FNT is used for both programming and erasing mechanisms. During the writing (programming) process, a high programming voltage V_{pp} is applied to the word-line of interest, say, *WL*[0], and V_{DD} is applied to all unselected word-lines. The drain select line (*DSL*) is

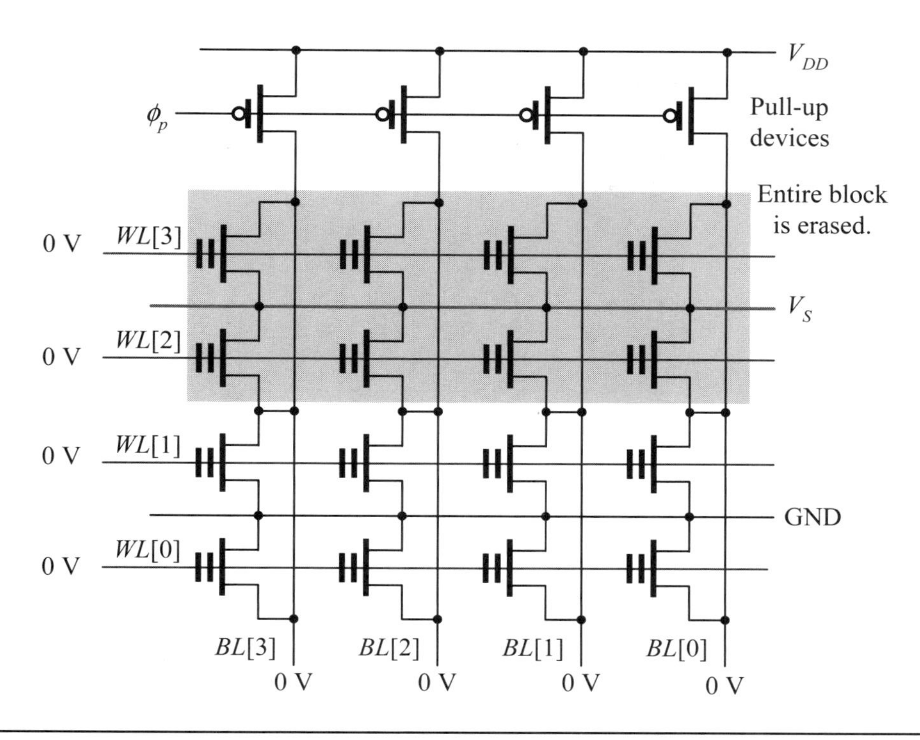

Figure 11.44: The erasing operation of NOR-type Flash memory.

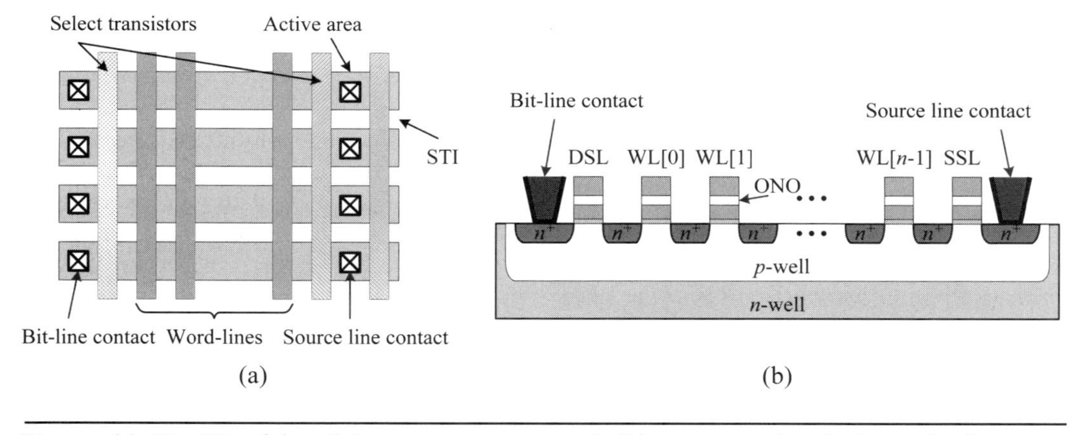

Figure 11.45: The (a) cell layout structure and (b) cross-sectional view of cell structure of NAND-type flash cells. (Both p^+ to p-well and n^+ to n-well connections are not shown.)

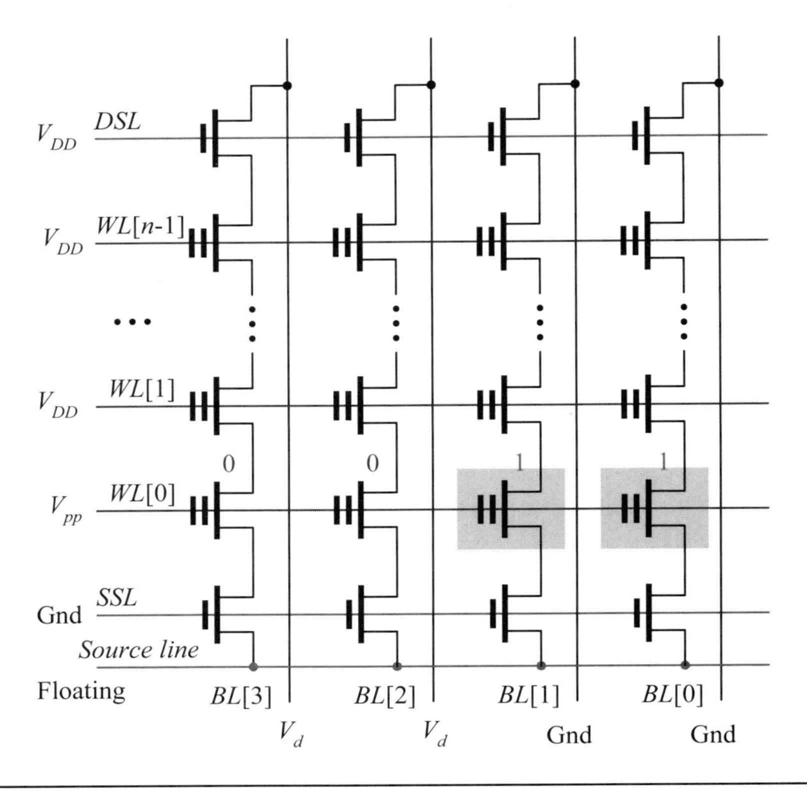

Figure 11.46: The write operation of NAND-type Flash memory.

connected to V_{DD}, while the p-well, n-well, and source select line (*SSL*) are grounded. For those bits to be programmed, their associated bit-lines are grounded; for those bits not to be programmed, their associated bit-lines are connected to a moderate voltage V_d. As illustrated in Figure 11.46, the shaded two cells are programmed and hence their threshold voltages are shifted to a higher value. During the reading process, these two cells are recognized as logic 1, and the other two cells are recognized as logic 0 because they are not programmed. The source line is floating to eliminate the possible DC-current path from the bit-line to ground.

During the reading process, all word-lines except the one of interest as well as both *DSL* and *SSL* are connected to V_{DD}, which is greater than the threshold voltage of programmed cells. The selected word-line, p-well, and n-well are grounded. In addition, the source line also needs to be grounded so that the related cell module can function as a NAND gate. As illustrated in Figure 11.47, bit-lines *BL*[3] and *BL*[2] will result in larger currents because these two cells are not programmed and turn on; *BL*[1] and *BL*[0] will result in smaller currents since these two cells are programmed and remain off.

■ Review Questions

Q11-47. Compare the features of both Flash memory architectures.

Q11-48. Compare the features of the multilevel cell and single-level cell.

Q11-49. Explain the read operation of NAND-based Flash memory.

Q11-50. How would you write data into a NAND-type Flash memory?

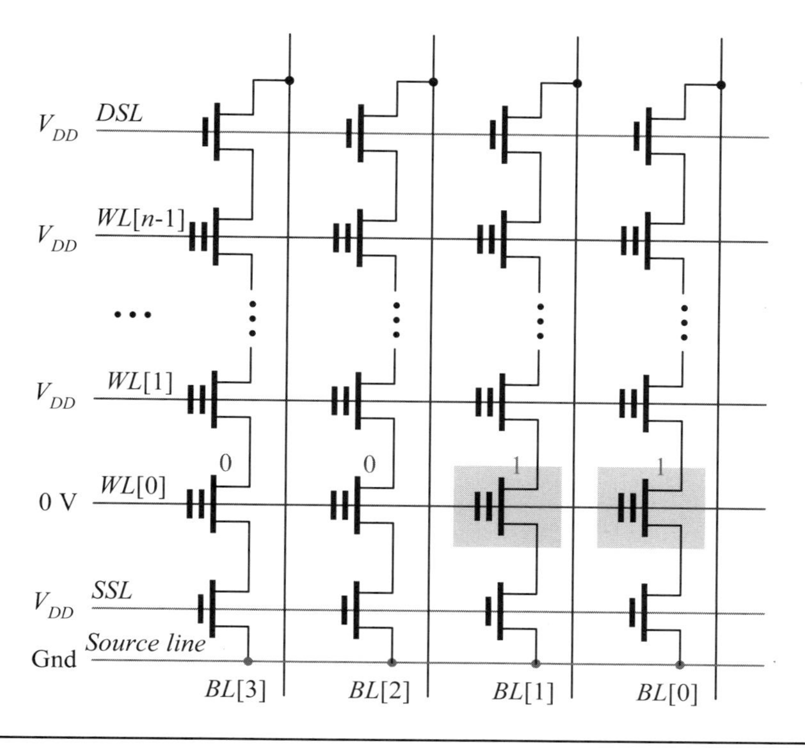

Figure 11.47: The read operation of NAND-type Flash memory.

11.5.2 Other Nonvolatile Memories

In addition to the aforementioned nonvolatile memory devices, magnetoresistive RAM (MRAM) and ferroelectric RAM (FRAM) are the other two widely used nonvolatile memories in the market. MRAM promises to become the universal memory to replace DRAM, SRAM and Flash memories and is the enabling technology for computer chip systems on a single chip. We briefly introduce the basics of these two new-type NVM devices.

11.5.2.1 Magnetoresistive RAM The important features of MRAM are that information is stored as magnetic polarization instead of a charge and the state of a bit is detected as a change in resistance. Most industrial magnetoresistive RAM (MRAM) devices are based on the *tunneling magnetoresistance* (TMR) effect.

The TMR effect exploits the quantum mechanical phenomenon between two ferromagnetic plates. As illustrated in Figure 11.48, in this structure two ferromagnetic plates are separated by a very thin insulator tunnel barrier, about 1.5 nm in thickness. The resistance is low if the magnetizations in two plates are in parallel because the majority spin-aligned electrons in two plates will tunnel more readily across the tunnel barrier and the resistance is high if the magnetizations in two plates are antiparallel since the majority spin-aligned electrons in one plate are inhibited from tunneling to the other plate. Hence, two different states are obtained. In practice, one ferromagnetic plate is fixed and the other is made free to store information. The resulting structure is often called a *magnetic tunnel junction* (MTJ).

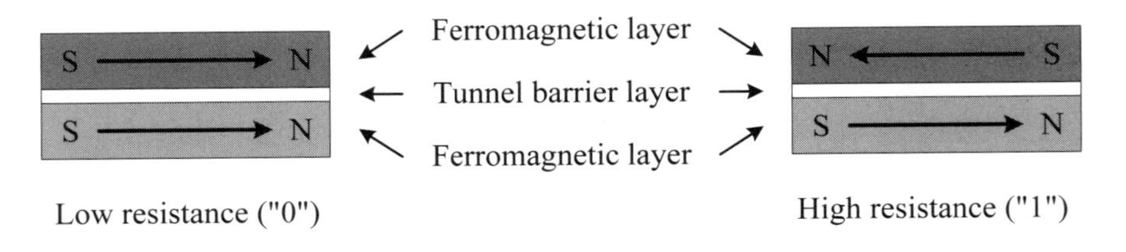

Figure 11.48: The TMR effect.

The TMR effect can be characterized by a metric, referred to as the *TMR ratio*, which is defined as

$$TMR = \frac{R_{AP} - R_P}{R_P} \tag{11.21}$$

where R_{AP} and R_P denote the resistances for antiparallel and parallel magnetization configurations between two ferromagnetic layers, respectively. For an MTJ with an MgO barrier layer and a dimension of 125 nm $\times$ 220 nm, the R_{AP} and R_P resistances are about 6.5 kΩ and 2.5 kΩ, respectively, leading to a typical TMR of 160%.

Most commercial MRAM devices are fabricated with standard CMOS processes as a "back-end" module, being integrated between the last two metal layers. Here, the "back-end" means that it is inserted after all of the associated CMOS circuits are fabricated. Therefore, it separates the specialized magnetic materials processing from the standard CMOS process and requires no alteration to the front-end CMOS process flow.

Depending on the ways of changing the MTJ magnetoresistance, many different MRAM structures are proposed over the past decade. We will introduce two of these: *toggled MRAM* and *spin-torque transfer MRAM* (STT-MRAM).

Toggled MRAM. The toggled MRAM cell has a structure shown in Figure 11.49. The MTJ is composed of a fixed magnetic layer, a thin dielectric tunnel barrier, and a free (storage) magnetic layer. During the read operation, as shown in Figure 11.49(a), the access transistor of the target bit is turned on to bias the MTJ, and the resulting current is compared to a reference to determine if the resistance state is low or high. During the write operation, as shown in Figure 11.49(b), the access transistor is turned off. A write operation involves driving a current unidirectionally down the digit line and bidirectionally along the bit-line. The hard magnetic field, controlled by the digit line current I_{HARD}, is used to reduce the strength of the soft magnetic field, controlled by the bit-line current I_{SOFT}, that is, required to switch between the high and low resistance states of the MTJ cell.

Spin-torque transfer MRAM (STT-MRAM). In STT-MRAM, data are written by applying spin-polarized currents through the MTJ element to change the magnetized orientation of the free magnetic layer. Because it does not use an external magnetic field, STT-MRAM consumes less power and is more scalable than toggled MRAM. Currently, STT-MRAM with an access time of 2 ns under a writing current of 100 μA is possible.

The cell structure of STT-MRAM is depicted in Figure 11.50(a), which is one transistor and one MTJ design (1T1J). The major challenge for implementing the STT writing mode in high-density and high-speed memory is the reduction of the switching current, which in turn reduces the size of the selection transistor in series with the MTJ element.

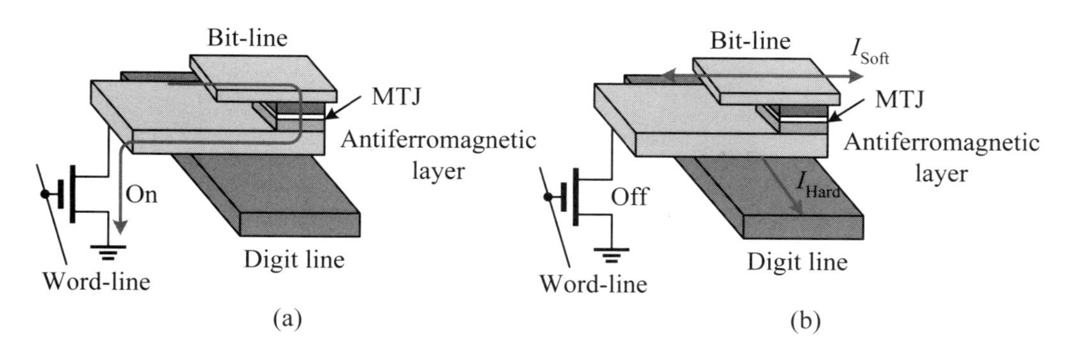

Figure 11.49: The (a) read and (b) write operations of toggled MRAM.

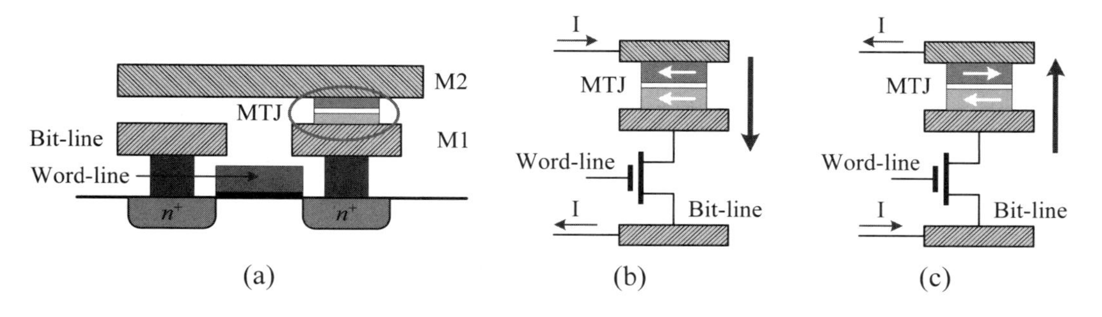

Figure 11.50: The (a) cell structure, (b) "0" write, and (c) "1" write of STT-MRAM.

The operation of STT-MRAM is as follows. The *spin-torque transfer effect* (STT effect) means that the magnetization direction in the MTJ free magnetic layer can be simply reversed by a spin-polarized current flowing through the junction. This switching mechanism occurs as the current density exceeds a critical density value, about 8×10^5 A/cm^2, corresponding to a critical current of a few hundred μA when the MTJ has a size of 125 nm $\times$ 220 nm. Hence, a simple minimum-sized CMOS current source can be used to write data into the MTJ element. Figures 11.50(b) and (c) show the current directions of writing data 0 and 1 into the MTJ element, respectively.

Since STT-MRAM retains all the good features of toggled MRAM, has no addressing errors, is capable of multibit (parallel) writing, has low power dissipation, and has a potentially high integration level and scalability, in particular, below 45 nm, it promises to replace existing DRAM, SRAM, and Flash memories and become a universal memory in the near future.

11.5.2.2 Ferroelectric RAM

Ferroelectric memory (FRAM) is a device that uses the *ferroelectric effect* as the charge (that is, information) storage mechanism. The ferroelectric effect is the ability of a material to store electric polarization in the absence of an applied electric field. The FRAM cell is similar to a DRAM cell except that the dielectric material of the capacitor is replaced by a ferroelectric material, such as *strontium bismuth titanate* (SBT) ($SrBi_2Ta_2O_9$) and $BaTiO_3$. Such a capacitor is usually called a *ferroelectric capacitor* and the material is known as a *Perovskite crystal material*, which can be polarized in one direction or the other.

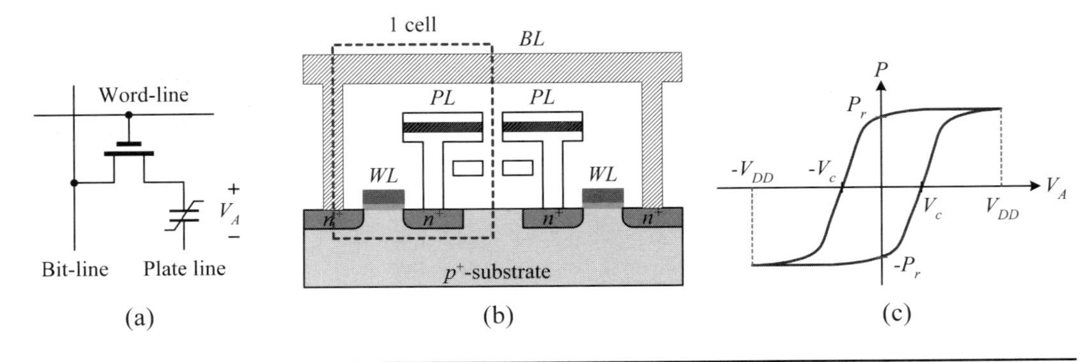

Figure 11.51: FRAM: (a) cell circuit; (b) cell layout; (c) polarization characteristics.

Table 11.2: Comparison of various nonvolatile memories.

	Cell Area	Program	Erase	Erase resolution	Cycles
ROM	Small	NA	NA	NA	0
EPROM	Small	HEI	UV exposure	Full memory	100
EEPROM	Large	FNT	FNT	Bit/Byte	10^4–10^5
Flash memory	Small	HEI/FNT	FNT	Block	10^4–10^5
FRAM	Small	Polarization	Polarization	Bit	Unlimited
MRAM	Small	Polarization	Polarization	Bit	Unlimited

Figure 11.51(a) gives the circuit symbol and circuit structure of using a ferroelectric capacitor as a memory cell. The cell layout is shown in Figure 11.51(b). The hysteresis loop of the polarization characteristics of the ferroelectric capacitor is shown in Figure 11.51(c). As the voltage is applied to a ferroelectric capacitor in one direction or the other, the crystals polarize and hold the polarization state after the voltage is removed, thereby creating a nonvolatile memory. During the write operation, the ferroelectric capacitor is polarized in one direction (V_{DD}) to store a logic 1 and the opposite direction ($-V_{DD}$) to store a logic 0.

During the read operation, the bit-line (BL) is precharged to a high-level voltage and the word-line (WL) is enabled. The current level is determined by the stored polarization direction, P_r or $-P_r$. A high-level current is obtained if the dipoles switch their directions and a low-level current is achieved otherwise. Like the DRAM cell, the read-out operation of a FRAM cell is destructive. Hence, a write-back operation is needed to write back the information of the cell that it has been read out.

11.5.2.3 Comparison of Nonvolatile Memories The features of various nonvolatile read/write memories (NVM) are summarized in Table 11.2 for comparison.

■ Review Questions

Q11-51. What is the tunneling magnetoresistance (TMR) effect?

Q11-52. Define the TMR ratio.

Q11-53. Describe the magnetic tunnel junction (MTJ) structure.

Q11-54. Describe the feature of toggled MRAM.

Q11-55. Describe the feature of STT-MRAM.

Q11-56. What is the ferroelectric capacitor?

Q11-57. What is the Perovskite crystal material?

11.6 Other Memory Devices

In addition to the aforementioned memory devices, many other types of memory have found their widespread use in various digital systems. In this section, we are concerned with some of them, including *content-addressable memory* (CAM), registers and register files, *dual-port RAM*, *programmable-logic array* (PLA), and first-in first-out (FIFO).

11.6.1 Content-Addressable Memory

A content-addressable memory (CAM) is a memory that compares an input data against all data stored in a table in parallel, and returns the index (address) of the matching data. CAMs find broad use in a wide variety of applications requiring high-speed search operations. These applications at least include the following: cache memory in computer systems, Huffman coding/decoding, Lempel-Ziv compression, and image coding. Nowadays, one primary commercial application of CAMs is to classify and forward *Internet protocol* (IP) packets in network routers.

11.6.1.1 The Operation of CAM The conceptual illustration of the operation of a CAM is given in Figure 11.52, where a table of stored data contains n items with each consisting of a tag and a data item. It returns the associated data item when it matches any tag with the interrogating keyword. To determine whether the keyword is in the table, the table may be searched at least in either of two ways. One approach is to search the table in a way such that the keyword is compared against tags in the table one by one. Such an approach is called a *sequential search* and is the usual way performed in most computer systems using a software algorithm as the unsorted data are stored in the table. Of course, this sequential search can also be realized with a RAM and a finite-state machine without the use of a more powerful microprocessor.

The other approach is to compare the keyword with all tags in the table at the same time. This results in a concept and technique known as a *parallel search*. To facilitate such a capability, a special hardware is needed. As indicated in Figure 11.52, the table is indeed composed of two parts: a tag part and a data item part. Both parts are a kind of memory array. However, the tag part needs the capability of a parallel search, whereas the data item part does not. The data item part simply outputs a data item whenever its associated tag matches the keyword. A memory array capable of a parallel search is referred to as a *content-addressable memory* (CAM). It is also called *associative memory* because the data associated with a tag is referenced whenever the stored tag matches the interrogating keyword.

11.6.1.2 CAM Organization The organization of a CAM is almost the same as that of a SRAM except that some minor modifications are made. As shown in Figure 11.53, a CAM comprises a row decoder, an array of memory cells, a write I/O and tag driver, and a priority encoder as well. Like SRAM, it needs to store data for looking up later; namely, it needs the write capability. Nevertheless, instead of a read-out operation, an interrogation is used to search the data. A match line for each word in the memory array is utilized to indicate whether the situation is matched. The match line is asserted as the word matches the interrogating keyword and is not asserted otherwise.

Keyword

Tag 1	Data item 1
Tag 2	Data item 2
Tag 3	Data item 3
•••	•••
Tag n-1	Data item n-1
Tag n	Data item n

Figure 11.52: The use of associative memory as a lookup table for performing a parallel search.

The matched signal, if any, is then decoded into a binary address, called the *matched address*, representing the matched word, along with the true *Valid* signal. If there is no matched word in the memory array, the valid signal *Valid* is set to false (logic 0) to invalidate the output matched address.

Because replica data might appear in the memory array, a situation of multiple matches may arise, leading to a wrong matched address. To solve this problem, a priority encoder is needed. With a predetermined priority, the matched address can then represent a unique matched word.

11.6.1.3 CAM Cells In addition to the regular write operation, each CAM cell needs the capability of interrogation. To facilitate this, some modifications of existing SRAM cell or a new type of memory cell must be made. The most widely used approaches at present are to extend the capability of the SRAM cell by adding some transistors. Figure 11.54 shows two CAM cell structures, being extended from the traditional 6T-SRAM cell. Figure 11.54(a) shows an *XOR-based CAM cell*, consisting of 10 transistors and Figure 11.54(b) is the *9T-CAM cell*.

The XOR-based CAM cell works as follows. To write data into the CAM cell, both bit-lines must start at a low-level voltage to keep the XOR gate off to avoid inadvertently pulling the match line to a low-level voltage. For instance, as shown in Figure 11.54(a), if the nMOS transistor M_7 is on and the $\bar{Q}$ is high, then the match line will be pulled to a low-level voltage. To complete the write operation, one of bit-lines rises to a high value.

The comparison is carried out as follows. The match line is precharged to a high-level voltage. The match line will be pulled to a low-level voltage if the bit-line values match the stored values and will remain at the high-level voltage otherwise. For example, let Q be 0 and $\bar{Q}$ be 1. If *Bit* and $\overline{Bit}$ have values of 0 and 1, respectively, then both nMOS transistors M_7 and M_8 will be on to pull the match line to a low-level voltage. On the other hand, if *Bit* and $\overline{Bit}$ have reverse values, the match line will remain at the high-level voltage because one of the stacked transistors associated with each side is off.

The 9T CAM cell works as follows. It has exactly the same write operation as the 6T-SRAM cell. The comparison logic is performed by the two nMOS transistors, M_8 and M_9. To interrogate the bit value, the match line is first precharged to a high-level voltage. The match line will be pulled to a low-level voltage if the bit-line values match the stored values and remain at the high-level voltage otherwise.

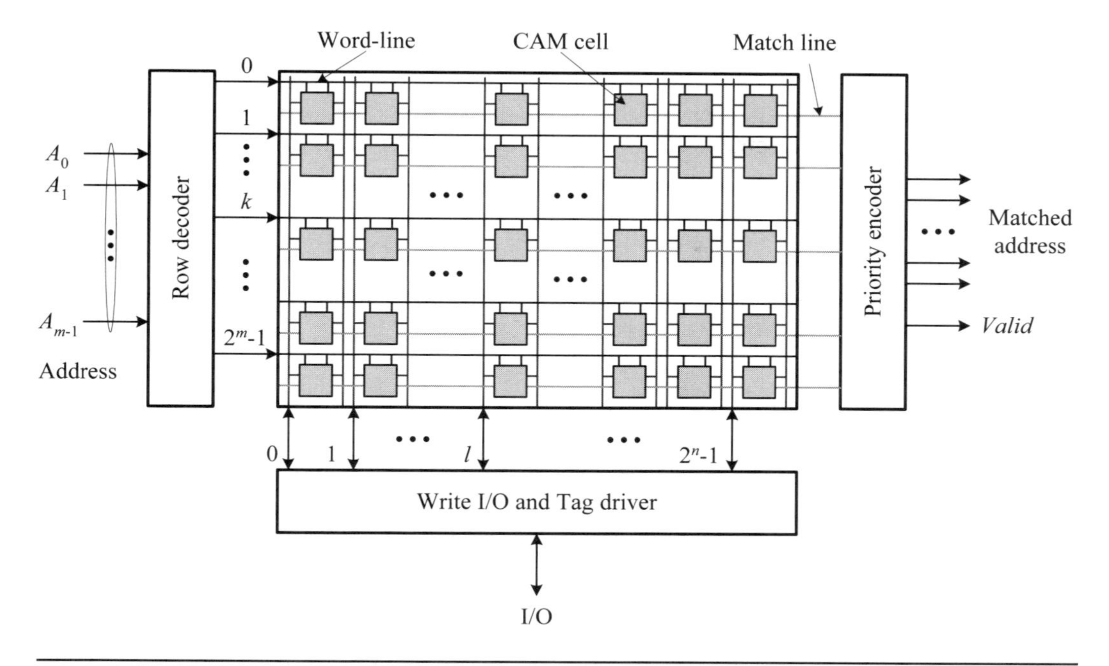

Figure 11.53: The organization of content-addressable memory (CAM).

11.6.1.4 A Combination of CAM and SRAM Figure 11.55 shows a combination of both CAM and SRAM memory arrays. This structure is widely used in many practical applications, such as the *translation-look-aside buffer* (TLB) used in the *memory management units* (MMUs) of computer systems for helping perform *virtual memory* or the *cache memory* used in the network routers for classifying and forwarding Internet protocol packets. This structure actually implements the lookup table described in Figure 11.52. The CAM memory array carries out the parallel search by comparing the incoming keyword with the stored tags. It generates a match-line output signal to drive the word-line of the data RAM memory array whenever a match occurs.

The selected word-line of the data RAM memory array needs to be enabled at the right time. This means that the word-line should not be activated until the CAM has completed its operation. To achieve this, a dummy row like the one described in Section 11.2.5.2, is used as a replica delay line to generate an enable signal with the worst-case timing of a match in the presence of PVT variations. This enable signal must be ANDed with each match line that drives the word-line.

Review Questions

Q11-58. What is content-addressable memory?

Q11-59. What is associative memory?

Q11-60. Explain why the priority encoder is needed in the CAM organization.

Q11-61. Explain why the *Valid* signal is needed in the priority encoder.

Q11-62. Explain the operation of the XOR-based and 9*T* cells of CAM.

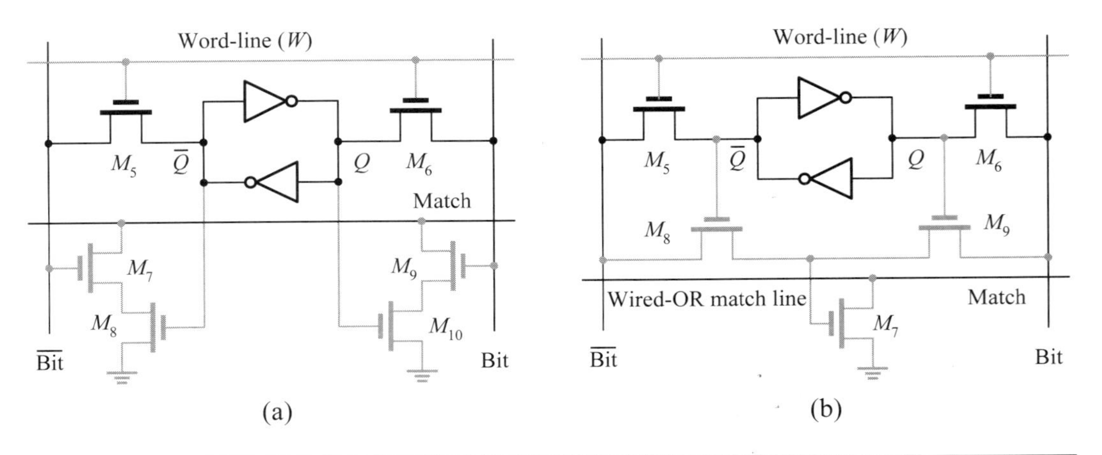

Figure 11.54: The (a) XOR-based and (b) $9T$-cell structures of CAM.

11.6.2 Register Files

In most digital systems, registers and register files are common in their datapaths. Hence, we introduce an example of a multiport register file and its associated register/memory cell.

11.6.2.1 Register Files One popular use of registers in datapaths is to construct a register file, which is a set of registers with multiple access ports. An example is given in Figure 11.56(a), which shows the role of a register file in a datapath consisting of an *arithmetic and logic unit* (ALU). The register file needs to provide two data operands at a time for the ALU and receives a result from the ALU after the ALU completes its operation. Therefore, the register file has to provide two concurrent read ports and one write port.

The general block diagram of a register file with two-read and one-write ports is illustrated in Figure 11.56(b). The core of register file is a set of n m-bit registers, in which each bit can be a flip-flop or a memory cell. There are three sets of address and data buses, two sets for two read ports and one set for the write port. To route data from one register of the register file to a read port, an m-bit n-to-1 multiplexer is required. Hence, two such multiplexers are required, one for each read port. The write port does not need any multiplexer because the destination register is controlled by the write enable signal, which is in turn driven by the write address decoder. In addition to the n m-bit registers, the m-bit n-to-1 multiplexers also contribute a large part to the cost of multiple-port register file, in particular, when the number of read ports is more than two. Hence, in practice the multiplexing operation is often incorporated into the register/memory cell to reduce the hardware cost and propagation delay.

11.6.2.2 Register/Memory Cells Since a flip-flop usually needs much more area than an SRAM cell, SRAM cells are usually used as the basic building blocks in most high-speed full-custom designs. Hence, in what follows we are concerned with two typical multiple-port memory cells.

Figure 11.57 shows the structure of a dual-port memory cell. It is an extension from the basic 6T-SRAM cell by adding an extra set of word-line and bit-lines as well as access transistors. The read and write operations for each set is exactly the same as that of a 6T-SRAM cell. Hence, we omit it here.

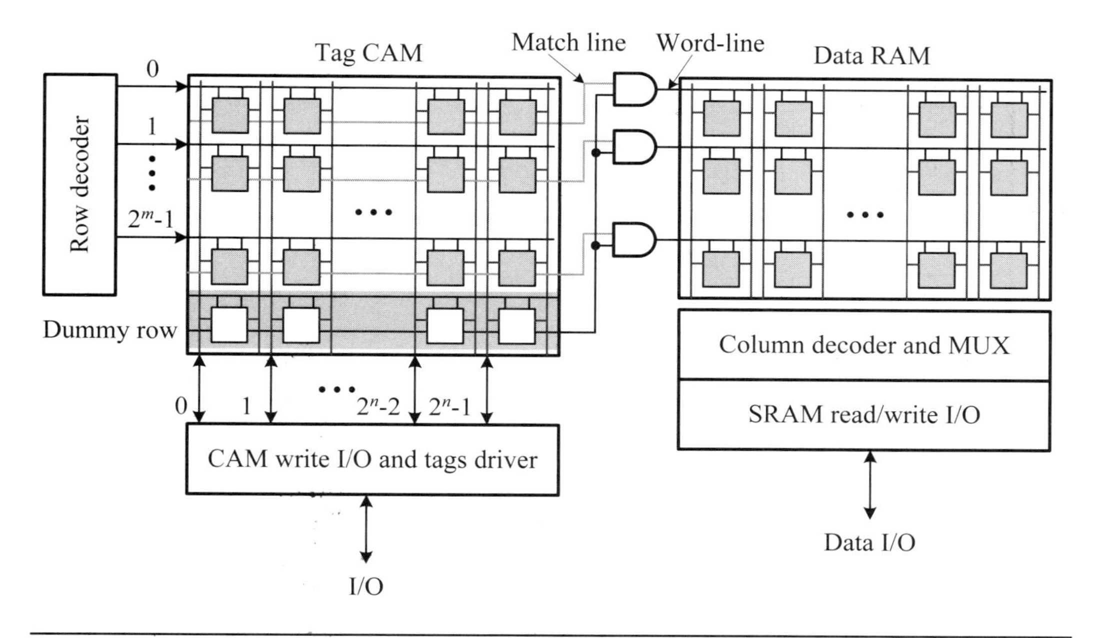

Figure 11.55: The combination of CAM and SRAM memory arrays.

The other circuit is illustrated in Figure 11.58. It is also extended from the basic 6T-SRAM cell. The nMOS transistors depicted on the lower right side will quickly write a logic 1 into the bit cell. In order to ensure the speedy latching of a logic 0, a stack of two nMOS transistors is used to allow data to be written into both sides of the cross-coupled inverters at the same time. The high-to-low transition is of particular importance because if a logic 1 value lingers too long on the latch, the associated bit-line will start a false transition.

■ Review Questions

Q11-63. Define a register file.

Q11-64. How would you implement an n m-bit register file with three read ports and one write port?

Q11-65. Explain the operation of the dual-port register cell structure shown in Figure 11.57.

Q11-66. Explain why an m-bit n-to-1 multiplexer will contribute a large amount of cost to a register file.

11.6.3 Dual-Port RAM

A dual-port RAM is a random-access memory that can be concurrently accessed by two independent sets of address, data, and control lines, as shown in Figure 11.59. The essential parts of a dual-port RAM are dual-port memory cells as well as control and address arbitration logic. The dual-port memory cells are used to store information while the control and address arbitration logic provides the concurrent access capability of the two ports.

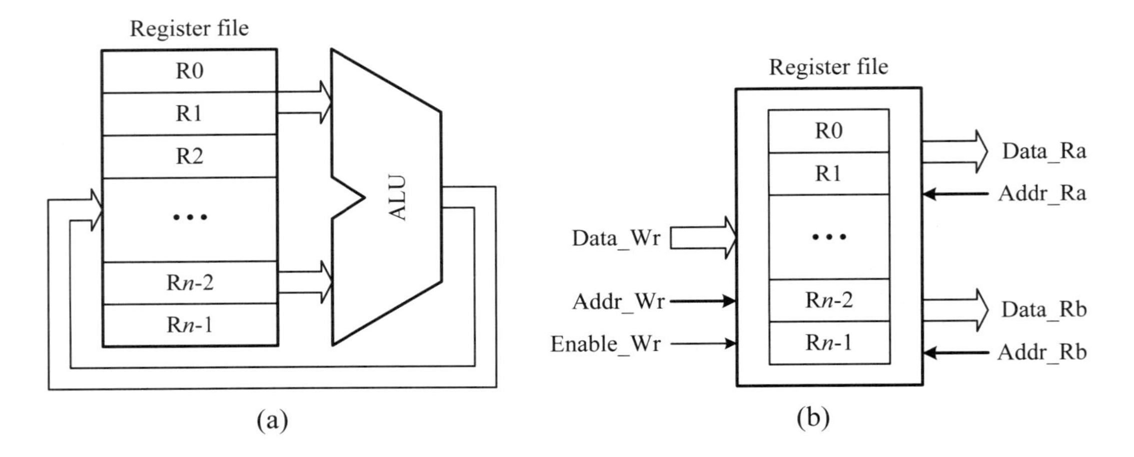

Figure 11.56: An application and the structure of a register file: (a) register and ALU; (b) block diagram of a register file.

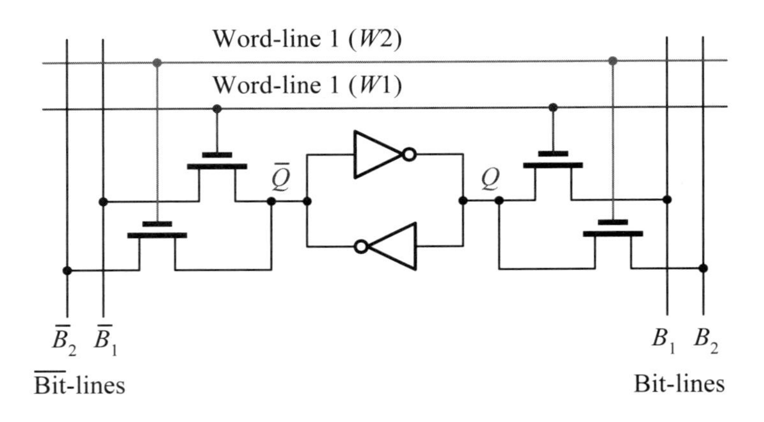

Figure 11.57: A dual-port register cell structure.

In designing a dual-port RAM with dual-port memory cells, care must be taken to cope with the possible conflict occurring when two ports are trying to access the same cell in the RAM memory array. These conflicts include the following three cases:

- *Both ports reading*: If both ports of a dual-port RAM are reading the same location at the same time, then they are reading the same data.
- *One port reading and the other port writing*: The result of arbitration logic will allocate priority to either the reading or the writing port. If the losing port is attempting to write data, the write is inhibited to prevent the data in memory from being corrupted. The $\overline{BUSY}$ flag to the losing port signals that the write was not performed. If the losing port is attempting to read data, the $\overline{BUSY}$ flag to the losing port indicates that the old data is still being read.
- *Both ports writing*: Only one port is allowed to proceed in its operation. The losing port determined by arbitration logic is prevented from writing so that the data would not be corrupted. $\overline{BUSY}$ is asserted to the losing port, indicating that the write operation was unsuccessful.

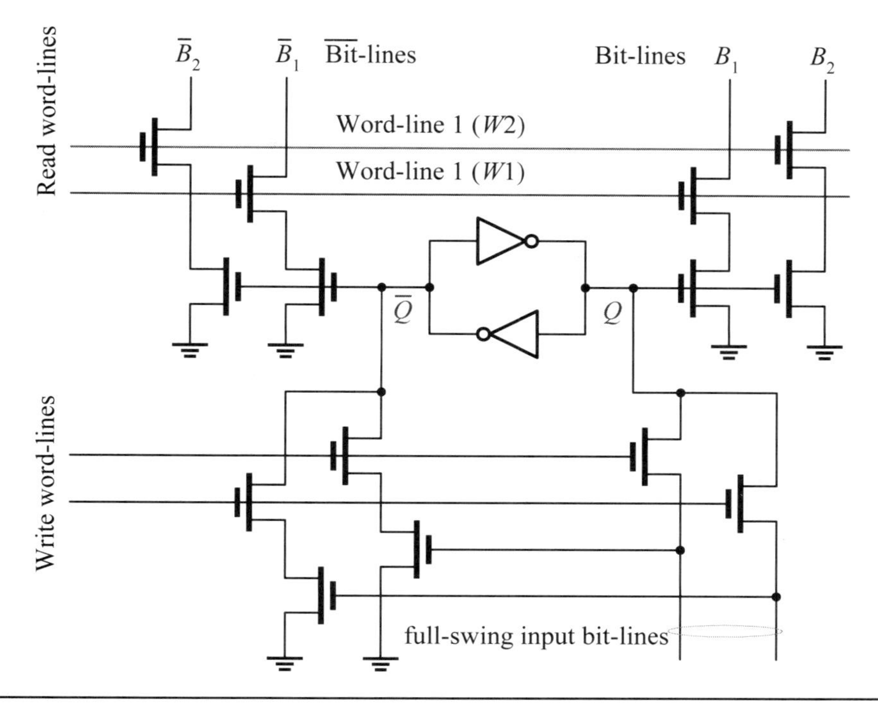

Figure 11.58: A dual-port register cell structure.

11.6.3.1 Arbitration Logic The *arbitration logic* is essential in a dual-port or multiport memory in which concurrent read and write operations are needed. The arbitration logic has three functions: to decide which port wins and which port loses if the addresses are identical at the same time, to prevent the losing port from writing, and to issue a busy signal to the losing port. As shown in Figure 11.60, the arbitration logic consists of two address comparators A and B, with their associated delay buffers; the arbitration latch formed by the cross-coupled, three-input NAND gates labeled L and R; the gates that generate the $\overline{BUSY}$ signals.

When the addresses of A and B ports are not identical, the outputs of the address comparators A and B are both low, while the outputs of gates L and R are both high. This condition results in all $\overline{BUSY}$ and internal write inhibit signals to be inactive. At this time, the arbitration latch does not function as a latch. On the contrary, when both ports access the same memory location simultaneously, the outputs of gates L and R go from low to high at exactly the same instant. The arbitration latch settles into one of two states and determines which port wins and which port loses. Hence, one set of $\overline{BUSY}$ and internal write inhibit signals is activated. The latch is designed in a way such that its two outputs are never low at the same time.

■ Review Questions

Q11-67. How would you realize a dual-port RAM with a single-port RAM?

Q11-68. How would you realize a dual-port RAM with dual-port memory cells?

Q11-69. Explain the operation of the arbitration logic shown in Figure 11.60.

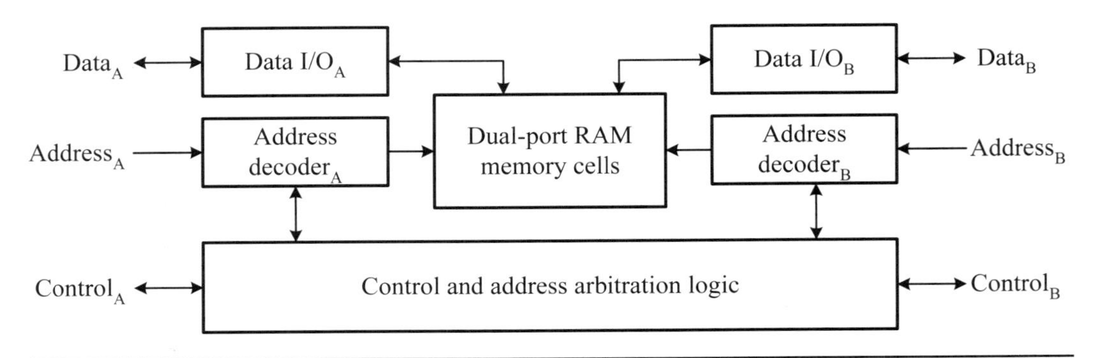

Figure 11.59: The block diagram of a dual-port RAM implemented with dual-port memory cells.

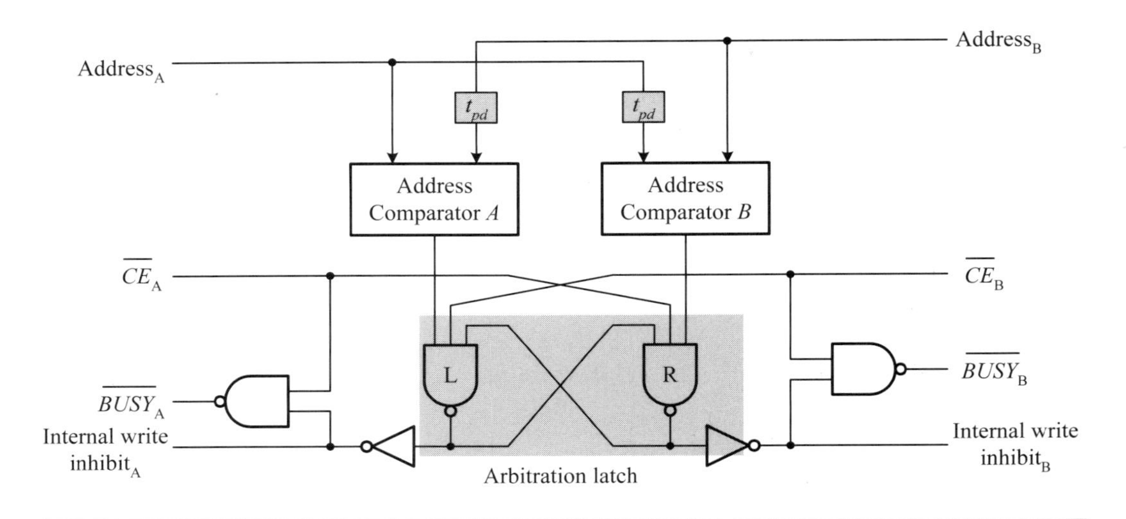

Figure 11.60: The arbitration logic of a dual-port RAM.

11.6.4 Programmable Logic Arrays

Like ROMs, a programmable logic array (PLA) is a two-level AND-OR logic structure and can be used to implement combinational logic in sum-of-product (SOP) form. If the feedback from the output is allowed, it can also be used to implement sequential logic.

The general structure of a typical $n \times k \times m$ PLAs is shown in Figure 11.61. There are n inputs and buffers/NOT gates, k AND gates, and m OR gates. There are $2n \times k$ programming points between inputs and AND gates, and $k \times m$ programming points between AND and OR gates. The programmable AND and OR arrays are referred to as AND and OR planes, respectively.

An illustration of using a PLA to implement combinational logic functions is demonstrated in the following example.

■ **Example 11-15: (PLA — Implementing switching functions.)** Figure 11.62 shows that two switching functions are implemented by using a PLA device.

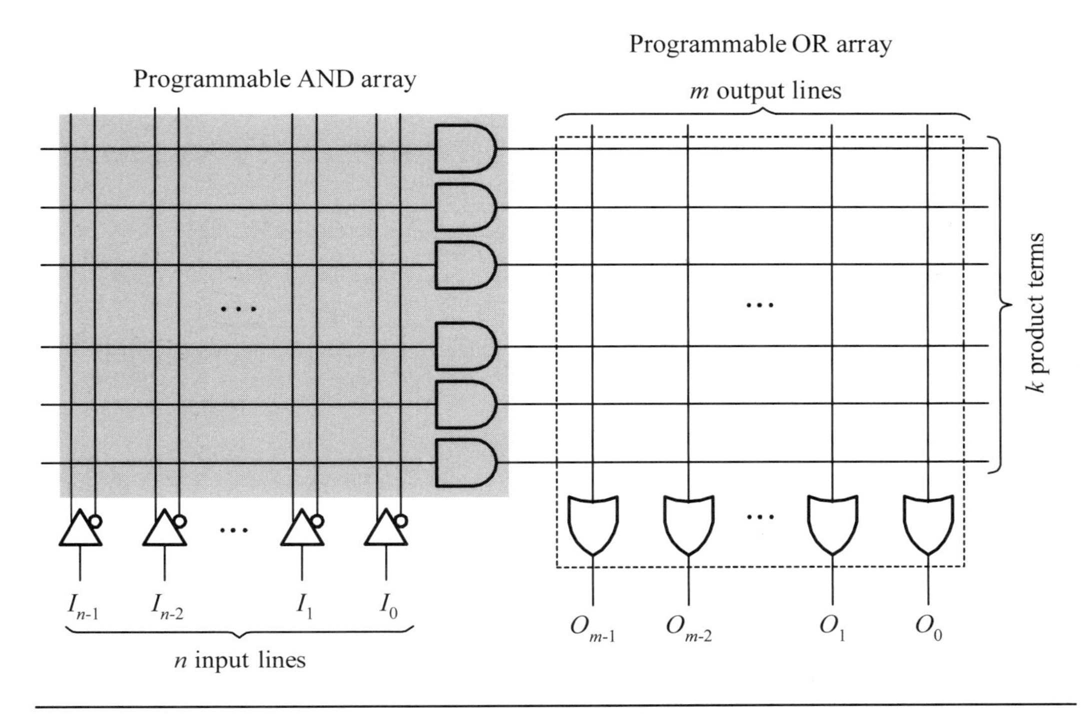

Figure 11.61: The block diagram of a typical PLA.

These two switching functions are as follows:

$$\begin{aligned} f_1(w,x,y) &= \Sigma(2,4,5,7) \\ f_2(w,x,y) &= \Sigma(1,5,7) \end{aligned}$$

Simplifying the above two switching functions using a multiple-output minimization process such as the one introduced in Lin [27], we obtain the following simplified switching expressions:

$$\begin{aligned} f_1(w,x,y,z) &= x\bar{y} + xz + \bar{x}y\bar{z} \\ f_2(w,x,y,z) &= xz + \bar{y}z \end{aligned}$$

These two switching functions can be expressed as the programming table shown in Figure 11.62(a). The product term xz is shared by both functions f_1 and f_2. There are four product terms, numbered from 0 to 3. The function f_1 consists of product terms: 0, 1, and 3. The function f_2 consists of product terms 1 and 2. The logic circuit that implements the programming table shown in Figure 11.62(a) is depicted in Figure 11.62(b). A symbolic diagram, as depicted in Figure 11.62(c), is often employed as a shortened notation of the logic circuit in practice.

■

11.6.4.1 Implementations of PLA A NOR-NOR implementation of the PLA logic circuit in Figure 11.62(b) using the pseudo-nMOS circuit style is depicted in Figure 11.63. Even though this results in a compact and fast structure, it consumes much power dissipation. Hence, for a larger PLA logic circuit, the dynamic logic is often

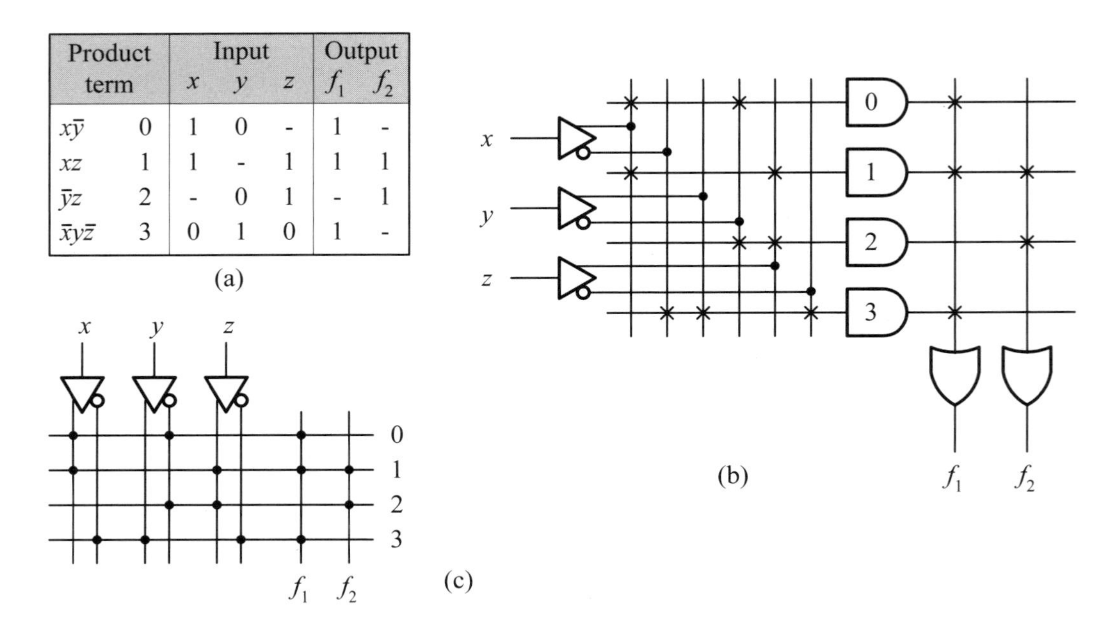

Product term		Input x	y	z	Output f_1	f_2
$x\bar{y}$	0	1	0	-	1	-
xz	1	1	-	1	1	1
$\bar{y}z$	2	-	0	1	-	1
$\bar{x}y\bar{z}$	3	0	1	0	1	-

Figure 11.62: An example of implementing two switching functions using a PLA: (a) programming table; (b) logic circuit; (c) symbolic diagram.

used. However, two basic dynamic logic circuits cannot be simply cascaded because of a possible partially discharged hazard described in Section 8.1.3. One way to overcome this difficulty is to use a delayed clock, like the clock-delayed domino logic described in Section 8.3.4.

Figure 11.64 shows the implementation of the PLA logic circuit described in Figure 11.62(b) using the dynamic logic style. Assume that the AND plane and OR plane are clocked with clocks ϕ_{AND} and ϕ_{OR}, respectively. The clock ϕ_{AND} starts to precharge the AND plane. After enough precharging time, the AND plane begins to evaluate its value by the rising ϕ_{AND} signal. At the same time, the OR plane starts to precharge by lowering its clock ϕ_{OR}. When the AND plane completes its evaluation, the OR plane begin to evaluate its value by rising its clock ϕ_{OR}.

Since the timing of the above clock events strongly depends on the PLA size, programming status of the AND plane, and PVT variations, a replica AND row with maximum loading is employed as the delay line to take into account the worst case. Based on this, the clock ϕ_{AND} can be derived from the input clock clk by NANDed the clock clk with a delayed version of itself. The OR plane clock ϕ_{OR} is derived from ϕ_{AND} in a similar way, as shown in Figure 11.64. The purpose of using inverters in the clock generators is to add some extra safe margins for both clocks, ϕ_{AND} and ϕ_{OR}.

■ Review Questions

Q11-70. Describe the features of PLA devices.

Q11-71. What do AND and OR planes mean?

Q11-72. What is the intention of the replica AND row?

Q11-73. How would you resolve the partially discharged problem in dynamic PLA?

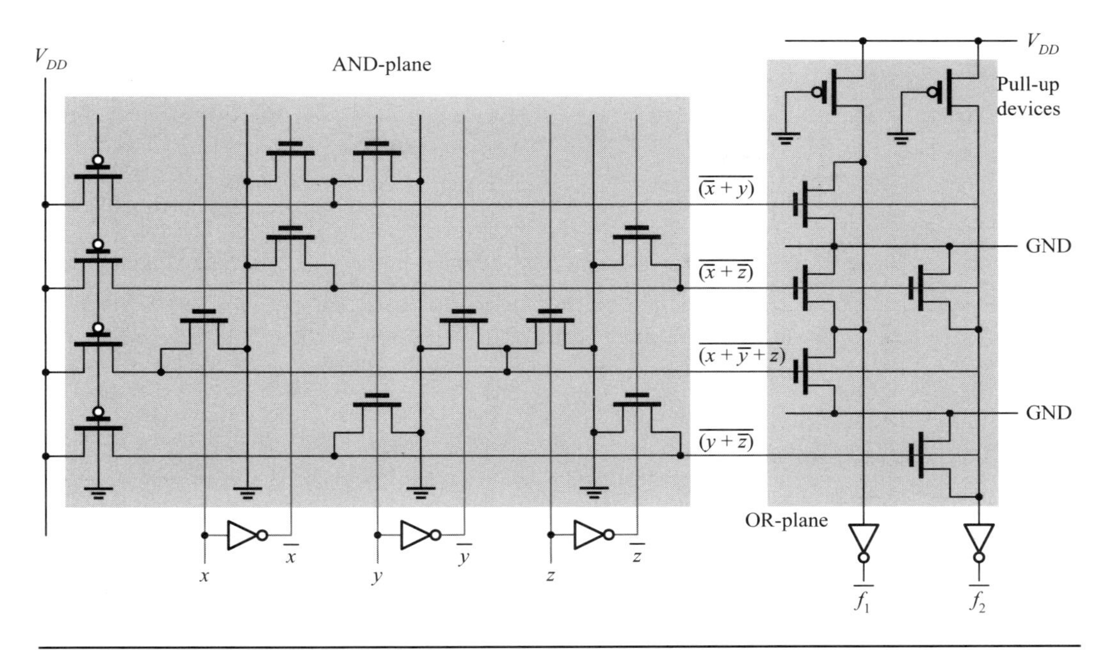

Figure 11.63: A pseudo-nMOS PLA for implementing the PLA logic circuit shown in Figure 11.62(b).

11.6.5 FIFO

First-in first-out (FIFO) is memory in which the data word comes in first and comes out first. It is often used in digital systems as an *elastic buffer* to harmonize two different data access rates, one for write and the other for read. As shown in Figure 11.65, the size of a FIFO is determined by the difference of data rates between the read and write operations.

FIFO is fundamentally a dual-port memory based on dual-port memory cells, in which one port is used to write into the memory, while the other port is employed to read the data out. Unlike RAMs and ROMs, FIFO does not require the address to access data; instead, two pointers associated with FIFO are employed implicitly to maintain the positions of write and read up to this point.

In general, there are two types of FIFOs: exclusive read/write FIFO and concurrent read/write FIFO. Exclusive read/write FIFO contains a variable number of stored data words, and hence, a necessary synchronism exists between the read and write operations. In such FIFOs, a timing relationship exists between the write and read clocks. A concurrent read/write FIFO also contains a variable number of stored data words and probably operates in an asynchronous way between the read and write operations.

The general block diagram of FIFO is shown in Figure 11.66. FIFO comprises an array of dual-port memory cells, and a read control and pointer, as well as a write control and pointer. In addition, two flags are used to indicate whether the status of FIFO is empty, partially full, or full.

The address of the front data word in FIFO is in the read pointer *head*, while the next address to be written is in the write pointer *tail*. After reset, both pointers, *head* and *tail*, are set to minus 1. After each write operation, the write pointer points to the next memory location to be written. After each read operation, the read pointer

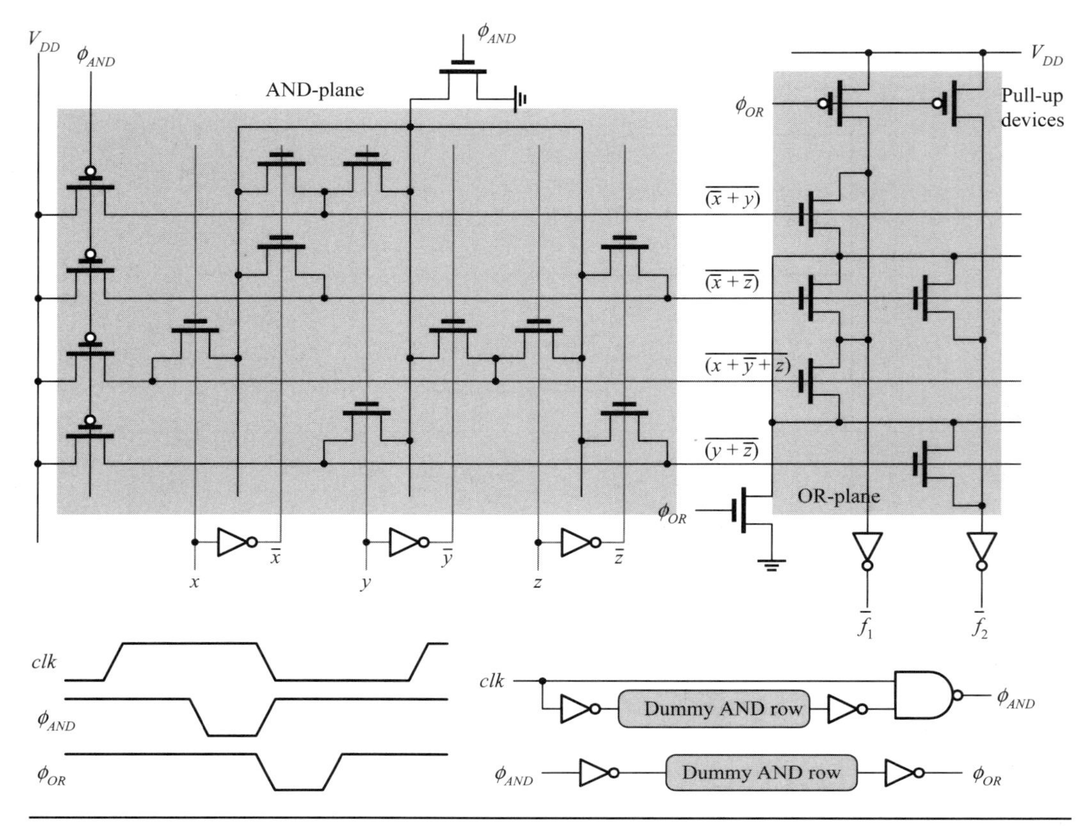

Figure 11.64: A dynamic PLA for implementing the PLA logic circuit shown in Figure 11.62(b).

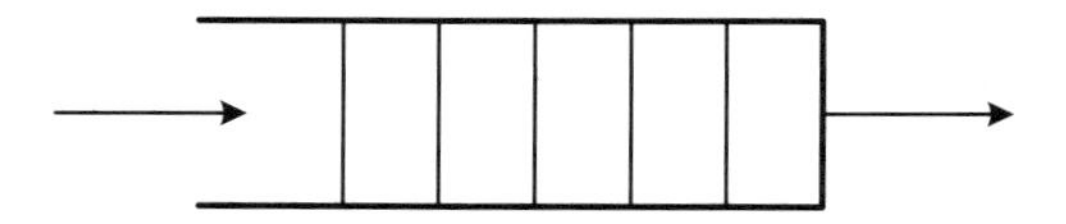

Figure 11.65: The concept of FIFO.

points to the next data word to be read out. The read pointer continuously follows the write pointer. When the read pointer catches up to the write pointer, FIFO is empty. On the other hand, if the write pointer reaches the read pointer, FIFO is full.

■ Review Questions

Q11-74. Describe the features of FIFO.

Q11-75. How would you design FIFO?

Q11-76. Describe the major components of FIFO.

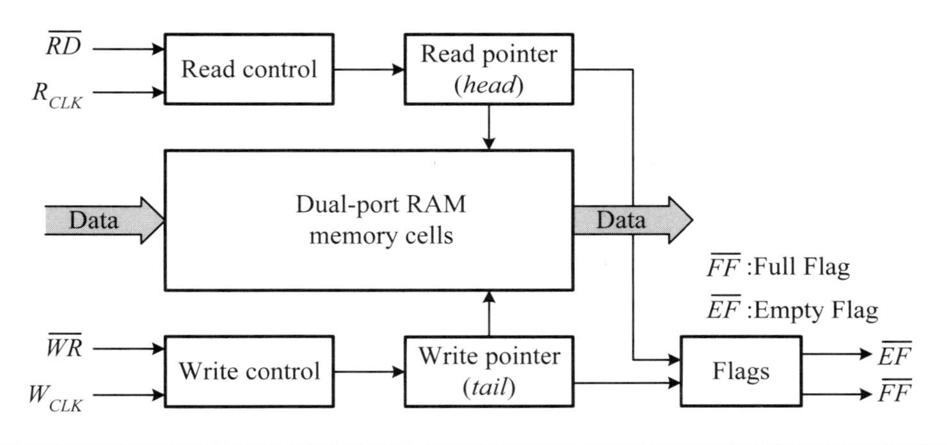

Figure 11.66: The block diagram of FIFO.

11.7 Summary

Semiconductor memory can be classified into serial access, content addressable, and random access according to the type of data access. Serial-access memory mainly contains shift registers and queues. The shift registers can be further subdivided into serial-in parallel-out (SIPO) and parallel-in serial-in (PISO), while the queues contain first-in first-out (FIFO) and first-in last-out (FILO). FIFO is generally called a queue or a buffer while FILO is known as a stack.

Content-addressable memory (CAM) is a device capable of searching all entries within it in parallel. CAM carries out the reverse operation of SRAM; namely, it receives as an input data and as outputs a matched address if the data match one in the CAM.

Random-access memory (RAM) is a memory device capable of accessing any word at random in a constant time. The random-access memory can be classified into read-/write memory and read-only memory. Read/write memory can be further subdivided into two types: static RAMs (SRAMs) and dynamic RAMs (DRAMs). The SRAM cell stores its information in the circuit state form; the DRAM cell retains its information in the charge form. Because of the leakage current of the access transistor, the capacitor (a soft node) in a DRAM cell is gradually losing its charge and eventually its information. To remedy this, a DRAM cell must be refreshed periodically.

Read-only memory (ROM) is only allowed to retrieve but not allowed to modify the stored information. The contents of ROM can be either programmed using photolithographic mask during manufacturing or programmed in laboratories. The former is called a mask ROM or ROM for short and the latter is called a programmable ROM (PROM). At present, many variations of PROM are available for use. These include fuse ROM, erasable PROM (EPROM), electrically erasable PROM (EEPROM), and Flash memory.

Semiconductor memory may also be categorized into volatile and nonvolatile in terms of the information retention capability once the power supply is interrupted. The volatile memories, such as static RAM and dynamic RAM, will lose their information when the power supply is removed, while the nonvolatile memories, such as the ROM family, ferroelectric RAM (FRAM), and magnetoresistance RAM (MRAM), still retain their information.

Content-addressable memory (CAM), registers and register files, dual-port RAM, programmable-logic array (PLA), and first-in first-out (FIFO) have also found broad use in digital systems.

References

1. B. Amrutur and M. Horowitz, "Speed and power scaling of SRAM's," *IEEE J. of Solid-State Circuits,* Vol. 35, No. 2, pp. 175–185, February 2000.
2. B. Amrutur and M. Horowitz, "Fast low-power decoders for RAMs," *IEEE J. of Solid-State Circuits,* Vol. 36, No. 10, pp. 1506–1515, October 2001.
3. B. Amrutur and M. Horowitz, "A replica technique for wordline and sense control in low-power SRAM's," *IEEE J. of Solid-State Circuits,* Vol. 33, No. 8, pp. 1208–1219, August 1998.
4. B. H. Calhoun and A. P. Chandrakasan, "Static noise margin variation for sub-threshold SRAM in 65-nm CMOS," *IEEE J. of Solid-State Circuits,* Vol. 41, No. 7, pp. 1673–1679, July 2006.
5. B. H. Calhoun, and A. P. Chandrakasan, "A 256-kb 65-nm sub-threshold SRAM design for ultra-low-voltage operation," *IEEE J. of Solid-State Circuits,* Vol. 42, No. 3, pp. 680–688, March 2007.
6. C. W. Chen et al., "A fast 32K $\times$ 8 CMOS static RAM with address transition detection," *IEEE J. of Solid-State Circuits,* Vol. 22, No. 4, pp. 533–537, August 1987.
7. M. K. Choi et al.,"A 0.25-μm 3.0-V 1T1C 32-Mb nonvolatile ferroelectric RAM with address transition detector and current forcing latch sense amplifier scheme," *IEEE J. of Solid-State Circuits,* Vol. 37, No. 11, pp. 1472–1478, November 2002.
8. Y. Chung, "FRAM design style utilizing bit-plate parallel cell architecture," *Electronics Letters,* Vol. 39, No. 24, pp. 1706–1707, 27th November 2003.
9. B. F. Cockburn, "Tutorial on magnetic tunnel junction magnetoresistive random-access memory," *The IEEE 2004 Int'l Workshop on Memory Technology, Design and Testing* (MTDT 04), pp. 46–51, 2004.
10. Cypress Semiconductor, *Application Note: Understanding Asynchronous Dual-Port RAMs,* Revised November 7, 1997.
11. M. Durlam et al., "A 1-Mbit MRAM based on 1T1MTJ bit cell integrated with copper interconnects," *IEEE J. of Solid-State Circuits,* Vol. 38, No. 5, pp. 769–773, May 2003.
12. B. N. Engel et al., "The science and technology of magnetoresistive tunneling memory," *IEEE Trans. on Nanotechnology,* Vol. 1, No. 1, pp. 32–38, March 2002.
13. Freescale Semiconductor, Inc., *Magnetoresistive Random Access Memory,* Revised June 23, 2006.
14. D. Gogl et al., "A 16-Mb MRAM featuring bootstrapped write drivers," *IEEE J. of Solid-State Circuits,* Vol. 40, No. 4, pp. 902–908, April 2005.
15. M. Golden and H. Partovi, "A 500 MHz, write-bypassed, 88-entry, 90-bit register file," *1999 Symposium on VLSI Circuits Digest of Technical Papers,* pp. 105–108, June 1999.

16. K. Grosspietsch, "Associative processors and memories: a survey," *IEEE Micro,* Vol. 12, No. 3, pp. 12–19, June 1992.

17. T. Hara et al., "A 146-mm^2 8-Gb multi-level NAND flash memory with 70-nm CMOS technology," *IEEE J. of Solid-State Circuits,* Vol. 41, No. 1, pp. 161–169, January 2006.

18. R. Heald and J. Holst, "A 6-ns cycle 256 kb cache memory and memory management unit," *IEEE J. of Solid-State Circuits,* Vol. 28, No. 11, pp. 1078–1083, November 1993.

19. H. Hidaka et al., "Twisted bit-line architectures for multi-megabit DRAMs," *IEEE J. of Solid-State Circuits,* Vol. 24, No. 1, pp. 21–27, February 1989.

20. T. Hirose, "A 20-ns 4-Mb CMOS SRAM with hierarchical word decoding architecture," *IEEE J. of Solid-State Circuits,* Vol. 25, No. 5, pp. 1068–1074, October 1990.

21. D. A. Hodges, H. G. Jackson, and R. A. Saleh, *Analysis and Design of Digital Integrated Circuits: In Deep Submicron Technology,* 3rd ed. New York: McGraw-Hill Books, 2004.

22. W. Hwang, R. Joshi, and W. Henkels, "A 500-MHz, 32-Word × 64-bit, eight-port self-resetting CMOS register file," *IEEE J. of Solid-State Circuits,* Vol. 34, No. 1, pp. 56–67, January 1999.

23. K. Imamiya et al., "A 125-mm^2 1-Gb NAND flash memory with 10-MByte/s program speed," *IEEE J. of Solid-State Circuits,* Vol. 37, No. 11, pp. 1493–1501, November 2002.

24. T. Kawahara et al., "2 Mb spin-transfer torque ram (SPRAM) with bit-by-bit bidirectional current write and parallelizing-direction current read," *Proc. of IEEE Int'l Solid-State Circuits Conf.*, pp. 480–481, February 2007.

25. C. H. Kim et al., "A forward body-biased low-leakage SRAM cache: device, circuit and architecture considerations," *IEEE Trans. on Very Large Scale Integration (VLSI) Systems,* Vol. 13, No. 3, pp. 349–357, March 2005.

26. M. B. Lin, "On the design of fast large fan-in CMOS multiplexers," *IEEE Trans. on Computer-Aided Design of Integrated Circuits and systems,* Vol. 19, No. 8, pp. 963–967, August, 2000.

27. M. B. Lin, *Digital System Design: Principles, Practices, and Applications,* 4th ed. Taipei, Taiwan: Chuan Hwa Book Ltd., 2010.

28. M. B. Lin, *Digital System Designs and Practices: Using Verilog HDL and FPGAs.* Singapore: John Wiley & Sons, 2008.

29. A. R. Linz, "A low-power PLA for a signal processor," *IEEE J. of Solid-State Circuits,* Vol. 26, No. 2, pp. 107–115, February 1991.

30. M. Matsumiya et al., "A 15-ns 16-Mb CMOS SRAM with interdigitated bit-line architecture," *IEEE J. of Solid-State Circuits,* Vol. 27, No. 11, pp. 1497–1503, November 1992.

31. T. C. May and M. H. Woods, "Alpha-particle-induced soft errors in dynamic memories," *IEEE Trans. on Electronic Devices,* Vol. 26, No. 1, pp. 2–9, January 1979.

32. H. P. McAdams, "A 64-Mb embedded FRAM utilizing a 130-nm 5LM Cu/FSG logic process," *IEEE J. of Solid-State Circuits,* Vol. 39, No. 4, pp. 667–677, April 2004.

33. J. I. Miyamoto, "An experimental 5-V-only 256-kbit CMOS EEPROM with a high-performance single-polysilicon cell," *IEEE J. of Solid-State Circuits,* Vol. 21, No. 5, pp. 852–860, October 1986.

34. H. Miyatake, M. Tanaka, and Y. Mori, "A design for high-speed low-power CMOS fully parallel content-addressable memory macros," *IEEE J. of Solid-State Circuits,* Vol. 36, No. 6, pp. 956–968, June 2001.

35. Y. Nakagome et al., "Circuit techniques for 1.5-3.6-V battery-operated 64-Mb DRAM," *IEEE J. of Solid-State Circuits,* Vol. 26, No. 7, pp. 1003–1010, July 1991.

36. H. Nambu et al., "A 1.8-ns access, 550-MHz, 4.5-Mb CMOS SRAM," *IEEE J. of Solid-State Circuits,* Vol. 33, No. 11, pp. 1650–1658, November 1998.

37. K. Nii et al., "Synchronous ultra-high-density 2RW dual-port 8T-SRAM with circumvention of simultaneous common-row-access," *IEEE J. of Solid-State Circuits,* Vol. 44, No. 3, pp. 977–986, March 2009.

38. M. Ohkawa et al., "A 98 mm^2 die size 3.3-V 64-Mb flash memory with FN-NOR type four-level cell," *IEEE J. of Solid-State Circuits,* Vol. 31, No. 11, pp. 1584–1589, November 1996.

39. N. Ohtsuka et al., "A 4-Mbit CMOS EPROM," *IEEE J. of Solid-State Circuits,* Vol. 22, No. 5, pp. 669–675, October 1987.

40. N. Otsuka and M. A. Horowitz, "Circuit techniques for 1.5-V power supply flash memory," *IEEE J. of Solid-State Circuits,* Vol. 32, No. 8, pp. 1217–1230, August 1997.

41. K. Pagiamtzis, and A. Sheikholeslami, "Content-addressable memory (CAM) circuits and architectures: a tutorial and survey," *IEEE J. of Solid-State Circuits,* Vol. 41, No. 3, pp. 712–727, March 2006.

42. J. M. Rabaey, A. Chandrakasan, and B. Nikolic, *Digital Integrated Circuits: A Design Perspective,* 2nd ed. Upper Saddle River, NJ: Prentice-Hall, 2003.

43. W. Regitz and J. Karp, "A three-transistor cell, 1,024-bit 500 ns MOS DRAM," *ISSCC Digest of Technical Papers,* pp. 42–43, 1970.

44. G. Samson, and L. T. Clark, "Low-power race-free programmable logic arrays," *IEEE J. of Solid-State Circuits,* Vol. 44, No. 3, pp. 935–946, March 2009.

45. E. Seevinck, F. List, and J. Lohstroh, "Static noise margin analysis of MOS SRAM cells," *IEEE J. of Solid-State Circuits,* Vol. 22, No. 5, pp. 748–754, October 1987.

46. E. Seevinck, P. J. van Beers, and H. Ontrop, "Current-mode techniques for high-speed VLSI circuits with application to current sense amplifier for CMOS SRAMs," *IEEE J. of Solid-State Circuits,* Vol. 26, No. 4, pp. 525–536, April 1991.

47. T. Seki et al., "A 6-ns 1-Mb CMOS SRAM with latched sense amplifier," *IEEE J. of Solid-State Circuits,* Vol. 28, No. 4, pp. 478–483, April 1993.

48. K. Seno et al., "A 9-ns 16-Mb CMOS SRAM with offset-compensated current sense amplifier," *IEEE J. of Solid-State Circuits,* Vol. 28, No. 11, pp. 1119–1124, November 1993.

49. A. K. Sharma, *Advanced Semiconductor Memories: Architectures, Designs, and Applications.* New York: Wiley-Interscience, 2003.

50. N. Shibata et al., "A 70 nm 16 Gb 16-level-cell NAND flash memory," *IEEE J. of Solid-State Circuits*, Vol. 43, No. 4, pp. 929–937, April 2008.

51. Texas Instrument, *Application Note: FIFO Architecture, Functions, and Applications*, July 1999.

52. J. P. Uyemura, *Introduction to VLSI Circuits and Systems.* New York: John Wiley & Sons, 2002.

53. C. Villa et al., "A 65 nm 1 Gb 2b/cell NOR Flash with 2.25 MB/s program throughput and 400 MB/s DDR interface," *IEEE J. of Solid-State Circuits*, Vol. 43, No. 1, pp. 132–140, January 2008.

54. J. Wang, C. Chang, and C. Yeh, "Analysis and design of high-speed and low-power CMOS PLAs," *IEEE J. of Solid-State Circuits*, Vol. 36, No. 8, pp. 1250–1262, August 2001.

55. B. Wicht, S. Paul, and D. Schmitt-Landsiedel, "Analysis and compensation of the bit-line multiplexer in SRAM current sense amplifiers," *IEEE J. of Solid-State Circuits*, Vol. 36, No. 11, pp. 1745–1755, November 2001.

56. W. Xu, T. Zhang, and Y. Chen, "Design of spin-torque transfer magnetoresistive RAM and CAM/TCAM with high sensing and search speed," *IEEE Trans. on Very Large Scale Integration (VLSI) Systems*, Vol. 18, No. 1, pp. 66–74, January 2010.

57. M. Yoshimoto et al., "A divided word-line structure in the static RAM and its application to a 64K full CMOS RAM," *IEEE J. of Solid-State Circuits*, Vol. 18, No. 5, pp. 479–485, October 1983.

Problems

11-1. Referring to Figure 11.2, assume that the memory capacity is 4 Mb.

(a) How many address lines are required if the data bus is 4 bits?

(b) How many address lines are required if the data bus is 8 bits?

11-2. Suppose that an embedded SRAM is physically arranged in a square fashion. Calculate the number of inputs of each column multiplexer under each of the following conditions:

(a) The embedded SRAM contains 512 8-bit words.

(b) The embedded SRAM contains 4096 16-bit words.

11-3. Referring to Figure 11.4, use the concepts of logical effort to show that the hierarchical word-line structure is better than single word-line structure.

11-4. Consider the 6T-SRAM cell shown in Figure 11.8. Assume that the following aspect ratios are used: M_1 and M_2 are $4\lambda/2\lambda$, and M_3 to M_6 are $3\lambda/2\lambda$.

(a) Draw a stick diagram.

(b) Design and construct a layout.

(c) Design an experiment with SPICE to measure the read-cycle and write-cycle times.

11-5. Referring to the 6T-SRAM cell shown in Figure 11.10, suppose that the cell is designed with $k_R = 1$. Is it possible to write to the cell when $W_3/L_3 = W_1/L_1$ and $W_4/L_4 = W_2/L_2$? Explain your answer in terms of the circuit.

11-6. Consider the 6T-SRAM cell shown in Figure 11.8. Assume that node $\bar{Q}$ can only tolerate a 0.1-V rise in voltage during read operation. Using 0.13-μm process parameters: $V_{T0n} = 0.35$ V, $v_{sat} = 1.5 \times 10^7$ cm/s, $\mu_{effn} = 298$ cm^2/V-s, and $E_{satn} = 9.5 \times 10^4$ V/cm, determine the ratio of W_1 to W_3. Suppose that $V_{DD} = 1.2$ V.

11-7. Assume that the bit-line capacitance associated with a 6T-SRAM cell is 500 fF. In the read cycle, if the desired transition time of the bit-line is 2 ns when its voltage is changed by 200 mV, determine the size of transistor M_3 using 0.13-μm process parameters.

11-8. Consider the 6T-SRAM cell shown in Figure 11.11. Assume that node Q must be pulled down below 0.2 V during the write operation. Using 0.13-μm process parameters: $V_{T0n} = |V_{T0p}| = 0.35$ V, $\mu_{effn} = 298$ cm^2/V-s, $\mu_{effp} = 97$ cm^2/V-s, $v_{sat} = 1.5 \times 10^7$ cm/s, $E_{satn} = 9.5 \times 10^4$ V/cm, and $E_{satp} = 42 \times 10^4$ V/cm, determine the ratio of W_6 to W_4. Suppose that $V_{DD} = 1.2$ V.

11-9. In a 64-kb SRAM, the memory core is composed of 256 × 256 cells. The transistors in each cell have their aspect ratios as follows: M_1 and M_2 are $4\lambda/2\lambda$, and M_3 to M_6 are $3\lambda/2\lambda$. The area of word-line is $26\lambda \times 2\lambda$ and the area of bit-line is $44\lambda \times 3\lambda$. The resistance of n^+/p^+ polysilicon layer is 5 $\Omega/\square$. Using 0.13-μm process parameters, calculate the capacitance and resistance of each word-line, and the word-line RC time constant as well.

11-10. Continued from the preceding problem, assume that each contact on the bit-line has a capacitance of 0.3 fF and the wire capacitance is 0.15 fF/μm. Using 0.13-μm process parameters, calculate the capacitance and resistance of the bit-line, and the bit-line RC time constant as well.

11-11. Assume that both the predecoder and final decoder are composed of two-input NAND gates and inverters.

(a) Design a 6-to-64 multilevel row decoder.

(b) Estimate the propagation delay using the logical effort technique.

(c) Compare with the single-level row decoder.

11-12. Assume that both the predecoder and final decoder are composed of two-input NAND gates and inverters.

(a) Design an 8-to-256 multilevel row decoder.

(b) Estimate the propagation delay using the logical effort technique.

(c) Compare with the single-level row decoder.

11-13. Consider the design of a column multiplexer in which 32 inputs need to be multiplexed to a single I/O output.

(a) Design the column multiplexer using a binary-tree structure. How many nMOS switches are required? What is the propagation delay of the resulting multiplexer?

(b) Design the column multiplexer using a heterogeneous-tree structure. How many nMOS switches are required? What is the propagation delay of the resulting multiplexer?

11-14. Consider the design of a column multiplexer in which 64 inputs need to be multiplexed to a single I/O output.

(a) Design the column multiplexer using a binary-tree structure. How many nMOS switches are required? What is the propagation delay of the resulting multiplexer?

(b) Design the column multiplexer using a heterogeneous-tree structure. How many nMOS switches are required? What is the propagation delay of the resulting multiplexer?

11-15. Show that the propagation delay t_{sense} of the latch-based sense amplifier can be expressed as a function of both input and output voltages and is equal to

$$t_{sense} = \left(\frac{C_{bit}}{g_{mn} + g_{mp}}\right) \ln\left(\frac{\Delta v_{out}(t_{sense})}{\Delta v_{in}}\right)$$

11-16. Assume that the bit-line capacitance C_{bit} is 800 fF. Using 0.18-μm-process parameters: $k'_n = 96\ \mu A/V^2$, $k'_p = 48\ \mu A/V^2$, and $V_{T0n} = |V_{T0p}| = 0.4$ V, design a latch-based sense amplifier to drive the voltage difference, about 200 mV, in bit-lines to 1.8 V in less than 2 ns.

11-17. Assume that each DRAM cell has a 40-fF storage capacitor and contributes 2 fF to the total bit-line capacitance. There are 512 bits on each bit-line that result in a bit-line capacitance of approximately 1024 fF. With the addition of replica cells and other parasitics, assume that the bit-line capacitance is 1200 fF. Using 0.18-μm parameters, calculate the bit-line voltage change when a logic 1 and logic 0 are read from a storage capacitor, respectively.

11-18. Assume that the storage capacitor C_S in a DRAM has a capacitance of 50 fF. The circuitry restricts the maximum voltage of the capacitor to a value of $V_{max} = 1.4$ V. The leakage current off the cell is estimated to be 50 nA.

(a) How many electrons can be stored on the storage capacitor C_S?

(b) Calculate the time needed to reduce the number of stored electrons to 200.

11-19. Assume that the storage capacitor C_S in a DRAM has a capacitance of 50 fF. It is used in a system in which $V_{DD} = 1.8$ V and $V_{T0n} = 0.4$. The bit-line capacitance is 500 fF. The leakage current off the cell is estimated to be 50 nA.

(a) Find the maximum amount of charge that can be stored on the storage capacitor C_S?

(b) If to detect a logic 1 state, the voltage on the bit-line must be at least 0.8 V, calculate the hold time.

11-20. Design an active-programming 4×4 ROM that contains the following specified data:

(a) 1000, 0110, 1010, 1110.

(b) 0100, 1011, 1001, 1111.

11-21. Design a dynamic CMOS AND-OR PLA based on NOR gates to realize the following four switching functions:

$$\begin{aligned}
d(w,x,y,z) &= x(y+z)+w \\
c(w,x,y,z) &= \bar{x}(y+z)+x\overline{(y+z)} \\
b(w,x,y,z) &= yz+\overline{(y+z)} \\
a(w,x,y,z) &= \bar{z}
\end{aligned}$$

where w, x, y, and z are inputs.

11-22. Design a dynamic CMOS AND-OR PLA based on NOR gates to realize the following three switching functions:

$$\begin{aligned}
f_1(w,x,y,z) &= w\bar{x}+xz+\bar{x}y \\
f_2(w,x,y,z) &= \bar{w}xz+y \\
f_3(w,x,y,z) &= xy+w\bar{x}\bar{y}+wxz
\end{aligned}$$

where w, x, y, and z are inputs.

12

Design Methodologies and Implementation Options

Once we have understood the basic features of a very-large-scale integration (VLSI) cell, datapath subsystems, and memory subsystems design, we are now in a position to understand the important system-level design issues. These issues include the logic module, interconnect network, power distribution network, clock generation and distribution network, input/output (I/O) module, and electrostatic discharge (ESD) protection network. The power management has been explored in Chapter 6. The important related design issues at the system level include signal integrity, clock integrity, and power integrity.

The logic module performs all desired functions in the VLSI system. In some applications, optional analog modules are also incorporated into the system to facilitate the required functions, thereby resulting in a mixed-signal system. In this chapter, we focus on the three closely related design issues: design methodology, synthesis flows, and implementation options. We begin with the introduction of design methodology, including system level and register-transfer level (RTL). Next, the synthesis-flow using hardware description language (HDL) for designing and implementing a digital system based on a field-programmable gate array (FPGA) device or a cell library is described concisely. Then, implementation options are covered. These options can be classified into three classes: platforms, application-specific integrated circuits (ASICs), and field-programmable devices. Finally, a case study is given to illustrate how a real-world system can be implemented with a variety of options, including the microprocessor/digital-signal processing (μP/DSP) system, a field-programmable device, and ASIC with a cell library.

12.1 Design Methodologies and Implementation Architectures

With the dramatically increasing integration density of VLSI devices, digital system designers have shifted their playgrounds from printed-circuit boards (PCBs) to silicon wafers by using various IPs instead of discrete devices. This trend not only reduces system cost and power dissipation but also increases system performance profoundly. In addition, it also makes the design style of a digital system significantly changed;

namely, computer-aided design (CAD) tools are extensively used to carry out a large amount of hard work. We briefly describe the related design approaches at both system and register-transfer levels, as well as implementation architectures.

12.1.1 Designs at System Level

A VLSI (digital) system can be designed at the *electronic system level* (ESL) by using one of the following methods: *function-based*, *architecture-based*, and the mixing use of above two methods.

12.1.1.1 Function-Based Method The function-based method uses a *computational model* to compose a variety of different components into a desired system. One of the most widely used computational models is Simulink® developed by MathWorks. Simulink can be used to simulate a continuous-time or discrete-time system. In Simulink, each design abstraction composes a block diagram. Each block diagram in turn contains two essential components: block and signal. Each block captures the behavior of a component, that is, the relationship between outputs and inputs; signals connect together related blocks so that the entire system can be solved to find the relationship between state variables and signals over the period of interest.

In the function-based method, the resulting design abstraction can be realized in one of the following ways:

- Direct translation into RTL
- High-level synthesis to RTL
- Direct mapping to IPs
- Compilation into software programs

Currently, the most widely used approaches for the implementation of a function-based design abstraction are to direct mapping to available IPs and hence to be implemented with a dedicated hardware, to transfer into software programs, and then to execute on a platform system.

12.1.1.2 Architecture-Based Method The architecture-based method combines the desired components from a predefined set into a desired system. This method is much like the traditional PCB-based design approach but instead uses IPs and a silicon wafer. In the architecture-based method, the processor component can be programmable; that is, it is not necessary to execute a fixed function.

In the architecture-based method, the resulting design abstraction can be realized through a number of ways.

- High-level synthesis to RTL
- Direct mapping to IPs
- Platform with configurable function units and instruction set

12.1.1.3 Hybrid Approach The more popular approach is to combine function- and architecture-based methods and proceed in a top-down, stepwise refinement fashion. As shown in Figure 12.1, the system designers capture the specifications and proceed with algorithm selection based on a functional model. The results are then refined into architecture components gradually through high-level synthesis based on an architecture model. These architecture components are in turn composed of *microarchitectures* and synthesized into gate-level netlists via RTL synthesis on the basis of the technology library.

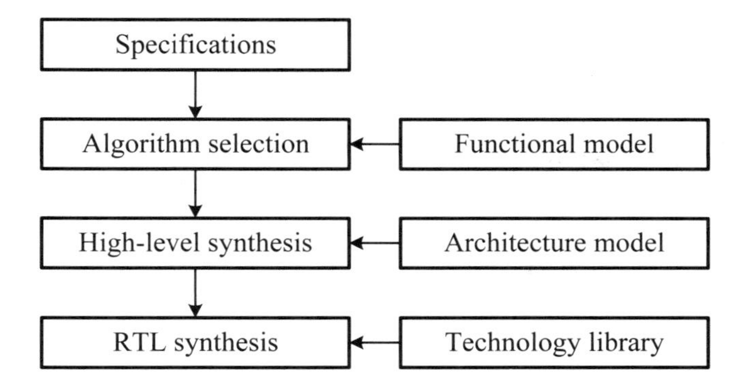

Figure 12.1: The function and architecture co-design method.

Before proceeding with the high-level synthesis, many related issues need to be considered. These include hardware/software partitioning, behavior mapping, architecture exploration, protocol generation, and topology synthesis. Since they are beyond the scope of this book, we would like to omit them here. The interested reader can refer to Wang et al. [11] for details.

12.1.1.4 Global Asynchronous and Local Synchronous Design Because of large parameter variations across a chip in the nanoscale era, it is prohibitively expensive to control delays in clocks and other global signals. Consequently, the *globally asynchronous and locally synchronous* (GALS) approach seems to be feasible and preferred in the near future, in particular, in the system-on-a-chip (SoC) realms. Nonetheless, the complexity of the numerous asynchronous/synchronous interfaces required in a GALS will eventually lead to entirely asynchronous solutions.

An SoC can be thought of as a complex distributed system in which a large number of parallel components communicate with one another and synchronize their activities by message exchange through an interconnect network or NoC (network on chip). These components may be synchronous or asynchronous. To communicate across different components, appropriate synchronization schemes must be applied. The communication across two asynchronous components is usually implemented as a handshaking protocol. The communications between a synchronous component and an asynchronous one may be realized by either synchronizers or stoppable clocks. In the synchronizer scheme, asynchronous signals are sampled by synchronizers so that they can be processed in the synchronous systems. In a stoppable clock scheme, the clock may be stopped by the asynchronous signal; thereby, all variables on the synchronous side are being treated as registers by the asynchronous part of the system. As long as the writes to the variables are complete before the asynchronous side releases the clock, there is no risk of misoperations.

The theme of a GALS is the implementation of send/receive communication, which is central to the methods of asynchronous logic. This form of communication is used at all levels of system design, from communication between two system components, such as a processor and a cache, down to the interaction between the control part and the datapath of an arithmetic and logic unit (ALU).

■ Review Questions

Q12-1. What is the function-based method?

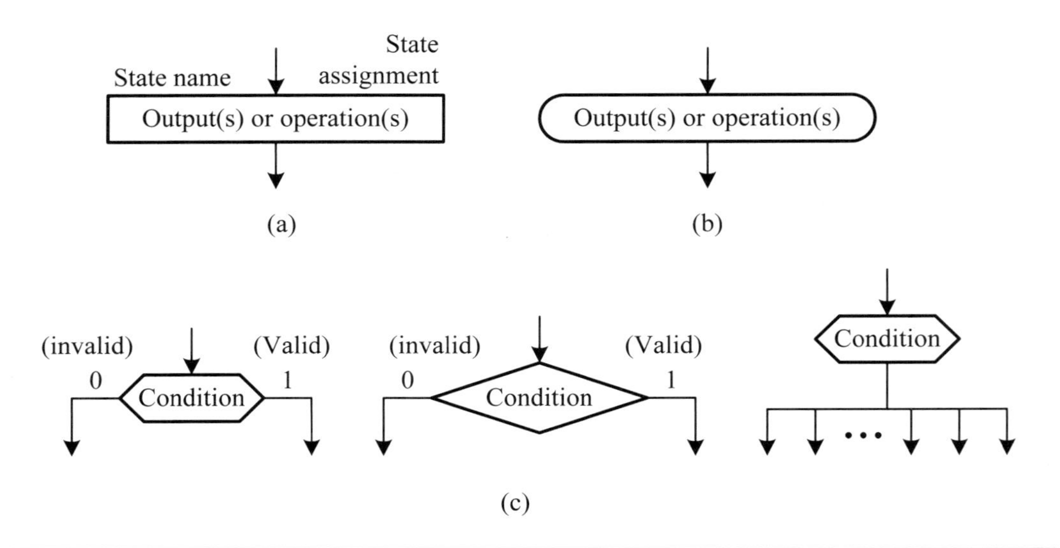

Figure 12.2: The ASM building blocks: (a) state block; (b) conditional output block; (c) decision block.

Q12-2. What is the architecture-based method?
Q12-3. What is the hybrid approach?
Q12-4. What is the essence of the GALS approach?

12.1.2 Designs at RTL

RTL design can be proceeded in either of the following: *algorithmic state machine chart* (ASM chart) and *finite-state machine with datapath* (FSMD). In this section, we only briefly describe their basic concepts. The details of these approaches can be referred to Lin [9] or other textbooks.

12.1.2.1 ASM Chart The ASM chart, sometimes called a *state machine* (SM) chart, is often used to describe a digital system design at the algorithmic level. Two essential features of ASM charts are as follows. First, they specify RTL operations since they define what happens on every cycle of operation. Second, they show clearly the flow of control from one state to another.

An ASM chart consists of three types of blocks: *state block*, *decision block*, and *conditional output block*, as shown in Figure 12.2. A state block, as illustrated in Figure 12.2(a), specifies a machine state and a set of unconditional RTL operations associated with the state. It may contain as many actions as desired and all actions in a state block occur in parallel. Each state along with its related operations occupy a clock period. In an ASM chart, state blocks are sequentially executed. In addition, a register may be assigned to only once in a state block. This is known as a *single-assignment rule.*

A conditional output block, as shown in Figure 12.2(b), describes the RTL operations that are executed under the conditions specified by one or more decision blocks. The input of a conditional output block must be from the output of a decision block or the other conditional output blocks. A conditional output block can only evaluate the present state or primary input value on the present cycle.

A decision block, as shown in Figure 12.2(c), describes the condition under which ASM will execute specific actions and select the next state based on the value of primary inputs and/or the present state. It can be drawn in either a two-way selection or multiway selection.

An ASM block contains exactly one state block, together with the possible decision blocks and conditional output blocks associated with that state. Each ASM block describes the operations to be executed in one state. An ASM block has the following two features. First, it has exactly one entrance path and one or more exit paths. Second, it contains one state block and a possible serial-parallel network of decision and conditional output blocks.

12.1.2.2 Finite-State Machine with Datapath (FSMD)

A finite-state machine with datapath (FSMD) is an extension of FSM introduced in Section 9.1.1 and is popular for use in high-level synthesis to generate an RTL result. However, it is also widely used in designing a complex system at RTL.

An FSMD is a quintuple $\mathcal{M}_{\mathcal{D}} = \langle \mathcal{S}, \mathcal{I} \cup S, \mathcal{O} \cup A, \delta, \lambda \rangle$, where $\mathcal{S} = \{s_0, \ldots, s_{n-1}\}$ is the set of n finite states, s_0 is the reset state, $\mathcal{I}$ is the set of input variables, $\mathcal{O} = \{o_k\}$ is the set of primary output values, $S = \{Rel(a, b) : a, b \in E\}$ is the set of statements specifying relations between two expressions from the set E, $E = \{f(x, y, z, \ldots) : x, y, z, ..., \in V\}$ is the set of expressions, V is the set of storage variables, and $A = \{x \Leftarrow e : x \in V, e \in E\}$ is the set of storage assignments.

The state transition function is defined as follows:

$$\delta : \mathcal{S} \times (\mathcal{I} \cup S) \rightarrow \mathcal{S}$$

which means that the δ function is determined by inputs and statements as well as state symbols.

The output function λ is defined as follows:

$$\begin{aligned} \lambda &: \mathcal{S} \times (\mathcal{I} \cup S) \rightarrow (\mathcal{O} \cup A) && \text{(Mealy machine)} \\ \lambda &: \mathcal{S} \rightarrow (\mathcal{O} \cup A) && \text{(Moore machine)} \end{aligned}$$

where in a Mealy machine, the output function is determined by both present input and statements as well as a state, whereas in a Moore machine the output function is solely determined by the present state.

The general circuit model of an FSMD is shown in Figure 12.3. It is also referred to as a *datapath and controller approach* or *control-point approach*. The circuit model of an FSMD is generally composed of three major parts, *datapath*, *memory*, and *control unit*. The major functions of each part are as follows:

- *Datapath* performs all operations needed in the system. A datapath usually consists of arithmetic and logic units, including adder, subtracter, multiplier, shifter, comparator, and so on, and a register file.
- *Memory* temporarily stores the data used and generated by the datapath unit. Memory can be one of random access memory (RAM), content-addressable memory (CAM), read-only memory (ROM), first-in, first-out (FIFO) buffer, and shift registers, or their mix.
- *Control unit* controls and schedules all operations performed by the datapath unit. The control unit is a finite state machine and can be implemented by ROM, programmable logic array (PLA), or random logic circuits. To adapt the control signals, a set of status signals is often fed back to the controller to indicate the status of datapath.

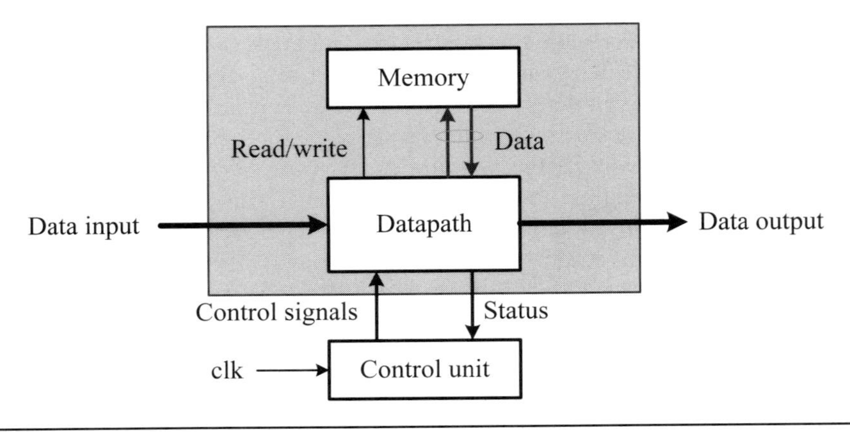

Figure 12.3: The general circuit model of FSMD.

In addition to the datapath, register file and interconnect network are the other two essential and difficult design issues for any digital system. The register file used in a datapath may need many read and write ports. For instance, a register file with two read ports and one write port is the most basic and common. Details of register file designs can be referred to in Section 11.6.2.

Interconnect network connects together all logic modules in an effective and efficient way. It mainly comprises switches, arbiters, multiplexers, and a bus. A bus is a set of conducting wires. A bus cannot support full connectivity between all datapath elements due to the inhibited cost and complexity. Thereby, some sort of time-division multiplexing has to be performed to enable the sharing of a bus among modules.

12.1.2.3 Relationship between ASM and FSMD For simple systems, we often begin to obtain an ASM chart from specifications and then derive both datapath and controller from the ASM chart if necessary. The datapath portion corresponds to the registers and function units in the ASM chart and the controller portion corresponds to the generation logic of control signals. For complex systems, the datapath and controller of a design are often derived from the specifications in a state-of-the-art manner.

■ Review Questions

Q12-5. Describe the general architecture of logic modules.

Q12-6. What are the three basic types of blocks used in ASM charts?

Q12-7. Describe the basic circuit model of FSMD.

Q12-8. What is the meaning of single-assignment rule?

Q12-9. What is a bus?

12.1.3 Implementation Architectures

Many options may be used to implement an RTL design. The rationale behind these options is a trade-off among performance (throughput), space (area), and time (operating frequency). The widely used implementation options in practical systems include: *single-cycle structure*, *multiple-cycle structure*, and *pipeline/parallelism structure*.

12.1.3.1 Single-Cycle Structures Generally speaking, a single-cycle structure only uses combinational logic to realize the desired functions. It may require a rather long propagation delay to finish a computation of the desired functions. A simple example of using the single-cycle structure is an n-bit adder, which has appeared many times in this book. More complex examples include array multipliers and dividers introduced in Chapter 10.

12.1.3.2 Multiple-Cycle Structures A multiple-cycle structure executes the desired functions in consecutive clock cycles. The actual number of needed clock cycles depends on the specific functions to be computed. The multiple-cycle structure may be further classified into two basic types: *linear structure* and *nonlinear structure*. The linear multiple-cycle structure simply cascades many stages together and performs the desired functions without sharing resources, and hence the same combinational logic circuit might be duplicated several times if two or more stages carry out the same function. The nonlinear multiple-cycle structure can be further subdivided into single-stage or multiple-stage. The single-stage nonlinear multiple-cycle structure performs the desired functions with sharing resources by using feedback. The multiple-stage nonlinear multiple-cycle structure carries out the required functions with sharing resources by using feedback or feed-forward connections; it may be a feedback or feed-forward structure or a combination of both. As a consequence, one important feature of nonlinear multiple-cycle structures is that they may reuse the same hardware many times.

12.1.3.3 Pipeline/Parallelism Structures A pipeline structure is a multiple-stage, multiple-cycle structure in which new data can be generally fed into the structure at a regular rate. Hence, it may output a result per clock cycle after the pipeline is fully filled. The pipeline structure may also have linear and nonlinear types. The difference between a pipeline structure and a regular multiple-stage, multiple-cycle structure is determined by whether the new data can be regularly fed into the structure.

■ Review Questions

Q12-10. What are the features of single-cycle architecture?

Q12-11. What are the features of multiple-cycle architecture?

Q12-12. What is the distinction between a multiple-stage, multiple-cycle structure and a pipeline structure?

12.2 Synthesis Flows

Once we have studied the various design options of a digital system, we introduce in this section the synthesis-flow using *hardware description language* (HDL) for designing and implementing a digital system based on an FPGA device or a cell library. When designing an FPGA-based system or an ASIC, we often follow a synthesis flow. A *synthesis flow* is a set of procedures that allows designers to progress from the specifications of a desired system to the final FPGA or chip implementation in an efficient and error-free way. We concisely describe the general synthesis flow of designing FPGA-based systems and ASICs.

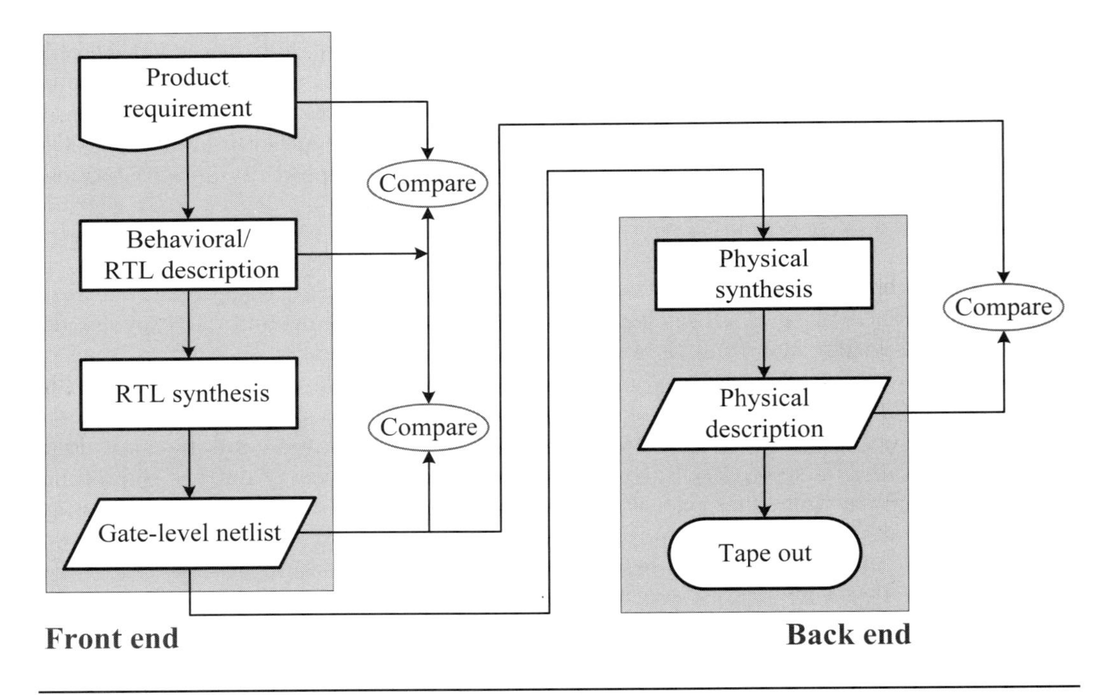

Figure 12.4: The general synthesis flow of an FPGA-based and ASIC design.

12.2.1 The General Synthesis Flow

An RTL synthesis is a mixed style combining both dataflow and behavioral styles with the constraint that the resulting description can be acceptable by synthesis tools. The general synthesis flow of an FPGA-based and ASIC design is shown in Figure 12.4.

From this figure, we know that the synthesis flow can be divided into two major parts: *front end* and *back end*. The front end is target-independent and contains three phases, starting from product requirement, behavioral/RTL description, and ending with RTL synthesis, and generates a gate-level netlist. The back end is target-dependent and mainly comprises the physical synthesis, which accepts the structural description of a gate-level netlist and generates a physical description. In other words, the RTL synthesis is at the heart of the front-end part and the physical synthesis is the essential component of the back-end part.

12.2.2 RTL Synthesis Flow

The general RTL synthesis flow is shown in Figure 12.5. The RTL synthesis flow begins with product requirement, which is converted into a *design specification*. The specification is then described with an RTL behavioral style in Verilog HDL or VHDL. The results are then verified by using a set of test benches written by HDL. This verifying process is called *RTL functional verification*. The functional verification ensures that the function of design entry is correct and conforms to the specification, in addition to performing some basic checks such as syntax error in HDL.

The RTL description is synthesized by a logic synthesizer after its function has been verified correctly. This process is denoted as *RTL synthesis* or *logic synthesis*. The essential operation of logic synthesizer is to convert an RTL description into generic gates and registers, and then optimize the logic to improve speed and area. In addition,

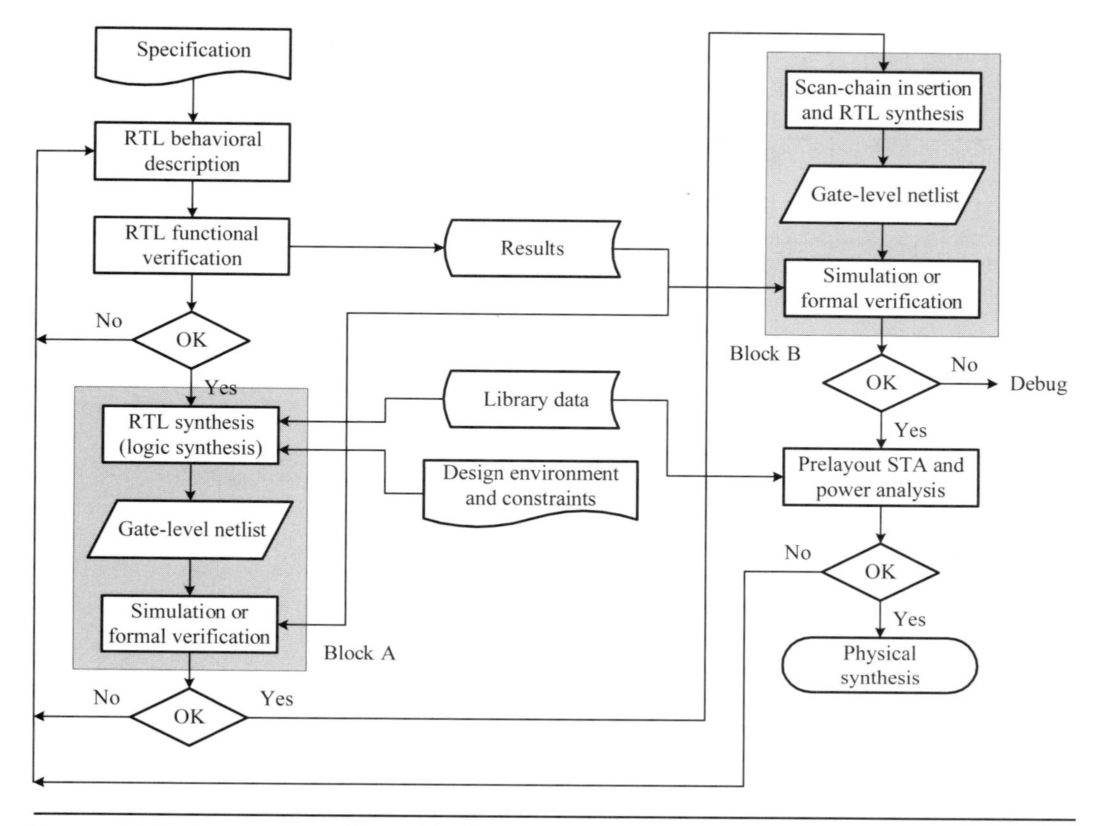

Figure 12.5: The general flow of RTL synthesis.

finite-state machine decomposition, datapath optimization, and power optimization may also be performed at this stage. Generally speaking, a logic synthesizer accepts three inputs, *RTL code*, *technology library*, and *design environment and constraints*, and generates a gate-level netlist.

After a gate-level netlist is generated, it is necessary to rerun the test benches used in the stage of RTL functional verification to check if they produce exactly the identical output for both behavioral and structural descriptions or to perform an RTL versus gate equivalence checking to ensure the logical equivalence between the two descriptions.

The next three steps often used in ASIC (namely, cell-based design) but not in FPGA-based designs are *scan-chain logic insertion*, resynthesis, and verification, as shown in the shaded block B. This block may be combined together with block A. The scan-chain (or test logic) insertion step is to insert or modify logic and registers to aid in the manufacturing test. *Automatic test pattern generation* (ATPG) and *built-in self-test* (BIST) are usually used in most modern ASIC designs. The details of these topics are addressed in Chapter 16.

The final stage of RTL synthesis flow is the prelayout *static timing analysis* (STA) and power dissipation analysis. Static timing analysis checks the temporal requirements of the design. The STA is a timing analysis alternative to the *dynamic timing analysis* (DTA), which is performed by simulation, by analyzing the timing paths of the design without carrying out any actual simulation. Through detailed STA, many timing problems can be corrected and system performance might also be optimized.

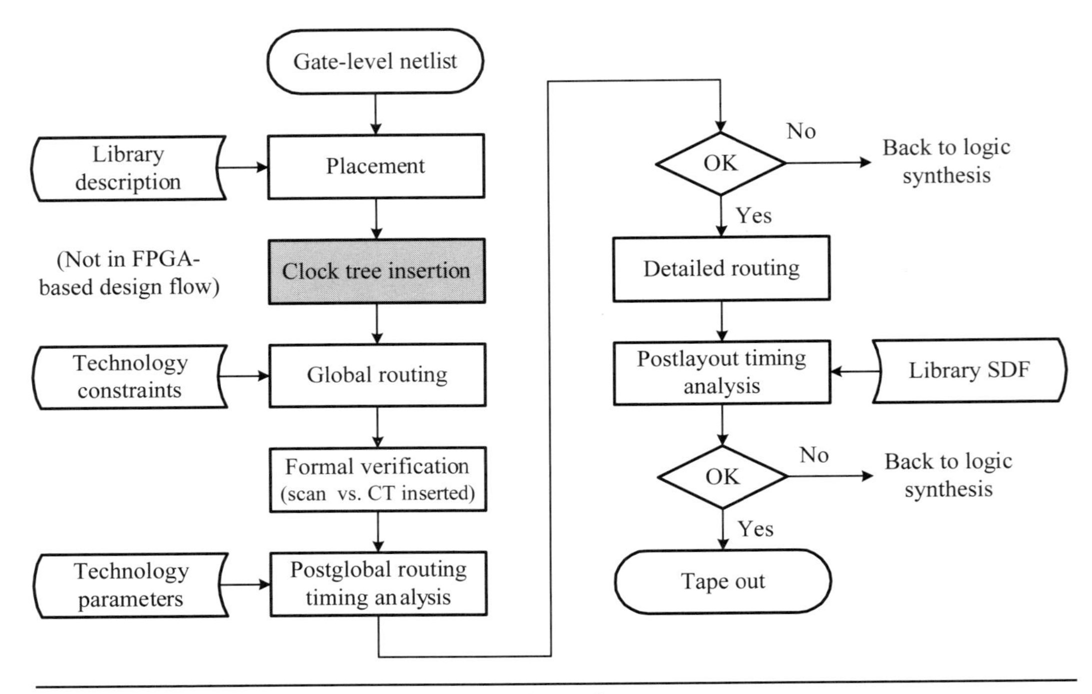

Figure 12.6: The general flow of physical synthesis.

The details of STA can be referred to Lin [9]. Power analysis estimates the power dissipation of the circuit. The power dissipation depends on the activity factors of the gates. Power analysis can be performed for a particular set of test vectors by running the simulator and evaluating the total capacitance switched at each clock transition of each node.

12.2.3 Physical Synthesis Flow

The second part of the synthesis flow of an FPGA-based or ASIC system is the physical synthesis. In this part, we have to choose a target, such as complex programmable logic device (CPLD), FPGA, or a cell library. Regardless of the FPGA-based or ASIC system, physical synthesis can be further subdivided into two major stages: *placement* and *routing*, as shown in Figure 12.6. Therefore, physical synthesis is usually called *place and route* (PAR) in CAD tools.

In the placement stage, logic cells (standard cells or building blocks) are placed at fixed positions to minimize the total area and wire lengths. In other words, the placement stage defines the location of logic cells (modules) on a chip and sets aside space for the interconnect of each logic cell (module). This stage is generally a mixture of three operations: *partitioning*, *floorplanning*, and *placement*.

Partitioning divides the circuit into parts such that the sizes of the components (modules) are within prescribed ranges and the number of connections between components is minimized. Floorplanning determines the appropriate (relative) location of each module in a rectangular chip area. Placement finds the best position of each module on the chip such that the total chip area is minimized or the total length of wires is minimized. Of course, not all CAD tools have their placement divided into the above three substeps. Some CAD tools may simply combine these three substeps into one big step known as *placement*.

After placement, a clock tree is inserted in the design. In this step, a clock tree is generated and routed coupled with the required buffers. A clock tree is often placed before the main logic placement and routing is completed to minimize the clock skew. This step is not necessary in an FPGA-based synthesis in which a clock distribution network is already fixed on the chip.

The next big stage is known as the *routing*, which is used to complete the connections of signal nets among the cell modules placed by placement. This stage is often further subdivided into two substages: *global routing* and *detailed routing*. Global routing decomposes a large routing problem into small and manageable subproblems (detailed routing) by finding a rough path for each net to reduce chip size, shorten the total length of wires, and evenly distribute the congestion over the routing area. Detailed routing carries out the actual connections of signal nets among the modules.

After both global and detailed routing steps, a separate static timing analysis for each of these two steps is performed. These timing analyses rerun the timing analysis with the actual routing loads placed on the gates to check whether the timing constraints are still valid.

The final tape-out stage has different meanings for cell-based (cell library or standard cells) and gate-array-based syntheses as well as CPLD/FPGA-based syntheses. For cell-based and gate-array-based syntheses, the tape-out stage generates the photomasks so that the resulting designs can be "programmed" in an IC (Integrated Circuit) foundry. For CPLD/FPGA-based syntheses, the tape-out stage generates the programming file to program the device. Of course, after fabricating a device or programming a CPLD/FPGA device, we often test the device in a real-world environment to see if it indeed works as expected.

■ Review Questions

Q12-13. What is a synthesis flow?

Q12-14. Which phases are included in the front-end part of the general synthesis flow?

Q12-15. Why is the front-end part often called logic synthesis?

Q12-16. Which phases are included in the back-end part of the general synthesis flow?

Q12-17. Why is the back-end part often called physical synthesis?

Q12-18. What are the two major stages of physical synthesis?

12.3 Implementation Options of Digital Systems

Design is a series of transformations from one representation of a system to another until a representation that can be fabricated exists. Implementation (or realization) is the process of transforming design abstraction into physical hardware components such as FPGAs or cell-based ICs (integrated circuits). In this section, we introduce a broad variety of options for digital system implementations available now. These options can be classified into the following three classes: *platforms*, *ASICs*, and *field-programmable devices*.

12.3.1 Platform-Based Systems

The choice of implementation options for a specific VLSI (digital) system depends on the following factors: power budget, performance requirements, time to market, and the cost of final product. Among these, probably the time to market and cost of final product are the two most important factors that often dominate the others and determine whether a product is successful despite how excellent the design is.

As shown in Figure 1.45, a platform-based system can exist in one of the following forms: *hardware* μ*P/DSP system*, *platform FPGA*, and *platform IP*. A platform FPGA is also called a *programmable system chip* (PSC) or a *system on a programmable chip* (SoPC) and is an FPGA device containing μP/DSP system modules in a variety of IP forms. A platform IP is a soft IP that contains μP/DSP system modules and can be synthesized into an FPGA or a cell library. In this subsection, we describe each of these in brief.

12.3.1.1 Hardware μP/DSP Systems Since the time to market and cost of final product are two vital factors that determine whether the product is successful, the use of the μP/DSP system to design a specific digital system is often the first attempt. Because of very large product volume of these systems, their amortized NRE cost is very low and can be ignored. In addition, the design and implementation of a given digital system using these systems are at a system level. Consequently, the time to market is very short. As for the consideration of power dissipation and performance requirements, it is necessary to choose an appropriate μP and/or DSP according to the actual requirements.

An electronic device that incorporates a computer (usually a microprocessor) or computers within the implementation is often referred to as an *embedded system*. In such a system, a computer is primarily used as a component like the others to simplify the system design and to provide flexibility; the user of the system is not even aware of the presence of a computer inside the system. In other words, an embedded system is the one that regards the computers as components for designing the system.

Nowadays, there are so many μP systems that can be used to design a digital system. These systems can be cast into low-end, medium-end, and high-end in terms of the computation power of the center processing unit (CPU), which is the heart of a μP system. The low-end systems include an 8-bit CPU along with some needed peripherals; the medium-end systems contain a 16-bit CPU with digital-signal processing enhancement coupled with some needed peripherals; the high-end systems are often targeted to multimedia applications and combine one or more 32-bit CPUs together with one or more DSPs as a possible option and some high-performance peripherals, in particular, having the networking capability.

12.3.1.2 Platform IPs At present, a lot of CPUs and a wide variety of peripherals are ready for use in both forms of software and hard modules. Such modules are referred to as macros or IPs. Because a broad variety of hard and soft IPs are available, it is possible to design a digital system at the system level using a cell-based approach or FPGA devices. Presently, the configuration of platform IPs can be automated or manual. To reduce the burden of designers, CAD tools with the capability of an *integrated-design environment* (IDE) are provided from vendors. By using these tools, the designers are able to design the desired system at the system level using a high-level programming language, such as C or C++. We briefly examine each of these two types of configurable platform IPs.

Automated configurable platform IPs. A popular automated configurable platform IP is Tensilica Xtensa, which is a user-configurable and user-extensible processor architecture on which an *application-specific instruction set processor* (ASIP) can be developed. To develop an application system based on this platform IP, designers first profile the software to be executed. Next, designers select the application-specific instruction set options, additional data types and instructions, execution units, memory hierarchy, and external interfaces, as well as other related building blocks. The CAD tool then configures the required processor and automatically generates the synthesizable RTL description accompanied with preverified control logic and its associated software development tools, including a C/C++ compiler, assembler, linker, simulator, and debugger. Finally, designers verify the processor hardware with an automatically generated test bench and checkers.

Manually configurable platform IPs. The representative of manually configurable platform IPs is ARM OptimoDE. To configure such a system, designers first manually define the instruction set processor architecture using DesignDE and OptimoDE resource libraries provided by the vendor. The RTL views of the individual blocks are preverified. Then, the tool automatically generates a simulator, the Verilog RTL or target FPGA implementation. Finally, designers manually develop all related softwares using DEvelop, a stand-alone version of the tool's C/C++ compiler.

A platform IP is the system designed using a cell library with CPU and peripheral IPs. A system designed with platform IPs is usually optimized and scalable since only the required function units are included into the system and the system function can be easily changed simply by adjusting some parameters at the system level.

12.3.1.3 Platform FPGAs

Recall that a platform FPGA combines features from both platform and field-programmable devices into a single device; it is an FPGA device containing one or more CPUs in a hard, soft, or hardwired IP form, some periphery modules, and field-programmable logic modules. The major distinction between a hard IP and a hardwired IP is that the former only comes as a predesigned layout while the latter is already fabricated along with FPGA fabrics. Hardwired IPs usually cover CPUs, memory blocks, and multipliers. The CPU is usually a type of 32-bit *reduced instruction set computer* (RISC). The features of such IPs are that the block size, performance, and power dissipation can be accurately measured. In addition to modules related to μP systems, platform FPGA devices also provide enough logic cells for customizing the logic function needed in application systems.

The choice of using which type of CPU, soft or hardwired, in a design is totally determined case by case, depending on the allowed cost and power dissipation as well as required performance.

12.3.1.4 Comparison of Various Platforms

Both platform IP and platform FPGA use system-level cells, such as μP/DSP, memory modules, and peripherals, to construct a desired system in much the same way as the hardware μP/DSP system. The hardware μP/DSP system uses hardwired modules or standard discrete modules, whereas platform IP and platform FPGA use hardwired, hard, and/or soft IPs. The distinction between platform IP and platform FPGA is that the former uses a cell-based approach, while the latter uses FPGA devices to build the desired system.

Because all of the three platform-based systems introduced above use computer(s) as components to build systems, they are called *embedded systems*. In addition, since these systems are founded on silicon, they are system-on-a-chip (SoC). In other words, an SoC is an embedded system on silicon. The design issues of platform-based systems

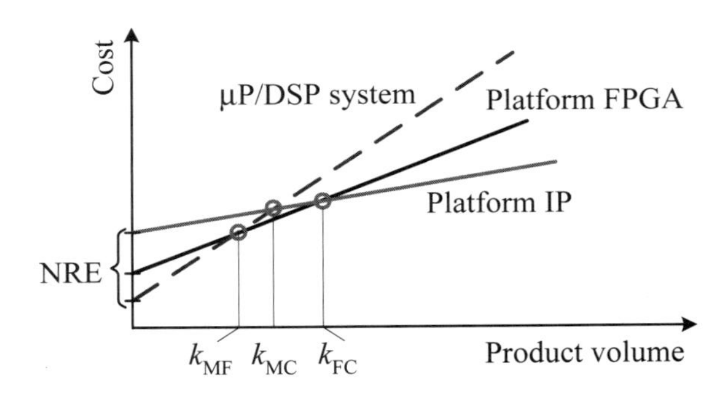

Figure 12.7: Comparison of various platform options in terms of cost versus product volume.

involve a system-level hardware and software co-design, which is beyond the scope of this book.

The aforementioned three implementation options are at the system level. They are able to accomplish the same function but at different costs and performances. The criteria for choosing an appropriate option to design a desired digital system are determined as follows. As shown in Figure 12.7, the comparison of various platform options is plotted in terms of cost versus product volume. The μP/DSP system has the lowest NRE cost but highest variable cost while the platform IP has the highest NRE cost but the lowest variable cost. The platform FPGA comes in between these two extremes. As a consequence, there is a cross-point k_{MF} of the cost versus product volume function between the μP/DSP system and platform FPGA, as illustrated in Figure 12.7. Below this point, it is more cost-effective to design and implement the system with the μP/DSP system. A cross-point k_{FC} also exists between the platform FPGA and platform IP. Above this point, it is more cost-effective to design and implement the system with platform IP. Between points k_{MF} and k_{FC}, the platform FPGA is the better choice. The cross-point k_{MC} between μP/DSP system and platform IP represents the situation that above this point using platform IP is much more inexpensive than the μP/DSP system. However, because of the higher NRE cost than platform FPGA, platform FPGA is usually used instead.

In summary, when designing a digital system needing a CPU, we first take into account the μP/DSP system because of its low NRE cost. If this type of system does not fit into the requirements or the resulting system requires some customized logic function, then we may shift to use platform FPGA before resorting to platform IP. Usually, we use the platform IP to design a system only when the market of the system has been successful and the product volume has reached the cross-point k_{FC}. Even though the final system is to be realized by platform IP, it often begins to prototype the system in platform FPGA because it has virtually no mask cost.

■ Review Questions

Q12-19. What are the three forms of platform-based systems?

Q12-20. What devices can be used in a hardware μP/DSP system?

Q12-21. What is an IP? What are the three common types of IPs?

Q12-22. What is the distinction between a hard IP and a hardwired IP?

Q12-23. What is the platform IP? What is the platform FPGA?

12.3.2 ASICs

ASICs refer to those integrated circuits that need to be processed in IC foundries. They can be designed and implemented with one of the following three methods: full-custom, cell-based, and gate-array-based. We briefly describe each of these.

12.3.2.1 Full-Custom Design Full-custom design starts from scratch and needs to design the layouts of every transistor and wire. It requires background knowledge at all levels, from system specification down to layout, as exhibited in Figure 1.27. The features of full-custom design are to design the specific digital system in a top-down manner, from the system level down to the physical level, and then realize it in a bottom-up fashion, from the physical level up to the system level.

Even though it provides the best performance among all options, the full-custom design is not an easy and a cost-effective way to design a digital system. The reasons are as follows. First, with the reduction of feature sizes, more expensive equipment is needed. Second, more challenging design issues arise with the reduction of feature sizes and the increased complexity of functionality in a VLSI system. Because of the parasitic resistance, capacitance, inductance, and their combined effects caused by closely spaced wires, to maintain power-supply integrity, clock integrity, and signal integrity, in such a wicked environment is much more challenging than before.

To tackle the above design challenges, a more complex synthesis flow, and more expensive CAD tools are required to verify the design. This means that the NRE cost is significantly increased with the reduction of feature sizes. This can be seen from the fact that nowadays the NRE cost is profoundly increased to a level where industry prototypes are also widely done using multiproject chips to amortize the mask cost over multiple designs.

In summary, because of the need of a complex synthesis flow, the design throughput is dramatically reduced and far behind the progress of feature-size reduction. Currently, the full-custom design is only reserved for designing the performance-critical devices, such as CPUs, graphics processing units (GPUs), FPGAs, and memory devices.

12.3.2.2 Cell-Based Design Cell-based design is a synthesizable flow using HDL, Verilog HDL or VHDL,[1] and is a simplified design flow of full-custom by using a set of predesigned layout cells. However, the resulting design still needs to be processed in an IC foundry in the same way as the full-custom design.

The standard cells in a typical cell library are shown in Table 12.1. They can be partitioned into three classes: combinational cells, sequential cells, and subsystem cells. The combinational cells cover inverters/buffers, basic gates, multiplexers/demultiplexers, decoders/encoders, Schmitt trigger circuits, and I/O pad circuits.

The inverters/buffers as well as tristate buffers are provided with various driving capabilities, called $1X$, $2X, \cdots$, and so on. The basic gates, including NAND/AND, NOR/OR, and XOR/XNOR, usually support 2 to 8 inputs and have three types of performance denoted as high, normal, and low power. Multiplexers and demultiplexers allow 2 to 16 inputs and outputs. Both cells support inverting and noninverting outputs. Encoders and decoders also provide 4 to 16 inputs and outputs, respectively, and have both inverting and noninverting output types. Schmitt trigger circuits with inverting and noninverting outputs are also provided. They are found in a variety of applications to reshape the input waveform with low-rise and low-fall times. The final set of combinational cells in the cell library is I/O pad circuits, containing input,

[1] It is an acronym of very-high-speed integrated circuit (VHSIC) hardware description language.

Table 12.1: The standard cells in a typical cell library.

Types of standard cells	Variations
• Combinational cells	
Inverters/buffers/tristate buffers	1X, 2X, 4X, 8X, 16X
NAND/AND gates	2 to 8 inputs (high, normal, low power)
NOR/OR gates	2 to 8 inputs (high, normal, low power)
XNOR/XOR gates	2 to 8 inputs (high, normal, low power)
MUX/DeMUX	2 to 16 inputs/outputs (inverting/noninverting output)
Decoders/Encoders	4 to 16 inputs/outputs (inverting/noninverting output)
Schmitt trigger circuits	Inverting/Noninverting output
I/O pad circuits	Input/Output (tristate, bidirectional, 1 to 16 mA)
• Sequential cells	
Latches	D type
Flip-flops/registers	D/JK-type (synchronous/asynchronous clear/reset)
Counters	BCD/binary (synchronous/asynchronous)
• Subsystem cells	
Adders/subtracters	4 to 32 bits (CLA/Brent-Kung/Kogge-Stone)
Multipliers	8/16/32 bits (signed/unsigned)
Barrel shifters	8/16/32 bits (arithmetic/logical, left/right)

output, and bidirectional cells. The output cells are often designed to provide different driving currents, ranging from 1 to 16 mA.

Sequential cells include D-type latches, D-type and JK flip-flops and registers, and binary-coded decimal (BCD) and binary counters. These counters can be either synchronous or asynchronous.

Subsystem cells include adders and subtractors, ranging from 4 to 32 bits in a step of 2^i, where i is a positive integer and in the range of 2 to 5. Three different adders are often provided in a typical cell library. These are carry-lookahead (CLA) adders, Brent-Kung adders, and Kogge-Stone adders. Sometimes a cell library also provides signed and unsigned multipliers with word widths of 8, 16, and 32 bits. Another type of widely used subsystem cells is barrel shifters, which may function as an arithmetic or a logical left/right shift with a word length of 8, 16, and 32 bits.

12.3.2.3 Gate-Array-Based Design Gate-array-based design combines a variety of soft IPs and builds the resulting design on a wafer with prefabricated transistors. Like the cell-based design, the gate-array-based design uses a synthesis flow to design a system and the resulting design also needs to be fabricated in an IC foundry.

The essential features of gate arrays are that standard transistors have been fabricated in advance using standard masks and only the metalization masks are left to users to define their final logic functions. Consequently, gate arrays (GAs) are also referred to as *uncommitted logic arrays* (ULAs) because their functions are left to be defined by users and are a type of semicustom design of ASICs. Here, the semicustom means that only a partial set of masks needs to be processed in an IC foundry. The basic element of gate arrays can be either NOR gate or NAND gates in CMOS technology. A particular subclass of gate arrays is known as *sea-of-gates* (SoGs). Sea-of-gates differ from gate arrays in that the array of transistors is continuous rather than segmented.

A typical structure of complementary metal-oxide-semiconductor (CMOS) gate arrays is shown in Figure 12.8, which consists of a set of pairs of CMOS transistors and routing channels. If the rows of n-type metal-oxide-semiconductor (nMOS) and p-type

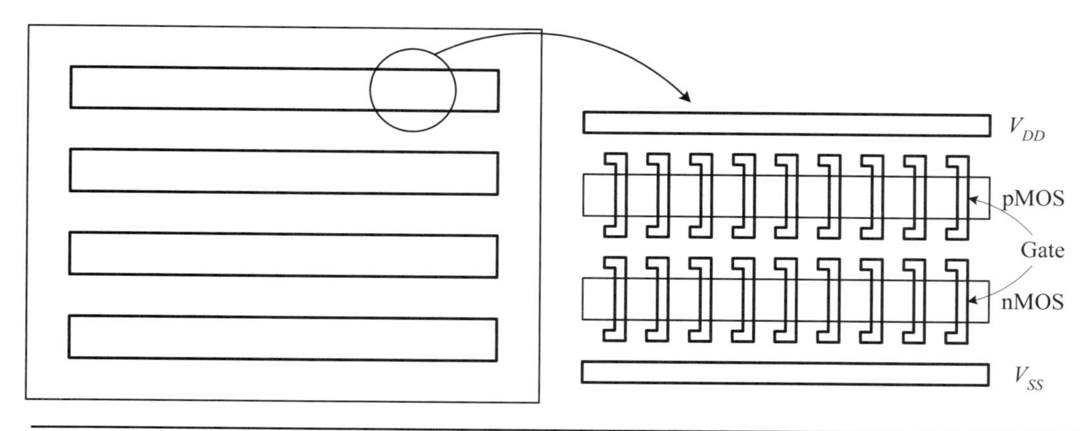

Figure 12.8: The basic structure of gate arrays and sea-of-gates.

metal-oxide-semiconductor (pMOS) transistors are broken into segments with each having two or three transistors, the result is the gate-array structure; otherwise, it is the sea-of-gates structure. The features of sea-of-gates are that they utilize multilayer metalizations and remove the routing channels. Hence, they have a higher density than gate arrays.

Like the full-custom design, it is quite clumsy and time consuming to start an ASIC design with gate arrays from scratch. Similar to the cell-based approach, macro libraries are usually provided by the vendors to save design time. The typical macro library associated with gate arrays is similar to the standard-cell library with only one major difference—that it contains only the metalization masks because the transistors on gate arrays are already fabricated in advance.

Compared to full-custom and cell-based designs, gate-array-based design has less functionality per unit area due to the overheads of interconnect and those unused prefabricated transistors. However, unlike full-custom and cell-based design, which need to process every mask, gate-array-based design only requires processing the final metalization masks. In addition, like cell-based design, gate-array-based design also uses the synthesis flow. Thus, gate-array-based design takes much less time to complete an ASIC design and prototyping.

Since all of full-custom, cell-based, and gate-array-based designs need their designs to be fabricated in IC foundries, these design approaches are time-consuming processes and need a great amount of time to prototype an ASIC. Hence, field-programmable devices have entered the market as alternatives to overcome this difficulty and provide faster prototyping. The field-programmable devices will be described in more detail in the following subsection.

■ Review Questions

Q12-24. What is the essential feature of ASICs?

Q12-25. What are the essential features of full-custom designs?

Q12-26. What is the distinction between cell-based and full-custom designs?

Q12-27. What is the uncommitted logic array?

Q12-28. What is the distinction between gate arrays and sea-of-gates?

12.3.3 Field-Programmable Devices

Fundamentally, field-programmable devices can be subdivided into two types, programmable logic device (PLD) and CPLD, and FPGA, according to their logic structures. The common feature of PLDs and CPLDs is that they use a two-level AND-OR logic structure to realize (or implement) switching functions. FPGA devices combine the features of gate arrays and the on-site programmability of PLDs and CPLDs.

The programmable options of PLDs/CPLDs and FPGAs can be either *mask-programmable* or *field-programmable*. When using mask-programmable devices, the designer needs to provide the vendors with the designed interconnect pattern of a given device for preparing the required masks to fabricate the final ASIC. Field-programmable devices can be personalized on site by the designer using appropriate *programming equipment*, referred to as a *programmer*. Field-programmable devices can be further subdivided into two types: *one-time programmable* (OTP) and *erasable*. OTP devices can only be programmed one time; erasable devices can be reprogrammed as many times as required. In summary, field-programmable devices are usually used at the start-up time of a design to gain flexibility or in low-volume production to save the NRE cost, whereas mask-programmable devices are used in high-volume production to reduce the cost.

12.3.3.1 Programmable Logic Devices When used to design a digital system, a PLD can replace many small-scale integration (SSI) and/or medium-scale integration (MSI) devices. Consequently, using PLDs to design digital systems allows us to considerably reduce the number of wires, the number of devices used, the area of the *printed-circuit board* (PCB), and the number of connectors. The hardware cost of the resulting system is then significantly lowered.

PLDs can be subdivided into the following three categories: *programmable logic array* (PLA), *read-only memory* (ROM), and *programmable array logic* (PAL). All of these three devices have a similar two-level AND-OR logic structure. The essential differences among these devices are the programmability of AND and OR arrays, as shown in Figure 12.9. For ROM devices, the AND array generates all minterms of inputs and hence is fixed, but the OR array is programmable to implement the desired functions. For PLA devices, both AND and OR arrays are programmable. Therefore, they provide the maximum flexibility among the three types of PLDs. For PAL devices, the AND array is programmable but the OR array is fixed to connect to some specified AND gates.

Nowadays, the discrete PLA devices have become obsolete. However, PLA structures and ROMs are often used in full-custom and cell-based designs to take advantage of their regular structures. Both PAL and ROM not only have commercial discrete devices but also are widely used in digital systems.

12.3.3.2 Programmable Interconnect (PIC) The programmable interconnect structures of field-programmable devices can be classified into three types: *static RAM* (SRAM), *Flash* (EEPROM), and *antifuse*, as shown in Figure 12.10. The interconnect structure based on SRAM cell is an nMOS or a TG switch controlled by an SRAM cell. The basic structure of an SRAM cell is a bistable circuit, as shown in Figure 12.10(a). Once programmed, the SRAM cell retains its state until it is reprogrammed or its supply voltage is removed. The Flash cell is like the cell used in Flash memory devices, as shown in Figure 12.10(b). Its basic structure is a floating-gate transistor, which can be programmed to store "1" or "0." Once programmed, the Flash cell retains its state permanently, even when the supply voltage is removed. However, it can be

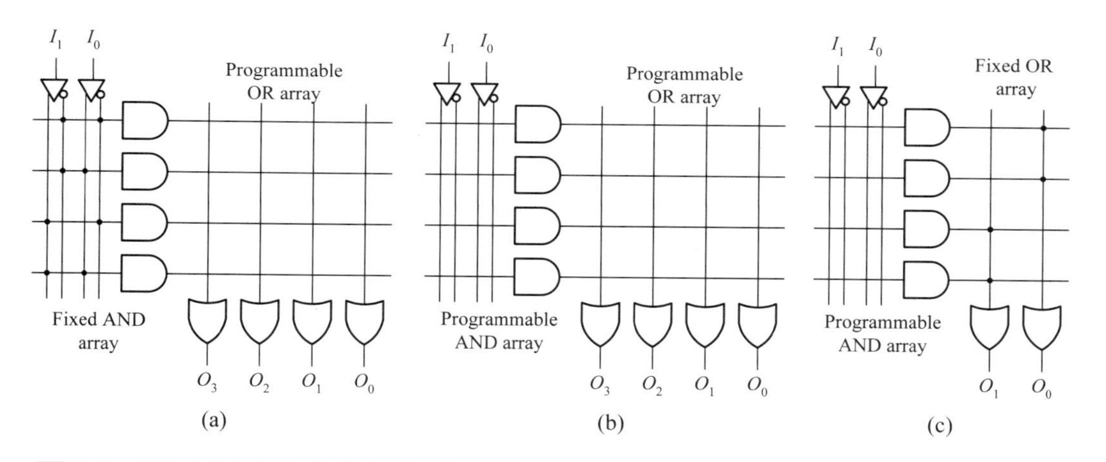

Figure 12.9: The basic structures of PLDs: (a) ROM; (b) PLA; (c) PAL.

Table 12.2: The basic features of programmable interconnect.

	SRAM	Flash	Antifuse
Process	Standard CMOS Standard CMOS	Standard two-level polysilicon	New type polysilicon
Programming approach	Shift register	FAMOS	Avalanche
Cell area	Very large	Large	Small
Resistance	≈ 2 kΩ	≈ 2 kΩ	≈ 500 Ω
Capacitance	50 fF	50 fF	10 fF

reprogrammed many times as needed. As shown in Figure 12.10(c), the antifuse is a device that operates in the reverse direction of a normal fuse; namely, it has high resistance in a normal condition but is changed to low resistance permanently when an appropriate voltage has been applied to it.

The basic characteristics of the above three types of programmable interconnect structures are summarized in Table 12.2. From the table, we can see that the antifuse structure has the best performance since it has the lowest resistance and capacitance. The other two structures have almost the same performance. In addition, the SRAM structure is volatile, but Flash (EEPROM) and antifuse structures are not.

12.3.3.3 Complex Programmable Logic Devices Because of the popularity of PAL devices along with the mature of VLSI technology, combining many PALs with a programmable interconnect structure into the same chip is feasible. This results in a device known as a complex PLD (CPLD). Recall that each output of a PAL device consists of an OR gate associated with a few AND gates, ranging from 5 to 8 gates. Such a circuit is used as the building blocks in CPLDs and is usually referred to as a *PAL macro* or a *macrocell* for short.

The basic structures of CPLDs consist of PAL macros and interconnect, as well as input/output blocks (IOBs). There are two basic types of CPLDs, which are classified according to the arrangement of PAL macros and the interconnect structures, as shown in Figures 12.11(a) and (b), respectively. The first type as depicted in Figure 12.11(a) is most widely used in commercial CPLDs, where PAL macros are placed on both sides of the programmable interconnect area. Another type of CPLD structure is depicted in Figure 12.11(b), where PAL macros are placed on all four sides and a programmable

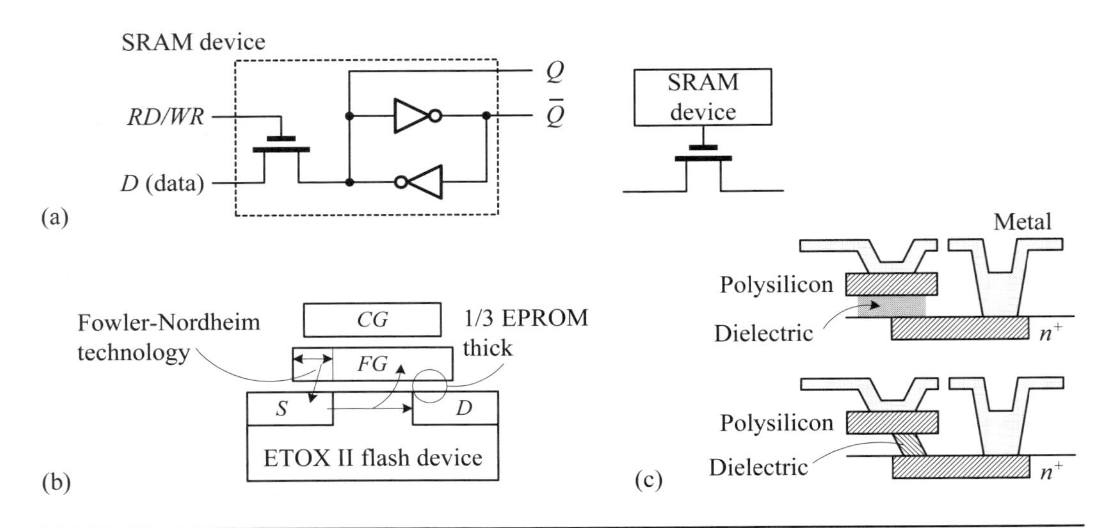

Figure 12.10: The basic structures of programmable interconnect: (a) SRAM cell; (b) Flash cell; (c) antifuse cell.

interconnect area is placed at the center region, called the *global routing area*. In addition, an *output routing area* is placed between the PAL macros and input/output blocks.

12.3.3.4 Field-Programmable Gate Arrays The basic structures of FPGAs are composed of *programmable logic blocks* (PLBs), interconnect, and input/output blocks (IOBs). These components are referred to as *fabrics* of FPGAs. The PLBs are called *configurable logic blocks* (CLBs) in Xilinx terminology and logic elements (LEs) in Altera terminology. The PLBs are usually used to implement combinational logic and sequential logic. Each PLB consists of a k-input *function generator* (namely, a *universal logic module*) and a D-type flip-flop, or other types of flip-flop, where k is usually set from 3 to 8. A circuit is known as a k-input function generator if it is capable of implementing any switching function with k variables at most. The most common k-input function generators used in FPGAs are implemented by using lookup tables or multiplexers. The lookup tables are usually constructed from SRAM or Flash memory.

A simplified PLB is shown in Figure 12.12, where two groups of a 4-input function generator (4-input LUT or 16-to-1 multiplexer) and a D-type flip-flop are contained in the PLB. The 4-input function generator in each group can realize any switching function with 4 variables at most. It can also be used to implement sequential logic when combined with the output D-type flip-flop. The D-type flip-flop is a universal flip-flop, with the capability of clock enable as well as asynchronous set and clear. The 2-to-1 multiplexer at the output of the PLB is used to bypass the D-type flip-flop when the PLB is employed to realize combinational logic. The multiplexer is set by the M bit, which is an SRAM bit, an antifuse cell, or a Flash cell, as we described before.

The k-input function generator can be realized by either a lookup table, composed of a $2^k \times 1$ RAM or Flash memory, or a 2^k-to-1 multiplexer. Regardless of which way is used, the essential idea is first to represent the switching function in the form of truth table and then store the truth table in the RAM/Flash memory or multiplexer (or multiplexer tree) to prepare for the look-up operations later.

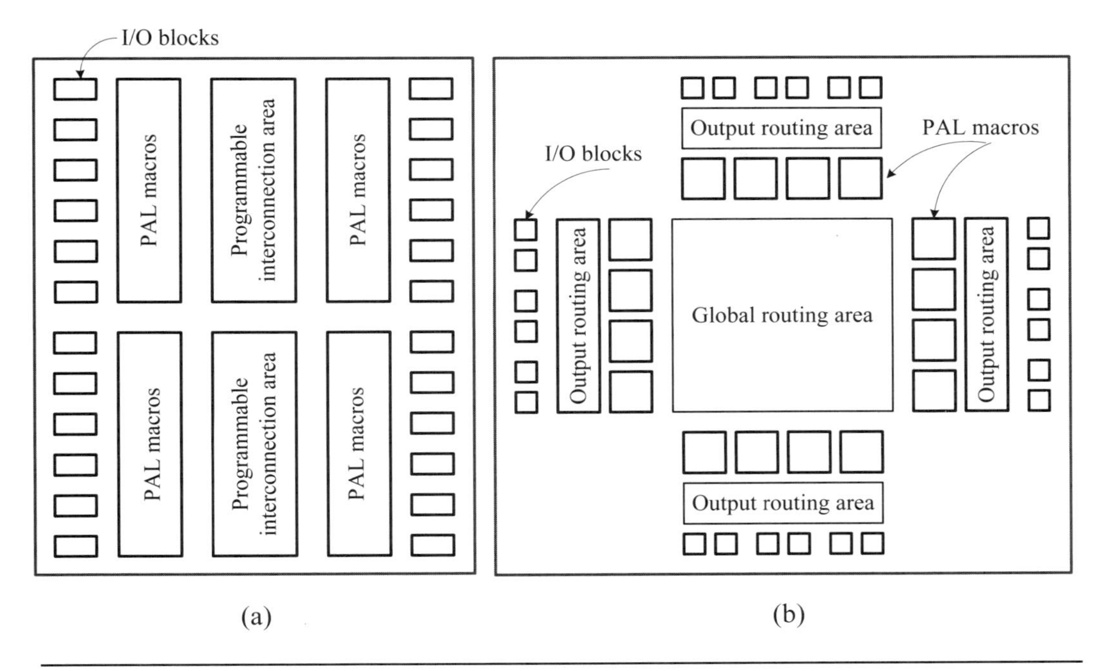

Figure 12.11: The basic structures of CPLDs: (a) CPLD; (b) pLSI.

According to the arrangements of PLBs on the chip, the basic structures of FPGA devices can be subdivided into two types: *matrix* and *row*. The matrix-type FPGA is shown in Figure 12.13(a), where PLBs are placed in a two-dimensional matrix manner. Between PLBs there are two types of interconnect, called *horizontal routing channels* and *vertical routing channels*, respectively. Figure 12.13(b) is a row-type FPGA, where the PLBs are placed intimately in a row fashion. The spaces between two rows are the routing channels.

Nowadays, many FPGAs also have been incorporated into specialized features for specific applications, including communications, multimedia, or consumer products. These features include carry-chain adders, multipliers, shifters, block and distributed RAMs, and even powerful 32-bit CPUs such as PowerPC and ARM CPU, as well as peripherals such as an Ethernet controller and USB controller. An FPGA device with such features is often called a platform FPGA, as previously defined.

12.3.4 Selection of Implementation Options

The criteria to choose an appropriate option for designing and implementing a specific digital system in a cost-effective way are usually based on the NRE cost and the easiness. The selection sequence is as follows in order:

- μP/DSP device(s)
- FPGA/SoPC device(s)
- GA/cell-based chip(s)
- Full-custom chip(s)

Nevertheless, it is not uncommon to mix use of many options in the same design.

When a system cannot be designed at the system level using a μP/DSP device (platform) due to the limitation of performance, it is natural to consider multiple

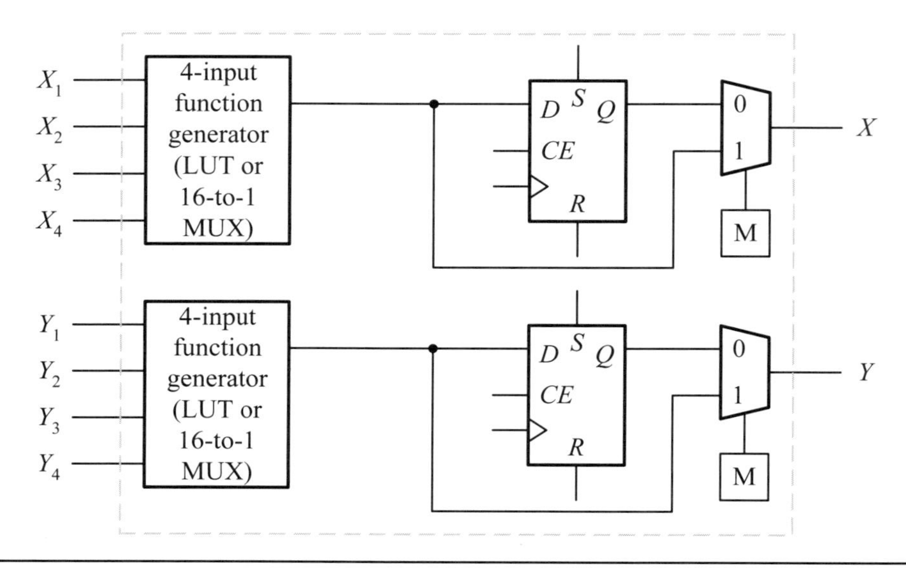

Figure 12.12: A simplified PLB showing the essential elements of typical PLBs.

μP/DSP devices or to design using FPGA, cell-based, or a full-custom approach. In addition to using a μP/DSP device, often an FPGA device is needed to realize some special logic functions required for the system. For such a case, a platform FPGA may be the appropriate choice. When the market scale is large enough, cell-based design is the proper selection to reduce the cost of the system.

For those systems not using microprocessors, they can be designed with FPGA, cell-based, or a full-custom approach. In general, cell-based design offers a 3 to 5 times performance over the FPGA design and full-custom design offers a 3 to 8 times frequency advantage over the cell-based design. However, the performance is not the unique criterion to determine whether an option is suitable for a specific system; the other two important factors, power dissipation and cost, also need to be taken into account.

As in the case of platform option, the product volume determines the proper implementation option for a specific digital system and the final cost of the system. The comparison of various options of ASICs in terms of cost versus product volume is illustrated in Figure 12.14. Among these three options, the NRE cost of FPGA is the lowest while that of full-custom is the highest. However, NRE cost alone does not decide the final cost of the system. The cost of a system is generally determined by the product volume, which amortizes the NRE cost; the higher product volume has the lower amortized NRE cost. As a result, as illustrated in Figure 12.14, using FGPA has the least cost among the three options when the product volume below the cross-point k_{FC} while using full-custom has the least cost when the product volume is above the cross-point k_{CF}.

Table 12.3 compares various promising options for implementing a digital system in terms of NRE cost, unit cost, power dissipation (PD), implementation complexity, performance, flexibility, and time to market.

In summary, except for those ICs needing high-performance and high product volume, such as dynamic random access memory (DRAM), SRAM, FPGAs, Flash memory devices, and so on, full-custom design is rarely used. Instead, a cell-based approach being combined with HDL synthesis flow is used to increase the design throughput to

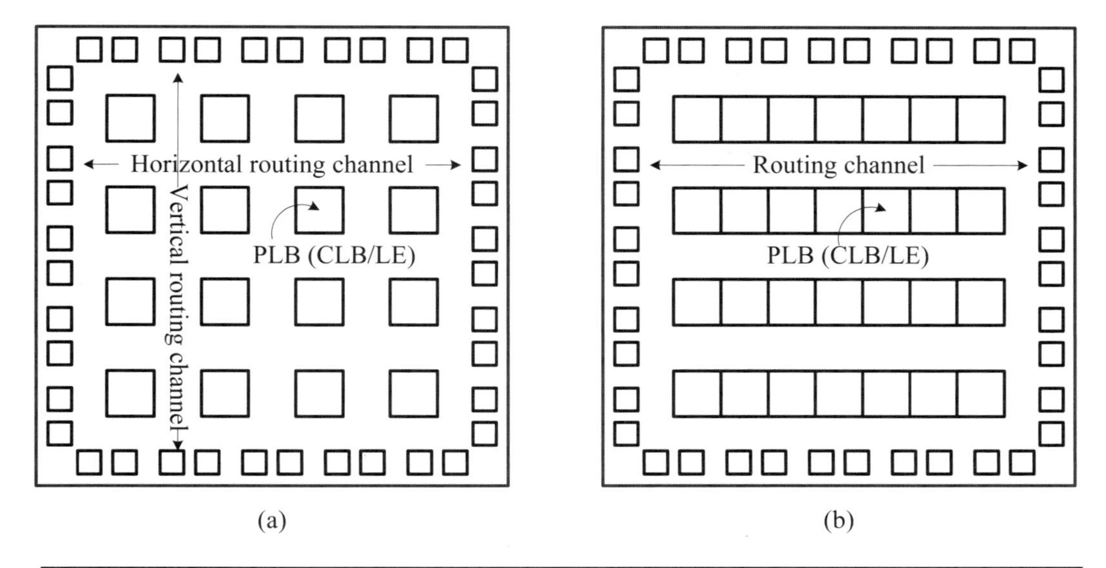

Figure 12.13: The basic structures of FPGAs: (a) matrix type; (b) row type.

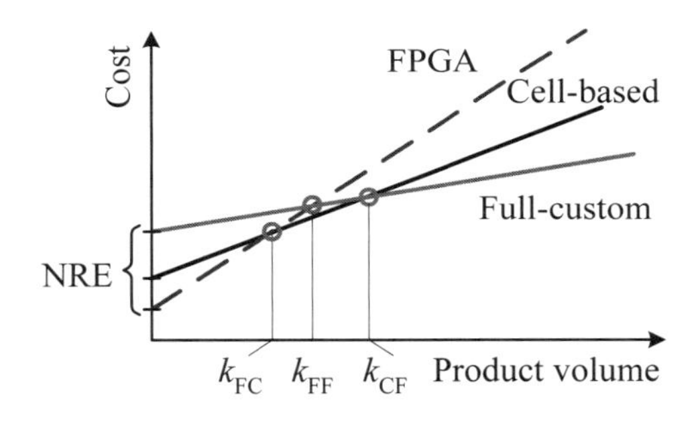

Figure 12.14: Comparison of various options in terms of cost versus product volume.

catch up the short time-to-market feature of modern digital products. Moreover, IP-based design approach is often used to design a very large system in much the same way as we have done on a PCB except that here the system is done on a silicon wafer rather than a PCB. For the ease of doing this, an *electronic system-level* (ESL) design is emerging and focuses on the high-level abstraction through which a large system can be designed, verified, as well as debugged, and the hardware and software of the system can be realized in a custom SoC, a platform FPGA, a system-on-board, or a multiboard system.

■ Review Questions

Q12-29. What is the difference between mask-programmable and field-programmable?

Q12-30. What is the essential feature of PLDs?

Q12-31. What is the distinction between CPLDs and FPGAs?

Q12-32. How would you distinguish ROM, PLA, and PAL?

Table 12.3: Comparison of various options for implementing a digital system.

		Field-programmable		ASIC		
Design method	μP/DSP	PLD	FPGA/CPLD	GA/SoG	Cell-based	Full-custom
NRE	Low	Low	Low	Medium	High	High
Unit cost	Medium	Medium	High	Medium	Low	Low
PD	High	Medium	Medium	Low	Low	Low
Complexity	Low	Low	Medium	Medium	High	High
Performance	Low	Medium	Medium	Medium	High	Very high
Flexibility	High	Low	High	Medium	Low	Low
Time to market	Low	Low	Low	Medium	High	High

12.4 A Case Study — A Simple Start/Stop Timer

As described previously, a specific digital system can be realized in a broad variety of options, including the μP/DSP system, a field-programmable device, and ASIC. In this section, we use a simple start/stop timer as an example to illustrate various possible design and implementation options of a given digital system so that the reader can understand the features of and the differences among these options.

12.4.1 Specifications

The start/stop timer has two buttons: *start* and *clear*. The start button controls an internal toggle switch, which in turn controls the operations of the start/stop timer in a toggle manner. That is, the start/stop timer will change its operation status from running to stop or vice versa each time when the start button is pressed. The clear button is used to clear the start/stop timer count while the start/stop timer is not running. In addition, suppose that the resolution of the timer is 10 ms and the maximum timer count is 99.99 seconds. The behavior of the start/stop timer is as follows:

```
// a start/stop timer with time resolution 10 ms
module start_stop_timer(
       input clk_10ms, reset_in, clear, start,
       output reg [15:0] count);
reg    start_flag;
// the body of the start/stop timer
assign reset = reset_in | (clear & ~start_flag);
always @(posedge start or posedge reset)
   if (reset) start_flag <= 0;
   else start_flag <= ~start_flag;
end
always @(posedge clk_10ms or posedge reset) begin
   if (reset) count <= 0;
   else if (start_flag) count <= count + 1;
end
endmodule
```

The display module used in the start/stop timer is shown in Figure 12.15. Figure 12.15(a) depicts the structure of typical common-anode seven-segment LED (light-emitting diode) display; Figure 12.15(b) shows the multiplexing-driven four-seven-

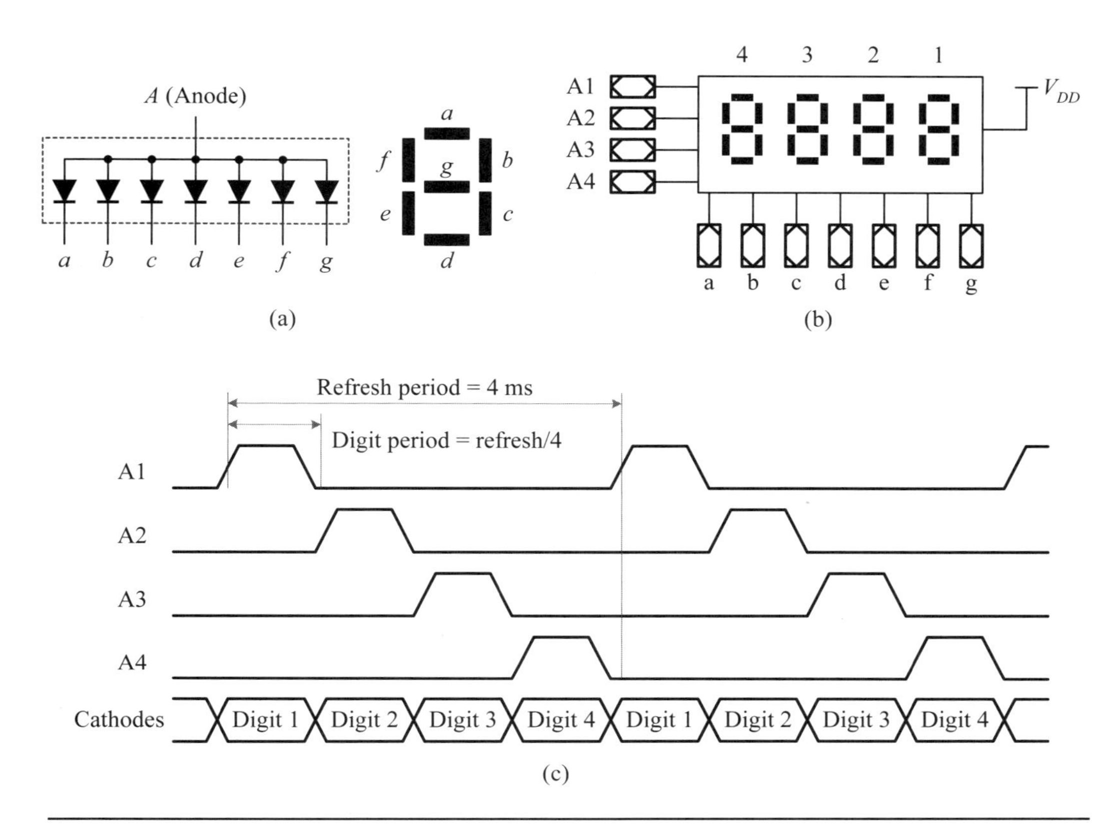

Figure 12.15: The multiplexing-driven 4-digit seven-segment LED display module: (a) common anode LED structure; (b) configuration of the display module; (c) timing diagram.

segment LED display module. The timing sequence for sending digits to the display module is portrayed in Figure 12.15(c). To display the ith digit, the seven-segment code of digit i is sent to the display module along with the activated signal Ai, where $i \in \{1, 2, 3, 4\}$, as shown in the figure. The digit 1 is on the rightmost position and digit 4 is on the leftmost position. Each digit is displayed about 1 ms in turn and repeated every 4 ms.

12.4.2 μP-Based Design

A possible embedded system, such as the AT89C51, a member of the MCS-51 family, can be used to design and implement the start/stop timer. The AT89C51 is a general-purpose microcomputer system (μC) and is composed of an 8-bit CPU, a serial port, four 8-bit I/O ports, two 16-bit timers, a two-level five-input interrupt controller, one 128-byte RAM, and one program memory (Flash memory) of 0 to 64 kbytes.

Of course, not all modules of the AT89C51 μC are needed in implementing the start/stop timer. To illustrate the use of a μC system to design and implement the start/stop timer, consider Figure 12.16. The μC system has to receive the status of two switches from outside and output the timer counts to the LED display module in the way of one digit at a time. Hence, two software functions are needed. One is to debounce the switches and the other is to display the timer counts on the LED display

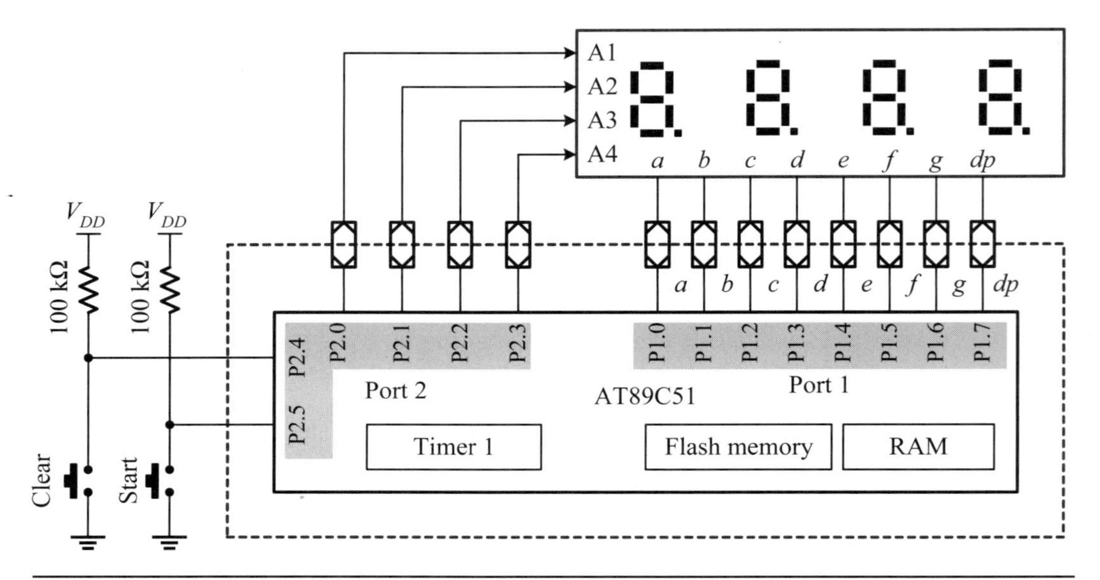

Figure 12.16: A possible embedded system for implementing the start/stop timer.

in a digit-by-digit manner at an interval of 1 ms. As a result, a 1-ms timing base is required for the display function.

The switch bouncing problem of a switch can be solved by reading the switch status twice in an interval of 10 ms apart. If both readings are the same, the switch status is stable; otherwise, it is unstable. Based on this idea, a 10-ms timing-base signal is needed for the switch debouncing function.

The display function is used to display the timer counts on the LED display module. This function is called once per milisecond to display a digit on the LED display. The digits are in turn displayed from leftmost (MSD) to rightmost (LSD), as shown in Figure 12.15. The display function has to implement the following two functions. First, it has to convert the BCD code of a digit into the seven-segment code before the BCD code is sent to the LED display module. Second, it also needs to send an enable signal in synchronism with the digit to be displayed.

From the above discussion, the resources required for the start/stop timer are the CPU, a timer, two 8-bit GPIOs, the Flash memory, and the RAM, as shown in Figure 12.16. The two GPIOs are used to drive the display module, where port 1 is used to send the seven-segment code and a half of port 2 is used to synchronize the seven-segment code being sent to the display module. The other two bits of port 2 are used to receive the status of two buttons: clear and start. To correctly receive the status of these two buttons, a switch debouncing software module is required as mentioned before.

As we can see from Figure 12.15 and the above discussion, the 1-ms timing base is indeed at the heart of the start/stop timer. In addition, a 10-ms timing base is also required since the resolution of the start/stop timer is 10 ms. However, this 10-ms timing base can be readily derived from the 1-ms timing base just by counting the 1-ms event 10 times. Consequently, only the 1-ms timing-base signal is indeed needed; it can be generated simply by using a 16-bit timer.

Another important use of the 1-ms timing-base signal is to schedule the operations of the entire start/stop timer. The main operation flow of start/stop timer is as follows. When the 1-ms generator (a 16-bit timer) is a time-out each time, an interrupt is raised. This interrupt notifies the CPU to call all related functions to process proper operations as stated in the following.

```
/* a start/stop timer with time resolution 10 ms */
void start_stop_timer(void)
{
   initialize the timer 1 to generate 1 ms delay;
   initialize start_flag to 0;
   for (;;) /* forever loop */
      while (interrupt) {
         initialize the timer 1 to generate 1 ms delay;
         if (count_10ms() == 9) {
            switch_debouncer(clk_10ms);
            if (clear && !start_flag) count = 0;
            else if (start) start_flag = ~start_flag;
            if (start_flag) count = count + 1;
         }
         display(count);
      }
}
```

The details of the use of the MCS-51 microcomputer system to design and realize a specific system is beyond the scope of this book. The interested reader can be referred to [8] or other μP- and/or μC-related textbooks for more details.

12.4.3 FPGA-Based Design

The start/stop timer can also be designed and realized with an FPGA device. Recall that from the behavior of the start/stop timer described at the beginning of this section the heart of the timer is the four-digit counter, which is a synchronous counter with asynchronous reset. The counter is controlled by two buttons and counts up once per 10 ms. As a result, three additional modules associated with the counter are needed. These modules include two switch debouncer modules and one timing-base generator. In addition, a display module is needed to interface the counter with the LED display module, as shown in Figure 12.17. The display module actually consists of four parts: a 2-to-4 decoder, a 2-bit binary counter, a four-bit 4-to-1 multiplexer, and a BCD-to-seven-segment decoder.

The synthesized result using Xilinx Virtex-II FPGA is shown in Figure 12.18. It needs 1004 equivalent gates corresponding to 100 4-input lookup tables (LUTs). The details of the use of FPGA devices to design and realize a specific system is beyond the scope of this book. The interested reader may refer to [9] or other FPGA-related textbooks for more details.

12.4.4 Cell-Based Design

A cell-based design proceeds in much the same way as an FPGA-based design except that now the target device is a cell library rather than an FPGA device. In most practical applications, prior to realizing in a cell library the design is often verified using one or more FPGA devices to take less risk of the failure of the design.

Like an FPGA-based design, a cell-based design entirely relies on the use of CAD tools. Figure 12.19(a) shows the placement result of the start/stop timer. From the figure, we can see that cells are placed in a row-by-row manner and cells in the same row

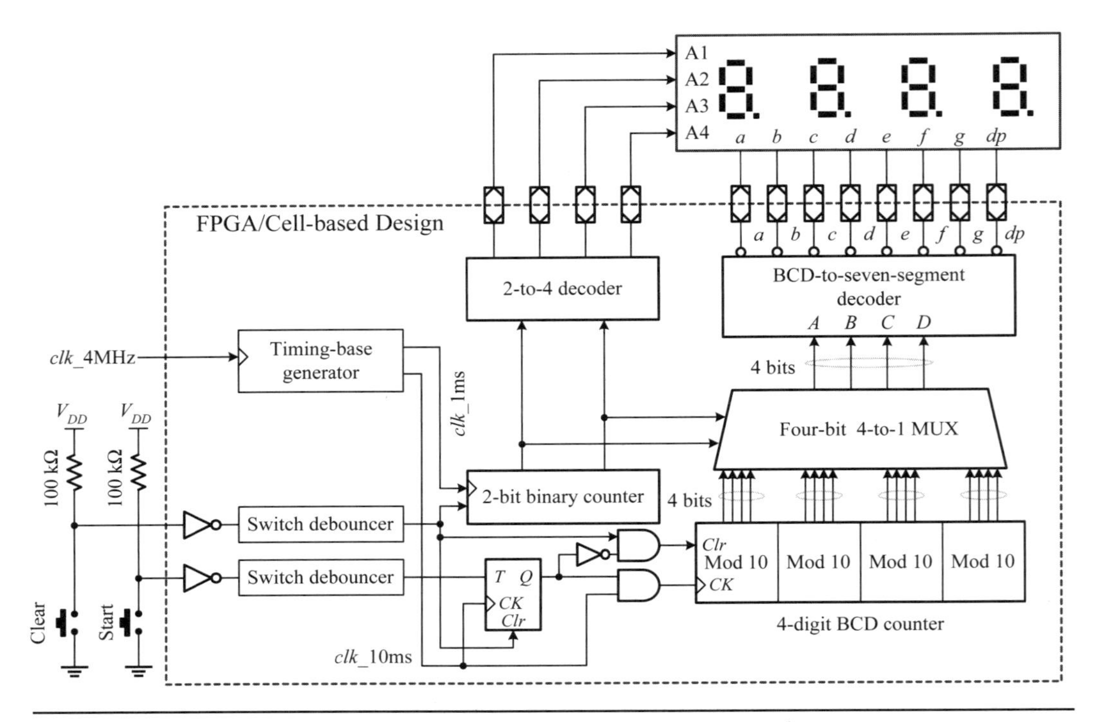

Figure 12.17: A possible FPGA implementation for the start/stop timer.

have the same height. Cells in different rows may have unequal heights to accommodate cell functions in different complexities. All rows are closely placed without the need of routing space between them since multilayer metals are used in such a cell library. Consequently, the routing between cells in the same row or different rows can be done in the higher-layer metals. As a rule of thumb, the first-layer metal is often reserved for the routing required in the cell itself, and the next one or two higher-layer metals are reserved for the interconnect between cells.

The place-and-routing result of the start/stop timer is shown in Figure 12.19(b), including both logic cells and the power distribution network. The complete chip layout, including core logic, power rings, I/O and power pads, is shown in Figure 12.20.

■ Review Questions

Q12-33. What is the bouncing problem associated with a mechanical switch?

Q12-34. How would you remove the bouncing problem of a mechanical switch?

Q12-35. Describe the operation of the embedded system shown in Figure 12.16.

Q12-36. Describe the operation of the FPGA system exhibited in Figure 12.17.

12.5 Summary

In this chapter, we focused on the three closely related design issues: design methodologies, synthesis flows, and implementation options. The design methodologies, including system level and register-transfer level (RTL), were introduced briefly. At the system

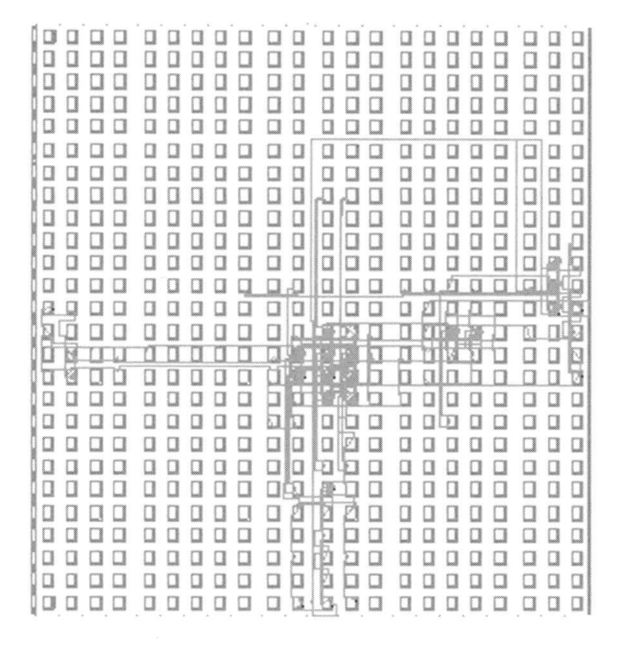

Figure 12.18: A synthesized result of the start/stop timer using Xilinx Virtex-II FPGA.

level, a design can proceed by using one of the following methods: function-based, architecture-based, and the mixing use of the above two methods. At RTL, a design can be done in either of the following: the algorithmic state machine chart (ASM chart) and finite-state machine with datapath (FSMD). In addition, the concept of the globally asynchronous and locally synchronous (GALS) approach was introduced. Furthermore, implementation architectures, including single-cycle structure, multiple-cycle structure, and pipeline/parallelism structure, were dealt with concisely.

The synthesis-flow using hardware description language (HDL) for designing and implementing a digital system based on an FPGA device or a cell library was described concisely and implementation options were covered. These options can be classified into the following three classes: platforms, application-specific integrated circuits (ASICs), and field-programmable devices. Finally, a case study was given to illustrate how a real-world system can be designed and implemented with a variety of options, including the μP/DSP system, a field-programmable device, and ASIC with a cell library.

References

1. N. Agarwal and N. Dimopoulos, "FSMD partitioning for low power using ILP," *IEEE Computer Society Annual Symposium on VLSI, 2008* (ISVLSI '08), pp. 63–68, April 2008.
2. B. Bailey, G. Martin, and A. Piziali, *ESL Design and Verification: A Prescription for Electronic System-Level Methodology.* San Francisco: Morgan Kaufmann, 2007.
3. S. Bhunia, A. Datta, N. Banerjee, and K. Roy, "GAARP: a power-aware GALS architecture for real-time algorithm-specific tasks," *IEEE Trans. on Computers,* Vol. 54, No. 6, pp. 752–766, June 2005.

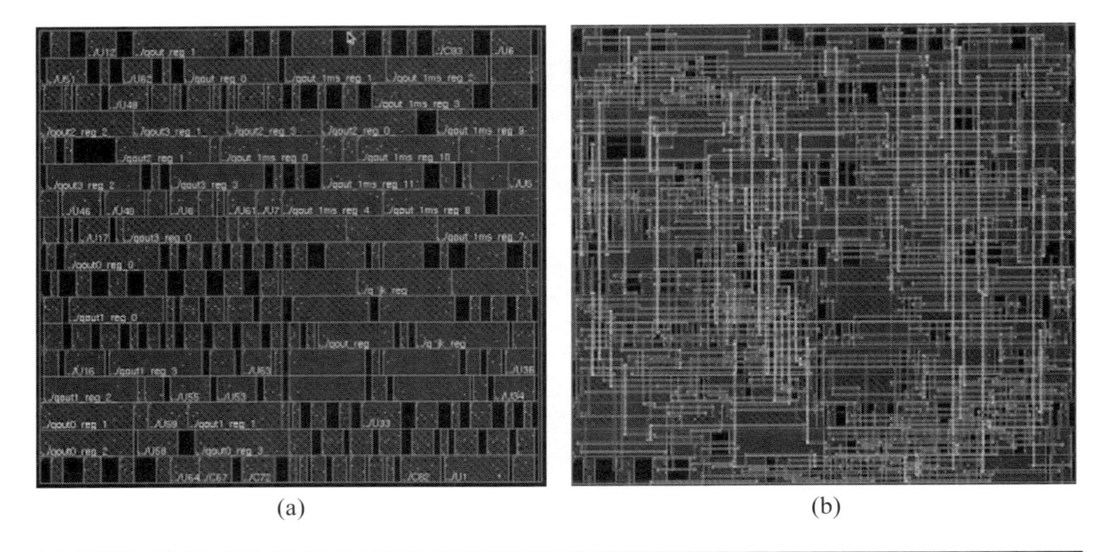

(a) (b)

Figure 12.19: The (a) cell-placement and (b) result of place and route of the start/stop timer using a cell library.

4. A. Hemani et al., "Lowering power consumption in clock by using globally asynchronous locally synchronous design style," *Proc. of the 36th Design Automation Conf.*, pp. 873–878, June 1999.
5. D. A. Hodges, H. G. Jackson, and R. A. Saleh, *Analysis and Design of Digital Integrated Circuits: In Deep Submicron Technology,* 3rd ed. New York: McGraw-Hill Books, 2004.
6. E. Hwang, F. Vahid, and Y. C. Hsu, "FSMD functional partitioning for low power," *Proc. of Design, Automation and Test in Europe Conf. and Exhibition,* pp. 22–28, March 1999.
7. M. B. Lin, *Digital System Design: Principles, Practices, and Applications,* 4th ed. Taipei, Taiwan: Chuan Hwa Book Ltd., 2010.
8. M. B. Lin, *Basic Principles and Applications of Microprocessors: MCS-51 Embedded Microcomputer System, Software, and Hardware,* 3rd ed. Taipei, Taiwan: Chuan Hwa Book Ltd., 2011.
9. M. B. Lin, *Digital System Designs and Practices: Using Verilog HDL and FPGAs.* Singapore: John Wiley & Sons, 2008.
10. A. J. Martin and M. Nystrom, "Asynchronous techniques for system-on-chip design," *Proc. IEEE,* Vol. 94, No. 6, pp. 1089–1120, June 2006.
11. L. T. Wang, Y. W. Chang, and K. T. Cheng, *Electronic Design Automation: Synthesis, Verification, and Test.* New York: Morgan Kaufmann, 2009.
12. N. H. E. Weste and D. Harris, *CMOS VLSI Design: A Circuit and Systems Perspective,* 4th ed. Boston: Addison-Wesley, 2011.

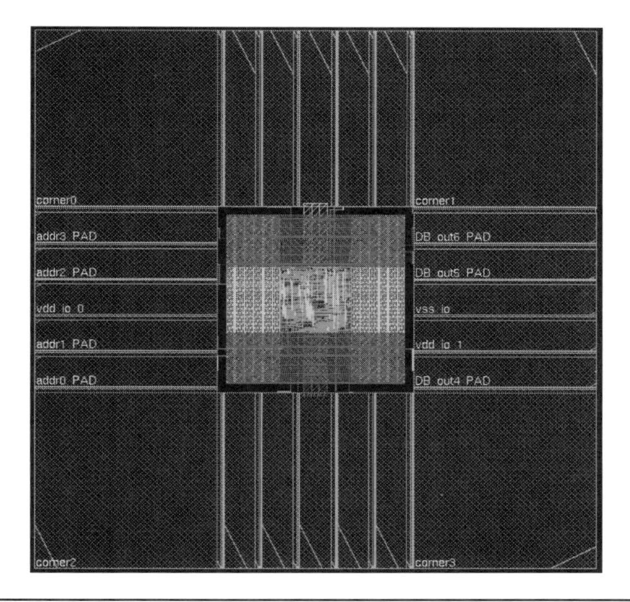

Figure 12.20: The complete chip layout of the start/stop timer.

Problems

12-1. A 4×16 multiplier. This problem involves the design of an unsigned multiplier that performs the multiplication of a 4-bit with 16-bit numbers to give a 20-bit product. To speed up the operation, a 4×4 unsigned array multiplier is used so that a multiplication of 4 bits rather than only 1 bit is performed at each step. Suppose that a 24-bit accumulator, an 8-bit adder, the 4×4 array multiplier, and a control unit are used to construct the system. Design this 4×16 multiplier.

(a) Draw the block diagram and label each component properly to indicate its functionality and size.

(b) Draw the ASM chart for the control unit.

(c) Write a Verilog HDL module to describe the 4×16 multiplier.

(d) Write a test bench to verify the resulting module.

(e) Using an available FPGA device, realize this multiplier and verify it.

(f) Using an available cell library, realize this multiplier and verify it.

12-2. A 16×16 signed multiplier. This problem considers the design of a signed multiplier that multiplies two 16-bit signed numbers to give a 32-bit product. To use as little hardware as possible, a multiple-cycle unsigned algorithm for multiplying two unsigned numbers based on the shift-and-add approach is used. Negative numbers are represented in two's complement form. The signed multiplier proceeds the multiplication of both input number as in the following steps. First, it converts the multiplier and multiplicand into unsigned numbers if they are negative. Next, it performs the multiplication on the unsigned numbers. Finally, it converts the product into two's complement form if necessary.

(a) Draw the block diagram and label each component appropriately to indicate its functionality and size.

(b) Draw the ASM chart for the control unit.

(c) Write a Verilog HDL module to describe the 16×16 signed multiplier.

(d) Write a test bench to verify the resulting module.

(e) Using an available FPGA device, realize this multiplier and verify it.

(f) Using an available cell library, realize this multiplier and verify it.

13

Interconnect

Interconnect means the wires that link together transistors, circuits, cell, modules, and systems as well. It plays an important role in any very-large-scale integration (VLSI) or digital system because it controls timing, power, noise, design functionality, and reliability. Interconnect in a VLSI or digital system mainly provides power delivery paths, clock delivery paths, and signal delivery paths. Power delivery paths distribute power to every element in the system and provides appropriate return paths, clock delivery paths deliver global or local clocks to storage elements, such as latches and flip-flops, or dynamic logic blocks, and signal delivery paths provide the communication capability among circuit elements and/or modules. For multiple-processor systems, a more complicated interconnect network referred to as a network-on-a-chip (NoC) technique is widely used.

To estimate and approximate the real behavior of a wire, a proper interconnect model is needed to represent the wire as a function of its parameters. For this purpose, a simple model is first used to capture the behavior of the wire and to simplify the simulation work. The more detailed considerations of interconnect parasitic effects then follow. These parasitic effects include RC delay, capacitive-coupling effects, and RLC effects. All of these influence the signal integrity and degrade the performance of the circuit.

Finally, the transmission-line terminations are dealt with examples. The two popular termination methods are series termination and parallel termination, depending on whether the reflected waves are absorbed (terminated) at the source or at the destination end. In addition, self-timed regenerators and network on a chip (NoC) as well as logical effort with interconnect are discussed concisely.

13.1 RLC Parasitics

We begin with the introduction of the resistance of conducting layers, including metal wires and semiconductor diffusion layers. Then, we are concerned with the capacitance between two conducting layers, such as a metal wire and a semiconductor, and two metal wires as well. Finally, we discuss the inductance of metal wires.

13.1.1 Resistance

In this subsection, we deal with the resistance of wires and diffusion layers of the semiconductor.

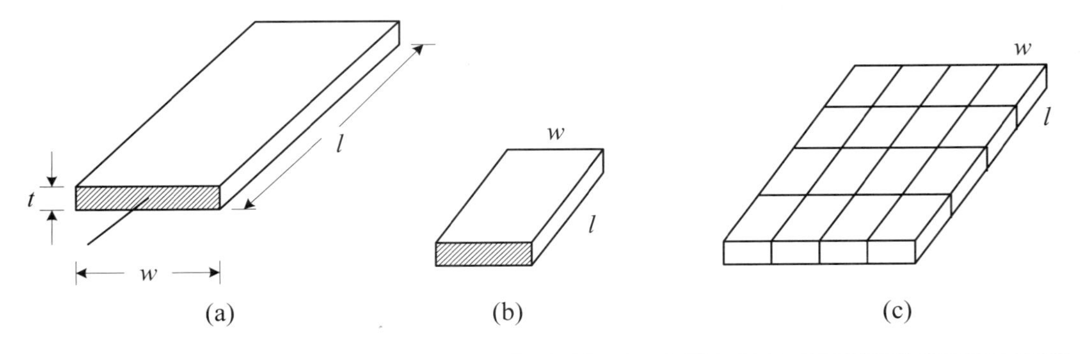

Figure 13.1: The resistance of typical uniform slabs: (a) uniform slab; (b) an example; (c) another example.

13.1.1.1 Resistance of Uniform Slabs A wire can be modeled as a uniform slab, as shown in Figure 13.1(a). For a uniform slab, the resistance R can be described as follows:

$$\begin{aligned} R &= \frac{\rho l}{wt} = \frac{l}{\sigma wt} \\ &= \frac{1}{e\mu n} \cdot \frac{l}{wt} = \left(\frac{1}{e\mu nt}\right)\frac{l}{w} \\ &= R_{sq}\frac{l}{w} \end{aligned} \tag{13.1}$$

where ρ is called the *resistivity* of conducting material and is the inverse of *conductivity* $\sigma(= e\mu n)$. The ratio of l/w is the number of squares of the conducting material. The square resistance R_{sq} is often referred to as *sheet resistance* and is defined as

$$R_{sq} = \frac{1}{e\mu nt} = \frac{\rho}{t} \tag{13.2}$$

The units for R_{sq} are in $\Omega/\square$ by convention.

According to the Equation (13.1), two uniform slabs of the same R_{sq} with different widths and lengths but with the same ratio of l/w will have the same resistance. The following example further manifests this.

■ **Example 13-1: (Resistance of two uniform slabs.)** The first example is depicted in Figure 13.1(b), where the total resistance is calculated as follows:

$$R = R_{sq}\left(\frac{l}{w}\right)\ \Omega$$

The second example is shown in Figure 13.1(c), where the total resistance is

$$R = R_{sq}\left(\frac{4l}{4w}\right) = R_{sq}\left(\frac{l}{w}\right)\ \Omega$$

Therefore, both examples have the same resistance even though their widths and lengths are different.

■

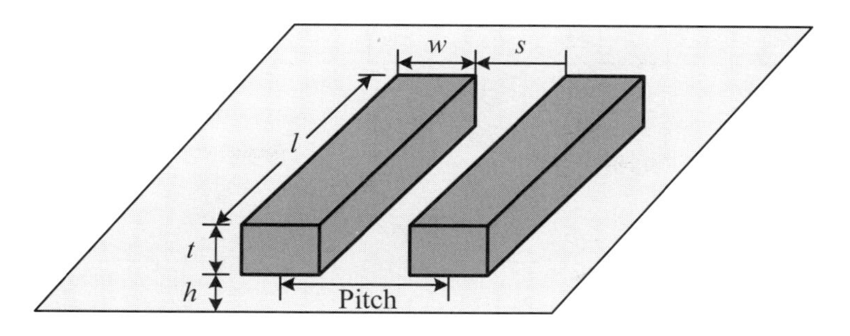

Figure 13.2: The metal wires of a 6-metal 0.18-μm process.

Table 13.1: The parameters of layer stack for a 6-metal 0.18-μm process.

Layer	t (nm)	h (nm)	w (nm)	s (nm)	AR
6	1720	1000	860	860	2.0
5	1600	1000	880	800	2.0
4	1080	700	540	540	2.0
3	700	700	320	320	2.2
2	700	700	320	320	2.2
1	480	800	250	250	1.9

Modern complementary metal-oxide-semiconductor (CMOS) processes provide many metal layers to reduce the chip area needed for systems with complex functions. Because higher metal layers are commonly used for routing supply voltages, global clocks, and global signals, they are intentionally made to be thicker than the lower metal layers. Figure 13.2 shows the *pitches* and *aspect ratio* (AR) of metal layers of a typical 6-metal 0.18-μm process. The related parameters are given in Table 13.1. Here, the pitch means the distance between the middle of two metal wires and is equal to the sum of the width of a metal wire and the separation of two such wires; that is, pitch is equal to $w + s$ and the aspect ratio (AR) denotes t/w. The wire width is equal to the separation of two wires. Furthermore, the wire height is almost two times the wire width.

Some widely used metal materials along with their resistivities are listed in Table 13.2. Silver metal has the lowest resistivity and its conductivity is the best of all metal materials. The next materials are copper, gold, aluminum, and tungsten in sequence. Except for silver, all of the other four metals are widely used in modern VLSI processes.

The following example gives the typical sheet resistance of metal wires commonly used in most CMOS processes.

Table 13.2: The resistivities of several metal materials.

Material	Resistivity ($\mu\Omega$-cm)
Silver (Ag)	1.62
Copper (Cu)	1.69
Gold (Au)	2.28
Aluminum (Al)	2.75
Tungsten (W)	5.25

Table 13.3: The thickness of metal layers for several common processes.

	M1	M2	M3	M4	M5	M6	M7	M8	M9	Unit
0.25 μm	0.61	0.88	0.88	1.73	2.43					μm
0.18 μm	0.48	0.70	0.70	1.08	1.60	1.72				μm
0.13 μm	0.40	0.40	0.40	0.40	0.40	0.40	0.80	0.80		μm
90 nm	150	256	256	320	384	576	972			nm
65 nm	170	190	200	250	300	430	650	975		nm
45 nm	144	144	144	216	252	324	504	720	7000	nm
32 nm	95	95	95	151	204	303	388	504	8000	nm

■ Example 13-2: (Sheet resistances of metal layers.) From Table 13.1, we know that metal1 has a thickness of 0.48 μm and metal5 has a thickness of 1.6 μm. Calculate the sheet resistances of these two metal wires.

Solution: The sheet resistances of these two metal wires are calculated as follows:

$$R_{sq} = \frac{\rho}{t} = \frac{2.75\ \mu\Omega\text{-cm}}{0.48\ \mu m} = 57\ m\Omega/\Box \quad \text{(metal1)}$$

$$R_{sq} = \frac{\rho}{t} = \frac{2.75\ \mu\Omega\text{-cm}}{1.6\ \mu m} = 17\ m\Omega/\Box \quad \text{(metal5)}$$

■

Although it seems that the sheet resistance is quite small, the resistance cannot be simply ignored if the wire is long enough, such as in the case of many system-on-a-chip (SoC) designs. The following example gives a feeling about the resistance of a long wire.

■ Example 13-3: (Resistance of a long metal1 wire.) Suppose that a metal1 wire is used. The length and width of the metal1 wire are 0.5 cm and 1 μm, respectively. Calculate the resistance of the metal1 wire.

Solution: The resistance of the metal1 wire is equal to

$$R_{Al} = \left(\frac{0.5 \times 10^4}{1}\right) \times 57 \times 10^{-3} = 285\ \Omega$$

It is a rather large value, which would possibly cause a long RC delay of the signal passing through it when the metal1 wire is used as a signal wire or a large IR drop on it when the metal1 wire is used as a power-supply wire.

■

Table 13.3 lists the typical thickness of metal layers for several common CMOS processes. Note that in 0.25- and 0.18-μm processes, all metals are aluminum and in 0.13 μm and beyond processes all metals are copper. Regardless of which process, the higher layers are made to be thicker, thereby reducing their resistivities so that they can be used to carry global signals or used as power-supply or ground wires.

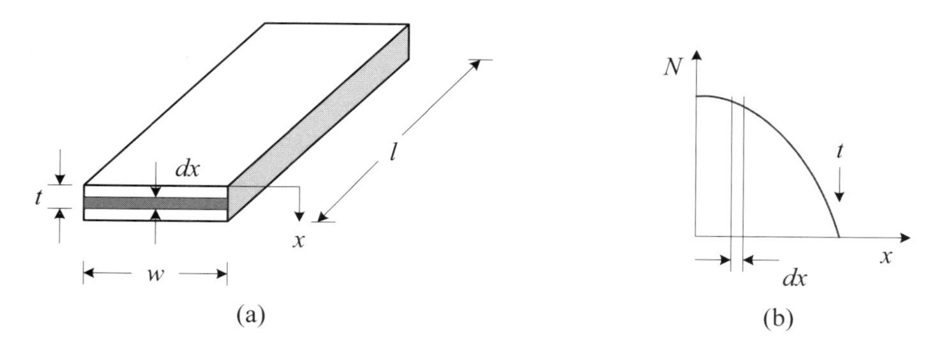

Figure 13.3: The typical (a) diffusion layer and (b) its resistance model.

13.1.1.2 Resistance of the Diffusion Layer For nonmetal conducting layers, such as heavily doped polysilicon layers or diffusion layers, the resistance is not uniform over the slabs due to the nonuniform doping within them. The resistance model of typical diffusion layers is depicted in Figure 13.3.

For a diffusion layer of thickness dx at depth x, the conducting sheet has the conductance of

$$dG = e\left(\frac{w}{l}\right)\mu N(x)dx \tag{13.3}$$

By integrating both sides, we obtain the conductance of the sheet as follows:

$$G = \int_0^{x_j} e\left(\frac{w}{l}\right)\mu N(x)dx = \left(\frac{w}{l}\right)\int_0^{x_j} e\mu N(x)dx \tag{13.4}$$

Consequently, the resistance, which is the reciprocal of conductance, can then be represented as the following equation:

$$R = \frac{1}{\int_0^{x_j} e\mu N(x)dx}\left(\frac{l}{w}\right) = R_{sq}\left(\frac{l}{w}\right) \tag{13.5}$$

where the sheet resistance R_{sq} is defined as

$$R_{sq} = \frac{1}{\int_0^{x_j} e\mu N(x)dx} \tag{13.6}$$

The sheet resistance is difficult to calculate analytically unless we known the distribution function $N(x)$ exactly. In practical applications, the sheet resistances of polysilicon and diffusion layers are measured from testing wafers or testing components rather than calculated theoretically. Heavily doped polysilicon with n^+ and p^+ have sheet resistance of about 200 and 440 $\Omega/\square$, respectively. It is quite large compared to metal wires. Consequently, polysilicon layers are only used for local interconnect to connect together adjacent metal-oxide-semiconductor (MOS) transistors. The following example gives a feeling about the resistance of a long polysilicon wire.

■ **Example 13-4: (Resistance of a long n$^+$ polysilicon wire.)** Calculate the resistance of an n^+ polysilicon wire, which is 0.5-cm long and 1-μm wide. The sheet resistance of the n^+ polysilicon is 200 $\Omega/\square$.

Table 13.4: The sheet resistances of various metal layers in a typical CMOS process.

Material	Sheet resistance (m$\Omega/\square$)	Material	Sheet resistance (m$\Omega/\square$)
Metal1	57	Metal4	25
Metal2	39	Metal5	17
Metal3	39	Metal6	16

Solution: By using the Equation 13.5, the resistance of the n^+ polysilicon wire is calculated as follows:

$$R = R_{sq}\left(\frac{l}{w}\right) = 200 \times \left(\frac{0.5 \times 10^4}{1}\right) = 1000 \text{ k}\Omega$$

The resistance is very large compared to that of the metal1 wire with the same length and width. Consequently, the polysilicon layer can be only used for local interconnect. ∎

The sheet resistances of metal layers in a typical 6-metal 0.18-μm process are summarized in Table 13.4. In most CMOS processes, the sheet resistances of lower metal layers are usually much larger than those of higher metal layers because as stated lower layers are usually made to be thinner than higher layers.

Recall that the basic building units or blocks in a VLSI system are cells. Within each cell, the available wires are polysilicon and metal1 layers. Because of the limited size of a cell, a thinner metal layer can easily satisfy the needs of signal, power supply, and ground wires. A unit or a large module consisting of many cells would have to use longer metal wires to connect together all required cells to perform the desired function. In this case, metal2 and metal3 are often used to reduce the RC delay and hence improve the performance. Metal4 and metal5 have smaller resistance and are often reserved for use as the paths for critical signals. Metal6 even has less resistance and is reserved for power-supply and ground distribution networks.

In summary, a sample usage of these six metal layers is as follows:

- Metal1 is used for interconnect and power-supply wires within a cell.
- Metal2 and metal3 are used for the wires between cells within units.
- Metal4 and metal5 are used to connect units and provide paths for critical signals.
- Metal6 serves as I/O pads, clock, power-supply, and ground paths.

Review Questions

Q13-1. Define the pitch and aspect ratio (AR) of metal layers.

Q13-2. Describe the meaning of sheet resistance R_{sq}.

Q13-3. Explain why polysilicon layers are only used for local interconnect.

13.1.2 Capacitance

In what follows, we are concerned with the capacitance of a parallel-plate capacitor and fringing-field effects.

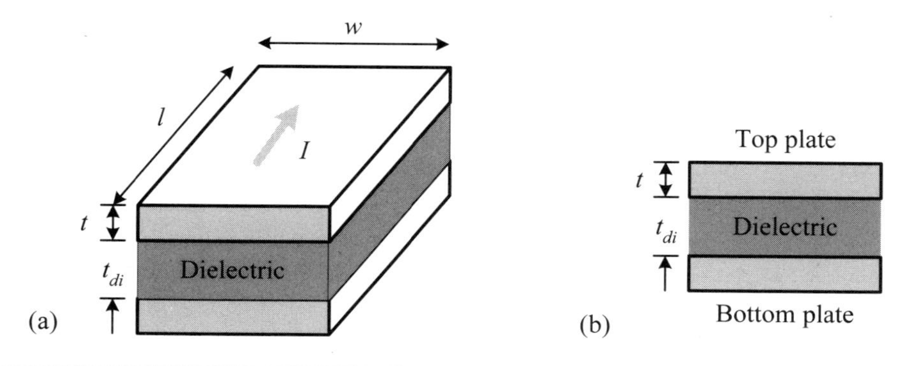

Figure 13.4: The (a) basic structure and (b) cross-sectional view of a parallel-plate capacitor.

Table 13.5: The relative permittivities of several common materials.

Material	Relative permittivity (ε_r)
Free space	1
Silicon	11.7
Silicon dioxide	3.9
Silicon nitride	6.5 to 7.5
Hafnium oxide	20.0
Zirconium oxide	23.0
Tantalum oxide	20 to 30

13.1.2.1 Parallel-Plate Capacitors Recall that a capacitor is formed when two conductors are separated by a dielectric (that is, an insulator). The simplest capacitor is formed by two equal-area parallel metal plates separated by a dielectric. Such a capacitor is referred to as an *area capacitor*, or a *parallel-plate capacitor*, as shown in Figure 13.4.

Assuming that the area of a metal plate is A and the distance between two metal plates is t_{di}, then the capacitance of an area capacitor can be expressed as follows:

$$C_{area} = \frac{\varepsilon_{di}}{t_{di}} A = \frac{\varepsilon_r \varepsilon_0}{t_{di}} A = \left(\frac{\varepsilon_r \varepsilon_0}{t_{di}} \right) wl \tag{13.7}$$

where $\varepsilon_{di} = \varepsilon_r \varepsilon_0$ is the permittivity of dielectric. The ε_r is the relative permittivity of dielectric and ε_0 is the free-space permittivity, equal to 8.854×10^{-14} F/cm. The w and l are the width and length of metal plates, respectively. Table 13.5 lists the relative permittivities of several common materials in typical CMOS processes.

13.1.2.2 Fringing-Field Effects Although the feature sizes of CMOS processes are continually reduced, the wire thickness is not scaled down with the same factor because it is necessary to minimize the sheet resistance by keeping the cross-sectional area of wires as large as possible. This results in a significant capacitance existing between both sidewalls and substrate. This capacitance is known as a *fringing capacitance* because it is caused by the fringing field between sidewalls and the substrate, as illustrated in Figure 13.5(a). Fringing capacitance cannot be simply ignored in deep submicron processes because it may contribute a significant part to the total wire capacitance.

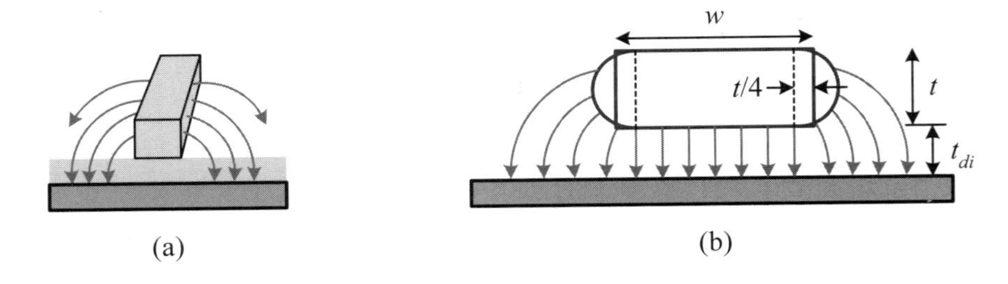

Figure 13.5: The wire capacitance model with the fringing-field effect: (a) fringing field; (b) capacitance model.

13.1.2.3 Single-Wire Capacitance Model A simple way to model both parallel-plate and fringing capacitors are depicted in Figure 13.5(b), where the fringing-field capacitor is modeled as a cylindrical wire with a diameter equal to the wire thickness, t. The area capacitance is still modeled as a parallel-plate capacitor. Area and fringing capacitances in terms of unit length can be described as the following two equations, respectively:

$$C_{area} = \varepsilon_{di}\left(\frac{w}{t_{di}}\right) \tag{13.8}$$

and

$$C_{fringe} = 2\varepsilon_{di}\ln\left(1 + \frac{t}{t_{di}}\right) \tag{13.9}$$

where 2 accounts for both sides of the wire.

The total unit-length wire capacitance is the sum of area and fringing capacitances. That is,

$$C_{int} = C_{area} + C_{fringe} \tag{13.10}$$

The following example illustrates the importance of fringing capacitance on the total unit-length capacitance of a narrower wire.

■ Example 13-5: (Area and fringing capacitance.) Assume that a metal1 wire is over the field oxide of a thickness of 0.52 μm. The width and thickness of the metal1 wire are 0.5 μm and 0.6 μm, respectively. Calculate the total unit-length capacitance per μm and the total capacitance of a 100-μm metal1 wire.

Solution: From Equations (13.8) and (13.9), the unit-length area and fringing capacitances of the metal1 wire can be calculated as follows:

$$\begin{aligned} C_{area} &= \varepsilon_{di}\left(\frac{w}{t_{di}}\right) \\ &= 3.9 \times 8.854 \times 10^{-14}\left(\frac{0.50 \times 10^{-4}}{0.52 \times 10^{-4}}\right) \\ &= 33.2 \text{ aF}/\mu\text{m} \\ C_{fringe} &= 2\varepsilon_{di}\ln\left(1 + \frac{t}{t_{di}}\right) \end{aligned}$$

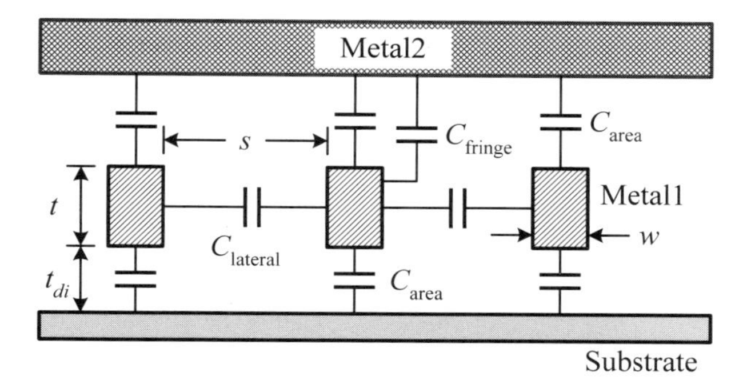

Figure 13.6: The interconnect capacitance of modern multilayer-metal processes.

$$
\begin{aligned}
&= 2 \times 3.9 \times 8.854 \times 10^{-14} \ln\left(1 + \frac{0.6 \times 10^{-4}}{0.52 \times 10^{-4}}\right) \\
&= 52.98 \text{ aF}/\mu\text{m}
\end{aligned}
$$

The total unit-length capacitance of the metal1 wire is the sum of area and fringing capacitances and is $33.2 + 52.98 = 86.18$ aF/μm. Hence, the total capacitance of the 100-μm metal1 wire is equal to

$$C_{wire} = C_{int} \times l = 86.18 \text{ aF}/\mu\text{m} \times 100\ \mu\text{m} = 8.62 \text{ fF}$$

■

13.1.2.4 Multilayer Interconnect Capacitance Model Multilayer metals are common in modern CMOS processes. As shown in Figure 13.6, in addition to both area and fringing capacitances, the capacitance between two adjacent wires on the same layer is included to reflect that the area capacitance formed between both sidewalls of two adjacent wires cannot be ignored if they are closely spaced. This capacitance is called *lateral capacitance*.

A simple way to estimate the unit-length wire capacitance in the multilayer-metal structure is to combine both individual area and lateral capacitances with fringing capacitance. The total wire capacitance comprises the area capacitance and lateral or fringing capacitance, depending on whether the wire is closely spaced to others. Lateral capacitance dominates when two wires are closely spaced, and fringing capacitance dominates when two wires are separated far apart. To illustrate this, an example is given to show how to account for the effects of lateral or fringing capacitance.

■ Example 13-6: (Wire capacitance.) For a 0.18-μm process, the upper layer of metal has the following parameters: $w = 0.5\ \mu$m, $t_{di} = 0.52\ \mu$m, $t = 0.6\ \mu$m, and $s = 0.5\ \mu$m.

1. Calculate C_{area}, $C_{lateral}$, and C_{fringe}.
2. Calculate the capacitance of the middle wire when wires are closely spaced.
3. Calculate the capacitance of the middle wire when wires are widely spaced.

Solution:

1. Using Equation (13.8), we can calculate C_{area} and $C_{lateral}$ as follows:

$$\begin{aligned} C_{area} &= \varepsilon_{ox}\frac{w}{t_{di}} \\ &= (3.9 \times 88.5 \times 10^{-4} \text{ fF}/\mu\text{m})\frac{0.5}{0.52} \\ &= 0.03 \text{ fF}/\mu\text{m} \end{aligned}$$

and

$$\begin{aligned} C_{lateral} &= \varepsilon_{ox}\frac{t}{s} \\ &= (3.9 \times 88.5 \times 10^{-4} \text{ fF}/\mu\text{m})\frac{0.6}{0.5} \\ &= 0.04 \text{ fF}/\mu\text{m} \end{aligned}$$

The fringe capacitance is calculated by using Equation (13.9) and is

$$\begin{aligned} C_{fringe} &= 2\varepsilon_{ox}\ln\left(1 + \frac{t}{t_{di}}\right) \\ &= 2 \times 0.035 \ln\left(1 + \frac{0.6}{0.52}\right) \text{ fF}/\mu\text{m} \\ &= 0.05 \text{ fF}/\mu\text{m} \end{aligned}$$

2. The capacitance of the middle wire when wires are closely spaced is

$$\begin{aligned} C_{int} &= 2C_{area} + 2C_{lateral} + C_{fringe} \\ &\approx 2(0.03) + 2(0.04) + 0.00 = 0.14 \text{ fF}/\mu\text{m} \end{aligned} \tag{13.11}$$

3. The capacitance of the middle wire when wires are widely spaced is

$$\begin{aligned} C_{int} &= 2C_{area} + 2C_{lateral} + C_{fringe} \\ &\approx 2(0.03) + 2(0.0) + 0.05 = 0.11 \text{ fF}/\mu\text{m} \end{aligned} \tag{13.12}$$

Regardless of whether wires are widely or closely spaced, the resulting capacitance is quite coincident with the result from the previous example.

■

For the purpose of timing calculation, a rule of thumb about the interconnect capacitance is $C_{int} = 0.1$ fF/μm for widely spaced wires and $C_{int} = 0.15$ fF/μm for closely spaced wires, assuming that the minimum width of metals is used. Here, "widely spaced" means that two wires are separated with a space much wider than the minimum spacing allowed by the layout design rules and "closely spaced" means that the minimum spacing is used. With this simple rule, the wire capacitance with a length of l can be estimated as

$$C_{wire} = C_{int}l \tag{13.13}$$

■ Example 13-7: (RC delay of a wire.) Assuming that the width and length of a metal1 wire are 0.4 μm and 200 μm, respectively, calculate the RC delay of this metal wire.

Solution: Using the unit-length capacitance 0.1 fF/μm, the total capacitance of this metal wire is equal to

$$C = 0.1 \text{ fF}/\mu\text{m} \times 200\ \mu\text{m} = 20 \text{ fF}$$

The resistance of the metal1 wire is

$$R = 57 \times 10^{-3}\ \Omega/\square \times \frac{200}{0.4} = 28.5\ \Omega$$

Hence, the RC delay of the wire is

$$t_{pd(wire)} = 0.69RC = 0.69 \times 28.5\ (\Omega) \times 20\ (\text{fF}) = 0.393\ \text{ps}$$

■

■ Review Questions

Q13-4. What is fringing capacitance?

Q13-5. What is area capacitance?

Q13-6. What is lateral capacitance?

13.1.3 Inductance

When a current flows through a wire, it sets up an electromagnetic field around it with a flux, $\phi = Li$, that stores energy. As the current changes, the magnetic flux acts on the wire itself and induces a voltage drop. This phenomenon is called *Faraday's law of induction* and can be described as follows:

$$\Delta V = L\frac{di}{dt} \tag{13.14}$$

In other words, the changing current induces a voltage proportional to the rate of change. This proportional constant is defined as the wire *inductance L*. Recall that we also use the symbol L to denote the channel length of a MOS transistor. However, it should be easy to distinguish one from another from the context.

The wire inductance can be expressed as follows:

$$L = \frac{\mu_0}{2\pi}\left[l \ln\left(\frac{2l}{w+t}\right) + \frac{l}{2} + 0.2235(w+t)\right] \tag{13.15}$$

where w, t, and l are the width, thickness, and length of the wire, respectively.

■ Example 13-8: (Wire inductance.) Consider the inductance of a metal1 wire over a field oxide of thickness 0.52 μm. Assume that the width and thickness of the metal1 wire are 0.5 μm and 0.6 μm, respectively. Calculate the total inductance of a 100-μm metal1 wire.

Solution: From Equation (13.15), we obtain

$$\begin{aligned} L &= \frac{\mu_0}{2\pi}\left[l \ln\left(\frac{2l}{w+t}\right) + \frac{l}{2} + 0.2235(w+t)\right] \\ &= \frac{4\pi \times 10^{-13}}{2\pi}\left[100 \ln\left(\frac{2 \times 100}{0.5+0.6}\right) + \frac{100}{2} + 0.2235(0.5+0.6)\right] \\ &= 114\ \text{pH} \end{aligned}$$

■

As in the case of capacitance, the wire inductance is usually estimated by a rule of thumb in practical applications. From electromagnetic theory, the unit-length wire inductance can be estimated from the unit-area capacitance of the wire according to the following equation:

$$C_{int}L_{int} = \varepsilon_{ox}\mu_{ox} \tag{13.16}$$

Substituting $C_{int} = 0.1$ fF/μm and both constants: $\varepsilon_{ox} = 3.9 \times 8.854 \times 10^{-12}$ F/m and $\mu_{ox} = 4\pi \times 10^{-7}$ H/m into above equation, we obtain the unit-length wire inductance as $L_{int} = 0.43$ pH/μm. Hence, we will use $L_{int} = 0.5$ pH/μm when calculating the wire inductance.

13.1.3.1 Inductance Effects The combination of both inductance and capacitance of a wire will make the wire behave as a transmission line. A simple criterion can be used to judge whether the wire inductance needs to be considered: If we measure different values at any two points along the wire, then the wire inductance is important and the wire should be regarded as a transmission line; otherwise, the wire inductance is unimportant. The details of transmission lines will be addressed in great detail in Section 13.4.

When a wire exhibits the transmission-line behavior, an abrupt changed signal passing through it will be reflected off the load and backed to the signal source along the wire, leading to a *ringing phenomenon*. Fortunately, this ringing phenomenon will gradually die out, and the signal on the wire will be settled down to a final steady value if the wire resistance R is large enough.

Remember that the wire resistance is proportional to the wire length. Thus, when the wire is long enough, the reflections will be insignificant and the wire can be simply modeled as an RC circuit. In other words, as the DC resistance R is greater than the characteristic impedance Z_0 of the wire, the wire will behave as an RC circuit. Consequently, the wire length has an upper bound below which the wire inductance should be taken into account.

On the other side, the rise time of the signal at the output of the driving buffer also determines whether the wire inductance should be taken into account. If the rise time t_{rise} is greater than the round-trip time of the signal, $2l\sqrt{L_{int}C_{int}}$, then the inductance effect is gone. As a consequence, the wire length has a lower bound above which the wire inductance should be considered.

Combining the above two extreme cases, a two-sided relationship determining the wire-length range in which the wire inductance is significant is as follows:

$$\frac{t_{rise}}{2\sqrt{L_{int}C_{int}}} \leq l \leq \frac{2}{R_{int}}\sqrt{\frac{L_{int}}{C_{int}}} \tag{13.17}$$

where $1/\sqrt{L_{int}C_{int}}$ is the signal velocity and the $\sqrt{L_{int}/C_{int}}$ is the characteristic impedance Z_0 of the wire. This range depends on the parasitic impedance of the wire per unit length as well as on the rise time of the signal at the output of the driving buffer. In some situations, this range can be nonexistent if the following condition is satisfied:

$$t_{rise} > 4\frac{L}{R} \tag{13.18}$$

which means that the wire inductance under consideration is not important for any length.

Example 13-9: (The inductance effect.) Suppose that the sheet resistance of a wire is 57 mΩ/□ and unit-length capacitance and inductance are 0.1 fF/μm and 0.5 pH/μm, respectively. The wire width is 0.5 μm. Determine the wire-length range in which the wire inductance needs to be taken into account if

1. The rise time of signal is 1.5 ns.
2. The rise time of signal is 15 ps.

Solution:

1. The resistance is $(57\ m\Omega/\square)\times(1\square/0.5\ \mu\text{m}) = 114\ m\Omega/\mu\text{m}$. From Equation (13.18), we obtain

$$1.5\text{ ns} > 4\frac{0.5\text{ pH}/\mu\text{m}}{114\ m\Omega/\mu\text{m}} = 0.018\text{ ns}$$

Hence, in this case, inductance is not important for any length of the interconnect under consideration.

2. From Equation (13.18), we obtain

$$0.015\text{ ns} > 4\frac{0.5\text{ pH}/\mu\text{m}}{114\ m\Omega/\mu\text{m}} = 0.018\text{ ns}$$

Hence, the inequality equation is not valid. This means that the inductance is important when the wire length falls between the two extremes.

$$\frac{15\text{ ps}}{2\sqrt{0.5\text{ pH}/\mu\text{m}\times 0.15\text{ fF}/\mu\text{m}}} \le l \le \frac{2}{114\ m\Omega/\mu\text{m}}\sqrt{\frac{0.5\text{ pH}/\mu\text{m}}{0.15\text{ fF}/\mu\text{m}}}$$

Thus, the wire-length range is $0.87\text{ mm} \le l \le 1.01\text{ mm}$.

Review Questions

Q13-7. Describe Faraday's law of induction.

Q13-8. Describe the upper bound of the wire length in which the inductance is significant and cannot be ignored.

Q13-9. Describe the lower bound of the wire length in which the inductance is significant and cannot be ignored.

13.2 Interconnect and Simulation Models

The interconnect model is used to estimate and approximate the real behavior of a wire as a function of its parameters. To simplify the simulation work for a wire, we often need a simple model to capture the behavior of the wire. We first introduce the interconnect models and then the widely used simulation models.

13.2.1 Interconnect Models

The interconnect model is used to estimate and approximate the real behavior of a wire as a function of its parameters. The most commonly used interconnect models include the *lumped-RC model*, *distributed-RC model*, and *transmission-line model*.

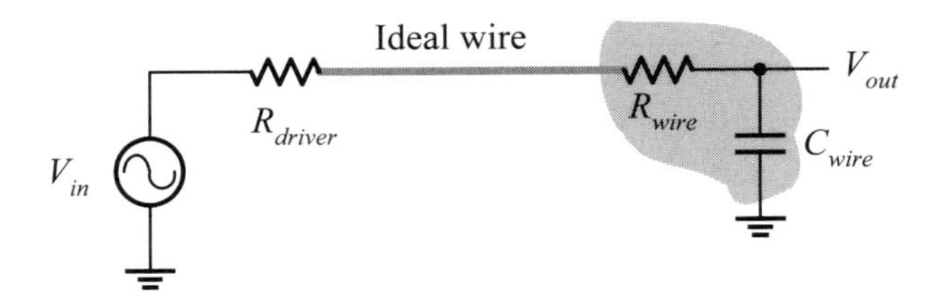

Figure 13.7: The lumped-RC model for short wires.

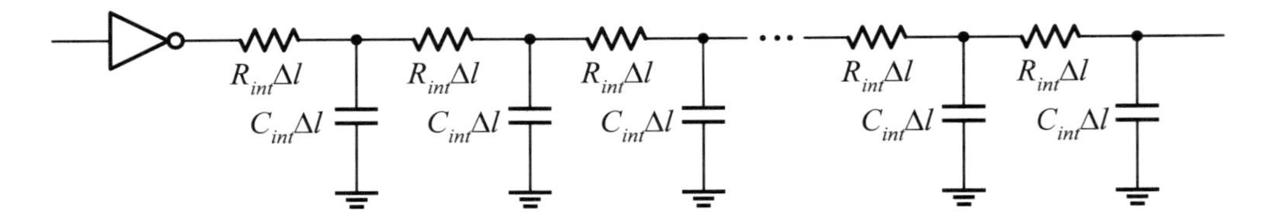

Figure 13.8: The distributed-RC model for long wires.

13.2.1.1 The Lumped-RC Model The lumped-RC model of a wire is the simplest model. It simply takes the entire wire as a single time constant RC circuit and as a combination of an *ideal wire* with a single resistor R_{wire} and capacitor C_{wire} being connected at the end of the ideal wire, as shown in Figure 13.7. An ideal wire is the one with zero value of resistance and capacitance. Assume that the wire length is l, then the wire resistance and capacitance can be separately estimated as follows:

$$R_{wire} = l \times R_{int} \tag{13.19}$$

$$C_{wire} = l \times C_{int} \tag{13.20}$$

where R_{int} and C_{int} are the unit resistance and capacitance of the wire, respectively.

If the output resistance of the driver is R_{driver}, the time constant (τ) of the lumped-RC model shown in Figure 13.7 is

$$\tau = (R_{driver} + R_{wire})C_{wire} \tag{13.21}$$

Based on this, the propagation delay is

$$t_{50\%} = 0.69\tau = 0.69(R_{driver} + R_{wire})C_{wire}$$

and rise and fall times are

$$t_{90\%} = 2.2\tau = 2.2(R_{driver} + R_{wire})C_{wire}$$

13.2.1.2 The Distributed-RC Model The *distributed-RC model* takes the wire as a cascading of simple RC networks, consisting of a set of segment resistance and segment capacitance, as depicted in Figure 13.8. Assume that each segment has a length of Δl. Then each segment has a resistance of $R_{int}\Delta l$ and a capacitance of $C_{int}\Delta l$. The total delay of the wire can then be estimated by using the Elmore delay model as follows:

$$\begin{aligned} t_{Elmore} &= (R_{int}\Delta l)(C_{int}\Delta l) + 2(R_{int}\Delta l)(C_{int}\Delta l) + \cdots + n(R_{int}\Delta l)(C_{int}\Delta l) \\ &= (\Delta l)^2 R_{int}C_{int}(1 + 2 + \cdots + n) \\ &= (\Delta l)^2 R_{int}C_{int}\frac{n(1+n)}{2} \\ &\approx (\Delta l)^2 R_{int}C_{int}\frac{n^2}{2} = \frac{1}{2}l^2 R_{int}C_{int} = \frac{1}{2}R_{wire}C_{wire} \end{aligned} \tag{13.22}$$

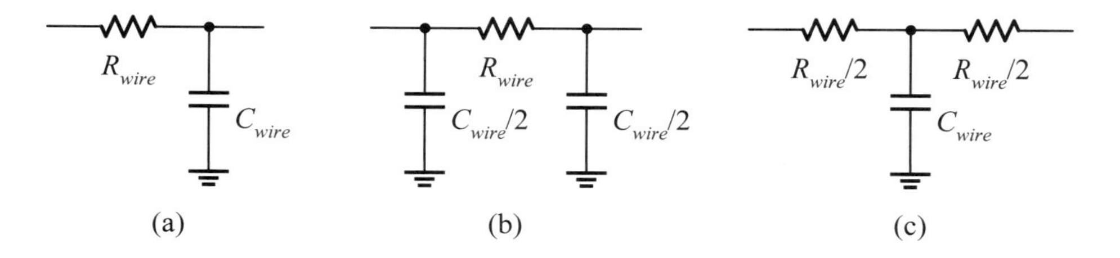

Figure 13.9: The most widely used simulation models of wires: (a) L-model; (b) Π-model; (c) T-model.

The actual delay is approximately to $0.38R_{wire}C_{wire}$. From the above equation, we can see that the delay of a wire is a quadratic function of its length.

13.2.1.3 Transmission-Line Model The transmission line is a distributed RLC wire, which includes not only distributed R and C, but also inductance L. In this model, a signal propagates along the wire as a wave. The details of the transmission-line model will be discussed in Section 13.4.

13.2.2 Simulation Models

To simplify the simulation work for a wire, we often need a simple model to capture the behavior of the wire. Figure 13.9 gives the three most widely used simulation models for wires. The first is the L-model, which produces a time constant of $R_{wire}C_{wire}$. The second is the Π-model, which produces a time constant of $R_{wire}C_{wire}/2$. The third is the T-model, which produces a time constant of $R_{wire}C_{wire}/2$. Both the Π-model and T-model produce an accurate model for long distributed-RC lines. But, in practice, the Π-model is the most popular.

■ Example 13-10: (An example of the Π model.) Assume that a metal4 wire segment with 10-mm long is driven by a 50X inverter. The wire width is 0.5 μm. Using the Π-model and 0.18-μm process parameters, estimate the delay of this metal wire with the Elmore delay model.

Solution: We begin to calculate the equivalent resistance and capacitance of the metal4 wire segment. Recall that 1 mm = 10^3 μm. The wire resistance is

$$R_{wire} = R_{sq}\left(\frac{l}{w}\right) = 0.025 \times \left(\frac{10000}{0.5}\right) = 500\ \Omega$$

and the wire capacitance is

$$C_{wire} = C_{int} \cdot l = 0.15 \times 10000 = 1.5 \text{ pF}$$

Hence, the Π-model for the metal4 wire segment is shown in Figure 13.10.

Next, we calculate the effective resistance and self-loading capacitance of the driving inverter. The on-resistance is equal to

$$R_{eqv} = R_{eqn}/50 = 8 \text{ k}\Omega/50 = 160\ \Omega$$

and the capacitance is

$$C_{self-load} = C_{para} \cdot (3W) \cdot 50 = 3.68 \times 3 \times 0.18 \times 50 = 99.4 \text{ fF (Ignored)}$$

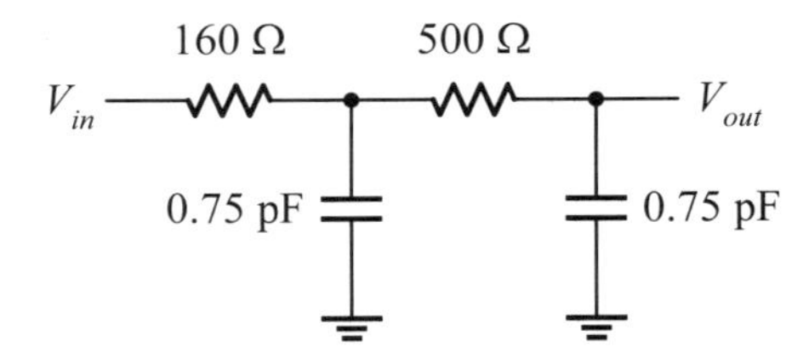

Figure 13.10: An example of using the Π-model to represent a long wire.

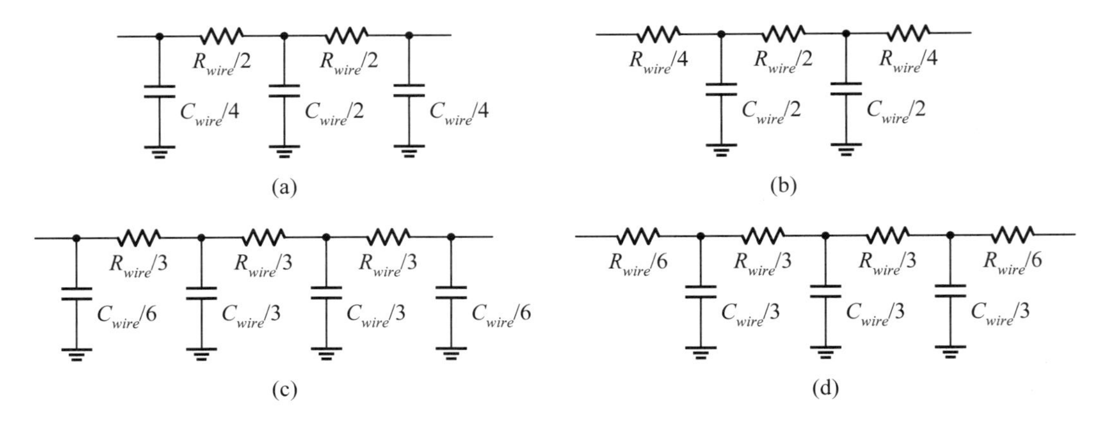

Figure 13.11: Some advanced simulation models of wires: (a) Π2-model; (b) T2-model; (c) Π3-model; (d) T3-model.

The resulting equivalent circuit is given in Figure 13.10. Using the Elmore delay model, the delay of the metal4 wire segment is found to be

$$t_{Elmore} = 160 \times 0.75 + (160 + 500) \times 0.75 = 615 \text{ ps}$$

■

Some more advanced interconnect models are given in Figure 13.11. These include the Π-2 and Π-3 models as well as the T-2 and T-3 models.

■ Review Questions

Q13-10. What are the most popular interconnect models?

Q13-11. Describe the lumped-RC model.

Q13-12. Describe the distributed-RC model.

13.3 Parasitic Effects of Interconnect

Recall that wires introduce three types of parasitic effects: resistive, capacitive, and inductive. All of these influence the signal integrity and degrade the performance of the circuits. The signal-integrity problem involves the quality of a signal being transmitted from one place to another through wires. We are concerned with each of these in more detail.

13.3.1 RC Delay

Resistive parasitics may cause an IR drop such that noise margins are reduced and the performance of logic circuits is deteriorated. The combination of resistive and capacitive parasitics induces RC delays of signals and clocks, which may severely impact on the performance of logic circuits. Hence, the reduction of RC delays is an important issue for designing high-performance logic circuits. For this purpose, the following three methods are proposed: *better interconnect materials, better interconnect strategies,* and *buffer insertion.*

13.3.1.1 Better Interconnect Materials Using better interconnect materials can effectively reduce the parasitic resistance and capacitance of wires, thereby, decreasing the RC delay significantly. General speaking, silicides and the use of copper wires are two popular approaches to reduce the resistance of polysilicon and metal wires, respectively. In addition, using low-permittivity (that is, low-k) dielectrics lowers the parasitic capacitance between two wires. Another popular method for reducing the resistance of a long polysilicon wire, such as the word-line of static random access memory (SRAM), is to use a metal bypass wire to tap the polysilicon word-line at many places such that both metal and polysilicon wires are effectively connected in parallel. As a result, it can effectively reduce the resistance of the word-line, referring to Figure 11.18 for more details.

13.3.1.2 Better Interconnect Strategies The use of a better routing strategy makes possible the reduction of the RC delay of wires. At present, most computer-aided design (CAD) tools only allow the wires to be routed horizontally and vertically, namely, in the Manhattan style. However, if the diagonal wiring strategy is allowed, the resulting chip would have above 20% interconnect length reduction and above 15% chip area saving plus above 30% via reduction. Unfortunately, this would complicate the parameter models and make the parameter extraction of layout much more difficult than the Manhattan style.

13.3.1.3 Buffer Insertion As described before, the delay of a wire is a quadratic function of its length. Hence, to reduce the propagation delay of a long wire, we need to break up the quadratic relationship associated with the length. To achieve this, probably the most straightforward way is to subdivide the long wire into many segments and then to insert a buffer between two segments so that the propagation delay of the resulting long wire becomes a linear function of the number of segments. The design approach based on this idea is referred to as *buffer insertion.* An illustration of buffer insertion is shown in Figure 13.12(a), where the long wire is evenly divided into n segments, each having a length of l/n. A buffer with the size of M times of the basic inverter is then inserted between two segments. The resulting equivalent RC circuit of a segment is then modeled by the Π-model, as revealed in Figure 13.12(b).

The design issues of buffer insertion are to find the optimal number of equal-sized buffers and to determine the optimal size of each buffer. To illustrate this, let us consider the basic block and equivalent circuit shown in Figure 13.12. Here, we suppose that the long wire is partitioned into the n segment and the buffer between two segments has the size of M times the basic inverter.

Recall that the input capacitance of a basic inverter can be expressed as

$$C_{in} = C_g(W_n + W_p) = C_G \tag{13.23}$$

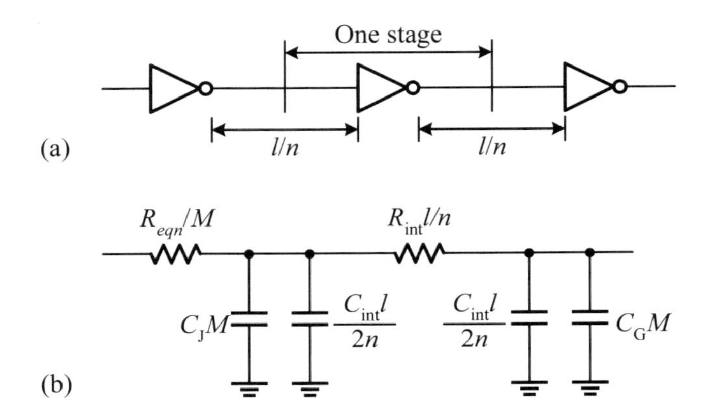

Figure 13.12: The reduction of the RC delay of a wire by inserting buffers: (a) basic block; (b) equivalent circuit.

and the output capacitance of a basic inverter is equal to its self-loading capacitance. That is,

$$C_{out} = C_{para}(W_n + W_p) = C_J \tag{13.24}$$

Now let M be the multiplier of the optimal buffer size. Then, the input capacitance of the optimal buffer is equal to MC_G and the output capacitance is MC_J. The effective resistance R_{eqv} of the optimal buffer is reduced by a factor of $1/M$ accordingly and is equal to $R_{eqv} = R_{eqn}/M$.

Remember that the wire resistance is equal to the product of unit resistance and the number of unit lengths of the wire. That is, the wire resistance is

$$R_{wire} = R_{sq}\left(\frac{l}{w}\right) = R_{int}l \tag{13.25}$$

where $R_{int} = R_{sq}/w$. Similarly, the wire capacitance is found to be

$$C_{wire} = C_{int}l \tag{13.26}$$

The resulting equivalent RC circuit of a wire segment represented in the Π-model is shown in Figure 13.12(b).

Using the Elmore delay model, the delay for each segment is equal to

$$t_{pd(segment)} = \frac{R_{eqn}}{M}\left(C_JM + \frac{C_{int}l}{2n}\right) + \left(\frac{R_{eqn}}{M} + \frac{R_{int}l}{n}\right)\left(C_GM + \frac{C_{int}l}{2n}\right) \tag{13.27}$$

The total delay is the sum of delays of all segments and can be expressed as the following equation:

$$\begin{aligned} t_{pd(total)} &= n \times t_{pd(segment)} \\ &= n(C_J + C_G)R_{eqn} + \left(R_{eqn}C_GM + \frac{R_{eqn}C_{int}}{M}\right)l + \left(\frac{R_{int}C_{int}}{2n}\right)l^2 \end{aligned} \tag{13.28}$$

To find the optimal number of segments, we take the partial derivative of $t_{pd(total)}$ with respect to the number of segments n and obtain

$$\frac{\partial t_{pd(total)}}{\partial n} = 0 = (C_J + C_G)R_{eqn} - \frac{R_{int}C_{int}}{2n^2}l^2 \tag{13.29}$$

Solving the above equation, the optimal number of segments n is found to be

$$n = \sqrt{\frac{R_{int}C_{int}l^2/2}{(C_J + C_G)R_{eqn}}} \tag{13.30}$$

To find the optimal buffer size, we take the partial derivative of $t_{pd(total)}$ relative to the multiplier M of buffers and have the following equation:

$$\frac{\partial t_{pd(total)}}{\partial M} = 0 = R_{int}lC_G - \frac{R_{int}C_{int}l}{M^2} \tag{13.31}$$

Solving the above equation, the multiplier M of the optimal buffer size is equal to

$$M = \sqrt{\frac{R_{eqn}}{C_G}\frac{C_{int}}{R_{int}}} \tag{13.32}$$

An illustration of the above design procedure by using equations for finding both n and M is given in the following example.

■ Example 13-11: (An example of buffer insertion.) Assume that a 10-mm metal4 wire driven by a 50X inverter is used in a circuit. The width of the metal4 wire is 0.5 μm. If the buffer insertion approach is used to reduce the wire delay, calculate the optimal number of buffers needed and the optimal buffer size, using 0.18-μm process parameters.

Solution: From the preceding example, we know that $R_{wire} = 500\ \Omega$ and $C_{wire} =$ 1.5 pF. Using Equation (13.30), the number of segments n is found to be

$$\begin{aligned} n &= \sqrt{\frac{R_{int}C_{int}l^2/2}{(C_J + C_G)R_{eqn}}} \\ &= \sqrt{\frac{25 \times 10^{-3}/0.5 \times 0.15 \times (10^4)^2/2}{(3.68 + 3.41) \times 3 \times 0.18 \times 8 \times 10^3}} = 3.5 \approx 4 \end{aligned}$$

Hence, the segment resistance is $500/4 = 125\ \Omega$, and segment capacitance is $C_{wire} =$ $1.5/4 = 0.375$ pF.

The optimal buffer size M can be calculated by using Equation (13.32) and equals

$$\begin{aligned} M &= \sqrt{\frac{R_{eqn}}{C_G}\frac{C_{int}}{R_{int}}} \\ &= \sqrt{\frac{8 \times 10^3}{3.41 \times 3 \times 0.18}\frac{0.15}{25 \times 10^{-3}/0.5}} = 114.16 \approx 115 \end{aligned}$$

The buffer on-resistance is

$$R_{eqv} = R_{eqn}/M = 8\ \text{k}\Omega/115 = 70\ \Omega$$

The buffer output capacitance $C_{self-load}$ is equal to

$$C_{self-load} = C_{para}(3W) \cdot M = 3.68 \times 3 \times 0.18 \times 115 = 229\ \text{fF}$$

The buffer input capacitance C_G equals

$$C_G = C_g(3W) \cdot M = 3.41 \times 3 \times 0.18 \times 115 = 212\ \text{fF}$$

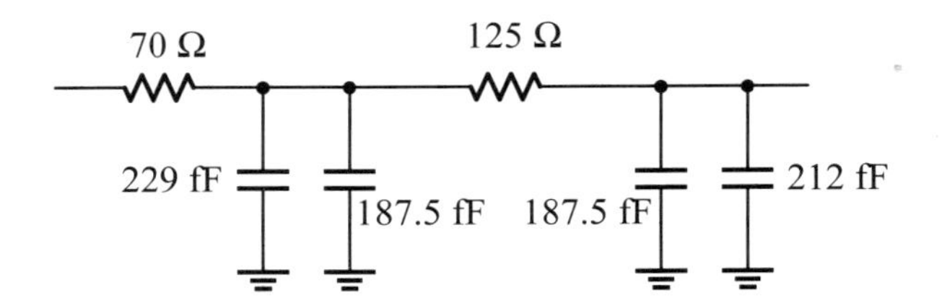

Figure 13.13: An example to illustrate the use of buffer insertion.

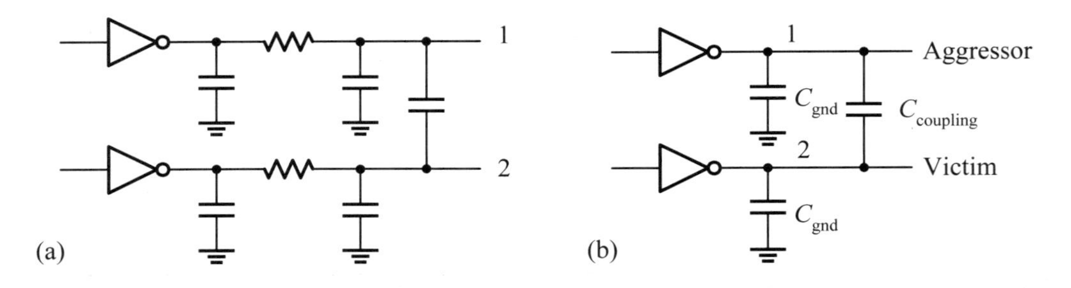

Figure 13.14: The capacitive-coupling effects: (a) *RC* model; (c) *C*-only model.

The resulting equivalent circuit of an interconnect segment is given in Figure 13.13. The total Elmore delay of the entire wire can be calculated as follows:

$$t_{Elmore} = 4 \times [70 \times (229 + 187.5) + (70 + 125) \times (187.5 + 212)] = 428.25 \text{ ps}$$

A value much less than the one without buffer insertion, which is 615 ps. Of course, the reduction of the propagation delay is at the cost of three large buffers, which need much hardware (chip area).

Review Questions

Q13-13. Define power-integrity and signal-integrity problems.

Q13-14. What are the three popular methods used to reduce the *RC* delay of interconnect?

Q13-15. What are the design issues of buffer insertion?

13.3.2 Capacitive-Coupling Effects

The capacitive-coupling effects may cause undesired crosstalk between different wires. These effects are illustrated in Figure 13.14. Figure 13.14(a) shows an equivalent circuit taking into account both resistive and capacitive parasitics and Figure 13.14 gives the equivalent circuit only considering the capacitive parasitics. For simplicity, we only consider the *C*-only model shown in Figure 13.14(b). The wire of interest is generally called the *victim*, while the neighboring wire is called the *aggressor*. In what follows, we begin with the discussion of the effective loading capacitance of the victim due to the coupling capacitance and then cope with the *crosstalk*.

13.3.2.1 Effective Loading Capacitance The effective loading capacitance of the victim depends on the switching state of the aggressor. When the aggressor is not switching, the effective loading capacitance of the victim is equal to

$$C_L = C_{gnd} + C_{coupling} \tag{13.33}$$

When both aggressor and victim are switching together, both ends of the coupling capacitor track each other. Hence, the effective loading capacitance of the victim is only its output capacitance.

$$C_L = C_{gnd} \tag{13.34}$$

However, when both aggressor and victim are switching oppositely, the effect of the coupling capacitor is doubled. Consequently, the effective loading capacitance of the victim is

$$C_L = C_{gnd} + 2C_{coupling} \tag{13.35}$$

13.3.2.2 Multilayer Interconnect Network For a complicated multilayer interconnect network, the coupling of multiple wires in such an environment can be generally described by a *capacitance matrix* as follows:

$$\mathbf{Q} = \mathbf{C} \times \mathbf{V} \tag{13.36}$$

For a two-wire system, the matrix equation is

$$\begin{bmatrix} Q_1 \\ Q_2 \end{bmatrix} = \begin{bmatrix} C_{11} & -C_{12} \\ -C_{21} & C_{22} \end{bmatrix} \times \begin{bmatrix} V_1 \\ V_2 \end{bmatrix} \tag{13.37}$$

where

$$C_{11} = \left.\frac{Q_1}{V_1}\right|_{V_2=0} \tag{13.38}$$

$$C_{22} = \left.\frac{Q_2}{V_2}\right|_{V_1=0} \tag{13.39}$$

$$C_{12} = C_{21} = -\left.\frac{Q_1}{V_2}\right|_{V_1=0} = -\left.\frac{Q_2}{V_1}\right|_{V_2=0} \tag{13.40}$$

In a symmetric system, C_{11} and C_{22} are the same as C (wire capacitance), and C_{12} and C_{21} are the coupling capacitance between two wires.

13.3.2.3 Crosstalk Crosstalk, also known as *capacitive noise* (note that noise is an undesired signal), is the problem of noise injection due to capacitive coupling arising when there is a nonswitching node (the victim) surrounded by switching aggressor nets. To quantify the amount of crosstalk, consider the equivalent circuit revealed in Figure 13.15. The amount of voltage change at node 1 is obtained by using the voltage-divider rule and can be expressed as follows:

$$\Delta V_1 = \frac{C_{coupling}\Delta V_2}{C_{gnd} + C_{coupling}} = \frac{C_{coupling}V_{DD}}{C_{gnd} + C_{coupling}} \tag{13.41}$$

when aggressor switches between power-supply rails; that is, $\Delta V_2 = V_{DD}$.

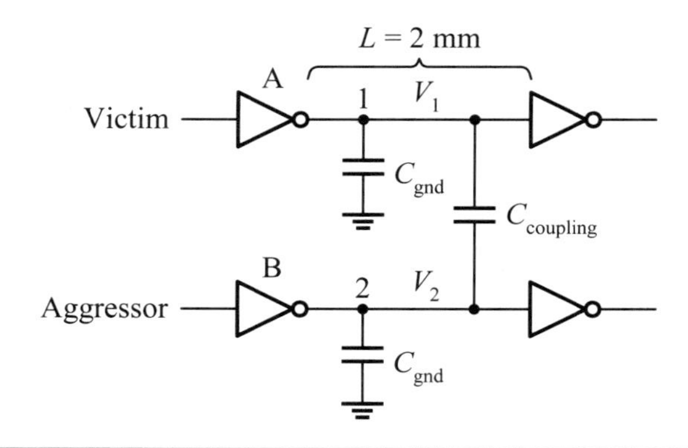

Figure 13.15: The crosstalk effects caused by capacitive coupling.

■ Example 13-12: (An example of crosstalk.) Suppose that the length l of two metal2 wires between two inverters is 2 mm and both victim and aggressor are $10X$ inverters, as depicted in Figure 13.15. Find the amount of crosstalk ΔV_1 using 0.18-μm process parameters. Assume that $C_{area} = 0.06$ fF/μm and $C_{lateral} = 0.08$ fF/μm.

Solution: The coupling capacitance is indeed the lateral capacitance $C_{lateral}$ and is

$$C_{coupling} = C_{lateral} \times l = 0.08 \times (2 \times 10^3) = 160 \text{ fF}$$

and the ground capacitance is

$$\begin{aligned} C_{gnd} &= C_{para}(W_n + W_p) + C_{area} \times l \\ &= 10 \times C_{para} \times 3W + 0.06 \times (2 \times 10^3) \\ &= 10 \times 3.68 \times 3 \times 0.18 + 120 = 139.87 \text{ fF} \end{aligned}$$

From Equation (13.41), the crosstalk is found to be

$$\begin{aligned} \Delta V_1 &= \frac{C_{coupling} V_{DD}}{C_{gnd} + C_{coupling}} \\ &= \frac{160 \times 1.8}{139.87 + 160} = 0.96 \text{ V} \end{aligned}$$

An amount of crosstalk seems to be intolerable in many systems. ■

13.3.2.4 Capacitive-Coupling Reduction Once we realized the capacitive-coupling effects caused by the capacitance between adjacent lines, we are now in a position to cope with these effects. As illustrated in Figure 13.16, two basic principles for reducing the capacitive-coupling effects are *spacing* and *shielding*. The former is based on the principle of spacing out the separation between wires and hence reduces the lateral capacitance between two adjacent wires significantly; the latter is by interleaving switching signals with nonswitching signals, such as powers and grounds, so that the coupling capacitance can be reduced profoundly.

Capacitive-coupling effects can also be reduced by a process using low-k dielectric material to reduce the coupling capacitance between adjacent layers or by a more aggressive layout strategy in which wires on adjacent layers are routed orthogonally to

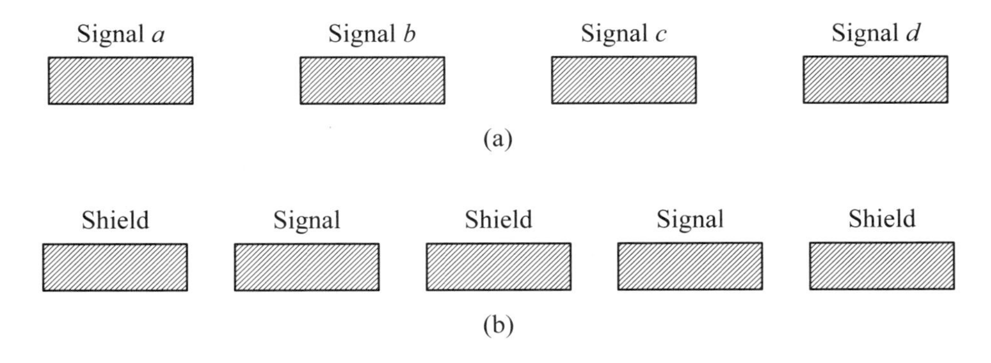

Figure 13.16: Two basic principles to reduce the capacitive-coupling effects: (a) spacing; (b) shielding.

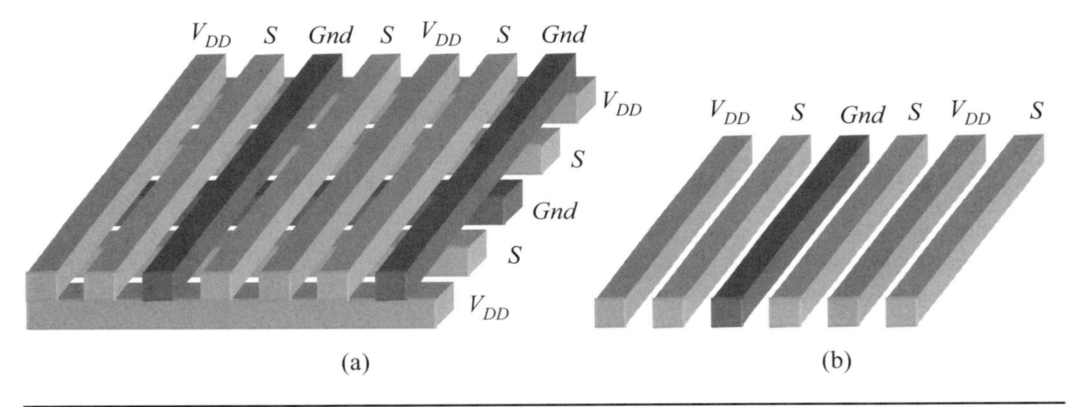

Figure 13.17: The layout strategy for reducing capacitive-coupling effects: (a) between adjacent layers; (b) in the same layer.

minimize the crosstalk and signal wires on the same layer are separated by V_{DD} and Gnd shields, as shown in Figure 13.17.

Another approach to reducing capacitive-coupling effects is based on the concept of *cancellation*, namely, using an inverting replica to cancel the original one. Based on this, there are three common approaches. Figure 13.18(a) depicts the situation when buffers (repeaters) are used in long wires. We can stagger the place of buffers properly so that injected noises can cancel out each other. Figure 13.18(b) illustrates the method of charge compensation in which an explicitly coupling capacitor injects phase-inverse noise to attack the unavoidable noise. The most effective method is the use of twisted, differential wires, as depicted in Figure 13.18(c). Both differential buffers and twisted wires are used in this case.

■ Review Questions

Q13-16. Define aggressor, victim, capacitive noise, and crosstalk.

Q13-17. What are the two basic principles to reduce capacitive-coupling effects?

Q13-18. What are the three approaches for reducing capacitive-coupling effects?

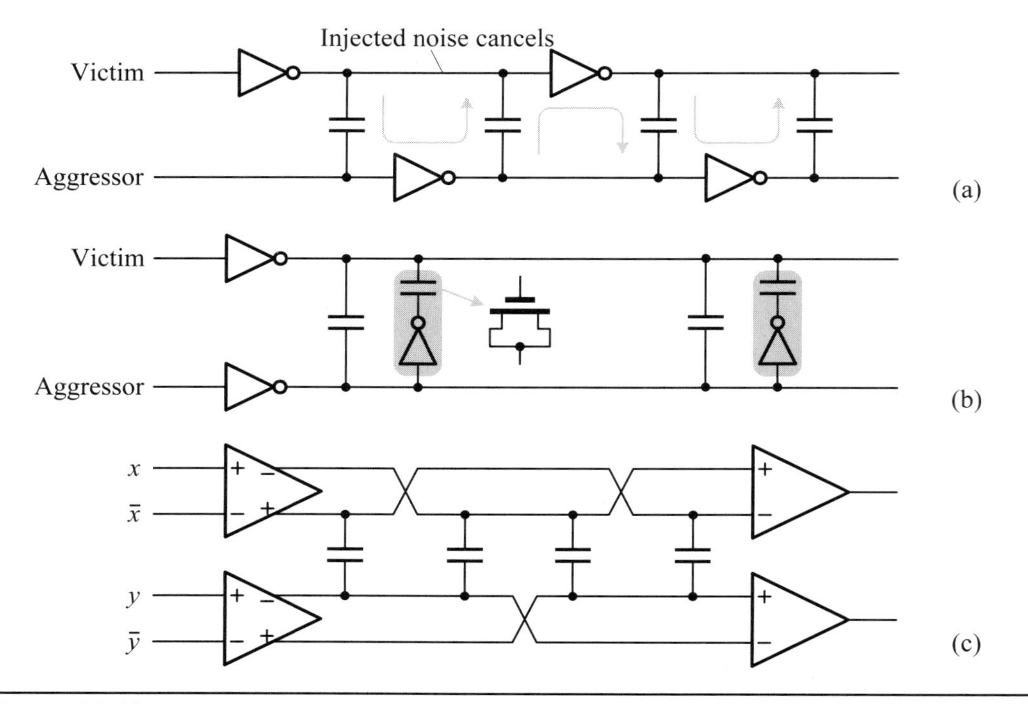

Figure 13.18: Three approaches for reducing capacitive-coupling effects: (a) staggering repeaters; (b) charge compensation devices; (b) twisted, differential wires.

13.3.3 RLC Effects

The inductance effect caused by a single wire alone may result in an Ldi/dt voltage drop (or called Ldi/dt noise). When it is combined with resistance and capacitance parasitics, a second-order RLC response will be resulted. A consequence of this RLC response is the *ringing effect* on the rising and falling edges during signal transitions. The ringing effect is caused by the interaction of forward and backward propagation signals.

A lumped RLC model of a wire is shown in Figure 13.19(a), where the $Z_{RL} = R + j\omega L$. From circuitry theory, as $R \gg j\omega L$, the inductance is not important and the RLC model is effectively reduced to a simple RC circuit. Nevertheless, as $R < j\omega L$, the Z_{RL} is a complex number and the inductance effect has a significant impact on the signal responses.

When a wire is being driven by a buffer with an output impedance of R_{buffer}, the self-inductance of the wire can be ignored if $R_{buffer} + R_{wire} \gg j\omega L$, as shown in Figure 13.19(b). For clock networks, the sizes of clock drivers are often intended to be very large and the wires are very long. Hence, $R_{buffer} + R_{wire} < j\omega L$ and the self-inductance would be noticeable and cannot be simply ignored.

Once we deal with the inductance effects on a single wire, we are now in a position to address the troublesome problem of mutual inductance. Mutual inductance can cause noise and delay effects on unsuspecting neighboring lines in unexpected ways. The inductance is so problematic is because inductance can only be defined for a closed current loop. Knowing the forward path of the current is not sufficient; the return paths must also be identified. Unfortunately, it is not easy to determine the return paths for most signals. In digital systems, clock wires are more likely prone

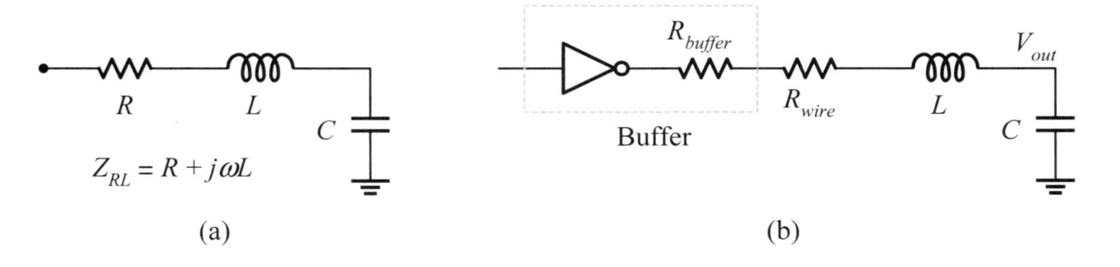

Figure 13.19: A lumped-RLC circuit model: (a) RLC model; (b) a buffer-driven interconnect wire.

to self-inductance effects and buses are more likely to experience mutual inductance effects. For clock wires, the wire length should be kept outside the boundary in which the inductance needs to be considered. For buses, shielding techniques can be used to limit the mutual inductance effects, as illustrated in Figure 13.16.

13.3.3.1 Multilayer Interconnect Network For a complicated multilayer-metal interconnect network, the inductive coupling of multiple wires can be generally described by an *inductance matrix* as follows:

$$\mathbf{\Phi} = \mathbf{L} \times \mathbf{I} \tag{13.42}$$

The induced voltage is then equal to

$$\frac{d\mathbf{\Phi}}{dt} = \mathbf{V} = \mathbf{L} \times \frac{d\mathbf{I}}{dt} \tag{13.43}$$

For a two-wire system, the matrix equation is

$$\begin{bmatrix} V_1 \\ V_2 \end{bmatrix} = \begin{bmatrix} L_{11} & -L_{12} \\ -L_{21} & L_{22} \end{bmatrix} \times \begin{bmatrix} \dfrac{dI_1}{dt} \\ \dfrac{dI_2}{dt} \end{bmatrix} \tag{13.44}$$

In a symmetric system, L_{11} and L_{22} are same as L (wire inductance) and L_{12} and L_{21} are the mutual inductance between two wires.

■ Review Questions

Q13-19. In what situation, can the inductance parasitic be ignored?

Q13-20. What is a ringing effect?

Q13-21. What is the meaning of return path?

13.4 Transmission-Line Models

The transmission-line model of a wire can be classified into the following two types:

- The *lossless transmission line*: When the wire resistance is small, the transmission line can be simplified to a capacitive/inductive model, which is called the lossless transmission line. It is applicable for the wires at the printed-circuit board (PCB) level.

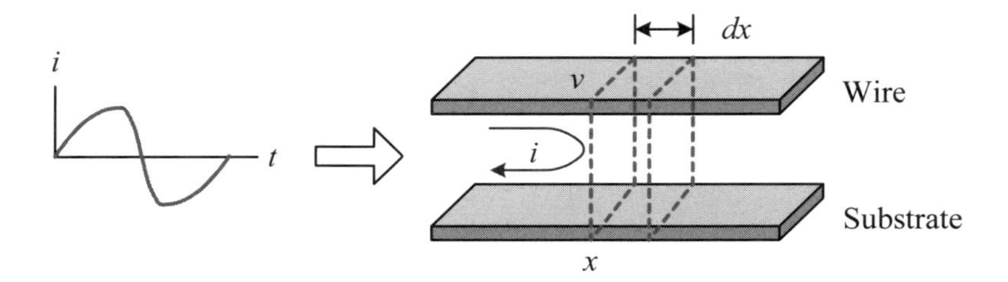

Figure 13.20: A conceptual illustration of the characteristic impedance of transmission line.

- The *lossy transmission line*: When the wire resistance plays an important role, the transmission line is called the lossy transmission line. It is applicable to on-chip wires and some thin-film package wires.

13.4.1 Lossless Transmission Lines

In the lossless transmission line, the signal velocity (u) is defined as

$$u = \frac{1}{\sqrt{L_{int}C_{int}}} = \frac{1}{\sqrt{\varepsilon\mu}} = \frac{c}{\sqrt{\varepsilon_r\mu_r}} \tag{13.45}$$

where c is the light speed in free space and L_{int} and C_{int} are unit-length inductance and capacitance of the transmission line, respectively.

Even though the values of L_{int} and C_{int} depend on the geometric shape of the wire, their product is constant and only a function of the surrounding media. The propagation time per unit length (t_p) is equal to the square root of the product of L_{int} and C_{int}:

$$t_p = \sqrt{L_{int}C_{int}} \tag{13.46}$$

13.4.1.1 Characteristic Impedance (Z_0) To illustrate the concept of characteristic impedance of a transmission line, consider Figure 13.20. Because the current i behaves as a wave when it reaches x, it regards the transmission line as an infinite long path. To make the current i progress a small distance dx, an additional capacitance $C_{int}dx$ must be charged, thereby yielding a differential current i equal to

$$i = \frac{dq}{dt} = C_{int}\frac{dx}{dt}v = C_{int}uv = \sqrt{\frac{C_{int}}{L_{int}}}v \tag{13.47}$$

The signal sees the remainder of the transmission line as a real impedance of Z_0.

$$Z_0 = \frac{v}{i} = \sqrt{\frac{L_{int}}{C_{int}}} = \frac{\sqrt{\varepsilon\mu}}{C_{int}} \tag{13.48}$$

The quantity Z_0 is the impedance at any location that looks toward an infinitely long transmission line without having reflections. It is called the *characteristic impedance* of the transmission line.

The characteristic impedance is a function of the dielectric medium and the geometry of the conducting wire and dielectric, and is independent of the length of wire and

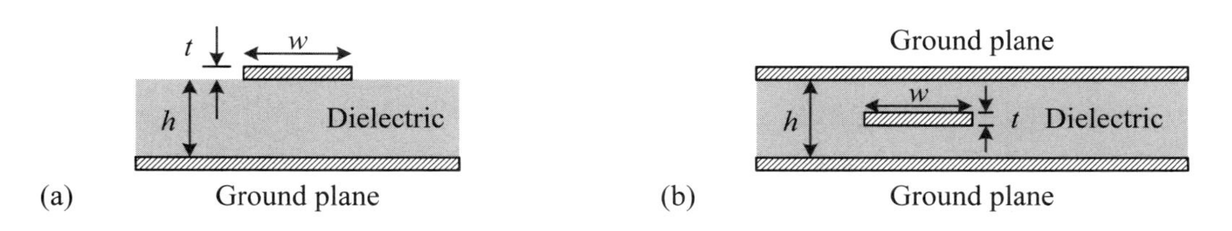

Figure 13.21: Examples of a semiconductor (a) microstrip and (b) stripline.

the frequency. Typical values of the characteristic impedance of wires in semiconductor circuits range from 10 to 200 Ω.

Generally speaking, a microstrip wire (or microstrip) is a metal wire on top of a dielectric setting on a ground plane, as shown in Figure 13.21(a), and has a characteristic impedance Z_0 being expressed as

$$Z_0 = 60\sqrt{\frac{\mu_r}{0.475\varepsilon_r} + 0.67 \ln\left(\frac{4h}{0.67t + 0.536w}\right)} \ \Omega \tag{13.49}$$

where μ_r and ε_r are the relative permeability and the permittivity of the dielectric, respectively.

A stripline is a flat strip of metal wire sandwiched between two ground planes, as shown in Figure 13.21(b). It has a characteristic impedance Z_0 of

$$Z_0 = 94\sqrt{\frac{\mu_r}{\varepsilon_r}} \ln\left(\frac{h+w}{t+w}\right) \ \Omega \tag{13.50}$$

Example 13-13: (Characteristic impedance examples.) Figure 13.21 gives examples of microstrip and stripline wires. The dielectric is silicon dioxide with $\mu_r = 1$ and $\varepsilon_r = 4.5$.

1. Assume that in Figure 13.21(a), $t = 1$ μm, $w = 2$ μm, and $h = 3$ μm, find the Z_0 of it.
2. Assume that in Figure 13.21(b), $t = 1$ μm, $w = 2$ μm, and $h = 3$ μm, find the Z_0 of it.

Solution:

1. From Equation (13.49), we have

$$\begin{aligned} Z_0 &= 60\sqrt{\frac{\mu_r}{0.475\varepsilon_r} + 0.67 \ln\left(\frac{4h}{0.67t + 0.536w}\right)} \ \Omega \\ &= 60\sqrt{\frac{1}{0.475 \times 4.5} + 0.67 \ln\left(\frac{4 \times 3}{0.67 \times 1 + 0.536 \times 2}\right)} \ \Omega \\ &= 123.5 \ \Omega \end{aligned}$$

2. Using Equation (13.50), we obtain

$$\begin{aligned} Z_0 &= 94\sqrt{\frac{\mu_r}{\varepsilon_r}} \ln\left(\frac{h+w}{t+w}\right) \ \Omega \\ &= 94\sqrt{\frac{1}{4.5}} \ln\left(\frac{3+2}{1+2}\right) \ \Omega \\ &= 22.6 \ \Omega \end{aligned}$$

■

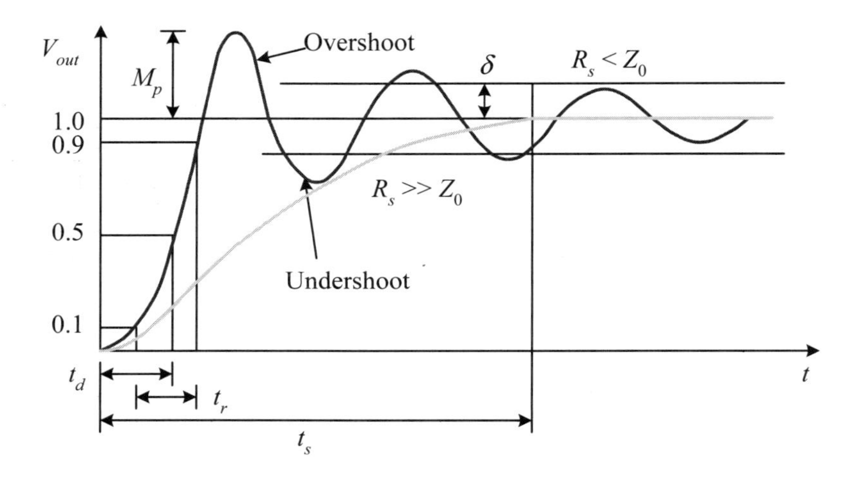

Figure 13.22: The typical responses of RLC interconnect.

13.4.1.2 RLC Responses Recall that the lossless transmission line only includes inductance and capacitance but no resistance. The characteristic impedance Z_0 of a lossless transmission line with unit inductance L_{int} and unit capacitance C_{int} is equal to

$$Z_0 = \sqrt{\frac{L_{int}}{C_{int}}} = \sqrt{\frac{0.45 \text{ pH}}{0.15 \text{ fF}}} = 54.8 \ \Omega \tag{13.51}$$

if $C_{int} = 0.2$ fF, Z_0 is equal to 47 Ω.

The lossless transmission line can be simply considered as an RC effect if $R_s \gg Z_0$, where R_s is the output resistance of the driver. When this is the case, the RC effect will dominate the transient response. However, the inductance effect will be important and the transient response is a second-order effect if $R_s < Z_0$. The general response of a lossless transmission line with respect to a unit-step input can be expressed as follows:

$$V_{out}(t) = \begin{cases} 1 - \exp\left(\dfrac{-t}{RC}\right) & R_s \gg Z_0 \\ 1 - r\exp\left(\dfrac{-t}{a}\right)\cos(bt + c) & R_s < Z_0 \end{cases} \tag{13.52}$$

where r, a, b, and c depend on the RLC values. The typical RLC responses with respect to a unit-step input are illustrated in Figure 13.22. From the figure, we can see that the response is similar to an RC response if $R_s \gg Z_0$ and is a second-order response if $R_s < Z_0$. Several related terms are also shown in the figure. The rise time t_r is the time interval for the output voltage rising from 10% to 90% of its final value. The *delay time* is the time for the output voltage to rise to 50% of its final value. The *settling time* t_s is the time interval for the output voltage to settle down to a limit δ within its final value. The *maximum overshoot* M_p is the maximum output voltage with respect to the final value.

13.4.1.3 Transmission-Line Behavior The transmission-line behavior is strongly influenced by its termination. The termination determines how much the electromagnetic wave will be reflected back to the source as the electromagnetic wave arrives at the

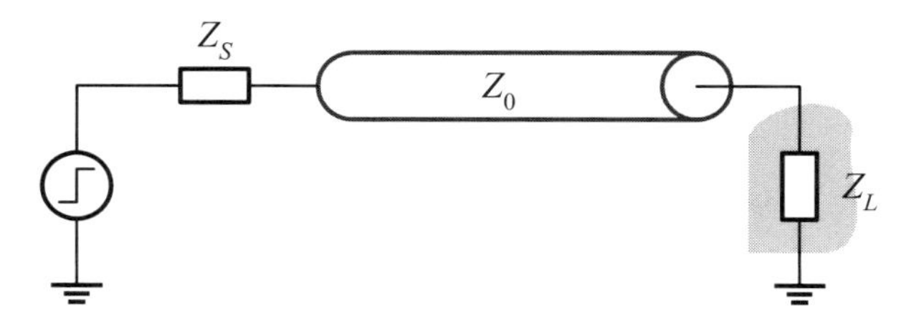

Figure 13.23: The definition of load and source reflection coefficients.

end of the transmission line. As shown in Figure 13.23, if the source impedance is Z_S and the termination impedance is Z_L, then we may define *load reflection coefficient* Γ_L and *source reflection coefficient* Γ_S as a function of these two impedances, respectively. More specifically, the load reflection coefficient Γ_L determines the amount of incident wave, which will be reflected by the load impedance. It is defined as

$$\Gamma_L = \frac{V_{reflection}}{V_{incident}} = \frac{I_{reflection}}{I_{incident}} = \frac{Z_L - Z_0}{Z_L + Z_0} \tag{13.53}$$

The *source reflection coefficient* Γ_S determines the amount of reflected wave which will be reflected by the source impedance and is defined as follows:

$$\Gamma_S = \frac{Z_S - Z_0}{Z_S + Z_0} \tag{13.54}$$

From Equation (13.53), we may sum up the transmission-line behavior as in the following:

- As the load impedance Z_L matches the characteristic impedance Z_0, there is no reflection wave. This condition is called *matched termination.*
- As the load impedance $Z_L = \infty$, namely, open-circuit termination, the incident wave is completely reflected without phase reversed.
- As the load impedance $Z_L = 0$, namely, short-circuit termination, the incident wave is completely reflected with the phase reversed.

■ Example 13-14: (An example of voltage transient diagrams.) Figure 13.24 shows how to draw voltage transient diagrams. To see this, assume that the load impedance $Z_L = 3Z_0$ and source impedance $Z_S = 2Z_0$. The load reflection coefficient Γ_L is

$$\Gamma_L = \frac{Z_L - Z_0}{Z_L + Z_0} = \frac{1}{2}$$

and the source reflection coefficient Γ_S

$$\Gamma_S = \frac{Z_S - Z_0}{Z_S + Z_0} = \frac{1}{3}$$

The initial voltage V_1^+ as the switch is turned on at time $t = 0$ is equal to the voltage across characteristic impedance Z_0 of the voltage divider formed by source impedance Z_s and characteristic impedance Z_0. It is equal to

$$V_1^+ = \frac{Z_0}{Z_S + Z_0} V_0 = \frac{1}{3} V_0$$

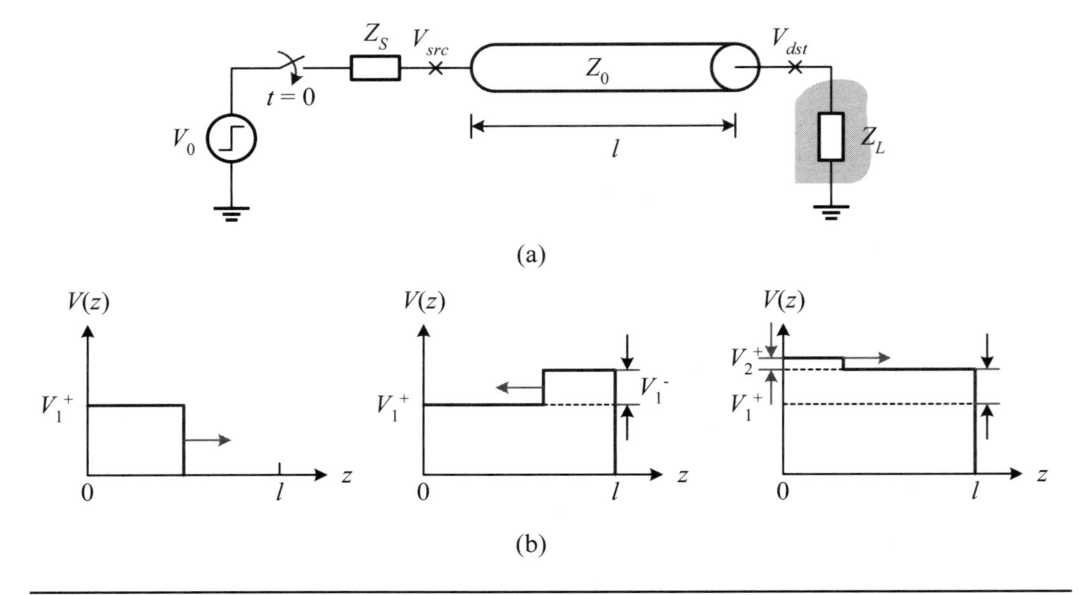

Figure 13.24: An example of voltage transient diagrams: (a) circuit; (b) voltage transient diagrams.

This voltage is reflected when it reaches the end of the transmission line. The magnitude of reflected voltage is equal to the product of incoming voltage and the load reflection coefficient Γ_L.

$$V_1^- = \Gamma_L V_1^+ = \frac{1}{6} V_0$$

Similarly, as the V_1^- reaches the source end, it will be reflected again as a scaled version with a factor of Γ_S.

$$V_2^+ = \Gamma_S V_1^- = \frac{1}{18} V_0$$

The above process continues forever.

■ Review Questions

Q13-22. Define delay time, settling time, and maximum overshoot (M_p).

Q13-23. What are the two types of transmission lines?

Q13-24. What is characteristic impedance?

Q13-25. Define the load reflection coefficient and source reflection coefficient.

13.4.2 Lossy Transmission Lines

As the resistance of a wire is large but less than the characteristic impedance Z_0 of the wire, the resistance plays an important role and the wire must be modeled as the lossy transmission line. When the (dc) resistance of the wire is larger than its characteristic

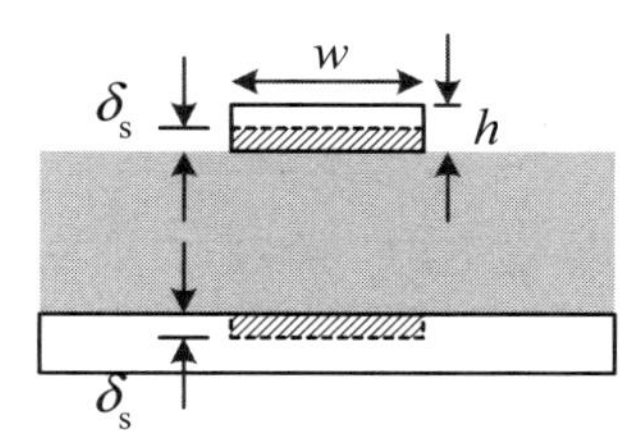

Figure 13.25: An illustration of the skin-effect of a conductor.

impedance, the resistance dominates the electrical behavior, and the inductance effects disappear. The wire should be modeled as a distributed-RC line. Generally, there are three major factors that cause attenuation in wires. These include *DC resistive loss*, *skin-effect loss*, and *dielectric loss*

DC resistive loss. All wires virtually have a finite resistance giving the electric field and the current to penetrate into the wires. This gives rise to resistive loss. When $R_{wire} \gg 2Z_0$, the fast-rising portion of the waveform is negligible, and the wire behaves like a distributed-RC line. Only 8% of the original voltage step may reach the end of the wire when the wire resistance R_{wire} is $5Z_0$.

Skin-effect loss. A steady current is distributed uniformly throughout the cross-section of a wire through which it flows; however, time-varying currents only concentrate near the surface of a wire, which is known as the *skin effect*, as shown in Figure 13.25. The depth of a wire from the surface into which the time-varying current can penetrate is defined as *skin depth* (δ_s) and can be expressed as

$$\delta_s = \sqrt{\frac{\rho}{\pi\mu f}} \tag{13.55}$$

where ρ and μ are wire resistivity and permeability, respectively, and f is the frequency of the incident time-varying current. For nonmagnetic materials, μ is approximately equal to $\mu_0 = 4\pi \times 10^{-9}$ H/cm.

For a microstrip wire, as shown in Figure 13.25, the low-frequency loss per unit length α_R is equal to

$$\alpha_R = \frac{R_{int}}{2Z_0} = \frac{\rho}{2whZ_0} \tag{13.56}$$

where $R_{int} = \rho/wh$, and w and h are the wire width and height, respectively, and Z_0 is the characteristic impedance. Because of the skin effect, the high-frequency components of the current are confined within the skip depth (δ_s). The high-frequency loss per unit length α_s can be expressed as

$$\alpha_s = \frac{2R_{skin}}{2Z_0} = \frac{\rho}{w\delta_s Z_0} = \frac{\sqrt{\pi\mu_0 f\rho}}{wZ_0} \tag{13.57}$$

where δ_s is the skin depth and is defined by Equation (13.55). The conductor (wire) loss α_C is equal to

$$\alpha_C = \max\{\alpha_R, \alpha_s\} \tag{13.58}$$

Dielectric loss. A common measure of the dielectric losses is called the *loss tangent*, $\tan\delta_D$, and is defined as

$$\tan\delta_D = \frac{G_{int}}{\omega C_{int}} = \frac{\sigma_D}{\omega\varepsilon_r} \tag{13.59}$$

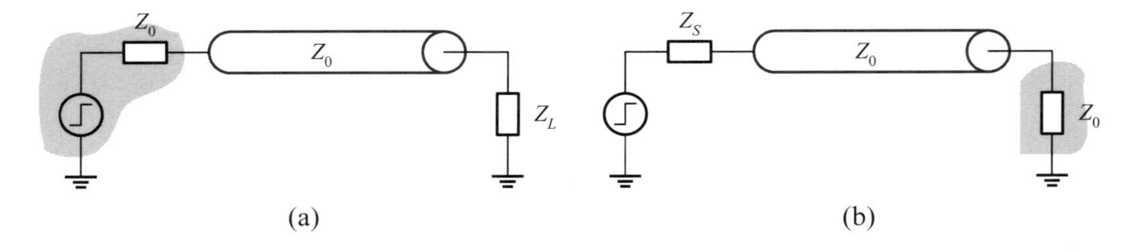

Figure 13.26: The two basic termination methods for transmission lines: (a) series termination; (b) parallel termination.

where σ_D is the conductivity of the dielectric and $\omega = 2\pi f$ is the angular frequency. ε_r is the relative permittivity of the dielectric.

The dielectric loss per unit length α_D is

$$\begin{aligned} \alpha_D &= \frac{G_{int}Z_0}{2} = \frac{2\pi f C_{int} \tan\delta_D}{2}\sqrt{\frac{L_{int}}{C_{int}}} \\ &= \pi f \tan\delta_D \sqrt{L_{int}C_{int}} = \frac{\pi f \sqrt{\mu_r \varepsilon_r}\tan\delta_D}{c} \end{aligned} \tag{13.60}$$

which can be significant in silicon or gallium arsenide. The constant c is the light speed in free space.

The RLC unit-step response. The RLC response of a lossy transmission line to a unit-step input is a combination of wave propagation and a diffusive component.

$$V_{step}(x) = V_{step}(0)\exp(-\alpha x) \tag{13.61}$$

where α is the attenuation constant composed of two components: conductor loss and dielectric loss.

$$\alpha = \alpha_C + \alpha_D = \frac{R_{int}}{2Z_0} + \frac{G_{int}Z_0}{2} \tag{13.62}$$

13.4.3 Transmission-Line Terminations

Recall that reflections occur in a transmission line when its load reflection coefficient or source reflection coefficient is not zero. Hence, to clean the ringing effects due to reflections, an appropriate termination should be made either at the source or destination end of the transmission line. We describe two popular termination methods: *series termination* and *parallel termination.*

13.4.3.1 Series Termination Series termination means that the reflected waves are absorbed (terminated) at the source end, as shown in Figure 13.26(a). The waves generated at the source propagate down the transmission line toward the destination end and then are reflected. The reflected waves are absorbed (namely, terminated) at the source end due to matched impedance. Series termination is popular with the slower CMOS-type I/O circuits and the situations where the wire impedance is small.

■ Example 13-15: (Examples of series termination.) Two basic examples of series termination are shown in Figures 13.27(a) and (b). The first one uses the

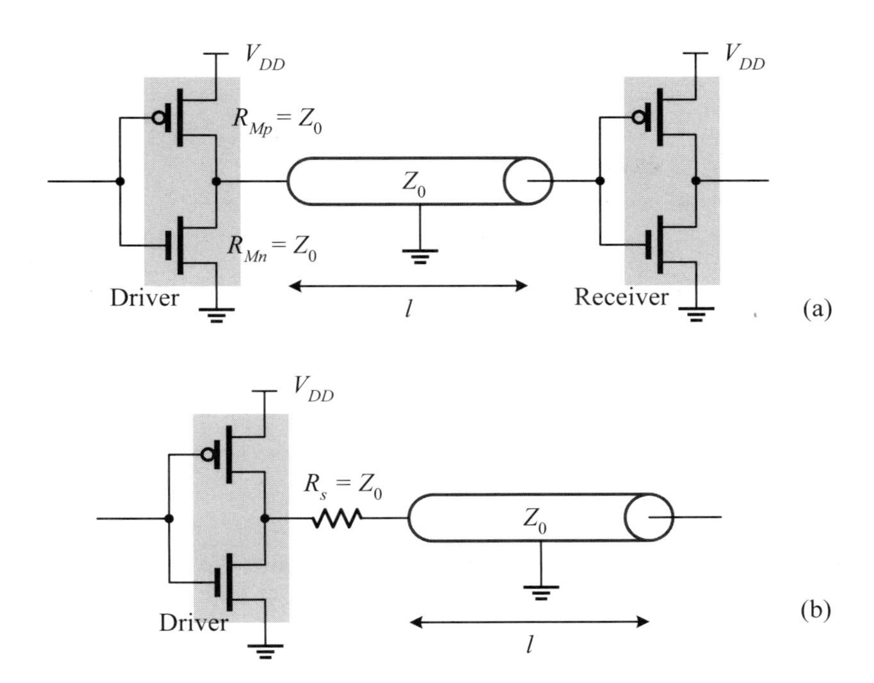

Figure 13.27: Two basic series termination approaches.

on-resistance of both nMOS and pMOS transistors to provide the desired series termination. The on-resistance of the driver is set equal to the Z_0 of the transmission line, as illustrated in Figure 13.27(a). The receiver behaves like a capacitive termination and the voltage step doubles with a time constant of $t = Z_0C_L$. In order to provide a matched impedance, a tunable output driver consisting of many stages with different sizes may be used. An illustration of this is discussed later in Section 15.3.4.

The second way provides the series termination by using a series resistor, being inserted between the output driver and the transmission line, as drawn schematically in Figure 13.27(b). The series resistance R_s is equal to Z_0 and the on-resistance of the driver is much less than Z_0. As compared to the above approach, it has better termination features across process, voltage, and temperature (PVT) variations. ■

13.4.3.2 Parallel Termination Parallel termination means that the reflected waves are absorbed (terminated) at the destination end, as shown in Figure 13.26(b). The waves are launched at the source end and transmitted down the transmission line toward the destination end and then absorbed (terminated) there due to matched impedance. Parallel termination is preferred in faster I/O circuits. A popular example is known as *Gunning transceiver logic* (GTL).

GTL is a standard for electrical signals in CMOS circuits used to provide high-speed data transfers with small voltage swings. The maximum signaling frequency is specified to be 100 MHz, but some applications use higher frequencies. The GTL signal swings between 0.4 and 1.2 V with a reference voltage of about 0.8 V. An illustration of GTL is explored in the following example.

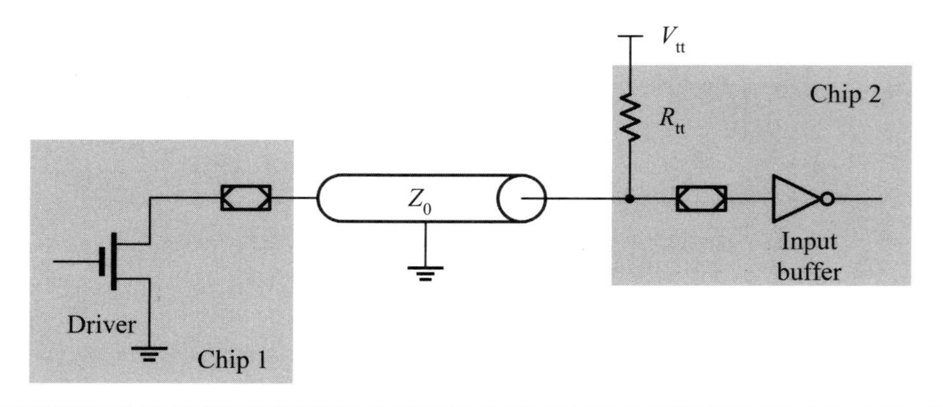

Figure 13.28: A simple GTL output driver.

■ **Example 13-16: (An example of parallel termination.)** The GTL scheme is similar to an open-drain scheme, as shown in Figure 13.28, where an active pull-down driver (a large n-type metal-oxide-semiconductor (nMOS) device) is required and a pull-up action is performed using a termination/pull-up resistor at the receiver end. Assume that the on-resistance of the pull-down driver is R_{on}. Then, both low- and high-level voltages can be expressed as

$$V_{OL} = \frac{R_{on}}{R_{on} + R_{wire} + R_{tt}} V_{tt}$$

and

$$V_{OH} = V_{tt}$$

Since both V_{OL} and V_{OH} are defined as 0.4 and 1.2 V, respectively, we may find the size of the pull-down driver by determining the value of R_{on} from the above V_{OL} equation if both resistances R_{tt} and R_{wire} are determined.

■

■ Review Questions

Q13-26. What does the open-circuit transmission line mean?

Q13-27. What is a lossless transmission line?

Q13-28. What is a lossy transmission line?

Q13-29. What is the skin effect?

Q13-30. Describe the two common methods of transmission-line terminations.

13.5 Advanced Topics

As manufacturing technology continues to be scaled down, the propagation delay of local interconnect decreases while the propagation delay of global interconnect remains the same or even increases. In addition, for deep submicron (DSM) processes, the global interconnect network, including repeaters and wires, may consume up to 40%

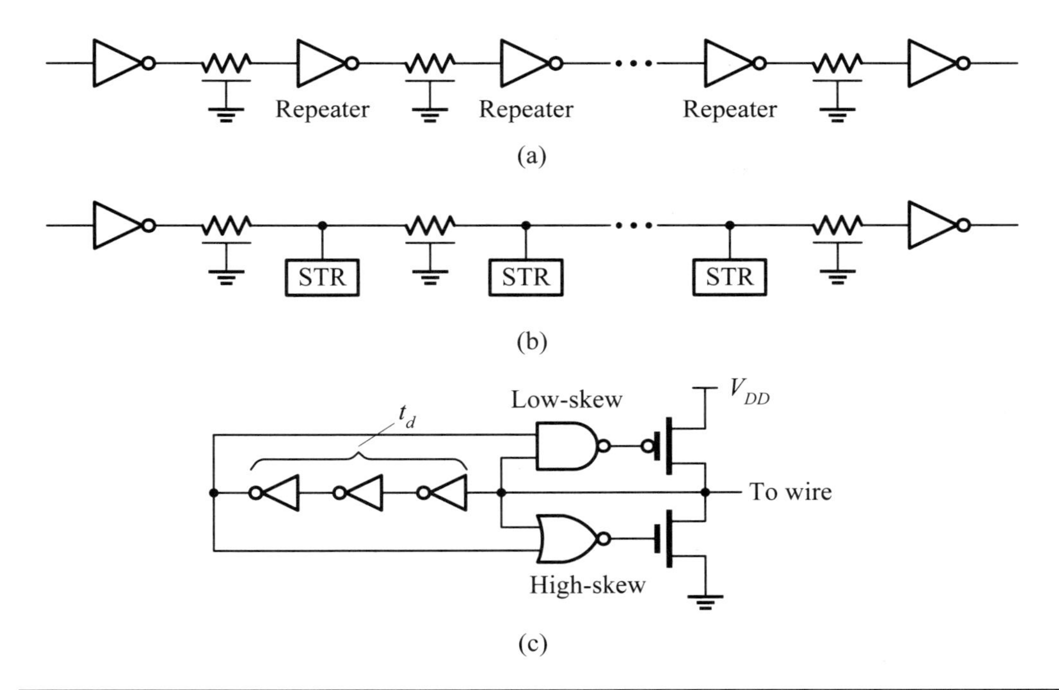

Figure 13.29: Repeater and STR implementations of a long interconnect signal wire: (a) repeater implementation; (b) STR implementation; (c) an example of a regenerator.

of the total power. As a consequence, on-chip global wires become a major bottleneck for circuit designs with respect to overall chip performance and power constraints.

13.5.1 Self-Timed Regenerators (STRs)

As described in Section 13.3, the use of *repeaters* (buffers) has been the widely used method to reduce the quadratic dependence of interconnect delay on wire length, as shown in Figure 13.29(a). Nevertheless, as CMOS technology continues to be scaled down, the number of repeaters increases dramatically, and power and area overhead due to repeaters is becoming a serious concern.

To overcome these difficulties, an alternative way is to use *self-timed regenerators* (STRs) or *boosters* to replace repeaters. An STR or a booster detects a transition earlier than a conventional inverter and then accelerates it to a full logic swing level. As shown in Figure 13.29(b), the STRs or boosters attach along the wire rather than interrupt it like the repeaters and can be used for driving bidirectional signals. As a consequence, STRs or boosters do not add any explicit propagation delay to the signals propagating along the wire as compared to repeaters. However, the design of STRs is not an easy task. An example of a STR is explained in what follows.

■ **Example 13-17: (An STR example.)** Figure 13.29(c) shows an example of STR. In a steady state, both pMOS and nMOS transistors are off. Hence, the wire is not actively driven. As the wire begins to rise, the low-skewed NAND gate detects the signal transition midway and turns on the *p*-type metal-oxide-semiconductor (pMOS) transistor to assist for a time t_d. After time t_d, the pMOS is eventually turned off. As

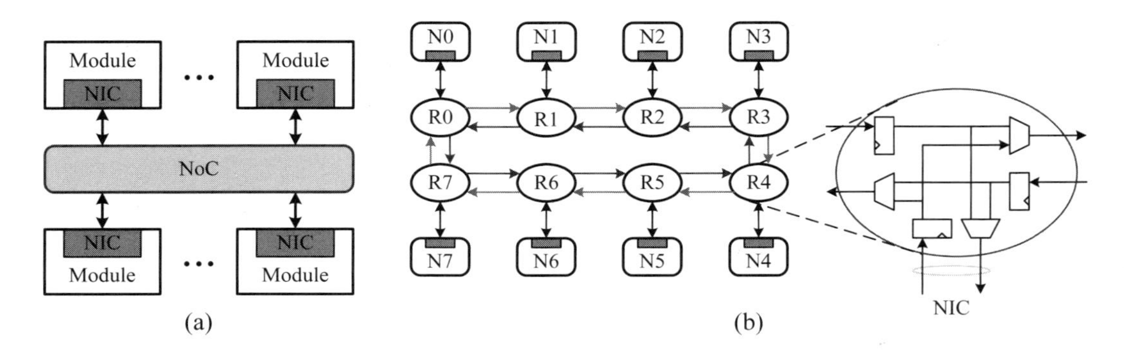

Figure 13.30: The (a) conceptual block diagram and (b) an 8-node ring topology of NoC.

the wire starts to fall, the high-skewed NOR gate detects the transition and turns on the nMOS transistor for a time t_d.

■

13.5.2 Network on a Chip

As the integration density continually increases to a level in which many processors and other hardware resources can be put together on the same chip, it is natural to migrate the design methodology of using platforms and computer systems, even networks, to a single chip (SoC) instead of using individual PCBs. The advantages of this new trend are that components, architectures, applications, and implementations can be reused.

For a large system containing many processors, memory modules, among other hardware resources, the interconnect network becomes the essential portion to effectively provide communicating paths between two different modules. The interconnect network used in such a system is usually based on the package-switching principle and is called a *network on a chip* (NoC). The system may scale from a few dozen up to a few hundred or even thousand of modules. Any module may be a processor core, a digital-signal processor (DSP) core, an field-programmable gate array (FPGA) block, a dedicated hardware block, a mixed signal block, or a memory block of any kind such as random access memory (RAM), read-only memory (ROM) or content-addressable memory (CAM). The conceptual block diagram is shown in Figure 13.30(a), where each hardware resource contains a *network interface circuit* (NIC), through which the hardware resource can communicate with the NoC.

■ Example 13-18: (An NoC example.) An 8-node ring topology without buffers is exhibited in Figure 13.30(b). The details of the router is also shown in the figure, where only pipeline registers are needed. The reason why buffers are not needed is as follows. Once a packet is injected into the ring network, the packet is guaranteed to make progress toward its destination by prioritizing those packets that are in flight. Hence, there are no contentions for network resources and do not need input buffers.

■

Many other interconnection topologies such as bus, crossbar, multistage interconnection networks (MINs), mesh, hypercube, and so on, may also be used as NoC. The interested reader can refer to Dally and Towles [6] and [8] for more details.

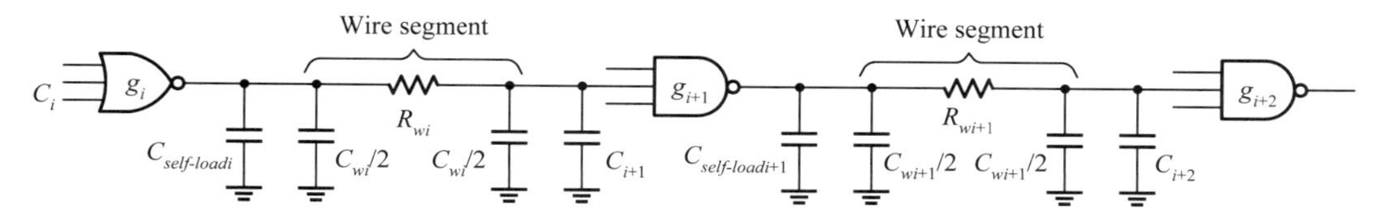

Figure 13.31: A partial logic chain with RC-wire segments.

Review Questions

Q13-31. What are the disadvantages of using repeaters as the CMOS technology is continually scaled down?

Q13-32. Describe the features of self-timed regenerators (STRs).

Q13-33. Describe the features of network on a chip (NoC).

13.5.3 Logical Effort with Interconnect

Recall that logical effort optimization for a logic chain without considering the RC-wire segments is equal to $g_i \cdot h_i = g_{i+1} \cdot h_{i+1}$; namely, the effort delay of all stages must be equal. We extend the logical effort to the situation of taking into account the interconnect delay.

13.5.3.1 Logical Effort of Interconnect To extend the logical effort definition from gate to interconnect, consider the partial logic chain with RC-wire segments shown in Figure 13.31. Here, the π-model is used to model the wire segment. The propagation delay of stage i can be computed by using the Elmore model. That is,

$$t_{pdi} = R_i(C_{self-loadi} + C_{w_i} + C_{i+1}) + R_{w_i}\left(\frac{1}{2}C_{w_i} + C_{i+1}\right) \tag{13.63}$$

where R_i and $C_{self-loadi}$ are the equivalent resistance and self-loading capacitance of gate i, respectively. C_{i+1} is the input capacitance of gate $(i+1)$.

Normalizing the propagation delay of stage i with respect to the intrinsic delay of inverter τ_{inv} $(= R_{eqv}C_{in})$, the above equation can be rewritten as in the following.

$$\begin{aligned} t_{Di} &= \frac{t_{pdi}}{\tau_{inv}} = \frac{R_iC_i}{\tau_{inv}}\left(\frac{C_{self-loadi} + C_{w_i} + C_{i+1}}{C_i}\right) + \frac{R_{w_i}}{\tau_{inv}}\left(\frac{1}{2}C_{w_i} + C_{i+1}\right) \\ &= g_i\left(h_i + \frac{C_{w_i}}{C_i}\right) + \frac{R_{w_i}}{\tau_{inv}}\left(\frac{1}{2}C_{w_i} + C_{i+1}\right) + p_i \end{aligned} \tag{13.64}$$

where $g_i = R_iC_i/\tau_{inv} = \tau_i/\tau_{inv}$ is the logical effort of gate i, h_i is the electrical effort of gate i and is defined as C_{i+1}/C_i, and p_i is the parasitic term and is defined as $g_i \times \gamma_i$.

For convenience, we now separately define *capacitive interconnect effort* h_w and *resistive interconnect effort* p_w as follows:

$$h_{w_i} = \frac{C_{w_i}}{C_i} \tag{13.65}$$

$$p_{w_i} = \frac{R_{w_i}}{\tau_{inv}}\left(\frac{1}{2}C_{w_i} + C_{i+1}\right) \tag{13.66}$$

The above equation can then be expressed as

$$t_{Di} = g_i(h_i + h_{w_i}) + (p_i + p_{w_i}) \tag{13.67}$$

13.5.3.2 Path-Delay Optimization The normalized path delay t_D with the RC-wire segment can be represented as

$$t_D = \sum_{i}^{n} g_i(h_i + h_{w_i}) + (p_i + p_{w_i}) = t_{Deffort} + t_{Dpara} \tag{13.68}$$

By using the same technique described in Section 5.3.3, the optimum condition is obtained.

$$(g_i + \frac{R_{w_i} C_i}{\tau_{inv}}) \cdot h_i = g_{i+1} \cdot (h_{i+1} + h_{w_{i+1}}) \tag{13.69}$$

For a logic chain without wires ($h_{w_i} = 0$ and $R_{w_i} = 0$), the optimum condition converges to the optimum condition: $g_i \cdot h_i = g_{i+1} \cdot h_{i+1}$.

To obtain a more intuitive interpretation of the optimum condition, the expression can be further simplified as follows:

$$(R_i + R_{w_i}) \cdot C_{i+1} = R_{i+1} \cdot (C_{i+2} + C_{w_{i+1}}) \tag{13.70}$$

The optimum size of gate $i + 1$ is achieved as the delay component $(R_i + R_{w_i} \cdot C_{i+1}$ due to the gate capacitance is equal to the delay component $R_{i+1} \cdot (C_{i+2} + C_{w_i})$ due to the effective resistance of the gate.

13.6 Summary

In this chapter, we studied parasitic resistance, capacitance, and inductance of interconnect wires, including metal wires and semiconductor diffusion layers. Interconnect mainly provides power delivery paths, clock delivery paths, and signal delivery paths. The most commonly used interconnect models include the lumped-RC model, distributed-RC model, and transmission-line model.

The parasitic effects of interconnect include the IR drop, RC delay, capacitive-coupling effect, and RLC effect. All of these influence the signal integrity and degrades the performance of the circuits. Resistive parasitics may cause an IR drop that reduces noise margins and deteriorates the performance of logic circuits. The combination of resistive and capacitive parasitics induces RC delays of signals and clocks. Three methods are proposed to reduce the RC delays: better interconnect materials, better interconnect strategies, and buffer insertion.

The capacitive-coupling effects may cause undesired crosstalk between different wires. A number of approaches used to overcome such effects include spacing, shielding, and cancellation. The inductance effect caused by a single wire alone may result in an Ldi/dt voltage drop (or called Ldi/dt noise). When it is combined with resistance and capacitance parasitics, a second-order RLC response is the result.

Finally, the transmission-line models and terminations were introduced with examples. The two popular termination methods are series termination and parallel termination, depending on whether the reflected waves are absorbed (terminated) at the source or at the destination end. In addition, self-timed regenerators and network on a chip (NoC) as well as logical effort with interconnect were discussed concisely.

References

1. N. Arora, K. V. Raol, R. Schumann, and L. Richardson, "Modeling and extraction of interconnect capacitances," *IEEE Trans. on Computer-Aided Design,* Vol. 15, No. 1, pp. 58–67, January 1996.

2. P. Bai et al., “A 65-nm logic technology featuring 35-nm gate lengths, enhanced channel strain, 8 Cu interconnect layers, low-k ILD and 0.57 μm^2 SRAM cell,” *2002 IEEE Int'l Electron Devices Meeting (IEDM)*, pp. 657–660, 2004.

3. H. Bakoglu, *Circuits, Interconnections, and Packaging for VLSI.* Reading, MA: Addison-Wesley, 1990.

4. E. Barke, “Line-to-ground capacitance calculation for VLSI: a comparison,” *IEEE Trans. on Computer-Aided Design,* Vol. 7, No. 2, pp. 295–298, February 1988.

5. D. K. Cheng, *Field and Wave Electromagnetics.* 2nd ed. Reading, MA: Addison-Wesley, 1989.

6. W. J. Dally and B. Towles, *Principles and Practices of Interconnection Networks.* Boston: Morgan Kaufmann, 2004.

7. I. Dobbalaere, M. Horowitz, and A. El Gamal, “Regenerative feedback repeaters for programmable interconnect,” *IEEE J. of Solid-State Circuits,* Vol. 30, No. 11, pp. 1246–1253, November 1995.

8. J. Duato, S. Yalamanchili, and L. Ni, *Interconnection Networks: An Engineering Approach.* Boston: Morgan Kaufmann, 2003.

9. C. Gauthier and B. Amick, “Inductance: implications and solutions for high-speed digital circuits: the chip electrical interface,” *Proc. of IEEE Int'l Solid-State Circuits Conf.,* Vol. 2, pp. 563–565, 2002.

10. P. R. Gray, P. J. Hurst, S. H. Lewis, and R. G. Meyer, *Analysis and Design of Analog Integrated Circuits,* 4th ed. New York: John Wiley & Sons, 2001.

11. R. Ho, K. Mai, and M. Horowitz, “The future of wires,” *Proc. IEEE,* Vol. 89, No. 4, pp. 490-504, April 2001.

12. R. Ho, K. Mai, and M. Horowitz, “Managing wire scaling: a circuit perspective,” *Proc. of IEEE Interconnect Technology Conf.,* pp. 177–179, 2003.

13. K. Mistry et al., “A 45-nm logic technology with high-k + metal gate transistors, strained silicon, 9 Cu interconnect layers, 193-nm dry patterning, and 100% Pb-free packaging,” *2007 IEEE Int'l Electron Devices Meeting (IEDM),* pp. 247–250, 2007.

14. D. A. Hodges, H. G. Jackson, and R. A. Saleh, *Analysis and Design of Digital Integrated Circuits: In Deep Submicron Technology,* 3rd ed. New York: McGraw-Hill Books, 2004.

15. X. Huang et al., “Loop-based interconnect modeling and optimization approach for multigigahertz clock network design,” *IEEE J. of Solid-State Circuits,* Vol. 38, No. 3, pp. 457–463, March 2003.

16. Y. Ismail, E. Friedman, and J. Neves, “Figures of merit to characterize the importance of on-chip interconnect,” *IEEE Trans. on Very Large Scale Integration (VLSI) Systems,* Vol. 7, No. 4, pp. 442–449, December 1999.

17. H. R. Kaupp, “Characteristics of microstrip transmission lines,” *IEEE Trans. on Electronic Computers,* Vol. 16, No. 2, pp. 185–193, April 1967.

18. J. Kim, “Low-cost router microarchitecture for on-chip networks,” *The 42nd Annual IEEE/ACM Int'l Symposium on Microarchitecture,* pp. 255–266, 2009.

19. J. Kim and H. Kim, “Router microarchitecture and scalability of ring topology in on-chip networks,” *The 2nd Int'l Workshop on Network-on-chip Architecture,* pp. 5–10, 2009.

20. S. Kumar et al., "A network on chip architecture and design methodology," *Proc. of the IEEE Computer Society Annual Symposium on VLSI* (ISVLSI 02), pp. 105–112, 2002.

21. A. Morgenshtein, E. G. Friedman, R. Ginosar, and A. Kolodny, "Unified logical effort - a method for delay evaluation and minimization in logic paths with *RC* interconnect," *IEEE Trans. on Very Large Scale Integration (VLSI) Systems,* Vol. 18, No. 5, pp. 689–696, May 2010.

22. P. Packan et al., "High performance 32-nm logic technology featuring 2nd generation high-k + metal gate transistors," *2009 IEEE Int'l Electron Devices Meeting (IEDM),* pp. 1–4, 2009.

23. J. M. Rabaey, A. Chandrakasan, and B. Nikolic, *Digital Integrated Circuits: A Design Perspective,* 2nd ed. Upper Saddle River, NJ: Prentice-Hall, 2003.

24. S. Rusu and G. Singer, "The first IA-64 microprocessor," *IEEE J. of Solid-State Circuits,* Vol. 35, No. 11, pp. 1539–1544, November 2000.

25. T. Sakurai, "Approximation of wiring delay in MOSFET LSI," *IEEE J. of Solid-State Circuits,* Vol. 18, No. 4, pp. 418–426, August 1983.

26. P. Singh, J. S. Seo, D. Blaauw, and D. Sylvester, "Self-timed regenerators for high-speed and low-power on-chip global interconnect," *IEEE Trans. on Very Large Scale Integration (VLSI) Systems,* Vol. 16, No. 6, pp. 673–677, June 2008.

27. S. Thompson et al., "A 90-nm logic technology featuring 50-nm strained silicon channel transistors, 7 layers of Cu interconnects, low k ILD, and 1 μm^2 SRAM cell," *2002 IEEE Int'l Electron Devices Meeting (IEDM),* pp. 61–64, 2002.

28. J. P. Uyemura, *Introduction to VLSI Circuits and Systems.* New York: John Wiley & Sons, 2002.

29. A. Vittal et al., "Crosstalk in VLSI interconnections," *IEEE Trans. on Computer-Aided Design of Integrated Circuits and systems,* Vol. 18, No. 12, pp. 1817–1824, December 1999.

30. N. H. E. Weste and D. Harris, *CMOS VLSI Design: A Circuit and Systems Perspective,* 4th ed. Boston: Addison-Wesley, 2011.

31. S. Yang et al., "A High Performance 180 nm Generation Logic Technology," *Proc. of IEEE Int'l Electron Devices Meeting (IEDM),* pp. 197–200, 1998.

32. C. P. Yuan and T. N. Trick, "A simple formula for the estimation of the capacitance of two-dimensional interconnects in VLSI circuits," *IEEE Electronic Device Letters,* Vol. 3, No. 12, pp. 391–393, December 1982.

Problems

13-1. Suppose that the dose $Q_d = 8 \times 10^{13}$ cm^{-2} of arsenic is implanted uniformly into a p-type substrate with the junction depth of 0.8 μm. The electron mobility μ_n is 550 cm^2/Vs. Calculate the sheet resistance of the n-type diffusion region.

13-2. Suppose that the dose $Q_d = 10^{14}$ cm^{-2} of boron is implanted uniformly into a polysilicon layer with the thickness of 1.0 μm. The hole mobility μ_p is 270 cm^2/Vs. Calculate the sheet resistance of the p-type polysilicon layer.

13-3. Suppose that a metal1 wire is used as the desired interconnect. The length and width of the metal1 wire are 0.5 cm and 1 μm, respectively.

(a) Calculate the resistance of the metal1 wire.

(b) Calculate the RC delay of the metal1 wire if the unit-length wire capacitance is 0.1 fF/μm.

13-4. An output buffer (an inverter) is used to drive a transmission line with a characteristic impedance Z_0 of 50 Ω. Assuming that a 0.18-μm process is used, calculate the sizes of nMOS and pMOS transistors.

13-5. Assume that the field oxide of 0.62 μm in thickness is inserted between metal2 and metal3 wires. The width and thickness of the metal2 wire are 0.6 μm and 0.7 μm, respectively. Using Equations (13.8) and (13.9), calculate the total unit-length capacitance per μm.

13-6. The unit-length wire capacitance can also be estimated by representing the rectangular line profile with an oval one, composed of a rectangle and two half cylinders, as depicted in Figure 13.5(b). The resulting capacitance is the sum as an area capacitance with width $w - t/2$ and a cylindrical one with radius $t/2$ and can be expressed as follows:

$$C_{int} = \varepsilon_{di}\left[\frac{\left(w - \frac{t}{2}\right)}{t_{di}} + \frac{2\pi}{\ln\left(1 + \frac{2t_{di}}{t} + \sqrt{\frac{2t_{di}}{t}\left(\frac{2t_{di}}{t} + 2\right)}\right)}\right] \quad \text{for } w \geq \frac{t}{2}$$

$$C_{int} = \varepsilon_{di}\left[\frac{w}{t_{di}} + \frac{\pi\left(1 - 0.0543 \cdot \frac{t}{2t_{di}}\right)}{\ln\left(1 + \frac{2t_{di}}{t} + \sqrt{\frac{2t_{di}}{t}\left(\frac{2t_{di}}{t} + 2\right)}\right)} + 1.47\right] \quad \text{for } w < \frac{t}{2}$$

Assume that a metal1 wire is over a field oxide of thickness of 0.52 μm. The width and thickness of the metal1 wire are 0.5 μm and 0.6 μm, respectively. Using the above equations, calculate the total unit-length capacitance per μm and the total capacitance if the length of the metal1 wire is 100 μm.

13-7. The following simple empirical expression can be used to estimate the total unit-length wire capacitance in the multilayer-metal structure.

$$C_{int} = \varepsilon_{di}\left[\left(\frac{w}{t_{di}}\right) + 0.77 + 1.06\left(\frac{w}{t_{di}}\right)^{0.25} + 1.06\left(\frac{t}{t_{di}}\right)^{0.5}\right]$$

where the first term denotes the area capacitance and the other three terms represent all sidewall effects, including fringing and lateral capacitances.

Assume that a metal1 wire is over a field oxide of thickness of 0.52 μm. The width and thickness of the metal1 wire are 0.5 μm and 0.6 μm, respectively. Calculate the total unit-length capacitance per μm and the total capacitance of a 100-μm metal1 wire.

13-8. Consider the inductance of a metal1 wire over a field oxide of thickness 0.52 μm. Assume that the width and thickness of the metal1 wire are 3.5 μm and 0.6 μm, respectively. Calculate the total inductance if the length of the metal1 wire is 100 μm.

13-9. Suppose that the sheet resistance of a wire is 38 m$\Omega/\Box$ and unit-length capacitance and inductance are 0.15 fF/μm and 0.5 pH/μm, respectively. The wire width is 2.0 μm. Determine the wire-length range in which the wire inductance needs to be considered if

(a) The rise time of signal is 2.0 ns.

(b) The rise time of signal is 25 ps.

13-10. Assume that a 5-mm long metal4 interconnect segment is driven by a 20X inverter. The width of the wire is 0.5 μm. Use the Π-model and Elmore delay model to estimate the delay of this metal wire.

13-11. Assume that an 8-mm metal4 wire driven by a 10X inverter is used in a circuit. The width of the metal4 wire is 0.4 μm. If the buffer insertion approach is used to reduce the delay, calculate the optimal number of buffers and the optimal buffer size, using 0.13-μm process parameters. Assume that the R_{sq} of metal4 is 0.025 Ω.

13-12. Referring to Figure 13.15, suppose that both victim and aggressor are $5X$ inverters. Using 0.13-μm process parameters, find the amount of crosstalk ΔV_1 under each of the following conditions. Assume that $C_{area} = 0.06$ fF/μm and $C_{lateral} = 0.08$ fF/μm.

(a) The length L of two wires between two inverters is 1 mm.

(b) The length L of two wires between two inverters is 5 mm.

13-13. Referring to Figure 13.21 and assuming that the dielectric is silicon dioxide with $\mu_r = 1$ and $\varepsilon_r = 3.9$.

(a) Assume that in Figure 13.21(a), $t = 1$ μm, $w = 5$ μm, and $h = 2$ μm, find the Z_0 of it.

(b) Assume that in Figure 13.21(b), $t = 1$ μm, $w = 6$ μm, and $h = 4$ μm, find the Z_0 of it.

13-14. A simple GTL output driver is shown in Figure 13.28. If V_{OL} and V_{OH} are defined as 0.4 and 1.2 V, respectively, determine the size of the pull-down driver. Assume that R_{wire} and R_{tt} are 5 Ω and 50 Ω, respectively.

14

Power Distribution and Clock Designs

In modern deep submicron chip design, power distribution and clock generation and distribution are the two essential issues that determine whether the chip may be successful on the market . For a system to be workable, each part of the logic module must be powered through a power distribution network and driven by a clock distribution network. The objective of the power distribution network is to evenly distribute the power supply to all individual devices in the system in an undisturbed fashion. We begin in this chapter with the introduction of the design issues of power distribution networks and then discuss power distribution networks. In addition, the decoupling capacitors and their related issues are discussed.

The goal of a clock system is to generate and distribute one or more clocks to all sequential devices or dynamic logic circuitry in the system with as little skew as possible. To reach this, we start to describe the clock system architecture and then address a variety of methods widely used to generate clocks. Finally, we consider a few general clock distribution networks.

In modern very-large-scale integration (VLSI) chip design, the frequency multiplier, clock-deskew circuit and clock recovery circuit are often needed. To reach this, phase-locked loops (PLLs) are widely used. A PLL is a circuit that uses feedback control to synchronize its output clock with the incoming reference clock, and to generate an output clock running at a higher rate of operation than the incoming reference clock. Like the PLL, a delay-locked loop (DLL) also uses feedback control to lock the output clock to the incoming reference clock in a constant phase. PLLs/DLLs can be pure analog, pure digital, or a hybrid of both, depending on the actual applications and requirements. When a PLL is constructed with all digital cells, it is referred to as an all-digital PLL (ADPLL).

14.1 Power Distribution Networks

The main goal of a power distribution network is to evenly distribute the power supply to all individual devices in the system in an undisturbed fashion. We begin with the introduction of the design issues of power distribution networks. Then, we discuss power distribution networks. Finally, we consider the *decoupling capacitors* and their related issues.

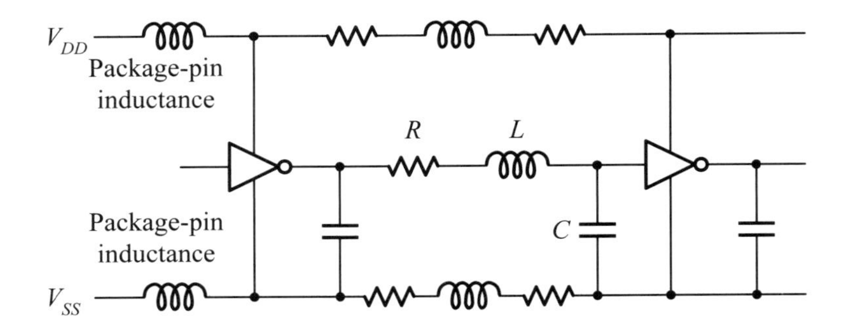

Figure 14.1: A circuit model incorporating package-pin inductance.

14.1.1 Design Issues of Power Distribution Networks

The important issues of power distribution networks are package-pin inductances, IR drop and Ldi/dt noise, electromigration, and power-supply noise. For a power distribution network to be completely designed, it is necessary to take all these issues into account. We deal with each of these in more detail.

14.1.1.1 Package-Pin Inductances A package is a necessary component used to hold a chip. The package body is a physical and thermal support for a chip. Leads (pins) in a package connect to pads and provide substrate connection to a chip. However, package pins can generally introduce significant inductance to the power-supply and signal paths of the logic circuitry. A circuit model incorporating package-pin inductance is given in Figure 14.1. Pads are library components required to be electrically designed with care. They allow the on-chip wires to be connected to the package.

Package pins have nontrivial inductance. This inductance is about 1 to 2 nH in the conventional dual-in-line package and is reduced to 0.1 nH in a BGA package. Power and ground nets typically require many package pins to supply the required amount of current due to the limited current capability of each power/ground pad. The induced voltage across a pin inductance of L with a current change rate di/dt is equal to $v_L = Ldi/dt$, according to Faraday's law of induction. For instance, if $L = 0.5$ nH and the power-supply current changes by 1 A in 1 ns, then the induced voltage is $v_L = 0.5$ V.

Regardless of digital or analog circuits, power supply is essential to keep them working normally. The digital transient signals may be coupled to analog circuits through common power-supply pads/pins. To see this, consider the equivalent circuit shown in Figure 14.2, where digital and analog blocks share the same power-supply pad and pin. Because of the existence of wire inductance, about 1 nH/mm, and resistance inherent in a bonding connection, the voltage drop ΔV, referred to as *ground* or *power bounce*, is equal to

$$\Delta V = R_1 I_{total} + R_3 I_{analog} + L\frac{dI_{total}}{dt} \tag{14.1}$$

where I_{total} is the sum of I_{analog} and $I_{digital}$. Consequently, the digital transient currents severely impact on the supply voltage of the analog block.

To minimize the effect of power-supply coupling, the following general rules can be applied: First, it should keep the separation between analog and digital blocks as large as possible. Second, it should keep the power-supply pins as close to the power pads as possible to reduce the parasitic inductance. Third, it should use separate

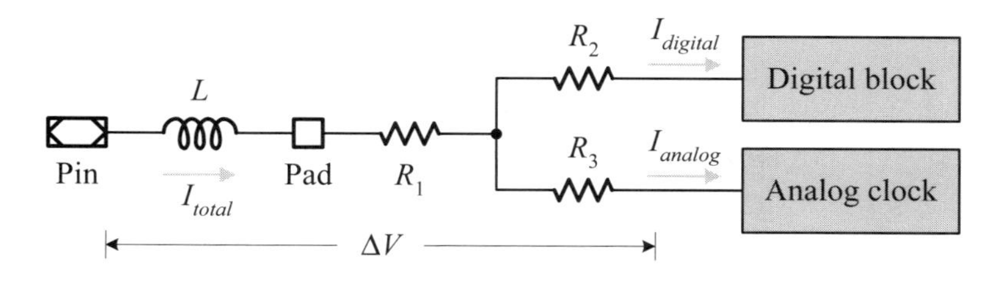

Figure 14.2: An illustration of the effect of power-supply coupling.

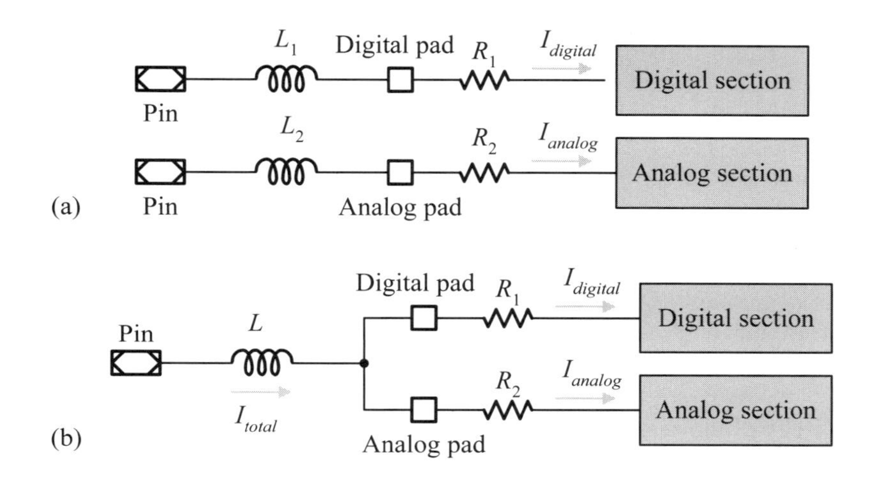

Figure 14.3: The minimization of the effect of power-supply coupling by using: (a) separate pads and pins; (b) separate pads and a sharing pin.

power-supply pads and pins for digital and analog blocks, as shown in Figure 14.3(a), whenever it is possible. If this is not the case, it should at least use separate pads for analog and digital blocks with relatively close bonding to the same pin, as shown in Figure 14.3(b).

14.1.1.2 IR Drop and Ldi/dt Noise The IR drop and Ldi/dt noise are two major factors that impact on the overall timing, noise margin, and functionality of a chip. The IR drop is due to the resistance appearing on V_{DD} power wires and the Ldi/dt is the voltage variation at package pins. The total voltage drop at any point in a power wire is the combination of the above two factors.

$$\Delta V = IR + L\frac{di}{dt} \tag{14.2}$$

As shown in Figure 14.4, the voltage at the connection to inverter n drops by ΔV. The voltage at the ground wire near the inverter will increase an amount of voltage ΔV when the output switches to low, thereby causing a ground bounce.

The ΔV drop impacts on logic circuits in terms of timing and noise margin. As illustrated in Figure 14.4, the IR drop reduces the drive capability of the gate and hence increases the propagation delay of the inverter n. In addition, because of the voltage drop and ground bounce on the power wires, the noise margin is reduced significantly. As a consequence, it is necessary to reduce the amount of ΔV drop. For this purpose,

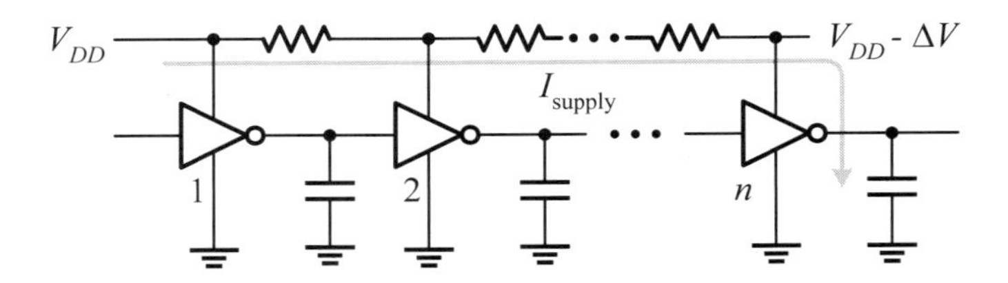

Figure 14.4: An illustration of the effects of the IR drop.

the most widely used approach is to place decoupling capacitors between power and ground wires. These decoupling capacitors can deliver the additional current needed temporarily from the power distribution network.

The general approaches for tackling the Ldi/dt noise are as follows. First, separate power pins to be used for I/O pads from those used for core logic because I/O pads may cause a large current surge. Second, multiple power and ground pins should be used and care must be taken to choose the positions of the power and ground pins on the package. Third, the rise and fall times of off-chip signals should be increased to the maximum extent allowable. Fourth, schedule current-consuming transitions carefully so that they do not occur simultaneously. Fifth, enough numbers of decoupling capacitors with proper capacitance should be added on the board and on the chip to limit the amount of the current surge.

14.1.1.3 Electromigration The phenomenon that metal molecules migrate from one area to another because of an amount of current in excess of its maximum allowable current density is called *electromigration* (EM), which would cause the metal wire eventually to be broken. Electromigration in the power wire is a function of the average current flowing in metal wires and vias. The electromigration effect may cause the reliability problem of a metal wire and generally is measured by *mean time to failure* ($MTTF$). The $MTTF$ of a metal wire can be expressed as follows:

$$MTTF = \frac{A}{J_{avg}^n} \exp(E_a/kT) \tag{14.3}$$

where A is an empirical scaling factor, depending on film structure such as grain size and processing, J_{avg} is the average current density, n is a constant close to 2, E_a is the activation energy, with a value ranging from 0.4 eV to 0.8 eV, determined by the material and its diffusion mechanism, k is Boltzmann's constant, and T is the temperature.

The typical current density of metal wires at which electromigration occurs is about 1 to 2 mA/μm^2 for aluminum wires and about 10 mA/μm^2 or better for copper wires.

Bidirectional wires. Even though bidirectional wires are less prone to electromigration, they may encounter another phenomenon called *thermal migration* caused by the *self-heating effect*. In a bidirectional wire such as a clock bus, the current flow is bidirectional and hence the average current is essentially zero. Nevertheless, the root-mean-square (rms) current is not zero, thereby leading to power dissipation and generating heat. Since the surrounding oxide or low-k dielectric is a thermal isolation, the temperature rise of a metal wire due to the self-heating effect may become significantly higher than the underlying substrate. This may in turn help the occurrence of electromigration since electromigration is a strong function of temperature, as stated. To limit the temperature rise in a tolerant range, the maximum root-mean-square (rms) current density should not be greater than 15 mA/μm^2 when aluminum wires are used.

In summary, the current density should not exceed 1 to 2 mA/μm^2 for unidirectional wires and 15 mA/μm^2 for bidirectional wires when aluminum wires are used. The current density limitation in bidirectional wires is generally about 10 times that in unidirectional wires.

14.1.1.4 Power-Supply Noise A power-supply distribution network contains power-wire resistance, power-wire capacitance, coupling capacitance, packaging inductance, and power-wire inductance. Power-wire resistance creates supply-voltage variations with current surge. The voltage drops on power wires depend on the dynamic behavior of circuits. These variations in supply voltage manifest themselves as noise in the logic circuits. The supply-voltage variations are more severe when many buffers are switched together, which may result in large IR drops (called *hot spots*) and ground bounce locally. Such a situation is often called *simultaneous switching noise* (SSN). The hot spots, if not solved appropriately, can cause timing closure problems or result in functional failures in extreme cases.

Methods used to tackle power-supply noise are as follows. It begins to measure and control the amount of current required by each block at varying times. At the worst case, it may need to move some activity from one clock cycle to another to reduce peak current. Even it may need to move the logic blocks, add or rearrange the power pins, and/or redesign power/ground network to reduce resistance at high current loads.

■ Review Questions

Q14-1. What are the design issues of a power distribution network?

Q14-2. Describe the electromigration (EM) phenomenon.

Q14-3. Define simultaneous switching noise (SSN).

14.1.2 Power Distribution Networks

With the ever-decreasing feature size of integrated circuits (ICs), the design of power and ground distribution networks has become a challenging task. These challenges are caused from shorter rise and fall times, lower supply voltages, lower noise margins, higher currents, and increased current densities. Power distribution networks, including power and ground distribution networks, require the interaction of system designers, thermal designers, system architects, board designers, and chip designers. A complete power distribution network needs to take into account the effects of IR drop (hot spots), Ldi/dt noises, ground bounce, and electromigration. It is necessary to demand global optimization rather than simply localized chip-level optimization. We first introduce power-tree and power-grid networks and then deal with decoupling capacitors.

14.1.2.1 Power-Tree Networks The *power-tree network* uses a single metal layer, as shown in Figure 14.5(a). Two trees are interdigitated to supply both sides of supply voltages. When designing such a power-tree network, it is necessary to properly size power-supply wires so that they can handle enough currents. The essence of designing a power-tree network is to design an appropriate topology for the V_{DD}/V_{SS} network. A planar wiring network is desired because all power-supply wires need to be routed in the same metal layer. More importantly, it needs to guarantee that each branch is able to supply the required current to all of its subsidiary branches.

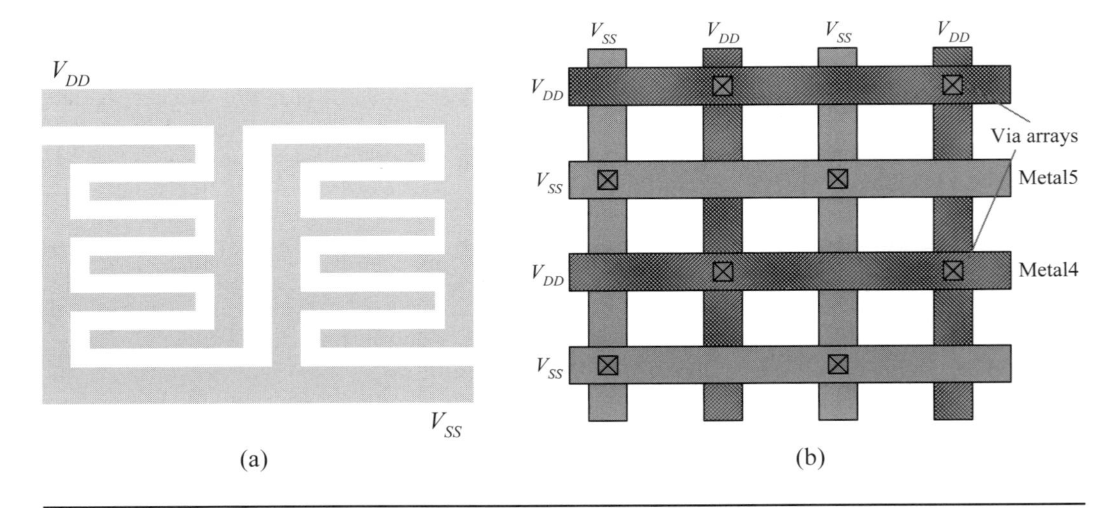

Figure 14.5: Examples of (a) power-tree and (b) power-grid distribution networks.

The key design challenges of the power-tree network are as follows. First, the power buses need to be properly sized to minimize the IR drop, while satisfying the required timing and area constraints at the same time. Second, since the same metal layer is also used for the routing clock and global signals, it is necessary to trade off between power and signal routing. This is often a difficult problem for system designers.

14.1.2.2 Power-Grid Networks In high-performance ICs, power distribution networks are usually organized as a multilayer grid network (or mesh array). In such a grid network, straight power and ground wires in each metal layer can span an entire die and are orthogonal to the wires in adjacent metal layers. Power and ground wires typically alternate in each layer. Vias connect a power (ground) wire to another at the overlap sites. An example of a power-grid network is shown in Figure 14.5(b), where two metal layers, Metal4 and Metal5, form a grid array and are properly connected by a via array. The resulting array is then tied to V_{DD} and ground in a way as indicated in the figure. Hence, the IR drop may be minimized.

14.1.2.3 Decoupling Capacitors As the clock frequency and current demands of a chip increase while the supply voltage decreases, it becomes more challenging to maintain the quality of power supply. To keep the power-supply variation within a certain percentage, 10% to 15%, of its nominal value, decoupling capacitors with proper size are often used to reduce IR drop and Ldi/dt noises.

A decoupling capacitor (decap) like a battery acts as a reservoir of charge, which is released when the supply voltage at a particular current load drops below some tolerable level. To be effective, care must be taken to deploy decoupling capacitors in proper locations in designing the power distribution networks in high-performance ICs, such as microprocessors. On the other hand, using decoupling capacitors is an effective way to reduce the impedance of power delivery systems operating at high frequencies.

An illustration of the use of decoupling capacitors is shown in Figure 14.6, where decoupling capacitors are placed beside the buffers to effectively suppress the current surge due to state changes of the buffers, thereby limiting the IR drop and Ldi/dt noise in the range of both nearby buffers.

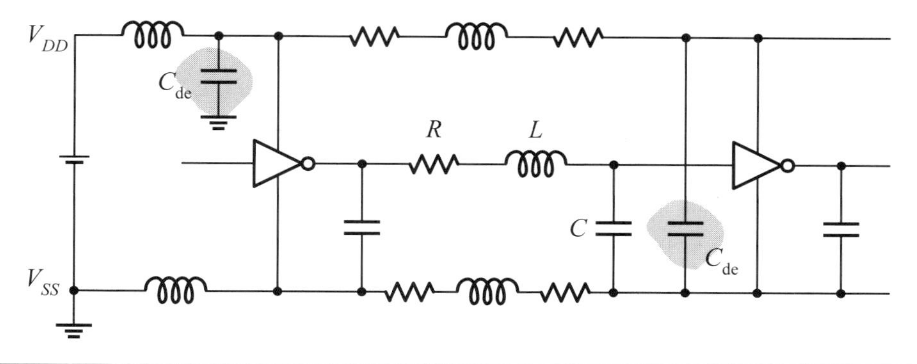

Figure 14.6: An illustration of the use of decoupling capacitors.

From the above discussion, we can see that two main design issues of decoupling capacitors are to decide how much decoupling capacitance should be included and where to place them. Generally speaking, the decoupling capacitance should be 10 times the switching capacitance to maintain the power supply within 10% of its nominal value. Nonetheless, because of some inherent parasitic capacitances already existing in the circuits that are not switched, such as gate and source/drain capacitance as well as wire capacitance, the actual desired decoupling capacitance is less than that of the above-mentioned. Moreover, the decoupling capacitance is also determined by the noise budget of the system.

The criteria for placing the decoupling capacitors are as follows: First, the decoupling capacitors should be placed nearby the large buffers that are switching during the peak demand periods. Second, the decoupling capacitors should be placed in as many open areas of the chip as possible. Third, the decoupling capacitors should be located near the power pins to offset any inductance effects due to solder bumps or bonding wires.

Decoupling capacitors can be implemented either off-chip or on-chip. The off-chip decoupling capacitors are added to the chip along with the package. The implementations of *on-chip decoupling capacitors* can be either passive or active. The passive on-chip decoupling capacitors are usually implemented using large n/p-type metal-oxide-semiconductor (nMOS/pMOS) transistors since gate capacitor offers a higher capacitance per area than other types of capacitors, such as *metal-insulator-metal* (MIM) structures.

■ **Example 14-1: (Decoupling capacitors.)** Figure 14.7(a) shows a conventional on-chip nMOS decoupling capacitor. One disadvantage of this simple decoupling capacitor is the large leakage power dissipation. To reduce the unnecessary leakage power dissipation when the associated subsystem is idle, the gated decoupling capacitor exhibited in Figure 14.7(b) can be used. In this structure, a sleep nMOS transistor M_{n2} is employed to enable the decoupling capacitor (that is, M_{n1}). When it is on, the nMOS transistor M_{n1} acts as a normal decoupling capacitor. The leakage power dissipation is equal to $V_{DD}(I_{G1} + I_{G2})$. When the sleep nMOS transistor M_{n2} is off, the nMOS transistor M_{n1} is deactivated and the voltage at node x is determined by the equivalent on-resistances of both nMOS transistors, M_{n1} and M_{n2}. At this point, the leakage power dissipation $V_{DD}(I_{G1})$ is exponentially decreased by the reduction of gate-to-source voltage due to the increase of voltage V_x above the ground level. ■

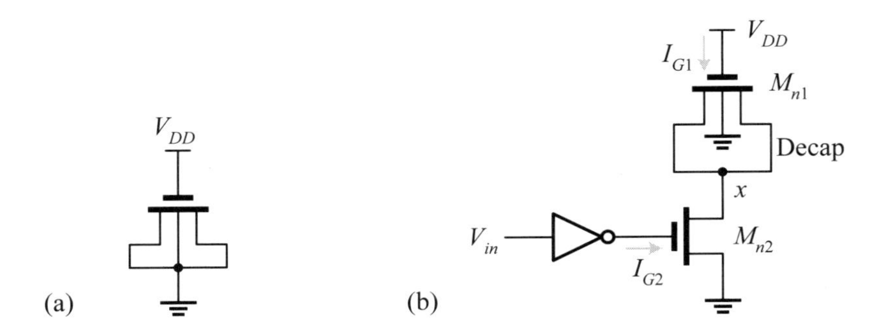

Figure 14.7: Examples of decoupling capacitors: (a) conventional decap; (b) gated decap.

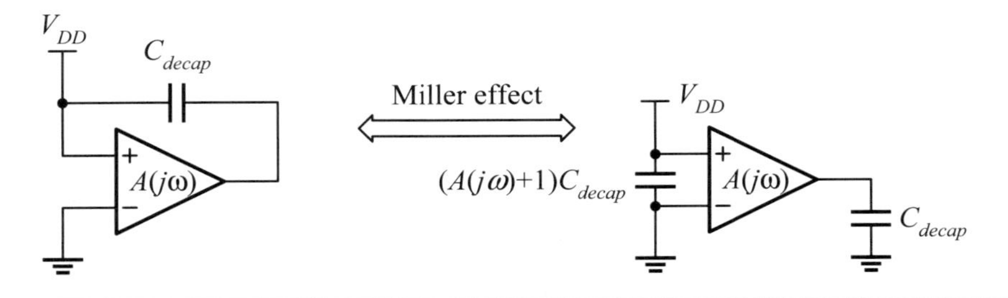

Figure 14.8: The concept of an active decoupling capacitor.

There are two major constraints that limit the usage of passive decoupling capacitors in scaled technologies. First, passive on-chip decoupling capacitors consume a large silicon area, about 20% of the total area in some high-end microprocessor chips. Second, passive on-chip decoupling capacitors introduce a large amount of gate leakage current and hence power dissipation.

The concept of *on-chip active decoupling capacitors* is based on the Miller effect, which uses the voltage gain of an active *operational amplifier* (op amp) to boost the effective capacitance. To get more insight into this, consider Figure 14.8, where a decoupling capacitance C_{decap} is connected between the positive input and the output of an op amp. The op amp has the voltage gain of $A(j\omega)$. Because of the Miller effect, the capacitance of decoupling capacitor seen from the power supply has been boosted by a factor of $(1 + A(j\omega))$, a value much greater than the original capacitance C_{decap} if $A(j\omega)$ is large. Of course, the design of the op amp needs some skills from analog circuitry, which might somewhat complicate the system design.

■ Review Questions

Q14-4. What are the features of power-tree networks?

Q14-5. What are the two main design issues of decoupling capacitors?

Q14-6. What are the features of passive on-chip decoupling capacitors?

Q14-7. What is the rationale behind active on-chip decoupling capacitors?

Q14-8. What effects will need to be taken into account in designing a power distribution network?

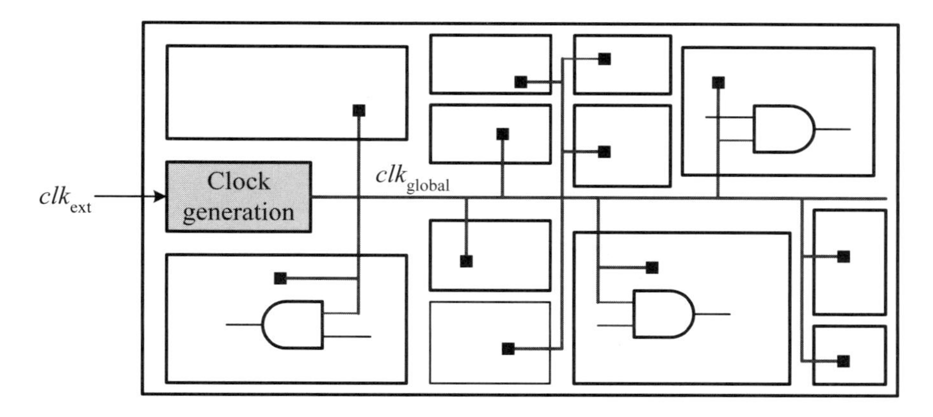

Figure 14.9: An architecture of a typical clock system in modern chips.

14.2 Clock Generation and Distribution Networks

To make the operations of a digital system progress continuously, each module in the system must be steered by one or more clocks appropriately. The main goal of a clock system is to generate and distribute one or more clocks to all sequential devices or dynamic logic circuitry in the system with as little skew as possible. In this section, we start to describe the clock system architecture. Then, we address a variety of methods widely used to generate clocks. Finally, we consider a few general clock distribution networks.

14.2.1 Clock System Architectures

A typical clock system on a chip is roughly composed of a *clock generation* and a *clock distribution network*, as illustrated in Figure 14.9. In addition to the clock generator, the clock generation may include a *phase-locked loop* (PLL) or a *delay-locked loop* (DLL) to adjust the frequency or phase of the global input clock.

Ideally, the clock should arrive at all flip-flops or latches, or synchronous memory modules, at the same time and have a fixed period T_{clk} as well as acceptable rise and fall times. Three important metrics associated with the clock are *clock skew*, *clock latency*, and *clock jitter*. Remember that the clock skew is defined as the maximum difference of the clock arrival times at the clock inputs of storage elements, such as flip-flops, latches, and so on. The clock latency is defined as the maximum delay from the clock source to the clock input of a flip-flop. The clock jitter is the variation of the clock period from one cycle to another. Both clock skew and clock jitter affect the effective clock cycle time. Nonetheless, the clock skew is constant for all clock cycles and clock jitter varies cycle by cycle.

In summary, two major design issues of a clock distribution network are clock skew and power dissipation. In order to minimize clock skew among all clocked elements, the clock distribution network of a system must be designed with care. In addition, it must be designed meticulously to minimize the unnecessary power dissipation since the clock distribution network may consume a large amount of power, up to 30% to 45% of total power in a typical system.

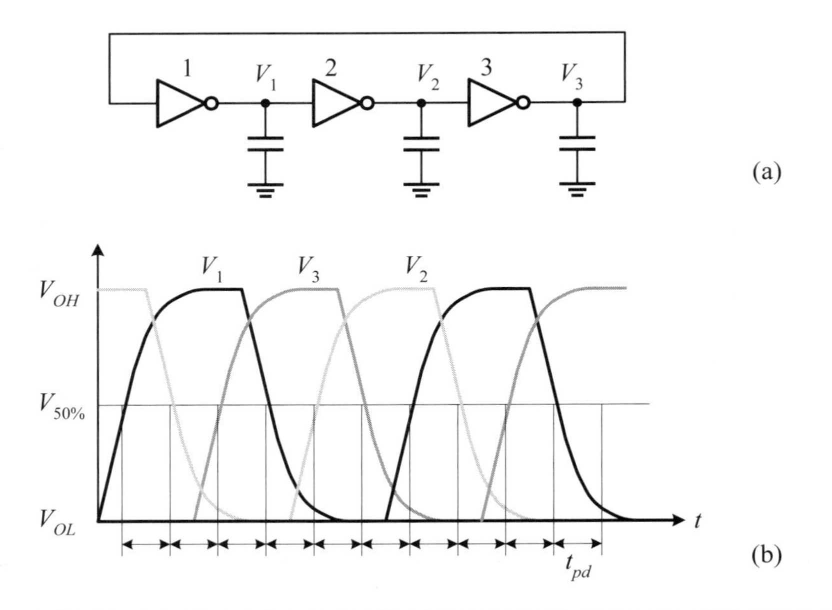

Figure 14.10: A three-stage ring clock generator: (a) circuit; (b) timing diagram.

14.2.2 Clock Generation Circuits

Clock generators can be categorized into two major types: *multivibrators* and *linear oscillators*. A multivibrator, also called a *function generator* or *relaxation oscillator*, generates a sequence of square waves, triangle waves, or pulses, with a controllable duty cycle and frequency. Two widely used multivibrators are the *ring oscillator* and *Schmitt-circuit-based oscillator*. A ring oscillator consists of a number of voltage-gain stages to form a closed-feedback loop. A Schmitt-circuit-based oscillator is based on the charge and discharge operations on one or more timing capacitors. Both period and duty cycle can be controlled through the time constants of charge and discharge paths. A *linear oscillator* (*resonator-based oscillator*) generates a single-frequency sinusoidal wave. It is usually built from either of the following two ways: an *RC-tuned circuit* and *LC-tuned circuit*. An *RC*-tuned oscillator uses an *RC*-frequency-selective feedback network to select the desired frequency. An *LC*-tuned oscillator uses an *LC*-frequency-selective feedback network to select the desired frequency. In what follows, we only consider multivibrators, including the ring oscillator and Schmitt-circuit-based oscillator.

14.2.2.1 Ring Oscillators A ring oscillator consists of n inverters cascaded into a loop, as displayed in Figure 14.10, where n is usually an odd integer. By properly arranging the inverters, an even-stage ring oscillator is also possible. In CMOS technology, at least three stages of inverters have to be used, as shown in Figure 14.10(a), to guarantee stable operations. The timing diagram of the three-stage ring oscillator is plotted in Figure 14.10(b).

The period of the clock generated by an n-stage ring oscillator can be expressed as

$$T = 2 \cdot n \cdot t_{pd} \tag{14.4}$$

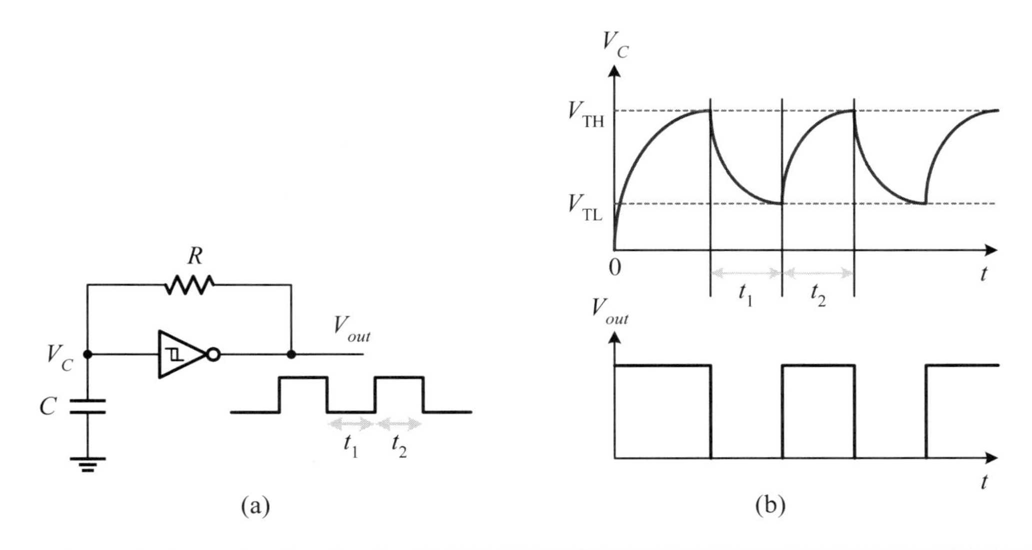

Figure 14.11: A Schmitt-circuit-based clock generator: (a) circuit; (b) timing diagram.

where t_{pd} is the propagation delay of one stage, supposing that the inverter is symmetric. The frequency f is equal to

$$f = \frac{1}{T} = \frac{1}{2 \cdot n \cdot t_{pd}} \tag{14.5}$$

For instance, the period of a three-stage ring oscillator is

$$T = 3 \cdot 2 \cdot t_{pd} = 6t_{pd}$$

For our generic 0.18-μm process, the intrinsic delay of the inverter is 15 ps (simulation result). As a result, the period is $6 \times 15 = 90$ ps, generating a clock with a frequency of 11.1 GHz. The frequency of a ring oscillator can be varied by controlling the number of stages and the propagation delay of each stage in the inverter loop. They can be made statically or dynamically.

14.2.2.2 A Schmitt-Circuit-Based Oscillator In addition to being used as a waveform shaper, a Schmitt (trigger) circuit can also be used as a pulse generator. To explore how it works as an oscillator, consider the circuit given in Figure 14.11. Once entering stable operations, the capacitor voltage V_c is periodically swung between V_{TH} and V_{TL}, where V_{TH} and V_{TL} are high- and low-threshold voltages of the Schmitt circuit, respectively. As the capacitor voltage V_C reaches V_{TH} from V_{TL}, the output voltage V_{out} falls from V_{DD} to 0 V and at the same time, the capacitor starts to discharge toward the ground level. When the capacitor voltage V_C reaches V_{TL}, the output voltage rises to V_{DD} and the capacitor reverses its operation and begins to charge toward V_{DD} again. The amounts of time in charging and discharging the capacitor C set the frequency of the oscillator.

The discharging and charging times t_1 and t_2 can be calculated by using the following general relation for *single time constant* (STC) circuits:

$$v(t) = V_f + (V_i - V_f)\exp(-t/\tau) \tag{14.6}$$

where V_i and V_f are the initial and final voltages of the node of interest, respectively. The τ is the time constant.

To calculate the time interval t_1, we substitute $\tau = RC$, $V_i = V_{TH}$ and $V_f = 0$ V into the above equation and obtain

$$V_C(t) = V_{TH} \exp(-t/RC) \tag{14.7}$$

The t_1 is the time when V_C reaches the low-threshold voltage V_{TL} of the Schmitt circuit and can be expressed as follows:

$$V_C(t_1) = V_{TL} = V_{TH} \exp(-t_1/RC) \tag{14.8}$$

Hence, the time interval t_1 is equal to

$$t_1 = RC \ln \frac{V_{TH}}{V_{TL}} \tag{14.9}$$

Similarly, t_2 can be obtained by substituting $V_i = V_{TL}$ and $V_f = V_{DD}$ and $V_C(t_2)$ into Equation (14.6) and is found to be

$$t_2 = RC \ln \frac{V_{DD} - V_{TL}}{V_{DD} - V_{TH}} \tag{14.10}$$

The period is the sum of $t_1$and t_2. Hence, the frequency can be expressed as

$$f = \frac{1}{t_1 + t_2} \tag{14.11}$$

■ **Example 14-2: (An example of a Schmitt-circuit-based clock generator.)** Referring to Figure 14.11, let the values of resistor R and capacitor C be 10 kΩ and 1 pF, respectively. Assume that the low- and high-threshold voltages of the Schmitt circuit are 1.0 and 1.8 V, respectively, and $V_{DD} = 2.5$ V. Calculate the discharging and charging times t_1 and t_2 of the capacitor C, and the clock frequency f.

Solution: Using Equation (14.9), the amount of discharging time t_1 of the capacitor C is

$$\begin{aligned} t_1 &= RC \ln \frac{V_{TH}}{V_{TL}} \\ &= 10^4 \times 10^{-12} \times \ln \frac{1.8}{1.0} = 5.87 \text{ ns} \end{aligned}$$

Using Equation (14.10), the amount of charging time t_2 of the capacitor C is

$$\begin{aligned} t_2 &= RC \ln \frac{V_{DD} - V_{TL}}{V_{DD} - V_{TH}} \\ &= 10^4 \times 10^{-12} \times \ln \frac{2.5 - 1.0}{2.5 - 1.8} = 7.62 \text{ ns} \end{aligned}$$

As a result, the clock frequency is $1/(t_1 + t_2) = 74.13$ MHz.

■

14.2.2.3 Crystal Oscillators Regardless of the ring oscillator or Schmitt-circuit-based oscillator, the generated clock can be quite sensitive to process, voltage, and temperature (PVT) variations and is unstable. Consequently, for those systems needing high accuracy and high frequency of operation, crystal oscillators are often used instead as the essential clock sources. For the time being, many commercial clock chips using crystal oscillators are available and can provide clock sources up to several hundred megahertz.

■ Review Questions

Q14-9. Define the terms: clock skew, clock latency, and clock jitter.

Q14-10. How would you construct a ring oscillator?

Q14-11. Explain the principle of a Schmitt-circuit-based clock generator.

14.2.3 Clock Distribution Networks

The design of a clock distribution network is one of the greatest challenges in designing a large chip. The objectives of a clock distribution network are as follows. First, the clock distribution network needs to deliver to all memory elements and dynamic circuitry as a clock with bounded skew and acceptable rise and fall times. Second, a good clock distribution network should also provide a controlled environment for the global clock buffers so that skew optimization and jitter reduction schemes can be easily accommodated to minimize clock inaccuracies.

To reduce the clock skew and to minimize delay, clock distribution networks are generally constructed with metal wires, especially the upper metal layers due to their low resistances. Along these clock wires, multiple drivers are also used to buffer and regenerate the clock. The path-delay optimization approach may be used to design such buffers. It is instructive to note that clock wires may create significant crosstalk to themselves and to its adjacent signal wires through capacitive coupling between clock and signal wires. We describe a few popular clock distribution networks.

14.2.3.1 Super-Buffer Trees and FO4 Trees The simplest approach for routing a clock to modules on a chip is to use a big super-buffer, as exhibited in Figure 14.12(a), to drive an interconnect network, which in turn feeds the clock to all modules. A super-buffer is usually employed to provide enough driving current for the rest of the clock trees. This approach is popular in small-scale modules. For large-scale modules, it is not easy to control the clock skew due to unbalanced RC propagation delays of different interconnect segments. In addition, the propagation delay of a clock will be proportional to the square of the segment length.

Another approach is by using an FO4 tree to distribute the clock to all modules. As shown in Figure 14.12(b), many basic buffers are distributed over the chip. All clock ports in a module are driven from a tree consisting of equal-sized buffers, with each having a fan-out of 4. To minimize clock skew, care must be taken to balance the propagation delays through the tree in designing such a clock tree.

14.2.3.2 Clock Grids and H-Trees (X-Trees) Other ways to reduce clock skew include clock grid, H-tree, and X-tree, as shown in Figure 14.13. A clock grid is a mesh of horizontal and vertical wires driven from the middle or edges, as shown in Figure 14.13(a). The features of the clock grid are as follows. First, the clock skew is low between clocked elements due to the low resistance between any two nearby

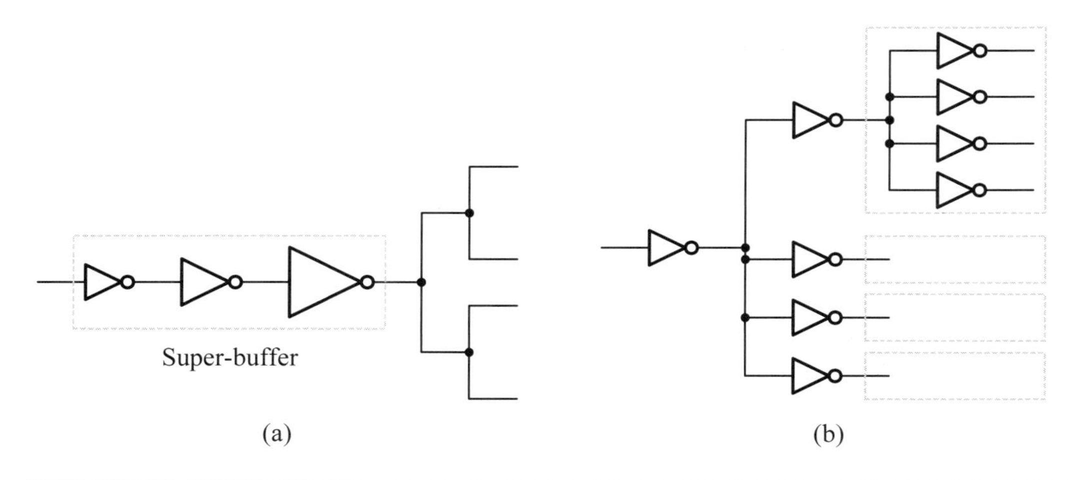

Figure 14.12: Examples of (a) super-buffer and (b) FO4-tree approaches for clock distribution.

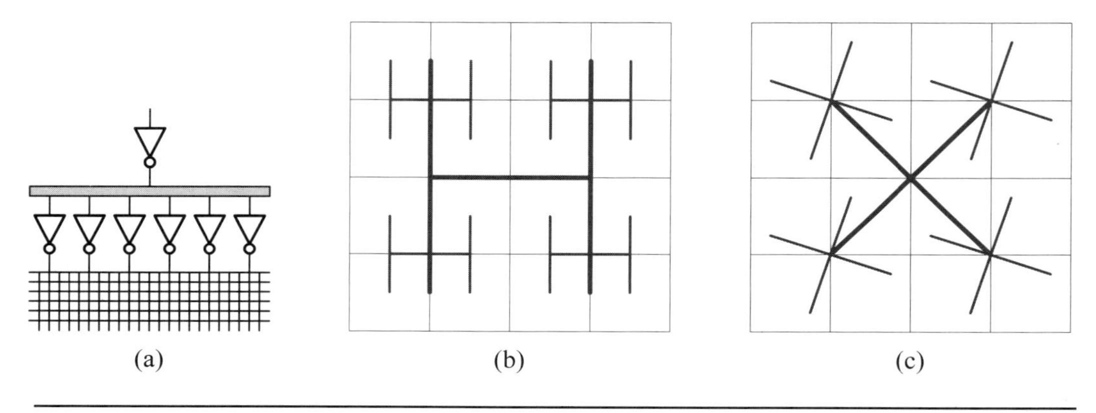

Figure 14.13: Examples of (a) clock grid, (b) H-tree, and (c) X-tree.

points in the mesh. Second, clock grid can be routed early in the design cycle without detailed knowledge of clocked-element placement. However, the clock-grid technique consumes much unnecessary power due to a lot of redundant interconnect segments existing in the grid. This might limit its use in those applications with a tight power budget.

Both H-tree and X-tree techniques are based on the concept of matched RC trees. Both H-tree and X-tree are a recursively structure built by drawing H and X shapes in which all elements are identical and can be distributed as binary and quad trees, respectively, as shown in Figures 14.13(b) and (c). They have the same feature of balancing path lengths at the leaf nodes in theory. Nevertheless, because of coupling capacitance and inductance from adjacent wires, it is very difficult to exactly balance the clock paths in practice.

14.2.3.3 Clock Spine and Hybrid Approach In the clock spine approach, many clock trunks (spines), that is, fat wires, are employed to deliver the clock from a common clock source to each individual clock tree. An example is illustrated in Figure 14.14, which is used in a Pentium-4 processor. Three clock spines are used to cover

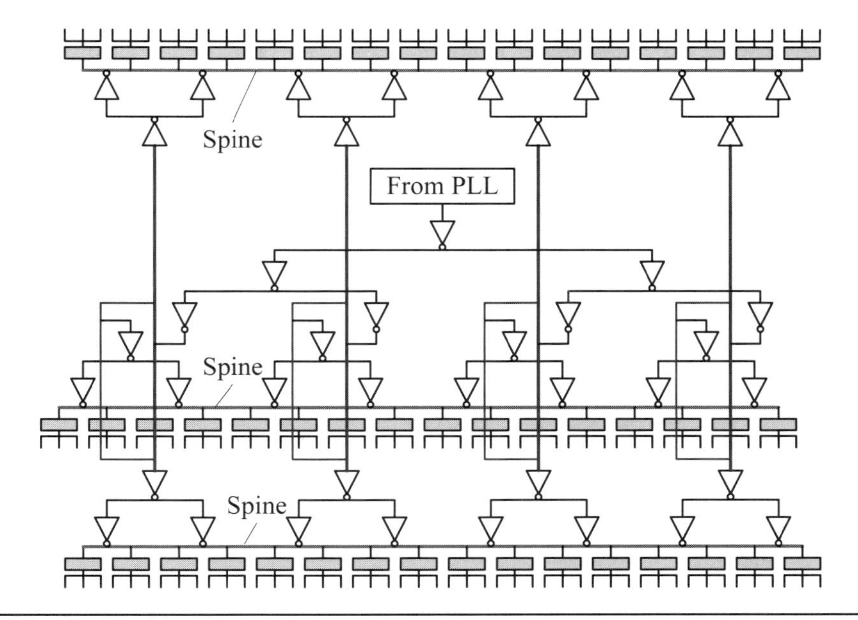

Figure 14.14: The clock spines used in a Pentium-4 processor.

the large die. Each clock spine contains a binary tree, with each leaf node supporting an independent clock domain. Because in the clock spine a controlled environment for the global clock buffers is provided, skew optimization and jitter reduction schemes can be readily accommodated to minimize clock inaccuracies and can be reached within 10% of the clock period.

The clock spine approach is indeed a hybrid of fat wires and binary trees. It is natural to mix the use of aforementioned approaches and others to form a new clock distribution network. The only concern is that the result should meet the goals of clock distribution networks mentioned above as closely as possible.

14.2.3.4 Clock Routers In cell-based synthesis flow, clock routers (that is, computer-aided design (CAD) tools) should be used to generate an appropriate clock tree, including clock-tree synthesis and clock-buffer insertion. The former automatically chooses the depth and structure of the clock tree according to the target system, while the latter tries to equalize the delay to the leaf nodes by balancing interconnect delays and buffer delays. An example of a clock tree generated automatically by a CAD tool is exhibited in Figure 14.15. Figure 14.15(a) is the original chip layout and Figure 14.15(b) shows the clock tree within the chip layout.

Gate arrays usually use a clock spine (or a regular grid) to eliminate the need for special routing and cell-based application-specific integrated circuits (ASICs) may use a clock spine, clock tree, or hybrid approach.

■ Review Questions

Q14-12. What is the basic concept on which H-tree and X-tree techniques are based?

Q14-13. Describe the essence of the clock spine technique.

Q14-14. What are the functions of clock routers?

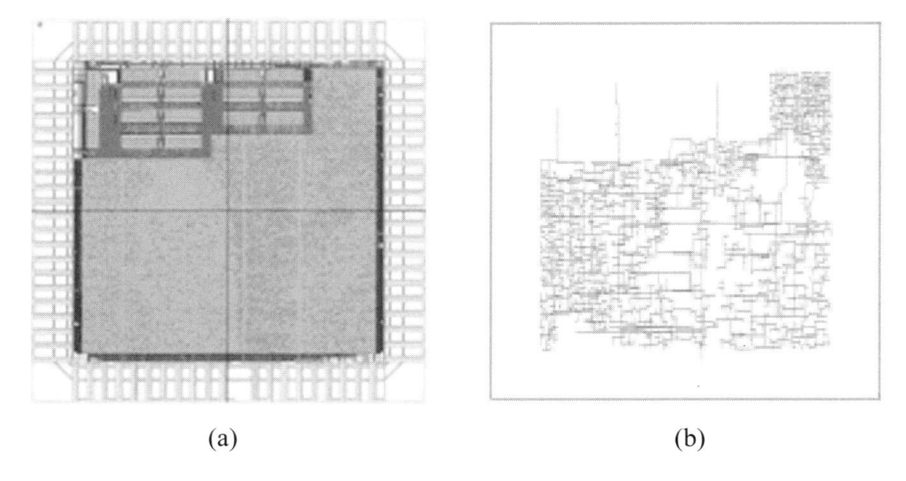

Figure 14.15: An example of a clock tree generated by a CAD tool: (a) an ECC chip layout; (b) the ECC clock tree.

14.3 Phase-Locked Loops/Delay-Locked Loops

A clock generator based on a *phase-locked loop* (PLL) is a circuit that uses feedback control to synchronize its output clock with the incoming reference clock, and to generate an output clock running at a higher rate of operation than the incoming reference clock. Like the PLL, a delay-locked loop (DLL) also uses feedback control to lock the output clock to the incoming reference clock in a constant phase.

Three important metrics of a PLL circuit are *lock-in range*, *capture range*, and *response time*. Lock-in (tracking) range means the range of input frequencies with respect to the center frequency for which the PLL remains locked. Capture (that is, acquisition or pull-in) range is the range of input frequencies for which the initially unlocked PLL will lock on an input signal. Response time denotes how fast the PLL can settle to its final state when a step-function signal is applied.

PLLs/DLLs can be pure analog, pure digital, or a hybrid of both, depending on the actual applications and requirements. In this section, we are concerned with the basic principles of *charge-pump PLLs* (CPPLLs), *all-digital PLLs* (ADPLLs), and DLLs.

14.3.1 Charge-Pump PLLs

A charge-pump PLL (CPPLL) is a closed-feedback loop consisting of a *phase-frequency detector* (PFD), a *charge pump* (CP), a *low-pass filter* (LPF), and a voltage-controlled oscillator (VCO), as shown in Figure 14.16(a). Its state diagram is given in Figure 14.16(b). We first address the basic principle of CPPLLs and then further deal with them in a more mathematical fashion.

14.3.1.1 Basic Principle The operation of CPPLL can be described as follows. The phase detector detects the phase difference between the output of VCO and reference clock, and yields two control signals *Up* and *Down* accordingly. These two signals are applied to a charge pump to switch current or voltage sources to charge or discharge its output. The output pulses of a charge pump are then filtered by the LPF to generate a more smooth control voltage, which in turn determines the oscillating frequency of

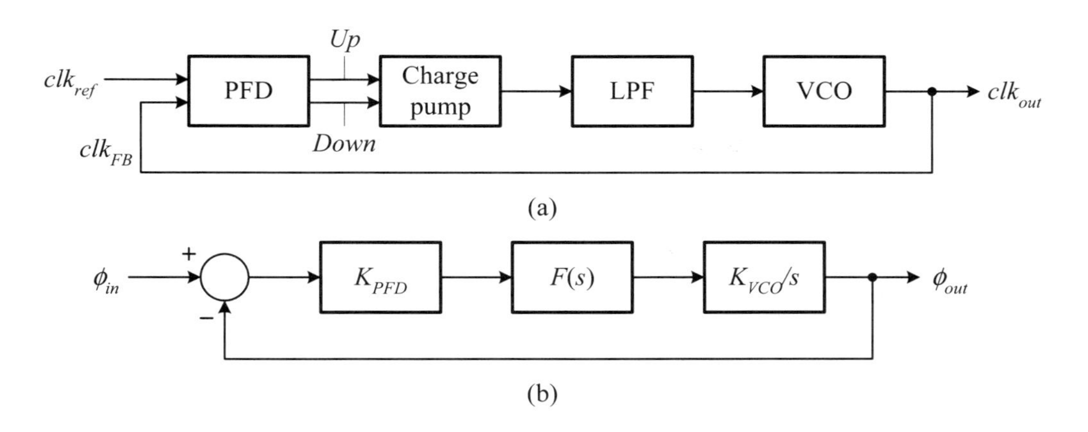

Figure 14.16: The (a) block and (b) state diagrams of typical charge-pump PLLs.

VCO. The output of VCO is then fed back to the phase detector, thereby forming a closed loop.

The PFD detects the phase difference between its two inputs and develops an amount of voltage accordingly. The output voltage v_{PFD} of PFD can be expressed as follows:

$$v_{PFD} = K_{PFD}(\phi_{in} - \phi_{out}) \tag{14.12}$$

where K_{PFD} is the voltage gain of PFD. This voltage is passed through a LPF with a transfer function $F(s)$ so that higher-frequency components are removed. The filtered signal is then applied to the VCO to create the output phase. Recall that the phase is the integral of frequency and the frequency of VCO is determined by the control voltage v_{LPF}, which is the output from the LPF. Mathematically, this can be expressed as

$$\phi = \int f dt = \int K_{VCO} v_{LPF} dt \tag{14.13}$$

namely, the VCO can be thought of as an integrator that converts the control voltage into the phase. Hence, the transfer function of it is K_{VCO}/s, which introduces a pole into the system transfer function and adds instability to the system.

Referring to the state diagram shown in Figure 14.16(b), the open-loop transfer function of the CPPLL can be expressed as follows:

$$G(s) = \frac{\phi_{out}(s)}{\phi_{in}(s)} = K_{PFD} \cdot F(s) \cdot \frac{K_{VCO}}{s} \tag{14.14}$$

where $F(s)$ is the transfer function of LPF.

14.3.1.2 Loop Filters As illustrated in Figure 14.17, there are three types of low-pass filters (LPFs) that can be used as the loop filter needed in a PLL to smooth the output from the PFD. The first is a simple capacitor, as shown in Figure 14.17(a). This capacitor adds a second pole at $s = 0$ into the system transfer function and makes the resulting system become a second-order PLL, thereby yielding an unstable PLL. The system transfer function is as follows:

$$G(s) = K_{PFD} \cdot \frac{1}{sC_1} \cdot \frac{K_{VCO}}{s} \tag{14.15}$$

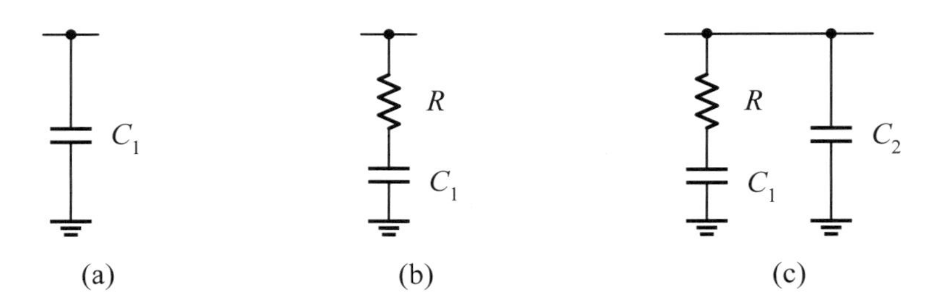

Figure 14.17: The low-pass filters for charge-pump PLLs.

The second type of LPF is shown in Figure 14.17(b), where a resistor is placed in series with a capacitor. Therefore, it inserts a zero into the system transfer function, which can be in turn expressed as follows:

$$G(s) = K_{PFD} \cdot \left(\frac{1 + sRC_1}{sC_1}\right) \cdot \frac{K_{VCO}}{s} \tag{14.16}$$

Because of the zero insertion, the overall frequency response of resulting PLL can be made to be stable by properly choosing the resistance R and capacitance C. One problem of this PLL is that small step discontinuities in the value of v_{LPF} occur as the LPF switches because of the capacitive feedthrough of both charge-pump switches. To compensate for this, a shunting capacitor is connected in parallel with the LPF shown in Figure 14.17(b). This results in the third LPF as depicted in Figure 14.17(c).

The third type of LPF is as shown in Figure 14.17(c), which is a second-order circuit. The resulting PLL system is a third-order one, and its system transfer function becomes

$$G(s) = K_{PFD} \cdot \left(\frac{1 + sRC_1}{s^2RC_1C_2 + s(C_1 + C_2)}\right) \cdot \frac{K_{VCO}}{s} \tag{14.17}$$

Although it is a third order, it can be made to be stable by choosing the values of R, C_1, and C_2 appropriately. Since capacitor C_2 functions much like the filtering capacitor of C_1, used to remove the step discontinuities, the value of C_2 is usually set to be $(1/10)C_1$.

Once we have the transfer functions of LPFs, the closed-loop transfer function can then be easily derived. To illustrate this, assuming that the LPF in Figure 14.17(b) is used in the system, the closed-loop transfer function $T(s)$ is as follows:

$$\begin{aligned} T(s) &= \frac{\phi_{out}(s)}{\phi_{in}(s)} = \frac{G(s)}{1 + G(s)} = \frac{K_{PFD} \cdot \left(\frac{1 + sRC_1}{sC_1}\right) \cdot \frac{K_{VCO}}{s}}{1 + K_{PFD} \cdot \left(\frac{1 + sRC_1}{sC_1}\right) \cdot \frac{K_{VCO}}{s}} \\ &= \frac{K_{PFD}K_{VCO}\left(\frac{1 + sRC_1}{C_1}\right)}{s^2 + K_{PFD}K_{VCO}Rs + \frac{K_{PFD}K_{VCO}}{C_1}} \end{aligned} \tag{14.18}$$

Therefore, it is a second-order system.

14.3.1.3 Voltage-Controlled Oscillators A voltage-controlled oscillator (VCO) is a circuit that generates an output signal with a frequency proportional to its input

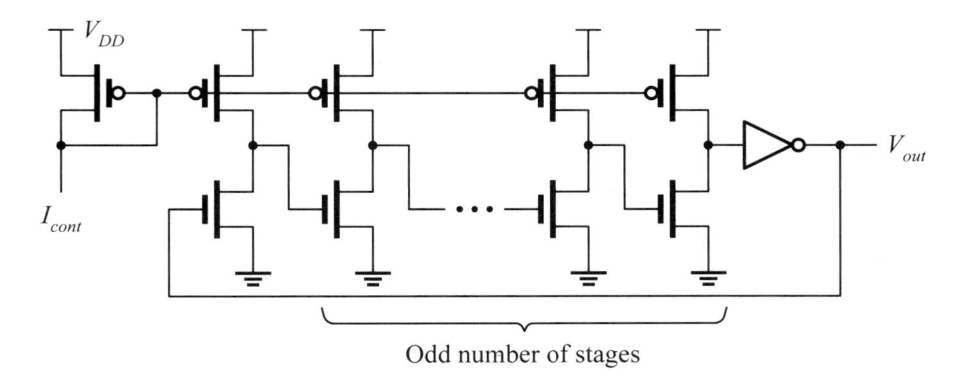

Figure 14.18: A multiple-stage current-controlled ring oscillator.

control voltage. The VCOs used in PLLs may be subdivided into two categories: *ring VCO* and *LC-tuned VCO*. Ring VCOs are widely used in a variety of applications due to their simplicity and low cost; LC-tuned VCOs are preferred in high-frequency applications, especially, in radio frequency (RF) systems where ultra-low phase noise is required. The detailed discussion of LC-tuned VCOs is beyond the scope of this book. Therefore, in what follows, we only consider a few simple ring VCO circuits.

A simple multiple-stage *current-controlled ring oscillator* is exhibited in Figure 14.18, where each controlled pMOS-transistor current source is served as the load of an nMOS transistor. Because pMOS transistors are operated in saturation regions, their gate-bulk capacitances have a negligible effect on the speed. This ring oscillator can achieve a wide-tuning range and a maximum speed relatively independent of pMOS transistors. In addition, with a small speed penalty, the transconductance of pMOS transistors can be minimized to reduce their contribution to the phase noise of the oscillator.

A single-stage ring oscillator based on the Schmitt circuit is illustrated in Figure 14.19, where a Schmitt circuit is used to generate a square wave output. The oscillation frequency can be controlled by adjusting the charging and discharging currents through transistors M_{p2} and M_{n1}, which are in turn controlled by an external input V_{cntl} through a current mirror composed of M_{p3} and M_{p2}.

The charging time t_1 is determined by the drain current I_{Mp2} of pMOS transistor M_{p2} and is equal to

$$t_1 = C\frac{V_{TH} - V_{TL}}{I_{Mp2}} \tag{14.19}$$

where V_{TH} and V_{TL} are high- and low-threshold voltages of the Schmitt circuit. The discharging time t_2 is controlled by the drain current I_{Mn1} of nMOS transistor M_{n1} and can be expressed as

$$t_2 = C\frac{V_{TH} - V_{TL}}{I_{Mn1}} \tag{14.20}$$

Both drain currents I_{Mp2} and I_{Mn1} are determined by the control voltage V_{cntl} of nMOS transistor M_{n3} and the current of the current mirror consisting of pMOS transistors M_{p3} and M_{p2}.

A multiple-stage ring oscillator with a current-starved inverter at each stage is shown in Figure 14.20. The control voltage adjusts the amount of current delivered to the inverter to charge/discharge the next stage. The operation of this ring oscillator

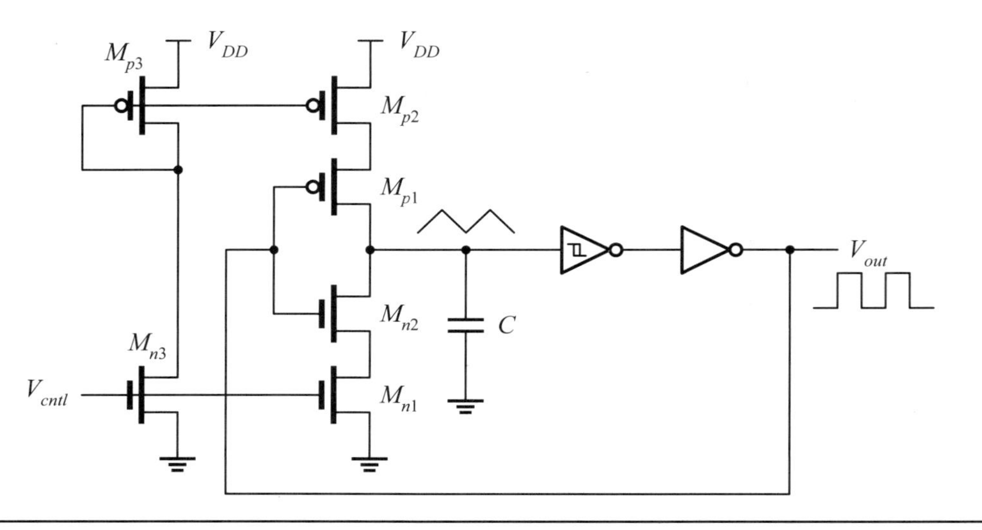

Figure 14.19: A single-stage ring oscillator using the Schmitt circuit.

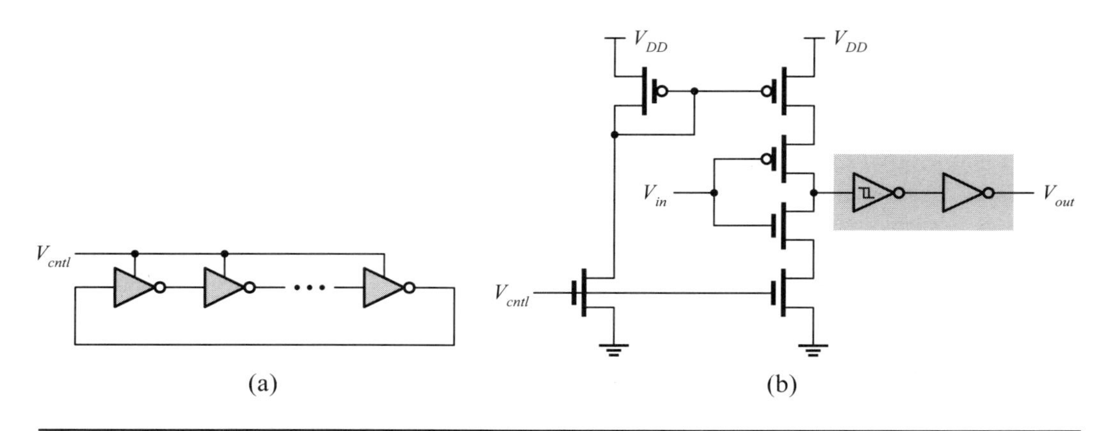

Figure 14.20: A multiple-stage ring oscillator with current-starved inverter: (a) block diagram; (b) a current-starved inverter.

is similar to Figure 14.10 except that here the propagation delay of each stage is controlled by an external voltage V_{cntl}. The frequency is determined by the number of stages and control voltage.

14.3.1.4 Phase Detector and Phase-Frequency Detector A *phase detector* (PD) is a circuit whose average output is linearly proportional to the phase difference (or time difference) between its two inputs. A *phase-frequency detector* (PFD) can detect both the phase and frequency differences between its two inputs. We first examine a simple XOR-gate PD and then address the PFD, a circuit widely used in modern PLLs.

An XOR gate is the simplest phase detector that produces error pulses on both rising and falling edges. As shown in Figure 14.21(c), when the two input signals, clk_{ref} and clk_{FB}, are completely in phase, the XOR gate will output a zero voltage. When both signals are 180° apart, the XOR gate will output a high-level signal. The phase characteristics of the XOR gate are shown in Figure 14.21(b). As the phase difference

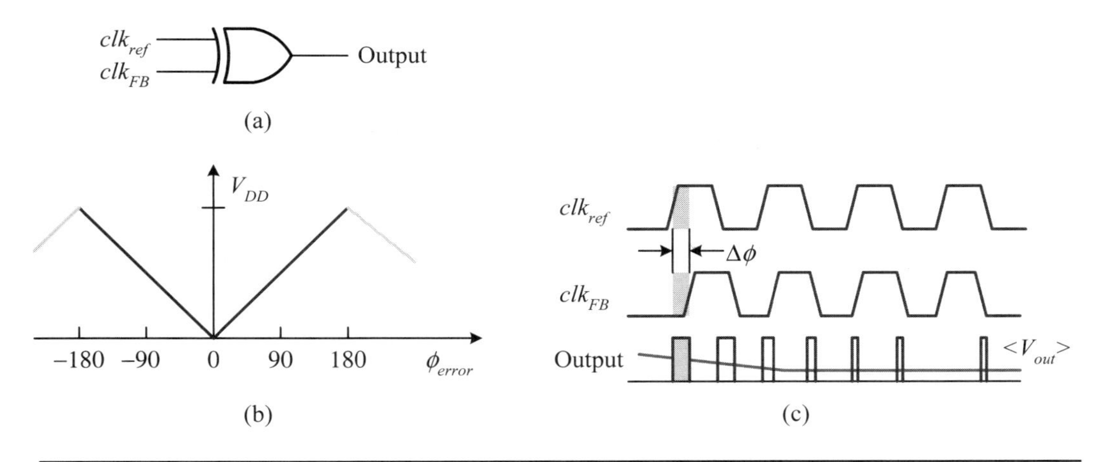

Figure 14.21: A simple phase detector using the XOR gate: (a) logic circuit; (b) phase characteristics; (c) timing diagram.

is between 0° and 180°, the XOR gate outputs a voltage value linearly proportional to the phase difference. Hence, if an out-phase condition occurs, the control loop would not know in which direction to correct. Because of this, it is usually locked on 90°. One problem with the XOR-gate phase detector is that it has a steady-state phase error if the input signal clk_{ref} or the VCO output clk_{FB} is asymmetric.

The phase-frequency detector (PFD) differs from the XOR-gate phase detector in that it not only can detect the phase difference but also can detect the frequency difference. As its name implies, the PFD output depends not only on the phase error but also on the frequency error. The PFD detector comprises an AND gate and two D-type flip-flops and can be in one of three states, *Down*, *Hold*, and *Up*, as illustrated in Figures 14.22 (a) and (b). The actual state of the PFD is determined by the positive-edges of both inputs signals, as shown in Figure 14.22(a). As the clk_{ref} switches ahead of clk_{FB}, the *Up* output would raise high to indicate that the frequency of VCO should be increased. On the other hand, as the clk_{FB} switches ahead of clk_{ref}, the *Down* output would raise high to indicate that the frequency of VCO should be decreased. As *Up* and *Down* outputs are both high, the PFD would be reset to zero. If *Up* and *Down* outputs are both zero, the PLL is locked in both the desired frequency and phase. The phase characteristics are exhibited in Figure 14.22(c), where the linear range is expanded to 4π compared to π of the XOR-gate phase detector. As the phase error is greater than 2π, the PFD works as a frequency detector.

A sample timing diagram is illustrated in Figure 14.22(d). For a small duration, both outputs are high. This duration is equal to the sum of the propagation delay of the AND gate and clear-to-Q delay of flip-flops.

14.3.1.5 Charge Pump The outputs from PFD must be combined into a single analog voltage so that it can be applied to a VCO for adjusting the output phase ϕ_{out} of the VCO. Two approaches are commonly used for this purpose: the *tristate inverter* and *charge pump*. As shown in Figure 14.23(a), the tristate inverter works as follows. As the *Up* signal is high, the pMOS pull-up transistor is turned on and pulls the output to V_{DD}; as the *Down* signal is high, the nMOS pull-down transistor is turned on and pulls the output down to the ground level. When both *Up* and *Down* signals are low, both pMOS and nMOS transistors are off and the output is at a high impedance. The

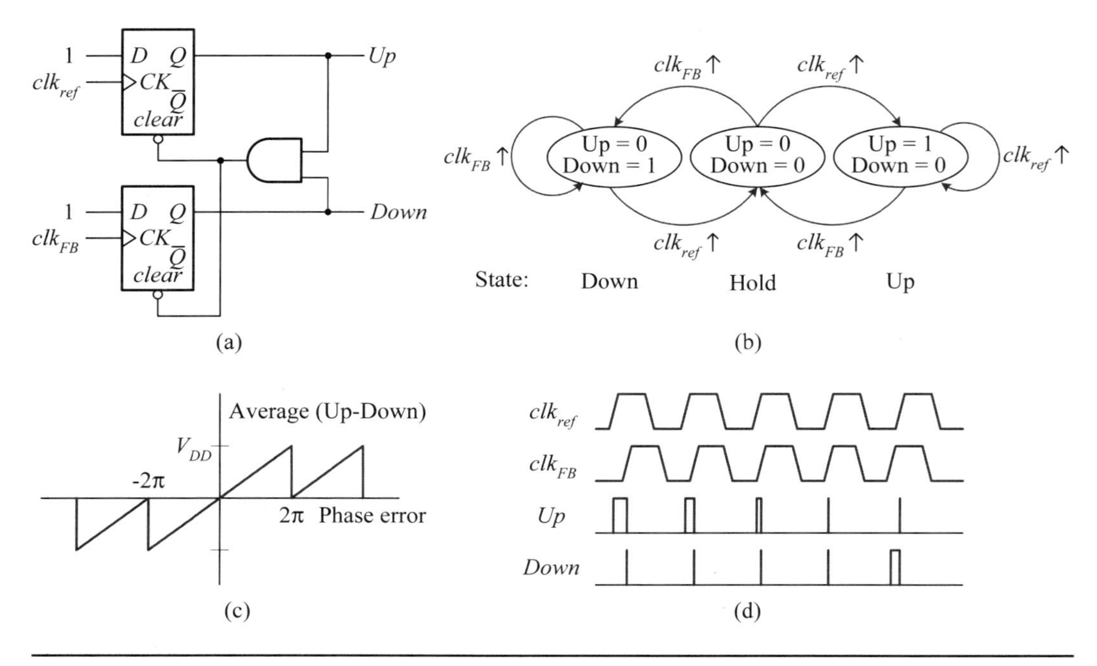

Figure 14.22: A typical phase-frequency detector used in PLLs: (a) logic circuit; (b) state transition diagram; (c) phase characteristics; (d) timing diagram.

always-on TG switch is used to balance the path delay of the *Down* signal relative to the *Up* path. The disadvantage of this circuit is that the VCO control voltage, and hence the phase of VCO output signal, might be modulated by the supply voltage when the pMOS transistor is on.

The charge-pump circuit is illustrated in Figure 14.23(b). This circuit comprises two constant current sources along with one pMOS and one nMOS switch. It is usually called a *charge pump* because of the use of constant current sources. As the *Up* signal goes high, the corresponding switch turns on, thereby connecting the current source to the LPF; as the *Down* signal goes high, the LPF is connected to the ground level. Because of the current sources made insensitive to power-suppply variations, the problem of VCO control voltage modulated by the power supply can be alleviated.

14.3.1.6 Applications Before leaving the discussion of PLLs, we are concerned with a few applications of PLLs. These include *frequency multiplier/synthesizer*, *clock-deskew circuit*, and *clock recovery circuit* (CRC).

Frequency multipliers/synthesizers. Although crystal oscillators can generate accurate, low-jitter clocks over a wide frequency range, from tens of megahertz to approximately 200 MHz, modern digital systems often need much higher frequencies than this. The most common way to provide an on-chip clock with a frequency higher than the input clock is by using a PLL. A PLL takes an external low-frequency crystal signal as a reference input f_{ref} and generates an output signal f_{out} with a frequency multiplied by a rational number M, namely, $f_{out} = Mf_{ref}$.

The general block diagram of a PLL-based frequency multiplier is shown in Figure 14.24. Since $f_D = f_{out}/M$ and f_D and f_{ref} must be equal under the locked condition, the output frequency of PLL is equal to f_{ref} multiplied by a factor of M.

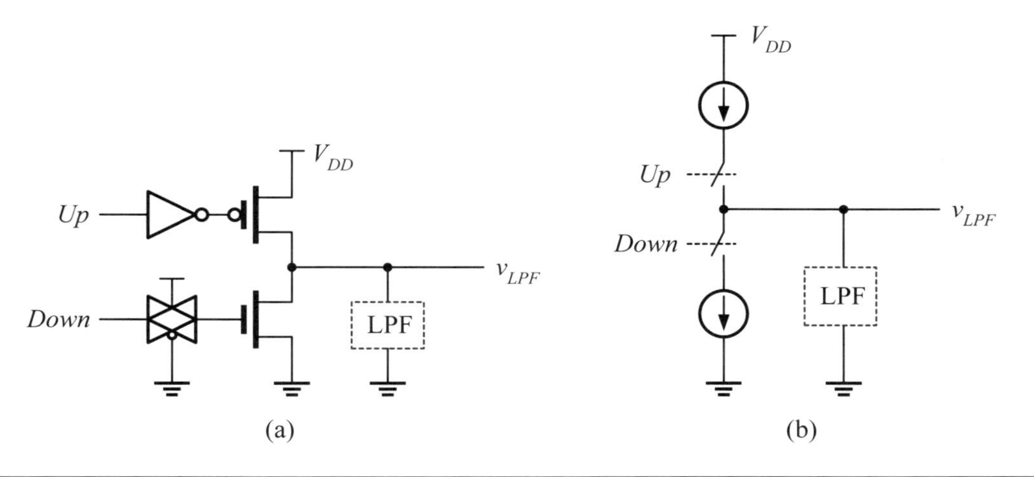

Figure 14.23: A typical phase-frequency detector used in PLLs: (a) tristate inverter; (b) charge pump.

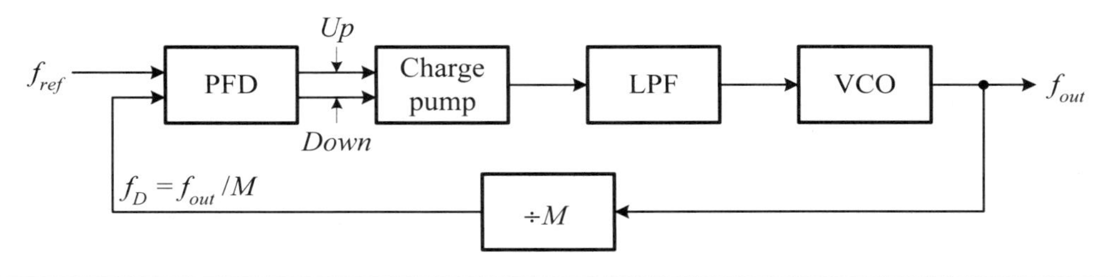

Figure 14.24: The PLL-based frequency multiplier.

The frequency divider circuit ($\div M$) is implemented by a counter that produces one output pulse for every M input pulse.

The frequency multiplier can also be used as a frequency synthesizer if the frequency divider circuit ($\div M$) is programmable. In this situation, the relative accuracy of f_{out} is equal to that of f_{ref} since $f_{out} = Mf_{ref}$. The output signal frequency f_{out} varies in steps equal to the reference signal frequency f_{ref} if M is changed by one each time. Nowadays, CMOS frequency synthesizers with gigahertz output frequencies have been widely used.

Clock-deskew circuits. The concept of applying PLL to reduce the clock skew can be illustrated by the block diagram shown in Figure 14.25. The clk_{in} is an off-chip clock and applies to a digital system, which in turn generates two internal clocks, clk_1 and clk_2. Both clk_1 and clk_2 are buffered before they are driving flip-flops, latches, or dynamic circuits. However, because of the propagation delay of clock buffers and wires, a phase difference is created between the internal and external clocks, as it can be seen from the clock clk_1 exhibited in Figure 14.25(b). The clock skew can be removed by placing a PLL between the external clock clk_{in} and the clock buffers, as the clock clk_2 shown in Figure 14.25(a). The clk_2 is fed back to the input of PLL to compare with the external clock clk_{in} so that clk_2 can track the phase of clk_{in} and both clocks can then be lined up with each other.

Clock recovery circuits. In digital systems, the simplest way to sample data is by way of using a D-type flip-flop, as illustrated in Figure 14.26(a). At each positive

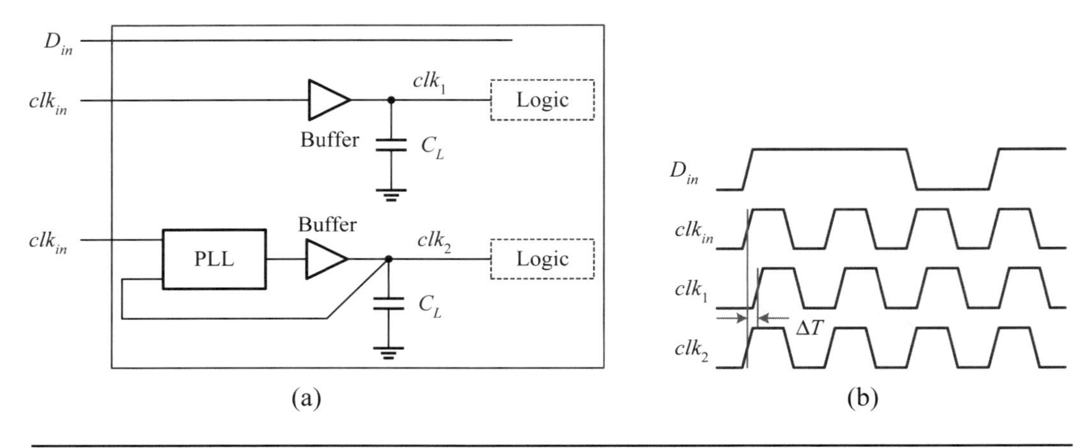

Figure 14.25: The PLL-based clock-skew reduction: (a) block diagram; (b) timing diagram.

edge of the clock being applied to the D-type flip-flop, the data input is sampled and latched in the flip-flop. The timing diagram is shown in Figure 14.26(b). As long as the clock is aligned with the center of or nearby the bit time of data, the data will be correctly sampled.

For a high-speed system, it is quite difficult to maintain the same propagation delay for both data and clock wires due to many factors such as propagation delay of buffers and wire parasitic resistance, capacitance, as well as inductance. As a consequence, the clock is often embedded into the data stream and transmitted to the receiver along with the data. At the receiver, the clock is first extracted from the data stream and then used to sample the data value, thereby obtaining the desired output data. The circuit used to extract the clock from a data stream is referred to as a *clock recovery circuit* (CRC).

A PLL-based CRC is illustrated in Figure 14.26(c). The PLL output signal serves as the sampling clock and at the same time feeds back to the PFD within the PLL so that it can be synchronized with the input clock embedded in the data stream. The design of a particular CRC strongly depends on the specific data encoding scheme, such as *nonreturn to zero* (NRZ) and *nonreturn to zero inverted* (NRZI). Hence, it should be designed in a case-by-case fashion.

■ Review Questions

Q14-15. Describe the principle of PLL-based clock-skew reduction.

Q14-16. What is the distinction between the phase detector and phase-frequency detector?

Q14-17. Describe the basic principles of CRC.

14.3.2 All-Digital PLLs

In an all-digital phase-locked loop (ADPLL), all function blocks of the system are entirely implemented by digital circuits. Hence, they are suitable for applications designed by synthesis flows and implemented with cell libraries. The benefits of using

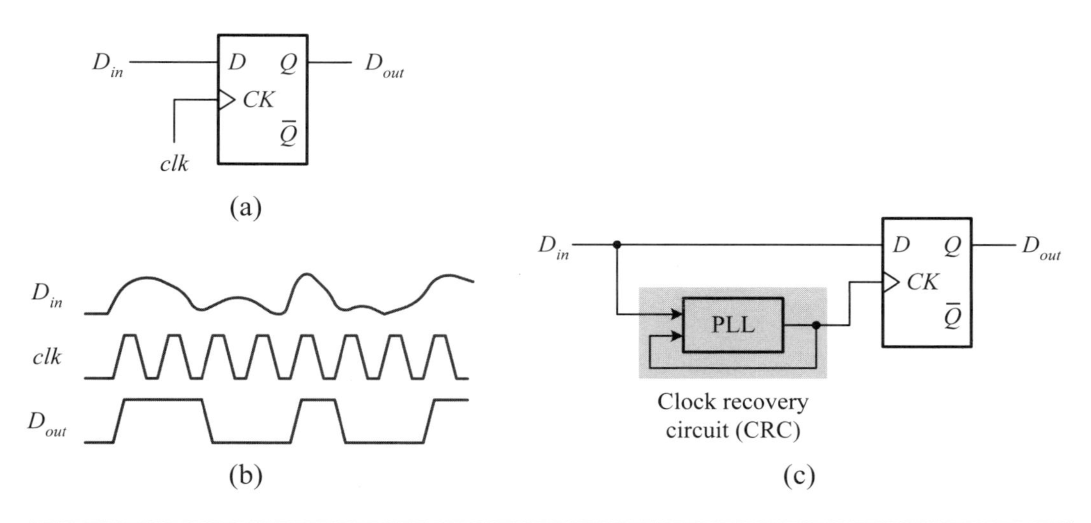

Figure 14.26: The PLL-based clock recovery cicruit: (a) sampling D-type flip-flop; (b) timing diagram; (c) clock recovery circuit.

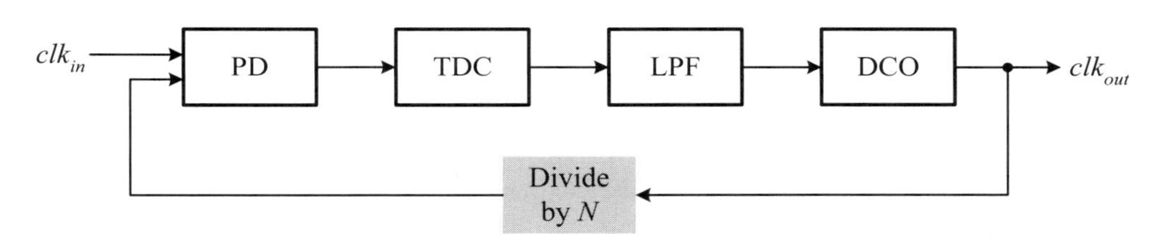

Figure 14.27: The general block diagram of ADPLLs.

ADPLLs over PLLs are as follows. First, they are easily scaled with process technology if they are made with only active components. Second, they are more robust against power-supply variations due to high control voltages compared to deviations.

14.3.2.1 Basic Principles The block diagram of an ADPLL is basically the same as that of a PLL except that now the oscillator is digital controlled, referred to as a *digital-controlled oscillator* (DCO). As shown in Figure 14.27, an ADPLL is also a closed-feedback system. The output clock of a DCO is compared to an input reference clock with a phase detector. The phase detector (PD) produces a pulse with a width equal to the phase error. This phase error is then translated into a digital signal through a *time-to-digital converter* (TDC). The digitalized time measurement output from the TDC is fed into an integrating filter (LPF), which generates a smooth output signal to control the DCO. After being divided by a *multiplication factor* (N), the output clock of the DCO is fed back to the phase detector to compare with the reference input clock. Eventually, it is locked on the frequency and phase of the input clock.

14.3.2.2 Phase Detectors The function of the phase detector (PD) in an ADPLL is to produce a pulse with a width equal to the phase error. The principle of PD can be illustrated by using an edge-triggered SR flip-flop depicted in Figure 14.28. The operation of the edge-triggered SR flip-flop is as follows. Both input signals, clk_{in} and

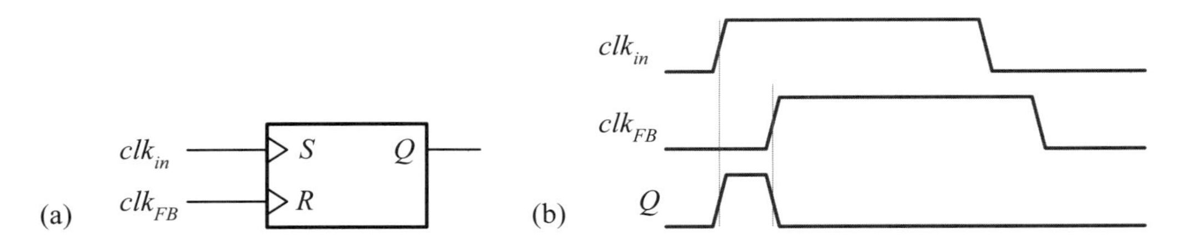

Figure 14.28: The edge-triggered SR flip-flop as a phase detector of ADPLLs: (a) logic symbol; (b) timing diagram.

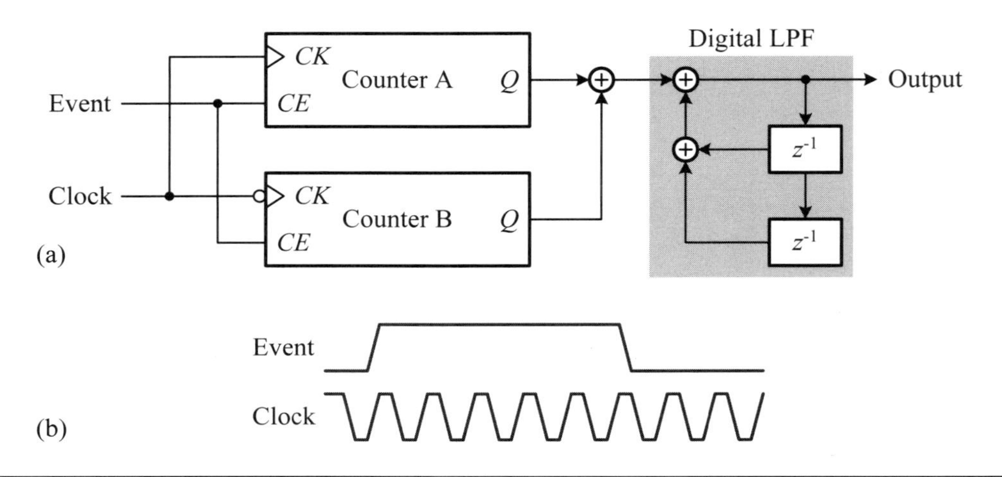

Figure 14.29: The loop filter based on the combination of TDC and LPF: (a) logic circuit; (b) timing diagram.

clk_{FB}, are used to set and reset the SR flip-flop, respectively. Hence, the duration of output Q of the SR flip-flop corresponds to the phase error of both inputs.

14.3.2.3 Time-to-Digital Converters and Loop Filters Like PLLs, the function of a loop filter in an ADPLL is to generate a stable control signal for the DCO according to the phase error generated by the phase detector. For this purpose, a TDC is employed to produce a control word with the value proportional to the phase error. The TDC can be simply implemented with an enable-controlled counter being driven by a high-frequency clock, which may be from the output of the DCO or other clock sources. An example of TDC is shown in Figure 14.29, where the TDC is composed of two counters to double the precision. One counter is positive edge-triggered and the other is negative edge-triggered. Both counters are enabled by the control signal *Event* and their outputs are added together. The output is then smoothed by a second-order integrator, that is, a recursive digital filter, before being sent to the DCO. To further increase the precision, a higher-frequency clock or a multiple-phase clock may be used.

14.3.2.4 Digital-Controlled Oscillators A DCO is a circuit that accepts a digital input word and adjusts its oscillation frequency accordingly. A wide variety of DCO structures have been proposed during the past decades. Here, we only introduce a DCO based on a ring oscillator because it is suitable for implemention with cell libraries.

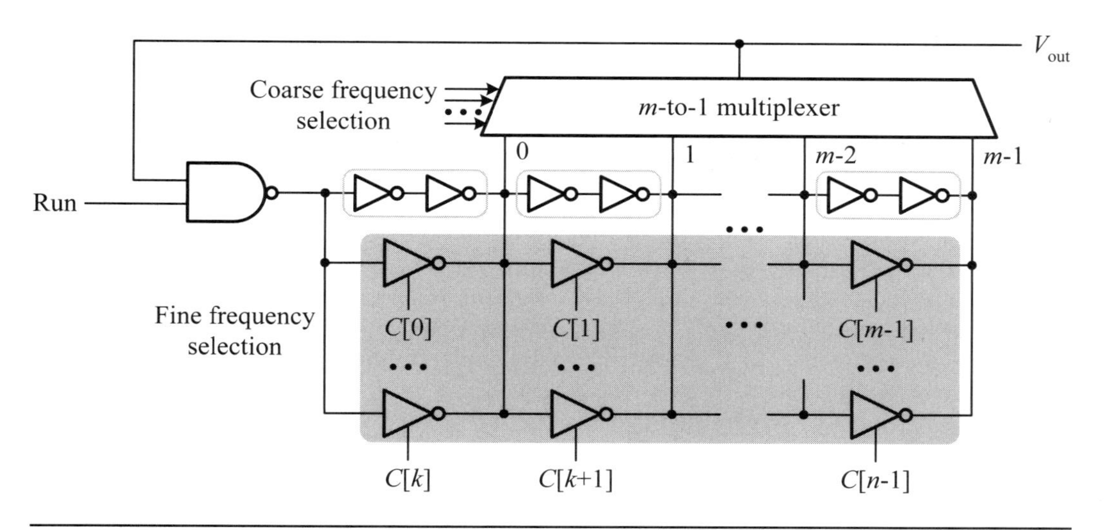

Figure 14.30: A DCO based on the structure of a ring oscillator.

As shown in Figure 14.30, the idea of using a ring oscillator to build a DCO is to use a variable number of inverters to implement a variable delay. Nevertheless, using only a variable number of inverters results in that the resolution of delay is at least equal to the propagation delay of an inverter, thereby giving an inaccurate and unstable phase lock for high-frequency applications.

To provide a higher resolution for the DCO, some mechanism must be used along with the inverters composing the DCO. As shown in Figure 14.30, the DCO is mainly a ring oscillator composed of an odd number of inverters. An m-to-1 multiplexer is employed to select the number of inverters in the ring oscillator and hence obtain the coarse frequency of the ring oscillator. To finely tune the oscillator frequency, each pair of inverters is paralleled with several tristate inverters to change the propagation delay of the stage by controlling the on-off status of individual tristate inverters. The control signals are decoded from the output of the loop filter. A NAND gate is used before the inverter chain to allow the ring oscillator to be shut down during idle mode to save power.

In summary, although the above DCO may be fully implemented with standard cells, it has several disadvantages. The DCO consumes relatively high power, over 50% of the ADPLL according to simulations, and has low maximum frequency due to high capacitance existing in the internal nodes of the ring oscillator. Hence, for applications demanding less power dissipation and high resolution, other approaches, even full-custom designs of the DCO are needed. Moreover, the nonlinear behavior of the DCO also makes it harder to find a proper loop gain for the ADPLL.

■ Review Questions

Q14-18. Describe the basic features of all-digital phase-locked loops.

Q14-19. Describe the operation of the edge-triggered SR flip-flop as a phase detector.

Q14-20. Find the transfer function of the digital LPF depicted in Figure 14.29.

Q14-21. How would you implement a DCO by using a digital-to-analog converter (DAC) and VCO?

14.3.3 Delay-Locked Loops

Like the PLL, a delay-locked loop (DLL) also uses feedback control to lock the output clock to the incoming reference clock within a constant phase. More specifically, a DLL is a circuit that adjusts the total delay of its output signal to be a multiple of the period of input signal. A DLL has a structure much like a PLL except that a *voltage-controlled delay line* (VCDL) is used instead of the VCO. In addition, like PLLs, the structure of a DLL can be analog or digital.

Unlike a PLL locking onto both frequency and phase of the input reference clock, a DLL only locks its output clock to a constant phase of its input reference clock. As a consequence, a DLL can only be used to synchronize a clock with the reference clock but cannot generate a clock with a multiple frequency of the reference clock. However, by adding extra circuits, a DLL can be modified into a clock multiplier, thereby providing a higher frequency than the reference clock. Such a circuit is referred to as a *multiplying DLL*. We are only concerned with the basic operation principle and structure of DLLs in brief.

14.3.3.1 Basic Principles The basic block diagram of a DLL is shown in Figure 14.31. The structure is similar to the PLL except that here a VCDL is used instead of the VCO. A DLL can be either analog or digital, according to whether the VCDL is controlled by an analog or a digital voltage. In an analog DLL, the VCDL is controlled by an analog voltage from the output of LPF as shown in Figure 14.31(a), whereas in a digital DLL, the VCDL is controlled by a digital signal from the output of a finite-state machine (FSM), as shown in Figure 14.31(b). In the digital DLL shown in Figure 14.31(b), the output of each stage of VCDL is connected to a phase selector (multiplexer) controlled by the digital control signals output from the FSM. The actual clock is output from the phase selector.

14.3.3.2 Voltage-Controlled Delay Lines The VCDL is at the heart of a DLL. Recall that a VCDL can be either analog or digital. An analog VCDL is usually implemented with the same circuit as that used in ring oscillators except that there is no feedback connection. The delay of a VCDL can be generally controlled by either the RC time constant or the charge and discharge currents of each stage in the VCDL. The former can be further subdivided into *capacitive tuning* and *resistive tuning*, whereas the latter is based on the fundamental relationship: $Q = It = CV$. The delay elements introduced in Section 8.3.4.2 can be used as components to control the RC time constant of VCDL. An illustration of the control of charge and discharge currents is exploited in the following example.

■ **Example 14-3: (An analog VCDL.)** An analog VCDL based on current-starved inverters is shown in Figure 14.32, which is basically the same as that used in the ring oscillator introduced in Figure 14.20 except that the Schmitt circuit is not used here and a reset circuit is added to facilitate the reset operation. The control voltage V_{cntl} is used to control the charge and discharge currents at the internal node of the current-starved inverter. By cascading an enough number of stages, the desired VCDL can be obtained.

■

14.3.3.3 Applications DLLs have many applications. The most popular one is the clock-skew reduction, serving the same function as PLLs. The second example is used

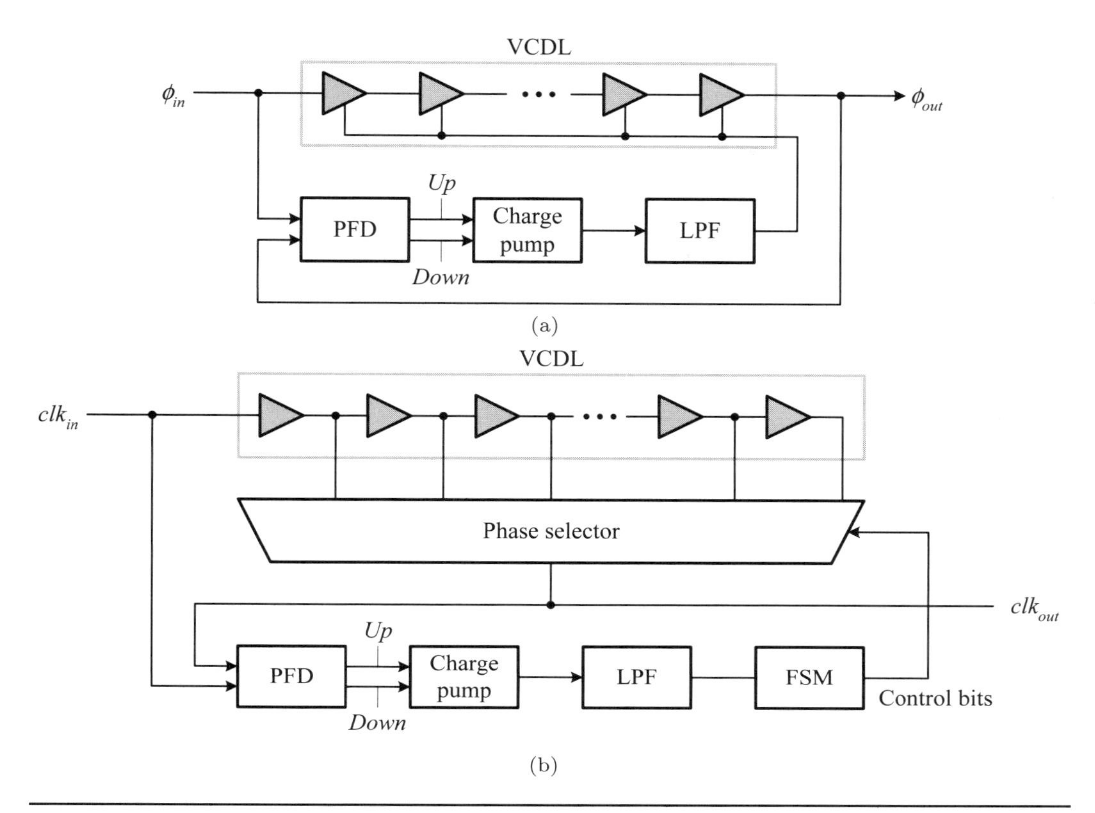

Figure 14.31: The basic block diagrams of (a) analog and (b) digital DLLs.

as a multiphase signal generator, generating a multiphase clock. The third example is served as a frequency multiplier. Nonetheless, to accomplish this, an extra circuit must be added.

■ Review Questions

Q14-22. Describe the basic features of delay-locked loops.

Q14-23. What is the distinction between analog and digital DLLs?

Q14-24. How would you implement a voltage-controlled delay line?

14.4 Summary

In modern deep submicron chip designs, power distribution and clock generation and distribution are the two essential issues that determine whether the chip may be successful on the market. For a system to be workable, each part of the logic module must be powered through a power distribution network and driven by a clock distribution network. The objective of the power distribution network is to evenly distribute the power supply to all individual devices in the system in an undisturbed fashion. The design issues of power distribution networks, clock distribution networks, and decoupling capacitors were discussed.

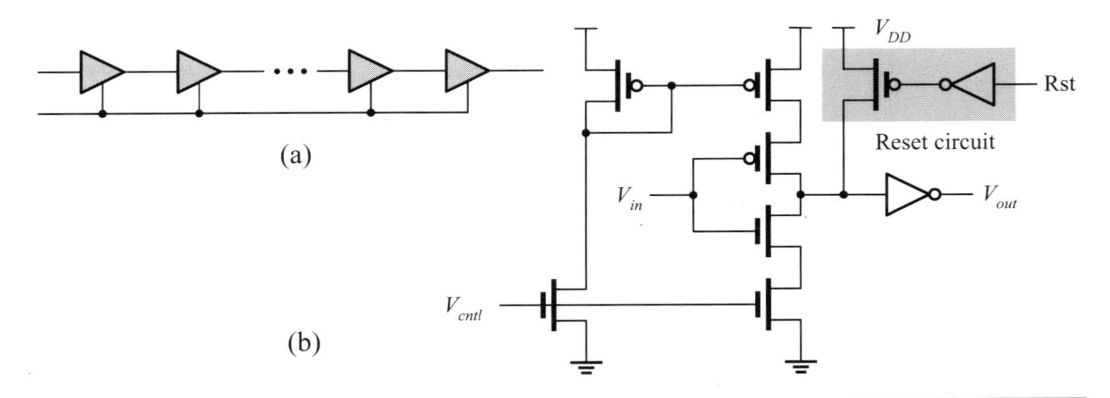

Figure 14.32: An analog VCDL based on current-starved inverters: (a) block diagram; (b) one-stage circuit.

The main goal of the clock generation and distribution network is to distribute the clocks to all individual sequential devices or dynamic logic circuitry in the system with as little skew as possible. The design issues include the clock system architecture, methods to generate clocks, and clock distribution networks. In addition, phase-locked loop (PLL) and delay-locked loop (DLL) and their applications were discussed.

References

1. F. Anderson, J. Wells, and E. Berta, "The core clock system on the next generation Itanium microprocessor," *Proc. of IEEE Int'l Solid-State Circuits Conf.*, pp. 146–147, 453, February 2002.

2. J. R. Black, "Electromigration—a brief survey and some recent results," *IEEE Trans. on Electronic Devices*, Vol. 16, No. 4, pp. 338–347, April 1969.

3. Y. Chen, H. Li, K. Roy, and C.-K. Koh, "Gated decap: gate leakage control of on-chip decoupling capacitors in scaled technologies," *IEEE Trans. on Very Large Scale Integration (VLSI) Systems*, Vol. 17, No. 12, pp. 1749–1752, December 2009.

4. C. C. Chung and C. Y. Lee, "A new DLL-based approach for all-digital multiphase clock generation," *IEEE J. of Solid-State Circuits*, Vol. 39, No. 3, pp. 469–475, March 2004.

5. D. Draper et al., "Circuit techniques in a 266-MHz MMX-enabled processor," *IEEE J. of Solid-State Circuits*, Vol. 32, No. 11, pp. 1650–1664, November 1997.

6. P. Gronowski et al., "High-performance microprocessor design," *IEEE J. of Solid-State Circuits*, Vol. 33, No. 5, pp. 676–686, May 1998.

7. J. Gu, R. Harjani, and C. H. Kim, "Design and implementation of active decoupling capacitor circuits for power supply regulation in digital ICs," *IEEE Trans. on Very Large Scale Integration (VLSI) Systems*, Vol. 17, No. 2, pp. 292–301, February 2009.

8. D. Harris and S. Naffziger, "Statistical clock skew modeling with data delay variations," *IEEE Trans. on Very Large Scale Integration (VLSI) Systems*, Vol. 9, No. 6, pp. 888–898, December 2001.

9. D. Harris, M. Horowitz, and D. Liu, "Timing analysis including clock skew," *IEEE Trans. on Computer-Aided Design of Integrated Circuits and Systems*, Vol. 18, No. 11, pp. 1608–1618, November 1999

10. D. A. Hodges, H. G. Jackson, and R. A. Saleh, *Analysis and Design of Digital Integrated Circuits: In Deep Submicron Technology*, 3rd ed. New York: McGraw-Hill Books, 2004.

11. J. Ingino and V. von Kaenel, "A 4-GHz clock system for a high-performance system-on-a-chip design," *IEEE J. of Solid-State Circuits*, Vol. 36, No. 11, pp. 1693–1698, November 2001.

12. D. K. Jeong, G. Borriello, D. A. Hodges, and R. H. Katz, "Design of PLL-based clock generation circuits," *IEEE J. of Solid-State Circuits*, Vol. 22, No. 2, pp. 255–261, April 1987.

13. N. Kurd et al., "A multigigahertz clocking scheme for the Pentium 4 microprocessor," *IEEE J. of Solid-State Circuits*, Vol. 36, No. 11, pp. 1647–1653, November 2001.

14. P. Larsson, "Parasitic resistance in an MOS transistor used as on-chip decoupling capacitance," *IEEE J. of Solid-State Circuits*, Vol. 32, No. 4, pp. 574–576, April 1997.

15. M. B. Lin, *Digital System Designs and Practices: Using Verilog HDL and FPGAs.* Singapore: John Wiley & Sons, 2008.

16. J. Maneatis, "Low-jitter process-independent DLL and PLL based on self-baised techniques," *IEEE J. of Solid-State Circuits*, Vol. 31, No. 11, pp. 1723–1732, November 1996.

17. J. Maneatis et al., "Self-biased high-bandwidth low-jitter 1-to-4096 multiplier clock generator PLL," *IEEE J. of Solid-State Circuits*, Vol. 38, No. 11, pp. 1795–1803, November 2003.

18. X. Meng, and R. Saleh, "An improved active decoupling capacitor for 'hot-spot' supply noise reduction in ASIC designs," *IEEE J. of Solid-State Circuits*, Vol. 44, No. 2, pp. 584–593, February 2009.

19. T. Olsson and P. Nilsson, "A digitally controlled PLL for SoC applications," *IEEE J. of Solid-State Circuits*, Vol. 39, No. 5, pp. 751–760, May 2004.

20. M. Popovich, M. Sotman, A. Kolodny, and E. G. Friedman, "Effective radii of on-chip decoupling capacitors," *IEEE Trans. on Very Large Scale Integration (VLSI) Systems*, Vol. 16, No. 7, pp. 894–907, July 2008.

21. M. Popovich, E. G. Friedman, M. Sotman, and A. Kolodny, "On-chip power distribution grids with multiple supply voltages for high-performance integrated circuits," *IEEE Trans. on Very Large Scale Integration (VLSI) Systems*, Vol. 16, No. 7, pp. 908–921, July 2008.

22. J. M. Rabaey, A. Chandrakasan, and B. Nikolic, *Digital Integrated Circuits: A Design Perspective*, 2nd ed. Upper Saddle River, NJ: Prentice-Hall, 2003.

23. B. Razavi, Kwing F. Lee, and R. H. Yan, "Design of high-speed, low-power frequency dividers and phase-locked loops in deep submicron CMOS," *IEEE J. of Solid-State Circuits*, Vol. 30, No. 2, pp. 101–109, February 1995.

24. P. Restle and A. Deutsch, "Designing the best clock distribution network," *1998 Symposium on VLSI Circuits Digest of Technical Papers*, pp. 2–5, 1998.

25. P. Restle et al., "A clock distribution network for microprocessors," *IEEE J. of Solid-State Circuits*, Vol. 36, No. 5, pp. 792–799, May 2001.

26. S.Rzepka, K. Banerjee, E. Meusel, and C. Hu, "Characterization of self-heating in advanced VLSI interconnect lines based on thermal finite element simulation," *IEEE Trans. on Components, Packaging, and Manufacturing Technology—Part A*, Vol. 21, No. 3, pp. 406–411, September 1998.

27. M. Shoji, "Elimination of process-dependent clock skew in CMOS VLSI," *IEEE J. of Solid-State Circuits*, Vol. 21, No. 5, pp. 875–880, October 1986.

28. S. Tam et al., "Clock generation and distribution for the first IA-64 microprocessor," *IEEE J. of Solid-State Circuits*, Vol. 35, No. 11, pp. 1545–1552, November 2000.

29. S. Tam, R. Limaye, and U. Desai, "Clock generation and distribution for the 130-nm Itanium 2 processor with 6-MB on-die L3 cache," *IEEE J. of Solid-State Circuits*, Vol. 39, No. 4, pp. 636–642, April 2004.

30. E. Temporiti et al., "A 3 GHz fractional all-digital PLL with a 1.8 MHz bandwidth implementing spur reduction techniques," *IEEE J. of Solid-State Circuits*, Vol. 44, No. 3, pp. 824–834, March 2009.

31. J. P. Uyemura, *Introduction to VLSI Circuits and Systems*. New York: John Wiley & Sons, 2002.

32. N. H. E. Weste and D. Harris, *CMOS VLSI Design: A Circuit and Systems Perspective*, 3rd ed. Boston: Addison-Wesley, 2005.

33. I. Young, J. Greason, and K. Wong, "A PLL clock generator with 5 to 110 MHz of lock range for microprocessors," *IEEE J. of Solid-State Circuits*, Vol. 27, No. 11, pp. 1599–1607, November 1992.

Problems

14-1. Referring to Figure 14.10, design a three-stage ring clock generator.

(a) Calculate the clock frequency of the clock generator using 0.18-μm parameters.

(b) Using SPICE, verify the functionality of the clock generator.

(c) Measure the clock frequency of the clock generator.

14-2. Referring to Figure 14.11, let the values of resistor R and C be 1 kΩ and 0.1 pF, respectively. Assume that the low- and high-threshold voltages of the Schmitt circuit are 0.4 and 0.8 V, respectively, and V_{DD} = 1.2 V. Calculate the discharging and charging times t_1 and t_2 of the capacitor, and the clock frequency f.

14-3. Referring to Figures 14.11 and 15.4, design a Schmitt-circuit-based clock generator.

(a) Calculate the clock frequency of the clock generator using 0.18-μm parameters.

(b) Using SPICE, verify the functionality of the clock generator.

(c) Measure the clock frequency of the clock generator.

14-4. Referring to Figure 14.19, design a single-stage ring oscillator using the Schmitt circuit.

(a) Calculate the clock frequency of the clock generator using 0.18-μm parameters.

(b) Using SPICE, verify the functionality of the clock generator.

(c) Measure the clock frequency of the clock generator.

14-5. This problem involves the design of a simple CPPLL circuit.

(a) Combine the four components, PFD, charge pump, LPF, and VCO, into a CPPLL circuit, as the standard block diagram shown in Figure 14.16(a).

(b) Design a SPICE experiment to explore the features of this simple CPPLL circuit.

15

Input/Output Modules and ESD Protection Networks

The input/output (I/O) module plays an important role for communicating with the outside of a chip or a very-large-scale integration (VLSI) system. The I/O module generally includes input and output buffers. Associated with I/O buffers are electrostatic discharge (ESD) protection networks used to create current paths for discharging the static charge caused by ESD events to protect the core circuits from being damaged.

Input buffers are circuits on a chip that take the input signals with imperfections, such as slow rise and fall times, and convert them into clean output signals for use on the chip. Based on this, the following circuits, including inverting and noninverting Schmitt circuits, level-shifting circuits, as well as differential buffers, are often used along with input buffers. An output driver or buffer is a big inverter that can be controlled as necessary in terms of the following features: transient or short-circuit current, slew rate, tristate, output resistance, and propagation delay. The issues related to output buffers include nMOS-only buffers, tristate buffer designs, bidirectional I/O circuits, driving transmission lines, as well as simultaneous switching noise (SSN) problems and reduction.

An ESD event means a transient discharge of the static charge arising from human handling or contact with machines. The ESD stress not only can destroy I/O peripheral devices but also can damage weak internal core circuit devices. Hence, proper electrostatic discharge (ESD) protection networks are necessary to create current paths for discharging the static charge caused by ESD events. In this chapter, the ESD problem and basic design issues about ESD protection networks are considered concisely.

15.1 General Chip Organizations

In this section, we begin with the introduction of general chip organization and then consider the types of pads, including power pads and I/O pads. The important issue, the SSN problem caused by simultaneous switching output (SSO), closely related to I/O pads and power pads is also introduced.

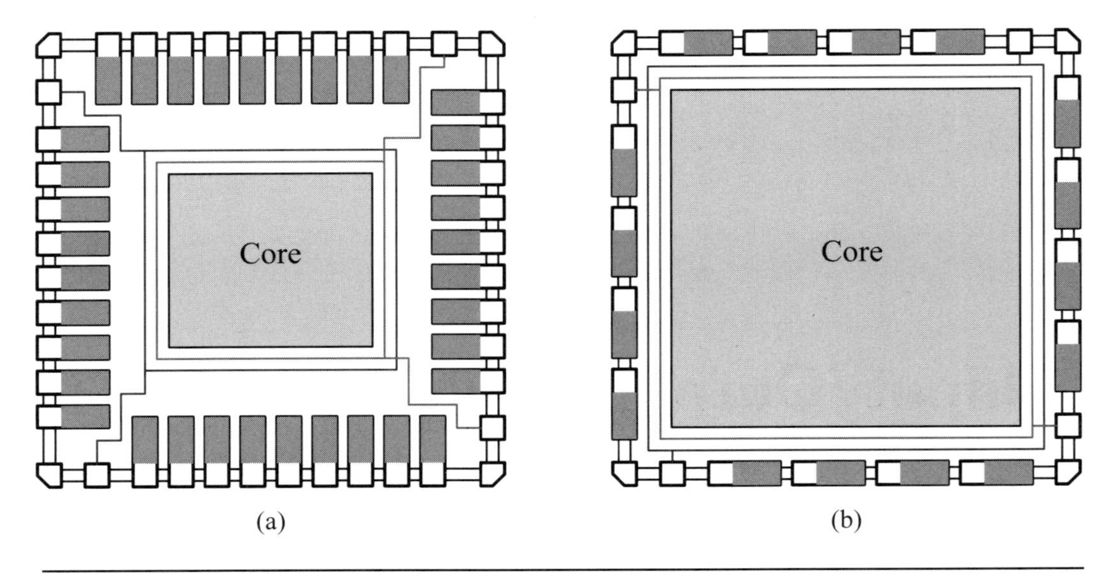

Figure 15.1: Two general types of dies: (a) I/O-limited die; (b) core-limited die.

15.1.1 General Chip Organizations

A die consists of a core circuit inside a pad ring and pads. Pads can be further categorized into two major types: *power pads* and *I/O pads*. The function of power pads is to provide power supplies to all I/O and internal core circuits. The function of I/O pads is to provide a bonding wire place along with related *electrostatic discharge* (ESD) protection networks and various necessary circuits for interfacing with the outside world.

Two general types of dies are *pad-limited* (*I/O-limited*) and *core-limited,* as shown in Figure 15.1. In a pad-limited die, the die is full of I/O pads. To fit the resulting chip into an area budget, we use tall, thin pads to maximize the number of pads that can be deployed around the outside of the chip, as shown in Figure 15.1(a). In a core-limited die, we use short, wide pads to maximize the space for the core circuit so that the resulting chip may fit into an area budget, as illustrated in Figure 15.1(b).

15.1.1.1 Power Pads Power pads are used to connect positive supply, V_{DD}, negative supply, V_{SS}, and the ground (GND/gnd). Power pads are further subdivided into *dirty* (*noisy*) *power pads* and *clean* (*quiet*) *power pads*. Dirty power pads are utilized to power I/O circuits and pads since they have to supply large transient currents to the outputs of MOS transistors. Clean power pads are used to power the internal core circuit.

When planing power pads, we have to keep all dirty power pads separate from clean power pads to avoid injecting noise into the power distribution network of the internal core circuit. In other words, we need to use one set of V_{DD}/V_{SS} pads to supply one power ring that runs around the pad ring and to power the I/O pads, and to use another set of V_{DD}/V_{SS} pads connected to a second power ring that supplies power to the internal core circuit.

15.1.1.2 I/O Pads The functions of I/O pads are for core circuit signals to communicate with the outside world. I/O pads can be further distinguished into input pads and output pads. Input pads receive signals from the outside world of the chip, while output pads send signals to the outside world. Input pads also contain special circuits to protect against electrostatic discharge (ESD) because the gates of metal-oxide-semiconductor (MOS) transistors in the input circuits are directly exposed to the pads, and these gates are very sensitive to high voltages and may be permanently damaged if they are not protected properly. Moreover, they may contain *level-shifting circuits* to be compatible with other logic families, such as transistor-transistor logic (TTL) or complementary metal-oxide-semiconductor (CMOS), with different supply voltages.

Features of output pads are that they generally do not need the ESD protection circuit because the gates of MOS transistors in the output circuits are not directly exposed to the pads. They must be able to drive large loading capacitance of pads and the outside world. Output pads can easily consume most of the power on a CMOS application-specific integrated circuit (ASIC) because of a much larger capacitance on a pad than that on the typical on-chip load. Like the input pads, they may also need level-shifting circuits to be compatible with other logic families.

In designing output pads, care must be taken to consider the noise caused by simultaneous switching outputs (SSO) and driving strengths. The noise caused by SSO is usually referred to as SSN. SSN problems are common in VLSI circuits. The driving strengths of TTL families are 2 mA, 4 mA, and 8 mA, while the drive strengths of CMOS logic are 1X, 2X, 3X, 4X, and 5X. If there is a choice, one should select the pins with the lowest inductance for ground pads because reducing the ground noise is more important than reducing the power noise, particularly, for TTL pads. It is also recommended that at least two power pads should be used for every three ground pads.

15.1.2 General Considerations

I/O and power pads are placed on the top metal layer to provide a place through which they can be bonded to the package pins. Some advanced packaging systems directly bond to the package without bonding wire; some allow pads across the entire chip surface, and such chips are called *flip chips*. Pads are typically placed around periphery of chip and collected as a *pad frame*. The design issues of the pad frame are as follows: First, appropriate power/ground pads must be supplied to each I/O pad as well as the core circuit. Second, positions of pads around the pad frame may be determined by pinout requirements on the package. Third, power/ground pads must be distributed as evenly as possible to minimize power distribution problems.

The other important issue closely related to I/O pads and power pads is the SSN problem caused by SSO. SSO occurs when an off-chip bus is being driven and may cause signal-integrity and power-integrity problems. To alleviate the SSN problems, it is necessary to provide dedicated V_{DD} and V_{SS} pads for every few SSO signals. The general rules for tackling SSN problems are as follows. First, one should not put the power and ground pads farther than 8 pad slots away from the SSO signals. Second, one should distribute power and ground pads as evenly as possible around the SSO signals, especially the ground pads.

■ Review Questions

Q15-1. What are the two types of dies in general?

Q15-2. What are the dirty and clean power pads?

Q15-3. How would you distinguish between SSN and SSO?

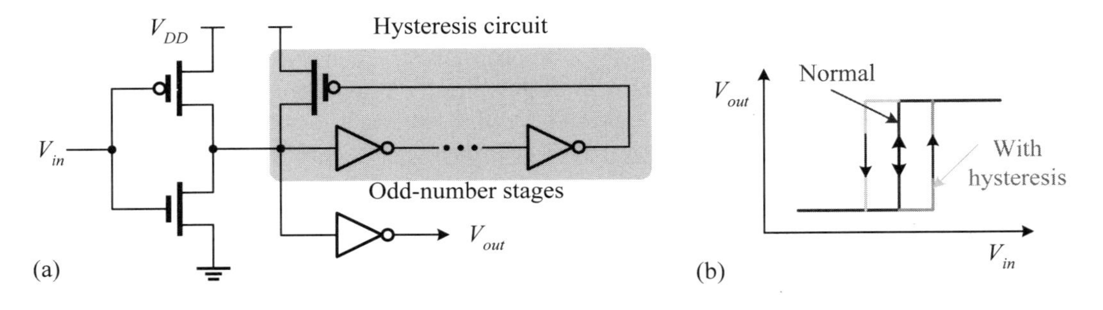

Figure 15.2: A CMOS input receiver with hysteresis: (a) circuit; (b) characteristics.

Q15-4. Describe the general rules for tackling SSN problems.

15.2 Input Buffers

Input buffers are circuits on a chip that take the input signals with imperfection, such as slow rise and fall times, and convert them into clean output signals for use on the chip. Based on this, the design issues of input buffers are as follows: First, input buffers shape the waveform of input signals to balance their rise and fall times. Second, input buffers reduce signal skew induced by logic circuitry. Third, if necessary, input buffers convert double-ended signals into single-ended signals and vice versa. Fourth, input buffers convert the input signal level to an appropriate one suitable for internal circuits.

15.2.1 Schmitt Circuits

In many applications, when an input signal changes its value very slowly and/or bounces between two voltage levels, we often need a circuit with hysteresis to shape the input waveform into one with fast transition times. We first look at a simple input receiver with hysteresis and then consider two Schmitt circuits in greater detail.

■ **Example 15-1: (An input receiver with hysteresis.)** Figure 15.2(a) shows an example of an input receiver with hysteresis. The hysteresis width is controlled by and proportional to the propagation delay of the inverter chain, consisting of an odd number of inverter stages. The characteristics of this circuit are given in Figure 15.2(b). ■

15.2.1.1 Schmitt Circuits A Schmitt (trigger) circuit is a waveform-shaping circuit that shapes a slowly changing input waveform into one with fast transition times. One feature of Schmitt circuits is that its voltage-transfer characteristics (VTC) exhibit a hysteresis loop. As stated previously, the basic principle used to design a CMOS Schmitt circuit is to make the k_R value different as the input voltage goes from 0 to V_{DD} and from V_{DD} to 0 V. Depending on how the circuit is designed, there are two types of Schmitt circuit: inverting and noninverting. The VTCs of both Schmitt circuits are shown in Figure 15.3. Figure 15.3(a) is an inverting type while Figure 15.3(b) is a

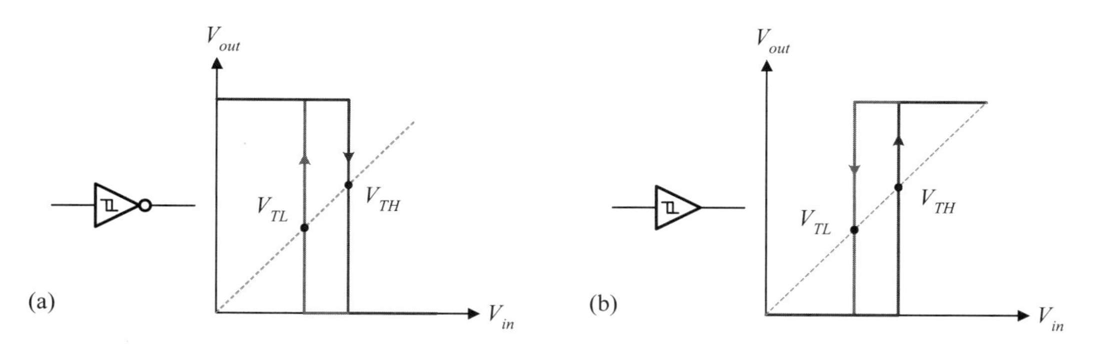

Figure 15.3: The two basic types of Schmitt circuits: (a) inverting; (b) noninverting.

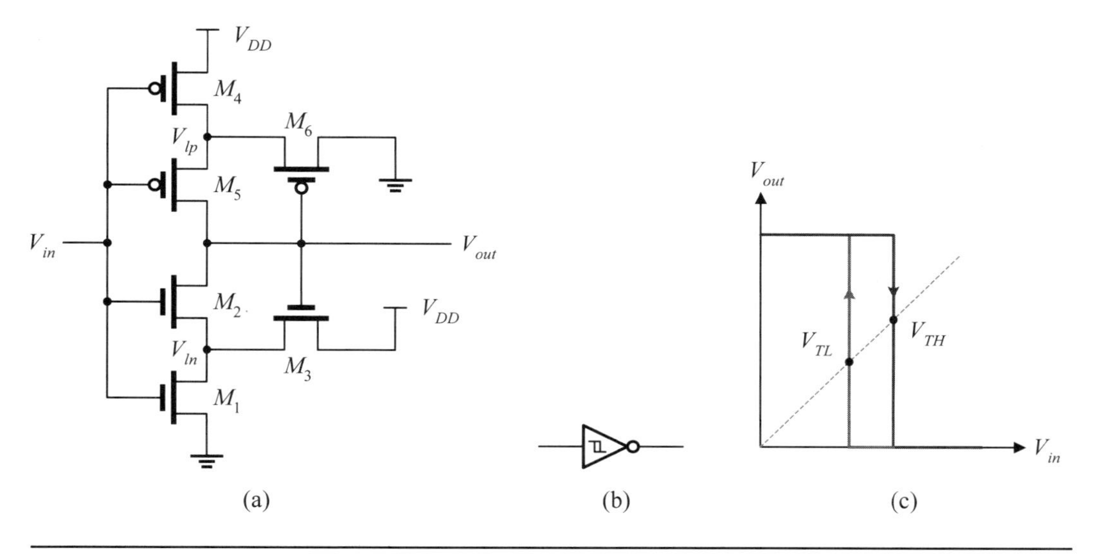

Figure 15.4: An inverting Schmitt circuit: (a) circuit; (b) logic symbol; (c) voltage-transfer characteristics.

noninverting type. To define the VTC of a Schmitt circuit, we need to determine the low- and high-threshold voltages, V_{TL} and V_{TH}, of the circuit, respectively.

15.2.1.2 Inverting Schmitt Circuit An inverting Schmitt circuit and its VTC are exhibited in Figure 15.4. Both transistors M_3 and M_6 are combined with transistors M_2 and M_5, respectively, to change k_R in accordance with the transition direction of output voltage. In what follows, we detail the analysis of both threshold voltages, V_{TL} and V_{TH}, of this Schmitt circuit.

We begin to find V_{TH}. At this point, the output is at a high voltage, say, V_{DD}. The input is raised from 0 V toward V_{DD}. At the nearby point of V_{TH}, both transistors M_1 and M_3 are in their saturation regions of operation. Hence, we have

$$\frac{k_1}{2}(V_{in} - V_{Tn})^2 = \frac{k_3}{2}(V_{DD} - V_{ln} - V_{Tn})^2 \tag{15.1}$$

The node voltage V_{ln} decreases with the increase of V_{in}. As $V_{ln} = V_{in} - V_{Tn}$, transistor M_2 is turned on and V_{out} drops to 0. Substituting V_{in} with V_{TH} into and solving the above equation for V_{TH}, we have

$$V_{TH} = \frac{V_{DD} + \sqrt{\frac{k_1}{k_3}} \cdot V_{Tn}}{1 + \sqrt{\frac{k_1}{k_3}}} \tag{15.2}$$

Next, we find V_{TL}. At this point, the output voltage is at the ground level, that is, 0 V. The input voltage decreases from V_{DD} toward 0 V. At the nearby point of V_{TL}, both transistors M_4 and M_6 are in their saturation regions of operation. Therefore, we have the following relation

$$\frac{k_4}{2}(V_{DD} - V_{in} - |V_{Tp}|)^2 = \frac{k_6}{2}(V_{lp} - |V_{Tp}|)^2 \tag{15.3}$$

The node voltage V_{lp} increases with the decrease of input voltage. As $V_{lp} = V_{in} + |V_{Tp}|$, transistor M_5 is turned on and V_{out} rises to V_{DD}. Substituting V_{in} with V_{TL} into and solving the above equation, we have

$$V_{TL} = \frac{\sqrt{\frac{k_4}{k_6}}(V_{DD} - |V_{Tp}|)}{1 + \sqrt{\frac{k_4}{k_6}}} \tag{15.4}$$

■ Example 15-2: (A design example.) Assume that the desired threshold voltages are $V_{TH} = 1.2$ V and $V_{TL} = 0.8$ V, respectively. Using 0.18-μm parameters and $V_{DD} = 1.8$ V, design an inverting Schmitt circuit.

Solution: Using Equation (15.2), we have

$$V_{TH} = \frac{V_{DD} + \sqrt{\frac{k_1}{k_3}} \cdot V_{Tn}}{1 + \sqrt{\frac{k_1}{k_3}}} = \frac{1.8 + \sqrt{\frac{k_1}{k_3}} \cdot 0.4}{1 + \sqrt{\frac{k_1}{k_3}}} = 1.2 \text{ V}$$

Hence, $k_1 = 0.56k_3$. $W_3 = 1.79W_1 \Rightarrow W_3 = 2.0W_1$, assuming that the feature size is used for the channel length of all MOS transistors. Using Equation (15.4), we obtain

$$V_{TL} = \frac{\sqrt{\frac{k_4}{k_6}}(V_{DD} - |V_{Tp}|)}{1 + \sqrt{\frac{k_4}{k_6}}} = \frac{\sqrt{\frac{k_4}{k_6}}(1.8 - 0.4)}{1 + \sqrt{\frac{k_4}{k_6}}} = 0.8 \text{ V}$$

Hence, $k_4 = 1.78k_6$. From this, we obtain $W_4 = 1.78W_6$ and take $W_4 = 2W_6$. ■

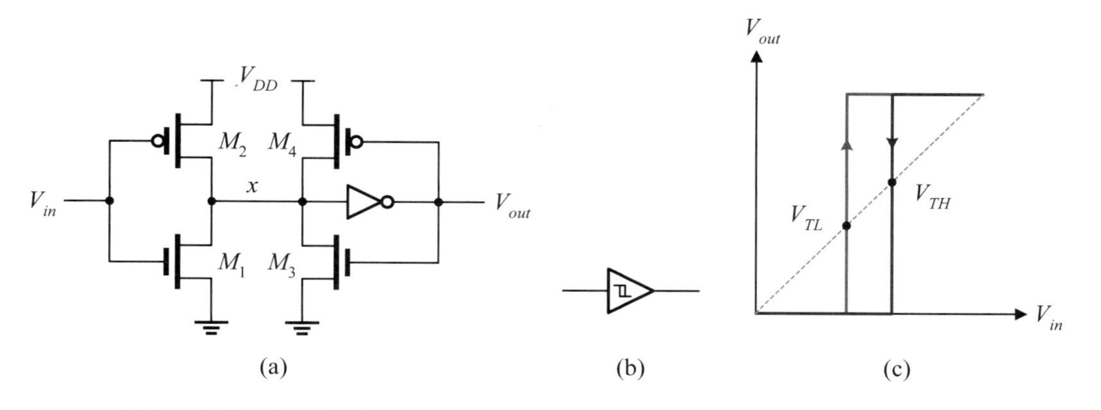

Figure 15.5: A noninverting Schmitt circuit: (a) circuit; (b) logic symbol; (c) voltage-transfer characteristics.

15.2.1.3 Noninverting Schmitt Circuit A noninverting Schmitt circuit and its VTC are shown in Figure 15.5. The operation of this Schmitt circuit is as follows. When the input V_{in} rises from 0 V to V_{DD}, transistor M_4 is effectively connected to M_2 in parallel, thereby reducing the ratio k_R of the input inverter. On the other hand, as the input V_{in} falls from V_{DD} to 0 V, transistor M_3 is effectively in-parallel connection with M_1, thereby increasing the ratio k_R of the input inverter. Consequently, the VTC of the circuit given in Figure 15.5(a) reveals a hysteresis loop and the circuit is a Schmitt circuit. In addition, since the input goes through two inverters to reach the output, the circuit is a noninverting type.

The analysis of this Schmitt circuit is as follows. Referring to Figure 15.6(a), by definition, at the high-threshold voltage V_{TH}, $V_{in} = V_x = V_{out} = V_{TH}$. Both p-type metal-oxide-semiconductor (pMOS) transistors are connected in parallel and $V_{SD4sat} = V_{SD2sat}$. Using the long-channel current equation, we have

$$k_1(V_{TH} - V_{T0n})^2 = k_2(V_{DD} - V_{TH} - |V_{T0p}|)^2 + k_4(V_{DD} - V_{TH} - |V_{T0p}|)^2 \tag{15.5}$$

Solving it, the high-threshold voltage V_{TH} is found to be

$$V_{TH} = \frac{V_{DD} - |V_{T0p}| + \sqrt{\dfrac{k_1}{k_2 + k_4}} \cdot V_{T0n}}{1 + \sqrt{\dfrac{k_1}{k_2 + k_4}}} \tag{15.6}$$

Similarly, referring to Figure 15.6(b), by definition, at the low-threshold voltage V_{TL}, $V_{in} = V_x = V_{out} = V_{TL}$. Both nMOS transistors are connected in parallel and $V_{DS1sat} = V_{DS3sat}$. Using the long-channel current equation, the following equation is obtained

$$k_2(V_{DD} - V_{TL} - |V_{T0p}|)^2 = k_1(V_{TL} - V_{T0n})^2 + k_3(V_{TL} - V_{T0n})^2 \tag{15.7}$$

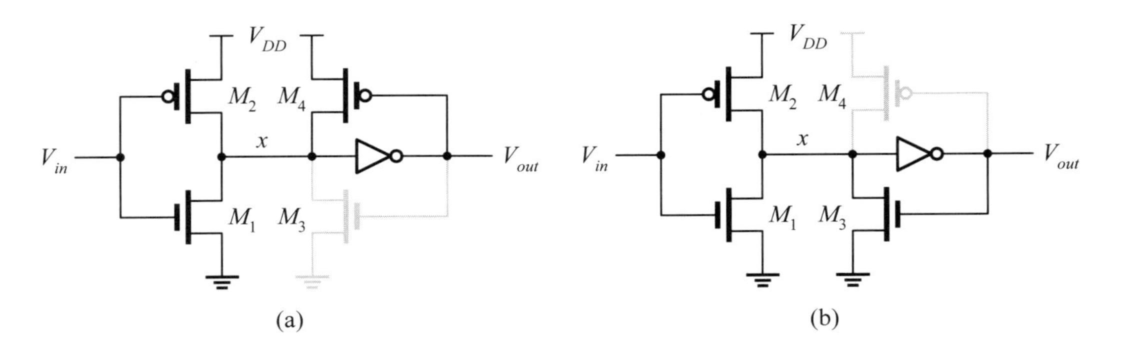

Figure 15.6: The operation of the noninverting Schmitt circuit: (a) calculation of V_{TH}; (b) calculation of V_{TL}.

Solving it, the low-threshold voltage V_{TL} can be expressed as follows:

$$V_{TL} = \frac{V_{T0n} + \sqrt{\dfrac{k_2}{k_1 + k_3}} \cdot (V_{DD} - |V_{T0p}|)}{1 + \sqrt{\dfrac{k_2}{k_1 + k_3}}} \tag{15.8}$$

Because of three parameters being involved in each equation for V_{TH} and V_{TL}, respectively, it is not easy to design the required noninverting Schmitt circuit using both equations. In practice, the following three-step procedure can be used to design such a Schmitt circuit:

1. Set the threshold voltage V_{th} of the inverter consisting of M_1 and M_2 to the midpoint of V_{TH} and V_{TL}.
2. Adjust the V_{TL} to the desired value by modifying the aspect ratio of nMOS transistor M_3.
3. Adjust the V_{TH} to the desired value by modifying the aspect ratio of pMOS transistor M_4.

The aspect ratios of both M_3 and M_4 transistors must be small enough to still allow the switching to occur. An example is given in the following to illustrate this approach.

■ **Example 15-3: (Noninverting Schmitt circuit.)** Refer to Figure 15.5(a) and suppose that the desired low- and high-threshold voltages V_{TL} and V_{TH} are 0.8 V and 1.0 V, respectively. If $V_{DD} = 1.8$ V, using 0.18-μm parameters, design the desired noninverting Schmitt circuit.

Solution: Referring to Figure15.6(a), we first design the inverter consisting of nMOS M_1 and pMOS M_2 with a threshold voltage at the midpoint of desired low- and high-threshold voltages; namely, $V_{th} = \frac{1}{2}(V_{TL} + V_{TH}) = 0.9$ V. From the inverter threshold voltage V_{th}, we obtain

$$\begin{aligned} V_{th} &= \frac{V_{DD} - |V_{T0p}| + \sqrt{k_R} \cdot V_{T0n}}{1 + \sqrt{k_R}} \\ &= \frac{1.8 - 0.4 + \sqrt{k_R} \cdot 0.4}{1 + \sqrt{k_R}} = 0.9 \end{aligned}$$

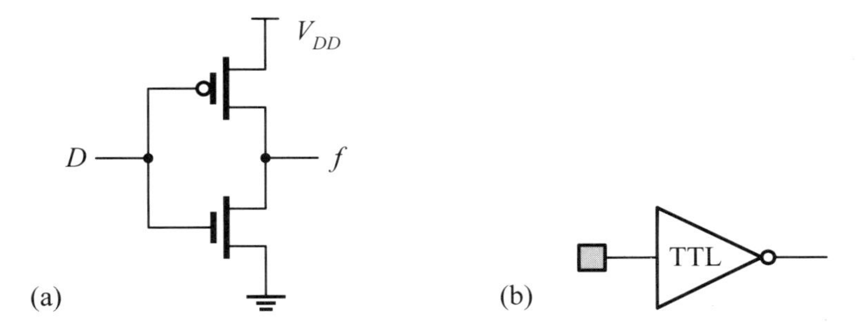

Figure 15.7: An inverting TTL-to-CMOS conversion circuit: (a) circuit; (b) logic symbol.

Solving it, the $k_R = k_1/k_2$ is found to be 1. Using $\mu_n = 287$ cm^2/V-s and $\mu_p = 88$ cm^2/V-s, we have $W_2 = 3.26W_1$. Setting $W_1 = 4\lambda$ and $W_2 = 16\lambda$, the simulation result shows that the V_{th} is about 0.7 V, less than 0.9 V. Hence, we need to decrease k_R value so that the V_{th} may be moved toward 0.9 V. By setting $W_2 = 20$, the threshold voltage V_{th} is about at 0.8 V with 0.15 V being expanded at both sides.

Next, we adjust the V_{TL} by setting the aspect ratio of nMOS transistor M_3. Because of V_{TL} below the desired value, we have to move it toward a higher value; namely, we need to decrease the total k_R, including M_3 into the inverter composed of M_1 and M_2. After setting the $L_3 = 7\lambda$ and $L_4 = 3\lambda$, both V_{TL} and V_{TH} fit into the desired values. The final aspect ratios of all transistors are $M_1 = M_6 = 4\lambda/2\lambda$, $M_2 = 20\lambda/2\lambda$, $M_3 = 3\lambda/7\lambda$, and $M_4 = 3\lambda/3\lambda$, $M_5 = 3\lambda/2\lambda$.

■

15.2.2 Level-Shifting Circuits

In many applications, an interface between TTL and standard CMOS circuits is often necessary. Both 3.3-V and 5-V CMOS logic families can be compatible with the TTL family if appropriate level-shifting circuits are used. In what follows, we give two examples to illustrate the approaches for designing such circuits.

15.2.2.1 Inverting TTL-to-CMOS Converter An inverting TTL-to-CMOS converter is nothing but a CMOS inverter with TTL-compatible voltage levels. To reach this, it is necessary to design a CMOS inverter with its threshold voltage other than $V_{DD}/2$. An illustration is explored in the following example.

■ **Example 15-4: (An inverting TTL-to-CMOS converter.)** Referring to Figure 15.7, suppose that the CMOS inverter is powered by a voltage of 3.3 V and wants to receive TTL signals. Design an inverter capable of receiving TTL voltage levels and calculate the corresponding noise margins: NM_L and NM_H. Assume that 0.35-μm process parameters are used; namely, $V_{T0n} = |V_{T0p}| = 0.55$ V, $\mu_n = 400$ cm^2/V-s, and $\mu_p = 120$ cm^2/V-s.

Solution: To be compatible with the input TTL voltage levels, the threshold voltage V_{th} of the CMOS inverter is set to 1.5 V.

$$\begin{aligned} V_{th} &= \frac{V_{DD} - |V_{T0p}| + \sqrt{k_R} \cdot V_{T0n}}{1 + \sqrt{k_R}} \\ &= \frac{3.3 - 0.55 + \sqrt{k_R} \cdot 0.55}{1 + \sqrt{k_R}} = 1.5 \text{ V} \end{aligned}$$

Hence, $k_R = 1.73$. Once we know this, we may use it to determine the channel widths of both nMOS and pMOS transistors.

$$k_R = \frac{k_n}{k_p} = \frac{\mu_n W_n}{\mu_p W_p} = \frac{400}{120}\left(\frac{W_n}{W_p}\right) = 1.73$$

As a result, $W_p = 1.93W_n$, taking $W_p = 2W_n$. Substituting the k_R into Equation (7.7), we obtain

$$\begin{aligned} V_{IL} &= \frac{2\sqrt{1.73}(3.3 - 0.55 - 0.55)}{(1.73 - 1)\sqrt{1.73 + 3}} - \frac{3.3 - 1.73 \times 0.55 - 0.55}{1.73 - 1} \\ &= 1.19 \text{ V} \end{aligned}$$

The V_{OH} can be found by using Equation (7.8) and is as follows:

$$\begin{aligned} V_{OH} &= \frac{(1 + 1.73) \times 1.19 + 3.3 + 0.55 - 1.73 \times 0.55}{2} \\ &= 3.07 \text{ V} \end{aligned}$$

To calculate V_{IH}, substituting the k_R into Equation (7.13), we obtain

$$\begin{aligned} V_{IH} &= \frac{2 \times 1.73(3.3 - 0.55 - 0.55)}{(1.73 - 1)\sqrt{1 + 3 \times 1.73}} - \frac{3.3 - 1.73 \times 0.55 - 0.55}{1.73 - 1} \\ &= 1.73 \text{ V} \end{aligned}$$

The V_{OL} can be found by using Equation (7.14) and is

$$\begin{aligned} V_{OL} &= \frac{(1 + 1.73) \times 1.73 - 3.3 + 0.55 - 1.73 \times 0.55}{2 \times 1.73} \\ &= 0.29 \text{ V} \end{aligned}$$

The noise margins NM_L and NM_H are calculated as follows:

$$\begin{aligned} NM_L &= V_{IL} - V_{OL} = 1.19 - 0.40 = 0.79 \text{ V} \\ NM_H &= V_{OH} - V_{IH} = 2.4 - 1.73 = 0.67 \text{ V} \end{aligned}$$

where V_{OL} and V_{OH} for standard TTL logic circuits are 0.4 V and 2.4 V, respectively. ∎

15.2.2.2 Noninverting TTL-to-CMOS Converter A simple noninverting TTL-to-CMOS level-shifting circuit is given in Figure 15.8. This converter consists of two stages. The first stage shifts the voltage level, while the second stage is a normal CMOS inverter. As the input voltage is low, namely, V_{OL}, nMOS transistor M_3 turns on while M_4 is turned off. pMOS transistor M_2 is on, causing the output voltage V_{out} to fall to the ground level. As the input voltage rises higher, that is, V_{OH}, nMOS transistor M_3 is turned off while nMOS transistor M_4 is on, thereby causing the output voltage V_{out} to rise to V_{DD}. As a result, it is a noninverting TTL-to-CMOS conversion circuit.

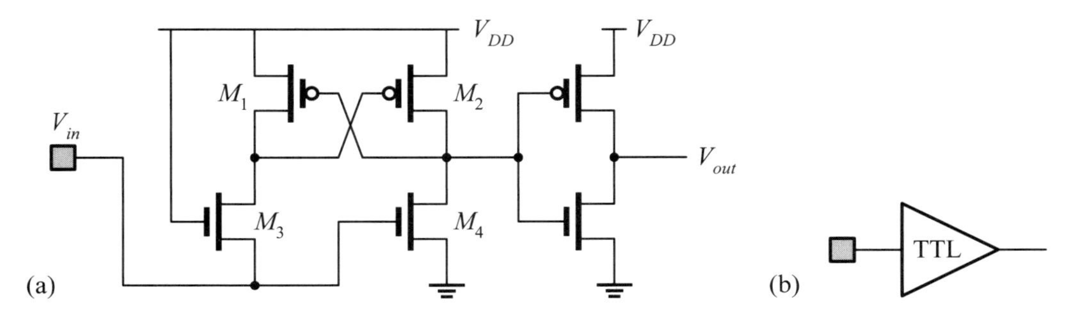

Figure 15.8: A noninverting TTL-to-CMOS conversion circuit: (a) circuit; (b) logic symbol.

15.2.3 Differential Buffers

Differential buffers (also amplifiers) are popular in designing high-speed digital receivers due to their good immunity on common-mode noise. A differential amplifier magnifies the difference of signals between its two inputs and outputs a pair of differential signals at its outputs.

15.2.3.1 nMOS-Input Differential Buffer A basic nMOS-input differential buffer is illustrated in Figure 15.9. The maximum and minimum common-mode voltages allowed by an nMOS-input differential buffer are determined by the condition that all transistors must be operated in their saturation regions. Hence,

$$V_{Tn} + V_{DS5sat} \leq V_{CM} \leq V_{DD} - |V_{Tp}| + V_{Tn} \tag{15.9}$$

where V_{CM} is the input common-mode voltage. As a consequence, the minimum common-mode input voltage of the nMOS-input differential amplifier is at least equal to $V_{Tn} + V_{DS5sat}$. The output voltage swings between $V_{in+} - V_{Tn}$ and $V_{DD} - V_{SD4sat}$. Namely,

$$V_{in+} - V_{Tn} \leq V_{out} \leq V_{DD} - V_{SD4sat} \tag{15.10}$$

because both transistors M_4 and M_2 must remain in their saturation regions.

15.2.3.2 pMOS-Input Differential Buffer Naturally, the input stage of a differential amplifier can also be built on the basis of pMOS transistors. A simple pMOS-input differential buffer is given in Figure 15.10. As in the case of an nMOS-input differential amplifier, all transistors in the amplifier must be operated in their saturation regions. Therefore, the range of common-mode input voltage of this differential amplifier is

$$V_{Tn} - |V_{Tp}| \leq V_{CM} \leq V_{DD} - |V_{Tp}| - V_{SD5sat} \tag{15.11}$$

The output voltage swings between V_{Tn} and $V_{in} + |V_{Tp}|$. That is

$$V_{Tn} \leq V_{out} \leq V_{in+} + |V_{Tp}| \tag{15.12}$$

because both transistors M_4 and M_2 must remain in their saturation regions.

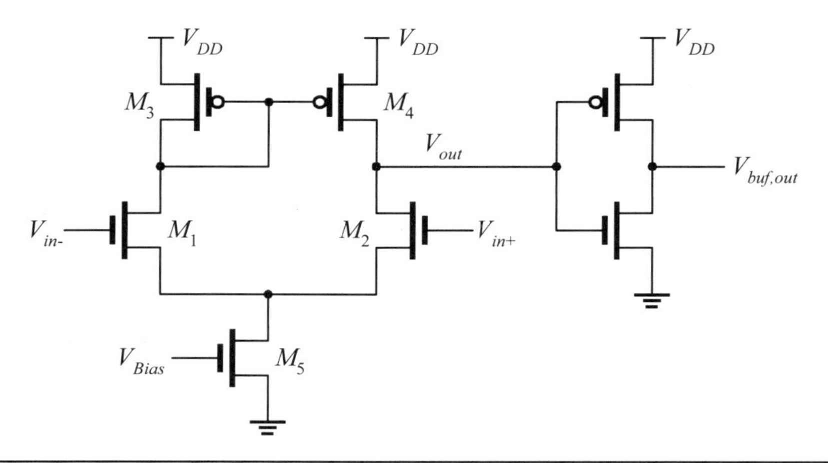

Figure 15.9: An nMOS-input differential buffer.

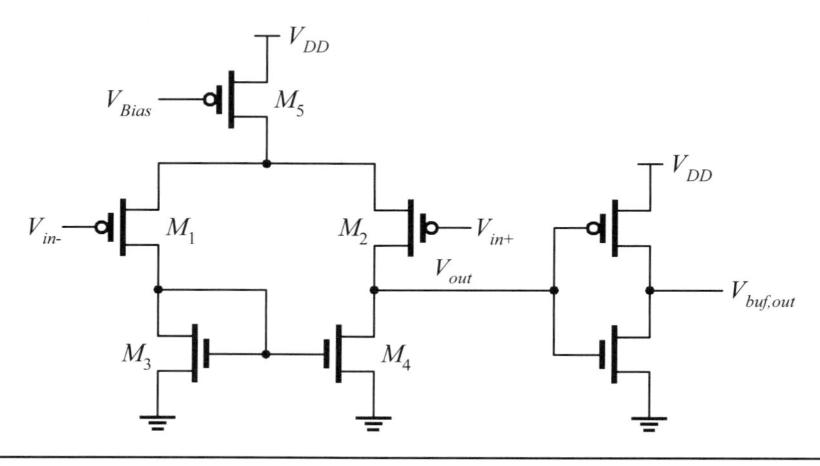

Figure 15.10: A pMOS-input differential buffer.

15.2.3.3 Full-Range Swing Buffer The shortcomings of nMOS-input and pMOS-input buffers are their output voltages cannot be fully ranged. By combining nMOS-input with pMOS-input buffers in parallel, a full-range swing buffer is obtained, as illustrated in Figure 15.11.

■ Review Questions

Q15-5. Describe the philosophy behind the noninverting Schmitt circuit.

Q15-6. Describe the operation of an inverting Schmitt circuit.

Q15-7. Describe the major functions of input buffers.

Q15-8. Explain why the output voltages of nMOS-input and pMOS-input buffers cannot be full-ranged.

Q15-9. Explain the operation of the full-range swing buffer shown in Figure 15.11.

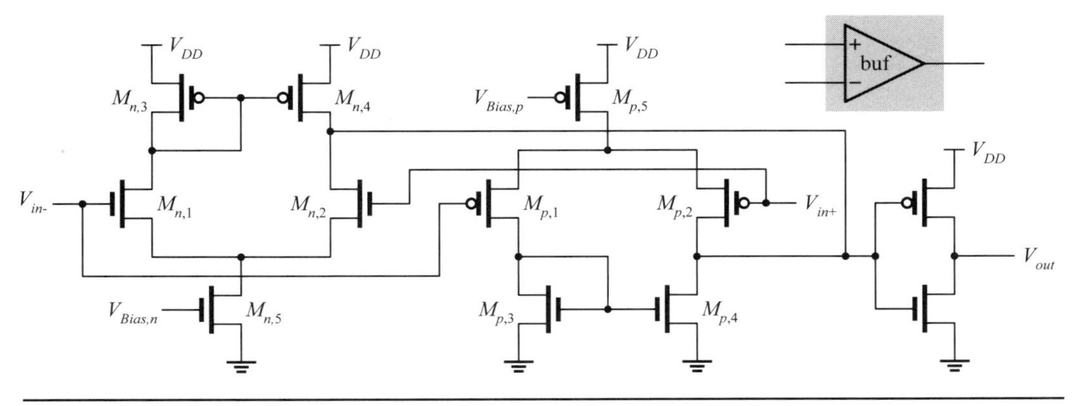

Figure 15.11: A full-range swing buffer.

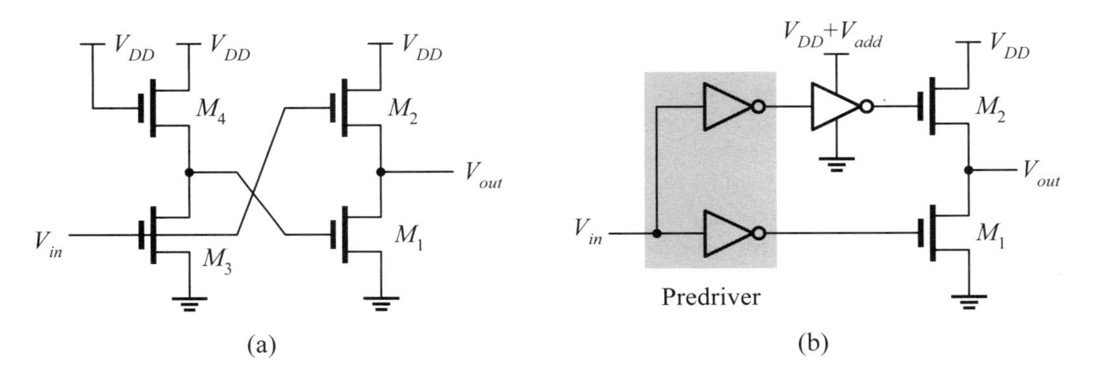

Figure 15.12: Two examples of nMOS-only buffers: (a) circuit; (b) circuit with a voltage booster.

15.3 Output Drivers/Buffers

An output driver or buffer is a big inverter that can be controlled as necessary in terms of the following features: transient or short-circuit current, slew rate, tristate, output resistance, and propagation delay. In what follows, we deal with nMOS-only buffers, tristate buffer designs, bidirectional I/O circuits, driving transmission lines, and SSN problems and reduction.

15.3.1 nMOS-Only Buffers

Because of the susceptibility of the basic CMOS inverter to latch-up, an nMOS-only buffer may be used for the output buffer, as shown in Figure 15.12(a). The nMOS-only buffer works as follows. When V_{in} is low, nMOS transistors M_2 and M_3 are off while nMOS transistors M_1 and M_4 are on. The output voltage V_{out} is pulled down to the ground level through the nMOS transistor M_1. When V_{in} rises high, nMOS transistors M_3 and M_2 are on while the nMOS transistor M_1 turns off. The output V_{out} is pulled up to $V_{DD} - V_{Tn}$ through M_2, assuming the input V_{in} is V_{DD}.

One disadvantage of the above nMOS-only buffer is a direct current (DC) from V_{DD} to the ground node, as transistor M_3 is on since M_4 is always on. Another disadvantage

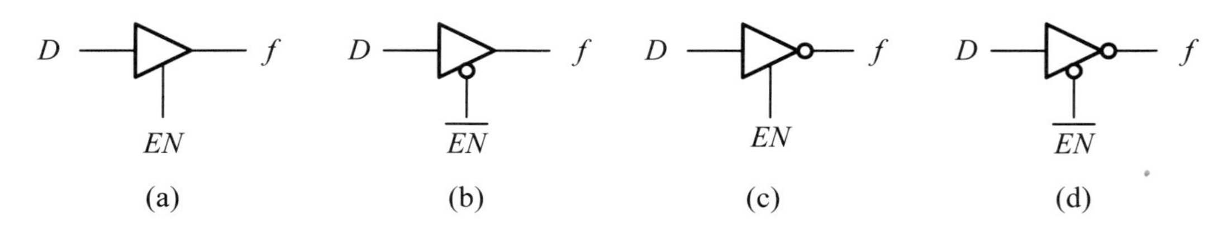

Figure 15.13: The four types of tristate buffers and inverters: (a) active-high enable buffer; (b) active-low enable buffer; (c) active-high enable inverter; (d) active-low enable inverter.

is that its maximum output voltage can only reach $V_{DD} - V_{Tn}$. Fortunately, this reduced output voltage can be improved by using an on-chip voltage booster circuit to provide a voltage greater than V_{DD} for driving the gate of nMOS transistor M_2. Hence, the output voltage may swing from 0 to V_{DD}, like the CMOS output buffer. The resulting nMOS-output buffer with a pumped voltage is depicted in Figure 15.12(b), where the inverter driving nMOS transistor M_2 is powered by a voltage greater than V_{DD}. The disadvantage of this approach is that it needs an extra power supply greater than V_{DD}.

One salient feature of the buffer shown in Figure 15.12(b) is that by adding enable control logic, the output of the buffer can be forced into a high impedance or can be with the same or opposite polarity of the input signal. This feature will be explored in exercises. We use a general paradigm of CMOS tristate buffers as an example to illustrate how to design a desired tristate buffer or inverter.

15.3.2 Tristate Buffer Designs

Tristate (three-state) buffers are often used in I/O circuits to route multiple signals into the same output, namely, serving as a multiplexer. Depending on circuit structures, there exists four types of tristate buffers and inverters: active-high enable buffer, active-low enable buffer, active-high enable inverter, and active-low enable inverter. They are shown in Figure 15.13.

Before presenting the general paradigm of CMOS tristate buffers, consider a simple but widely used tristate buffer shown in Figure 15.14(a). The two common implementations of it are shown in Figures 15.14(b) and (c), respectively. One is based on the cascading of an inverter and a transmission gate (TG) switch, and the other is based on a clocked CMOS (C^2MOS) inverter. Because two transistors are cascaded at the output of either one, they have longer propagation delays compared to standard inverters.

To be more precise, we can consider their logical effort. The logical effort of the combination of inverter and TG switch shown in Figure 15.14(b) can be calculated as follows:

$$g_{inv+TG} = \frac{\tau_{inv+TG}}{\tau_{inv}} = \frac{(2R)3WC_g}{3RWC_g} = 2 \tag{15.13}$$

The logical effort of the C^2MOS inverter shown in Figure 15.14(c) is as follows:

$$g_{C^2MOS} = \frac{\tau_{C^2MOS}}{\tau_{inv}} = \frac{(R)6WC_g}{3RWC_g} = 2 \tag{15.14}$$

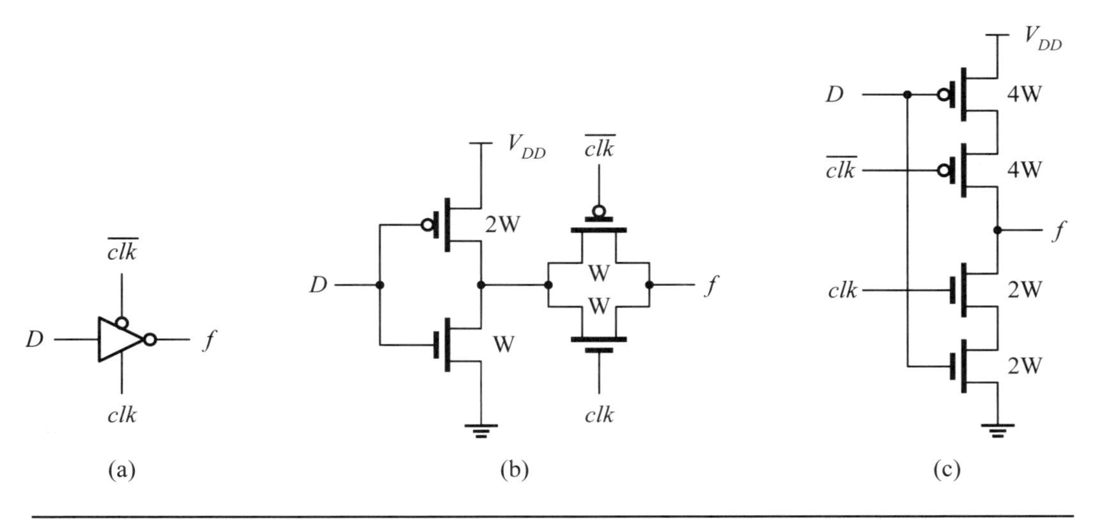

Figure 15.14: The general implementations of tristate buffers: (a) logic symbol; (b) TG-based circuit; (c) C^2MOS circuit.

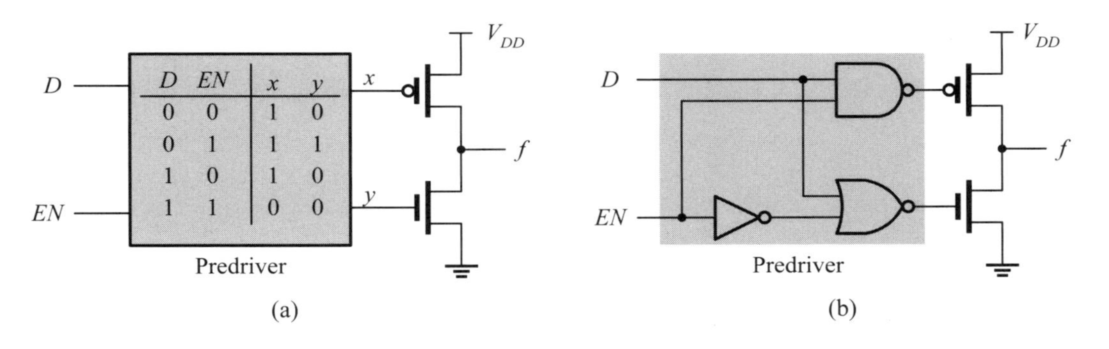

Figure 15.15: The general paradigm of tristate buffers: (a) design principle; (b) the resulting circuit.

Both have the same logical effort but are much larger than that of standard inverters. Therefore, more popular tristate buffers are designed on the basis of standard inverters; namely, they use the standard inverter at the output stage and control the on-off states of both transistors separately.

The general paradigm for designing a tristate buffer is illustrated in Figure 15.15, where pMOS and nMOS transistors are connected as a totem-pole just like an inverter circuit. However, instead of connecting both gates of pMOS and nMOS transistors together to the input, both gates are controlled by an enable logic circuit (namely, predriver), which has two inputs and two outputs. The design procedure is to derive the truth table of the enable logic circuit, simplify it, and then implement the results in CMOS gates. To illustrate this design procedure, an example is given in the following.

■ **Example 15-5: (A tristate buffer design.)** Assume that an active-high enable tristate buffer is desired. Using the paradigm depicted in Figure 15.15(a), design such a buffer.

Solution: As shown in Figure 15.15(a), both gates, x and y, of the pMOS and nMOS transistors composing the output stage are controlled by the inputs, EN and D. From the specifications, the circuit functions as a buffer if the enable input EN is high and its output is in a high impedance otherwise. As a result, the truth table is obtained as displayed in Figure 15.15(a). Simplifying it, we have

$$x = \overline{EN} + \bar{D} = \overline{EN \cdot D}$$

and

$$y = \bar{D} \cdot EN = \overline{\overline{EN} + D}$$

The resulting logic circuit is shown in Figure 15.15(b).

■

The other three types of tristate buffer and inverter circuits can be designed in a similar way. Hence, we omit them here and leave them as exercises to the reader.

15.3.3 Bidirectional I/O Circuits

Bidirectional input/output (I/O) circuits are popular in digital systems because of the significant reduction of I/O ports (that is, pads) needed in a chip. As an output buffer and an input buffer connect to the same port (or pad/pin), the resulting I/O circuit and port are called a *bidirectional I/O circuit* and *port*, respectively. To avoid the interference of the output signal to the input signal, the output buffer should place its output at a high impedance when the I/O port functions as an input port. As a consequence, the output buffer should be a tristate circuit.

■ **Example 15-6: (A bidirectional I/O circuit.)** Figure 15.16(a) shows a bidirectional I/O circuit, which combines the noninverting TTL-to-CMOS conversion circuit depicted in Figure 15.8 with the active-high enable tristate buffer given in Figure 15.15(b). In the circuit, we also show an ESD protection network (ESD PN), which is necessary for the input circuit and will be addressed later in a dedicated subsection briefly. The logic symbol is given in Figure 15.16(b).

■

15.3.4 Driving Transmission Lines

In many applications, an output circuit often needs to drive a transmission line. As stated, a wire may behave like a transmission line if its length is in a specific range as described in Section 13.1.3. In addition, as shown in Figure 15.17, the output buffer usually has an impedance other than the characteristic impedance (Z_0) of the transmission line. Hence, some special circuit must be designed to solve this problem.

Because of the difficulty of matching the output impedance of an output driver with the Z_0 of a transmission line, an output driver with tunable or selectable impedance is often used. One way to design such a circuit is to combine a number of selectable segmented drivers in a parallel-connected fashion, thus yielding a tunable or selectable impedance. The following example illustrates how this idea is realized in practice.

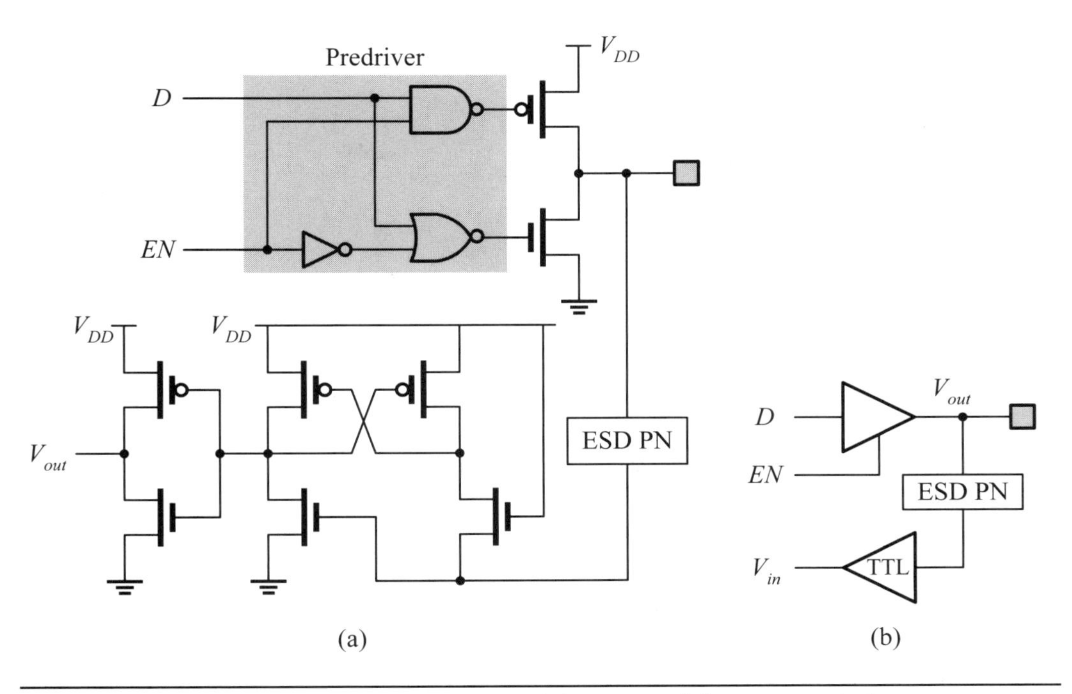

Figure 15.16: A bidirectional I/O circuit: (a) circuit; (b) logic symbol.

■ Example 15-7: (A selectable segmented output driver.) Figure 15.18 shows an example to adaptively control the output impedance by a number of output drivers connected in parallel. Each of the output drivers has a different impedance by shaping its channel width a factor S_i. For example, S_i may be set to 2^i, where i is a positive integer or zero. Each output driver S_i is switched in and out by control line C_i to make the output impedance of a selectable segmented output driver match the Z_0 of the transmission line as closely as possible. Of course, it is not possible to exactly match the output impedance of an output driver with the Z_0 of a transmission line. ■

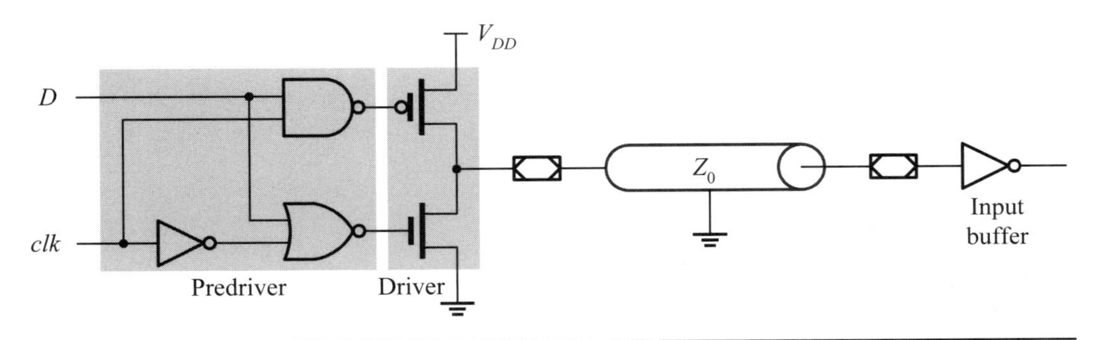

Figure 15.17: A tristate buffer used to drive a transmission line.

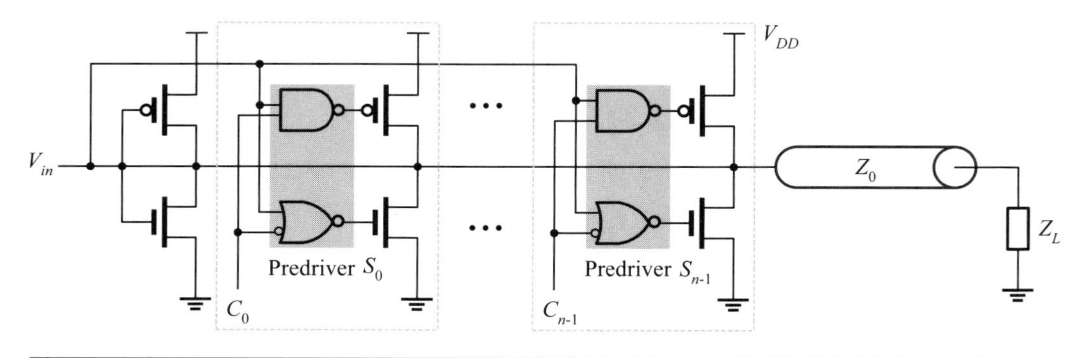

Figure 15.18: An example of selectable segmented output driver.

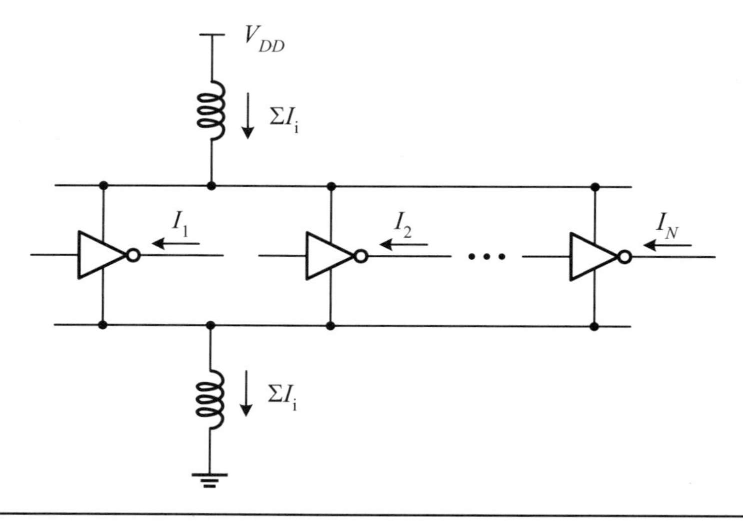

Figure 15.19: The basic circuit used to illustrate the formation of SSN.

15.3.5 Simultaneous Switching Noise

Recall that SSN or Δi noise, is an inductive noise caused by several outputs switching at the same time. An illustration of SSN is shown in Figure 15.19, where N inverters work in parallel.

Recall that the induced voltage v_L due to the changing current of a wire with a self-inductance L is equal to

$$v_L = L\frac{di}{dt}$$

Assuming that each inverter has an amount of current I_i passing through it, the total induced voltage is then equal to

$$v_{SSN} = NL_{total}\frac{dI_i}{dt} \tag{15.15}$$

where v_{SSN} is the SSN voltage, N is the number of drivers that are switching, L_{total} is the equivalent inductance in which the total current must pass, and I_i is the current per driver. Since the current must pass through an inductance L_{total}, the noise of the v_{SSN} value will be introduced onto the power supply, which in turn manifests itself at

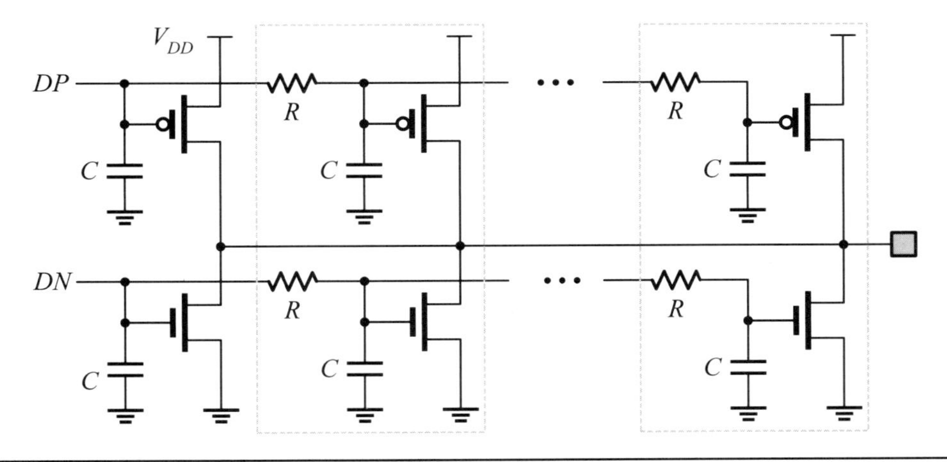

Figure 15.20: The use of a predriver to reduce SSN.

the driver output. This may result in the malfunction or cause a timing error in the logic circuits. Therefore, in practical circuit design, we have to reduce the amount of SSN as much as possible.

15.3.5.1 Designs for SSN Reduction Because the amount of SSN is determined by the inductance and the changing rate of total current in the output buffers, to reduce the SSN the inductance and/or changing rate or total current needs to be reduced. Based on this idea, many options are possible. The reduction of package inductance by using more advanced packaging technology is the most straightforward method. The use of a low-weight coding scheme in the output buffers to prevent all bits from toggling simultaneously is another approach. We are concerned with two circuit design options: *slew-rate control* and *differential-signaling scheme*.

Slew-rate control. In order to provide a large current to a heavy capacitive load, an output buffer is often designed with a number of inverter stages connected in parallel. Because of this, an approach to reducing SSN is to skew the turn-on times of these inverter stages to prevent all inverters from being switched on at the same time. One intuitive way to achieve this is to make an RC-delay line using the C inherent in the inverter and the R formed by a polysilicon or diffusion layer or a pass transistor. An example is illustrated in Figure 15.20, where n inverters are connected together to drive an output pad. The second and later stages from left to right are turned on by an RC delay relative to its preceding stage. Hence, the switching current rate di/dt and hence the Ldi/dt noise can be significantly reduced if the RC values are properly set.

In addition to controlling the resistance, it is possible to control the slew rate of the output buffer with the loading capacitance C on the predriver. An example is shown in Figure 15.21, where capacitors $C_{i(p)}$ and $C_{i(n)}$, are connected to the output nodes, DP and DN, of the predriver, under the control of digital control bits $\bar{c}_i$ andc_i. Both capacitors $C_{i(p)}$ and $C_{i(n)}$ are separately connected to DP and DN, if the digital control bits c_i are 1 (hence, $\bar{c}_i$ are 0), and they are bypassed otherwise.

Differential-signal scheme. The SSN is severe and acute when all buffers are simultaneously switched from one state to another. This generates the maximum Ldi/dt noise. To reduce the SSN for such a situation, a differential-signaling scheme can be used. As illustrated in Figure 15.22, every bit has its complement transmitted

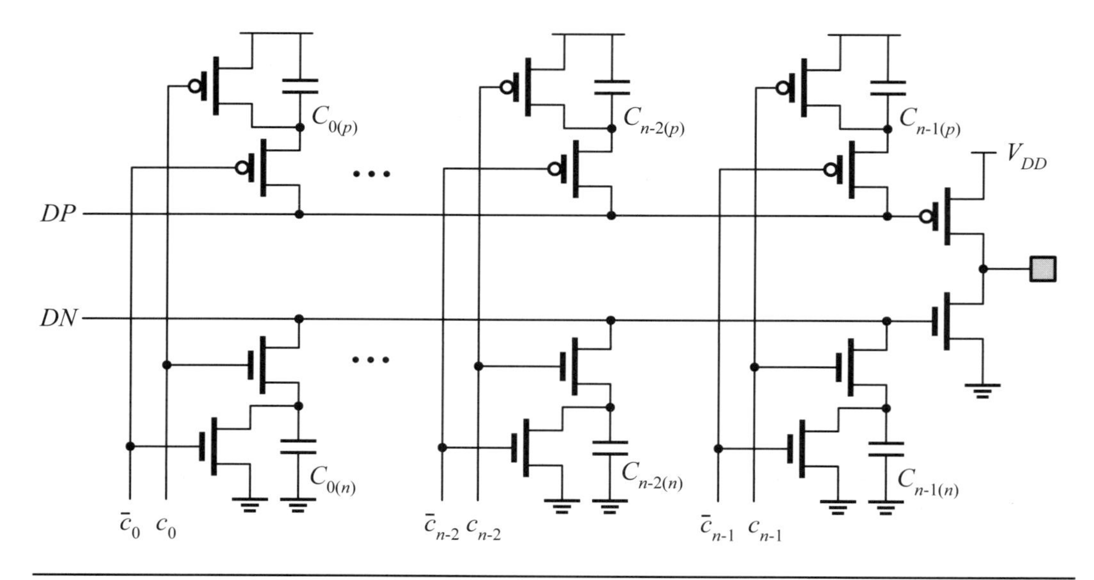

Figure 15.21: A simple approach used to adjust the slew rate of an output buffer.

besides its true signal. Hence, the total switching current is zero in theory. In addition, if both true and complementary signals are closely routed, crosstalk and ground bounce effects are virtually reduced or even eliminated. Differential-signaling schemes are widely used in modern high-speed serial input/output buses, such as low-voltage differential signal (LVDS) and *serial advanced technology attachment* (*serial ATA* or SATA for short). LVDS provides a data rate up to 655 Mb/s over twisted-pair copper wire and the maximum data rate of SATA can reach 6 Gb/s.

■ Review Questions

Q15-10. What are the features of output buffers?

Q15-11. What are the four types of tristate buffers?

Q15-12. Describe the advantages of nMOS-only buffers.

Q15-13. Show the differences between single-ended and differential signaling schemes in terms of SSO noise.

Q15-14. What are the features of LVDS?

Q15-15. What are the objectives of slew-rate control?

15.4 Electrostatic Discharge Protection Networks

The basic principle of electrostatic discharge (ESD) protection is to create current paths for discharging the static charge caused by ESD events. An ESD event means a transient discharge of the static charge arising from human handling or contact with machines. The potentially destructive nature of ESD becomes apparent as the feature sizes are shrunk smaller and smaller. The ESD stress not only can destroy I/O peripheral devices but also can damage weak internal core circuit devices. In this section, we briefly introduce the ESD problem and the basic design issues about ESD protection networks. The effective ESD protection networks are strongly process-related and always an active research area since the onset of integrated circuits. The interested

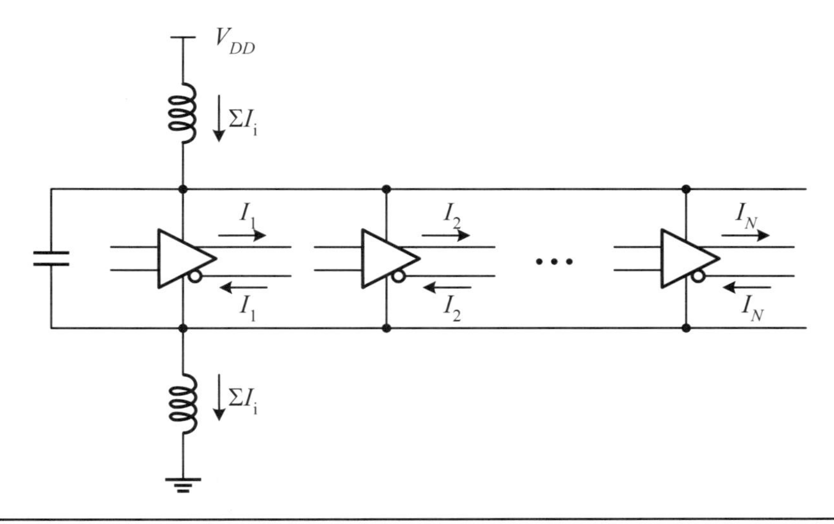

Figure 15.22: An illustration of a differential-signal scheme.

reader should consult related textbooks, such as Amerasekera and Duvvury [1] and Dabral and Maloney [4], or papers for more details.

15.4.1 ESD Models and Design Issues

Electrical overstress (EOS) refers to the events that encompass time scales in the microsecond and millisecond ranges and includes lighting and *electromagnetic pulses* (EMP). EOS events can occur due to electrical transients at the board level of systems, during the burn-in test or device characterized by product engineers. ESD is a subset of EOS and refers to the events encompassing time scale in 100 ns. ESD events are characterized by small distances (less than a centimeter) and low voltages (about a few kV). ESD damage can occur during wafer processing, dicing, packaging, testing, board assembly, transportation, and in the field.

Some typical ESD-related failures are as follows:

- *Junction breakdown*: Junctions may melt and cross diffuse or lattice damage may occur.
- *Oxide breakdown*: Oxides may have void formation, vaporization, and filament formation, leading to shorts or opens and causing transistors to fail.
- *Metal/via damage*: Metals and contacts may be melted and vaporized, leading to shorts or opens and causing circuits to malfunction.

15.4.1.1 ESD Models Three ESD models associated with integrated circuits (ICs) are *human body model* (HBM), *charge-device model* (CDM), and *machine model* (MM). The equivalent circuits for these models are illustrated in Figure 15.23. A high-voltage pulse source (HVPS) is applied to all circuits.

The human body model (HBM) represents the situation when a charged human discharges into an IC. It was standardized by the ESD Association standard in 1996, called HBM96. The circuit components to simulate a charged people are a 100-pF capacitor in series with a 1.5-Ω resistor, as illustrated in Figure 15.23(a). For commercial IC products, the ESD level is generally required to be higher than 2 kV in human-body-model (HBM) ESD stress.

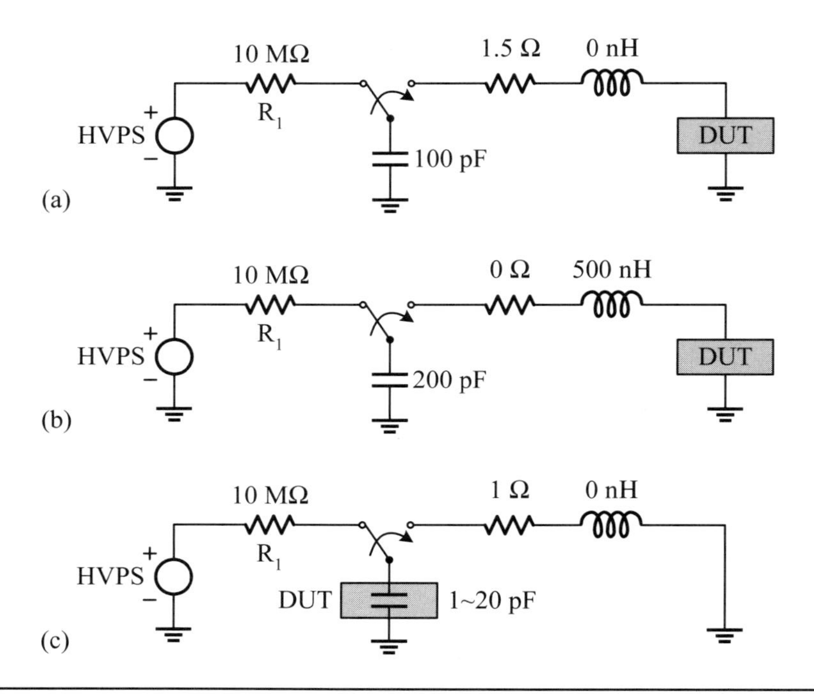

Figure 15.23: The equivalent circuits of ESD models: (a) HBM; (b) MM; (c) CDM.

The machine model (MM) represents the charging phenomenon due to machine handling. It was standardized by the ESD Association standard in 1998, denoted MM98. The circuit component is a 200-pF capacitor with no resistive component. However, in a real ESD tester, the resistance is always greater than zero. Its equivalent circuit is given in Figure 15.23(b).

The charge-device model (CDM) denotes the case during ICs self-charging and then self-discharging. It was standardized by the ESD Association standard in 1999, called CDM99. CDM discharge occurs at less than 5 ns where typically the rise time of the event is of the order of 250 ps. The equivalent circuit is exhibited in Figure 15.23(c).

15.4.2 General ESD Protection Network

A good ESD protection design should be capable of surviving various ESD events and protect the internal circuits connected to the IC pins from being damaged. As stated, the basic principle of ESD protection is to create current paths to power-supply rails for discharging the static charge caused by ESD events. Based on this, the general ESD protection network for typical CMOS ICs can be modeled as Figure 15.24. It comprises three parts: input ESD protection network, output ESD protection network, and power-rail ESD clamp network. Both input and output ESD protection networks can be clamped or shunted to V_{SS}, V_{DD}, or both. The diode from V_{SS} to V_{DD} represents the p-substrate to n-well diode inherent in any CMOS process. The power-supply clamp is used to limit the supply voltage in a safe range that the internal circuits can tolerate.

Depending on the strategy and circuitry used, the input ESD protection networks can be further subdivided into primary and secondary stages and a resistor R_s is placed between them. The secondary stage is often a scaled-down version of the primary stage.

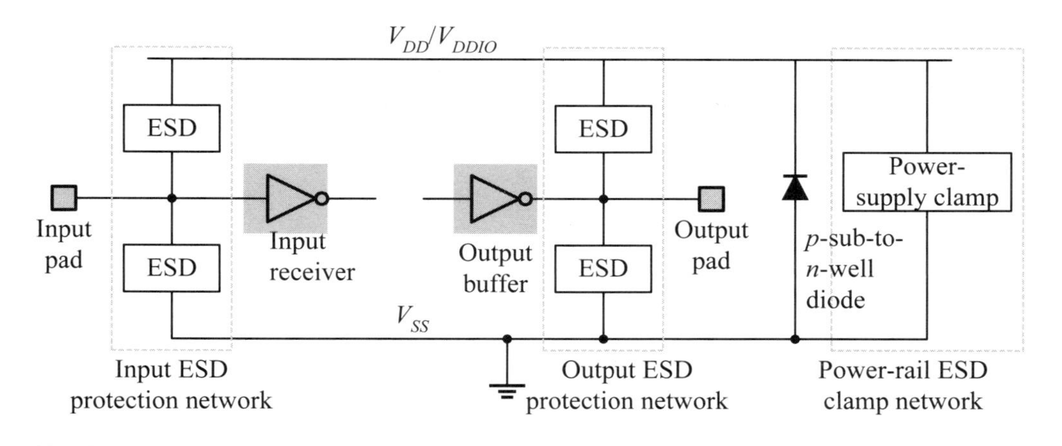

Figure 15.24: The general ESD protection network for typical CMOS processes.

The function of resistor R_s is to limit the peak current flowing into the secondary stage during ESD events. The resistor R_s is usually formed by the polysilicon (Poly) layer.

The devices used to create current paths for an ESD protection network can be either nonbreakdown (NBD) oriented or breakdown (BD) oriented. The basic non-breakdown component is a diode while the basic breakdown components include gate-grounded nMOS (GGnMOS)/gate-V_{DD} pMOS (GGpMOS) and silicon-controlled rectifier (SCR) devices (MVTSCR and LVTSCR[1]) among others. In the rest of this section, we will briefly address the features of these components along with their related ESD circuits.

15.4.2.1 Quality Metrics of ESD Protection Networks

To protect ICs from being damaged by ESD events, a variety of protection networks are proposed during the last decades. To evaluate the quality of an ESD protection network, the following four major metrics are used: *robustness*, *effectiveness*, *speed*, and *transparency*.

Robustness describes the ability of an ESD protection network to handle ESD events by itself. Usually, it is proportional to the size of ESD devices. Effectiveness describes the ability of the ESD protection network to limit the voltage to a safe level to prevent the protected circuits from being damaged by ESD events. Speed means that a robust and effective ESD protection network must activate fast enough to clamp the ESD event at a safe level. Transparency means that the ESD protection network must not interfere with the normal operations of I/O and internal circuits.

■ Review Questions

Q15-16. What is the distinction between EOS and ESD?

Q15-17. Describe the basic principle of ESD protection.

Q15-18. Describe the four quality metrics of an ESD protection network.

Q15-19. Describe the basic components of the general ESD protection network.

[1] The MVTSCR and LVTSCR mean the medium-voltage triggered SCR and low-voltage triggered SCR, respectively.

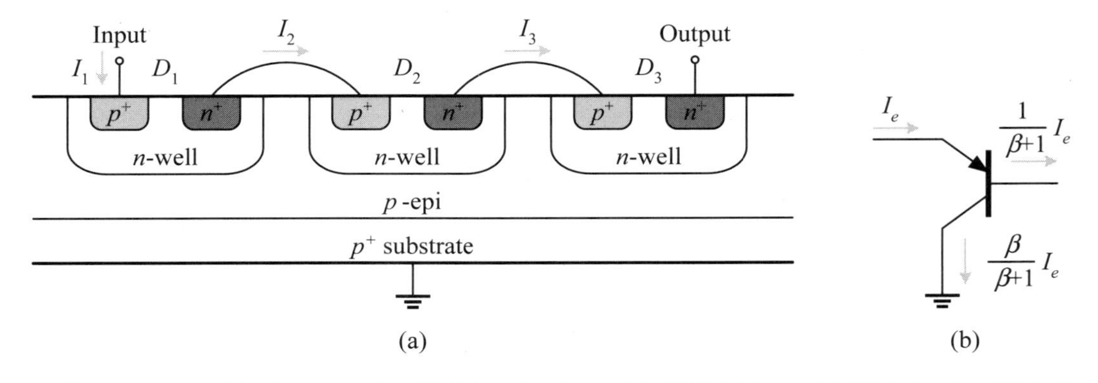

Figure 15.25: The (a) general structure and (b) equivalent circuit of a diode chain used in the ESD protection network.

15.4.3 ESD Protection Networks

In this subsection, we are only concerned with a few typical ESD protection devices and their related ESD networks. More about ESD devices and networks can be found in Amerasekera and Duvvury [1], Dabral and Maloney [4], or related research and technical papers.

15.4.3.1 Diode-Based ESD Protection Networks Diodes are the simplest and earliest semiconductor devices. In most n-well CMOS processes, a diode is indeed a *pnp* transistor with its collector grounded. If ideal diodes are cascaded, the turn-on voltage V_{on} of a diode chain will linearly increase with the number of diodes. However, for a diode chain constructed using *pnp* transistors, the turn-on voltage is no longer linearly increased with the number of diodes. To see this, referring to Figure 15.25, we know that the current I_2 for the second diode is only a small portion of that for the first diode in the chain because the second diode is connected to the base of the first diode.

From the ideal-diode equation described in Chapter 2, we have

$$\ln \frac{I_1}{I_s} = \frac{eV_{EB1}}{nkT} \tag{15.16}$$

and the current in diode 2 is

$$\ln \frac{I_2}{I_s} = \frac{eV_{EB2}}{nkT} = \ln \frac{I_1}{(\beta+1)I_s} = \ln \frac{I_1}{I_s} - \ln(\beta+1) \tag{15.17}$$

As a consequence, the voltage across diode 2 can be expressed as

$$V_{EB2} = V_{EB1} - \frac{nkT}{e}\ln(\beta+1) = V_{EB1} - \ln(10)\frac{nkT}{e}\log(\beta+1) \tag{15.18}$$

If all m diodes in the chain are of the same size, then the total turn-on voltage is equal to

$$V_{total} = mV_{EB1} - \frac{1}{2}m(m-1)V_0\log(\beta+1) \tag{15.19}$$

where $V_0 = \ln(10)nkT/e = 60$ mV for an ideal diode at room temperature.

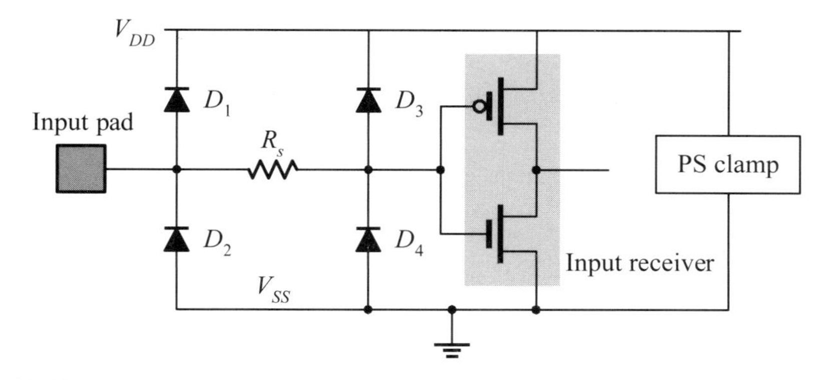

Figure 15.26: The structure of a four-diode-based input protection network.

Example 15-8: (The turn-on voltage of a diode chain.) Referring to Figure 15.25, suppose that there are 5 diodes connected in series. Calculate the total turn-on voltage of the diode chain if the current gain β of *pnp* transistors is 20 and $V_{EB1} = 0.7$ V.

Solution: Using Equation (15.19), we obtain

$$\begin{aligned} V_{total} &= mV_{EB1} - \frac{1}{2}m(m-1)V_0 \log(\beta + 1) \\ &= 5 \times 700 - \frac{1}{2}5(5-1)60 \log(20+1) \ \ (\text{mV}) \\ &= 2.7 \text{ V} \end{aligned}$$

which is less than 3.5 V for five ideal diodes connected in series.

In CMOS technology, n^+ and p^+ source/drain diffusions of nMOS and pMOS devices accompanied with the substrate and n-well provide diodes that can be used to clamp the input node and protect the internal devices. A widely used four-diode-based input protection network is given in Figure 15.26. Clamp diodes D_2 and D_4 are formed by the n^+ diffusions of an nMOS transistor, whereas clamp diodes D_1 and D_3 are formed by the p^+ diffusions of a pMOS transistor within a n-well. The diode D_2 turns on when the input voltage goes 0.7 V below the ground level. The diode D_1 turns on when the input voltage goes 0.7 V above V_{DD}. Resistor R_s is used to limit the peak current flowing into diodes D_3 and D_4 in ESD events. Diodes D_3 and D_4 are used to filter out the smaller energy surge in the input that is beyond the sensitivity of D_1 and D_2.

15.4.3.2 GGnMOS and GDpMOS Transistors The thin-oxide nMOS or pMOS transistor when used in ESD protection network with gate grounded (GGnMOS) or gate connected to V_{DD} (GDpMOS) is usually called a *field-plated diode* (FPD) or *gated diode* due to the effect of the gate on the diode breakdown voltage (avalanche breakdown). The structure of a GGnMOS transistor is shown in Figure 15.27(a), where a parasitic *npn* transistor is displayed explicitly in lighter lines. Figure 15.27(b) gives an example of using the GGnMOS transistor and GDpMOS transistor as an input ESD protection network. To sustain the desired ESD level, both GGnMOS and

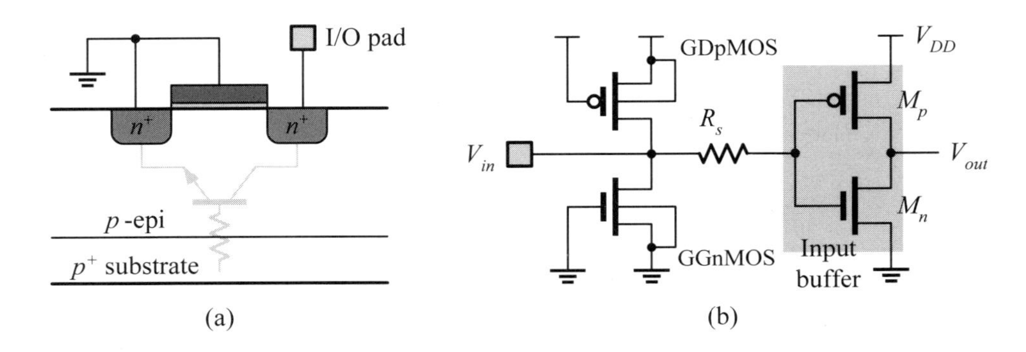

Figure 15.27: GGnMOS device: (a) structure; (b) application to ESD circuit.

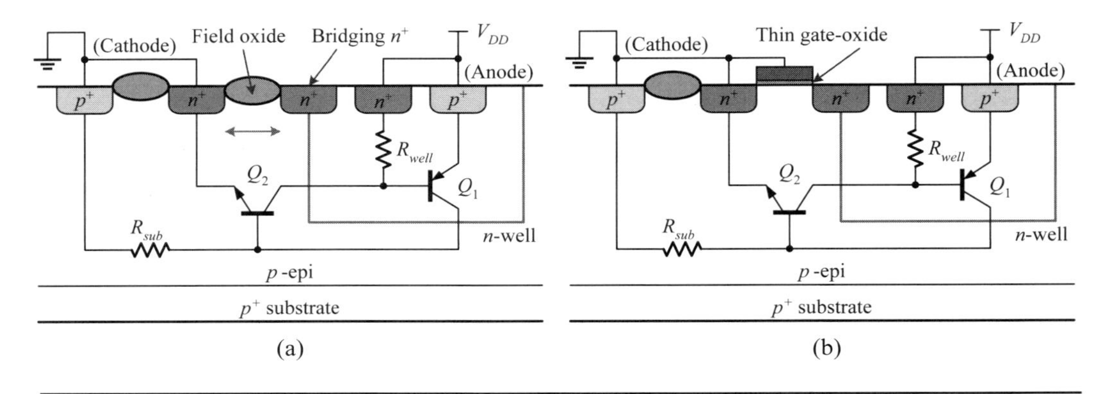

Figure 15.28: The structures of (a) MVTSCR and (b) LVTSCR devices.

GDpMOS transistors should be designed with a large device dimension and a wider drain-contact-to-polygate layout spacing.

15.4.3.3 SCR Devices As stated before, the parasitic *pnpn* device, also called the *silicon controlled rectifier* (SCR), plays an essential role in CMOS latch-up, and we often do our best to avoid its conduction. Nevertheless, when designing an ESD protection circuit, we intentionally build an SCR device to fit some requirements. An SCR device consists of a *pnp* transistor formed by a p^+, the n-well of the pMOS, and p-epi substrate, and an *npn* transistor formed by an n^+ source of the nMOS, the p-epi substrate, and the n-well of the pMOS, as shown in Figure 15.28.

The normal SCR device has a high trigger voltage, leading to poor ESD performance. To improve the ESD performance, we have to lower the trigger voltage of the SCR device. Recall that the breakdown voltage of a *pn* junction is inversely proportional to the doping concentration. Hence, by adding a bridging n^+ area between the n-well and p-epi substrate, as shown in Figure 15.28(a), the trigger voltage of the SCR device can be effectively reduced because now the junction formed by the n^+ and p-epi substrate has a lower breakdown voltage than the junction formed by the n-well and p-epi substrate. The resulting device is called a *medium-voltage triggered SCR* (MVTSCR). It is shown that for typical 0.35-μm CMOS processes, the trigger voltage is about 10 V.

To further reduce the trigger voltage of the SCR device, the field oxide between bridging n^+ and cathode n^+ regions in the MVTSCR device may be replaced by

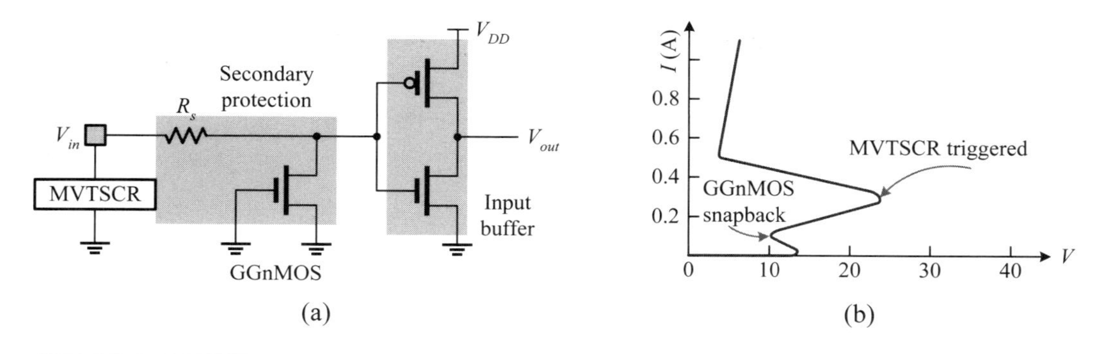

Figure 15.29: An MVTSCR-based input protection network: (a) circuit; (b) SCR characteristics.

thin gate oxide and the nMOS transistor formed is then connected in a grounded-gate fashion, as shown in Figure 15.28(b). This results in a *low-voltage triggered SCR* (LVTSCR) device and it has been shown that a trigger voltage of about 8 V for typical 0.35-μm CMOS processes can be obtained.

Figure 15.29 depicts an application of the MVTSCR device to an input ESD protection network. The MVTSCR device serves as the primary protection and a polysilicon resistor R_s combined with a GGnMOS transistor composes the secondary protection of the input buffer. After GGnMOS transistor snapback, the IR drop across the resistor R_s triggers the MVTSCR. As long as the SCR trigger occurs below the failure current, the protection scheme would be effective.

Review Questions

Q15-20. Show that Equation (15.19) is valid.

Q15-21. Explain the principle of the four-diode-based input protection network.

Q15-22. What are LVTSCR and MVTSCR?

Q15-23. What is GGnMOS transistor?

15.5 Summary

The input/output (I/O) module plays important roles for communicating with the outside of a chip or a system. I/O buffers include input buffers and output buffers. Input buffers are circuits on a chip that take the input signals with imperfection, such as slow rise and fall times, and convert them into clean output signals for use on the chip. For these purposes, the following circuits, including inverting and noninverting Schmitt circuits, and level-shifting circuits, as well as differential buffers, are often used along with input buffers. An output driver or buffer is a big inverter that can be controlled as necessary in terms of the following features: transient or short-circuit current, slew rate, tristate, output resistance, and propagation delay. Design issues related to output buffers are nMOS-only buffers, tristate buffer designs, bidirectional I/O circuits, driving transmission lines, and SSN problems and reduction.

An ESD event means a transient discharge of the static charge arising from human handling or contact with machines. The ESD stress not only can destroy I/O peripheral devices but also can damage weak internal core circuit devices. Hence, proper electrostatic discharge (ESD) protection networks are necessary to create current paths for discharging the static charge caused by ESD events. In this chapter, the ESD problem and basic design issues about ESD protection networks were considered thoroughly.

References

1. A. Amerasekera and C. Duvvury, *ESD in Silicon Integrated Circuits.* New York: John Wiley & Sons, 2002.
2. R. J. Baker, *CMOS Circuit Design, Layout, and Simulation.* New York: John Wiley & Sons, 2005.
3. A. Chatterjee and T. Polgreen, "A low-voltage triggering SCR for on-chip protection at output and input pads," *Electronic Device Letters,* Vol. 12, pp. 21–22, 1991.
4. S. Dabral and T. J. Maloney, *Basic ESD and I/O Design.* New York: John Wiley & Sons, 1998.
5. M. J. Declercq, M. Schubert, and F. Clement, "5V-to-75V CMOS output interface circuits," *IEEE Int'l Solid-State Circuits Conf.* (ISSCC 1993), pp. 162–163, 1993.
6. T. J. Gabara, et al., "Forming damped *LRC* parasitic circuits in simultaneously switched CMOS output buffers," *IEEE J. of Solid-State Circuits,* Vol. 32, No. 7, pp. 407–418, March 1997.
7. H. I. Hanafi et al., "Design and characterization of a CMOS off-chip driver/receiver with reduced power-supply disturbance," *IEEE J. of Solid-State Circuits,* Vol. 27, No. 5, pp. 783–791, May 1992.
8. D. A. Hodges, H. G. Jackson, and R. A. Saleh, *Analysis and Design of Digital Integrated Circuits: In Deep Submicron Technology,* 3rd ed. New York: McGraw-Hill Books, 2004.
9. M. D. Ker, C. Y. Wu, T. Cheng, and H. H. Chang, "Capacitor-coupled ESD protection circuit for deep-submicron low-voltage CMOS ASIC," *IEEE Trans. on VLSI Systems,* Vol. 4, No. 9, pp. 307–321, September 1996.
10. M. D. Ker and S. L. Chen, "Design of mixed-voltage I/O buffer by using nMOS-blocking technique," *IEEE J. of Solid-State Circuits,* Vol. 41, No. 10, pp. 2324–2333, October 2006.
11. M. D. Ker and C. C. Yen, "Investigation and design of on-chip power-rail ESD clamp circuits without suffering latchup-like failure during system-level ESD test," *IEEE J. of Solid-State Circuits,* Vol. 43, No. 11, pp. 2533–2545, November 2008.
12. J. M. Rabaey, A. Chandrakasan, and B. Nikolic, *Digital Integrated Circuits: A Design Perspective,* 2nd ed. Upper Saddle River, NJ: Prentice-Hall, 2003.
13. C. N. Wu and M. D. Ker, "ESD protection for output pad with well-coupled field-oxide device in 0.5-μm CMOS technology," *IEEE Trans. on Electronic Devices,* Vol. 44, No. 3, pp. 503–505, March 1997.

Problems

15-1. Assume that the desired threshold voltages of an inverting Schmitt circuit are $V_{TH} = 0.8$ V and $V_{TL} = 0.4$ V, respectively.

(a) Using 0.13-μm parameters and $V_{DD} = 1.2$ V, design the inverting Schmitt circuit.

(b) Using SPICE, verify the functionality of the inverting Schmitt circuit.

(c) Adjust the sizes of transistors so that the resulting Schmitt circuit can meet the desired specification.

15-2. Referring to Figure 15.12(b), answer the following questions:

(a) Plot the general paradigm of an nMOS-only buffer.

(b) Design an active-high enable inverter.

(c) Design a SPICE experiment to measure the rise and fall times of this tristate circuit.

(d) Design a SPICE experiment to measure the turn-on and turn-off times of this tristate circuit.

15-3. This problem describes the design of an nMOS-only active-low enable inverter.

(a) Design an active-low enable inverter.

(b) Design a SPICE experiment to measure the rise and fall times of this tristate circuit.

(c) Design a SPICE experiment to measure the turn-on and turn-off times of this tristate circuit.

15-4. This problem involves the design of an nMOS-only active-high enable buffer.

(a) Design an active-high enable buffer.

(b) Design a SPICE experiment to measure the rise and fall times of this tristate circuit.

(c) Design a SPICE experiment to measure the turn-on and turn-off times of this tristate circuit.

15-5. This problem involves the design of an nMOS-only active-low enable buffer.

(a) Design an active-low enable buffer.

(b) Design a SPICE experiment to measure the rise and fall times of this tristate circuit.

(c) Design a SPICE experiment to measure the turn-on and turn-off times of this tristate circuit.

15-6. This is a problem involving the design of an active-low enable buffer using the general paradigm shown in Figure 15.15.

(a) Design an active-low enable buffer.

(b) Design a SPICE experiment to measure the rise and fall times of this tristate circuit.

(c) Design a SPICE experiment to measure the turn-on and turn-off times of this tristate circuit.

15-7. This is a problem involving the design of an active-high enable inverter using the general paradigm shown in Figure 15.15.

(a) Design an active-high enable inverter.

(b) Design a SPICE experiment to measure the rise and fall times of this tristate circuit.

(c) Design a SPICE experiment to measure the turn-on and turn-off times of this tristate circuit.

15-8. This is a problem involving the design of an active-low enable inverter using the general paradigm shown in Figure 15.15.

(a) Design an active-low enable inverter.

(b) Design a SPICE experiment to measure the rise and fall times of this tristate circuit.

(c) Design a SPICE experiment to measure the turn-on and turn-off times of this tristate circuit.

15-9. Referring to Figure 15.25, calculate the total turn-on voltage of the diode chain under each of the following specifications:

(a) There are 6 diodes connected in series, supposing that the current gain β of *pnp* transistors is 25 and $V_{EB1} = 0.7$ V.

(b) There are 10 diodes connected in series, supposing that the current gain β of *pnp* transistors is 10 and $V_{EB1} = 0.6$ V.

16

Testing, Verification, and Testable Designs

Testing is an essential step in any very-large-scale integration (VLSI) system, including cell-based, field-programmable-gate-array-based (FPGA-based), and full-custom implementations, even in a simple digital logic circuit because the only way to ensure that a system or a circuit may function properly is through a careful testing process. The objective of testing is to find any existing faults in a system or a circuit.

The test of any faults in a system or a circuit must be based on some predefined fault models. The most common fault models found in complementary metal-oxide-semiconductor (CMOS) technology include stuck-at faults, bridge faults, and stuck-open faults. Based on a specific fault model, the system or circuit can then be tested by applying stimuli, observing the responses, and analyzing the results. The input stimulus special for use in a given system or a circuit is often called a test vector, which is an input combination. A collection of test vectors is called a test set. Except in an exhaustive test, a test set is usually a proper subset of all possible input combinations. A test vector generation for combinational logic circuits based on path sensitization is described concisely.

Nowadays, the most effective way to test sequential circuits is by adding some extra circuits to the circuit under test to increase both controllability and observability. Such a design is known as a testable circuit design or as a design for testability. The widely used approaches include the ad hoc approach, scan-path method, and built-in self-test (BIST). The scan-path method is also extended to the system-level testing, such as static random access memory (SRAM), core-based system, and system-on-a-chip (SoC).

16.1 An Overview of VLSI Testing

Generally speaking, VLSI testing include three types: *verification testing*, *manufacturing testing*, and *acceptance testing*. The verification testing verifies the correctness of the design and correctness of the test procedure. Manufacturing testing is a factory testing by which all manufactured chips for parametric and logic faults, and analog specifications are verified to see whether they meet the desired specifications. Acceptance testing, also called *incoming inspection*, is carried out by the user (customer) to test purchased parts to ensure quality before they are integrated into the target

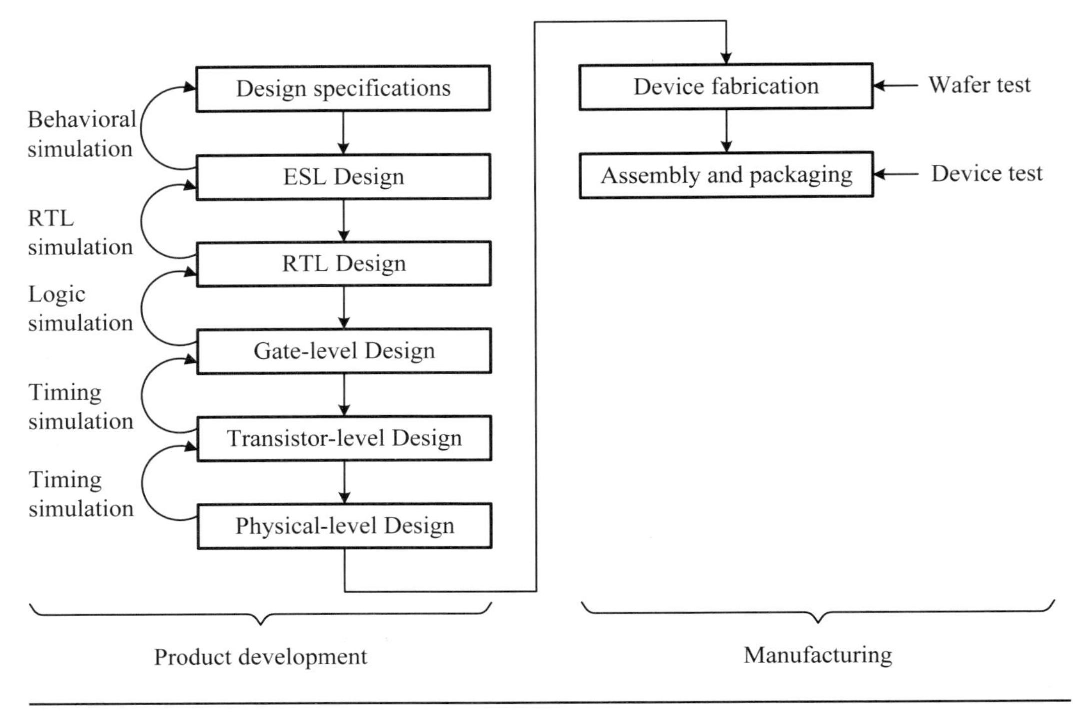

Figure 16.1: The typical VLSI development process.

systems. We are concerned with verification testing and manufacturing testing. Manufacturing testing includes the *wafer test* and *device test*.

16.1.1 Verification Testing

A VLSI device must go through several phases before it emerges into the market for being integrated into application systems. The typical VLSI development process is shown in Figure 16.1, where each phase involves some form of verification testing. The development cycle of a VLSI device can be partitioned into two major parts: *product development* and *manufacturing*. The former begins with product specifications and ends in the physical-level design; the latter consists of device fabrication as well as assembly and packaging. The overall VLSI testing can be categorized into two stages: *design verification* (also *preproduction verification*) and manufacturing testing. Design verification belongs to the product development phase and manufacturing testing is associated with the manufacturing phase. We describe the overview of VLSI testing in more detail.

After a design is completed, design verification is performed to verify the function and timing of the design to ensure that they meet the desired specifications. Before fabrication, the design has to be verified in functionality and/or timing at each abstraction level through various dynamic simulations, static-timing analysis (STA), and/or formal verification.

16.1.1.1 Verification Testing After the completion of the design cycle, *verification testing*, also known as *characterization testing*, *design debug*, or *silicon debug*, is performed at the wafer level to verify the design on the basis of silicon before it is in production. During the course of characterization, we usually test for the worst case

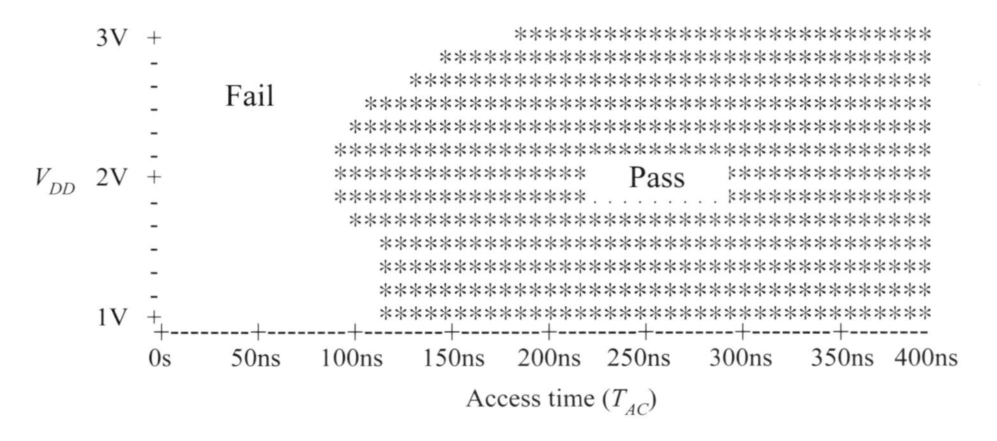

Figure 16.2: An example of the Shmoo plot.

by selecting a test that results in a chip pass/fail decision. The test conditions may be voltage, access time, propagation time, frequency, temperature, or a combination of above or among others. After a test condition is determined, test vectors are applied to a statistically significant sample of devices and the success of the test is recorded. This process is then repeated enough times so that enough information can be collected. The results are then plotted as a graph, known as a *Shmoo plot*.

In general, a Shmoo plot is a graph of test results represented as a function of test conditions. A Shmoo plot gives a more complete picture of the performance of the device and can tell where the device is most susceptive to failure. An example of the Shmoo plot is shown in Figure 16.2, where access time is on the X-axis and supply voltage (V_{DD}) is on the Y-axis. It shows the results of test vectors that are applied to each combination of voltage and access time. From Shmoo plot, we can see the exact limits of operating values of the device. Once the device weaknesses are recognized, the device design, fabrication process, and test program can then be improved to maximize production yield.

Review Questions

Q16-1. What are the three types of VLSI testing?

Q16-2. What are the two parts of the development cycle of a VLSI device?

Q16-3. Why are the features of a Shmoo plot?

16.1.2 Wafer Test

After a wafer has finished all of its process steps, it is ready for a wafer test, assembly and packaging, as well as a final test to be finished into a set of good, usable integrated circuits. The wafer test includes the *in-line parameter test* and *wafer sort*. In-line parameter test, also known as the *wafer electrical test* (WET) or direct current (DC) test, is carried out right after the first metal layer etch to monitor information about process and device performance. Wafer sort is done at the completion of wafer fabrication to determine which dies on the wafer meet the desired specifications and are acceptable for assembly and packaging.

16.1.2.1 In-Line Parameter Test During wafer fabrication, it is necessary to monitor and control the process in an optimal condition. For this purpose, a variety of test structures, also called *process control monitors* (PCMs), are designed and arranged at specific locations on the wafer. For verification testing, test structures may be a test pattern on an entire chip; for production testing, test structures are usually located in the scribe line regions to avoid wasting useful area.

Examples of test structures include discrete transistors, various line widths, box in a box, resistor structure, contact or via chain, among others. The discrete transistors are utilized to monitor device features such as leakage current, breakdown voltage, threshold voltage, and effective channel length. Various line widths and box in a box are used to measure the *critical dimensions* (CDs) and overlap registration, respectively. Register structure is employed to monitor the resistivity of film thickness. Contact or via chain is utilized to measure contact resistance and connections.

It is essential to obtain the process parameter as soon as possible during fabricating a wafer. Hence, the parameter test is carried out right after the first metal layer etch so that contact probes of *automatic test equipment* (ATE) can make electrical contact with the test structures to measure and monitor information about process and transistor performance. The results are used to determine whether the wafer should be continued in the fabrication process or aborted based on pass/fail criteria, to assess process trends to make process improvement, to identify process problems, and to evaluate wafer-level reliability.

16.1.2.2 Wafer Sort/Probe Wafer sort, also called *wafer probe*, is the first time that chips are tested to see whether they function as they should. It is performed at the end of wafer fabrication to knock out bad dies and to sort the good ones according to their performance. Hence, it is named. To accomplish this, each die on a wafer is tested against all product specifications in both alternating current (AC) and DC parameters.

To carry out wafer sort, three basic tools used as a set are required. These are a *wafer prober*, a *probe card*, and an ATE. The wafer prober is a material-handling system that takes wafers from their carriers, loads them to a flat chuck, aligns and positions them precisely under a set of fine contacts on a probe card. The probe card provides fine electrical probes for contacting each input-output or power pad on the die to translate the small individual die pad features into connections to the tester. The ATE is used to functionally exercise all of the chip's designed features under software control.

Any failure to meet the desired specification is identified by the tester and the device is cataloged as a reject. The tested results are then categorized with a bincode number, with each presenting a class of dies. A *bin map* is used for each wafer to provide a visual image that highlights the bincode number of each die on the wafer. Special analysis softwares, such as *spatial signature analysis* (SSA), can be used to analyze, recognize, and pinpoint the source of the wafer defects from the unique bin map failure distributions.

The good dies can be further classified into several degrees according to their performance. This is accomplished by applying wafer sort tests, also known as *AC tests*, because of the use of a system clock and high-frequency input signal. Before performing longer AC tests, some DC tests are applied first to check continuity, short/opens, leakage current, and I_{DDQ} tests. Continuity means that each pin is properly connected to an appropriate pad.

I_{DDQ} is the quiescent current from the drain to the source of the transistor when it is off. The rationale behind this test is that the current should be on the order that can be neglected when the transistor is off. However, when some physical defects exist, the magnitude of I_{DDQ} can be increased to a level that can no longer be ignored. The most common faults that can be detected by the I_{DDQ} test include physical shorts (bridging), power supply to ground shorts, gate-oxide shorts, and punchthrough. One drawback of the I_{DDQ} test is the difficulty in determining the root cause of the fault. In addition, it is being eroded in the deep submicron process due to the increased subthreshold current.

The major purpose of wafer sort is to ensure that only those ICs meet the data-sheet specifications are sent to the next stage, assembly and packaging. To achieve this, a test practice called *guardbanding* is performed by tightening the test limits to account for equipment and process variations, including instrument error and measurement error, and product variability. The tightened test limits are a set of more stringent requirements than those specified in the data-sheet specifications. The typical tightened test limits could be in the form of reduced electrical requirements, such as leakage current from a 10 pA to 8 pA at final test and 7 pA at wafer sort, and elevated temperature from 75° to 85°.

At the wafer sort stage, wafers have completed the entire fabrication process. The *wafer sort yield* is defined as the percentage of acceptable dies that passes the wafer sort test and denotes the overall stability and cleanliness of the wafer fabrication processes. The wafer sort yield can be expressed as

$$\text{Wafer sort yield} = \frac{\text{No. of good dies}}{\text{Total no. of dies on wafers}} \tag{16.1}$$

Wafer sort yields are typically approximated 60% for the first year of production and 80% to 90% for several years later.

16.1.3 Device Test

After a device passes wafer sort, the next step is to saw between each die in both directions and separate or "dice" out the good dies. The good dies are then assembled and packaged. Device test or package test is the last time that chips are tested to see whether they function as they were designed before leaving the factory. To test a device, three basic tools are needed. They are a handler, a socket, and an ATE. The handler is a material-handling system that takes packaged devices from their carriers, loads them into sockets, and sets the environmental temperature as specified. The socket on a custom-designed printed-circuit board, known as a *device under test board* (*DUT board*) or a *load board*, is employed to contact each pin on the chip's package. The ATE is utilized to functionally exercise all of the chip's designed features under software control. The device test includes *burn-in test*, or called *stress test*, and *final test*.

16.1.3.1 Burn-in or Stress Test Burn-in or stress test is a process by which chips are placed in a high temperature and over-voltage supply environment while production tests are running. Burn-in test is used to isolate *infancy failures* and *freak failures*. Infancy failures are often caused by manufacturing process defects. A process may not be tuned correctly or a process step may be inadvertently skipped. These failures are likely to be screened out in the first few days of operation before they are shipped to customers. Freak failures, on the other hand, usually refer to component defects within the assembly. Examples are weak wire bonding, poor die attachment, and induced defects that work properly but fail when put into a mild stress scenario. Freak failures usually can be screened out with a moderate amount of burn-in time.

16.1.3.2 Final Test Final test includes functional tests and parameter tests. The functional tests use test patterns to verify the functionality of the device, and parameter tests measure the electrical characteristics and determine whether they meet the desired specifications.

DC parameter test. DC parameter tests use Ohm's law to test steady-state electrical characteristics of a device. They include contact test, power dissipation test, output short current test, output drive current test, and threshold voltage test. The contact test verifies that there are no opens or shorts on a chip's pin. The power dissipation test measures the worst-case power dissipation for static and dynamic cases. The output short current test measures the output currents at both high- and low-output voltages. The output drive current test verifies the output voltage that is maintained for a specified output drive current. The threshold voltage test determines the low- and high-level input voltages, V_{IL} and V_{IH}, needed to cause the output to switch from low to high and high to low, respectively.

AC parameter tests. In AC parameter tests, alternating voltages with a specified set of frequencies are applied to a device and measure the terminal impedance or dynamic resistance or reactance. AC parameter tests include rise- and fall-time tests, setup- and hold-time tests, and propagation-delay tests as well.

16.1.3.3 Data Sheet A data sheet usually contains a feature summary and description, principles of operation, absolute maximum ratings, electrical characteristics, timing diagrams, application information, circuit schematic, and die layout. Some data sheets also include a few basic application examples of the device.

■ Review Questions

Q16-4. Describe the goal of the wafer test.
Q16-5. Define the wafer sort yield.
Q16-6. Describe I_{DDQ} testing.
Q16-7. Define burn-in and stress tests.
Q16-8. What are included into the final test?
Q16-9. What are infancy failures and freak failures?
Q16-10. What should be included into the data sheet of a device?

16.2 Fault Models

Any test must be based on some kind of fault models. Even though there are many fault models that have been proposed over the past decades, the most widely used fault model in logic circuits is the *stuck-at fault* model. In addition, two additional fault models, bridge fault and stuck-open fault, are also common in CMOS technology. Furthermore, in deep submicron CMOS processes, delay faults caused by the capacitive-coupling effects become more prevalent with the reduction of feature sizes. The delay fault models covers gate-delay fault, transition fault, and path-delay fault. We first introduce some useful terms along with their definitions and then deal with the fault detection.

16.2.1 Fault Models

A *fault* is the physical defect in a circuit. The physical defect may be caused by process defects, material defects, age defects, or even package defects. When the fault

Figure 16.3: The stuck-at fault models: (a) stuck-at-0 fault; (b) stuck-at-1 fault.

manifests itself in the circuit, it is called a *failure*. When the fault manifests itself in the signals of a system, it is called an *error*. In other words, a defect means the unintended difference between the implemented hardware and its design. A fault is a representation of a defect at an abstract function level and an error is the wrong output signal produced by a defective circuit.

16.2.1.1 Stuck-at Faults The most common fault model used today is still the stuck-at fault model due to its simplicity and that it can model many faults arising from physical defects, such as broken wire, opened diode, shorted diode, as well as a short-circuit between the power supply and the ground node. The stuck-at fault model can be *stuck-at-0* or *stuck-at-1* or both. "Stuck-at" means that the net will adhere to a logic 0 or 1 permanently. A stuck-at-0 fault is modeled by assigning a fixed value 0 to a signal wire or net in the circuit. Similarly, a stuck-at-1 fault is modeled by assigning a fixed value 1 to a signal wire or net in the circuit.

■ **Example 16-1: (Examples of stuck-at faults.)** Figure 16.3(a) shows the case of stuck-at-0 fault. Because the logic gate under consideration is NAND, its output is always 1 whenever one of its inputs is stuck-at-0, regardless of the other input value. Figure 16.3(b) shows the case of stuck-at-1 fault. Because the underlying logic gate is NOR, its output is always 0 whenever one of its inputs is stuck-at-1, regardless of the other input value.

■

16.2.1.2 Equivalent Faults Some stuck-at faults will cause the same output values regardless of what inputs are. Such indistinguishable faults are referred to as *equivalent faults*. More precisely, an equivalent fault means that two or more faults of a logic circuit transform the circuit in a way such that two or more faulty circuits have the same output function. The size of the test set can be considerably reduced if we use equivalent faults. A process used to select one fault from each set of equivalent faults is known as *fault collapse*. The set of selected equivalent faults is called the *set of equivalent collapsed faults*. The metric of this is the *collapse ratio*, which is defined as follows:

$$\text{Collapse ratio} = \frac{|\text{Set of equivalent collapsed faults}|}{|\text{Set of faults}|}$$

where $|x|$ denotes the cardinality of the set x.

■ **Example 16-2: (Examples of equivalent faults.)** In Figure 16.3(a), the output of the NAND gate is always 1 whenever one or both inputs a and b are stuck-at-0, or output f is stuck-at-1. These faults are indistinguishable and hence are equivalent faults. Similarly, in Figure 16.3(b), the output of the NOR gate is always 0 whenever one or both inputs a and b are stuck-at-1, or output f is stuck-at-0. Hence, they are equivalent faults.

■

When a circuit has only one net occurring at the stuck-at fault, it is called a *single fault*; when a circuit has many nets occurring at stuck-at faults at the same time, it is known as *multiple faults*. For a circuit with n nets, it can have $3^n - 1$ possible stuck-at net combinations but at most $2n$ single stuck-at faults. Of course, a single fault is a special case of multiple faults. In addition, when all faults in a circuit are either stuck-at-0 or stuck-at-1 but cannot be both, the fault phenomenon is called a *unidirectional fault*.

16.2.1.3 Bridge and Stuck-Open/Stuck-Closed Faults In a CMOS circuit, there are two additional common fault models: *bridge fault* and *stuck-open fault*. A bridge fault is a short-circuit between any set of nets unintentionally and often change the logic function of the circuit. In general, the effect of a bridge fault is completely determined by the underlying technique of the logic circuit. In CMOS technology, a bridge fault may evolve into a stuck-at fault or a stuck-open fault, depending on where the bridge fault occurs.

■ Example 16-3: (Examples of bridge faults.) Figure 16.4(a) shows two examples of a bridge fault. The pMOS block functions as $\overline{y+z}$ rather than its original function $\overline{x(y+z)}$ when the bridge fault S_1 occurs. The nMOS block functions as $(x+y)$ rather than its original function $(x+yz)$ when the bridge fault S_2 occurs. Consequently, bridge faults often change the logic function of the circuit.

■

A stuck-open fault is a unique feature of CMOS circuits. It means that some net is broken during manufacturing or after a period of operation. A major difference between it and a stuck-at fault is that the circuit is still a combinational logic as a stuck-at fault occurs but the circuit will be converted into a sequential circuit as a stuck-open fault occurs. To make this point more apparent, consider the Figure 16.4(b), which is a two-input NOR gate. When the nMOS transistor Q_{1n} is at a stuck-open fault, the output f will change its function from $\overline{x+y}$ into a sequential logic function (Why? Try to explain it).

Sometimes a *stuck-closed* (*stuck-on*) fault is also considered in CMOS technology. A stuck-open fault is modeled as a switch being permanent in the open state. A stuck-closed (stuck-on) fault is modeled as a switch being permanent in the shorted state. The effect of a stuck-open fault is to produce a floating state at the output of the faulty logic circuit, while the effect of a stuck-closed fault is to yield a conducting path from power supply to ground.

16.2.1.4 Delay Faults For a logic circuit to be fault-free, both of its logic function and timing must be correct. More precisely, the logic circuit not only has to carry out its logic function correctly but also has to propagate the correct logic signals within a specified time limit. However, with the decreasing feature sizes, the capacitive-coupling effects may severely prolong the propagation delay of signals along a path to fall outside the specified limit, thereby resulting in an incorrect logic value. Such a

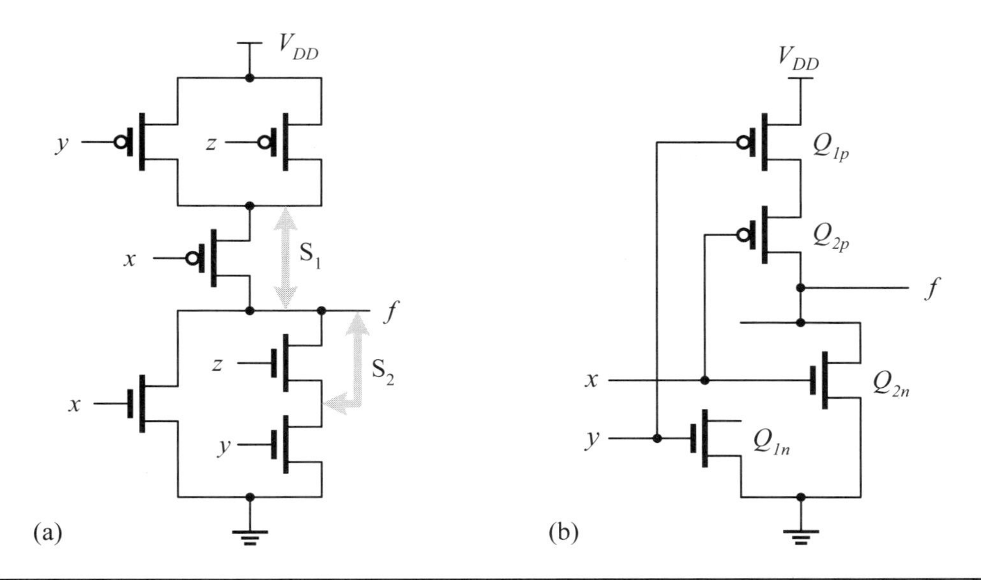

Figure 16.4: The (a) bridge and (b) stuck-open fault models.

result is referred to as a *delay fault*. Delay faults have become more prevalent with the reduction of feature sizes.

Three different delay fault models are *gate-delay fault*, *transition fault*, and *path-delay fault*. The gate-delay fault and transition fault models assume that the delay fault affects only one gate in the circuit. The gate-delay fault means that a delay fault occurs when the time interval taken for a signal propagating from the gate input to its output exceeds its specified range. The transition fault means that a delay fault occurs when the time interval taken for a rising or falling transition exceeds its specified range. Two transition faults are associated with each gate, a slow-to-rise fault and a slow-to-fall fault. The path-delay fault, as its name implies, considers the cumulative propagation delay along a signal path through the *circuit under test* (CUT). In other words, it calculates the sum of all gate delays and interconnect delays along a signal path and determines whether the result is within the specified limit. As a consequence, the path-delay fault is more useful than gate-delay fault in practice.

■ Review Questions

Q16-11. Define stuck-at-0 and stuck-at-1 faults.

Q16-12. Define bridge fault and stuck-open fault.

Q16-13. What is the meaning of a fault model?

Q16-14. How many single stuck-at faults can an m-input logic gate have?

Q16-15. Define delay fault and path-delay fault.

Q16-16. Distinguish between gate-delay fault and transition fault.

16.2.2 Fault Detection

To begin with, we define three related terms: *test*, *fault detection*, and *fault location*. A test is a response generated by a procedure that applies proper stimuli to a specified

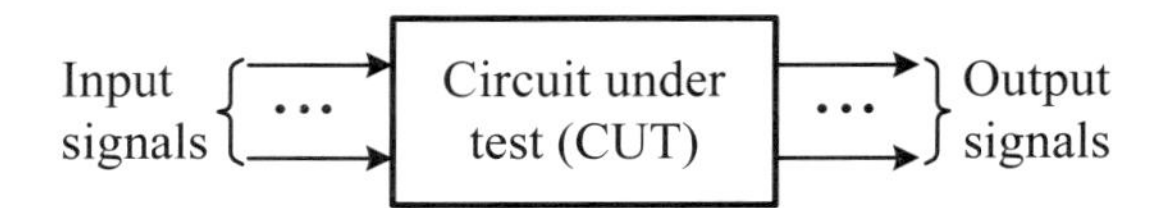

Figure 16.5: The basic model of a logic circuit under test (CUT).

fault-free circuit. Fault detection is a procedure that determines whether the circuit under test (CUT) is faulty by applying a test. The fault location means locating the exact position, or nearby, of a fault. However, before a fault can be located, it is generally necessary to detect whether the fault has occurred. As a result, locating a fault often requires more resources and is more difficult than detecting a fault.

The basic model of a logic CUT is shown in Figure 16.5. The CUT is a black box with a known logic function, even the logic circuit, but all stimuli must be applied at the (primary) inputs and all responses must be detected at the (primary) outputs. To test whether a fault has occurred, we need to apply an appropriate stimulus from the inputs to set the net with a fault to be detected to an opposite logic value and to propagate the net value to the outputs so that we can observe the net value and determine whether the fault has occurred. More precisely, the capability of fault detection is based on the two features of the CUT: *controllability* and *observability*.

- The controllability of a particular node in a logic circuit is a measure of the ease of setting the node to a 1 or a 0 from the inputs.
- The observability of a particular node in a logic circuit is the degree to which you can observe the node value at the outputs.

In a combinational logic circuit, if at least one test can be found to determine whether a specified fault has occurred, the fault is said to be *detectable* or *testable*; otherwise, the fault is said to be *undetectable*. In other words, a detectable fault means that at least one input combination can be found to make the outputs different between the fault-free and faulty circuits. An undetectable fault means that there exists no input combination to make the outputs different between the fault-free and faulty circuits, namely, a fault for which no test can be found. The metric of faults that can be detected is known as *fault coverage*, which is defined as follows:

$$\text{Fault coverage} = \frac{\text{No. of detected faults}}{\text{Total no. of faults}}$$

which means what fraction of faults can be detected from all possible faults. Commercial computer-aided design (CAD) tools such as *Sandia controllability and observability analysis program* (SCOAP) can be employed to analyze the fault coverage in a logic circuit. To achieve a world-class quality level, the fault coverage of a circuit has to be in excess of 98.5%. In theory, the condition that all stuck-at faults of a combinational logic circuit can be detected is the combinational logic circuit must be irredundant (or irreducible); namely, its logic function must be an irredundant switching expression.

■ Example 16-4: (Detectable and undetectable stuck-at faults.) Figure 16.6 is a logic circuit with a redundant gate; that is, it is not an irredundant logic circuit. From the basic features of the NAND gate, to make the output of a NAND gate equal to the complementary value set by a specified input, all the other inputs of the gate must be set to 1. As a result, nets α and β can be independently set to 0 or 1. Both α_{s-a-0} and β_{s-a-0} are unobservable faults but α_{s-a-1} and β_{s-a-1}

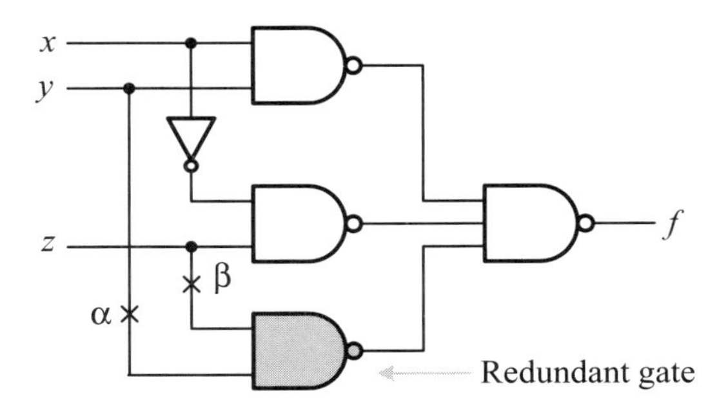

Figure 16.6: A redundant gate and undetectable faults.

are observable faults. Why? Hence, α_{s-a-0} and β_{s-a-0} are undetectable faults but α_{s-a-1} and β_{s-a-1} are detectable faults.

From the above example, we can draw a conclusion. A fault of a CUT that is detectable or undetectable depends on the following two parameters of the fault: controllability and observability. As mentioned before, when a combinational logic circuit is irredundant, any physical defect occurring on any net of the logic circuit will cause the output to deviate from its normal value. Hence, it can be detected by a test.

Review Questions

Q16-17. Define test, fault detection, and fault location.

Q16-18. Define detectable and undetectable faults.

Q16-19. Define controllability and observability.

Q16-20. How many single stuck-at faults does the Figure 16.6 have, once the redundant AND gate is removed? Can they all be detectable?

16.3 Automatic Test Pattern Generation

To test a combinational logic circuit, it is necessary to derive a test set with a minimal number of test vectors[1] to save test cost and speed up the test process. There are many ways that can be used to derive the set of test vectors. In this section, we only consider the *path sensitization* approach. The interested reader can refer to Abramovici et al. [1], Bushnell and Agrawal [6], and Wang et al. [38] for further details.

16.3.1 Test Vectors

In testing a logic circuit, it is necessary to find a simple set of input signals that can be applied to the input of the logic circuit. This set of input signals is called a *test set*. A combination of input signals used to test a specified fault is called the *test vector* (also called the *test pattern*) for the fault. A test set of a logic circuit is a union of test

[1] In the industry and literature, the terms of test vector and test pattern are often used interchangeably.

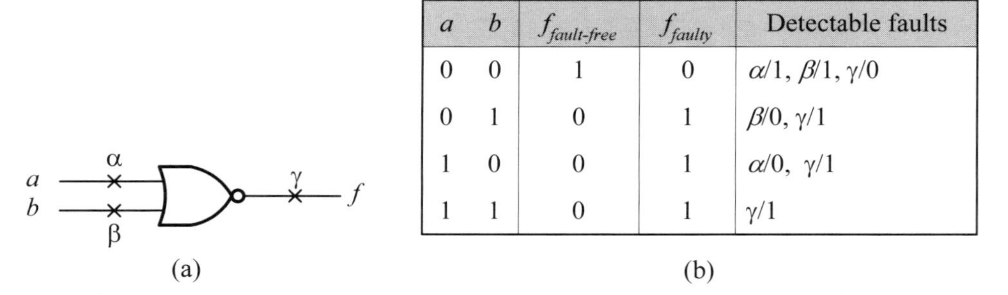

a	b	$f_{fault\text{-}free}$	f_{faulty}	Detectable faults
0	0	1	0	$\alpha/1, \beta/1, \gamma/0$
0	1	0	1	$\beta/0, \gamma/1$
1	0	0	1	$\alpha/0, \gamma/1$
1	1	0	1	$\gamma/1$

Figure 16.7: An example of a complete test.

vectors for the circuit. If a test set can test all testable (detectable) faults of a logic circuit, the test set is known as a *complete test set*.

In general, a truth table of a logic circuit is a complete test set of the logic circuit. However, there are 2^n possible input combinations of an n-input combinational logic circuit. Consequently, it is not possible or ineffective for a large n in practice. Fortunately, we do not generally need to use the entire truth table as a complete test set. Through carefully examining the logic circuit, we usually can find a complete test set with a size much smaller than the truth table. An example to illustrate this is explored in the following.

■ Example 16-5: (An example of a complete test set.) Figure 16.7 shows a two-input NOR gate. As mentioned above, the simplest way to test this circuit is to apply the four combinations of inputs a and b one by one and then compare the result of the output with its truth table to determine whether it is faulty.

When the combination of inputs a and b is 00, the output is 1 if it is fault-free but is 0 if net α or β is stuck-at 1 or γ is stuck-at 0. When the combination of inputs a and b is 01, the output is 0 if it is fault-free but is 1 if net β is stuck-at 0 or γ is stuck-at 1. When the combination of inputs a and b is 10, the output is 0 if it is fault-free but is 1 if net α is stuck-at 0 or γ is stuck-at 1. Since the above three combinations have completely tested all possible single stuck-at faults, they constitute a complete test set. That is, the complete test set of the two-input NOR gate is $\{00, 01, 10\}$. It saves 25% test cost compared to an exhaustive test with the entire truth table.

■

In summary, the technique of using all possible input combinations of a combinational logic circuit to test the logic circuit is known as an *exhaustive test*, which belongs to a kind of complete test. However, it can usually find a complete test set of smaller size than the truth table if we examine the combinational logic circuit under test more carefully.

■ Review Questions

Q16-21. Define test vector, test set, and complete test set.

Q16-22. Define the exhaustive test.

Q16-23. Why is the size of a complete test set usually smaller than that of the truth table of the combinational logic circuit?

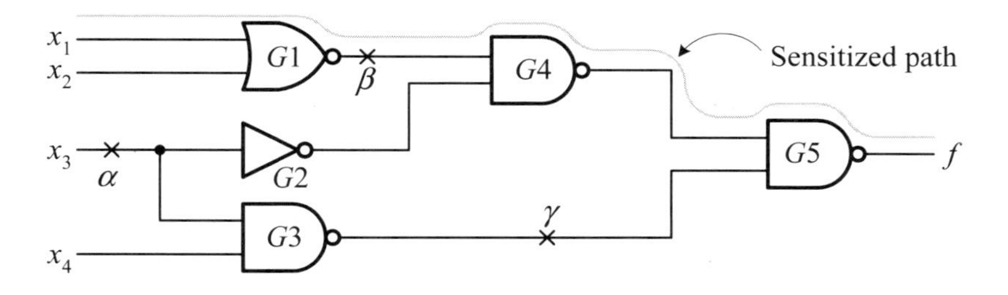

Figure 16.8: An example of path sensitization.

16.3.2 Path Sensitization

Recall that the essence of detecting a fault is to apply a proper set of input values (test vector) to set the net of interest with an opposite value of its fault value, propagate the fault effect to the primary outputs, and observe the output value to determine whether the fault at the net has occurred. The path-sensitization approach is a direct application of this idea and has the following three major steps:

1. *Fault sensitization:* This step sets the net to be tested (namely, test point) to an opposite value from the fault value. Fault sensitization is also known as *fault excitation.*
2. *Fault propagation:* This step selects one or more paths, starting from the test point (that is, the specified net) to the outputs, to propagate the fault effect to the output(s). Fault propagation is also known as *path sensitization.*
3. *Line justification:* This step sets the inputs to justify the internal assignments previously made to sensitize a fault or propagate its effect. Line justification is also known as a *consistency operation.*

An illustration of explaining how the path-sensitization approach works is explored in the following example.

■ Example 16-6: (The path-sensitization approach.) As mentioned above, the stuck-at-1 fault at net β of Figure 16.8 is tested by first setting the logic value of net β to 0, which in turn requires either input x_1 or x_2 to be set to 1. Then, to propagate the logic value at net β to the output, a path from the net β to the output should be established. This means that input x_3 has to be set to 0, which also sets net γ to 1. As a result, the value of input x_4 is irrelevant. Therefore, the test vectors are the set: $\{(1, \phi, 0, \phi), (\phi, 1, 0, \phi)\}$.

■

A path is called a *sensitizable path* for a stuck-at-fault net if it can propagate the net value to the primary outputs with a consistency operation. Otherwise, the path is called an *unsensitizable path* for the stuck-at-fault net. The following example illustrates how the sensitizable and unsensitizable paths occur in an actual logic circuit.

■ Example 16-7: (The difficulty of the path-sensitization approach.) As the stuck-at-0 fault at net h shown in Figure 16.9 is to be tested, both inputs x_2 and x_3 have to be set to 0 to set the logic value at net h to 1. Two possible paths from net h could be set as the sensitization paths to the output. If path 1 is used, both inputs

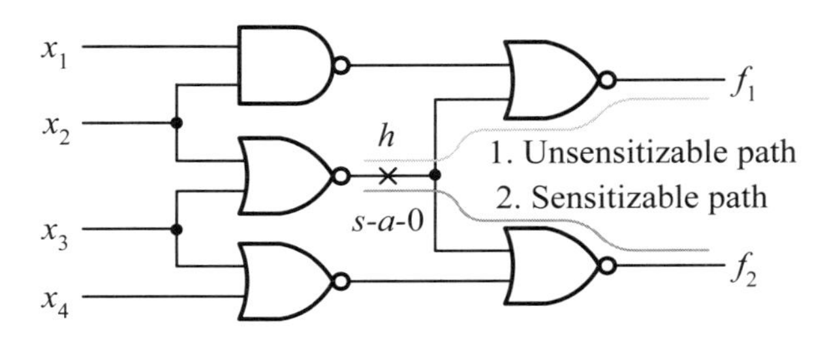

Figure 16.9: Unsensitizable path in the path-sensitization approach.

x_1 and x_2 need to be set to 1 to propagate the logic value at net h to the output. As a result, we have an inconsistent setting of input x_2 and the path 1 is an unsensitizable path. If path 2 is used, input x_4 needs to be set to 1 to propagate the logic value at net h to the output. This is a consistent operation. Therefore, the test vectors are the set: $\{(\phi, 0, 0, 1)\}$, which is independent of input x_1. The path 2 is a sensitizable path for the stuck-at-0 fault at net h.

In the path-sensitization approach, when only one path is selected, the result is known as a *one-dimensional* or *single-path sensitization* approach; when multiple paths are selected at the same time, the result is called a *multiple-dimensional* or *multiple-path sensitization* approach. The following example demonstrates the difficulty of single-path sensitization and how the multiple-path sensitization can be used to overcome it.

Example 16-8: (Multiple-path sensitization.) As shown in Figure 16.10(a), we obtain an inconsistent setting of the input values x_3 and $\bar{x}_3$ when testing the stuck-at-0 fault at input a. In this case, both inputs x_3 and $\bar{x}_3$ must be set to 0, leading to an inconsistent setting. Nonetheless, when all paths related to the input x_1 are sensitized at the same time, as shown in Figure 16.10(b), both inputs x_3 and $\bar{x}_3$ are separately set to 1 and 0, which is a consistent operation. Consequently, the stuck-at-0 fault at input a is detectable.

In general, when a signal splits into several ones, which pass through different paths, and then reconverges together later at the same gate, the signal is called a *reconvergent fan-out signal*. For such a reconvergent fan-out signal, as illustrated above, the single-path sensitization approach cannot be generally applied to detect the stuck-at faults along it because of the difficulty or impossibility of finding a consistent input combination. Fortunately, the multiple-path sensitization approach can be applied to sensitize all paths related to the signal at the same time to solve this difficulty.

Review Questions

Q16-24. Describe the basic operations of the path-sensitization approach.

Q16-25. What are the sensitizable and unsensitizable paths?

Q16-26. What is the major drawback of the single-path sensitization approach?

Q16-27. What is a reconvergent fan-out signal?

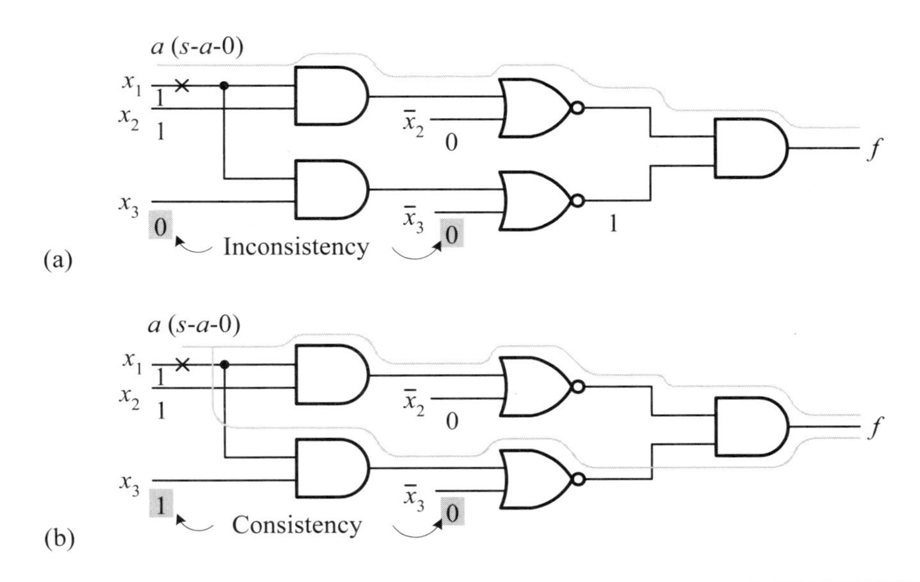

Figure 16.10: Comparison between (a) single and (b) multiple sensitized paths in the path-sensitization approach.

Q16-28. What are the basic differences between the single-path and multiple-path sensitization approaches?

16.4 Testable Circuit Designs

As stated, the controllability and observability of a net determine whether a fault at that net can be detected. As a consequence, the best way to facilitate the testability of a logic circuit is to increase the controllability and observability of the logic circuit by adding to it some extra logic circuits or modifying its structure appropriately. Such an approach by adding extra logic circuits to reduce the test difficulty is known as a *testable circuit design* or *design for testability* (DFT). In this section, we consider the three widely used DFT approaches: *ad hoc approach*, *scan-path method*, and *built-in self-test* (BIST).

16.4.1 Ad Hoc Approach

The basic principles behind the ad hoc approach are to increase both the controllability and observability of the logic circuits by using some heuristic rules. The general guideline rules are listed as follows:

1. *Providing more control and test points:* Control points are used to set logic values of some selected signals and test points are used to observe the logic values of the selected signals.
2. *Using multiplexers to increase the number of internal control and test points:* Multiplexers are used to apply external stimuli to the logic circuit and take out the responses from the logic circuit.

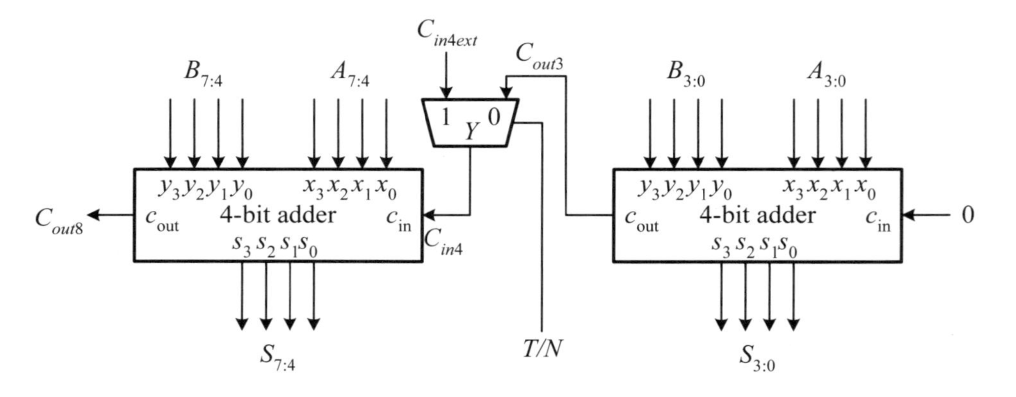

Figure 16.11: An illustration of an exhaustive test method.

3. *Breaking feedback paths:* For sequential logic circuits with feedback paths, multiplexers, AND gates, or other appropriate gates may be applied to disconnect these feedback paths to transform sequential logic circuits into combinational logic circuits during testing.
4. *Using state registers:* Reduce the additional input/output (I/O) pins required for testing signals.

Moreover, the exhaustive test is feasible when the number of input signals of a logic circuit is not too large. For a logic circuit with a large number of input signals, the exhaustive test might still be feasible if the logic circuit can be partitioned into several smaller modules. These modules can then be separately tested in a sequential or parallel fashion.

The following example demonstrates how it is possible to perform an exhaustive test by partitioning a logic circuit into two smaller modules and using multiplexers to increase the number of internal control and test points.

■ Example 16-9: (An example of the exhaustive test method.) Figure 16.11 shows an 8-bit binary adder consisting of two 4-bit binary adders. By simply using the exhaustive test method, we can continuously apply each input combination of 16 inputs to the adder, which requires 65536 tests in total.

In order to reduce the test time, namely, the required number of tests, a multiplexer is used to cut off the connection between the two 4-bit binary adders, as depicted in Figure 16.11. When the source selection signal T/N (Test/Normal) of the multiplexer is set to 1, the circuit is in test mode. The two 4-bit binary adders are allowed to operate independently. They can be tested in parallel and only require 256 tests. Two additional tests are required for testing the carry propagation, one for C_{in4ext} and the other for C_{out3}. When the source selection signal T/N (Test/Normal) of the multiplexer is set to 0, the two 4-bit binary adders are cascaded into an 8-bit binary adder and operate in its normal mode.

■

■ Review Questions

Q16-29. Describe the essence of testable circuit designs.

Q16-30. Describe the basic rules of the ad hoc approach.

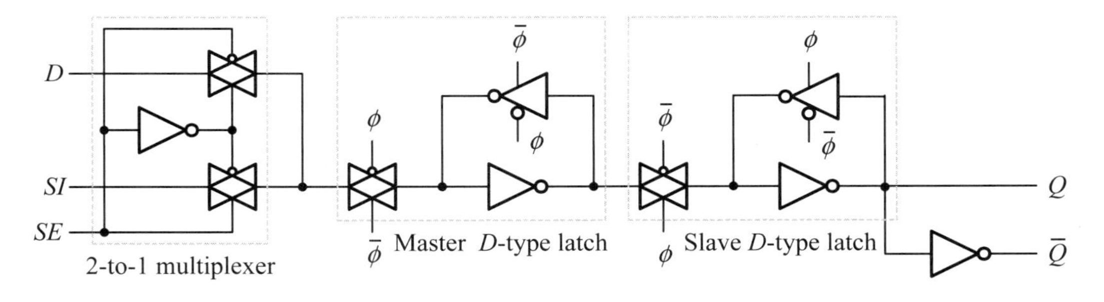

Figure 16.12: A muxed-D scan cell.

Q16-31. In what conditions can the exhaustive test be applied to test a circuit?

16.4.2 Scan-Path Method

It is very difficult to test a sequential logic circuit without the aid of any extra logic circuits. The scan-path method is an approach to removing this difficulty by adding a 2-to-1 multiplexer at the data input of each D-type flip-flop to allow the access of the D-type flip-flop directly from an external logic circuit. As a result, both controllability and observability increase to the point where the states of the underlying sequential logic circuit can be easily controlled and observed externally. In what follows, we first discuss some typical scan cells and then address scan architectures.

16.4.2.1 Scan Cells We describe three widely used scan cell structures: *muxed-D scan cell*, *clocked-D scan cell*, and *level-sensitive scan design* (LSSD) cell.

Muxed-D scan cells. A muxed-D scan cell comprises a D-type flip-flop and a 2-to-1 multiplexer, as shown in Figure 16.12. The 2-to-1 multiplexer uses a scan enable (SE) input to select between the normal data input (D) and scan input (SI). In normal mode, the scan enable SE is set to 0, the muxed-D scan cell operates as an ordinary D-type flip-flop. In scan mode, the scan enable SE is set to 1, the scan input SI is used to shift data into the D-type flip-flop. At the same time, the current data in the cell is shifted out. Even though each muxed-D scan cell adds a multiplexer delay to the functional path, it is widely used in modern designs due to its compatibility to cell-based designs and support by existing CAD tools.

Clocked-D scan cells. One way to remove the 2-to-1 multiplexer delay from the function path is to use the clocked-D scan cell. As shown in Figure 16.13, a clocked-D scan cell also has a data input (D) and a scan input (SI). However, instead of using a 2-to-1 multiplexer, two independent clocks are used to select the data input to be fed into the D-type flip-flop. In normal mode, the data clock ϕ_d is used to capture the input value from the D input. In scan mode, the shift clock ϕ_s is used to capture the input value from the scan input SI. Meanwhile, the current data in the cell are shifted out. Compared to the muxed-D scan cell, the clocked-D scan cell does not cause the performance degradation but needs an additional clock. The two clocks, ϕ_d and ϕ_s, can be easily generated with a 1-to-2 demultiplexer to demultiplex the clock ϕ by using the scan enable SE as the selection signal.

Level-sensitive scan design (LSSD) cells. Both above-mentioned muxed-D and clocked-D scan cells are generally used for replacing edge-triggered flip-flops in digital designs. For level-sensitive latch-based designs, it is also possible to design a

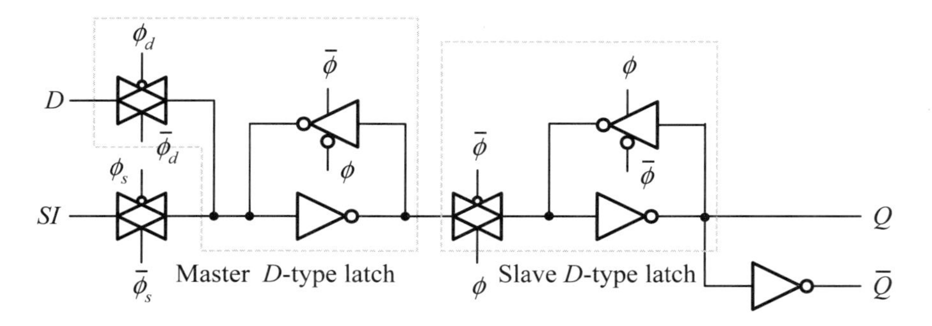

Figure 16.13: A clocked-D scan cell.

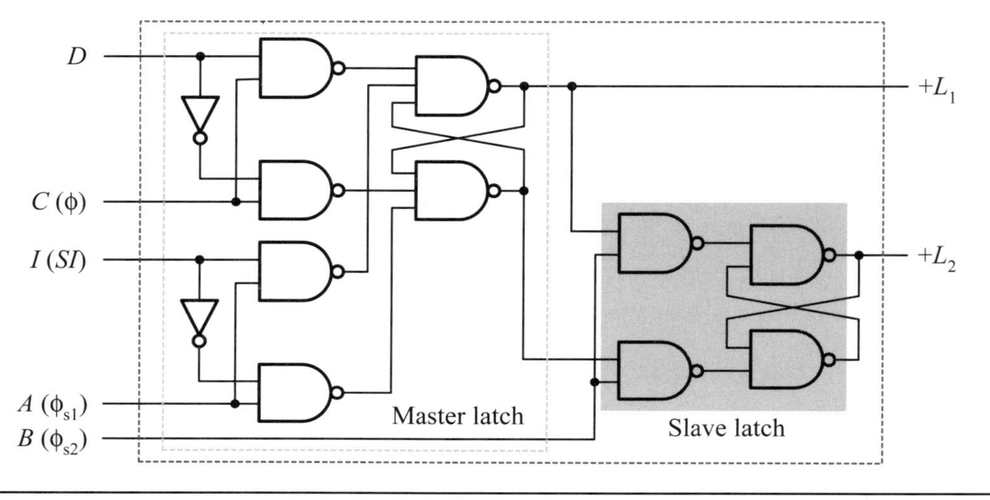

Figure 16.14: A level-sensitive scan design (LSSD) cell.

scannable latch cell. As shown in Figure 16.14, this scan cell consists of two latches, a master dual-port latch and a slave latch. In normal mode, only the master latch is used while in the scan model, both master and slave latches are used to form a flip-flop. This is the basic principle for constructing a scannable level-sensitive latch-based scan cell. During normal mode, clock C (ϕ) is used to capture the data D into the master latch. During scan mode, clock A (ϕ_{s1}) is utilized to capture the scan input I (SI) into the master latch. Clock signal B (ϕ_{s2}) is employed to transfer the data from the master latch into the slave latch. To guarantee a race-free operation, clocks $A(\phi_{s1})$, $B(\phi_{s2})$, and $C(\phi)$ are applied in a nonoverlapping fashion.

Unlike the cases of using muxed-D and clocked-D scan cells, the technique of LSSD cells can guarantee a race-free operation. In addition, using LSSD scan cells may allow us to insert a scan path into a latch-based design. The major disadvantage of using the LSSD technique is that it requires a little complex clock generation and routing.

16.4.2.2 Scan Architectures There are three widely used scan architectures: *full-scan structure*, *partial-scan structure*, and *random-access scan structure*. In the full-scan structure, during scan mode all storage elements are replaced by scan cells, which are in turn configured as one or more shift registers, called *scan chains* or *scan paths*. In the partial-scan structure, only a subset of storage elements is replaced by scan cells.

These scan cells are also configured into one or more scan chains. In the random-access scan structure, instead of using serial scan designs in the full-scan and partial-scan structures, all scan cells are organized into a two-dimensional array, in which each scan cell can be individually accessed in a random fashion. We only deal with the full-scan structure briefly. The details of scan architectures are beyond the scope of this book and can be referred to Wang et al. [38].

Muxed-D full-scan structure. In the muxed-D full scan structure, the 2-to-1 multiplexer at the data input of each D-type flip-flop operates in either of two modes: *normal* and *test*. In the normal mode, the data input of each D-type flip-flop connects to the output of its associated combinational logic circuit to carry out the normal operation of the sequential logic circuit. In the test mode, all D-type flip-flops are cascaded into a shift register (that is, scan chain) through the use of those 2-to-1 multiplexers associated with the D-type flip-flops. Consequently, all D-type flip-flops can be set to specific values or read out to examine their values externally.

Generally, the test approach using the scan-path method for a sequential logic circuit works as follows:

1. Set the D-type flip-flops into the test mode to form a shift register. Shift a sequence of specific 0 and 1 into the shift register and then observe whether the shift-out sequence exactly matches the input sequence.
2. Use either of the following two methods to test the sequential logic circuit.
   (a). Test whether each state transition of the state diagram is correct.
   (b). Test whether any stuck-at fault exists in the combinational logic part.

An illustration of how the muxed-D full-scan structure can be applied to a practical sequential logic circuit is explored in the following example.

■ Example 16-10: (An example of the scan-path method.) Figure 16.15 shows a muxed-D scan chain embedded in a sequential logic circuit. In the original circuit, the 2-to-1 multiplexers at the inputs of all D-type flip-flops do not exist. The objective of adding 2-to-1 multiplexers is to form a scan chain. As the source selection signal T/N of all 2-to-1 multiplexers is set to 1, the sequential logic circuit is in the test mode. All 2-to-1 multiplexers and D-type flip-flops form a 2-bit shift register with *ScanIn* as input and *ScanOut* as output, as indicated by the lighter lines in the figure. Therefore, any D-type flip-flop can be set to a specific value by the shift operation and the output of any D-type flip-flop can be shifted out as well. That is, each D-type flip-flop has the features of perfect controllability and observability. All D-type flip-flops are operated in their original function desired in the sequential logic circuit as the source selection signal T/N of all 2-to-1 multiplexers is set to 0.

■

■ Review Questions

Q16-32. Describe the basic principles of the scan-path method.
Q16-33. What are the three types of scan cells?
Q16-34. What are the features of the level-sensitive scan design (LSSD) cell?
Q16-35. What is the meaning of the muxed-D full-scan structure?
Q16-36. How would you apply the scan-path method to test a sequential logic circuit?

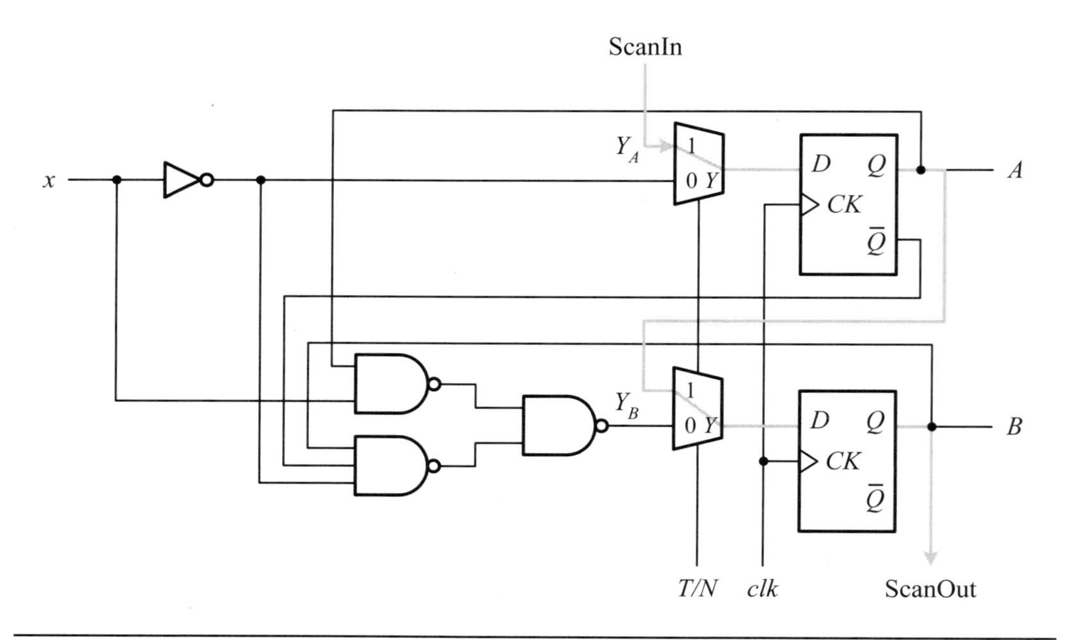

Figure 16.15: An example of a muxed-D scan chain embedded in a sequential logic circuit.

16.4.3 Built-in Self-Test

The built-in self-test (BIST) relies on augmenting logic circuits to allow them to carry out operations to prove the correct operations of the logic circuit. The BIST principle can be illustrated by using Figure 16.16, which includes four major blocks: test vectors (ATPG), *circuit under test* (CUT), *output response analyzer* (ORA), and the BIST controller. The required test vectors for the BIST system are generated automatically by a logic circuit and directly applied to the CUT. The response signals from the CUT are compressed into a signature by a response-compression circuit. The signature is then compared with the fault-free signature to determine whether the CUT is faulty. These three components compose the output response analyzer. The BIST controller schedules the operations of the BIST circuit. Since all of the above test vector generation and response-compression circuits are embedded into the system, they have to be simple enough so as not to incur too much cost. The logic circuit used to generate the test vectors is known as an *automatic test pattern generator* (ATPG) and the response-compression logic circuit is called a *signature generator*.

16.4.3.1 Random Test For a complicated logic circuit, much time may be needed to generate the test set, and it may not be feasible to embed a BIST logic circuit into the CUT due to too much hardware required for storing test vectors. In such a situation, an alternative way, referred to as *random test*, may be used instead. The random test only generates enough test vectors randomly.

The concept of a random test is to use a maximum-length sequence generator, that is, a PR-sequence generator (PRSG), to generate the desired stimuli for testing the target circuit. The random test is not an exhaustive test because it does not generate all input combinations, and it does not guarantee that the logic circuit passing the test is completely fault-free. However, the detection rate of detectable faults will approach

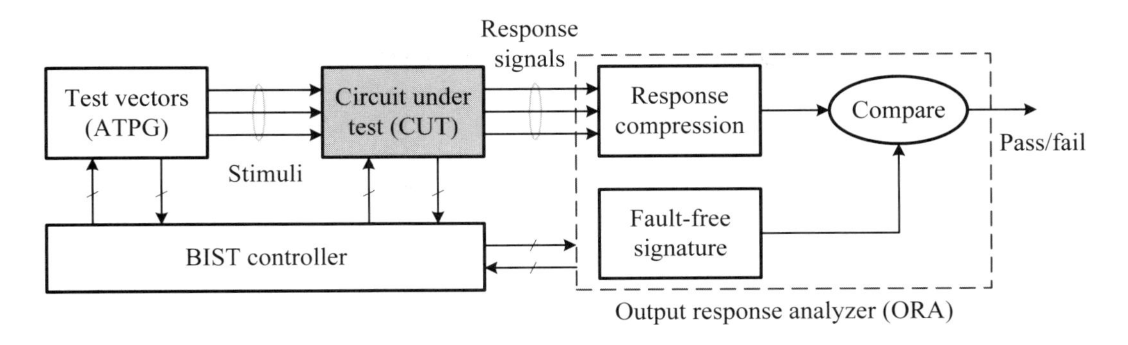

Figure 16.16: The BIST principles of digital systems.

100% when enough test vectors are applied. The implementations of a PR-sequence generator and sample primitive polynomials have been discussed in Section 10.2.4. The interested reader may refer to that section again for details.

The output signals from a PRSG can be an output in one of two modes: *serial* and *parallel*. In the serial output mode, the output signal may be taken from any D-type flip-flop; in the parallel output mode, the output signals are taken from all or a portion of all D-type flip-flops.

16.4.3.2 Signature Generator/Analysis As illustrated in Figure 16.16, the response signals in BIST have to be compressed into a small amount of data, called a *signature*, and then compared with the expected result to determine whether the CUT is faulty. At present, the most widely used signature generator is a circuit that uses an n-stage linear feedback shift register (LFSR).

Signature generator/analysis can be classified into two types: *serial signature generator/analysis* and *parallel signature generator/analysis*. The n-stage *serial-input signature register* (SISR) is a PRSG with an additional XOR gate at the input for compacting an m-bit message M into the standard-form LFSR. The n-bit *multiple-input signature register* (MISR) is a PRSG with an extra XOR gate at each input of the D-type flip-flop for compacting an m-bit message M into the standard-form LFSR. Examples of SISR and MISR are shown in Figures 16.17(a) and (b), respectively. Except for the data inputs, the feedback functions of both circuits are the same as that of the PRSG introduced in Section 10.2.4.

■ Example 16-11: (An application of signature analysis.) An application of signature analysis is shown in Figure 16.18. The inputs x, y, and z are generated by a PRSG. The output response f of the logic circuit is sent to a 4-stage SISR for compressing into a 4-bit signature, as shown in Figure 16.18(a). Figure 16.18(b) lists the outputs and signature values of fault-free, stuck-at-0 at both nets α and β, and stuck-at-1 at both nets α and β under six input combinations. Consequently, whenever it is needed to test the logic circuit, the test vectors are generated in sequence and applied to the logic circuit. Then, the signature of output responses is compared with that of the fault-free circuit to determine whether the logic circuit is faulty. If both signatures are equal, the logic circuit is fault-free; otherwise, the logic circuit is faulty. ■

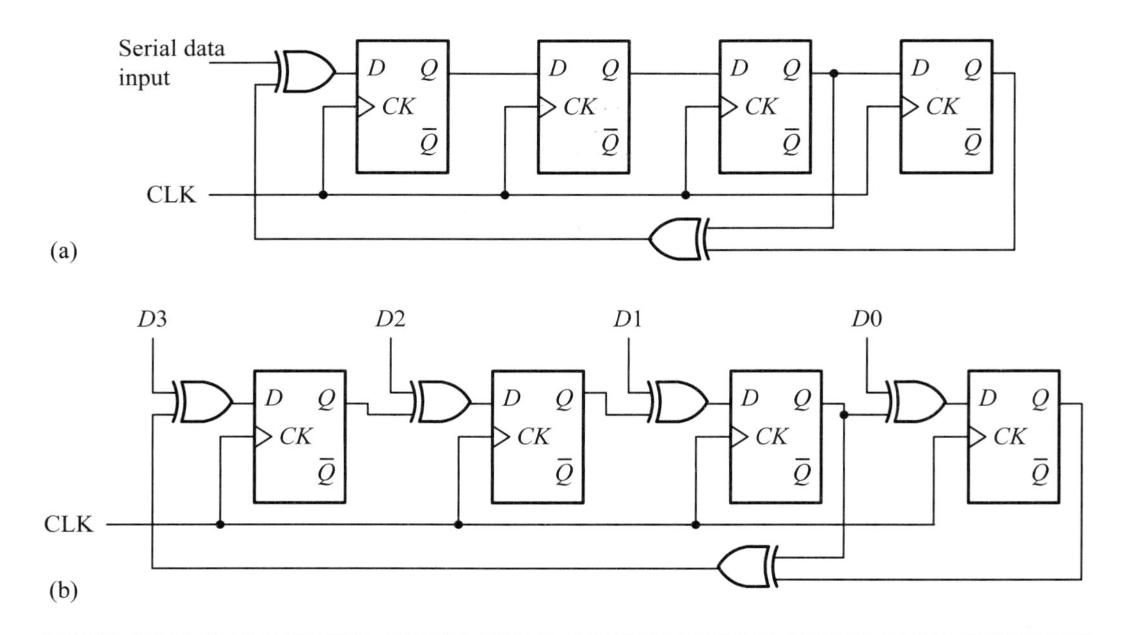

Figure 16.17: Two examples of signature generators: (a) SISR circuit; (b) MISR circuit.

The basic principle of signature analysis is to compress a long stream into a short one (that is, signature). Hence, it is possible to map many streams onto the same signature. In general, if the length of the stream is m bits and the signature generator has n stages and hence the signature is n bits, then there are $2^{m-n} - 1$ erroneous streams that will yield the same signature. Since there are a total of $2^m - 1$ possible erroneous streams, the following theorem follows immediately [1].

Theorem 16.4.1 *For an m-bit input data stream, if all possible error patterns are equally likely, then the probability that an n-bit signature generator will not detect an error is equal to*

$$P(m) = \frac{2^{m-n} - 1}{2^m - 1} \tag{16.2}$$

which approaches 2^{-n}, for $m \gg n$.

Consequently, the fault detection capability of an n-stage signature generator will asymptotically approach to be perfect if n is large enough.

16.4.3.3 BILBO As mentioned above, both PRSG and signature generator need n-stage shift registers and use primitive polynomials. Consequently, both logic circuits can be combined together and use only one register. This results in a logic circuit known as a *built-in logic block observer* (BILBO).

Figure 16.19 shows a typical BILBO logic circuit. The BILBO has four modes as described in the following:

1. *Scan mode*: When mode selection signals $M_1M_0 = 00$, it supports the scan-path method.

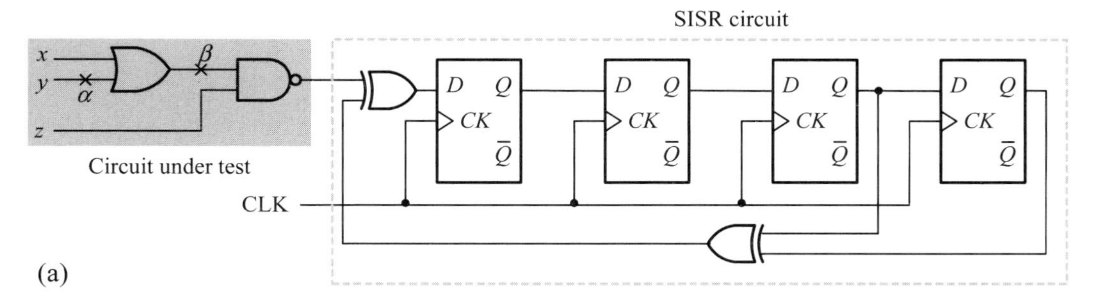

Input			Fault-free output	Faulty output		
x	y	z	f	$f_{\alpha/0}$	$f_{\beta/0}$	$f_{\alpha/1}$ and $f_{\beta/1}$
0	0	1	1	1	1	0
0	1	0	1	1	1	1
0	1	1	0	1	1	0
1	0	0	1	1	1	1
1	0	1	0	0	1	0
1	1	0	1	1	1	1
			0000	1001	1101	0110

(b)

Figure 16.18: An example of the application of signature analysis: (a) logic circuit; (b) a numerical example.

2. *Multiple-input signature register* (MISR) *mode*: When mode selection signals $M_1M_0 = 01$, it supports the signature analysis.
3. *Clear mode*: When mode selection signals $M_1M_0 = 10$, it clears the content of the register.
4. *Parallel-load mode*: When mode selection signals $M_1M_0 = 11$, it functions as a register with the capability of parallel load.

The function selections of the logic circuit shown in Figure 16.19(a) are summarized in Figure 16.19(b). A simple application is illustrated in Figure 16.19(c). In normal mode, the PRSG at the input of the combinational logic circuit is effectively isolated and removed by a multiplexer, not shown in the figure. The registers of the signature generator are used as the state registers and combined with the combinational logic circuit to function as sequential logic. In the test mode, the inputs of the combinational logic circuit are the random test vectors generated by the PRSG and the responses from the combinational logic circuit are sent to the signature generator to carry out the signature analysis.

Review Questions

Q16-37. Describe the essential idea of BIST.

Q16-38. Describe the basic components of a BIST.

Q16-39. Describe the basic concept of a random test.

Q16-40. Describe the basic principles of BILBO.

Q16-41. What is the probability that an n-stage signature generator fails to detect a fault?

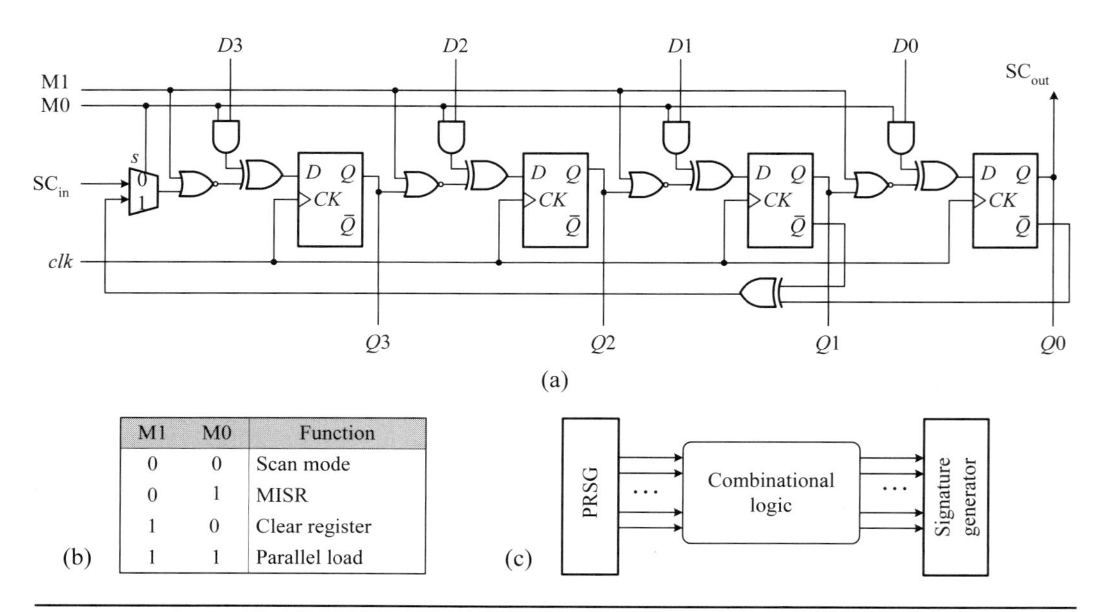

M1	M0	Function
0	0	Scan mode
0	1	MISR
1	0	Clear register
1	1	Parallel load

Figure 16.19: An example of a BILBO logic circuit: (a) logic circuit; (b) function table; (c) an application example.

16.4.4 Boundary-Scan Standard—IEEE 1149.1

The test methods described previously are limited to a single chip or device. It is much more difficult to test the entire system on a printed-circuit board (PCB). To decrease the test cost of PCB systems, the *Joint Test Advisory Group* (JTAG) proposed a testable bus specification in 1988, and then it was defined as a standard known as the *IEEE 1149.1* by the IEEE [15, 38]. IEEE 1149.1, also referred to as the *boundary scan standard*, has become a standard that most integrated circuits must follow.

The goals of the boundary scan standard are to provide a data transfer standard between the *automatic test equipment* (ATE) and the devices on a PCB, a method of interconnecting devices on a PCB, and a way of using test bus standard or BIST hardware to find the faulty devices on a PCB as well. An example to show the boundary scan architecture is given in Figure 16.20.

Figure 16.20 shows a PCB-based system containing four devices facilitated with the boundary scan standard. The boundary scan cells of all four devices are cascaded into a single scan chain through connecting the test data output (TDO) of one device into the test data input (TDI) of another device. The resulting system can carry out the following test functions: *interconnection test*, *normal operation data observation*, and *each device test*.

The essential idea of the boundary scan standard is to extend the scan-path method into the entire PCB system. The boundary scan standard provides the following two major operation modes: *noninvasive mode* and *pin-permission mode*. In the noninvasive mode, the outside world is allowed to serially write in test data or instructions or serially read out test results in an asynchronous way. In the pin-permission mode, the boundary scan standard takes control of the IC input/output pins, thus disconnecting the system logic from the outside world. All ICs adhering to the boundary scan standard must be designed to power-up in the noninvasive mode. Moreover, the boundary scan standard allows delivery of BIST mode commands (e.g., RUNBIST)

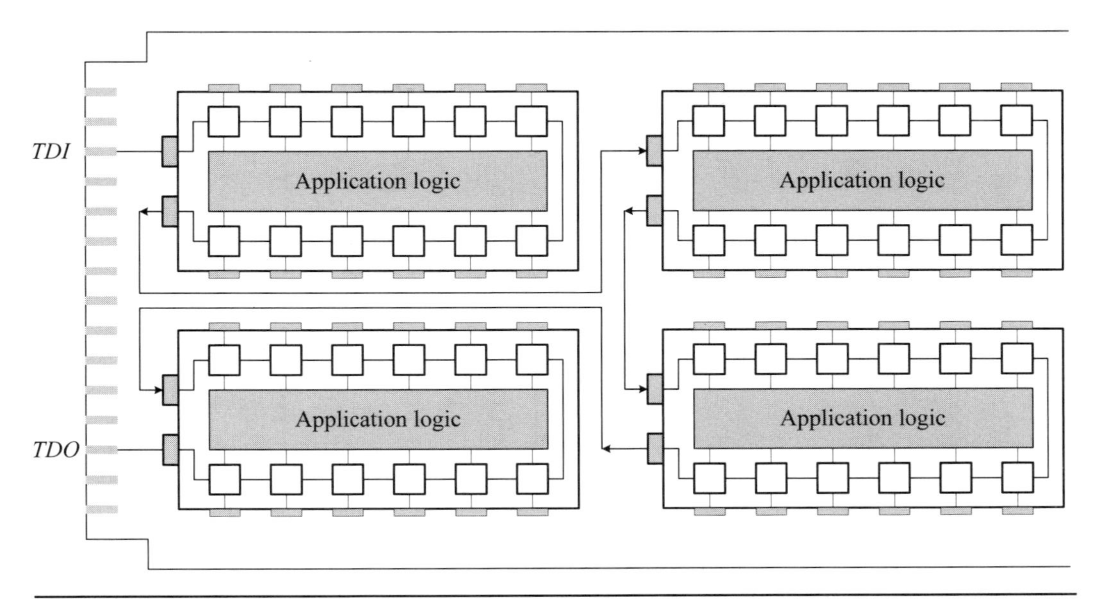

Figure 16.20: A boundary scan architecture used to test the entire board-level module.

through JTAG hardware to the device under test.

■ Review Questions

Q16-42. What is the goal of the boundary scan standard?

Q16-43. What is the essential idea of the boundary scan standard?

Q16-44. What are the two major operation modes of the boundary scan standard?

16.5 System-Level Testing

In this section, we are concerned with system-level testing, including SRAM BIST and the March test, core-based testing, and SoC testing. Moreover, the IEEE 1500 standard is introduced in brief.

16.5.1 SRAM BIST and March Test

As mentioned before, SRAM is an important component or module in most digital systems. In practice, SRAM modules with a variety of capacities are often embedded into a digital system. Therefore, it is necessary to test them appropriately; otherwise, they might become the faulty corners of the systems.

16.5.1.1 SRAM BIST At present, there are many schemes that can be used to test a memory device or module, such as those reported in Bushnell and Agrawal [6], and Wang et al. [38]. Among these, testing a memory at system level with BIST is probably the most widely used approach. A simple example of using BIST is shown in Figure 16.21. Here, a 10-bit counter generates the test data and an MISR is used to record the signature. In the test mode, the 10-bit counter takes over the control of the

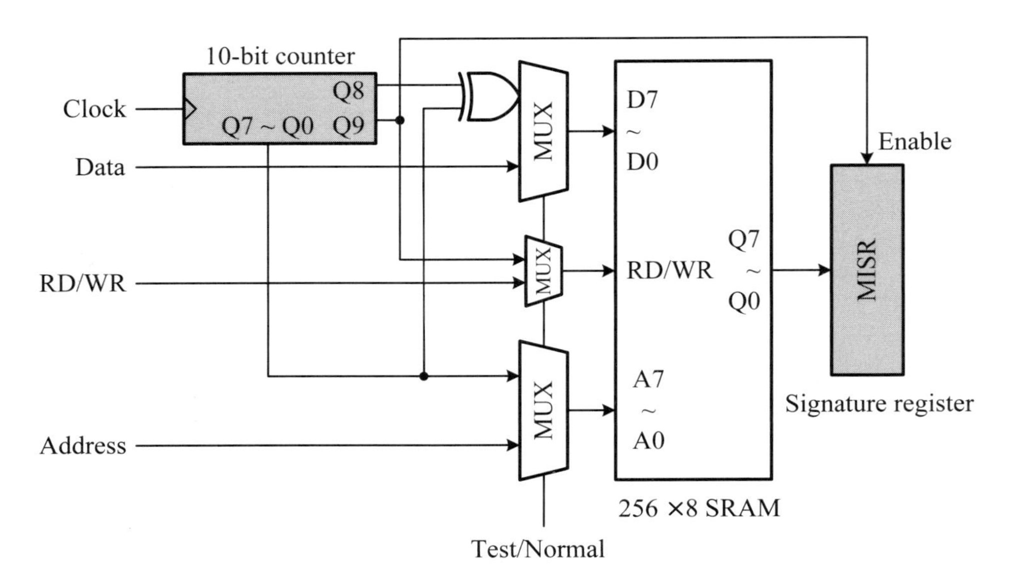

Figure 16.21: SRAM BIST with an MISR.

SRAM and generates all required data for testing the SRAM, including both the read and write modes. The test data are first written into the SRAM and then read out and captured by the MISR to compute the signature. In the normal mode, all BIST hardware are effectively disabled or removed by using multiplexers.

16.5.1.2 March Test Another famous algorithm widely used to test memory with BIST is known as the *March algorithm* or *March test*. As shown in Figure 16.22, the required BIST hardware for the March algorithm includes the following:

- A BIST controller controls the test procedure.
- An address counter generates the required address during the testing procedure.
- A set of multiplexers (MUXs) is used to feed the memory with the required data during self-test from the controller.
- A comparator checks the response from memory by comparing it with data patterns generated by the read pattern generator under the control of the BIST controller.
- A test pattern generator is used to generate the required test patterns for writing into memory.
- A read pattern generator is used to generate the required test patterns for comparing with the data read from memory.

Notations for the March+ test are as follows: **r0** and **r1** denote the read of a 0 and 1 from the specific memory location, respectively, while **w0** and **w1** represent the write of a 0 and 1 to the specific memory location, respectively. The arrows are used to the address-related operations. $\Uparrow$ and $\Downarrow$ represent that the addressing orders are increasing and decreasing, respectively; $\Updownarrow$ denotes that the addressing order can be either increasing or decreasing.

The March test is applied to each cell in memory before proceeding to the next, which means that if a specific pattern is applied to one cell, then it must be applied to all cells. For example, **M0:** $\Updownarrow$(**w0**); **M1:** $\Uparrow$(**r0**, **w1**); **M2:** $\Downarrow$(**r1**, **w0**). The detailed operations of the above March test are summarized as the following algorithm.

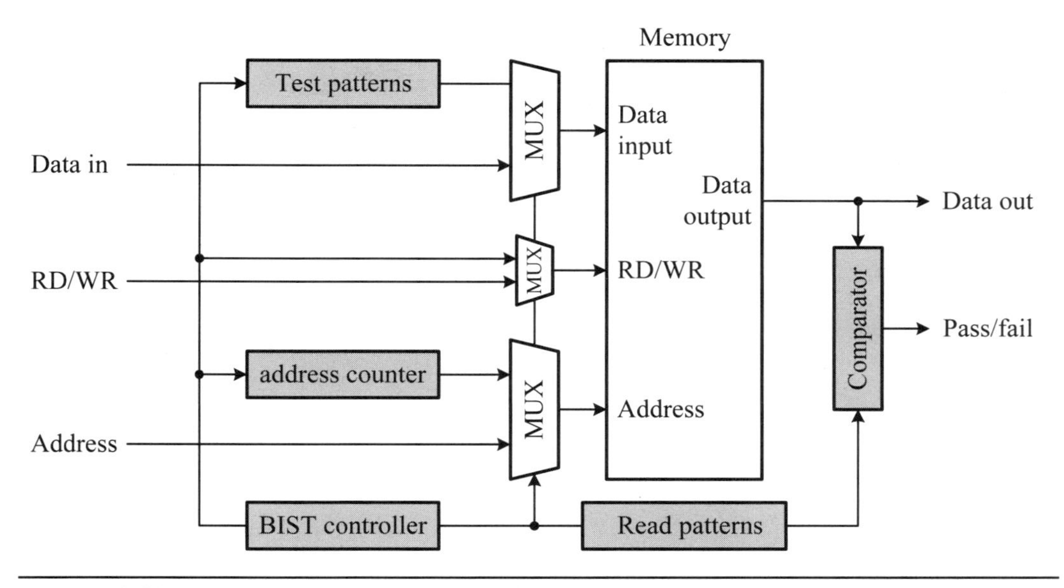

Figure 16.22: The BIST hardware architecture for testing SRAM based on the March test.

Algorithm: MATS+ march test

M0: /* March element $\Updownarrow$ (w0) */
  for(i = 0; i <= n - 1; i++)
    write 0 **to** A[i];
M1: /* March element $\Uparrow$ (r0, w1) */
  for (i = 0; i <= n - 1; i++) **begin**
    read A[i]; /* expected value = 0 */
    write 1 **to** A[i];
  end
M2: /* March element $\Downarrow$ (r1, w0) */
  for (i = n - 1; i >= 0; i − −) **begin**
    read A[i]; /* expected value = 1 */
    write 0 **to** A[i];
  end ∎

Even though memory BIST circuitry may be used to test embedded memory blocks, in the scan mode, the outputs of these memory blocks are still unknown. To overcome this difficulty and increase the fault coverage of the design, a bypass logic is often used, as shown in Figure 16.23. The fault coverage is low due to unknown data appearing at the outputs of memory block when the memory block is not bypassed. However, the fault coverage is increased much more after implementing the bypass logic at the output of the memory block.

16.5.2 Core-Based Testing

A core is an intellectual property (IP) of the vendor, which is a predesigned and verified functional block included on a chip. In general, the vendor provides tests for the core, but the SoC designer must provide the boundary scan chain to a core embedded on

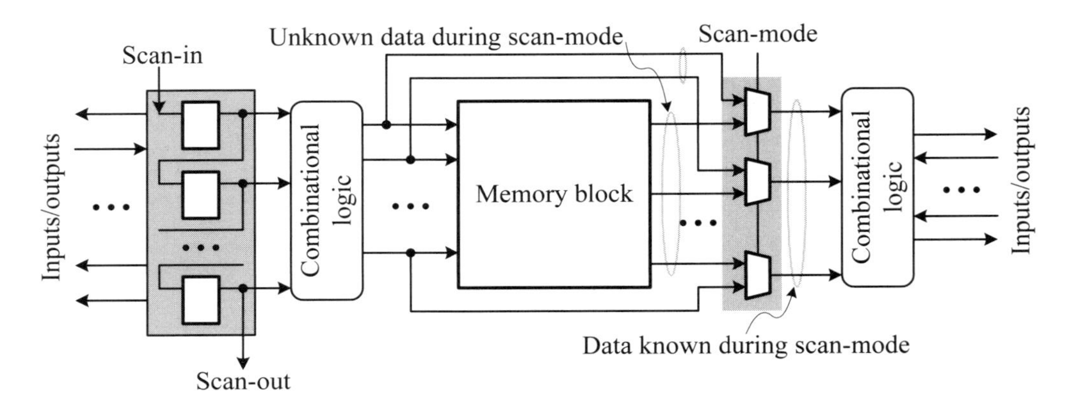

Figure 16.23: Bypassing BIST memory for increasing fault coverage.

the chip. It will become difficult to test the core and its surrounding logic if the core does not embed into the boundary-scan logic.

The most widely used approach to overcoming the above-mentioned difficulty is by surrounding the core with a test logic circuit known as a *test wrapper*, as shown in Figure 16.24. Indeed, this is an application of the concept and technique of the scan-path method.

The test wrapper includes a cell for each core I/O port. For each input port of the core, the test wrapper has to provide three modes of operation: *normal mode*, *external test mode*, and *internal test mode*. The external test mode allows the test wrapper to observe the core input port for the interconnect test, and the internal test mode allows the test wrapper to test the core function.

For each output port of the core, the test wrapper also has to provide three modes of operation: *normal mode*, *external test mode*, and *internal test mode*. The external test mode allows the test wrapper to observe the core output port for the interconnect test, and the internal test mode allows the test wrapper to observe the core output for testing the core function.

16.5.3 SoC Testing

The test of SoC is an extension of the core-based test; that is, it uses the technique that is a combination of the test wrapper concept and the boundary-scan approach. Also, the ATPG and signature analysis or other response checking techniques are often incorporated into the test of SoC. The overall architecture is defined by IEEE in 2005 as a standard known as the *IEEE 1500 standard*. We only briefly describe the basic concepts of the standard. The interested reader may refer to the IEEE Std. 1500-2005 Standard [16] and Wang et al. [38] for further details.

A system overview of the IEEE 1500 standard is shown in Figure 16.25, where the system is composed of n cores, with each being wrapped with a 1500 wrapper. The 1500 standard provides both serial and parallel test modes. The serial test mode is facilitated with the *wrapper serial port* (WSP), which is a set of I/O terminals of the wrapper for serial operations. It comprises the *wrapper serial input* (WSI), the *wrapper serial output* (WSO), and several *wrapper serial control* (WSC) signals. Each wrapper has a *wrapper instruction register* (WIR) used to store the instruction to be executed in the corresponding core.

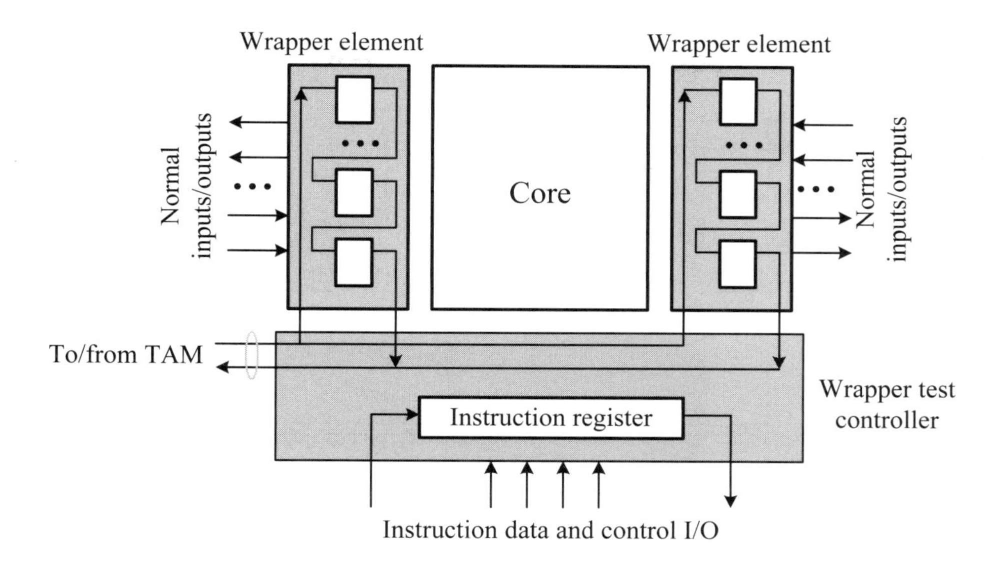

Figure 16.24: The general concept of core-based testing.

The parallel test mode is achieved by incorporating a user-defined, parallel *test access mechanism* (TAM), as shown in Figure 16.25. Each core can have its own TAM-in and TAM-out ports, composed of a number of data or control lines for parallel test operations. The test source includes counters, PRSGs, or test vectors stored in ROMs and the test sink contains the signature analysis or other response compressions.

■ Review Questions

Q16-45. How would you test a core that does not have the test access port?

Q16-46. Explain the operations of Figure 16.21.

Q16-47. Explain the operation of the March test.

Q16-48. What is the required BIST hardware needed in supporting the March test?

Q16-49. What is the meaning of a test wrapper used in core-based testing?

16.6 Summary

Testing is an essential step in any PCB-based, FPGA-based, cell-based, or full-custom system, even in a simple digital logic circuit because the only way to ensure that a system or a circuit may function appropriately is through a careful testing process. The objective of testing is to find the faults in a system or a logic circuit.

There are three types of VLSI testing, including verification testing, manufacturing testing, and acceptance testing. The verification testing verifies the correctness of the design and test procedure. Manufacturing testing is a factory testing by which all manufactured chips for parametric and logic faults and analog specifications are verified to see whether they meet the desired specifications. It includes the wafer test and device test. Acceptance testing, also called incoming inspection, is carried out by

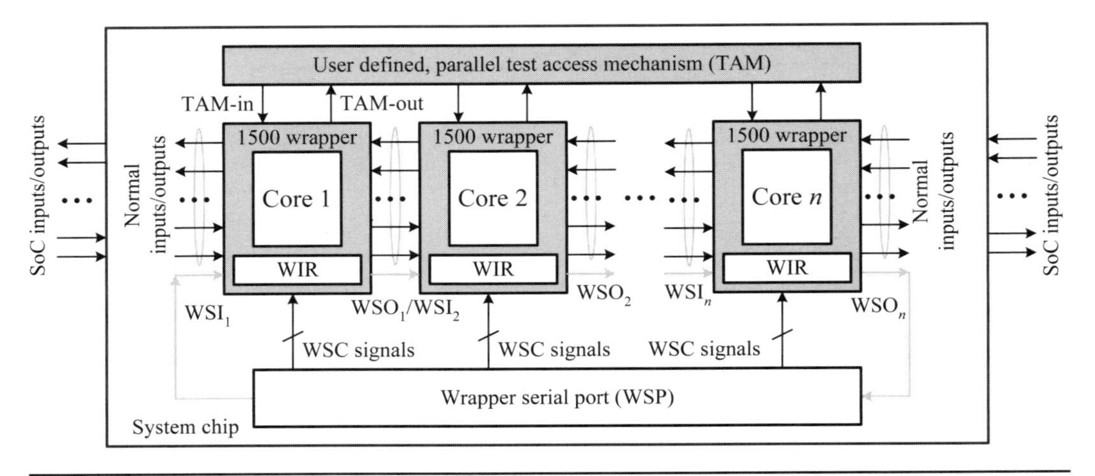

Figure 16.25: A system view of the IEEE 1500 standard for testing an SoC.

the user (customer) to test purchased parts to ensure quality before they are integrated into the target systems.

The most common fault models found in CMOS technology include stuck-at faults, bridge faults, and stuck-open faults. The stuck-at means that the net will adhere to a logic 0 or 1 permanently. The bridge fault means that a short-circuit exists between any set of wires (nets) unintentionally. A stuck-open fault is a feature of CMOS circuits and means that some net is broken during manufacturing or after a period of operation. In deep submicron CMOS processes, delay faults caused by the capacitive-coupling effects become more prevalent with the reduction of feature sizes. The delay fault models covers gate-delay fault, transition fault, and path-delay fault.

To test a fault in a VLSI system or a logic circuit, we have to input stimuli, observe the responses, and analyze the results. The input stimulus special for use in a given VLSI system or a logic circuit is often referred to as a test vector. A test vector is an input combination used to detect a fault in the logic circuit. The collection of test vectors for testing detectable faults in a system or a combinational logic circuit is called a test set. A widely used test vector generation, path sensitization, was described.

The testability of a logic circuit is around two basic features: controllability and observability. To increase these two features of a logic circuit, some extra logic circuits are intended to be added to the design. This results in a technique known as design for testability or a testable circuit design. At present, there are three approaches that are widely used to achieve this purpose. These are the ad hoc approach, scan-path method, and built-in self-test (BIST). The boundary scan standard (namely, IEEE 1149.1) was also introduced.

The final section was concerned with system-level testing based on the scan-path method, including the SRAM BIST and March test, core-based testing, and SoC testing. The most widely used approach to testing a core (or IP) is by surrounding the core with a test wrapper through which test access can be made. The test of SoC is a combination of the core-based test, which uses the test wrapper concepts and techniques, and the boundary-scan approach. Also, the ATPG and signature analysis or other response checking techniques are often incorporated into the test of SoC. The overall architecture of the SoC test was defined by the IEEE in 2005 as a standard known as the IEEE 1500 Standard.

References

1. M. Abramovici, M. A. Breuer, and A. D. Friedman, *Digital Systems Testing and Testable Design*, 2nd ed. Piscataway: IEEE Press, 1996.

2. J. M. Acken, "Testing for bridging faults (shorts) in CMOS circuits," *20th Design Automation Conf.*, pp. 717–718, 1983.

3. R. A. Allen, M. W. Cresswell, and L. W. Linholm, "Junction-isolated electrical test structures for critical dimension calibration standards," *IEEE Trans. on Semiconductor Manufacturing*, Vol. 17, No. 2, pp. 79–83, May 2004.

4. K. Baker and J. van Beers, "Shmoo plotting: the black art of IC testing," *IEEE Design and Test of Computers*, Vol. 14, No. 3, pp. 90–97, July-September 1997.

5. Himanshu Bhatnagar, *Advanced ASIC Chip Synthesis: Using Synopses Design Compiler and Prime Time.* Boston: Kluwer Academic Publishers, 1999.

6. M. L. Bushnell and V. D. Agrawal, *Essentials of Electronic Testing for Digital, Memory & Mixed-Signal VLSI Circuits.* Boston: Kluwer Academic Publishers, 2000.

7. A. Cabrini et al., "A test structure for contact and via failure analysis in deep-submicrometer CMOS technologies," *IEEE Trans. on Semiconductor Manufacturing*, Vol. 19, No. 1, pp. 57–66, February 2006.

8. R. Dekker, F. Beenker, and L. Thijssen, "A realistic fault model and test algorithms for static random access memories," *IEEE Trans. on Computer-Aided Design*, Vol. 9, No. 6, pp. 567–572, June 1990.

9. E. Eichelberger and T. Williams, "A logic design structure for LSI testability," *Journal of Design Automation and Fault Tolerant Computing*, Vol. 2, No. 2, pp. 165–178, May 1978.

10. R. Frowerk, "Signature analysis: a new digital field service method," *Hewlett Packard Journal*, Vol. 28, pp. 2–8, May 1977.

11. J. Galiay, Y. Crouzet, and M. Verginiault, "Physical versus logical fault models MOS LSI circuits: impact on their testability," *IEEE Trans. on Computers*, Vol. 29, No. 6, pp. 527–531, June 1980.

12. C. T. Glover and M. R. Mercer, "A method of delay fault test generation," *Proc. of the 25th ACM/IEEE Design Automation Conf.*, pp. 92–95, 1988.

13. D. Harris, *Skew-Tolerant Circuit Design.* San Francisco, CA: Morgan Kaufmann, 2001.

14. J. P. Hayes, *Introduction to Digital Logic Design.* Reading, MA: Addison-Wesley, 1993.

15. IEEE Std 1149.1-2001 Standard, *IEEE Standard Test Access Port and Boundary-Scan Architecture.* New York: IEEE Press, 2001.

16. IEEE Std 1500-2005 Standard, *IEEE Standard for Embedded Core Test.* New York: IEEE Press, 2005.

17. B. W. Johnson, *Design and Analysis of Fault Tolerant Digital Systems.* Reading, MA: Addison-Wesley, 1989.

18. D. Josephson, "The manic depression of microprocessor debug," *Proc. of the Int'l Test Conf.*, pp. 657–663, 2002.

19. B. Koenemann, J. Mucha, and G. Zwiehoff, "Built-in logic block observation techniques," *Proc. of IEEE Int'l Test Conf.*, pp. 37–41, October 1979.

20. Z. Kohavi and N. K. Jha , *Switching and Finite Automata Theory,* 3rd ed. Cambridge: Cambridge University Press, 2010.

21. P. K. Lala, *Practical Digital Logic Design and Testing.* Upper Saddle River, NJ: Prentice-Hall, 1996.

22. K. Lee and M. Breuer, "Design and test rules for CMOS circuits to facilitate IDDQ testing of bridging faults," *IEEE Trans. on Computer-Aided Design,* Vol. 11, No. 5, pp. 659–670, May 1992.

23. M. B. Lin, *Digital System Design: Principles, Practices, and Applications,* 4th ed. Taipei, Taiwan: Chuan Hwa Book Ltd., 2010.

24. M. B. Lin, *Digital System Designs and Practices: Using Verilog HDL and FPGAs.* Singapore: John Wiley & Sons, 2008.

25. E. J. McCluskey, "Built-in self-test techniques," *IEEE Design and Test of Computers,* Vol. 2, No. 2, pp. 21–28, 1985.

26. E. J. McCluskey, "Built-in self-test structures," *IEEE Design and Test of Computers,* Vol. 2, No. 2, pp. 29–36, 1985.

27. E. J. McCluskey, *Logic Design Principles.* Englewood Cliffs, NJ: Prentice-Hall, 1986.

28. H. Nadig, "Signature analysis—concepts, examples and guidelines," *Hewlett Packard Journal,* Vol. 28, No. 9, pp. 15–21, May 1977.

29. S. Naffziger et al., "The implemetnation of the Itanium 2 microprocessor," *IEEE J. of Solid-State Circuits,* Vol. 37, No. 11, pp. 1448–1460, November 2002.

30. R. Nair, S. Thatte, and J. Abraham, "Efficient algorithms for testing semiconductor random-access memories," *IEEE Trans. on Computers,* Vol. 27, No. 6, pp. 572–576, June 1978.

31. V. P. Nelson, H. T. Nagle, B. D. Carroll, and J. D. Irwin, *Digital Circuit Analysis & Design.* Upper Saddle River, NJ: Prentice-Hall, 1995.

32. T. Oshawa et al.,"A 60-ns 4-Mbit CMOS DRAM with built-in self-test function," *IEEE J. of Solid-State Circuits,* Vol. 22, No. 5, pp. 663–668, October 1987.

33. B. Paul and K. Roy, "Testing cross-talk induced delay faults in static CMOS circuit through dynamic timing analysis," *Proc. of Int'l Test Conf.*, pp. 384–390, October 2002.

34. J. P. Roth, "Diagnosis of automata failures: a calculus and a method," *IBM J. of Research and Development,* Vol. 10, No. 4, pp. 278–291, July 1966.

35. D. S. Suk and S. M. Reddy, "A march test for functional faults in semiconductor random-access memories," *IEEE Trans. on Computers,* Vol. C-30, no. 12, pp. 982–985, 1981.

36. L. T. Wang and E. J. McCluskey, "Hybrid designs generating maximum-length sequences," *IEEE Trans. on Computer-Aided Design,* Vol. 7, No. 1, pp. 91–99, January 1988.

37. L. T. Wang and E. J. McCluskey, "Circuits for pseudoexhaustive test pattern generation," *IEEE Trans. on Computer-Aided Design,* Vol. 7, No. 10, pp. 1068–1080, October 1988.

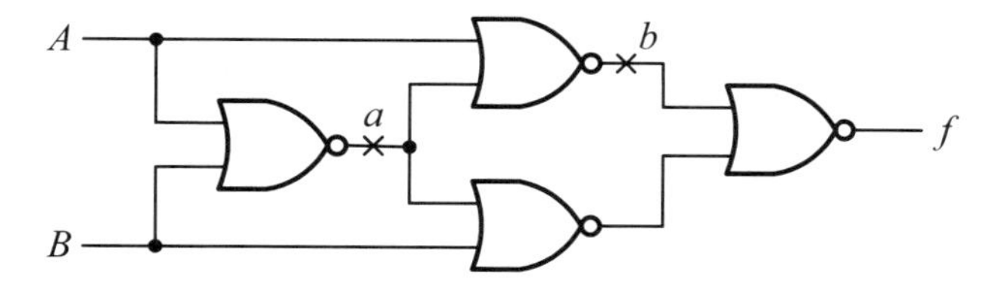

Figure 16.26: The logic circuit for problem 16-2.

38. L. T. Wang, C. W. Wu, and X. Wen, *VLSI Test Principles and Architectures: Design for Testability.* New York: Morgan Kaufmann Publishers, 2006.

39. N. H. E. Weste and D. Harris, *CMOS VLSI Design: A Circuit and Systems Perspective*, 3rd ed. Boston: Addison-Wesley, 2005.

40. T. Williams and K. Parker, "Design for testability: a survey," *Proc. of IEEE*, Vol. 71, No. 1, pp. 98–112, January 1983.

Problems

16-1. Supposing that the stuck-at fault model is used, answer the following questions.

(a) List all equivalent faults of a two-input NAND gate.

(b) List all equivalent faults of a two-input NOR gate.

16-2. Consider the logic circuit shown in Figure 16.26.

(a) Using the single-path sensitization approach, find the test vector(s) for the stuck-at-0 fault at net a.

(b) Using the multiple-path sensitization approach, find the test vector(s) for the stuck-at-0 fault at net a.

(c) Using the path-sensitization approach, find the test vector(s) for the stuck-at-1 fault at net b.

16-3. Consider the logic circuit shown in Figure 16.27.

(a) Using the path-sensitization approach, find the test vector(s) for the stuck-at-1 fault at net a.

(b) Using the path-sensitization approach, find the test vector(s) for the stuck-at-0 fault at net b.

(c) Using the path-sensitization approach, find the test vector(s) for the stuck-at-0 fault at net c.

16-4. The following are problems when using the exhaustive test:

(a) How many test vectors are required for testing an 8-bit binary counter?

(b) Refer to the 8-bit adder example shown in Figure 16.11. Divide the 8-bit counter into two 4-bit counters and draw the resulting circuit.

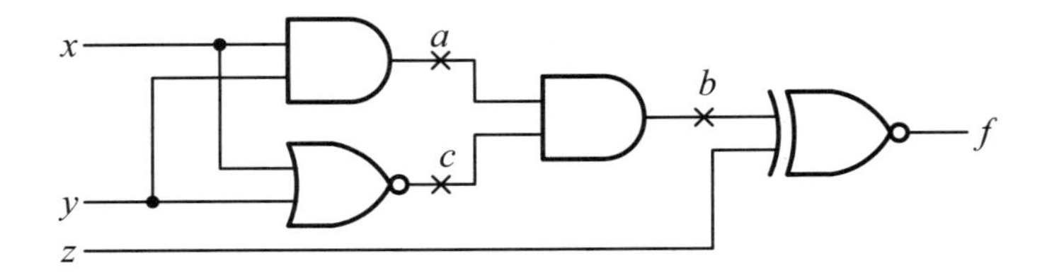

Figure 16.27: The logic circuit for problem 16-3.

(c) How many test vectors are required for testing the above 8-bit counter constructed by cascading two 4-bit counters?

16-5. Let $f(x) = 1 + x + x^5 + x^6 + x^8$; design a BILBO to function as follows:

$m_1m_0 = 00$: Autonomous LFSR mode $m_1m_0 = 10$: Parallel-load mode

$m_1m_0 = 01$: Scan mode $m_1m_0 = 11$: MISR mode

16-6. Let $f(x) = 1 + x^3 + x^{10}$; design a BILBO to function as follows:

$m_1m_0 = 00$: Autonomous LFSR mode $m_1m_0 = 10$: Parallel-load mode

$m_1m_0 = 01$: Scan mode $m_1m_0 = 11$: MISR mode

A

An Introduction to Verilog HDL/SystemVerilog

Using hardware description languages (HDLs) to design digital systems has become an essential part of modern electronic engineering. A major feature inherent in HDLs is that it has the capability of modeling a digital system at many levels of abstraction, ranging from the algorithm-level to the gate-level, even to the switch-level. The two most commonly used HDLs in industry are Verilog HDL (some texts call it Verilog for short) and very high-speed integrated circuit (VHSIC) hardware description language (VHDL). The essence of HDL is to describe a hardware module in text. Hence, we must have a hardware module in mind whenever writing a Verilog HDL module. In other words, we have to think the module in hardware mind rather than software mind. In this appendix, we focus on a synthesizable subset of Verilog HDL.

Verilog HDL originated in 1983 at Gateway Design Automation,[1] as a hardware modeling language associated with their simulator products. Since then, Verilog HDL has gradually gained popularity in industry. To increase its popularity, Verilog HDL was placed in the public domain in 1990. In 1995, it was standardized by IEEE as IEEE Std. 1364-1995, which has been promoted by Open Verilog International (OVI) since 1992. An updated version of the language was standardized by IEEE in 2001 as IEEE Std. 1364-2001. The new version covered many new features, including the port-list declaration style, configurations, and generate statements. The revision proposed in 2005, referred to as IEEE Std. 1364-2005, does not include new features but instead corrects and clarifies features ambiguously described in the 1995 and 2001 editions. In 2005, a superset of Verilog HDL, known as SystemVerilog, was defined by IEEE as a new standard, IEEE Std. 1800-2005, revised in 2009 as IEEE Std. 1800-2009. SystemVerilog added features that have proven useful in verification. Some basic extended features of SystemVerilog relating to synthesizable structures are also introduced in this appendix.

A.1 Introduction

In this section, we first give a simple example and then deal with module structure. Both the interface and body of modules are addressed concisely.

[1] Gateway Design Automation has been acquired by Cadence Design Systems.

A.1.1 A Simple Example of Verilog HDL

Because of the nature of hardware, a Verilog HDL module needs to capture the *timing* and *concurrency* features inherent in the underlying hardware module. The following example illustrates how to describe a hardware module.

■ Example A-1: (A simple Verilog HDL module.) In this example, a module is designed to perform a three-input majority gate. A delay #5 is employed to model the operations of actual hardware. The resulting **assign** continuous statement is to carry out the three-input majority function and produces a result in 5 time units.

```
module majority_gate(
       input  x, y, z,  // I/O ports declaration
       output f);
// specify the three-input majority gate
   assign #5 f = x & y | y & z | x & z;
endmodule
```

■

A Verilog HDL module usually begins with the keyword **module** followed by the module name and a port list (if any), declarations of each item in the port list (if any), an **assign** continuous statement or other statements, and ends with the keyword **endmodule**. The text after // are comments and are ignored by Verilog HDL compilers and simulators. Another way to denote comments is block comments, beginning with /* and ending with */.

The objectives of a Verilog HDL module are synthesis and simulation. The synthesized result of the above module is shown in Figure A.1(a) and its simulation results are plotted in waveform as depicted in Figure A.1(b).

A.1.1.1 Value Set Usually, the logic value has only two values: 0 (logic 0) and 1 (logic 1). However, to describe a real-world hardware circuit, a tristate (or three state) value z is needed to indicate a high-impedance condition of a node or net. In addition, another logic value called x is used to specify an unknown logic value of a net or node. Hence, the resulting logic system is a *four-value logic* system with the value set {0, 1, x, z}.

A.1.1.2 Constants Three types of constants provided in Verilog HDL are *integer*, *real*, and *string*. Examples of constants are as follows:

```
-1234    // is decimal -1234
2011     // is decimal 2011
3.14159  // legal
2.34E6   // the exponent symbol can be e or E
4'b0010  // a 4-bit binary number 0010
16'habcd // a 16-bit number abcd in hexadecimal
```

A.1.2 Module Concepts

The basic unit of a digital system is a *module*, consisting of a *body* (also called an *internal* or *core circuit*) and an *interface* (called *ports*). The body performs the required

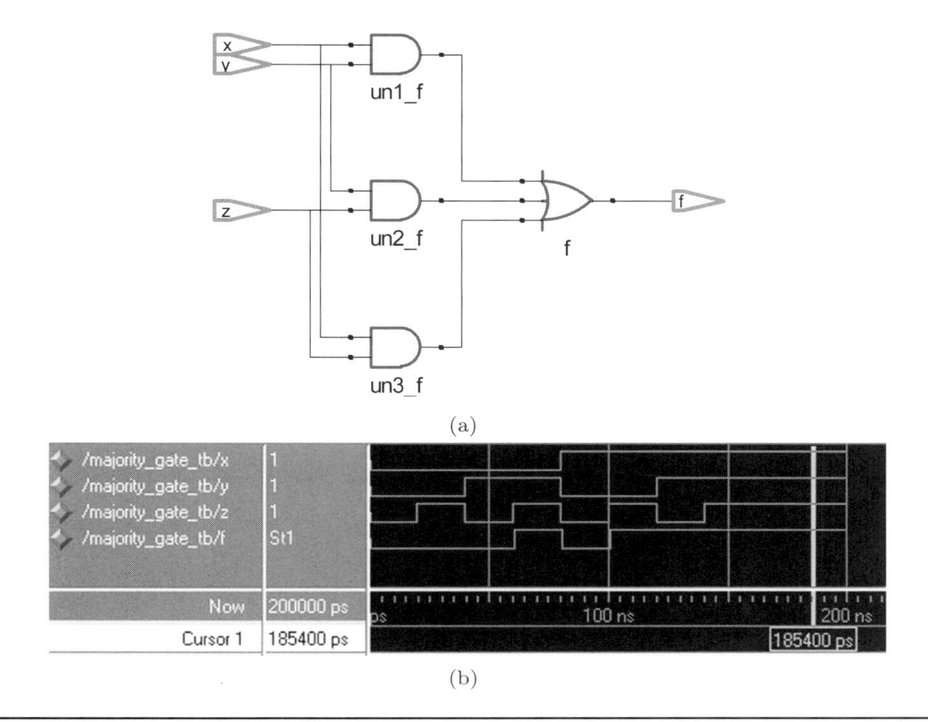

Figure A.1: The (a) synthesized and (b) simulation results of the three-input majority gate.

function of the module while the interface carries out the needed communication between the body and the outside world. In addition, power and ground also need to be provided by the interface. Like a hardware module, a Verilog HDL module can be an element or a collection of lower-level design blocks (modules). The operation to "call" a built-in gate primitive, or another module, is called *instantiation* and each copy of the called primitive or module is called an *instance*.

A.1.2.1 Port Declaration The interface signals (including supply and ground) of any Verilog HDL module can be cast into one of the following three types:

- *Input ports* are a group of input signals declared by the keyword **input**.
- *Output ports* are a group of output signals declared by the keyword **output**.
- *Bidirectional ports* are a group of bidirectional signals declared by the keyword **inout**.

The complete interface of a module can be described in a way like a C function as follows:

```
// port-list declaration style
module adder(input [3:0] x, y,
             input c_in,
             output reg [3:0] sum,
             output reg c_out
); // sometimes called ANSI-C style
```

The net data type **wire** after keywords **input** and **output** is often omitted. Nevertheless, the variable data type **reg** should be declared explicitly. In SystemVerilog, the keyword **logic** may be used to replace **wire** and **reg**.

A.1.2.2 Port Connection Rules Connecting ports to external signals can be done by one of the following two methods:

- **Named association:** In this method, the ports are connected by listing their names. The port identifiers and their associated port expressions are explicitly specified as follows:

  ```
  .port_id1(port_expr1),..., .port_idn(port_exprn)
  ```

  An unconnected port is just skipped or places no signal, such as "`.port_id()`".
- **Positional association:** In this method, the ports are connected by the ordered list of ports, with each corresponding to a port. The form is as follows:

  ```
  port_expr1, ..., port_exprn
  ```

  An unconnected port is just skipped, such as "`x, ,y`", where a port is skipped between `x` and `y`.

However, these two methods cannot be mixed in the same module.

■ Review Questions

QA-1. What are the major differences between a software program and a hardware module?

QA-2. How do we introduce a comment into a Verilog HDL module?

QA-3. How many possible values may a **wire** net or **reg** variable have?

QA-4. What is the difference between named association and positional association?

A.1.3 Module Modeling Styles

We briefly deal with the approaches used to describe the body of a module.

A.1.3.1 Modeling the Body of a Module For each module in Verilog HDL, the body can be modeled as one of the following styles: *structural style*, *dataflow style*, *behavioral or algorithmic style*, and *mixed style*. In industry, the term *register-transfer level* (RTL) is often used to mean a structure that combines both behavioral and dataflow constructs and can be acceptable by logic synthesis tools.

A.1.3.2 Structural Modeling The structural modeling of a design is by connecting required instantiations of built-in primitives or other (user-defined) modules through nets.

■ Example A-2: (A 2-to-1 multiplexer.) By combining two tristate buffers with opposite polarities of enable control: **bufif0** and **bufif1**, a 2-to-1 multiplexer is obtained. The tristate buffer **bufif0** is enabled if `s` is 0 while the tristate buffer **bufif1** is enabled if `s` is 1. Note that because the output f is driven by two tristate buffers, it needs to be declared as a **tri** net.

```
// a 2-to-1 multiplexer
module mux21 (
       input  x, y, s,
       output tri f);
// data selector body
   bufif0 b1 (f, x, s);   // enabled if s = 0
   bufif1 b2 (f, y, s);   // enabled if s = 1
endmodule
```

■

■ Example A-3: (A 4-to-1 multiplexer.) By combining three 2-to-1 multiplexers, a 4-to-1 multiplexer is obtained. The resulting module is as follows:

```
// A 4-to-1 multiplexer
module mux41 (
       input  i0, i1, i2, i3, s1, s0,
       output out);
wire y0, y1;
// the body of 4-to-1 multiplexer
   mux21 mux_a (i0, i1, s0, y0);
   mux21 mux_b (i2, i3, s0, y1);
   mux21 mux_c (y0, y1, s1, out);
endmodule
```

■

A.1.3.3 Dataflow Modeling The **assign** continuous assignment is the essential structure used to model a design in dataflow style. In an **assign** continuous assignment, a value is assigned onto a net. It must be a net because **assign** continuous assignments are used to model the behavior of combinational logic circuits. All **assign** continuous assignments in a module execute concurrently regardless of the order they appear.

■ Example A-4: (A full adder modeled in dataflow style.) In this example, we assume that the adder requires 5 time units to complete its operations. The delay will be ignored by the synthesis tools when the module is synthesized because the delay will be replaced by the actual delays of gates used to realize the adder.

```
module full_adder_dataflow(
       input  x, y, c_in,
       output sum, c_out);

// the body of full adder
assign #5 {c_out, sum} = x + y + c_in;
endmodule
```

■

Like C programming language, Verilog HDL has a rich set of operators, including arithmetic, shift, bitwise, case equality, reduction, logic, relational, equality, and mis-

Table A.1: The summary of operators in Verilog HDL.

Arithmetic	Bitwise	Reduction	Relational
+: add	~: negation (not)	&: and	>: greater than
−: subtract	&: and	|: or	<: less than
∗: multiply	|: or	~&: nand	>=: greater than or equal to
/: divide	^: xor	~|: nor	<=: less than or equal to
%: modulus	^~, ~^: xnor	^: xor	
∗∗: power (exponent)		^~, ~^: xnor	
Logical equality	**Logical**	**Miscellaneous**	**Shift**
==: equality	&&: and	{, }: concatenation	<<: logical left shift
! =: inequality	||: or	{const{}}: replication	>>: logical right shift
Case equality	!: negation (not)	? : : conditional	<<<: arithmetic left shift
===: equality			>>>: arithmetic rigth shift
! ==: inequality			

Table A.2: The precedence of operators in Verilog HDL.

Operators	Symbols	Operation
Unary (plus, minus)	+ −	Highest
Unary (logical negation, bitwise negation)	! ~	
Reduction (and, nand, or, nor, xor, xnor)	& ~& | ~| ^ ^~ (~^)	
Power (exponent)	∗∗	
Multiply, divide, modulus	∗ / %	
Add (binary plus), subtract (binary minus)	+ −	
Logical shift (left, right)	<< >>	
Arithmetic shift (left, right)	<<< >>>	
Relational (less than, less than, or equal to)	< <=	
Relational (greater than, greater than, or equal to)	> >=	
Logical equality (equality, inequality)	== ! =	
Case equality (equality, inequality)	=== ! ==	
Bitwise (and, xor, xnor, or)	& ^ ^~ (~^) |	
Logical (and, or)	&& ||	
Conditional	? :	
Concatenation and replication	{ } {{}}	Lowest

cellaneous operators, as listed in Table A.1. The precedence of operators in Verilog HDL is listed in Table A.2.

■ Example A-5: (A big-endian to little-endian converter.) The following example converts a big-endian representation into its corresponding little-endian one and vice versa. In the big-endian representation, the low-order byte is on the high address while in the little-endian representation, the low-order byte is on the low address.

```
module swap_bytes (
       input  [31:0] in,
       output [31:0] out);
// use part-select
assign out [31:24] = in [7:0],
```

```
        out [23:16] = in [15:8],
        out [15:8]  = in [23:16],
        out [7:0]   = in [31:24];
endmodule
```

■

■ Example A-6: (A 9-bit parity generator.) The operation of reduction operators is to reduce a vector operand into a single-bit result. In the following module, the reduction operator ^ (xor) is employed to evaluate the input vector `x` into a 1-bit result, thereby resulting in a 9-bit parity generator.

```
module parity_gen_9b(
        input  [8:0] x,
        output ep, op);
// the body of 9-bit parity generator
assign ep = ^x;   // even parity generator
assign op = ~ep;  // odd parity generator
endmodule
```

■

Another simple but useful application of reduction operators is to detect whether all bits in a byte are zeros or ones. In the following example, we assume that both results are needed.

■ Example A-7: (An all-bit-zero/one detector.) The outputs `zero` and `one` are assigned to 1 if all bits of the input vector `x` are zeros and ones, respectively. These two detectors are readily implemented by the reduction operators: | (or), and & (and). The resulting module is as follows:

```
module all_bit_01_detector (
        input   [7:0] x,
        output  zero, one);
// the body of zero and one detectors
assign zero = ~|x;  // all-bit zero detector
assign  one = &x;   // all-bit one detector
endmodule
```

■

■ Example A-8: (A 4-to-1 multiplexer.) In this example, a 4-to-1 multiplexer is described by a nested conditional operator in which both true and false expressions also contain their own conditional operators. Although a construct like this is quite concise, it is not easy to understand for a naive reader.

```
module mux41_conditional (
        input  i0, i1, i2, i3,
        input  s1, s0,
```

```
        output out);
// use a nested conditional operator (?:)
assign out = s1 ? ( s0 ? i3 : i2) : (s0 ? i1 : i0);
endmodule
```

■

A.1.3.4 Behavioral Modeling The behavioral style uses the following two procedural constructs: **initial** and **always** statements. The **initial** statement can only be executed once and therefore is usually used to set up initial values of variable data types (such as **reg** and **integer**), whereas the **always** statement, as the name implies, is executed repeatedly. The **always** statements are used to model combinational or sequential logic. Each **always** statement corresponds to a piece of logic. All other statements used to model a logic in behavioral style must be placed in **initial** and/or **always** statements. All **initial** and **always** statements begin their execution at simulation time 0 concurrently and are independent of their order in the module.

■ **Example A-9: (A full adder modeled in behavioral style.)** Basically, the expression used to describe the operations of a 1-bit full adder in behavioral style is the same as that in dataflow style except that it needs to be put inside an **always** statement. In addition, a *sensitivity list* `@(x, y, c_in)` is used to sense the changes of input signals. The resulting module is as follows:

```
module full_adder_behavioral(
        input      x, y, c_in,
        output reg sum, c_out);

// the body of full adder
always @(x, y, c_in) // or always @(x or y or c_in)
   #5 {c_out, sum} = x + y + c_in;
endmodule
```

■

In SystemVerilog, **always @(...)** may be replaced by **always_comb** to emphasize that the piece of logic described by the **always** block is a combinational logic circuit in nature.

A.1.3.5 Mixed-Style Modeling The mixed-style modeling is usually used to construct a hierarchical design in large systems. Nevertheless, we are still able to model a simple design in mixed style. The following example illustrates this idea by using the full adder depicted in Figure 1.31. This full adder consists of two half adders and an OR gate.

■ **Example A-10: (A full adder modeled in mixed style.)** The first half adder is modeled in structural style, the second half adder in dataflow style, and the OR gate in behavioral style.

```
module full_adder_mixed_style(
        input      x, y, c_in,
        output     s,
```

```
        output reg c_out);
// internal nets
wire   s1, c1, c2;
   // model HA 1 in structural style
   xor xor_ha1 (s1, x, y);
   and and_ha1(c1, x, y);
   // model HA 2 in dataflow style
assign s = c_in ^ s1;
assign c2 = c_in & s1;
   // model the output OR gate in behavioral style
always @(c1, c2) // or use always @(*)
   c_out = c1 | c2;
endmodule
```

■

■ Review Questions

QA-5. Describe the features of structural style.
QA-6. What is the essential structure of dataflow style?
QA-7. Explain the operations of an **initial** block.
QA-8. Explain the operations of an **always** block.
QA-9. What are the basic statements used in behavioral style?

A.2 Behavioral Modeling

Like any other programming languages, such as C, Verilog HDL also provides the three basic structures: assignments, selection statements, and iterative (loop) statements.

A.2.1 Assignments

Assignment statements contain **assign** continuous assignment and procedural assignments, including blocking assignment (=) and nonblocking assignment (<=). In the following, we deal with procedural assignments.

As mentioned previously, an **assign** continuous assignment is used to continuously assign values onto a net (such as a **wire**) in a manner similar to the way that a logic gate drives a net. In contrast, a procedural assignment puts values in a variable (such as a **reg**), which holds the value until the next procedural assignment updates that variable.

■ Example A-11: (A 4-to-1 multiplexer.) A procedural assignment within the **always** statement is used to implement the operations of the 4-to-1 multiplexer. The procedural assignment is the same as the **assign** continuous assignment except that the keyword **assign** is absent.

```
module mux41_behavioral(
       input  i0, i1, i2, i3, // data input
       input  s1, s0,         // source selection
       output reg out);       // data output
```

```
// use bitwise operators
always@(*) // or always @(i0, i1, i2, i3, s1, s0)
   out = (~s1 & ~s0 & i0)|
         (~s1 &  s0 & i1)|
         ( s1 & ~s0 & i2)|
         ( s1 &  s0 & i3);
endmodule
```

■

A.2.1.1 Blocking versus Nonblocking Assignments Suppose that we want to model the 4-bit shift register depicted in Figure 10.18(a). For simplicity, we ignore the reset function here. An incorrect module shown in the following example might be written by the naive reader.

■ Example A-12: (An incorrect version of shift register module.) According to the logic diagram shown Figure 10.18(a), it is quite straightforward to directly employ four blocking assignments, with each describing one flip-flop. However, the resulting module fails to work properly. The reason is that blocking assignment statements are executed in the order specified, one after the other, and after they are optimized by a logic synthesizer, the final result is exactly the same as the one shown in Figure A.2(a).

```
// a shift register module --- an incorrect version
module shift_reg_blocking(
       input clk,  // clock input
       input sin,  // serial data input
       output reg [3:0] qout);
// the body of 4-bit shift register
always @(posedge clk)
   begin              // using blocking assignments
      qout[0] = sin;
      qout[1] = qout[0];
      qout[2] = qout[1];
      qout[3] = qout[2];
   end
endmodule
```

■

The essential operation of shift register is that at each positive edge of the clock, each flip-flop is assigned a new value, which is the output of its previous stage. Thus, the value to be assigned to a flip-flop at each positive edge of the clock is like the one taken from a temporary variable, which stores the output of its previous stage in the previous clock cycle. This is exactly the operation that nonblocking assignments perform.

■ Example A-13: (A 4-bit shift register.) In this example, we replace the four blocking assignments in the preceding example with nonblocking assignments. The synthesized result is correct and shown in Figure A.2(b). Of course, an even better approach is to use a single nonblocking assignment: `qout <= {qout[2:0], sin}`.

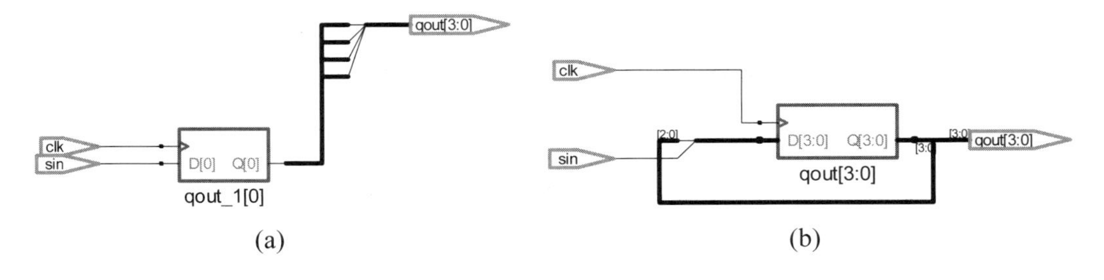

Figure A.2: The synthesized results of a shift register described by using (a) blocking and (b) nonblocking assignments, respectively.

```
// a shift register module --- a correct version
module shift_reg_nonblocking(
       input  clk,  // clock input
       input  sin,  // serial data input
       output reg [3:0] qout);
// the body of 4-bit shift register
always @(posedge clk)
   begin
      qout[0] <= sin;
      qout[1] <= qout[0];
      qout[2] <= qout[1];
      qout[3] <= qout[2];
   end
endmodule
```

■

In SystemVerilog, the **always @(posedge...)** statement may be replaced by the **always_ff @(posedge...)** statement to emphasize the fact that the piece of logic described by the **always** block is a flip-flop and its associated logic.

A.2.2 Selection Statements

Selection statements are used to make a selection according to a given condition. They contain two statements: **if-else** and **case**, including **case**, **casex**, and **casez**.

A.2.2.1 The if-else Statement Basically, an **if-else** statement is used to perform a two-way selection according to a given condition. Nonetheless, a nested **if-else** statement is also allowed so that it can carry out a multiway selection as well.

■ Example A-14: (A 4-bit synchronous binary counter.) This example describes a modulo-16 counter. The counter is cleared whenever a positive-edge `clear` signal is activated; otherwise, the counter counts up by 1 at every negative edge of the `clock` signal. The clear operation is asynchronous since the `clear` signal is put inside the sensitivity list to serve as a trigger source.

```
module counter_4b (
       input  clock, clear,
```

```
        output reg [3:0] qout);
// the body of 4-bit binary counter
always @(negedge clock or posedge clear) begin
   if (clear) qout <= 4'd0;
   else       qout <= (qout + 1);
end
endmodule
```

■

A.2.2.2 The case Statement A **case** statement is used to perform a multiway selection according to a given input condition and equivalent to a nested **if-else** statement. The **case** statement acts like a multiplexer. Consequently, it is often used to model a multiplexer, such as the following 4-to-1 multiplexer.

■ Example A-15: (A 4-to-1 multiplexer.) In this example, a **case** statement is used to model a 4-to-1 multiplexer. Based on this, it is quite straightforward to extend to the case of an n-to-1 multiplexer.

```
// a 4-to-1 multiplexer using a case statement
module mux41_case (
        input i0, i1, i2, i3,
        input [1:0] s,     // two-bit selection signal
        output reg  y);
// the body of 4-to-1 multiplexer
always @(*) // or always @(i0, i1, i2, i3, s)
   case (s)
      2'b00: y = i0;
      2'b01: y = i1;
      2'b10: y = i2;
      2'b11: y = i3;
   endcase
endmodule
```

■

A.2.2.3 The casex and casez Statements Both **casex** and **casez** statements are used to perform a multiway selection like that of the **case** statement except that the **casez** statement treats all `z` values as "don't cares" while the **casex** treats all `x` and `z` values as don't cares. These two statements only compare non-`x` or `z` positions in the `case_expression` and the `case_item` expressions.

■ Example A-16: (Counting the trailing zeros in a nibble.) Because the **casex** statement only compares non-`x` or `z` positions in the `case_expression` and the `case_item` expressions, by properly arranging each `case_item` and its associated expression in the order as shown in the module, the job is done.

```
// count the trailing zeros in a nibble
module trailing_zero_4b(
        input  [3:0] data,
        output reg [2:0] out);
```

```
// the body of trailing-zero counting module
always @(data)
   casex (data) // treat both x and z as don't care
      4'bxxx1: out = 0;
      4'bxx10: out = 1;
      4'bx100: out = 2;
      4'b1000: out = 3;
      4'b0000: out = 4;
      default: out = 3'b111;
   endcase
endmodule
```

■

■ Review Questions

QA-10. What is the distinction between blocking and nonblocking assignments?

QA-11. What are the differences between **case** and **casex** statements?

QA-12. What are the differences between **casex** and **casez** statements?

A.2.3 Iterative (Loop) Statements

Iterative (loop) statements are used to repeatedly execute a set of statements. There are four iterative (loop) statements: **while**, **for**, **repeat**, and **forever**.

A.2.3.1 The while Statement The **while** loop is used to perform a procedural statement until the given condition becomes false.

■ Example A-17: (Counting the zeros in a byte.) A **while** loop is used to control the number of iterations, and an **if** statement is used to check and add up the number of nonzero bits. The loop is terminated when the maximum number of iterations is reached. The bit value is checked in sequence from bit 0 to bit 7, which is controlled by an **integer** variable `i`. Note that the **if** statement may be replaced with `out = out + ~data[i]`. Why? Give the reasons for this.

```
// count the zeros in a byte
module zero_count_while (
       input      [7:0] data,
       output reg [3:0] out);
// the body of zero-counting module
integer i;                 // loop counter
always @(data) begin
   out = 0; i = 0;
   while (i <= 7) begin  // simple condition
      if (data[i] == 0) out = out + 1;
      i = i + 1; end
end
endmodule
```

■

A.2.3.2 The for Statement The **for** loop is used to perform a procedural statement a fixed number of times. The behavior of the **for** loop is much like the **for** statement in C programming language.

■ Example A-18: (Counting the trailing zeros in a byte.) This example counts the trailing zeros in a byte. A **for** loop is used to check the bit value from bit 0 to 7 in sequence and add up the number of zero bits. The loop is terminated whenever a nonzero bit is found or the maximum number of iterations is reached.

```
// count the trailing zeros in a byte
module trailing_zero_for(
       input      [7:0] data,
       output reg [3:0] out);
// the body of trailing-zero counting module
integer i;              // loop counter
always @(data) begin
   out = 0;
   // using a complex condition
   for (i = 0; data[i] == 0 && i <= 7; i = i + 1)
       out = out + 1;
end
endmodule
```

■

A.2.3.3 The repeat Statement The **repeat** loop is used to perform a procedural statement a specified number of times. The following program segment repeatedly executes the **begin-end** block 12 times.

```
   i = 0;
   repeat (12) begin
      state[i] = 0;        // initialize to zeros
      i = i + 1;           // next item
   end
```

A.2.3.4 The forever Statement The **forever** loop continuously performs a procedural statement until the **$finish** system task is encountered or is got out by a **disable** statement within the procedural statement.

In the following, a simple **forever** loop is used to generate a clock signal with a period of 10 time units and a duty cycle of 50%:

```
   initial begin
      clock <= 0;
      forever #5 clock <= ~clock;
   end
```

■ Review Questions

QA-13. What are the four types of loop statements?

QA-14. Describe the operations of the **while** loop statement.

QA-15. Describe the operations of the **for** loop statement.

QA-16. Describe the operations of the **repeat** loop statement.

A.3 Hierarchical Structural Modeling

In this section, we consider two closely related issues of hierarchical structural modeling: instantiations and generate statements.

A.3.1 Parameterized Modules

Parameters are constants, which can be used throughout the module defining them, and are often used to specify delays and widths of variables. Parameters are defined by the keyword **parameter**. Some examples are as follows:

```
parameter SIZE = 8;
parameter WIDTH_BUSA = 16, WIDTH_BUSB = 8;
parameter signed [3:0] mux_selector = 4'b0;
parameter integer CNT_SIZE = 12;
```

■ **Example A-19: (A parameterized adder.)** In this example, an n-bit adder is defined by using the keyword **parameter**. The value defined by the **parameter** is called a *default value*, which can be redefined or overridden whenever the module is instantiated.

```
module adder_nbits
       #(parameter N = 4)(  // default size
         input    [N-1:0] x, y,
         input    c_in,
         output   [N-1:0] sum,
         output   c_out);
// the body of n-bit adder
assign {c_out, sum} = x + y + c_in;
endmodule
```

■

A.3.2 Instantiation of Modules

The hierarchical structure is formed by embedding modules into others. Higher-level modules create instances of lower-level modules and communicate with them through input, output, and bidirectional ports, which can be scalar or vector. The two ways that can be used to override parameter values defined in lower-level modules are as follows:

1. **defparam** statement
2. module instance parameter value assignment

A.3.2.1 Using the defparam Statement A **defparam** statement is used to redefine the values of parameters defined by the keyword **parameter** in any module instance throughout the design using the hierarchical path names of the parameters. This approach is particularly useful for grouping together the parameter overriding assignments in the same module.

■ Example A-20: (Parameter overriding using defparam statement.) This example demonstrates how the **defparam** statement is used to change the parameter values defined within the lower-level module, `counter_nbits`. The default value of `N` within the module `counter_nbits` is 4. The result is that two instances of `counter_nbits` with bits of 4 and 8, respectively, are created.

```
// define top level module
module two_counters(
       input clock, clear,
       output [3:0] qout4b,
       output [7:0] qout8b);
// instantiate two counter modules
defparam cnt_4b.N = 4, cnt_8b.N = 8;
counter_nbits cnt_4b (clock, clear, qout4b);
counter_nbits cnt_8b (clock, clear, qout8b);
endmodule

module counter_nbits
       #(parameter N = 4)( // default counter size
         input       clock, clear,
         output reg [N-1:0] qout);
always @(negedge clock or posedge clear) begin
   if (clear) qout <= {N{1'b0}};
   else       qout <= (qout + 1) ;
end
endmodule
```

■

A.3.2.2 Module Instance Parameter Value Assignment In this approach, the parameters defined with the keyword **parameter** within a module are overridden by the parameters passed through parameter ports whenever the module is instantiated.

■ Example A-21: (Module instance parameter value assignment.) In this example, two instances of module `counter_nbits` are instantiated and the new parameters are passed to the instantiated modules through module instance parameter value assignments with positional association. One of them is a 4-bit counter and the other is an 8-bit counter.

```
// define top level module
module two_counters(
       input clock, clear,
       output [3:0] qout4b,
       output [7:0] qout8b);
// instantiate two counter modules
counter_nbits #(4) cnt_4b (clock, clear, qout4b);
counter_nbits #(8) cnt_8b (clock, clear, qout8b);
endmodule
```

■

Review Questions

QA-17. Describe the two methods that can be used to override module parameter values defined by **parameter**.

QA-18. What does a parameterized module mean?

QA-19. Explain the meaning of the **defparam** statement.

A.3.3 The generate Statement

The **generate** statement may conditionally generate declarations and instantiations into a design. Almost what can be put inside a module can also be placed within **generate** statements. The **generate** statement can be employed to build a large regular design, such as n-bit ripple carry adders, $m \times n$ array multipliers, and $m \times n$ array dividers, among others.

There are three kinds of statements that can be used to create **generate** statements: *generate-loop*, *generate-conditional*, and *generate-case*.

A.3.3.1 Generate-Loop Statement A generate-loop is formed by using a **for** statement within a **generate** statement. The generate-loop allows modules, gate primitives, **assign** continuous assignments, **initial** blocks, and **always** blocks to be duplicated at elaboration time.

Example A-22: (The generate-loop statement.) This example uses an **assign** continuous assignment within a generate-loop for converting Gray code into binary code. To convert a Gray code into its equivalent binary code, we may count the number of "1s" from most-significant bit (MSB) to the current position, i. If it is odd, the binary bit is 1, otherwise the binary bit is 0. Therefore, the reduction operator ^ is applied to an indexed subrange, counting from $N-1$ to i, of the **gray** vector and reduces it into a single binary bit. By repeating this process from bit 0 to bit $N-1$, the conversion is completed.

```
// an example of converting Gray code into binary code
module gray2bin
       #(parameter N = 8)( // set default value
         input  [N-1:0] gray,
         output [N-1:0] bin);
genvar i;                 // define generate-loop index
generate for (i = 0; i < N; i = i + 1) begin: bit_convert
   assign bin[i] = ^gray[N-1:i];
end endgenerate
endmodule
```

An **always** statement can also be used to replace the **assign** continuous assignment.

A.3.3.2 Generate-Conditional Statement A generate-conditional statement allows modules, gate primitives, **assign** continuous assignments, **initial** blocks, and **always** blocks to be instantiated into another module based on an **if-else** conditional expression. An illustration is given in the following example.

■ **Example A-23: (A parameterized n-bit ripple-carry adder.)** In this example, a conditional expression **if-else** is employed to set up the boundary cells, the least-significant bit (LSB) and the MSB, as well as the other bits. The LSB cell needs to accept the external carry input `c_in` and the MSB cell needs to send the carry `c_out` out of the module. Each of the other cells accepts the carry from its preceding cell and sends the carry to its succeeding cell.

```
// using module instantiations inside generate block
module adder_nbits_generate
       #(parameter N = 4)(// set default value
         input  [N-1:0] x, y,
         input  c_in,
         output [N-1:0] sum,
         output c_out);
// the body of N-bit adder using generate statement
genvar i;
wire   [N-2:0] c;      // internal carries declared as nets
generate
for (i = 0; i < N; i = i + 1) begin: adder
   if (i == 0)          // specify LSB
      full_adder fa (x[i], y[i], c_in, sum[i], c[i]);
   else if (i == N-1) // specify MSB
      full_adder fa (x[i], y[i], c[i-1], sum[i], c_out);
   else                 // specify other bits
      full_adder fa (x[i], y[i], c[i-1], sum[i], c[i]);
end endgenerate
endmodule

module full_adder (
       input  x, y, c_in,
       output sum, c_out);

// the body of full adder
assign #5 {c_out, sum} = x + y + c_in;
endmodule
```

■

A.3.3.3 Generate-Case Statement Like the generate-conditional statement, a generate-case statement allows modules, gate primitives, **assign** continuous assignments, **initial** blocks, and **always** blocks to be conditionally instantiated into another module.

■ **Example A-24: (The generate-case statement.)** This example demonstrates how the parameter port can be used to control the actual implementation of the n-bit adder. The n-bit adder is implemented by instantiating the `adder_nbit` when the `WIDTH` is 4 and 8, and by instantiating the `adder_cla` when the `WIDTH` has other values.

```
module adder_nbits
       #(parameter WIDTH = 4)(  // set default value
         input  [WIDTH-1:0] x, y,
```

```
        input  c_in,
        output [WIDTH-1:0] sum,
        output c_out);
// the body of n-bit adder using generate statement
generate // instantiate a proper instance based on WIDTH
   case (WIDTH)
           4: adder_nbits #(4)  adder4(x,y,c_in,sum,c_out);
           8: adder_nbits #(8)  adder8(x,y,c_in,sum,c_out);
     default: adder_cla   #(WIDTH) cla(x,y,c_in,sum,c_out);
   endcase
endgenerate
endmodule
```

■

■ Review Questions

QA-20. What components can be used within a **generate** statement?

QA-21. Describe the features of generate-loops.

QA-22. Describe the features of generate-conditional statements.

QA-23. Describe the features of generate-case statements.

QA-24. Can generate-loops be nested?

A.4 Combinational Logic Modules

The widely used combinational logic modules are decoders, priority encoders, multiplexers, demultiplexers, comparators, and tristate buffers.

A.4.1 Decoders

A decoder is a code converter that converts an input code into a desired output code. To describe such a combinational logic circuit, a convenient way is to use the **case** statement, such as the following example.

■ Example A-25: (A 2-to-4 decoder.) This example uses a **case** statement to code the function table shown in Figure 10.2(b). The output vector y is all 1s if the enable control is high and depends on the input vector x otherwise.

```
// a 2-to-4 decoder with both active-low output and
// enable control
module decoder_2to4_low(
       input  [1:0] x,
       input  enable_n,
       output reg [3:0] y);
// the body of the 2-to-4 decoder
always @(x or enable_n)
   if (enable_n) y = 4'b1111; else
      case (x)
         2'b00 : y = 4'b1110;
```

```
      2'b01 : y = 4'b1101;
      2'b10 : y = 4'b1011;
      2'b11 : y = 4'b0111;
   endcase
endmodule
```

The function of ROM is fundamentally a decoder. Hence, the **case** statement can be employed to describe a ROM device as well.

A.4.2 Priority Encoders

Because of the inherent priority feature of the **if-else** statement, it is possible to describe a priority encoder using such a feature. The following example illustrates this idea.

■ Example A-26: (A 4-to-2 encoder.) This example uses a nested **if-else** statement to build a 4-to-2 encoder. Each conditional expression in the nested **if-else** statement compares the input code sequentially with a constant that determines the corresponding output value, as specified in the function table shown in Figure 10.4(b). For simplicity, here we omit the `valid` output signal.

```
// a 4-to-2 priority encoder using if-else statements
module priority_encoder_4to2_ifelse(
       input   [3:0] in,
       output reg [1:0] y);
// the body of 4-to-2 encoder
always @(in) begin
   if (in[3] == 1'b1) y = 2'b11; else
   if (in[2] == 1'b1) y = 2'b10; else
   if (in[1] == 1'b1) y = 2'b01; else
   if (in[0] == 1'b1) y = 2'b00; else
                      y = 2'bx;
end
endmodule
```

The priority encoder may also be described with a **casex** statement. Recall that the **casex** statement only compares non-x positions in the `case_expression` and `case_item`. As a consequence, the function table shown in Figure 10.4(b) can be coded into `case_item` directly.

■ Example A-27: (A 4-to-2 priority encoder.) This example uses a **casex** statement to model a 4-to-2 priority encoder by coding the function table directly into `case_item`. Nevertheless, because of the **casex** statement being incompletely specified, a **default** statement is needed to avoid inferring a latch.

```
// a 4-to-2 priority encoder using a casex statement
module priority_encoder_4to2_casex(
```

```
        input   [3:0] in,
        output reg [1:0] y,
        output valid_in);
// the body of 4-to-2 priority encoder
assign valid_in = |in;
always @(in) casex (in)
   4'b1xxx: y = 3;
   4'b01xx: y = 2;
   4'b001x: y = 1;
   4'b0001: y = 0;
   default: y = 2'bx;
endcase
endmodule
```

■

A.4.3 Multiplexers

Like decoders or encoders, multiplexers are also ready to be described by the **case** statement. Many examples have been introduced before. We only further consider a 3-to-1 multiplexer, which needs the use of the **default** statement.

■ **Example A-28: (A 3-to-1 multiplexer.)** In this example, a **default** statement is needed to completely specify the **case** statement to avoid the unwanted latch inferred by synthesis tools. The procedural assignment used to assign a value to the output y when the **default** statement is executed must consider the actual requirement, such as to leave the output an unknown x, a z, or a definite known value 0 or 1.

```
// a 3-to-1 multiplexer using a case and default statements
module mux31_case (
        input i0, i1, i2,
        input [1:0] s,     // two-bit selection signal
        output reg  y);
// the body of 3-to-1 multiplexer
always @(i0 or i1 or i2 or s) // or use always @(*)
   case (s)
      2'b00: y = i0;
      2'b01: y = i1;
      2'b10: y = i2;
      default: y = 1'b0;
   endcase
endmodule
```

■

A.4.4 Demultiplexers

The following example models an n-bit 1-to-4 demultiplexer without an enable control in behavioral style by using a nested **if-else** statement.

■ **Example A-29: (An n-bit 1-to-4 demultiplexer.)** This example uses a nested **if-else** statement to describe an n-bit 1-to-4 demultiplexer.

```
// an N-bit 1-to-4 demultiplexer using if-else statements
module demux_1to4_ifelse
       #(parameter N = 4)(  // set default width
         input   [1:0] select,
         input   [N-1:0] in,
         output reg [N-1:0] y3, y2, y1, y0);

// the body of N-bit 1-to-4 demultiplexer
always @(select or in) begin
   if (select == 3) y3 = in; else y3 = {N{1'b0}};
   if (select == 2) y2 = in; else y2 = {N{1'b0}};
   if (select == 1) y1 = in; else y1 = {N{1'b0}};
   if (select == 0) y0 = in; else y0 = {N{1'b0}};
end
endmodule
```

■

A.4.5 Magnitude Comparators

A cascadable n-bit magnitude comparator can be easily described by using relational and logical equality operators, such as the following example.

■ **Example A-30: (A cascadable n-bit magnitude comparator.)** Because the operation of a magnitude comparator is to compare and indicate the relative magnitude of the two input numbers, logical equality (==) and relational operators, greater than (>) and less than (<), are suitable for describing such a circuit. As stated, three inputs used to indicate the input relationship from its preceding stage are also provided to design a cascadable module.

```
// a cascadable N-bit comparator example
module comparator_cascadable
       #(parameter N = 4)(   // set default size
         input  Iagtb, Iaeqb, Ialtb,
         input  [N-1:0] a, b,
         output reg Oagtb, Oaeqb, Oaltb);
// the body of N-bit comparator
always @(*) begin
   Oaeqb = (a == b) && (Iaeqb == 1);          // equality
   Oagtb = (a > b)||((a == b)&&(Iagtb == 1));// greater than
   Oaltb = (a < b)||((a == b)&&(Ialtb == 1));// less than
end
endmodule
```

■

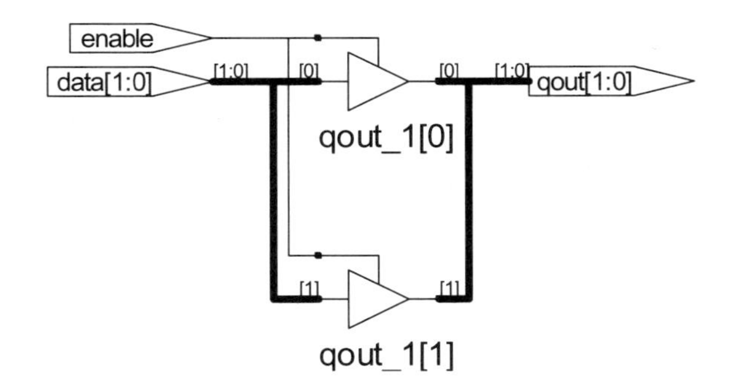

Figure A.3: The block diagram of a 2-bit tristate buffer.

A.4.6 Tristate Buffers

Tristate buffers can be used instead of multiplexers to construct a bus, although it is not good practice. We first give an n-bit tristate buffer and then describe the construction of a tristate bus.

■ Example A-31: (An n-bit tristate buffer.) Suppose an n-bit tristate buffer is to be modeled. When the enable control is asserted, the buffer is enabled and places its input data on the output. Otherwise, the output of the buffer is in a high-impedance state.

```
// a tristate buffer
module tristate_buffer
      #(parameter N = 2)(    // define buffer width
        input  enable,
        input  [N-1:0] data,
        output [N-1:0] qout);

// the body of tristate bus
assign qout = enable ? data : {N{1'bz}};
endmodule
```

■

The synthesized result is depicted in Figure A.3. You may try it on your system and check the result. The following example further explores how to use the tristate buffers to construct a tristate bus.

■ Example A-32: (An n-bit tristate bus.) In this example, four n-bit source modules are assumed to be connected on the same bus through tristate buffers. Each tristate buffer has its own enable control to control its operation independently. When the enable control is asserted, the tristate buffer is enabled and its input data are placed onto the bus. Otherwise, the output of the tristate buffer is in a high-impedance state. To avoid the data conflict on the bus, at any time at most one tristate buffer can be enabled. Because `qout` is driven by multiple drivers, it needs to be declared as a **tri** net.

```
// a tristate bus with four drivers
module tristate_bus
       #(parameter N = 2)(    // define bus width
         input  [3:0] enable,
         input  [N-1:0] data0, data1, data2, data3,
         output tri [N-1:0] qout);
// the body of tristate bus
assign qout = enable[0] ? data0 : {N{1'bz}},
       qout = enable[1] ? data1 : {N{1'bz}},
       qout = enable[2] ? data2 : {N{1'bz}},
       qout = enable[3] ? data3 : {N{1'bz}};
endmodule
```

■

■ Review Questions

QA-25. Describe the operation of an m-to-n decoder.

QA-26. Describe an m-to-n priority encoder by using the **casex** statement.

QA-27. Describe the operation of a multiplexer.

QA-28. Describe the operation of a demultiplexer.

A.5 Sequential Logic Modules

Latches and flip-flops are basic sequential elements. Synchronizers, counters, registers, shift registers, and register files are composed of flip-flops. We cope with these with examples. In addition, synchronous RAM and FSM modeling are also addressed with examples.

A.5.1 Latches

As mentioned, the major difference between latches and flip-flops is the transparent property. Latches are level-sensitive devices and have a transparent property, whereas flip-flops are edge-sensitive and have no transparent property.

■ Example A-33: (A D-type latch.) Because of the inherent transparent property associated with a latch, both enable and data input need to be put inside the sensitivity list of the **always** block.

```
// a D-type latch
module Dlatch (
       input  enable,
       input       [3:0] d,
       output reg  [3:0] q);
// the body of D-type latch
always @(enable or d)  // always_latch in SystemVerilog
   if (enable) q <= d;
endmodule
```

■

In SystemVerilog, the keyword **always** used in an **always** block being used to describe a piece of logic associated with a latch may be replaced by **always_latch** to emphasize this fact explicitly.

A.5.2 Flip-Flops

Flip-flops are edge-sensitive devices and have no transparent property. A flip-flop can often be reset (that is, cleared) synchronously or asynchronously. To model an asynchronous reset signal of a D-type flip-flop, it needs to put the reset signal within the sensitivity list of the **always** block being used to describe the flip-flop.

■ Example A-34: (A D-type flip-flop with asynchronous reset.) A complete description of a D-type flip-flop with asynchronous reset is as follows. The output `q` is cleared to 0 whenever the reset signal `reset_n` is active, that is, 0, regardless of the status of the clock signal.

```
// an asynchronous reset D-type flip-flop
module DFF_async_reset(
       input  clk, reset_n, d,
       output reg q);
// the body of flip flop
always @(posedge clk or negedge reset_n)
   if (!reset_n) q <= 0; // active-low reset_n signal
   else          q <= d;
endmodule
```

■

To model a synchronous reset signal of a D-type flip-flop, it needs to put the signal outside the sensitivity list of the **always** block being used to describe the flip-flop.

■ Example A-35: (A D-type flip-flop with synchronous reset.) At each positive edge (identified by the keyword **posedge**) of the clock signal `clk`, the reset signal `reset` is checked to see whether it is active. If it is active, then the output `q` is cleared to 0; otherwise, the input data `d` is sampled into the flip-flop.

```
// A synchronous reset D-type flip-flop
module DFF_sync_reset(
       input  clk, reset, d,
       output reg q);
// the body of flip flop
always @(posedge clk)
   if (reset)  q <= 0; // active-high reset signal
   else        q <= d;
endmodule
```

■

A.5.3 Synchronizers

Synchronizers are essential to sample asynchronous signals into a synchronous system. As discussed, a simple synchronizer is basically a two-stage shift register. Consequently, it can be readily modeled by using the same style of shift register.

■ Example A-36: (A simple synchronizer.) An asynchronously resettable two-stage synchronizer is as follows. The output `q_out` is cleared to 0 whenever the reset signal `reset_n` is active, that is, 0, regardless of the status of the clock signal.

```
// a two-stage synchronizer with asynchronous reset
module synchronizer(
       input  clk, reset_n, d,
       output q_out);
reg    [1:0]  q;
// the body of synchronizer
assign q_out = q[1];
always @(posedge clk or negedge reset_n)
   if (!reset_n) q <= 0;
   else          q <= {q[0], d};
endmodule
```

■

A.5.4 Counters

Counters are widely used to count events or to provide desired timing signals. Counters can be cast into enable-controlled counters and autonomous counters. Regardless of which types, a counter can usually be reset asynchronously or synchronously if necessary.

■ Example A-37: (An n-bit binary counter.) The following module is an enable-controlled synchronous modulo 2^n binary counter with synchronous reset. It consists of n flip-flops. At each positive edge of clock `clk`, if `reset` signal is asserted, the counter is cleared; otherwise, if the `enable` signal is active, the counter counts up by one.

```
// an N-bit controlled binary counter with synchronous reset
module binary_counter
       #(parameter N = 4)( // set default size
         input clk, enable, reset,
         output reg [N-1:0] qout);

// the body of N-bit binary counter
always @(posedge clk)
   if (reset)         qout <= 0;
   else if (enable) qout <= qout + 1;
endmodule
```

■

A.5.5 Registers

In most practical applications, a data register often has the capability of clearing its outputs asynchronously. The following example illustrates such a data register.

■ **Example A-38: (An n-bit data register.)** This example simply expands the single bit D-type flip-flop with asynchronous reset to an n-bit register by declaring both `din` and `qout` as vectors of n bits.

```
// an N-bit data register with asynchronous reset
module register_reset
      #(parameter N = 4)( // set default size
        input  clk, reset_n,
        input  [N-1:0] din,
        output reg [N-1:0] qout);
// the body of N-bit data register
always @(posedge clk or negedge reset_n)
   if (!reset_n) qout <= {N{1'b0}};
   else          qout <= din;
endmodule
```

■

■ **Example A-39: (A parallel-load n-bit data register.)** The output `qout` is cleared whenever the reset signal `reset_n` is active. At each positive edge of the clock signal `clk`, a new input data `din` is sampled and loaded into the data register `qout` when load control `load` is enabled; otherwise, the output remains unchanged.

```
// an N-bit parallel-load data register with asynchronous reset
module register_load_reset
      #(parameter N = 4)(  // set default size
        input  clk, load, reset_n,
        input  [N-1:0] din,
        output reg [N-1:0] qout);
// the body of an N-bit data register
always @(posedge clk or negedge reset_n)
   if    (!reset_n) qout <= {N{1'b0}};
   else if (load)  qout <= din;
endmodule
```

■

A.5.6 Shift Registers

The following example describes an n-bit shift register with an asynchronous active-low reset signal.

■ **Example A-40: (An n-bit shift register.)** The outputs `qout` are cleared to 0 if the reset signal `reset_n` is active; otherwise, they are assigned the concatenation of serial input `sin` and the portion of `qout`, `qout[N-1:1]`. Hence, the result is a shift-right operation.

```
// a shift-right register module
module shift_register
      #(parameter N = 4)( // default number of bits
        input  clk, reset_n,
        input  sin,
        output reg [N-1:0] qout);
// the body of an N-bit shift-rigth register
always @(posedge clk or negedge reset_n)
   if (!reset_n) qout <= {N{1'b0}};
   else          qout <= {sin, qout[N-1:1]};
endmodule
```

■

A.5.7 Register Files

In most register-file-based datapaths, three operands are often required for carrying out an operation, and thus, the register file needs to have two-read and one-write ports.

■ **Example A-41: (An n-word register file.)** In this module, suppose that the access to each read port is unconditional and the access of the write port is controlled by a write enable signal `wr_enable` in synchronism with the clock signal, `clk`. Each port regardless of read or write has its own access address.

```
// an N-word register file with one-write and two-read ports
module register_file
      #(parameter M = 4,  // default number of address bits
        parameter N = 16, // N = 2**M
        parameter W = 8)( // default word width in bits
        input  clk, wr_enable,
        input  [M-1:0] rd_addra, rd_addrb, wr_addr,
        input  [W-1:0] din,
        output [W-1:0] douta, doutb);

reg     [W-1:0] reg_file[N-1:0];
// the body of the N-word register file
assign douta = reg_file[rd_addra],
       doutb = reg_file[rd_addrb];
always @(posedge clk)
   if (wr_enable) reg_file[wr_addr] <= din;
endmodule
```

■

A.5.8 Synchronous RAM

Synchronous RAM devices are widely used in synthesis-flow designs. To describe such a device, a **reg** array is declared. The array can be one or two dimensions.

■ **Example A-42: (A synchronous RAM module.)** This example demonstrates how to model a synchronous RAM with one read and one write port. Since

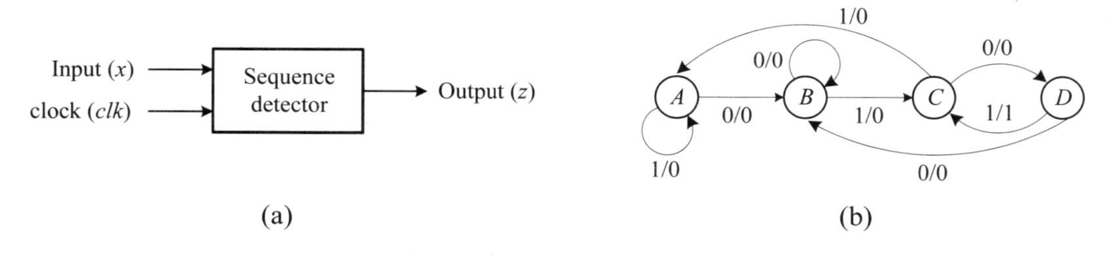

Figure A.4: The (a) block and (b) state diagrams of a sequence 0101 detector.

both access operations are in synchronism with the clock, they are placed inside the **always** block under the event control of the clock edge.

```
// a synchronous RAM module
module syn_ram
      #(parameter N = 16, // default number of words
        parameter A = 4,  // default number of address bits
        parameter W = 4)( // default word width in bits
        input  [A-1:0] addr,
        input  [W-1:0] din,
        input  cs, wr, clk,
        output reg [W-1:0] dout);
// define an N * W memory array
reg [W-1:0] ram [N-1:0];
// the body of synchronous RAM
always @(posedge clk)
   if (cs) if (wr) ram[addr] <= din;
           else    dout <= ram[addr];
endmodule
```

A.5.9 FSM Modeling

Two major modeling issues of a finite-state machine (FSM) are the declaration and update of the state register, and the computation of both next state and output functions. The state register is often declared as two registers such as `present_state` and `next_state` in accordance with the Huffman model. Output and next state functions can be computed by one of **assign** continuous assignment, function, and **always** block.

To describe an FSM, it is necessary to derive the state diagram. Then, the Huffman model is employed to describe the FSM. This idea is demonstrated by the 0101-sequence detector describe in Figure A.4.

■ Example A-43: (The 0101 sequence detector.) This example declares two state registers `present_state` and `next_state` and uses an **always** block to initialize and update the state register. The next state is computed by another **always** block. The output function is simply computed by the third **always** block. Both next state and output functions are separately computed by using a **case** statement and its related **if-else** statements.

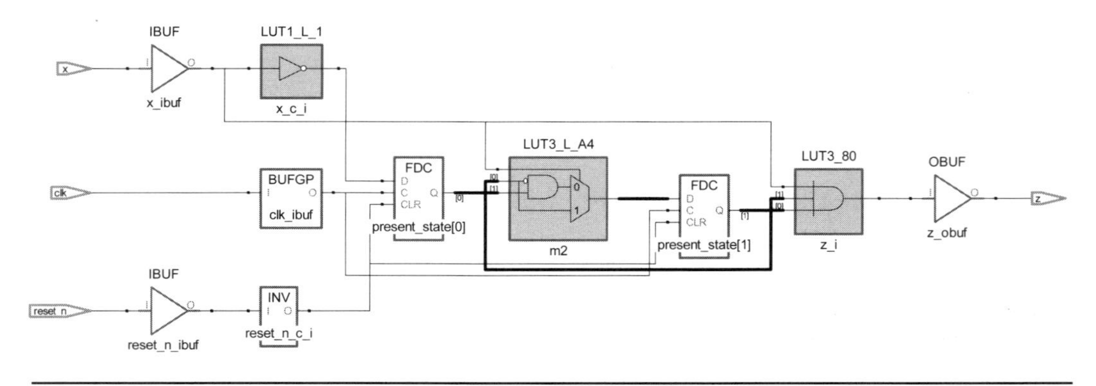

Figure A.5: The synthesized result of the sequence 0101 detector.

```
// a 0101 sequence detector in behavioral style
module sequence_detector_mealy(
       input  x, clk, reset_n,
       output reg z);
// declare state registers: present state and next state
reg [1:0] present_state, next_state;
localparam A = 2'b00, B = 2'b01, C = 2'b10, D = 2'b11;
// part 1: Initialize to state A and update state register
always @(posedge clk or negedge reset_n)
   if (!reset_n) present_state <= A; // A is the start state
   else present_state <= next_state; // update present state
// part 2: compute the next state function
always @(present_state or x)
   case (present_state)
      A: if (x) next_state = A; else next_state = B;
      B: if (x) next_state = C; else next_state = B;
      C: if (x) next_state = A; else next_state = D;
      D: if (x) next_state = C; else next_state = B;
   endcase
// part 3: evaluate output function z
always @(present_state or x)
   case (present_state)
      A: if (x) z = 1'b0; else z = 1'b0;
      B: if (x) z = 1'b0; else z = 1'b0;
      C: if (x) z = 1'b0; else z = 1'b0;
      D: if (x) z = 1'b1; else z = 1'b0;
   endcase
endmodule
```

■

The synthesized result is shown in Figure A.5. The use of two registers, `present_state` and `next_state`, does not mean the synthesized result will result in two state registers. It only provides a convenient way to write the RTL code.

Review Questions

QA-29. How would you describe an asynchronous reset D-type flip-flop?

QA-30. How would you describe a synchronous reset D-type flip-flop?

QA-31. How would you describe a modulo-10 synchronous counter?

QA-32. What are the features of register files?

QA-33. Explain the approach for describing a FSM with three **always** blocks.

A.6 Synthesis

In this section, we address how the synthesis tools deal with the language structures, including the **if-else** statement, **case** statement, and **always** statement.

A.6.1 General Considerations of Language Synthesis

The most straightforward language structures in Verilog HDL are assignment statements, including **assign** continuous and procedural assignments. Almost all operators in Verilog HDL are synthesizable. The exponent and modulus operators are generally the exceptions. The operands can be **wire** and **tri** net data types, **reg** variable data type, parameter, as well as functions.

Module instances, primitive gate instances, and tasks are synthesizable when the timing constructs are ignored. Procedural statements **always**, **if-else**, **case**, **casex**, and **casez** are all synthesizable but the **initial** statement is not supported. Procedural blocks **begin-end**, named blocks, as well as the **disable** statement are also synthesizable.

Loop statements, **for**, **while**, and **forever**, are synthesizable except that **while** and **forever** statements must contain timing control `@(posedge clk)` or `@(negedge clk)`.

A.6.2 Synthesis of Selection Statements

For many applications, one often encounters the case that only the **if** part is enough to describe a piece of hardware without the need of the **else** part, thereby leading to an incomplete **if-else** statement. An incomplete **if-else** statement may be inferred as a latch by synthesis tools when it describes a piece of combinational logic, such as the following statement:

```
always @(enable or data)
   if (enable) y = data;   // infer a latch
```

The synthesis tools will by default assume that the value of variable `y` is unchanged otherwise. As a result, a latch is inferred to keep the value of `y` unchanged. This problem can be solved by either of the following two ways:

- Complete the **if-else** statement by adding the **else** part. For example,

```
always @(enable or data)
   if (enable) y = data;
   else y = 1'b0;
```

- Initialize the outputs of the combinational logic circuit before setting their values conditionally. For example,

```
always @(enable or data)
   y = 1'b0;    // initialize y
   if (enable) y = data;
```

The **case** statement infers a multiplexer. An incomplete specified **case** statement will be inferred a latch when it describes a piece of combinational logic. To solve this problem, a **default** statement should be used as the last case item.

■ Example A-44: (Latch inference—an incomplete case statement.) This example demonstrates the situation of latch inference due to an incomplete **case** statement. It intends to model a 3-to-1 multiplexer and hence three case items are naturally used to select and route the input data to the output of the multiplexer. However, it leaves a case without being specified, which is then the reason to be inferred a latch.

```
// an example of latch inference
module latch_infer_case(
       input  [1:0] select,
       input  [2:0] data,
       output reg y);
// the body of 3-to-1 MUX
always @(select or data)
   case (select)
      2'b00: y = data[select];
      2'b01: y = data[select];
      2'b10: y = data[select];
   endcase
endmodule
```

■

■ Review Questions

QA-34. Explain why the exponent operator is generally not synthesizable.

QA-35. Explain why the modulus operator is in general not synthesizable.

QA-36. Explain the meaning of an incomplete **if-else** statement.

QA-37. Explain the meaning of an incomplete **case** statement.

A.6.3 Delay Values

The delay values are often used in source description to mimic the delays of logic gates and wires during simulation. Nevertheless, the synthesis tools ignore the delay values because the ultimate delays of the logic circuits will be determined by the actual delays of the gates used to implement the gate-level netlist.

■ Example A-45: (Ignored delay values—a not synthesizable version.) This example intends to model a four-phase clock generator. The phase 0 starts at 0 time unit from the positive edge of the clock signal. Each phase lasts for 5 time units. Although this program may generate the desired timing during the simulation time, it is not synthesizable because all delay values are ignored by the synthesis tools. The synthesized result is only the last statement; namely, it assigns `4'b0001` to `phase_out` after logic optimization.

```
// a four phase clock example --- an incorrect hardware
module four_phase_clock_wrong(
       input  clk,
       output reg [3:0] phase_out);   // phase output
// the body of the four phase clock
always @(posedge clk) begin
   phase_out <= #0  4'b0001;
   phase_out <= #5  4'b0010;
   phase_out <= #10 4'b0100;
   phase_out <= #15 4'b1000;
   phase_out <= #20 4'b0001;
end
endmodule
```

■

To generate a four-phase clock signal, the simplest way is to use a ring-counter structure, such as the following example.

■ **Example A-46: (A four-phase clock generator.)** At each positive edge of the clock signal `clk`, output signals `phase_out` change one time; namely, the value 1 is shifted to the next higher significant bit. The vector `phase_out` returns to its original status after four clock cycles through the **default** case item. As a result, it generates the desired four-phase clock signals.

```
// a four phase clock example --- a synthesizable version
module four_phase_clock(
       input  clk,
       output reg [3:0] phase_out);   // phase output
// the body of the four phase clock
always @(posedge clk)
   case (phase_out)
      4'b0000: phase_out <= 4'b0001;
      4'b0001: phase_out <= 4'b0010;
      4'b0010: phase_out <= 4'b0100;
      4'b0100: phase_out <= 4'b1000;
      default: phase_out <= 4'b0001;
   endcase
endmodule
```

■

A.6.4 Synthesis of Positive and Negative Signals

Most synthesis tools support the mixed use of two or more different edge-triggered signals but cannot accept the mixed use of edge-triggered with level-sensitive signals in the same sensitivity list of an **always** block. The following example illustrates this.

■ **Example A-47: (The mixed use of edge-triggered and level-sensitive signals.)** The **always** block uses both edge-triggered signal `clk` and level-sensitive signal `reset`. When synthesizing this module, we generally obtain an error message from the synthesis tool: "Error: Can't mix posedge/negedge use with plain signal references."

```
// an example of mixed use of posedge and level signals.
module DFF_bad (
       input  clk, reset, d,
       output reg q);
// the body of D-type flip-flop
always @(posedge clk or reset)
   if (reset) q <= 1'b0;
   else       q <= d;
endmodule
```

■

However, the mixed use of **posedge** and **negedge** signals, such as in the following example, is generally synthesizable.

■ **Example A-48: (The mixed use of posedge and negedge signals.)** In this example, a positive-edge event of clock signal `clk` is used to trigger the sampling operation of D-type flip-flop and a negative-edge event of reset signal `reset_n` to reset the flip-flop. This is a common way used to describe a D-type flip-flop with an asynchronous reset.

```
// an example of mixed use of posedge and negedge signals
module DFF_good(
       input  clk, reset_n, d,
       output reg q);
// the body of D-type flip-flop
always @(posedge clk or negedge reset_n)
   if (!reset_n) q <= 1'b0;
   else          q <= d;
endmodule
```

■

A.7 Verification

Verification is a necessary process to ensure that a design can meet its specifications. The verification of a design can be subdivided into two parts: functional verification and timing verification.

A.7.1 Related Compiler Directive and System Tasks

We introduce in the following the most basic and widely used compiler directive, **'timescale**, display system tasks, **$display** and **$monitor**, simulation time system functions, **$time** and **$realtime**, as well as simulation control system tasks, **$finish** and **$stop**.

A.7.1.1 'timescale Compiler Directive When carrying out simulations, we need to specify the physical unit of measure, or time scale, of a numerical time delay value. It also needs to specify the resolution of the time scale, that is, the minimum step size of the scale during simulation. In Verilog HDL, this is accomplished by using a **'timescale** compiler directive:

```
`timescale time_unit / time_precision
```

where " ' " is the back quote. The `time_unit` specifies the unit of measure for times and delays; the `time_precision` specifies how delay values are rounded during simulation. Only the integers 1, 10, and 100 may be used to specify `time_unit` and `time_precision`; the valid units are s, ms, us, ns, ps, and fs.

A.7.1.2 Display System Tasks During simulation, we need to display information about the design for debugging or other useful purposes. Verilog HDL provides two widely used system tasks: **$display** and **$monitor**, for displaying information on the standard output. The **$display** system task displays information only when it is called but **$monitor** system task continuously monitors and displays the values of any variables or expressions.

A.7.1.3 Simulation Time System Functions There are two system functions that provide access to current simulation time: **$time** and **$realtime**. The **$time** system function returns a 64-bit integer of time and the **$realtime** system function returns a real number of time.

A.7.1.4 Simulation Control System Tasks There are two simulation control system tasks: **$finish** and **$stop**. The **$stop** system task suspends the simulation, whereas the **$finish** system task terminates the simulation.

A.7.2 Test Bench Designs

The basic principles of test bench design are the test bench should generate stimuli and check responses in terms of test cases, the test bench should employ reusable verification components whenever possible, rather than coding from scratch each time, and the response checking must be automatic (self-checking test bench).

■ Example A-49: (Test bench example 1: Exhaustive test.) In this example, deterministic test signals are generated on the fly during simulation and the end-of-test checking is used. Of course, generating stimuli exhaustively can only be applied to a small design with small input space.

```
// test bench design example 1: exhaustive test.
`timescale 1 ns / 100 ps
module adder_nbits_tb1;
parameter N = 4;
reg  [N-1:0] x, y;
reg  c_in;
wire [N-1:0] sum;
wire c_out;
// Unit Under Test port map
   adder_nbits UUT (
      .x(x), .y(y), .c_in(c_in), .sum(sum), .c_out(c_out));
```

```
reg [2*N-1:0] i;
initial
   for (i = 0; i <= 2**(2*N)-1; i = i + 1) begin
      x[N-1:0] = i[2*N-1:N]; y[N-1:0] = i[N-1:0];
      c_in =1'b0;
      #20;
   end
initial
   #1280 $finish;
initial
   $monitor($realtime,"ns %h %h %h %h", x, y, c_in,
           {c_out, sum});
endmodule
```

The following is an example of using random stimulus generation on the fly during simulation.

■ Example A-50: (Test bench example 2: Random test.) In this example, the test signals are generated randomly on the fly during simulation and the on-the-fly checking is used.

```
// Testbench design example 2: Random test.
`timescale 1 ns / 100 ps
module adder_nbits_tb2;
parameter N = 4;
reg  [N-1:0] x, y;
reg  c_in;
wire [N-1:0] sum;
wire c_out;
// Unit Under Test port map
   adder_nbits UUT (
      .x(x), .y(y), .c_in(c_in), .sum(sum), .c_out(c_out));
integer i;
reg [N:0] test_sum;
initial
   for (i = 0; i <= 2*N ; i = i + 1) begin
      x = $random % 2**N;
      y = $random % 2**N;
      c_in =1'b0;
      test_sum = x + y;
      #15;
      if (test_sum != {c_out, sum})
         $display("Error iteration %h", i);
      #5;
   end
initial
   #200 $finish;
initial
   $monitor($realtime,"ns %h %h %h %h", x, y, c_in,
```

```
        {c_out, sum});
endmodule
```

In SystemVerilog, the above **if** statement

```
if (test_sum != {c_out, sum})
        $display("Error iteration %h", i);
```

can be replaced with the following **assert-else** statement

```
assert (test_sum == {c_out, sum}) else
        $error("Error iteration %h", i);
```

The following example combines the use of golden vectors and on-the-fly checking.

■ Example A-51: (Test bench example 3: Golden vectors.) In this example, the golden vectors are read in through using the **$readmemh** system task and stored in memory. The stimuli are applied to the device under test (DUT) one by one and the outputs from the DUT are then compared with the golden vectors on the fly during simulation. If any error is detected, then a message is displayed on the standard ouput.

```
// Testbench design example 3: Using golden vectors.
`timescale 1 ns / 100 ps
module adder_nbits_tb3;
//Internal signals declarations:
parameter N = 4;
parameter M = 8;
reg  [N-1:0] x, y;
reg  c_in;
wire [N-1:0] sum;
wire c_out;
// Unit Under Test port map
   adder_nbits UUT (
      .x(x), .y(y), .c_in(c_in), .sum(sum), .c_out(c_out));
integer i;
reg [N-1:0] x_array [M-1:0];
reg [N-1:0] y_array [M-1:0];
reg [N:0] expected_sum_array [M-1:0];
initial begin  // reading verification vector files
   $readmemh("inputx.txt", x_array);
   $readmemh("inputy.txt", y_array);
   $readmemh("sum.txt", expected_sum_array);
end
initial
   for (i = 0; i <= M - 1 ; i = i + 1) begin
      x = x_array[i];
      y = y_array[i];
      c_in =1'b0;
      #15;
      if (expected_sum_array[i] != {c_out, sum})
```

```
        $display("Error iteration %h", i);
      #5;
    end
initial
    #200 $finish;
initial
    $monitor($realtime,"ns %h %h %h %h", x, y, c_in,
             {c_out, sum});
endmodule
```

The contents of files: `inputx.txt`, `inputy.txt`, and `sum.txt`, are as follows:

inputx.txt	inputy.txt	sum.txt
4	1	05
9	3	0c
d	d	1a
5	2	07
1	d	0e
6	d	13
d	c	19
9	6	0f

Review Questions

QA-38. What are the three major components of a test bench?

QA-39. What is the function of the **‘timescale** compiler directive?

QA-40. What is the distinction between system tasks **$display** and **$monitor**?

QA-41. What is the distinction between system functions **$time** and **$realtime**?

QA-42. What is the function of system task **$finish**?

A.8 A Start/Stop Timer

As illustrated in Figure 12.17, the start/stop timer comprises a four-digit counter, two switch debouncers, a timing-base generator, and a display module.

A.8.1 Top Module

The top module of the start/stop timer is as follows. All modules except the 2-to-4 decoder and the 2-bit binary counter are easily recognized with reference to Figure 12.17. In writing a Verilog HDL module, it is easy to combine both the 2-to-4 decoder and 2-bit binary counter into one module, called the `mux_timing_generator` to reduce the number of intermodule ports.

```
// the top module of start_stop_timer
module start_stop_timer_top(
       input clk_4MHz, reset, clear, start,
       output wire [3:0] addr,
       output wire [6:0] DB_out);
```

```
wire [3:0] qout3, qout2, qout1, qout0, current_data_out;
wire clk_1ms, clk_10ms, clear_out, start_out;
wire [1:0] mux_sel;

// instantiate various modules of start_stop_timer
timing_base_generator basic_timing(
      clk_4MHz, reset, clk_1ms, clk_10ms);
switch_debouncer      button_clear(
      clk_10ms, reset, ~clear, clear_out);
switch_debouncer      button_start(
      clk_10ms, reset, ~start, start_out);
mux_timing_generator  mux_select  (
      clk_1ms, reset, mux_sel,addr);
four_bit_4_to_1_mux   mux_BCD  (
      mux_sel, qout3, qout2, qout1, qout0, current_data_out);
BCD_to_seven_segment_decoder segment_seg_LED(
      current_data_out,DB_out);
four_digit_counter    timer    (
      clk_10ms, reset, clear_out, start_out, qout3, qout2,
      qout1, qout0);
endmodule
```

A.8.2 Timing-Base Generator Module

The timing-base generator module receives a 4-MHz clock and generates two clocks with periods of 1 ms and 10 ms, respectively. These two clocks are denoted `clk_1ms` and `clk_10ms`, respectively. The timing-base generator module contains two binary counters. One is a 12-bit counter to divide the input 4-MHz clock into a 1-kHz clock, `clk_1ms`. The other is a 4-bit counter that further divides the 1-kHz clock into a 100-Hz clock, `clk_10ms`.

```
// time base generator --- generate all reference clocks
module timing_base_generator(
       input clk_4MHz, reset,
       output wire clk_1ms, clk_10ms);
reg [11:0] qout_1ms;
reg [3:0]  qout_10ms;

// generate clk_1ms clock
always @(posedge clk_4MHz or posedge reset) begin
   if (reset) qout_1ms <= 0;
   else if (qout_1ms == 3999) qout_1ms  <= 0;
   else qout_1ms  <= (qout_1ms + 1);
end
assign clk_1ms = qout_1ms[11];
// generate clk_10ms clock
always @(posedge clk_1ms or posedge reset) begin
   if (reset) qout_10ms <= 0;
   else if (qout_10ms == 9) qout_10ms <= 0;
   else qout_10ms <= (qout_10ms + 1);
end
```

```
assign clk_10ms = qout_10ms[3];
endmodule
```

A.8.3 Switch Debouncer Module

Every mechanical switch has an inherent bouncing phenomenon due to its inertial effect. To prevent this effect from being read many times and recognized as many switch presses by an electronic circuit, a switch debouncing circuit is needed. A simple but effective means to debounce a mechanical switch is based on the assumption that any valid switch press must be sustained at least 10 ms. Any switch signal shorter than this is deemed to be an invalid one and should be ignored. Based on this idea, a cascaded two-stage shift register clocked by a 10-ms clock is used to detect the switch signal, as described in the first `always` block. The second `always` block using a JK flip-flop to generate a single pulse when the switch is pressed each time.

```
//a switch debouncing logic
module switch_debouncer(
       input clk_10ms, reset, sw_in,
       output wire sw_out);
wire d;
reg  [1:0] q_sample;
reg  q_jk;
// debounce the bouncing effect associated with an
// input mechanical switch.
always @(posedge clk_10ms or posedge reset) begin
   if (reset) q_sample <= 0;
   else begin
      q_sample[1] <= q_sample[0];
      q_sample[0] <= sw_in;
   end
end
// using a JK flip-flop to generate a single pulse
// when the switch is pressed each time.
assign d = (((q_sample[0] & q_sample[1]) & ~q_jk) |
           (~(~q_sample[0] & ~q_sample[1])& q_jk));
always @(posedge clk_10ms or posedge reset)
   if (reset)   q_jk <= 1'b0;
   else         q_jk <= d;
assign sw_out = q_jk;
endmodule
```

A.8.4 Timer Module

The timer is a four-digit BCD counter, which is formed by cascading four binary-coded decimal (BCD) counters. The clock input of the counter is gated by a T-type flip-flop, which is toggled by the switch *start*. This logic is described by the first `always` block. The other part of the module is the body of the four-digit BCD counter. This counter may be cleared by either the switch *clear* or the global reset control signal. The reset control signal is also applied to all other sequential-related modules.

```
// the four_digit counter
```

```
module four_digit_counter(
       input  clk_in, reset_in, clear, start,
       output reg [3:0] qout3, qout2, qout1, qout0);
wire clk;
reg  qout;
// generate the controlled clock from start and clk_in
assign reset = reset_in | (clear & ~qout);
always @(posedge start or posedge reset)
   if (reset) qout <= 0;
   else qout <= ~qout;
assign clk = clk_in & qout;
// the LSD of counter
always @(posedge clk or posedge reset)
   if (reset) qout0 <= 0;
   else if (qout0 == 9) qout0 <= 0;
   else qout0 <= (qout0 + 1);
// the second digit of counter
always @(posedge clk  or posedge reset)
   if (reset) qout1 <= 0;
   else if (qout1 == 9 && qout0 == 9) qout1 <= 0;
   else if (qout0 == 9) qout1 <= (qout1 + 1);
// the third digit of counter
always @(posedge clk or posedge reset)
   if (reset) qout2 <= 0;
   else if (qout0 == 9 && qout1 == 9 && qout2 == 9)
           qout2 <= 0;
   else if (qout0 == 9 && qout1 == 9) qout2 <= (qout2 + 1);
// the MSD of counter
always @(posedge clk or posedge reset)
   if (reset) qout3 <= 0;
   else if (qout0 == 9 && qout1 == 9 && qout2 == 9
            && qout3 == 9) qout3 <= 0;
   else if (qout0 == 9 && qout1 == 9 && qout2 == 9)
            qout3 <= (qout3 + 1);
endmodule
```

A.8.5 Display Module

The display module actually consists of four parts: a 2-to-4 decoder, a 2-bit binary counter, a four-bit 4-to-1 multiplexer, and a BCD-to-seven-segment decoder. In the following, we briefly describe the operation of each module.

A.8.5.1 The `mux_timing_generator` Module As we have described previously, the 2-to-4 decoder and the 2-bit binary counter are combined into one module, called the `mux_timing_generator`. This module is composed of two major parts: the 2-bit binary counter described by the first `always` block and the 2-to-4 decoder described by the second `always` block.

```
// binary counter with decoder serve as a timing generator
module mux_timing_generator(
       input  clk, reset,
```

```
        output wire [1:0] mux_sel,
        output reg  [3:0] addr);
// the body of binary counter
reg    [1:0] qout;

always @(posedge clk or posedge reset)
  if (reset) qout <= 0;
  else       qout <= qout + 1;
assign mux_sel = qout;
// decode the output of the binary counter
always @(qout)
   case(qout)
      2'b00: addr = 4'b0001;
      2'b01: addr = 4'b0010;
      2'b11: addr = 4'b1000;
      default: addr = 4'b0000;
   endcase
endmodule
```

A.8.5.2 Four-Bit 4-to-1 Multiplexer The four-bit 4-to-1 multiplexer is described by a `case` statement in Verilog HDL. The multiplexer is selected by the source selection inputs coming from the output of the 2-bit binary counter in the `mux_timing_generator` module.

```
// selecting one out of four digits for display
module four_bit_4_to_1_mux(
        input [1:0] mux_sel,
        input [3:0] data_in3, data_in2, data_in1, data_in0,
        output reg [3:0] data_out);
// counter_mux --- multiplexing output the counter
always @(*)
   case (mux_sel)
      2'b00: data_out = data_in3;
      2'b01: data_out = data_in2;
      2'b10: data_out = data_in1;
      2'b11: data_out = data_in0;
   endcase
endmodule
```

A.8.5.3 BCD-to-Seven-Segment Decoder As we have seen from Figures 12.15(a) and (b), only one BCD digit is displayed on the light-emitting diode (LED) display module at any time and all four BCD digits are displayed repeatedly in a time-multiplexing manner. However, the selected BCD digit must be converted into a seven-segment code before it is sent to the LED display module. The following module carries out this conversion. The seven-segment code for the LED display module can be easily derived from Figure 12.15(a), thereby leaving it to the interested reader as an exercise.

```
// Convert input data into seven_segment_display code with
// active-low outputs.
// The input data is assumed in the range of 0 to 9.
```

```
module BCD_to_seven_segment_decoder(
       input [3:0] data_in,
       output reg [6:0] data_out);
// the default case is assumed to be blank.
always @(data_in)
   case (data_in)
      0: data_out = 7'b100_0000;    // h40
      1: data_out = 7'b111_1001;    // h79
      2: data_out = 7'b010_0100;    // h24
      3: data_out = 7'b011_0000;    // h30
      4: data_out = 7'b001_1001;    // h19
      5: data_out = 7'b001_0010;    // h12
      6: data_out = 7'b000_0010;    // h02
      7: data_out = 7'b111_1000;    // h78
      8: data_out = 7'b000_0000;    // h00
      9: data_out = 7'b001_0000;    // h10
      default: data_out = 7'b111_1111;
   endcase
endmodule
```

■ Review Questions

QA-43. What does the top module mean?

QA-44. What is the function of the switch debouncer module?

QA-45. What is the function of the timing-base generator module?

QA-46. Explain the operation of the display module.

QA-47. Explain the rationale behind the design of the **mux_timing_generator** module.

A.9 Summary

This appendix simply surveyed some synthesizable features with examples of Verilog and SystemVerilog HDL. Through the illustrations of these examples, readers can readily write their own hardware modules. In addition, basic design approaches of test benches were also dealt with thoroughly. Finally, the complete description of the start/stop timer described in Chapter 12 was presented in the context of Verilog HDL.

References

1. IEEE 1364-2001 Standard, *IEEE Standard for Verilog Hardware Description Language*, 2001.
2. IEEE 1364-2005 Standard, *IEEE Standard for Verilog Hardware Description Language*, 2005.
3. IEEE 1800-2005 Standard, *IEEE Standard for SystemVerilog-Unified Hardware Design, Specification, and Verification Language*, 2005.

4. IEEE 1800-2009 Standard, *IEEE Standard for SystemVerilog-Unified Hardware Design, Specification, and Verification Language*, 2009 (a revision of IEEE 1800-2005 Standard).

5. M. B. Lin, *Digital System Design: Principles, Practices, and Applications*, 4th ed. Taipei, Taiwan: Chuan Hwa Book Ltd., 2010.

6. M. B. Lin, *Digital System Designs and Practices: Using Verilog HDL and FPGAs*. Singapore: John Wiley & Sons, 2008.

Problems

A-1. Model the following switching expression at the gate level in the structural style:

$$f(x, y, z) = \overline{xy + x\bar{z}}$$

Write a test bench to verify whether the module behaves correctly.

A-2. Model the following switching expression at the gate level in the structural style:

$$f(w, x, y, z) = \overline{(wx + \bar{y}z) + \overline{xy}}$$

Write a test bench to verify whether the module behaves correctly.

A-3. Write a module to convert the BCD code into an excess-3 code. Also, write a test bench to verify the functionality of the module.

A-4. A majority circuit is a device that outputs 1 whenever more than half of its inputs are 1.

(a) Describe the behavior of a majority circuit, assuming that there are 8 inputs.

(b) Model the majority circuit in the behavioral style.

(c) If the inputs are checked only when `ready` signal is 1, redo part(b).

A-5. A minority circuit is a device that outputs 1 whenever less than half of its inputs are 1.

(a) Describe the behavior of a minority circuit, assuming that there are 8 inputs.

(b) Model the minority circuit in the behavioral style.

(c) If the inputs are checked only when `ready` signal is 1, redo part(b).

A-6. An excess-3 code checker is a circuit whose output is 1 if its input is a valid excess-3 code and is 0 otherwise. Describe the behavior of this circuit and model it in the behavioral style. Write a test bench to verify it.

A-7. Assume that we want to design a module that counts the number of ones in a byte. Describe the module in the behavioral style and write a test bench to verify its functionality.

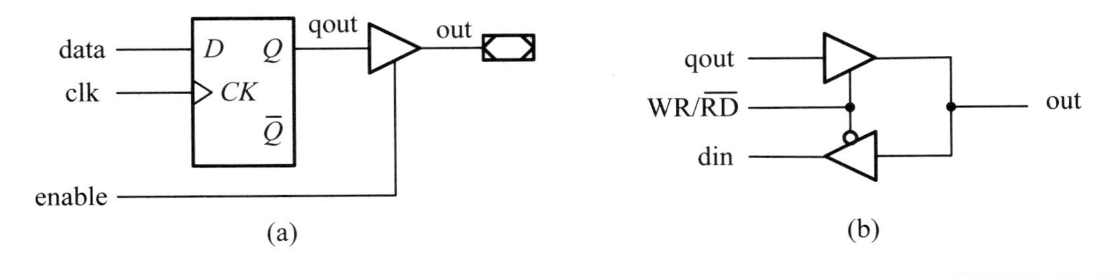

Figure A.6: Figure for problem **A-16**.

A-8. Assume that we want to design a module that counts the number of trailing ones in a byte. Describe the module in the behavioral style and write a test bench to verify its functionality.

A-9. Design a 3-to-8 decoder with the following specified output polarities and then write Verilog HDL modules to describe them.

(a) Active-high output

(b) Active-low output

A-10. Design a 4-to-16 decoder using at most two enable-controlled 3-to-8 decoders and write a Verilog HDL module to describe it.

A-11. Design an n-bit 4-to-1 multiplexer with active-low enable control and write a Verilog HDL module to describe it. The output is at high-impedance when the enable control is deasserted.

A-12. Use two 2-to-1 multiplexers to design a 3-to-1 multiplexer and write a Verilog HDL module to describe it.

A-13. Use two 4-bit magnitude comparators to design an 8-bit magnitude comparator. Also write Verilog HDL modules to describe them.

A-14. Write a Verilog HDL module to describe a D-type flip-flop with both asynchronous clear and preset.

A-15. Write a Verilog HDL module to describe a D-type flip-flop with both synchronous clear and preset.

A-16. Consider the simple logic circuit shown in Figure A.6:

(a) Write a Verilog HDL module to describe the circuit shown in Figure A.6(a).

(b) Replace the output tristate buffer of the circuit shown in Figure A.6(a) with the bidirectional buffer shown in Figure A.6(b) and redo (a).

A-17. Consider the shift register module shown in the following:

```
// an example of shift register with computation
module shift_reg_blocking(clk, sin, qout);
input  clk;
input  sin;  // serial data input
output reg [3:0] qout;
```

```
// the body of a 4-bit shift register with computation
always @(posedge clk)begin
   qout[0] = sin;
   qout[1] = qout[0];
   qout[2] = qout[1] ^ qout[0];
   qout[3] = qout[1] & qout[2];
end
endmodule
```

(a) Can it be operated as a 4-bit shift-left register with the required computation?

(b) Synthesize it and examine the result.

(c) Explain why the synthesized result is like what you have obtained.

(d) Correct the statements within the module so that it can function correctly.

A-18. This problem concerns the design of an n-bit program counter (PC) that has the following functions:

(a) PC is cleared to 0 when the asynchronous reset is asserted.

(b) PC is loaded a new value in parallel when the PCload is asserted.

(c) PC is incremented by 1 when the PCinc is asserted.

Design the program counter and write a Verilog HDL module to describe it.

A-19. Suppose that a 4-bit programmable counter is needed. The counter can be set to modulo n, where n may be set from 0 to 15 externally by the control signals $C3$ to $C0$. Write a Verilog HDL module to describe this counter and write a test bench to verify it.

A-20. Suppose that an 8-bit programmable counter is required. The counter can be set to modulo $4n$, where n can be set from 1 to 15 externally by control signals $C3$ to $C0$. Write a Verilog HDL module to describe this counter and write a test bench to verify it.

A-21. Assume that a clock generator is required. The clock output `clk` is initialized to 0 and toggles its value every 20 time units. Describe this clock generator without using the **always** statement.

A-22. Assume that a clock generator with burst output is required. When the `enable` control signal is raised to high, the clock generator produces 25 pulses with a period of 10 time units and a duty cycle of 50%. The output pulses should start from 0. Model this clock generator.

A-23. Assume that a clock generator is required. The clock output `clk` is initialized to 0 and toggles its value every 20 time units. Describe this clock generator using a **while** loop.

A-24. Assume that a clock generator is required. The clock output `clk` is initialized to 0 and has a period of 40 time units and a duty cycle of 25%. Describe this clock generator using a **for** loop.

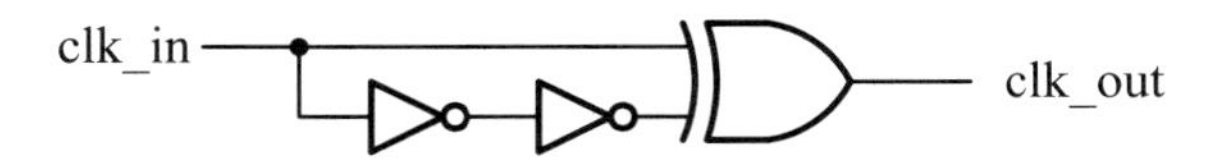

Figure A.7: A simple frequency doubler.

A-25. If we want to use four inverters cascaded together in some applications, write a Verilog HDL to model it. Synthesize your design with an available field-programmable gate array (FPGA) device and see what happens.

A-26. A simple frequency doubler is a circuit that uses XOR gates to extract the edge information from an input clock signal, such as the one shown in Figure A.7.

Index